Ecoregions

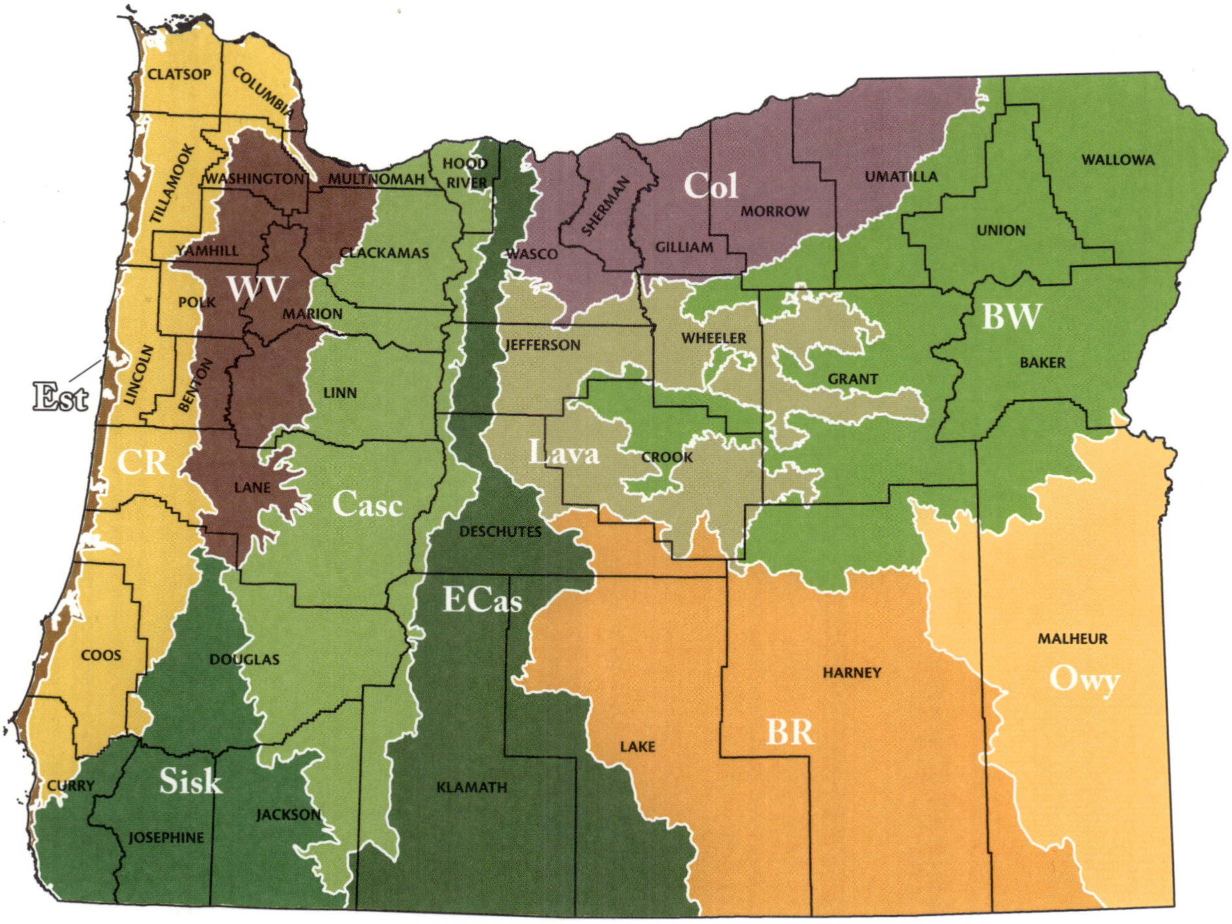

BR	Basin and Range
BW	Blue and Wallowa Mountains
Casc	Cascades
CR	Coast Range
Col	Columbia Basin
ECas	East Cascade Slopes and Foothills
Est	Estuarine Coast
Lava	Lava Plains
Owy	Owyhee Uplands
Sisk	Siskiyous
WV	Willamette Valley

Flora of Oregon

Oregon Flora Project
Dept. of Botany & Plant Pathology
Oregon State University
2082 Cordley Hall
Corvallis, OR 97331-2902, U.S.A.
www.oregonflora.org

John Myers, principal illustrator
Tanya Harvey, layout and design

OSU Oregon State University
College of Agricultural Sciences

Native Plant Society of Oregon
Founded 1961

National System of Public Lands
U.S. Department of the Interior
Bureau of Land Management

FNA

OCF The Oregon Community Foundation
Here for Oregon. Here for Good.

Flora of Oregon

Volume 1: Pteridophytes, Gymnosperms, and Monocots

Stephen C. Meyers, Thea Jaster,
Katie E. Mitchell, & Linda K. Hardison, eds.

BRIT
PRESS
Botanical Research Institute of Texas
Fort Worth, Texas
2015

FLORA OF OREGON

Printed in Korea

Please cite this book as follows:

Meyers, S.C., T. Jaster, K.E. Mitchell & L.K. Hardison. 2015. Flora of Oregon. Volume 1: Pteridophytes, Gymnosperms, and Monocots. Botanical Research Institute of Texas, Fort Worth, Texas, U.S.A.

Published and distributed by:
Botanical Research Institute of Texas Press
1700 University Drive
Fort Worth, Texas 76107-3400, U.S.A.
817-332-4441
Fax: 817-332-4112
www.britpress.org
shop.brit.org

ISBN: 978-1-889878-46-1

Library of Congress Cataloging-in-Publication Data

Flora of Oregon. Volume 1, Pteridophytes, gymnosperms, and monocots / Stephen C. Meyers, Thea Jaster, Katie E. Mitchell, Linda K. Hardison, editors. — First edition.
volumes cm
Includes bibliographical references and index.
ISBN 978-1-889878-46-1 (alk. paper)
1. Plants—Oregon. 2. Plant ecology—Oregon. 3. Pteridophyta—Oregon. 4. Gymnosperms—Oregon. 5. Monocotyledons—Oregon. I. Meyers, Stephen C., 1966– editor. II. Title: Pteridophytes, gymnosperms, and monocots.
QK182.F56 2015
581.709795—dc23 2015011057

BRIT
PRESS

MIX
Paper from responsible sources
FSC
www.fsc.org
FSC® C023083

Flora of Oregon

is dedicated to

Scott Sundberg

(1954–2004)

Scott's knowledge, his passion for the flora of this state, and his vision for a comprehensive botanical reference continue to inspire us and guide the Oregon Flora Project to the fulfillment of its mission.

The mission of the Oregon Flora Project is to serve as a comprehensive resource about the vascular plants of Oregon that grow in nature without cultivation, and to foster effective use of this knowledge by all citizens.

DIGITAL RESOURCES

Additional information is presented digitally through the Oregon Flora Project's website, www.oregonflora.org. Current resources available include the following:

The Oregon Vascular Plant Checklist

OFP's synonymized Checklist, with cross-references to nine key references. It is published annually with updates.

The Oregon Plant Atlas

An interactive plant mapping program with a database of more than 540,000 records from Flora vouchers (specimens of the OSU Herbarium and other selected records which serve as documentation for all taxa treated in the *Flora of Oregon*), 35 other herbaria from the Pacific Northwest, type specimens of Oregon taxa from multiple herbaria, and unvouchered observations provided by governmental agencies, the Oregon Biodiversity Information Center, the Native Plant Society of Oregon, and individuals.

The Photo Gallery

Field photographs of each plant featuring its habitat, general features, and details, as well as images of representative herbarium specimens.

The Rare Plant Guide

Field-oriented fact sheets and a searchable database about some of Oregon's rare and threatened species.

Oregon Wildflowers

A mobile app for rapid identification of over 1000 wildflowers, vines, and shrubs statewide.

CONTENTS

OREGON FLORA PROJECT

Staff

Director
Dr. Linda K. Hardison

Taxonomic Director
Dr. Stephen C. Meyers

Database Managers, Botanists
Thea Jaster
Katie E. Mitchell

Senior Graphic Artist, Principal Illustrator
John Myers

Advisory Board (2011–2015)

Dr. Gail Baker, Eugene
Lynda Boyer, Salem
Jeff Cook, Eugene
Sam Friedman, Roseburg
Dr. Aaron Liston, Corvallis
Dr. Daniel Luoma, Corvallis
Carol Savonen, Philomath
Joan Seevers, Ashland
Paul Slichter, Gresham
Dr. Robert Soreng, Washington, DC
José Vega, Corvallis

Contributors to Flora of Oregon, Volume 1

Authors

Dennis A. Albert
Oregon State University, Corvallis, OR

Edward R. Alverson
Eugene, OR

Richard E. Brainerd
Carex Working Group, Corvallis, OR

Frank T. Callahan
Central Point, OR

Bridget Chipman
Jefferson, OR

Thea Jaster
Oregon Flora Project, Corvallis, OR

Linda K. Hardison
Oregon Flora Project, Corvallis, OR

Susan R. Kephart
Willamette University, Salem, OR

Stephen C. Meyers
Oregon Flora Project, Corvallis, OR

Katie E. Mitchell
Oregon Flora Project, Corvallis, OR

Rachel E. Newton
Corvallis, OR

Rory Nichols
Silverton, OR

Nick Otting
Carex Working Group, Corvallis, OR

Tamra Dickson Prior
Portland, OR

James P. Riser II
Washington State University, Pullman, WA

Emily B. Sessa
University of Florida, Gainesville, FL

Lareina Shipler
Portland, OR

Mark W. Skinner
US Department of Agriculture, Forest Service, Pacific Northwest Region, Portland, OR

Robert J. Soreng
Smithsonian Institution, Washington, DC

Kathryn E. Theiss
Willamette University, Salem, OR

Duncan Thomas
Washington State University, Vancouver, WA

Barbara L. Wilson
Carex Working Group, Corvallis, OR

Ann Willyard
Hendrix College, Conway, AR

Peter Zika
University of Washington, Seattle, WA

Artists

Photographers

ACKNOWLEDGMENTS

The *Flora of Oregon* is inextricable from its organizational parent, the Oregon Flora Project. These acknowledgments attempt to capture my immense gratitude for the support the entire project has received in addition to the contributions made on behalf of this first volume.

The Department of Botany & Plant Pathology of Oregon State University (OSU) and the OSU Herbarium have given us a physical home since our inception. The Department has contributed the indirect costs of our program, and Herbarium Director Aaron Liston has provided unlimited access to the collections of the OSU Herbarium. These resources have given the Oregon Flora Project (OFP) the opportunity to focus our energies on the development of our content and, in doing so, expand the impact of our work. An early collaboration with the Northwest Alliance for Computational Science & Engineering (NACSE) at OSU launched the digital presentation of our data. I thank NACSE Director Cherri Pancake, Sherry Pittam, Joe Hanus, and colleagues for sharing their expertise. The Agricultural Research Foundation has served as our fiscal sponsor since 2008 and greatly facilitated our ability to interact efficiently with donors and foundation-based funders.

The Native Plant Society of Oregon is an enthusiastic sponsor of the OFP and provided the original core of financial donors and volunteers through the formation of a Friends of the OFP Committee. Beyond the sustained financial assistance and the engagement given to our program, I am touched by the generosity of spirit and personal support I have received from many individuals of this group and the Society as a whole.

The Oregon/Washington Bureau of Land Management (BLM) provided early financial support to develop the Oregon Plant Atlas. Through the diligence of state botanists Joan Seevers (retired) and Mark Mousseaux, the BLM continues to be a key funder of the OFP.

The John and Betty Soreng Environmental Fund of The Oregon Community Foundation championed our cause at the bleakest time in our project's history. Their financial support in 2008 reinstated operations after we had exhausted all funding, and their sustained generosity has allowed us to bring the published *Flora of Oregon* and many web-based tools to completion.

Considering the importance of our online presence and the communication of data through our website, it is especially noteworthy that the coding for the original Oregon Plant Atlas and subsequent website development was accomplished by volunteers. Clayton Gautier wrote the original Atlas software, and Jeff Cook developed Version 2 of the Atlas in addition to providing major updates to our website. Undergraduates Kit Hoffman and David Fontenot assisted in website design. We are deeply indebted to the hundreds of hours of effort each provided.

The content of our website (www.oregonflora.org) augments the information presented in the paper-published *Flora.* I extend my thanks to the many individuals and groups that have provided distribution data and images for our web-based tools, and suggest that readers visit the website to view their acknowledged contributions.

The development of a synonymized checklist created the foundation for the *Flora of Oregon.* Two advisory bodies, the Checklist Advisory Board and the Checklist Project Leaders, were critical in its formation. Upon Scott Sundberg's death in 2004, these groups rallied to ensure the completion of the Checklist. Advisory Board members were Ed Alverson, Karen Antell, Henrietta Chambers, John Christy, Tom Kaye, Susan Kephart, Frank Lang, Don Mansfield, Kareen Sturgeon, and Barbara Wilson. The Checklist Project Leaders were Kenton Chambers, Richard Halse, Jimmy Kagan, Aaron Liston, Rhoda Love, Robert Meinke, David Wagner, and Peter Zika. Of the many contributors to the Checklist, Henrietta Chambers, Kenton Chambers, and Richard Halse receive special thanks for having prepared Checklist treatments for 877 genera, representing 86% of the flora.

I am especially grateful to the hundreds of volunteers that have shared their time and talents over the twenty years of our project. Individuals have assisted in every aspect of work and can rightfully claim membership on the Oregon Flora Project team. I would like to acknowledge a few for their exceptional help over many years: Wilbur Bluhm, Native Plant Society of Oregon liaison for Atlas observations; Henrietta Chambers, author and researcher of significant portions of the synonymous Checklist; Rhoda Love, editor of the *Oregon Flora Newsletter* for 18 years; Gene Newcomb, photographer of herbarium specimens and long-term volunteer; Don Roberts, photographer of herbarium specimens for the Photo Gallery; and Charlene Simpson, coordinator for the production of the Lane County Checklist and long-term volunteer. For the many whose names I cannot list here, please know that your contributions are deeply appreciated.

Over 140 students have been employed by the Oregon Flora Project, assisting in almost every aspect of our program. It has been a privilege to play a role in their educational experience, especially to those who have pursued careers or graduate studies in botany.

Many hands have helped to craft the details that are captured between the covers of this book. The authors of floristic treatments have graciously shared their expertise and time. I have greatly enjoyed interacting with those who have visited the OSU Herbarium in the course of their participation. James Hall has kindly allowed us to use the artwork of his late wife Bonnie on the book jacket. We hope this serves to honor her memory, and we express our appreciation for the generosity Jim and Bonnie have shown to the OFP. Thanks go to Dennis Albert, author of the Ecology chapter, and Edward Alverson, contributor of the chapter detailing places to experience the Oregon flora. The color images in these two chapters bring the content to life; we thank the photographers listed on page ix for allowing us to use their images. I acknowledge Rhoda Love, Cindy Roché, Susan Kephart, Nancy Weber, and Henrietta Chambers for permission to adapt their essays on historic botanical figures, and Cindy Roché, editor of *Kalmiopsis*, for assistance in acquiring the manuscripts. Associated images are used with permission and assistance from the Special Collections and Archives units at OSU, University of Oregon, Willamette University, and the Leach Botanical Garden. Illustrations from the publications of the Flora of North America Association are published here through a reciprocal agreement; many thanks to John Myers and James Zarucchi for facilitating this exchange. The illustrations of Juncaceae (excluding *Luzula subsessilis*) are reprinted with permission of the University of Washington Press from: Hitchcock, C. Leo and Arthur Cronquist. *Flora of the Pacific Northwest: An Illustrated Manual*, ©1973. We also acknowledge Mary Barkworth and the Utah State University Herbarium for use of Poaceae illustrations. We appreciate the reviewers who have critically read the manuscripts and tested keys: Edward Alverson, Jerry Black, Anna Bradford, Kelly Burnett, Melissa Carr, Kim Clark, Mike Faden, Christian Feuillet, Brandon Fraser, Emily Gillespie, Matt Groberg, Kirsten Hall, Tina Huckabee, Troy Maddux, Eli Piper, Stormy Scharzenberger, Lareina Shipler, Matthew Silbernagel, Charlene Simpson, Carla Wise, and others. Tanya Harvey has provided the design and layout for the *Flora of Oregon*, as well as editing of all texts. She has transformed reams of manuscripts into this beautiful *Flora* with an apparent ease that belies the hard work and attention to detail she provides. Barney Lipscomb and the publications staff of BRIT Press have guided us through the myriad steps to produce this volume; their care and support are appreciated.

Any number of superlatives can be used to describe the Oregon Flora Project staff: passionate, talented, insightful, dedicated. All of them are true and reflect the character that imbues our program. Former staff members Ann Willyard, Rena Schlachter, and Jennifer Sackinger, along with our current staff of Thea Jaster, Stephen Meyers, Katie Mitchell, and John Myers have been friends and colleagues to me. After the death of Scott Sundberg, the founding director and my husband, they exhibited leadership and dedication as they sustained the project activities and patiently helped me grow into the role of OFP director. I am deeply grateful for their compassion and their commitment. I hold each of these people in the highest regard and feel fortunate to have worked with them on such a meaningful project.

All salaries and direct operating costs for the Oregon Flora Project are derived from competitive grants and charitable donations. I gratefully acknowledge the financial support we have received over the lifetime of the program from these organizations and individuals. Over 970 donors have provided financial support for the Oregon Flora Project. Donations have ranged from the large to the small; some are given on a regular basis and others as a one-time gift. I am grateful for each contribution and the confidence in our ability to achieve the OFP mission that such support denotes. The first list of names represents donors who have given $1,000 or more since 2000. I also extend my thanks to our sustaining donors: those who have made ten or more donations since the formation of the OFP. A list acknowledging these contributors who are not represented in the first category is given.

Every effort has been made to present correct listings; I assume responsibility and sincerely apologize for any errors or omissions.

Linda K. Hardison

BENEFACTORS

KEY SUPPORTERS

John and Betty Soreng Environmental Fund of The Oregon Community Foundation

Oregon/Washington Bureau of Land Management

Native Plant Society of Oregon

Department of Botany & Plant Pathology, Oregon State University

SUPPORTING ORGANIZATIONS

Benton County Master Gardener Association

H.J. Andrews Experimental Forest

Jackson Foundation

Mountaineers Foundation

National Science Foundation

Natural History Museum Fund of Oregon State University

North American Rock Garden Society

Oregon State University General Research Fund

US Forest Service Pacific Northwest Research Station Forest Inventory and Analysis Program

US Forest Service Region 6

Willamette Industries, Inc.

DONORS

Donors who have given $1,000 or more since 2000

Dennis Albert
Norma Alberthal
Karl Anderson
John Annear
Betty Bahn
David Barts
Dan Bates
Connie Battaile
Wilbur Bluhm
Bert Brehm
Kay Butler
Gerald & Barbara Carr
George & Fanny Carroll
Kenton & Henrietta Chambers
Nancy Chapman
Glen Cole
Thea Jaster & Jeff Cook
M.F. Crane
Ellen A. Crumb
Geraldine Custer
David & Janet Dobak
Lorena Dornfeld
Janet Dorow
Jim Duncan & Elaine Plaisance
Michael & Nancy Fahey
Robert & Elizabeth Frenkel
Linda Friedman
Clayton Gautier & Gail Baker
Edwin Gerow
Sylvia Giustina
Jim Hall
Glenn & Barbara Halliday
Michael Hartman
Cliff Heselton
Noel & Patricia Holmgren
Mary Hough
Lawrence Janeway
Anne Jaster
Albert Johnson
Jesse Jonas & Steven Cantor
Karol Kochsmeier
John Koenig
Jennifer Lippert
Aaron & Sara Liston
Kenneth & Robin Lodewick
Cynthia Lord
Rhoda & Glen Love
Jay & Ann Lunn
Dan Luoma & Joyce Eberhart
Alex & Lillian Maksymowicz
Louise Marquering
Vernon Marttala
Sara Mauritz
Cheryl McCaffrey
Bruce McCune & Patricia Muir
Peter & Esther Gruber McEvoy
Kathy Merrifield
Andy Moldenke
Native Plant Society of Oregon:
Blue Mountain
Cheahmill
Corvallis
Emerald
High Desert
Mid-Columbia
Portland
Siskiyou
Umpqua Valley
Willamette Valley
Julie Kierstead Nelson
Bonnie Brunkow Olson
Theodore & Laramie Palmer
Carol & Kurt Putnam
Torrance Raymond
Phyllis Reynolds

Donors who have given $1,000 or more since 2000 (continued)

Dell Rhodes & Will Bloch
Don Roberts
Georgie Robinett
Cindy Roché & Robert Korfhage
Robert A. Ross
John Roth
Sandy Roth
Barbara Roy
Freeman Rowe
Ruth & Delbert Shirley
Charlene Simpson
Paul Slichter
Deanna Sterett
David Steward
Leland Steward
Kent Sundberg & Beth de la Fuente
Norman & Donna Sundberg
Ron & Pam Swisher
Tetronix Foundation
The Nature Conservancy
Charles Trainer
Robert Trask
George Van Vechten
Jody Waugh
Nancy Weber
Carolyn Wright
Sunia Yang
Desi & Karen Zamudio
Donald Zobel

Donors who have made ten or more donations to OFP

Sara Barnum
Julie Biddle & Tom Robertson
Paula Brooks
Nancy Broshot
Rob Castleberry
James Conklin
Alan Curtis
Robyn Darbyshire
Heidi E.M. Dobson
Nona Donahue
Nancy Eid
Jenifer & Rodger Ferriel
D. Ann Gahr
Dave Garcia
Janice Gerdemann
Marvel Gillespie
Louise Godfrey
Andrew Gray
Candice Guth
Tanya Harvey
Kenneth Himes
Charlene Holzwarth
Kathryn Jernstedt
Keith Karoly
Doramay Keasbey
Sarah Malaby
Elizabeth Marantz
Stanley & Margaret Meierhenry
Greg & Gail Miles
Marilyn Miller
Dan & Pat Montague
Jeanne & Frank Moore
Linda Mullens
Barbara Mumblo
Julie & Mark Nichols
Nick Otting
Warren Pavlat
Kathy Pendergrass
Sandy Poinsett
Robert & Marna Porath
Ginny & Jim Post
Ray & Peg Prag
Richard Prusz
Dana Abel & Charles Quinn
Julie Reilly
Wes Reynolds
Doug Ripley
Mary Ann Roberts
Wayne Rolle
Ben Ross
Rachel Schwindt
Thomas Seddon
Corrine Sherton & Dennis Martin
Alice Smith
James P. Smith
Veva Stansell
Kareen Sturgeon
Mildred Thiele
Karl & Elaine Urban
Kelli Van Norman & Jack Maynard
Josephine Von Hippel
Jeff Walker
Luise Walker
Faith Wilkins
Margi Willowmoon
Mary Alice Willson
L.K. Wright
Jo Ann Yeager
Walt Yungen

Erythronium oregonum

INTRODUCTION

HISTORY OF THE OREGON FLORA PROJECT

Linda K. Hardison

"Thomas Jefferson told Lewis and Clark to go out there and identify everything. Well, it's been two hundred years, and it is high time we had an inventory. In a way, it is fitting that this should be such a grassroots sort of thing. Instead of a huge agency that could have put a few million into a Flora Project, but hasn't and won't, it is heartening to see that many dedicated people will get the job done."

~Jerry Igo, naturalist; Mosier, OR

The Oregon Flora Project (OFP) was created in 1994 to meet a long-standing need for a comprehensive, modern flora for the state of Oregon. At that time, the most recent synopsis was Morton Peck's *A Manual of the Higher Plants of Oregon*, originally published in 1941 and updated in 1961. The *Flora of Northwest America* (1897–1903) by Thomas Jefferson Howell was the first floristic work that treated the entire state of Oregon.

In 1994, the closing of the herbarium at the University of Oregon (UO) and the transfer of specimens from Eugene to the Oregon State University (OSU) Herbarium in Corvallis was the catalyst for initiating a flora-writing effort. A grant from the National Science Foundation (NSF) funded the move and integration of the UO collection, which created the largest collection of Oregon vascular plant specimens in the state. However, it was the personal initiative of Scott Sundberg to leverage the centralized resource created by the reorganization of Oregon's two largest herbaria into an opportunity to prepare a new flora. As he combined the 66,831 UO vascular plant specimens with the OSU Herbarium holdings, he began to update the nomenclature and prepare a current checklist of all native and exotic vascular plants of the state. He concurrently sought collaborations with cartographers and computer scientists at OSU to develop the digital capacities of a contemporary flora. In this setting, the Oregon Flora Project was formed.

MARK SUNDBERG

Scott Sundberg

INSTITUTIONAL ORGANIZATION

The Oregon Flora Project is based in the Department of Botany & Plant Pathology at OSU. Since its inception, the OFP has been intimately aligned with, but independent of, the OSU Herbarium. The Department of Botany & Plant Pathology finances indirect costs and provides office space, infrastructure, and access to the Herbarium. Competitive grants and charitable donations fund all salaries and direct operating expenses. Recognizing the need for broad public engagement in the project, Sundberg initially approached the Native Plant Society of Oregon (NPSO) for support in launching the OFP. The Society enthusiastically responded by partnering with OFP and promoting their shared mission to expand knowledge of Oregon's plants. The NPSO continues to provide consistent financial backing and engages its membership in supporting the activities of the OFP.

Sundberg recognized that for a contemporary flora to maintain its relevance and reflect ongoing taxonomic research, it would need to present information in a digital format. He saw that online communication of floristic data could eliminate many of the constraints of paper publication and allow customized data retrieval and presentation. Making data available online would also help disseminate the information to a wider audience, which has always been an underlying goal. The OFP, therefore, was charged with producing dynamic digital content in addition to a traditional paper-bound flora.

Troy Maddux

Linda Hardison

Beginnings

A synonymized checklist serves as the taxonomic framework for Oregon plants and provides a foundation for the entire project. The checklist was initiated by drawing from existing datasets. The most substantive contribution was a database of the vascular plants of US Forest Service Region 6 (Pacific Northwest) compiled by Karl Urban; in 1989 Urban submitted it to Kenton Chambers as part of his master's thesis. Smaller, regional checklists from individuals and governmental agencies were also assessed.

An early partnership with the Northwest Alliance for Computational Science & Engineering (NACSE) at OSU was instrumental in developing a robust database structure that could accommodate the large amounts of information anticipated during the project. In collaboration with NACSE, the OFP received an NSF grant (2001–2004) to develop a "Personal Digital Field Guide," designed to enable users to select and download subsets of project data to a personal computer or mobile device. This forward-thinking proposal presaged the technological capacity that now exists. Though the Personal Digital Field Guide as initially conceived was not completed, the partnership developed components of the Field Guide and a dataset of morphological characters that would support interactive web tools.

The Oregon Vascular Plant Checklist

To develop the Oregon Vascular Plant Checklist, Sundberg organized a team of participants including eight project leaders, an advisory board composed of ten professional and amateur botanists from throughout the state, invited contributors, student assistants, and volunteers. For each accepted taxon, the contributor designated the location of a voucher specimen, preferably the OSU Herbarium. An estimated 75% of the Oregon specimens in the OSU Herbarium have been annotated since 1994 as a result of the preparation of the Checklist and *Flora* manuscripts. The Checklist reflects the decisions of checklist treatment authors regarding the delimitation of taxonomic concepts, with special attention to synonymy. Together with the accepted name of each terminal taxon, the Checklist also lists the nomenclature used in nine floras covering parts or all of Oregon (referred to as *key references*; see page 4). The Checklist was completed and published online in January 2011, and updated versions of the Checklist are published annually, serving as mileposts in the dynamic field of botanical nomenclature.

Digital Components of the Flora

The initiative to produce an online atlas of Oregon vascular plants began in early 1995 as a partnership with the NPSO. Sundberg's intention was to integrate the contributions of amateur botanists and to promote citizen science. Complementing the *Flora of Oregon* text, the interactive Oregon Plant Atlas lets users select taxa and display their distribution on a map, with the ability to view map overlays and details of each occurrence. Data sources include herbarium specimen labels, unvouchered observations from lists prepared for academic and governmental agency studies, other checklists, and the Oregon Biodiversity Information Center (formerly Oregon Natural Heritage Information Center) database of rare plant occurrences.

The Atlas received a significant boost from a Bureau of Land Management grant to database a specimen record for each taxon found in each Oregon county. This funding also allowed the OFP to hire its first staff members, Ann Willyard and Thea Jaster Cook. Significant effort went into designing data entry forms to minimize error and immediately reflect taxonomic decisions. Once those forms were in place, the OFP launched a large data entry effort in 2002, and an enthusiastic crew of 20 student and volunteer workers entered more than 30,000 records. Further adding to the specimen data, the OSU Herbarium and OFP collaborated on an NSF proposal (2003–2006) to database and georeference the Oregon herbarium specimens not yet included in the Atlas. Katie Mitchell was hired in 2004 to assist with this effort, which culminated in the databasing of all Oregon

Thea Jaster

Stephen Meyers

vascular plant specimens—now approximately 149,400—in the OSU herbarium (OSC, ORE, and WILLU). In 2005, the Oregon Plant Atlas was launched on the OFP website as the first major digital component of the Flora.

The Photo Gallery was the second major digital component of the Flora. The gallery started with field photographs of plants from the OSU Herbarium slide collection, which were digitized, catalogued, and databased. The OFP then solicited public contributions and in 2002 also began photographing representative herbarium specimens for each taxon. Rena Schlachter was employed as OFP illustrations editor from 2004–2006 to manage the images dataset and provide other artwork. The Photo Gallery was launched on the OFP website in June 2009.

In 2007, the *Oregon Rare Plant Guide*, which previously existed in paper form, was converted to digital format and published on the OFP website. The US Forest Service funded the work to digitize and expand the number of taxa treated in the guide, which was first developed in 2002 for Willamette Industries to help with rapid identification of sensitive plant taxa in the field. Jennifer Sackinger was hired as a staff member to assist with this initiative.

In all, five databases were created at various times to manage all the OFP data needed for a searchable digital flora: checklist, specimen, observation, image, and morphology. In 2011, all five were consolidated into a single PostgreSQL database.

Transitions

The OFP suffered a great loss in December 2004 with the death of its director and founder, Scott Sundberg. Existing staff members maintained their essential roles in the functioning of the Project, and Linda Hardison assumed the position of director in 2005. Some of the taxonomic decisions that had previously been made by Sundberg (e.g., in the preparation of the Checklist) were now accomplished by consensus of project participants.

In June 2008, the OFP was halted due to lack of funds. Through the timely and generous support of the John and Betty Soreng Environmental Fund of The Oregon Community Foundation, all staff members were rehired and the Project resumed operations in the autumn of 2008. The sustained support of this fund has enabled the OFP to extend public access to every digital facet of the project, including the Photo Gallery in 2009, version 2.0 of the Atlas in January 2010, and the Checklist in January 2011.

An eleven-member advisory board was formed in 2011, representing the diverse stakeholders interested in OFP's offerings. Representatives from governmental agencies, academia, K–12 education, and computer science participate.

The *Flora of Oregon* Book

With all major digital components of the Flora publicly available and the taxonomic foundation of the Checklist complete, work began in earnest on the manuscript of the *Flora of Oregon*. In July 2010, Dr. Stephen Meyers was hired as taxonomic director. His association with OFP began in 2001 as a post-baccalaureate student worker. His understanding of Sundberg's vision for the OFP and his intimate knowledge of the dataset made him well suited for the position. Artist John Myers joined the OFP in August 2012 to provide original illustrations for the *Flora*, and to implement an agreement that permits the exchange of artwork with the Flora of North America Association.

Looking Ahead

While progress on the *Flora of Oregon* text has recently been a priority for OFP, numerous other initiatives are concurrently underway. A 21st century implementation of the Personal Digital Field Guide concept was accomplished with the release in spring 2014 of a wildflower identification app for mobile devices. A project promoting the use of native plant taxa in the horticultural and landscaping industry was funded by the Oregon Department of Agriculture in autumn 2014.

Finding improved ways to meaningfully present data on the OFP website presents continuous challenges, and ensuring the sustainability of the OFP and its ability to keep data current and accessible are key priorities. As Scott Sundberg's vision is realized, it will be exciting to witness the future trajectory and accomplishments of the Oregon Flora Project.

L to R: Tanya Harvey, David Henken, Stephen Talbot

Thea Jaster

Katie Mitchell

John Myers

PHILOSOPHY OF TAXONOMIC APPROACH

Stephen C. Meyers

Volume 1 of the *Flora of Oregon* has been written with the goal of being accessible to a wide array of botanists and plant enthusiasts. The users of this manual include academic botanists, employees of government agencies, ecologists, gardeners, native plant growers, environmental consultants, amateur botanists, and students. To meet the needs of this diverse audience, we have made a concerted effort to present floristic treatments that are scholarly and scientific yet also readable.

The authors and reviewers of the floristic treatments and accompanying chapters range from well-established academic scholars and post-doctoral researchers to graduate students and skilled amateur botanists. Though this diversity has the potential to create problems with consistency across the *Flora*, we feel that these discordances ultimately provide value by presenting varied perspectives.

Broadly speaking, practitioners of taxonomy have always been cast into one of two opposing philosophical camps: splitters or lumpers. In reality, most competent taxonomists fall somewhere in the middle of these two extremes, sometimes splitting taxonomic concepts and sometimes lumping them, according to the logical course suggested by the data.

Some plant groups suffer from a history of inadequate or poor taxonomic work. Compounding this problem, some modern taxonomists have been reluctant to revise the conclusions of the "experts" who preceded them. Seeking to partially rectify this, we have strongly encouraged treatment authors to revise the taxonomy of their groups to reflect their research results, using synonymization, the application of previously described taxonomic concepts, and the description of new taxa.

Some authors recognize taxa that are either cryptic or have few or seemingly vague defining morphological characteristics. Although this can make a plant group appear unduly complex, it is helpful to remember that the characteristics that place a taxon on a distinct evolutionary path are not necessarily evident to human senses.

Whenever possible, we recruit authors with expertise in specific plant groups. Unfortunately, over the past several decades the number of active, university-trained taxonomists has vastly declined. As a result, we have occasionally used authors whose assigned plant group is not the focus of their work. While this unfamiliarity sometimes requires the author to "catch up" with the relevant literature, it also brings the benefit of a fresh perspective on the group. Some authors expressed a desire to devote more time and effort to the study of their assigned plant groups. This was particularly the case with complex groups and questionable taxonomic concepts for which 21st century molecular techniques could help answer longstanding questions. Unfortunately, a dearth of time and financial resources often constrained the authors' admirable enthusiasm. As a result, a hint of some authors' desire is often reflected in the *Flora* by the statement "more study is needed."

The Oregon Flora Project refers to the nine floras and field guides that cover all or parts of the state as *key references*. Morton Peck's *A Manual of the Higher Plants of Oregon* (1961) is the most recent and complete regional flora. The remaining key references are: *Flora of North America North of Mexico* (Flora of North America Editorial Committee, eds. 1993+), *Flora of the Pacific Northwest* (Hitchcock and Cronquist 1973), *Guide to the Plants of the Wallowa Mountains of Northeastern Oregon* (Mason 1975), *Illustrated Flora of the Pacific States* (Abrams and Ferris 1923–1960), *Intermountain Flora: Vascular Plants of the Intermountain West, USA* (Cronquist et al. 1972–2013), *The Jepson Manual: Higher Plants of California* (Hickman 1993), *The Jepson Manual: Higher Plants of California, second edition* (Baldwin et al. 2012), and *Vascular Plants of the Pacific Northwest* (Hitchcock et al. 1955–1969).

Ann Willyard

Rena Schlachter

Jennifer Sackinger

L to R: Gene Willyard, Stephen Meyers, Doug Sackinger

DESIGN AND CONTENT

SCOPE OF WORK

The *Flora of Oregon* is a three-volume reference publication circumscribing the vascular plants of Oregon that occur in Oregon's uncultivated, natural habitat. It covers Oregon's 254,799 km^2 (98,379 mi^2), bounded on its southern border by the 42nd parallel; its eastern and northern borders predominately by the Snake and Columbia Rivers, respectively; and its western border by the Pacific Ocean. Additional information supporting this publication can be found at the Oregon Flora Project (OFP) website (www.oregonflora.org).

Volume 1 includes the ferns and fern allies (pteridophytes), gymnosperms, and monocots; *Volume 2a* and *Volume 2b* will cover the dicots. The nomenclatural status of dicots referenced in this volume is based on currently existing floristic treatments for *Volume 2a* and *Volume 2b* of the *Flora of Oregon* or, if not available, on the Oregon Flora Project Checklist Version 1.4, dated April 2014.

The *Flora of Oregon* includes both native and exotic (non-native) plant taxa. A *native* plant taxon is defined as a floristic component of Oregon prior to European settlement, or a newly described taxon found in the state. An *exotic* or non-native plant is a floristic component that established in Oregon post-European settlement. The OFP recognizes eleven categories of exotic plants (described on the OFP website) and has assigned each non-native taxon to one of these categories. The categories fall into two broad groups: 1) naturalized species (exotic taxa that have persisted in the wild for at least 3–5 years and have spread beyond the area of their original introduction); and 2) non-naturalized species that are persisting or spreading within a humanly altered environment. Examples include a) weeds and b) ephemerals such as waifs (solitary or small groups of alien plants not documented as having persisted beyond one season) and ballast plants (waifs that have appeared on ship ballast).

Native and naturalized exotic taxa accepted by the OFP receive full descriptions. An *accepted taxon* is one with documented occurrences in Oregon and at least one *flora voucher*, that is, a herbarium specimen deposited in the Oregon State University Herbarium (OSC, ORE, and WILLU) or in some cases at another institution (Figure 1). Frequently occurring interspecific hybrids and those that form persistent, self-maintaining populations also receive full descriptions. Rarely occurring hybrids and historic or waif taxa are generally discussed but usually not given full treatments. Waifs discussed in the *Flora of Oregon* include ballast plants, taxa recently escaped from cultivation or otherwise introduced, and taxa from adjacent states that have been reported from a single Oregon location or plant population.

Table 1 lists statistics for all accepted vascular plants of Oregon. A *terminal taxon* refers to the narrowest taxonomic category recognized by the OFP; a terminal taxon may be a species, subspecies, variety, or forma. An *endemic* is a native terminal taxon whose occurrence has not been documented in adjacent states. At the time of publication of *Volume 1*, there are 4,653 terminal taxa that are presently occurring within the state or that are presumed extirpated. This represents a 6% increase from the 4,378 taxa listed in the initial drafts of the Oregon Flora Project Checklist produced in 1995. The increase is due primarily to the range extension of plants from adjacent states into Oregon, the introduction of exotic species, and taxonomic revisions. *Volume 1* contains full descriptions of 1,054 pteridophyte, gymnosperm, and monocot terminal taxa.

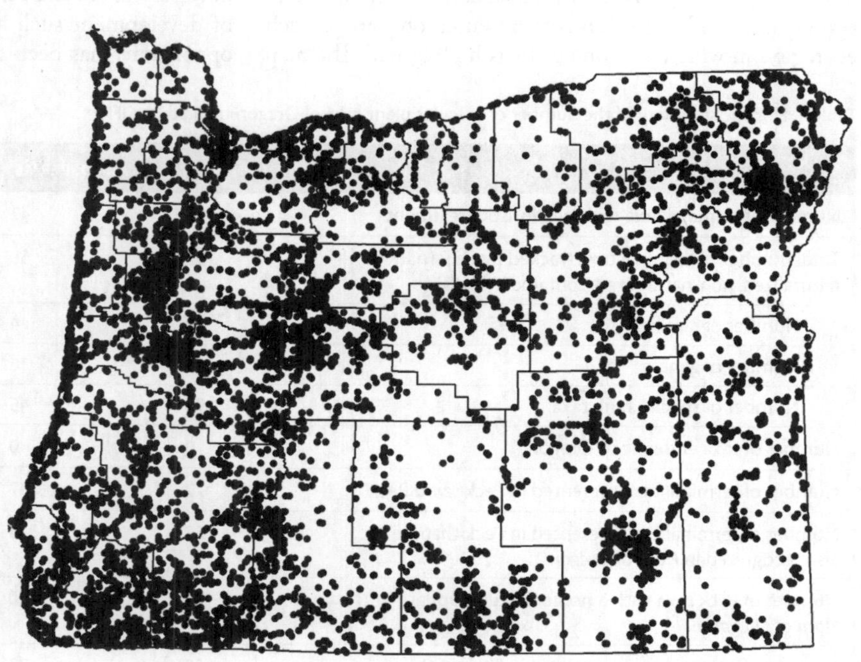

Figure 1. Map depicting the distribution of flora vouchers for all accepted vascular plant taxa.

Before the publication of the *Flora of Oregon*, the most recent comprehensive treatment of the state's vascular plants was Morton Peck's *A Manual of the Higher Plants of Oregon*, which was last updated in 1961. Table 1 compares the scope of Peck's *Manual* to that of the *Flora of Oregon*, serving as a reminder of the advances in botanical studies that have transpired in the last half-century.

CONVENTIONS USED

Book Design

Organization

The introductory material is followed by identification keys and detailed descriptions of the three major groups of vascular plants (pteridophytes, gymnosperms, and monocots) treated within this volume. There is a key to the families within each major group. The content for each major group—organized by family, the genera within each family, the species within each genus, and the infraspecific taxa within each species—is arranged alphabetically.

Illustrations

Each genus includes an illustration of at least one species, with few exceptions. For genera with more than four species, typically 20–25% of the species are illustrated. In most cases, illustrations represent the plant's habit, flower, and specific details that aid identification. If a taxon is illustrated, the page number for the drawing appears in parentheses to the right of the scientific name in the description.

Maps

The distribution of each terminal taxon within each Oregon county and ecoregion is represented on a map. Each ecoregion in which a taxon occurs is highlighted. The map also shows plant occurrence records reflecting flora vouchers. The identification of the majority of these specimens was confirmed for the OFP. Details of each record are available from the Oregon Plant Atlas on the OFP website.

Measurements

Measurements are given in metric units, with metric conversion scales at the end of each volume.

Index

The index is a combined alphabetical listing of all common and scientific names of families, genera, species, infraspecific epithets, and synonyms within the floristic treatments.

Keys

Dichotomous keys are provided to identify unknown plants to the level of family, genus, and terminal taxon. A *dichotomous key* is a series of paired, parallel statements. A pair of statements is referred to as a couplet; each statement is referred to as a lead. Of the paired statements, the user selects the statement that most closely matches the characters of the unknown plant and rejects the alternative statement. The selected statement ends either in a scientific name—the identity of the unknown plant—or directs the user to another couplet. The user repeats the selection process until the plant's name is obtained. The user can then compare the plant's features to those in the floristic description to confirm that the identification deduced from keying the specimen is correct.

The characters within each lead are predominately morphological features discernible with the unaided eye or a hand lens. Readily evident features are listed early in the leads. Habitat and distribution information is used when necessary, as are character suites that require multiple phases of development such as mature fruits and flowers. Every opportunity has been taken to develop lead pairs that are

Table 1. Floristic summary of the number of vascular plant taxa of Oregon tracked by OFP.

	Pteridophytes	Gymnosperms	Monocots	Dicots	Total
Number of terminal taxa with full floristic treatment	92	42	920	—	—
Total number of terminal taxa tracked by OFP (native, naturalized, non-naturalized, sporadic hybrids)	102	50	1,000	3,501	4,653
Number of families	18	4	27	111	160
Number of genera	33	16	204	780	1,033
Number of species rank taxa	96	45	915	3,040	4,096
Number of taxa endemic to Oregon	0	0	20	95	115
Number of terminal taxa presented in Peck *sensu* Peck	72	37	808	3,057	3,974
Number of terminal taxa presented in Peck that are also recognized in *Flora of Oregon*	67	34	721	2,865	3,687
Number of Peck taxa with a nomenclatural update in *Flora of Oregon*	31	10	360	1,554	1,955
Number of terminal taxa presented in *Flora of Oregon* that are also recognized in Peck *sensu* OFP	65	32	617	2,424	3,138

parallel; that is, both leads ask questions about the same characters. When a critical character is unique to one of the leads in a couplet, that character is put at the end of the lead and all other elements remain in parallel. In highly diverse taxonomic groups, multiple leads may direct users to the same group in different sections; such sections of a taxonomic group will be designated "in part" in the key. All characters used in the keys appear in the descriptions.

Descriptions

All Descriptions

Descriptions at all taxonomic levels delimit the features of each taxon as it occurs in Oregon. The content of each description applies to all taxonomic groups hierarchically below the one discussed; that is, the characters used to describe a family are exhibited by all described genera of that family, and a genus description applies to all described species within the genus. The general format described below is followed for all levels of descriptions.

Each treatment begins with the accepted scientific name in boldface type, followed by the common name. If there are multiple common names, the first name listed is designated by the OFP as the *primary common name*. The primary common name is preferentially a published regional convention; alternatively it may be a translation of the scientific name or a name provided by the treatment author that describes a distinctive morphological feature. Each description begins with the general habit of the plant and its persistence (annual, biennial, or perennial). Characters generally are listed in the following order: from the ground up; as they appear during the plant's development; from proximal to distal; abaxial before adaxial; and staminate before pistillate. Each major structure or organ is in boldface type and followed by specific details in the following order: number, odor, shape, dimensions, color, texture, base, peripheral region, central area, apex, surfaces, internal parts, and exudates.

The expected quantity or measurement of a feature is described by a discrete value or range of values. Any extreme values that are uncommonly encountered are placed in parentheses. Two-dimensional measurements are first cited by length, then by width. For example, (5)7–9.5 × 2–3 mm indicates the length is normally 7–9.5 mm but may be as small as 5 mm, while the range of the width is 2–3 mm.

The morphological description is followed by information about range, habitat, and number of subsidiary taxa that are fully treated in the *Flora*. General discussion may follow.

Family Descriptions

Each family treatment begins with the family's scientific name in all caps and boldface type, followed by its common name and acknowledgment of the floristic treatment's contributor. Family names reflect the nomenclature established by the Angiosperm Phylogeny Group III, or "APG3" (Stevens 2001 onwards) with some exceptions. If members of the family were previously classified within a different family, the original family name is also referenced.

The range and global geographic distribution of the family, its habitat, number of genera worldwide, and the number of genera that are fully treated in the *Flora* are given. When the habitat of family member taxa varies widely, details are given at a lower taxonomic level. When a family has a single genus worldwide, the family description serves as the generic description.

Genus Descriptions

The conventions found in genus descriptions follow those used in family treatments. If a genus is comprised of a single species, a genus description is not included, as the details in the species description suffice.

Species Descriptions

Each species description begins with the scientific name in boldface type, followed by the publishing author and the common name(s). Below this are listed the synonyms, if any. The morphological description is followed by a paragraph describing the plant's habitat and range. Flowering time and elevation range are based on specimen label data from OSU Herbarium vouchers or information provided by the treatment contributor. Minimum elevation values are rounded down and maximum values are rounded up in 100 m increments; 50 m increments are used if the species is limited to areas with an elevation less than 100 m. The Oregon ecoregions in which the taxon has a documented occurrence are listed; ecoregion abbreviations are defined below. This is followed by information about the species' distribution beyond Oregon, first listing adjacent states and then North American and global occurrences. The species' origin (native or exotic) concludes this section. Taxa designated as endemic are indicated with a ❧ symbol; also see Appendix 2.

When multiple infraspecific taxa are recognized, a key to the subtaxa follows the species description. If a single subspecies or variety occurs in Oregon, it appears in the text as if a full species.

ABBREVIATIONS & SYMBOLS

General

2*n*	diploid chromosome number
ca.	circa; approximately
cm	centimeter
ft	feet
in	inch
km	kilometer
m	meter
mi	mile
mm	millimeter
µm	micron or micrometer (0.001 mm)
sp.	species (singular)
spp.	species (plural)
ssp.	subspecies (singular)
sspp.	subspecies (plural)
var.	variety

Ecoregions

BR	Basin and Range
BW	Blue and Wallowa Mountains
Casc	Cascades
Col	Columbia Basin
CR	Coast Range
ECas	Eastern Cascade Slopes and Foothills
Est	Estuarine Coast
Lava	Lava Plains
Owy	Owyhee Uplands
Sisk	Siskiyou Mountains
WV	Willamette Valley

Institutions

BLM	Bureau of Land Management
CAS	California Academy of Sciences Herbarium. San Francisco, CA
GH	Gray Herbarium, Harvard University. Cambridge, MA
MO	Missouri Botanical Garden Herbarium. St. Louis, MO
NPSO	Native Plant Society of Oregon
ODFW	Oregon Department of Fish and Wildlife
ODSL	Oregon Department of State Lands
OFP	Oregon Flora Project
OPRD	Oregon Parks and Recreation Department
ORBIC	Oregon Biodiversity Information Center
ORE	University of Oregon, Oregon State University Herbarium. Corvallis, OR
OSC	Oregon State University Herbarium. Corvallis, OR
OSU	Oregon State University. Corvallis, OR
RNA	Research Natural Area
RSA	Rancho Santa Ana Botanic Garden Herbarium. Claremont, CA
SOC	Southern Oregon University Herbarium. Ashland, OR
TNC	The Nature Conservancy
UO	University of Oregon. Eugene, OR
USFWS	United States Fish and Wildlife Service
USFS	United States Forest Service
US	Smithsonian Institution Herbarium. Washington, DC
UTC	Utah State University Herbarium. Logan, UT
WILLU	Willamette University, Oregon State University Herbarium. Corvallis, OR
WS	Washington State University Herbarium. Pullman, WA
WTU	University of Washington Herbarium. Seattle, WA

Symbols

❦	taxon is endemic to Oregon
♀	female
♂	male
°	degrees
×	multiplication; used for range of measurements and to indicate hybrid taxon
×	multiplication; used for amount of magnification

States

AL	Alabama
AK	Alaska
AZ	Arizona
AR	Arkansas
CA	California
CO	Colorado
CT	Connecticut
DC	District of Columbia
DE	Delaware
FL	Florida
GA	Georgia
HI	Hawaii
ID	Idaho
IL	Illinois
IN	Indiana
IA	Iowa
KS	Kansas
KY	Kentucky
LA	Louisiana
ME	Maine
MD	Maryland
MA	Massachusetts
MI	Michigan
MN	Minnesota
MS	Mississippi
MO	Missouri
MT	Montana
NE	Nebraska
NV	Nevada
NH	New Hampshire
NJ	New Jersey
MN	New Mexico
NY	New York
NC	North Carolina
ND	North Dakota
OH	Ohio
OK	Oklahoma
PA	Pennsylvania
RI	Rhode Island
SC	South Carolina
SD	South Dakota
TN	Tennessee
TX	Texas
US	United States
UT	Utah
VT	Vermont
VA	Virginia
WA	Washington
WV	West Virginia
WI	Wisconsin
WY	Wyoming

NOTABLE OREGON BOTANISTS

Linda K. Hardison

INTRODUCTION

Oregon Country has long been a destination for explorers. Unlike earlier forays that focused primarily on trade and imperialism, exploration in the 19th century placed a greater emphasis on scientific discovery. Plants new to the scientific record were catalogued and described, and the names given to many of these taxa honored the botanists involved in their discovery. During the first half of the 1800s, Lewis and Clark (1804–1805), David Douglas (1825–1827), and Thomas Nuttall (1834–1836) led expeditions in Oregon that were relatively brief visits of intense exploration. In contrast, the individuals highlighted in this chapter lived in Oregon for much of their lives. Their explorations spanned decades and benefited from an intimate knowledge of place that can only be acquired over time. Notables William Cusick, Thomas Jefferson Howell, and Louis Henderson were also farmers, an occupation that demands awareness of one's natural surroundings.

Most of the contemporary botanists (and several of the historic figures) have been teachers and mentors. Their ability to inspire students has impacted botanical research and informed plant conservation in Oregon and beyond. All of them have shown dedication and perseverance in their pursuit of knowledge, even as some overcame obstacles of financial duress and failing health.

The lives of the men and women featured in this chapter span 173 years. In addition to any written records they produced, the plant specimens each collected serve as a valuable documentation of the plant communities that were encountered. Many of these historic specimens were deposited in Oregon's prominent regional herbaria—those at Oregon State College (which in 1961 became Oregon State University), University of Oregon, and Willamette University. Specimen accession numbers continue to reflect the originating herbarium (OSC, ORE, or WILLU, respectively), but currently all collections are housed within the Oregon State University Herbarium in Corvallis. The WILLU collection is on permanent loan to Oregon State University (OSU) and is filed independently; the University of Oregon (UO) specimens were integrated with those of OSU in 1994.

Whether farmer or academic, celebrated or obscure, all the men and women profiled share an abiding curiosity about the natural world. Each of them has pursued their studies with admirable commitment and in doing so they have influenced the course of botanical knowledge in Oregon. We are indebted to their contributions of intellect and passion.

HISTORIC

William C. Cusick (1842–1922)

William Conklin Cusick was a pioneer in his personal and professional life. As an eleven-year-old, he walked from Illinois to Oregon's Willamette Valley with a wagon train. At the age of thirty, William was one of the earliest settlers of the Powder River Valley in eastern Oregon. During the next forty years, despite nearing complete loss of his hearing and eyesight, Cusick undertook the pioneering botanical explorations of the remote Wallowa and Blue Mountains.

William Cusick, the oldest child of Robert G. and Sarah H. Cusick, moved with his extended family to farm a 320-acre claim in the Willamette Valley, near Kingston, Linn County. Even as a boy, Cusick was interested in botany and noted flowers he saw during his westward journey. He attended public schools in Illinois and Oregon, receiving his higher education through the (now closed) La Creole Academy in Polk County and later at Willamette University in Salem, where he studied math, physics, and geology.

In 1865, near the end of the Civil War, 23-year-old Cusick joined the Union Army as a volunteer and was originally stationed in Lapwai, Idaho, as sergeant in the 1st Oregon Infantry. During this time, Cusick sent away to Portland for his first botanical text—Asa Gray's *First Lessons in Botany*. Cusick was discharged from the Army with a small pension in 1866 and settled near Salem to teach. However, at the age of 24, his hearing began to fail, and he was forced to abandon his profession.

Cusick and his younger brother, Samuel Franklin, moved in 1872 to the Powder River Valley to start adjoining ranches in Thief Valley of Baker County. There, Cusick encountered Dr. Reuben D. Nevius, a minister-botanist who taught him how to collect and document scientifically useful specimens and where to send specimens for identification. Nevius undoubtedly told Cusick that pressed plants could be sold to eastern US and European herbaria and wealthy private collectors. Records at the Gray Herbarium at Harvard show that Cusick began sending plants, including many undescribed taxa, to Asa Gray for identification. This piqued the interest of Gray who sent his assistant, Sereno Watson, to visit Cusick in November of 1880.

Cusick's time was largely consumed with maintaining his ranch. However, between 1880 and 1885, he took brief botanical forays whenever possible to the Blues, Wallowas, Imnaha River, and Steens Mountain. In 1886, he dedicated a full season to collecting in the southern Wallowa Mountains and Malheur County—a notable accomplishment considering he traveled alone while dealing with hearing loss and the lingering effects of pleurisy. During these trips, "Cusick never carried fire arms or fishing tackle as he felt that when unmolested the wild animals would not trouble him, and that he was too busy with his botanical collecting to waste any time in fishing."

For unknown reasons, in 1887 the Cusick brothers moved from Thief Valley to land on Jimmy Creek in Union County. In 1892, William married Emma A. Alger, a widow, who died only four months later in February 1893. Cusick took responsibility for her two sons, aged 9 and 18, and did almost no botanizing from 1892 through 1895. His vision may have begun to fail about this time.

William Cusick at around age 35 when he first began sending plant specimens to Asa Gray for identification. Courtesy Albert and Carrie Sweetser Papers, Ax 075, Box 1, Special Collections and University Archives, University of Oregon Libraries.

In 1896, Charles Vancouver Piper, recently hired botany professor at Washington State College, initiated correspondence with Cusick, and the two men made a week-long collecting trip into the Wallowas. Piper seemed to have motivated Cusick to resume collecting, and during the subsequent five years, Cusick botanized much of eastern Oregon and portions of Idaho and Nevada. In 1902, Cusick, at 60 years old, undertook the longest collecting trip of his career accompanied by his nephew, George. The three-month itinerary in the Klamath region, Siskiyous, Cascades, and central Oregon was well documented in George's daily journal.

Contrary to contemporary scientific thought, Cusick's study of geology and knowledge of plants of the Blue and Wallowa Mountains convinced him that the ranges had distinct origins and unique floras. He continued to collect in those regions, gathering information to test his theory. In 1910, Cusick, at age 68, traveled to the University of Oregon Herbarium in Eugene to conduct further research.

Shortly after this visit, Cusick's herbarium, consisting of 10,000 sheets, was purchased by UO in 1911, doubling their collection. While the purchase price is unknown, it no doubt provided much-needed income for the impoverished botanist.

Cusick moved in 1913 to the Oregon State Soldiers' Home in Roseburg and began studying the flora of Douglas County. This was his final major botanical undertaking. He wrote, "I am 71 years old, nearly blind with cataracts of the eyes and quite deaf." He spent part of three years collecting hundreds of new specimens, which he eventually sold to Washington State College. In 1915, Cusick returned to eastern Oregon to live with his brother's family on the outskirts of Union. Despite several surgeries, his vision was reduced to partial sight in one eye. In 1916, perhaps not realizing the extent of Cusick's handicap, Piper asked for Cusick's help in compiling a flora of the Blue Mountains. Over the next five years, Cusick twice accompanied Piper and a colleague into the Blues, where he drew upon his unerring memory to recall plant localities for the project.

Continued failing health left Cusick unable to study his botanical collections, and he sold approximately 6,000 sheets to Washington State College in 1922. He died in October of that year at his brother's home and was buried in Union County. His contributions to the knowledge of the flora of the Blue Mountains were appreciated both nationally and internationally, and over 60 taxa were named in his honor.

~Adapted from Love, Rhoda M. 2007. *Kalmiopsis* 14:8–16.

Thomas J. Howell (1842–1912)

Thomas Jefferson Howell was born in Cooper County, Missouri, on October 8, 1842, the youngest of five children of Benjamin and Elizabeth (Matthews) Howell. The family did not wish to live in a slave state, so in 1850 they joined others in a small wagon train that left Missouri in April and arrived in Oregon in October. The family first settled in Hillsboro, then moved in 1851 to Sauvie Island. Although Thomas's father was trained as a physician, he did not practice medicine but instead assumed possession of a 240-acre land claim on the island in 1853, which he and his three sons further cleared and farmed.

Thomas Howell's formal education consisted of only three months in 1855 at the first school built on the island. Otherwise, he and his brothers were self-taught via reading, with help from their father. As a youngster, Thomas became interested in learning the names of plants that grew wild on Sauvie Island, and he began collecting and describing plants. He developed a strong interest in the science of botany, at the same time losing enthusiasm for farming. In 1877, he published a 22-page *Catalogue of the Flora of Oregon, Washington, and Idaho*. The work also served as advertisement for the pressed and living plants that he sold. Over the next ten years he published additional plant listings, including the 28-page *A Catalogue of the Known Plants (Phaenogamia and Pteridophyta) of Oregon, Washington, and Idaho*, which listed 2,379 taxa.

Howell was impoverished throughout his life and took various jobs in his struggle to support his family: he was a farmer, postmaster, grocer, and pieceworker who assembled teamsters' mittens with a sewing machine. The sale of pressed specimens and living plants was an important source of income for Howell. Depending upon the quantity ordered, he sold specimens for four to ten cents apiece. Correspondence in the early 1890s with E.L. Greene, head of the Botany Department at University of California at Berkeley, indicates Greene sought to buy herbarium specimens from Howell; his requests were countered with Howell's insistence on fair compensation for his effort and expertise. Because replies from Greene to Howell have not survived, it is not known if Greene contracted with Howell to collect plants.

Howell lacked references and herbarium specimens with which to identify his collections and sent his specimens to botanists elsewhere for identification. His coterie of identifiers included George Vasey of the U.S. Department of Agriculture in Washington, D.C., L.H. Bailey at Cornell University, and Asa Gray and Sereno Watson, both at Harvard University. Through time, Howell developed an intimate knowledge of the Northwest flora through his extensive field work and from his interactions with fellow botanists. He collected with Louis Henderson and Portland botanist Martin Gorman; he also corresponded with Gray, Watson, W.N. Suksdorf, E.L. Greene, Willis Jepson,

Thomas Jefferson Howell in 1910, proudly displaying a copy of his complete, bound *Flora of Northwest America*. Courtesy Albert and Carrie Sweetser Papers, Ax 075, bound papers 7, Special Collections and University Archives, University of Oregon Libraries.

and others. By 1887, Howell had begun naming plants on his own: "I had these on hand and could not distribute them until they were named; and I have to sell all I can to pay the very heavy expense of collecting in this country."

After years of gathering information for a flora, Howell began writing in 1882. The first fascicle (a literary term meaning material published separately as well as a botanical term referring to a cluster of flowers or leaves) was published in 1897. Financial hardships dogged Howell throughout his efforts to publish, and a breach of contract delayed printing of the first fascicle: "This leaves me on the verge of despair for the manuscript that I have represents ten years work of the best part of my life and to loose [*sic*] it now looks to me like throwing away life itself." Portland printers apparently were unwilling to cope with the technical terminology of the flora and experienced difficulty with Howell's often-illegible handwriting, so Howell himself set the type at home in sets of eight pages, which he then took to a printer.

The second fascicle was published April 1, 1898, and subsequent fascicles followed at irregular intervals. The seventh and last section was published August 10, 1903, with a print run of 1,000 copies. In total, Howell's magnum opus, *A Flora of Northwest America* (1897–1903), consisted of 792 pages and described 3,150 species, of which 89 were newly described by Howell.

Louis Henderson collecting along the Yukon River, Alaska, 1932. His journal comments on the voracious mosquitoes and flies he encountered. Courtesy University Photograph Collection, UA REF 3, Box 61, Special Collections and University Archives, University of Oregon Libraries.

On December 3, 1912, Thomas Howell died at the age of 70 in Woodstock, Oregon, after a long illness. He is buried in the family plot in Vancouver, Washington. After Howell's death, Greene wrote that Howell "accomplished the greatest amount of meritorious and valuable scientific work that was ever done by any man of any epoch, on so very rudimentary an education in letters." Jepson added that Howell had "organized diagnoses of genera and species scattered in the works of many writers into a pioneer flora which, considering the circumstances of its production, is balanced, judicious, and highly useful. Few men leaving so durable a contribution to American botany have led so obscure an existence as did Howell."

Howell will be remembered as one who advanced the study of botany despite conditions of extreme hardship. Self-educated, forever impoverished, and lacking access to reference material, he nevertheless made outstanding contributions to the botanical knowledge of the Northwest.

~Adapted from Ornduff, Robert. 2008. *Kalmiopsis* 15:32-41.

Louis F. Henderson (1853–1942)

Louis Forniquet Henderson was one of Oregon's most important early botanists. During his 65-year career, he collected plants throughout all of Oregon as well as Washington and Idaho. At the time, he may have understood the flora of the Northwest better than any living person. Henderson's tens of thousands of meticulously annotated specimens—now at Oregon State University, the University of Washington, the National Herbarium, the Gray Herbarium, and elsewhere—provide a detailed record of the changing plant communities of the Northwest from the 1870s to World War II.

Henderson was born in Roxbury, Massachusetts, on September 17, 1853, the grandson of U.S. Senator John Henderson of Mississippi. His mother, Catharine Leland, was a teacher from a prominent Massachusetts family, and his father, John Henderson Jr., was an attorney in New Orleans. When young Louis's father, a supporter of Lincoln, was murdered shortly after the end of the Civil War, his mother took Louis and his older brother, Leland, north to New England.

Directly after completing his degree at Cornell University, Henderson went west. He arrived in Oregon in 1877 and secured a position as a high school teacher in Portland. Six years later, he married fellow teacher Kate Robinson; the couple had two daughters.

Henderson's life in botany passed through many phases. Almost from the day of his arrival in Oregon, he began collecting botanical specimens, spending many spring and summer weekends botanizing near Portland and riding trains to his favorite wildflower areas in such mountain ranges as the Blues and the Siskiyous. His early explorations also took him into Washington State, where he collected near the mouth of the Columbia and at Mount Adams.

In 1889, Henderson, ill with typhoid fever, resigned his school position and moved his family to Olympia, Washington, to recuperate near his brother's home. By 1890, he had recovered sufficiently to explore the interior of the

Olympic Peninsula with the O'Neil Expedition. He then accepted a job with the State of Washington to collect plants and set up the state's display at the 1893 Chicago World's Fair. After returning from Chicago, Henderson took a position at the University of Idaho (UI) in Moscow as its first botany professor, teaching and botanizing throughout the state. He founded the UI Herbarium, which grew to 85,000 specimens, all of which, coupled with his complete personal collection, burned in the Administration Building fire of 1906. Henderson remarked after the fire that he had lost "one of the joys of my life." Disheartened, he retired from teaching in 1908.

Three years later, when his daughters had finished college at the University of Idaho, Henderson moved his family to an 80-acre homestead in Hood River bequeathed to him by his mother. Here, he and Kate planted a 40-acre commercial apple orchard. Henderson remained on the orchard until he was 70 years old. His correspondence, however, makes it clear that he missed the botanical life.

On September 8, 1923, just days before his seventieth birthday, Henderson celebrated by swimming across the Columbia River from Hood River to Washington. The feat received statewide newspaper coverage and may have attracted notice by Albert Raddin Sweetser, head of the botany department at the University of Oregon. Shortly thereafter, Sweetser offered Henderson the curatorship of the Oregon plant collection at the University's herbarium. Early the following year, Henderson moved to Eugene, delighted to leave the orchard in Kate's capable hands and return to academia. For the next decade and a half, Henderson spent every summer in the field collecting plant specimens. During his tenure at UO, he enriched the herbarium by nearly 20,000 sheets. During the winters, he labeled and curated the specimens and sent exchanges to herbaria throughout the United States. Of this time in Eugene he wrote, "These past years have been one of the pleasantest periods of my life."

In 1939, Henderson retired to Tacoma to live with his daughter, Margaret (Kate died in 1934). Margaret reported that even in his mid-80s he was physically active. Throughout his life, he was "an avid outdoorsman, explorer, mountain climber, fly fisherman, oarsman, and swimmer. He was also said to have been a keen baseball player, a fascinating raconteur, an excellent singer, and a fine bridge player." On June 14, 1942, at the age of 88, Henderson "went to sleep at last," as his daughter later wrote. He is buried in the Oakwood Hill Cemetery in Tacoma.

Henderson is credited with naming and describing 64 species, subspecies, and varieties of plants. The most famous of these plants is arguably *Kalmiopsis leachiana*, the Oregon endemic shrub first collected in the Siskiyou Mountains by Henderson's friends Lilla and John Leach. Sixteen Pacific Northwest plant species still bear the Henderson name. Among them are *Agrostis hendersonii, Angelica hendersonii, Carex hendersonii, Dodecatheon hendersonii, Erythronium hendersonii, Phlox hendersonii, Sidalcea hendersonii,* and *Triteleia hendersonii.*

~Adapted from Love, Rhoda M. 2001. *Native Plant Society of Oregon Occasional Paper* 2:1-58.

Morton Peck, Willamette University, ca. 1918. Courtesy Paulus Glass Plate Negative Collection, PLS90A216, Archives and Special Collections, Willamette University.

Morton E. Peck (1871–1959)

A native of La Porte, Iowa, Morton Eaton Peck was born on March 12, 1871, to Clara Eaton and George D. Peck. In an unpublished, handwritten autobiography that was probably solicited by Albert Sweetser, Peck connects his earliest interest in the out-of-doors to "birds, insects, and fishing" and to his father. According to Peck, his father preferred living on the outskirts of civilization unlike his mother who "liked neighborhoods instead." Thus, in compromise, his birthplace and childhood home was a log cabin somewhat distant from town, where he attended a country school. He first collected organisms at age 11, and his literary aspirations began before he left the country school at 16 to attend La Porte City High School. Peck reflects that his first literary attempts yielded "abundant and very bad verses." He left high school without graduating but authored his first published work, *Webster's Ghosts and Other Verses*, prior to his junior year.

After high school, Peck taught in country schools for several years before entering Cornell College in Mt. Vernon, Iowa. He initially gravitated toward literature and the classics, but when he began to study plants more intimately in his junior year, "he knew he had found his field." Peck had an interest in the slime molds (myxomycetes), and Thomas Macbride, a botany professor and myxomycete specialist at the nearby University of Iowa, became a valued mentor and colleague.

Upon graduation from Cornell College in 1895, Peck became a biology instructor at the Marionville (Missouri) Collegiate Institute. In 1897, he accepted a professorship in biology at Ellsworth College in Iowa Falls, where he spent the next eight years. In 1905, Peck married Jessie Grant, a botany student at Ellsworth, and the two immediately departed for a wilderness excursion in Belize. Their brief visit lengthened to two years, during which they collected specimens and discovered dozens of plants and birds new to science.

Once back in the United States, Peck accepted a faculty position at Iowa Wesleyan University. Severe rheumatism led him to move to Salem, Oregon, in 1908, where his parents were already residing. Here he became the sole biology faculty member at Willamette University. Correspondence in the early 1920s between University of Oregon president P.L. Campbell and botanist Albert Sweetser suggests Peck may have considered moving southward; however, the move never transpired. In his handwritten 1936 autobiography, Peck comments, "I have enjoyed my teaching work thoroughly and have never felt that I would like to give it up." It is evident from published accounts and personal anecdotes that Morton and Jessie Peck were held in high regard by both students and colleagues, and the couple formed many strong friendships while at Willamette.

Peck nurtured his love of literature throughout his life. He was part of a faculty book club, he figured prominently in the writers section of the Salem Arts League in the mid-1920s, and in 1923 he published a volume of poetry entitled *The Book of Bardons and Other Poems*.

This specimen was collected by William Cusick on July 26, 1898, in Union County, Oregon. Cusick's notes say "at very low alt. for Zygadenus the (elegans) species." It was subsuquently annotated with additional notes by later botanists, including Kenton Chambers and Richard Halse. The designation "TYPE" indicates that this specimen was used as a reference when the species was first named. This specimen will require a new annotation now that this species has been renamed *Anticlea elegans* (see p. 312). Herbarium specimens are verifiable documentation of plant communities from a given time and place and thus provide a record of the changes that occur to a region's flora through time. Properly preserved, herbarium specimens can serve as research material for hundreds of years. Not just of interest to botanists, they are valued by historians, artists, and scientists in many other fields.

Whereas Peck pursued his literary interests "just for the enjoyment of it," he studied botany as a serious researcher and considered the pleasure it brought him as an added benefit. His two greatest achievements were the development of the Peck Herbarium of Willamette University (WILLU) and the writing of two editions (1941, 1961) of *A Manual of the Higher Plants of Oregon*. By the time of his retirement in 1941, Peck and his collaborators—including his wife Jessie—had collected about 40,000 sheets of plants from Oregon and adjacent western states. The herbarium was placed on permanent loan to the Oregon State University Herbarium, where it is maintained as a collection distinct from the combined UO and OSU specimens. Today, type specimens and duplicates of his many plant collections can be found in major world herbaria, including Harvard, Kew, Stanford, the Missouri and New York botanical gardens, the California Academy of Sciences, the U.S. National Herbarium, and the University of Washington Herbarium.

Peck's influence on taxonomic botany is evident in the many names, author citations, and collections that bear his mark. Namesakes include plants and fungi as diverse as *Astragalus peckii*, *Castilleja peckiana*, *Cirsium eatonii* var. *peckii*, *Lomatium peckianum*, *Phacelia peckianum*, *Penstemon peckii*, and *Sanicula peckiana*.

Peck's health began to fail as early as 1952, when, by letter, his California colleague John Thomas Howell expressed hope that his condition would improve. The cause of his death on December 4, 1959, is unclear. What we do know from Willamette University President Doney's comments, however, is that Peck's home was likely "in the hearts of some three thousand men and women whom he instructed and inspired." His wife published the second edition of his *Manual* posthumously in 1961, and it has stood until 2015 as the sole comprehensive flora of Oregon plants. Today, students and professionals alike are reminded of Peck's contributions via the Peck Herbarium, his manual of Oregon plants, and the Peck Scholarship, which has been awarded annually since 1973 to Willamette undergraduate students of botany.

~Adapted from Kephart, Susan R. 2001. *Kalmiopsis* 7:1-7.

HELEN M. GILKEY (1886–1972)

Helen Margaret Gilkey was born March 6, 1886, to John Augustus and Fannie Barnes Gilkey in Montesano, Washington Territory. Helen was the third of five children who survived to adulthood. In 1903, the family moved to Corvallis, Oregon, where John was employed as supervising horticulturalist of the grounds at Oregon Agricultural College (now OSU). The same year, Gilkey enrolled at the College; Professor Edward Ralph Lake was her first mentor in botany and sparked her interest in local mushrooms. She graduated in 1907 with a B.S. in Household Science, the standard degree for undergraduate women at the time. Gilkey continued her studies and received her M.S. degree in 1911; her thesis, "*Oregon Mushrooms*," contained brief descriptions and watercolor illustrations of several species of mushrooms, most of which were collected in or near Corvallis.

Gilkey began her doctoral studies at the University of California, Berkeley in 1912 with Dr. William Albert Setchell as her supervising professor. He introduced her to what became a lifelong passion: the underground fungi called truffles in the group Tuberales. Her dissertation, "*A Revision of the Tuberales of California,*" was illustrated with her own drawings. She received her degree in 1916 and was the first woman to receive a Ph.D. in botany from U.C. Berkeley. Gilkey remained there the next two years, working in the Botany Department as a scientific illustrator. She anonymously illustrated Willis Linn Jepson's 1925 *Manual of the Flowering Plants of California*, the first comprehensive floristic work for the state.

Helen Gilkey in 1952, one year after her retirement from Oregon State College. Courtesy Harriet's Collection, HC-3022, Special Collections & Archives Research Center, Oregon State University Libraries.

In 1918, Gilkey returned to Corvallis to accept a position as assistant professor and curator of the Herbarium at Oregon State College. Her responsibilities included Herbarium management, public outreach, and teaching. During her 33-year tenure, she oversaw and participated in eight moves of the Herbarium around campus, while the number of vascular plant specimens increased from 25,000 to 75,000. She sought to diversify and increase OSC's holdings of Oregon plants for use in research, teaching, and extension work. Gilkey herself contributed 2,109 specimens to the Herbarium. They represent plants she collected in every county of Oregon except Gilliam, and for most of them she was the sole collector listed on the specimen label. Gilkey described and illustrated two species of flowering plants: *Fritillaria gentneri* (Gentner's fritillary) and *Trifolium owyheense* (Owyhee clover).

Gilkey was a gracious ambassador of both the Herbarium and Oregon Agricultural College through her role as curator of the Herbarium. She responded to inquiries from a diverse audience seeking botanical help, she spoke and wrote for regional news media, and she gave presentations to garden clubs and nature study groups and at wildflower festivals. She took pleasure in serving as an instructor and mentor and recounted in her retirement that "...one of the greatest joys is that of hearing from our former students, particularly those of whose lives and careers we can be very proud." Gilkey was an engaged citizen and outspoken advocate for humanitarian and social causes, women's equality, and environmental concerns.

After her retirement in 1951, Helen Gilkey pursued numerous research and writing projects that she lacked the time to complete while a faculty member. In addition to technical articles and biographical sketches of botanists, she authored a taxonomic treatment of the fungi Tuberales in *North American Flora.* In 1967, at age 81, she published the highly popular *Handbook of Northwestern Plants* in collaboration with La Rea Dennis. She contributed the artwork illustrating many of her scholarly works.

In the course of her career, Helen Gilkey received numerous honors and awards, including in 1967 the OSU Distinguished Service Award, the university's highest honor. She was recognized as an accomplished scholar, an artist, and a thoughtful friend and mentor; she was a gentlewoman whose life and work impacted colleagues in botanical circles as well as her community. Gilkey died August 8, 1972, in Corvallis.

~ Adapted from Weber, Nancy S. 2014. Unpublished manuscript.

Lilla I. Leach (1886–1980)

Lilla Irvin Leach was an independent field botanist who, with her husband John, botanized throughout Oregon and other western states. The couple extensively explored the rugged Siskiyou Mountains of Curry County and spent nine summers there between 1928 and 1938, travelling more than 1,000 miles of primitive trails with their burros Pansy and Violet. Lilla was not directly affiliated with a university and received no monetary compensation for her botanical work nor for her expedition expenses. In 1963, Leach donated her private collection of approximately 3,000 pressed plant specimens to the University of Oregon.

Lilla Irvin, whose family had immigrated to Oregon over the Oregon Trail, was born on March 13, 1886, on her parents' farm in Barlow, Clackamas County. She attended school in Barlow and Canby and then the Tualatin Academy in Forest Grove. There she met John Leach from the bunchgrass country of eastern Oregon, who began (unsuccessfully at first) to woo her. Lilla attended the University of Oregon, where she studied with botanist Albert Sweetser. She received her B.A. degree in 1908. For the next five years she taught high school science and botany in Eugene.

Meanwhile, John Leach finally won her hand in marriage by assuring her he knew how to handle pack animals and "could pack her back where the flowers were different and cake-eating botanists could never get." They were married in an outdoor ceremony in 1913 on her parents' farm rather than in a more formal setting because, according to John, he was "…too wild to be in a church." After their wedding, the Leaches moved to Portland, where John established a pharmacy. They botanized together every summer for 32 years, traveling to Yellowstone Park, the Olympic Peninsula, the Cascade Mountains of Washington, and throughout Oregon.

A favorite destination of Lilla's and John's was southern Oregon's Siskiyou Mountains. Prior to their efforts over nearly a decade, this area had not been systematically botanized. Their explorations in remote and rugged terrain led to Lilla's discovery of a striking pink-flowered shrub in the heath family that was unknown to science. She recounted that on June 14, 1930, "I was in the lead where I usually walk in order to get the first chance over the burros to anything of interest that might be growing, when suddenly I beheld a small patch of beautiful, low growing, deep rose-colored plants. Because of their beauty, I started running and dropped to my knees… I had never seen anything so beautiful before. John came and started taking pictures while I examined the plant with a hand lens. I believed it was new." The specimen was first described by Morton Peck as a new species of *Kalmia,* which he named in her honor: *K. leachiana*. Lilla also sent a specimen to Louis Henderson at the UO Herbarium. Henderson originally determined the plant to be a novel species of *Rhododendron*. With further study, he declared it to be a variety of *R. lapponicum* but in 1931 returned to Peck's original determination and formally described the plant as *R. leachianum*. New plant material collected by the Leaches in 1931 was

Lilla Leach in 1953, at her microscope. Image accompanying a March 29, 1953, article in *The Oregonian*. Photograph by Carl Vermilya, courtesy Leach Botanical Garden.

sent to Alfred Rehder, curator of the Arnold Arboretum Herbarium of Harvard University. He identified the small shrub as a new genus and named it *Kalmiopsis leachiana,* the name it is known by today.

The genus *Kalmiopsis* is endemic to southwestern Oregon and occurs nowhere else in the world; Oregon's Kalmiopsis Wilderness is named for the plant. In addition to *Kalmiopsis leachiana,* Lilla is credited with the discovery of over a dozen additional Oregon species, including *Sophora leachiana*, *Iris innominata, Sedum moranii, Allium cascadense* (=*A. crenulatum*), and *Penstemon deustus* var. *suffrutescens*.

In 1931, Lilla and John bought five acres along Johnson Creek in southeast Portland and named the property Sleepy Hollow. Over the following five years they built their home and lived there until John's death in 1972. Lilla deeded the property to the City of Portland for preservation as a botanical area; it was designated as the Leach Botanical Garden and now includes approximately 16 acres open to the public. Lilla Leach died September 10, 1980, in Portland at the age of 94. Her ashes and John's were scattered in the Kalmiopsis Wilderness.

~Adapted from Love, Rhoda M. 1991. *Kalmiopsis* 1:3–8.

Georgia Mason (1910–2007)

Georgia Mason was originally from New Jersey and had lived in Oregon for only three years when she set out in 1961 to compile a flora of the Wallowa Mountains. Mason, whose birth name was Mavropoulos, was born in West Orange, New Jersey, on March 16, 1910, the second of two daughters of Greek immigrants Peter and Bessie Mavropoulos. She attended high school and college in her home state. From 1931 to 1958, she taught elementary and middle school-aged children in the New Jersey Public School System. It is said that she changed her name from Mavropoulos to Mason in order to make it easier for her pupils to pronounce. Seizing an opportunity to pursue a graduate degree in science to strengthen her teaching credentials, she left New Jersey in 1958 and moved to Corvallis to enroll at Oregon State College. Her passion for fieldwork and the lure of Oregon's surrounding wilderness proved strong. Although she was expected to return to the East Coast and resume teaching, she remained in Oregon for the remainder of her life. Mason became one of Oregon's major plant collectors and published two local floras before she died in Eugene at the age of 97.

In September 1958, Mason (now 48) left Clifton, New Jersey, for Oregon. In the Botany Department of Oregon State College, she soon met Dr. Helen Gilkey and her students. Gilkey, professor emerita and retired curator of the Herbarium, remained active in the Herbarium. She encouraged other women to seek careers in botany as she had done and was likely an inspiration to Mason.

Another important connection Mason made upon arrival at Oregon State College was that of fellow graduate student George Van Vechten. Van Vechten was researching the geology and ecology of the Three Sisters Mountains of Central Oregon and introduced Mason to hiking and plant collecting as she accompanied him in field work in the Three Sisters.

Mason received her Master of Science degree from Oregon State College in August 1959 and began collecting plant specimens in Western Oregon. In June of 1960, Georgia visited her friend Van Vechten, who himself had taken a teaching position at Eastern Oregon College in La Grande (located in the foothills of the Wallowa Mountains). Their hikes into the Wallowa high country were likely the inspiration for Mason to pursue her serious study of the Wallowa flora. It seems reasonable that the example of Dr. Helen Gilkey coupled with the encouragement of Dr. Kenton Chambers (newly hired curator of what was now the OSU Herbarium) persuaded Mason to undertake this project. During the spring and summer of 1961, she collected 796 specimens in Wallowa, Union, and Baker Counties. At the age of 51, she was most likely living on savings.

In the introduction to her flora of the Wallowas, Mason described the vast and rugged territory she covered during her eleven summers of botanizing in the Wallowas. She wrote that she concentrated mainly on the area designated as Eagle Cap Wilderness, where she botanized numerous high peaks, declaring "though rugged, all these trails are accessible to the hiker," despite her being a rather diminutive woman in her 50s and carrying collecting gear!

Mason was hired as acting assistant curator of the University of Oregon Herbarium during the sabbatical leave of Curator LeRoy Detling for the 1961–1962 academic year. After this position, she was unemployed. However, she chose to remain in Eugene, which would be her home for the remainder of her life. Mason traveled to the Wallowa Mountains each year between 1961 and 1971, spending entire summers there until she was over the age of 60. During this period she corresponded with Chambers and donated 1,417 specimens to OSU, expressing her gratitude to Chambers for his encouragement.

She returned to the UO Herbarium as acting curator from 1970 to 1976. During her tenure as curator, Mason accessioned the collections of Lilla Leach, Orlin Ireland, and others, adding approximately 7,000 specimens to the Herbarium.

Mason's *Guide to the Plants of the Wallowa Mountains of Northeastern Oregon* was published in December 1975. During her final two years as Herbarium curator, she collected and catalogued wetland and weedy species of Eugene and Lane County. She reluctantly retired in 1976 at the age of 66. She self-published *Plants of Wet to Moist Habitats in and Around Eugene Oregon* in 1982 and taught adult education field botany courses at Lane Community College in Eugene for several years.

Georgia Mason's contributions toward the understanding of the flora of the Wallowa Mountains were significant. During her eleven years of solitary exploration in that rugged landscape, she collected and contributed 4,858 specimens to UO and OSU, making her one of Oregon's most prolific collectors. She died in Eugene on October 8, 2007.

~Adapted from Love, Rhoda M. 2010. *Kalmiopsis* 17:9–16.

Georgia Mason, as she appeared when she entered Oregon State College in 1958. This is the only photo known to exist from her time in Oregon. Courtesy Student Academic Records (RG 053 – SG 2), Special Collections & Archives Research Center, Oregon State University Libraries.

CONTEMPORARY

Kenton L. Chambers (1929–)

When seeking an expert on the plants of Oregon and the Pacific Northwest, most trails lead—directly or indirectly—to Kenton Chambers. Stanford-trained in classical plant taxonomy, Ken's expertise in the aster family and numerous diverse genera has brought clarity to the details of Oregon's botanical resources as well as an increased appreciation of their intrinsic value.

Kenton Lee Chambers was born in Los Angeles, California, on September 27, 1929, to Edna (Miller) and Maynard Macy Chambers. Ken spent his early childhood in Pasadena, where his brother Derrell was born in 1934. Ken studied music in his youth; he played piano, sang, and played in the marching band through his high school and college years. Upon graduating from Paso Robles High School, he received a four-year scholarship to attend Whittier College and began classes in 1946. A memorable classmate in Ken's first-year zoology class was upperclassman Henry "Harry" Thompson, who graduated and returned to Whittier in 1948 as a biology instructor. Ken excelled in Thompson's classes and developed an interest in the aster family (Asteraceae) during Thompson's field botany class.

After graduating from Whittier College in 1950, Ken attended Stanford University, following in the footsteps of his friend and mentor, Harry Thompson. Ken's Ph.D. research on the genus *Microseris* included field work in the California Great Valley and the Coast Ranges. He received his degree in 1955 under the direction of Dr. Ira Wiggins. After an extended field trip to botanize in Baja California, Ken began a National Science Foundation (NSF) postdoctoral fellowship with Dr. Harlan Lewis at University of California at Los Angeles, counting chromosomes of Asteraceae and other families.

In 1956, the native Californian relocated to the east coast when he accepted a faculty position in the Department of Botany at Yale University. Ken developed a research program focusing on eastern genera in the chicory tribe of Asteraceae. With his colleague and good friend, William L. Stern, Ken travelled to Panama and the Florida Keys to collect plants. In 1957, Ken met Henrietta Laing, a Ph.D. student studying the taxonomy of the mint *Pycnanthemum*. Their friendship blossomed into romance, and they married in 1958.

Ken returned to the west coast in 1960 to accept an associate professorship at Oregon State College. He succeeded Dr. Albert Steward, who died in 1959, and worked with the acting curator of the Herbarium, La Rea Dennis. In addition to directing the Herbarium, Ken taught agrostology, plant evolution, systematics, and general biology. His research interests continued to include *Microseris* and *Krigia*, which he grew in the greenhouse for hybridization studies.

Some of Ken's scholarly activities led him to distant places and promoted interactions with diverse colleagues. He lived in Dominica for three months during 1966, collecting plants in preparation for a regional flora by the Smithsonian Institution. Ken and his family spent 1967 in the Washington, D.C. area when he was selected by NSF to direct the Systematic Biology grants section. In 1979, he hosted three Soviet botanists, sponsored by the USA/USSR Botanical Exchange Program, on an extended trip throughout Oregon and northern California as part of their month-long visit. A long-standing collaboration with Dr. Konrad Bachmann from the University of Heidelberg focused on the genetic basis of variation in species and hybrids of *Microseris*. Bachmann visited OSU in 1978, and, in 1983, Ken and Henrietta spent three months working in his laboratory in Heidelberg.

Ken's influence as a botanist in the Pacific Northwest is significant. In addition to teaching undergraduates, he served as major professor for 16 Ph.D. students and 16 M.S. students at OSU. Ken was the local organizer of several gatherings for the botanical community. In 1968, he chaired the Biology Colloquium Committee at OSU, selecting the topic "Biochemical Coevolution," and the following year he led a field trip for International Botanical Congress participants prior to their gathering in Seattle. In 1988, he organized and chaired a symposium entitled "The Future of Endangered Plant Species Studies in the Pacific Northwest" for the University of California at Los Angeles Pacific Division annual meeting.

Rare and endangered plant species have long been an interest of Ken's. Shortly after his arrival at OSU, he began to compile a list of Oregon species whose reported distribution in scientific literature and herbarium records

COURTESY OF KLC

Kenton Chambers in 1989, professor of botany and curator of the OSU Herbarium.

suggested they were rare in the state. Jean Siddall, an amateur botanist from Lake Oswego, was doing similar work using field observation data. The two joined efforts, and in 1975 they produced the provisional list of Rare, Threatened and Endangered Plants in Oregon—568 species—which was published by the Oregon State Land Board as part of a study outlining regional needs for conservation and land use planning. Ken asserted "...a solid taxonomic foundation is essential for any legitimate conservation action." Their efforts contributed to the passing of the Oregon Endangered Species Act in 1987.

Although Ken retired in 1990, it did little to alter his activity level. He taught agrostology for another four years, and he continues to collect plants and write articles for amateur botanists in the *Oregon Flora Newsletter* and for native plant society publications. He has played a critical role in the Oregon Flora Project as an advisor and a prolific author of hundreds of treatments for both the Checklist and the *Flora of Oregon*. In 2006, he coauthored a monograph of the genus *Claytonia*. Additionally, Ken pursues the describing of new plant taxa from the early Cretaceous and Tertiary periods that are preserved in amber with colleague George Poinar. Retirement also gave Ken the opportunity to reengage with his music interests; he plays baritone horn in the Corvallis Community Band and has performed in several University productions of Gilbert and Sullivan operettas.

The sphere of Kenton Chambers' influence on botany in Oregon, the Pacific Northwest, and beyond is substantial, and it continues to expand as his protégés advance botanical studies in the tradition of this notable scholar and mentor.

~ Adapted from Chambers, Henrietta L. 2001. *Kalmiopsis* 8:2–9.

Rhoda M. Love (1932–)

"As a scholar, Rhoda is untiring in her pursuit of the answer to an important question," says Glen Love of his wife, Rhoda Mae Love. This perseverance is evident in Rhoda's many activities as an instructor, environmental activist, historian, and citizen science educator. Rhoda's interest in the natural world began during childhood in Seattle, Washington. She was born on August 16, 1932, the only child of Mary Biehn Moore, a nurse, and James Moore, a police officer. After an education in the public schools, she enrolled at the University of Washington in Seattle. She entered as an English major, but botany became her emphasis after taking a class taught by Dr. C. Leo Hitchcock. Hitchcock was a highly influential figure in Rhoda's life. He recognized her keen interest and scholarly talent as a young adult student and provided constant encouragement during her professional development. In 1952, Rhoda participated in a summer-long field botany course led by Hitchcock and Arthur Kruckeberg and traveled throughout the American West. Rhoda earned a B.S. in Biology in 1955.

Soon after completing her bachelor's degree, Rhoda married Glen Love. To finance their graduate education, the two scholars took turns working and enrolling in graduate classes. In this cooperative way they both finished advanced degrees: Rhoda earned a master's degree in Botany in 1959, and Glen earned a Ph.D. in English in 1964.

Rhoda began her teaching career in 1959 as a high school biology teacher in Seattle. Employment for Glen took them to California, and she taught biology at San Diego State College from 1963 to 1965. In 1966, her husband accepted a faculty position at the University of Oregon, and the couple, with their infant son, moved to Eugene. In the same year, Rhoda was hired as a biology instructor at the newly formed Lane Community College (LCC). Classes were held in buildings throughout Eugene; Rhoda's first classroom was next to a railroad, and her evening lectures were often interrupted with the clatter of connecting rail cars. Rhoda taught 23 years at LCC, between 1966 and 1997. She gained several insights during this period. She acknowledged how hard-working many of her students were and the challenges they overcame to pursue their education, notably those who took classes at night while working during the day. She also witnessed the impact of gender inequity and consequently strived to be a mentor and role model to young female science students. Rhoda was the only female faculty member of the LCC Biology Department for many years. She also was grateful to a department chair who accommodated her desire to balance employment with child-rearing through a part-time appointment for most of her career.

As a community college teacher, Rhoda also assumed her role as an environmental steward. She has said "When

Bryan Hayne

Rhoda Love in 2011, holding a tin vasculum, a device historically used for temporarily storing plant material before pressing it into herbarium specimens.

you teach ecology, you have to become an activist because you see what is happening to the world as well as opportunities to stop some of the things that you know are going to be adverse to the environment. I felt it was my duty to get out there and do something about it." Both Rhoda and her husband were energized by the environmental issues arising in the early 1960s. One of Rhoda's first significant literary efforts was *Ecological Crisis: Readings for Survival* (1970), an anthology of environmental writing she co-authored with Glen. The book was widely adopted as a primer for ecological studies. She also worked effectively to oppose the construction of a regional Oregon nuclear power plant in the late 1960s. Under the auspices of the Native Plant Society of Oregon (NPSO), Rhoda and Esther McEvoy led the passage of the Oregon Endangered Species Act.

Beginning in 1975, Rhoda took a leave of absence from teaching at LCC to pursue Ph.D. studies at the University of Oregon. She researched the ecology of Willamette Valley hawthorns (*Crataegus*) under the direction of Stanton Cook and completed her Ph.D. in 1980.

Several organizations benefited from the focus of Rhoda's attention and efforts. Rhoda became an enthusiastic proponent of the development of Mount Pisgah Arboretum and served the Arboretum in many capacities. Beginning in 1980, she was an administrative assistant and grant writer, she designed the Arboretum Nature Trail, established the education program, edited the newsletter, and served on the board of directors. She became involved with the NPSO in 1980, and frequently was an officer at both state and local chapter levels. In recognition of her contributions, Rhoda, along with her friends Charlene Simpson and Veva Stansell, received the NPSO Fellows award in 2001, the organization's highest honor. In 1994, Rhoda began volunteering for the newly formed Oregon Flora Project, serving as a Checklist Project leader, checklist treatment author, and editor of the *Oregon Flora Newsletter*, a responsibility she held for 18 years. She says of her volunteerism: "I liked the idea of being part of something. I was very enthusiastic about bringing some of these organizations along, letting people know about them by writing newsletters for them."

Rhoda has researched and published numerous biographies of Northwest botanists. She is co-editor in chief, with Arthur Kruckeberg, of *Plant Hunters of the Pacific Northwest: Botanical Explorers of Oregon, Washington, Montana, Idaho, and British Columbia, 1790–1970* (in preparation). It is notable that she shares authorship on this historical opus with one of her first university botany professors, one who inspired her and supported her academic pursuits. Likewise, through her own efforts, Rhoda has influenced countless individuals and has ushered a new generation of instructors and environmentalists to action—a befitting legacy for this Pacific Northwest native.

Gerald Carr

Richard Halse in 2014, collecting a specimen of *Alisma lanceolata*.

Richard R. Halse (1947–)

Richard Ray Halse was born on March 15, 1947, in Watertown, South Dakota. He was raised in the nearby small town of Clear Lake, where his parents owned and operated a grocery store. Both maternal and paternal relatives owned farmland nearby. Richard began his higher education as a math major at Northern State College in Aberdeen, South Dakota. He "quickly learned the error of his ways" and, with the influence of a good biology teacher, decided to pursue the study of biology. One of his early mentors was Dr. Gertrude Miller, a plant taxonomist and the instructor of his first upper division biology class. In addition to enthusiasm for his new course of study, Richard also discovered he enjoyed teaching; he was an instructor of undergraduate biology labs between 1968 and 1970.

Upon receiving his B.S. in Biology in 1970, Richard moved to Tucson to begin graduate school at the University of Arizona. Choosing a project that had a significant component of field work, Richard wrote a flora of Canyon de Chelly National Monument for his master's thesis. His information about the extant flora of the region was used to help identify plant fragments found in an archaeological dig of the Antelope House Ruin in the Monument. Richard spent the summer of 1971 living at the field site, collecting and documenting approximately 470 plant species from the area.

It was at OSU that Richard embraced a third professional interest, that of herbarium management. He began his graduate studies in the Botany and Plant Pathology Department in 1973 with Dr. Kenton Chambers, curator of the Herbarium. The duties of his graduate research assistantship exposed him to curation and collections management skills. His dissertation addressed the systematics of *Phacelia* sect. *Miltitzia* (Hydrophyllaceae), a group of plants that grow in southeastern Oregon and the Great

Basin, and involved field work as well as teaching. Richard received his Ph.D. in 1980.

For the next ten years, Richard worked for a pharmaceutical company in Monmouth, Oregon, and as a botanical consultant. In 1990 he returned to OSU, first as a botany instructor and next as assistant curator of the Herbarium. With the retirement of Dr. Chambers in 1991, the leadership of the Herbarium was divided between a Herbarium director (now held by Dr. Aaron Liston) and a Herbarium curator, the position Richard has held since 1992.

Richard's enjoyment of field work has brought significant benefit to the OSU Herbarium: he has deposited approximately 10,000 specimens (7,127 from the Pacific Northwest) in the OSU Herbarium, making him the 3rd most prolific contributor of specimens to the Herbarium. During his annual collecting trips, he often gathers enough material for multiple specimen sheets and contributes the duplicates to an international herbarium specimen exchange program.

The Herbarium has also benefited from Richard's contributions to the Oregon Flora Project Checklist. In the course of preparing checklist treatments for 240 genera (representing 17% of the flora), he examined and annotated thousands of herbarium specimens to reflect the latest nomenclature.

Twenty percent of Richard's position is funded by OSU Extension Service for the purpose of providing plant identification services to university extension agents, state and federal agencies, and the public. Richard estimates that over 50 percent of the identification requests he receives are from the general public; they range from native species to houseplants. He is most frequently asked to identify grasses due to their complexity and their prevalence. More unusual requests venture into forensic botany, requiring Richard to examine the rumen contents of animals to search for poisonous plant fragments, or to assist police in deducing the particulars of a death that occurred outdoors. This aspect of Richard's work requires knowledge of taxonomic groups found worldwide. "Identifying some plants is like figuring out a mystery novel," he says.

Teaching and mentoring continues to be an element of Richard's work, coupled with undergraduate student advising. He has instructed classes in plant taxonomy, anatomy, aquatic botany, and agrostology as well as general biology and botany.

Much of the success of Richard's work within the Herbarium reflects the personal knowledge he brings to his work. His familiarity with the flora of the state has come from decades of plant collecting throughout Oregon, identifying plants unfamiliar to him, and curating the specimens of the state's largest and most comprehensive herbarium. "I feel that my most important contribution to Oregon botany is my collecting of plants. It expands our knowledge of the plant species in the state, their distribution, and their variability," he says. "I have in my head a fairly good idea of Oregon's flora." Through his service as OSU Herbarium curator, instructor, and plant identifier, Richard's knowledge has been instrumental in maintaining and expanding the botanical resources of Oregon.

Barbara L. Wilson (1951–)

Barbara Lynn Wilson was born on August 2, 1951, in Ann Arbor, Michigan, the oldest of seven children. Her father, Lewis Williams, was an attorney, and her mother, Jeanette Sprung Williams, was a homemaker. The family lived for ten years in Paw Paw, Michigan, adjacent to her paternal grandparents; in 1961 they moved to a home in the country nearby. One of Barbara's first memories is standing at her grandmother's kitchen window, watching large yellow birds (she now knows they were evening grosbeaks) at a bird feeder. Her grandmother's interest in nature, especially plants, was an inspiration to Barbara. As a child, she spent much of her time outdoors, exploring bogs and ponds. She and her siblings frequently brought home snakes and toads, which sometimes escaped in the house, much to her mother's dismay. Barbara nurtured her interest in the natural world by attending classes and volunteering at the Kalamazoo Nature Center. Beginning at age 12, every Saturday during the school year and throughout the summers, her mother would drive her the hour's journey to the Nature Center.

After high school, Barbara attended Kalamazoo College in Michigan, where she studied biology and pursued her interest in birds. She received a B.A. in Biology in 1972. Encouraged by her parents' recent move to Coos Bay, Oregon, Barbara went the following year to the University of Oregon and obtained a teaching certificate. She then joined the Peace Corps and taught high school biology and general science in the Kingdom of Tonga from 1974–1976. While there, she met John Wilson, a fellow Peace Corps volunteer.

Richard Brainerd

Barbara Wilson in 2014, discussing *Festuca roemeri* for a video presentation, Marys Peak, Oregon.

Upon returning to the United States, they married and moved to the Wilson family farm in southwestern Iowa.

During her first year back in Iowa, Barbara attempted to become a farm wife. She says of that effort, "It was a good life, but not for me." The details of the natural world beyond the farm called for her attention, and she soon enrolled at the University of Nebraska at Omaha (UNO) to pursue graduate research in bird behavior. Following the receipt of her M.A. in biology in 1982, Barbara worked for the University of Nebraska State Museum preparing bird and mammal specimens for study and conducting small research projects. She became increasingly interested in the plants that formed the habitats for birds and shifted her focus to botany under the mentorship of taxonomist Dr. David Sutherland. Barbara was particularly intrigued with sedges (*Carex*) because they were diverse and were also one of the very few groups that Dr. Sutherland found challenging.

Barbara worked as an instructor in the UNO Biology Department from 1988 through 1992. She greatly enjoyed teaching and found it was an opportunity to share her passion for the natural world with others. One of her favorite classes to teach was an evening environmental science class for non-majors because "it was often the only science class these students would take, and it served as an important opportunity to get people interested and thinking critically about science."

In 1991, Barbara's husband died, and the following year she returned to Oregon to begin graduate school at OSU in the Department of Botany & Plant Pathology. Her dissertation topic was genetic variation in the grass genus *Festuca* (fescue) with Dr. Aaron Liston as her advisor. She completed her Ph.D. in 1999. During this time, Barbara maintained her interest in sedges and discovered there were others in the region that shared this interest. In 1993, with the financial support of the Native Plant Society of Oregon, Barbara and a small group of other botanists organized to produce *The Atlas of Oregon Carex* (1999). Pleased with the process, they decided to meet weekly to study taxonomic questions in *Carex*.

Developing the sedge atlas demonstrated the need for a more effective way to manage data. A collaboration with the Northwest Alliance for Computational Science and Engineering (NACSE) at OSU resulted in the design of a database that both captured biological data and tracked nomenclatural changes. This work led to the basic database structure that was subsequently used by the Oregon Flora Project.

The original group of "sedgeheads" eventually narrowed to Richard Brainerd, Danna Lytjen, Bruce Newhouse, Nick Otting, Peter Zika, and Barbara. In 2002, they incorporated as the *Carex* Working Group LLC, with the purpose of studying sedge taxonomy and providing botanical consultation. Barbara explained her mission as a taxonomist this way: "Many people do not even 'see' plants, much less know their names. Providing the correct plant name is important to me because with it, you can then gather more knowledge about that plant." For Barbara, plant taxonomy is both enjoyable and important; "Working with plant names is a fun great game, but it isn't the end goal. Nomenclature is an important step toward understanding the biology of the plants and providing sustainable plant management."

In 2008, The *Carex* Working Group published the *Field Guide to the Sedges of the Pacific Northwest*. The Group, now comprised of Barbara, Nick Otting, and Richard Brainerd, continues to provide botanical consulting and publish field guides to challenging plant groups of the Pacific Northwest. They have also authored taxonomic treatments of Poaceae, Cyperaceae, Salicaceae, and other taxa for the *Flora of Oregon*.

Through her efforts and through the creation of the *Carex* Working Group, Barbara has brought clarity to many difficult plant groups in Oregon and the Pacific Northwest and has made that knowledge accessible to a wide audience.

Donald H. Mansfield (1951–)

Remote southeastern Oregon is home to Steens Mountain and the Owyhee basin; these are landscapes that would remain poorly understood and underappreciated today without the exploration and research conducted by Don Mansfield. Donald H. Mansfield was born on October 30, 1951, in Salem, Oregon, to Oliver T. Mansfield, an FBI employee, and Helen R. Mansfield, a homemaker and school volunteer. His family moved frequently because of his father's work. The first two years of Don's life were spent in Salem, and it was not until the mid-1960s that the Mansfields returned to the Pacific Northwest, specifically

Emma George

Don Mansfield in 2011, teaching undergraduate field botany in Death Valley, California.

to Seattle. Don became involved in Explorer Scouts and spent extensive time mountain climbing and hiking in the Cascades. He graduated from high school in 1969 and enrolled in Willamette University in Salem. During his third year, a biology class taught by Dr. Martha Springer prompted Don to major in biology. Don completed his undergraduate studies at Colorado College in Colorado Springs, where he was inspired by Dr. Jack L. Carter and focused intensely on botany, especially on the mechanisms of evolution and systematics. He graduated in 1973 with a B.A. in biology. Don completed his M.S. in botany at the University of British Columbia in 1977; it was here that he discovered his passion for teaching. He then moved to Idaho State University to pursue a D.A. (Doctor of Arts) degree with an emphasis in undergraduate biology education. Don was interested in combining his love of teaching with his commitment to scholarly research. Toward that end, ecologist Dr. Jay Anderson became an influential mentor to Don by demonstrating how those pursuits are compatible. Don received his doctorate in 1979, and spent the next ten years in research and teaching positions at the University of California at Davis, Colorado College, and Rollins College in Florida.

In 1989, Don accepted a tenure-track position teaching at The College of Idaho, a small liberal arts college in Caldwell, Idaho, near the Oregon border. For his research program to successfully incorporate undergraduates, he realized it would need to be low budget and compatible with student research capabilities. A study of the flora of Steens Mountain seemed an ideal topic. "In that way I could meet several objectives: contribute meaningful science, engage and involve students, and have fun exploring landscapes that I loved."

Don began his research by searching for checklists and plant collections from Steens Mountain and quickly realized there was little information available about the flora of this remote area. Throughout the 1990s, he collected extensively; a record of his botanizing is preserved in specimens deposited in the College of Idaho Herbarium. Noting the discrepancies between existing data and his field observations, Don felt strongly that the work he was pursuing added important knowledge to our understanding of the region. He relished the opportunity to introduce his students to field work, as he found it an effective way to teach and to emphasize the relevance and value of their efforts.

The twelve-plus years of Don's endeavors have produced both intangible and tangible results. He has inspired and informed a generation of students about the value of natural places and the resources they contain, and he has trained undergraduates for careers as scientists and researchers. Don has published *Flora of Steens Mountain* (2000) based on his and his students' work, and he has also identified opportunities for further research in evolution, biodiversity, and the elucidation of new plant taxa. The centuries-old process of botanical exploration has revealed plants—such as members of the genus *Lomatium*—with classical taxonomic ambiguities that Don plans to investigate using modern genetic techniques.

Since 2002, Don has focused on the floristics of the Owyhee drainage in the southeastern corner of Malheur County. In the process of writing his notes on the exploration of the Owyhee drainage, he noticed the geographic connections to the Steens. At the same time, there are distinctive differences in some of the plant taxa found in the two areas despite their geographic similarities. "I have collected lupines on a transect between the Steens and Jarbidge mountains, and by looking at the plants, I can tell you where they were collected. The plant forms are specific to their geographies, but their taxonomy isn't yet clear." Many of the projects that Don pursues result from the recognition of patterns that become evident only over several decades of field observation. Floristic work requires long-term study and benefits from the synthesis of knowledge a researcher can apply only after many years of dedication to such a project.

More than thirty years of exploration have given Don an appreciation of the importance of cataloguing existing biodiversity and for the necessity of conservation. "Conservation biology must keep pace with the challenges we face as a culture," he says. His hope is that sound science will influence policy and prevent species from perishing so that we can learn from them. Don's contributions to the botanical knowledge of the Steens and Owyhee regions serve as an important foundation for a more comprehensive understanding of this unique area in Oregon.

Scott D. Sundberg (1954–2004)

Assembling and organizing collections was a lifelong pastime for Scott, so it comes as no surprise that the child who collected cones, fossils, shells, and other found objects channeled his focus to the gathering of information about his life's passion, the plants of Oregon.

Scott Donald Sundberg was born on February 10, 1954, in Eugene, Oregon. He was the third of four sons of Norman Sundberg, a psychology professor, and Donna Varner Sundberg, a homemaker. From a very young age, Scott was fascinated by the natural world and spent much of his time exploring the outdoors. He attended public schools in Eugene. At age 11, Scott and his family lived in Delhi, India, for a year. This experience and other international travels both individually and with his family gave Scott an abiding appreciation of global diversity. After graduating from high school in 1972, Scott enrolled in the University of Oregon as a biology major. His interest in plant systematics began when he took a taxonomy course taught by Dr. George Carroll. Scott also pursued undergraduate research in Carroll's lab, studying nutrient cycling in the canopies of old-growth Douglas-fir trees. Scott collected his study material from 400-year-old trees rigged with climbing ropes in the H.J. Andrews Experimental Forest. During this time, he collected the type specimen of the rare lichen, *Nephroma occultum*.

Scott's interest in vascular plant systematics led him to the UO Herbarium, where he benefited from working with two herbarium curators: recently-retired Georgia Mason and her successor, David Wagner. He researched ferns and grew them from spores in a home greenhouse. Scott graduated in 1978 with a B.S. (honors) in biology.

Camping and backpacking were important activities for Scott as a youth and adult. He was adept at route finding with map and compass and frequently explored the backcountry of the Cascades and Oregon Coast Range. He botanized throughout western Oregon and made numerous noteworthy collections. At 6'5", Scott was a big man, yet he delighted in finding obscure and tiny plants.

From 1978 to 1980, Scott worked summers as a botanist for the Bureau of Land Management (BLM), Coos Bay District. Scott's close friend John Christy held the comparable position in an adjacent BLM District, and the two continued their botanizing in the Coast Range as friends and professionals. In 1981, Scott began graduate school at the University of Texas at Austin under the guidance of Dr. Billie L. Turner. His cohorts included Matt Lavin, Alan Whittemore, Tina Ayers, and Kevin Nixon. Scott's dissertation was on the systematics of a section of the genus *Aster*, and he travelled extensively throughout Mexico and the southwestern U.S. for his field work. In 1983, he met Linda Hardison, a botany and marine biology major; she often accompanied him on plant collecting trips. Scott received his Ph.D. in 1986, and he and Linda were married in Jakarta, Indonesia, later that year.

Scott spent the following year at Ohio State University as a postdoctoral fellow, working with Dr. Tod Stuessy on species-isolating mechanisms in the aster family and the *Flora de Nicaragua*. In 1987, he moved to Seattle, where Linda began her Ph.D. studies in botany at the University of Washington. Scott did post-doctoral research there on *Sedum oreganum*, using molecular biological techniques, and was an instructor of undergraduate botany and plant taxonomy courses. In 1991, he began work for Ebasco Environmental, Inc. as a botanical consultant. He was happy to return to field work, doing wetland and rare plant surveys throughout the Pacific Northwest. His first project as a consultant was conducting wetland delineations on property proposed for development; Scott's exacting assessment of the botanical and ecological resources present resulted in the developer ceding much of the site for preservation.

In 1994, Scott moved to Corvallis to coordinate the integration of the UO Herbarium into the OSU Herbarium. He viewed this as an opportunity to be in the place where he could best accomplish his life-long aspiration to write a new flora for the state of Oregon. In his job application, he noted that it was his intention to pursue writing a flora as a personal endeavor outside the responsibilities of the job. Within the year, Scott initiated the Oregon Flora Project (OFP).

Scott developed the OFP through partnering with the Native Plant Society of Oregon, engaging collaborators and hundreds of volunteers, and ultimately receiving grant funding from the BLM and the National Science Foundation. He had a vision for the OFP to not only produce a much-needed flora for the state but also to be an effective and relevant resource through time by managing and presenting data in a digital format. His early focus on database structure reinforced this and led to the development of effective ways to track scientific name changes and synonymy. In 1998, Scott and Kenton Chambers coauthored a *Checklist of the Asteraceae*; this served as a model of the taxonomic framework to be used for the entire *Flora*.

The year 1999 brought significant events to Scott's life: he was promoted to research assistant professor, his son Matthew was born, and Scott was diagnosed with multiple sclerosis. Though the disease caused him to struggle with pain and chronic fatigue, he continued to advance the work of the OFP. He hired the first OFP staff members in 2001, promoted public involvement in the gathering of plant observation data and field photos, produced a rare plant guide, and developed a database of plant morphological characters. He also contributed to floristics on a larger scale by co-authoring taxonomic treatments of 99 taxa in several genera of Asteraceae for the *Flora of North America*.

As Scott's health declined, his wife Linda assumed a greater role in the OFP, working first as a volunteer, and later as a part-time employee to aid in project management. In the summer of 2004, after seven months of severe pain attributed to multiple sclerosis, he was diagnosed with cancer. Scott died December 30, 2004.

Scott's life brought him full circle: from the genesis of his passion for botany and nature in Eugene, to his scholarly development with outstanding mentors, then back to the Pacific Northwest, and ultimately to Oregon and the flora he loved. Through the formation of the Oregon Flora Project and his vision that guides the production of the *Flora of Oregon*, Scott's contributions are a lasting legacy to the study and appreciation of Oregon botany.

Norman Sundberg

Scott Sundberg in 1983 at Golden Lake, Three Sisters Wilderness, Oregon.

ECOLOGY OF OREGON

Dennis A. Albert
with contributions from Bridget Chipman

Introduction

The ecology of Oregon is discussed within several frameworks and scales that are recognized as important for understanding the distribution of plant species. Climate and substrate are recognized as important physical factors affecting vegetation, and these are often shaped by geology and landform at both regional and local scales.

Oregon's vegetation is briefly described as it relates to the phytogeographical and regional vegetation classifications presented by Barbour and Christensen (1993) and by Takhtajan (1986). Vegetation and phytogeographic types are large-scale frameworks that often span multiple ecoregions and cross geopolitical boundaries. The state of Oregon is subdivided here into eleven ecoregions, in which major geological features, climate, habitats, and plant communities are described.

The common plant habitats of Oregon are identified and described; some of the plant species common to each habitat are also listed. Habitats encompass a mix of plant communities and physical features such as cliffs, talus slopes, or seeps. Habitats and plant communities are described at a relatively local scale, with multiple habitats occurring within each ecoregion.

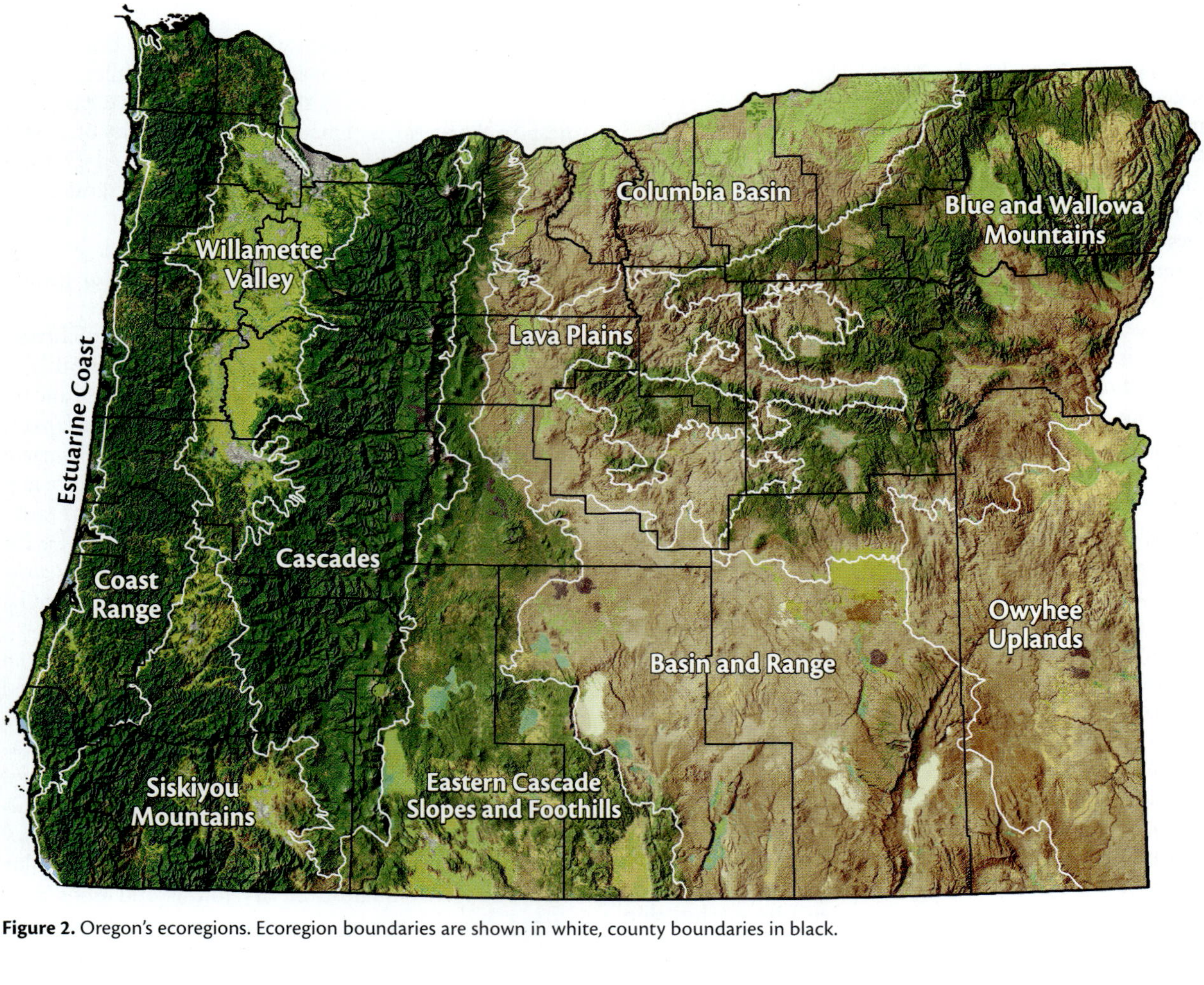

Figure 2. Oregon's ecoregions. Ecoregion boundaries are shown in white, county boundaries in black.

Physical Factors Affecting Plant Distribution

Climate and Substrate

Climatic factors strongly influence the flora of Oregon and the Pacific Northwest, with a distinct gradient of precipitation decreasing from the Pacific Ocean eastward to the border with Idaho. Annual precipitation ranges between 940 and 5,080 mm (37 and 200 in) from the Coast Range to the Cascades. Abundant precipitation is further augmented by regularly occurring fog along the Pacific Coast. This moisture contributes to the conifer forests of western Oregon being some of the most productive in the world. There is an abrupt decrease in precipitation to the east of the Cascade crest, an effect called rain shadow that is common on the leeward side of most mountain ranges. The result is an average precipitation as low as 150 mm (6 in) in parts of the Owyhee Uplands. While the amount of precipitation decreases strongly from west to east, precipitation increases in parallel with an increase in elevation in all of the mountain ranges in Oregon. The high elevation habitats in the Blue and Wallowa Mountains receive up to 2,000 mm (80 in) annually, largely as snow during the winter months.

The timing of precipitation is another critical factor impacting plant distribution and phenology. With the exception of the Coast Range, most of Oregon experiences precipitation from autumn to spring, with almost no summer rainfall except for the occasional thunderstorm. Consequently, many plants flower in the spring and early summer, rather than in late summer or early autumn. This is especially true for low elevation habitat in eastern Oregon.

Temperature variation also impacts vegetation. The climate of the Coast Range and Willamette Valley is moderated by the Pacific Ocean. Temperatures are seldom below freezing for more than a few days, and the frost-free period is 160–280 days annually. Temperatures in the Cascades are much lower in high elevation areas, and the frost-free period is as short as 10 days in the alpine zone of the Blue Mountains.

Some of the most extreme climatic conditions occur along the coastal estuaries, where the vegetation is regularly exposed to a combination of tidal inundation, sediments, saline water, and extreme winds. The nearby coastal dunes are equally severe, with salt spray, strong winds, and excessively dry, unstable, blowing sands. The internally draining playa lakes of the Basin and Range ecoregion experience similar challenging growth conditions of high salinity, seasonally fluctuating water levels, strong winds, and extreme summer temperatures.

Soil type and depth varies greatly across Oregon. The deep, rich silts and clays of the Willamette Valley and the rich and productive loess soils along the rolling plateau of the Columbia Basin have resulted in broad-scale conversion of these landscapes to agriculture. In contrast, the marine sedimentary bedrock of the Coast Range breaks down to produce clay soils, but these soils are acid and remain in forest production. Much of eastern Oregon has thin, coarse-textured, droughty soil derived from basalt, pumice, or volcanic ash. These soils support grazing, forest production, and wildlife habitat and continue to host sagebrush steppe or grassland plant communities.

Exposed bedrock is common in all Oregon ecoregions except for the Willamette Valley and can limit plant growth. Rock type varies from easily eroded marine sandstone and siltstone of the Coast Range to volcanic deposits of pumice, ash flow, and basalt over most of eastern Oregon. Diverse igneous and metamorphic rock characterizes both the Blue and Wallowa Mountains and the Siskiyou Mountains. The nutrient-poor, heavy metal-rich, ultramafic rock of the Siskiyou Mountains is most limiting to plant diversity, but these extreme conditions have resulted in some of the highest levels of plant endemism in the United States. Similarly, the diverse rocky, thin soils of eastern Oregon support many plant species with localized distributions.

Regional Vegetation Classification

In the *Flora of North America*, Barbour and Christensen (1993) identify three of the major vegetation types of North America as occurring in Oregon: *Pacific Coast conifer forest*, *western montane conifer forests*, and *intermountain grassland* or *shrub steppe*.

Pacific Coast conifer forest, which stretches from Monterey County, California, northward to Cook Inlet, Alaska, is considered to be one of the most productive, luxuriant forests in the world, with net annual productivity of 1,500–2,500 g C/m^2 (Daubenmire 1978, Franklin and Dyrness 1973). Conifers thrive in the moderate temperatures and high rainfall of the region, which allow them to photosynthesize and grow throughout the year, especially in the moister fall, winter, and spring; they account for greater than 99% of timber volume (Küchler 1946). The central section of the Pacific Coast conifer forest vegetation type is occupied by Oregon's Coast Range (Daubenmire 1978), which supports a low elevation *Picea sitchensis* zone and an extensive *Tsuga heterophylla* zone. The *T. heterophylla* zone often contains greater amounts of *Pseudotsuga menziesii* than *Tsuga heterophylla* following logging or disturbance, and *Thuja plicata* is another important dominant (Franklin and Dyrness 1973). All of the dominants in the Pacific Coast conifer forest are long-lived, with *Pseudotsuga menziesii* living as long as 1,000 years before being replaced by *Tsuga heterophylla*.

The *western montane conifer forests*, found growing in the Cascades, Eastern Cascade Slopes and Foothills, and the Blue and Wallowa Mountain ecoregions, represent a complex group of forest types determined by both elevation and precipitation. In the moist Cascades, *Pseudotsuga menziesii* and *Tsuga heterophylla* are replaced by *Abies amabilis*, *A. procera*, and *T. mertensiana* as elevation increases. In the Eastern Cascade Slopes and Foothills and the Blue and Wallowa Mountains, the dry forests and woodlands proceed from *Juniperus occidentalis* at low elevation to *Pinus ponderosa*, *P. contorta*, *Pseudotsuga menziesii*, and *Abies grandis* at mid-elevation, and *A. lasiocarpa* and *T. mertensiana* at high elevation.

The third major vegetation type in Oregon, *intermountain grassland or shrub steppe*, is prevalent in the dry Columbia Basin, Lava Plains, Basin and Range, and Owyhee Uplands ecoregions. The dominant species of the shrub steppe remain *Artemisia tridentata* and *Pseudoroegneria spicata*, along with less common grasses, including *Festuca idahoensis, Poa secunda, and Stipa thurberiana*. An outlier in this region of dry grasslands is the Willamette Valley ecoregion, where the original prairies were a mix of dry savannas, prairies and wet prairies, with *Carex* spp., *Danthonia californica, Deschampsia cespitosa, Festuca roemeri, Hordeum brachyantherum, Juncus* spp., and many forbs (Christy and Alverson 2011).

Barbour and Christensen also briefly discuss the most common vegetation types along the Oregon coast: *salt marsh, beach*, and *dune*. Plant diversity, low in the salt marsh due to salinity and regularly fluctuating water levels, includes *Carex lyngbyei, Deschampsia caespitosa, Distichlis spicata, Glaux maritima, Jaumea carnosa, Plantago maritima, Puccinellia nutkaensis, P. pumila, Schoenoplectus pungens, Sarcocornia perennis*, and *Triglochin maritima*. Beach and dune vegetation types are similarly low in diversity because of the salt spray, blowing sand, and droughty conditions. Common dune plants include *Abronia latifolia, Cakile edentula, Carex macrocephala, Elymus mollis, Lathyrus littoralis, L. japonicus, Poa macrantha, Polygonum paronychia*, and the introduced beach grasses *Ammophila breviligulata* and *A. arenaria*. Moist depressions between the dunes contain *Carex obnupta, Juncus falcatus, J. nevadensis, Potentilla pacifica, Ranunculus repens, Salix hookeriana, and Symphyotrichum chilense.*

Phytogeography of Oregon

Floristic or *phytogeographic provinces* provide a large-scale categorization of the landscape based on plant communities and physical parameters. Multiple ecoregions can be found within any one phytogeographic province; likewise

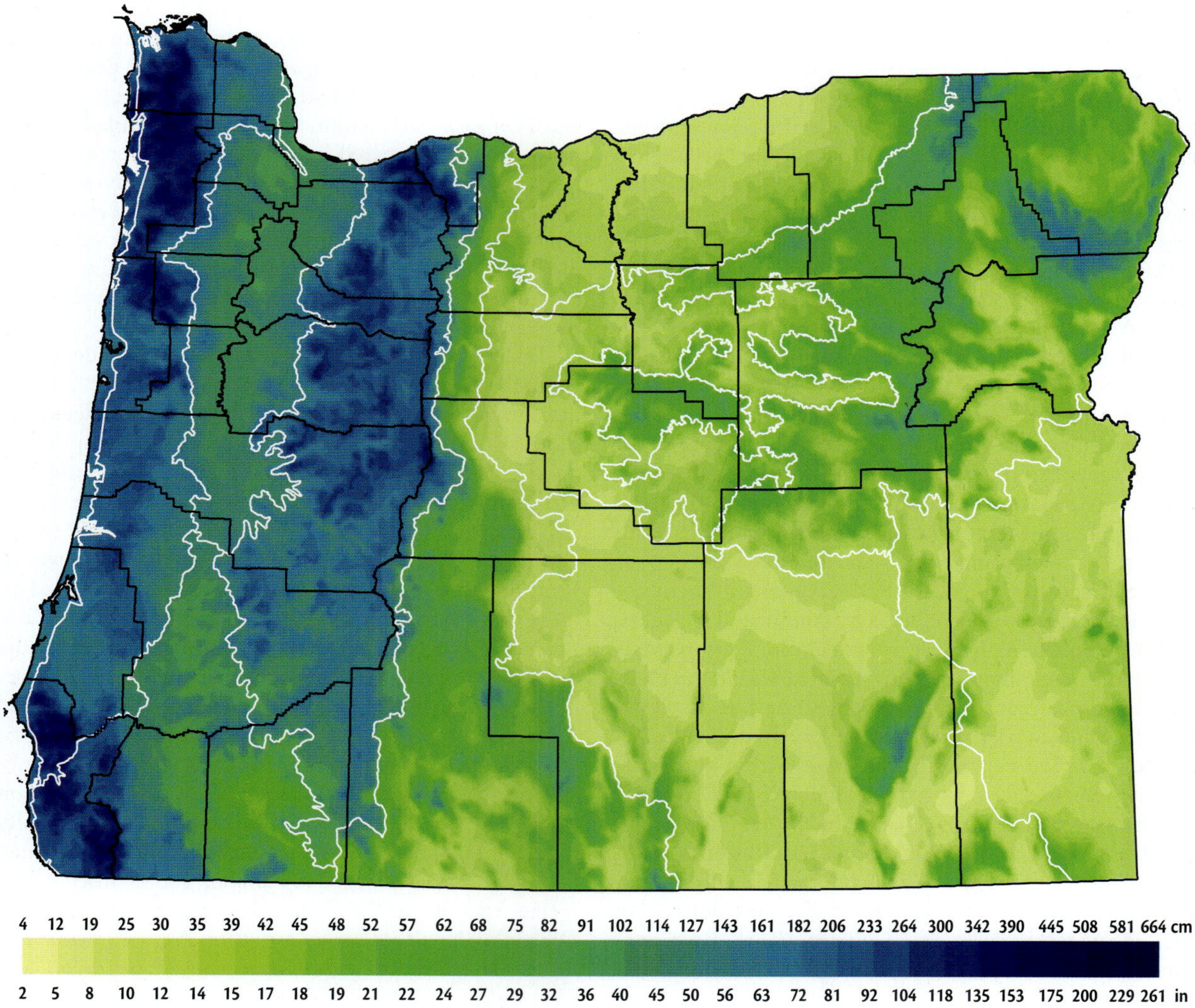

Figure 3. Average annual precipitation. Precipitation is greatest in the areas of dark shading and least in the lightest areas. County boundaries are shown with black lines and ecoregion boundaries with white lines. ©2015, PRISM Climate Group, Oregon State University, http://prism.oregonstate.edu. Map created January 2015.

several phytogeographic provinces can exist in the more-inclusive category of *phytogeographic regions*. Thorne (1993) maps and identifies four floristic provinces in Oregon based on the work of Takhtajan (1986). The Coast Range, Willamette Valley, Cascades, and northern portion of the Siskiyou Mountains ecoregions are roughly included in the *Vancouverian Province* (4a) of the Rocky Mountain Region, while the Columbia Basin, Lava Plains, and Blue and Wallowa Mountains ecoregions are included in the *Rocky Mountain Province* (4b) of the same region. The Basin and Range and Owyhee Uplands ecoregions are part of Takhtajan's *Great Basin Province* (9a) of the Madrean Region. Within the Siskiyou Mountains ecoregion, only the southern edge of the ecoregion is mapped as part of Takhtajan's *Californian Province* (9b), and the much greater northern portion is included in the Vancouverian Province. Takhtajan's treatment of the Siskiyou Mountains ecoregion does not take into account the extreme diversity and high number of regional endemic species shared between northern California and the ultramafic portion of the Siskiyou Mountains, nor other shared genera that are considered characteristic of the Californian Floristic Province, as is discussed below. The Rocky Mountain Floristic Region has no endemic plant families, but four genera have their major center of diversity in this region: *Arnica*, *Castilleja*, *Erigeron*, and *Lomatium*. The genus *Carex* is also represented by approximately 200 species in the Rocky Mountain Region.

In the broad coastal and Vancouverian Floristic Province, endemic or near-endemic genera sharing a small degree of overlap with the Californian Province include *Columbiadoria*, *Raillardella*, *Rainiera* (Asteraceae), and *Sequoia sempervirens*. Also found are species of *Cascadia*, *Elmera*, *Nothochelone*, *Oemleria*, *Romanzoffia*, *Scoliopus*, *Tolmiea*, and *Whipplea*, and five genera of achlorophyllous herbs within the family Ericaceae (*Allotropa*, *Hemitomes*, *Pityopus*, *Pleuricospora*, and *Sarcodes*). The province is also noted for high levels of conifer diversity and dominance, both in the Coast Range and the Cascades. The northern part of the Siskiyou Mountains ecoregion, also included in this phytogeographic province, contains numerous endemic plants restricted to the ultramafic soils and bedrock.

The Oregon portion of the Rocky Mountain Phytogeographic Province was unaffected by continental ice sheets and therefore has higher levels of endemism than areas farther to the north that were fully glaciated during the Pleistocene. Vertical zonation characterizes the forests, which, like the Vancouverian Province, are largely dominated by conifers. Vertical zonation is discussed within our ecoregion descriptions. Endemic species identified by Takhtajan for the sagebrush lands, which include Oregon's Columbia River, John Day Valley, and other portions of the Lava Plains, include *Astragalus arthurii*, *Camassia cusickii*, *Castilleja xanthotricha*, *Chaenactis nevii*, *Erigeron disparipilus*, *Hackelia hispida* var. *hispida*, *Lomatium rollinsii*, *L. serpentinum*, *Penstemon elegantulus*, *P. triphyllus*, *Phlox colubrina*, and *P. viscida*.

The Madrean Phytogeographic Region is extremely different from the Rocky Mountain Region, having a locally derived flora and four provinces that continue south into Mexico (Takhtajan 1986). Within the Madrean Region is the Great Basin Province; it encompasses most of the Great Basin, which is surrounded by high mountain ranges in Oregon, Idaho, California, Arizona, and the edges of Utah and Nevada. The basin contains north-south trending mountain ranges separated by internally draining basins with alkaline conditions, and the climate is extremely dry during the summer and cold during the winter. These features restrict migration of plants from surrounding floristic provinces that have more moderate climates. However, there are shared species, with low elevation species common to adjacent arid provinces and high elevation species shared with the Rocky Mountain Province. *Artemisia tridentata* is a dominant shrub, and native bunchgrasses are common as well, with alkaline-tolerant wetland plants in the basins. Endemism within the Great Basin Province is not considered high; approximately 25% of the species are unique to this province (Takhtajan 1986). The three genera Takhtajan (1986) considers best represented in the entire province are *Astragalus*, with 175 species, and *Eriogonum* and *Penstemon* with more than 100 species each. In Oregon's ecoregions within the Great Basin Province, these three genera are well represented, totaling 151 species and an additional 60 varieties, but only *Astragalus* species occur primarily in the province, with 86% within, while *Eriogonum* has 51% and *Penstemon* 58% within the province. Similarly, of the other genera considered by Takhtajan to be concentrated in the Great Basin Province—*Castilleja*, *Chrysothamnus*, *Cryptantha*, *Cymopterus*, *Erigeron*, *Gilia*, *Lomatium*, and *Phacelia*—none has more than 62% of its taxa within the province. However, many of these genera, as well as *Astragalus*, *Eriogonum*, and *Penstemon*, did have endemics within the province.

Takhtajan (1986) maps only the southern edge of Oregon's Siskiyou Mountains ecoregion within the Californian Province of the Madrean Phytogeographic Region, in contrast to the more inclusive delineation of the province given by Raven and Axelrod (1978). Our review of the taxa present indicates there is a greater affinity of the northern portion of the Siskiyou Mountains ecoregion to the flora of the Californian Province than to the Vancouverian Province. This is consistent with Takhtajan's acknowledgement that the Siskiyou or Klamath area is transitional between the two provinces. Several genera common in the Siskiyou Mountains ecoregion are centered in the Californian Province, for example *Arctostaphylos*, *Calochortus*, and *Ceanothus*; however, these and other genera do not exhibit the high levels of endemism in the Oregon Siskiyous as occurs farther south in California. Genera that are common throughout the ecoregion and whose center of diversity is in the Californian Province include *Brodiaea*, *Clarkia*, *Collinsia*, *Plagiobothrys*, *Streptanthus*, and *Triteleia*. Tree species endemic to the Californian Province that occur in Oregon include *Lithocarpus densiflora*, *Pinus balfouriana*, *P. jeffreyi*, *P. sabiniana*, *Quercus × morehus*, *Q. chrysolepis*, and *Umbellularia californica*. Endemic vernal pool taxa include eight species of *Plagiobothrys*, *Isoëtes howellii*, *Navarretia leucocephala*, and *Pogogyne zizyphoroides*.

Ecoregions

The authors of Oregon's earliest floras attempted to identify physical conditions of soil, geology, and climate that resulted in regional concentrations of distinctive groups of plant species. Morton E. Peck (1961), in his *Manual of the Higher Plants of Oregon*, identified nine such regions, which he called *plant areas*. These plant areas captured the coarse boundaries of many of the floristic or ecological regions that we now call *ecoregions*. With the assistance of aerial photos and satellite images, ecoregional boundaries have been refined in recent years. Concurrently, the ability to capture precise plant locations has further refined our understanding of the relationships between plant distributions and ecological or ecoregional boundaries. Franklin and Dyrness (1973, 1988) created a revised map of the *Physiographic and Geological Provinces of Oregon and Washington*, identifying ten provinces that maintained most of Peck's major map units but increased the number east of the Cascades. The current map of Oregon's ecoregions builds on more detailed work by the United States Environmental Protection Agency (Pater et al. 1998, Thorson et al. 2003) and the United States Forest Service (Bailey 1995). Current boundaries are similar to those used for the identification of statewide conservation priorities (Defenders of Wildlife 1998, Natural Heritage Advisory Council 1998) and development of a statewide management plan (Oregon Department of Fish and Wildlife 2006). There are minor alterations to parts of the boundaries of the Basin and Range, the Lava Plains, and the Owyhee Uplands. The Estuarine Coast ecoregion was separated from the higher-elevation headlands and mountains, which are treated as the Coast Range ecoregion.

Estuarine Coast

The Estuarine Coast forms a narrow low-lying band of sand dunes, beaches, marshes, swamps, and brackish (salty) tidal estuaries at the mouths of rivers that drain the adjacent Coast Range mountains. The estuarine coast has been considered elsewhere a part of the Coast Range but is separated in our treatment due to the low elevations and specialized harsh environments of the dunes and estuaries. Elevations range from sea level along the Pacific Ocean to 130 m (400 ft) inland, with heavy annual precipitation averaging 1,520–2,150 mm (60–85 in) and an ocean-moderated climate of cool summers and relatively frost-free winters. Along the shoreline, soft siltstone and sandstone create low, easily eroded cliffs broken at irregular intervals by steep headlands of volcanic (basalt) rock. The zone is 18–20 km (11–12 mi) at its widest near Tillamook and Florence in the north and North Bend and Bandon in the south, becoming less than 2 km (1 mi) wide south of Bandon. The Estuarine Coast continues south into northern California as a narrow disjunct strip and much farther north into Washington and British Columbia.

The most common type of vegetation in the estuaries is salt marsh, adapted to the harsh combination of salt or brackish water and extreme tidal water-level fluctuations. Broadly distributed plants of the salt marsh include *Carex lyngbyei*, *Deschampsia cespitosa*, *Distichlis spicata*, *Juncus balticus* ssp. *ater*, *Sarcocornia perennis*, *Schoenoplectus pungens*, and *Triglochin maritima*. *Picea sitchensis* swamps, once common in the upper estuary, have been widely converted to pasture. Sand dunes are equally harsh habitats due to a combination of salt spray and continually shifting sand, but are populated with *Abronia latifolia*, *Calystegia soldanella*, *Festuca rubra*, *Glehnia littoralis*, *Lupinus littoralis*, *Poa* spp., and other plants tolerant of burial. More stable back dunes support dense shrub thickets of *Gaultheria shallon* and wind-bent forests of *Pinus contorta* var. *contorta* and *Picea sitchensis*. The coastal dunes have been stabilized by the intentional planting of exotics *Ammophila breviligulata*, *A. arenaria*, *Cytisus scoparius*, and *Ulex europaeus*, resulting in reduced habitat for many native dune plants. Introduced species on the dunes have set in motion major changes in dune ecology/succession that are still playing out. Small bogs once occupied depressions among old dunes, providing habitat for the rare *Lilium occidentale* and *Darlingtonia californica*, but many of these bogs have been converted to commercial cranberry production.

Figure 4. Estuarine Coast: Salt marsh forms in the lower reaches of most major rivers, where protection is provided from the full energy of the Pacific Ocean's waves by narrow sand spits. At the mouth of the Salmon River, a sand spit has formed along the river's south shore, while a bluff of resistant basalt, here seen shrouded in fog, blocks wind and waves along the river's north shore.

More than 90 plant species, subspecies, or varieties are restricted to the Estuarine Coast in Oregon, although many of these taxa of estuaries and sand dunes also extend their range to the south into northern California and to the north into Washington and British Columbia. These species are found primarily in salt marsh, sand dunes, meadows on coastal bluffs, or in wet swales or small bogs growing within coastal dune fields. Three rare plants occur in Oregon's salt marshes, *Chloropyron maritimum* ssp. *palustre*, *Sidalcea hendersonii*, and *Stellaria humifusa*. Another showy, more common plant growing only in the estuaries is *Castilleja ambigua* ssp. *ambigua*. Three additional rarities are found growing on sand dunes of northern California

Figure 5. Estuarine Coast: European and American dune grasses were introduced to reduce sand movement along the Oregon coast, resulting in the stabilization of sand dunes and loss of native plant diversity.

and southern Oregon, *Gilia millefoliata*, *Oenothera wolfii*, and *Phacelia argentea*, while *Abronia umbellata* continues as far north as British Columbia. *Lilium occidentale* inhabits wet shrublands and boggy coastal wetlands near the coast. (Figures 4, 5, 6, 59).

Coast Range

The Coast Range is a 25–75 km (15–45 mi) wide band of steeply eroded ridges of soft, marine siltstone and sandstone underlain by volcanic rock, mostly formed in an offshore ocean environment during the late Eocene (40 million years ago) and then elevated as the oceanic platform merged with the continent (Bishop 2003, Orr and Orr 2012). Most of the steep headlands along the northern coast, such as Tillamook Head, Cape Falcon, Cape Meares, Cape Lookout, Cape Foulweather, and Yaquina Point, are volcanic rock of the Miocene age (15 million years ago), Columbia River basalt that erupted from sources in eastern Oregon and Idaho. The coastal mountains are low, with the highest elevation of 1,249 m (4,097 ft) at the summit of Marys Peak west of Corvallis. The region is the wettest in the state, with 1,270–5,080 mm (50–200 in) of annual precipitation compounded with the strong influence of marine fog. Major rivers, including the Umpqua, Rogue, and Columbia, transect the north-south mountain range.

The combination of high rainfall and mild winter temperatures provides conditions for one of the most productive temperate conifer forests in the world, with deciduous trees restricted largely to river valleys. Dominant conifers of the Coast Range include *Abies grandis*, *Pseudotsuga menziesii*, *Thuja plicata*, and *Tsuga heterophylla*, with *Picea sitchensis* restricted to near the Pacific shoreline. *Acer macrophyllum* and *Alnus rubra* are locally dominant along streams. A diverse and often dense forest understory of shrubs includes *Berberis nervosa*, *Gaultheria shallon*, *Holodiscus discolor*, *Oplopanax horridus*, *Rhododendron* spp., *Ribes sanguineum* var. *sanguineum*, *Rubus parviflorus*, and *Vaccinium parvifolium*.

Figure 6. Estuarine Coast: Impenetrable shrub thickets and herb-rich grassy meadows carpet the steep, foggy headlands and rocky bluffs along the Pacific, backed by dense rain forests of *Picea sitchensis*, *Pseudotsuga menziesii*, and other conifers.

Figure 7. Coast Range: Moss-covered *Alnus rubra* forms dense stands along the rocky margins of most Coast Range rivers.

Chest-high *Polystichum munitum* is dense on steep slopes, while *Adiantum aleuticum* inhabits streams and rock faces, and *Polypodium glycyrrhiza* grows within deciduous tree crowns. Two other impressive aspects of the flora are mushroom diversity on the forest floor and the predominance of mosses and lichens within tree canopies and on soil and rotting logs. In recent decades, extensive portions of the conifer forest have been replaced by plantations of *Pseudotsuga menziesii*, greatly reducing ecological and plant diversity.

There are only eleven plant species that are known to occur only in Oregon's Coast Range, while a similar number are known from both the Estuarine Coast and Coast Range. The majority of plants in the Coast Range are shared with moist conifer forests of the Western Cascades. Two rare plants occupy the grassy coastal bluffs at the forest's edge, *Sidalcea hirtipes* and *Silene douglasii* var. *oraria*. *Castilleja chambersii* grows on open talus and shaded cliffs on peaks in Clatsop County. *Erythronium elegans*, an Oregon endemic, and *Dodecatheon austrofrigidum* and *Filipendula occidentalis*, found in both Washington and Oregon, grow in higher elevation forests. *Dodecatheon austrofrigidum* and *F. occidentalis* occur on moist river edges, cliffs, and seeps in shaded, conifer forests, while *E. elegans* grows on rocky slopes, cliffs, meadows, and bog edges. *Erythronium revolutum*, although not as rare as *E. elegans*, is also found on moist forested sites. Another showy plant that grows primarily in the shrubby forest edge and bluffs is *Castilleja affinis* var. *litoralis*. Some of the higher peaks in the Coast Range, including Saddle Mountain and Onion Peak, are considered glacial refugia, and support disjunct northern plants such as *Elliottia pyroliflora*, and northern species at the southern extent of their range, such as *Douglasia laevigata* and *Synthyris schizantha* (Chambers 1973). (Figures 7, 8, 9, 60).

Figure 8. Coast Range: Because of high rainfall for seven or more months beginning in October, mosses and lichens drape branches and trunks of *Pseudotsuga menziesii*, even as far inland as Green Mountain near the eastern margin of the Coast Range.

Figure 9. Coast Range: Much of the Coast Range is logged as short-rotation *Pseudotsuga menziesii* plantation. The several clear cuts seen in this photo occurred between 2000 and 2010.

Willamette Valley

The relatively flat alluvial terraces and floodplain of the Willamette River Valley are sharply bordered by foothills of the Coastal Mountain Range to the west and the Cascade Range to the east. Rounded knobs of volcanic rock scattered the length of the valley on its eastern margin support the few remnants of native upland prairie and white oak savanna that once characterized the region. Elevations range from near sea-level along the Columbia River at Portland to 152 m (500 ft) at the southern valley limit. Elevations reach 640 m (2,100 ft) in the foothills along the edges of both the Cascade Mountains and the Coast Range, and much higher at the upland margins of the watershed. The combination of abundant rainfall, averaging 940–1,520 mm (37–60 in) annually, and rich alluvial silt and heavy clay soils has resulted in early settlement and almost complete conversion of the region to agriculture.

Beneath the present surface of the Willamette Valley, marine sediments were deposited in a shallow tropical sea between 24 and 30 million years ago during the Eocene and Oligocene epochs, followed by dramatic cooling of the climate and regional uplift of the area in the latest Oligocene. Miocene basalts began to flow from eastern Oregon into the northern basin around 15 million years ago, decomposing into an irregular, hilly landscape. Much later, beginning 2 million years ago, volcanic activity in the Portland area continued until 50,000 years ago, stippling the area with small cones. Thick deposits of silt, sand, and gravel cut by Cascade glaciers were carried into the subsiding valley by west-flowing streams off the volcanic range. But the rich, fine-textured surface deposits of the Willamette Valley, up to 100 m (300 ft) thick, are the product of scores of great Ice-Age floods between 14,500 and 19,000 years ago. These "Missoula" floods originating from a huge lake in western Montana repeatedly turned the entire Willamette Valley into a muddy body of water, leaving the thick silt/clay soils that today foster agriculture. Recent detailed studies show that some of the surface clay deposits are derived from the fluvial redistribution of volcanic ash deposits from Mount Mazama approximately 7,000 years ago (James and Baitis 2003; William Orr, personal communication).

Up until the mid–20th century, the Willamette and its many tributaries regularly flooded in the spring, greatly limiting land use near the river. Dam construction on Cascade streams has slowed these events, but major floods still occur at an average ten-year interval. Reduced flooding has

Figure 10. Willamette Valley: Diverse wet prairies characterized the flat, poorly drained alluvial plain of the Willamette River and its tributaries prior to the extensive agricultural development that followed flood control and drainage.

Figure 11. Willamette Valley: Curving meander loops mark abandoned channels in the farm fields along Muddy Creek, one of the many small tributary streams that eventually joins the Willamette River. The rich alluvial soils of the Willamette Valley in the foreground are heavily farmed, while conifer forests cloak the steep ridges of the Coast Range to the west.

allowed the region to become the most intensively farmed and most densely populated area in the state. The cultivation of wheat, grass seed, hazelnuts, wine grapes, Christmas trees, and ornamental nursery stock has largely eliminated natural vegetation from the region in all but steep, rocky, and poorly drained sites. Along with the agricultural development of the valley came the introduction of plant species from elsewhere, especially Europe. The Willamette Valley has over 700 introduced plants, more than in any other ecoregion in the state.

Prior to European settlement, the upland ridges were covered with a mix of forest, woodlands, and savannas, while the more regularly flooded alluvial deposits along the Willamette River and tributaries supported a complex matrix of dry and wet prairie, marshes, and riparian forest. Native American use of fire for resource management maintained the open prairies, savannas, and woodlands. Over 31% of the landscape was dominated by prairie, with 67% drier upland prairie and 33% wet prairie (Christy and Alverson 2011). Among the common upland prairie grasses were *Achnatherum lemmonii*, *Danthonia californica*, *Elymus trachycaulus*, *Festuca roemeri*, *Koeleria macrantha*, and *Poa secunda.* Forbs of the drier prairie included species of *Balsamorhiza*, *Brodiaea*, *Lomatium*, *Madia*, *Perideridia*, and *Triteleia*, along with *Fritillaria affinis*—all important food crops for the Native American residents of the valley. Other prairie forbs include *Calochortus tolmiei*, *Lupinus oreganus*, *Sidalcea virgata*, and *Wyethia angustifolia.* Wet prairie in most of the valley was dominated by *Carex* spp., *Deschampsia cespitosa*, and *Juncus* spp., while *Camassia leichtlinii* ssp. *suksdorfii*, *C. quamash* ssp. *maxima*, *Madia* spp., and *Perideridia* spp. were important food plants that often dominated large areas of the wet prairie. Over 98% of the prairies have been converted to agricultural and residential use.

Savanna of *Quercus garryana* (and *Q. kelloggii* in Lane County, where it is at its northern range limit) and mixed oak-conifer (particularly *Pseudotsuga menziesii* and *Pinus ponderosa*), covered another 18% of the ecoregion's uplands, followed by 14% of semi-open woodlands of the same trees. Savannas supported many of the already mentioned upland prairie grasses and forbs, as well as others that include *Agrostis hallii*, *Cynoglossum grande*, *Elymus glaucus*, *Erythronium oregonum*, *Eucephalis vialis*, *Festuca californica*, *Iris tenax*, *Lathyrus holochlorus*, *Ligusticum apiifolium*, and *Sidalcea campestris.* Shrub cover was high in the woodlands, including *Corylus cornuta* var. *californica*, *Holodiscus discolor*, *Rosa nutkana*, *Symphoricarpos albus*, and *Toxicodendron diversilobum*. Prairies, savannas, and woodlands in the Willamette Valley are thought to have largely been maintained by Native American fire management practices.

In the mid-1800s, 26% of the Willamette Valley was upland forests, primarily on the low ridges and foothills at margins of the valley, with *Pseudotsuga menziesii-Tsuga heterophylla*, *Pseudotsuga menziesii*, and *Pseudotsuga menziesii-Quercus garryana* forests making up the majority (Christy and Alverson 2011). While large, old growth conifers were present in the Willamette Valley in the mid-1800s, much of the upland forest represented younger seral stages, perhaps reflecting a frequent fire return interval.

Figure 12. Willamette Valley: Vernal pools with *Plagiobothrys nothofulvus* or *scouleri* were much more prevalent in the Willamette Valley prior to creation of flood-control dams on the Willamette River and its larger tributaries.

Figure 13. Willamette Valley: Open upland prairie and oak savanna occupied the rolling ridges along the margins and within the Willamette Valley. Fire control allowed *Quercus garryana* and *Pseudotsuga menziesii* forest to replace prairies and savannas, resulting in reduced herbaceous diversity.

Riparian areas along the Willamette River and tributaries occupied seven percent of the ecoregion and were once populated with *Acer macrophyllum*, *Fraxinus latifolia*, *Populus trichocarpa*, and *Salix* spp., along with a variety of understory shrubs. These forests supported a diverse spring flora, particularly on the higher terraces, including *Delphinium trolliifolium*, *Dicentra formosa*, *Erythronium oregonum*, *Hydrophyllum* spp., *Maianthemum racemosum*, *Mertensia platyphylla*, *Thalictrum occidentale*, *T. polycarpum*, *Trillium albidum*, and *Viola glabella*. Other wetlands were scattered throughout the ecoregion, including beaver ponds, marshes, shrub swamps, and seasonally flooded vernal pools. The bottomlands of the Columbia River in the vicinity of Portland were historically a rich mosaic of floodplain prairie, marshes, swamps, and forest but are now largely modified by agriculture and flood control.

Only twelve plant species have distributions that are limited to the Willamette Valley ecoregion in Oregon, and several of these are rare species. *Castilleja levisecta* once occupied wet prairies and meadows from the Willamette Valley, where it is being reintroduced, to British Columbia, while *Delphinium pavonaceum* is a Willamette Valley endemic found on drier prairies. *Delphinium leucophaeum* occurs on rocky cliffs and open ground and is now mainly restricted to ditches and fencerows; its range extends into southern Washington. *Lupinus oreganus* is an upland prairie species occurring primarily in the Willamette Valley, with outlying populations in the Umpqua Valley and in SW Washington. *Erigeron decumbens* is a Willamette Valley endemic of wet prairie, dry prairie, and savanna on heavy clay soils. *Horkelia congesta* ssp. *congesta* is a wet prairie endemic that also grows in the Siskiyou Mountains. *Lomatium bradshawii* is a Willamette Valley and southern Washington endemic of wet prairies. *Sidalcea nelsoniana* grows in wet prairies, meadows, and *Fraxinus latifolia* riparian forests in the Willamette Valley and southern Washington and occasionally along the edge of the Coast Range. *Howellia aquatilis* is currently known from only a couple of ephemeral ponds, lakes, and sloughs in the Willamette Valley and southern Washington but was previously collected in Idaho, Montana, and California. (Figures 10, 11, 12, 13).

Siskiyou Mountains

The Siskiyou and Klamath Mountains form a rugged, geologically complex block of mountains best known for its ultramafic rock and unique flora. The Siskiyous continue south into California, with the Pacific Ocean to the west, the Coast Range to the north, and the higher Cascade Mountains to the east. The Klamath Mountains were formed by the accretion of some 15 terranes or ocean crustal slabs of volcanic plateaus and islands onto the west coast of the continent between 200 and 100 million years ago (Orr and Orr 2012). Because these terranes were constructed over an extended period of time and in diverse settings across the Pacific basin, they display a remarkable diversity of rocks and minerals, many of which are of economic importance. Gold, platinum, silver, and copper have

Figure 14. Siskiyou Mountains: Vast expanses of golden *Lasthenia californica* blanket the vernal pools atop Upper Table Rock in early spring. As the pools dry down throughout the spring and early summer, an ever-changing display of wildflowers emerges.

Figure 15. Siskiyou Mountains: Fens along rocky streams and seeps support *Darlingtonia californica* and several other rare endemic plants. *Pinus jeffreyi* and *Chamaecyparis lawsoniana* occupy the rocky uplands bordering the fen.

Figure 16. Siskiyou Mountains: Few plant species can tolerate the high levels of heavy metals found in the ultramafic rock of the Siskiyou Mountains, resulting in an open, steep landscape of exposed bedrock and scattered small trees and shrubs, with abundant signs of past fires.

Figure 17. Siskiyou Mountains: *Rhododendron occidentale* blooms along the Illinois River. *Pinus jeffreyi* and *Chamaecyparis lawsoniana* grow on the surrounding serpentine slopes.

been continuously mined here since the mid 19th century. The distinctive ultramafic rocks of the region, which include serpentinite, dunite, and peridotite, were formed as an oceanic floor which was later altered by heat and pressure before being forced to the surface and joined to the western edge of the North American continent. Only a few plant species can grow on the nutrient-poor soils derived from these rocks, which are typically low in silica and high in magnesium and iron, as well as the heavy metals copper, chromium, cobalt, zinc, and nickel. The highest peak in the region is Mt. Ashland at 2,280 m (7,533 ft), while the remainder of the province varies from 600 m (1,968 ft) near the coast to 1,060 m (3,477 ft) to the east. Annual precipitation averages from 3,300 mm (130 in) near the coast to 500 mm (20 in) in the east. Historically, mining of gold, chromium, and nickel was a major regional industry, and signs of mining are widespread. Logging remains a major commercial activity in the productive non-serpentine areas of the region, and orchards and vineyards are locally important to the economy in eastern valleys and foothills.

The nutrient-poor soils formed from serpentine, peridotite, and dunite support over 140 species of specialized native endemic plants able to tolerate heavy metals and are found only in this ecoregion. Open woodlands are composed of a limited number of trees capable of growing on serpentine soils, including *Chamaecyparis lawsoniana*, *Pinus jeffreyi*, and *Umbellularia californica*. Woodland understory is diverse in shrubs that include *Arctostaphylos viscida*, *Ceanothus pumilus*, *C. integerrimus*, *C. thyrsiflorus*, *Notholithocarpus densiflorus*, and *Rhododendron occidentale*, as well as perennial grasses, including *Achnatherum occidentale* ssp. *californicum*, *Danthonia californica*, and *Koeleria macrantha*. Distinctive fens, common along small rocky streams and seepages, are home to *Darlingtonia californica*, *Gentiana setigera*, and *Rudbeckia glaucescens*, while a common fern on the dry serpentine rock is *Aspidotis densa*. Forests on non-serpentine rock and soils are characterized by *Pseudotsuga menziesii* and *Quercus garryana* and have understory and groundcover diversity similar to that of the adjacent Coast Range and Cascades ecoregions. Both *Quercus chrysolepis* and *Q. kelloggii* are equally as common as *Q. garryana* on rocky non-serpentine soils in the ecoregion.

Rare plants in the region are concentrated on serpentinite and dunite and soils derived from these heavy-metal rich rocks. Many of these plants are narrow endemics of only southwestern Oregon, but several have ranges that extend into adjacent northwestern California. Endemics that grow in the *Darlingtonia*-dominated fens include *Carex klamathensis*, *C. serratodens*, *Castilleja elata*, *Epilobium oreganum*, *Gentiana setigera*, *Hastingsia alba*, *H. bracteosa* var. *atropurpurea*, *H. bracteosa* var. *bracteosa*, and *Viola primulifolia* ssp. *occidentalis*. Several species also grow in vernal pools or seasonally moist meadows in valley bottoms, including *Limnanthes alba* ssp. *gracilis*, *L. pumila* ssp. *grandiflora*, *L. pumila* ssp. *pumila*, *Lomatium cookii*, *Plagiobothrys figuratus* ssp. *corallicarpus*, *P. hirtus*, and *P. lamprocarpus*. *Picea breweriana*, a regional endemic of high elevations in both Oregon and California, is considered one of the earliest spruce species, linking the genus' origin to the North American continent. *Quercus sadleriana*, which occupies the understory of the ecoregion's open conifer forests, is also considered a relict oak related to eastern North American and Asian chestnut oaks (Muth 1976). Endemic plants that grow on dry, rocky soils of the woodlands include *Arabis mcdonaldiana*, *Calochortus coxii*, *C. greenei*, *C. howellii*, *Carex scabriuscula*, *C. serpenticola*, *Castilleja brevilobata*, *C. schizotricha*, *Diplacus congdonii*, *D. kelloggii*, *Erythronium citrinum* var. *citrinum*, *Fritillaria gentneri*, *Microseris howellii*, *Ranunculus austrooreganus*, *Sedum oblanceolatum*, *Sophora leachiana*, *Streptanthus howellii*, and *Triteleia hendersonii* var. *leachiae*. *Horkelia hendersonii*, *Lupinus lepidus* var. *ashlandensis*, and *Tauschia howellii* grow on gravelly, granitic soils in open, dry forests. *Epilobium siskiyouense* colonizes open talus, while *Sedum moranii* inhabits cliffs and rock faces. (Figures 14, 15, 16, 17, 61, 62).

Figure 18. Cascades: Where moist Western Cascades forests escape logging and fires, they can develop impressive old growth trees.

Cascades

The Cascades ecoregion consists of two ranges, the Western Cascades and the High Cascades. Collectively known as the Cascades, they form a north-south trending chain of extinct and active volcanoes 30–120 km (19–75 mi) wide that stretches across Oregon and continues north into Washington and British Columbia and south into northern California.

The High Cascades contains the state's highest peaks, including Mount Hood, Mount Jefferson, and the Three Sisters, which range from 3,062 to 3,427 m (10,043 to 11,240 ft) in elevation. These steep volcanic edifices formed less than a million years ago, and most display flanks deeply eroded by glaciation. Modest to tiny glaciers persist on the highest peaks, and their size continues to diminish with the warming climate of the present interglacial phase. The Western Cascades comprises the western two-thirds of the region and consists of older volcanoes and lava flows,

Figure 19. Cascades: The Western Cascades are characterized by rocky ridges with steep, mostly forested slopes.

Figure 20. Cascades: Fire plays a big part in the forests of the ecoregion. Here, the remnants of an old fire can be seen on Grizzly Peak in Jackson County, while smoke billows above from another fire burning in the distance.

Figure 21. Cascades: In the Western Cascades, steep, rocky meadows on low- to mid-elevation, south-facing slopes support *Quercus garryana*, and abundant annuals follow seeps.

dating back some 40 million years ago. Weathering and heavy erosion have sculpted this region into steep ridges dissected by numerous stream valleys. Ridge crests average 1,500 m (4,900 ft) but reach just over 2,000 m (6,560 ft) in the south.

The rocks are predominantly dark volcanic basalts or lighter-colored andesites, but volcanic eruptions in the last few thousand years have blanketed large areas of the High Cascades with pumice, cinders, and ash. Soils are commonly thin and rocky, with thicker soils associated with glacial moraines. Lava flows in fresh condition are still exposed in some areas, notably at McKenzie and Santiam passes.

Annual precipitation ranges from 1,140 to 3,550 mm (45 to 140 in), greatest at high elevations. Rainfall declines rapidly to the east in the rain shadow of the north-south crest of the Cascades. Small glacial (cirque) tarn lakes are clustered around the bases of the high peaks, and narrow streams flow off both the east and west flanks of the range. West-flowing rivers tend to have narrower valleys with higher gradients in contrast to the younger, less mature, east-flowing streams. Virtually all of the large west-flowing streams of any size have been dammed, creating reservoirs that generate electricity, control flooding, and provide water for municipalities and the irrigation of crops in the adjacent Willamette Valley.

The moist climate supports extensive conifer forests. The tree species of the forests are arranged zonally by elevation. Below 1,200 m (3,940 ft), forest dominants include *Pseudotsuga menziesii* and *Tsuga heterophylla*, with *Calocedrus decurrens* and *Thuja plicata* and understories of *Acer circinatum*, *Berberis nervosa*, and *Polystichum munitum*, and a diverse assortment of perennials. *Chrysolepis chrysophylla* is a frequent sub-canopy tree on rocky soils and upper slopes, and drier meadows associated with stands of *Arbutus menziesii* and *Quercus garryana* occur on south-facing slopes. Between elevations of 1,200–2,000 m (3,940–6,560 ft), *Tsuga heterophylla* gives way to *T. mertensiana*, and *Abies procera* and *A. amabilis* become common, along with *Vaccinium* spp. and *Rhododendron macrophyllum*. *Quercus garryana* in dry openings tends to be dwarfed. Commercial forestry is the primary land use, especially at middle and lower elevations. The Western Cascades region is a highly productive tree-growing area, and, as a result, the region is riddled with logging roads, which are often conduits of non-native, invasive species, including *Centaurea × moncktonii*, *Cytisus scoparius*, and *Lotus corniculatus*, which was purposely planted following logging.

The highest peaks above 2,000 m (6,560 ft) are characterized by open moraines, glacial lakes, pumice flats, and permanent snowfields. Vegetation is a mosaic of *Carex* spp., *Phyllodoce empetriformis*, *Vaccinium* spp., low-growing *Salix* spp., and tufted-hairgrass meadows. *Xerophyllum tenax* and *Vaccinium* spp. grow beneath open woodlands of *Tsuga mertensiana* and *Abies lasiocarpa* var. *lasiocarpa*. *Pinus albicaulis* grows in open parkland and rocky areas. In poor soil, so common in pumice-laden areas along the Cascade crest, extensive areas may be covered with forests of *Pinus contorta* var. *latifolia* with little understory other than *Vaccinium scoparium*. *Abies concolor × A. grandis*, *Abies magnifica × A. procera*, and *Calocedrus decurrens* are more common to the south where the Cascades meet the Siskiyou Mountains, while *Callitropsis nootkatensis* becomes more common to the north.

The Cascades cross from the northerly Vancouverian Floristic Province to the Californian Province in southern Lane and northern Douglas counties. This is in part due to an east-west arm of the Western Cascades known as the Calapooya Mountains, which reaches as high as 1,860 m (6,100 ft) at its east end and tapers down as it winds its way

Figure 22. Cascades: Wildflowers line a meltwater creek on the south side of Broken Top, a typical peak in the High Cascades.

west toward the Willamette Valley, forming somewhat of a barrier to the migration of many plant species. A number of species typical of the Vancouverian Floristic Province, such as *Castilleja rupicola* and *Douglasia laevigata*, reach their southern limit here, often finding refuge only on the cool, shaded, north-facing sides of ridges, even though they may grow in open areas farther north. The area to the south tends to be warmer and drier with a higher percentage of meadows, and typical Western Cascades plants are joined by some species more common east of the Cascades or from the Siskiyous, including *Fritillaria glauca* and *Ribes cereum*. Conversely, the northern half of the Western Cascades is generally cooler and moister with more snow cover in the winter and a larger number of wetlands. Plants of the Coastal ecoregions, which favor cool, damp conditions, may find suitable habitat at the northern end of the Cascades, *Trientalis europaea* ssp. *arctica*, for example.

Forty-nine plant taxa are found growing only in the Cascades ecoregion. The southern portion of the ecoregion contains the only known populations of four Cascades endemic plants that occur only in Oregon: *Kalmiopsis fragrans*, a shrub of rocky outcrops within conifer forests; *Eriogonum villosissimum*, a perennial of crevices and ledges on exposed cliff faces; *Boechera horizontalis*, a cespitose subshrub of high elevation pumice slopes; and *Collomia mazama*, a perennial of high-elevation meadows and woodlands of lodgepole pine and true fir. *Iris tenuis* is another narrow endemic that grows within *Pseudotsuga menziesii* stands almost entirely in Clackamas County. This ecoregion also supports two endemic plants, *Erigeron howellii* and *E. oreganus*, which are restricted to rock outcrops along the forested portion of the Columbia River Gorge in Washington and Oregon, and *Eucephalus gormanii*, which also grows on rocky slopes and cliffs only in the central part of the Cascades. *Cimicifuga laciniata* and *Corydalis aquae-gelidae* are herbaceous perennials that occupy relatively cool riparian habitats along small streams that are submerged during spring high water conditions. Adjacent to streams and waterfalls in the Columbia River Gorge within the Cascades ecoregion is the rare grass, *Agrostis howellii*. The Cascades is also the only Oregon ecoregion with populations of *Asplenium septentrionale*, a rare but widely distributed rock-outcrop fern found throughout the western US, and for *Sisyrinchium sarmentosum*, which occupies mid-elevation mountain meadows

Figure 23. Cascades: *Pinus contorta* var. *latifolia* woodlands grow on lava north of North and Middle Sister that flowed less than a thousand years ago. Little soil development is evident on these recent lava flows.

Figure 24. Cascades: In the High Cascades, mountain glaciers have scoured the bedrock, forming an abundance of small cirque lakes such as Bays Lake at Jefferson Park.

Figure 25. Eastern Cascade Slopes and Foothills: Open stands of *Pinus ponderosa* border the meandering Deschutes River, which has cut through the region's characteristic ash and lava beds.

Figure 26. Eastern Cascade Slopes and Foothills: Rowena Plateau, renowned for its wildflower diversity, characterizes the transition from open *Pinus ponderosa* and *Quercus garryana* woodland to shrub and grassland farther east in the Columbia River Gorge. *Lupinus polyphyllus* and *Balsamorhiza sagittata* blanket the grassland in late spring.

in northern Oregon and southern Washington. Other rare Oregon endemics occurring mainly in the Western Cascades include *Romanzoffia thompsonii*, which grows on seepy slopes; *Phacelia verna*, an annual of gravelly washes found only in Douglas and southern Lane counties; and *Frasera umpquaensis*, a tall perennial of meadows and open forest in southwestern Oregon. (Figures 18, 19, 20, 21, 22, 23, 24, 54, 56, 57, 63, 64, 65, 66).

Eastern Cascade Slopes and Foothills

The Eastern Cascade Slopes and Foothills ecoregion, located in the rain shadow of the Cascades, consists of an extensive volcanic plateau with a surface veneer of volcanic ash and pumice. The southern third of the ecoregion contains large playa lakes and wetlands formed by block faulting, similar to that in the Basin and Range ecoregion to the east, and the area between Klamath Lake and Goose Lake is mapped by geologists as part of the Basin and Range ecoregion (Orr and Orr 2012). Elevations and precipitation are both lower than in the Cascade Mountains, with elevations 760–2,530 m (2,500–8,300 ft) and annual precipitation of 380–1,400 mm (15–55 in). Winter temperatures are cold with significant snow accumulation. Soils derived from ash and pumice are excessively drained and prone to drought. Springs and perennially flowing streams are common throughout the region. Historically Klamath and Goose lakes were part of a much larger glacial (pluvial) lake that extended well to the south into California and Nevada.

Open, fire-dependent ponderosa pine and lodgepole pine woodlands and forests dominate the broad ridges of much of the region, with an open understory of shrubs including *Arctostaphylos patula*, *Ceanothus velutinus*, *Purshia tridentata*, and *Symphoricarpos albus*, and several grasses including *Achnatherum occidentale*, *Festuca idahoensis*, and other bunchgrasses. East of Mount Hood, precipitation is

Figure 27. Eastern Cascade Slopes and Foothills: Recent lava flows dating from 8,000 to 12,000 years ago characterize portions of the ecoregion, as seen on Paulina Peak. Portions of these young flows remain unvegetated, but *Pinus albicaulis* has established locally in the foreground and *Pinus contorta* var. *latifolia* in large areas below.

Figure 28. Eastern Cascade Slopes and Foothills: Open *Pinus ponderosa* woodlands, as seen in this photo, and lodgepole pine forests occupy the greater part of the rolling slopes of the ecoregion.

Figure 29. Columbia Basin: Steep fescue grasslands with scattered *Ericameria nauseosa* and *Artemisia tridentata* blanket the slopes above the Columbia River near Wallula Gap.

Figure 30. Columbia Basin: Open scree or talus slopes below cliffs of basaltic caprock are locally stabilized by *Festuca idahoensis* and other native grasses.

Figure 31. Columbia Basin: Rolling loess (windblown silt) hills form a broad plateau above the Columbia River. Fescue grassland, as seen in the foreground, has been largely replaced by dryland wheat farming, except for the steep ravines and canyons.

higher, and the forests contain *Abies grandis*, *Pseudotsuga menziesii*, and *Quercus garryana* in places. Riparian areas were originally inhabited by *Alnus incana*, *Cornus sericea*, *Populus tremuloides*, and *Salix* spp. Klamath Lake was populated with large expanses of *Schoenoplectus acutus* var. *occidentalis* and *Typha latifolia*, but the saline-alkaline soils of the playa lakes support *Distichlis stricta* and *Sarcobatus vermiculatus*. Forests are currently managed for timber, while riparian areas and lake margins have typically been converted to either pasture or grass production. East of Mount Hood, forest land has been converted to pear and apple orchards.

While there are six native plant species whose range in Oregon is restricted to this ecoregion, only two are endemic to the region. One of these species, *Astragalus applegatei*, is very localized in alkali wetlands near Klamath Lake, while *Penstemon glaucinus* occupies open lodgepole pine stands. *Botrychium pumicola*, also found in California, grows on pumice and loose volcanic ash south of Bend, with several populations near Newberry crater, Broken Top, Mount Bachelor, and surrounding high-elevation pumice and ash deposits, as well as a single population from open *Pinus contorta* var. *latifolia* forest at low elevation. *Penstemon peckii* is scattered in open *Pinus ponderosa* stands near Sisters on gravelly volcanic soils, while *Astragalus peckii* occurs in pumice openings in *Juniperus occidentalis* and pine woodlands. *Penstemon barrettiae* is restricted to rock outcrops and cliffs, while *Lomatium columbianum* grows on rocky slopes along the Columbia River and nearby canyons in northern Oregon and southern Washington. Several species are restricted to open forests in the southern portion of the ecoregion and nearby California and/or Nevada, including *Carex davyi* and *Phacelia hydrophylloides*. (Figures 25, 26, 27, 28, 67).

Columbia Basin

The Columbia Basin region encompasses the basin of the Columbia River in both Oregon and Washington with a deep, spectacular gorge through the Cascades and a rolling, broad plateau that covers northeastern Oregon and southeastern Washington. The region borders three other Oregon ecoregions: the Blue and Wallowa Mountains on the south and east, the Lava Plains on the south, and the Eastern Cascade Slopes and Foothills on the west. The deep, 2–3 km (1–2 mi) wide gorge was re-cut several times after successive flows of the Miocene age Columbia River Basalt Group plugged the canyon. Catastrophic Pleistocene floods from glacial Lake Missoula in western Montana flushed the gorge of sediments over 100 times.

Wind-blown silt (loess) from glacial outwash streams in eastern Washington contributed fertile soils to the broad plateau along the south side of the river during the Pleistocene. The plateau is up to 110 km (65 mi) wide, and while the loess forms deep, fertile soils, annual precipitation is low, 170–650 mm (7–25 in). The region is underlain by a very thick sequence of lava flows, collectively known as the Columbia River Basalts, with local interbedded soil layers and volcanic ash deposits (Bishop 2003, Orr and Orr 2012). These intermittent widespread Miocene lava flows spanned a period of approximately 11 million years, between 17 and 6 million years ago. Elevations range from 60 m (200 ft) along the river to 1,340 m (4,400 ft) on the steep upper terraces abutting the Blue and Wallowa Mountains. The Deschutes, John Day, and Umatilla rivers and several large creeks cut deep, narrow canyons through the rolling loess topography and basalt plateau. The upper plateau is dissected by numerous small streams, creating a high-relief, mature landscape along its southern boundary.

The majority of the region in Oregon was originally a mix of sagebrush and grasslands. The arid sagebrush steppe is dominated by *Artemisia tridentata* ssp. *wyomingensis* and *A. tridentata* ssp. *tridentata* and interspersed with grasslands of *Festuca idahoensis*, *Poa secunda* ssp. *secunda*, and *Pseudoroegneria spicata*. There were also showy wildflowers in the grass and shrublands, including species of *Balsamorhiza*, *Eriogonum*, *Lomatium*, and *Lupinus*. The large sandy areas along the Columbia River were dominated by *Purshia tridentata* with *Elymus trachycaulus*, *Hesperostipa comata* ssp. *comata*, and *Poa secunda*. Forests were limited to the dissected upper plateau, where open stands of *Pinus ponderosa* and *Pseudotsuga menziesii* grew, and narrow riparian forests in the steep ravines. Natural vegetation is now limited to the steep ravines, bluffs, and cliff faces, as the plateau is grazed and extensively farmed for dryland wheat, while the Columbia River bottomlands are irrigated for grass, alfalfa, and a variety of other crops.

The ecoregion shares many plant species with other high desert ecoregions east of the Cascades. Only 15 plant species are restricted to the Columbia Basin ecoregion, and of these, five species are rare or endemic. *Achnatherum hendersonii* grows in Oregon, Washington, and Idaho on shallow soils derived from basalt, along with *Poa secunda* (Binney and Bradfield 2000). *Artemisia campestris* var. *wormskioldii* historically grew at sites in Washington and Oregon, occupying rocky shoreline within the flood zone of the

Figure 32. Columbia Basin: Dissected Columbia River plateau converted to wheat farming.

Figure 33. Lava Plains: Twickenham. The rolling, eroded ash beds of the ecoregion are vegetated with open grassland, sage steppe, or woodlands of *Pinus ponderosa*, *P. contorta* var. *latifolia*, or *Juniperus occidentalis* var. *occidentalis*.

Figure 34. Lava Plains: As seen along the deeply incised canyon of the Deschutes River at Cove Palisades State Park, parallel layers of volcanic ash and lava are interbedded with layers of stream gravel between one and seven million years old.

Figure 35. Lava Plains: While land on most of the droughty volcanic soils and exposed rock of the Lava Plains is used as range for cattle, the richer alluvial soils along the John Day River and other rivers is either more intensively grazed or irrigated for fodder or row crops.

Columbia River and tributaries. Alteration of the flood regime by dams has eliminated northern wormwood from all but two Washington sites, although restoration is underway on Oregon islands in the Columbia River. *Astragalus collinus* var. *laurentii*, a narrow endemic from four counties, occupies a diversity of sites ranging from loess to sandy or rocky soils overlying basalt, where it grows with native bunchgrasses. *Astragalus tyghensis*, another local endemic, grows on rocky sandy to clay soils in sage steppe and native grasslands of the Tygh Valley. *Erythranthe jungermannioides* is restricted to moist crevices and seeps in basalt cliff faces and canyon walls along the Columbia River and both small and large tributaries in Washington and Oregon. (Figures 29, 30, 31, 32, 52).

Figure 36. Lava Plains: Clarno Butte, the core of one of many volcanos that deposited massive amounts of lava and ash upon the landscape 40–50 million years ago, has been deeply eroded by wind and water. Droughty soils and limited precipitation restrict trees (*Juniperus occidentalis* var. *occidentalis*) to ravines and canyons over large areas.

Figure 37. Lava Plains: Eocene and Oligocene ash flows have weathered to montmorillonite clays at Lands Inn. Rapid shrinking and swelling of these clays greatly limits vegetation establishment.

Lava Plains

The Lava Plains is an elevated plateau lying east of the Cascade Range and southwest of the Blue Mountains that includes the Deschutes, Crooked, and John Day river valleys, as well as all three of the ash units of the John Day Fossil Beds National Monument. Elevations on most of the plateau range from 365–1,370 m (1,200–4,500 ft), with isolated low mountains reaching 1,890 m (6,200 ft). A resistant basalt cap from Newberry volcano has reduced erosion and maintained the flat character of the landscape along the southern Deschutes River and locally elsewhere. While there are local loess soils, most of the ecoregion's soils are volcanic, derived from flows, ash, and pyroclastic debris. Sediments here have been widely transported and reworked by water. The long history of volcanic ash deposition, beginning in the Eocene over 40 million years ago, is preserved in such features as the John Day Fossil Beds, while as recently as 5–7 million years ago in the late Miocene sediments were deposited along the Deschutes River at the western edge of the ecoregion. The nature of the volcanic deposits is equally as variable, ranging from resistant welded tuff and basalt lava flows to more easily weathered ash fall deposits. Weathering of ash deposits has created colorful alkaline montmorillonite clays, noted for their shrink-swell characteristics (Garrett and Youtie 1992). These clays, formed from late Eocene and early Oligocene ash (24–44 million years old), create the barren landscape of the Painted Hills and habitat for some Lava Plains endemics. The entire region is in the rain shadow of the Cascade Mountains, with average annual precipitation of only 200–300 mm (8–12 in).

Characteristic vegetation is sagebrush grassland, with *Achnatherum thurberianum*, *Artemisia tridentata* var. *wyomingensis*, *A. tridentata* var. *vaseyana*, *Festuca idahoensis*, *Leymus cinereus*, and

Figure 38. Basin and Range: Diamond Craters is one of several volcanic craters within the ecoregion. Pumice, basalt, and ash limit vegetation to a few species tolerant of droughty conditions.

Figure 39. Basin and Range: Christmas Valley is a broad basin seen from Black Hills, an uplifted thrust-fault mountain. Center pivot irrigation has allowed the basin to be extensively farmed.

Pseudoroegneria spicata. Rocky outcrops are common and support *Juniperus occidentalis*; open stands of old-growth juniper are found northeast of Bend. *Pinus ponderosa* occurs in scattered, open stands on coarse-textured soils within the lava plain. Riparian areas provide habitat for *Alnus rhombifolia*, *Pinus ponderosa*, *Populus trichocarpa*, and *Salix* spp. Currently, sagebrush grassland and open ponderosa pine forests are used for livestock grazing, and overgrazing has resulted in widespread invasion by *Bromus tectorum*. Irrigated agricultural crop production is common along the Deschutes River, and includes wheat, alfalfa, and potato.

Only eight native plants are restricted to the Lava Plains ecoregion. *Chaenactis nevii* is an endemic that grows on nearly bare hills of montmorillonite clay, along with the more widely distributed *Peritoma platycarpa*. Most other endemics in the region are found on shallow, rocky soil derived from basalt or welded tuff, including the rare *Astragalus misellus* var. *misellus*, *Lomatium ochocense*, and *Eriogonum microthecum* var. *microthecum*, which is also found in Idaho. *Astragalus diaphanus* var. *diurnus* and *Eremothera pygmaea* occupy seasonally dry streambeds, while *Thelypodium eucosmum* grows in or near springs. *Castilleja xanthotricha* is a rare hemiparasite associated with sagebrush, *Gutierrezia sarothrae*, and bunchgrasses. (Figures 33, 34, 35, 36, 37).

Basin and Range

The Basin and Range, a small portion of a region that covers parts of California, Nevada, Idaho, Utah, Arizona, New Mexico, and Mexico, is characterized by a series of parallel northwest-southeast trending fault-block mountains that enclose internally draining basins. Between 16.5 and 15.5 million years ago, the Basin and Range ecoregion was a level plain formed from nearly a thousand meters (thousands of feet) of basalt flowing from the volcanic vents of Steens Mountain (Bishop 2003, Orr and Orr 2012). At about 12 million years ago, the offshore tectonic plate rotated to the northwest beneath the edge of the North American

Figure 40. Basin and Range: Extensive wetlands have formed within the shallow Malheur Lake basin, an important site for waterfowl migration. Malheur Lake would have been a much larger and deeper lake during the Pleistocene.

continent, triggering formation of the Basin and Range province, as the thin crust began stretching, faulting, and breaking into huge, flat-topped mountain blocks, separated by down-dropping basins. The escarpments of these fault-block mountains rise dramatically from the valley floors, more than 600 m (2,000 ft) at Abert Rim and Winter Rim. The flat summits of the fault-block mountains, called rim rock, are formed by the most resistant of the basalt flows. Elevations range from 1,200 m (4,000 ft) in the basins to 2,350–2,967 m (7,700–9,733 ft) at the summits of Hart Mountain, Warner Peak, and Steens Mountain.

Internal drainage creates shallow, saline lakes, surrounded by broad salt flats and low dune fields. Summer Lake, Warner Lake, Silver Lake, Alvord Lake, and Lake Abert are among the largest of these playa lakes, which, during the wetter Pliocene and Pleistocene periods, were as much as 115 m (375 ft) deep. The volume of these playa lakes varies seasonally, with many lakes drying completely in some years. The region is one of the driest of the state, with average annual precipitation of 150–350 mm (6–14 in). Precipitation is greater on the highest ridges, reaching 660 mm (26 in) on Hart and Steens mountains.

Upland plant communities are adapted to drought and consist largely of sagebrush steppe with bunchgrasses. *Artemisia arbuscula*, *A. tridentata* ssp. *wyomingensis*, and *Chrysothamnus viscidiflorus* ssp. *viscidiflorus* are common shrubs of the steppe, along with *Festuca idahoensis*, *Poa secunda*, and other grasses. Invasive *Bromus tectorum* has replaced most native plants in overgrazed areas. The dune fields and margins of alkali lakes are harsh environments that often support a depauperate flora of *Artemisia spinescens*, *Atriplex canescens* var. *canescens*, *Distichlis spicata*, *Sarcobatus vermiculatus*, and *Sporobolus airoides*. The area also has numerous non-alkaline playas which support *Artemisia canescens* and grasses, including *Leymus cinereus* and *Poa secunda*. Lakes that seasonally flood provide important habitat for migratory birds and have a diversity of sedges, rushes, *Deschampsia cespitosa*, and *Leymus triticoides*. Low-density cattle grazing occurs throughout the region in the sagebrush steppe and in stream and lake margins planted to pasture. Irrigation allows the flat basins and stream margins to be planted to grass and alfalfa which is exported throughout the state.

Twenty-six native plant species are restricted to the Basin and Range ecoregion in Oregon, although many of these are also found in adjacent states within the Basin and Range. Common genera in the ecoregion are *Antennaria*, *Astragalus*, *Cryptandra*, *Downingia*, *Epilobium*, *Eriogonum*, *Lomatium*, *Penstemon*, *Phacelia*, and *Solidago*. Many genera and species common in the ecoregion are also found

Figure 41. Basin and Range: Steens Mountain, the highest fault-block mountain in Oregon, is the home of *Eriogonum ovalifolium* and other plants able to survive the short growing season and extreme low temperatures of the site.

Figure 42. Basin and Range: Borax Lake, a saline playa lake, with the fault-block escarpment of Steens Mountain in the background.

growing in adjacent Owyhee Uplands, Lava Plains, and Blue and Wallowa Mountains ecoregions, where eroded volcanic rock, ash, and derived soils are common.

Rare species endemic to Oregon and the ecoregion are *Eriogonum cusickii*, which grows on open, gravelly flats with volcanic soils, and *Cirsium eatonii* var. *peckii* and *Draba cusickii* var. *cusickii*, both found on gravelly slopes at high elevation on Steens Mountain. Other Basin and Range endemics that occur in adjacent states as well include *Carex tiogana*, which grows on lake margins and in meadows at high elevations on Steens Mountain and in the Sierra Nevada Mountains of California. In southern Oregon and northern Nevada, *Eriogonum crosbyae* var. *crosbyae* inhabits dry ridges of welded tuffs and pumice outcrops, *E. crosbyae* var. *mystrium* populates gravelly eroded volcanic soils, *Ivesia rhypara* var. *shellyi* occupies rock cracks and crevices in pumice and tuff, and *Lomatium roseanum* grows on dry basalt scree over clays. Four other endemic species are shared by adjacent ecoregions: *Castilleja pilosa* var. *steenensis*; *Eriogonum chrysops* and *Penstemon glaucinus*, which are found on dry, gravelly soils; and *Thelypodium howellii* ssp. *howellii*, which inhabits meadows. (Figures 38, 39, 40, 41, 42, 51, 55, 68, 69).

Blue and Wallowa Mountains

The Blue and Wallowa ranges comprise one of the most geologically complex landscapes in Oregon, representing almost 400 million years of geologic history with multiple tectonic collision events between volcanic island arc terranes and North America. Island arc rock suites represent different diverse ages and settings (Bishop 2003, Orr and Orr 2000, Orr and Orr 2012). Isolated mountain ranges of substantial elevation characterizes much of the ecoregion, and while the mountains are not as high as those of the Cascade Range, there are 21 peaks over 2,744 m (9,000 ft) and 36 over 2,439 m (8,000 ft) tall. The highest, Mount Sacajawea, reaches 3,000 m (9,838 ft) in elevation. Each of the ranges bears a distinctive geologic setting varying from granitic to silica-rich sedimentary rock of deep-sea origin and metamorphics overlain by more recent acidic volcanics and basalts. The highlight of the distinctive topography is the impossibly steep, deep canyon lands along the eastern edge of the ecoregion, best displayed in Hells and Imnaha River canyons, where multiple flows of Columbia River basalt and even much earlier Permian rock were deeply eroded by the Bonneville flood 15,000 years ago.

Average annual precipitation ranges from 220 mm (9 in) at lower elevations to 2,000 mm (80 in) at high elevations. Most precipitation at high elevations falls as snow.

The combination of ancient and highly dissected topography and extreme geologic diversity results in a flora rich with an abundance of endemic species. The dissected hills and low mountains of the region are dominated by *Pinus ponderosa* forests mixed with *Pseudotsuga menziesii* and diverse understories that include *Artemisia tridentata*, *Festuca idahoensis*, *Pseudoroegneria spicata*, and *Symphoricarpos albus*. Droughty soils support increased levels of lodgepole pine forests with lesser amounts of western larch and grand fir. At higher elevations, *Abies lasiocarpa* var. *lasiocarpa* and *Picea engelmannii* are the typical dominant species where either soils retain more moisture or precipitation is elevated. *Juniperus occidentalis* woodlands occur on many rocky and xeric sites. A diverse matrix of soils, slope, and aspect creates a complex mosaic of grassland, shrubland, and forest types. This suite of vegetation typically includes species such as *Artemisia rigida*, *A. arbuscula*, *Cercocarpus ledifolius*, and *Purshia tridentata*. Important grasses include *Danthonia unispicata*, *Festuca idahoensis*, *Poa secunda* ssp. *secunda*, and *Pseudoroegneria spicata*. The peaks of the Wallowa and Idaho's adjacent Seven Devils ranges are divided by steep forested

Figure 43. Blue and Wallowa Mountains: Eagle Lake. Eagle Cap Wilderness contains many small, clear lakes formed by melting of mountain glaciers. The flora of these tarn or cirque lakes is limited by low temperatures, low nutrient levels, and short growing season.

Figure 44. Blue and Wallowa Mountains: The Imnaha Canyon has been cut through thousands of feet of Columbia River lava deposits before joining the Snake River at the northern edge of the ecoregion.

Figure 45. Blue and Wallowa Mountains: At the base of Matterhorn Peak, a seasonal stream meanders across an alpine meadow in a cirque created when a mountain glacier plucked away underlying rock as it moved slowly down slope. Despite the short growing season, plant diversity is high in these protected alpine meadows.

Figure 46. Blue and Wallowa Mountains: The Logan Valley, more than 1.5 km wide at its lower end, is fed by several mountain streams. The rich alluvial soils originally supported a diverse prairie flora, as seen in this photo, but most of the valley has been converted to pasture or fodder production.

Figure 47. Owyhee Uplands: Thick deposits of tuff, a form of lightweight volcanic ash, were eroded by wind and rain to form the Pillars of Rome. Few plant species grow on the droughty, easily eroded tuff, but tuff provides habitat for some regional endemic plants.

canyons thousands of feet deep formed by streams that often flow along pre-existing geological fault lines. Above the canyons, high-elevation forests of *Abies lasiocarpa* var. *lasiocarpa*, *Picea engelmannii*, *Pinus albicaulis*, and *P. contorta* var. *latifolia* form a mosaic with alpine meadows of *Festuca viridula* and a high diversity of *Carex* species. Within many of the open forests and grasslands, there is also a diversity of drought-tolerant flowering plants, including 36 taxa of *Astragalus*, 29 of *Erigeron*, 29 of *Eriogonum*, 25 of *Lomatium*, and 20 of *Lupinus*. The summits of the taller mountains include largely unvegetated rock outcrops, rubble, and snowfields. Forest logging is important to the economy, as is cattle grazing in the valleys and at high elevations.

Over 120 of Oregon's native plant species grow only in the Blue and Wallowa Mountains ecoregion, and although the number of endemic plants is much lower, the number of recently discovered rare endemics point to the need for greater botanical study of this region. Rare species are found in three distinct zones: high elevation alpine and subalpine habitats, mid- to low-elevation valley grasslands, and the rugged canyon lands (Defenders of Wildlife 1998). Among the plants that inhabit rocky, high-elevation alpine and subalpine habitat—the location for the greatest number of endemics in the Blue and Wallowa Mountains—are *Castilleja fraterna*, *C. rubida*, *Eriogonum scopulorum*, *Lomatium erythrocarpum*, *L. cusickii*, *L. oreganum*, *L. greenmanii*, *Luina serpentina*, and *Penstemon pennellianus*. *Astragalus tegetarioides* and *Lupinus sabinianus* are found in lower elevation grasslands and forest openings.

Hells Canyon and similar canyon lands support populations of rare species requiring or tolerant of open, gravelly or rocky soils and steep slopes. Among these plants are *Achnatherum wallowaense*, *Boechera hastatula*, *Calochortus macrocarpus* var. *maculosus*, *Erythranthe hymenophylla*, *Mirabilis macfarlanei*, *Phacelia minutissima*, *Pyrrocoma radiata*, and *P. scaberula*. Another Snake River endemic in Oregon and Idaho, *Rubus bartonianus*, forms thickets along protected stream banks in canyons near and immediately along the river, while *Camassia cusickii*, *Diplacus clivicola*, and *Erythranthe patula* grow in open seeps. *Lomatium serpentinum* occupies open crevices among rocks along the Snake River just above the water line. (Figures 43, 44, 45, 46, 70, 71, 72).

Owyhee Uplands

The Owyhee Uplands region of far southeastern Oregon is a geologically unique, dissected plateau with a youthful stream drainage system and few playa lakes. The immature stream system and paucity of playas, along with only limited evidence of block fault features, separate the Owyhee Uplands from the Basin and Range region to the west. Elevations of the plateau and irregular uplands range from 760–1,980 m (2,500–6,500 ft), while the Owyhee River creates a deep canyon whose base elevation is 610 m (2,000 ft) where it meets the Snake River. The region is predominantly volcanic, including thick basaltic and rhyolitic lava flows, ash flows, and associated ash deposits and sediments. Its long, violent volcanic history is reflected by many surface features, such as the Miocene-age (17 million years ago) Three Fingers Caldera and Mahogany Mountain. Extensive deposits of ash and welded tuff are abundant, and the much more recent Pleistocene-age (1.9 million years ago) Jordan Crater produced rope-like, glassy basalt flows. Volcanic activity continued as recently as 4,000 years ago in the southeastern corner of the ecoregion (Orr and Orr 2012).

Many of the smaller headwater streams are intermittent due to the region's arid climate—average annual precipitation is only 150–350 mm (6–14 in). Such arid conditions support sagebrush steppe, with *Artemisia tridentata* ssp. *wyomingensis* and *Chrysothamnus viscidiflorus* ssp. *viscidiflorus*. Grasses of the sagebrush steppe include *Achnatherum thurberianum*, *Festuca idahoensis*, *Leymus cinereus*, *Poa secunda* ssp. *secunda*, and *Pseudoroegneria spicata*. Saline soils sustain *Atriplex canescens* var. *canescens*, *Distichlis spicata*, and *Sarcobatus vermiculatus*. Cattle grazing is a common land-use, and overgrazing often results in *Bromus tectorum* replacing native vegetation.

Thirty-one plant species have distributions restricted to the Owyhee Uplands in Oregon, and several of these are Oregon endemics. Endemics with local distribution in Oregon include *Senecio ertterae* which grows on ash beds. *Amsinckia carinata* occurs on talus on low volcanic cones at the edge of the Malheur Valley, while *Eriogonum chrysops* grows locally on rhyolite flows. Both *Astragalus mulfordiae* and *Hackelia cronquistii* are found in sandy sagebrush steppe at the edges of the Malheur and Snake Rivers (Defenders of Wildlife 1998). Several other regional endemics also occur in adjacent Idaho or Nevada on volcanic ash, including *Artemisia packardiae*, *Lomatium cronquistii*, *Mentzelia mollis*, *M. packardiae*, *Phacelia lutea* var. *calva*, *Trifolium leibergii*, *T. owyheense*, and the recently discovered *Lomatium bentonitum*. (Figures 47, 48, 49, 50, 53, 73).

Figure 48. Owyhee Uplands: Leslie Gulch is a botanically diverse area of highly eroded basalt and ash flows along the canyons of the Owyhee River and its many small tributaries.

Figure 49. Owyhee Uplands: The drainage system of Owyhee River has carved deep canyons and ravines through much of the ecoregion, eventually entering the Snake River 25 km (15 mi) south of Ontario, OR. The rock of the Owyhee Uplands is largely soft ash flow (tuff) that erodes to form the region's canyons and strange land forms.

Figure 50. Owyhee Uplands: Aerial view of intensive agriculture in the Snake River portion of the ecoregion, which has been largely converted to row crops.

Habitats

Upland

Alpine Zone

At high elevations above the timberline, conditions are too harsh to support trees, and only low prostrate or cushion-shaped shrubs (*Cassiope mertensiana* ssp. *mertensiana* and ssp. *gracilis*, and *Phyllodoce empetriformis*), forbs (*Antennaria media*, *Hulsea nana*, *Oreostemma alpigenum* var. *alpigenum* and var. *haydenii*), grasses (*Festuca viridula* and *Trisetum spicatum*), and sedges (*Carex nigricans* and *C. spectabilis*) can survive. Soil is thin or absent, and snow persists late into the summer. Plants survive among rocks and scree. Cold, desiccating winds blow almost continually, and sunlight is extreme at high elevations.

Figure 51. Cushion plants in the alpine zone atop Steens Mountain.

Cliff, Rock Outcrop, and Talus Slope

The mountainous landscapes of Oregon contain expansive areas of cliff, rock outcrop, and talus slope, all features that are characterized by shallow, well-drained to droughty soils and often extreme temperatures. This habitat is found in all of Oregon's ecoregions and frequently hosts regionally endemic species such as *Eriogonum villosissimum*, *Erythronium elegans*, and *Penstemon barrettiae*. The serpentine-laden Siskiyous, with their extensive areas of rock surface and shallow soils, support the highest levels of endemism in the state. All of these habitats can receive abundant seasonal surface water, which produces moist microhabitats that support genera such as *Erythranthe* and *Saxifraga*.

Conifer Forest

Conifers dominate the forests of Oregon on both the west and east sides of the crest of the Cascade Mountains. Oregon has 35 conifer species, 31 occurring in and west of the Cascades and 22 east of the Cascades. At the time of European settlement, most of western Oregon was forested with conifers up to the tree line. Exceptions are the Willamette Valley and the Siskiyou Mountains, where a combination of excessive or poor drainage conditions and Native American land management resulted in wet prairie, dry prairie, and oak savanna dominance. Forests in eastern Oregon are also largely dominated by conifers. The lower amount of precipitation restricts forests to mid-elevations, and many are better described as open woodlands, especially those dominated by *Pinus ponderosa*. Species dominance in conifer forests is determined by several factors including elevation, precipitation, successional history, soils, length of growing season, and extreme minimum temperature. Some of the more common forest types are discussed within individual ecoregions.

Deciduous Forest

Deciduous forests are much less ecologically important than coniferous forests in Oregon. The Willamette Valley hosts small areas of *Quercus garryana* var. *garryana* forest and narrow corridors of *Acer macrophyllum*, *Alnus rubra*, *Populus trichocarpa*, and *P. tremuloides* along stream courses.

Disturbed

This class is quite general and includes both naturally occurring and human-generated sites such as road cuts and margins, fallow fields, gravel pits, railroad right-of-ways, dumps, and cracks in sidewalks. Disturbed habitats frequently host exotic taxa further delimited as invasives or waifs. Some of the species that establish in disturbed habitats are ephemeral due to inappropriate soils or climatic conditions, while others become naturalized and spread into adjacent habitats to grow among or replace native plants.

Lava and Ash Flow

Lava and ash flows characterize the young volcanic landscapes of the Cascades as well as older volcanic regions throughout eastern Oregon. Deep, droughty ash deposits and dark lava flows create extremely harsh habitats, but deep-rooted perennial species of *Astragalus*, *Eriogonum*, and *Penstemon* are able to establish and thrive.

Meadow, Prairie, and Grassland

At the time of European settlement in Oregon, there were large areas dominated by grasses and herbaceous plants, with few shrubs or trees. Such open, treeless landscapes were often the result of several factors, including regularly occurring fire, grazing, and insufficient soil or precipitation to support trees. Because many grasslands are the result of some type of disturbance, they often occur within a matrix of shrublands, woodlands, or forests. Some of the largest grasslands were historically found along steep, rocky slopes or rolling hills of the Columbia Basin, within a matrix of oak savanna and wet prairie in the Willamette Valley, and among the *Pinus ponderosa* woodlands of the Eastern Cascades Slopes and Foothills ecoregions. In mountainous areas, dry meadows are found mainly on south-facing or steep slopes and are more common in the drier regions east of the Cascade crest. Grasslands persist throughout Oregon. Currently non-native grasses and forbs may be prevalent components of their flora, but undisturbed areas, such as ungrazed mountain slopes, support a large diversity of wildflowers and native grasses including *Festuca idahoensis*, *F. viridula*, *Phleum alpinum*, and *Trisetum spicatum*. These open areas are important to wildlife; restoration efforts focus on controlling non-natives and encroaching trees that are no longer kept at bay by naturally occurring wildfires.

Savanna and Woodland

Prior to European settlement, oak forest in the Willamette Valley and portions of the Siskiyou Mountains had only scattered trees and an open, grassy groundcover due to fire management by Native Americans. Such savanna or woodland habitat is dependent on regular, human-initiated fires. The subsequent reduction of fire activity has resulted in the conversion of savannas and woodlands to closed-canopy forests, thereby diminishing species diversity within the understory and groundcover. In contrast, *Pinus jeffreyi* forests in the Siskiyou Mountains and *Pinus ponderosa* forests in eastern Oregon ecoregions are the result of natural fire activity, droughty soils, and low levels of precipitation; these conifer stands remain open and are thus designated as woodlands.

Figure 52. Meadow of flowers and bunchgrasses in the Columbia Basin.

Figure 53. Shrub steppe in Owyhee Uplands.

Shrub Steppe

Open shrub steppe characterizes the dry lands of eastern Oregon, where rainfall is insufficient to support trees. *Artemisia tridentata* ssp. *tridentata* dominates the shrub steppe, typically with lesser amounts of *Chrysothamnus viscidiflorus*, *Ericameria nauseosa*, and *Purshia tridentata*. Grasses and drought-tolerant forbs can be locally abundant; these include species of *Astragalus*, *Balsamorhiza*, *Eriogonum*, *Lupinus*, and *Penstemon*.

Subalpine Meadow

Subalpine meadows are a transition zone between forested habitats and the timberline, where snow can persist late into the summer and limit the growing season. Grasses and sedges dominate, often with a rich diversity of flowering forbs that include *Antennaria*, *Astragalus*, *Castilleja*, *Erigeron*, and *Potentilla*. Trees and shrubs are scattered and small, and rock outcrops are common. Dwarf shrubs include species of *Cassiope*, *Phyllodoce*, *Sorbus*, and *Vaccinium*.

Wetland

Alkaline Wetland

Alkaline wetlands are restricted to ecoregions east of the Cascades, where low rainfall and high summer temperatures result in evaporation of lakes and ponds to create alkaline soils and salt flats. These wetlands are most common in the Basin and Range, and to a lesser degree the Owyhee Uplands, where many basins are internally draining and form playa lakes that dry down each summer. In the most extreme conditions, this habitat cannot support vegetation. Among the plants that have adapted to the saline or alkaline conditions, including many that also occur in tidal marshes, are *Distichlis spicata*, *Lepidium davisii*, *Suaeda moquinii*, and species of *Eleocharis* and *Schoenoplectus*. Shrubs populating the playas of this habitat include *Atriplex confertifolia*, *Nitrophila occidentalis*, and *Sarcobatus vermiculatus*.

Figure 54. *Sphagnum* mound within a bog in the Cascades.

Bog

Bogs are wetlands characterized by extremely poor drainage, acid (pH less than 5.5) peat accumulation, and dominance by species of *Sphagnum* moss. Extremely wet or cold conditions reduce the rate of vegetation decomposition, creating the ideal habitat for bog formation. Bogs vary greatly in structure, with *Sphagnum* spp. and herbaceous plants dominant in some, while shrubs and trees dominate others. Ecoregions where one commonly encounters bogs include the Estuarine Coast, Cascades, and the Blue and Wallowa Mountains. Bogs occupy depressions among the dunes along the coast; shallow cirque lakes support bogs in mountainous areas. The species diversity within this habitat in Oregon is much lower than what is found in previously glaciated landscapes to the north and in the midwestern and eastern U.S. and Canada. Bog herbs include *Drosera rotundifolia*, *Pedicularis groenlandica*, and *Platanthera* spp., while the shrubs include numerous representatives of the Ericaceae.

Fen

Fens are wetlands fed by groundwater and dominated by mosses, herbs, and shrubs. Fens are distinguished from bogs by having a pH generally higher than 5.5 and lacking domed *Sphagnum* peat accumulation. However, localized areas within fens do occur where *Sphagnum* has accumulated as hummocks or lawns and created a pH of 4.0. Fens occur throughout Oregon, with the greatest concentration in dunal troughs or swales along the Pacific Ocean and along small streams and seeps in the Siskiyou Mountains. These coastal and Siskiyou fens are well known for supporting the showy species *Darlingtonia californica* and *Lilium occidentale*. Fens often occur in springs or ponds that form along upland margins. They are found in most ecoregions of the state, although they have not been documented in the Basin and Range, Columbia Basin, or Owyhee Uplands.

Freshwater Marsh

This habitat includes wetlands that are flooded most of the year and occupy margins of lake basins and streams. They often include open water. The dominant vegetation is emergent or submergent herbaceous species (see Lake and Pond habitat); marshes differ from swamps in their lack of woody species. Freshwater marshes are found within all ecoregions of Oregon and can support regionally endemic species.

Lake and Pond

This habitat encompasses lakes, ponds, pools, beaver channels, and other areas of open, standing, or slow-moving bodies of water. Clear lakes protected from strong winds can have plants blanketing their entire basin, sometimes to depths of ten or more meters. Given sufficient water levels, shallow ponds and channels also host aquatic plant taxa. Aquatic species affiliated with wetland habitats are often subdivided into three classes: emergent plants that grow partially above the water, floating plants that rest on the surface of the water, and submergent plants that grow predominately below the water's surface. Common emergent genera are *Carex*, *Eleocharis*, *Juncus*, and *Schoenoplectus*, as well as diverse forbs such as *Alisma*, *Polygonum*, *Sagittaria*, and *Sparganium*. Common floating plants include both rooted (*Nuphar*, *Nymphaea*) and free-floating (*Lemna* and *Spirodela*) genera. Submergent genera include *Ceratophyllum*, *Elodea*, and *Myriophyllum*. *Potamogeton* has both submergent and floating species.

Riparian

The vegetation along the margins of streams varies greatly across Oregon as a result of differences in stream size and gradient, as well as substrate. Trees, shrubs, and herbaceous plants along stream margins differ from those of adjacent uplands because of increased moisture availability and shade. Floating, emergent, and submergent plants increase within streams as stream velocity decreases, as long as water clarity remains high.

Figure 55. *Nuphar polysepala* in a shallow lake in the Basin and Range.

Figure 56. *Cascadia nuttallii* on steep-sloping seepage in the Cascades.

Seepage

Seepages or seeps are places where ground water reaches the surface, often at the lower edge of slopes or where bedrock and soil surfaces meet. They can be quite small or localized and, like vernal pools, seeps can be seasonal. Water in seeps is flowing and is often oxygen-rich, allowing a flora to establish that is distinct from wetlands, where oxygen levels are too low for most upland plants to survive. The flora of seeps can include *Cascadia nuttallii*, *Erythranthe guttata*, *Plectritis congesta*, *Romanzoffia thompsonii*, and others.

Shrub Swamp

Shrub swamps occur along streams as a type of riparian wetland, in depressions between coastal dunes, and along lake margins, including playa lakes. Genera common to this habitat include *Alnus*, *Betula*, and *Salix*.

Swamp Forest

Swamp forests are those that are flooded for extended periods during the growing season. They commonly occur along streams or in depressions within the landscape and are found in many ecoregions. Deciduous trees such as *Alnus rubra*, *Fraxinus latifolia*, *Populus balsamifera* ssp. *trichocarpa*, and *P. tremuloides* are common dominants, with conifers (*Abies lasiocarpa*, *Picea engelmannii*, *P. sitchensis*, *Pinus contorta* var. *latifolia*) occurring more frequently in some ecoregions.

Tidal Marsh (Estuary)

Tidal marshes occur in the lower reaches of coastal rivers regularly flooded by ocean tides. The salinity of tidal marshes decreases with distance from the ocean, gradually becoming freshwater upstream. Freshwater tidal marsh along the Columbia River extends more than 160 km (100 mi) inland. Salt and brackish marshes support plants adapted to high salinity, and taxa tolerant of highly variable salinities, such as *Eleocharis macrostachya*, *Juncus balticus*, *Schoenoplectus pungens*, and *Triglochin maritima* occur throughout this habitat.

Vernal Pool

Vernal pools are seasonal wetlands that are flooded during the fall, winter, and early spring but are dry during the late spring and summer. Most vernal pools are small, but they often occur as complexes in some environments. They share many of the characteristics of alkali wetlands, especially in the Basin and Range ecoregion. Species diversity varies greatly by ecoregion. These pools are well recognized as home to numerous rare taxa such as *Juncus uncialis*, *Limnanthes* spp., *Lomatium cookii*, and *Plagiobothrys hirtus* and are threatened by drainage for agricultural and urban development.

Wet Meadow

Wet meadows are a diverse group of wetlands associated with seasonally moist montane habitats. Typically saturated in the spring and early summer, wet meadows are often dry by late summer and fall, resulting in seasonally oxygenated conditions that allow organic material to rapidly decompose. Wet meadows are common in most ecoregions, with the exception of the Willamette Valley and Columbia Basin. Grasses and sedges are the dominant meadow plants, but forb diversity can also be high. Species composition varies greatly by ecoregion. The large, white-flowered spikes of *Veratrum californicum* are conspicuous in wet meadows throughout the state.

Wet Prairie

Wet prairies are grasslands characterized by annual flooding early in the growing season, typically by large rivers such as the Willamette, Columbia, and Umpqua. Flood control on Oregon's rivers has eliminated regular inundation; coupled with extensive conversion to agricultural use, most of this habitat has been lost. The Willamette Valley ecoregion historically supported the largest area of wet prairie, and currently its prairie remnants are greatly diminished. *Camassia quamash*, *Carex aperta*, *Deschampsia caespitosa*, *Eryngium petiolatum*, and *Triteleia hyacinthina* are taxa characteristic of this habitat. Drier conditions have allowed exotic species to replace native prairie plants, and regularly recurring fires that maintained prairie diversity are suppressed.

Figure 57. *Gentiana sceptrum* in a wet meadow in the Cascades.

EXPLORING OREGON'S BOTANICAL DIVERSITY

Edward R. Alverson

Oregon is blessed with an extensive network of lands designated to promote the conservation, research, and exploration of our native flora and fauna. Of the many possible locations where one can experience the diversity of our flora, the fifty sites listed here are highlighted because they provide good examples of native habitats and plant communities, support a diverse native flora, host endemic or rare species, or have impressive wildflower displays. Preference has been given to sites that are readily accessible and that provide a broad geographic and ecological representation.

Each site lists ownership and location by ecoregion and county; this should be sufficient for the reader to identify additional resources on the internet. Directions and trailhead locations for most of the hikes (as well as numerous other sites of interest) can be found in the following references: Franklin et al. (1972); Horn (2006); Houck and Cody (2011); Jolley (1988); Jones (1992); MacKinnon (2014); Malaby (2007); Ross and Chambers (1988); Sullivan (2009–2013); Wuerthner (2001). The best times of year to visit individual sites varies greatly depending upon ecoregion, elevation, and year-to-year fluctuations in weather and snowpack, but native plants can usually be seen blooming somewhere in the state from February through October. Field trips led by the Native Plant Society of Oregon and other conservation organizations are also an excellent way to learn more about Oregon's flora.

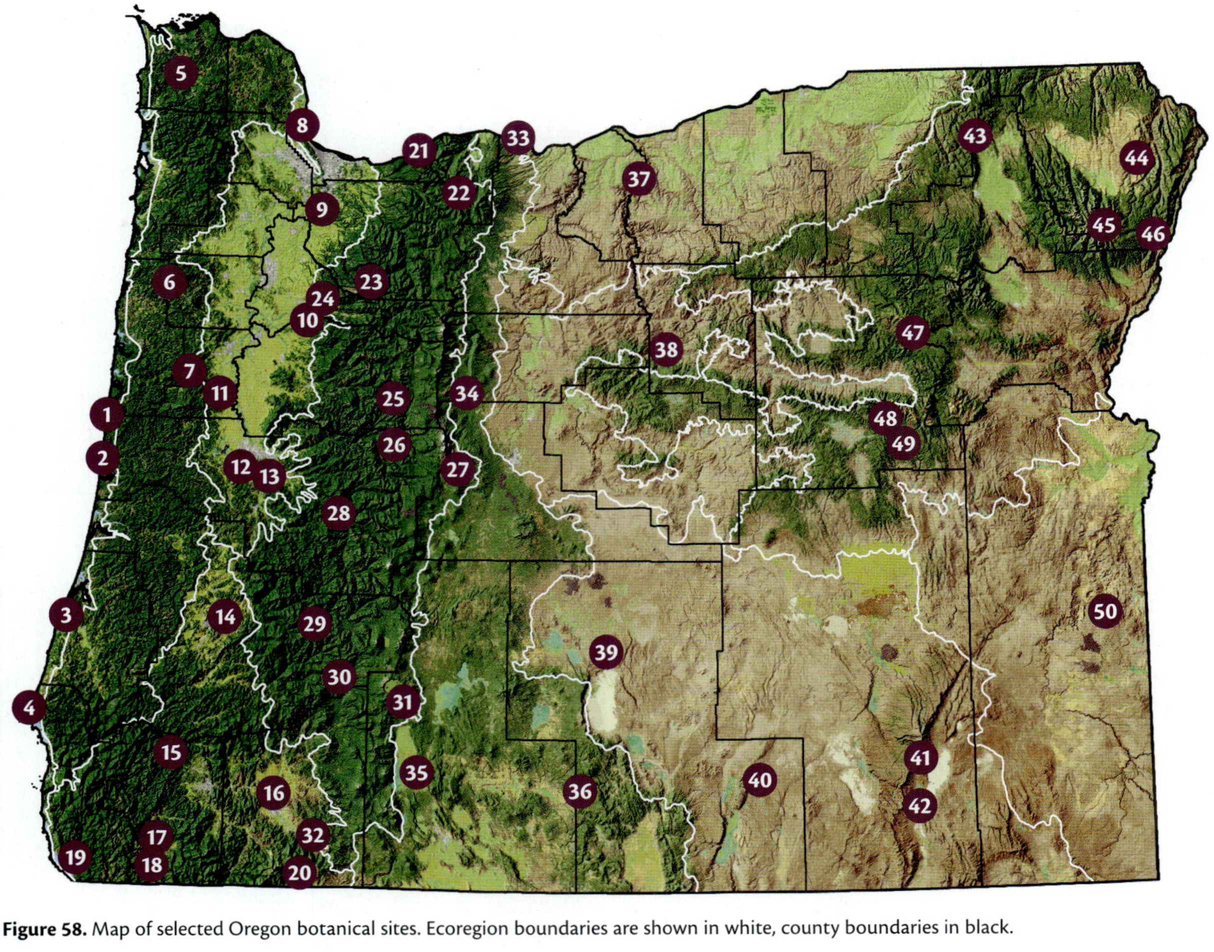

Figure 58. Map of selected Oregon botanical sites. Ecoregion boundaries are shown in white, county boundaries in black.

Table 2. List of selected Oregon botanical sites. Numbering follows Figure 58.

Site Number & Name		Ecoregion	Counties
1	Cape Perpetua Scenic Area	Est	Lane & Lincoln
2	Sutton Creek	Est	Lane
3	South Slough National Estuarine Research Reserve	Est	Coos
4	Cape Blanco State Park	Est	Coos & Curry
5	Saddle Mountain State Park	CR	Clatsop
6	Valley of the Giants Outstanding Natural Area	CR	Polk
7	Marys Peak Scenic Botanical Area	CR	Benton
8	Sauvie Island	WV	Columbia & Multnomah
9	Willamette Narrows	WV	Clackamas
10	Kingston Prairie	WV	Linn
11	William L. Finley National Wildlife Refuge	WV	Benton
12	West Eugene Wetlands	WV	Lane
13	Mt. Pisgah & Howard Buford Recreation Area	WV	Lane
14	North Bank Habitat Management Area	Sisk	Douglas
15	Rogue River Trail	Sisk	Josephine
16	Lower & Upper Table Rocks	Sisk	Jackson
17	Eight Dollar Mountain & T.J. Howell Botanical Drive	Sisk	Josephine
18	Rough and Ready Creek	Sisk	Josephine
19	Chetco River Redwood Groves	Sisk	Curry
20	Mt. Ashland-Siskiyou Peak Botanical Area	Sisk	Jackson
21	Western Columbia River Gorge: McCord Creek & Elowah Falls	Casc	Multnomah
22	Mt. Hood Subalpine Zone	Casc	Clackamas & Hood River
23	Table Rock Wilderness	Casc	Clackamas
24	Silver Falls State Park	Casc	Marion
25	Iron Mountain & Cone Peak	Casc	Linn
26	Forests of the Upper McKenzie River Valley	Casc	Lane & Linn
27	Broken Top Area, Eastern Side of the Three Sisters Wilderness	Casc	Deschutes
28	Tire Mountain	Casc	Lane
29	North Umpqua River	Casc	Douglas
30	Rogue-Umpqua Divide Wilderness	Casc	Douglas & Jackson
31	Crater Lake National Park	Casc	Douglas, Jackson & Klamath
32	Grizzly Peak	Casc	Jackson
33	Eastern Columbia River Gorge: Rowena Plateau	ECas	Wasco
34	Metolius River & Black Butte	ECas	Jefferson
35	Wood River Wetland	ECas	Klamath & Lake
36	Gearhart Mountain	ECas	Lake
37	Cottonwood Canyon State Park	Col	Gilliam & Sherman
38	Painted Hills	Lava	Wheeler
39	Fort Rock & Christmas Valley	BR	Lake
40	Warner Basin & Hart Mountain National Antelope Refuge	BR	Lake
41	Steens Mountain	BR	Harney
42	Alvord Desert & Pueblo Valley	BR	Harney
43	Skyline Road	BW	Umatilla & Union
44	Zumwalt Prairie	BW	Wallowa
45	Eagle Cap Wilderness High Country	BW	Wallowa
46	Wallowa Mountain Loop Highway	BW	Baker & Wallowa
47	Greenhorn Mountains	BW	Grant
48	Strawberry Mountains	BW	Grant
49	Logan Valley	BW	Grant
50	Leslie Gulch	Owy	Malheur

Estuarine Coast Ecoregion

Along the immediate coast, where the continent meets the Pacific Ocean, a number of specialized habitats are home to species with ranges that extend only a short distance inland. Estuarine habitats, coastal dunes, and wind-buffeted Sitka spruce (*Picea sitchensis*) forests support plant communities unique to this narrow coastal strip.

1 Cape Perpetua Scenic Area

Lane & Lincoln counties
US Forest Service (USFS)

Cape Perpetua is the Oregon coast in a microcosm. Wind-pruned thickets hug the rocky shore, coastal headlands rise steeply above the ocean, and old growth forests occupy sheltered valleys. A short trail follows Cape Creek through the forest to a giant Sitka spruce tree, and the open south-facing headland below Cape Perpetua proper harbors a great diversity of grassland wildflowers, including camas, paintbrush, and lupine. This is a good place to see the winter-flowering shrub, wavy leaf silk tassel (*Garrya elliptica*), close to its northern range limit.

2 Sutton Creek

Lane County
Oregon Parks and Recreation Dept. (OPRD), The Nature Conservancy (TNC), USFS

Just north of Florence, Sutton Creek flows out of the Coast Range into Sutton Lake and then winds its way through the dunes to the Pacific Ocean. Much of the landscape is stabilized dune forest dominated by shore pine (*Pinus contorta* var. *contorta*), often with a dense understory of rhododendrons and many other species of broadleaf evergreen shrubs. Closer to the ocean, a range of open, early successional dune and deflation plain habitats are found and are home to several rare pteridophytes. An unusual northern population of the carnivorous California pitcher plant (*Darlingtonia californica*) is found at the Darlingtonia Wayside, and the rare moss *Limbella fryei* grows at Sutton Lake.

3 South Slough National Estuarine Research Reserve

Coos County
Oregon Department of State Lands (ODSL)

Estuaries are highly productive ecosystems that support unique plant communities and species. Much of Oregon's estuarine habitat has been lost or altered, but 5,000-acre South Slough Reserve is an excellent place to experience this ecosystem. Intertidal salt marsh plants such as tufted hairgrass (*Deschampsia cespitosa*), *Carex lyngbeyi*, common silverweed (*Potentilla anserina* ssp. *anserina*), and Douglas' aster (*Symphyotrichum subspicatum*) typify the high marsh, and saltgrass (*Distichlis spicata*), pickleweed (*Sarcocornia perennis*), and fleshy jaumea (*Jaumea carnosa*) typify lower, more inundated sites. The rare saltmarsh birdbeak (*Chloropyron maritimum* ssp. *palustre*) is found in this area, as is the more common aquatic eel-grass (*Zostera marina*).

Figure 59. Cape Perpetua Scenic Area (#1)

4 Cape Blanco State Park

Coos & Curry counties
Bureau of Land Management (BLM), OPRD

Cape Blanco, the westernmost point in Oregon, is one of a number of scenic headlands on Oregon's south coast. A portion of the state park has been designated as a State Natural Area and supports shore pine forest, shrublands of crowberry (*Empetrum nigrum*), and western azalea (*Rhododendron occidentale*), along with coastal headland grassland. Many wildflowers can be found on the headland, including the endemic large-flowered goldfields (*Lasthenia ornduffii*). Here, a number of rare lichens grow on Sitka spruce trees. Nearby Floras Lake State Park has beach and forest habitats worth visiting, including a pygmy shore pine forest growing on infertile soils. Immediately north in Coos County is the New River Area of Critical Ecological Concern. Its dune communities have been seeded with the rare pink sandverbena (*Abronia umbellata* var. *breviflora*), following removal of European beach grass and other introduced species.

Coast Range Ecoregion

Though lacking high peaks, the topography of the Coast Range is extremely rugged. High rainfall supports lush forest vegetation. Endemic species are primarily associated with rocky sites or meadows, and disjunct populations of taxa suggest that this ecoregion served as a refugium for northern species during Pleistocene glaciations.

5 Saddle Mountain State Park
Clatsop County
OPRD

Saddle Mountain is a basalt monolith that rises in sheer cliffs above the surrounding Coast Range forest and is buffeted by incoming Pacific storms. Saddle Mountain's natural rock garden is reminiscent of traditional Chinese landscape paintings depicting mountains and pines. The flora is showy and diverse and includes a number of endemics and northern disjuncts. Endemic or noteworthy species here include Saddle Mountain bittercress (*Cardamine pattersonii*), frigid shooting star (*Dodecatheon austrofrigidum*), copper bush (*Cladothamnus pyroliflorus*), fringed kittentails (*Synthyris schizantha*), and Saddle Mountain saxifrage (*Micranthes hitchcockiana*).

6 Valley of the Giants Outstanding Natural Area
Polk County
BLM

This 51-acre old growth forest remnant is a classic example of Coast Range old growth forest with trees over 400 years old. Located at the confluence of Warnicke Creek and the North Fork Siletz River, access to this site requires a lengthy drive along bad logging roads. A loop trail through the forest highlights the Douglas-fir-western hemlock forest and its understory. The infrequently found queen-of-the-forest (*Filipendula occidentalis*) also grows in the area.

7 Marys Peak Scenic Botanical Area
Benton County
BLM, USFS

Marys Peak is a unique and easily accessed high point in the Oregon Coast Range, supporting an isolated noble fir (*Abies procera*) forest, extensive montane meadows, and a natural rock garden on the exposed summit. At an elevation of 1,250 m (4,100 ft), Marys Peak is high enough to accumulate a substantial snowpack in many winters, so that the early meadow plants such as *Anemone lyallii* and glacier-lily (*Erythronium grandiflorum*) may not begin blooming until late spring. Meadow plants bloom successively into summer; the rock garden at the summit is typically at its peak in late June and early July, with harsh Indian paintbrush (*Castilleja hispida*), spreading phlox (*Phlox diffusa*), and Cascade desert parsley (*Lomatium martindalei*) present.

Willamette Valley Ecoregion

The Willamette Valley is home to the majority of Oregon's population, and much of its native habitat has been lost or altered by development, agriculture, or invasive species. However, significant conservation areas support remnant habitats as well as populations of rare or endemic plant species.

8 Sauvie Island
Columbia & Multnomah counties
Oregon Dept. of Fish and Wildlife (ODFW), OPRD

Located at the confluence of the Willamette and Columbia rivers, Sauvie Island was the home of T.J. Howell, one of Oregon's first resident botanists. His botanical collections document the diverse flora of the Columbia River bottomlands as they existed in the late 19th century. Although subsequent agriculture and hydrologic changes have greatly diminished the native flora, important wetland communities such as wapato (*Sagittaria latifolia*)-dominated marsh can still be found. In late summer especially, Sauvie Island is an excellent place to explore the distinctive hardwood forests, swamps, marshes, and exposed river banks along the Columbia River.

9 Willamette Narrows
Clackamas County
Metro, OPRD, TNC

The area immediately above and below Willamette Falls in Oregon City supports a number of both common and unusual habitats. Of particular note are herbaceous balds and associated oak woodlands, such as at the Camassia Natural Area, with showy spring wildflower displays of camas, plectritis, monkeyflowers (*Erythranthe*), and others. The endemic pale larkspur (*Delphinium leucophaeum*) has its last stronghold in this area. Peach Cove Fen is a wetland with a floating vegetation mat, unusual in the area. Several state parks in this area have second growth forest habitats inhabited by typical forest understory flora.

Figure 60. Marys Peak Scenic Botanical Area (#7)

10 Kingston Prairie
Linn County
TNC

Kingston Prairie exists as an island of prairie surrounded by agricultural lands because its shallow soils over basalt bedrock preclude successful farming. Impressive displays of darkthroat shooting star (*Dodecatheon pulchellum* ssp. *monanthum*) and two species of camas (*Camassia leichtlinii* and *C. quamash*) are found in mid-spring, and the prairie supports a number of rare or uncommon plants, including Willamette Valley larkspur (*Delphinium oreganum*), Pacific fleabane (*Erigeron decumbens*), and *Lomatium bradshawii*. Numerous small native annuals adapted to seasonally wet soils over bedrock can be found here. The prairie is also a stronghold for declining grassland birds such as Western Meadowlark.

11 William L. Finley National Wildlife Refuge
Benton County
US Fish and Wildlife Service (USFWS)

Finley NWR is one place where visitors can immerse themselves in the historic Willamette Valley landscape of prairies, savannas, and riparian and upland forests. The Willamette Floodplain Research Natural Area, a National Natural Landmark, is the largest remaining contiguous tract of wet prairie in the Willamette Valley and supports an impressive sequence of spring and summer wildflowers. The refuge is the last stronghold for the endemic peacock larkspur (*Delphinium pavonaceum*); three species of *Sidalcea*, including the threatened Nelson's checkermallow (*S. nelsoniana*), also occur here. Extensive recovery efforts for listed plant and animal species occur here, including the reintroduction of golden paintbrush (*Castilleja levisecta*), which had been extirpated from Oregon.

12 West Eugene Wetlands
Lane County
BLM, City of Eugene, McKenzie River Trust, ODFW, TNC, US Army Corps of Engineers

The heavy clay soils of the southern end of the Willamette Valley precluded the extensive agricultural alteration that most of the Willamette Valley's native prairies experienced following pioneer settlement. A conservation partnership has been in existence for over three decades to conserve and restore remnants of wet prairies and associated habitats on the western edge of the city of Eugene. This area supports excellent displays of spring wildflowers such as camas, Oregon saxifrage (*Micranthes oregana*), and narrowleaf mule's ears (*Wyethia angustifolia*), and is an important recovery area for rare or endemic plants such as Pacific fleabane, Bradshaw's lomatium, shaggy horkelia (*Horkelia congesta* ssp. *congesta*), Kincaid's lupine (*Lupinus oreganus*), and goldenweed (*Pyrrocoma racemosa*). Several prairie species are at or near their northern range limit here, including Monterey mariposa lily (*Calochortus uniflorus*), timwort (*Cicendia quadrangularis*), and *Sidalcea cusickii*.

13 Mt. Pisgah & Howard Buford Recreation Area
Lane County
Lane County Parks, Mt. Pisgah Arboretum

Rising 305 m (1,000 ft) above the adjacent floodplain of the Willamette River, Mt. Pisgah has perhaps the largest remaining contiguous block of upland prairie and savanna in the Willamette Valley. Over 400 native vascular plant taxa have been documented in the park. Mt. Pisgah supports the only upland oak/buckbrush savanna in this ecoregion. Forested floodplain habitats also provide an excellent spring wildflower display featuring trilliums, violets, and poison larkspur (*Delphinium trolliifolium*).

Siskiyou Mountains Ecoregion

The Siskiyou Mountains are an area of global biodiversity significance and represent a northern extension of the California Floristic Province. Great topographic variation and geological diversity, including numerous areas of serpentine rocks, along with a lack of major glaciation during the Pleistocene, make the Siskiyous home to both ancient and newly evolved endemic species. Seven species of Fagaceae, including five species of oaks (*Quercus*), grow here. The lily family is also well represented by numerous species of *Calochortus*, *Erythronium*, *Fritillaria*, and *Lilium*, some of which are endemics. Much of the ecoregion, including numerous designated botanical areas, is under federal ownership, which provides varying degrees of protection and abundant opportunities for botanical exploration.

Figure 61. North Bank Habitat Management Area (#14)

14 North Bank Habitat Management Area

Douglas County
BLM

Originally established as a refuge for the rare Columbian white-tail deer, North Bank is an ideal example of the Umpqua Valley oak savanna habitat. California black oak (*Quercus kelloggii*), Oregon white oak (*Q. garryana*), and Pacific madrone (*Arbutus menziesii*) dominate the rolling hills, which resemble a typical Californian landscape; this is logical, considering that the Umpqua Valley is, under some interpretations, the northern-most extension of the California Floristic Province. The habitat of the western half of the North Bank site is well preserved and supports a diverse native prairie flora, including the endemic great white camas (*Camassia leichtlinii* ssp. *leichtlinii*), Hooker's pink (*Silene hookeri* ssp. *hookeri*), an interesting hybrid swarm of *Iris tenax* × *I. chrysophylla*, and numerous native grasses. The entire stretch of North Bank Road, though largely in private ownership, makes a good spring wildflower drive, and the annual wildflower show held the last weekend of April in nearby Glide is a must-see event.

15 Rogue River Trail

Josephine County
BLM, USFS

Downstream from Grants Pass, the Rogue River enters a narrow canyon through the Siskiyou Mountains that is accessible only by trail or boat. The canyon is easily accessed for day hikes via the road ending at Grave Creek, with trails on both banks of the river. The Rogue River National Recreation Trail traverses the north, more xeric bank 64 km (40 mi) westward to the downstream trailhead of Foster Bar. Dominant trees here are Douglas-fir and canyon live oak, but numerous other conifers, broadleaf evergreens, and deciduous hardwoods also grow here. Wildflower diversity is high due to the great variation in soil moisture and exposure, and ferns are also diverse. Hikers may see the endemic Rogue River stonecrop (*Sedum moranii*) and the red-flowered larkspur (*Delphinium nudicaule*), which is close to its northern range limit.

16 Lower & Upper Table Rocks

Jackson County
BLM, TNC

The Table Rocks are andesite mesas ringed by cliffs that rise 240 m (800 ft) above the center of the Rogue Valley near Medford. Dry mixed woodlands of Oregon white oak, California black oak, Pacific madrone, and ponderosa pine skirt the slopes, and the plateaus support mounded grasslands and vernal pools. Native annuals such as Cascade downingia (*Downingia yina*), slender goldfields (*Lasthenia californica* ssp. *californica*), bracted popcornflower (*Plagiobothrys bracteatus*), and the narrow endemic dwarf woolly meadowfoam (*Limnanthes pumila* ssp. *pumila*) populate the vernal pools, blooming as water levels recede. The spring flora also highlights regional specialties such as Henderson's fawnlily (*Erythronium hendersonii*) and scarlet fritillary (*Fritillaria recurva*).

Figure 62. Rogue River Trail (#15)

17 Eight Dollar Mountain & T.J. Howell Botanical Drive

Josephine County
USFS

A 13-km (8-mi) section of Eight Dollar Mountain Road has been designated the "T.J. Howell Botanical Drive," to commemorate the pioneering botanist who was the first to botanically explore the Illinois Valley (see p. 11 for biography of T.J. Howell). The drive, which highlights both serpentine and non-serpentine vegetation areas, includes a stop at a boardwalk through a serpentine fen on the south side of Eight Dollar Mountain where California pitcher plant and other serpentine fen plants can be easily seen alongside blooming western azalea in late spring. Further to the west, at the Days Gulch Botanical Area, serpentine soils support a Jeffrey pine savanna. The road continues to the boundary of the Kalmiopsis Wilderness where Brewer's spruce (*Picea breweriana*) and deer oak (*Quercus sadleriana*) can be seen in the Babyfoot Lake Botanical Area. The endemic *Kalmiopsis leachiana* grows nearby but can be seen more easily downstream along the Illinois River in the York Creek Botanical Area. Much of the Kalmiopsis Wilderness burned in the 2002 Biscuit fire, the largest wildfire in Oregon history.

18 Rough and Ready Creek

Josephine County
BLM, OPRD, TNC, USFS

One of the easiest and most rewarding places to experience the serpentine country of the Illinois Valley is along Rough and Ready Creek south of Cave Junction. The alluvial terrace comprised of boulders washed down from the adjacent serpentine slopes supports a xeric shrubland of *Arctostaphylos*, *Ceanothus*, and *Garrya* interspersed with diverse native grasses, spring bulbs, and wildflowers. Higher elevations in the drainage support a variety of serpentine forest communities. The extensive plant list includes regional endemics such as purple-flowered rush-lily (*Hastingsia bracteosa* var. *atropurpurea*), *Lomatium cookii*, and Siskiyou butterweed (*Packera hesperia*). A portion of the trail is wheelchair accessible.

19 Chetco River Redwood Groves

Curry County
OPRD, USFS

A few groves of coast redwood (*Sequoia sempervirens*), the world's tallest tree species, are found in the very southwestern part of Oregon at the northern limit of the species' natural range. The Redwood Nature Trail, located about 13 km (8 mi) northeast of Brookings, passes through a stand of old growth redwoods and is a good place to see the associated flora. An old growth grove of Oregon myrtle (*Umbellularia californica*) is found in the adjacent Alfred Loeb State Park.

20 Mt. Ashland-Siskiyou Peak Botanical Area

Jackson County
USFS

Located about 8 km (5 mi) north of Oregon's southern border, the granite ridge extending from Mt. Ashland (2,300 m [7,560 ft]) west to Siskiyou Peak is home to a unique flora that is transitional between California and Oregon floras. This subalpine zone supports a mix of conifer stands, brush fields, wet meadows, and open rocky habitats. The Ashland lupine (*Lupinus lepidus* var. *ashlandensis*) is endemic to Mt. Ashland; *Horkelia hendersonii* and *Tauschia howellii* are other rarities that can be seen here. One can explore the site while driving along Forest Service Rd. 20 or hiking on the Pacific Crest Trail, which parallels the road in this area. West of Siskiyou Peak, serpentine endemics are found on peridotite soils of the Red Mountain Botanical Area.

Cascades Ecoregion

Spanning the length of the state and ranging from sea level to mountain peaks, the Cascades is perhaps Oregon's most diverse ecoregion. Important habitats include old growth coniferous forests, montane wetlands, meadows, and rock outcrops. Topographic diversity and the limited extent of past glaciation suggest that many areas provided refugia during times of climate extremes, resulting in significant regional and local endemism. Heavy snowfall at high elevations makes for an extended flower viewing season. Much of the land is in public ownership, so habitats are readily accessible. Note that this ecoregion also includes higher elevations just east of the Cascade crest, where habitats are more similar to habitats just west of the crest than to more arid regions at lower elevations to the east.

Figure 63. Table Rock Wilderness (#23)

21 Western Columbia River Gorge: McCord Creek & Elowah Falls

Multnomah County
OPRD, USFS

The region on the south side of the Columbia River from Crown Point to Cascade Locks is well known for its waterfalls and viewpoints. Each of the waterfalls and side streams provides an opportunity to experience the Gorge flora. A particular concentration of Gorge endemics and other specialties can be seen along the trails around McCord Creek and Elowah Falls. The trail to the upper falls traverses basalt cliffs, providing close access to Gorge specialties such as western mountain kittentails (*Synthyris missurica* ssp. *stellata*), Howell's daisy (*Erigeron howellii*), and gorge daisy (*Erigeron oreganus*). Other taxa found in the area include bolandra (*Bolandra oregana*), white shooting star (*Dodecatheon dentatum* ssp. *dentatum*), cliff dwarf-primrose (*Douglasia laevigata*), and *Selaginella douglasii*. Note that this trail is not for small children or people with fear of heights, and other trails in the area can be just as rewarding botanically.

22 Mt. Hood Subalpine Zone

Clackamas & Hood River counties
USFS

At 3,430 m (11,240 ft), Mt Hood is the highest peak in Oregon and is one of the best places to see wildflowers of the subalpine and alpine zones of the Cascades. The rugged 64-km (40-mile) Timberline Trail circles the mountain, and numerous spur trails provide day hikers access to higher elevations. On the north side, trails to McNeil Point, Vista Ridge, and Cooper Spur make good day hikes into the subalpine zone, with its mountain hemlock (*Tsuga mertensiana*) groves, fell fields, and flowery meadows.

23 Table Rock Wilderness

Clackamas County
BLM

It is unfortunate so many places in Oregon are named "Table Rock," as they lose their distinctiveness. This is a most distinctive botanical destination: a rugged area of forests, talus, and cliffs located in the Molalla River drainage. The flora has nearly 240 native species and includes the rare: Hall's false rue anemone (*Enemion hallii*); the endemic: Clackamas iris (*Iris tenuis*); the stunning: Cascade lily (*Lilium washingtonianum* ssp. *purpurascens*) and cliff penstemon (*Penstemon rupicola*); and the mundane but important: six species of blueberries (*Vaccinium*).

Figure 64. Broken Top (#27)

24 Silver Falls State Park

Marion County
BLM, OPRD

This is Oregon's largest state park and one of the largest remaining tracts of mature, low-elevation conifer forest in Oregon. Numerous waterfalls cascade over basalt ledges, and the lush forests and cliffs support a diverse flora of over 300 native plant species, including a good representation of Saxifragaceae and ferns.

25 Iron Mountain & Cone Peak

Linn County
USFS

Iron Mountain, along with nearby Cone Peak and Tombstone Prairie, is well known as a botanical hot spot due to the diverse habitats present, from moist forest to dry meadows and rock outcrops, with significant disjunct species from east of the Cascades as well. In late spring and early summer, the open slopes are carpeted with species such as Cascade mariposa lily (*Calochortus subalpinus*), harsh Indian paintbrush, Menzies' larkspur (*Delphinium menziesii*), and skyrocket (*Ipomopsis aggregata* ssp. *formosissima*). Also of particular note here are Hells Canyon rockcress (*Boechera hastatula*), Gordon's ivesia (*Ivesia gordonii* var. *alpicola*), and five species of *Penstemon*.

26 Forests of the Upper McKenzie River Valley

Lane & Linn counties
USFS

The pure water of the McKenzie River comes from a distinctive forested landscape. The McKenzie River National Recreational Trail parallels the river for 43 km (27 miles) from McKenzie Bridge to the wetland of Fish Lake and is easily accessed at many points. Nearby, the Delta Nature Trail passes through a rich old growth forest with a diverse understory and impressively large trees. The trail to Proxy Falls passes through open forest of scattered conifers and the broadleaf evergreen golden chinquapin (*Chrysolepis chrysophylla* var. *chrysophylla*) that grow among the jumbled lava flows. The under-appreciated fall color provided by vine maple and other deciduous shrubs lights up the lava flows in October.

27 Broken Top Area, Eastern Side of the Three Sisters Wilderness

Deschutes County
USFS

The subalpine country just east of South Sister provides numerous wildflower hikes for all abilities. The short trail along Todd Lake (elevation 1,860 m [6,100 ft]) passes through extensive wildflower meadows and wetlands. The Green Lakes, set at the base of South Sister, are a popular day hike destination. Timberline habitats such as pumice flats, wildflower-lined creeks, rocky ridges, and parklands dotted with whitebark pine (*Pinus albicaulis*) can be seen on Broken Top and reached either by trail from the south or via Tam McArthur Rim from the east. Noteworthy species on Broken Top include great alpine whitlow grass (*Draba aureola*) and alpine mountain trumpet (*Collomia debilis*).

28 Tire Mountain
Lane County
USFS

Located on a south-facing ridge at 1,220 m (4,000 ft) in the Western Cascades, Tire Mountain's flora nonetheless has many similarities with the Willamette Valley. An easy trail leads to a series of open rocky slopes that are covered with a diverse array of wildflowers in late spring. Particularly prominent are hues of yellow from species of monkeyflower, woolly sunflower (*Eriophyllum lanatum*), and deltoid balsamroot (*Balsamorhiza deltoidea*); pink from rosy plectritis (*Plectritis congesta*); and blue from blue-eyed Mary (*Collinsia grandiflora*), Menzies' larkspur (*Delphinium menziesii*), and bluehead gilia (*Gilia capitata*).

Figure 66. Tire Mountain (#28)

29 North Umpqua River
Douglas County
USFS

The forests, cliffs, and waterfalls of the North Umpqua River provide habitats for a noteworthy flora, including species at their northern range limit such as marbled wild ginger (*Asarum marmoratum*), canyon live oak (*Quercus chrysolepis*), and snow plant (*Sarcodes sanguinea*), as well as the endemic *Kalmiopsis fragrans*. Much of the Boulder Creek wilderness has been burned in recent years by wildfires and provides an opportunity to compare the floras of different natural successional stages. Numerous trails provide ready access to the region; less accessible but protected sites such as the Limpy Rock Research Natural Area also warrant exploration.

30 Rogue-Umpqua Divide Wilderness
Douglas & Jackson counties
USFS

This out-of-the-way corner of the Western Cascades offers a mix of montane forest, meadows, and rocky slopes. Several research natural areas—Cougar Butte and Abbott Creek—contain representative forest and meadow communities. A 50-km (31-mi) trail traverses the wilderness from north to south with numerous entry points. Visit Abbott Butte soon after the snow melts to see steer's head (*Dicentra uniflora*), glacier-lily, Klamath fawn-lily (*Erythronium klamathense*), and three-leaf lewisia (*Lewisia triphylla*) in bloom; the meadows display continuous blooms as the season progresses. In the northern portion of the wilderness, the meadows of Fish Creek Valley have late-blooming species of gentian, Cascade grass-of-Parnassus (*Parnassia cirrata* var. *intermedia*), the curious umbel western oxypolis (*Oxypolis occidentalis*), and the rare orchid starry ladies' tresses (*Spiranthes stellata*). A trail leads across winding Fish Creek up to the rocky cliffs and high meadows of 2,029-m (6, 656-ft) Rattlesnake Mountain.

Figure 65. Rattlesnake Mountain in the Rogue-Umpqua Divide (#30)

31 Crater Lake National Park
Douglas, Jackson & Klamath counties
National Park Service

Most visitors are drawn to Crater Lake, Oregon's only national park, by the stunning views of the lake formed in the caldera of the collapsed Mt. Mazama. However, numerous short trails located along Rim Drive provide an easy opportunity to explore the high-elevation flora. Look for the rust-colored flowers of endemic Crater Lake currant (*Ribes erythrocarpum*) among the rugged mountain hemlock trees. Mount Mazama collomia (*Collomia mazama*) is another endemic of Crater Lake and nearby parts of the Cascade crest. The 8-km (5-mi) round-trip hike to the summit of Mt. Scott leads to rocky, high-elevation (2,720 km [8,930 ft]) habitats with a diverse subalpine flora. Five species of rock cress (*Boechera*, formerly *Arabis*) have been collected on Mt. Scott.

32 Grizzly Peak
Jackson County
BLM

Reaching nearly 1,830 m (6,000 ft), Grizzly Peak is a high plateau above Ashland and the Bear Creek Valley. The conifer forest on the north slope—mostly true firs that are intermediate between Sierra white fir and grand fir (*Abies concolor* and *A. grandis*)—hosts tall bugbane (*Cimicifuga elata*) as well as dense stands of mountain larkspur (*Delphinium glaucum*). Open meadows and natural rock gardens occupy much of the top of the plateau. The southwestern end of the loop trail passes through outcrops of volcanic rock and formerly forested areas now open after a 2002 wildfire. Nearly 300 vascular plant taxa have been found here, including a number that are known from few localities in Oregon.

Figure 67. Rowena Plateau (#33)

Eastern Cascade Slopes and Foothills Ecoregion

In the lee of the Cascade Crest, the pine-dominated landscape of the Eastern Cascade Slopes and Foothills is the region that attracts westsiders seeking a dose of sunshine. In addition to the ponderosa pine stands (with Oregon white oak at the north and south ends), there are areas of old growth Oregon juniper (*Juniperus occidentalis*) woodlands and montane wetlands.

33 Eastern Columbia River Gorge: Rowena Plateau

Wasco County
OPRD, TNC, USFS

The meadows and open woodlands of ponderosa pine and white oak of the eastern Columbia River Gorge are renowned for their wildflower displays. The Rowena Plateau is a classic locality for eastern Gorge flora. The wildflower display begins in late winter with yellow bells (*Fritillaria pudica*), Piper's desert parsley (*Lomatium piperi*), and grass widows (*Olsynium douglasii*) and continues into early summer. The mid-spring display of lupine and balsamroot has graced many scenic calendars, and at least eight species of *Lomatium* grow here, including the statuesque *L. columbianum*. Vernal pools formed on the bedrock plateau provide an additional unique habitat.

34 Metolius River & Black Butte

Jefferson County
USFS

The Metolius River emerges fully formed from a spring on the north side of Black Butte, and the forest and streamside habitats of the region can be explored on a number of short trails. This is ponderosa pine savanna country, and the US Forest Service has been working over the years to restore these pine habitats with thinning and controlled burning. The trail to the summit of Black Butte provides rewarding views and an interesting flora. Peck's penstemon (*Penstemon peckii*) is one of the endemic specialties of this region.

35 Wood River Wetland

Klamath County
BLM

The wetlands of the Klamath Basin are well known for their value to birds, fish, and other wildlife. This wetland, located along the Wood River on the northern side of Agency Lake, provides an easily accessible introduction to Klamath Basin wetland communities. Species of Cyperaceae are particularly diverse in these wetlands, including river bulrush (*Bolboschoenus fluviatilis*), *Carex aquatilis*, common spikerush (*Eleocharis palustris*), tule (*Schoenoplectus acutus*), and softstem bulrush (*S. tabernaemontani*). Other common or noteworthy wetland plants here include Baltic rush (*Juncus balticus* ssp. *ater*), broadfruit bur reed (*Sparganium eurycarpum*), common cattail (*Typha latifolia*), common bladderwort (*Utricularia vulgaris* ssp. *macrorhiza*), and several species of willow. Other wetlands in the Klamath Basin, such as the Upper Klamath and the Klamath Marsh National Wildlife Refuges, Sycan Marsh, and Tunnel Creek wetland are also noteworthy but more difficult to access (some require a canoe). Klamath County is a paradise for sedge enthusiasts; so far, 70 *Carex* taxa have been documented from the county.

36 Gearhart Mountain

Klamath & Lake counties
USFS

In south-central Oregon, the Eastern Cascade Slopes and Foothills ecoregion expands eastward toward the Great Basin. Gearhart Mountain is the high point in this region, rising to 2,550 m (8,360 ft). Pleistocene glaciers carved dramatic cliffs in the volcanic rocks, and a series of meadows follow the drainages downslope. Forests are dominated by ponderosa pine and Oregon juniper on the lowest and driest sites, with Sierra white fir, lodgepole pine (*Pinus contorta* var. *latifolia*), and sugar pine (*P. lambertiana*) at middle elevations, and whitebark pine (*P. albicaulis*) at the highest elevations. However, the forests have been extensively altered by tree mortality due to pine beetles in recent years. Both rocky habitats and wet meadows here are floristically rich, with good representation of genera such as *Eriogonum*, *Leptosiphon*, *Lupinus*, and *Penstemon*, including the endemic blue-leaved penstemon (*Penstemon glaucinus*).

Columbia Basin Ecoregion

The Columbia Basin is an arid low elevation region of dry grassland and shrub steppe and is an extension of the larger Columbia Basin ecoregion across the Columbia River in eastern Washington. Though much of Oregon's portion has been converted for agriculture, areas of rugged topography along the two major river canyons, the Deschutes and John Day, invite exploration.

37 Cottonwood Canyon State Park

Gilliam & Sherman counties
OPRD

Lacking major dams, the John Day River is the longest free-flowing river in Oregon. Cottonwood Canyon State Park is a good central location for exploring the canyon country of the Columbia Basin. This is a landscape of sagebrush and bunchgrasses and is a hot spot for certain genera of the dry interior. For example, sixteen species of milkvetch (*Astragalus*), ten species of buckwheat (*Eriogonum*), and seventeen species of *Lomatium* have been reported from Gilliam and Sherman counties. Look also for the infrequent deciduous tree, western hackberry (*Celtis reticulata*), growing on rocky slopes above the river. Several nearby BLM Wilderness Study Areas are also worth exploring.

Figure 69. Fort Rock (#39)

Lava Plains Ecoregion

The Lava Plains ecoregion is transitional between the Columbia Basin and the northern Basin and Range province. It is arid country of sagebrush (*Artemisia*) and juniper, and because much of the ecoregion is managed by BLM, there are numerous BLM designated natural areas to explore.

38 Painted Hills

Wheeler County
National Park Service

The Painted Hills are named for the colorful sedimentary deposits associated with the Oligocene John Day Formation, and are protected by a unit of the John Day Fossil Beds National Monument. At a relatively low elevation of 610 m (2,000 ft), sagebrush and other shrub-steppe communities dominate the vegetation. This area supports a number of endemic plants, including yellow-haired paintbrush (*Castilleja xanthotricha*) and John Day pincushion (*Chaenactis nevii*). The extent of the wildflower display varies substantially from year to year depending upon precipitation but can be impressive in a wet year. Two other units of the John Day Fossil Beds National Monument, the Clarno and Sheep Rock Units, are also good places to learn about the present day flora as well as the area's fossil history.

Figure 68. Wildhorse Lake on Steens Mountain (#41)

Basin and Range Ecoregion

The remote valleys and mountains of Oregon's Basin and Range country are deservedly popular destinations for botanical exploration. Diversity is surprisingly high, and many endemics are found in this ecoregion, though some are shared with adjacent parts of Nevada or California. For the optimal botanical experience, timing is everything and depends upon elevation and precipitation patterns. The sites described below are more broadly defined than for Oregon's other ecoregions, thereby allowing each site to encompass many more features.

39 Fort Rock & Christmas Valley

Lake County
BLM, OPRD

This remote, arid region of northern Lake County has much of interest, particularly in a wet spring. Numerous BLM botanical areas provide access to a variety of habitats, from basalt buttes to lava flows, sand dunes, and alkaline basins. Sagebrush and juniper dominate the vegetation, but an isolated ponderosa pine stand is found in the Lost Forest Research Natural Area. Fort Rock State Park provides botanical interest but is also known for its archaeology, with over 10,000 years of documented human presence.

40 Warner Basin & Hart Mountain National Antelope Refuge

Lake County
BLM, USFWS

Though perhaps better known for wildlife viewing, the Warner Basin and adjacent Hart Mountain are also botanically significant—nearly 500 native plant species have been documented in this area. Species diversity is a function of the substantial range of habitats found here, from wetlands and riparian corridors to upland shrub steppe and rock outcrops at elevations ranging from 1,340 m (4,400 ft) to over 2,440 m (8,000 ft). There is strong representation of grasses and sedges here, as well as a dozen species each of buckwheat and milkvetch, nine species of lupine, and six species of onion (*Allium*).

41 Steens Mountain

Harney County
BLM, ODSL

The uplifted fault block of Steens Mountain, rising to 2,970 m (9,730 ft), is a world unto itself, an island of montane and subalpine habitats above the sagebrush desert. The valleys of Kiger, Little Blitzen, and Big Indian Gorges are classic examples of glacially carved valleys. Steens Mountain has few trees other than Oregon juniper, poplar, willow, and a few groves of Sierra white fir. Still, nearly 1,100 vascular plant taxa have been documented here, and the unique geography and ecology of Steens Mountain provides habitat for over 60 rare plant species. Regional or local endemics found on Steens Mountain include Cusick's horsemint (*Agastache cusickii*), Steens Mountain whitlow grass (*Draba cusickii* var. *cusickii*), Steens Mountain thistle (*Cirsium eatonii* var. *peckii*), and Steens Mountain paintbrush (*Castilleja pilosa* var. *steenensis*). All elevation zones are readily accessible via the Steens Loop Road; visits to Fish Lake, Wildhorse Lake, and the actual summit provide access to the full spectrum of landscapes and species.

42 Alvord Desert & Pueblo Valley

Harney County
BLM

The Alvord Desert, located in the lee of Steens Mountain, is Oregon's largest internally draining alkaline basin. The Pueblo Valley is a southward extension located east of the Pueblo Mountains. This is one of the driest parts of Oregon, and the climate and alkaline soils result in distinctive vegetation. Mickey Hot Springs Research Natural Area supports an uncommon shrub community of winter fat (*Krascheninnikovia lanata*) and other halophytic plants. A trail up Pike Creek provides views of the Alvord desert and, if the timing is right, a diverse selection of shrub steppe wildflowers. South of Fields, the Pueblo Foothills RNA supports stands of black cottonwood (*Populus trichocarpa* and possible intergrades with other species) along Cottonwood Creek, as well as Nevada ephedra, an uncharacteristic gymnosperm for Oregon and one at the northern limits of its range.

Blue and Wallowa Mountains Ecoregion

The Blue and Wallowa Mountains extend over a large area of eastern and central Oregon. The climate is generally continental, with substantial snow in the winter. Geographically, this area provides a connection to the Middle Rocky Mountains, which is reflected in the flora of the ecoregion. While much of the area is volcanic in origin, there are actually many rock types represented, which makes for a particularly diverse flora, with many endemic species and spectacular wildflower habitats.

Figure 70. Zumwalt Prairie (#44)

Figure 71. Strawberry Mountains (#48)

43 Skyline Road

Umatilla & Union counties
USFS

Following along the crest of the Blue Mountains at about 1,520 m (5,000 ft) elevation, the 64-km (40-mi) drive along Skyline Road provides easy access to the high country of the Blue Mountains. Habitats range from montane forest to meadows and open scablands. Try to find the locally restricted *Lupinus sabinianus*, a yellow lupine discovered here by David Douglas during his botanical explorations in the 1820s. Another plant unique to the region is Blue Mountain penstemon (*Penstemon pennellianus*).

44 Zumwalt Prairie

Wallowa County
TNC, USFS

Unlike most areas of Palouse Prairie, Zumwalt Prairie's elevation is too high to support agriculture and thus has been managed as rangeland. Because of this, extensive grasslands remain in Wallowa County, providing habitat for a high diversity of bunchgrasses as well as rarities such as *Silene spaldingii*. Most of this landscape is privately owned, but The Nature Conservancy provides trail access to the portions it owns (approximately 10%) along Camp Creek and Harsin Butte.

45 Eagle Cap Wilderness High Country

Wallowa County
USFS

Peaks over 3,050 m (10,000 ft), diversity of bedrock types (basalt, greenstone, limestone, and granitic rocks), and proximity to the Rocky Mountains to the east make the Wallowa Mountains a noteworthy locality for plants found nowhere else in Oregon. Access to the botanical riches of this area is best achieved via its numerous trails. Many of the most popular destinations, such as the Lake Basin under the summit of Eagle Cap, require a multi-day backcountry trip. The East Fork Wallowa River Trail leads to the high country around Aneroid Lake and Polaris Pass, home to eight species of paintbrush (*Castilleja*), including the endemic *Castilleja fraterna*. A less strenuous route to the south leads from the Tenderfoot trailhead to Bonny Lakes. The Hurricane Creek Trail passes through calcareous meadows with a diversity of moonwort (*Botrychium*) species. The easiest way to take in the high country, however, is to ride the Mt. Howard tram above Wallowa Lake, which carries visitors to an elevation of over 2,440 m (8,000 ft), where the endemic *Lomatium greenmanii* can be found growing on the subalpine ridgetop.

Figure 72. Greenhorn Mountains (#47)

46 Wallowa Mountain Loop Highway

Baker & Wallowa counties
USFS

The Wallowa Mountain Loop Highway provides access to the upper part of the rugged Imnaha River drainage, between the Wallowa Mountains and Hells Canyon. The landscape is a mosaic of mixed coniferous forest including ponderosa pine and open rocky slopes. Composites (Asteraceae) are particularly well represented in the flora. *Camassia cusickii* and Wallowa primrose (*Primula cusickiana* var. *cusickiana*), both regional endemics named in honor of the early botanist William C. Cusick, are also found in this area. Follow the side road up the Imnaha River or take the short detour off the loop highway to the Hells Canyon overlook, where summer wildflowers are likely to distract from the views across the Snake River canyon into Idaho. Hat Point, farther north and reached by a rugged side road, is another noteworthy botanical locality overlooking Hells Canyon.

47 Greenhorn Mountains

Grant County
USFS

Named for their serpentine formations, the Greenhorn Mountains support a variety of bedrock types, which in turn influence the flora. Despite a long history of mining, the area has been protected within the Vinegar Hill-Indian Rock Scenic Area. Several trails provide access to a long, rounded subalpine ridge crest at about 2,440 m (8,000 ft), where one can find extensive sagebrush communities and numerous species of wildflowers and ferns, including the easternmost populations of the serpentine-loving Shasta sword fern (*Polystichum lemmonii*).

Figure 73. Leslie Gulch (#50)

48 Strawberry Mountains

Grant County
USFS

The Strawberry and Aldrich mountains south of John Day have many areas of scenic and botanical interest. In addition to the classic hike to Strawberry Lake, the Canyon Mountain Trail, located in the northwest corner of the Strawberry Wilderness, passes through areas of serpentine bedrock. The endemic colonial luina (*Luina serpentina*) can be seen along this trail, though the name is a misnomer, as the species usually does not grow on serpentine. The Canyon Mountain Trail continues eastward along the crest of the Strawberry Mountains and can be experienced as a 60-km (37-mi), one-way, backcountry trip. On the southwest side of the wilderness, the East Fork Canyon Creek Trail leads up a long forested valley, providing access to unmanaged ponderosa pine stands in the Canyon Creek Research Natural Area. The crest of the Strawberry Mountains and its wildflower meadows and unusual rock formations can be reached from a southerly approach via the Buckhorn Meadows and Indian Butte Trails.

49 Logan Valley

Grant County
USFS

Located immediately southeast of the Strawberry Mountains are the mile-high meadows of Logan Valley. Unlike many of the other similar large montane meadows in the Blue Mountains, much (but not all) of Logan Valley is in public ownership. The habitat is a mix of wet and dry meadows abundant with native plants—nearly 300 species—with good representation among meadow familiars such as paintbrushes, delphiniums, fleabanes, biscuitroot, buttercups, asters, sedges, and rushes.

Owyhee Uplands Ecoregion

Hugging the Oregon-Idaho border, the Owyhee Uplands put a unique spin on the Basin and Range flora. This is due in part to the endemic flora associated with volcanic ash and tuff deposits, which are still yielding botanical novelties.

50 Leslie Gulch

Malheur County
BLM

Amid Leslie Gulch's canyons and rock pinnacles, ancient volcanic ash deposits with unusual soil chemistry support an array of interesting plants. A number of local or regional endemics grow here, and five of these species—*Artemisia packardiae,* grimy mousetail (*Ivesia rhypara* var. *rhypara*), Packard's blazingstar (*Mentzelia packardiae*), Owyhee phacelia (*Phacelia lutea* var. *lutea*), and Ertter's groundsel (*Senecio ertterae*)—were first described as new to science as recently as the 1970s. Nearly forty species of composites have been documented here as well.

VASCULAR PLANTS OF OREGON

Key to Plant Groups:

1. Plants without seeds, flowers, or cones; plants with an evident alternation of generations . . . . **PTERIDOPHYTES** (p. 66)

1' Plants with seeds produced by flowers or cones; plants without an evident alternation of generations.

2. Seeds naked, not enclosed in carpels, borne between scales of cone; plants without fruits or typical flowers . . . . **GYMNOSPERMS** (p. 108)

2' Seeds enclosed in carpels; plants with fruits and typical flowers.

3. Floral parts in 3's or 6's, rarely in 2's or 4's, never in 5's; leaves usually parallel-veined; vascular bundles scattered throughout pith; embryo with 1 cotyledon . . . . **MONOCOTS** (p. 127)

3' Floral parts usually in 4's or 5's; leaves pinnate- or palmate-veined, rarely parallel-veined; vascular bundles arranged concentrically around a central pith; embryo with 2 cotyledons . . . . **DICOTS** (Volumes 2a & 2b)

FAMILY INDEX

PTERIDOPHYTES

Duncan Thomas (except where noted)

Herbs (ours), perennial, with an evident alteration of generations, and both generations independent at maturity. **Leaves** typically "fern"-like, moss-like (Lycopodiaceae, Selaginellaceae), rush-like (Equisetaceae) or quill-like (Isoëtaceae). **Sporophytes** reproducing by minute, numerous, 1-celled spores borne in sporangia; sporangia borne on vegetative or modified leaves, or in their axils.

Vascular plants without seeds are lumped together to form the Division Pteridophyta. As an ancient plant group, great diversity exists within this Division. This is reflected in the morphological and genetic differences between the ferns, club-mosses, spike-mosses, and quillworts.

Key to the Pteridophyte families:

1. Plants not fern-like; leaves small and scale-like or long, narrow and not dissected.
  2. Plants small free-floating aquatics of ponds and calm water . . . . **SALVINIACEAE**
  2' Plants not free-floating aquatics.
    3. Leaves scale-like, whorled, and usually fused into a sheath around the nodes of a stem; branches (when present) whorled; sporangia on peltate scales clustered into terminal cones . . . . **EQUISETACEAE**
    3' Leaves not scale-like, whorled, or fused into a sheath; branches not whorled; sporangia not on peltate scales.
      4. Rooted aquatics or emergent plants of (seasonally) wet places with clover- or grass-like leaves and sporangia in rounded/oblong nut-like sporocarps . . . . **MARSILEACEAE**
      4' Plants lacking above combination of characters.
        5. Leaves grass-like, clustered on short erect stems; sporangia embedded in the leaf bases; plants submerged in lakes or emergent/terrestrial in (seasonally) wet places . . . . **ISOËTACEAE**
        5' Leaves not grass-like; sporangia not embedded in the leaf bases; plants not submerged in lakes or emergent/terrestrial in (seasonally) wet places.
          6. Leaves usually 1, more than 2 mm wide; plants not resembling robust mosses or conifer seedlings . . . . **OPHIOGLOSSACEAE** (in part)
          6' Leaves many, slender, less than 2 mm wide; plants resembling robust mosses or conifer seedlings.
            7. Megaspores absent; stems generally more than 3 mm wide; fertile leaves not ranked; leaves generally more than 3 mm long (2–3.5 mm in *Diphasiastrum*) . . . . **LYCOPODIACEAE**
            7' Microspores and megaspores present; stems generally less than 3 mm wide; fertile leaves 4-ranked; leaves generally less than 3 mm long . . . . **SELAGINELLACEAE**

1' Plants fern-like; leaves usually broad and dissected (i.e., feathery or leafy fronds).
  8. Sporangia 1 mm or more, on separate bladeless stalk; new leaves not coiled . . . . **OPHIOGLOSSACEAE** (in part)
  8' Sporangia less than 0.5 mm, not easily visible without magnification, generally under leaves; new leaves coiled.
    9. Leaves commonly over 75 cm long; stems arising singly from a far-creeping rhizome; petiole and rachis erect to arching, bearing pairs of pinnae decreasing in size distally, lowest pair (2)3(4)-pinnate; sporangia marginal and covered by recurved leaf margin . . . . **DENNSTAEDTIACEAE**
    9' Plants without above combination of characters.
      10. Indusium well developed, peltate or reniform, not elongate; sori on veins in the mid-area of the segment.
        11. Abaxial leaf surface bearing 1-celled colorless hairs; indusia reniform . . . . **THELYPTERIDACEAE**
        11' Abaxial leaf surface lacking 1-celled hairs; indusia reniform or peltate . . . . **DRYOPTERIDACEAE**
      10' Indusium (if present) not peltate or reniform; sori abaxial or marginal or on veins, or sporangia not aggregated into sori.
        12. Mid-region of leaf blade pinnatifid or pinnate; pinnae not deeply divided into more than 2 segments.
          13. Leaves pinnatifid; pinnae widest at the base.
            14. Leaves clustered, strongly dimorphic; plants usually on soil; indusia present . . . . **BLECHNACEAE** (in part)
            14' Leaves not clustered, not dimorphic; plants usually on trees or rocks; indusia absent . . . . **POLYPODIACEAE**
          13' Leaves pinnate; pinnae narrowed at the base, usually to a distinct petiolule.
            15. Sporangia in elongate sori along veins, covered by flap-like indusia . . . . **ASPLENIACEAE**
            15' Sporangia in submarginal bands on the abaxial leaf surface, partly covered by the recurved leaf margin (false indusia) or not . . . . **PTERIDACEAE** (in part)

12' Mid-region of leaf blade bipinnate or if pinnate then pinnae deeply lobed.
16. Sori absent or linear and marginal/submarginal/along veins or covered by recurved segment margins (false indusia) . . . . **PTERIDACEAE** (in part)
16' Sori round to oval or J-shaped; indusia present or not.
17. Pinnules bearing 2 lines of elongate sori parallel to the costa; leaves clustered with deeply pinnatifid pinnae . . . . **BLECHNACEAE** (in part)
17' Pinnules and sori not as above; leaves clustered or not.
18. Widest pair of pinnae at or near bottom of leaf; stems creeping; leaves not clustered . . . . **CYSTOPTERIDACEAE** (in part)
18' Widest pair of pinnae in mid-region of leaf; stems not creeping; leaves clustered.
19. Older leaf bases persistent, numerous, and wiry; indusium composed of filaments resembling strings of beads under the sporangium; small plants of dry habitats . . . . **WOODSIACEAE**
19' Older leaf bases sparse, not wiry; indusium not as above or none; small to large plants of dry to wet habitats.
20. Stems creeping; indusia hood-like; usually small plants of moist to dry sites; stems easily broken. . . . . **CYSTOPTERIDACEAE** (in part)
20' Stems erect or ascending; indusia elongate to J-shaped or none; robust plants of wet to moist sites; stems not easily broken . . . . **ATHYRIACEAE**

ASPLENIACEAE Spleenwort family

Plants perennial, herbaceous. **Stems** short and erect, ascending or rarely creeping, mostly unbranched; leaf bases persistent; scales dense, usually lattice-like. **Leaves** monomorphic, mostly evergreen. **Blades** simple to 3- or 4-pinnate; indumentum of narrow scales and glandular hairs or none. **Sori** on abaxial veins, not round, slightly elongated to linear; indusia usually present on 1 side of the sori, rarely absent. **Spores** with a simple linear scar, bilateral, usually narrowly winged; wall ornamentation various. **Gametophytes** thin, heart-shaped, green.

Cosmopolitan. Approximately 24 genera; 1 genus treated in *Flora*.

ASPLENIUM Spleenwort

Stems perennial, mostly short, ascending, occasionally long-creeping, usually unbranched, densely covered with lattice-like scales and persistent leaf bases. **Leaves** monomorphic, evergreen or wintergreen. **Petioles** green or brown. **Blades** with or without bulbils, often linear and 1-pinnate or broader and more divided; veins free, ending in a hydathode gland and not reaching the margin. **Sori** elongated, borne along a vein; indusium hyaline, present on 1 side of the sorus and partially covering it.

Cosmopolitan. Approximately 360 species; 3 species treated in *Flora*.

Asplenium bulbiferum, a Chinese native, has been collected once in Coos County, escaped from cultivation. This species has a large three- to four-pinnate frond with bulbils on some pinnae, thus it cannot be confused with any of our native species. Other *Asplenium* species including *A. scolopendrium* (Hart's tongue fern) are widely cultivated and likely to persist in abandoned gardens.

Key to *Asplenium* species:

1. Fronds very narrow, graminoid, forked distally. . . . . ***A. septentrionale***
1' Fronds linear, 1-pinnate, not graminoid, not forked.
2. Mature rachis dark brown, tip sometimes green . . . . ***A. trichomanes*** ssp. ***trichomanes***
2' Mature rachis green . . . . ***A. viride***

Asplenium septentrionale (Linnaeus) Hoffmann

Northern spleenwort, forked spleenwort

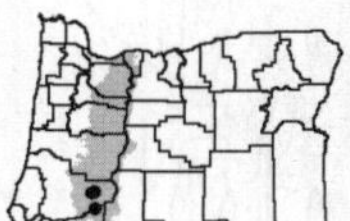

Stems thick, ascending or shortly creeping in rock crevices, usually unbranched, densely covered with non-bicolored, long (approximately 5 mm) lattice-like scales and long persistent leaf bases. **Leaves** graminoid, forked, evergreen, or at least overwintering, often numerous in a dense cluster along the stem, 3–10 cm long; petioles, rachises and blades poorly differentiated. **Petioles** glossy yellow-brown basally with sparse narrow scales, green and flattened and glabrous distally, rachises (above the first fork, if present) green and flattened. **Blades** linear, forked, branched up to 3 times more or less dichotomously or irregularly. **Pinnae** 1–2 mm wide, linear, elliptic to spathulate, often with a few small tooth-like lobes. **Sori** elongate, 1 to several on the abaxial surface, parallel to the midrib, often poorly delineated and anastomosing when mature; indusia present, pale, attached along veins and partially covering the sori on 1 side. **2*n***=144.

Exposed or shady rocky crevices. 400–1700 m. Casc. CA; scattered locations in western US and WV; Africa, Eurasia. Native.

Asplenium trichomanes Linnaeus ssp. ***trichomanes*** (p. 68)

Maidenhair spleenwort

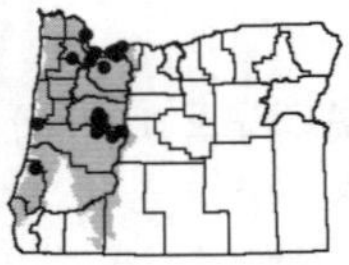

Stems erect or ascending, usually unbranched, densely covered with scales and persistent long leaf bases; scales more or less bicolored, usually with a darker central stripe. **Leaves** up to 30 cm long, evergreen, old leaves shedding leaflets leaving a twig-like petiole and rachis. **Petioles** of mature leaves glossy yellowish brown, usually short, up to 25% the leaf length, semicircular in cross section with 2 narrow wings bordering the flat adaxial side; rachis similar in color to the petiole, sometimes green near the tip; petiole and rachis with sparse narrow lattice-like scales and glandular hairs or glabrescent. **Blades** linear, 1-pinnate, to approximately 20 × 1.8 cm. **Pinnae** ovoid or oblong, to 8 × 6 mm, sparsely glandular-pubescent to glabrous; margins crenate with broad teeth; apex rounded; base unequal. **Sori** linear to oblong, to 2.5 mm long, on lateral veins diverging from the costa at an acute angle, usually 4–8 sori per segment; indusia pale, borne on 1 side of the sorus and partially covering it. ***2n***=72.

Moist rock crevices, rarely on soil. 0–1600 m. Casc, CR, WV. CA, WA; western and northeastern Canada and much of US; Eurasia. Native.

Asplenium trichomanes ssp. *quadrivalens* has been collected once in Oregon. It is a polyploid (2*n*=144) with larger spores than spp. *trichomanes* (37–43 µm, vs. 27–32 µm). It is reported to grow in crevices in limestone and other calcareous substrates.

Asplenium viride Hudson

Green spleenwort

Asplenium trichomanes-ramosum L. (illeg.)

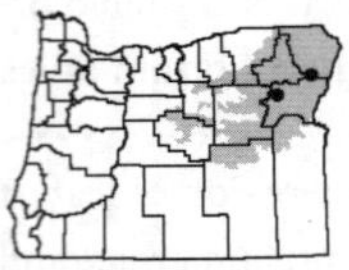

Stems short, ascending, usually unbranched, densely clad with scales and persistent long leaf bases; scales with very prominent cell walls, darker central stripe absent. **Leaves** often wintergreen but not fully evergreen, up to approximately 15 cm long. **Petioles** short, up to 33–50% the leaf length, glossy yellowish brown at the base, green above, adaxial side often grooved, not winged; rachis green; petiole and rachis with sparse narrow lattice-like scales and glandular hairs. **Blades** linear, 1-pinnate, to approximately 10 × 1.5 cm. **Pinnae** ovate or deltate, to 7 × 5 mm, sparsely glandular-hairy; margins with few broad rounded or obtuse teeth; apex rounded; base unequal. **Sori** linear to oblong, to 2 mm long, on lateral veins diverging from the costa at an acute angle, usually 2–6 sori per segment; indusia pale, borne on 1 side of the sorus and partially covering it. ***2n***=36.

Moist crevices in shaded limestone outcrops. 2000–2400 m. BW. CA, ID, NV, WA; western and northeastern North America; Eurasia. Native.

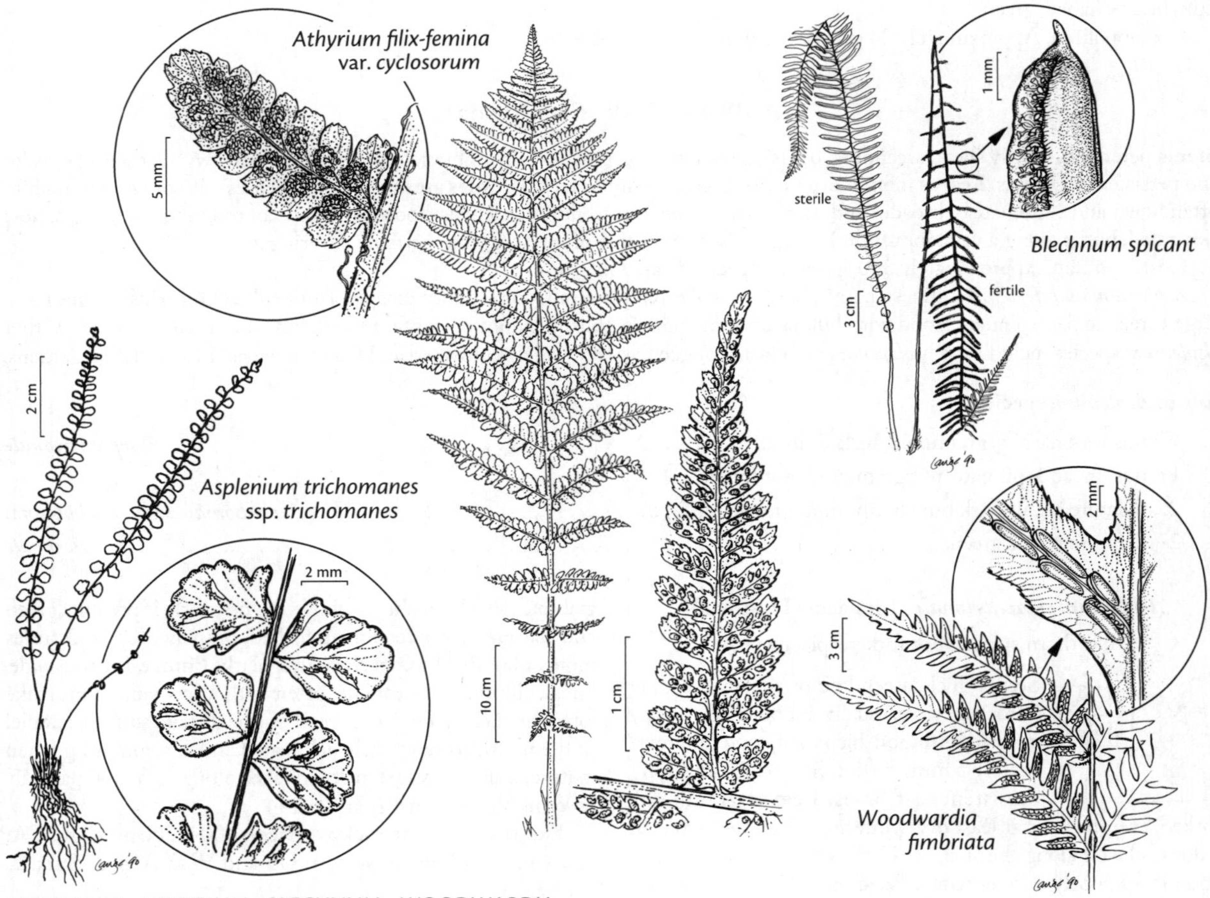

ASPLENIUM • ATHYRIUM • BLECHNUM • WOODWARDIA

ATHYRIACEAE Lady fern family

Plants herbaceous, perennial. **Stems** unbranched. **Leaves** monomorphic. **Petioles** straw-colored, bases usually persistent and thickened, basal scales resembling those on the rhizomes, reduced in size distally. **Sori** borne on veins. **Indusia** present or absent, when present flap-like and attached along the veins, opening laterally. **Spores** monolete, ornamentation more or less smooth to tuberculate or ridged. **Gametophytes** flat, heart-shaped, green.

Cosmopolitan. 16 genera; 1 genus treated in *Flora*.

The members of this family have been formerly included in Woodsiaceae. Recent molecular studies support a narrow view of Woodsiaceae, with Athyriaceae and Cystopteridaceae as separate families.

ATHYRIUM Lady fern

Stems erect or ascending, with abundant narrow scales. **Leaves** annual, in an apical cluster on the stem; leaf bases persistent. **Petioles** straw-colored distally, with a dark swollen slightly flattened base with dark brown lanceolate scales. **Blades** 1–3-pinnate-pinnatifid, lanceolate or narrowly elliptic, widest at or below mid-region; pinnae reduced in size to the acute apex and to the base; pinnules with 1 to several non-spinose teeth; indumentum glabrous or of scales, hairs and/or glands. **Veins** simple or forked, not anastomosing. **Sori** round or elongate to comma-shaped. **Spores** yellow to brown. **2*n*=80.**

Cosmopolitan. 215 species; 2 species treated in *Flora*.

Key to *Athyrium* species:

1. Sori round, submarginal; indusia absent or small; rachis with scales, lacking glands . . . . . . ***A. distentifolium* var. *americanum***

1' Sori elongate or comma-shaped, medial; indusia well developed; rachis lacking scales, usually with glands . . . . ***A. filix-femina* var. *cyclosorum***

Athyrium distentifolium Tausch ex Opiz var. ***americanum*** (Butters) Cronquist

Alpine lady fern, American alpine lady fern

Athyrium alpestre (Hoppe) Clairv. var. *americanum* Butters, *Athyrium americanum* (Butters) Maxon, *Athyrium distentifolium* Tausch

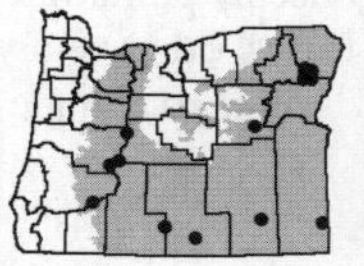

Petioles 10–30 cm long. **Blade segments** deeply divided, lanceolate to narrowly elliptic, 10–60 × 5–25 cm; rachis and costae with sparse pale brown scales, glands absent. **Pinnae** with a distinct stalk, narrowly triangular; acroscopic lobe scarcely enlarged. **Sori** round, borne on veins near the margins of the pinnules. **Indusia** absent or minute; pinnule teeth cupping the sorus on 1 side.

Wet montane meadows, rocky hillsides, talus slopes. 1000–3000 m. BR, BW, Casc, ECas, Owy. CA, ID, NV, WA; north to British Columbia, disjunct populations in Yukon, Quebec, Newfoundland, and Greenland. Native.

Athyrium filix-femina (Linnaeus) Roth var. ***cyclosorum*** Ruprecht (p. 68)

Lady fern, northwestern lady fern

Athyrium filix-femina (L.) Roth var. *californicum* Butters

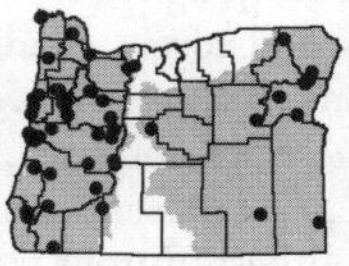

Petioles 15–50 cm long. **Blade segments** lobed, elliptic, 25–110 × 10–50 cm; rachis and costa with sparse linear scales, glands usually present, occasionally glabrous. **Pinnae** sessile or very shortly stalked, narrowly oblong; acroscopic lobe usually enlarged/auriculate. **Sori** elongate or comma-shaped, borne on veins distant from the margins. **Indusia** conspicuous, flap-like, ciliate to dentate, attached along the vein on 1 side of the sorus; pinnule teeth not curved over the sorus.

Wet forests, meadows, stream banks. 0–2300 m. BR, BW, Casc, CR, Est, Lava, Owy, Sisk, WV. CA, ID, NV, WA; north to AK, east to SD, disjunct populations in Ontario and Quebec. Native.

BLECHNACEAE Chain Fern family

Plants herbaceous, perennial. **Stems** creeping or ascending, scales dense. **Leaves** monomorphic or dimorphic, often large. **Blades** not much divided, deeply pinnatifid to 2-pinnate (rarely simple or more divided), glabrous or glabrescent, often reddish when young. **Veins** some anastomosing on fertile leaves, forming at least a secondary vein parallel to the costa and enclosing a line of elongated areas of leaf (areolae); non-fertile leaves usually with free veins. **Sori** elongate, oval to linear, borne on the areolae parallel to the costa; indusium present, conspicuous, attached to the secondary vein and usually covering the sorus, open on the costal side. **Spores** monolete, gibbous, wall sculpting various, almost smooth to coarsely ridged or tuberculate. **Gametophytes** thin, heart-shaped, green.

Cosmopolitan. Approximately 20 genera; 2 genera treated in *Flora*.

Key to Blechnaceae genera:

1. Leaves strongly dimorphic, more than 6 times longer than wide, pinnate with entire pinnae; sori long and linear, 2 per pinna . . . . **BLECHNUM**

1' Leaves monomorphic, less than 6 times longer than wide, pinnate with deeply divided pinnae; sori short, forming chain-like lines on the pinna segments . . . . **WOODWARDIA**

BLECHNUM Hard fern

Stems long-creeping to short and ascending or erect; brown or black scales sparse to dense. **Leaves** monomorphic or dimorphic, in cespitose clusters or produced further apart. **Blades** mostly pinnatifid or pinnate. **Veins** usually free and forked in vegetative leaves, anastomosing to form a secondary vein parallel to the costa and a line of areolae in fertile leaves. **Sori** continuous along the secondary vein, forming a linear structure on each side of the costa, at least partly covered by an indusium attached along the secondary vein, arched over the sori and open on the costal side of the sori.

Cosmopolitan but mostly tropical. Approximately 100 species; 1 species treated in *Flora*.

Blechnum spicant (Linnaeus) Smith (p. 68)

Deer fern, hard fern

Struthiopteris spicant (L.) Weiss

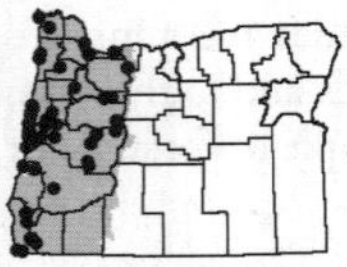

Stems shortly creeping or ascending, usually unbranched, densely scaly, with short, persistent leaf bases. **Leaves** densely cespitose, strongly dimorphic, fertile leaves rarely exceeding 100 cm, vegetative leaves to 80 cm, all leaves less than 10 cm wide. **Petioles** and proximal rachises dark purple-brown, rachis green distally, with sparse narrow hair-like scales abaxially; petiole scaly at base with brown scales which are smaller and sparser distally; petiole of vegetative leaves short, nearly absent in young plants, up to about 20% of leaf length; petiole of fertile leaves longer, up to 33(50)% of leaf length. **Blades** glabrous, pinnate, narrow, leathery, evergreen, sometimes reddish when young; vegetative blades very narrowly elliptic, widest in the mid-region; fertile blades more oblong, narrowed more abruptly to the base and apex. **Pinnae** vegetative pinnae oblong, to about 4.5 × 0.6 cm, spreading, often curved slightly forwards, apex rounded, base slightly widened, margins entire or rarely toothed or lobed; fertile pinnae linear, to about 4.5 × 0.2 cm, usually angled forwards, widened into a triangular base, apex blunt and sterile, margins parallel and entire. **Sori** linear, paired on each pinna, extending from near the base to just short of the apex, borne between the costa and the secondary vein; indusium pale and more or less continuous along the pinna, intramarginal with a distinct but narrow green stripe between the indusium and the margin. **2*n***=68.

Shaded moist to fairly dry sites, forests, road banks, stream banks, bogs. 0–2000 m. Casc, CR, Est, Sisk, WV. CA, ID, WA; western North America, north to AK; interrupted circumboreal distribution, Europe. Native.

WOODWARDIA Chain fern

Plants cespitose or with distant leaves. **Stems** creeping to erect, with brown scales and persistent leaf bases. **Leaves** usually monomorphic, scales present on the petiole, rachis, and costae. **Blades** not much dissected, 1-pinnate with entire to deeply pinnatifid pinnae, rarely with bulbils on the rachis. **Veins** anastomosing, forming narrow areolae along the costae of all leaves, other veins open and forking or forming additional areolae. **Sori** elongated, discrete, not anastomosing into lines, each sorus limited to a single areola, clearly visible on both surfaces as a chain of pale indusia abaxially and a line of raised bumps adaxially.

North and Central America, East Asia, southern Europe. 24 species; 1 species treated in *Flora*.

Woodwardia fimbriata Smith (p. 68)

Chain fern, giant chain fern

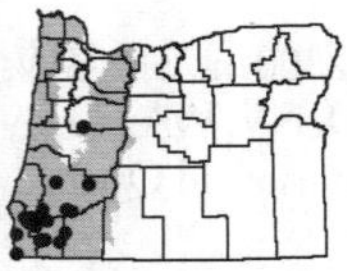

Stems shortly creeping, ascending, densely clothed in brown scales and persistent leaf bases, branched sparingly, producing a clump of large arching leaves. **Leaves** monomorphic, to approximately 1.2 m long, vegetative fronds (if present) resembling the fertile, bulbils absent. **Petioles** and rachises pale yellow-brown, densely red-brown scaly at the base, sparse smaller scales above. **Blades** evergreen, sub-herbaceous, lanceolate, 1-pinnate with deeply pinnatifid pinnae, dissected to the winged costa; costa and blade with hair-like glandular scales when young, then glabrescent. **Pinnae** shortly petiolate, lanceolate-elliptic, to approximately 25 cm long, deeply pinnatifid, up to 20 pairs of segments with a less divided usually sterile apex; segments oblong-lanceolate, up to 3 × 0.5 cm, distal 33% narrowed to an acute apex; margins parallel to the base, merging with the broadly winged costa, margin finely and regularly toothed. **Sori** conspicuous, oval or oblong, to 1.5–3 mm long, borne in depressions in the frond between the costa and the secondary vein, visible as lines of raised bumps on the adaxial surface, 1 sorus per areola; indusium pale, attached along the secondary vein, covering the developing sporangium, open side along the costa. **2*n***=68.

Road banks, shaded woods and forests, mostly riparian. 50–1000 m. Casc, CR, Est, Sisk. CA, NV, WA; north to British Columbia, south to AZ; Mexico. Native.

This species has a disjunct distribution in the Pacific Northwest. It is present in California and adjacent southwestern Oregon but also occurs around the Puget Sound. It is apparently absent in between. Oregon plants tend to have smaller leaves than California specimens.

CYSTOPTERIDACEAE Bladder fern family

Plants herbaceous, perennial. **Stems** slender, often branched, rarely hairy; scales dense. **Leaves** annual, monomorphic; bulbils occasionally present. **Petioles** slender and pale, glabrous above; base persistent, scaly. **Blades** 1–3-pinnate, glabrous or glandular. **Sori** rounded to ovoid, on veins. **Indusia** present or absent, narrowly ovate to rounded, attached proximally on the vein, often arching distally over the sporangia. **Spores** monolete, wall ornamentation various. **Gametophytes** flat, heart-shaped, green.

Cosmopolitan. 4 genera; 2 genera treated in *Flora*.

Formerly included in Woodsiaceae (or Dryopteridaceae), recent molecular studies support a narrow view of Woodsiaceae, with Athyriaceae and Cystopteridaceae as separate families.

Key to Cystopteridaceae genera:

1. Blades broadly triangular, widest near base. . . . . **GYMNOCARPIUM**
1' Blades lanceolate to elliptic, widest near middle. . . . . **CYSTOPTERIS**

CYSTOPTERIS Bladder fern, brittle fern

Stems shortly creeping; scales lanceolate. **Leaves** annual, clustered or distant on the stems, monomorphic. **Petioles** shorter or equaling the blade in length; bases thickened. **Blades** lanceolate to elliptic, 1–2-pinnate-pinnatifid, glandular or glabrous. **Indusia** attached at the vein on the proximal side of the sorus, ovate to lanceolate, cupped over the developing sporangia and partially covering them. **Sori** round, on the veins of the pinnules. **Spore** walls rough with tubercles or spines.

Cosmopolitan. 23 species; 1 species treated in *Flora.*

Cystopteris fragilis (Linnaeus) Bernhardi (p. 71)

Fragile fern, brittle fern

Cystopteris fragilis (L.) Bernh. var. *fragilis*

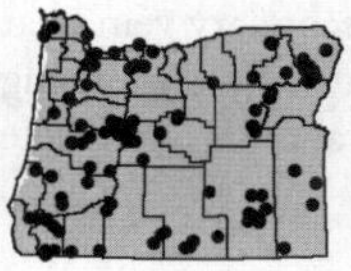

Stems scales tan to brown. **Leaves** to approximately 40 cm long but usually much shorter. **Petioles** dark at base, lighter colored distally; base sparsely scaly. **Blades** lanceolate, widest near middle, tapered to an acute tip, tapered slightly to base. **Pinnae** with several pairs of lobed/toothed pinnules. **Sori** usually abundant; indusia ovate to lanceolate. $2n$=168, 242.

Cliffs and rocky sites. 50–3000 m. All ecoregions except Est. CA, ID, NV, WA; throughout North America; nearly worldwide. Native.

Plants occurring at low elevations west of the Cascades go dormant in early summer but may produce new leaves with fall rains and persist through the winter.

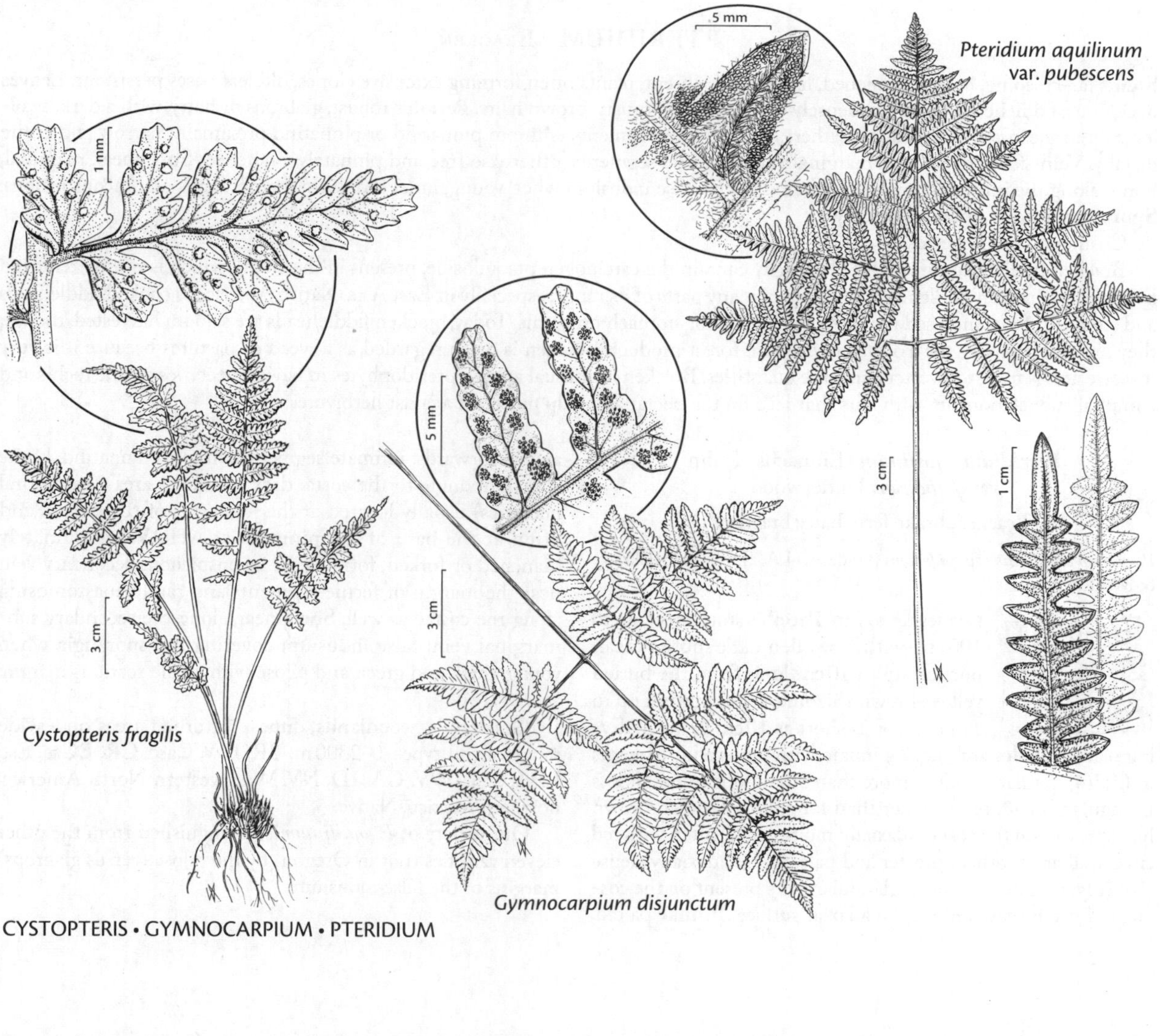

CYSTOPTERIS • GYMNOCARPIUM • PTERIDIUM

GYMNOCARPIUM Oak fern

Stems far-creeping; scales sparse, short. **Leaves** annual, monomorphic. **Petioles** long and slender; bases not swollen. **Blades** broadly triangular, 3-pinnate-pinnatifid; small glands often present. **Indusia** absent. **Sori** round, on veins on the lamina. **Spores** brown, wrinkled.

Temperate regions of the Northern Hemisphere. 8 species; 1 species treated in *Flora*.

Gymnocarpium disjunctum (Ruprecht) Ching (p. 71)

Western oak fern

Gymnocarpium dryopteris (L.) Newman var. *disjunctum* (Rupr.) Ching

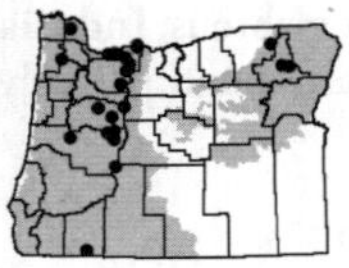

Stems branching dichotomously; scales 2–4 mm. **Leaves** 20–70 cm. **Petioles** 10–45 cm, proximal sparse scales to 6 mm, with sparse glandular hairs distally; bases slender. **Blades** 10–25 cm; basal pinnae much larger than the others, asymmetric with the downward-pointing pinnules; 2nd pair of pinnae with asymmetric proximal pinnules, toothed to shallowly lobed margins; 3rd pair of pinnae less divided and scarcely asymmetric. **Sori** in 1 row between midrib and margin. $2n$=80.

Shady rocky habitats, woods, riparian areas. 100–1700 m. BW, Casc, CR, ECas, Sisk, WV. ID, WA; north to AK, east to WY; northeastern Asia. Native.

DENNSTAEDTIACEAE Bracken family

Plants perennial, herbaceous. **Stems** long-creeping to ascending, often branched, usually hairy. **Leaves** monomorphic. **Blades** simple to 4-pinnate, indumentum of hairs and glands or absent, veins forked or pinnate, free or forming a secondary vein near the margin in fertile leaves. **Sori** at or close to the margin, short and semi-lunar to long and linear; indusium usually comprising a false indusium formed by the recurved leaf margin and an inconspicuous (absent) true indusium. **Spores** tetrahedral and trilete or monolete and gibbous. **Gametophytes** thin, heart-shaped, green.

Cosmopolitan. 20 genera; 1 genus treated in *Flora*.

PTERIDIUM Bracken

Stems far-creeping, buried, branched, hairy, scales absent, plants often forming extensive clones, old leaf bases persistent. **Leaves** deciduous, fiddleheads (croziers) densely covered by long rusty-brown hairs. **Petioles** robust, glabrous or hairy, with a dark, swollen subterranean portion. **Blades** leathery, 2–4 pinnate, segments entire to pinnatifid or pinnatifid proximally, narrow and entire distally. **Veins** forming an intramarginal vein in fertile segments, otherwise free and pinnately-branched. **Sori** linear, marginal, borne along the intramarginal vein, covered by the false indusium when young, inner indusium membranous and inconspicuous. **Spores** tetrahedral, rounded distally.

Cosmopolitan. 1 species.

Bracken fiddleheads are edible. They may contain the carcinogen ptaquiloside, present in the stems, which is mostly removed by cooking. Bracken fiddleheads are eaten in many parts of its range, especially in East Asia. Native Americans cooked fiddleheads and stems for food and used the root medicinally for stomach problems. Today, bracken fiddleheads are seldom harvested, though they are still listed by the Forest Service as a forest product. Bracken is often regarded as a weed of pastures because it is very invasive and is toxic to livestock in large quantities. Bracken is unusual among pteridophytes in having nectaries on the rachis and a mutualistic relationship with ants that feed on the nectar and help protect it against herbivores.

Pteridium aquilinum (Linnaeus) Kuhn var. ***pubescens*** Underwood (p. 71)

Bracken, brake fern, hairy bracken

Pteridium aquilinum ssp. *pubescens* (Underw.) J.A. Thomson, Mickel & Mehltreter

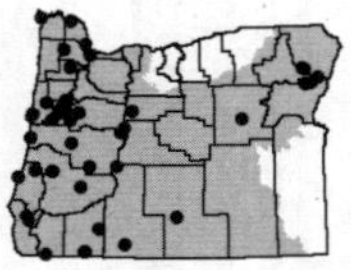

Leaves up to 2 m. **Petioles** stout, erect, up to 100 cm, with a swollen dark subterranean portion up to 10 cm long from the buried yellow-brown rhizome, aerial portion up to 1 m long, or as short as 10 cm in open, drying places. **Blades** arching, 2-pinnate with pinnatifid segments to (2)3(4)-pinnate, seldom more than 50 cm long, more or less triangular, broad, often wider than long; long reddish brown hairs present on the costae adaxially, multi-celled, often with red cross wall areas, others shorter and pale, sparse adaxially, dense abaxially; minute red tipped glandular hairs present on the costae and often scattered on the adaxial surface. **Pinnae** paired, angled forwards; ultimate segments entire to pinnatifid, borne at a right angle to the costa; dark glandular areas (extrafloral nectaries) usually located at the junctions of the pinnae and often at the base of the pinnules as well. **Veins** pinnately branched or forked, forming an inconspicuous secondary vein near the margin of fertile segments and rarely anastomosing along the costae as well. **Sori** linear, along the secondary submarginal vein; false indusium covering the sporangia when young, thick and green and pilose with a pale scarious margin. $2n$=104.

Grasslands, woodlands, dunes, disturbed areas on a wide range of soil types. 0–2300 m. BR, BW, Casc, CR, ECas, Est, Lava, Sisk, WV. CA, ID, NV, WA; western North America, northern Mexico. Native.

Our variety of *P. aquilinum* is distinguished from the other eleven varieties (not in Oregon) by the pilose (versus glabrous) margins of the false indusium.

DRYOPTERIDACEAE Wood Fern family

Plants herbaceous. **Stems** often stout and densely scaly. **Leaves** monomorphic or dimorphic, often forming a crown at the stem apex; indumentum usually scaly, occasionally hairy or glandular; veins pinnate or forked or anastomosing. **Sori** round, discrete or rarely covering the leaf surface, with a peltate or reniform indusium (rarely absent). **Spores** monolete and gibbous/reniform, surrounded by inflated flanges or wings. **Gametophytes** heart-shaped, green.

Cosmopolitan. Approximately 34 genera; 2 genera treated in *Flora*.

Key to Dryopteridaceae genera:

1. Indusium peltate .... **POLYSTICHUM**
1' Indusium attached at a marginal sinus, reniform .... **DRYOPTERIS**

DRYOPTERIS Wood fern, shield fern

Emily B. Sessa

Plants herbs, terrestrial, perennial. **Stems** short-creeping or erect. **Leaves** monomorphic, remaining green or dying back in winter; petioles 25–33% of leaf length, with scattered light brown to brown scales which may be hair-like (in 1 species) or have a dark stripe (in another species). **Leaf blades** deltate-ovate to ovate-lanceolate, pinnate-pinnatifid to 3-pinnate-pinnatifid at leaf base, with or without glands. **Sori** round, on ultimate segments midway between midvein and margin; indusia reniform, glands present or not; spores brown.

Cosmopolitan. 225 species; 3 species treated in *Flora*.

There are 13 species of wood ferns in North America, most of which are restricted to the eastern parts of the continent. Three species occur in the western United States and Canada, and all of these can be found in Oregon. These three are easily distinguishable from one another based on a number of characters, including scale color, level of leaf division, and relative length of the petiole compared to the rest of the frond. All three prefer wooded habitats, though with varying degrees of openness, and generally with rocky or sloping substrates.

Key to *Dryopteris* species:

1. Petiole 25–33% of leaf length; scattered scales of 1 kind.
   2. Scales on petiole light brown; leaf blade at base pinnate-pinnatifid to 2-pinnate .... ***D. arguta***
   2' Scales on petiole brown with a dark brown central stripe; leaf blade 3-pinnate throughout .... ***D. expansa***
1' Petiole less than 25% of leaf length; scattered scales of 2 kinds, 1 broad and 1 thin; leaf blade at base pinnate-pinnatifid to 2-pinnate .... ***D. filix-mas***

Dryopteris arguta (Kaulfuss) Maxon (p. 74)

Coastal wood fern, California wood fern, marginal wood fern

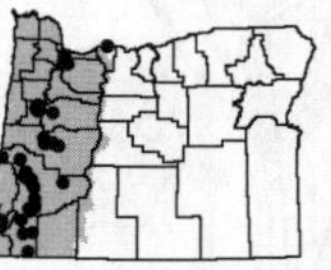

Stems short-creeping or erect. **Leaves** up to 90 × 30 cm, monomorphic, remaining green throughout winter; petioles 25–33% of leaf length, with scattered light brown scales. **Blades** ovate-lanceolate, pinnate-pinnatifid to 2-pinnate at leaf base, with conspicuous glands. **Pinnae** deltate, with basal basiscopic and basal acroscopic pinnules approximately equal in length; pinnule margins serrate and may be bristle-tipped. **Sori** round, on ultimate segments midway between midvein and margin; indusia reniform, glands present when young and fading with maturity; spores brown.

Open woodlands and shaded slopes. 0–1600 m. Casc, CR, Est, Sisk, WV. CA, WA; north to British Columbia, southeast to AZ. Native.

Dryopteris arguta occurs in several variable, unnamed morphological forms. Only the most common of these is found in Oregon, and it is described here. *Dryopteris arguta* is somewhat similar morphologically to the eastern species *D. marginalis*, to which it may be closely related. Unlike most North American *Dryopteris*, it is not known to form any hybrids with other species.

Dryopteris expansa (C. Presl) Fraser-Jenkins & Jermy

Spreading wood fern, northern wood fern

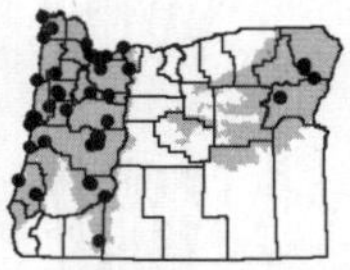

Stems short-creeping or erect. **Leaves** up to 60(90) × 30 cm, monomorphic, dying back in winter; petioles 33% of leaf length with scattered light brown scales, each with a dark central stripe. **Blades** deltate-ovate, 3-pinnate-pinnatifid, usually without glands. **Pinnae** deltate, with basal basiscopic pinnule longer than basal acroscopic pinnule; pinnule margins serrate but not bristle-tipped. **Sori** round, on ultimate segments midway between midvein and margin; indusia reniform, usually without glands; spores brown.

Cool forests or rocky slopes. 0–2100 m. BW, Casc, CR, Est, WV. CA, ID, WA; western and northeastern North America; Eurasia, Greenland. Native.

Dryopteris expansa is a member of the reticulate complex of the North American wood ferns. It is one of the diploid progenitors of the eastern allotetraploid species, *D. campyloptera*, whose other progenitor is the eastern species *D. intermedia*. In Europe, *D. expansa* is a diploid progenitor of the allotetraploid *D. dilatata*, and historically the names *D. dilatata* and *D. expansa* have been used interchangeably in North America. It is now understood that these are separate taxa and *D. dilatata* does not occur outside of Europe. *Dryopteris expansa* will not likely be confused with other *Dryopteris* species in Oregon due to differences in leaf division and the presence of a distinctive dark central stripe on its petiole scales, but it is morphologically similar to *D. carthusiana*, which occurs in Washington and southwestern British Columbia. The latter can be distinguished by its tan scales that lack stripes and its bristle-tipped pinnule margins.

Dryopteris filix-mas (Linnaeus) Schott
Male fern

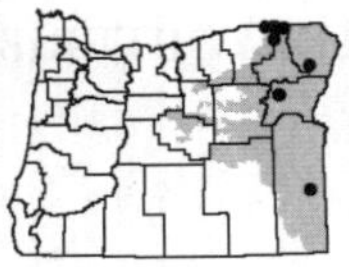

Stems short-creeping or erect. **Leaves** up to 120 × 30 cm, monomorphic, dying back in winter; petioles 25% or less of leaf length, with 2 kinds of scattered brown scales, 1 broad and 1 thin and hair-like. **Blades** ovate-lanceolate, pinnate-pinnatifid to 2-pinnate at leaf base, without glands. **Pinnae** ovate-lanceolate, with basal basiscopic and basal acroscopic pinnules equal in length; pinnule margins serrate or lobed. **Sori** round, on ultimate segments midway between midvein and margin; indusia reniform, usually without glands; spores brown.

Dense forests and talus slopes. 500–1800 m. BW, Owy. CA, ID, NV, WA; western and northeastern North America; Eurasia, Greenland. Native.

At least two variable, unnamed morphological forms of *D. filix-mas* occur in North America, but only the most common occurs in Oregon, and it is described here. *Dryopteris filix-mas* is widely distributed in the northern hemisphere, and it is thought to be an allopolyploid between two European species. It may be confused with *D. arguta* due to similar leaf division, but *D. filix-mas* is generally larger, and it is the only species in the genus with two distinct scale types along the petiole.

POLYSTICHUM Sword fern, holly fern

Plants herbaceous. **Stems** often short, stout, erect or ascending; chaffy scales usually dense on the stem and petiole base, often extending to the upper petiole and rachis. **Leaves** forming a crown, mostly erect/ascending to arching, overwintering at least into a second year; petiole bases persistent. **Petioles** with chaffy scales proximally, continuing to the rachis or becoming sparse or glabrescent distally. **Blades** mostly 1-pinnate, 1-pinnate-pinnatifid or rarely 2-pinnate, often rather leathery in texture; leaf blade indumentum with scales (microscales) much smaller and narrower than the chaffy scales; microscale base wider than the long narrow apex, sometimes with basal appendages, microscales present and often abundant on the abaxial leaf blade, sparse to absent on the adaxial surface. **Pinnae** narrow, ovate, often falcate, in the same plane and appearing 2-dimensional or twisted out of 1 plane or rarely folded into 3-dimensional structures; margins often with spinose teeth, spreading or incurved; pinnae with a

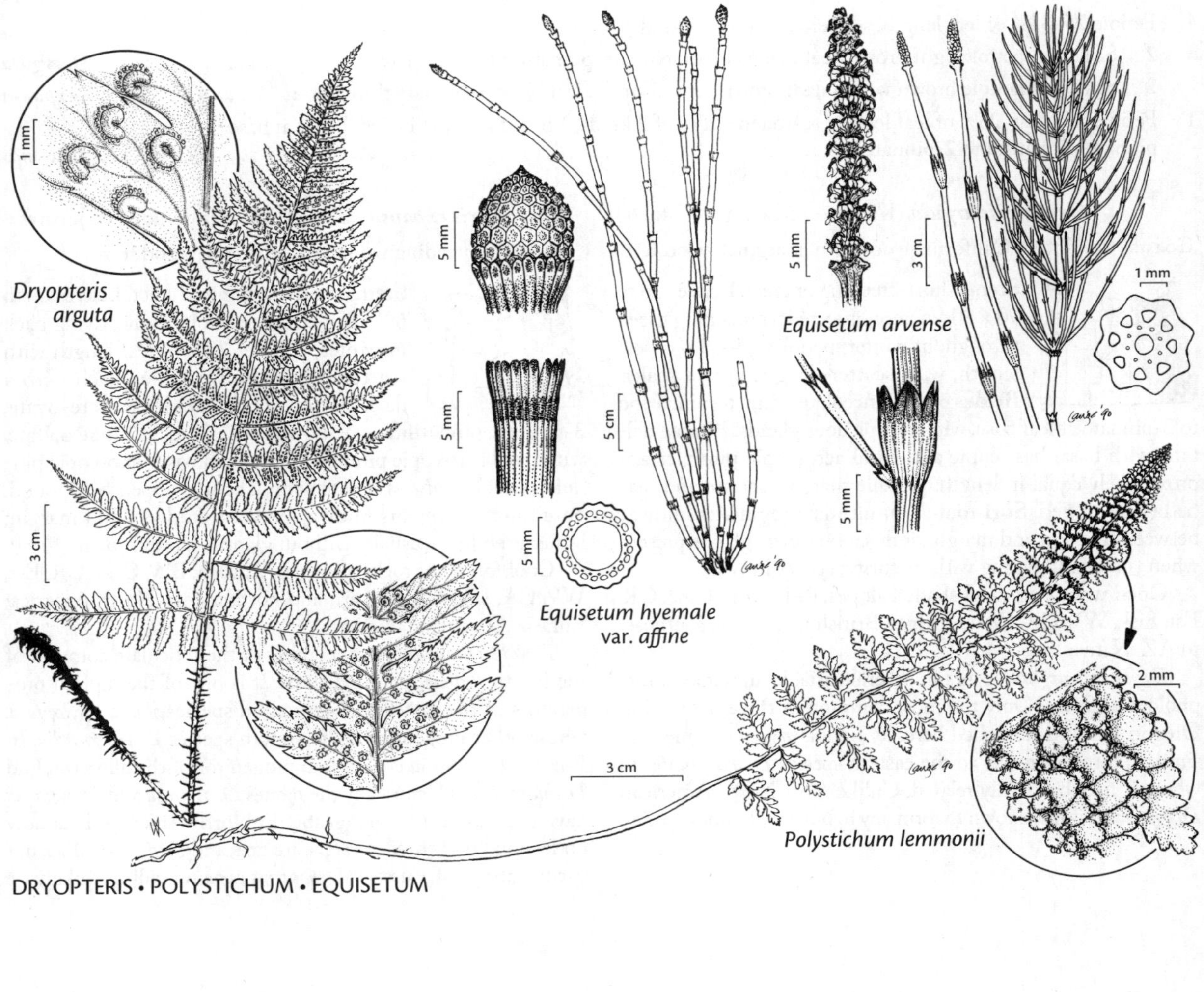

DRYOPTERIS • POLYSTICHUM • EQUISETUM

forward-pointing auricle or the lowest lobe/pinnule enlarged, base usually unequal; veins free, simple to sparingly-branched. **Sori** round, on laminal veins; indusia peltate; margins variously entire, ciliate, or dentate; in 1-pinnate species, the sori are borne in 2 neat lines along each fertile pinna, with more complex patterns in pinnatifid species. **2*n***=82.

Cosmopolitan. Approximately 260 species; 8 species treated in *Flora*.

Two hybrids have been reported from Oregon: *P. andersonii* × *P. munitum* and *P. californicum* × *P. munitum*.

Key to *Polystichum* species:

1. Most leaves with a bulbil on the distal third of the rachis; pinnae pinnatifid . . . . *P. andersonii*

1' Bulbils lacking; pinnae various.

  2. Leaves 1-pinnate; pinnae toothed but not deeply lobed.

    3. Petiole short, blade narrowed to base, margins spreading-spinulose . . . . *P. lonchitis*

    3' Petiole longer, blade not narrowed to base, margins mostly incurved-spinulose.

      4. Indusia ciliate; pinna base cuneate, apex acuminate; distal petiole scales wide . . . . *P. munitum*

      4' Indusia entire or dentate; pinna base obtuse, apex apiculate; distal petiole scales narrow or absent . . . . *P. imbricans* ssp. *imbricans*

  2' Leaves 2-pinnate or with at least proximal pinnae deeply lobed.

    5. Leaves 2-pinnate; pinnae petiolate; pinnules not spine-tipped . . . . *P. lemmonii*

    5' Leaves with pinnatifid pinnae; if appearing fully 2-pinnate then pinnules spine-tipped.

      6. Longest pinnae over 4 cm . . . . *P. californicum*

      6' Longest pinnae less than 4 cm.

        7. Pinna apex with spreading teeth . . . . *P. kruckebergii*

        7' Pinna apex with incurved teeth . . . . *P. scopulinum*

Polystichum andersonii Hopkins

Anderson's sword fern, Anderson's holly fern

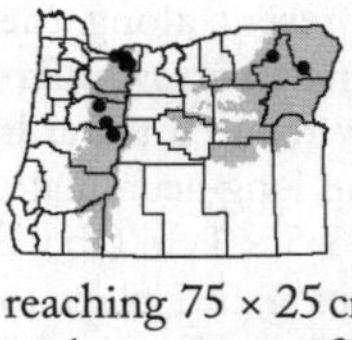

Stems erect. **Leaves** robust, to approximately 1 m long. **Petiole** to 25 cm, densely clothed with brownish scales which continue along the rachis. **Blades** 1-pinnate-pinnatifid, narrowly elliptic to lanceolate, reaching 75 × 25 cm, narrowing to apex and base; abaxial veins with numerous filiform microscales; scales sparse to nearly absent adaxially distal 33% of the adaxial rachis bearing 1 or more scaly bulbils. **Pinnae** narrowly ovate to triangular, slightly falcate, to approximately 9 cm long, borne in 1 plane and not overlapping, deeply pinnatifid in mid-region with sinuses progressively reduced distally; distal lobes in the form of a triangular tooth with an incurved apical spine; proximal pinnules oblong with terminal and lateral spines; acroscopic pinnule enlarged, sometimes free to the costa (2-pinnate) on proximal pinnae. **Sori** with sparsely ciliate indusia margins. **2*n***=164.

Riparian and moist woods, thickets, rocky areas. 700–1600 m. BW, Casc. ID, WA; north to AK, east to MT. Native.

Polystichum andersonii differs from *P. californicum* in its petiole, which bears chaffy scales throughout versus scales which are sparse distally. In addition, *P. andersonii* is more divided at the middle of the pinna with pinnules free for most of the way to the costa versus about halfway to the costa in *P. californicum*.

Polystichum californicum (D.C. Eaton) Diels

California sword fern

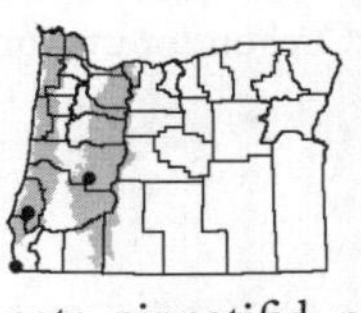

Stems erect or ascending. **Leaves** robust to approximately 70 cm; bulbils absent. **Petiole** up to 33% leaf length; scales long, brown and densely chaffy proximally, becoming smaller and sparser distally. **Blades** 1-pinnate-pinnatifid, narrowly lanceolate, narrowed apically, not much narrowed to the base, up to 55 × 10 cm, filiform microscales abundant on abaxial veins, sparse to glabrescent adaxially. **Pinnae** lanceolate, falcate, to approximately 9 × 1.5 cm, borne in 1 plane, overlapping or not, pinnatifid to about 50% of pinna width in mid-region with sinuses progressively reduced distally, distal lobes in the form of a triangular tooth with an incurved apical spine, proximal pinnules oblong with a terminal and lateral spines incurved or spreading, acroscopic lobe enlarged, nearly incised to costa on proximal pinnae. **Sori** with ciliate indusial margins. **2*n***=164.

Cliffs and shaded rocky sites. 0–1100 m. Casc, CR. CA, WA; north to British Columbia. Native.

Rarely, *P. andersonii* may lack bulbils, in which case it will key out here.

Polystichum imbricans (D.C. Eaton) D.H. Wagner ssp. *imbricans*

Imbricate sword fern

Polystichum munitum (Kaulf.) C. Presl var. *imbricans* (D.C. Eaton) Maxon, *Polystichum munitum* (Kaulf.) C. Presl ssp. *imbricans* (D.C. Eaton) Munz

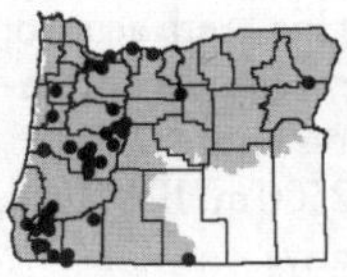

Stems short, erect, with a crown of leaves. **Leaves** to 80 cm long; bulbils absent. **Petioles** well developed, up to 33% leaf length, with dense chaffy brown scales proximally, becoming smaller and sparse distally, largely absent from the rachis. **Blades** 1-pinnate, narrowly lanceolate, to 70 × 8 cm, tapering to apex, not much reduced in width at base; microscales narrow, linear to lanceolate, abundant abaxially, absent above. **Pinnae** lanceolate to narrowly oblong, to 4 × 1 cm, usually overlapping and twisted out of plane; apex cuspidate; margins regularly dentate with forward-pointing spine-tips; acroscopic auricle enlarged, broadly triangular or ovate, cuspidate, sinus mostly lacking but occasionally deep. **Sori** with irregularly lobed indusial margins.

On rocks and cliffs, frequently in sunny locations. 50–2300 m. BW, Casc, Col, CR, ECas, Lava, Sisk, WV. CA, WA; north to British Columbia. Native.

Polystichum kruckebergii W.H. Wagner

Kruckeberg's sword fern, Kruckeberg's holly fern

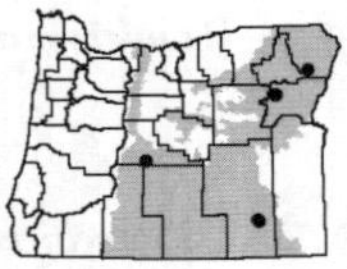

Stems ascending. **Leaves** small, mostly under 20 cm; bulbils absent. **Petiole** variable in length from very short up to almost 50% of the leaf length, chaffy at base; scales sparser distally and present on at least the proximal part of the rachis. **Blades** 1-pinnate-pinnatifid (sometimes 1-pinnate), narrowly lanceolate, not much reduced basally; microscales narrow, confined to the abaxial costa. **Pinnae** more or less ovate, small, to about 1.5 cm, overlapping, usually twisted out of plane; distal part strongly spiny-dentate with spreading spines; a few proximal segments more deeply incised, with several spreading spines; acroscopic lobe enlarged, sometimes forming a distinct pinnule. **Sori** with entire indusium. ***2n***=164.

Rocky places. 800–3000 m. BR, BW, ECas. CA, ID, NV, WA; most of western North America. Native.

Polystichum kruckebergii can appear pinnate, with even the proximal pinna segments not deeply lobed. It then resembles *P. lonchitis,* from which it differs in its more rounded pinnae with broader teeth. It can also be hard to distinguish from small specimens of *P. scopulinum*. The spores of P. *kruckebergii* are darker in color than the brown or yellow/brown spores of *P. scopulinum*. According to Wagner (1979), both species are allopolyploids that share one common parent (*P. lemmonii* × *P. lonchitis* = *P. kruckebergii*; *P. lemmonii* × *P. munitum* = *P. scopulorum*).

Polystichum lemmonii Underwood (p. 74)

Shasta sword fern, Lemmon's holly fern

Polystichum mohrioides (Bory) C. Presl var. *lemmonii* (Underw.) Fernald

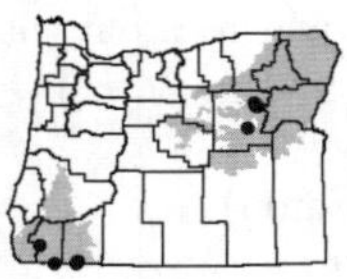

Stems ascending or decumbent. **Leaves** small, up to 35 cm; bulbils absent. **Petioles** up to 25% leaf length, chaffy near the base; scales very sparse distally, small and sparse along the rachis. **Blades** 2-pinnate, narrowly lanceolate, not much narrowed at base; microscales linear, sparse on the ab- and adaxial costa. **Pinnae** ovate, up to 2 cm long; lowest pinnae often remote, not overlapping; distal pinnae overlapping, twisted out of plane with pinnules folded again creating a complex 3-dimensional structure; pinnules ovate, pinnatifid with irregularly lobed or toothed margins, teeth acute or obtuse, not spine-tipped; basal (acroscopic) pinnule not enlarged. **Sori** with indusial margins more or less entire. ***2n***=82.

Rocky (often serpentine) slopes. 1300–2500 m. BW, Sisk. CA, ID, WA; north to British Columbia. Native.

The combination of its small size and two-pinnate leaves with complex folding makes this species distinctive among all other Oregon members of the genus.

Polystichum lonchitis (Linnaeus) Roth

Northern holly fern

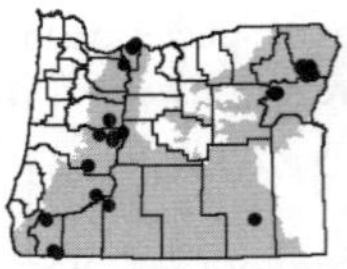

Stems erect. **Leaves** to 50 cm but usually smaller, 10 cm or even shorter in specimens from exposed sites; bulbils absent. **Petiole** short, less than 20% of the leaf length; leaves sometimes subsessile; chaffy scales sparse to dense on the petiole and rachis. **Blades** 1-pinnate, very narrow, more or less linear, to approximately 45 × 5 cm, widest in the mid-region and tapered to apex and base, microscales narrow on abaxial surface, adaxial surface glabrous. **Pinnae** oblong-lanceolate, slightly falcate, up to 4 × 1.5 cm, lowest pinnae often broadly triangular, rarely overlapping, in 1 plane, margins regularly spiny-toothed, spines conspicuous, spreading or angled forward, acroscopic auricle large, broadly triangular, ovate, bearing several spreading spines. **Sori** with irregularly toothed indusia. ***2n***=82.

Shaded, rocky or occasionally wet places. 300–3000 m. BR, BW, Casc, ECas, Sisk. CA, ID, NV, WA; north to AK, circumboreal. Native.

Polystichum munitum (Kaulfuss) C. Presl

Common sword fern, western sword fern

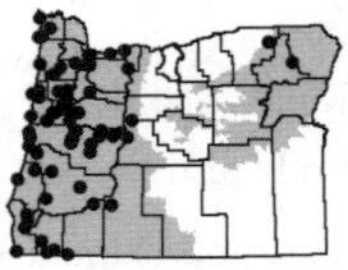

Stems typically short and erect. **Leaves** robust, to 150 cm, in dense crowns; bulbils absent. **Petioles** up to 25% leaf length, densely chaffy near base, becoming sparser distally; chaffy scales present along the length of the rachis and costae. **Blades** 1-pinnate, narrowly lanceolate, tapered to the apex, not or scarcely reduced towards base; microscales lanceolate or narrowly ovate, long-acuminate. **Pinnae** linear-lanceolate, to approximately 9.5 × 1 cm (above the anterior lobe), often falcate, not overlapping, in 1 plane to twisted; margins serrulate with small, regular spine-tipped teeth, the spines bent forward and parallel to the margin; acroscopic auricle large, broadly triangular to ovate, with a prominently cuspidate apex and lateral teeth similar to those on the pinna margin. **Sori** with fimbriate/ciliate indusial margins. ***2n***=82.

Understories of moist forests, meadows, rocky sites. 0–2200 m. BW, Casc, CR, ECas, Est, Sisk, WV. CA, ID, WA; north to AK, east to SD. Native.

This species is so abundant in the Pacific Northwest that one might forget to look for other pinnate sword ferns. At higher elevations, *P. lonchitis* is identified by narrowing blade bases and spreading spines on the pinna margins versus scarcely reduced blade bases and incurved spines in *P. munitum*. On moist, sunny, rocky sites the pinnae of *P. munitum* may be twisted like venetian blinds, making it look much like *P. imbricans* ssp. *imbricans*. Besides the characters listed in the key, these species also differ in the distribution of chaffy scales. In *P. imbricans* ssp. *imbricans*, these are concentrated on the proximal portion of the petiole, while in *P. munitum* they occur all along the length of the petiole and rachis. *Polystichum munitum* is widely cultivated as an ornamental, and leaves are harvested from Oregon forests for sale to florists.

Polystichum scopulinum (D.C. Eaton) Maxon

Rock sword fern, mountain holly fern

Polystichum mohrioides (Bory) C. Presl var. *scopulinum* (D.C. Eaton) Fernald

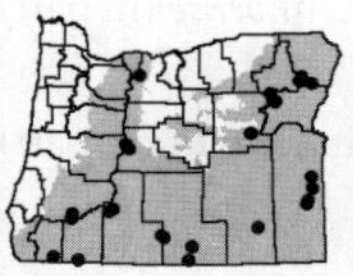

Stems erect. **Leaves** small, to about 40 cm; bulbils absent. **Petioles** with chaffy scales near the base, sparse and smaller distally and on the proximal part of the rachis. **Petiole** highly variable in length, from short to over 33% length of the leaf. **Blades** 1-pinnate-pinnatifid to more or less bipinnate proximally, linear-lanceolate, to 25 × 6 cm, widest below mid-length and narrowed to the apex and base; microscales narrow, confined to the abaxial costa. **Pinnae** ovate to narrowly ovate, to approximately 3 cm long, overlapping and twisted out of plane; distal part of pinna toothed, incised less than halfway to costa, teeth acute but not spine-tipped; proximal part (at least on the lower pinnae) deeply incised with 1–several pairs of pinnules, each entire to lobed/toothed; acroscopic pinnule somewhat enlarged. **Sori** with irregularly toothed indusial margins. **2*n***=164.

Rock crevices. 700–3000 m. BR, BW, Casc, ECas, Owy, Sisk. CA, ID, NV, WA; western and northeastern North America. Native.

This species is very similar to the rarer *P. kruckebergii*; see comments under that species.

EQUISETACEAE Horsetail family

Herbs perennial, with creeping often deeply-buried and dark-colored rhizomes; roots and shoots produced at the nodes. **Aerial stems** annual or persisting over winter, clustered, erect to decumbent, surface rough with silicified cell walls; main aerial stem unbranched or few-branched or with whorls of horizontal branches. **Stems** with well-differentiated solid nodes and hollow internodes, internodes in the form of a cylinder around a central cavity, 2 alternating whorls of smaller canals in the cylinder tissue; longitudinal ribs present on stems, often with a central furrow, stem ribs and leaves usually equal in number at each node, with the rib continuing into the leaf as its midrib. **Lateral branches**, when present, borne at nodes, growing through the base of the leaf sheath, spreading or ascending. **Leaves** borne in whorls at the nodes of stems and branches, fused into a sheath with the leaf tips forming a ring of teeth; teeth free or fused in groups, persistent or early-deciduous. **Cones** terminal on the main stems or ascending branches, composed of whorls of peltate sporophylls; cone apices blunt to sharply pointed; mature cone axis elongating to separate the sporophylls and allow the spores to be shed. **Sporangia** 5–10 per sporophyll, attached to the inner surface, elongate, dehiscing longitudinally. **Fertile spores** round, green, outer wall layer spitting into spiral strap-like sections (elaters) during dehiscence. **Gametophytes** green, terrestrial, unisexual; male gametophytes smaller than female.

Cosmopolitan. 1 genus.

EQUISETUM Horsetail, scouring rush

Family comprised of one genus, *Equisetum*.

15 species; 8 species treated in *Flora*.

Infertile hybrids between species of the same subgenus are likely to be found where both parents are present and may persist through vegetative reproduction from the rhizomes. Hybrids have intermediate characters and misshapen, pale spores.

Equisetum pratense has been reported from Union County. It is similar in appearance to *E. arvense* and resembles that species in having sharply three-angled branches with the first internode exserted from the sheath on the non-fertile stems. Like *E. arvense* and *E. telmateia*, *E. pratense* has dimorphic stems with the pale fertile frond produced early in the growing season. Unlike the other two species, however, the fertile stem of *E. pratense* is not ephemeral but develops chlorophyll and short branches after the spores are shed. *Equisetum arvense* and *E. telmateia* occasionally produce cones on stems otherwise identical to the vegetative stems. *Equisetum pratense* differs in having persistent fertile stems with shorter branches, obviously dissimilar to the more profusely branched non-fertile stems.

Key to the *Equisetum* species:

1. Stems simple or with few often ascending branches.
   2. All parts lacking chlorophyll and stomata.
      3. Stem ribs 16 or more; teeth on sheaths in clumps, not separating individually .... ***E. telmateia*** var. ***braunii*** (in part)
      3' Stem ribs 14 or fewer; teeth on sheaths mostly separated individually .... ***E. arvense*** (in part)
   2' Aerial parts mostly green; stomata present.
      4. Stems slender, under 2 mm in diameter .... ***E. variegatum***
      4' Stems wide, over 2 mm in diameter.
         5. Sheath teeth not articulated at the base; blade persistent.......................... ***E. fluviatile*** (in part)
         5' Sheath teeth articulated at the base; blade early-deciduous leaving a ring of incurved bases.
            6. Stems annual, green; aerial sheaths green with a single apical dark band; cone blunt .... ***E. laevigatum***
            6' At least the lower part of stems persisting over winter, dark green; aerial sheaths green or gray, dark bands 1 or 2; cones apiculate.

7. Most of the stem persisting over winter; mature spores round and green. . . . . ***E. hyemale*** var. ***affine***
7' Only lower part of stem persisting over winter; spores misshapen and pale . . . . ***E. ×ferrissii***

1' At least some stems with regular whorls of spreading branches.
8. Branches sharply 3(4)-angled; stem ribs 14 or fewer, teeth on sheaths mostly separated individually . . . . ***E. arvense*** (in part)
8' Branches not sharply angled though often deeply grooved; ribs and teeth various.
9. Teeth on sheaths in clumps; stem ribs 16 or more . . . . ***E. telmateia*** var. ***braunii*** (in part)
9' Teeth on sheaths distinct, not fused in clumps; stem ribs various.
10. Stems shallowly grooved; central cavity very large comprising approximately 90% of stem diameter. . . . . ***E. fluviatile*** (in part)
10' Stems deeply grooved; central cavity small comprising less than 33% of stem diameter . . . . ***E. palustre***

Equisetum arvense Linnaeus (p. 74)

Common horsetail

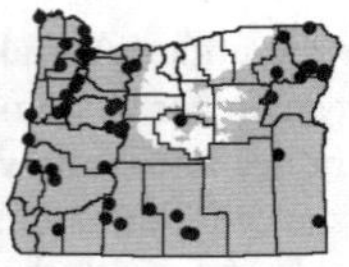

Aerial stems annual, dimorphic, green or lacking chlorophyll and stomata; vegetative stems erect and bushy, up to 40 cm or semi-decumbent, the stem apex extended into a long narrow branchless tip; up to 14 prominent ribs lacking a furrow; central cavity 33–67% of the stem diameter. **Sheaths** nearly square, up to 5 × 6 mm (rarely larger); teeth up to 3 mm, dark, firm-textured, persistent, with simple midribs continuous with the stem ribs, free or united in pairs. **Branches** simple, whorled, solid, sharply 3(4)-angled; branch bases dark; first internode distinctly longer than its subtending sheath. **Fertile stems** shorter than the vegetative stems, unbranched, pale brown, lacking chlorophyll and stomata, and with larger sheaths, produced in spring before the vegetative stems. **Cones** exserted from last sheath when mature; green stem usually lacking cones. $2n=216$.

Moist to dry open places, grasslands, meadows, disturbed sites, along roads; 0–2600 m. BR, BW, Casc, CR, ECas, Est, Owy, Sisk, WV. CA, ID, NV, WA; throughout North America; Eurasia. Native.

For identification hints for plants with fertile, green stems, see comments under *E. palustre*.

Equisetum ×ferrissii Clute

Ferriss' horsetail

Equisetum hyemale L. var. *elatum* (Engelm. ex A. Br.) C.V. Morton

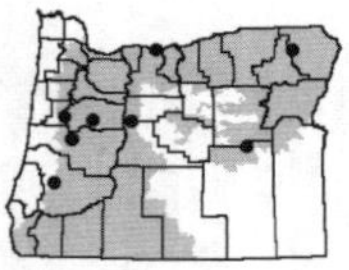

Aerial stems persisting over winter, usually unbranched, to 50 cm tall, green; central cavity approximately 75% of stem diameter. **Sheaths** rectangular, up to 6 × 12 mm, green, with single dark terminal band around the bases of teeth or 2 dark bands and a pale section between bands; teeth articulated with an early-deciduous blade and persistent dark incurved bases. **Branches** usually absent, where present ascending, hollow, and resembling narrow main stems. **Cones** subsessile or shortly pedunculate, pale and misshapen; spores infertile.

Wet places, sandy shores, roadsides. 0–2000 m. BW, Casc, Col, ECas, Sisk, WV. CA, ID, NV, WA; throughout North America. Native.

This species is a hybrid between *E. hyemale* var. *affine* and *E. laevigatum*. It is usually found where both parents occur. The species is sterile but persists by vegetative growth.

Equisetum fluviatile Linnaeus

River horsetail

Equisetum limosum L.

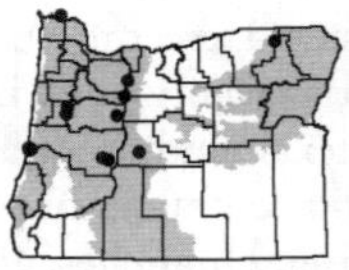

Aerial stems annual, monomorphic, simple or branched, to 110 cm tall, shallowly grooved, green; ribs 9–24; central cavity approximately 90% of the stem diameter. **Sheaths** square, to approximately 10 × 10 mm, green; teeth dark, narrow, persistent, not articulated at base, hyaline margin lacking or very narrow. **Branches** (when present) rough, spreading or sometimes ascending and fertile; first internode of the branch shorter than the lowest sheath and not exserted; ridges 4–6. **Cones** terminal on the main stem or occasionally on branches, apiculate. $2n=216$.

Bogs, lakes, and marshes. 0–1500 m. BW, Casc, CR, ECas, WV. ID, WA; throughout northern North America; Eurasia. Native.

Equisetum hyemale Linnaeus var. ***affine*** (Engelmann) Calder & Roy L. Taylor (p. 74)

Scouring rush

Equisetum hyemale L. ssp. *affine* (Engelm.) Calder & Roy L. Taylor, *Equisetum hyemale* L. var. *californicum* Milde, *Equisetum praealtum* Raf.

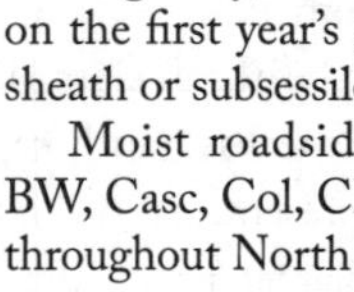

Aerial stems persisting over winter, monomorphic, simple or with a few hollow, ascending branches, up to 2 m tall, dark green; ribs 18–40, not furrowed; central cavity 70% or more of the stem diameter. **Sheaths** nearly square when flattened, up to 15 × 16 mm, green proximally; 2 dark bands usually present above the green base, with a pale gray section between them; teeth articulated, tooth blade detaching early, leaving the dark incurved base. **Cones** terminal on the first year's growth, shortly exserted from the terminal sheath or subsessile, apiculate. $2n=216$.

Moist roadsides, woods, riparian areas. 0–2300 m. BR, BW, Casc, Col, CR, ECas, Est, Sisk, WV. CA, ID, NV, WA; throughout North America. Native.

Equisetum laevigatum A. Braun

Smooth scouring rush

Equisetum funstonii A.A. Eaton, *Equisetum kansanum* J.H. Schaffn.

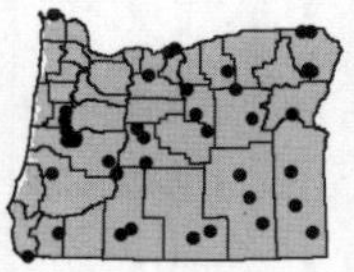

Aerial stems annual, monomorphic, up to 120 cm tall, usually simple, occasionally with a few hollow ascending branches, green; ribs 16–30, not furrowed; central cavity up to 75% of the stem diameter. **Sheaths** rectangular, 7–12 × 5–6 mm (uppermost sheath shorter), same color as stem and continuous with it, except for a terminal black band formed by the inturned persistent bases of the teeth; teeth early-deciduous, articulated and campanulate. **Branches** (when present) with a dark base; first internode very short. **Cones** shortly exserted from the terminal sheath or subsessile, blunt. ***2n***=216.

Moist meadows, roadsides, and river banks. 50–2000 m. All ecoregions except Est. CA, ID, NV, WA; throughout North America. Native.

Equisetum palustre Linnaeus

Marsh horsetail

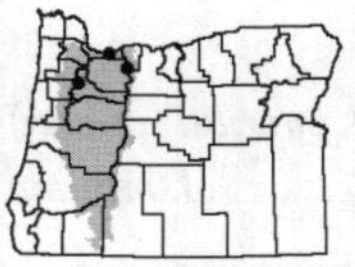

Aerial stems annual, monomorphic, up to 80 cm tall, deeply grooved, green; ribs 5–10; central cavity less than 33% of the stem diameter. **Sheaths** rectangular, 5–10 mm, green; teeth 3–7 mm, persistent with a dark center and pale hyaline margins. **Branches** in whorls, from the mid-region of the stem, numerous to sparse (rarely absent); ridges 4–6; first internode shorter than the sheath and not exserted. **Cones** pedunculate. ***2n***=216.

Swamps, marshes. 0–1200 m. Casc, WV. Northern North America; Asia, Europe. Native.

Damaged green stems of *E. arvense* can sometimes produce cones and resemble *E. palustre.* However, the latter differs from *E. arvense* in the size of its central cavity (greater than 33% stem diameter vs. less than 33% in *E. arvense*) and the first branch internode (shorter than sheath vs. longer than sheath in *E. arvense*).

Equisetum telmateia Ehrhart var. *braunii* (J. Milde) Hauke

Giant horsetail

Equisetum maximum Lam., *Equisetum telmateia* Ehrh. ssp. *braunii* (J. Milde) Hauke

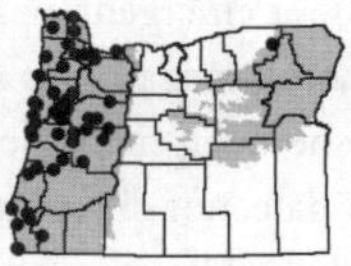

Aerial stems annual, dimorphic, green or lacking chlorophyll and stomata; vegetative stems to 100(175) cm, green to pale yellow-green; central cavity approximately 70% of the stem diameter; ribs 16–30, usually with furrows extending into the sheaths and teeth. **Sheaths** rectangular, approximately 12 × 15 mm; teeth long acuminate, 4–8 mm, connate into groups of 3–5. **Branches** whorled or simple, rough; ridges 4–5, not sharply angled; first internode short, concealed by the sheath. **Fertile stems** shorter and thicker than green stems, unbranched, pale brown, lacking stomata; sheaths larger than those of the green stems, up to approximately 4 cm long. **Cones** well exserted from the last sheath; green stems usually lacking cones.

Stream banks, marshes, roadsides. 0–1900 m. BW, Casc, CR, Est, Sisk, WV. CA, ID, WA; north to British Columbia. Native.

Equisetum variegatum Schleicher ex F. Weber

Variegated scouring rush

Equisetum variegatum Schleich. ex F. Weber & D. Mohr ssp. *variegatum*, *Equisetum variegatum* Schleich. ex F. Weber & D. Mohr var. *variegatum*

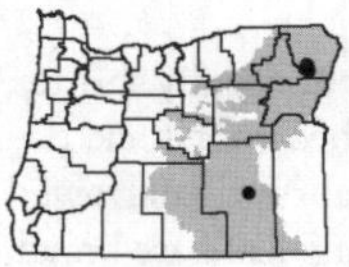

Aerial stems persisting over winter, slender and under 2 mm in diameter, simple or occasionally branched near the base, erect to decumbent, 10–30 cm tall, green; central cavity small, up to 33% of the stem diameter; ribs 4–8, furrow usually absent on the internode, present on the midribs of the sheath and teeth. **Sheaths** to approximately 3 × 3 mm, green; teeth 1–2 mm, free, with black, 2-ribbed midstripe and hyaline margins, base not articulated, not early-deciduous, distal part often damaged with age. **Cones** shortly pedunculate. ***2n***=216.

River banks, lakeshores, wet woods, roadsides. 100–2500 m. BR, BW. ID, WA; northern North America; Eurasia. Native.

ISOËTACEAE Quillwort family

Herbs perennials, monoecious, aquatic or terrestrial. **Rootstock** short, erect, globose. **Roots** produced from 2 or 3 vertical grooves in the rootstock. **Leaves** spirally arranged, deciduous or evergreen; base more or less triangular, narrowed abruptly or gradually into the grass-like blade with a vestigial leaf or tongue at the junction of the blade and base on the adaxial surface. **Sporangia** solitary, embedded in the leaf base, exposed on the adaxial leaf surface, partially or completely covered by a thin layer of leaf tissue (velum), interior divided into chambers by several transverse septae. **Megasporangia** and **microsporangia** in zones among the leaves, megasporangia forming the external zone in deciduous species, with microsporangia in the center; megasporangia with white (rarely black) megaspores, diameter 0.3–0.7 mm, with 3 prominent sutures from an equatorial ridge. **Megaspore** wall ornamentation nearly smooth to rugulose, tuberculate, echinate or cristate; ornamentation of the equatorial girdle is often different from the rest of the surface. **Microspores** 20–45 µm, usually gray to pale brown, wall ornamentation smooth, papillose, or echinate. **Gametophytes** endosporic, female producing archegonia with brown necks in the white gametophyte tissue exposed by spore germination, males producing 4 multiflagellate motile gametes.

Oregon species of this family are diploid (2*n*=22) with the exception of *Isoëtes occidentalis,* which is hexaploid. *Isoëtes maritima,* from coastal Washington, is tetraploid.

Cosmopolitan. 1 genus.

ISOËTES Quillwort

Family comprised of a single genus, *Isoëtes*.

Approximately 150 species; 7 species treated in *Flora*.

Key to *Isoëtes* species:

1. Plants aquatic, submerged in lakes, ponds, or emergent on lakeshore sediments.
    2. Leaves narrowed abruptly to a fine point; megaspores rugulate or tuberculate . . . . ***I. bolanderi***
    2' Leaves tapered to the tip, without a shoulder; megaspores various.
        3. Megaspores echinate; girdle echinate, not differentiated . . . . ***I. tenella***
        3' Megaspores nearly smooth to tuberculate, rugulate or cristate; girdle smooth.
            4. Sporangium wall unpigmented; megaspores greater than 0.5 mm; plants in lakes, seldom emergent . . . . ***I. occidentalis***
            4' Sporangium wall brown-streaked; megaspores less than 0.5 mm; plants in marshes and shallow water, often emergent . . . . ***I. howellii*** (in part)

1' Plants terrestrial, in seepage or seasonally wet places.
    5. Velum covering sporangium; megaspores smooth to lightly rugulate, white or gray.
        6. Leaves stiff, up to 20 cm long; plants on wet ground; megaspores averaging more than 0.4 mm . . . . ***I. nuttallii***
        6' Leaves rather flexuous, less than 10 cm long; plants of vernal pools; megaspores averaging less than 0.35 mm . . . . ***I. orcuttii***
    5' Velum incomplete; megaspores rugulate to echinate, white or black.
        7. Megaspores rugulate or tuberculate, white; amphibious plants up to 30 cm long; rootstock 2-lobed . . . . ***I. howellii*** (in part)
        7' Megaspores echinate, white or black; minute terrestrial plants to 6 cm long; rootstock 3-lobed . . . . ***I. minima***

Isoëtes bolanderi Engelmann

Bolander's quillwort

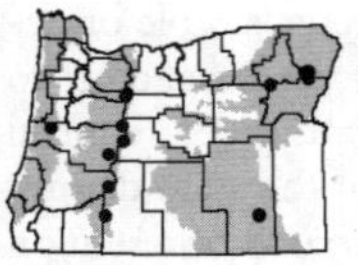

Leaves up to 17 cm long but usually shorter, bright green, from a 2-lobed rootstock, deciduous, rather soft; apex abruptly narrowed to a fine tip several mm long; leaf base broad, more or less triangular around the sporangium; hyaline leaf margins extending less than 1 cm above the sporangium, abruptly contracted into the blade. **Sporangia** wall pigmented with brownish streaking. **Velum** covering less than half of the sporangium. **Megaspores** 0.3–0.5 mm, white, rugulate to tuberculate; girdle differentiated, smooth. **Microspores** 20–30 µm, brown, spinulose.

Submerged in mountain lakes, sometimes emergent on lakeshore sediments. 100–2800 m. BR, BW, Casc, CR. CA, ID, NV, WA; western US, north to British Columbia and Alberta. Native.

In suitable lakes, populations of this species may be very large. Sterile hybrids are possible between *I. bolanderi* and *I. occidentalis* or *I. tenella* where the species grow together in the same lake. Hybrids have misshapen megaspores, but neither hybrid has been confirmed from Oregon. The unique shape of the leaf tip distinguishes this species from all other western *Isoëtes*, but care is needed to find an undamaged leaf tip and ensure that the abrupt contraction is present. Tapering leaf tips on *I. occidentalis* and *I. tenella* occasionally have irregularities but lack the "shoulders" of *I. bolanderi*.

Isoëtes howellii Engelmann

Howell's quillwort

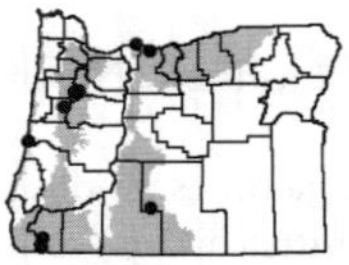

Leaves up to 30 cm long, bright green, from a rounded, 2-lobed rootstock, deciduous, fairly rigid; leaf base broad; hyaline margins extending up to 5 cm above the sporangium, narrowed gradually into the leaf blade. **Sporangia** ovoid; wall pigmented with brown streaking or mostly brown. **Velum** covering less than half of the adaxial surface. **Megaspores** white, 0.3–0.5 mm, rugulate to tuberculate. **Microspores** 25–35 µm, brown, spinulose.

Vernal pools, lake margins. 0–1600 m. Col, ECas, Est, Sisk, WV. CA, ID, WA; north to British Columbia, east to MT and UT. Native.

Isoëtes howellii is more aquatic than the three Oregon terrestrial species and is found in shallower and smaller bodies of water than the three fully aquatic species, such as ponds or grassy creeks that dry out in late summer. This species can occur around the marshy margins of lakes that contain fully aquatic species; therefore, it cannot be separated from the latter on habitat alone.

Isoëtes minima A.A. Eaton

Midget quillwort

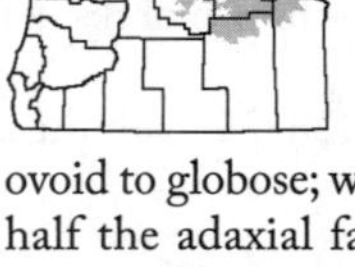

Leaves up to 6 cm long, yellow-green with a white base, from a rounded 3-lobed rootstock, rather rigid, deciduous, narrowing abruptly to the apex; leaf base broad, abruptly narrowed into the leaf blade. **Sporangia** ovoid to globose; wall pale, unpigmented. **Velum** covering about half the adaxial face of the sporangium. **Megaspores** 0.35–0.4 mm, black, covered with blunt spines; girdle not differentiated. **Microspores** approximately 30 µm, brown, rugulate.

Seepage areas in dry prairie (often with sagebrush) that dry out by midsummer. 1700–1800 m. BW. WA; north to British Columbia. Native.

Isoëtes minima is globally rare, known only from a few sites, and has only recently been found (2009) in northeastern Oregon. This species is easily overlooked and should be sought in seasonal seepage areas with very short vegetation in dry prairies. Washington material of this species (including the type specimen) has white megaspores. The only known Oregon population has black megaspores and may be taxonomically distinct.

Isoëtes nuttallii A. Braun ex Engelmann (p. 82)

Nuttall's quillwort

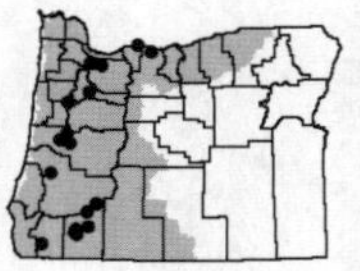

Leaves up to 20 cm long but usually shorter, yellow-green, from a 2- or 3-lobed rootstock, rigid, slightly brittle, deciduous, abruptly tapered at the tip; leaf bases dark, scale-like, often persistent. **Sporangia** wall unpigmented. **Velum** complete. **Megaspores** 0.36–0.6 mm, lustrous white, smooth to rugulose; girdle not differentiated. **Microspores** 20–30 µm, brown, papillose.

Wet ground with short vegetation. 0–1700 m. Casc, Col, CR, ECas, Sisk, WV. CA, WA; north to British Columbia. Native.

This species can be distinguished from all other *Isoëtes* in Oregon (except *I. orcuttii*) by the combination of a terrestrial habit and a velum completely covering the sporangia. It differs from *I. orcuttii* in the larger size of the plants and megaspores and also in its habitat preference.

Isoëtes occidentalis L.F. Henderson

Western quillwort

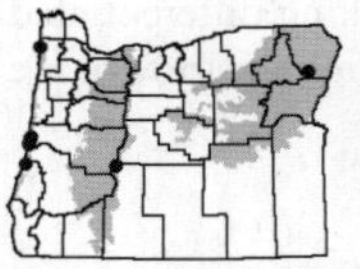

Leaves up to 20 cm long, dark green, from a 2-lobed rootstock, tough, rigid, evergreen, distal 10 mm tapered gradually to the apex. **Sporangia** wall unpigmented. **Velum** partial, covering approximately half of the sporangium. **Megaspores** 0.5–0.7 mm, white, wall sculpting reportedly variable from rugulate to tuberculate (echinate) and cristate; girdle smooth. Megaspores in examined Oregon specimens have nearly smooth walls, consequently the girdle is scarcely differentiated. **Microspores** 35–45 µm, brown, papillose to spinulose.

Submerged in coastal and inland lakes, lowlands, mountains; sometimes occurring at depths of 2 m or more, occasionally emergent in summer. 0–2200 m. BW, Casc, Est. CA, ID, WA; north to AK, east to MT and WY. Native.

Hybrids are possible with *I. bolanderi* or *I. tenella* where species are growing together; none are confirmed from OR. Unique among *Isoëtes* species in Oregon, more than one pair of zones of micro- and megasporangia can occur in *I. occidentalis*.

Isoëtes orcuttii A.A. Eaton

Orcutt's quillwort

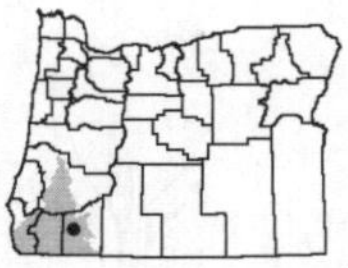

Leaves up to 7 cm long, yellow-green, from a 3-lobed rootstock, deciduous, tapered abruptly to apex, rather soft. **Sporangia** unpigmented. **Velum** completely covering sporangium. **Megaspores** 0.2–0.38 mm, white or gray, lustrous, ridges very prominent, surface nearly smooth; girdle not differentiated. **Microspores** 20–30 µm, gray, smooth to papillose.

Vernal pools. 0–1500 m. Sisk. CA; south to Baja California. Native.

Isoëtes orcuttii is similar to *I. nuttallii* but differs in its smaller size, smaller megaspores, and habitat (vernal pools vs. seasonally to perennially wet soil).

Isoëtes tenella Leman

Spiny-spored quillwort

Isoëtes echinospora Durieu, *Isoëtes setacea* Lam.

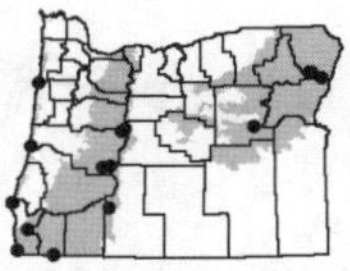

Leaves up to 20 cm long, bright green to reddish-green, from a 2-lobed rootstock, rather soft and flexuous, narrowed to the apex in the distal 1 cm; base broad and triangular around the sporangium, gradually tapered for about 2 cm above the sporangium. **Sporangia** rounded to ovoid; wall usually brown-streaked. **Velum** partial, covering approximately half the sporangium. **Megaspores** 0.4–0.55 mm, white, echinate; girdle not differentiated. **Microspores** 20–30 µm, gray or brown, smooth to spinulose.

Submerged in coastal and inland lakes. 0–2300 m. BW, Casc, Est, Sisk. CA, ID, WA; circumboreal. Native.

Isoëtes tenella is the only aquatic Oregon quillwort with echinate megaspores. *Isoëtes maritima* also has echinate megaspores, but it does not occur south of Puget Sound. Sterile hybrids are possible between *I. tenella* and *I. occidentalis* or *I. bolanderi* where the species grow together. Neither hybrid has been reported from Oregon.

LYCOPODIACEAE Clubmoss family

Plants herbs; roots emerging from the stems, basally or along the length of trailing stems, sometimes produced apically and migrating downwards through the stem tissues to emerge near the base. **Stems** erect, pendant or trailing, sometimes buried, dichotomous branching; gemmae formed from modified branches present or absent. **Leaves** simple with a single vein, small, needle-like to ovate, often narrowly lanceolate or linear. **Cones** differentiated from the stems or not, appearing sessile or stalked. **Sporophylls** identical to or very different from the vegetative leaves. **Gametophytes** free-living, variously subterranean and achlorophyllous or exposed and green.

Cosmopolitan. 3–10 genera; 4 genera treated in *Flora*.

There is wide variation in genus concepts within Lycopodiaceae. Genera counts vary from three or four up to ten or more. Genera with indeterminate shoot growth (including *Huperzia*) are sometimes separated into family Huperziaceae.

There is much disagreement about generic limits in *Lycopodium*, with some authors combining it with *Diphasiastrum* and *Lycopodiella*. *Diphasiastrum* species have a different appearance, however, with shorter leaves in fewer ranks, resembling cedar or microphyllous Ericaceae. *Lycopodiella* is morphologically very similar to *Lycopodium*, but its leaves and stems start to senesce after one year, in contrast to the long-lived stems and leaves of our *Lycopodium* species.

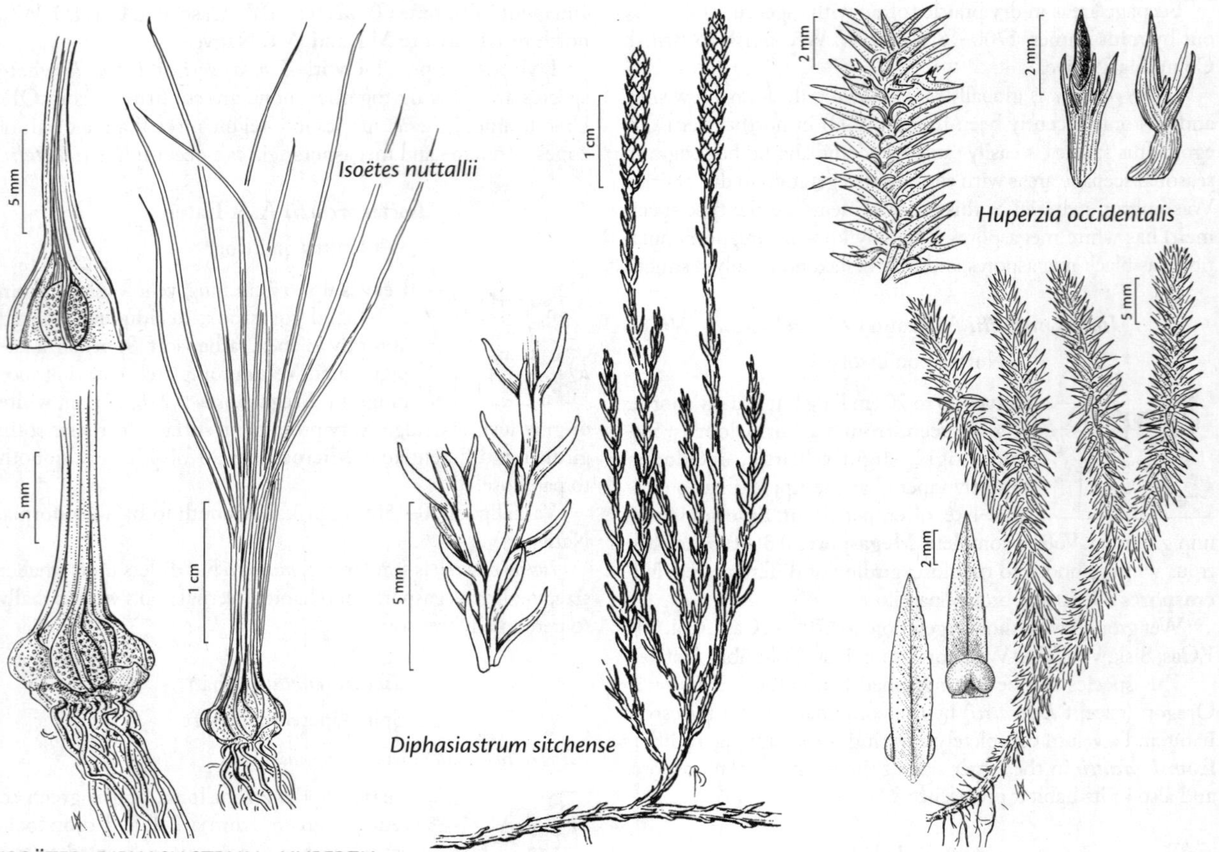

ISOËTES • DIPHASIASTRUM • HUPERZIA

The common name, clubmoss, refers to the club-like, differentiated, erect cone of *Lycopodium.* Genera lacking differentiated cones are called firmoss or fir clubmoss, while pendant epiphytes (tropical) with long slender cones of indeterminate growth are called tassel moss.

Key to Lycopodiaceae genera:

1. Sporangia borne along the stems; cones not distinct; sporophylls resembling leaves; stems erect; short gemmiferous branches present . . . . *Huperzia*

1' Sporangia borne in distinct terminal cones; sporophylls unlike leaves; some stems trailing; gemmae absent.

  2. Stems distinctly flattened; leaves scale-like, those on the underside smaller . . . . *Diphasiastrum* (in part)

  2' Stems not flattened; leaves not scale-like.

    3. Leaves less than 4 mm, in fewer than 6 ranks; cones sessile on leafy stems. . . . . *Diphasiastrum* (in part)

    3' Leaves greater than 4 mm, in 6 ranks or more; cones sessile or with distinct peduncle with short sparse leaves.

      4. Erect stems simple, usually bearing a single cone; distinct peduncle lacking . . . . *Lycopodiella*

      4' Some erect stems branched, cones 1–several per branch; peduncle differentiated or not. . . . . *Lycopodium*

DIPHASIASTRUM Clubmoss

Plants perennial; trailing, with long-creeping horizontal stems, sometimes buried, from which arise at fairly regular intervals roots and clusters of erect stems, some fertile. **Stems** densely to sparsely leafy; annual growth sections on the stems often separated by a constriction with very short leaves. **Leaves** on erect branches borne in 4 or 5 ranks, either distinct and linear, or scale-like. **Cones** single, from end of erect, leafy fertile branch, or 1–several in a simple to branched peduncle. **Gemmae** absent.

Cosmopolitan, but mostly temperate to arctic. 15–20 species; 2 species treated in *Flora*.

Key to *Diphasiastrum* species:

1. Branches distinctly flattened and cedar-like; leaves on the dorsal/underside much reduced in size, in 4 ranks, free part of leaf less than 2 mm . . . . *D. complanatum*

1' Branches not flattened; leaves all similar, in more than 4 ranks, greater than 3 mm long. . . . . *D. sitchense*

Diphasiastrum complanatum (Linnaeus) Holub

Ground cedar, groundcedar, Northern running-pine

Lycopodium complanatum L.

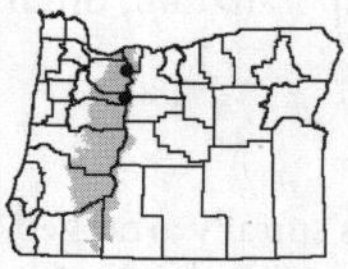

Herbs perennial, glabrous. **Stems** main stem trailing, 30 cm or more in length, often buried, rooting irregularly; ascending stems up to 20 cm tall, arising at intervals from the trailing stem, dichotomously branched, composed of vertical and layered spreading branches, some vertical stems turning into peduncles and bearing cones; conspicuous constrictions present between annual growth segments. **Leaves** narrow, approximately 2 mm long, sparse, in 4 ranks; lateral pair opposite, flattened, with a free acute tip 1–2.5 mm long, base strongly decurrent; upper leaves resembling lateral leaves but more strongly appressed to the stem; lower leaves triangular, scarcely decurrent. **Peduncles** distinct, 2–5 cm long, simple or branched. **Cones** 1–4 per peduncle, 10–25 × 2 mm. **Sporophylls** pale, ovate; margins wavy to irregularly lobed, abruptly narrowed into a decurrent base on which the reniform sporangium is seated; sporangia approximately 1.5 mm wide, completely covered by the sporophyll. **2*n***=46.

Moist coniferous woods. 1000–1700 m. Casc. ID, WA; northern US, Canada; circumboreal. Native.

This species is found in northern Oregon, where it reaches its southern limit in the western US.

Diphasiastrum sitchense (Ruprecht) Holub (p. 82)

Alaska clubmoss, Sitka clubmoss

Lycopodium sitchense Rupr., *Lycopodium sabinaefolium* Willd. var. *sitchense* (Rupr.) Fernald

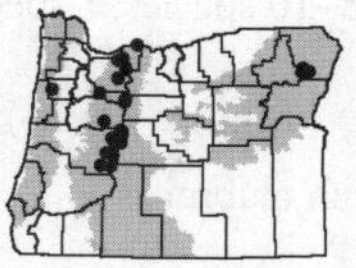

Herbs perennial, glabrous. **Stems** main stem slender, trailing or occasionally buried, rooting irregularly, sparsely to densely leafy, slightly sulcate stem visible; erect dichotomous branches arising at intervals along the main stem, dichotomously branched near the base, forming shrubby clumps of upright branches, up to 10 cm tall, a few stems fertile and taller, with cones borne above the vegetative growth. **Leaves** narrowly lanceolate, stiff, dark green, 2–3.5 mm, ascending and incurved, concave adaxially, borne in (4)5 ranks, leaves of horizontal and principal erect stems shorter and sparser than those of the branches. **Peduncles** with reduced or absent leaves; cone appearing sessile on the leafy stem. **Cones** distinct, 10–15 × 3 mm long, borne singly on longer erect stems. **Sporophylls** pale, up to 3.6 × 2.8 mm, deltate, triangular or occasionally ovate, margins irregularly toothed to subentire; sporangia reniform, narrower than the sporophylls. **2*n***=46.

Open to wooded areas in the mountains, often on peaty soil, often mixed with low ericaceous shrubs or on rocks. 100–2500 m. BW, Casc, CR, ECas. ID, WA; northern US, Canada; East Asia. Native.

HUPERZIA Firmoss, fir clubmoss

Stems clustered, erect, simple or dichotomous, of indeterminate growth; roots initiated near the stem apex, emerging near the base. **Leaves** linear to lanceolate, arranged irregularly around the stems, margin often with small teeth formed by papillae on the walls of the leaf margin cells. **Cones** not differentiated. **Sporophylls** producing sporangia in poorly defined zones along the stems. **Gemmae** present on short branches. **2*n***=134, 136.

Cosmopolitan in temperate to arctic regions. 10–15 species; 1 species treated in *Flora*.

Huperzia differs from the other genera in the family by its lack of horizontal stems and differentiated cones, the indeterminate growth of the branches, and the presence of gemmiferous branches.

Huperzine A is a sesquiterpene alkaloid isolated from *Huperzia* species that increases neural activity by inhibiting the enzyme acetylcholinesterase. It is used as a dietary supplement to combat memory loss.

Huperzia miyoshiana has been collected twice in Oregon. Both occurrences were in Clatsop County: once on Saddle Mountain (1978) and again on Sugarloaf Mountain (2007). This species differs from *H. occidentalis* by its narrowly lanceolate or slightly triangular leaves.

Huperzia occidentalis (Clute) Kartesz & Gandhi (p. 82)

Western firmoss, western clubmoss

Lycopodium selago L.

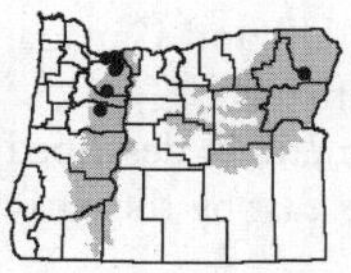

Herbs glabrous. **Stems** in a cluster of erect, simple to dichotomous branches of indeterminate growth, sometimes decumbent in age, not rooting above the base, up to 25 cm long but usually much shorter; roots emerging basally from stems. **Leaves** reflexed to spreading, dull yellow-green when dry, darker when fresh, 6–12 × approximately 0.15 mm, narrowly oblanceolate, broadest at mid-length or above, apex acute, base shortly decurrent, smallest leaves sometimes narrowly triangular and broadest at base, margins with tiny teeth or papillae. **Cones** not differentiated. **Sporophylls** similar in size and shape to the leaves or slightly smaller; reniform sporangia wider than the sporophylls and borne in the axils, golden yellow when mature, pale when dehisced. **Gemmae** approximately 4.5 × 3.5 mm, composed of a very short stem and 4 leaves, 2 large and 2 small, produced singly on short lateral upturned branches about 1 mm long with a terminal cluster of short narrow leaves of varying lengths, produced in a small cluster or rarely scattered more diffusely. **2*n***=134.

Moist sites in forests, stream banks, on old mossy logs. 300–1900 m. BW, Casc. ID, WA; north to Yukon Territory, northeast to MT. Native.

LYCOPODIELLA Bog clubmoss, marsh clubmoss

Herbs perennial, creeping, rooting sparsely from the underside of the stems. **Stems** branching dichotomously 1 to several times per year, 1 branch continuing horizontal, the other usually ascending and producing a cone. **Leaves** all similar, often toothed. **Peduncles** not differentiated. **Sporophylls** differing widely from the leaves with a wide base, covering the sporangium, often toothed. **Gemmae** absent. **2*n*=156.**

North temperate and neotropic regions. 8–10 species; 1 species treated in *Flora*.

Lycopodiella inundata (Linnaeus) Holub (p. 85)

Northern bog clubmoss, northern marsh clubmoss

Lycopodium inundatum L., *Lycopodium inundatum* L. var. *inundatum*

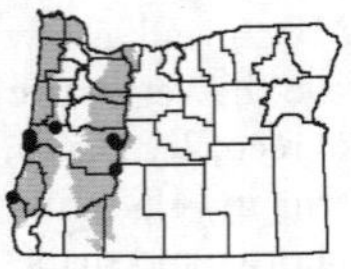

Herbs perennial, creeping, glabrous. **Roots** few. **Stems** decumbent ones densely leafy; erect stems simple, usually 1–2 per year, 3–10 cm tall. **Leaves** healthy-looking only on current year's growth; leaves of decumbent stems falcate, swept upwards, lanceolate, approximately 5 mm long, entire, acute; leaves of erect stems spirally arranged, resembling those of the horizontal branches but becoming shorter above, often appearing whorled below cone. **Cones** 1–5 × 0.3–0.5 cm. **Sporophylls** approximately 5 mm long, becoming straw-colored, ovate at the base with several irregular teeth and a long acumen.

Coastal or inland peat bogs, wet sandy soil, or other wet habitats with short vegetation, sometimes on logs and stumps. 0–1700 m. Casc, CR, Est. CA; northwestern and northeastern US into Canada; Europe. Native.

LYCOPODIUM Clubmoss

Herbs perennial, trailing; roots at regular intervals from horizontal or buried stems. **Stems** densely leafy; annual growth sections often separated by a leafless constriction. **Leaves** linear to narrowly lanceolate, entire or minutely toothed; leaves on erect branches borne in 6 or more distinct ranks. **Cones** at ends of erect, leafy, fertile branches, or in a simple to branched peduncle with reduced leaves bearing 1–several cones. **Gemmae** absent. **2*n*=68.**

Temperate to subarctic regions. 15–25 species; 2 species treated in *Flora*.

Spores of *Lycopodium clavatum* and other common species are harvested and sold as *Lycopodium* powder, used for a variety of purposes including ignition and lubrication.

Key to *Lycopodium* species:

1. Leaves with caducous tip hairs; peduncles well developed . . . . ***L. clavatum***

1' Leaves acute with no tip hairs; peduncles not differentiated . . . . ***L. annotinum***

Lycopodium annotinum Linnaeus (p. 85)

Stiff clubmoss, bristly clubmoss

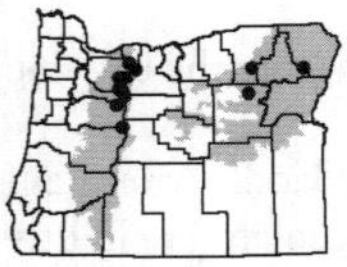

Herbs perennial, often forming loose mats; roots produced infrequently along trailing stems. **Stems** densely leafy; main stem long-creeping; ascending stems produced dichotomously from main stem at regular intervals, branching to form clumps of erect, densely leafy stems, to approximately 10 cm tall, with prominent annual bud constrictions. **Leaves** of trailing stems 5–6 × 0.9 mm, falcate, not obscuring the sulcate stem, leaves of ascending stems dense, arranged in 6–8 ranks, sometimes appearing whorled, pale yellow-green when dry, darker when fresh, 5–7 mm long, lanceolate and broadest just above the base, margin with small teeth, apex acute, lacking a tip hair. **Peduncles** not differentiated. **Cones** 15–25 × approximately 4 mm. **Sporophylls** buff-colored when mature, approximately 3 × 2 mm, ovate or triangular and long-acuminate.

Moist coniferous forests, stream banks, and seepage, occasionally in open boggy meadows. 0–1900 m. BW, Casc. ID, WA; western and northeastern US, Canada, Greenland; circumboreal. Native.

Lycopodium clavatum Linnaeus (p. 85)

Common clubmoss, stag's horn clubmoss

Lycopodium clavatum L. var. *clavatum*

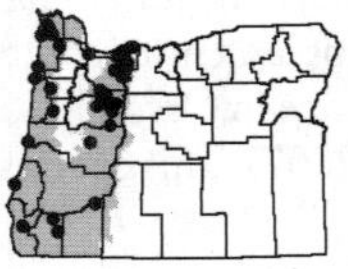

Herbs perennial, forming loose mats with main stems up to 1 m long. **Stems** sparsely to densely leafy, lateral branches arising from the main stem, forming clumps of erect, densely leafy stems, 10–25 cm tall. **Leaves** linear, incurved, in 6 or more ranks, 7–8 mm long, usually entire, occasionally with sparse teeth, narrowed at the apex into a long caducous hair tip, tip 1–2 mm long. **Peduncles** well developed. **Cones** 2–5 cm long. **Sporophylls** straw colored, approximately 3.5 × 2 mm, ovate, apex with a caducous tip hair.

Moist to dry sites in forests. 0–1700 m. Casc, CR, Est, Sisk. CA, ID, WA; mainly western and eastern US and Canada, south to South America; Africa, Eurasia. Native.

Plants from western North America, sometimes described as var. *integrifolium*, differ from those in the east by the early shedding of the hairs on the leaf tips.

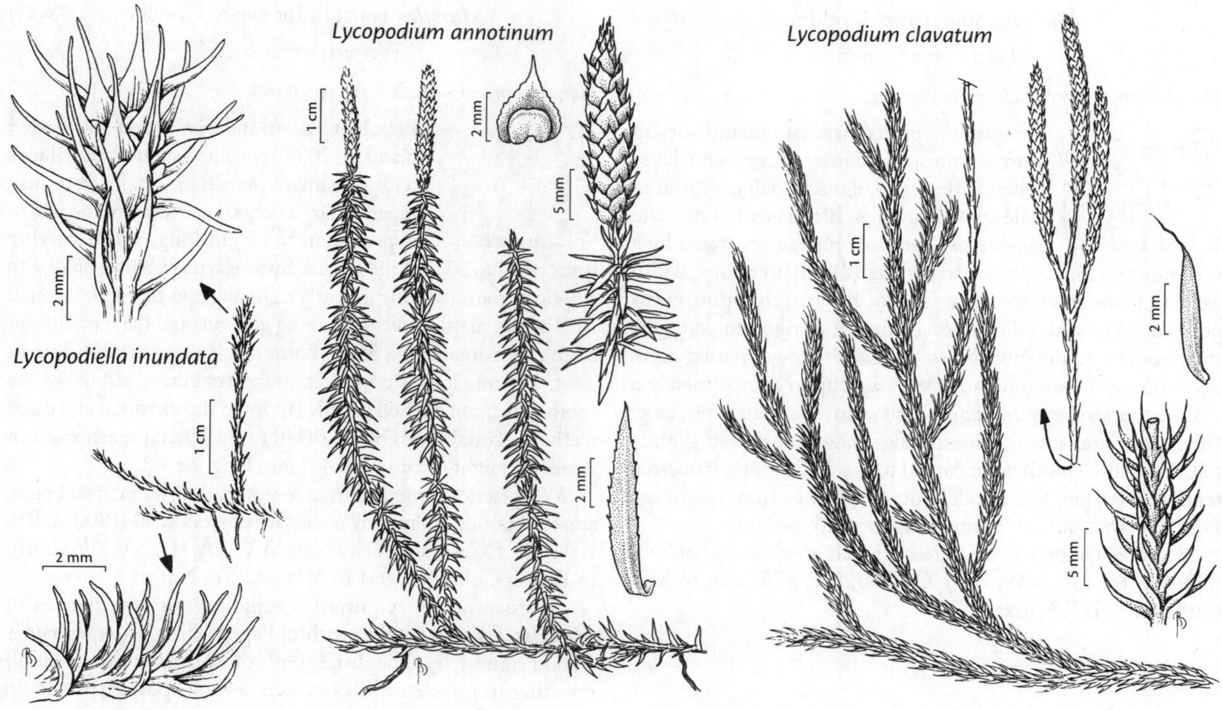

LYCOPODIELLA • LYCOPODIUM

MARSILEACEAE Waterclover family

Plants herbaceous, perennial, heterosporous, monoecious, aquatic or terrestrial. **Stems** horizontal, creeping, stolon-like; indumentum of long hairs or none. **Leaves** long-petiolate, expanded pinnae present or absent, hairy or glabrous. **Sporocarps** nut-like, globose or ovoid and flattened in cross section, with matted hairs, wall eventually splitting releasing a gelatinous mass in which the sori are embedded, produced on stalks at leaf bases; stalks erect or reflexed and simple with a single sporocarp. **Gametophytes** largely endosporic; female producing a single archegonium; male producing motile gametes.

Cosmopolitan. 3 genera; 2 genera treated in *Flora*.

Key to Marsileaceae genera:

1. Leaves consisting of a petiole with 4 clover-like pinnae . . . . **MARSILEA**

1' Leaves simple, grass or rush-like, lacking pinnae . . . . **PILULARIA**

MARSILEA Waterclover

Stems horizontal, submerged in shallow water or emergent, creeping stolon-like on mud or sand, rooting from the nodes, sparsely hairy or glabrescent, or with short leafy lateral branches usually with dense tufts of rusty hairs; nodes either simple or with a single leaf. **Leaves** in 2 ranks along the stems, clover-like when floating or emergent, often small and linear when submerged; floating leaves with lax petiole, pinnae forming a floating circle with 4 quadrants, often with reddish stripes between the veins below, sparsely hairy to glabrescent; emergent leaves with shorter wiry/stiff petioles and unpigmented pinnae, smaller on average than floating pinnae. **Blades** of floating and emergent leaves palmately divided into 4 triangular pinnae; lateral margins normally plane or slightly concave; distal margin curved, usually entire. **Veins** numerous, anastomosing to form a net of narrow areolae. **Sporocarps** hard and seed-like, shaggy with long matted hairs, glabrescent with age, oblong, elliptic or nearly rounded in lateral view, more or less compressed in cross section, each containing up to 25 sori; sporocarp stalks unbranched, shortly fused to the sporocarp at the apex, fused portion straight or curved; megasporangia and microsporangia borne within the same sorus.

Cosmopolitan. Approximately 45 species; 2 species treated in *Flora*.

The portion of the sporocarp stalk that is fused to the sporocarp usually bears two woody teeth. These are referred to in the key and descriptions below as the *proximal* and *distal* teeth.

Key to *Marsilea* species:

1. Distal teeth 0–0.4 mm long, straight, blunt, or absent; sporocarps oblong or elliptic laterally . . . . *M. oligospora*

1' Distal teeth 0.4–1 mm long, straight or curved, acute; sporocarps oblong to more or less rounded laterally . . . . . . *M. vestita*

Marsilea oligospora Goodding (p. 87)

Pacific waterclover

Marsilea vestita Hook. & Grev. in part

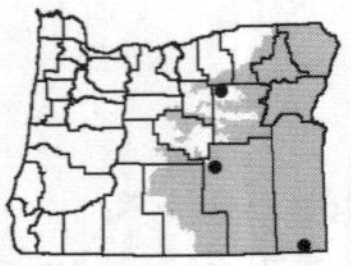

Petioles to approximately 6 cm and wiry in non-submerged plants, longer and laxer when submerged, densely hairy with long pale spreading hairs. **Blades** with 4 palmate pinnae or 2 pairs of pinnae separated by a very short rachis; segments triangular, (2)5–10 mm long, width varying from equal to about 50% of the length. **Sporocarps** nut-like, oblong to elliptic, long axis not at right angle to the stalk but angled downwards, covered with brown matted hairs, becoming glabrous and darker with age, up to approximately 6 × 4 mm, borne singly on unbranched stems less than 8 mm long; stems erect and straight, continuing straight or curved slightly to the proximal tooth, then curved to the distal tooth. **Proximal teeth** 0.2–0.6 mm long, straight or curved away from sporocarp. **Distal teeth** blunt, to 0.4 mm long or nearly absent.

Seasonal ponds, wet areas in sagebrush or other dry habitats. 600–1900 m. BR, BW, Owy. CA, ID, NV, WA; east to MT, southeast to UT. Native.

Marsilea vestita Hooker & Greville (p. 87)

Western waterclover

Marsilea vestita Hook. & Grev. ssp. *vestita*

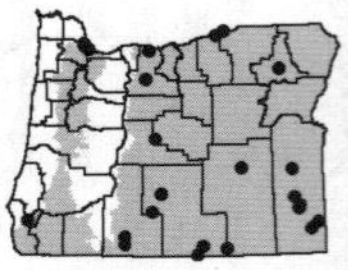

Petioles to 10 cm and wiry in non-submerged plants, to 20(60) cm when submerged. **Blades** with 4 palmate pinnae or 2 pairs of pinnae separated by a very short rachis, triangular, to approximately 1.5 cm long, width varying from equal to 33% of the length. **Sporocarps** nut-like, oblong to elliptic to rounded, long axis at a right angle to the stalk, covered with brown matted hairs, becoming glabrous and darker with age, up to approximately 7 × 5 mm, borne singly on unbranched stems less than 6 mm long; stems erect, straight or curved just below the sporocarp, fused portion usually curves to the proximal and distal teeth. **Proximal teeth** blunt, 0.3–0.6 mm. **Distal teeth** stout or slender, straight or curved, 0.4–1 mm long. **2*n***=40.

Wet places, submerged to exposed, shallow water, river banks, seasonal ponds, on muddy or sandy substrates. 50–1800 m. BR, BW, Col, ECas, Lava, Owy, Sisk, WV. CA, ID, NV, WA; north to British Columbia, east to MN and FL. Native.

Sporocarps are required to separate our two species of *Marsilea*. Care is needed with old sporocarps. Loss of hair is a good sign of age. The distal tooth of *M. vestita* can break off resulting in misidentification, thus several sporocarps should be examined.

PILULARIA Pillwort

Stems horizontal, creeping stolon-like over mud or sand, green, rooting at the nodes. **Leaves** glabrous or sparsely pubescent, rolled up when young, graminoid, lacking expanded pinnae, borne in 2 ranks along the stem, short leafy lateral branches often present in the leaf axils. **Sporocarps** globose, not flattened, borne on a short reflexed stalk, divided internally into 3–4 chambers, each with a single sorus containing both micro- and megasporangia; megasporangia arranged basally on the elongated sorus receptacle.

North and South America, Europe, Oceania. Approximately 6 species; 1 species treated in *Flora*.

Pilularia americana A. Brown (p. 87)

American pillwort

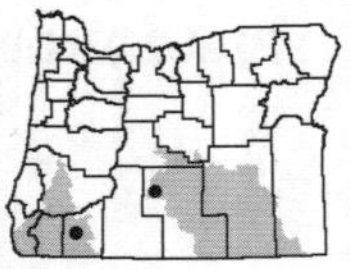

Stems creeping on mud or other substrates, submerged in shallow water or exposed, forming loose mats, glabrous. **Leaves** small, 1.5–6(10) cm long, bright green, borne singly or in small clumps at fairly regular intervals along the stem usually a few millimeters apart. **Sporocarps** reddish brown, globose, to approximately 2.2 mm in diameter, matted-hairy, losing much of the hair when old, borne on or just under the surface of the substrate; stalks less than 2 mm long, reflexed. **2*n***=20.

Seasonal ponds. 400–1700 m. BR, Sisk. CA; scattered sites throughout the US and Mexico. Native.

Due to its inconspicuous morphology, *P. americana* is seldom collected. As a result, this species may be more widespread and common than shown by current collection records.

OPHIOGLOSSACEAE Adder's tongue family

Ben Legler

Herbs perennial, terrestrial, forming apparently obligatory mycorrhizal relationships. **Roots** mostly unbranched, lacking root hairs. **Stems** solitary, subterranean, erect, usually unbranched, enclosed by sheathing leaf base; gemmae present or absent, formed from axillary meristems. **Leaves** in our species 1(2) per stem, usually erect, divided into a sterile, foliaceous portion (trophophore) and a fertile, spore-bearing portion (sporophore), these united at base to form a common stalk. **Trophophores** simple to pinnately compound, rarely absent, venation reticulate or free and fan-like. **Sporophores** simple or pinnately branched, rarely absent. **Sporangia** exposed or embedded, globose, 2-valved, without an annulus, 0.5–1.5 mm diameter, containing thousands of spores. **Spores** all of 1 kind. **Gametophytes** subterranean, non-chlorophyllous.

Cosmopolitan. 7 genera; 4 genera treated in *Flora*.

Ophioglossaceae is an ancient lineage of ferns sister to Psilotaceae. The circumscription of genera within Ophioglossaceae has been debated. *Botrypus* and *Sceptridium* are treated here as distinct from *Botrychium*.

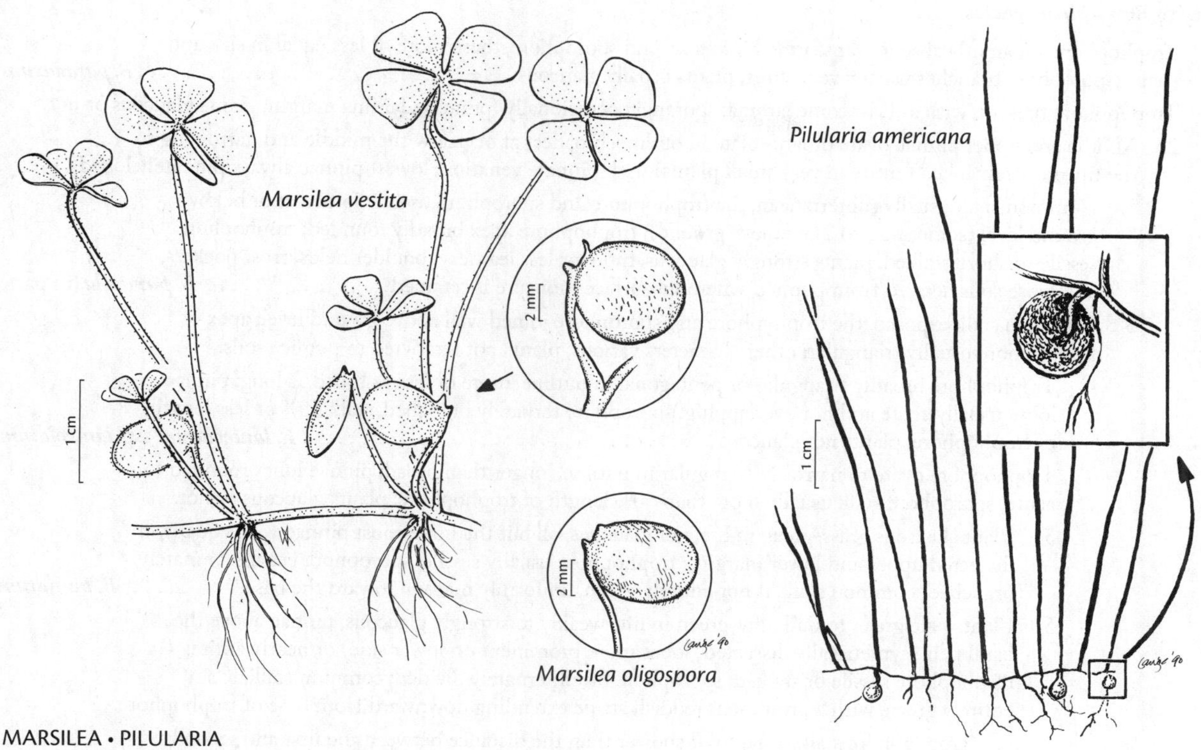

MARSILEA • PILULARIA

Key to Ophioglossaceae genera:

1. Trophophore (leaf-like portion) simple, entire; veins reticulate, not extending to leaf margin; sporophore (spore-bearing portion) unbranched, with sporangia sunken in axis and confluent in 2 linear rows . . . . **OPHIOGLOSSUM**

1' Trophophore compound or pinnately lobed (occasionally simple or absent in *Botrychium*); veins forked, terminating near leaf margin; sporangia exposed in sessile to stalked clusters, not sunken in axis.

   2. Trophophore 1–2-pinnate (occasionally simple or absent), narrowly lanceolate to triangular or deltate, less than 5(9) cm wide; sporophore nearly always present; plants glabrous . . . . **BOTRYCHIUM**

   2' Trophophore 3–4-pinnate, triangular to deltate, (5)10–30 cm wide; sporophore often absent in smaller plants; plants sparsely hairy when young, becoming glabrous.

      3. Trophophore sessile, thin and herbaceous, deciduous; lobes acute; trophophore and sporophore diverging well above ground level. . . . . **BOTRYPUS**

      3' Trophophore long-stalked, thick and more or less leathery, often wintergreen; lobes rounded to broadly acute; trophophore and sporophore diverging near or below ground level . . **SCEPTRIDIUM**

BOTRYCHIUM Moonwort

Plants glabrous; gemmae present or absent. **Leaves** 1 per stem above ground. **Trophophores** pinnately compound, sometimes simple, rarely absent, venation free and fan-like. **Sporophores** simple or pinnately branched. **Sporangia** exposed.

Cosmopolitan, mostly in boreal, temperate, and montane areas. Approximately 35 species; 13 species treated in *Flora*.

Botrychium is a difficult genus for which diagnostic characters are limited, particularly in small plants, and not all individuals can be confidently identified. Further confusion arises from the presence of numerous allotetraploids whose morphologies are intermediate between the diploid species. Identification is aided by sampling multiple plants and carefully pressing them to clearly show pinnae shape and margins.

Botrychium species often grow intermixed in genus communities. Most species in our area occur at montane to subalpine elevations, where they are often found in well-drained, mesic meadows or open, historically disturbed sites (natural or man-made) with a low, sparse to dense, herbaceous cover. Good indicator species are *Fragaria virginiana*, *Antennaria* spp., *Taraxacum officinale*, *Achillea millefolium*, low *Carex* spp., Poaceae, and, in disturbed sites, recruiting conifer saplings. Other characteristic habitats include the margins of small seeps or streams, moist duff beneath mature *Thuja plicata*, and in saturated, mossy soils around fens or wetlands.

Key to *Botrychium* species:

1. Trophophore apparently absent, transformed into a second sporophore, these more or less equal in size and shape; sporophore branches usually very short; plants usually glaucous . . . . ***B. paradoxum***

1' Trophophore present, with at least some lamina; sporangia occasionally found on lamina margin; plants glaucous or not.

2. Middle and upper pinnae ovate or lanceolate in outline, broadest at or below the middle and narrowing to the tip, pinnately lobed (entire in very small plants) with pinnate venation; lowest pinnae always pinnately lobed.

3. Common stalk usually subterranean, the trophophore and sporophore usually joined at or below ground level (sometimes to 2 cm above ground); trophophore apex broadly rounded; trophophore sessile to short-stalked; plants strongly glaucous, more or less leathery; boulder fields, frost pockets, and loose soils derived from pumice, where subsurface moisture is retained . . . . ***B. pumicola*** (in part)

3' Common stalk exposed, the trophophore and sporophore joined well above ground level; apex of trophophore usually triangular; other characters various; plants not restricted to pumice soils.

4. Trophophore broadly triangular or pentagonal in outline, more or less as broad as long; pinnae lobes usually acute and not overlapping; sporophore ternately branched, stalk 50% or less length of trophophore; plants not glaucous . . . . ***B. lanceolatum*** ssp. ***lanceolatum***

4' Trophophore ovate to narrowly triangular in outline, longer than broad; pinnae lobes rounded to acute; sporophore stalk usually more than 50% length of trophophore; plants glaucous or not.

5. Plants lustrous grass-green in life, not glaucous; all but the uppermost pinnae with equally dissected upper and lower margins; trophophore usually sessile; sporophore usually pinnately branched; common stalk, if not entirely green, uniformly maroon toward the base . . . . ***B. pinnatum***

5' Plants pale green to dull gray-green in life, weakly to strongly glaucous; pinnae above the basal pair asymetrically dissected (lobes more prominent on lower side) or nearly entire; trophophore sessile or stalked; sporophore often ternately divided; common stalk, if not entirely green, with a prominent reddish stripe extending downward from base of trophophore.

6. Trophophore stalk equal to or shorter than the distance between the first and second pairs of pinnae; sporangia rarely present on margins of lower pinnae; pinnae tips and lobes usually broadly rounded, more or less overlapping; basal pinnae asymmetrically enlarged on lower side. . . . . ***B. hesperium***

6' Trophophore stalk equal to or exceeding the distance between the first and second pairs of pinnae; sporangia usually present on margins of lower pinnae (often absent in small or shade-grown plants); pinnae tips and lobes generally narrower than the previous, scarcely overlapping . . . . ***B. pedunculosum***

2' Middle and upper pinnae broadly to narrowly fan-shaped, wedge-shaped, or linear in outline (or trophophore simple), broadest at the outer margin or parallel-sided, entire or palmately lobed with palmate or parallel venation; lowest pinnae usually similar to the middle and upper, or sometimes elongate and pinnately lobed.

7. Trophophore and sporophore joined at or below ground level (occasionally 1–2 cm above ground level in *B. pumicola* or in small plants of *B. simplex*); junction of pinnae outer and side margins usually broadly rounded; lowest pinnae on large plants often elongate and pinnately lobed, replicating the central axis; trophophore apex broadly rounded.

8. Sporophore stalk at spore release shorter than trophophore, stout with numerous branches; trophophore sessile to short-stalked (stalk to 10 mm); common stalk well developed but usually hidden below-ground; plants strongly glaucous in life, more or less leathery; middle and upper pinnae usually lobed; boulder fields and loose soils derived from pumice, where subsurface moisture is retained . . . . ***B. pumicola*** (in part)

8' Sporophore stalk at spore release much longer than trophophore, slender and lax, with branches in small flag-like cluster at tip; trophophore stalked (stalk to 30 mm); common stalk usually very short, scarcely extending past the leaf sheath; plants grass-green to glaucous, herbaceous; middle and upper pinnae sometimes lobed; moist meadows, fens, stream banks, and lake shores . . . . ***B. simplex***

7' Trophophore and sporophore joined well above ground level on an exposed common stalk; junction of pinnae outer and side margins often angular; lowest pinnae rarely elongate or pinnately lobed; trophophore apex rounded to triangular.

9. Basal pinnae broadly fan-shaped, spanning an arc of (120)150–180° as measured from the point of attachment on the rachis to the outer edge of the pinnae sides.

10. Pinnae texture paper-thin and delicate; margins finely toothed, crenulate, or sometimes lacerate; pinnae in 2–5 pairs, usually spreading at ca. right angle to rachis, well spaced; trophophore sessile or with stalk to 17 mm long; basal sporophore branches often twisted downwards; plants usually in or near saturated substrate. . . . . ***B. crenulatum*** (in part)

10' Pinnae texture fingernail-thick and firm; margins entire to undulate, rarely toothed or cleft; pinnae in 3–7 pairs, the middle and upper pinnae ascending, usually overlapping; trophophore usually sessile; basal sporophore branches spreading to ascending; plants not growing in saturated substrate .... ***B. lunaria***

9' Basal pinnae parallel-sided, wedge-shaped, or fan-shaped, spanning an arc of less than 120°, or the trophophore simple.

11. Sporophore stalk at spore release ca. 50% or less the length of the trophophore (occasionally longer in *B. ascendens*), generally shorter than the spore-bearing portion; trophophore sessile or, if stalked, the stalk usually shorter than the distance between the first and second pairs of pinnae; pinnae mostly ascending, linear to narrowly fan-shaped.

12. Span of unlobed pinnae usually greater than 45°; pinnae and pinnae lobes mostly narrowly fan-shaped; outer margins of pinnae and pinnae lobes often coarsely toothed; basal pinnae often bearing a few marginal sporangia; trophophore sessile to short-stalked; sporophore stalk approximately 50% the length of the trophophore, occasionally longer .... ***B. ascendens***

12' Span of unlobed pinnae less than 45°; pinnae lobes linear to narrowly wedge-shaped; outer margins of pinnae and pinnae lobes not toothed; basal pinnae usually lacking marginal sporangia; trophophore strictly sessile; sporophore stalk 33% or less the length of the trophophore .... ***B. lineare***

11' Sporophore stalk at spore release 0.5–2 times the length of the trophophore, generally equal to or longer than the spore-bearing portion; trophophore sessile or more often with a stalk equal to or longer than the distance between the first and second pairs of pinnae; pinnae spreading or ascending.

13. Middle and upper pinnae rectangular or rhombic with parallel sides, broadly adnate to the rachis, outer margin truncate, often toothed; pinnae often irregularly spaced with adjacent pinnae often confluent (or trophophore simple); trophophore stalk 25–50% or more of the total trophophore length .... ***B. montanum***

13' Pinnae wedge-shaped or fan-shaped, the sides rarely parallel, outer margin rounded, lobed, toothed, or entire; pinnae more or less evenly spaced, not confluent; trophophore sessile to stalked, the stalk rarely greater than 25% of the total trophophore length.

14. Pinnae texture paper-thin and delicate; margins finely toothed, crenulate, or sometimes lacerate; pinnae in 2–5 pairs, usually spreading at ca. right angle to rachis; basal sporophore branches usually spreading or twisted downwards; plants usually growing in or close to saturated substrate. ***B. crenulatum*** (in part)

14' Pinnae texture fingernail-thick and firm; margins entire, undulate, or with 3–5 rounded lobes; pinnae in 3–7 pairs, mostly ascending; sporophore branches usually ascending; plants usually not growing in saturated substrate .... ***B. minganense***

Botrychium ascendens W.H. Wagner

Upswept moonwort

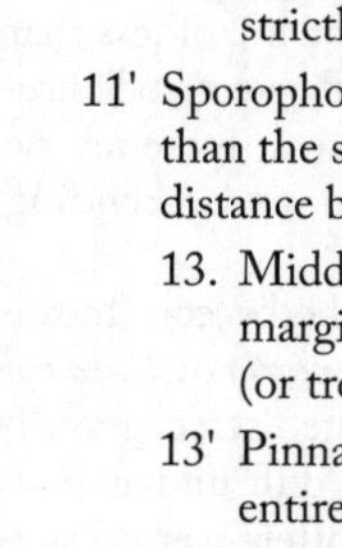

Plants 5–12(20) cm tall above ground. **Common stalks** well developed, greater than 2 cm, green. **Trophophores** sessile or short-stalked, the stalk occasionally to 2 cm long; blade 1-pinnate, narrowly oblong-triangular, to 6 × 2 cm, yellow-green and shiny in life; pinnae to 6 pairs, strongly ascending, narrowly fan-shaped, spanning an arc of less than 90°, broadest at the outer margin, usually coarsely toothed and with 2–5 lobes; lower pinnae often bearing marginal sporangia. **Sporophores** stiffly erect, stalk 0.5–1 times as long as the trophophore at spore release, rarely longer, branches mostly simple, strongly ascending. **Spores** (40)44–47(54) µm. **2*n***=180 (allotetraploid, apparently derived from *B. crenulatum* and *B. lineare*).

Open, well-drained areas and historically disturbed sites with a sparse to dense cover of grasses, forbs, and small shrubs or conifer saplings, such as montane or subalpine meadows, avalanche meadows, road shoulders, ski areas, and mining disturbances. 1500–2000 m. BW. CA, ID, NV, WA; scattered in western and northeastern North America. Native.

Botrychium ascendens is easily confused with *B. crenulatum*, *B. lineare*, and *B. minganense*.

Botrychium crenulatum W.H. Wagner

Crenulate moonwort, dainty moonwort

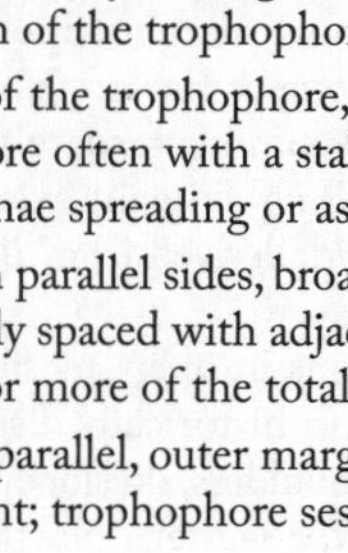

Plants 6–10(16) cm tall above ground. **Common stalks** well developed, greater than 3 cm, green. **Trophophores** stalk (1)5(17) mm long; blade 1-pinnate, ovate-oblong to linear-oblong, to 6.5 × 1.8 cm (averaging 2 cm long), yellow-green, usually with a thin, herbaceous texture; pinnae in (2)3(5) pairs, overlapping or not, spreading at nearly a right angle to rachis, fan-shaped, spanning an arc of 60–150(180)°, broadest at the outer margin, usually finely crenulate and sometimes with 1–5 incisions. **Sporophores** erect, stalk 0.5–2 times as long as the trophophore at spore release, branches well developed, usually spreading or twisted downwards. **Spores** 45(40–48) µm. **2*n***=90.

Moist to saturated soils in or near seeps, fens, or on the margins of small streams, usually among low, dense herbaceous cover, from mid-montane to subalpine elevations. 1200–2100 m. BR, BW. CA, ID, NV, WA; western North America. Native.

Botrychium crenulatum grows in more saturated substrates than most other moonworts. It is frequently confused with *B. minganense*, from which it can usually be distinguished by the two to five pairs of fan-shaped, spreading pinnae with finely crenulate margins, thinner texture, and spreading or downward-pointing sporophore branches. However, many plants do not exhibit all of these characters.

Botrychium hesperium (Maxon & Clausen) W.H. Wagner & Lellinger

Western moonwort

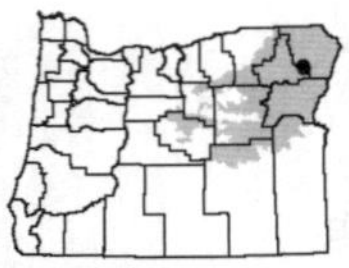

Plants 5–20 cm tall above ground. **Common stalks** well developed, greater than 3 cm, with a reddish brown stripe below the trophophore. **Trophophores** short-stalked (stalk usually shorter than distance between first and second pairs of pinnae); blade pinnate-pinnatifid, subdeltate, to 5 cm long, dull gray-green and usually appearing glaucous; pinnae ascending, usually overlapping, ovate to lanceolate, broadly attached to the rachis, broadest at or below the middle, asymmetrically lobed with basiscopic lobes enlarged; pinnae lobes with rounded apices, the margins slightly rolled; lowest pinnae often exaggerated, up to twice as long as the adjacent pinnae; sporangia rarely present on pinnae margins. **Sporophores** erect with stalk up to 2 times as long as the trophophore at spore release, in larger plants becoming ternately branched with lateral branches nearly as long as main axis. **Spores** 37(29–50) µm. ***2n***=180 (allotetraploid, derived from *B. lanceolatum* and an unknown diploid).

At upper montane to subalpine elevations in open, mesic meadows with low, herbaceous cover. Also in historically disturbed areas with gravelly soil, such as road shoulders, ski slopes, or mine tailings. 1400–1700 m. BW. ID, WA; western North America. Native.

Unpublished forms with deeply lobed pinnae (*B. hesperium* var. '*fenestratum*' and *B.* '*venulosum*') do not warrant taxonomic recognition.

Botrychium lanceolatum (S.G. Gmelin) Angström ssp. ***lanceolatum*** (p. 92)

Triangle moonwort

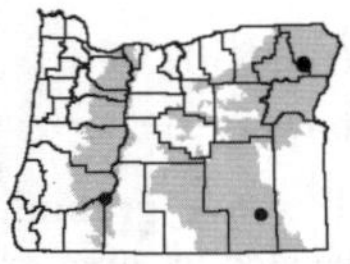

Plants 5–20 cm tall above ground. **Common stalks** well developed, greater than 4 cm, green or sometimes with a reddish brown stripe below trophophore. **Trophophores** sessile or nearly so (stalk less than 1 mm); blade 1–2-pinnate, broadly deltate, with the basal pinnae nearly as long as entire trophophore, to 9 cm wide and long, yellow-green to dull or shiny green; pinnae to 5 pairs, narrowly to broadly lanceolate, broadest at or below the middle, with acute to rounded tips. **Sporophores** erect, stalk 0.5–2 cm, shorter than length of trophophore at spore release, ternately branched with lateral branches nearly as long as main axis and again branched. **Spores** 35(29–42) µm. ***2n***=90.

Mesic meadows, gravelly road shoulders, forest openings, under shrubs, and in moist duff beneath conifers, at upper montane to subalpine elevations. 1400–2800 m. BR, BW, Casc. ID, WA; throughout northern North America; Eurasia. Native.

2 subspecies; 1 subspecies in Oregon.

Two morphologically distinct genotypes can be recognized within ssp. *lanceolatum*. The green genotype is characterized by an entirely green common stalk, a glossy, deep green color in life, and relatively broad, closely spaced pinnae lobes with the basal pinnae straight and symmetrically lobed. The red genotype shows a prominent reddish brown stripe extending up the common stalk and into the base of the trophophore, a dull yellow-green color in life, and relatively narrow, widely spaced pinnae lobes with the basal pinnae curved upwards and asymmetrically lobed. Only the green genotype has been confirmed in Oregon, but the red genotype should be expected.

Botrychium lineare W.H. Wagner

Slender moonwort

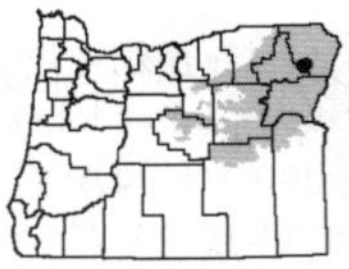

Plants 6–18 cm tall above ground. **Common stalks** well developed, usually greater than 2 cm, green. **Trophophores** sessile or rarely with a stalk to 1 cm long; blade 1-pinnate, oblong, to 4 × 1.3 cm, pale green in life, not glaucous; pinnae to 5 pairs, usually ascending, linear with nearly parallel sides, sometimes narrowly fan-shaped or bifid with linear lobes, unlobed pinnae spanning an arc of less than 45°, outer margin truncate. **Sporophores** stiffly erect, stalk usually 33% or less the length of the trophophore at spore release, branches usually numerous, closely spaced, and ascending. **Spores** 36(35–39) µm. ***2n***=90.

In Oregon, known from mesic, grassy or herbaceous meadows and avalanche meadows among *Pinus contorta* or *Picea engelmannii*; elsewhere also in sparsely vegetated scree, gravelly road shoulders, fen-like seeps, and shaded duff under *Thuja plicata*, at montane to subalpine elevations, often over calcareous substrates. 1500–1700 m. BW. CA, ID, NV, WA; north to AK, east rarely to Quebec, southeast to NM. Native.

Botrychium lineare is among the rarest of our moonworts. In Oregon, it is known only from the Wallowa Mountains.

Botrychium lunaria (Linnaeus) Swartz (p. 92)

Common moonwort

Botrychium lunaria (L.) Sw. var. *lunaria*

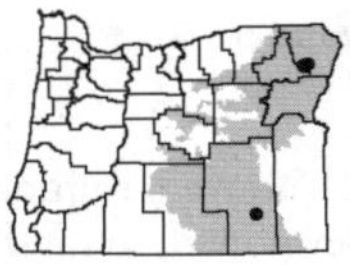

Plants 4–20 cm tall above ground. **Common stalks** well developed, greater than 2 cm, green. **Trophophores** sessile or occasionally with a stalk to 1 cm long; blade 1-pinnate, oblong, to 10 × 4 cm, grass-green, not glaucous; pinnae to 9 pairs, usually overlapping except in shaded plants, lower pinnae spreading at nearly a right angle to rachis, fan-shaped, spanning an arc of (120)150–180°, broadest at the outer margin, entire to undulate, upper pinnae fan-shaped, usually with upper margin ascending and lower margin spreading to form an arc of 90–150°. **Sporophores** erect, stalk (0.5)1–2 times as long as the trophophore at spore release, branches spreading to ascending, usually arranged on 1 side of stalk to form a triangular flag. **Spores** 33–39 µm. ***2n***=90.

Mesic meadows at upper montane to subalpine elevations, occasionally on sparsely vegetated scree. 1500–2800 m. BR, BW. CA, ID, NV, WA; throughout most of Canada and the northern and western US; Australia, Eurasia, New Zealand. Native.

North American plants of *B. lunaria* are genetically distinct from Eurasian plants. The tetraploid between these two lineages, *B. yaaxudakeit* Stensvold & Farrar, is known from northeastern Oregon. It has larger spores (43–48 µm) but is otherwise morphologically indistinguishable from *B. lunaria*.

Botrychium minganense Victorin (p. 92)

Mingan moonwort

Botrychium lunaria (L.) Sw. var. *minganense* (Vict.) Dole

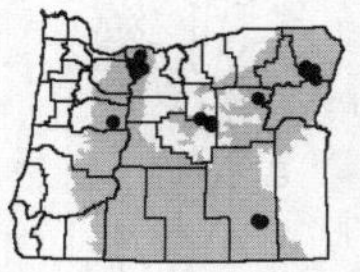

Plants 3–20 cm tall above ground. **Common stalks** well developed, greater than 2 cm except in very small plants, green. **Trophophores** with stalk to 2 cm long, occasionally sessile; blade 1-pinnate, oblong to linear, to 10 × 2.5 cm, grass-green to dull green, sometimes slightly glaucous; pinnae to 10 pairs, slightly overlapping to widely spaced, spreading to ascending, wedge-shaped to narrowly fan-shaped or rarely with nearly parallel sides, spanning an arc of 45–120°, broadest at the outer margin, entire, toothed, or with 3–5 rounded lobes, the lower pinnae occasionally elongate and pinnately lobed. **Sporophores** erect, sometimes lax, stalk (0.5)1–2 times as long as the trophophore at spore release, branches spreading to ascending, sometimes much reduced. **Spores** 45(38–50) µm. **2*n*=**180 (allotetraploid, derived from *B. lunaria* and an undescribed diploid that resembles *B. montanum*).

Mesic meadows, gravelly road shoulders, ski slopes, avalanche meadows, margins of seeps and wetlands, and in moist duff under *Thuja plicata*, from mid-montane to subalpine elevations. 800–2800 m. BR, BW, Casc, ECas. CA, ID, NV, WA; western and northern North America; Iceland. Native.

Botrychium minganense is among the most widespread, abundant, and highly variable moonworts, and can be easily confused with *B. ascendens*, *B. crenulatum*, *B. lunaria*, and sometimes *B. simplex*.

Botrychium montanum W.H. Wagner

Western goblin, mountain moonwort

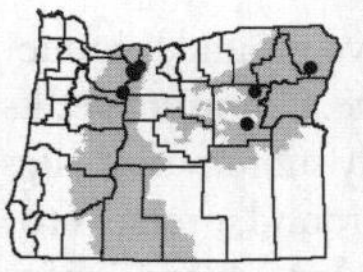

Plants 2–12 cm tall above ground. **Common stalks** well developed, greater than 2 cm except in very small plants, green. **Trophophores** with stalk 0.3–2 cm long, usually comprising 25–50% or more the total trophophore length; blade 1-pinnate, oblong to linear, to 4 × 1 cm, dull green, glaucous, often appearing fleshy; pinnae to 3(5) pairs, often irregularly spaced with adjacent pinnae often confluent, sometimes completely fused to form a spoon-shaped blade, spreading to ascending, wedge-shaped with nearly parallel sides and a truncate or slightly rounded, often lacerate outer margin. **Sporophores** stiffly erect, stalk (0.5)1–3 times as long as the trophophore at spore release, branches much reduced, often appearing absent. **Spores** 54(47–57) µm. **2*n*=**90.

Moist duff around seeps or streams under mature *Thuja plicata* or, rarely, *Picea engelmannii*, at mid-montane elevations. Occasional in fen-like seeps or in moist to wet meadows with moss or lush herbaceous cover at subalpine elevations. 900–2500 m. BW, Casc, ECas. CA, ID, WA; north to AK, northeast to MT. Native.

Botrychium montanum is our smallest moonwort.

Botrychium paradoxum W.H. Wagner (p. 92)

Paradox moonwort

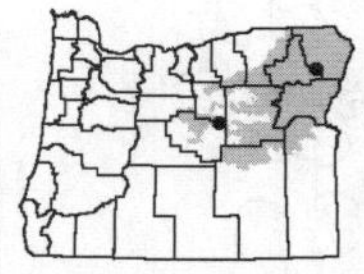

Plants 7–15 cm tall above ground, dull green, glaucous in life. **Common stalks** well developed, greater than 2 cm, green. **Trophophores** appearing absent, converted entirely to a second sporophore without any trace of lamina. **Sporophores** stiffly erect, slightly unequal in length, nearly sessile or with stalks up to 50% the entire sporophore length at spore release, branches much reduced and usually less than 5 mm long, often appearing absent. **Spores** 36–43 µm. **2*n*=**180 (allotetraploid, derived, in part, from an undescribed diploid that resembles *B. montanum*).

Mesic subalpine meadows, usually among grasses, sedges, and low forbs, but occasionally in denser herbaceous cover or under trees. 1500–1800 m. BW. CA, ID, WA; northwestern North America. Native.

Botrychium paradoxum is unique in having two sporophores and no trophophore. However, other species occasionally produce forms with the trophophore partly or completely transformed into a second sporophore. These usually have sporophores of clearly unequal lengths or with elongate branches, and often occur as isolated individuals among normal, trophophore-bearing plants. As currently circumscribed, *B. paradoxum* contains two distinct genotypes.

Botrychium pedunculosum W.H. Wagner (p. 92)

Stalked moonwort

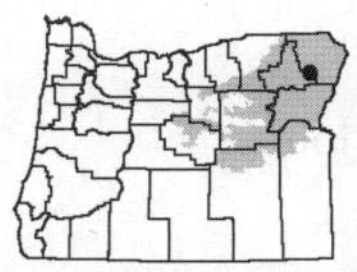

Plants 5–13(20) cm tall above ground. **Common stalks** well developed, greater than 3 cm, with a prominent reddish brown stripe below trophophore. **Trophophores** stalk 0.8–2.5 cm long, the stalk usually longer than distance between first and second pairs of pinnae; blade pinnate-pinnatifid, ovate-oblong to deltate-oblong, to 4.5 × 2.5 cm, dull gray-green and glaucous; pinnae ascending, approximate, the basal pinnae asymmetrically lobed, angular, broadest at or below the middle, upper pinnae becoming spatulate; pinnae lobes and apices usually acute, the margins slightly rolled; sporangia usually present on margins of basal pinnae. **Sporophores** erect with stalk 0.5–1.5 times as long as the trophophore at spore release, in larger plants becoming ternately branched with lateral branches nearly as long as main axis. **Spores** 36–43 µm. **2*n*=**180 (allotetraploid, derived from *B. lanceolatum* and an unknown diploid).

Mesic meadows with low, herbaceous or grassy cover, gravelly soil, brushy areas, and in duff under mature *Thuja plicata*, at upper montane to subalpine elevations. 1200–1800 m. BW. CA, ID, WA; north to AK, northeast to MT, disjunct in Quebec. Native.

Shade-grown plants often lack marginal sporangia on the trophophore. Conversely, occasional individuals of other pinnate-pinnatifid species may possess marginal sporangia.

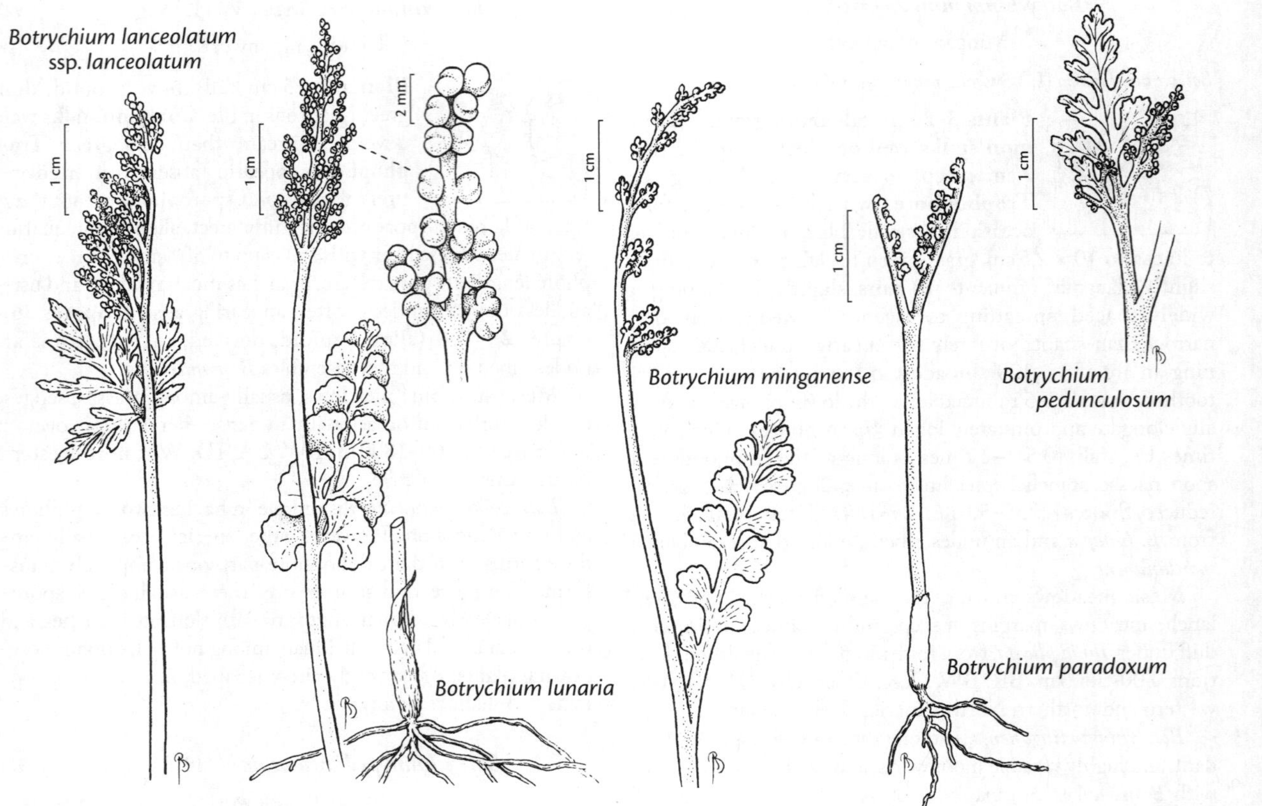

BOTRYCHIUM

Botrychium pinnatum H. St. John

Northwestern moonwort

Botrychium boreale Milde var. *obtusilobum* (Rupr.) M. Broun

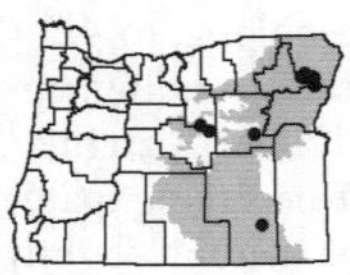

Plants 3–15 cm tall above ground. **Common stalks** well developed, greater than 2 cm except in very small plants, sometimes with a reddish brown stripe below trophophore. **Trophophores** sessile or rarely with stalk to 2 mm long; blade pinnate-pinnatifid, oblong to subdeltate, to 8 × 5 cm, grass-green and lustrous, not glaucous; pinnae spreading or slightly ascending, approximate to overlapping, ovate to lanceolate-oblong, broadest at or below the middle, symmetrically lobed; pinnae lobes with rounded to broadly acute apices; sporangia rarely present on pinnae margins. **Sporophores** erect, usually pinnately branched, stalk 0.5–1 times as long as the trophophore at spore release. **Spores** 36–45 µm. ***2n***=180 (allotetraploid, derived from *B. lanceolatum* and *B. lunaria)*.

Moist meadows, forest openings, along small streams under shrubs or conifers, gravel road shoulders, and old clearcuts, at upper montane to subalpine elevations. 1400–3000 m. BR, BW. CA, ID, NV, WA; north to AK, southeast to AZ and NM. Native.

All reports of *Botrychium boreale* from North America outside of Greenland are *B. pinnatum.*

Botrychium pumicola Coville

Pumice moonwort

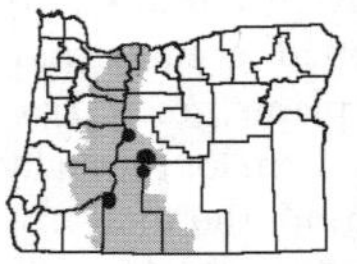

Plants 2–10 cm tall above ground, producing underground gemmae. **Common stalks** well developed but mostly subterranean, less than 2 cm high above ground, green. **Trophophores** stalk 0–1 cm long; blade 1-pinnate to pinnate-pinnatifid, oblong to deltate with rounded apex, to 6 × 4 cm, whitish green, strongly glaucous and somewhat leathery; pinnae to 6 pairs, overlapping, ascending, broadest at or above the middle (rarely below), often palmately lobed, broadly attached to rachis with 2–4 veins entering rachis; lower pinnae often elongate and pinnately lobed, replicating the central axis; pinnae lobes rounded, margins entire or rarely crenate. **Sporophores** erect, pinnately branched with crowded branches, stalk shorter than trophophore. **Spores** averaging 34 µm, released in tetrads. ***2n***=90.

Sparsely vegetated boulder fields and soils derived from pumice on ridge crests or in depressions with more or less permanent subsurface moisture, generally where snow melts early, at mid-montane to subalpine elevations. 1200–2800 m. Casc, ECas. CA. Native.

Botrychium pumicola is the only species to release its spores in tetrads. All other species release spores singly. Small plants do not develop the characteristic elongate basal pinnae; these can usually be recognized by the hidden common stalk, glaucous, leathery texture, broadly rounded trophophore apex, strongly overlapping pinnae with the lower usually lobed, and distinctive habitat.

Botrychium simplex E. Hitchcock

Least moonwort

Plants 4–15 cm tall above ground. **Common stalks** very short, less than 2 cm long, scarcely extending past the leaf sheath and usually not rising above ground level, green. **Trophophores** stalk to 3 cm long; blade 1-pinnate to pinnate-pinnatifid, oblong to deltate, to 6 × 6 cm, dull to bright green, sometimes glaucous, fleshy to herbaceous; pinnae to 7 pairs, spreading to ascending, overlapping or widely spaced, cuneate to fan-shaped, broadest above the middle, broadly to narrowly attached to rachis (with 1–2 veins entering rachis), in small plants confluent to form a simple, spoon-shaped blade; lower pinnae in larger plants usually elongate and pinnately lobed, replicating the central axis; pinnae lobes broadly rounded, margins entire to shallowly undulate. **Sporophores** lax, often recurved, pinnately branched, stalk 1–4 times as long as trophophore at spore release. **Spores** averaging 46 µm. **2*n***=90.

2 varieties.

Smaller plants of the two varieties tend to look the same and are difficult to identify except in association with larger, better developed plants. Very small plants may have an elongate common stalk and may resemble *B. minganense* or *B. montanum*, both of which generally differ in their smaller, narrower, more angular pinnae, thicker texture, and stouter sporophore.

Key to *Botrychium simplex* varieties:

1. Middle and upper pinnae narrowly attached to rachis, scarcely decurrent; terminal pinnae (and terminal pinnules of basal pinnae) not or only partly fused; plants grass-green, sometimes glaucous . . . . ***B. s.*** var. ***compositum***

1' Middle and upper pinnae broadly attached to rachis, strongly decurrent; terminal pinnae (and terminal pinnules of basal pinnae) often partly or wholly fused; plants usually glaucous . . . . ***B. s.*** var. ***simplex***

Botrychium simplex E. Hitchcock var. *compositum* (Lasch) Milde

Yosemite moonwort

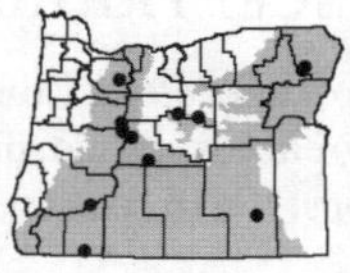

Trophophores green or sometimes glaucous; pinnae above the basal pair narrowly attached to rachis, slightly or not at all decurrent; basal pinnae often secondarily divided (except in small plants); terminal pinnae not or only partly fused.

Wet, mountain meadows and stream side meadows, where at least seasonally wet but not inundated. 1300–2200 m. BR, BW, Casc, ECas, Sisk. CA, ID, WA; north to British Columbia and Alberta, east to CO. Native.

Botrychium simplex E. Hitchcock var. *simplex*

Least moonwort

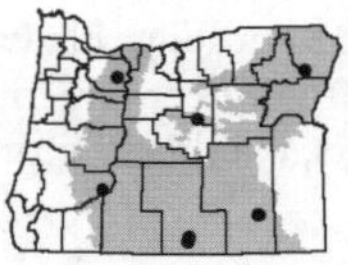

Trophophores usually glaucous; pinnae above the basal pair broadly attached to rachis, strongly decurrent; basal pinnae mostly undivided, spoon-shaped; terminal pinnae often partly or wholly fused.

Wet meadows, fens, hardwater seeps, and stream sides, where at least seasonally wet but not inundated. 800–2700 m. BR, BW, Casc, ECas. CA, ID, NV, WA; western and northeastern North America; Europe, Japan. Native.

BOTRYPUS Rattlesnake-fern

Plants sparsely pubescent when young. **Leaves** 1 per stem. **Common stalks** erect, the trophophore and sporophore diverging well above ground level. **Trophophores** sessile, pinnately compound, thin and herbaceous; venation free and fan-like; midribs prominent. **Sporophores** stalked, twice-pinnately branched, often absent. **Sporangia** exposed.

Cosmopolitan. Approximately 7 species; 1 species treated in *Flora.*

Botrypus virginianus (Linnaeus) Michaux (p. 94)

Rattlesnake-fern

Botrychium virginianum (L.) Sw.

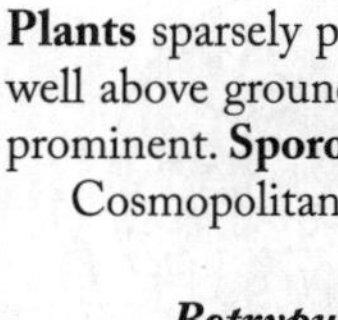

Plants to 40 cm tall. **Trophophores** 3–4-pinnate, to 30 × 30 cm; grass-green, neither glaucous nor glossy; pinnae to 12 pairs, spreading to slightly ascending, flat, slightly overlapping or well spaced, lobes lanceolate, tips acute, margins entire to serrate. **Sporophores** stalk (0.5)1–2 times as long as trophophore at spore release. **2*n***=184.

Moist soil and duff in coniferous or deciduous forest or under shrubs, often near streams or on margins of wetlands and fens, rarely on meadow margins. 100–1700 m. BW, Casc, ECas. CA, ID, NV, WA; throughout North America, South America; Eurasia. Native.

Very small plants in exposed habitats could be confused with *Botrychium lanceolatum*, from which it can be distinguished by the thinner, herbaceous texture, prominent midribs, sharply acute pinnae lobes, and longer sporophore stalk.

OPHIOGLOSSUM Adder's tongue

Plants glabrous. **Stems** solitary; gemmae absent. **Leaves** mostly solitary. **Common stalks** subterranean or exposed. **Trophophores** simple, sessile or stalked, lanceolate to ovate; margins entire; venation reticulate. **Sporophores** simple. **Sporangia** in 2 rows, deeply sunken into rachis.

Cosmopolitan. 25–30 species; 1 species treated in *Flora.*

Ophioglossum pusillum Rafinesque (p. 94)
Northern adder's tongue

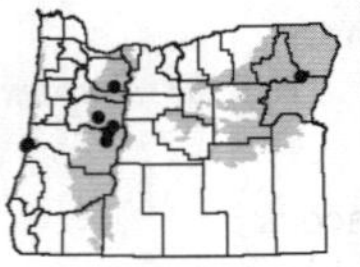

Plants to 30 cm tall. **Common stalks** usually exposed above ground level. **Trophophores** stalked, blade oblanceolate to obovate or ovate, widest at the middle, tapering to the base, rounded at tip, to 10 × 3.5 cm, pale green, soft; venation forming free veinlets within the larger reticulations. **Sporophores** stalk 1–3 times as long as trophophore at spore release. **2*n***=960.

Moist meadows, fens, brushy marsh edges, and vegetated depressions among dunes. 0–1500 m. BW, Casc, Est. CA, ID, WA; northern North America. Native.

Ophioglossum pusillum is inconspicuous and easily overlooked among grasses and herbaceous vegetation.

SCEPTRIDIUM Grape-fern

Plants pubescent. **Stems** solitary. **Leaves** 1 or 2 per stem. **Common stalks** short, mostly subterranean. **Trophophores** pinnately compound; venation free and fan-like. **Sporophores** stalked, pinnately branched, often absent. **Sporangia** exposed.

Cosmopolitan, mostly in temperate regions. 10–13 species; 1 species treated in *Flora*.

Sceptridium multifidum (S.G. Gmelin) M. Nashida ex Tagawa (p. 94)
Leather grape-fern

Botrychium multifidum (S.G Gmel.) Trevis, *Botrychium multifidum* (S.G. Gmel.) Trevis. ssp. *silaifolium* (C. Presl) R.T. Clausen, *Botrychium silaifolium* C. Presl

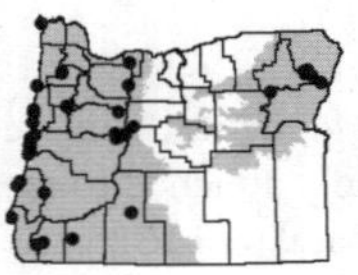

Plants to 40 cm tall. **Common stalks** short, usually not arising above ground level. **Trophophores** stalk to 15 cm long; blade 2–3-pinnate, to 30 × 30 cm, deep green, glossy; pinnae to 10 pairs, overlapping or well spaced; lobes ovate, flat or curled, tips rounded, margins entire to shallowly crenate. **Sporophores** pinnately branched, stalk 1–2 times as long as trophophore at spore release. **2*n***=90.

Moist to wet meadows, wetland margins, lake shores, vegetated depressions among dunes, forest edges, moist ditches. 0–2300 m. BW, Casc, CR, ECas, Est, Sisk, WV. CA, ID, NV, WA; throughout most of western and northern US and Canada; western Asia, Europe. Native.

Sceptridium multifidum is the most widespread and common species of Ophioglossaceae in Oregon.

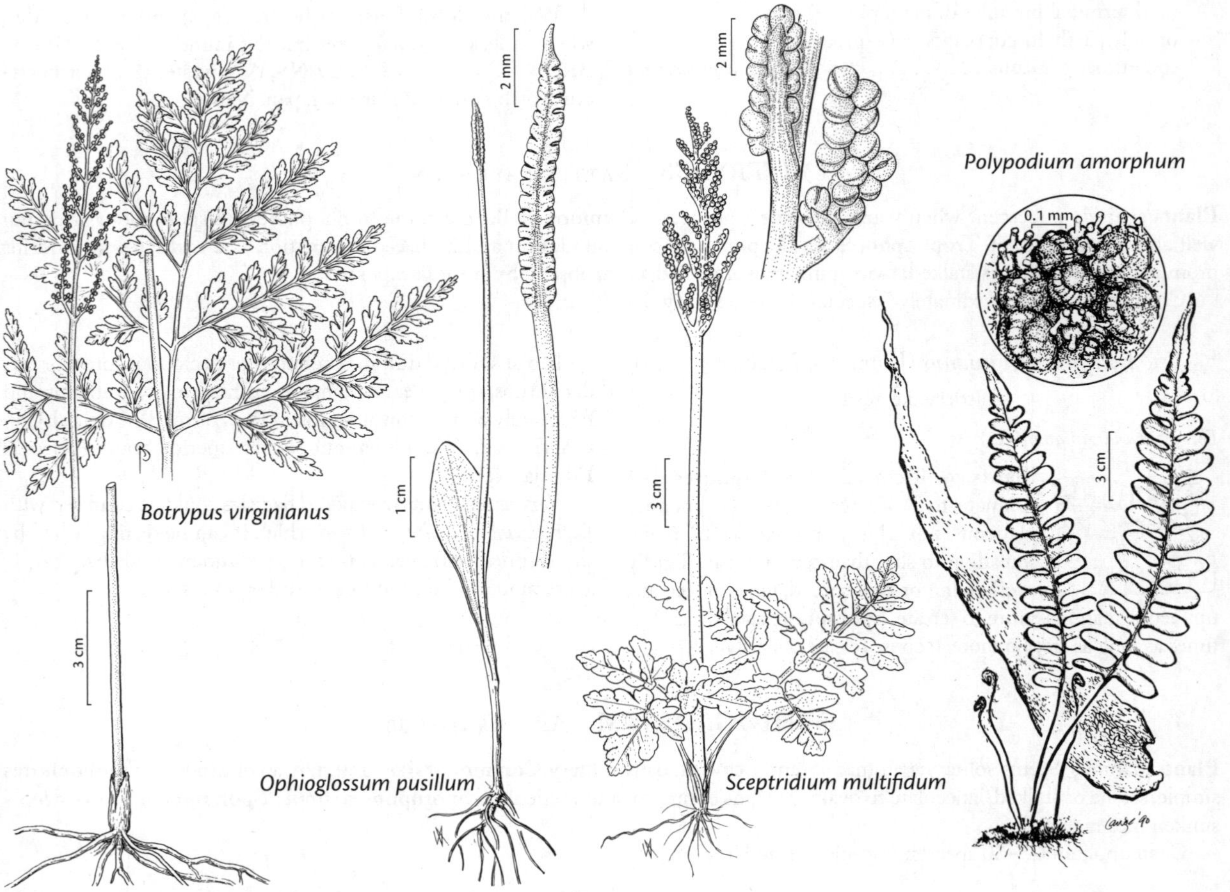

BOTRYPUS • OPHIOGLOSSUM • SCEPTRIDIUM • POLYPODIUM

POLYPODIACEAE Polypody family

Plants herbs, perennial. **Stems** creeping over soil or rock or epiphytic, bearing scales several cells wide. **Leaves** monomorphic or dimorphic; petiole usually articulated at the base and lacking scales. **Blades** mostly simple to pinnatifid or pinnate; indumentum of scales and occasionally hairs usually sparse. **Sori** produced on veins on the abaxial surfaces of the leaves, round to elongate, rarely marginal or covering the abaxial surface; indusia absent. **Spores** bilateral, pale to transparent.

Cosmopolitan. Approximately 90 genera; 1 genus treated in *Flora*.

POLYPODIUM Polypody

Plants usually growing on rocks or trees. **Stems** creeping, branched, often white-pruinose, densely scaly; leaf bases not persistent; taste sweet, bitter/acrid, licorice-flavored, or a combination of these. **Leaves** monomorphic, petioles articulated at the base; scales absent; leaf bases not persistent on the rhizome. **Blades** pinnatifid to pinnate; rachis with sparse scales or glabrescent abaxially, hairy to glabrous adaxially. **Pinnae** opposite or offset, entire to toothed, rounded to acuminate; veins free or anastomosing to produce a line of areolae each side of the pinna midrib, veins often ending in hydathodes. **Sori** on abaxial veins, often limited to the distal part of the blades, mostly round; indusia absent; modified sporangia interspersed with the normal sporangia or not. $2n$=74.

Cosmopolitan. Approximately 160 species; 5 species treated in *Flora*.

Key to *Polypodium* species:

1. Leaves thick, dark green, and leathery; sori greater than 3 mm; coastal . . . . ***P. scouleri***

1' Leaves not leathery, paler green; sori less than 3 mm; coastal or inland.

  2. Veins of adaxial leaf surface pubescent with whitish multicellular hairs, often sparse.

    3. Brown scales on abaxial rachis narrow, 2–3 cells wide . . . . ***P. glycyrrhiza***

    3' Brown scales on abaxial rachis wider, usually more than 3 cells wide . . . . ***P. calirhiza***

  2' Hairs absent, adaxial leaf surface glabrous.

    4. Sporangiasters absent; scales on abaxial rachis less than 6 cells wide . . . . ***P. hesperium***

    4' Sporangiasters present; scales on abaxial rachis more than 6 cells wide . . . . ***P. amorphum***

Polypodium amorphum Suksdorf (p. 94)

Pacific polypody

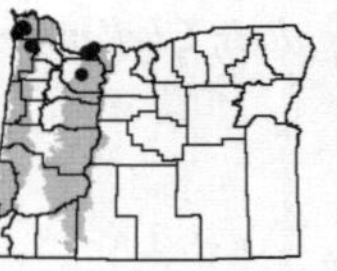

Stems slender, usually white-pruinose, scales dense, lanceolate, toothed, reddish brown, usually with paler margins; taste acrid/bitter. **Leaves** to approximately 30 cm long; petiole well developed, up to 10 cm, slender, approximately 1 mm thick. **Blades** pinnatifid, slightly leathery, glabrous above, 6–12 cm, occasionally longer, 2–4 cm wide, narrowly oblong or slightly wider in the mid-region; hairs absent; scales narrowly ovate, more than 6 cells wide, often sparse, mostly confined to the abaxial rachis. **Pinnae** short, oblong, up to 10 × 2.5 cm, apex rounded or obtuse; margins subentire to entire; veins free. **Sori** to 2.3mm; sporangiasters present, interspersed with the fertile sporangia. $2n$=74.

Moist cliffs and rocky outcrops. 0–1000 m. Casc, CR. WA; north to British Columbia. Native.

Polypodium calirhiza S.A. Whitmore & A.R. Smith

Nested polypody

Polypodium vulgare L. var. *occidentale* Hook.

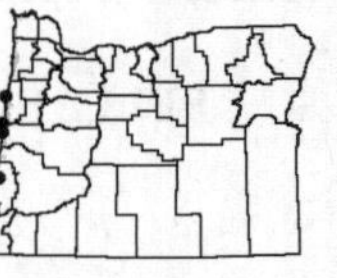

Stems moderately stout, often white-pruinose, with dense brown lanceolate scales; taste slightly bitter and lacking strong licorice flavor. **Leaves** to 30 cm long; petiole slender, to 13 × 0.2 cm, smooth. **Blades** pinnatifid, to 18 × 8 cm, narrowly ovate to oblong, often slightly leathery; adaxial surface pubescent with whitish multicellular hairs (at least along rachis); lanceolate scales usually present on abaxial surface of rachis, usually more than 3 cells wide. **Pinnae** narrow, lanceolate to oblong, to 3 × 0.6 cm; margin serrate; apex usually acute, not long-attenuate; veins anastomosing or not, forming areolae on some pinnae. **Sori** less than 3 mm; sporangiasters absent. $2n$=148.

On cliffs and rocks/gravel near the coast. 0–200 m. Est. CA; south to northern Mexico. Native.

This species can produce sterile hybrids with *P. glycyrrhiza*.

Polypodium glycyrrhiza D.C. Eaton

Licorice fern

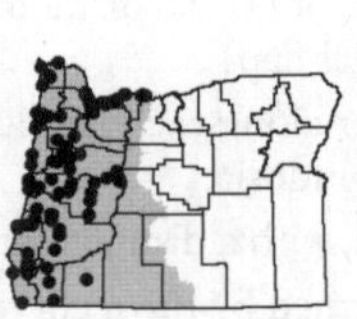

Stems moderately stout, not white-pruinose; scales dense, lanceolate, brown; sweet licorice taste. **Leaves** to 60 cm; petiole slender, to 20 × 0.2 cm. **Blades** to 40 cm, pinnatifid, narrowly ovate to oblong, not leathery, adaxial surface with whitish multicellular hairs (at least along the rachis); scales sparse on the rachis abaxially, brown, narrow, and hair-like, mostly less than 3 cells wide. **Pinnae** narrowly oblong, to 4 × 0.8 cm; margins serrate; apex mostly acute to long-acuminate; veins free. **Sori** round, less than 3 mm; sporangiasters absent. $2n$=74.

Epiphytic and on cliffs and rocks. 0–2300 m. Casc, CR, ECas, Est, Sisk, WV. CA, ID, WA; north to AK; northeastern Asia. Native.

Polypodium hesperium Maxon

Western polypody

Polypodium vulgare L. var. *columbianum* Gilbert

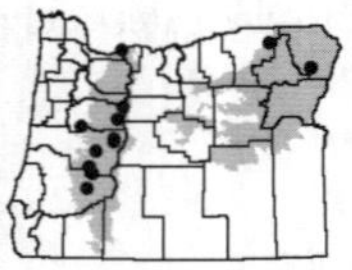

Stems slender, white pruina often present; scales dense, brown, lanceolate, toothed or entire; taste acrid/bitter. **Leaves** to 30 × 5 cm; petiole slender, to 12 × 0.1 cm. **Blades** oblong to narrowly ovate, pinnatifid, slightly leathery, up to 18 × 5 cm, glabrous above; scales on abaxial rachis sparse, brown, less than 6 cells wide. **Pinnae** small, oblong, to 2.5 × 1 cm; margin entire to serrulate; apex rounded to acute; veins free. **Sori** rounded, less than 3 mm, often oval when immature; sporangiasters absent, but aborted purplish sporangia often present. **2*n***=148.

Cliffs, crevices, and talus slopes. 50–1800 m. BW, Casc. CA, ID, WA; north to British Columbia, south to northwestern Mexico. Native.

This is the only *Polypodium* species in Oregon found east of the Cascade range. The lack of sporangiasters distinguishes it from *P. amorphum*, but the frequent sterile sporangia can cause confusion where the ranges of the two species overlap in northwestern Oregon.

Polypodium scouleri Hooker & Greville

Leather-leaved polypody, coast polypody

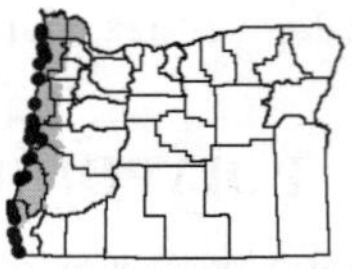

Stems stout, white-pruinose, densely clothed in scales; scales brown, long-toothed, margins often pale; taste bland. **Leaves** to 60 cm; petiole to 25 × 0.2 cm. **Blades** pinnatifid, narrowly ovate, stiff and leathery, dark green, to 35 × 8 cm, with relatively few pairs of robust pinnae, adaxial surface glabrous; scales on abaxial surface scattered, broad, convex, brown, with pale margins. **Pinnae** oblong, up to 9 × 2 cm; margin usually subentire; apex rounded to acute; some veins anastomosing, forming indistinct areolae along the pinnae. **Sori** round, large, over 3 mm, often absent on proximal pinnae; sporangiasters absent. **2*n***=74.

Epiphytic and on rocks. 0–600 m. CR, Est. WA, CA; north to British Columbia, south to northern Mexico. Native.

PTERIDACEAE Brake family

Plants herbaceous, mostly small and perennial. **Stems** stout and erect/ascending or more slender and creeping, often branched; scales and/or hairs present. **Leaves** monomorphic or dimorphic, circinate or non-circinate. **Blades** with a wide range of variation in dividedness and indumentum. **Sporangia** borne abaxially on simple, forked or anastomosing veins, not aggregated into distinct sori, often appearing to cover the leaf undersurface or form (sub)marginal lines; indusia absent; false indusia formed from recurved leaf margins present or not. **Spores** trilete, tetrahedral often with a hemispherical distal section, wall ornamentation various. **Gametophytes** mostly reniform or heart-shaped, flat, green.

Cosmopolitan. At least 50 genera and 1000 species; 6 genera treated in *Flora*.

Pteridaceae is a large family in the Polypodiales, currently with five fairly well-defined subfamilies (Christenhusz and Chase 2014). Oregon's three subfamilies with genera are: Cryptogrammoideae (*Cryptogramma*), Cheilanthoideae (*Aspidotis, Cheilanthes, Pellaea, Pentagramma*), Vittarioideae (*Adiantum,* formerly placed in Adianteae).

Key to Pteridaceae genera:

1. Indusia absent; leaf margins plane or slightly recurved.
   2. Leaves broadly triangular to pentagonal; pinnae deeply divided with lobed segments .... **PENTAGRAMMA**
   2' Leaves linear, pinnate, with entire, rounded segments .... ***Pellaea bridgesii***

1' Indusia marginal or inframarginal, covering some or all of the sporangia.
   3. Indusia along a single lateral margin or on the distal margin of the broad ultimate leaf segments; some sporangia borne on the indusial flap .... **ADIANTUM**
   3' Indusia bilateral on leaf segments, or along each lateral margin of irregularly shaped lobes; sporangia not borne directly on the indusia.
      4. Indusia linear and inframarginal, with a distinct narrow flange between the indusia and the margin ... **ASPIDOTIS**
      4' Indusia linear or not, clearly formed by the recurved leaf margin.
         5. Leaves strongly dimorphic; non-fertile leaves shorter and spreading; fertile leaves longer and more erect .... **CRYPTOGRAMMA**
         5' Leaves monomorphic, most leaves of mature plants fertile.
            6. Ultimate leaf segments small, less than 4 mm long, convex and bead-like adaxially, densely hairy abaxially; scales present or not .... **CHEILANTHES**
            6' Ultimate leaf segments greater than 4 mm, not bead-like, glabrous or sparsely hairy abaxially; scales present .... **PELLAEA**

ADIANTUM Maidenhair fern

Plants perennial, herbaceous, in moist to wet habitats. **Stems** creeping to ascending, with dense narrow reddish or yellowish scales, bicolored or monochromatic. **Leaves** monomorphic, deciduous, usually thin and herbaceous. **Petioles** long, brittle, usually dark-colored, often equal to or longer than blades, glabrescent to sparsely hairy/scaly. **Blades** narrow to wide, 1–4 pinnate, ultimate segments often broad and leaf-like, fan-shaped to strongly asymmetric, false indusia present, formed from the recurved margin, variously linear to crescent-shaped. **Sporangia** borne on and below the false indusium. **Spores** tetrahedral-globose, with a rounded distal section, rough-textured.

Cosmopolitan. 126 species; 2 species treated in *Flora.*

Adiantum capillus-veneris is a widespread native species of the southern U.S. that can also escape from cultivation. It is reported from California, including Siskiyou County, and is likely to occur in Oregon. It resembles *A. jordanii* and can be distinguished by the shape of the leaf segments. In *A. capillus-veneris,* some segments are lobed more than 25% of the way to the cuneate base, while *A. jordanii* is more shallowly lobed, and the base is obtuse or truncate.

Key to *Adiantum* species:

1. Leaf palmate; petiole splitting into 2 subequal branches just below the blade . . . . *A. aleuticum*
1' Leaf pinnate; pinnae produced on a rachis continuous with the unbranched petiole. . . . . *A. jordanii*

Adiantum aleuticum (Ruprecht) C.A. Paris (p. 97)

Northern maidenhair fern, western maidenhair fern

Adiantum pedatum L. var. *aleuticum* Rupr.

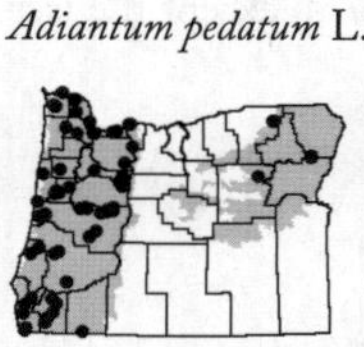

Stems shortly creeping, branched; scales abundant, to about 6 mm long, with thick, dark cell walls and pale yellow-brown lumina, appearing uniformly yellow-brown or chestnut-brown overall; petiole bases persistent. **Leaves** few to numerous, loosely clustered, to about 90 cm, herbaceous and pale/mid green, often reddish when young. **Petioles** long and rather brittle, up to about 60 cm, often longer than the blade, purplish black, scaly and paler at the base, otherwise glabrous or with sparse narrow scales, appearing to branch dichotomously at the base of the blade into (sub)equal branches that resemble the petiole. **Blades** rounded/reniform, borne more or less parallel to the ground, length and width subequal or slightly wider than long, to 35 × 40 cm, composed of a pair of pinnate sections each with an arching rachis; first pinna on each section longest and pointing forwards, up to 35 cm, subsequent pinnae shorter and spreading laterally. **Ultimate segments** oblique, oblong or wedge-shaped up to 3 × 0.6 cm, coarsely divided to about halfway into a maximum of 7 truncate or emarginate segments, purple-black or green; base narrowed to short (less than 1 mm) stalk. **False indusia** flap-like, broadly crescent-shaped, up to 3 mm long, green with a pale margin, formed from the reflexed distal mid-section of each lobe margin, often causing the lobe to appear emarginate. **Sporangia** borne on and under the false indusium. **2*n***=56.

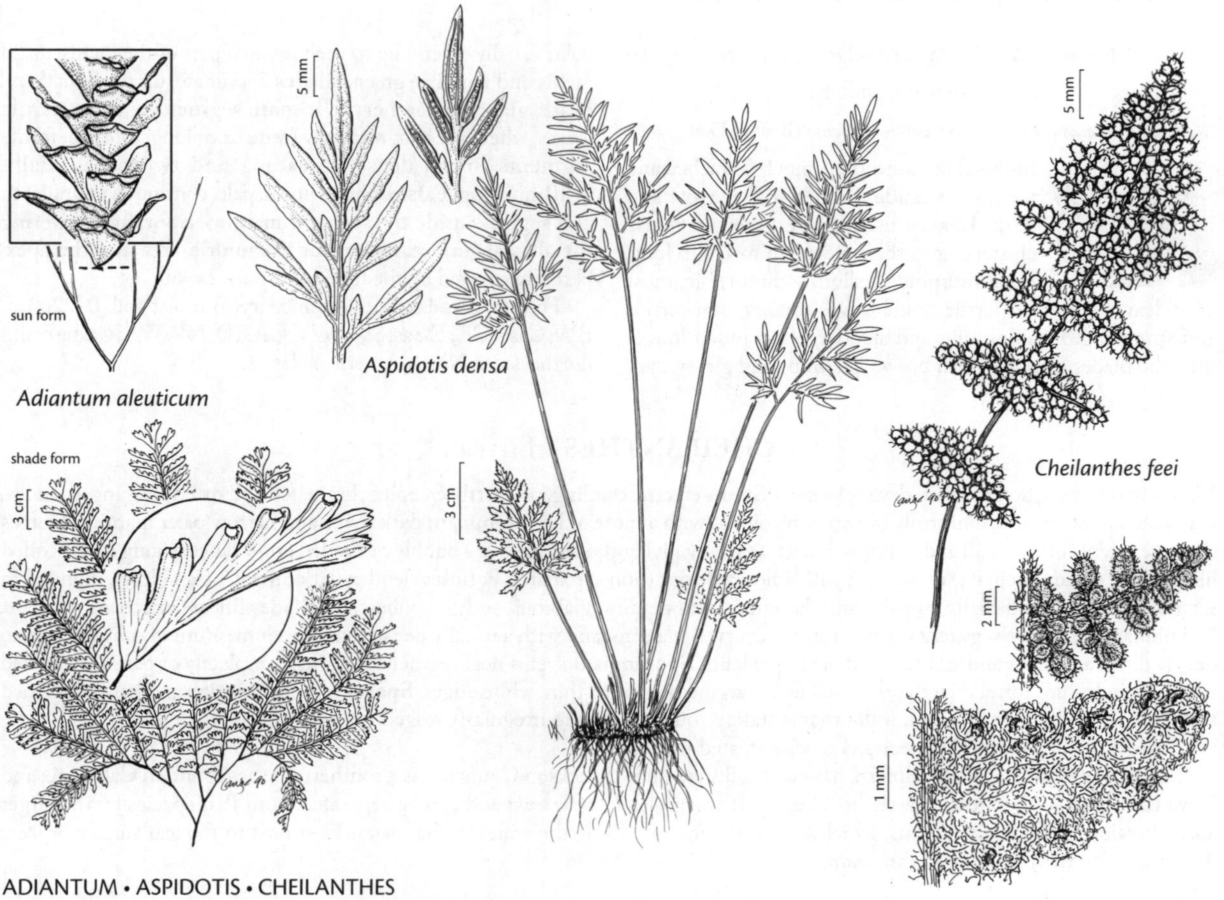

ADIANTUM • ASPIDOTIS • CHEILANTHES

Wet/riparian, usually shaded sites, often rocky. 0–2300 m. BW, Casc, CR, Sisk, WV. CA, ID, NV, WA; western and northeastern North America. Native.

Adiantum aleuticum is abundant in and west of the Cascades but is largely absent from the more arid parts of Oregon.

Adiantum jordanii C.H. Müller

California maidenhair fern

Stems short creeping and branched; scales abundant, brown, to about 3 mm long. **Leaves** few to numerous, loosely clustered, to about 80 cm long but usually much shorter, herbaceous, pale to medium (yellow-) green. **Petioles** variable in length and rather brittle, up to about 50 cm and much longer than the blade to much shorter than the blade, purplish black, scaly at the base, otherwise glabrous or with a few scales; rachis similar to and continuous with the petiole. **Blades** (broadly) lanceolate, pinnate (bipinnate), to 28 × 10 cm, arching or pendulous; lower pinnae (2)3(4)-pinnate. **Ultimate segments** fan-shaped, leaf-like, up to 1.5 × 2 cm; distal margins of fertile segments more or less regular, shallowly and sparsely incised between the false indusia less than 25% of the way to base, margins of non-fertile segments more lobed or toothed; base truncate to obtuse; stalk purple-black, 1–4 mm. **False indusia** linear, up to about 1 cm long, pale, often forming a white band along the curved distal leaf margin. **Sporangia** borne on and under the false indusium. **2*n***=60.

Rocky steam banks, seepage areas. 100–500 m. Sisk. CA; Baja California. Native.

In California, *A. jordanii* hybridizes with *A. aleuticum* to produce a sterile hybrid, *A.* × *tracyi*, and this hybrid is likely to be found in southwestern Oregon where the ranges of the two species overlap. The appearance of the leaf segments is intermediate between the parents, and the distal half of the leaf blade is unbranched in *A.* × *tracyi*, while in *A. jordanii* all except the last few pinnae are at least bipinnate.

ASPIDOTIS Lacefern

Plants perennial, in dry to moist sites. **Stems** ascending or creeping, branched, often forming a caudex; scales lanceolate and (dark) brown, monochromatic or bicolored with narrow, paler margins. **Leaves** clustered, long-petiolate, monomorphic to dimorphic; plants often with a few spreading vegetative leaves with broader segments in addition to the longer, more or less erect fertile leaves. **Petioles** dark colored, glabrous above the base, often much longer than the blade. **Blades** leathery to subherbaceous, glabrous, more or less ovate, 3–4 pinnate; fertile segments linear or lanceolate with an acute sterile tip. **Sporangia** with false indusia borne inframarginal ly inside the adaxial leaf margin, and therefore having the appearance of true indusia, continuous or discontinuous along the 2 sides of the segment; sori borne on and under the false indusium. **Spores** trilete and tetrahedral with a hemispherical distal quarter.

North America. 5 species; 1 species treated in *Flora*.

Aspidotis densa (Brackenridge) Lellinger (p. 97)

Indian's dream, podfern

Cheilanthes siliquosa Maxon, *Cryptogramma densa* (Brack.) Diels

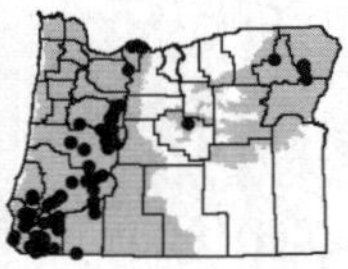

Stems short, ascending, much-branched and forming a caudex, scales stiff, narrow and dark. **Leaves** usually evergreen in dense clusters from the caudex, up to 25 cm long, monomorphic to slightly dimorphic, most or all leaves erect and fertile, some leaves smaller, non-fertile, and spreading. **Petioles** long and slender, often much longer than the blades, dark purplish brown, glabrous and glossy, dark color usually extending to the proximal part of the rachis, distal rachis and rachillae green. **Blades** 3-pinnate, up to about 6 cm, ovate, glabrous, leathery. **Ultimate segments** of non-fertile leaves, where present, narrowly ovate or oblong, toothed, fertile segments linear, dense, spreading and twisted adaxially, 3–10 mm long. **False indusia** linear, pale, continuous, parallel to and slightly inside the segment margins. **Sporangia** covering the abaxial surface, except for the midrib area and the apex, partially covered by the false indusium. **2*n***=60.

Forests, grasslands, rocks and dry to moist soil. 0–2200 m. BW, Casc, CR, ECas, Sisk, WV. CA, ID, NV, WA; western and northeastern North America. Native.

CHEILANTHES Lipfern

Plants herbaceous, perennial, in dry rocky sites. **Stems** erect, ascending, or shortly creeping, branched and often forming a caudex, scales abundant, narrow, uniformly brown or bicolored with a more or less prominent dark stripe and narrow paler margins. **Leaves** short; segments often small and strongly convex adaxially, giving the leaf surface a bubbly or beaded appearance; young leaves coiled like a fiddle-head (circinate) or straight with a hooked apex (non-circinate). **Petioles** slender, dark brown, often glossy, rounded in section or flattened adaxially, to paler and flattened; indumentum glabrous, scaly or pubescent. **Blades** linear-lanceolate to ovate, 2–4 pinnate. **Ultimate segments** with strongly recurved margins, and with a usually dense abaxial indumentum of ovate, ciliate, to deeply dissected scales and/or branched or simple hairs. **Sporangia** on veins near segment margins, completely or partially covered by a false indusium formed by the recurved leaf margin, often with thin, white edges. **Spores** tetrahedral-globose with 3 flattened sides and the distal side rounded; wall ornamentation rough, rugose to irregularly ridged.

Cosmopolitan. About 155 species; 3 species treated in *Flora*.

Cheilanthes covillei, Coville's lipfern, has been collected once in Jackson County. It is a southern species found in California and Nevada but not in the areas adjacent to Oregon. It resembles *C. intertexta* and can be separated from that species by the larger abaxial scales on the leaf segments, which exceed the margin, while the scales in the lowest layer next to the leaf surface are less deeply divided than those of *C. intertexta*.

Key to *Cheilanthes* species

1. Abaxial leaf segments lacking scales; long hairs abundant; leaves circinate when young . . . . *C. feei*

1' Abaxial leaf segments bearing small scales; branched hairs abundant; leaves non-circinate when young.

  2. Leaf blades lanceolate, scales laciniate . . . . *C. gracillima*

  2' Leaves ovate; scales graduated, outer scales ciliate, inner scales laciniate . . . . *C. intertexta*

Cheilanthes feei T. Moore (p. 97)

Fee's lipfern, slender lipfern

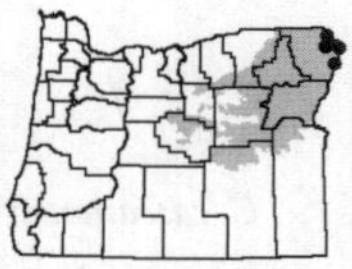

Stems ascending or shortly creeping in rock crevices, much-branched and forming a caudex; scales narrow, dense and persistent, some monochromatic brown, mostly bicolored with a well-defined dark central stripe and pale, scarious margins, some entirely pale brown. **Leaves** short, up to 20 cm, produced in clusters from the caudex, circinate when young. **Petioles** slender, up to about half of the leaf length, dark brown, with basal scales, pubescent with long, more or less spreading, brownish, multicellular hairs; indumentum and appearance of rachis similar to the petiole. **Blades** lanceolate to 12 × 3 cm, 3(4)-pinnate, scaleless, widest near the base; lowest pinnae not or slightly shorter; distal half of blade with pinnae gradually reduced in length apically. **Ultimate segments** rounded, convex, mid-green, and bead-like adaxially, concave and red-brown abaxially with the leaf margin recurved but not or hardly covering the widely dispersed sporangia; abaxial indumentum very dense, of long, red-brown, tangled hairs that spread beyond the lobe margins, hairs sparse to absent adaxially. **Sporangia** submarginal, often covering the underside of the leaf segment except for the midrib area. ***2n***=90 (apogamous triploid).

Cliffs, rocky areas, basalt or limestone. 300–900 m. BW. CA, ID, NV, WA; north to British Columbia, east to WI and TX, south to northern Mexico. Native.

Cheilanthes gracillima D.C. Eaton

Lace lipfern, lace fern

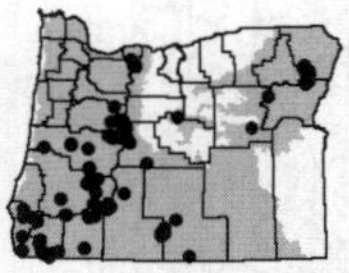

Stems ascending or shortly creeping in rock crevices, much-branched and forming a caudex; scales dense and persistent, variously brown and monochromatic to bicolored with a well-defined, dark, central stripe and pale margins. **Leaves** short, seldom over 20 cm, produced in clumps from the caudex, non-circinate when young. **Petioles** slender, up to about half of the leaf length, reddish brown, glabrous or with a sparse indumentum of hairs and/or scales; rachis similar in appearance to the petiole and more scaly. **Blades** linear up to about 10 × 2 cm, 2(3)-pinnate, parallel-sided proximally, distal half gradually narrowed to the apex. **Ultimate segments** oblong to rounded, convex, dark green, bead-like, and more or less glossy adaxially, concave and almost bowl-like abaxially with the leaf margin recurved and arching, the edge thin and white, forming a false indusium where sporangia are present; abaxial indumentum very dense but not covering the leaf margin, composed of multilayered small scales and branched hairs; scales not graduated, those appressed to the leaf surface similar to the outer scales, all deeply divided into narrow segments, adaxial leaf surface glabrous or with sparse hairs and scales. **Sporangia** submarginal.

Dry sites, cliffs, talus, dry soil, scrubland, occasionally in forests. 50–2700 m. BR, BW, Casc, CR, ECas, Sisk, WV. CA, ID, NV, WA; north to British Columbia, northeast to MT. Native.

This is the only common, widespread *Cheilanthes* species in Oregon.

Cheilanthes intertexta (Maxon) Maxon

Coastal lipfern

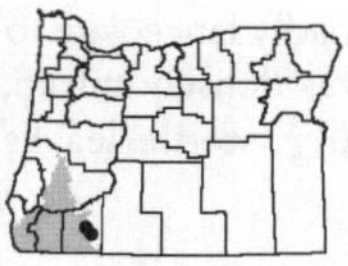

Stems ascending or shortly creeping in rock crevices, much-branched and forming a caudex; scales dense and persistent, mostly bicolored with a well-defined, dark, central stripe and pale narrow margins. **Leaves** short, seldom over 15 cm, produced in clusters from the caudex, non-circinate when young. **Petioles** slender, up to about half of the leaf length; petiole and rachis dark brown, with a sparse indumentum of hairs and scales. **Blades** (broadly) lanceolate, up to about 8 × 3 cm, 3-pinnate, pinnae on proximal half of blade more or less equal in length, distal pinnae reduced in size apically. **Ultimate segments** oblong, convex, dark green, bead-like, and more or less glossy adaxially, concave, red-brown, and almost bowl-like abaxially with the leaf margin strongly recurved and arching, the edge thin and white, forming a false indusium; abaxial indumentum very dense, red-brown, extending to the leaf margin, composed of multilayered small scales and branched hairs; scales graduated, those appressed to the leaf surface deeply divided into narrow segments, outer scales narrowly ovate, margins ciliate near the base; adaxial leaf surface glabrous or with sparse hairs and scales. **Sporangia** submarginal.

Rocky places. 700–1400 m. Sisk. CA, NV. Native.

CRYPTOGRAMMA Parsley fern, rock-brake

Edward R. Alverson

Plants on rock. **Stems** decumbent to erect, or creeping, few to many branched; scales colorless or brownish, concolored or bicolored, ovate, lanceolate or linear, margins entire. **Leaves** dimorphic, scattered or densely tufted; fertile leaves 5–25 cm; sterile leaves 3–20 cm, shorter than fertile leaves. **Petioles** dark brown proximally, light brown to green distally, with single longitudinal groove adaxially, scaly, with single vascular bundle. **Blades** deltate, lanceolate to elliptic, 2–4-pinnate, somewhat leathery or herbaceous to membranaceous, abaxially glabrous, adaxially glabrous or sparsely pubescent, dull to somewhat lustrous, not striate; rachis straight. **Ultimate segments** of blade short-stalked or sessile, free or adnate to midrib; stalks dull, green; segments of sterile leaves ovate, elliptic, obovate, or fan-shaped, usually less than 4 mm wide, margins plane, dentate or shallowly to deeply cut; fertile segments strongly differentiated from sterile segments, lanceolate to linear, usually less than 2 mm wide, margins reflexed to form false indusia extending over entire length of segments, at first covering young sporangia, often becoming plane at maturity. **Veins** of ultimate segments usually obscure, free, pinnately branched, and divergent distally. **False indusia** greenish to brown, broad, clearly marginal, usually concealing sporangia. **Sporangia** scattered along veins on abaxial leaf surface, often intermixed with farina-producing glands, containing 64 spores. **Spores** yellow, tetrahedral, trilete, verrucose, equatorial flange absent. **Gametophytes** glabrous.

Temperate regions of North America, South America, and Eurasia. 8–11 species; 3 species treated in *Flora*.

Key to *Cryptogramma* species:

1. Stems 1–1.5 mm diameter, creeping, few-branched; leaves scattered along stems, delicate and ephemeral; petioles mostly dark brown in proximal 50–67%, greenish distally; often in calcareous habitats . . . . ***C. stelleri***

1' Stems 4–20 mm diameter, decumbent to erect, many branched from base; leaves strongly tufted, herbaceous to somewhat leathery, present throughout growing season; petioles dark brown in proximal 12% or less; mostly of noncalcareous habitats.

  2. Blades herbaceous, thin, translucent when dried, shed in autumn, not persistent; hydathodes on dried leaves superficial; hairs absent from adaxial leaf surface. . . . . ***C. cascadensis***

  2' Blades somewhat leathery, opaque when dried, green and persistent over winter; hydathodes on dried leaves sunken below surface; hairs small, appressed, cylindric, present in grooves of petiole, costae, and costules of adaxial leaf surface . . . . ***C. acrostichoides***

Cryptogramma acrostichoides R. Brown (p. 100)
American parsley fern

Cryptogramma crispa (L.) R. Br. ex Hook. var. *acrostichoides* (R. Br.) C.B. Clarke

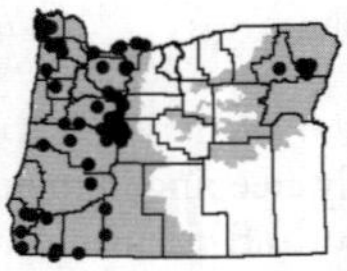

Stems decumbent to erect, much branched from base, stout, 10–20 mm diameter (including hardened, persistent leaf bases); scales bicolored, dense, broadly lanceolate to linear, to 6 × 2 mm. **Leaves** densely tufted, green over winter, persistent after withering; sterile leaves spreading, 3–17 cm; fertile leaves erect, 5–25 cm; hairs small, appressed, cylindric, scattered along grooves of petiole and along costae and costules of adaxial blade surface. **Petioles** 1–2 mm wide, green to straw-colored, dark brown on proximal 12% or less, firm and straw-like, not collapsed; scales bicolored or more or less concolored, becoming sparse distally. **Blades** deltate to ovate-lanceolate, all 2–3-pinnate, somewhat leathery, opaque; hydathodes sunken below leaf surface. **Segments** of sterile leaves oblong to ovate-lanceolate, bases cuneate, distal 50–67% of segments crenate to dentate, often somewhat more deeply incised

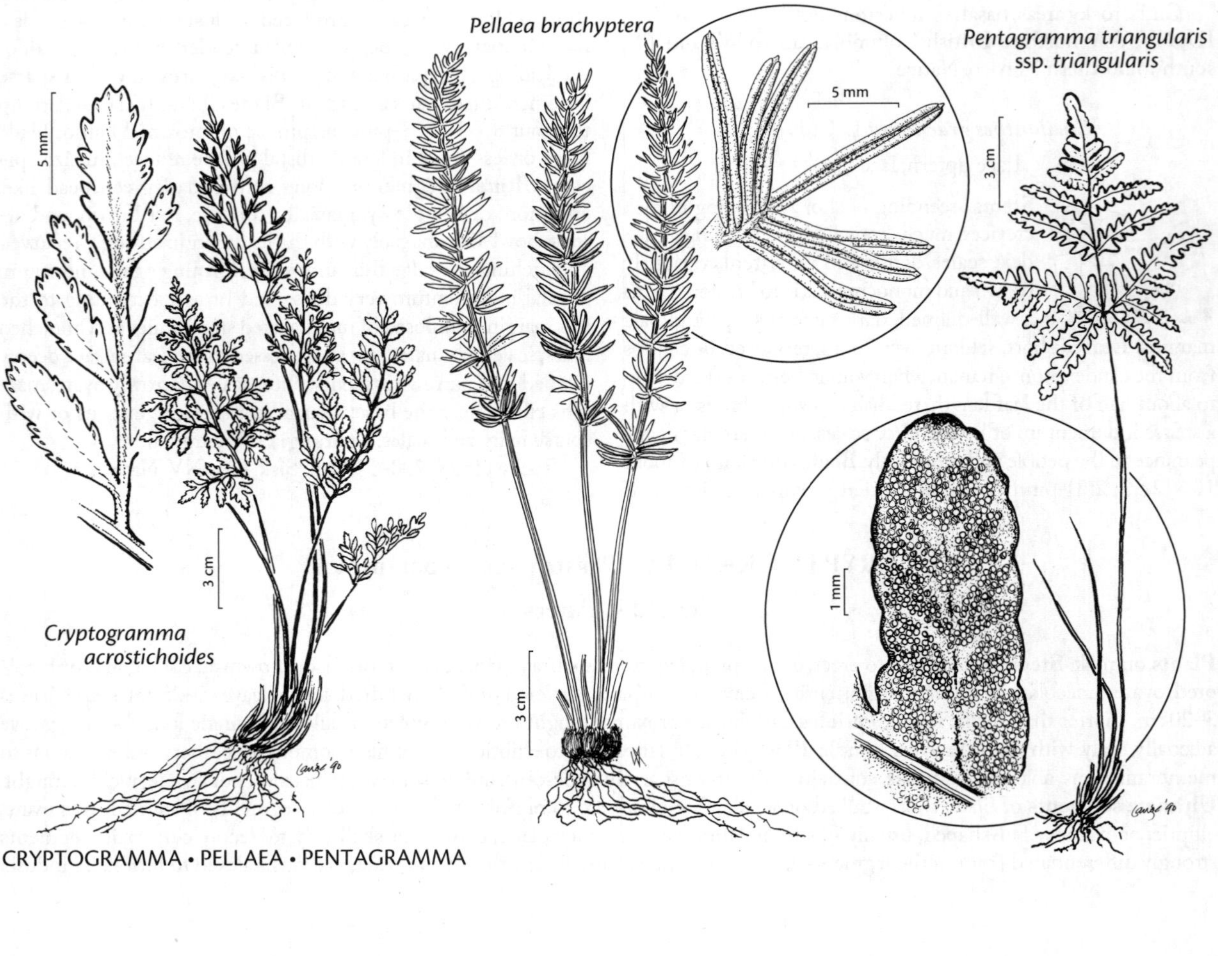

CRYPTOGRAMMA • PELLAEA • PENTAGRAMMA

every second tooth; segments of fertile leaves horizontal to ascending, strongly differentiated from those of sterile leaves, linear, 3–12 × 1–2 mm; margins of fertile segments revolute, covering sporangia. **Sporangia** in sori that coalesce at maturity. **2*n*=60.**

Noncalcareous cliff crevices, rock outcrops, and talus, often relatively dry habitats. 0–3000 m. BW, Casc, CR, ECas, Est, Sisk, WV. CA, ID, NV, WA; western North America; Asia. Native.

Cryptogramma acrostichoides has often been treated as a variety or subspecies of the strictly European *Cryptogramma crispa*, which has a chromosome number of **2*n*=120.**

Cryptogramma cascadensis E.R. Alverson

Cascade parsley fern

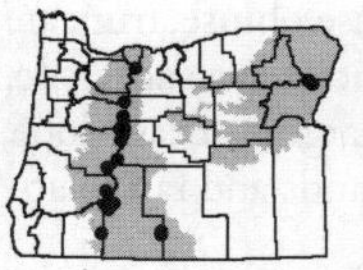

Stems decumbent to erect, much branched from base, stout, 4–8 mm diameter (including hardened, persistent leaf bases); scales often bicolored, dense, broadly lanceolate to linear, to 6 × 2 mm. **Leaves** strongly tufted, deciduous; sterile leaves spreading, 3–20 cm; fertile leaves erect, 5–25 cm; petioles, costae, and costules glabrous. **Petioles** approximately 1 mm wide when dry, green to straw-colored, dark brown only on proximal 12% or less,collapsing and strongly furrowed; scales bicolored or more or less concolored, becoming sparse distally. **Blades** deltate to ovate-lanceolate, all 2–3-pinnate, herbaceous, thin and translucent when dried, hydathodes superficial. **Segments** of sterile leaves oblong to fan-shaped, bases cuneate, distal 33–50% of each segment regularly dentate and often more deeply incised every 2nd and 4th tooth; segments of fertile leaves ascending to erect, strongly differentiated from those of sterile leaves, linear, 3–12 × 1–2 mm; fertile segments revolute, covering sporangia. **Sporangia** in sori that coalesce at maturity. **2*n*=60.**

Talus slopes and cliff crevices, often on igneous rocks, typically in relatively mesic subalpine habitats. 1400–2600 m. BW, Casc, ECas. CA, ID, WA; north to British Columbia, east to MT. Native.

Populations of *Cryptogramma cascadensis* were previously identified as *C. acrostichoides*.

Cryptogramma stelleri (S.G. Gmelin) Prantl

Slender rock-brake, Steller's rock-brake, fragile rock-brake

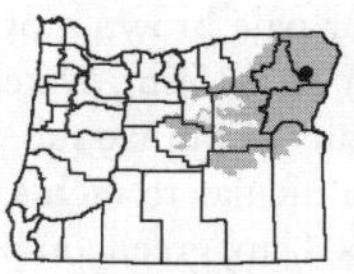

Stems creeping, few branched, slender, 1–1.5 mm diameter, succulent, brittle; scales colorless, sparse, transparent-reticulate, ovate, 0.4 × 0.3 mm; stems shriveling in second year following emergence of leaves. **Leaves** scattered along stems, ephemeral (dying by late summer), soon shed; sterile leaves erect, 3–15 cm; fertile leaves erect, 5–20 cm; petioles, costae, and costules glabrous. **Petioles** approximately 1 mm wide when dry, dark brown in proximal 50% or less, becoming greenish distally,only slightly furrowed, glabrous. **Blades** broadly lanceolate to ovate-lanceolate, all pinnate-pinnatifid to 2-pinnate, herbaceous to membranous, thin; hydathodes superficial, often poorly developed or absent. **Segments** of sterile leaves ovate-lanceolate to fan-shaped, distal 33–50% shallowly lobed; segments of fertile leaves horizontal to ascending, often only partially differentiated from sterile leaves, lanceolate to linear, 8–25 × 2–4 mm; margins reflexed, forming continuous false indusia. **Sporangia** often in discrete sori. **2*n*=60.**

Sheltered calcareous cliff crevices and rock ledges in coniferous forest or other boreal habitats. 1500–3000 m. BW. NV, WA; western and northeastern North America; Asia. Native.

PELLAEA Cliff-brake

Plants of dry places, perennial, herbaceous. **Stems** either short, ascending, branched and forming a caudex or (shortly) creeping and branched; scales more or less dense, long and very narrow, or shorter and lanceolate (ovate), uniformly brown or bicolored with a darker mid-stripe; dead leaves and petiole bases often persistent. **Leaves** monomorphic, evergreen, growing in clumps on the multiple caudex apices, or more distant and in loose tufts. **Petioles** dark and more or less glossy, or paler and matte, glabrous in Oregon species except for basal scales, more or less rounded in section, with a twiggy appearance, with or without several annular grooves (articulation lines) near the base. **Blades** 1–3-pinnate; rachis resembling the petiole, at least proximally. **Segments** linear to rounded, leathery to subherbaceous, green or glaucous, sessile or shortly stalked; margin entire to crenate, usually recurved over the sporangia abaxially to form a false indusium the same color as the blade or the margins paler, leaf margin rarely plane. **Sporangia** scattered on the abaxial leaf surfaces, nor aggregated into sori, in submarginal lines, glands producing a waxy substance are often interspersed among the sporangia.

Cosmopolitan. About 40 species; 4 species treated in *Flora*.

Genus limits in *Pellaea* are controversial, and there is some molecular evidence that the genus as currently constituted may be polyphyletic. Our four species are all distinctive and easily separated from one another.

Key to *Pellaea* species:

1. Leaves mostly 3-pinnate; petioles yellowish/brownish, not dark or glossy . . . . *P. andromedifolia*

1' Leaves at most 2-pinnate; petioles dark (reddish) brown and more or less glossy.

  2. Leaves 1-pinnate with broad segments.

    3. Pinnae all entire . . . . *P. bridgesii*

    3' Most pinnae deeply 2-lobed . . . . *P. breweri*

  2' Leaves 2-pinnate; pinnules narrow and needle-like . . . . *P. brachyptera*

Pellaea andromedifolia (Kaulfuss) Fée

Coffee fern

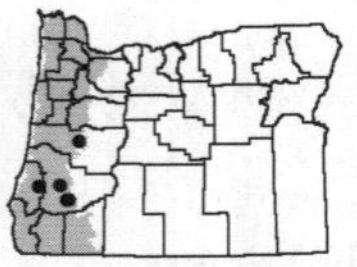

Stems rather slender, creeping, branched, not forming a caudex; scales moderately dense, lanceolate, toothed, mostly bicolored. **Leaves** large, to about 55 × 15 cm, more or less distant on the stem (roughly 1 cm apart), forming loose clumps. **Petioles** well developed, up to half the leaf length, annular basal grooves lacking, petioles and rachises all similarly colored, straw-colored or pale brown, not glossy. **Blades** 3-pinnate proximally, to about 30 × 15 cm, ovate or triangular, widest near base, lowest 2–3 pairs of pinnae subequal in length, pinnae ascending or perpendicular to rachis. **Segments** broad, oblong-elliptic, up to 1.5 × 1 cm, green, only slightly leathery, entire, glabrous or sparsely pubescent; scales absent; apex rounded, often notched; base obtuse, truncate or subcordate; margins recurved, entire to slightly crenate; veins visible, free and branching. **Sporangia** submarginal along the 2 sides of each segment, partially or completely covered by the false indusium, waxy glands and farina absent. ***n*=29** (in fertile diploids), asexual triploids, tetraploids also occur.

Dry, open sites with rocky soil. 100–900 m. CR, Sisk, WV. CA; Baja California. Native.

Pellaea brachyptera (T. Moore) Baker (p. 100)

Sierra cliff-brake

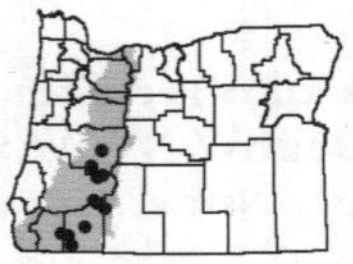

Stems ascending, branched and forming a caudex; scales dense, long and narrow, bicolored. **Leaves** short and narrow, to about 30 × 3 cm. **Petioles** well developed, up to half the leaf length, petioles and rachises dark reddish brown, more or less glossy, basal annular grooves lacking. **Blades** linear, bipinnate to simply pinnate distally; proximal pinnae with short rachillae resembling the rachis, twisted out of plane; leaves with bunches of pinnules superficially resembling a branch of rosemary or larch. **Segments** linear and needle-like, leathery, glabrous and glaucous; margins strongly recurved, more or less crenate, forming a false indusium and covering the sporangia and most of the abaxial surface; veins free, not visible. **Sporangia** completely covered by the recurved leaf margin, waxy glands and farina abundant.

Dry, open sites with rocky soil. 300–1900 m. Casc, Sisk. CA, WA. Native.

Pellaea breweri D.C. Eaton

Brewer's cliff-brake

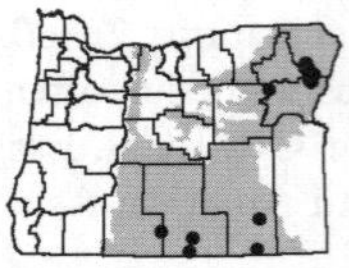

Stems ascending, stout, much-branched and forming a caudex; scales dense, long and very narrow, uniformly brown. **Leaves** in more or less dense clumps from multiple apices, small, to about 20 × 4 cm. **Petioles** up to half the leaf length, dark brown and glossy, color and texture extending into proximal part of rachis, distal rachis tip usually green, prominent annular articulation lines present near the petiole base. **Blades** linear oblong lanceolate, pinnate-pinnatifid; pinnae twisted out of plane, curved forwards and upwards. **Segments** broad, most deeply divided into 2(3) unequal lobes, acroscopic lobe largest, up to 1.5 × 1 cm, green, leathery, glabrous or glaucous; apex rounded to acute; base obtuse, truncate or subcordate; veins free, visible; shortly petiolate to subsessile. **Sporangia** in submarginal lines along the margins of each lobe, mostly covered by a false indusium, waxy glands and farina absent. **2*n*=58**.

Rocky soil and cliffs in the mountains. 1500–3100 m. BR, BW, ECas. CA, ID, NV, WA; most mountain ranges in the western US. Native.

Pellaea bridgesii Hooker

Bridges' cliff-brake

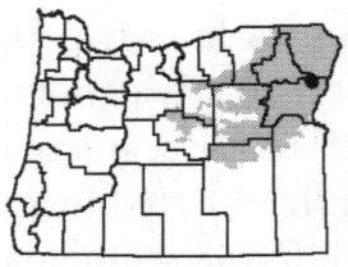

Stems ascending, branched and forming a caudex; scales dense, long, narrow and bicolored. **Leaves** in more or less dense tufts from the caudex, small, to about 25 × 4 cm. **Petioles** well developed, up to half the leaf length, petiole and rachis medium to dark brown, glossy, basal annular lines absent. **Blades** simply pinnate and linear, at least the proximal pinnae twisted out of plane. **Segments** broadly ovate, leathery, glabrous; apex rounded; base rounded or subcordate, slightly falcate; margins glaucous, margin entire or slightly crenate, not or only slightly recurved, not forming a false indusium, glaucous; veins anastomosing but rather obscure. **Sporangia** forming a submarginal line on each side of the pinna, absent from the base, extending into the rounded apex, often almost to the tip, not covered by the leaf margin, waxy glands and farina abundant. **2*n*=58**.

Dry, rocky sites in the mountains. 1500–2400 m. BW. CA, ID, NV. Native.

PENTAGRAMMA Goldback fern, silverback fern

Plants herbaceous, perennial, in dry to moist rocky/stony sites. **Stems** shortly creeping or ascending, usually unbranched; scales dense, narrow, stiff, bicolored with a black center and narrow brown margins. **Leaves** monomorphic, evergreen, in few-leaved clusters. **Petioles** long, dark often black and glossy. **Blades** broad, triangular and appearing somewhat pentagonal because of enlarged basal pinnules, and ternate-pinnate because of the greatly enlarged basal pinnae, adaxial surface glabrous or glandular, abaxial surface with abundant glands producing waxy white or yellow farina. **Sporangia** borne along veins on the leaf segments; indusia and false indusia absent. **Spores** 3-cornered, ornamented with large tubercles.

North America. 2 species; 1 species treated in *Flora*.

Pentagramma triangularis (Kaulfuss) Yatskievych, Windham, & E. Wollenweber ssp. ***triangularis*** (p. 100)

Goldback fern

Pityrogramma triangularis (Kaulf.) Maxon, *Pityrogramma triangularis* (Kaulf.) Maxon var. *triangularis*

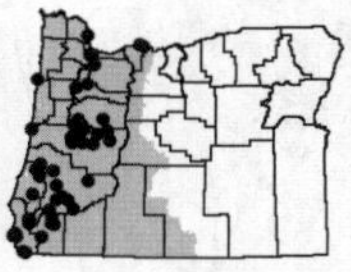

Stems usually unbranched; scales dense, stiff, narrow, dark. **Leaves** evergreen, to about 30 cm long, borne in a few-leaved cluster from the stem apex. **Petioles** long and slender, glabrous except for the sparse scales at the base, dark, glossy, purplish brown or black, often longer than the blade; rachis similar to the petiole but with narrow green wings, sometimes completely immersed in green tissue distally. **Blades** triangular-pentagonal, to about 10 × 10 cm; adaxial surface green, glossy and glabrous, abaxial surface glandular, coated with yellow (white) waxy farina; distal part of frond pinnate with deeply pinnatifid pinnae, rapidly reduced in length apically; proximal pair of pinnae much larger than the others, giving a ternate appearance to the blade; first downward-pointing (basiscopic) lobe longer than the others. **Segments** deeply pinnatifid, lobes ovate; apex more or less rounded; margins subentire, narrowly recurved, not forming a false indusium. **Sporangia** borne in lines along the forking veins of the pinnules, yellowish green, becoming straw-colored, surrounded by the yellow or golden farina. **$2n$**=60.

Dry to moist sites, often on rocks or rocky soils, mainly in open places, sometimes shaded places. 0–1800 m. Casc, CR, ECas, Est, Sisk, WV. CA, ID, WA; north to British Columbia, south to Baja California. Native.

SALVINIACEAE Floating Fern family

Plants aquatic, perennial, herbaceous, heterosporous, monoecious, usually floating on still water. **Stems** horizontal, branching (sub-) dichotomously, dying back to produce clonal colonies as the successive dichotomies separate, roots long or reduced to hairs. **Leaves** monomorphic, alternate with 2 dissimilar lobes. **Sori** enclosed, sac-like, of 2 types, mega- and microsporocarps. **Megasporocarps** with 1–30 megasporangia, each with a single megaspore, megaspores attached to a complex episporic structure. **Microsporocarps** with numerous microsporangia, each producing 32–64 microspores, dehiscing and releasing spore masses with anchor-shaped hooks (*Azolla*).

Cosmopolitan. 2 genera; 1 genus treated in *Flora*.

Salviniaceae is comprised of two genera (*Azolla* and *Salvinia*). *Salvinia* has not been reported in the wild from Oregon, although it is sometimes grown in aquaria. *Salvinia molesta* is highly invasive in warm temperate parts of the United States and is listed by the Oregon Department of Agriculture as potentially invasive. *Salvinia* leaves are larger than those of *Azolla*, rounded, and the upward facing surface is covered with long stiff hairs with branched tips.

AZOLLA Duckweed fern, mosquito fern, water fern

Plants aquatic, floating, fan-shaped, green and/or red. **Roots** long, unbranched, produced on the underside of the stem at branching points. **Stems** horizontal, with regular dichotomous branching after every 3rd leaf on the stem, strongly dorsi-ventral with the ventral side in contact with the water (or mud), dying back and releasing successive dichotomies that become free plants, individuals seldom more than 1 cm diameter. **Leaves** sessile, 2-lobed; ventral lobe flat and rounded; dorsal lobe inflated, more or less erect and angled forwards. **Sporocarps** of 2 kinds producing megasporangia or microsporangia, produced in pairs from the ventral leaf lobe of the first leaf after a stem dichotomy, microsporocarps more conspicuous and usually more abundant than megasporocarps. **Microsporocarps** globose with a terminal blunt dark tip, up to 2 mm in diameter; the numerous stalked, globose sporangia visible through the translucent sporocarp wall; sporangia approximately 0.25 mm in diameter, with up to 64 approximately 3-µm spores which are aggregated into several masses, each surrounded by anchor-shaped structures (glochidia); glochidia septate or not; male gametophytes and gametes are produced inside the spores. **Megasporocarps** approximately 0.5 mm in diameter, bearing a single megasporangium in which a single megaspore develops; megaspore walls variously ornamented with pits, ridges, and hairs and surmounted by an episporic structure composed of a collar enclosing the proximal end of the spore.

Cosmopolitan. 3–9 species; 2 species treated in *Flora*.

The *Azolla* species present in Oregon are difficult to separate. The diagnostic characters often require at least 40× magnification. The dorsal lobes of *Azolla* species are inflated with a central cavity filled with mucous and with filaments of *Anabaena azollae*, a symbiotic nitrogen-fixing cyanobacterium.

Key to the *Azolla* species:

1. Papilla-like hairs on the upper surface of the dorsal leaf lobe straight, forward-pointing, and uniformly 1-celled (the hair is part of the epidermal cell from which it protrudes); megaspore walls with sparsely hairy irregular ridges and tubercles, pits forming irregular channels . . . . ***A. filiculoides***

1' Papilla-like hairs on the upper surface of the dorsal leaf lobe (1)2(3)-celled (including the epidermal cell from which it protrudes), straight or curved apically, forward-pointing; megaspores with a pitted surface, pits not anastomosing, ridges between the pits sparsely hairy . . . . ***A. microphylla***

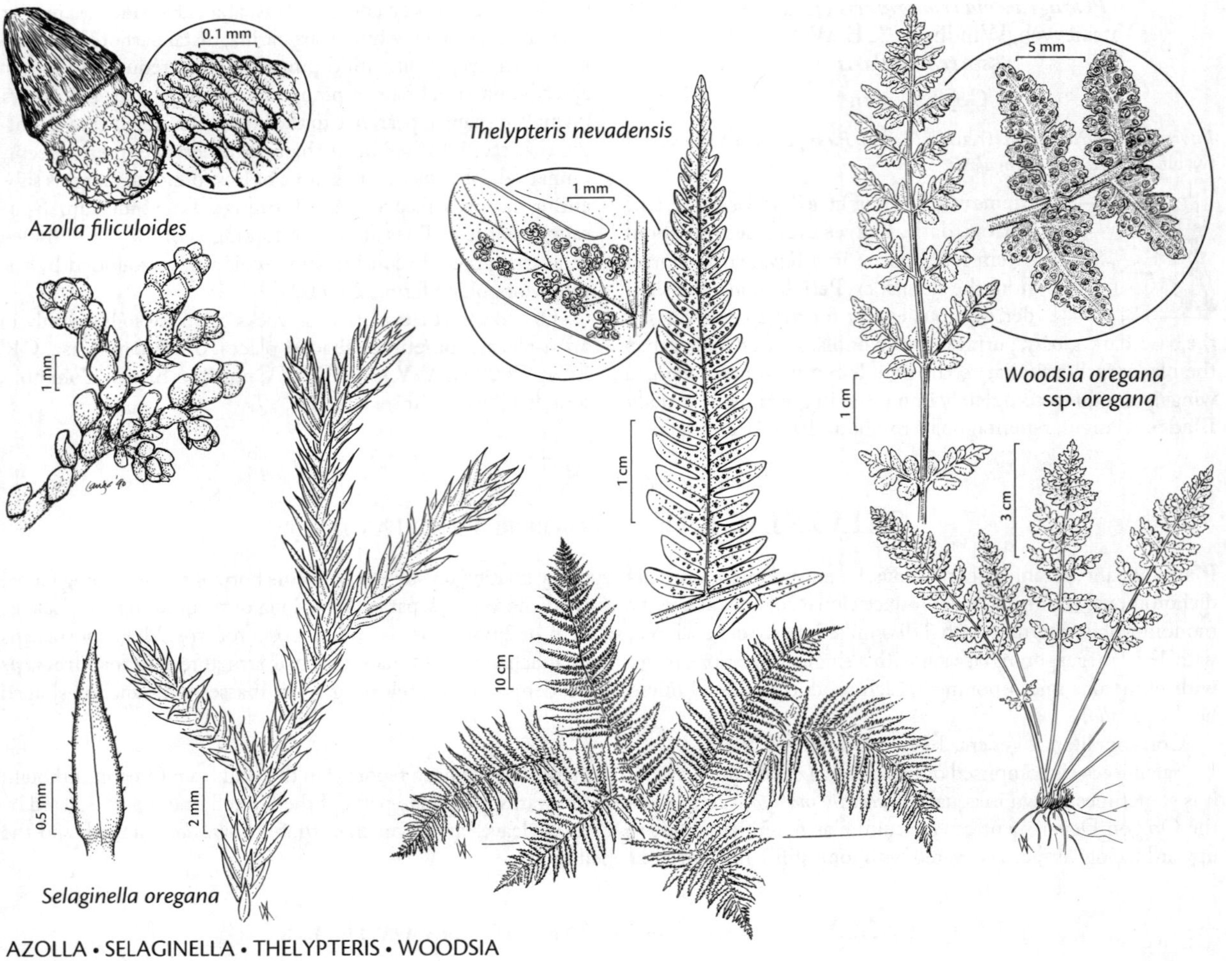

AZOLLA • SELAGINELLA • THELYPTERIS • WOODSIA

Azolla filiculoides Lamarck

(p. 104)

Duckweed fern, Pacific mosquito fern

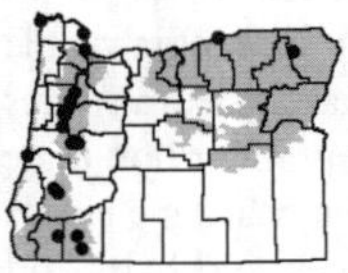

Plants forming red mats, especially on ponds with full exposure, greener where shaded. **Stems** horizontal, becoming ascending in summer when growth is very dense. **Leaves** 2-lobed; ventral lobe ovate-rounded and often submerged, with wide thin margins; dorsal lobe thick, red or green, margin often pale, many epidermal cells forming an erect, straight, forward-pointing non-septate papilla or hair. **Sporocarps** produced in late spring/summer in pairs from the ventral leaf lobe, mostly male, pairs sometimes male/female. **Megasporangia** each producing a single megaspore approximately 0.4 mm in diameter; megaspore walls with sparsely hairy irregular ridges and tubercles, pits between the ridges forming irregular channels. **2*n***=44.

Floating on (seasonal) ponds, shallow lakes, sloughs, persisting on mud. 0–1700 m. BW, Col, Est, Sisk, WV. CA, WA; north to British Columbia, southeast to AZ, Mexico; widespread in temperate (subtropical) regions. Native.

Azolla microphylla Kaulfuss

Mexican water fern, Mexican mosquito fern

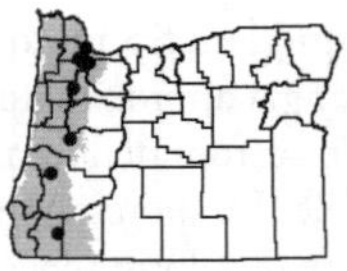

Plants forming red mats, especially on ponds with full exposure, greener where shaded. **Stems** horizontal, becoming ascending in summer when growth is very dense. **Leaves** 2-lobed; ventral lobe ovate-rounded and often submerged, with wide thin margins; dorsal lobe thick, red or green, margin often pale, many epidermal cells forming an erect, forward-pointing septate hair, with the distal cell often curved. **Sporocarps** produced in late spring/summer in pairs from the ventral leaf lobe, mostly male, pairs sometimes male/female. **Megasporangia** each producing a single megaspore; megaspore walls with a pitted surface resembling a golf ball, pits not anastomosing, ridges between the pits sparsely hairy. **2*n***=44.

Floating on (seasonal) ponds, shallow lakes, sloughs, persisting on mud. 0–400 m. CR, Sisk, WV. CA, NV, WA; west and central US, Mexico, Central America, South America. Native.

Azolla microphylla is often found growing in the same habitats as *A. filiculoides*.

SELAGINELLACEAE Spikemoss family

Rory Nichols

Herbs perennial, forming dense to open mats or cushions. **Roots** generally adventitious on aerial shoots. **Stems** leafy, branched, few to many, mostly irregular, sometimes dichotomous. **Leaves** simple, monomorphic or dimorphic, single-veined, petiole absent. **Strobili** conspicuous (or not), single or paired at branch terminal, sessile; sporophylls (fertile leaves) unlike sterile leaves on stems but generally tightly arranged in 4 distinct rows. **Sporangia** 1 per sporophyll, axile. **Spores** of 2 types; male typically distal, tiny, hundreds; female (1)4, large, yellow-orange.

Cosmopolitan, primarily in tropical and subtropical regions. 1 genus.

Out of the three subgenera found in the U.S., two are represented in Oregon—subgenus *Stachygynandrum* (includes *Selaginella douglasii*) and subgenus *Tetragonostachys* (*S. oregana*, *S. scopulorum*, *S. wallacei*, and *S. watsonii*)—subgenus *Selaginella* is not represented here. The chief morphological difference between the two represented subgenera is that *Stachygynandrum* has dimorphic sterile leaves (of two types, two lateral rows and two median rows), while *Tetragonostachys* has generally monomorphic sterile leaves (not ranked, spirally arranged).

SELAGINELLA Spikemoss

Herbs terrestrial or epiphytic, moss-like. **Stems** decumbent or creeping to long-pendent. **Leaves** dimorphic or monomorphic, if monomorphic, spirally arranged, typically linear-lanceolate; if dimorphic, 4-ranked (2 sets of lateral leaves and 2 sets of median leaves), typically ovate-oblong; base decurrent or not, auricles present or not; margins entire to dentate or short-ciliate; apices rounded or cuspidate to long-bristled. **Strobili** 4–60 mm long, sporophylls ovate to lanceolate.

Approximately 700 species; 5 species treated in *Flora*.

Selaginella, in general, has been under-collected in Oregon. This is likely due to the inconspicuous, moss-like appearance of most species. A minimum of 20× magnification is best for viewing leaf characters. Leaf lengths include bristle, if present.

Key to *Selaginella* species:

1. Sterile leaves dimorphic, oval to ovate, mostly 1–2 times longer than wide . . . . **S. douglasii**

1' Sterile leaves monomorphic, linear-lanceolate, mostly 4–6 times longer than wide.

  2. Plants epiphytic or terrestrial, hanging in appearance with pendent stems . . . . **S. oregana**

  2' Plants strictly terrestrial, in rocky habitats, forming relatively tight mats.

    3. Main stem internodes generally greater than 5 mm long (giving a loosely branched appearance); leaf base generally not decurrent; strobili not tightly clustered, often inconspicuous . . . . **S. wallacei**

    3' Main stem internodes generally less than 5 mm long; leaf base typically decurrent; strobili dense, distinct, numerous.

      4. Bristles toothed; leaf abaxial grooves extending to base of bristle or nearly so . . . . **S. scopulorum**

      4' Bristles smooth; leaf abaxial grooves not extending to base of bristle (giving leaf a general boat shape) . . . . **S. watsonii**

Selaginella douglasii (Hooker & Greville) Spring

Douglas selaginella

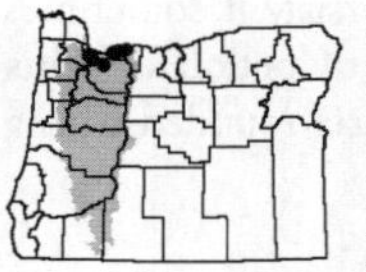

Stems loosely mat-like, long-creeping; branches dichotomous or irregular. **Leaves** dimorphic, distant, fragile, papery, rugose (when dry); base oblique, unequally auriculate. **Lateral leaves** ovate or ovate-oblong to oval, 1.5–3 × 0.5–2 mm; apices rounded. **Median leaves** ovate-obovate, 1.5–2.5 × 1–1.5 mm; apices cuspidate to short-bristled. **Strobili** distinct, 6–11(15) mm long.

On moist, rocky slopes, cliffs, rock, on moss in partial shade in mixed conifer-hardwood forest. 50–700 m. Casc, WV. ID, WA. Native.

This species is very distinct among *Selaginella* in the Pacific Northwest due to its two sets of differentiated sterile leaves on the same stem. It is most commonly found in the western Columbia River Gorge and disjunct in northern Idaho.

Selaginella oregana D.C. Eaton in S. Watson (p. 104)

Oregon selaginella, Oregon spikemoss

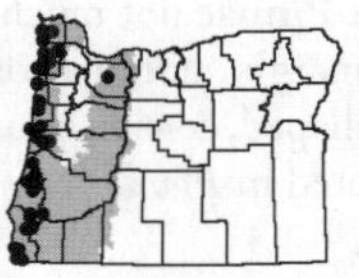

Stems epiphytic or terrestrial, loose mat-forming, often hanging, long-pendent, up to 50 cm long; branches irregular, branch tips curled when dry. **Leaves** monomorphic, distant (on main stem), linear-lanceolate, 2–3 × 0.4–0.6 mm; base decurrent; bristle inconspicuous (blending in with leaf as if apiculate), greenish, brittle, 0.1–0.4 mm long, entire. **Strobili** (5)10–25(60) mm long.

In shade on moss-covered tree trunks, branches, or sometimes on ground, boulders, logs, vertical cliffs, and overhangs in mixed conifer-hardwood forest. 0–1500 m. Casc, CR, Est, Sisk. CA, WA; north to British Columbia. Native.

Selaginella oregana is epiphytic and found growing mostly on *Acer macrophyllum* (occasionally on *Picea sitchensis, Alnus rubra,* and *Umbellularia californica.*) This species primarily grows west of the Coast Range crest; rarely farther inland.

Selaginella scopulorum Maxon

Rocky Mountain spikemoss

Selaginella densa Rydb., *Selaginella densa* Rydb. var. *scopulorum* (Maxon) R.M. Tryon

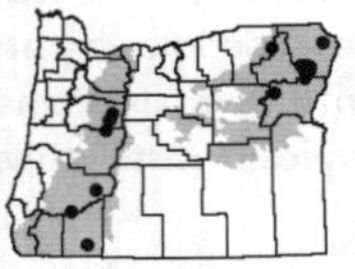

Stems dense to open mats, decumbent to creeping; branches irregular, with fertile branches dense and prominent; main stem internodes less than 5 mm long. **Leaves** lower and upper leaves of main stem slightly differentiated, linear-lanceolate, 2.5–4 × 0.3–0.5 mm; lower leaves often falcate, decurrent; upper leaves not decurrent; abaxial groove extending to leaf apex; base decurrent or occasionally not; bristle 0.3–0.8 mm long, toothed. **Strobili** prominent, dense, (6)8–25(35) mm long.

In conifer forest on exposed rocky slopes, outcrops, and cliffs. 700–2800 m. BW, Casc, Sisk. CA, ID, WA; throughout the Rocky Mountains. Native.

Selaginella scopulorum can be confused with the relatively uncommon *S. watsonii*. The latter, however, has a lower density/amount of strobili and boat-shaped leaves with smooth bristles.

Selaginella wallacei Hieronymus

Wallace's selaginella, Wallace's spikemoss

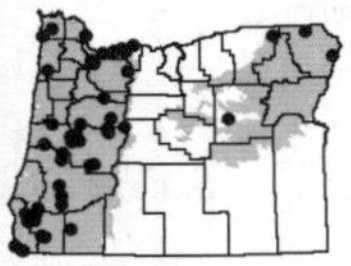

Stems compact to short-creeping cushion in sun, or larger open, decumbent to long-creeping mats in shade; loosely branched; main stem internodes greater than 5 mm long. **Leaves** monomorphic, distant, linear-lanceolate, 1.5–3.5 × 0.3–0.7 mm; base generally not decurrent; bristle 0.2–0.5 mm long, entire or sometimes toothed. **Strobili** often inconspicuous, 4–11(17) mm long.

On dry, exposed to moist, shady rocky outcrops, stream banks, cliffs in conifer or mixed conifer-hardwood forest. 50–1800 m. BW, Casc, CR, Est, Sisk, WV. CA, ID, WA; north to central British Columbia, northeast to MT. Native.

This species is the most common *Selaginella* in Oregon. In sunny, exposed habitats, this plant grows smaller and relatively compact, resembling *S. scopulorum*, but the strobili are never as prominent and tightly packed. Larger specimens of *S. wallacei* in moist, shady habitats can resemble smaller *S. oregana* specimens, but *S. wallacei* does not have a hanging habit and very rarely grows on tree trunks.

Selaginella watsonii Underwood

Alpine spikemoss, Watson's selaginella

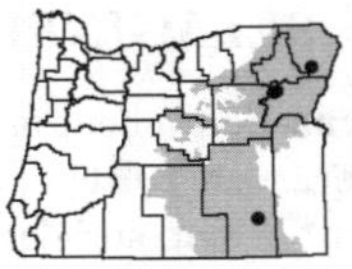

Stems cushion-like, dense to somewhat open, decumbent to creeping. **Leaves** monomorphic, boat-shaped, linear-lanceolate, 2.5–4 × 0.4–0.7 mm; abaxial groove prominent, not extending to leaf apex; base decurrent; bristle 0.2–0.4 mm long, entire. **Strobili** distinct, 5–13(23) mm long.

Open, dry rocky outcrops, talus slopes and gravelly soils in conifer forests. 900–2700 m. BR, BW. CA, ID, NV; east to WY, southeast to AZ. Native.

Selaginella watsonii is the least common of our spikemosses. Thus far it is confirmed from the Wallowa and Elkhorn ranges and Steens Mountain, but it should be looked for anywhere *S. scopulorum* occurs in eastern Oregon.

THELYPTERIDACEAE Marsh fern family

Plants perennial, herbaceous. **Stems** long-creeping to ascending or erect, scaly at least at the apex. **Leaves** mostly monomorphic with stout petioles, forming a crown on a short stem, or produced at intervals along a creeping stem and then often forming extensive stands. **Blades** mostly pinnate, with entire to deeply pinnatifid pinnae; pale hairs usually present, simple or variously branched; scales usually absent. **Spores** monolete, gibbous; wall ornamentation various, often with conspicuous ridges or spines. **Gametophytes** thin, green, reniform.

Cosmopolitan. 1–35 genera (depending on circumscription); 1 genus treated in *Flora*.

THELYPTERIS Marsh fern, maiden fern

Stems long-creeping to short and ascending, simple or branched; scales dense to sparse around the stem apex. **Blades** usually hairy with simple or branched pale hairs at least on the costae; glands often present; scales usually absent; leaf bases persistent, sometimes large and contiguous and covering the stem. **Pinnae** not much divided, entire to deeply pinnatifid, often shortly petiolate. **Veins** free, usually unbranched, extending to the margin, not anastomosing. **Sori** round or slightly elongated; indusia rounded with a sinus at the point of attachment or kidney-shaped, conspicuous to small, persistent to ephemeral.

Cosmopolitan. 875 species; 1 species treated in *Flora*.

Thelypteris nevadensis (Baker) Clute ex C.V. Morton (p. 104)

Sierra wood fern, Sierra marsh fern

Dryopteris oregana C. Chr.

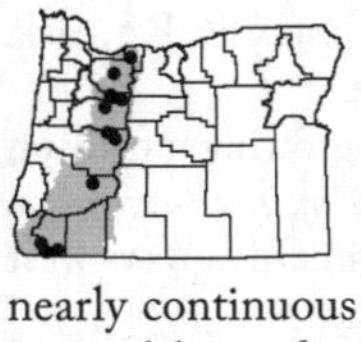

Stems shortly creeping and ascending, bearing a crown of deciduous arching monomorphic leaves. **Leaves** deciduous, monomorphic, to 70 × 15 cm; leaf bases short and persistent, thick and numerous, forming a nearly continuous layer on the scaly stem. **Petiole** short and stout, glabrous distally, strongly grooved adaxially, straw-colored, base darker with sparse scales and small white tubercles; rachis similar to the petiole. **Blades** herbaceous, elliptic, narrowed from the mid-region to the apex and gradually narrowed to the base, pale green; abaxial surface glandular; pale hairs present on the costae. **Pinnae** linear, to 10 × 2 cm; lowest pinna sometimes less than 1 cm long, deeply and regularly incised; segments oblong, angled slightly forwards, rounded; margins plane or slightly recurved, (sub-)entire. **Sori** round, borne on veins equidistant between the margin and costa; indusia thin, reniform. **2*n*=54.**

Wet places, seeps, bogs, and creek banks. 300–1500 m. Casc, Sisk. CA, ID, WA; north to British Columbia. Native.

WOODSIACEAE Cliff Fern family

(also see Athyriaceae, Cystopteridaceae)

Plants herbaceous, perennial. **Stems** ascending, usually unbranched; scales dense, lanceolate, entire or dentate. **Leaves** annual, clustered, monomorphic, bulbils absent. **Petioles** slender, pale to dark, glabrous above, persistent, scaly at the base, base not thickened. **Blades** 1–2-pinnate-pinnatifid, linear to ovate, mostly lanceolate, widest near mid region, narrowed to apex, narrowed basally; proximal pinnae often short and remote, glandular or glandular-hairy; veins free, ending submarginally. **Pinnae** in 1 plane, not overlapping; margins incised/toothed, not spiny; veins simple or forked, not anastomosing, ending before the leaf margin. **Sori** laminar, round, on veins. **Indusia** forming a basal ring of hair- or scale-like filaments under the sporangia. **Spores** bilateral, brownish, winged, wall ornamentation various. **Gametophytes** heart-shaped, flat, green. **2*n***=76.

Northern temperate regions and high elevations in the tropics. 1 genus.

Woodsiaceae formerly included the taxa now included in Athyriaceae and Cystopteridaceae. Members of all three families have, in the past, been included in Dryopteridaceae.

WOODSIA Cliff fern

Family comprised of single genus, *Woodsia*.
36 species; 2 species in Oregon.

Key to the *Woodsia* species:

1. Pinnae with non-glandular hairs in midribs; petioles pliable . . . . ***W. oregana*** ssp. ***oregana***
1' Pinnae without non-glandular hairs on midribs; petioles brittle and easily broken . . . . ***W. scopulina***

Woodsia oregana D.C. Eaton ssp. *oregana* (p. 104)

Oregon cliff fern

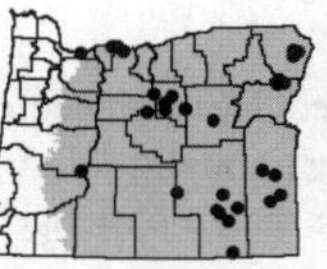

Stems with a crown of leaves and persistent petiole bases. **Leaves** 5–25 × 1–4 cm. **Petioles** brittle and easily broken. **Blades** sparse to moderately glandular; glandular hairs on thin stalks. **Pinnae** ovate, with non-glandular hairs in midribs, apex narrowed to rounded. **Pinnules** with blunt teeth. **Indusia** often concealed by the mature sporangia. **2*n***=76.

Cliffs and rocky slopes. 100–2600 m. BR, BW, Casc, Col, ECas, Lava, Owy. CA, ID, NV, WA; north to British Columbia, east to WY. Native.

Woodsia scopulina D.C. Eaton

Mountain cliff fern

Stem with persistent petiole bases and densely covered in lanceolate scales. **Leaves** 4–25 × 1–4 cm; petioles pliable; blades moderately glandular; rachis with glandular and non-glandular hairs; most glandular hairs on thick stalks. **Pinnae** lanceolate, without non-glandular hairs on midribs, apex rounded. **Pinnules** with lobed or toothed margins. **Indusia** often largely concealed by the mature sporangia.

3 subspecies; 2 subspecies treated in *Flora*.

Key to *Woodsia scopulina* subspecies:

1. Spores 50–57 µm; some stem and petiole scales with black, shiny, discontinuous (occasionally continuous) mid-stripe . . . . ***W. s.*** ssp. ***laurentiana***
1' Spores 42–50 µm; scales 1 color, or with only a few dark cells . . . . ***W. s.*** ssp. ***scopulina***

Woodsia scopulina D.C. Eaton ssp. *laurentiana* Windham

Laurent's cliff fern, Laurentian cliff fern

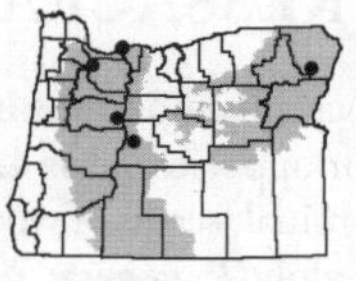

Stem scales of 2 types, brown and bicolored brown/black; bicolored scales with reddish brown margins and a discontinuous (occasionally continuous) mid-stripe of glossy black cells. **Spores** 50–57 µm. **2*n***=152.

Cliffs and rocky slopes. 50–1600 m. BW, Casc, ECas, WV. CA, ID, NV, WA; north to British Columbia, east to SD, disjunct in AZ, Quebec, and Ontario. Native.

Woodsia scopulina D.C. Eaton ssp. *scopulina*

Rocky Mountain woodsia, mountain cliff fern

Stem scales generally reddish brown or brown; dark scales if present few. **Spores** 42–50 µm. **2*n***=76.

Cliffs and rocky slopes. 0–1800 m. BW, Casc, Lava. CA, ID, NV, WA; north to AK, southeast to CO. Native.

GYMNOSPERMS

Stephen C. Meyers

Trees or shrubs monoecious or dioecious, perennial, without a distinct alternation of generations; wind pollinated. **Leaves** usually evergreen (deciduous in *Larix*), needle-like, slender, or scale-like, usually photosynthetic (non-photosynthetic in Ephedraceae). **Ovules** and seeds not enclosed in an ovary, female gametophyte becoming the food-storage tissue of the seed; male and female stroboli (cones) dissimilar.

Key to the Gymnosperm families:

1. Shrubs without functional leaves; stems photosynthetic, green . . . . . **EPHEDRACEAE**
1' Shrubs or trees with green functional leaves; stems generally not photosynthetic.
  2. Leaves awl-like or scale-like . . . . . **CUPRESSACEAE** (in part)
  2' Leaves needle-like.
    3. Seed cones soft, red arils; 1 seed per cone; leaves alternate . . . . . **TAXACEAE**
    3' Seed cones woody or blue to bluish black and berry-like; seeds per cone 1–many; leaves alternate, opposite, or whorled.
      4. Seed cones blue to bluish black and berry-like. . . . . **CUPRESSACEAE** (in part)
      4' Seed cones woody.
        5. Seed cone scales peltate; leaves of two kinds, needle-like and awl-like; trees giant. . . . . **CUPRESSACEAE** (in part)
        5' Seed cones imbricate; leaves of one kind; trees not giant. . . . . **PINACEAE**

CUPRESSACEAE Cypress family

Trees or **shrubs** evergreen, monoecious (but usually dioecious in *Juniperus*). **Bark** smooth to fibrous and furrowed. **Leaves** needle-like, awl-like, or scale-like, simple, alternate or opposite, 2- or 4-ranked or in whorls of 3. **Pollen cones** maturing annually, solitary, terminal. **Seed cones** woody or berry-like, terminal; scales of woody cones imbricate or peltate. **Seeds** 1–many per scale, not winged.

Temperate regions worldwide. Approximately 25 genera; 7 genera treated in *Flora.*

Although some authorities segregate Taxodiaceae from Cupressaceae, most current researchers unite the families. Among the conifers, Cupressaceae has the widest distribution, occurring on all continents except Antarctica.

Key to Cupressaceae genera:

1. Seed cones fleshy, berry-like; scales fused; seeds retained. . . . . **JUNIPERUS**
1' Seed cones woody, not berry-like; scales distinct; seeds shed.
  2. Leaves alternate, of 2 kinds, linear or awl-like . . . . . **SEQUOIA**
  2' Leaves opposite in 4 ranks, of 1 or 2 kinds, scale-like or awl-like.
    3. Seed cones oblong or elongated; scales not peltate.
      4. Seed cones pendent, 1.8–2.5 cm long, scales in 3 pairs, middle pair fertile, leaves appearing 4-whorled; seed wings unequal. . . . . **CALOCEDRUS**
      4' Seed cones erect to reflexed, 1–1.5 cm long, scales in 4–6 pairs, middle 2–3 pairs fertile, leaves opposite in 4 ranks; seed wings equal. . . . . **THUJA**
    3' Seed cones round; scales peltate.
      5. Branchlets in nearly cylindrical 3-dimensional clusters; seeds 5–20 per scale, wing shorter than the width of body. . . . . **HESPEROCYPARIS**
      5' Branchlets in flat clusters; seeds 2–4 per scale, wing equal to or greater than the width of body.
        6. Branchlets glaucous; underside of branchlets with a white pattern of x's; seed cones with 7–10 scales; bark fibrous . . . . . **CHAMAECYPARIS**
        6' Branchlets not glaucous; underside without a white pattern of x's; seed cones with 4–6 scales; bark scaly . . . . . **CALLITROPSIS**

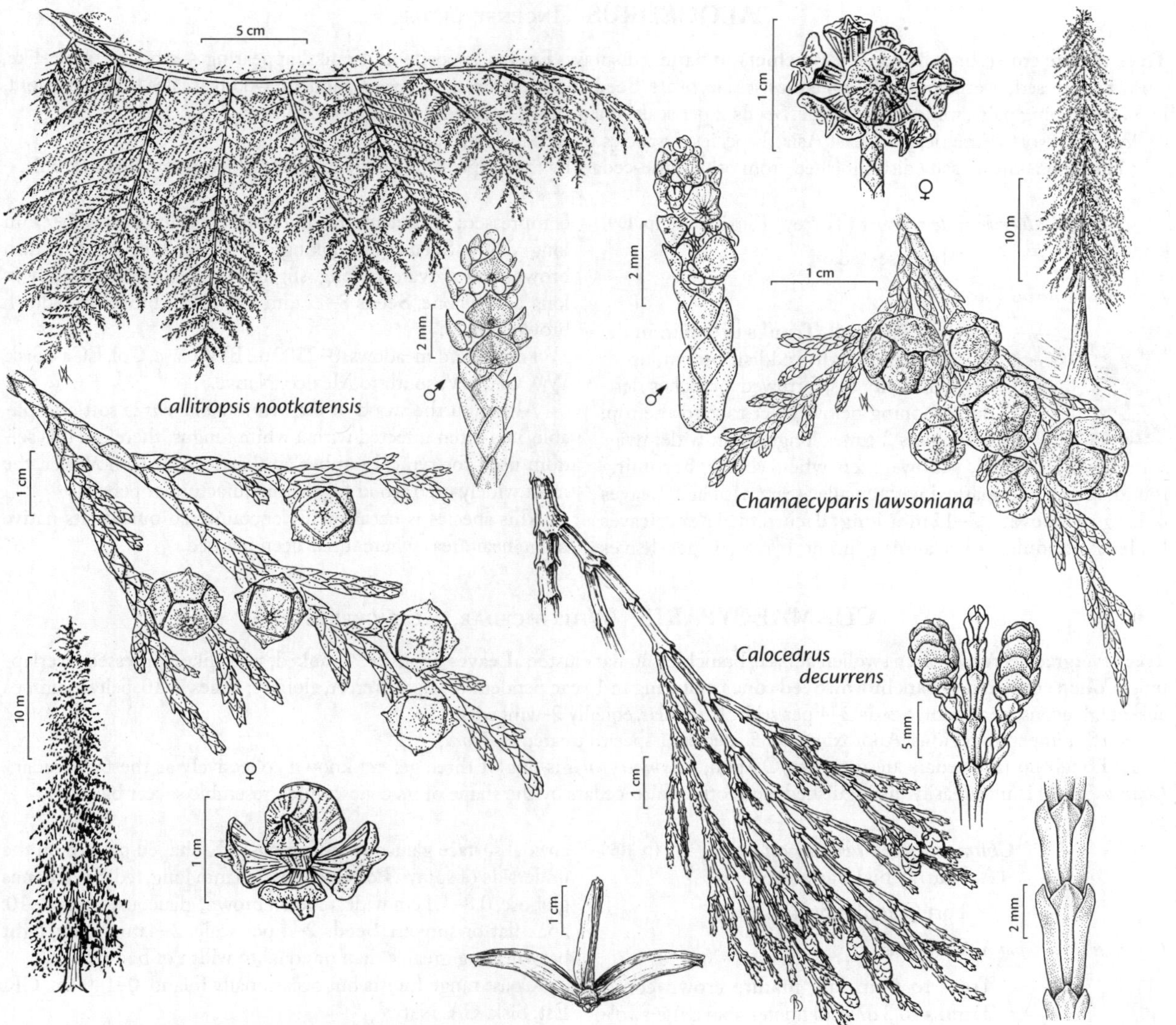

CALLITROPSIS • CALOCEDRUS • CHAMAECYPARIS

CALLITROPSIS Alaska-cedar

Genus comprised of one species, *C. nootkatensis*.

Once considered a member of *Cupressus*, this monotypic genus is now segregated from all other North American "cedar" taxa by many modern authorities.

Callitropsis nootkatensis (D. Don) D.P. Little (p. 109)

Alaska-cedar, Alaska yellow-cedar, yellow-cypress

Chamaecyparis nootkatensis (D. Don) Spach, *Cupressus nootkatensis* D. Don

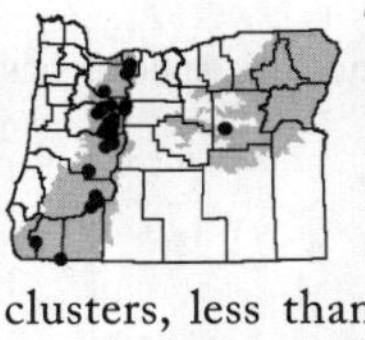

Trees or **shrubs** to 40 m tall; mature crown conic. **Trunks** to 2 m in diameter; bark grayish brown, fissured to expose a bright brown inner bark, up to 2 cm thick; branches spreading and drooping; branchlets in flat clusters, less than 3.5 mm in diameter. **Leaves** opposite, 4-ranked, scale-like, 1.5–2.5 mm long, green to bluish green, not glaucous, rarely resinous. **Pollen cones** 2–5 mm long. **Seed cones** nearly globose, 0.6–1.2 cm wide, dark red-brown, glaucous, resinous; scales 4–6, umbos present. **Seeds** 2–4 per scale, 2–6 mm, reddish brown, sharp pointed, somewhat flattened, wing greater than or equal to width of body.

Mid-elevation mountainous slopes. 1000–2100 m. BW, Casc, Sisk. CA, WA; north to AK. Native.

Callitropsis nootkatensis is not normally harvested for timber in Oregon but is of commercial importance in other states where it is more numerous. This species is distinguished from somewhat similar *Hesperocyparis bakeri* by its flat clusters of branchlets.

CALOCEDRUS Incense-cedar

Trees mature crown broadly conic. **Branchlets** in flattened sprays. **Leaves** opposite in 4 ranks, appearing 4-whorled, scale-like, closely appressed, overlapping. **Pollen cones** sessile, ovate. **Seed cones** maturing in 1 year, pendent, woody; scales thin, persistent, in 3 overlapping pairs, middle pair fertile. **Seeds** 2 per scale, unequally 2-winged.

Western North America and East Asia. 3 species; 1 species treated in *Flora.*

Calocedrus is most easily distinguished from other false-cedars by the shape of its pendent cones, which resemble duck bills.

Calocedrus decurrens (Torrey) Florin (p. 109)

Incense-cedar

Libocedrus decurrens Torr.

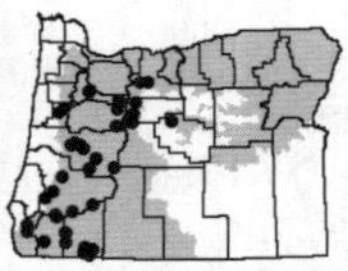

Trees to 50 m tall. **Trunks** to 3–5 m in diameter; bark bright reddish brown, up to 2 cm thick, fibrous, furrowed; branches slender, drooping below, erect toward the top; branchlets 2 times longer than wide; twigs somewhat flattened, yellow-green when young, becoming round, brown and marked with circular scars with age. **Leaves** oblong or obovate, 3–12 mm long, decurrent; lateral leaves keeled, glandular, overlapping inner leaves; inner leaves compressed; apex acute to mucronate. **Pollen cones** 5–7 mm long, yellow. **Seed cones** oblong, 1.8–2.5 cm long, light reddish brown; scales ovate-oblong, slightly convex, approximately as long as the cone. **Seeds** 8–12 mm long, light brown to reddish brown. **2*n***=22.

Forests and meadows. 0–2300 m. BW, Casc, Col, ECas, Sisk, WV. CA, NV; south to Mexico. Native.

Although the wood of *Calocedrus decurrens* is soft and pliable, it is often infected with a white fungus; therefore it is seldom used for typical lumber applications. It is, however, the most widely used wood for the manufacture of pencils.

This species is occasionally encountered outside its native range near areas where it has been planted.

CHAMAECYPARIS White-cedar, false-cypress

Trees evergreen. **Trunks** often swollen at base; branchlets in flat clusters. **Leaves** opposite, 4-ranked, scale-like, appressed, overlapping. **Pollen cones** red to dark brown. **Seed cones** maturing in 1 year, pendent to slightly erect, globose; scales 7–10, peltate, umbos absent, glaucous, persistent. **Seeds** 2–4 per scale, lenticular, equally 2-winged.

North temperate regions. Approximately 5 species; 1 species treated in *Flora.*

Although no true cedars are native to Oregon, *Chamaecyparis* is one of three genera known collectively as the false-cedars. *Chamaecyparis* is most easily distinguished from other false-cedars by the shape of its cones, which resemble soccer balls.

Chamaecyparis lawsoniana (A. Murray bis) Parlatore (p. 109)

Port Orford cedar

Cupressus lawsoniana A. Murray bis

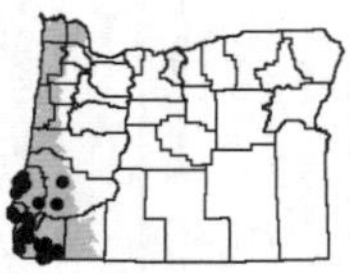

Trees to 60 m tall; mature crown conic. **Trunks** to 3 m in diameter above the swollen base; bark up to 25 cm thick with age, reddish brown, smooth when young, ridged and splitting into long thin shreds; branches horizontal to somewhat drooping; branchlets in flattened clusters. **Leaves** approximately 1.5–3 mm long, light green, acute, abaxial surface glaucous producing an X-shaped pattern on the underside of sprays. **Pollen cones** 2–4 mm long, red. **Seed cones** globose, 0.8–1.1 cm wide, reddish brown, glaucous; scales 7–10, apex flat or sunken. **Seeds** 2–4 per scale, 2–4 mm wide, light brown, wing greater than or equal to width of body. **2*n***=22.

Coast range forests but occasionally inland. 0–1400 m. CR, Est, Sisk. CA. Native.

This species is closely related to the *Chamaecyparis obtusa*, which is considered sacred to some in Japan. As a result, the wood of this species is highly prized by Japanese wood merchants. Overharvesting and a rot caused by *Phytophthora* are current threats to this species.

HESPEROCYPARIS Western cypress

Trees evergreen. **Branchlets** in nearly cylindrical 3-dimensional or flat clusters. **Leaves** opposite, 4-ranked, scale-like, appressed, overlapping. **Pollen cones** yellow. **Seed cones** maturing in 1–2 years, pendent, globose to nearly globose; scales 4–8, peltate, umbos present, glaucous or not, persistent. **Seeds** 2–20 per scale, lenticular, 2-winged.

North temperate regions. Approximately 7 species; 2 species treated treated in *Flora.*

The nomenclatural history of this genus is complex, confusing, and controversial. Traditionally, most species in this genus had been assigned to *Cupressus*, while more recently proposed names include *Xanthocyparis* and *Callitropsis.* At the time of this writing, the general consensus is to assign the generic name *Hesperocyparis* to the former North American *Cupressus* taxa, with the exception of Alaska-cedar (*Callitropsis nootkatensis*).

Hesperocyparis sargentii (Sargent cypress) has recently been collected in Josephine County (2010). Whether this species has historically been present in Oregon or this represents a recent range extension needs further research.

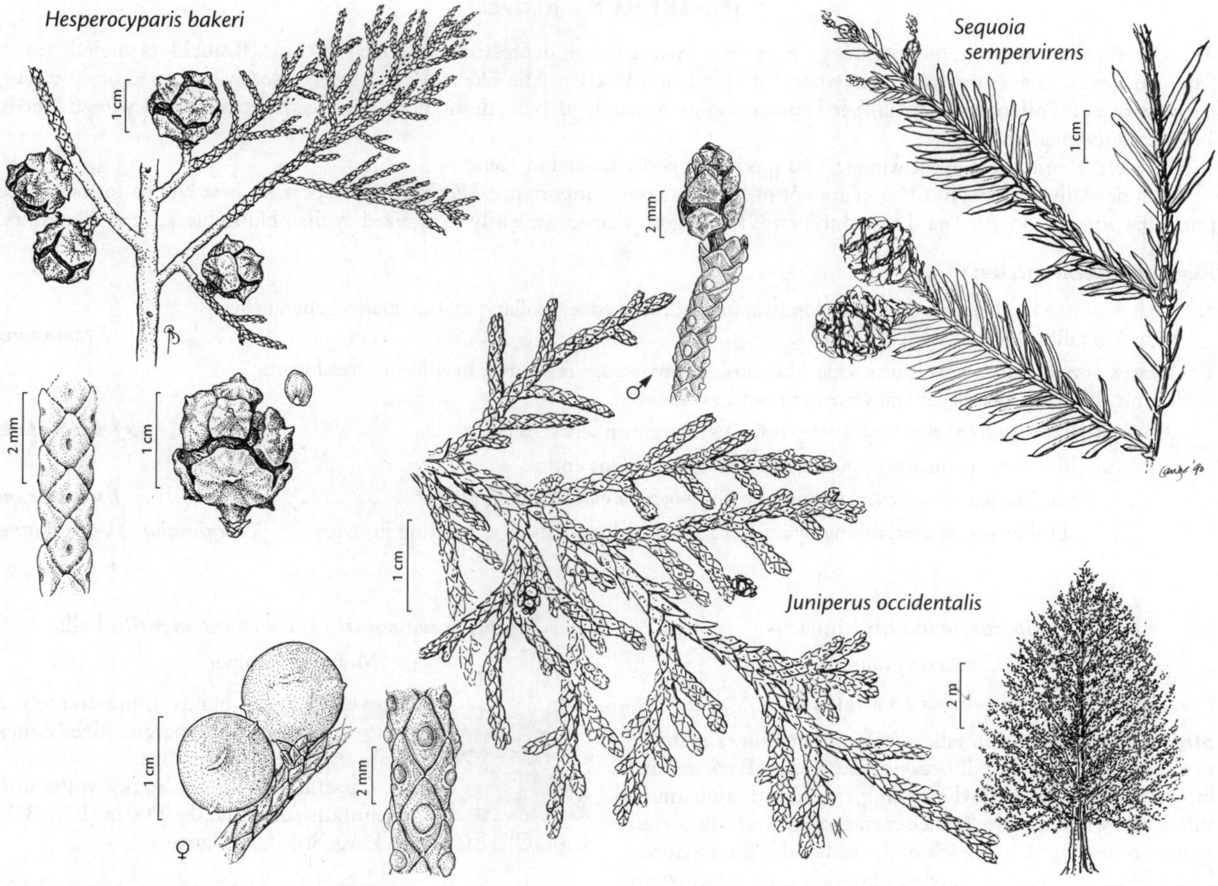

HESPEROCYPARIS • JUNIPERUS • SEQUOIA

Key to *Hesperocyparis* species:

1. Most leaves with a conspicuous, pit-like, abaxial gland; resin copious and sticky. . . . . ***H. macnabiana***
1' Most leaves without a conspicuous, pit-like, abaxial gland; resin not copious or sticky . . . . ***H. bakeri***

Hesperocyparis bakeri (Jepson) Bartel (p. 111)

Modoc cypress

Callitropsis bakeri (Jeps.) D.P. Little, *Cupressus bakeri* Jeps., *Cupressus bakeri* Jeps. ssp. *matthewsii* C.B. Wolf

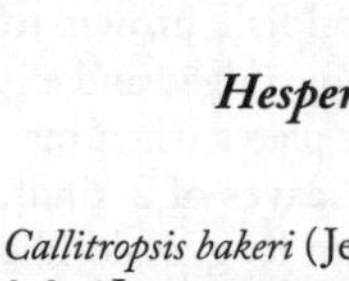

Trees to 25 m tall but usually much shorter; mature crown in a column. **Trunks** to 0.4 m in diameter; bark dark reddish brown, smooth when young, separating into long thin scales with age; branches often crooked; branchlets in nearly cylindrical 3-dimensional clusters, less than 2.5 mm in diameter. **Leaves** approximately 1.5 mm long, gray-green to dark green, glaucous, slightly resinous. **Pollen cones** approximately 2.5 mm long. **Seed cones** silvery, globose, 1–2 cm wide, not glaucous; scales 6–8, umbos present. **Seeds** 5–20 per scale, 2–4 mm, tan to brown, wing less than width of the body.

Mixed evergreen forests. 1100–1700 m. Sisk. CA. Native.

Hesperocyparis macnabiana (A. Murray bis) Bartel

MacNab cypress

Callitropsis macnabiana (A. Murray bis) D.P. Little, *Cupressus macnabiana* A. Murray bis

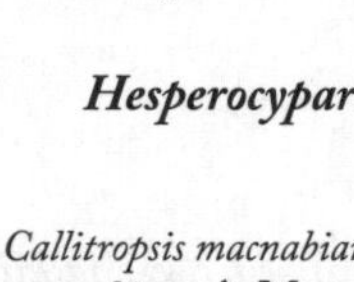

Trees to 10 m; crown broadly conical. **Trunks** to 1 m in diameter; bark rough, furrowed, fibrous; branchlets comb-like, to 1 mm in diameter. **Leaves** ovate, approximately 1.5 mm long, dark green, occasionally glaucous, with conspicuous pit-like abaxial gland, resin copious and sticky, sometimes glaucous. **Pollen cones** 15–25 mm long. **Seed cones** globose, 1.5–2.5 cm long, brown to gray, not glaucous; scales 6–8, umbos present, 2–4 mm long. **Seeds** 2–5 mm, light to medium brown, wing less than width of the body.

Lower elevation woodland, often on serpentine substrates. 300–1400 m. Sisk. CA. Native.

While long rumored to occur in Oregon, it was not until 2010 that botanist Frank Callahan located and documented this species in a remote area of Jackson County.

JUNIPERUS Juniper

Trees or **shrubs** evergreen, monoecious or dioecious; mature crown depressed to rounded or conic. **Branchlets** angled, not in flattened sprays. **Leaves** opposite or in whorls of 3, of 1 or 2 kinds, needle-like and/or scale-like, closely appressed or divergent, resinous or not. **Pollen cones** small. **Seed cones** globose to nearly globose, fleshy, berry-like, seeds retained; scales fused. **Seeds** 1–3 per cone, wingless.

Northern Hemisphere. Approximately 60 species; 4 species treated in *Flora*.

Commercially, junipers in Oregon are not of high economic importance. Worldwide, junipers are best known for being the prime flavoring in gin. Among the gymnosperms in Oregon, junipers are easily recognized by their bluish black, berry-like cones.

Key to *Juniperus* species:

1. Leaves of one kind, needle-like, not closely appressed; seed cones axillary; mature plants generally less than 1 m tall . . . . *J. communis*

1' Leaves of two kinds, most leaves scale-like, closely appressed, a few leaves needle-like; seed cones terminal; mature plants generally greater than 1 m tall.

  2. Scale-like leaves in whorls of 3, very resinous, margins finely serrated . . . . *J. occidentalis*

  2' Scale-like leaves generally opposite, not resinous, margins entire.

    3. Scale-like leaves not overlapping or barely so; bark exfoliating in plates . . . . *J. scopulorum*

    3' Scale-like leaves overlapping by at least 25% their length; bark exfoliating in strips . . . *J. virginiana* var. *virginiana*

Juniperus communis Linnaeus

Common juniper

Juniperus communis L. var. *montana* Aiton (in part)

Shrubs dioecious, to 1 m tall, spreading, sometimes mat-like; mature crown depressed, occasionally conic. **Bark** reddish brown to brown, smooth when young, ridged and exfoliating in thin strips when mature; branches spreading; branchlets erect. **Leaves** of one kind, in whorls of 3, needle-like, linear-lanceolate, 0.5–1.5 cm × approximately 2 mm wide, divergent, margins entire; abaxial surface dark green, keeled; adaxial surface white, glaucous, slightly concave, stomatal bands 2–4 times as wide as green margin; apex acute, rarely obtuse. **Pollen cones** 3–6 mm long. **Seed cones** maturing in 2–3 years, globose to elongate, 6–9 mm wide, bluish black, glaucous, short stalked. **Seeds** 1–3 per cone, 3–5 mm wide, brown. ***2n***=22.

North America; Eurasia. 3–5 varieties; 2 varieties treated in *Flora*.

Key to *Juniperus communis* varieties:

1. Stomatal bands 3–4 times as wide as green margins . . . . *J. c.* var. *jackii*

1' Stomatal bands approximately 2 times as wide as green margins . . . . *J. c.* var. *saxatilis*

Juniperus communis Linnaeus var. *jackii* Rehder

Jack's juniper

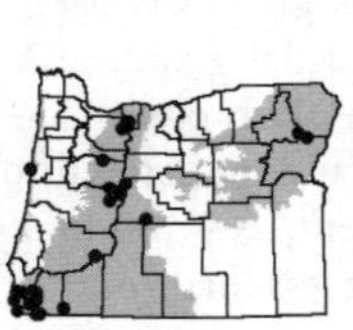

Leaves stomatal bands 3–4 times as wide as green margins. **Seed cones** globose to elongate. **Seeds** 1(2) per cone.

Coastal bluffs, dry rocky soils, and mountain summits. 0–2600 m. BW, Casc, ECas, Est, Sisk. CA. Native.

Juniperus communis Linnaeus var. *saxatilis* Pallas

Mountain juniper

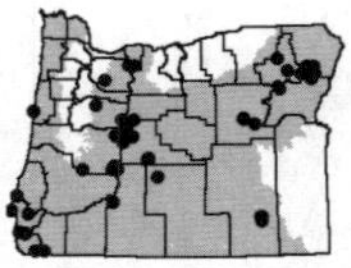

Leaves stomatal bands approximately 2 times as wide as green margins. **Seed cones** globose. **Seeds** 1–3 per cone.

Coastal bluffs, dry rocky soils, and mountain summits. 0–3000 m. BR, BW, Casc, CR, ECas, Est, Lava, Sisk. CA. Native.

Juniperus occidentalis W.J. Hooker (p. 111)

Western juniper

Juniperus californica Carrière var. *siskiyouensis* L.F. Hend., *Juniperus occidentalis* Hook. var. *occidentalis*, *Sabina occidentalis* (Hook.) A. Heller

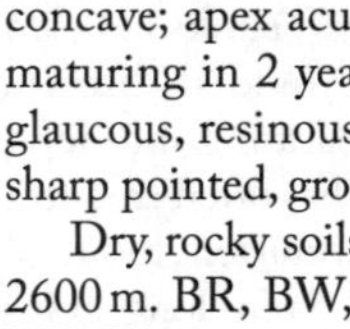

Trees monoecious or dioecious, to 20 m tall; mature crown rounded. **Trunks** to 1 m in diameter; bark light reddish brown to brown, smooth when young, ridged and exfoliating in thin strips or plates when mature; branches spreading; branchlets erect. **Leaves** of 2 kinds, the less numerous leaves needle-like, 3–6 mm long; scale-like leaves in whorls of 3, 1–3 mm long, light green, closely appressed, margins finely serrated; adaxial surface very resinous, concave; apex acute. **Pollen cones** 2–4 mm long. **Seed cones** maturing in 2 years, globose to nearly globose, bluish black, glaucous, resinous, 2–3-seeded. **Seeds** 2–4 mm wide, brown, sharp pointed, grooved. ***2n***=22.

Dry, rocky soils, foothills, mountain slopes and bases. 100–2600 m. BR, BW, Col, ECas, Lava, Owy, Sisk. CA, ID, NV, WA. Native.

Juniperus occidentalis is the most common *Juniper* species in Oregon. This taxon is not widely harvested for its timber, however, it is used occasionally for fence posts, firewood, and carved, novelty artwork.

Juniperus scopulorum Sargent

Rocky Mountain juniper

Sabina scopulorum (Sarg.) Rydb.

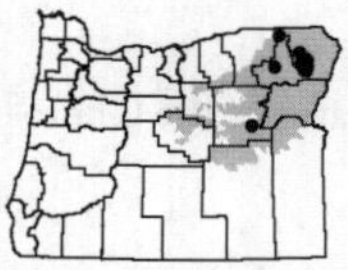

Trees dioecious, to 20 m tall; mature crown rounded. **Trunks** to 1 m in diameter, often branched; bark reddish brown to reddish gray, smooth when young, becoming narrowly ridged and covered with long fibrous scales when mature, exfoliating in plates; branches spreading to ascending. **Leaves** of 2 kinds; needle-like leaves 3–6 mm long; scale-like leaves opposite, 1–3 mm long, light green to blue-gray, margins entire, not or barely overlapping, closely appressed, sometimes glaucous; adaxial surface somewhat concave, not resinous; apex generally acute. **Pollen cones** 2–3 mm long. **Seed cones** maturing in 2 years, globose, 6–8 mm wide, light blue to bluish black, glaucous, resinous to fibrous, usually 2-seeded. **Seeds** 4–5 mm wide, brown, sharp pointed, angled.

Rocky soils on hills and slopes. 800–2000 m. BW. ID, NV, WA; north to AK, east to ND, southeast to TX and Mexico. Native.

In Oregon, this species is found only in the Wallowa Mountains. Morphologically it is very similar to *Juniperus virginiana*, but it can be distinguished from the latter by its non-overlapping or barely overlapping leaves.

Juniperus virginiana Linnaeus var. *virginiana*

Eastern red-cedar

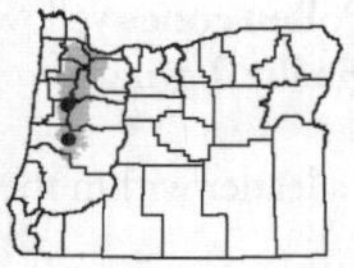

Trees dioecious, to 30 m tall; mature crown conic or dome-shaped. **Trunks** to 1.5 m in diameter; bark brown to reddish brown, smooth when young, fissured and exfoliating in long strips when mature; branches spreading to ascending. **Leaves** of 2 kinds; needle-like leaves 3–6 mm long; scale-like leaves green, usually opposite, rarely in whorls of 3, 2–4 mm long, overlapping by more than 25% of their length, closely appressed, margins entire; adaxial surface slightly concave, not resinous; apex acute. **Pollen cones** 2–3 mm long. **Seed cones** maturing in 1 year, globose to nearly globose, 4–7 mm wide, bluish black, glaucous, resinous, 1–2(3) seeded. **Seeds** round, 2–4 mm wide, brown, sharp pointed.

Wet remnant prairies (in Oregon). 0–200 m. WV. Eastern North America. Exotic.

Juniperus virginiana is a recent introduction to Oregon, having first been reported to escape cultivation in 1988. Currently this taxon has only been recorded as established in a few localities in the Willamette Valley and one unconfirmed locality in Union County.

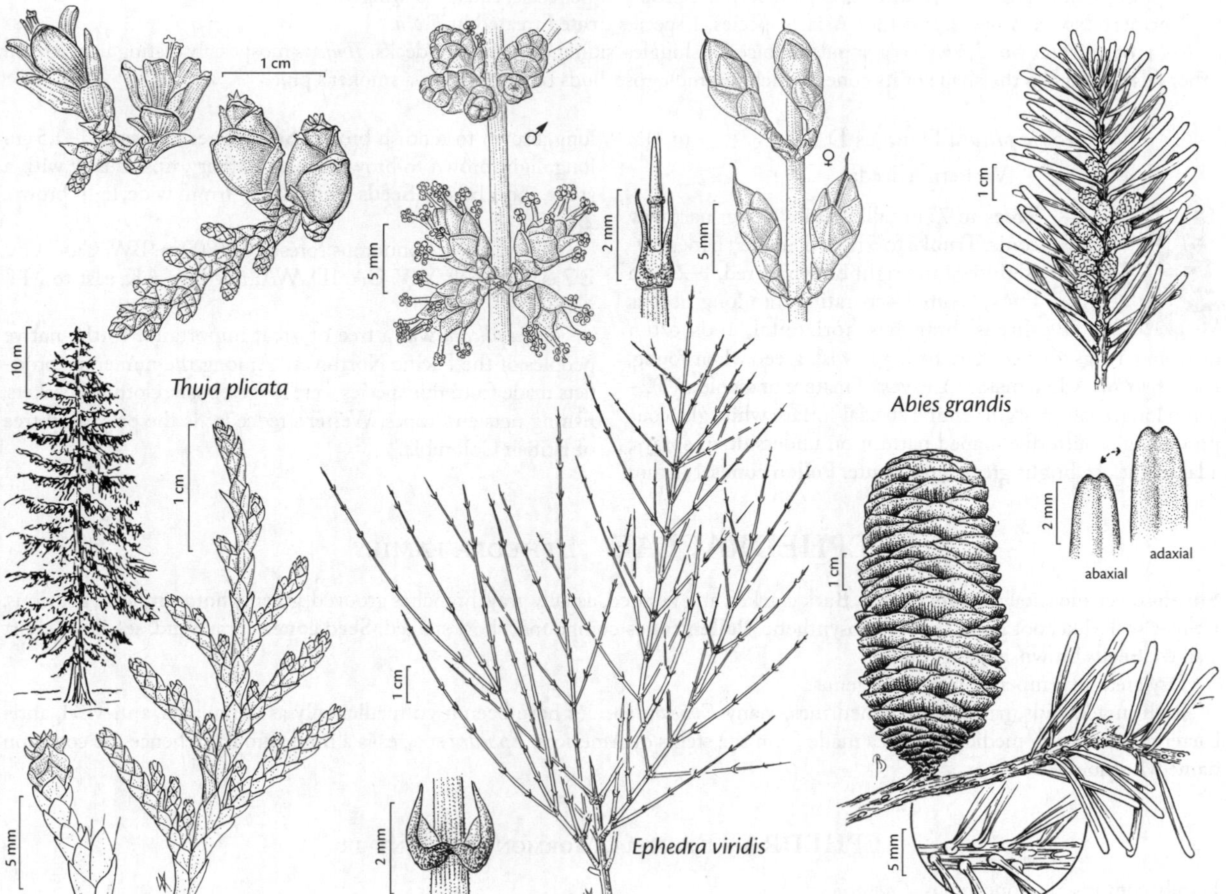

THUJA • EPHEDRA • ABIES

SEQUOIA Redwood

Trees giant, evergreen; mature crown cylindrical. **Branches** slender, round in cross section. **Leaves** alternate, in 2 ranks, of 2 kinds, linear and awl-like, divergent to appressed. **Pollen cones** yellow to brown. **Seed cones** maturing in 1 year, pendent, woody; scales persistent, peltate. **Seeds** 2–7 per scale, marginally 2-winged.

Western North America. 1 species.

This genus is named for Chief Sequoyah, a leader within the Cherokee tribe in the early 19th century and creator of the Cherokee alphabet.

Sequoia sempervirens (D. Don) Endlicher (p. 111)
Redwood, coast redwood

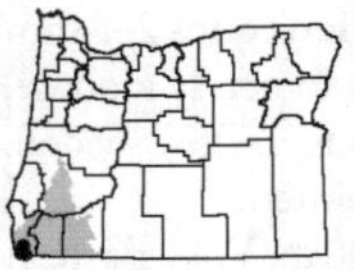

Trees to 110 m tall. **Trunks** to 9 m in diameter, bases swollen, buttressed and somewhat fluted; bark reddish brown, up to 35 cm thick, deeply furrowed with broad round ridges; branches drooping below, erect toward the top, lower portion of trunk usually devoid of branches. **Leaves** of higher branchlets awl-like, appressed, less than 6 mm long; leaves of mature branchlets linear, spreading, 6–20 mm long, somewhat curved; abaxial surface slightly revolute, ribbed, with prominent white stomatal bands, margins entire; adaxial surface dark green, shiny; apex acute, often sharply stiff pointed. **Pollen cones** ovoid, 2–5 mm long, short stalked. **Seed cones** oblong, 2–3.5 cm long, reddish brown; scales 15–25. **Seeds** lenticular, 3–6 mm wide. **2*n***=66.

Coastal forests. 50–200 m. Sisk. CA. Native.

Redwoods are one of the few conifers that are able to regrow vegetatively from their stumps, an ability that helps them regenerate after fires. Due to its color and resistance to decay, lumber from this species is popular for both interior and exterior building applications. Collectively, *Sequoia sempervirens* and *Sequoiadendron giganteum* are the state trees of California.

THUJA Arborvitae

Trees monoecious, evergreen. **Branches** spreading, horizontal; branchlets in flat clusters, generally horizontal. **Leaves** opposite in 4 ranks, closely appressed, scale-like. **Pollen cones** ovate. **Seed cones** maturing in 1 year, erect to reflexed, woody; scales persistent, overlapping, oblong, basifixed, middle 2–3 pairs fertile. **Seeds** 1–3 per scale, equally 2-winged.

Northern North America and East Asia. 5 species; 1 species treated treated in *Flora*.

As a rot resistant wood, *Thuja* is a popular choice for shingles, siding, and outdoor decks. *Thuja* is most easily distinguished from other false-cedars by the shape of its cones, which resemble rose-buds or the bowl of a smoker's pipe.

Thuja plicata Donn ex D. Don (p. 113)
Western redcedar

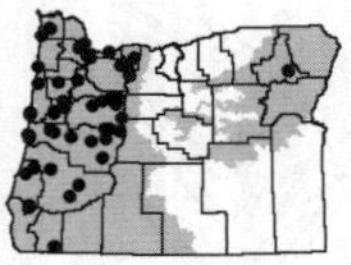

Trees to 75 m tall; mature crown narrowly conic. **Trunks** to 5 m in diameter; bark grayish brown to bright brownish red, 1–2.5 cm thick, fissured, separating into long fibrous shreds; branches horizontal, ends often drooping; twigs 4-sided, flattened, yellowish green when young, reddish brown when mature. **Leaves** of mature branchlets (1)2–6 mm long, ovate, margins entire; abaxial surface white glaucous producing a butterfly-shaped pattern on underside of sprays; adaxial surface bright green; apex acute. **Pollen cones** 1–3 mm long, brown to reddish brown. **Seed cones** ellipsoid, 1–1.5 cm long, light brown to brown; scales leathery, tips usually with a stout, short bristle. **Seeds** lenticular, 4–6 mm wide, light brown. **2*n***=22.

Moist, mixed coniferous forests. 0–1800 m. BW, Casc, CR, ECas, Est, Sisk, WV. CA, ID, WA; north to AK, east to MT. Native.

Thuja plicata was a tree of great importance to the native peoples of the Pacific Northwest. Among the numerous products made from this species were lodges, boats, clothing, baskets, fishing nets and ropes. Western redcedar is the provincial tree of British Columbia.

EPHEDRACEAE Ephedra family

Shrubs erect, dioecious. **Roots** fibrous. **Bark** cracked and fissured, usually gray; branches grooved, green, photosynthetic, glaucous. **Leaves** scale-like, not green, not photosynthetic. **Pollen cones** compound, short stalked. **Seed cones** compound, sessile or short stalked. **Seeds** brown, angled at top.

Dry areas in temperate regions. 1 genus.

Rich in alkaloids, particularly ephedrines, many *Ephedra* species have been used medicinally as stimulants, anti-viral, antibacterial, and cough medicines. A tea made from the stems of American *Ephedra* species is a mild stimulant, hence the common name Mormon tea.

EPHEDRA Ephedra, Mormon tea, jointfir

Family comprised of one genus, *Ephedra*.

Approximately 60 species; 2 species treated in *Flora*.

Key to *Ephedra* species:

1. Twigs pale green to bluish green; leaves deciduous. . . . . **E. nevadensis**
1' Twigs bright green to yellowish green; leaves persistent. . . . . **E. viridis**

Ephedra nevadensis S. Watson

Nevada ephedra

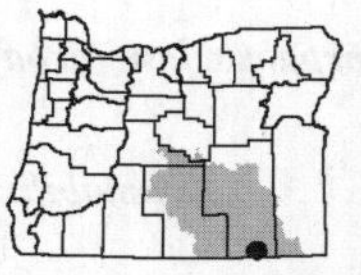

Shrubs 0.25–1 m. **Branches** alternate or whorled; angle of divergence about 45°, twigs pale green to bluish green, fading to gray, glaucous when young. **Buds** conic, apex obtuse. **Leaves** opposite, rarely in whorls of 3, 2–6 mm long, deciduous, leaf bases gray. **Pollen cones** 1–5 per node, ellipsoid, 4–8 mm. **Seed cones** 1–several per node, globose, 5–10 mm; bracts opposite, 3–5 pairs, circular, margins entire, brown to green center. **Seeds** 1–2, globose to ellipsoid, 6–9 mm, smooth. **2*n***=28.

Dry, rocky slopes and sandy flats. 0–1900 m. BR. CA, NV; southeast to AZ. Native.

Typically coning in late winter to early spring, *Ephedra nevadensis* is most easily distinguished from our only other *Ephedra* species, *E. viridis*, by the bluish green color of its twigs.

Ephedra viridis Coville (p. 113)

Green ephedra, green Mormon tea

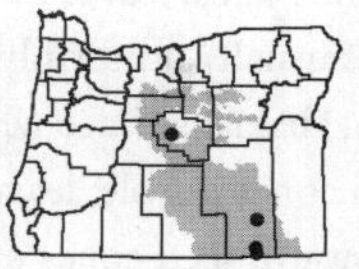

Shrubs 0.5–1 m. **Branches** alternate or whorled; angle of divergence about 30°; twigs bright green to yellowish green, fading to yellow. **Buds** conic, apex obtuse. **Leaves** opposite, 2–5 mm long, persistent, leaf bases brown to black. **Pollen cones** 2–6 per node, ovoid, 5–7 mm. **Seed cones** 2–several per node, ovoid, 6–10 mm; bracts opposite, 6–10 pairs, ovate, margins entire, yellow center. **Seeds** 2, ellipsoid, 5–8 mm, smooth. **2*n***=14, 28.

Dry, rocky slopes, canyons, and sagebrush scrub. 900–1900 m. BR, Lava. CA, NV; east to CO, southeast to NM. Native.

Ephedra viridis typically cones in spring and is easily distinguished by its yellowish green twigs.

PINACEAE Pine family

Trees evergreen (deciduous in *Larix*), monoecious. **Bark** smooth to scaly or furrowed. **Leaves** needle-like, simple, alternate or spirally arranged. **Pollen cones** maturing annually, axillary. **Seed cones** woody, maturing in 1–3 seasons; scales imbricate. **Seeds** 2 per scale, winged or wingless.

Primarily Northern Hemisphere. 10 genera; 6 genera treated in *Flora*.

Of major economic importance, the members of this family are used as ornamentals and as a source of turpentine, tars, essential oils, and food. Additionally, most of the world's softwood timber is harvested from Pinaceae species. Primarily a Northern Hemisphere family, only one species, *Pinus merkusii*, is found south of the equator.

Key to *Pinaceae* genera:

1. Leaves in bundles of 2 or more, base scaly sheathed.
   2. Leaves deciduous, in clusters of 10 or more. . . . . **LARIX**
   2' Leaves evergreen, in clusters of 2–5 . . . . **PINUS**
1' Leaves borne singly, base not scaly sheathed.
   3. Leaf tips sharp pointed or acute, more or less square-shaped, triangular, or somewhat flattened in cross section. . **PICEA**
   3' Leaf tips rounded or notched, flattened in cross section (but square-shaped in *Abies procera*).
      4. Seed cones erect, scales and bracts deciduous . . . . **ABIES**
      4' Seed cones pendent, scales and bracts persistent.
         5. Twigs with peg-like projections after leaves fall; seed cone bracts shorter than scales . . . . **TSUGA**
         5' Twigs without peg-like projections; seed cone bracts longer than scales . . . .**PSEUDOTSUGA**

ABIES Fir

Trees large to medium sized, evergreen. **Bark** initially smooth, generally becoming furrowed with age; branches short, horizontal to drooping, those bearing cones may be ascending; leaf scars prominent. **Leaves** borne singly, persisting at least 5 years, appearing 1- or 2-ranked, sessile, sheath absent. **Pollen cones** grouped. **Seed cones** deciduous, maturing in 1 season, erect, cylindrical or oblong-cylindrical, not falling as a whole; scales fan-shaped. **Seeds** winged.

North temperate regions. Approximately 44 species; 4 species and 2 hybrids treated in *Flora*.

Firs are a major component of Pacific Northwest coniferous forests. In addition to providing food and habitat for many animals, several species are cut for lumber. Firs are also grown on a large scale for use as ornamentals and Christmas trees. Although most major fir species are readily identifiable, hybrid taxa may pose a challenge in distinguishing them from their parental taxa.

When identifying specimens using leaf characteristics, leaves from the lower branches (as opposed to those near the top of the tree) should be used.

Key to *Abies* species:

1. Seed cones with bracts longer than the scales; leaves somewhat to fully 4-sided.
   2. Bracts of seed cones covering more than 90% of the cone surface, strongly reflexed; seed cones 10–15 cm long . . . . . ***A. procera***
   2' Bracts of seed cones covering less than 85% of the cone surface (often much less), usually reflexed; seed cones 15–20 cm long . . . . . ***A. magnifica* × *A. procera***

1' Seed cones with bracts no longer than the scales; leaves flat.
   3. Adaxial and abaxial surface of leaves usually uniformly grayish to bluish green . . . . . ***A. lasiocarpa* var. *lasiocarpa***
   3' Adaxial surface of leaves dark green, abaxial surface white except midrib.
      4. Leaves forwardly directed; seed cones purple; leaves 1.5–2 cm long . . . . . ***A. amabilis***
      4' Leaves spreading or curving upward; seed cones usually green, rarely purple; leaves 1.5–6 cm long.
         5. Leaves spreading, those on lower branches often notched at tip; scales of seed cones equal in length and width . . . . . ***A. grandis***
         5' Leaves spreading to curving upward, those on lower branches rarely notched at tip; scales of seed cones smaller in length than width . . . . . ***A. concolor* × *A. grandis***

Abies amabilis Douglas ex J. Forbes

Pacific silver fir, silver fir

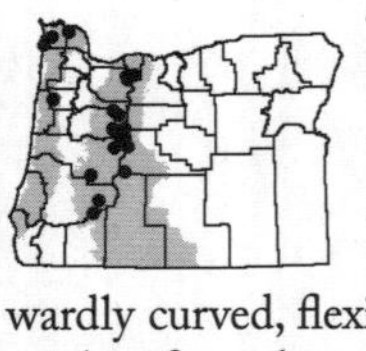

Trees up to 75 m tall; mature crown steeple-like. **Trunks** to 2.6 m in diameter; bark gray, smooth and scaly with age; branches at right angles to trunk; twigs generally opposite, pubescence tan. **Leaves** usually 2-ranked, forwardly curved, flexible, 1–2.5 × 1–3 mm, cross section flat; abaxial surface white, dark midrib with 5–6 stomatal rows on each side; adaxial surface dark green, grooved; apex notched to round. **Pollen cones** red to reddish yellow. **Seed cones** cylindrical, 8–15 cm long, purple; scales puberulent; bracts included, approximately 33% the length of the scales. **Seeds** 10–12 × 4 mm, tan, wings nearly as long as body.

Montane to subalpine forests. 300–2300 m. Casc, CR, ECas, WV. CA, WA; north to southeastern Alaska. Native.

The white, glaucous coating on the abaxial side of *Abies amabilis* leaves gives this species a "silver" appearance, hence the common name, silver fir.

Abies concolor (Gordon & Glendinning) Lindley ex. Hildebrand × *Abies grandis* (Douglas ex D. Don) Lindley

California white fir, Sierra white fir

Abies lowiana (Gordon) A. Murray bis

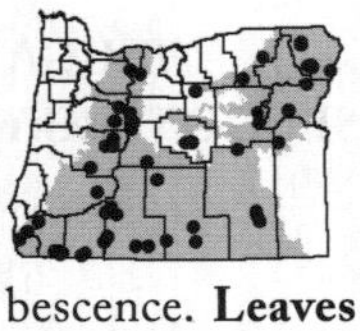

Trees up to 75 m tall; mature crown steeple-like. **Trunks** to 1.5 m in diameter; bark white to gray, smooth, furrowed; branches spreading, drooping; twigs generally opposite, glabrous or with yellow to brown pubescence. **Leaves** 2-ranked or if 1-ranked only on upper branches, spreading to erect and curving upward, flexible, cross section flat, 1.5–6 cm × 1.5–3 mm; abaxial surface grayish to bluish green, often with white glaucous bands, 4–7 stomatal rows on each side of midrib; adaxial surface grayish to bluish green, grooved, with 0–12 stomatal rows; apex generally rounded, rarely notched on lower branches. **Pollen cones** red, purple, yellow or green. **Seed cones** cylindrical, 6–12 cm long, usually green, rarely purple; scales pubescent; bracts included, shorter than the scales. **Seeds** 7–12 × 3 mm, tan, wings 1.5–2 times long as body.

Mid-elevation forests. 900–2500 m. BR, BW, Casc, ECas, Sisk, WV. CA. Native.

DNA evidence indicates that pure *Abies concolor* may not exist in Oregon. Rather, all previously known populations of *A. concolor* are now assigned as *A. concolor* × *A. grandis*. These populations vary in their degree of morphological intermediacy between *A. concolor* and *A. grandis*. However, specimens of *A. concolor* × *A. grandis* can be distinguished from pure *A. grandis* by leaf morphology and the seed cone scales.

Abies grandis (Douglas ex D. Don) Lindley (p. 113)

Grand fir

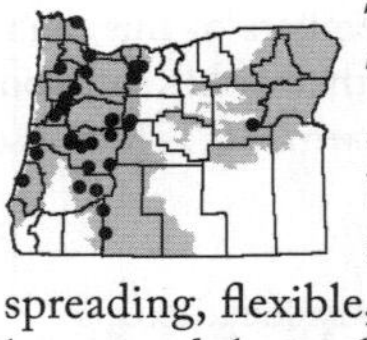

Trees up to 75 m tall; mature crown conic. **Trunks** to 1.5 m in diameter; bark gray, becoming brown with age, shallow-furrowed; branches spreading, drooping; twigs generally opposite, pubescent. **Leaves** 2-ranked, spreading, flexible, cross section flat, 1.5–5 cm × 1.5–3 mm, longer and shorter leaves intermixed; abaxial surface white with 5–7 stomatal rows on each side of midrib; adaxial surface light to dark green, grooved; apex generally notched. **Pollen cones** red, purple, yellow or green. **Seed cones** cylindrical, 8–15 cm long, usually green, rarely purple or gray; scales pubescent; bracts included, shorter than the scales. **Seeds** 6–8 × 3–4 mm, tan with red tinge, wings 2 times as long as body. $2n$=24.

Low to mid elevations in mixed evergreen forests. 0–2000 m. BW, Casc, CR, ECas, WV. CA, WA; north to British Columbia, northeast to MT. Native.

In southern Oregon, the Cascades, and the Blue and Wallowa mountains, *A. concolor* × *A. grandis* individuals are not uncommon.

Abies lasiocarpa (Hooker) Nuttall var. *lasiocarpa*

Subalpine fir

Abies bifolia A. Murray bis

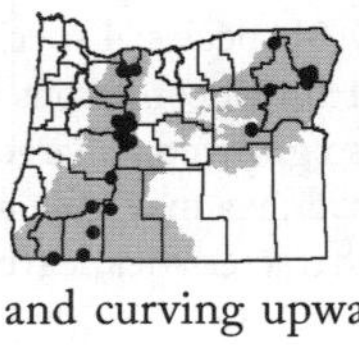

Trees up to 30 m tall; mature crown steeple-like. **Trunks** to 1 m in diameter; bark gray, smooth, furrowed with age; branches straight; twigs opposite to whorled, puberulent for several years. **Leaves** spiraled, erect and curving upward, flexible, cross section flat, 2–3.5 cm ×

1.5–2 mm; abaxial surface grayish to bluish green, often with white glaucous bands, 4–5 stomatal rows on each side of midrib; adaxial surface grayish to bluish green, glaucous, grooved, 4–6 stomatal rows; apex rounded to notched. **Pollen cones** bluish to purplish green. **Seed cones** cylindrical, 6–12 cm long, purple; scales pubescent; bracts included, shorter than the scales. **Seeds** 5–6 × 2–3 mm, tan to light brown, wings 1.5 times as long as body. **2*n***=24.

Subalpine forests, wetlands. 1000–2600 m. BW, Casc, ECas, Sisk, WV. CA, WA; north to AK. Native.

All *Abies lasiocarpa* specimens found in our state are variety *lasiocarpa*. The single other variety, *A. lasiocarpa* var. *arizonica* is endemic to Arizona, New Mexico, and Colorado. While *Abies lasiocarpa* is commonly recognized by its narrow, steeple-like crown, in exposed areas and near tree-line, it often forms "skirts" of low-lying branches that, pressed down by snow, are able to take root and sprout leaders, forming clusters of dwarfed trees. This is our only fir capable of vegetative reproduction.

Abies magnifica A. Murray bis × *Abies procera* Rehder

Shasta red fir

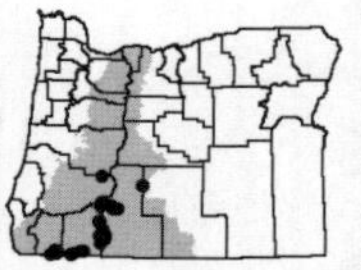

Trees to 60 m tall; mature crown narrow and conic. **Trunks** to 2.5 m in diameter; bark gray, smooth, bark of lower trunk deeply furrowed with reddish plates; branches ascending in crown, descending lower; twigs opposite to whorled, pubescent 1st year, glabrous after 1st year. **Leaves** usually 1-ranked, occasionally 2-ranked on lower branches, curving upward, flexible, cross section 4-sided, occasionally 3-sided, 2–3.5 cm × 2–2.5 mm; abaxial surface bluish green, glaucous, with 8–10 stomatal rows; adaxial surface bluish green, glaucous, often with 2 whitish bands, not or weakly grooved, 8–10 stomatal rows; apex rounded to acute. **Pollen cones** purple. **Seed cones** oblong-cylindrical, 15–20 cm long, purple to greenish brown; scales pubescent; bracts exserted and reflexed over scales, covering less than 85% (often much less) of cone at maturity. **Seeds** 14–16 × 5–7 mm, red, wings approximately same length as body.

Mid-elevation to subalpine forests. 1100–2400 m. Casc, ECas, Sisk. CA. Native.

Morphological and DNA evidence indicates that no pure *Abies magnifica* is likely to exist in Oregon. As a result, all previously described populations of *A. magnifica* are now assigned to *A. magnifica* × *A. procera*. Specimens of *A. magnifica* × *A. procera* can be distinguished from pure *A. procera* by seed cone and seed cone bract morphology.

Abies procera Rehder (p. 118)

Noble fir

Abies nobilis Lindl.

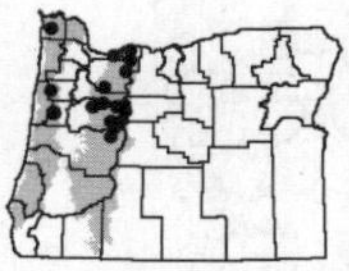

Trees to 80 m tall; mature crown steeple-like, the top rounded. **Trunks** to 2.5 m in diameter; bark gray, smooth when young, furrowed and reddish brown with age; branches at right angles to trunk; twigs opposite to whorled, pubescent for several years. **Leaves** 1-ranked, curving upward, flexible to stiff, 1–3.5 cm × 1.5–2 mm, cross section 4-angled, occasionally weakly 4-angled; abaxial surface bluish green, glaucous, with 2–4 whitish bands, 12–28 stomatal rows; adaxial surface bluish green, glaucous, often with 2 whitish bands, 0–14 stomatal rows; apex rounded to notched. **Pollen cones** reddish. **Seed cones** oblong-cylindrical, 10–15 cm long, purple to green; scales pubescent; bracts exserted and reflexed over scales, covering over 90% of cone at maturity. **Seeds** 11–13 × 2–4 mm, tan to brown, wings slightly longer than body. **2*n***=24.

Mid-elevation to subalpine forests. 300–2000 m. Casc, CR. CA, WA. Native.

Abies procera grows in isolated populations on mountains. As a result of this isolation, morphology may vary greatly between populations. In addition, stands of *A. magnifica* × *A. procera* are known to exist in Oregon. However, *A. procera* is most easily distinguished by its seed cone bracts which cover over 90% of the cones.

LARIX Larch, tamarack

Trees deciduous. **Trunks** straight, slender; bark initially smooth, becoming furrowed with age; branches slender, ascending to drooping; leaf pegs prominent. **Leaves** in dense clusters of 10–30, turning yellow before dropping, soft, sheath absent. **Pollen cones** solitary. **Seed cones** maturing in 1 season, erect, oblong-ovoid, falling as a whole; scales orbicular, persistent. **Seeds** winged.

Cold north temperate regions. 10 species; 1 species treated in *Flora*.

The deciduous needle-like leaves of *Larix* species are unique among the gymnosperms. The hard, heavy, rot resistant wood of *Larix* species is often utilized for outdoor applications such as rail road ties, pilings and fence poles.

Larix occidentalis Nuttall (p. 118)

Western larch

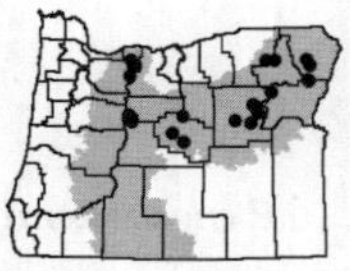

Trees to 40 m tall; mature crown conic. **Trunks** to 2 m in diameter; bark reddish brown, thin when young, thick and furrowed with age, scaly, flaky; branches somewhat ascending on upper branches to drooping on lower branches, in forests often without branches for most of height; twigs stout, orange-brown pubescent when young, becoming glabrous, with prominent leaf pegs. **Leaves** 2–4.5 cm × 0.5–1 mm, soft, flexible, triangular in cross section; abaxial surface pale green, keeled; adaxial surface pale green, with a convex midrib; apex sharp pointed. **Pollen cones** ovate-oblong, less than 1 cm long, yellow, sessile or on short stalks. **Seed cones** oblong-ovoid, 2.5–4 cm long, reddish brown, on short curved stalks; scales ovate; bracts exserted several millimeters beyond the scales, tipped with awns to 3 mm. **Seeds** 3–6 mm long, reddish brown, wings twice as long as body. **2*n***=24.

Mid-elevation forests. 700–2200 m. BW, Casc, ECas, Lava, WV. ID, WA; north to British Columbia, northeast to MT. Native.

The largest of the American *Larix* species, this species produces one of the most valuable timbers in the western United States. Trees of this species may live over 500 years.

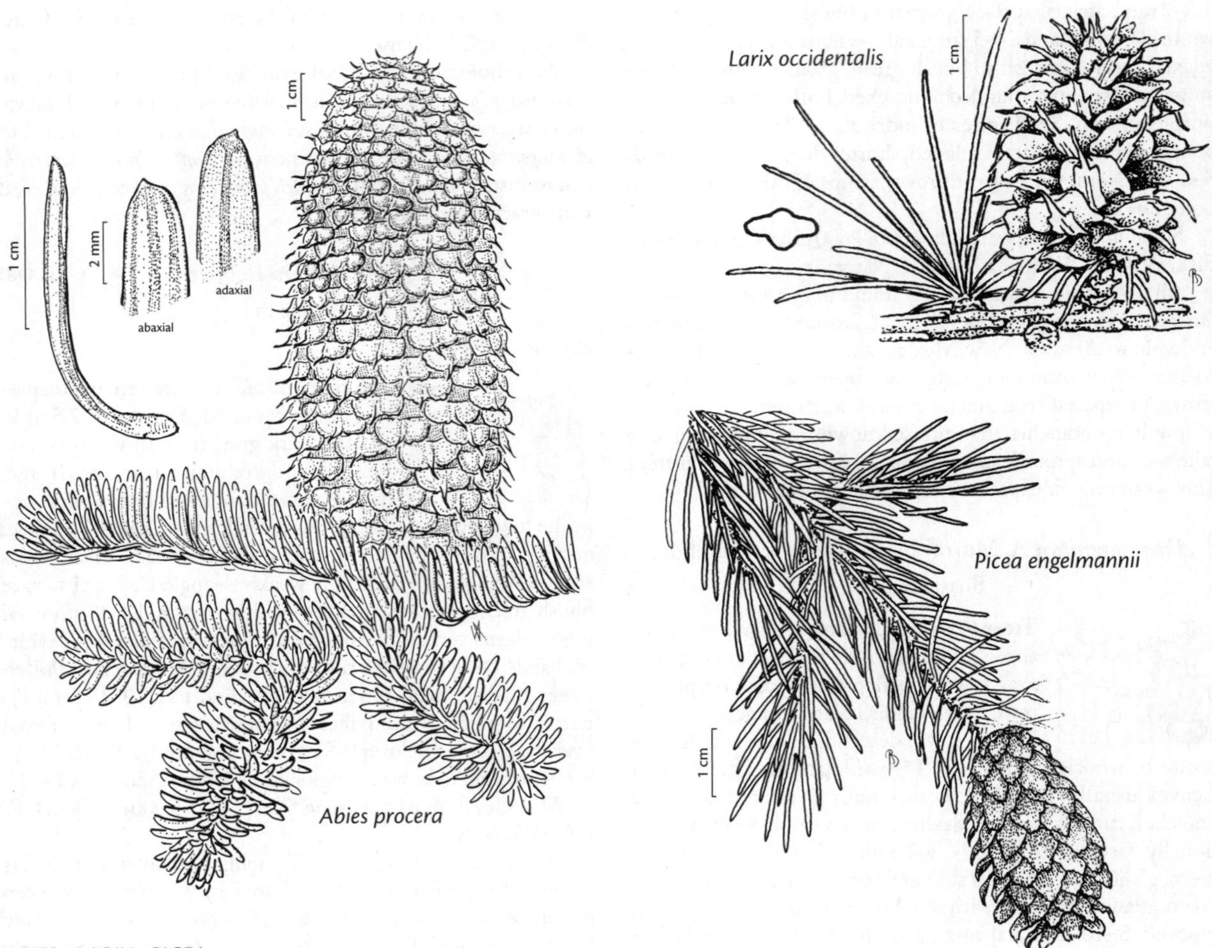

ABIES • LARIX • PICEA

PICEA Spruce

Trees evergreen; crown conic to narrowly conic; top most shoots erect. **Bark** whitish gray to reddish brown, thin, scaly, becoming furrowed with age; branches whorled, horizontal to drooping, leaf scars prominent. **Leaves** rigid, borne singly, persisting 4–10 years, spreading in all directions, 4-sided to flattened or 3-sided, apex sharp to bluntly pointed, sheath absent. **Pollen cones** grouped. **Seed cones** maturing in 1 season, pendent, oblong to cylindrical; scales elliptic or diamond-shaped to fan-shaped, persistent; bracts included.

North temperate regions. Approximately 35 species; 3 species treated in *Flora*.

As a result of their symmetrical conic growth habit, spruce species are popular as ornamentals and Christmas trees. In the past, spruce woods were highly valued in aircraft manufacturing. Notably, the first Wright Brothers aircraft and the Hughes H-4 Hercules (nicknamed the "Spruce Goose") were constructed from the wood of this genus.

Key to *Picea* species:

1. Branches conspicuously lax and drooping; leaves with blunt apexes; seed cone scales with usually entire margins, fan-shaped . . . . *P. breweriana*

1' Branches generally spreading, not lax and drooping or if drooping only somewhat; leaves with sharp apexes; seed cone scales with usually erose margins, elliptic to diamond-shaped.

   2. Leaves 4-sided or nearly so, stomatal bands conspicuous on both adaxial and abaxial surfaces; seed cone bracts less than 25% the length of the scales . . . . *P. engelmannii*

   2' Leaves flattened or 3-sided, stomatal bands conspicuous on adaxial surface only (stomatal bands on abaxial surface, if present, weak); bracts greater than 25% the length of the scales . . . . *P. sitchensis*

Picea breweriana S. Watson

Weeping spruce, Brewer spruce

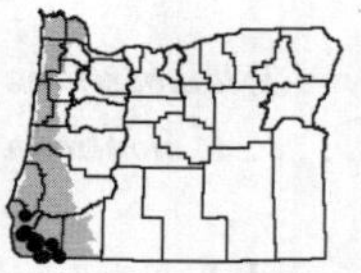

Trees to 40 m tall; mature crown conic. **Trunks** to 1.3 m in diameter; bark whitish gray to brown, scaly, becoming furrowed with age; branches lax and drooping; twigs grayish brown, slender, pendent, pubescent. **Leaves** spreading in all directions, flattened, 1.5–3 cm × 1–2 mm, triangular or rounded triangular in cross section, somewhat rigid; abaxial surface dark green, shining, angled or rounded, without stomatal bands; adaxial surface flattened, whitish green, glaucous, ridged, with conspicuous stomatal bands; apex blunt. **Pollen cones** 1–2 cm long, purple to reddish. **Seed cones** cylindrical and narrowed at both ends, 6–12 cm long, purple to orange-brown, rigid, glaucous when young; scales fan-shaped, 1.5–2 × 1.5–2 cm, margins usually entire; bracts less than 25% the length of the scales. **Seeds** 2.5–3.5 mm long, brown, wings 4 times as long as body.

Montane and subalpine ridges. 1000–1800 m. CR, Sisk. CA. Native.

The rarest of the North American spruces, this species is not generally harvested for lumber. However, *Picea breweriana* is utilized as an ornamental, particularly in northern Europe, where it is popular because of its drooping branches and twigs, unique among the spruces.

Picea engelmannii Parry ex Engelmann (p. 118)

Engelmann spruce

Picea engelmannii Parry ex Engelm. var. *engelmannii*, *Picea engelmannii* Parry ex Engelm. var. *glabra* Goodman

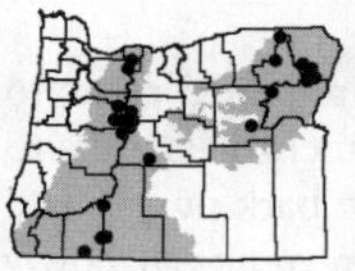

Trees to 55 m tall; mature crown narrowly conic. **Trunks** to 2 m in diameter; bark gray to reddish brown, scaly, deeply furrowed with age; branches horizontal to somewhat drooping; twigs greenish yellow to dark yellow-brown, stout, not pendent, pubescent. **Leaves** spreading in all directions, 4-angled in cross section, 2–3 × 1–1.5 cm, rigid; abaxial and adaxial surfaces blue-green, glaucous when young, ridged, with 2 obvious white stomatal bands; apex sharply pointed. **Pollen cones** 1–2 cm long, purplish to reddish. **Seed cones** ovoid-oblong, broadest near middle, 3–7 cm long, flexible, light green when young, light brown when mature; scales elliptic to diamond-shaped, 1–2 × 1–1.5 cm wide, margins usually erose; bracts less than 25% the length of the scales. **Seeds** 2.5–3.5 mm long, brown, wings twice as long as body. ***2n***=24.

Wetlands, montane and subalpine forests. 900–2400 m. BW, Casc, ECas, Sisk. CA, ID, NV, WA; north to British Columbia, northeast to MT, southeast to NM, south to Mexico. Native.

As a weak, knotty wood *Picea engelmannii* is utilized mainly for pulp and low grade lumber. Slow-growing, high altitude trees are used occasionally for musical instruments. Among the Oregon spruces, *P. engelmannii* is the only species with four-sided leaves.

Picea sitchensis (Bongard) Carrièrre

Sitka spruce

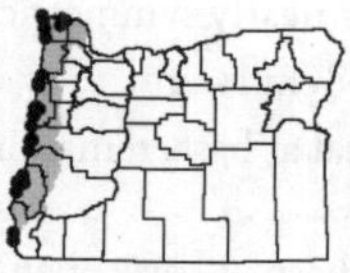

Trees to 70 m tall; mature crown narrowly conic. **Trunks** to 5 m in diameter; bark grayish brown to reddish brown, thin, scaly; branches horizontal to somewhat drooping; twigs yellowish to reddish brown, stout, not pendent, glabrous. **Leaves** spreading in all directions, triangular or rounded triangular in cross section, flattened, 1–3 cm × 1–1.5 mm, rigid; abaxial surface yellow-green to blue-green, angled or rounded, with faint or absent stomatal bands; adaxial surface whitish green, flattened, glaucous, ridged, with conspicuous stomatal bands; apex sharp pointed. **Pollen cones** 2–4 cm long, reddish. **Seed cones** oblong, 5–10 cm long, yellow to reddish brown, rigid; scales elliptic to diamond-shaped, 1.5–2 × 1–1.5 cm, margins usually erose; bracts greater than 25% the length of the scales. **Seeds** 1.5–2.5 mm long, red-brown, wings 6 times as long as the body. ***2n***=24.

Coastal forests. 0–1000 m. CR, Est. CA, WA; north to AK. Native.

As a result of its high strength to weight ratio, *Picea sitchensis* is valued as lumber for musical instruments, boats, and, in the past, aircraft. Only found in wet coastal forests, *P. sitchensis* is distinguished from other spruces in our area by the combination of its flat to three-sided leaves and spreading branches.

PINUS Pine

Trees evergreen; mature crowns conic to irregular. **Bark** smooth when young, becoming fissured with age; branches whorled, ascending to descending; twigs generally resinous. **Leaves** simple, alternate, in clusters of 2–5, square in cross section or 2–3-angled and rounded on abaxial surface, apex acute, sheath deciduous or persistent. **Pollen cones** in clusters. **Seed cones** maturing in 2(3) years, pendent to erect, conic to cylindric; scales persistent; bracts included.

North temperate and mountain regions. Approximately 100 species; 11 species treated in *Flora*.

In Oregon, *Pinus* species are readily identified by their leaves in clusters of two to five. Only one other gymnosperm genus in Oregon (*Larix*) has clustered leaves. The leaves of *Larix*, however, are in clusters of ten or more and are deciduous.

Pinus muricata (Bishop pine) has been collected once in Oregon, in Curry County (1965). Whether this species has historically been present in Oregon or this represents a relatively recent range extension needs further research. It is a two-needled pine, distinguished from *P. contorta* and *P. sylvestris* by its larger leaves and cones (both over 5 cm long).

Pinus radiata (Monterey pine) is known to occasionally escape near tree plantations where the species is cultivated.

Key to *Pinus* species:

1. Leaves in clusters of 5.
    2. Leaves straight; seed cones usually 15 cm long or longer, stalks greater than 2 cm long.
        3. Seed cones generally greater than 25 cm long; conspicuous stomatal lines on all surfaces of leaves . . . ***P. lambertiana***
        3' Seed cones shorter than 25 cm long, conspicuous stomatal lines only on adaxial surface of leaves . . . . . ***P. monticola***
    2' Leaves curved; seed cones less than 15 cm long; stalks absent or less than 2 cm long.
        4. Leaves 1.5–4 cm long; seeds wingless or nearly so; umbos central . . . . ***P. balfouriana***
        4' Leaves 3–7 cm long; seed wings as long or longer than body; umbos terminal.
            5. Mature seed cones 9 cm long or less, ovoid to nearly globose, remaining closed at maturity; scales often broken off by animals . . . . ***P. albicaulis***
            5' Mature seed cones 7 cm long or greater, oblong to lance-cylindrical, opening at maturity; scales attached . . . . ***P. flexilis***

1' Leaves in clusters of fewer than 5.
    6. Leaves in clusters of 2, less than 8 cm long.
        7. Base of seed cones asymmetric . . . . ***P. contorta***
        7' Base of seed cones symmetric or nearly symmetric. . . . . ***P. sylvestris***
    6' Leaves in clusters of 3, greater than 8 cm long.
        8. Seed cones strongly asymmetrical at base, remaining closed for many years after maturity; leaves 8–14 cm long. . . . . ***P. attenuata***
        8' Seed cones symmetrical or nearly so at base, opening at maturity; leaves 12–32 cm long.
            9. Seed cones less than 18 cm long . . . . ***P. ponderosa***
            9' Seed cones usually greater than 15 cm long.
                10. Seed cone apophyses continuous with umbos forming long, upcurving claws; buds resinous . . . ***P. sabiniana***
                10' Seed cone apophyses not continuous with umbos and not forming long, upcurving claws; buds not resinous . . . . ***P. jeffreyi***

Pinus albicaulis Engelmann

Whitebark pine

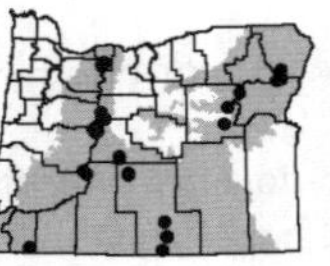

Trees to 20 m tall; mature crown compact and conic or rounded, sometimes deformed by wind. **Trunks** to 1.5 m in diameter, straight to contorted and twisted; bark whitish to gray, smooth when young, fissured into scales with age; inner bark reddish; branches horizontal to ascending, usually persistent to base; twigs orange to dark reddish brown when young, gray to gray-brown when mature, slightly roughened by scars. **Buds** ovoid, light red-brown, not resinous. **Leaves** in clusters of 5, usually upcurved, 3–7 cm × 1–2 mm, dark green to yellow-green, margins entire; adaxial surfaces whitened by stomatal bands; apex sharply stiff-pointed; sheaths deciduous. **Pollen cones** ovoid, 10–15 mm long, dark red. **Seed cones** ovoid to sub-globose, 4–9 cm long, gray to dark purple, erect, symmetric, sessile to short stalked, maturing in 2 years, remaining closed, generally opened by animal activity; umbos terminal, short, tip acute, prickled. **Seeds** obovoid, sharp pointed, 7–12 mm long, dark brown, wingless. ***2n***=24.

Montane forests, rocky soils near timberline. 1300–2800 m. BR, BW, Casc, ECas, Sisk. CA, ID, NV, WA; north to British Columbia, northeast to Alberta, east to WY. Native.

The cones of this species remain closed and are generally ripped open by animals, and in this manner the seeds are dispersed.

Pinus attenuata Lemmon (p. 122)

Knobcone pine

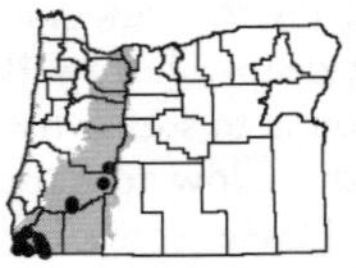

Trees to 30 m tall; mature crown conic to rounded, often many branched. **Trunks** to 1 m in diameter, straight; bark purplish to dark brown, smooth when young, shallowly fissured into scaly plates when mature; branches ascending; twigs dark yellow brown, roughened by bud-scales. **Buds** oblong-ovoid, dark brown, resinous. **Leaves** in clusters of 3, straight to slightly curved, twisted, 8–14 cm × 1–2 mm, yellowish green, margins serrulate; all surfaces with white stomatal bands; apex sharp callous-tipped; sheaths persistent. **Pollen cones** ellipsoid, 10–15 mm long, light brown to orange brown. **Seed cones** oblique-ovoid, very asymmetric, 6–15 cm long, light brown to reddish brown, reflexed to recurved, stalks less than 1 cm long, maturing in 2 years, remaining closed for several decades or opening at burning; umbos central, pyramidal, sharp pointed, prickled. **Seeds** oblong-oval, sharp pointed, 6–7 mm long, black, wings 3 times as long as body. ***2n***=24.

Chaparral. 300–1600 m. Casc, Est, Sisk, WV. CA; south to Mexico. Native.

The cones of this species will generally not open unless exposed to the heat of fire. *Pinus attenuata* × *radiata* plantations have been planted in Oregon.

Pinus balfouriana Greville & Balfour

Foxtail pine

Pinus balfouriana Grev. & Balf. ssp. *balfouriana*

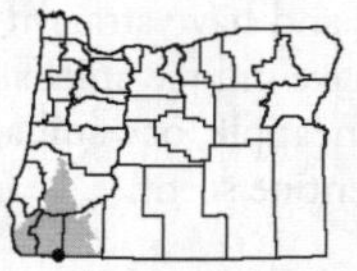

Trees to 22 m tall; mature crown compact conic to irregular. **Trunks** to 2 m in diameter, straight; bark gray to reddish brown, somewhat smooth when young, irregularly fissured and often divided into irregular square, scaly plates when mature; branches ascending or descending, often contorted; twigs reddish brown to gray, occasionally puberulent, roughened by scars. **Buds** ovoid-acuminate, reddish brown, resinous. **Leaves** in clusters of 5, upcurved, 1.5–4 cm × 1–1.5 mm, bluish green to yellow-green, margins entire; abaxial surfaces with resin bands; adaxial surfaces with white stomatal bands; apex acute and sharp pointed; sheaths deciduous. **Pollen cones** ellipsoid, 6–10 mm long, reddish. **Seed cones** ovoid-cylindrical, symmetric, 6–12 cm long, purple to reddish brown, pendent, stalks less than 1.5 cm long, maturing in 2 years, opening at maturity; umbos central, depressed or less than 1 mm, prickles very short or absent. **Seeds** ovoid, rounded at apex, 7–10 mm long, brown with purple or red spots, wings approximately equal to body. **2*n***=24.

Subalpine to alpine forests. 1800–1900 m. Sisk. CA. Native.

To date, *Pinus balfouriana* has only been collected once in Oregon, in Josephine County, within one mile of the California/Oregon border.

Pinus contorta Loudon

Lodgepole pine, shore pine

Trees to 45 m tall; mature crown variable. **Trunks** to 0.8 m in diameter, straight to contorted; bark orange-brown to reddish brown, fissured to plated; branches descending to ascending; twigs yellowish orange when young, dark reddish brown with age, roughened by bud scales. **Buds** ovoid, reddish brown, resinous. **Leaves** in clusters of 2, straight to slightly curved, twisted, 2–8 cm × 0.7–2.5 mm, dark green to yellow-green, margins finely serrulate; all surfaces with fine stomatal bands; apex usually acute; sheaths persistent. **Pollen cones** ellipsoid to cylindrical, 0.5–1.5 cm long, orange-red. **Seed cones** ovoid to sub-globose, asymmetric, 2–5 cm long, light brown to reddish brown, pendent, stalks less than 1 cm long, maturing in 2 years, opening at maturity; umbos central, triangular, less than 5 mm, prickles less than 5 mm long. **Seeds** obovoid, oblique and tapering at apex and base, 2–6 mm long, brown or reddish brown with black spots, wings at least twice as long as body. **2*n***=24.

Western North America. 3 varieties; 2 varieties treated in *Flora*.

With the exception of *Pinus sylvestris*, which one is unlikely to encounter in the wild, *P. contorta* is the only two-needled pine species found in Oregon.

Key to *Pinus contorta* varieties:

1. Leaves dark green, usually less than 1 mm wide; mature bark furrowed; seed cones strongly asymmetrical, often in whorls . . . ***P. c.* var. *contorta***

1' Leaves yellow-green, usually 1 mm wide or greater; mature bark not furrowed; seed cones asymmetrical, usually solitary or in 2's . . . . ***P. c.* var. *latifolia***

Pinus contorta Loudon var. ***contorta*** (p. 122)

Shore pine

Pinus contorta Douglas ex Loudon ssp. *contorta*

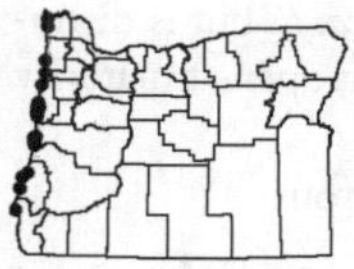

Trees to 10 m tall, sometimes **shrubs** formed by wind and salt spray; mature crown irregular or flat. **Trunks** to 0.5 m in diameter, straight inland, contorted and bent near ocean; bark furrowed and divided into small scaly plates; branches often contorted. **Leaves** 2–7 cm × 0.7–1 mm, dark green. **Seed cones** often in whorls.

Maritime forests. 0–600 m. Est. CA, WA; north to AK. Native.

This subspecies is found only in coastal forests. Often specimens of this subspecies have a distorted and contorted appearance due to the harsh, windy environment in which they grow.

Pinus contorta Loudon var. ***latifolia*** Engelmann

Lodgepole pine

Pinus contorta Douglas ex Loudon ssp. *murrayana* (Grev. & Balf.) Critchf., *Pinus contorta* Douglas ex Loudon var. *murrayana* (Grev. & Balf.) Engelm.

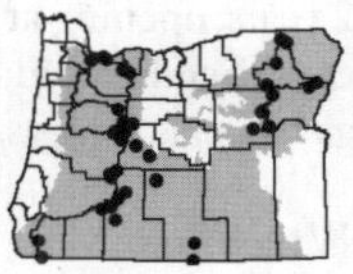

Trees to 45 m tall; mature crown conic. **Trunks** to 0.8 m in diameter, straight; bark not furrowed, separated into loose plates; branches usually horizontal. **Leaves** 5–8 cm × 1–2.5 mm, dark green. **Seed cones** solitary or in pairs.

Montane forests from low to high elevations, wetlands. 0–2400 m. BR, BW, Casc, ECas, Sisk, WV. ID, WA; north to AK, east to SD, southeast to CO. Native.

In the High Cascade Mountains and to the east, *P. c.* var. *latifolia* is often found in pure stands. Conversely, in the Western Cascades, the species is typically found as individuals in marginal habitats such as wetlands and ridgetops. The straight slender trunks of this subspecies were popular with Native Americans for use in lodges and teepees, hence the common name.

Pinus flexilis E. James

Limber pine

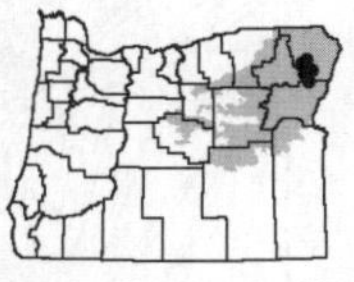

Trees to 20 m tall; mature crown conic to rounded. **Trunks** to 2 m in diameter, straight, sometimes contorted; bark gray, smooth, becoming ridged with rectangular scaly plates with age; branches horizontal to ascending, usually persistent to base; twigs reddish brown and pubescent when young, becoming gray with age, smooth or nearly so with some bud scars. **Buds** ovoid, reddish brown, resinous. **Leaves** in clusters of 5, upcurved, 3–6 cm × 1–1.5 mm, dark green, margins entire to finely serrulate; abaxial surfaces with faint stomatal bands; adaxial surfaces with obvious stomatal bands; apex acute; sheaths deciduous. **Pollen cones** ellipsoid-cylindrical, 1.2–1.6 cm long, red or yellow. **Seed cones** ovoid to ovoid-cylindrical, symmetric, 7–15 cm long, light brown, horizontal to pendent, sessile to less than 2 cm long, maturing in 2 years, opening at maturity; umbos terminal, less than 2 mm, prickles absent. **Seeds** oval, compressed, 6–12 mm long, brown to reddish brown, sometimes with spots, wingless or nearly so. **2*n***=24.

High altitude montane forests. 1500–2500 m. BW. CA, ID, NV, WA; north to Alberta, east to NE, southeast to NM. Native.

In Oregon, *Pinus flexilis* is found only in the Wallowa Mountains. Morphologically, it is similar to *P. albicaulis* but is distinguished from the latter by its larger cones that open at maturity.

Pinus jeffreyi Greville & Balfour

Jeffrey pine

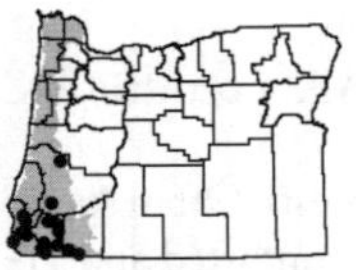

Trees to 60 m tall; mature crown conic to rounded. **Trunks** to 2.3 m in diameter, straight; bark yellow-brown to reddish brown, deeply furrowed, forming large irregular scaly plates; branches horizontal to ascending; twigs purplish brown, usually glaucous, roughened by bud scars. **Buds** ovoid, light brown to reddish brown, not resinous. **Leaves** in clusters of 3, straight, slightly twisted, 12–25 cm × 1.5–2 mm, bluish green to yellow-green, margins finely serrulate; all surfaces with fine stomatal lines; apex acute; sheaths persistent. **Pollen cones** lance-cylindrical, 2–3.5 cm long, yellow to brown. **Seed cones** oblong, slightly asymmetric, (12)15–30 cm long, brown to reddish brown, horizontal to pendent, stalks less than 3 cm long, maturing in 2 years, opening at maturity; umbos central, raised, prickles reflexed. **Seeds** ovoid, 8–12 mm long, brown with black spots, wings twice as long as body. **2*n***=24.

Dry forests. 50–2000 m. CR, Sisk. CA, NV; south to Mexico. Native.

Morphologically, *Pinus jeffreyi* is very similar to *P. ponderosa*, and the two species are known to hybridize. *Pinus jeffreyi* can be differentiated from *P. ponderosa* by its cones, which are generally longer than 15 cm and have reflexed prickles. Conversely, *P. ponderosa* cones are less than 15 cm long and have straight prickles. Some claim that the two species have unique smells. Allegedly, *P. jeffreyi* trees have a banana, pineapple, or vanilla scent, whereas *P. ponderosa* trees have a turpentine scent.

Pinus lambertiana Douglas

Sugar pine

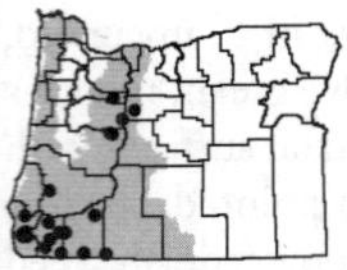

Trees to 70 m tall; mature crown conic to broad and flattened with horizontal branches. **Trunks** to 3 m in diameter, straight; bark smooth and gray when young, brown to reddish brown, furrowed with long scaly plates when mature; branches horizontal to ascending, lower branches sometimes drooping; twigs green to brown and pubescent when young, becoming glabrous and gray with age, smooth with bud scars flush to surface or nearly so. **Buds** ovoid, reddish brown, resinous. **Leaves** in clusters of 5, straight, slightly twisted, 5–9 cm × 1–2 mm, bluish green, margins finely serrulate; abaxial surfaces with faint stomatal lines; adaxial surfaces with obvious white stomatal lines; apex acute; sheaths deciduous. **Pollen cones** ellipsoid-cylindrical, 10–15 mm long, yellow. **Seed cones** cylindrical, symmetric, 30–45 cm long, pendent, stalks 5–16 cm long, maturing in 2 years, opening at maturity; umbos terminal, slightly excurved; prickles absent. **Seeds** ovoid, 10–20 mm long, brown, wings twice as long as body. **2*n***=24.

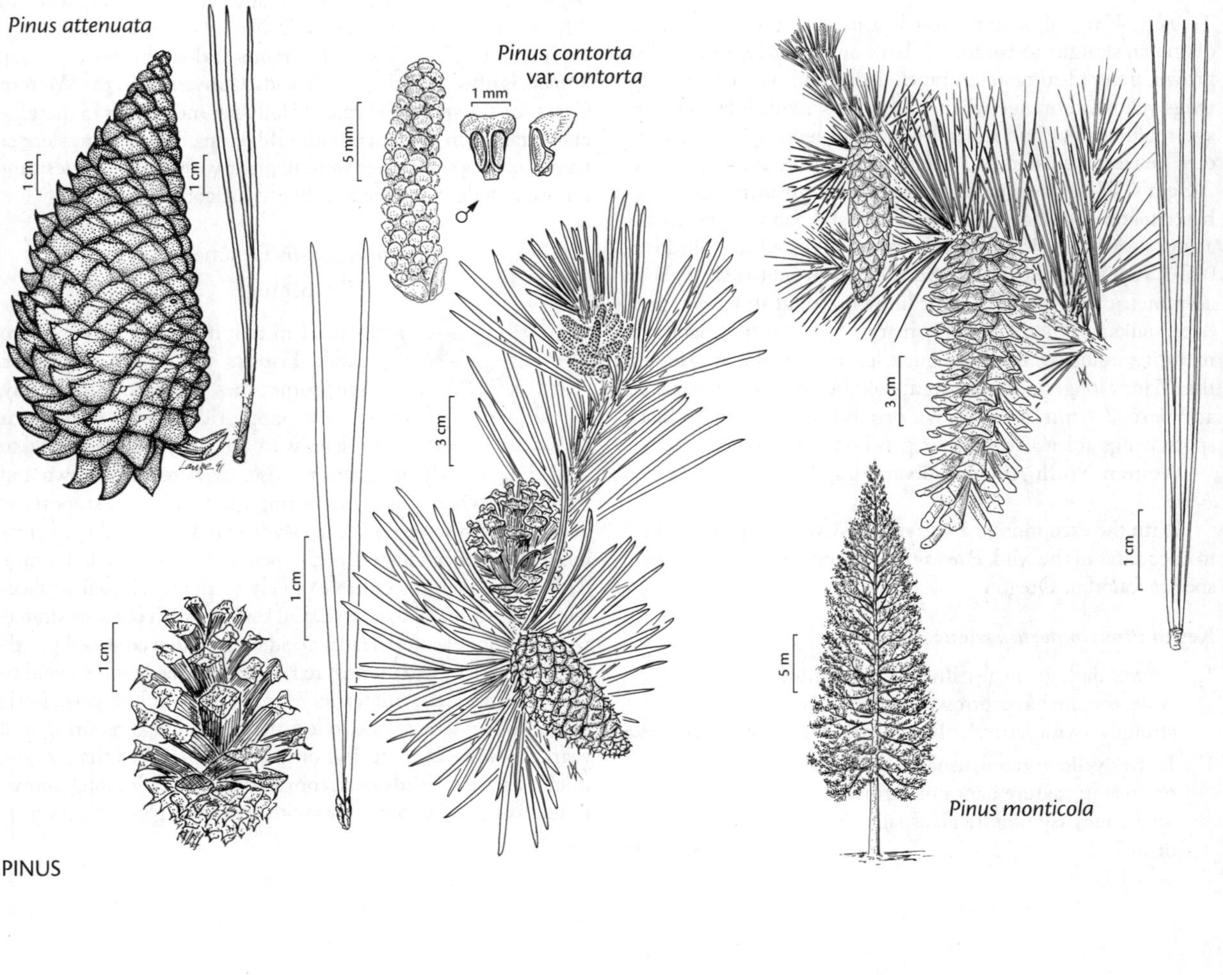

PINUS

Montane forests. 200–2100 m. Casc, CR, ECas, Sisk. CA, NV; south to Mexico. Native.

Pinus lambertiana is the largest of all pine species. Its large cones, often weighing several pounds, in combination with leaves in clusters of 5, distinguish it from all other pine species in Oregon.

Pinus monticola Douglas ex D. Don (p. 122)
Western white pine

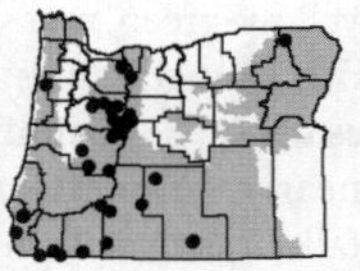

Trees to 70 m tall; mature crown conic. **Trunks** to 2.5 m in diameter, straight; bark gray, forming scaly plates; branches spreading and ascending; twigs reddish brown, smooth. **Buds** ellipsoid to cylindric, reddish brown, resinous. **Leaves** in clusters of 5, straight, slightly twisted, 4–10 cm × 0.7–1 mm, blue-green, margins finely serrulate; abaxial surface without stomatal lines; adaxial surface with stomatal lines; apex acute; sheaths deciduous. **Pollen cones** ellipsoid, 10–15 mm long, yellow. **Seed cones** lanceoloid to cylindric, symmetric, 10–25 cm long, brown to yellowish, pendent, stalks less than 2 cm long, maturing in 2 years, opening at maturity; umbos terminal, depressed. **Seeds** obovoid, 5–7 mm long, reddish brown, wings about 4 times as long as body. ***2n***=24.

Montane forests, wetlands. 400–2500 m. BR, BW, Casc, CR, ECas, Sisk. CA, ID, NV, WA; north to British Columbia, northeast to MT. Native.

Pinus monticola is the state tree of Idaho.

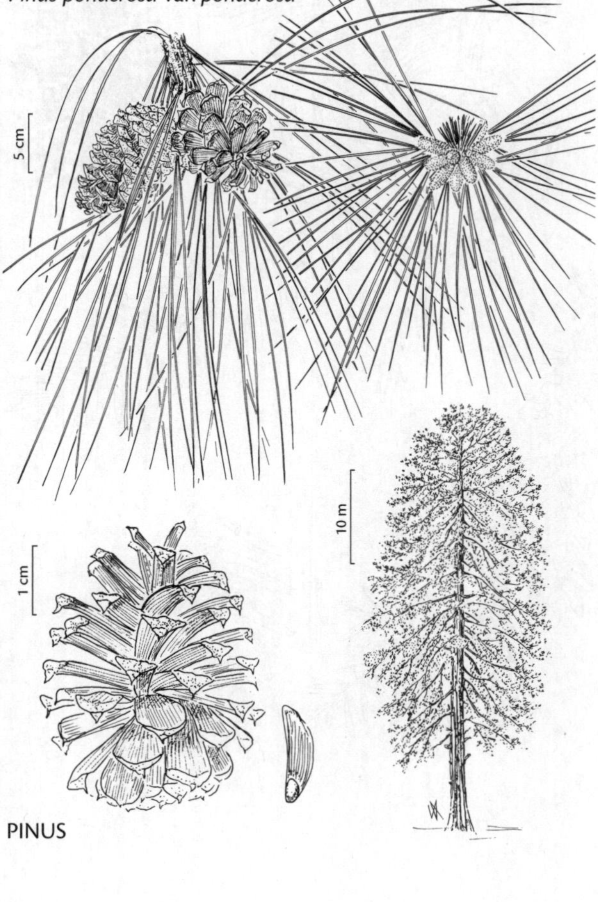

Pinus ponderosa var. *ponderosa*

PINUS

Pinus ponderosa Douglas ex Lawson
Ponderosa pine

Pinus ponderosa Douglas ex C. Lawson var. *washoensis* (H. Mason & Stockw.) J.R. Haller & Vivrette, *Pinus washoensis* H. Mason & Stockw.

Trees to 70 m tall; mature crown conic to rounded. **Trunks** to 2.5 m in diameter, straight; bark light yellow-brown to reddish brown, deeply furrowed, forming large irregular scaly plates; branches descending to ascending; twigs orange-brown, roughened by bud scars. **Buds** ovoid, light brown to reddish brown, resinous. **Leaves** in clusters of 3, straight, slightly twisted, 12–25 cm × 1.5–2 mm, grayish green to deep green, margins finely serrulate; all surfaces with fine stomatal lines; apex acute; sheaths persistent. **Pollen cones** lance-cylindrical, 2–3.5 cm long, yellow to reddish brown. **Seed cones** oblong, slightly asymmetric, 5–18 cm long, brown to reddish brown, horizontal to pendent, stalks less than 2 cm long, maturing in 2 years, opening at maturity; umbos central, raised; prickles straight. **Seeds** ovoid, 6–9 mm long, brown with black spots, wings about 4 times as long as body.

Western North America. 3–5 varieties. 2 varieties treated in *Flora*.

The number of, and characteristics within, varieties or subspecies of *Pinus ponderosa* varies depending on the authority. Conservatively, only two varieties of *P. ponderosa* are present in Oregon. Our *P. ponderosa* varieties are mainly distinguished from other varieties by the number of needles in their clusters (3), as opposed to varieties *scopulorum* and *arizonica* (2–3 and 4–5, respectively).

Key to *Pinus ponderosa* varieties:

1. Leaves deep green; bark light or moderate yellowish brown .... ***P. p.*** var. ***benthamiana***

1' Leaves grayish green; bark moderate yellowish brown to dark red .... ***P. p.*** var. ***ponderosa***

Pinus ponderosa Douglas ex Lawson var. ***benthamiana*** (Hartweg) Vasey
Bentham's ponderosa pine

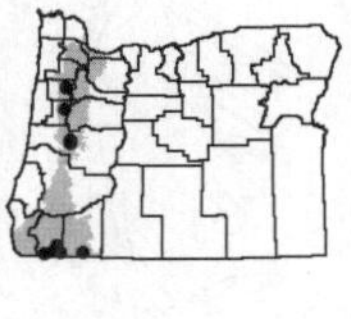

Bark light or moderate yellowish brown. **Leaves** deep green.

Hillsides and gravelly floodplains, usually mixed with oaks or other conifers. 50–1700 m. Sisk, WV. CA, WA. Native.

Pinus ponderosa Douglas ex Lawson var. ***ponderosa*** (p. 123)
Ponderosa pine

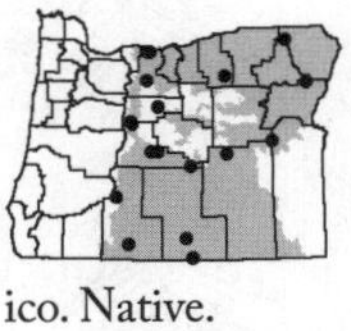

Bark moderate yellowish brown to dark red. **Leaves** grayish green.

Dry forests. 100–2200 m. BR, BW, Col, ECas. CA, ID, NV, WA; north to British Columbia, northeast to MT, south to Mexico. Native.

The morphological characteristics separating our two varieties are sometimes unreliable. This unsatisfactorily forces the identifier to use geography as the determining "character." More taxonomic study may be warranted.

Pinus sabiniana Douglas ex D. Don

Gray pine, foothill pine

Trees to 24 m tall; mature crown sparse, open and round. **Trunks** to 1.2 m in diameter, straight to contorted or forked; bark dark brown to black, irregularly furrowed with rounded ridges; inner bark orangish; branches usually ascending; twigs brown to black, glaucous, slightly roughened by bud scars. **Buds** ovoid, reddish brown, resinous. **Leaves** in clusters of 3, dull, straight, slightly twisted, 15–30 cm × 1–2 mm, bluish green, margins serrulate; all surfaces with stomatal lines; apex acute; sheaths persistent. **Pollen cones** ellipsoid, 10–15 mm long, yellow. **Seed cones** oblong-ovoid, nearly symmetric, 15–20 cm long, light reddish brown, pendent, stalks 1–5 cm long, maturing in 2 years, opening at maturity; umbos continuous with apophyses to form long upcurving claws. **Seeds** oblong, 1.5–2 cm long, dark brown to black, wings about 50% the length of the body. ***2n***=24.

Dry mixed forests. 200–1400 m. Sisk. CA. Native.

Pinus sylvestris Linnaeus

Scots pine

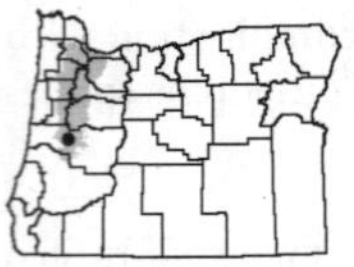

Trees to 40 m tall; mature crown conic to flattened. **Trunks** to 0.5 m in diameter, usually straight; bark reddish brown to gray, furrowed, scaly; branches horizontal to ascending; twigs green to gray-brown, not glaucous, roughened by bud scars. **Buds** ovoid, reddish brown, resinous. **Leaves** in clusters of 2, straight to slightly curved, twisted, 3–6 cm × 1.5–2 mm, bluish green to yellow-green, margins entire to finely serrulate; all surfaces with stomatal lines; apex acute; sheaths deciduous. **Pollen cones** ovoid, 5–10 mm long, yellow to pink. **Seed cones** ovoid, nearly symmetric, 3–6 cm long, yellow-brown to gray-brown, stalks less than 1 cm long, maturing in 2 years, opening at maturity; umbos central, less than 3 mm; prickles absent. **Seeds** obovoid, 3–5 mm long, gray to black, wings 3 times as long as body.

Various habitats, escaped from cultivation. 0–200 m. WV. Throughout North America; Europe. Exotic.

Pinus sylvestris is widely planted in North America for pulpwood and Christmas trees and has escaped cultivation in many states and provinces. This species is native to Europe.

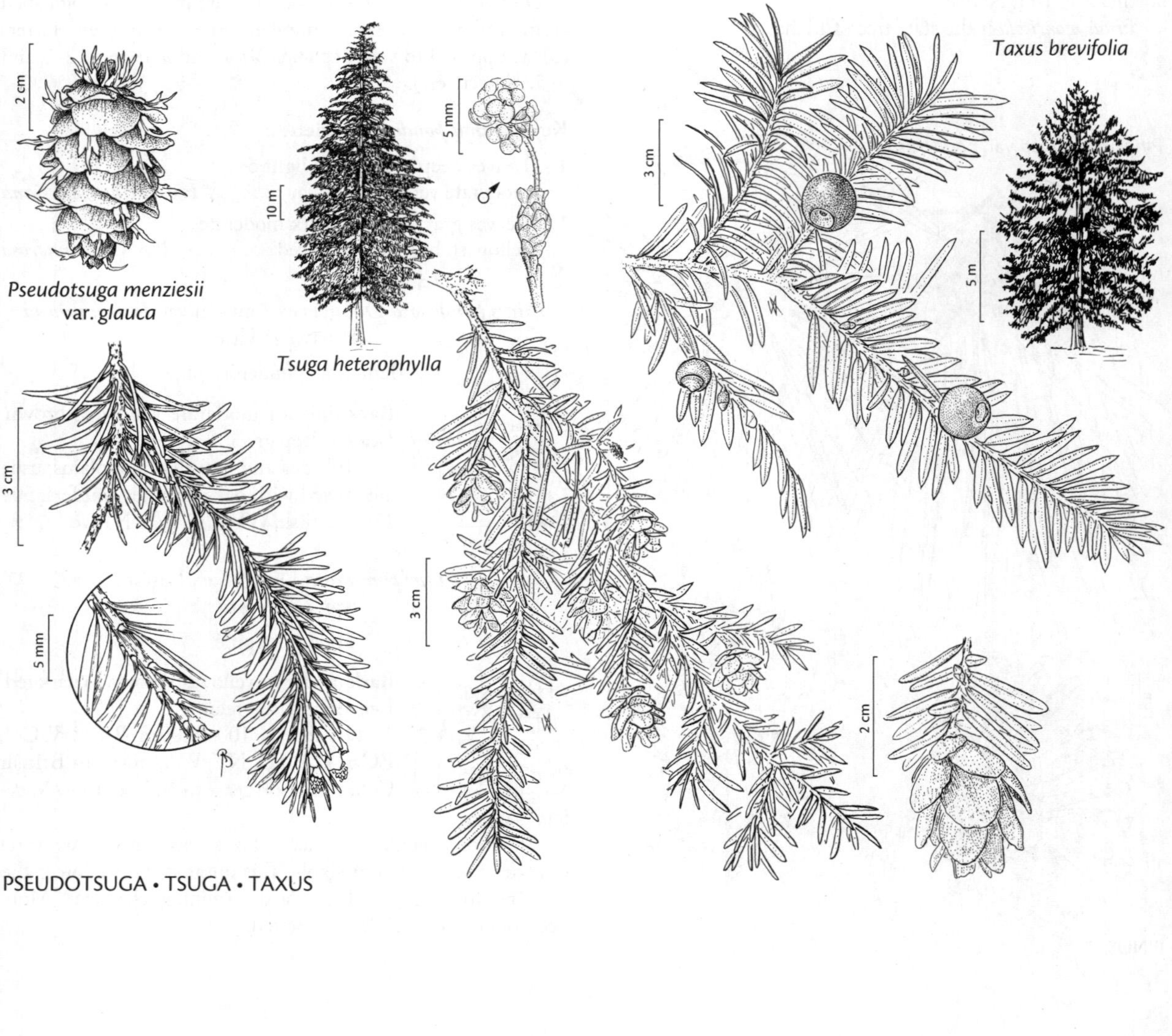

PSEUDOTSUGA • TSUGA • TAXUS

PSEUDOTSUGA Douglas-fir

Trees evergreen; crown conic; topmost shoots generally erect. **Bark** reddish brown, with resin blisters, smooth when young, becoming furrowed with age; branches whorled, ascending to slightly drooping, leaf scars flush with twigs. **Leaves** alternate, flattened, apex bluntly pointed, borne singly, persisting up to 8 years, sheath absent. **Pollen cones** axillary. **Seed cones** oblong-ovoid, maturing in 1 season, pendent; scales orbicular, persistent; bracts exserted.

North temperate region. 5 species; 1 species treated in *Flora*.

Commercially, *Pseudotsuga* is the most important tree genus in Oregon. In the past, this genus has been considered a part of *Abies*, *Picea*, *Pinus*, and *Tsuga*. This taxonomic confusion is reflected in both its common name (Douglas-fir) and scientific name, which means false hemlock.

Pseudotsuga menziesii (Mirbel) Franco

Douglas-fir

Trees to 100 m; mature crown conic. **Trunks** to 4 m diameter; bark reddish brown, smooth when young, becoming thick and furrowed with age; twigs pubescent and orange-yellow when young, becoming glabrous and reddish brown to gray with age, slender. **Leaves** linear, 1.5–3 cm × 1–1.5 mm, yellow-green to bluish green, flattened in cross section, usually spreading; abaxial surface ridged and somewhat glaucous; adaxial surface rounded or grooved; apex bluntly pointed, rarely acute. **Pollen cones** 1–2.5 cm long, yellow-red to red. **Seed cones** 4–10 × 3–3.5 cm; scales orbicular, about 2 cm long, margins slightly irregularly toothed; bracts inversely arrow-shaped, 2.5–3.5 cm long, conspicuously exserted. **Seeds** 5–6 mm long, reddish brown to dark brown, wings twice as long as seed body. **2*n***=26.

Western North America. 2 varieties.

Pseudotsuga menziesii is the most commercially important tree in North America. Uses range from pulp, lumber, and mulch to Christmas trees. The species is most easily distinguished from other coniferous species by the bracts of its seed cones which resemble the tail and feet of a mouse diving into a hole. This species is the state tree of Oregon.

Key to *Pseudotsuga menziesii* varieties:

1. Seed cones 6–10 cm long; leaves yellow-green . . . . ***P. m.*** var. ***menziesii***

1' Seed cones 4–7 cm long; leaves blue or gray-green to dark green . . . . ***P. m.*** var. ***glauca***

Pseudotsuga menziesii (Mirbel) Franco var. ***glauca*** (Mayr) Franco (p. 124)

Rocky Mountain Douglas-fir

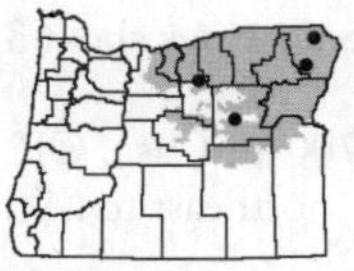

Trees to 50 m. **Trunks** to 1.1 m diameter. **Leaves** blue or gray-green to dark green. **Seed cones** 4–7 cm long; bracts spreading.

Coniferous forests east of the Cascade Mountains. 500–2200 m. BW, Col. ID, NV, WA; north to British Columbia, east to WY, southeast to TX and Mexico. Native.

Pseudotsuga menziesii (Mirbel) Franco var. ***menziesii***

Coast Douglas-fir

Pseudotsuga taxifolia (Lamb.) Britton

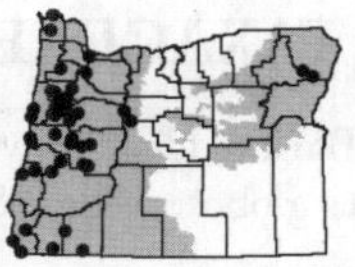

Trees to 100 m. **Trunks** to 4 m diameter. **Leaves** yellow-green. **Seed cones** 6–10 cm long; bracts appressed.

Coniferous forests. 0–2500 m. BW, Casc, CR, ECas, Est, Sisk, WV. CA, NV, WA; north to British Columbia. Native.

The two varieties of *Pseudotsuga menziesii* are sympatric in British Columbia, Washington, and eastern Oregon.

TSUGA Hemlock

Trees evergreen; mature crowns pyramidal; uppermost branches usually drooping. **Bark** gray to reddish brown, scaly, furrowed; branches horizontal and arching downward; twigs usually with rough peg-like projections from previous leaf bases. **Leaves** borne singly, appearing 2-ranked or radiating in all directions, flattened to 4-sided, set in peg-like projections, sheath absent. **Pollen cones** solitary. **Seed cones** maturing in 1 season, pendent, sessile or on very short stalks. **Seeds** winged.

North America and East Asia. Approximately 10 species; 2 species treated in *Flora*.

Hemlocks are most easily identifiable from other coniferous trees in Oregon by their short leaves and conspicuously drooping uppermost branches. In addition, because hemlocks do not need full sun, trees within forests retain their lower branches as they mature, as opposed to most Douglas-firs and firs.

Key to *Tsuga* species

1. Leaves appearing 2-ranked; upper branches of trees very lax and drooping; seed cones less than 2.5 cm long . . ***T. heterophylla***

1' Leaves radiating in all directions; upper branches of trees moderately drooping; seed cones over 3 cm long. . . . ***T. mertensiana***

Tsuga heterophylla (Rafinesque) Sargent (p. 124)

Western hemlock

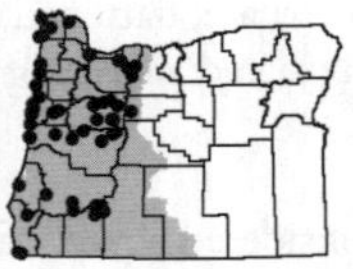

Trees to 50 m tall; mature crown narrowly pyramidal. **Trunks** to 2.5 m in diameter; bark gray-brown to reddish brown, smooth, scaly, becoming furrowed with age; branches horizontal, upper branches obviously drooping; twigs yellow-brown, pubescent. **Leaves** appearing 2-ranked, flattened, 0.8–2 cm × 1.5–2 mm; abaxial surface light green, glaucous, with 2 whitish stomatal bands; adaxial surface bright green, grooved; margins minutely dentate; apex blunt to rounded. **Pollen cones** globose, 3–4 mm long on a stalk of equal length, brown. **Seed cones** oblong-cylindrical while immature, ovoid when mature, 1–2.5 × 1–2.5 cm, scales ovate, thin and flexible, longer than bracts. **Seeds** 2–4 mm long, wings about 3 times as long as body. **2*n***=24.

Coastal to mid-elevation forests. 0–1700 m. Casc, CR, ECas, Est, Sisk, WV. ID, WA; north to AK, northeast to MT. Native.

Tsuga heterophylla can resemble *T. mertensiana*, especially at higher elevations. However, the two-ranked appearing leaves and short seed cones of *T. heterophylla* are reliable for identification. *Tsuga heterophylla* is the state tree of Washington.

Tsuga mertensiana (Bongard) Carrière

Mountain hemlock

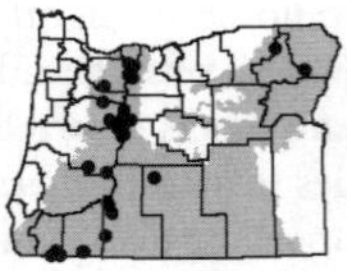

Trees to 40 m tall; mature crown very narrowly pyramidal. **Trunks** to 1.5 m in diameter; bark dark gray to reddish brown, scaly, fissured; branches horizontal, upper branches moderately drooping; twigs yellow-brown, glabrous to puberulent. **Leaves** radiating in all directions, curved near apex, usually flattened, occasionally rounded to 4-sided in cross section, 1–2.5 cm × 1–2 mm; abaxial and adaxial surfaces green, glaucous, with faint stomatal bands; margins entire; apex rounded. **Pollen cones** brown, 6–8 mm long on a stalk of equal or slightly greater length, globose. **Seed cones** oblong-cylindrical, 3–7 × 1.5–3 cm, scales obovate, stiff, longer than bracts. **Seeds** 3–5(6) mm long, wings about 4 times as long as body. **2*n***=24.

Subalpine to alpine forests. 800–2600 m. BR, BW, Casc, ECas, Sisk, WV. CA, ID, NV, WA; north to AK, northeast to MT. Native.

This species exhibits varied morphology, depending on the elevation at which it is found. For example, at high, exposed elevations, this species may exhibit a short, prostrate morphology (Krummholz form). *Tsuga mertensiana* is always readily identifiable, however, by its radiating leaves and cones over 3 cm long.

TAXACEAE Yew family

Small trees or **shrubs** evergreen, dioecious. **Bark** reddish brown, scaly. **Leaves** needle-like, simple, alternate, appearing 2-ranked, not in bundles; apex mucronate. **Pollen cones** globose, yellow. **Seed cones** arils, open at top, maturing in 1 season, axillary. **Seeds** partially enclosed by a red aril.

Northern Hemisphere. 5 genera; 1 genus treated in *Flora*.

Torreya californica (California nutmeg) has been collected once in Oregon, in Curry County (1949). Whether this species has historically been present in Oregon or this represents a range extension needs further research. This species, if found again, could easily be distinguished from *Taxus brevifolia* by its soft, green, purple-striped seed cones (versus red).

TAXUS Yew

Small shrubs or **trees**. **Branches** horizontal to drooping; twigs alternate. **Leaves** flexible, persisting for 4–5 years; apex not sharp to touch. **Seed cones** cup-shaped, soft, fleshy, mucilaginous. **Seeds** brown.

Northern temperate regions. 10 species; 1 species treated in *Flora*.

North American *Taxus* species, until recently, were of conservation concern as the result of over harvesting. The reason being the bark of the North American members of this genus is a source of paclitaxel, a chemotherapeutic drug used to treat breast and lung cancer. However, this drug is now produced synthetically, eliminating the need to harvest wild stands of *Taxus*.

Taxus brevifolia Nuttall (p. 124)

Pacific yew

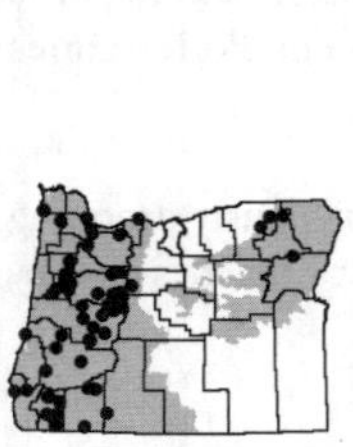

Small trees or **shrubs** to 20 m tall. **Trunks** to 1 m in diameter, usually much smaller, often irregular in shape and lacking a distinctive form; bark scaly, shredding; branches horizontal to drooping; twigs green, becoming bright brown with age. **Leaves** linear, flexible, flattened in cross section, 1–2.5 cm long, 1–3 mm wide; abaxial surface pale green, with a yellowish midrib, usually revolute; adaxial surface yellow-green; apex mucronate but soft to touch. **Pollen cones** less than 1 cm long. **Seed cones** 0.5–1.5 cm long, red, thick, fleshy. **Seeds** ovoid, 0.5–1 cm long, brownish red.

Rocky ridges, along streams, valley bottoms, and lower slopes under the shade of other coniferous trees. 0–2000 m. BW, Casc, CR, ECas, Est, Sisk, WV. CA, ID, WA; north to AK, northeast to Alberta and MT. Native.

Mucronate leaf tips and red, fleshy arils distinguish *Taxus brevifolia* from other Oregon gymnosperms. Similar to redwoods, Pacific yews are one of the few conifers that are able to regrow vegetatively from their stumps.

MONOCOTS

Herbs or rarely vines (ours); stems with vascular bundles scattered throughout pith. **Leaves** usually alternate and entire, often sheathing the stem at base, generally parallel veined; usually without petioles. **Flowers** with parts in 3's or 6's, rarely in 2's or 4's, never in 5's. **Ovules** borne within a closed ovary, ripened ovary becoming a fruit. **Embryo** with 1 cotyledon.

Key to the Monocot families:

1. Flowers on a spadix.
   2. Plants in marine aquatic (intertidal and subtidal) or alkaline habitats, usually fully submersed.
      3. Leaves ribbon-like .... **ZOSTERACEAE**
      3' Leaves linear and thread-like .... **RUPPIACEAE** (in part)
   2' Plants aquatic or terrestrial but not marine, usually not fully submersed.
      4. Leaves less than 3 cm wide .... **ACORACEAE**
      4' Leaves greater than 3 cm wide .... **ARACEAE** (in part)

1' Flowers not on a spadix.
   5. Plants thalloid, not obviously differentiated into stems, leaves and roots; plants less than 2 cm long; aquatic and free floating .... **ARACEAE** (in part)
   5' Plants not thalloid, obviously differentiated into stems, leaves and roots; plants usually more than 2 cm long; habits various.
      6. Stems flaccid, floating in or on water (inflorescence may be vertical and above water).
         7. Leaves less than 4 cm long .... **HYDROCHARITACEAE** (in part)
         7' Most leaves over 4 cm long.
            8. Leaves ribbon-like .... **HYDROCHARITACEAE** (in part)
            8' Leaves linear, thread-like or broadly elliptic.
               9. Plants monoecious.
                  10. Inflorescences usually with 2 flowers (1 staminate, 1 pistillate) .... **ZANNICHELLIACEAE**
                  10' Inflorescences with more than 2 flowers .... **TYPHACEAE** (in part)
               9' Plants with bisexual flowers.
                  11. Peduncles usually coiled; plants in marine or alkaline habitats .... **RUPPIACEAE** (in part)
                  11' Peduncles not coiled; plants in freshwater habitats .... **POTAMOGETONACEAE**
      6' Stems stiff, vertical (or nearly so), all or part of stem above water or terrestrial.
         12. Flowers naked and/or subtended by chaffy or bristle-like bracts or papery scales, not radially or bilaterally symmetric; perianth parts generally not in 3's.
            13. Inflorescences in globose heads.
               14. Plants monoecious .... **TYPHACEAE** (in part)
               14' Plants with bisexual flowers .... **CYPERACEAE** (in part)
            13' Inflorescences various but not in globose heads.
               15. Inflorescences dense, cylindrical, 1000+-flowered spike-like heads .... **TYPHACEAE** (in part)
               15' Inflorescences not as above.
                  16. Leaves elliptical or round in cross section; plants acaulescent .... **JUNCAGINACEAE** (in part)
                  16' Leaves flat; plants caulescent.
                     17. Leaves 2-ranked; stems usually hollow, not triangular; nodes usually prominent and swollen .... **POACEAE**
                     17' Leaves 3-ranked; stems solid or pithy, usually triangular; nodes not prominent or swollen .... **CYPERACEAE** (in part)
         12' Plants with radially or bilaterally symmetric flowers; perianth parts commonly in 3's.
            18. Perianth parts less than or equal to 2 mm long.
               19. Tepals in 2 alternating whorls .... **JUNCACEAE** (in part)
               19' Tepals not alternating .... **JUNCAGINACEAE** (in part)
            18' Perianth parts greater than 2 mm long.

20. Perianth parts less than or equal to 4 mm long.
    21. Flowers with morphologically distinct petals and sepals . . . . **ALISMATACEAE** (in part)
    21' Flowers without morphologically distinct petals and sepals (tepals present).
        22. Pistils 3–6; stems in a Z-like pattern. . . . . **SCHEUCHZERIACEAE**
        22' Pistil 1; stems generally not in a Z-like pattern . . . . **JUNCACEAE** (in part)
20' Perianth parts greater than 4 mm long.
    23. Styles 6 or more . . . . **ALISMATACEAE** (in part)
    23' Styles 1 or 3.
        24. Styles 3.
            25. Plants aquatic; rooting in mud or occasionally free floating. . . . . **PONTEDERIACEAE** (in part)
            25' Plants terrestrial.
                26. Inflorescences umbels . . . . **SMILACACEAE**
                26' Inflorescences solitary, racemes, or panicles.
                    27. Leaves 2-ranked . . . . **TOFIELDIACEAE**
                    27' Leaves not 2-ranked . . . . **MELANTHIACEAE**
        24' Style 1.
            28. Plants aquatic; rooting in mud or occasionally free floating. . . . . **PONTEDERIACEAE** (in part)
            28' Plants terrestrial.
                29. Flowers bilaterally symmetric; stamens 1 or 2 . . . . **ORCHIDACEAE**
                29' Flowers radially symmetric; stamens 3, 4, or 6.
                    30. Stamens 4 . . . . **ASPARAGACEAE** (in part)
                    30' Stamens 3 or 6.
                        31. Stamens 3 and staminodia 0.
                            32. Tepals with prominent purple-brown lines; ovary superior . . . . **LILIACEAE** (in part)
                            32' Tepals without prominent lines; ovary inferior . . . . **IRIDACEAE**
                        31' Stamens 6, or stamens 3 and staminodia 3, or stamens 6 and staminodia 6.
                            33. Stamens 6 and staminodia 6 . . . . **TECOPHILAEACEAE**
                            33' Stamens 6, or stamens 3 and staminodia 3.
                                34. Filaments densely woolly; plants of southwestern Oregon . . . . . . **NARTHECIACEAE**
                                34' Filaments glabrous to pubescent, but never densely woolly; range various.
                                    35. Plants smelling of onion. . . . . **AMARYLLIDACEAE** (in part)
                                    35' Plants not smelling of onion.
                                        36. Perianths conspicuously crowned (daffodils) and ovary inferior . . . . **AMARYLLIDACEAE** (in part)
                                        36' Perianths not conspicuously crowned (but crowned by perianth appendages in *Dichelostemma*) and ovary superior.
                                            37. Leaves less than 1 mm wide . . . . **ASPARAGACEAE** (in part)
                                            37' Leaves greater than 1 mm wide.
                                                38. Plants from rhizomes or fleshy roots.
                                                    39. Plants usually over 10 cm tall . . . . **LILIACEAE** (in part)
                                                    39' Plants less than or equal to 10 cm tall . . . . **ASPARAGACEAE** (in part)
                                                38' Plants from bulbs.
                                                    40. Leaves cauline or primarily cauline.
                                                        41. Inflorescences more than 20-flowered . . . . **ASPARAGACEAE** (in part)
                                                        41' Inflorescences less than 20-flowered . . . **LILIACEAE** (in part)
                                                    40' Leaves basal.
                                                        42. Inner and outer perianth segments dissimilar . . . . **LILIACEAE** (in part)
                                                        42' All perianth segments similar.
                                                            43. Flowers not nodding. . . . . . **ASPARAGACEAE** (in part)
                                                            43' Flowers nodding.

44. Perianth appendages protruding from base of perianth lobes and forming a crown .... **ASPARAGACEAE** (in part)
44' Perianth appendages 0.
   45. Perianth parts less than 1 cm long .... **ASPARAGACEAE** (in part)
   45' Perianth parts 1 cm long or greater.
      46. Flowers blue, bluish purple, or rarely white or pink .... **ASPARAGACEAE** (in part)
      46' Flowers not blue, bluish purple, white, or pink .... **LILIACEAE** (in part)

ACORACEAE Sweet-flag family

Bridget Chipman

Herbs perennial, aquatic, aromatic (particularly in rhizomes); rhizomes horizontal, creeping at or near surface, branched, with aromatic oil. **Stems** prostrate, branched. **Leaves** linear, bright green, not differentiated into petiole and blade, equitant, sheathing at base; 1–6 prominent veins parallel along the length of the leaf. **Inflorescences** solitary, arising from sympodial leaf spadices, peduncle fused with proximal portion of sympodial leaf that extends beyond spadix, true spathe absent. **Flowers** bisexual, tepals 6, brownish; stamens 6, distinct, anthers yellow; ovaries 3-locular, sometimes 1-locular, green, sessile; stigma sessile, styles small and appearing absent. **Fruits** berries, oblong, broader at apex, brown to reddish, longitudinally striped, with thin, almost leathery pericarp. **Seeds** 1–6.

Temperate areas worldwide, higher mountains in the tropics. 1 genus.

Based on morphological and molecular evidence, Acoraceae is now treated as a separate family from Araceae. The same evidence also suggests that Acoraceae is the oldest extant family of monocotyledons (Grayum 1987; Duvall et al. 1993).

ACORUS Sweet-flag

Family comprised of 1 genus, *Acorus*.

3–6 species; 1 species treated in *Flora*.

There are two species of *Acorus* in North America, *A. calamus* and *A. americanus*. Traditionally these species were collectively cited as a singular species, *A. calamus*. Current taxonomy, however, treats all fertile North American plants as *Acorus americanus* and presumes they are native. Putatively introduced, sterile triploids are referred to as *A. calamus* (Thompson 1995). Further study to clarify the taxonomy and biogeography of this genus is in order.

Acorus calamus Linnaeus (p. 129)

Sweet-flag

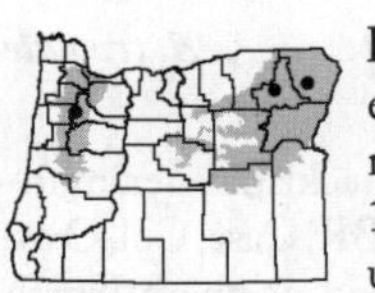

Leaves basally white with pink or red, otherwise bright green, with a single prominent midvein. **Vegetative leaves** up to 1.75 m × 0.5–2 cm, occasionally margins undulate or crisped. **Sympodial leaves** 30–160 × 0.5–2 cm, usually equal to vegetative leaves. **Inflorescences** 10–16 cm × 12–22 mm. **Flowers** 3–4 mm wide; pollen not staining in aniline blue. **Fruits** not produced in North American plants.

Wet areas, marshes, swamps, or swales; often found on gravel beds. Flowering May–Sep. 0-1200 m. BW, WV. CA; disjunct populations in western US, widespread in eastern US. Exotic.

The sterile *A. calamus* in North America tends to be larger than the North American native *A. americanus* (found in the northern United States and Canada). *Acorus americanus* is also distinguished from *A. calamus* by producing fruit, pollen staining in aniline blue, and the presence of multiple prominent veins on the leaves. The ethnobotanical uses of *A. calamus* are many. The root stock was and continues to be used as a flavoring agent, a candy, a breath sweetener, a toothache treatment, and to increase appetite and aid in digestion. Calamus oil is used for flavoring beer, gin, and bitters and also in perfumes and as a moth repellent.

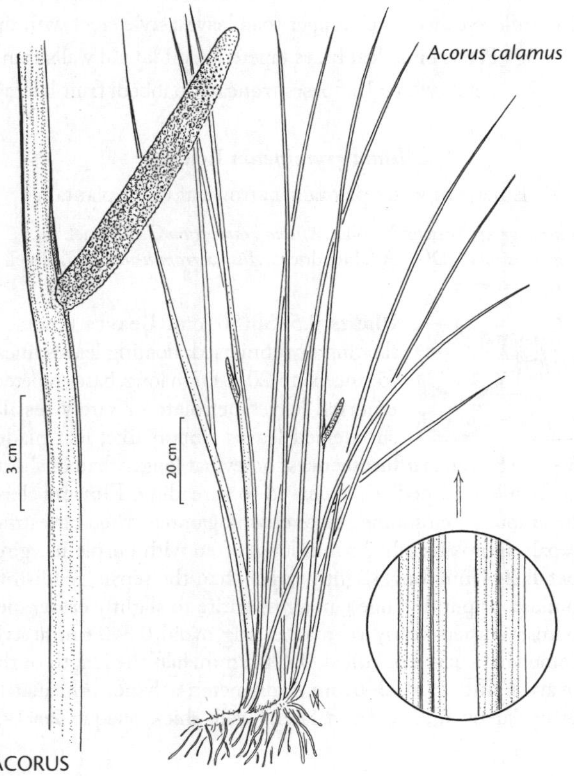

ACORUS

ALISMATACEAE Water plantain or Arrowhead family

Tamra Dickson Prior

Herbs annual or perennial, aquatic or on mud; from caudices, corms, stolons, rhizomes, or tubers. **Roots** fibrous, septate or not. **Stems** short caudices. **Leaves** basal, emersed, floating or submersed, simple, linear to sagittate, palmately veined. **Inflorescences** scapose, umbel-, raceme-, or panicle-like; branches whorled. **Flowers** monoecious, dioecious, or bisexual, radial; sepals 3, generally green, generally persistent; petals 3, generally bigger than sepals, white or pink; stamens 6–many; pistils 6–many, free or fused at the base. **Fruits** achenes, compressed, beaked. **Seeds** endosperm absent in mature seed.

Cosmopolitan, mainly tropics and subtropics. 12 genera worldwide; 3 genera treated in *Flora*.

Members of this family are grown in Asia for both human food and animal fodder. Wild plants in North America are a valuable food for wildlife.

Key to Alismataceae genera:

1. Pistils and fruits fused at base into a star-like aggregation; flowers bisexual; stamens 6 . . . . **DAMASONIUM**

1' Pistils and fruits free, in 1 whorl on margin of flat receptacle, or spiraled on convex receptacle; flowers bisexual or unisexual; stamens 6 or 7–30.

2. Pistils and fruits in 1 whorl on margin of flat receptacle; stamens 6; flowers bisexual. . . . . **ALISMA**

2' Pistils and fruits spiraled on convex receptacle; stamens 7–30; proximal flowers unisexual. . . . . **SAGITTARIA**

ALISMA Water plantain

Herbs annual, glabrous. **Roots** rhizomes often present. **Leaves** sessile or petiolate; leaf blades linear to ovate, base attenuate to rounded, apex obtuse to acute, margins entire; petioles (if present) triangular. **Inflorescences** erect, rarely submersed; panicles in 2–10 whorls; peduncle generally smooth; pedicels ascending, subtended by smooth delicate bracts; bracts lanceolate, shorter than pedicels. **Flowers** bisexual, chasmogamous or cleistogamous; receptacle flattened; sepals erect, herbaceous, 1–4 mm; petals entire to slightly erose, white or pink; stamens 6, filaments filiform; pistils many, free, in 1 whorl on margin of receptacle; ovule 1; styles lateral. **Fruits** abaxially 2–3 ribbed, lateral walls opaque to translucent; beaks erect or nearly so.

Temperate regions nearly worldwide. 9 species; 3 species treated in *Flora*.

Key to *Alisma* species:

1. Inflorescence somewhat longer than or shorter than leaves; style coiled; pedicels recurved to spreading. . . . . . ***A. gramineum***

1' Inflorescence much longer than leaves; style erect with tip recurved; pedicels ascending to erect.

2. Petals pink; leaf bases tapered; fruit lateral walls thin, translucent. . . . . ***A. lanceolatum***

2' Petals white; leaf bases truncate to lobed; fruit lateral walls thick, opaque . . . . ***A. triviale***

Alisma gramineum Lejeune

Emergent water plantain, narrowleaf water plantain

Alisma geyeri Torr. ex Nicollet, *Alisma gramineum* C.C. Gmel. var. *angustissimum* (DC.) A.J. Hendricks, *Alisma gramineum* C.C. Gmel. var. *gramineum*

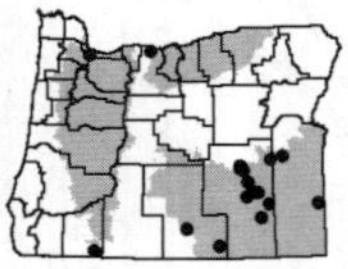

Plants 0.5–30(50) cm. **Leaves** emersed, floating or submersed; floating leaves linear to lanceolate, 20-100 cm long, bases tapered; emersed leaves petiolate or rarely sessile; submersed leaves ribbon-like; leaf blades 3–7 × 0.8–2 cm. **Inflorescences** somewhat longer than or shorter than leaves; pedicels recurved to spreading. **Flowers** cleistogamous when submersed or chasmogamous when terrestrial; sepals approximately 2.5 mm long, often with purple margins; petals 2–4 mm long, slightly larger than the sepals, reddish to pinkish or purple tinged, margins entire to slightly erose, apex obtuse; anthers nearly as broad as long, ovoid, 0.3–0.6 mm; style somewhat curved or coiled, 0.4–0.5 mm, half the length of the ovary. **Fruits** heads 3–6 mm in diameter; achenes orbicular to orbicular-ovoid, 2–2.7 mm, lateral walls thick, opaque. **2*n***=14.

Marshy shallow places, either fresh or brackish water, muddy shores. Flowering Jun–Aug. 50–1700 m. BR, Casc, Col, Owy, WV. CA, NV, ID, WA; most of western and northern North America; Africa, Eurasia. Native.

Alisma lanceolatum Withering

Lanceleaf water plantain

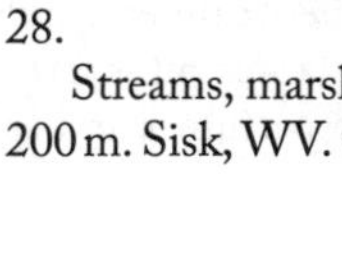

Plants to 70 cm. **Leaves** emersed, petiolate, 12–40 cm; leaf blades lanceolate, 6–15 × 1–3 cm, bases tapered. **Inflorescences** much longer than leaves; pedicels ascending to erect. **Flowers** chasmogamous; sepals 1.6–3 mm; petals 4.4–6.5 mm, pink, margins erose, apex acuminate; anthers ellipsoid, 0.6–1.1 mm; style straight, erect, 0.4–0.6 mm, equaling ovary in length. **Fruits** heads 4–8 mm in diameter; achenes obovoid, 2–3 mm, lateral walls thin, translucent. **2*n***=26, 28.

Streams, marshes, ditches, ponds. Flowering Jun–Aug. 50–200 m. Sisk, WV. CA; British Columbia; Eurasia. Exotic.

Alisma triviale Pursh (p. 131)

Northern water plantain, American water plantain

Alisma plantago-aquatica L. var. *americanum* Schult. & Schult. f.

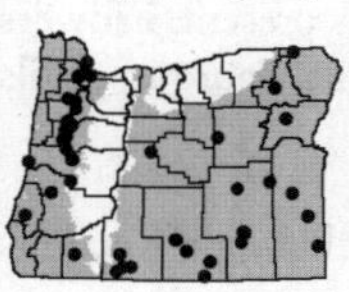

Plants to 100 cm. **Leaves** emersed, petiolate, 7-40 cm; leaf blades linear-lanceolate to ovate, 5.5–15 × 1.5–10 cm, bases truncate to lobed. **Inflorescences** much longer than leaves; pedicels ascending to erect. **Flowers** chasmogamous; sepals 3–6 mm; petals 3.5–6 mm, white, margins erose, apex obtuse; anthers ovoid, 0.6–1 mm; style 0.4–0.6 mm erect, tip curved, equal to ovary in length. **Fruits** heads 4.1–7 mm in diameter; achenes 2.1–3 mm, lateral walls thick, opaque. **2*n***=28.

Marshes, stream margins, shallow muddy ponds and ditches. Flowering Jun–Sep. 0–1500 m. BR, BW, CR, ECas, Est, Lava, Owy, Sisk, WV. CA, NV, ID, WA; throughout North America. Native.

DAMASONIUM Damasonium

Herbs perennial, glabrous. **Roots** corms present. **Leaves** floating or emersed; petiole triangular; leaf blades linear to ovate, tapered to base or sometimes rounded, apex acute, margins entire. **Inflorescences** emersed racemes or rarely panicles, in 1–9 whorls, erect; peduncles generally smooth; pedicels ascending to spreading; bracts lanceolate to ovate-lanceolate, much shorter than pedicels, apex acuminate. **Flowers** bisexual; receptacle flat; sepals 3.5–6 mm, erect; petals white to pink, erose; stamens 6, filaments dilated, glabrous; pistils 6–15, in a single whorl, radiating in a star-like pattern, fused at base, styles terminal; ovules 1(2). **Fruits** spreading, laterally compressed, sides opaque; beak terminal, greater or equal in length to body. **2*n***=14.

North America, Australia, Europe. 5 species; 1 species treated in *Flora*.

Damasonium californicum Torrey (p. 131)

California damasonium, fringed water plantain

Machaerocarpus californicus (Torr.) Small

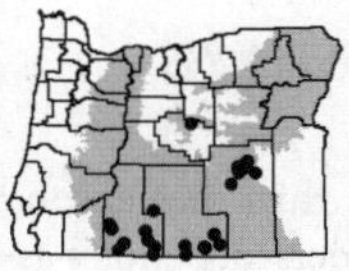

Plants to 35 cm. **Leaves** erect, linear to narrowly ovate, 5–35 cm; leaf blades lanceolate to ovate, 3–9 × 0.5–3 cm, margins entire; petioles 3–15 cm. **Inflorescences** racemes or panicles, generally longer than leaves; bracts broadly ovate-lanceolate, thin, dry. **Flowers** sepals oblong or ovate, 4–55 mm, green, hooded or not; petals triangular- to square-shaped, 6–10 mm, white with basal yellow spot, rarely pink, margins erose distally; stamens 2–3 mm, anthers red; pistils 6–10. **Fruits** 3–5.5 mm, horned by enlarged styles; beak 3–6 mm.

Ponds, vernal pools, intermittent streams, on mud. Flowering May–Jul. 1200–1800 m. BR, BW, Casc, ECas. CA, NV, ID, WA. Native.

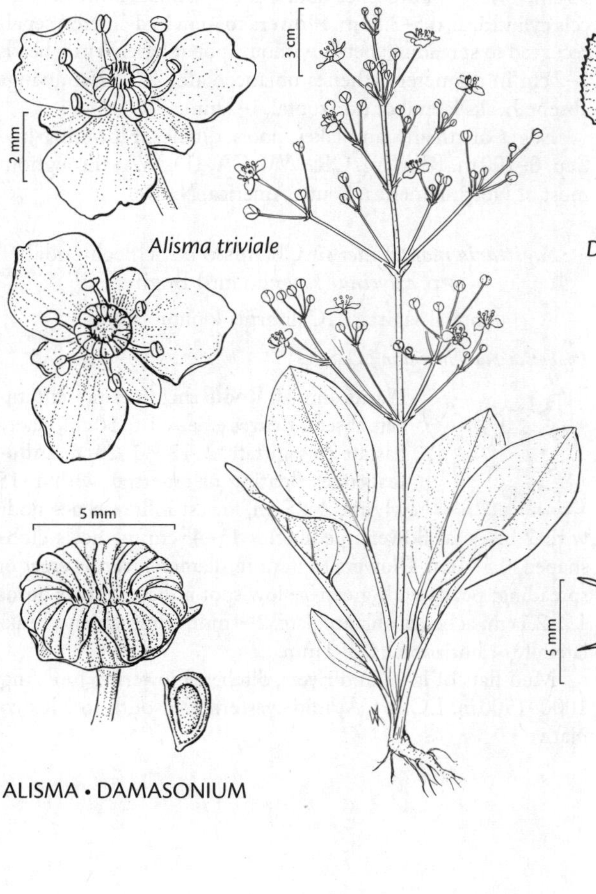

Alisma triviale

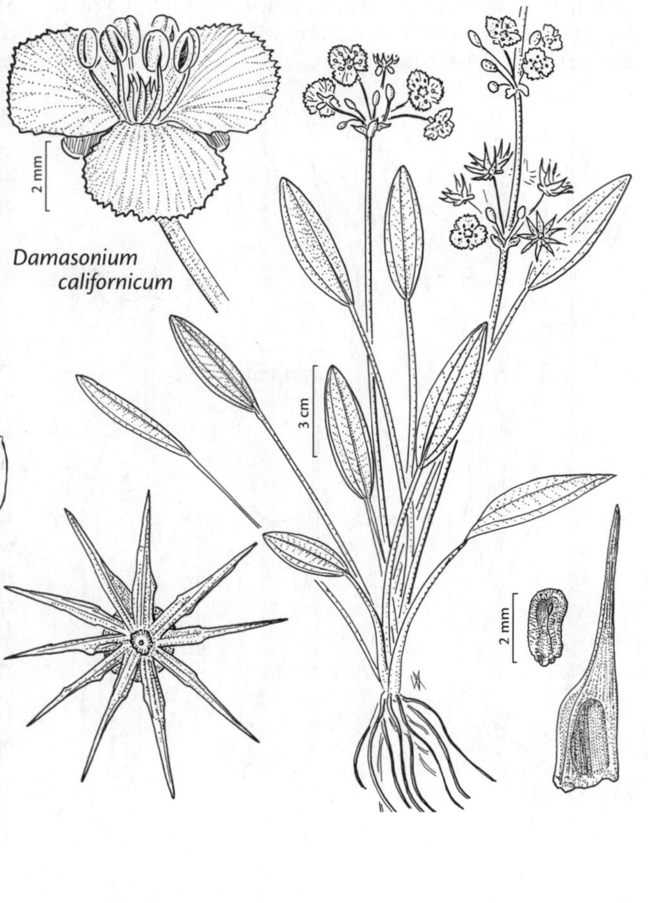

Damasonium californicum

ALISMA • DAMASONIUM

SAGITTARIA Arrowhead

Herbs perennial or annual, to 100 cm, glabrous to sparsely pubescent. **Roots** rhizomes present or not; stolons and corms present, tubers present or not. **Leaves** emersed, floating or submersed; blades linear to ovate, apex acute to rounded, margins entire. **Inflorescences** racemes or panicles; lower nodes with pistillate or bisexual flowers, staminate flowers above; bracts present. **Flowers** sepals 3–10 mm, reflexed to appressed; petals white, with or without spot at base, entire; stamens 7–30, filaments glabrous; pistils many, arranged spirally on receptacle. **Fruits** compressed, up to 4 mm long, winged or ridged. **2*n*=22.**

Cosmopolitan. Approximately 30 species; 3 species treated in *Flora.*

This genus is important as food for wildlife. Beavers, muskrats, and porcupines feed on the whole plants, including the starchy tubers of some species. Ducks and other waterfowl eat the seeds. Some species are deliberately planted in hunting areas.

Sagittaria platyphylla has recently (2014) been reported in Portland. How widespread it is and how persistent it may be are currently unknown.

Key to *Sagittaria* species:

1. Tubers 0; lowest inflorescence node with 2 bisexual flowers. . . . . **S. montevidensis** ssp. **calycina**
1' Tubers present; lowest inflorescence node with 3 pistillate flowers.
  2. Petioles of emersed leaves recurved; fruit beak terminal, erect, less than 1 mm long . . . . **S. cuneata**
  2' Petioles of emersed leaves ascending to erect; fruit beak lateral, spreading, 1 mm long or greater. . . . . **S. latifolia**

Sagittaria cuneata E. Sheldon

Arumleaf arrowhead, wapato

Sagittaria arifolia Nutt. ex J.G. Sm.

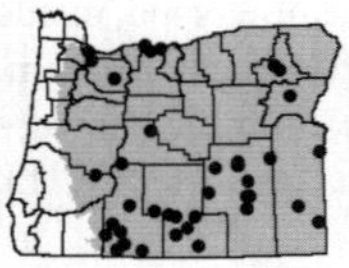

Plant perennial. **Roots** rhizomes absent; tubers oblong. **Leaves** emersed leaves to 45 cm, petioles recurved, blades linear to sagittate, 2–18 × 2–10 cm; floating leaves to 100 cm, blades usually sagittate or cordate, 7–8 × 3–4 cm; submersed leaves flat, to 40 cm. **Inflorescences** emersed, with 2–10 whorls, 10–20 cm, lowest inflorescence node with 3 pistillate flowers; peduncle 10–50 cm; pedicels cylindrical, 0.5–2 cm. **Flowers** to 25 mm in diameter; sepals recurved; petals without spot at base. **Fruits** heads 0.5–1.5 cm in diameter; achenes ovoid, 1–2.5 mm, wing 1 or absent; beaks terminal, erect, 0.1–0.5 mm.

Ponds, lake shores, ditches. Flowering Jun–Sep. 0–1800 m. BR, BW, Casc, Col, ECas, Lava, Owy, WV. CA, NV, ID, WA; throughout North America. Native.

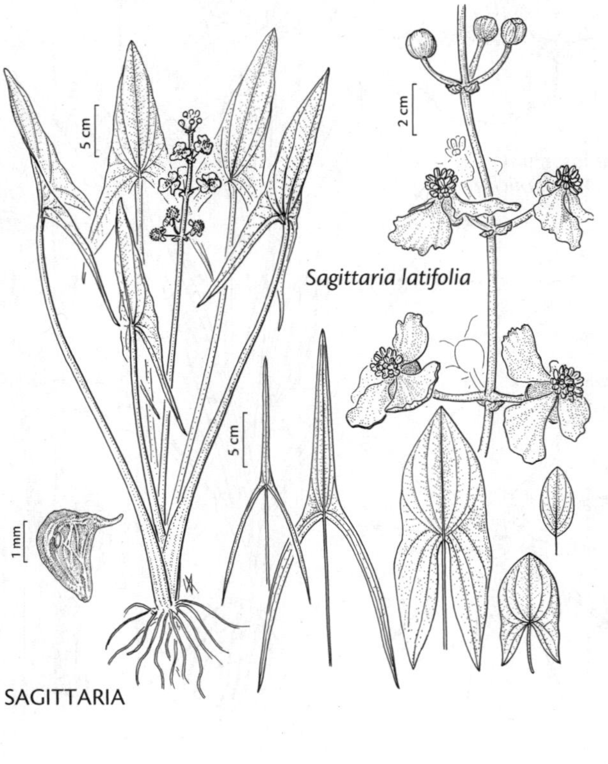

SAGITTARIA

Sagittaria latifolia Willdenow (p. 132)

Wapato, broadleaf arrowhead, duck potato

Sagittaria latifolia Willd. var. *latifolia*

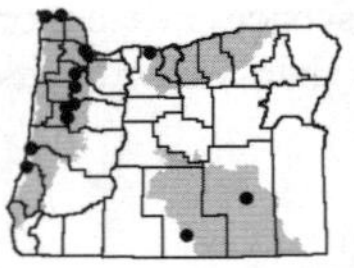

Plants perennial. **Roots** rhizomes absent; tubers present. **Leaves** emersed, petioles ascending to erect, to 50 cm, blades sagittate, 1–30 × 2–15 cm. **Inflorescences** emersed, with 3–9 whorls, 4–25 cm; peduncles 10–55 cm; lowest inflorescence node with 3 pistillate flowers; pedicels cylindrical, 0.5–3.5 cm. **Flowers** to 4 cm in diameter; sepals recurved to spreading; petals without spot at base. **Fruits** heads 1–2 cm in diameter; achenes oblanceolate, 2–3.5 mm, wings absent; beaks lateral or horizontal, 1–2 mm.

Edges of streams and lakes, pools, ditches. Flowering Jul–Sep. 0–100 m. BR, Col, CR, WV. CA, ID, WA; throughout most of North America; South America. Native.

Sagittaria montevidensis Chamisso & Schlechtendal ssp. *calycina* (Engelmann) Bogin

Annual sagittaria, California lophotocarpus

Lophotocarpus californicus J.G. Sm.

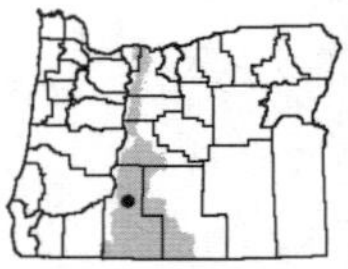

Plants annual. **Roots** rhizomes present, tubers absent. **Leaves** emersed, to 50 cm, blades hastate to sagittate, 2–17 × 1–20 cm. **Inflorescences** floating or emersed, with 1–15 whorls, 2–25 cm, lowest inflorescence node with 2 bisexual flowers; peduncles 15–45 cm; pedicels club-shaped, 0.5–4 cm. **Flowers** to 5 cm in diameter; sepals erect or spreading; petals with green-yellow spot at base. **Fruits** heads 1.2–2.1 cm; achenes oblanceolate, 2–4 mm, wings absent; beaks laterally or horizontal, 0.5–1 mm.

Mud flats of lakes and rivers, ditches. Flowering Jun–Aug. 1000–1500 m. ECas. CA; mid-western US south to Mexico. Native.

AMARYLLIDACEAE Amaryllis family

Herbs perennial, from membranous or scaly bulbs; stems scapose, cylindric. **Leaves** 1–several, linear, terete, channeled or flat. **Inflorescences** solitary or umbellate. **Flowers** radially symmetrical, erect or pendent, perianth parts fused or not; tepals 6, in 2 whorls; stamens 6; ovary superior or inferior; style 1, stigma capitate or 3-lobed. **Fruits** capsules, dehiscence loculicidal. **Seeds** black, smooth to roughened.

Cosmopolitan. 82 genera; 2 genera treated in *Flora.*

The genera currently within Amaryllidaceae were formerly included in Liliaceae. Under a narrower concept of Liliaceae, some genera, including *Allium*, have been delineated (split) into Alliaceae. Here, we take a more middling and current approach that recognizes a concept of Amaryllidaceae which includes all Alliaceae genera.

Key to Amaryllidaceae genera:

1. Perianths with tubular crowns . . . . . **NARCISSUS**

1' Perianths without crowns . . . . . **ALLIUM**

ALLIUM Onion

Nick Otting, Richard E. Brainerd, and Barbara L. Wilson

Plants herbs, perennial, scapose, from tunicate bulbs, with onion odor and taste; bulbs solitary or clustered and dividing at base, or on rhizomes and forming again each year; outer coats, smooth, fibrous, or with cellular reticulations; inner coats membranous. **Leaves** usually withering from tip by anthesis, usually persistent, 1–12, basal; blade usually linear, terete, channeled, or flat (keeled in *A. triquetrum* and *A. unifolium*), straight or more or less falcate, usually not petiolate. **Scapes** usually persistent, terete, flattened, or triangular. **Inflorescences** umbellate, flowers sometimes replaced by bulbils; pedicels erect or spreading (lax in *A. triquetrum*); subtending spathe bracts conspicuous, more or less fused, membranous. **Flowers** erect (pendent in *A. triquetrum*); tepals 6, in 2 similar whorls, more or less distinct, petal-like, usually persistent; stamens 6, filaments usually broad at base, fused into ring, glabrous (papillose to ciliate proximally); ovary superior, 3-lobed, sometimes crested, crests obscure to prominent; style 1, stigma usually narrowly capitate to more or less 3-lobed. **Fruits** loculicidal capsules. **Seeds** black, smooth and shiny or minutely roughened.

Mostly Northern Hemisphere. 550–700 species; 34 species treated in *Flora.*

In the recent splitting of Liliaceae *sensu lato* (Stevens 2012), *Allium* was classified within the family Amaryllidaceae, in the order Asparagales. All members of *Allium* smell more or less like onions or garlic and are more or less edible, though not always pleasant tasting. Bulb coat reticulations are frequently important in identification; they are best viewed on the inner surface of the outer coats. Descriptions of reticulations assume 10× magnification.

Allium membranaceum (papery onion) was collected in Jackson County in 2010. It may be a waif. This species is known to occur in California. *Allium cepa* (cultivated onion) occurs as a waif near onion fields and along dirt roads on which they are transported. It is also known from the wild in California and Washington.

Key to *Allium* species:

1. Inflorescence mostly consisting of bulbils; flowers few.
   2. Leaf blades solid, 6–30 cm long; range east of Cascades . . . . . ***A. geyeri*** (in part)
   2' Leaf blades hollow below middle, 20–60 cm long; range west of Cascades . . . . . ***A. vineale***

1' Inflorescence mostly consisting of flowers; bulbils absent or few.
   3. Inflorescence nodding at anthesis and thereafter; ovary strongly crested; stamens exserted . . . . . ***A. cernuum***
   3' Inflorescence erect at anthesis and thereafter; ovary crested or not; stamens exserted or not.
      4. Leaf blades terete to subterete, or folded and narrowly channeled in cross section, not falcate.
         5. Leaves 1 per scape.
            6. Scapes 5–15 cm long; stamens shorter than tepals; range southeastern Oregon . . . . . ***A. nevadense***
            6' Scapes 18–54 cm long; stamens equaling or longer than tepals; range southwestern Oregon . . . . . ***A. sanbornii*** var. ***sanbornii***
         5' Leaves usually 2 or more per scape.
            7. Tepals 5–8(9) mm long, spreading from base; ovary crests more or less conspicuous, triangular . . . . . ***A. amplectens***
            7' Tepals (7)9–15 mm long, erect; ovary crests absent or obscure.
               8. Umbel compact; pedicels much shorter than flowers; range Willamette Valley and Columbia Basin . . . . . ***A. schoenoprasum***
               8' Umbel more open; pedicels longer than flowers; range various.

9. Tepals more or less equal in length; outer bulb coats with obscure reticulum; umbel 10–20-flowered; range Klamath Mts. . . . . *A. bolanderi*
9' Tepals unequal, the outer longer than the inner; outer bulb coats with distinct reticulum; umbel (5)15–40-flowered; range various.
10. Outer bulb coat reticulations square to polygonal, with thickened walls; range various . . *A. acuminatum*
10' Outer bulb coat reticulations transversely elongate and forming a herringbone pattern, without thickened walls; uncommon; range southern Cascades . . . . *A. peninsulare* var. *peninsulare* (in part)

4' Leaf blades flat or broadly channeled in cross section, sometimes falcate.
11. Scapes sharply 3-angled in cross section; range near the coast . . . . *A. triquetrum*
11' Scapes round to flattened in cross section, not sharply 3-angled, sometimes winged; range various.
12. Scapes round or only slightly compressed in cross section, sometimes very narrowly winged.
13. Umbel 75–500-flowered; scapes (45)60–180 cm tall; uncommon introductions in Willamette Valley and Oregon Coast.
14. Tepals 6–9 mm; leaves 3–6. . . . . *A. nigrum*
14' Tepals 4–5.5 mm; leaves 6–9 . . . . *A. ampeloprasum*
13' Umbel 10–75-flowered; scapes 1.6–80 cm tall; range various.
15. Stamens more or less equaling tepals or exserted.
16. Leaf blades usually more than 5 mm wide, flat; spathe bracts 3 . . . . *A. douglasii*
16' Leaf blades 2–3 mm wide, flat to channeled; pathe bracts 2.
17. Outer bulb coat with quadrate to polygonal reticulations; leaf blade less than to more or less equaling scape; range north-central Oregon . . . . *A. nevii*
17' Outer bulb coat without reticulations or with 2–3 rows adjacent to roots; leaf blade usually exceeding scape; range widespread east of Cascades. . . . . *A. macrum*
15' Stamens much shorter than tepals.
18. Stigma usually with 3 short lobes; outer bulb coat reticulations distinct, forming a herringbone pattern; range southern Cascades . . . . *A. peninsulare* var. *peninsulare* (in part)
18' Stigma not, or slightly lobed at 10×; outer bulb coat reticulations (if evident) hexagonal to transversely elongate, sometimes contorted; range various.
19. Ovary conspicuously crested with 6 triangular processes; tepals usually spreading from base.
20. Tepals usually less than 7 mm long, usually papery in fruit, the tips rarely with strongly involute margins or a pronounced keel; leaves commonly withering by anthesis . . . . *A. campanulatum*
20' Tepals usually greater than 7 mm long, rigid in fruit, tips strongly involute, usually with a pronounced keel; leaves usually green at anthesis . . . . *A. bisceptrum*
19' Ovary crests absent or inconspicuous; tepals usually not spreading from base.
21. Tepals usually at least 11 mm long; pedicels 15–40 mm long; western Oregon . . . *A. unifolium*
21' Tepals usually less than 11 mm long; pedicels usually less than 15(25) mm long; range Blue Mts. and SW Oregon.
22. Leaves 3–5; tepal midveins usually inconspicuous. . . . . *A. geyeri* (in part)
22' Leaves 1–2; tepal midveins prominent.
23. Tepals oblong-elliptic, apices obtuse; leaves 1.5–4 times the length of the scape; range southwestern Oregon. . . . . *A. obtusum* var. *obtusum*
23' Tepals lanceolate to elliptic, apices acute to acuminate; leaves usually less than twice the length of the scape; range Blue Mts.
24. Larger bulbs with a cluster of basal bulbils; outer bulb coat reticulations often absent, or if present, obscure and contorted . . . . . . *A. madidum*
24' Bulbs without basal bulbils or rarely with 1 or 2; outer bulb coat reticulations usually evident, contorted to hexagonal or transversely elongate.
25. Outer bulb coat reticulations contorted; ovary obscurely crested. . *A. fibrillum*
25' Outer bulb coat reticulations usually distinct, more or less hexagonal to transversely elongate; ovary rarely at all crested . . . *A. brandegeei*

12' Scapes flattened or moderately compressed, often winged.

26. Scapes 30–70 cm tall; leaves 3–6; strongly rhizomatous; plants of swampy meadows . . . . *A. validum*

26' Scapes 3–25 cm tall; leaves 1–2; plants may or may not have rhizomes; diverse, seasonally wet to moist habitats.

27. Umbel compact; pedicels usually much shorter than flowers; leaves 1–3(4) mm wide; sand and gravel deposits and sandy slopes along the Columbia River . . . . *A. robinsonii*

27' Umbel open; pedicels usually equal to or greater than flowers; leaves 1–18 mm wide; diverse habitats, not sand and gravel deposits and sandy slopes along the Columbia River.

28. Leaves (6)8–18 mm wide, strongly falcate; umbels 30–90-flowered; stamens exserted; south-central Oregon . . . . *A. platycaule*

28' Leaves 1–8(10) mm wide, falcate or not; umbels 5–60-flowered; stamens included to exserted; range various.

29. Stamens equaling tepals or exserted; range southeastern Oregon.

30. Tepals linear to lance-linear, greater than 4 times as long as wide; scapes 5–15 cm tall . . . . . . *A. anceps*

30' Tepals lanceolate, less than 4 times as long as wide; scapes 6–20 cm tall . . . . *A. lemmonii*

29' Stamens much shorter than tepals; range various.

31. Range in and west of Cascade Mountains.

32. At least inner tepal margins finely denticulate; range southwestern Oregon.

33. Tepals 9–15 mm long; scapes 5–25 cm tall. . . . . *A. falcifolium*

33' Tepals 8–11 mm long; scapes 3–8 cm tall. . . . . *A. siskiyouense* (in part)

32' Tepal margins entire; range various.

34. Bulbs oblique or oblique-ovoid, renewal bulbs borne on rhizomes; tepals usually spreading from base at anthesis; leaf and scape margins sometimes crenulate-denticulate . . . . *A. crenulatum*

34' Bulbs ovoid to globose, rhizomes absent, renewal bulbs formed within coats of parent bulb; tepals ascending to erect, sometimes spreading from tips at anthesis; leaf and scape margins not crenulate-denticulate.

35. Tepals often spreading at mid-length, narrowly lanceolate to ovate, apices obtuse to acute, keeled in fruit; umbels 10–35-flowered . . . . *A. siskiyouense* (in part)

35' Tepals erect, broadly lanceolate to ovate, apices obtuse, not keeled in fruit; umbels 5–30-flowered . . . . *A. parvum* (in part)

31' Range east of Cascade Mountains.

36. Pedicels 5–25 mm; umbels 10–40-flowered; tepals lanceolate, apices acute to acuminate, spreading, the margins involute and the midrib prominent . . . . . . *A. tolmiei* var. *tolmiei*

36' Pedicels 3–12 mm; umbels 5–30-flowered; tepals widely lanceolate to ovate, apices obtuse to acute, usually not spreading; midrib not prominent.

37. Outer bulb coats with transverse elongate reticulations . . . . *A. punctum*

37' Outer bulb coats not cellular-reticulate or with 2–3 rows adjacent to roots . . *A. parvum* (in part)

Allium acuminatum W.J. Hooker (p. 137)

Tapertip onion, Hooker's onion

Allium acuminatum Hook. var. *acuminatum*

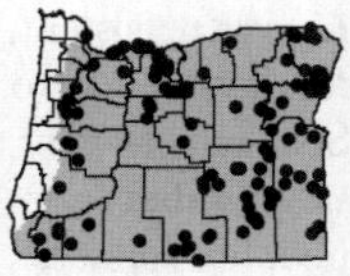

Bulbs 1–12+, not basally clustered, ovoid to more or less globose; outer coats enclosing 1 or more renewal bulbs, prominently cellular-reticulate, membranous; cells square or polygonal, walls thick, obscurely sinuous, without fibers. **Leaves** persistent, withering from tip by anthesis, 2–3; blades solid, subterete or more or less channeled, 7–30 cm × 1–3 mm. **Scapes** persistent, solitary, erect, solid, terete, 10–35 cm × 1–3 mm. **Umbels** persistent, erect, loose, 10–40-flowered, hemispheric; pedicels 6–25 mm; spathe bracts 2. **Flowers** 8–15 mm; tepals erect, lanceolate to lance-ovate, unequal, pink to rose-purple or white, becoming rigid and keeled in fruit, margins finely denticulate (inner tepals more prominently so), apex acuminate; outer tepals longer and wider than inner, spreading to recurved at tip; inner tepals with strongly recurved tips; stamens included; ovary obscurely crested with 3 minute, 2-lobed processes; stigma capitate, scarcely thickened, obscurely 3-lobed.

Dry to seasonally moist meadows, prairies, sagebrush steppe, and open forests on clay to sandy soils. Flowering May–Jul. 0–2500 m. BR, BW, Casc, Col, ECas, Lava, Owy, Sisk, WV. CA, ID, NV, WA; north to British Columbia, east to WY, southeast to AZ. Native.

Allium dictuon, an extremely rare onion of the Blue Mountains of southeastern Washington, superficially resembles *A. acuminatum* although it is more closely related to *A. bolanderi*. It differs from *A. acuminatum* in its rhizomatous habit and its bulb coat reticulations, which are polygonal to oblong with thick, obscurely sinuous walls. Its new bulbs form outside the bulb coats of the parent bulb, which disappears by anthesis, except for the persistent root pad. Additionally, its bulbs are oblique.

Allium ampeloprasum Linnaeus

Wild leek, elephant garlic

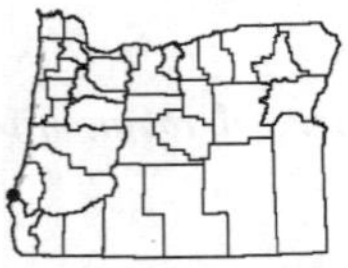

Bulbs 1–3+, variable, some poorly developed, others ovoid with 1–2 large bulbs and several bulbils at base; outer coat enclosing 1 or more bulbs, membranous. **Leaves** withering from tips by anthesis, 6–9; blades solid, flat, channeled, 1–5 × 2–20(30) mm. **Scapes** persistent, solitary, erect, hollow, terete, 45–180 cm × 3–7 mm. **Umbels** persistent, erect, compact, to 500-flowered, few-flowered in variants with bulbils, globose; pedicels 15–50 mm; spathe bracts 3–5. **Flowers** 4–5.5 mm; tepals erect, white, pink, or dark red; outer tepals oblong-lanceolate, margins entire, apex obtuse, sometimes mucronate; inner tepals narrowly ovate to spatulate, margins entire, apex obtuse; stamens equaling perianth or exserted; ovary crestless; stigma scarcely thickened, unlobed.

Roadsides and other disturbed areas. Flowering Apr–Jul. 0–50 m. Est. Occasionally escaped in North America; North Africa, Asia, Europe. Exotic.

This species may be conspecific with the cultivated leek *A. porrum*.

Allium amplectens Torrey

Slim-leaf onion

Allium serratum S. Watson

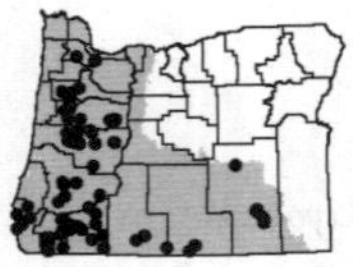

Bulbs 1–15+, increase bulbs absent or more or less equaling parent bulbs, never appearing as a basal cluster, ovoid to more or less globose; outer coats enclosing 1 or more bulbs, prominently cellular-reticulate, membranous; cells in more or less vertical rows, forming irregular herringbone pattern, transversely elongate, without fibers. **Leaves** persistent, withering from tip at anthesis, 2–4; blades solid, subterete or more or less channeled, 10–36 cm × 0.5–2 mm. **Scapes** persistent, solitary, erect, solid, terete, 15–50 cm × 3–5 mm. **Umbels** shattering after seeds mature, each flower deciduous with its pedicel as a unit, erect, compact, 10–50-flowered, hemispheric; pedicels 4–16 mm; spathe bracts 2–3. **Flowers** 5–9 mm; tepals spreading, lanceolate, more or less equal, white to pink, margins entire, apex acute; stamens included; ovary crested with 6 prominent processes; stigma scarcely thickened, unlobed. **2*n***=14, 21, 28.

Moist seeps, clay soils including dry meadows, savanna, and serpentine barrens. Flowering May–Jul. 0–2500 m. BR, Casc, CR, ECas, Est, Sisk, WV. WA, CA; north to British Columbia. Native.

Allium anceps Kellogg

Two-edged onion

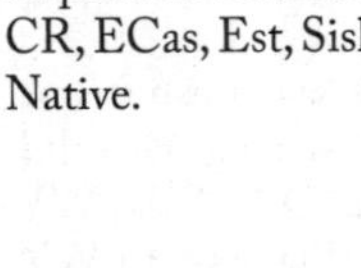

Bulbs 1–5, ovoid; outer coats enclosing 1 or more bulbs, membranous, more or less prominently cellular-reticulate, cells transversely elongate, in regular vertical rows, narrowly rectangular, without fibers. **Leaves** usually deciduous with scape, withering from tip at anthesis, 2; blades solid, flat, falcate, 7–26 cm × 4–6 mm. **Scapes** usually forming abscission layer and deciduous with leaves after seeds mature, solitary, erect, solid, flattened, narrowly winged, 5–15 cm × 1–3 mm. **Umbels** persistent, erect, more or less compact, 15–35-flowered, hemispheric; pedicels 7–17(20) mm; spathe bracts 2. **Flowers** 8–12 mm; tepals spreading, linear-lanceolate, more or less equal, light pink with diffuse greenish midveins, margins entire, apex acute; stamens more or less equaling tepals; ovary crested with 6 inconspicuous processes; stigma scarcely thickened, unlobed. **2*n***=14.

Heavy, usually barren, clay soils, rocky openings in sagebrush steppe. Flowering May–Jun. 700–1900 m. BR, BW, Owy. CA, ID, NV. Native.

Allium bisceptrum S. Watson

Twincrest onion

Allium bisceptrum S. Watson var. *bisceptrum*

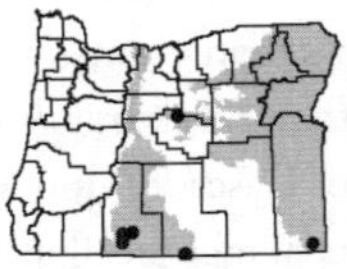

Bulbs 1–7+, commonly producing either a cluster of stalked, basal bulbils or filiform rhizomes to 10 cm, terminated by bulbils; rhizomes usually lost when specimens are collected; ovoid, outer coats enclosing 1 or more bulbs, membranous, obscurely cellular-reticulate, cells rectangular, walls minutely sinuous, vertical, varying to irregular, all sinuous, without fibers. **Leaves** persistent, green at anthesis, 2–5; blades solid, flat, broadly channeled, 8–30 cm × 1–13 mm. **Scapes** persistent, solitary or clustered, 1–3, erect, solid, terete, 10–30(40) cm × 1–5 mm. **Umbels** persistent, erect, loose, 15–40-flowered, globose; pedicels 10–20 mm, often becoming flexuous and deflexed in fruit; spathe bracts 2. **Flowers** 7–10 mm; tepals spreading, lanceolate, more or less equal, lilac to white, becoming papery in fruit, not keeled, margins entire, apex acuminate, not involute; stamens included; ovary conspicuously crested with 6 triangular processes; stigma unlobed. **2*n***=14, 28.

Meadows and aspen groves, less commonly on open slopes in mountains, grasslands with juniper, pumice barrens. Flowering Jun–Jul. 1400–2000 m. BW, ECas, Owy. CA, ID, NV; southeast to NM. Native.

Allium bolanderi S. Watson

Bolander's onion

Bulbs 1–6+, not basally clustered, new bulbs borne terminally on rhizomes; rhizomes 1–3, slender, less than 2 cm long including renewal bulb; parent bulbs disappearing by anthesis except for still-functional roots and bulb coat, oblique-ovoid to more or less oblong; outer coats not enclosing bulbs, more or less obscurely reticulate, membranous, reticulum delicate, cells transversely elongate, V-shaped or more or less wavy, forming obscure herringbone pattern, without fibers. **Leaves** persistent, withering from tip at anthesis, 2–3; blades solid, subterete to more or less channeled, 9–30 cm × 1–2 mm. **Scapes** persistent, solitary, erect, solid, terete, 10–35 cm × 1–3 mm. **Umbels** persistent, erect, compact to loose, 10–20-flowered, hemispheric; pedicels 10–20 mm; spathe bracts 2. **Flowers** 7–14 mm; tepals erect, narrowly lanceolate to lance-ovate, more or less equal, reddish purple, rarely white, margins finely denticulate (inner tepals more prominently so), apex acute to obtuse, becoming involute at tip and appearing acuminate; stamens included; ovary obscurely crested with 3 minute processes; stigma scarcely thickened, obscurely 3-lobed. **2*n***=14.

Clay soils, usually on serpentine, dry openings and savanna. 2 varieties.

The varieties of *A. bolanderi* are extremely difficult to distinguish in the absence of bulb material. Although great care must be taken to obtain good bulbs whenever collecting any *Allium* taxon, it is especially critical within this species.

Key to *Allium bolanderi* varieties:

1. Tepals lance-ovate, 7–12 mm long; bulbs ovoid; outer bulb coat breaking with serrate edge. . . . . ***A. b.* var. *bolanderi***

1' Tepals narrowly lanceolate, 9–14 mm long; bulbs oblong, irregular, resembling small tubers; bulb coat breaking with irregular edge . . . . ***A. b* var. *mirabile***

Allium bolanderi S. Watson var. ***bolanderi***

Bolander's onion

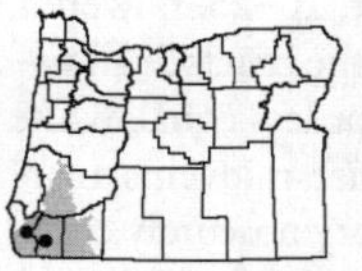

Bulbs ovoid, 0.7–1.2 cm; outer coats breaking with serrated edges, cells of reticulum V-shaped forming obscure herringbone pattern. **Flowers** 7–12 mm; tepals lance-ovate, apex acute to obtuse.

Flowering May–Jul. 400–1300 m. Sisk. CA. Native.

Allium bolanderi S. Watson var. ***mirabile*** (L.F. Henderson) McNeal

Slender-stemmed onion, potato bulb, Bolander's onion

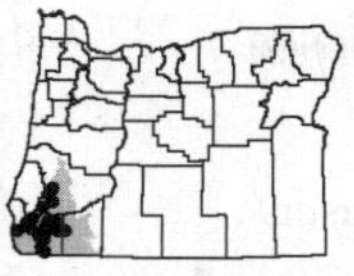

Bulbs oblong, more or less irregularly shaped, suggesting small tuber, 0.9–1.4 cm; outer coats breaking with irregular edges, cells of reticulum wavy. **Flowers** 9–14 mm; tepals narrowly lanceolate, apex acute.

Flowering May–Jun. 50–1200 m. Sisk. CA. Native.

Allium brandegeei S. Watson (p. 137)

Brandegee's onion

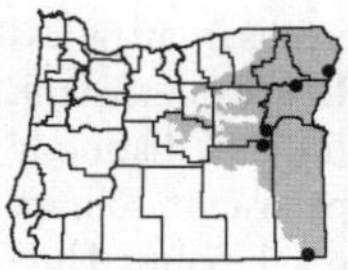

Bulbs 1–5, ovoid to globose; outer coats enclosing 1 or more bulbs, membranous, usually distinctly cellular-reticulate, cells isodiametric or more or less narrowly hexagonal, transversely elongate, without fibers. **Leaves** usually persistent, green at anthesis, 2; blades solid, flat or channeled, more or less falcate, 8–27 cm × 1–3 mm. **Scapes** persistent, solitary, erect, solid, terete, rarely slightly compressed, narrowly or not winged, 3–10(20) cm × 1–3(4) mm. **Umbels** persistent, erect, compact, 8–25(30)-flowered, hemispheric; pedicels 5–15 mm; spathe bracts 2. **Flowers** 5–8 mm; tepals erect, lanceolate to elliptic, more or less equal, white with prominent green to purplish midvein, becoming involute at tip in

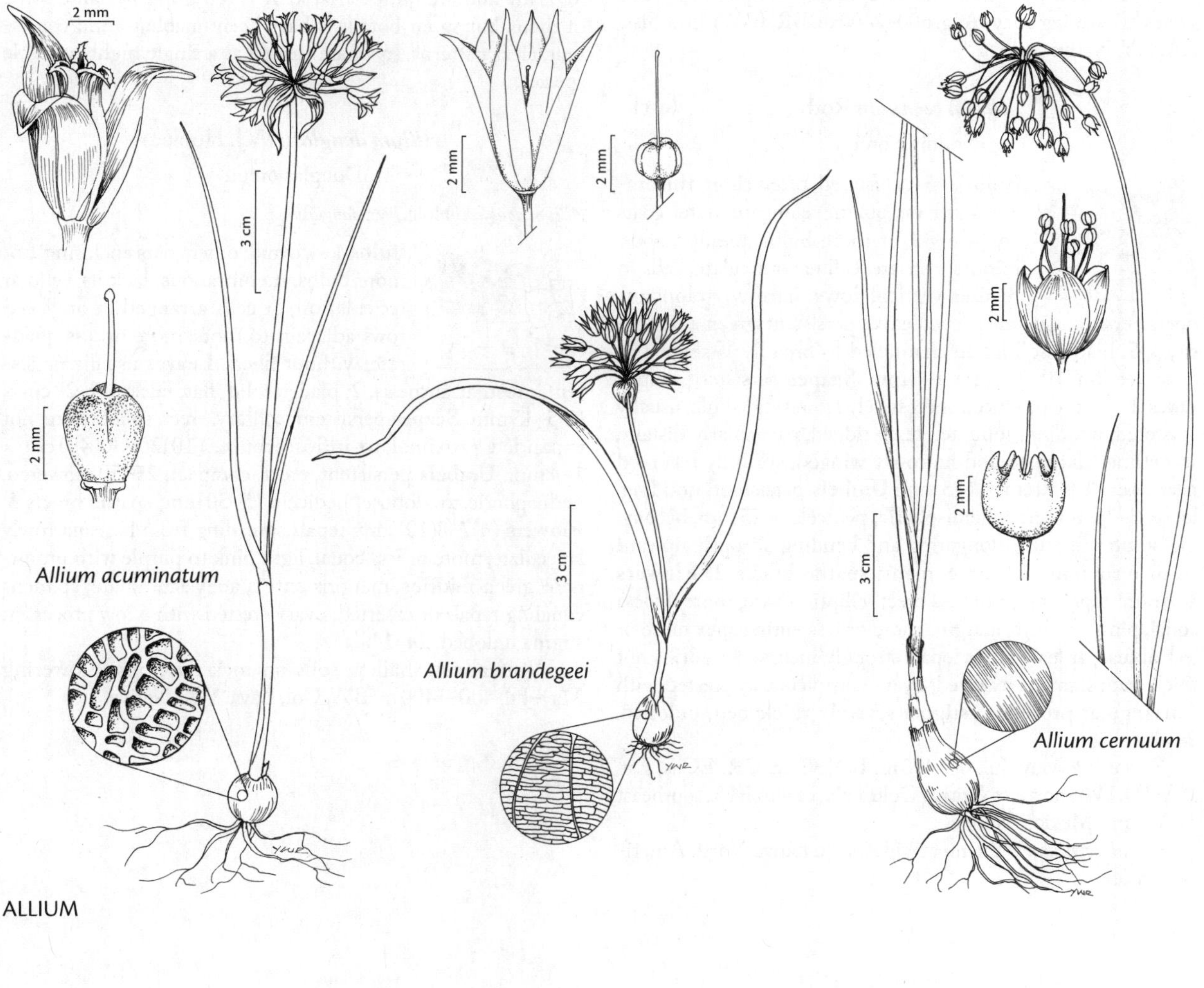

Allium acuminatum

Allium brandegeei

Allium cernuum

ALLIUM

fruit, margins entire, apex obtuse to acuminate; stamens included; ovary crestless or obscurely crested with 3 processes; stigma scarcely thickened, unlobed. **2*n*=14.**

Heavy soils in meadows and seasonally wet areas. Flowering Apr–Jun. 1300–1900 m. BW, Owy. ID, NV; east to MT and CO. Native.

Allium campanulatum S. Watson

Sierra onion

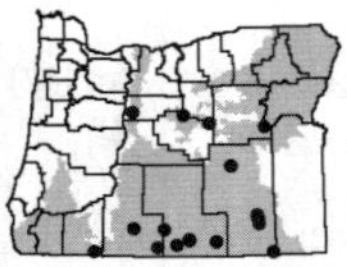

Bulbs 1–3, producing cluster of stalked basal bulbils above roots or filiform rhizomes to 10 cm and terminated by bulbils, ovoid; outer coats enclosing 1 or more bulbs, membranous, cellular-reticulate, cells more or less quadrate, walls very sinuous, without fibers. **Leaves** persistent, withering from tip by anthesis, 2; blades solid, flat, distinctly concave-convex, 8–40 cm × 1–5 mm. **Scapes** persistent, solitary or clustered, 1–3, erect, solid, terete, 10–30 cm × 1–5 mm. **Umbels** persistent, erect, loose, 10–50-flowered, more or less globose; pedicels 10–20 mm, becoming flexuous and mostly strongly deflexed in fruit; spathe bracts 2–3. **Flowers** 5–8 mm; tepals spreading, lanceolate to ovate, more or less equal, rose to purple (rarely white) with darker purple basal crescent on ventral side, keeled in fruit, becoming erect, more or less shiny, rigid, margins entire, apex acuminate, strongly involute at tip; stamens included; ovary crested with 6 prominent processes; stigma scarcely thickened, unlobed. **2*n*=14, 28.**

Usually sandy soils and rocky uplands, on open or shaded slopes. Flowering May–Aug. 600–2400 m. BR, BW, ECas, Sisk. CA, NV. Native.

Allium cernuum Roth (p. 137)

Nodding onion

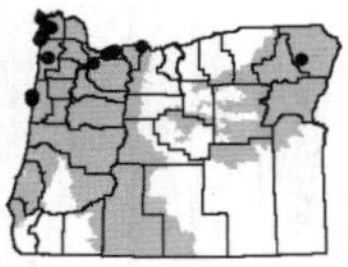

Bulbs 2–5+, clustered, often short-rhizomatous at base, oblong, elongate; outer coats enclosing 1 or more bulbs, membranous, minutely striate, cellular-reticulate, cells in regular vertical rows, narrowly elongate, fibers persistent, parallel, few. **Leaves** persistent, green at anthesis, 3–5; blades solid, flat, channeled to broadly V-shaped in cross section, 10–25 cm × 1–6 mm. **Scapes** persistent, sometimes 2 or more produced successively from single bulb, usually clustered, nodding, solid, terete or ridged, particularly distally, sometimes flattened and narrowly winged, abruptly recurved near apex, 10–50 cm × 1–3 mm. **Umbels** persistent, nodding, loose, 8–35-flowered, hemispheric; pedicels 6–25 mm, becoming stouter in fruit, elongating and bending abruptly upward from near point of attachment; spathe bracts 2. **Flowers** 4–6 mm; tepals more or less erect, elliptic-ovate, more or less equal, pink or white, margins more or less entire, apex more or less obtuse, at least outer tepals strongly incurved, midribs not thickened; stamens exserted; ovary conspicuously crested with 6 triangular processes; stigma scarcely thickened, unlobed. **2*n*=14.**

Flowering May–Jul. 0–1400 m. BW, Casc, CR, ECas, Est, WV. ID, WA; north to British Columbia, east to MA, southeast to AZ and Mexico. Native.

Allium cernuum is the most widespread native North American onion.

Allium crenulatum Wiegand

Scalloped onion, Olympic onion

Allium cascadense M. Peck, *Allium watsonii* Howell

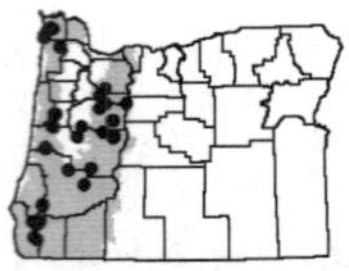

Bulbs 1–6+, replaced annually with new bulbs borne terminally on short, secondary rhizomes; parent bulb disappearing by anthesis except for still-functional roots and shriveled bulb coat, oblique-ovoid; outer coats not or only partially enclosing bulbs, membranous, lacking cellular reticulation or cells arranged in only 2–3 rows adjacent to roots, more or less quadrate, without fibers. **Leaves** usually deciduous with scape, withering from tip at anthesis, (1)2; blades solid, flat, falcate, 10–33 cm × 1.5–10 mm. **Scapes** usually forming abscission layer and deciduous with leaves after seeds mature, solitary, more or less erect, solid, flattened, winged distally, wings frequently crenulate proximal to umbel, 5–15 cm × 1–5 mm. **Umbels** persistent, erect, compact, 10–25-flowered, conic to hemispheric; pedicels 6–16 mm; spathe bracts 2. **Flowers** 6–12 mm; tepals erect, lanceolate, more or less equal, white to pinkish with deeper pink, purple or greenish midveins, margins entire, apex acute; stamens included; ovary obscurely crested with 3 minute, 2-lobed processes; stigma scarcely thickened, unlobed. **2*n*=14.**

Talus slopes and clay soils, including serpentine, on bald summits and ridges. Flowering May–Jul. 0–1400 m. BW, Casc, CR, ECas, Est, WV. WA; north to British Columbia. Native.

Oregon's *Allium crenulatum* populations differ morphologically and are quite variable. It is tempting to name some of them, but so far botanists have been unable to find consistent, clear patterns, so all are treated as a single highly variable species.

Allium douglasii W.J. Hooker

Douglas onion

Allium douglasii Hook. var. *douglasii*

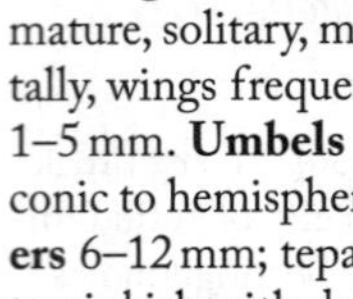

Bulbs 1–4, ovoid; outer coats enclosing 1 or more bulbs, membranous, lacking cellular reticulation, or cells arranged in only 2–3 rows adjacent to roots, more or less quadrate, without fibers. **Leaves** usually persistent, green at anthesis, 2; blades solid, flat, falcate, 9–28 cm × (2)5–15 mm. **Scapes** persistent, solitary, erect, solid, terete, not expanded proximal to inflorescence, (10)20–30(40) cm × 1–4 mm. **Umbels** persistent, erect, compact, 25–50-flowered, hemispheric to globose; pedicels 15–30 mm; spathe bracts 3. **Flowers** (6)7–8(10) mm; tepals spreading from base, narrowly lanceolate, more or less equal, light pink to purple with prominent green midribs, margins entire, apex acuminate; stamens equaling tepals or exserted; ovary crested with 6 low processes; stigma unlobed. **2*n*=14.**

Winter-wet, shallow soils on rock outcrops. Flowering May–Jul. 400–1400 m. BW, Col, Lava. WA. Native.

Allium falcifolium W.J. Hooker & J.A.W. Arnott (p. 140)

Sickle-leaf onion

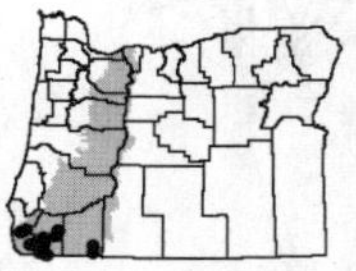

Bulbs 1–5+, rhizomes absent, increase bulbs absent or more or less equaling parent bulbs, never appearing as a basal cluster, ovoid; outer coats enclosing renewal and increase bulbs, membranous, lacking cellular reticulation or cells arranged in only 2–3 rows adjacent to roots, more or less quadrate, without fibers. **Leaves** usually deciduous with scape, withering from tip at anthesis, 2; blades solid, flat, falcate, 8–21 cm × 2–8 mm. **Scapes** usually forming abscission layer and deciduous with leaves after seeds mature, solitary, erect, solid, strongly flattened, winged distally, 5–25 cm × 1–4 mm. **Umbels** persistent, erect, compact to more or less loose, 10–30-flowered, hemispheric; pedicels 8–15 mm; spathe bracts 2. **Flowers** 9–15 mm; tepals erect, lanceolate, more or less equal, reddish purple or dingy white, at least inner margins denticulate, apex long-acuminate; stamens included; ovary crested with 3 low processes; stigma unlobed. **2*n***=14.

Heavy, rocky, clay soils, usually serpentine, rocky openings to savanna. Flowering Apr–Jun. 200–1700 m. Casc, Sisk. CA. Native.

Allium fibrillum Abrams

Fringed onion

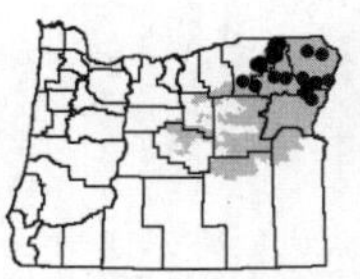

Bulbs 1–5+, without cluster of basal bulbils, more or less globose; outer coats enclosing 1 or more bulbs, brownish, membranous, reticulate; cells very fine, often obscure, narrow, more or less transversely elongate, intricately contorted, without fibers. **Leaves** usually persistent, withering from tip by anthesis, 2; blades solid, flat, more or less falcate, linear, channeled, 7–24 cm × 0.5–3 mm. **Scapes** persistent, solitary, erect, solid, terete or slightly flattened to very narrowly winged, 3–15 cm × 0.5–2 mm. **Umbels** persistent, erect, compact, 10–20(40)-flowered, hemispheric to globose; pedicels 3–10 mm; spathe bracts 2. **Flowers** 5–8 mm; tepals erect, lanceolate, more or less equal, white with prominent greenish or pink midribs, margins entire, apex obtuse to acuminate, more or less involute at tip; stamens included; ovary crestless or obscurely crested with 3 low processes; stigma unlobed. **2*n***=14.

Moist, shallow soils in rocky openings and grasslands in conifer woodlands. Flowering May–Jul. 800–2600 m. BW. WA; northeast to MT. Native.

Morphologically, *A. fibrillum* is similar to *A. madidum*. The latter, however has a basal cluster of bulbils.

Allium geyeri S. Watson

Geyer's onion

Bulbs 2–10+, ovoid or slightly elongate; outer coats enclosing 1 or more bulbs, reticulate, cells rather coarse-meshed, open, fibrous. **Leaves** persistent, usually green at anthesis, usually 3–5; blades solid, more or less straight, flat, channeled, (6)12–30 cm × 1–3(5) mm. **Scapes** persistent, solitary, erect, terete or somewhat 2-angled, 10–50 cm × 1–3 mm. **Umbels** persistent, erect, compact, 10–25-flowered, hemispheric to globose, not producing bulbils, or 0–5-flowered, largely replaced by ovoid, acuminate bulbils; pedicels becoming rigid and stiffly spreading in fruit, 8–13 mm; spathe bracts 2–3. **Flowers** (4)6–8(10) mm; tepals erect or spreading, ovate to lanceolate, more or less equal, pink to white, margins often obscurely toothed, apex obtuse to acuminate; stamens included; ovary when present, inconspicuously crested with 3–6 low processes; stigma unlobed or obscurely lobed.

2 varieties.

Key to *Allium geyeri* varieties:

1. Inflorescence mostly consisting of flowers, bulbils absent or few . . . . ***A. g.*** var. ***geyeri***

1' Inflorescence mostly consisting of bulbils, flowers few . . . . ***A. g.*** var. ***tenerum***

Allium geyeri S. Watson var. *geyeri*

Geyer's onion

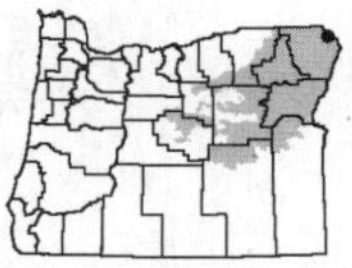

Umbels 10–25-flowered, bulbils absent or few. **Flowers** tepals white to pink; pedicels 8–13 mm. **Seeds** commonly produced. **2*n***=14, 28.

Moist, open slopes, meadows, or stream banks in mountains. Flowering Apr–May. 300–400 m. BW. WA, NV, ID; north to Saskatchewan, east to NE and TX, south to NM. Native.

Allium geyeri var. *tenerum* M.E. Jones

Bulbil onion

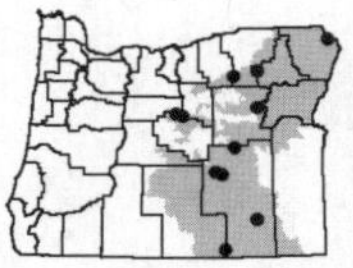

Umbels 0–5-flowered, flowers mostly replaced by bulbils; bulbils sessile or stalk less than 1 mm. **Flowers** tepals white to pink; flowering pedicels, if present, 8–13 mm. **Seeds** rarely produced. **2*n***=28, 35, 42.

Meadows and along streams. Flowering May–Jul. 300–2300 m. BR, BW. ID, NV, WA; north to British Columbia and Alberta, east to WY, southeast to AZ. Native.

Allium lemmonii S. Watson (p. 140)

Lemmon's onion

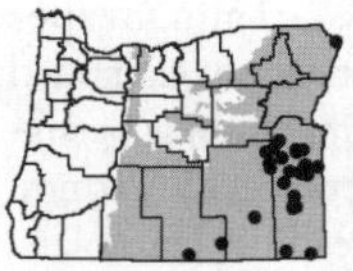

Bulbs 1–5+, ovoid; outer coats enclosing 1 or more bulbs, membranous, more or less prominently cellular-reticulate, cells in more or less regular vertical rows, narrowly rectangular, transversely elongate, without fibers. **Leaves** usually deciduous with scape, green or withering only at tip at anthesis, 2; blades solid, flat, falcate, 8–30 cm × 3–5 mm. **Scapes** usually forming abscission layer and deciduous with leaves after seeds mature, solitary, erect, solid, flattened, narrowly winged, 6–20 cm × 1–4 mm. **Umbels** persistent, erect, compact to more or less loose, 10–40-flowered, hemispheric; pedicels 7–20 mm; spathe bracts 2–3. **Flowers** 6–9 mm; tepals erect, lance-ovate, more or less equal, pink to whitish, margins entire, apex acute to acuminate; stamens more or less equaling tepals; ovary obscurely crested with 6 low processes; stigma scarcely thickened, unlobed. **2*n***=14.

Heavy clay soils often derived from volcanic ash, typically barren, open sites. Flowering Apr–Jun. 600–2000 m. BR, BW, ECas, Owy. CA, ID, NV; east to Utah. Native.

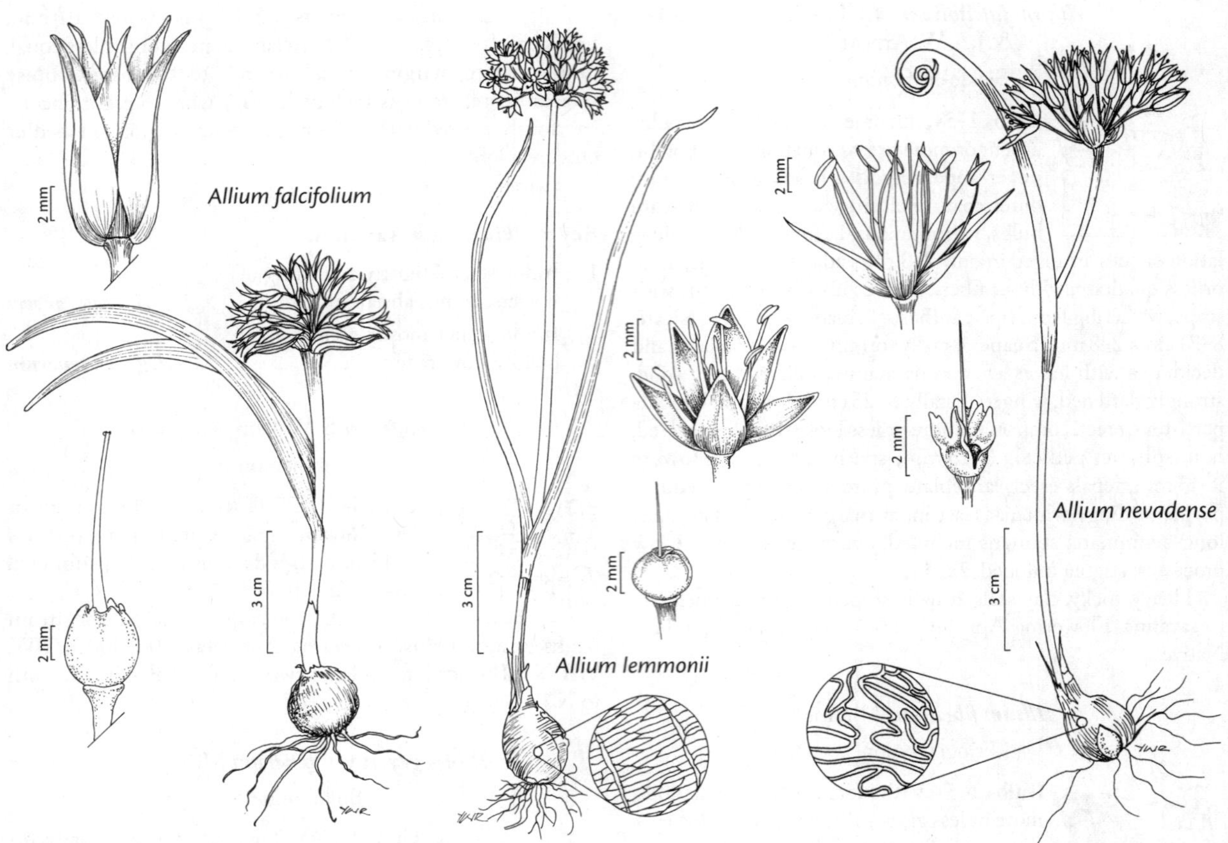

ALLIUM

Allium macrum S. Watson

Rock onion

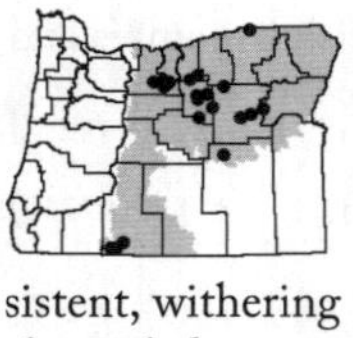

Bulbs 1–3, globose to ovate; outer coats enclosing 1 or more bulbs, membranous, lacking cellular reticulation or cells arranged in only 2–3 rows adjacent to roots, more or less quadrate, without fibers. **Leaves** usually persistent, withering from tip at anthesis, 2; blades solid, flat or channeled, more or less falcate, 12–38 cm × 2–3 mm. **Scapes** persistent, solitary, erect, solid, more or less terete, not expanded proximal to inflorescence, 9–32 cm × 0.5–4 mm. **Umbels** persistent, erect, loose, 10–20-flowered, globose; pedicels 5–20 mm; spathe bracts 2. **Flowers** 5–7 mm; tepals spreading, lanceolate, more or less equal, white or pinkish with prominent red midveins, margins entire, apex acute to acuminate; stamens equaling tepals or exserted; ovary crested with 3 obtuse processes; stigma scarcely thickened, unlobed. **2*n***=14.

Open, dry meadows to vernally moist rocky, clay swales. Flowering Apr–May. 400–1800 m. BW, Col, ECas, Lava. WA. Native.

Allium madidum S. Watson

Swamp onion

Bulbs 1–3, larger bulbs each usually with cluster of 10–30 easily detached bulbils to one side of base, globose to ovoid; outer coats enclosing 1 or more bulbs, membranous, without fibers. **Leaves** persistent, green at anthesis, 2; blades solid, flat or channeled, 10–25 cm × 1–4 mm. **Scapes** persistent, solitary, erect, solid, more or less terete, sometimes ridged, 10–20 cm × 1–2 mm. **Umbels** persistent, erect, compact, 10–20(40)-flowered, hemispheric; pedicels 5–12 mm; spathe bracts 2. **Flowers** 6–10 mm; tepals erect or more or less spreading, white with prominent green or pink midveins, lanceolate, more or less equal, margins entire, apex obtuse to acuminate, becoming involute at tip; stamens included; ovary crestless or obscurely crested with 3 minute processes; stigma unlobed or slightly lobed. **2*n***=28, 42.

Open mixed conifer forests, gravelly seeps, seasonally wet meadows and scablands. Flowering May–Jul. 1000–2200 m. BW. ID. Native.

Allium nevadense S. Watson (p. 140)

Nevada onion

Allium nevadense S. Watson var. *nevadense*

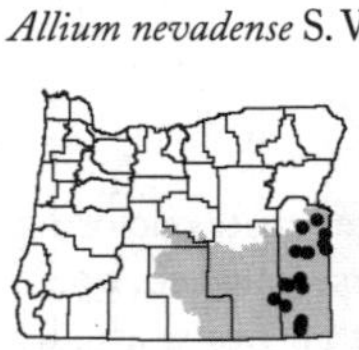

Bulbs 1–3, with 0–2 stalked, basal bulbils, ovoid; outer coats enclosing 1 or more bulbs, membranous, usually prominently cellular-reticulate, cells more or less transversely elongate, intricately contorted, without fibers. **Leaves** persistent, 1; blades solid, terete, 10–20 cm × 1–3 mm. **Scapes** persistent, solitary, erect, more or less solid, terete, 5–15 cm × 1–2.5 mm. **Umbels** persistent, erect, compact, 5–25-flowered, hemispheric; pedicels 8–16 mm; spathe bracts 2–3. **Flowers** 7–12 mm; tepals spreading, lanceolate to ovate, more or less equal, white or pinkish white with dark pink midveins, margins entire, apex acute to acuminate; stamens included; ovary crested with 6 prominent processes; stigma scarcely thickened, unlobed. **2*n***=14, 28.

Rocky soils in shrub-steppe, seasonally moist clay soils in scablands. Flowering May–Jun. 800–1600 m. BR, Owy. CA, ID, NV; east to CO, southeast to AZ. Native.

Allium nevii S. Watson (p. 143)

Nevius' onion

Allium douglasii Hook. var. *nevii* (S. Watson) Ownbey & Mingrone

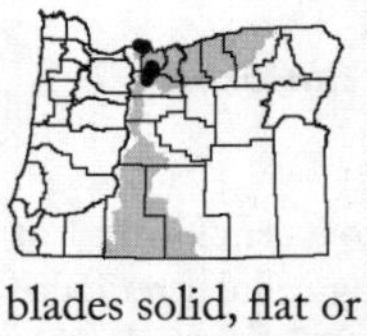

Bulbs 1–5+, ovoid; outer coats enclosing 1 or more bulbs, membranous, more or less reticulate, cells quadrate to polygonal, vertically oblong, without fibers. **Leaves** usually persistent, withering from tip at anthesis, 2; blades solid, flat or more or less channeled, falcate, 12–25 cm × 2–3 mm. **Scapes** persistent, 1–3, erect, solid, terete or somewhat flattened, 2-edged, not expanded proximal to inflorescence, 15–30 cm × 1–3 mm. **Umbels** persistent, erect, compact, 10–30-flowered, hemispheric; pedicels 8–12 mm; spathe bracts 2. **Flowers** 6–8 mm; tepals spreading, lanceolate, more or less equal, rose-colored, margins entire, apex acuminate; stamens more or less equaling tepals, or exserted; ovary distinctly crested with 6 low processes; stigma unlobed. **2*n***=14.

Wet meadows and along streams, in rocky clay soils, seasonally moist scablands. Flowering Apr–Jun. 0–1200 m. Col, ECas. WA. Native.

Allium nigrum Linnaeus

Black garlic, Homer's garlic

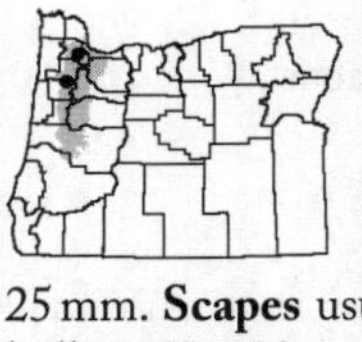

Bulbs 3–4, ovoid; outer coats enclosing 1 or more bulbs, thin, papery, reticulate, cells narrow, vertically elongate, without fibers. **Leaves** usually persistent, 3–6; blades solid, flat, broadly channeled, 10–60 cm × 10–25 mm. **Scapes** usually persistent, solitary, erect, solid, terete, hollow, 60–100 cm × 5–10 mm. **Umbels** persistent, erect, compact, 75–150-flowered, hemispheric-globose; pedicels 25–40 mm; spathe bracts 2–4. **Flowers** 6–9 mm; tepals more or less spreading, narrowly oblong, more or less equal, white with green midvein, margins entire, apex obtuse; stamens included; ovary crestless; stigma scarcely thickened, obscurely 3-lobed.

Disturbed roadsides. Flowering Jun. 50–100 m. WV. Mediterranean. Exotic.

Oregon has the only known naturalized population of *A. nigrum* in North America. It seems highly probable, however, that this species has escaped elsewhere.

Allium obtusum Lemmon var. ***obtusum***

Red Sierra onion

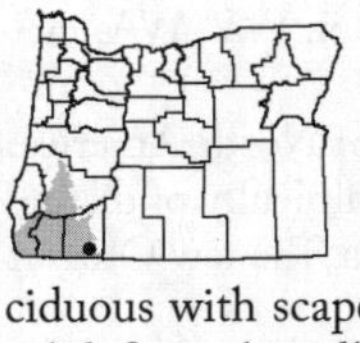

Bulbs 1–5, ovoid; outer coats enclosing 1 or more bulbs, brown, membranous, cellular-reticulate, cells quadrate, polygonal, or more or less rectangular, more or less transversely elongate, without fibers. **Leaves** usually deciduous with scape, withering from tip at anthesis, 1; blades solid, flat or broadly channeled, more or less falcate, 9–22 cm × 2–14 mm. **Scapes** usually forming abscission layer and deciduous with leaves after seeds mature, solitary, erect, solid, terete, 2.5–17 cm × 0.5–2 mm. **Umbels** persistent, erect, compact, 10–65-flowered, hemispheric; pedicels 3–14(20) mm; spathe bracts 2–3(5). **Flowers** 4–11.5 mm; tepals erect, lanceolate, more or less equal, pink with dark purplish midveins, margins entire, apex acute; stamens included; ovary obscurely to prominently crested with 3 rounded processes; stigma scarcely thickened, obscurely 3-lobed. **2*n***=14.

Moist, sandy soils on open slopes and flats. Flowering May–Jun. 1500–3000 m. Sisk. CA, NV. Native.

Currently rare in Oregon and Nevada, this taxon is more common in California.

Allium parvum Kellogg

Dwarf onion

Allium parvum Kellogg var. *parvum*

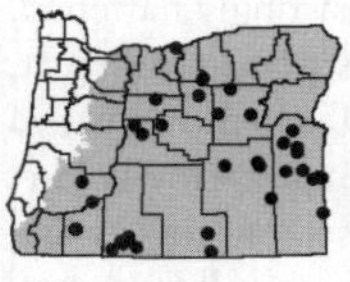

Bulbs 1–5+, renewal bulbs formed within coats of parent bulb, ovoid to more or less globose; outer coats enclosing 1 or more bulbs, membranous, lacking cellular reticulation or cells arranged in only 2–3 rows adjacent to roots, more or less quadrate, without fibers. **Leaves** usually deciduous with scape, withering from tip at anthesis, 2; blades solid, flat, more or less falcate, 8–15 cm × 1–5(6) mm. **Scapes** usually forming abscission layer and deciduous with leaves after seeds mature, solitary, erect, solid, flattened, frequently more or less winged distally, or, in smaller specimens, more or less terete, 3–12 cm × 0.5–2 mm. **Umbels** persistent, erect, compact, 5–30-flowered, hemispheric to conic; pedicels 3–12 mm; spathe bracts 2. **Flowers** 6–9 mm; tepals erect, slightly spreading toward tips, oblong to elliptic, unequal, white or tinged with pink, with broad, dark, reddish purple or brown midvein, outer longer, wider than inner, margins entire, apex obtuse to more or less acute; stamens included; ovary obscurely crested with 3 low processes; stigma scarcely thickened, unlobed or obscurely 3-lobed. **2*n***=14.

Rocky, clay slopes and talus, seasonally moist scablands and rocky slopes. Flowering Apr–Jun. 300–2200 m. BR, BW, Casc, Col, ECas, Lava, Owy, Sisk. CA, ID, NV; southeast to Utah. Native.

Allium peninsulare Lemmon ex Greene var. ***peninsulare***

Peninsular onion

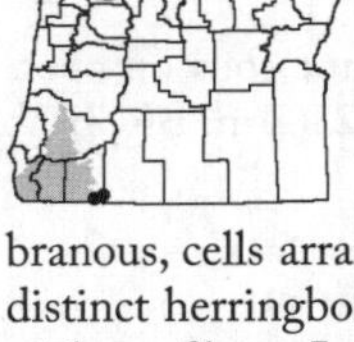

Bulbs 1–5+, increase bulbs absent or more or less equaling parent bulbs, never appearing as a basal cluster, ovoid to more or less globose; outer coats enclosing 1 or more bulbs, prominently cellular-reticulate, membranous, cells arranged in more or less vertical rows, forming distinct herringbone pattern, transversely elongate, V-shaped, without fibers. **Leaves** withering from tip by anthesis, 2–3; blades solid, subterete or more or less channeled, straight or nearly so, 8–30 cm × 1–3 mm. **Scapes** persistent, solitary, erect, solid, terete, 12–45 cm × 1–3 mm. **Umbels** persistent, erect, loose, 5–35-flowered, hemispheric to conic; pedicels 10–40 mm; spathe bracts 2. **Flowers** 10–15 mm; tepals erect, lanceolate to elliptic, unequal, reddish purple, outer tepals longer and wider than inner, margins entire or inner minutely denticulate, apex acute or short-acuminate, inner tepals spreading at tip, outer more or less recurved; stamens included; ovary obscurely crested with 3 minute processes; stigma distinctly 3-lobed. **2*n***=14.

Clay soils including serpentine. Flowering Apr–Jul. 1600–1800 m. Sisk. CA; south to Mexico. Native.

Allium platycaule S. Watson (p. 143)

Flatstem onion, broadstem onion

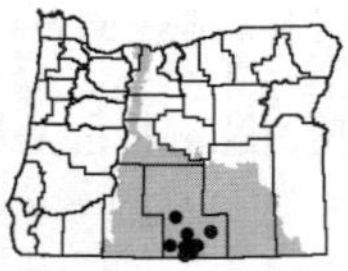

Bulbs 1–5+, ovoid, renewal bulbs formed within coats of parent bulb, more or less equaling parent bulbs; outer coats enclosing 1 or more bulbs, membranous, lacking cellular reticulation or cells arranged in only 2–3 rows adjacent to roots, more or less quadrate, without fibers. **Leaves** usually deciduous with scape, withering from tip at anthesis, 2; blades solid, flat, falcate, 10–30 cm × (6)8–18 mm. **Scapes** usually forming abscission layer and deciduous with leaves after seeds mature, solitary, erect, solid, strongly flattened, winged distally, 7–25 cm × 2–7 mm. **Umbels** persistent, erect, loose, 30–90-flowered, globose; pedicels 10–25 mm; spathe bracts 2–5. **Flowers** 8–15 mm; tepals spreading, narrowly lanceolate, more or less equal, bright pink to rose, margins entire, apex long-acuminate; stamens exserted; anthers yellow; ovary crestless; stigma scarcely thickened, unlobed. **2*n***=14.

Rocky, sandy slopes. Flowering May–Jul. 1300–2500 m. BR, ECas. CA, NV. Native.

Allium punctum L.F. Henderson

Punctate onion

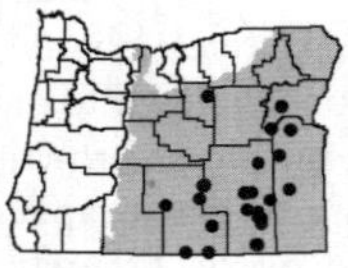

Bulbs 1–3, ovoid to subglobose; outer coats enclosing 1 or more bulbs, membranous, more or less prominently cellular-reticulate, cells in more or less regular vertical rows, narrowly rectangular, transversely elongate, without fibers. **Leaves** usually deciduous with scape, green or withering only at tip at anthesis, 2; blades solid, flat, falcate, 9–18 cm × 2–5 mm. **Scapes** usually forming abscission layer and deciduous with leaves after seeds mature, solitary, erect, solid, more or less flattened, sometimes slightly winged, 3–10 cm × 1–2 mm. **Umbels** persistent, erect, more or less compact, 6–20-flowered, hemispheric; pedicels 5–11 mm; spathe bracts 2. **Flowers** 6–13 mm; tepals erect, oblong-lanceolate, more or less equal, white to pinkish with narrow to very broad, purple midvein, margins entire, apex obtuse; stamens included; ovary crested with 3 low processes; stigma scarcely thickened, unlobed. **2*n***=14.

Sandy, rocky soil, clay, sometimes alkaline soils on open slopes and flats. Flowering Apr–Jun. 500–2300 m. BR, BW, ECas, Lava, Owy. CA, NV. Native.

Allium robinsonii L.F. Henderson

Robinson's onion

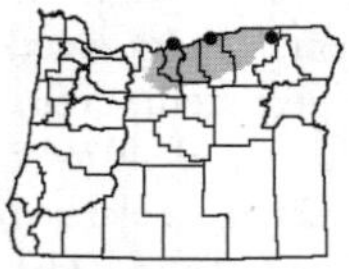

Bulbs 1–3+, renewal bulbs formed within coats of parent bulb, increase bulbs absent or 1–2, more or less equaling parent bulbs, never appearing as basal cluster, ovoid; outer coats membranous, lacking cellular reticulation or cells arranged in only 2–3 rows adjacent to roots, more or less quadrate, without fibers. **Leaves** usually deciduous with scape, 2; blades solid, flat, strongly falcate, 5–18 cm × 1–3(4) mm. **Scapes** usually forming abscission layer and deciduous with leaves after seeds mature, solitary, erect, solid, flattened, slightly 2-edged, 3–7 cm × 1–3 mm. **Umbels** deciduous with scape, erect, compact, 5–12-flowered, hemispheric; pedicels 3–12 mm; spathe bracts 2. **Flowers** 7–9 mm; tepals erect, oblong, more or less equal, white to pale pink with red midribs, margins entire, apex obtuse; stamens included; ovary more or less distinctly crested with 3 or 6 low processes; stigma unlobed. **2*n***=14.

Sand and gravel deposits and sandy slopes. Flowering Apr–May. 50–800 m. Col. WA. Native.

Allium sanbornii A.W. Wood var. ***sanbornii***

Sanborn's onion

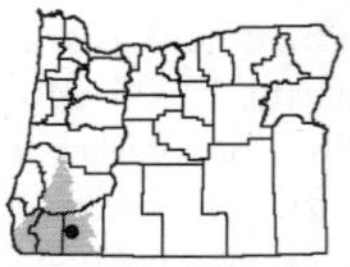

Bulbs 1–3, ovoid; outer coats enclosing 1 or more bulbs, papery, lacking cellular reticulation or cells arranged in 2–3 rows adjacent to roots, more or less quadrate, without fibers. **Leaves** persistent, withering from tip by anthesis, 1; blades solid, terete, 30–45 cm × 2–4 mm. **Scapes** persistent, solitary, erect, more or less solid, terete, 18–54 cm × 2–3 mm. **Umbels** persistent, erect, compact, 18–190-flowered, hemispheric to globose; pedicels 8–22 mm; spathe bracts 4. **Flowers** 5–9 mm; tepals erect, unequal, white to pink with darker midveins, margins entire or more or less irregular, apex acute; outer tepals lanceolate to ovate, reflexed at tip, 3–4 mm long; inner 25–33% longer than outer, 4.5–6.5 mm, ovate to broadly ovate; stamens exserted; ovary crested with 6 prominent processes; stigma obscurely 3-lobed. **2*n***=14.

Heavy serpentine clay. Flowering Jun–Aug. 400–600 m. Sisk. CA. Native.

Allium schoenoprasum Linnaeus (p. 143)

Chives

Allium sibiricum L.

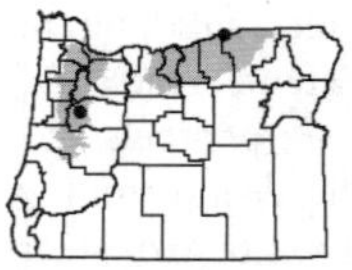

Bulbs 1 or more, clustered, short-rhizomatous at base, cylindric, elongate; outer coats enclosing bulbs, persisting as fibrous reticulum but often appearing membranous because outer coats are often not collected, cells minutely striate. **Leaves** persistent, green at anthesis, usually 2; blades hollow, terete, 20–60 cm × 2–7 mm. **Scapes** persistent, 2–12+, clustered, erect, terete, hollow, 20–50 cm × 3–5 mm. **Umbels** persistent, erect, compact, 30–50-flowered, more or less subglobose; pedicels 2–6 mm; spathe bracts 2. **Flowers** 8–12 mm; tepals erect, elliptic to lanceolate, more or less equal, pale purple to deep lilac, drying pink, or white, permanently enclosing capsule, margins entire, apex acute to acuminate, tips more or less recurved, midribs not thickened; stamens included; ovary crestless; stigma unlobed or obscurely lobed. **2*n***=16.

Wet meadows, rocky or gravelly stream banks and lake shores. Flowering May–Aug. 50–200 m. Col, WV. WA; circumboreal. Exotic?

Allium schoenoprasum is native to parts of North America. Its native range is unknown because of the difficulty of distinguishing native from introduced populations. The few Oregon populations are probably introduced.

Allium siskiyouense Traub

Siskiyou onion

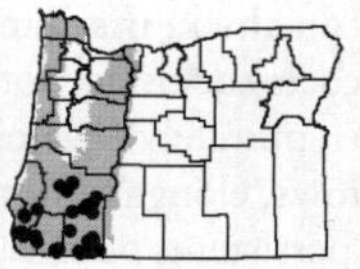

Bulbs 1–5, increase bulbs absent or more or less equaling parent bulbs, never appearing as basal cluster, ovoid; outer coats enclosing renewal bulbs, membranous, lacking cellular reticulation or cells arranged in only 2–3 rows adjacent to roots, more or less quadrate, without fibers. **Leaves** usually deciduous with scape, withering from tip at anthesis, 2; blades solid, flat, falcate, 8–18 cm × 3–6 mm. **Scapes** usually forming abscission layer and deciduous with leaves after seeds mature, solitary, erect, solid, flattened, winged distally, wing margins commonly minutely serrulate proximal to umbel, 3–8 cm × 1–4 mm. **Umbels** persistent, erect, compact, 10–35-flowered, hemispheric; pedicels 5–16 mm; spathe bracts 2. **Flowers** 8–11 mm; tepals erect, lanceolate to elliptic, more or less equal, pink, usually with darker pink midveins, outer margins entire, inner margins usually minutely denticulate distally, apex obtuse, margins becoming more or less involute, appearing acuminate in age; stamens included; ovary obscurely crested with 3 minute processes; stigma scarcely thickened, unlobed. ***2n***=14.

Rocky, clay, or sandy soils, usually on serpentine. Open ridgetops, slopes, and forest openings. Flowering May–Jul. 300–2100 m. Casc, CR, Sisk. CA. Native.

Allium siskiyouense is known only from SW Oregon and NW California.

Allium tolmiei Baker var. ***tolmiei***

Tolmie's onion

Allium cusickii S. Watson, *Allium pleianthum* S. Watson, *Allium tolmiei* Baker var. *platyphyllum* (Tidestr.) Ownbey

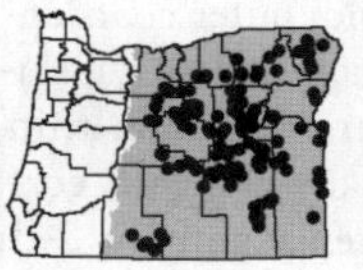

Bulbs 1–10+, renewal bulbs formed within coats of parent bulb, or borne terminally on short rhizomes outside coats of parent bulb, parent bulb disappearing by anthesis except for still-functional roots and shriveled bulb coat, ovoid; outer coats enclosing 1 or more bulbs, membranous, lacking cellular reticulation or cells arranged in only 2–3 rows adjacent to roots, more or less quadrate, without fibers. **Leaves** forming abscission layer at soil surface and deciduous, withering at tip at anthesis, 2; blades solid, flat, more or less falcate, 10–35 cm × 1–10 mm. **Scapes** deciduous, solitary, erect, solid, strongly flattened and winged, 5–28 cm × 1–5 mm. **Umbels** persistent, erect, compact to more or less loose, 10–40-flowered, globose to hemispheric; pedicels 5–25 mm; spathe bracts 2. **Flowers** 6–12 mm; tepals erect to more or less spreading, lanceolate, more or less equal, almost white to pink, with prominent, darker midribs, margins entire, apex acute to acuminate; stamens included; ovary crestless or crested with 3 or 6 rounded to triangular processes; stigma unlobed or obscurely lobed.

Rocky, gravelly, or clayey soils in dry to moist meadows, seasonally wet scablands, and shrub-steppe. Flowering Apr–Jul. 400–2800 m. BR, BW, Col, ECas, Lava, Owy. CA, ID, WA. Native.

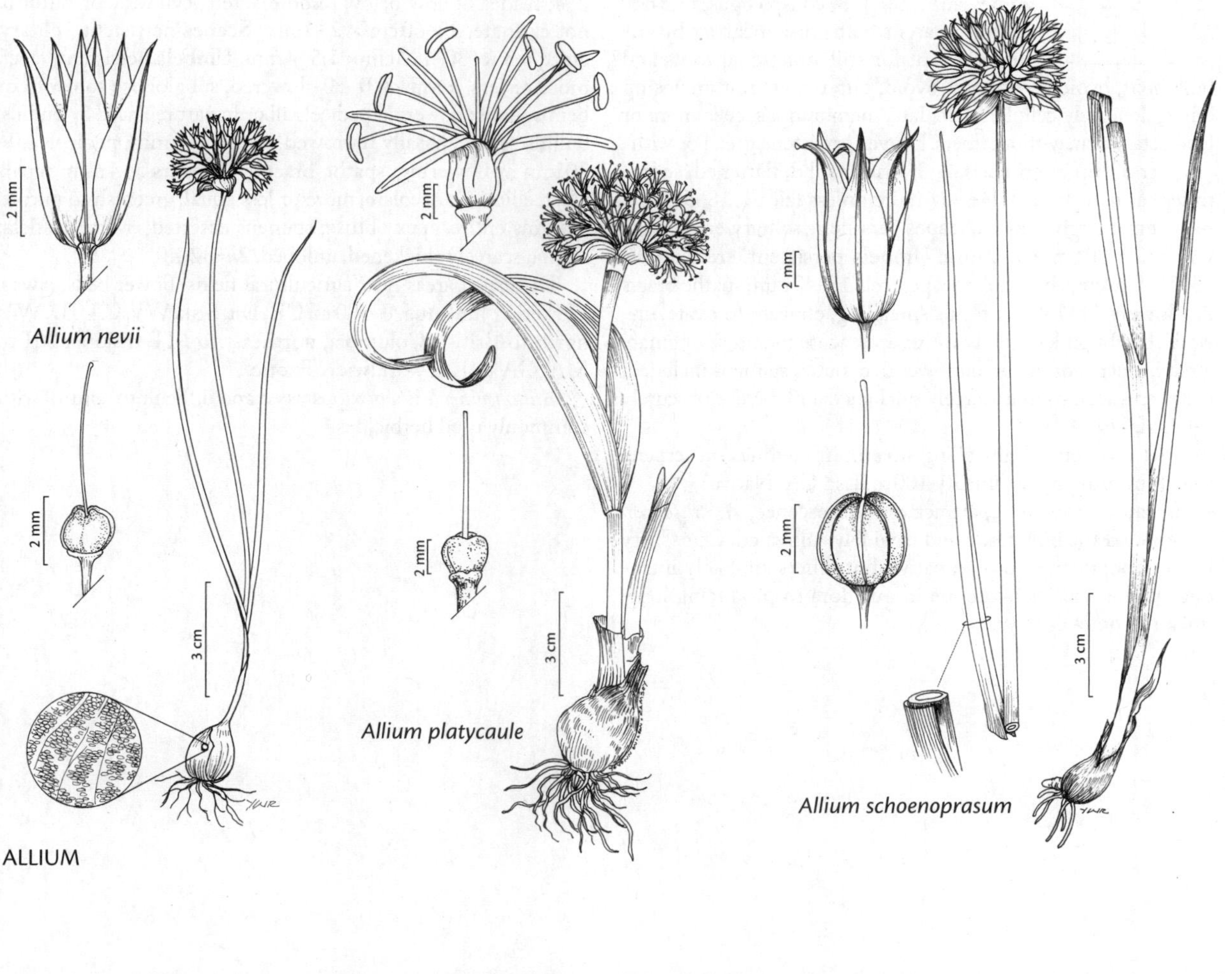

ALLIUM

Allium triquetrum Linnaeus

Three-corner leek

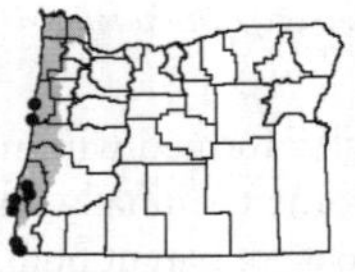

Bulbs 5–20+, increase bulbs absent or more or less equaling parent bulbs, never appearing as basal cluster, ovoid; outer coats enclosing renewal bulbs, more or less translucent, obscurely cellular-reticulate, thin, membranous, meshes delicate, cells vertically elongate, contorted, without fibers. **Leaves** persistent, green at anthesis, 2–3; blades solid, flat, not falcate, more or less keeled, 15–50 cm × 3–15 mm, margins entire. **Scapes** persistent, clustered, erect, solid, sharply 3-angled, 10–40 cm × 1–10 mm. **Umbels** persistent, lax, loose, 3–15-flowered, more or less 1-sided; pedicels 15–25 mm; spathe bracts 1–2. **Flowers** becoming pendent, 10–18 mm; tepals erect to spreading, lanceolate, more or less equal, white with prominent green midrib, margins entire, apex acute; stamens included; ovary crestless; stigma scarcely thickened, unlobed. ***2n***=18.

Road ditches, lawns, and other disturbed moist sites. Flowering Apr–Jul. 400–2800 m. BR, BW, Col, ECas, Lava, Owy. CA; Europe. Exotic.

Allium triquetrum is cultivated for its attractive flowers. It readily escapes in south coastal Oregon and California and has potential to become invasive.

Allium unifolium Kellogg

One-leaf onion

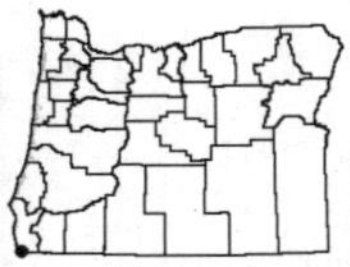

Bulbs solitary, replaced annually by new bulbs borne terminally on secondary rhizome; rhizomes 1–3, conspicuous, to 5 cm, smooth; parent bulbs disappearing by anthesis except for still-functional roots and bulb coat, ovoid to oblique-ovoid; outer coats not enclosing bulbs, delicately cellular-reticulate, membranous, cells more or less rectangular, without fibers. **Leaves** persistent, green or withering from tip at anthesis, 2–3; blades solid, flattened, sometimes carinate, keeled dorsally, more or less falcate, 18–50 cm × 4–10 mm, margins entire. **Scapes** persistent, solitary, erect, solid, terete, 20–80 cm × 2–7 mm. **Umbels** persistent, erect, loose, 15–35-flowered, hemispheric; pedicels 15–40 mm; spathe bracts 2. **Flowers** 11–15 mm; tepals spreading, obovate to ovate, unequal, bright pink or rarely white, apex acute to obtuse or emarginate, inner shorter and narrower than outer; stamens included; ovary crestless; stigma scarcely thickened, unlobed or obscurely 3-lobed. ***2n***=14.

Wet clay soil, usually along streams, sometimes on serpentine. Flowering May–Jun. 50–100 m. Est. CA. Native?

Because of its large, attractive inflorescence, *A. unifolium* has entered the bulb trade and is widely cultivated. Our Curry County population may be native, but others are likely introduced. This species is unique in our flora in producing long, thick rhizomes each year.

Allium validum S. Watson

Pacific swamp onion

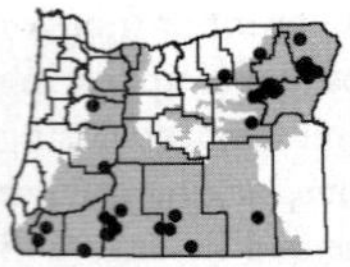

Bulbs 2–20+, clustered on thick, iris-like rhizome, elongate; outer coats enclosing 1 or more bulbs, membranous, minutely striate, cells in regular vertical rows, elongate, not fibrous-reticulate, fibers persistent, parallel, few, coarse. **Leaves** persistent, green at anthesis, 3–6; blades solid, flat, 20–70(80) cm × 4–15 mm. **Scapes** persistent, solitary, erect, solid, flattened and narrowly winged distally, 30–70 cm × 2–7 mm. **Umbels** persistent, erect, compact, 15–30-flowered, hemispheric; pedicels 10–15 mm, elongating and becoming stout in fruit; spathe bracts 2. **Flowers** 8–10 mm; tepals erect to more or less spreading, narrowly lanceolate, more or less equal, pink, margins entire, apex acuminate, midribs scarcely thickened; stamens exserted; ovary crestless; stigma unlobed. ***2n***=28, 56.

Wet meadows, fens, springs, and stream banks in mountains. Flowering Jun–Sep. 500–2500 m. BR, BW, Casc, ECas, Sisk. CA, ID, NV, WA. Native.

Allium vineale Linnaeus

Wild garlic, cow garlic, field garlic

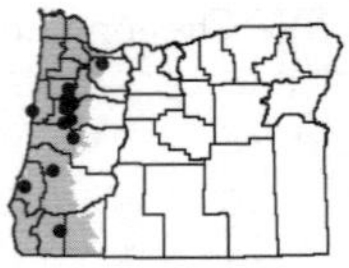

Bulbs 5–20, clustered, stipitate, hard-shelled, asymmetric, ovoid; outer coats enclosing bulbs, membranous, vertically striate, splitting into parallel strips and fibers, cellular reticulations vertical, arranged in more or less wavy rows. **Leaves** persistent, green at anthesis, 2–4; blades hollow below middle, terete, cylindric or filiform, not carinate, 20–60 cm × 2–4 mm. **Scapes** persistent, solitary, erect, terete, 30–120 cm × 1.5–4 mm. **Umbels** persistent, erect, more or less compact, 0–50-flowered, subglobose to ovoid or hemispheric; flowering pedicels all or in part replaced by bulbils; bulbils sessile, basally narrowed, 4–6 × 2–3 mm; pedicels 10–20 mm at flowering; spathe bract 1. **Flowers** 3–4 mm; tepals erect, elliptic-lanceolate, more or less equal, greenish to purple, margins entire, apex obtuse; stamens exserted; ovary crestless; stigma scarcely thickened, unlobed. ***2n***=32, 40.

Disturbed areas near agricultural fields, flower beds, lawns. Flowering Jun–Aug. 0–300 m. CR, Est, Sisk, WV. CA, ID, WA; north to British Columbia, northeast to MT, disjunct MI to MA, GA to LA; worldwide. Exotic.

Allium vineale is a noxious weed and difficult to control with commonly used herbicides.

NARCISSUS Daffodil

Stephen C. Meyers

Herbs perennial; from scaly, ovoid bulbs. **Leaves** 2–4, linear, flat, 20–45 cm × 8–12 mm. **Inflorescences** solitary, or umbellate in 2–15-flowered clusters. **Flowers** perianth parts tubular and fused at base, spreading in middle, conspicuously crowned above, 2–7 cm wide; tepals 6, yellow and/or white, margins usually wavy; stamens 6; ovary inferior; style 1; stigma 3-lobed. **Fruits** capsules. **Seeds** many, subglobose, black.

Cosmopolitan. Approximately 26 species; 2 species treated in *Flora.*

All populations of *Narcissus* outside the native range of the genus (Europe and northern Africa) are escaped. Several centuries of selective breeding, hybridization, and cultivation have produced a vast and complex array of morphology within the genus, and subsequently among escaped populations. As a result of the highly reticulated genetic history of most exotic populations, it may be pragmatically best to consider all escaped plants as one species. Here, however, we describe two of the most common and distinct forms, cognizant that hybrids between these "species" (*N.* × *incomparabilis*) and other variations within the genus are known to be occasionally naturalized (*N. tazetta*).

Key to *Narcissus* species:

1. Tepals white; crown cup-shaped, yellow with red margin, much shorter in length than subtending tepal lobes . . . ***N. poeticus***

1' Tepals yellow; crown tube-shaped, yellow without a colored margin, equal to or slightly longer in length to subtending tepal lobes . . . . ***N. pseudonarcissus***

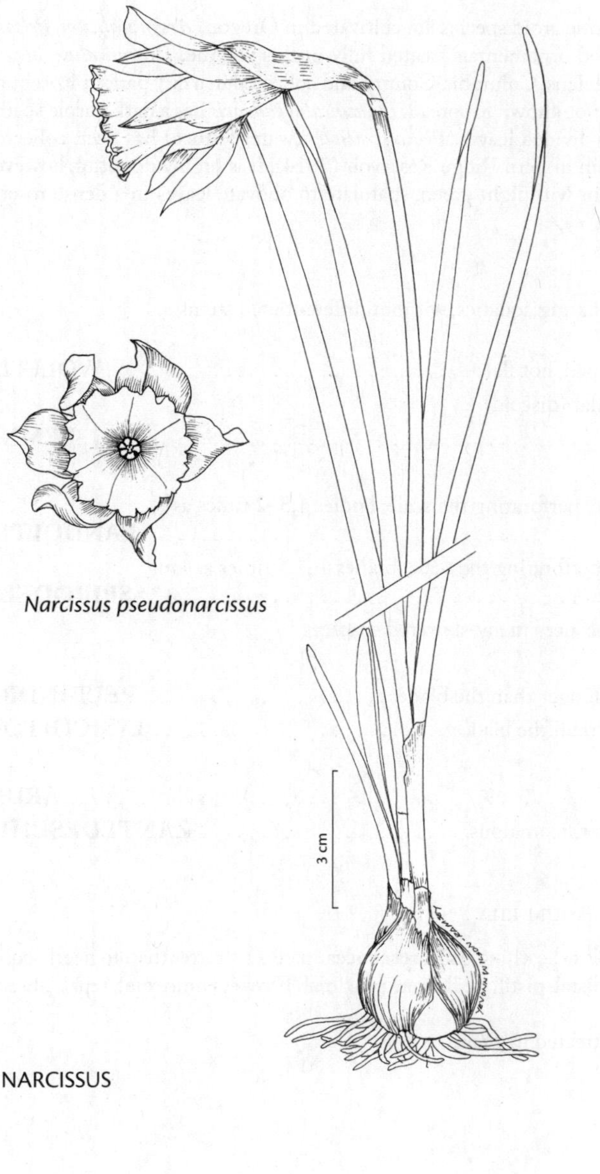

Narcissus pseudonarcissus

NARCISSUS

Narcissus poeticus Linnaeus

Poet's daffodil

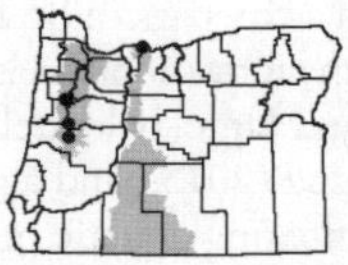

Plants 25–50 cm tall. **Flowers** perianth tube 2–3.5 cm; perianth 4–7 cm wide; crown cup-shaped, 3–5 mm long, yellow with a red margin; tepals white, lobes 1.5–3.5 cm long, apex mucronate; stamens arranged in 2 rows, 3 stamens shorter and included in perianth tube, 3 stamens longer and exserted.

Meadows, roadsides. Flowering Apr–May. 50–400 m. ECas, WV. WA; scattered throughout North America; worldwide. Exotic.

Hybrids between *N. poeticus* and *N. pseudonarcissus* (*N.* × *incomparabilis*) are occasionally encountered. These plants have white to yellow tepals and yellow crowns, with or without a red margin, that are approximately half as long as the tepal lobes.

Narcissus pseudonarcissus Linnaeus (p. 145)

Daffodil

Plants 20–50 cm tall. **Flowers** perianth tube 1–2.5 cm; perianth 4–7 cm wide; crown tube-shaped, 3–4 cm long, yellow without a colored margin; tepals yellow, lobes 2.5–3.5 cm long, apex acute; stamens arranged in 1 row, equal in length and exserted. **2*n***=14.

Meadows, roadsides. Flowering Mar–Apr. 0–400 m. Est, WV. CA, WA; scattered throughout North America; worldwide. Exotic.

Although frequently encountered in many locations each spring, this taxon is rarely collected. As a result, the true range of this taxon is poorly documented. See also discussion for *N. poeticus.*

ARACEAE Arum family

Katie E. Mitchell

Herbs aquatic or terrestrial; annual or perennial; generally monoecious. **Leaves** basal, simple to pinnately or palmately compound, with a membranous basal sheath, or lacking true leaves and plant body highly reduced. **Inflorescences** indeterminate spikes with many unisexual or bisexual flowers clustered on a fleshy axis (spadix) that is subtended by a large, leaf-like or petaloid bract (spathe), or reduced to a single bisexual flower (interpreted by some to be 2 unisexual flowers) in a pouch or cavity. **Flowers** unisexual or bisexual; tepals usually absent, sometimes 4–6, fleshy and inconspicuous; stamens 1–6 and free or 2–12 and connate to form a synandrium; pistil 1; ovary 1, superior to half inferior and embedded in the spadix, 1(4)-locular; style 1. **Fruits** berries or follicles.

Cosmopolitan, mostly tropical. Approximately 115 genera; 8 genera treated in *Flora*.

Two different types of inflorescences are characteristic of this family. The typical aroid inflorescence is a spadix, a fleshy spike with many flowers, subtended by a spathe, a large, leaf- or petal-like bract. Plants in subfamily Lemnoideae (formerly family Lemnaceae) have inflorescences consisting of one or two stamens and a single pistil, which has been interpreted as either a single, highly reduced flower or a small inflorescence with two or three unisexual flowers.

The minute, green, herbaceous plant bodies of the Lemnoideae lack true leaves and stems and have been referred to by various terms including thalli, thalloid stem/leaf units, and fronds. The phrase 'plant body' is used in this treatment. Reduction of both reproductive and vegetative parts leaves few characteristics to distinguish taxa within Lemnoideae. Reproduction by budding is common. Plants have one or two budding pouches or a basal cavity from which daughter plants arise. Turions (sometimes called winter buds) are smaller than the mother plant, usually darker in color, and filled with starch grains. They are produced in budding pouches in the fall, eventually detach, and then sink to the bottom of the water body. In the spring, they float to the surface and resume growth.

Grown for showy inflorescences and glossy leaves, several exotic aroid species are cultivated in Oregon. *Arum italicum*, *Peltandra virginica*, and *Zantedeschia aethiopica* are occasionally escaped ornamentals treated fully in this volume. *Dracunculus vulgaris* (dragon arum or voodoo lily) was observed in a city park in St. Helens, Columbia County, and collected in a city park in Roseburg, Douglas County. It is probably persisting from cultivation and is not known to spread. *Dracunculus vulgaris* has a dark purple spadix and spathe with a fetid, rotting flesh odor, and large, palmately divided leaves. *Pistia stratiotes* (water lettuce) has been collected twice, once in the Willamette River in Portland (2013), and again in Fern Ridge Reservoir (2014). It is highly doubtful, however, that these plants have persisted. It is a free-floating aquatic plant with light green, spatulate to obovate leaves in a dense rosette 15–30(40) cm in diameter.

Key to Araceae genera:

1. Plants minute (each plant body less than 2 cm long), free-floating aquatics, without differentiated stems and leaves; inflorescences single or few-flowered or absent.
    2. Plant bodies without roots; globular, ovoid, or boat-shaped, not flat . . . . **WOLFFIA**
    2' Plant bodies with usually one or more roots; generally flat (discoid).
        3. Plant bodies with 1 or no roots . . . . **LEMNA**
        3' Plant bodies with 2 or more roots.
            4. Plant bodies with usually 2–7 roots, 1 or 2 roots perforating the scale; bodies 1.5–2 times as long as wide . . . . **LANDOLTIA**
            4' Plant bodies with usually 7–21 roots, all roots perforating the scale; bodies 1–1.5 times as long as wide . . . . **SPIRODELA**

1' Plants larger, if in wet places then not free-floating; inflorescences many-flowered spadices.
    5. Wetland plants, growing in shallow water or mud.
        6. Leaf bases hastate to sagittate; petioles equal to or longer than the blades . . . . **PELTANDRA**
        6' Leaf bases cuneate to subtruncate; petioles shorter than the blades . . . . **LYSICHITON**
    5' Upland plants.
        7. Peduncle shorter than the petiole; plants tuberous . . . . **ARUM**
        7' Peduncle equal to or longer than the petiole; plants rhizomatous . . . . **ZANTEDESCHIA**

ARUM Arum lily

Herbs tuberous. **Leaves** basal, simple, sagittate, net-veined, petioles sheathing. **Inflorescences** spadices shorter than to nearly equal to the subtending spathes, upper 50% sterile, staminate flowers distal, pistillate flowers proximal. **Flowers** unisexual; tepals absent; stamens 3–4; pistil 1. **Fruits** obovoid, orange to red berries.

Northern Hemisphere. Approximately 25 species; 1 species treated in *Flora*.

Arum italicum Miller (p. 147)

Italian arum, Italian lords-and-ladies

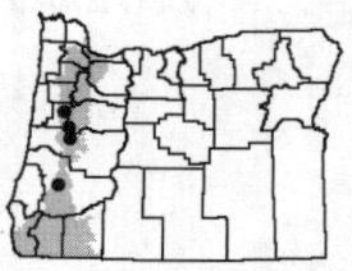

Herbs 10–50 cm. **Leaves** deciduous, 15–45 cm long, 5–25 cm wide, often withering before fruits develop; veins white. **Spathes** pale green, tubular. **Spadices** white to yellow.

Moist, shady, forested areas near habitations. Flowering Apr–Jun. 0–200 m. Sisk, WV. CA, WA; LA, NC; Europe. Exotic.

Seeds of this species are often spread by birds and mammals, and tubers spread in contaminated soil and compost.

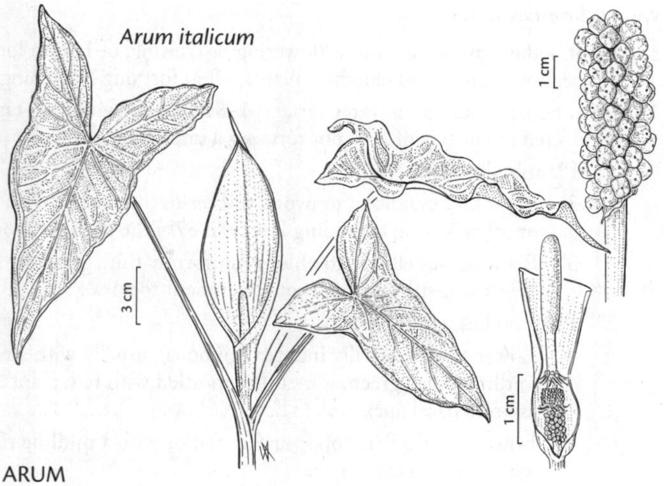

ARUM

LANDOLTIA Duckweed

Plants free-floating aquatics, on the surface of the water; small, green, flat plant bodies without true stems and leaves; ovate to lanceolate. **Roots** (1)2–7(12) per plant body. **Buds** formed in 2 lateral pouches; turions absent. **Inflorescences** a single flower in a membranous spathe within the budding pouch, rarely produced. **Flowers** bisexual; tepals 0; stamens 2. **Fruits** 1–2-seeded follicles.

Landoltia punctata is the only species within its genus. It is recognized as its own clade most closely related to *Lemna* and *Spirodela* but distinct from both (Les and Crawford, 1999).

1 species.

Landoltia punctata (G. Meyer) Les & D.J. Crawford (p. 147)

Dotted duckweed

Spirodela punctata (G. Mey.) C.H. Thomps., *Spirodela oligorrhiza* (Kurz) Hegelm.

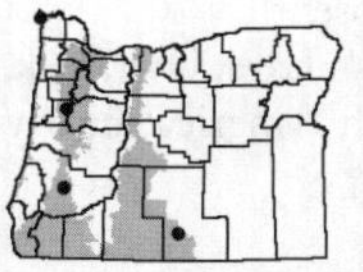

Plant bodies 1.5–2 times as long as wide; 3–5(7) veins; lower surface smooth and reddish purple; upper surface green with papillae along the midline. **Roots** up to 7 cm long.

Ponds, lakes, slow moving or stagnant water. 0–100 m. ECas, Est, Sisk, WV. CA, WA; throughout southern North America; temperate to subtropical regions worldwide. Exotic.

This species was first collected in Oregon in 1993, with several more recent, scattered collections at low elevations west of the Cascades.

Brown pigment cells in the epidermis, visible in dead plants, give a punctate, or dotted, appearance.

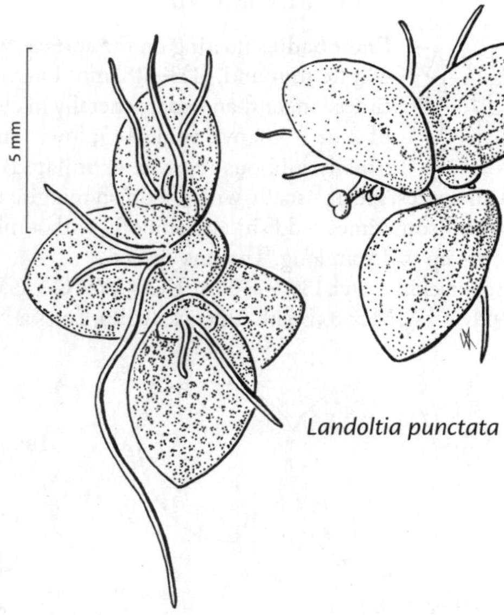

LANDOLTIA

LEMNA Duckweed

Plants free-floating aquatics, floating on the surface or submersed; small, green, sometimes reddish, flat plant bodies without true stems and leaves; veins 1–5 per plant body. **Roots** (0)1. **Buds** formed in 2 lateral pouches; turions present in some species. **Inflorescences** a single flower in a membranous spathe within the budding pouch, rarely produced. **Flowers** bisexual; tepals 0; stamens 2. **Fruits** 1–several seeded follicles.

Cosmopolitan. 13 species; 6 species treated in *Flora*.

Lemna rarely flower, but reproduction by budding is common. Daughter plants usually appear sessile but are connected to the mother plant by a short, thin, white stipe that decays after the daughter plant is grown. The exception is *Lemna trisulca*, in which the stipe is long and green and usually remains connected, resulting in a network of plant bodies. Vein morphology is best determined on fresh material using backlighting.

Key to *Lemna* species:

1. Plant bodies submersed unless flowering or fruiting; 6–10 mm long with a green stipe 2–20 mm long connecting mother and daughter plants, often forming branching networks of about 10–30 . . . . ***L. trisulca***

1' Plant bodies floating on water surface; 1–8 mm long without a tapered stipe, plant bodies solitary or connected in clusters of 2–8, not forming a branching network.

  2. Plant bodies with 1 vein.

    3. Plant bodies oblong to ovoid, thicker in the middle than along the edge, generally in clusters of 2 or solitary; vein extending less than 67% the distance from root attachment to apex, often faint . . . . ***L. minuta***

    3' Plant bodies elliptic to oblong, uniformly thin, generally in clusters of 4–8; vein extending at least 75% the distance from root attachment to apex . . . . ***L. valdiviana***

  2' Plant bodies with 3–5 veins.

    4. Lower surface usually inflated (gibbous), usually with a reddish margin; upper surface with a midline ridge, green, sometimes mottled with red; plant bodies orbicular-ovate with a slightly asymmetrical apex . . . . ***L. gibba***

    4' Lower surface flat; upper surface flat or with a midline ridge, green; plant bodies obovate-ovate with a symmetrical apex.

      5. Upper surface with a midline row of 3–7 papillae on a slight, broad ridge; plant bodies obovate-ovate; lower surface green but sometimes reddish (especially near the root); often forming dark-colored turions . . . . ***L. turionifera***

      5' Upper surface smooth and flat; plant bodies broadly oblong to elliptic, lower surface green; never forming turions . . . . ***L. minor***

Lemna gibba Linnaeus (p. 148)

Inflated duckweed

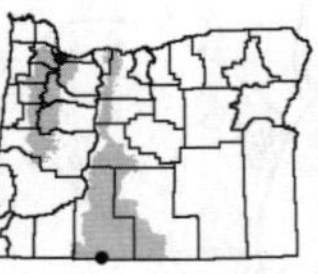

Plant bodies floating on the surface, widely elliptic to round, (1)3–6(8) mm long, 1–1.5 times longer than wide, generally in clusters of 2 or 3 (many), veins 4–5; lower surface usually gibbous (convex or inflated) with enlarged air spaces, green, usually with a reddish margin; upper surface green (sometimes reddish) and smooth, with a midline ridge. **Roots** up to 15 cm long. **Turions** absent.

Fresh or brackish water. 1300–2100 m. ECas, WV. CA, NV; east to NE, south to TX, Mexico; South America; Africa, Eurasia. Native.

Lemna minor Linnaeus (p. 148)

Common duckweed

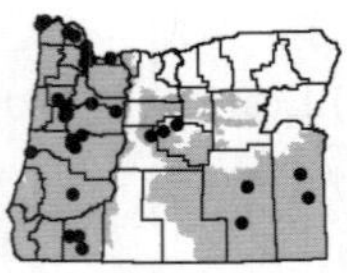

Plant bodies floating on the surface, elliptic to obovate, (1)2–5(8) mm long, 1.3–2 times longer than wide, generally in clusters of 2 or 3 (sometimes many), veins 3(5), if more than 3 then the outer veins branching from inner veins; lower surface flat, green, rarely reddish near the root attachment; upper surface flat, green (sometimes reddish), and smooth. **Roots** up to 15 cm long. **Turions** absent.

Freshwater. 0–1600 m. BR, Casc, CR, Est, Lava, Owy, Sisk, WV. CA, ID, NV, WA; throughout North America; nearly worldwide. Native.

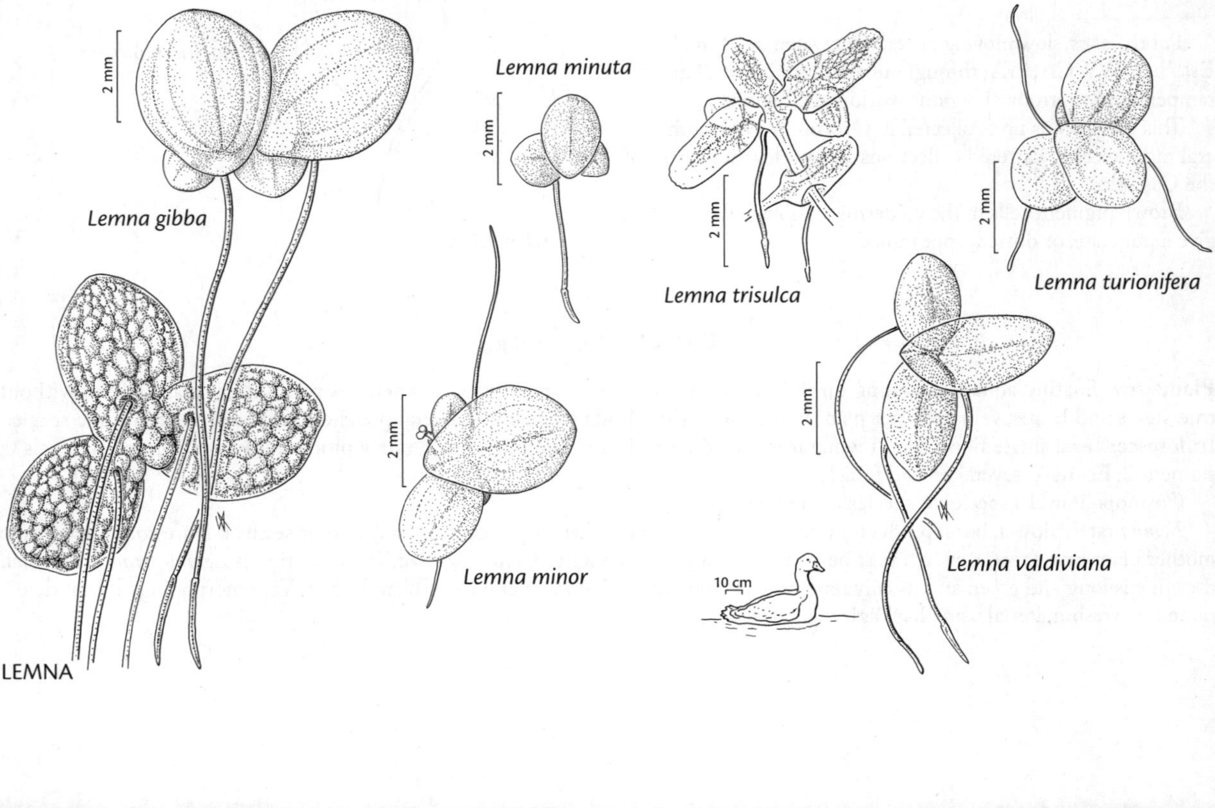

Lemna minor is distinguished from similar species with more than one vein (*L. gibba* and *L. turionifera*) by a lack of distinctive features such as a midline ridge or papillae on the upper surface, reddish or inflated lower surface, turions, and an asymmetrical apex.

Lemna minuta Kunth (p. 148)

Least duckweed

Lemna minima Phil. ex Hegelm.

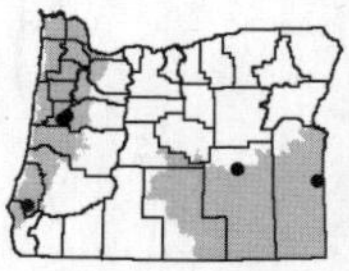

Plant bodies floating on the surface, widely elliptic to oblong or obovate, (0.8)1–2.5(4) mm long, 1–2 times longer than wide, slightly thicker in the middle with thinner margins, 1 vein, often faint, less than 67% the distance from root attachment to apex; lower surface flat, green; upper surface flat, green, and usually smooth, sometimes with papillae along the midline; generally in clusters of 2 (sometimes 4). **Roots** up to 1.5 cm long. **Turions** absent.

Freshwater. 50–1400 m. BR, CR, Owy, WV. CA, NV, WA; throughout North America; South America; Eurasia. Native.

This is the smallest *Lemna* species in Oregon.

Lemna trisulca Linnaeus (p. 148)

Star duckweed

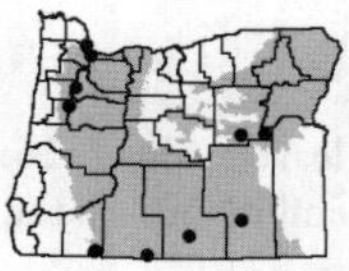

Plant bodies submersed unless flowering or fruiting, oblong to lanceolate, (3)6–10(15) mm long, 2–3.5 times longer than wide, apex obtuse, and base narrowed to a stipe, (1)3 veins, with the middle vein more prominent; lower surface flat and thin, green, often reddish; in branching networks of about 10–30 plant bodies connected by persistent green stipes, 2–20 mm long. **Roots** up to 2.5 cm long, but often absent. **Turions** absent.

Freshwater. 200–2100 m. BR, BW, Casc, ECas, WV. CA, ID, NV, WA; throughout North America; Africa, Australia, Eurasia. Native.

Lemna turionifera Landolt (p. 148)

Turion duckweed

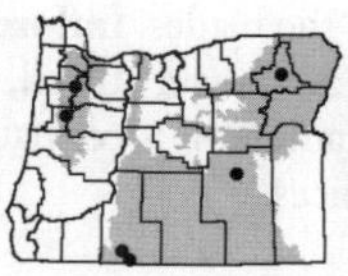

Plant bodies floating on the surface, orbicular to widely elliptic or obovate, (1)2–4 mm long, 1–1.5 times longer than wide, 3 veins; lower surface flat, usually reddish especially near root attachment; upper surface green, sometimes with red spots, with a row of 3–7 papillae along a midline ridge; generally in clusters of 2 or 3. **Roots** up to 15 cm long. **Turions** 0.8–1.6 mm diameter, greenish brown, formed in the fall and sink to the bottom.

Freshwater. 50–1400 m. BR, BW, ECas, WV. CA, ID, NV, WA; throughout North America; Africa, Australia, Eurasia. Native.

Lemna turionifera is the only Oregon *Lemna* species that produces distinct turions.

Lemna valdiviana Philippi (p. 148)

Valdivia duckweed

Lemna cyclostasa C.H. Thomps.

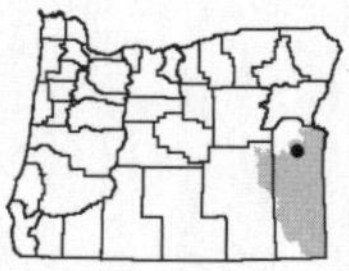

Plant bodies floating on the surface, rarely submersed, elliptic to oblong, (1)2–4(5) mm long, 1.3–3 times longer than wide, thin, 1 vein, greater than 75% the distance from root attachment to apex; lower surface flat, green; upper surface flat, green, and usually smooth, sometimes with papillae along the midline; generally in clusters of 4–8. **Roots** up to 1.5 cm long. **Turions** absent.

Freshwater. 800–900 m. Owy. CA, ID, NV; throughout North and South America in temperate to tropical climates. Native.

LYSICHITON Skunk cabbage

Herbs perennial. **Roots** fleshy and upright. **Leaves** basal, simple, elliptic to oblanceolate, pinnately-veined, leaf bases cuneate to subtruncate, petiole shorter than the blade, malodorous when crushed. **Inflorescences** spathes boat-shaped and open at maturity. **Flowers** bisexual; tepals 4; stamens 4; pistil 1. **Fruits** green, berry-like, embedded in spadix.

Western North America and northeastern Asia. 2 species; 1 in Oregon.

Lysichiton americanus Hultén & H. St. John (p. 149)

Skunk cabbage

Lysichiton camtschatensis (L.) Schott

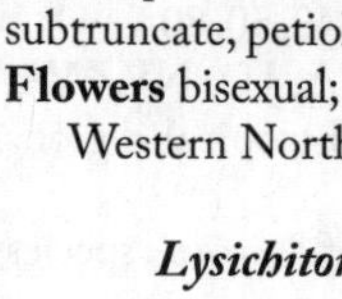

Plants to 150 cm. **Leaves** deep green with a prominent midvein, large, blades 30–100(150) × 8–45 cm, somewhat fleshy. **Inflorescences** spadices 4–12 cm long, stipitate; spathes 8–20(40) cm, yellow. **Flowers** yellowish green. **Fruits** 1–2-seeded. $2n$=28.

Wet areas. Flowering Feb–Jul. 0–1800 m. Casc, CR, Est, Sisk, WV. CA, ID, WA; north to AK, northeast to MT; Europe. Native.

The spadix of this species is initially shorter than the spathe but by fruiting is well exserted due to elongation of the stipe.

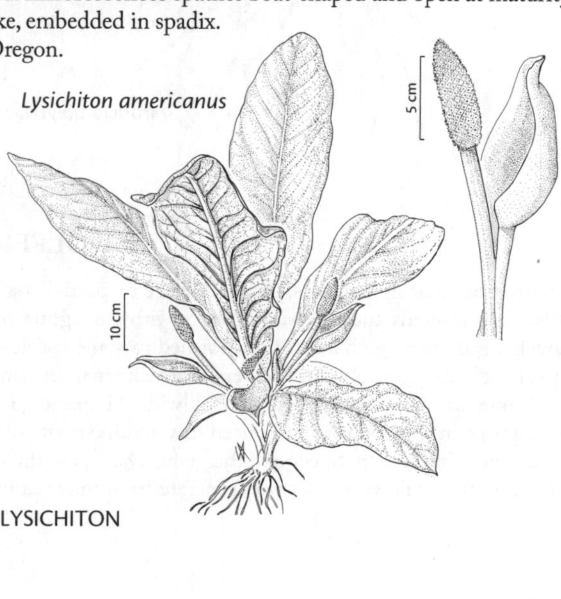

LYSICHITON

PELTANDRA Arrow arum

Herbs with vertical rhizomes. **Leaves** basal, simple, bases hastate to sagittate, pinnately veined, petioles equal to or longer than the blades. **Inflorescences** spadices shorter than the subtending spathes, staminate flowers distal, pistillate flowers proximal. **Flowers** unisexual; tepals absent; stamens 4–5, connate; pistil 1. **Fruits** 1–2(4) seeded berries, not embedded in spadices.

North America. 2 species; 1 in Oregon.

Peltandra virginica (Linnaeus) Schott (p. 150)

Green arrow arum

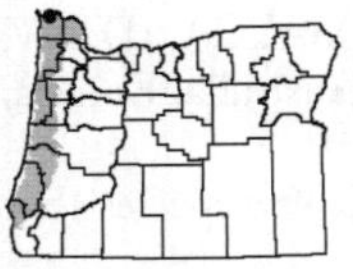

Herbs to 150 cm. **Leaves** up to 50 cm long, 5–20 cm wide, with lateral veins of 2 different thicknesses. **Inflorescences** spathes uniformly green or green with white or yellow-green margins; spadices 50% to almost as long as spathes. **Fruits** green to purple-green berries. **2*n***=28.

Freshwater and low salinity wetlands. Flowering Apr–Sep. 0–50 m. CR. CA, WA; eastern North America. Exotic.

This species is native to eastern North America and was likely introduced to Oregon as an ornamental.

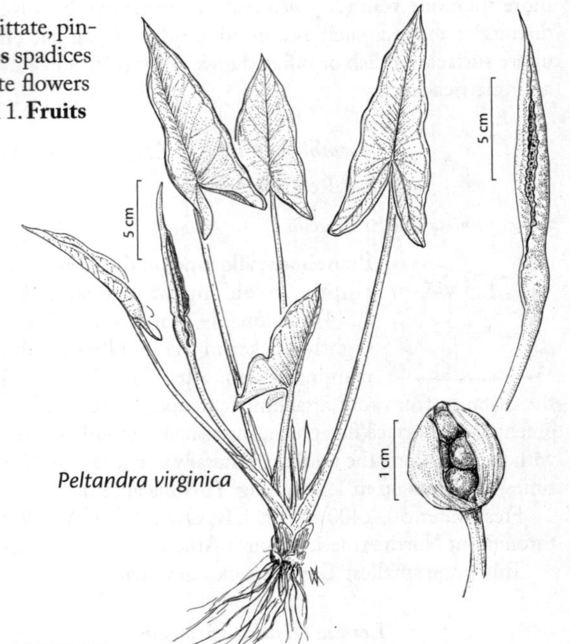

PELTANDRA

SPIRODELA Duckweed

Plants free-floating aquatics, on the surface; green, sometimes reddish, flat plant bodies without true stems and leaves; veins 7–16. **Roots** 7–21. **Buds** formed in 2 lateral pouches; turions present or absent. **Inflorescences** a single flower in a membranous spathe within the budding pouch, rarely produced. **Flowers** bisexual; tepals 0; stamens 2. **Fruits** 1–2(5) seeded follicles.

Freshwater. Cosmopolitan. 2 species; 1 species treated in *Flora*.

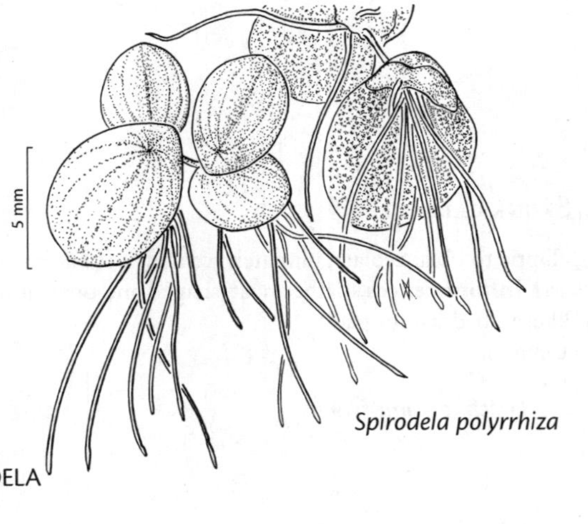

SPIRODELA

Spirodela polyrrhiza (Linnaeus) Schleiden (p. 150)

Greater duckweed

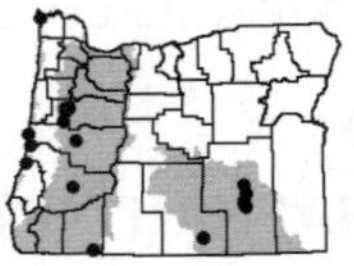

Plant bodies (3)5–10 mm long, 1–1.5 times longer than wide, orbicular to obovate; 7–16 veins; lower surface flat, reddish purple; upper surface green, usually with a reddish purple spot above the root attachment; generally in clusters of 2–5 or solitary. **Roots** up to 3 cm long. **Turions** 1–2 mm diameter, greenish brown. **2*n***=30, 38, 40, 50, 80.

0–1700 m. BR, Casc, Est, Sisk, WV. CA, ID, NV, WA; throughout North America; temperate and tropical regions worldwide. Native.

There are pigment cells in the epidermis of *Spirodela* species that are visible in dead plants as brown dots.

WOLFFIA Water meal

Plants free-floating aquatics, on the surface or partially submersed; small, green, globular, ovoid or boat-shaped plant bodies without true stems and leaves; solitary or with a daughter plant attached at basal end; veins 0. **Roots** 0. **Buds** formed in a basal pouch, light green, globular turions formed in some species. **Inflorescences** a single flower in a cavity along the midline of the upper surface; spathe absent. **Flowers** bisexual; tepals 0; stamen 1. **Fruits** single seeded follicles.

Temperate and tropical regions worldwide. 11 species; 3 species treated in *Flora*.

Flowers in this genus are produced in a midline cavity of the upper surface, not within the budding pouch like other Lemnoideae genera in Oregon. Species of the genus *Wolffia* are the smallest of all angiosperms. *Wolffiella*, not yet known from Oregon, is similar in that it flowers in a cavity separate from the budding pouch, but the floral cavity is to one side of the midline.

Key to *Wolffia* species:

1. Apex (the end opposite the bud, if present) pointed and curved upward; upper surface flattened; dead plants with brown spots (pigment cells) . . . . ***W. borealis***

1' Apex rounded and not curved upward; upper surface rounded, or flattened with a central, conical papule; dead plants with or without brown spots.

   2. Upper surface rounded; dead plants without brown spots . . . . ***W. columbiana***

   2' Upper surface flattened with a central, conical papilla; dead plants with brown spots (pigment cells). . . . . ***W. brasiliensis***

Wolffia borealis (Engelmann ex Hegelmaier) Landolt (p. 151)

Northern water meal

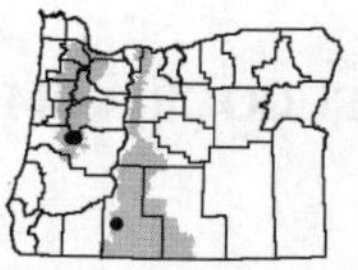

Plant bodies floating with entire upper surface above water, boat-shaped, 0.7–1.2(1.5) mm long, 1.3–2 times longer than wide, in clusters of 2 or solitary, apex pointed and curved upward; lower surface rounded, green; upper surface dark green, flattened, smooth, stomata 50–100. **Pigment cells** present and visible in dead plants as brown dots. **2*n***=20, 22, 30, 40.

Freshwater. 100–1300 m. ECas, WV. CA, ID, WA; scattered throughout northern North America. Native.

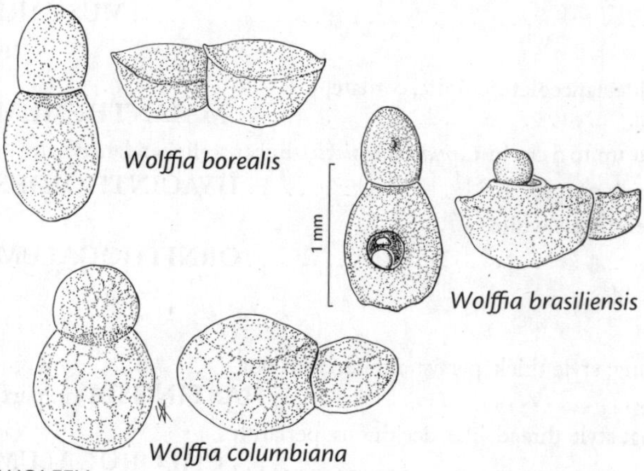

WOLFFIA

Wolffia brasiliensis Weddell (p. 151)

Brazilian water meal

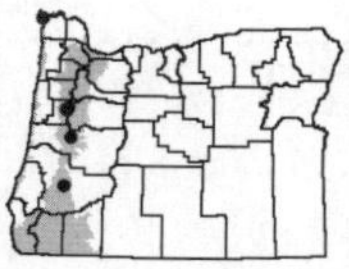

Plant bodies floating with entire upper surface above water, boat-shaped to ovoid, (0.5)0.7–1.2(1.6) mm long, 1–1.5 times longer than wide, in clusters of 2 or solitary, apex rounded, lower surface rounded, green; upper surface green, flattened with a central, conical papilla, stomata 50–100. **Pigment cells** present and visible in dead plants as brown dots. **2*n***=20, 40, 42, 50, 60, 80.

Freshwater. 0–300 m. Est, Sisk, WV. CA, WA; eastern North America; South America. Exotic.

Wolffia columbiana H. Karsten (p. 151)

Columbia water meal

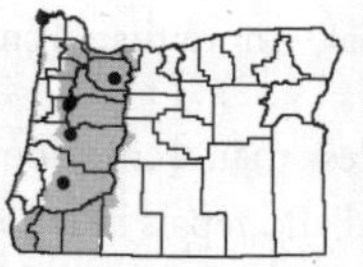

Plant bodies floating with only the central portion of the upper surface above water, nearly globular or spherical, (0.5)0.8–1.3(1.4) mm long, 1–1.3 times longer than wide, in clusters of 2 or solitary; lower surface green, rounded; upper surface rounded, green, smooth, stomata 1–10(30). **Pigment cells** absent.

Freshwater. 0–500 m. Casc, Est, Sisk, WV. CA;CA; throughout much of North America, south to South America; western Europe. Native.

Wolffia columbiana has recently been discovered as an introduced species in the Netherlands and Germany (Schmitz, et al. 2014).

ZANTEDESCHIA Calla-lily

Herbs rhizomatous. **Leaves** basal, simple, hastate or sagittate, pinnately-veined, with long petioles. **Inflorescences** spadices usually shorter than the subtending spathes, staminate flowers distal, pistillate flowers proximal. **Flowers** unisexual; tepals absent; stamens 2–3; pistil 1. **Fruits** berries.

Native to Africa, introduced worldwide. Approximately 8 species; 1 species treated in *Flora*.

Zantedeschia aethiopica (Linnaeus) Sprengel

Calla-lily

Herbs to 100 cm. **Leaves** deciduous, 15–45 cm long, 10–25 cm wide, uniformly green. **Peduncles** equal to or longer than the petioles. **Spathes** white or pale green, funnel-shaped, tip often recurved. **Spadices** shorter than the spathes, yellow. **Fruits** orange berries. **2*n***=26.

Near streams and ponds, ditches, wet areas near cultivation. Flowering Mar–Jun. 0–50 m. Est. CA; escaped in scattered locations worldwide. Exotic.

This species is usually not far escaped from where it was cultivated.

ASPARAGACEAE Asparagus family

Herbs perennial; from bulbs, rhizomes, or fleshy roots. **Stems** leafy or scapose. **Leaves** basal or cauline, 1–many, linear, linear-lanceolate, lanceolate, elliptic, cordate, or ovate; petioles present or not. **Inflorescences** solitary, panicles, racemes, or umbels. **Flowers** radially symmetrical (but sometimes bilateral, irregular or biradial in *Camassia*), erect or pendent; perianth parts 4 or 6, in 2 whorls (sometimes not in *Maianthemum*), fused or not; stamens 3, 4, or 6, staminodia present or not; ovary superior; style 1, stigmas 3-lobed. **Fruits** capsules or berries. **Seeds** few to many.

Cosmopolitan. 118 genera; 12 genera treated in *Flora*.

As a result of the taxonomic classification system proposed by Stevens (2012), many genera formerly contained within Liliaceae have been placed in the formerly monogeneric family Asparagaceae.

Key to Asparagaceae genera:

1. Leaves less than 1 mm wide . . . . **ASPARAGUS**
1' Leaves more than 1 mm wide.
  2. Plants without bulbs . . . . **LEUCOCRINUM**
  2' Plants with bulbs.
    3. Inflorescences scapose umbels.
      4. Perianth appendages protruding from base of perianth lobes and forming a crown . . . **DICHELOSTEMMA**
      4' Perianth appendages 0.
        5. Staminodia present, alternate fertile stamens. . . . . **BRODIAEA**
        5' Staminodia not present, all stamens fertile . . . . **TRITELEIA**
    3' Inflorescences various but not scapose umbels.
      6. Perianth parts fused . . . . **MUSCARI**
      6' Perianth parts free.
        7. Stems leafy; leaves usually more than 4 cm wide, lanceolate, elliptic, cordate, or ovate. . . . . **MAIANTHEMUM**
        7' Stems scapose; leaves less than 4 cm wide (but up to 5 cm in *Camassia cusickii*), linear to linear-lanceolate.
          8. Tepal tips recurved and tepals blue to bluish purple . . . . **HYACINTHOIDES**
          8' Tepal tips not recurved, or if recurved then tepals white.
            9. Filaments flat and broad. . . . . **ORNITHOGALUM**
            9' Filaments narrow, round, and thread-like.
              10. Inflorescences panicles.
                11. Tepals usually under 1 cm long; style thick, persistent; perianth not twisted in fruit. . . . . **HASTINGSIA** (in part)
                11' Tepals usually over 1 cm long; style thread-like, deciduous; perianth twisted in fruit. . . . . **CHLOROGALUM**
              10' Inflorescences racemes.
                12. Tepals usually under 1 cm long . . . . **HASTINGSIA** (in part)
                12' Tepals 1 cm long or greater . . . . **CAMASSIA**

ASPARAGUS Asparagus

Stephen C. Meyers

Plants perennial; erect, from rhizomes. **Stems** photosynthetic. **Flowers** tepals 6; stamens 6, distinct, equal; ovary superior; style 1. **Fruits** with tepals persisting at base. **Seeds** round to angled.

Cosmopolitan. 200–300 species; 1 species treated in *Flora*.

Asparagus officinalis Linnaeus (p. 153)

Asparagus officinalis L. ssp. *officinalis*

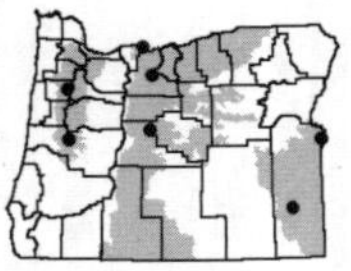

Herbs 1–2 m tall; rhizomes thick, matted. **Stems** branched, green. **Leaves** whorled, 4–18 mm long, less than 1 mm wide, linear. **Inflorescences** axillary racemes; pedicels 1–2(3)-flowered, 5–10 mm long, jointed in middle. **Flowers** bell-shaped, yellow to greenish yellow to greenish white; tepals 3–7 × 1–2 mm, connate basally, in 2 whorls; filaments slender; anthers oblong; style slender, stigmas 3, short. **Fruits** berries, red, 6–10 mm, round. **Seeds** 2–6, black. **2*n***=20, 40.

Disturbed areas, roadsides, fields. Flowering May–Jul. 50–1300 m. Col, ECas, Lava, Owy, WV. CA, ID, NV, WA; throughout North America; worldwide in temperate regions. Exotic.

Technically, but unintuitively, the leaves of this species are actually modified branches. Here, however, we treat them as "leaves" as this is the structure most observers will assume when viewing this plant part.

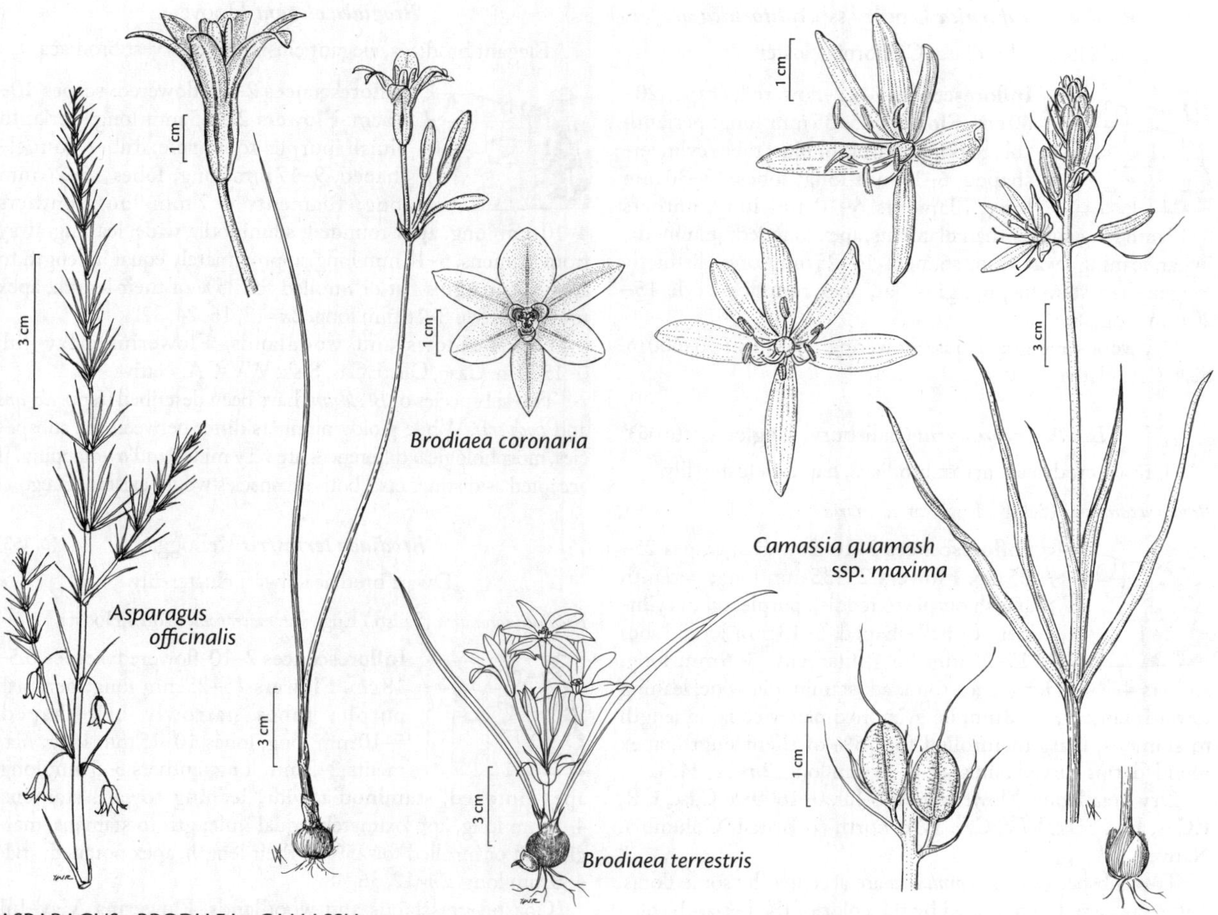

ASPARAGUS • BRODIAEA • CAMASSIA

BRODIAEA Brodiaea, cluster-lily

Stephen C. Meyers

Herbs perennial; from corms. **Leaves** 1–6, basal, linear, glabrous, C-shaped in cross section, entire. **Inflorescences** umbellate; scapes solitary, slender, straight; bracts present. **Flowers** perianth with 6 tepals, perianth tubes cylinder-, bell- or funnel-shaped, 3 outer tepals narrower than 3 inner tepals; stamens 3, opposite and attached to the 3 inner tepals; staminodia (sterile stamens) 3, opposite outer tepals; ovary superior; style 1, stigma lobes 3, distinct, reflexed. **Fruits** capsules, ovoid. **Seeds** round to oblong, black.

Western North America. 14–17 species; 4 species treated in *Flora*.

In the past, *Brodiaea*, *Dichelostemma*, and *Triteleia* have been treated as a single genus. Although the morphological and molecular evidence to split the larger, encompassing genus into three smaller units is only weakly compelling, we do so here for the sake of uniformity with other recent floristic works. Species and subspecific delimitation within *Brodiaea* is often based on slight differences in staminode morphology. More basic study is needed to determine if many of the described taxa merit scientific recognition.

Brodiaea minor and *B. nana* have each been collected once in Oregon (2007 and 2001 respectively). Whether these small populations remain is unknown.

Key to *Brodiaea* species:

1. Staminodia linear, distinctly longer than stamens, margins inrolled for 25–50% of their length . . ***B. californica*** ssp. ***californica***

1' Staminodia wide, approximately equal in length to stamens, margins inrolled for up to 25% of their length, 75% or more of their length, or flat.

  2. Perianth tubes funnel-shaped; staminodia leaning away from stamens. . . . . ***B. elegans***

  2' Perianth tubes cylinder- or bell-shaped; staminodia leaning toward stamens.

    3. Apex of anthers and staminodia rounded; staminodia margins inrolled 75% of their length; staminodia tips curved out at apex . . . . ***B. coronaria***

    3' Apex of anthers and staminodia notched; staminodia margins flat or inrolled 25% of their length; staminodia tips not curved out at apex . . . . ***B. terrestris***

Brodiaea californica Lindley ssp. ***californica***

California brodiaea, California cluster-lily

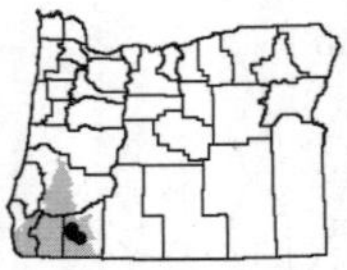

Inflorescences 3–12-flowered; scapes 20–60 cm. **Flowers** 25–35 mm long; perianth pale to bright purple or blue; tubes cylinder-shaped, 6–12 mm long; lobes 18–30 mm long; filaments 6–10 mm long; anthers 8–12 mm long, longer than filaments, apex rounded; staminodia linear, leaning away from stamens 15–25 mm long, distinctly longer than stamens, margins flat, apex rounded; style 15–20 mm long. **2*n***=12.

Dry woods and grasslands. Flowering Apr–Jun. 500–600 m. Sisk. CA. Native.

Brodiaea coronaria (Salisbury) Engler (p. 153)

Crown brodiaea, harvest brodiaea, harvest cluster-lily

Brodiaea coronaria (Salisb.) Engl. var. *coronaria*

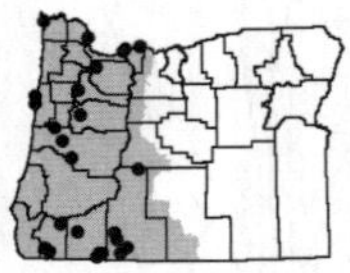

Inflorescences 3–10-flowered; scapes 25–35 cm. **Flowers** 25–35 mm long; perianth bluish purple to reddish purple; tubes cylinder- to bell-shaped, 5–13 mm long; lobes 15–25 mm long; filaments 3–6 mm long; anthers 4–7 mm long, apex rounded; staminodia wide, leaning toward stamens, 8–12 mm long, approximately equal in length to stamens, margins inrolled for 75% of their length, apex rounded, tips curved out; style 6–11 mm long. **2*n***=12, 24, 42.

Dry grasslands. Flowering Apr–Jul. 0–1600 m. Casc, CR, ECas, Est, Sisk, WV. CA, WA; north to British Columbia. Native.

Two subspecies of *B. coronaria* are accepted by some floras. These taxa are differentiated by the color of the perianth (bluish violet vs. rosy purple). Our plants are presumably *B. c.* ssp. *coronaria.*

Brodiaea elegans Hoover

Elegant brodiaea, elegant cluster-lily, harvest brodiaea

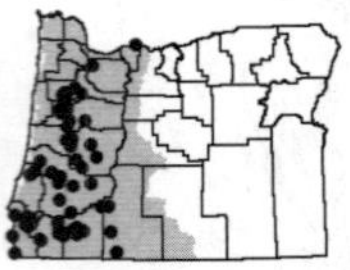

Inflorescences 2–10-flowered; scapes 10–45 cm. **Flowers** 25–35 mm long; perianth bluish purple to purple; tubes funnel-shaped, 9–17 mm long; lobes 15–26 mm long; filaments 4–7 mm long; anthers 4–10 mm long, apex rounded; staminodia wide, leaning away from stamens, 6–10 mm long, approximately equal in length to stamens, margins flat or inrolled for 25% of their length, apex rounded; style 8–16 mm long. **2*n***=12, 16, 24, 32.

Dry meadows and woodlands. Flowering May–Jul. 0–1500 m. Casc, CR, ECas, Sisk, WV. CA. Native.

Two subspecies of *B. elegans* have been described (sspp. *elegans* and *hooveri*). While ploidy numbers differ between the subspecies, morphological differences are very minor and overlapping. If accepted as distinct taxa, both subspecies would exist in Oregon.

Brodiaea terrestris Kellogg (p. 153)

Dwarf brodiaea, dwarf cluster-lily

Brodiaea coronaria (Salisb.) Engl. var. *macropoda* (Torr.) Hoover

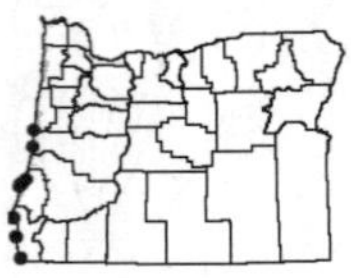

Inflorescences 2–10-flowered; scapes 0.5–18 cm. **Flowers** 15–25 mm long; perianth purple; tubes narrowly bell-shaped, 5–10 mm long; lobes 10–15 mm long; filaments 2–3 mm long; anthers 3–5 mm long, apex notched; staminodia wide, leaning toward stamens, 4–8 mm long, approximately equal in length to stamens, margins flat or inrolled for 25% of their length, apex notched; style 4–8 mm long. **2*n***=12, 36, 48.

Coastal grasslands and woodlands. Flowering May–Jul. 0–400 m. Est. CA. Native.

Plants with slightly larger flowers in central and southern California are referred to as *B. t.* ssp. *kernensis* by some authorities. Our plants, with smaller flowers, are referred to as *B. t.* ssp. *terrestris.*

CAMASSIA Camas

Susan R. Kephart

Herbs perennial, 10–130+ cm tall; bulbs solitary or rarely clustered; tunics brown or black. **Stems** scapose, erect, glabrous. **Leaves** basal, 3–20, lanceolate to linear, 15–60 × 0.5–5 cm, glabrous, margins entire, keeled. **Inflorescences** solitary, racemes bracteate or rarely with small side branches; sterile bracts 0–7, present along entire scape below raceme. **Flowers** 2–130 or more, symmetry radial, bilateral, irregular, or rarely biradial; sepals often longer and narrower than petals; tepals free, 6, blue, blue-violet to nearly purple, or white, withering connivently or separately, deciduous or persisting on raceme; veins 3–9; ovary superior, locules 3, nectaries septal; stamens 6; stigma lobes 3. **Fruits** capsules, globose to oblong, round to triangular in cross section, sometimes strongly veined at dehiscence. **Seeds** ovate-obovate, 6–36, black, lustrous. **2*n***=30.

North America. 6 species; 4 species treated in *Flora.*

Although small, the genus *Camassia* is diverse in Oregon, which supports all western species and all but two of ten traditionally recognized subspecies. Multiple broad phylogenetic studies confirm placement of camas in Agavaceae (=Agavoideae, Asparagaceae; APG 2009, Bogler et al. 2006) as does the presence of a similar bimodal karyotype (McKain et al. 2012, Ranker & Hogan 2002). *Camassia* forms a strongly supported monophyletic clade sister to the rush lilies, *Hastingsia* (Archibald et al. 2015, Halpin & Fishbein 2013), yet species delimitation remains challenging given high morphological and genetic variability, habitat diversity, and potentially recent radiation (Fishbein et al. 2010, Sultany et al. 2007, Uyeda & Kephart 2006).

Spring-blooming camas meadows host a diverse assemblage of bee pollinators and are now part of restoration efforts where accurate identification is critical.

Key to *Camassia* species:

1. Leaves usually 10 or more; shoots clonal; racemes densely flowered . . . . *C. cusickii*

1' Leaves mostly 3–8; shoots solitary; racemes with dense to non-overlapping flowers.

  2. Flowers radially symmetric; tepals twisting connivently; old flowers deciduous from top of intact pedicels.

    3. Flowers small, tepals 1–2 cm long; fruits subglobose, shiny when young . . . . *C. howellii*

    3' Flowers large, tepals 2–4 cm long; fruits oblong-ovoid, surface dull . . . . *C. leichtlinii*

  2' Flowers bilateral, less often radially symmetric; tepals withering individually or connivently until separated by enlarging ovaries; old flowers persistent on stem . . . . *C. quamash*

Camassia cusickii Watson
Cusick's camas

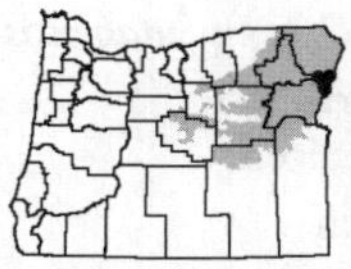

Plants diurnal; 50–90 cm tall; bulbs closely clustered to form evident "colony" of 2–20 or more basal shoots. **Leaves** (7)10–20; broadly linear, 25–70 × (1)2–5 cm. **Inflorescences** nodes 20–70, with 2–30 flowers open at a time; sterile bracts 0–7, most longer than pedicel, tan to blue. **Flowers** corollas bilateral, radial or irregular, varying in symmetry on a single raceme; tepals 15–35 × 3–5 mm, pale blue to blue-violet, withering separately, persistent along with old flowers, veins 3(5). **Fruits** erect, appressed to stem, oblong, 15–25 mm. **Seeds** 5–10 per locule.

Open meadows, gentle to steep rocky slopes and basalt cliffs, full sun along wet seeps. Flowering May–Jun. 600–1800 m. BW. ID. Native.

Cusick's camas is most readily distinguished from all other *Camassia* species in Oregon by its high basal leaf number (10 or more) and clonal habit.

Camassia howellii Watson
Howell's camas

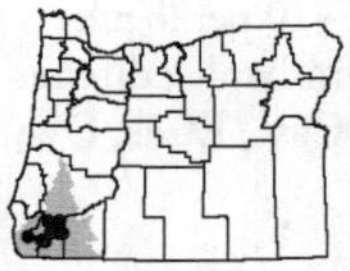

Plants vespertine; 20–60 cm tall; bulbs solitary. **Leaves** 4–9(10), lanceolate, 10–70 cm × 3–12 mm. **Inflorescences** nodes 5–80, with up to 30 flowers open at once; sterile bracts 0–3, most equal or shorter than pedicels, tan to pale green; pedicels often form broad angle with stem, but may curve inward. **Flowers** corollas radially symmetric; tepals 10–20 × 3–5 mm, blue to blue-violet, withering connivently and twisting off top of fruit or deciduous from pedicels that remain on stem, veins 3(5). **Fruits** globose, 5–10 × 5–9 mm, shiny when young but dulling with age. **Seeds** 2–5 per locule.

Open grassy or gravelly meadows, Jeffrey pine slopes, oak-madrone savanna, usually on serpentine soils. Flowering Apr–Jun. 200–700 m. Sisk. CA. Native.

The initially globose, shiny fruits and vespertine flowering found in *C. howellii* are unusual. In most camas species, new flowers open late afternoon to early evening and remain open, but flowers of Howell's camas have a short floral life span (tepals close between dusk and dawn or shortly thereafter without reopening; Howell 1903, Kephart et al. unpublished). This feature may link *Camassia* to vespertine species within *Chlorogalum* (Jernstedt 2002). The small-flowered *C. howellii* sometimes grows in mesic habitats near great camas (*C. leichtlinii),* which flowers earlier in wetter, rocky seeps. Despite these differences, putative hybrids occur, as supported by morphological and phylogenetic evidence (Fishbein et al. 2010).

Camassia leichtlinii (Baker) Watson
Great camas, Leichtlin's camas

Plants diurnal or rarely vespertine; 20–130 cm tall; bulbs usually solitary. **Leaves** 4–8(9), lanceolate, 20–70 cm × 5–30 mm. **Inflorescences** 6–60-flowered, often only 1–4 flowers open at a time; sterile bracts 0–3(5), most equal or shorter than pedicels, tan to pale green; pedicels spreading, forming broad angle with stem. **Flowers** corollas radially symmetric; tepals 2–4 cm × 4–12 mm, blue to deep blue-purple or creamy white, withering connivently and twisting off top of fruit or deciduous from pedicels that remain on stem, veins 5–9. **Fruits** oblong-ovoid, 1–3 cm × 5–15 mm, dull. **Seeds** 6–12 per locule.

Western North America. 2 subspecies.

Subspecies *suksdorfii* is the most broadly distributed of the two *C. leichtlinii* taxa. In Douglas County, it occurs near *C. l.* ssp. *leichtlinii* but usually in separate populations. These subspecies are potentially interfertile based on a historically known site near Roseburg where a creamy-white form dominated, yet plants showed the full color range. Further population-level study is needed.

The tall racemes, strictly radial symmetry, and connivently twisted, deciduous tepals distinguish *C. leichtlinii* from all subspecies of *C. quamash*. Plants in populations of *C. cusickii, C. quamash* ssp. *breviflora*, and ssp. *walpolei* can produce radial, near radial, or slightly bilateral flowers, but these taxa have three tepal veins in contrast to five to nine in *C. leichtlinii*. In *Camassia*, anthesis length varies, but flower buds usually open from mid-afternoon to dusk. Relative to *C. quamash* and *C. cusickii*, the mostly diurnal flowers of *C. leichtlinii* remain open for a shorter time, sometimes closing as early as late morning or early afternoon, leaving racemes with only buds above the withered flowers (Gould 1942).

Key to *Camassia leichtlinii* subspecies:

1. Tepals creamy-white; range Umpqua Valley . . . . *C. l.* ssp. ***leichtlinii***

1' Tepals blue-purple; throughout western Oregon . . . . *C. l.* ssp. ***suksdorfii***

Camassia leichtlinii (Baker) Watson ssp. ***leichtlinii***
Leichtlin's camas, great white camas, large camas

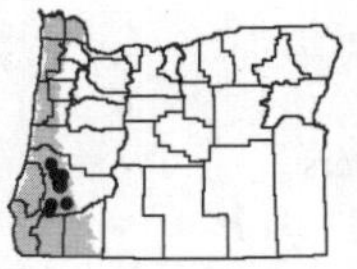

Flowers tepals creamy-white.

Open wet swales, grassy slopes, and shaded moist woodlands. Flowering Apr–May. 50–300 m. CR, Sisk. Native.

The creamy-white corollas and geographic range most reliably characterize this subspecies. Other traits lie within the range of variation in size measured for *C. l.* ssp. *suksdorfii*, but often in the largest categories. This may reflect the moist habitats and low elevations at which ssp. *leichtlinii* occurs. Except for flower color, the two subspecies show little differentiation, morphologically or phylogenetically.

Near Wilbur, plants of *C. l.* ssp. *leichtlinii* grow intermixed (with some habitat differentiation) in the same meadow as *C. quamash* ssp. *intermedia*. Putative hybrids are present, but infrequent.

Camassia leichtlinii (Baker) Watson ssp. ***suksdorfii*** (Greenman) Gould

Great camas

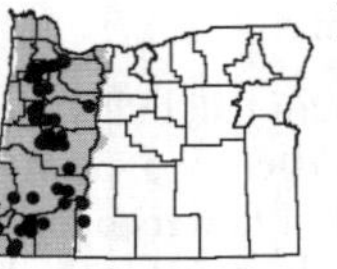

Flowers tepals blue-purple (rarely, white individuals are also present).

Wet meadows and ditches, oak savannas, forest openings, and partially shaded riparian woodlands. Flowering Mar–Jul. 0–1800 m. Casc, CR, Sisk, WV. CA, WA; north to British Columbia. Native.

Camassia leichtlinii ssp. *suksdorfii* and *C. quamash* ssp. *maxima* often grow in close proximity, yet remain distinct despite some hybridization (Uyeda & Kephart 2006, Fishbein et al. 2010). Plants of *C. quamash* ssp. *maxima* flower earlier, occurring in open, often more mesic habitats than the riparian woods or prairie-forest ecotones of *C. leichtlinii* ssp. *suksdorfii*. In serpentine mosaics, great camas occurs in pine-oak woodlands near *C. howellii*, which grows in sun or partly shaded meadow openings.

Camassia quamash (Pursh) Greene

Common camas

Plants diurnal; 10–80 cm tall; bulbs solitary. **Leaves** 3–5, lanceolate to linear, 25–70 cm × 5–15 mm. **Inflorescences** nodes 3–45, with 2–5 or more flowers open at once; sterile bracts 0–2, most longer than pedicels, tan to blue; pedicel-stem angle broad or less often narrow. **Flowers** corollas bilateral, rarely radial; tepals 10–35 × 3–5 mm, pale blue to deep blue-violet, initially withering individually or connivently but separating in fruit and persisting on stem, veins 3–9. **Fruits** oriented away from or appressed to stem, ovoid-oblong, 15–25 mm. **Seeds** 5–10 per locule.

Western North America. 8 subspecies; 6 subspecies treated in *Flora*.

Camassia quamash includes a confusing array of morphologically and geographically diverse subspecies. Differentiating them remains challenging, yet genetic data have revealed a detectable molecular signature between groups that grow "east" (*breviflora, quamash, utahensis*) and "west" of the Cascades (*intermedia, maxima, walpolei*). Taxa within these groups are still unresolved (Fishbein et al. 2010).

Key to *Camassia quamash* subspecies:

1. Tepal veins 3(5); flowers bilateral to radial; fruits held close to stem.
    2. Tepals withering separately as flower closes and remaining so in fruit.
        3. Leaves glaucous adaxially; anthers blue-violet to brown; flowers distinctly bilateral; stems to 70 cm tall; plants east of Cascades mountains . . . . ***C. q.*** ssp. ***quamash***
        3' Leaves not glaucous (uniformly green on both sides); anthers bright yellow; flowers radial to bilateral; stems to 45 cm tall; plants in southwestern Oregon . . . . ***C. q.*** ssp. ***walpolei***
    2' Tepals withering connivently but later separating if ovary enlarges to form fruit.
        4. Leaves 10–30 mm long; stems to 50 cm tall; flowers slightly to strongly bilateral or nearly radial . . . . ***C. q.*** ssp. ***breviflora***
        4' Leaves 20–50 mm long; stems to 70 cm; flowers distinctly bilateral . . . . ***C. q.*** ssp. ***utahensis***

1' Tepal veins 5–9; flowers distinctly bilateral; fruits borne on spreading or incurved pedicels.
    5. Leaves not glaucous (uniformly green on both sides); fruits close to stem on incurved pedicels . . . . ***C. q.*** ssp. ***intermedia***
    5' Leaves glaucous adaxially; fruit to stem distance variable but pedicels often slightly incurved. . . . . ***C. q.*** ssp. ***maxima***

Camassia quamash (Pursh) Greene ssp. ***breviflora*** Gould

Small camas

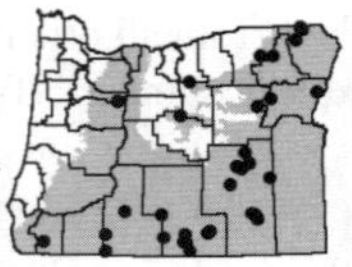

Plants 10–50 cm tall. **Leaves** 10–30 cm × 7–20 mm, glaucous adaxially. **Inflorescences** sterile bracts 0–2. **Flowers** corollas variably bilateral, sometimes radial; tepals 15–25(30) × 3–6 mm, pale blue to deep blue-violet, withering connivently and separating in fruit, veins 3(5); anthers usually bright yellow. **Fruits** erect on pedicels and positioned close to stem, 10–20 mm. **Seeds** 5–10 per locule.

Full sun in wet, marshy meadows, stream banks, ditches. Flowering May–Jul. 100–2300 m. BR, BW, Casc, ECas, Owy, Sisk. CA, ID, NV, WA. Native.

Camassia quamash (Pursh) Greene ssp. ***intermedia*** Gould

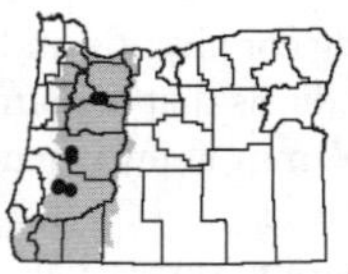

Plants 30–80 cm tall. **Leaves** 20–55 cm × 6–20 mm, not glaucous but uniformly green on both sides. **Inflorescences** sterile bracts 0–1. **Flowers** corollas distinctly bilateral; tepals pale blue to blue-violet, 20–35 × 3–7 mm, withering separately, veins 5–7(9); anthers usually bright yellow. **Fruits** erect on pedicels and positioned close to stem, 10–15 mm. **Seeds** 6–10 per locule.

Wet meadows, swales, and ash-dominated woodlands. Flowering Apr–May. 100–500 m. Sisk, Casc, WV. Native.

The evident bilateral symmetry, larger flowers, and 5–7-veined tepals clearly distinguish *Camassia quamash* ssp. *intermedia* from *C. q.* ssp. *walpolei*. *Camassia quamash* ssp. *intermedia* and ssp. *maxima* differ reliably only in adaxial leaf surface coating (*C. q.* ssp. *maxima* is glaucous, while *C. q.* ssp. *intermedia* is not). Many plants of *C. q.* ssp. *maxima* also bear fruits on pedicels that spread away from the raceme, but the capsules can also mature closely appressed to the stem as in *C. q.* ssp. *intermedia*. Ongoing morphological and molecular studies are needed to resolve these two taxa.

Camassia quamash (Pursh) Greene ssp. *maxima* Gould (p. 153)

Common camas

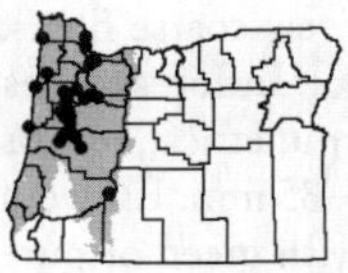

Plants 20–80 cm tall. **Leaves** 20–60 cm × 6–25 mm, glaucous adaxially. **Inflorescences** sterile bracts 0–1. **Flowers** corollas distinctly bilateral; tepals pale blue to deep blue-violet, 15–35 × 3–8 mm, withering separately, veins 5–9; anthers yellow to blue-violet to brown. **Fruits** positioned close to or distant from stem, often on slightly incurved pedicels, 9–17 mm. **Seeds** 5–10 per locule.

In or near wet springs, seeps on rocky bluffs, open mesic prairies, and oak savannas. Flowering Apr–Jul. 0–1600 m. Casc, CR, Est, WV. CA, WA; north to British Columbia. Native.

From its eastern Oregon counterpart, *C. q.* ssp. *breviflora*, western *C. q.* ssp. *maxima* differs in both its tepal venation and withering; both taxa range widely, yet their distributions are largely distinct with no known intergradation among populations.

Camassia quamash (Pursh) Greene ssp. *quamash*

Common camas

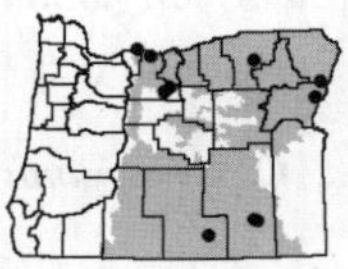

Plants 20–70 cm tall. **Leaves** 18–50 cm × 6–30 mm, glaucous adaxially. **Inflorescences** sterile bracts 0–1. **Flowers** corollas bilateral; tepals pale blue to deep blue-violet, 10–35 × 5–8 mm, withering separately, veins 3(5); anthers blue-violet to brown. **Fruits** erect on pedicels and positioned close to stem, 8–16 mm. **Seeds** 5–10 per locule.

Flowering Apr–Jul. 50–2200 m. BR, BW, Col, ECas. ID, WA; north to British Columbia, northeast to Alberta and MT. Native.

In Oregon, *C. q.* ssp. *quamash* is present only east of the Cascades and is more abundant in Idaho and Montana. In the past, some authorities misapplied this name to plants of *C. leichtlinii* outside this range, leading to misidentification, particularly in California, and southern and western Oregon.

Camassia quamash (Pursh) Greene ssp. *utahensis* Gould

Utah camas

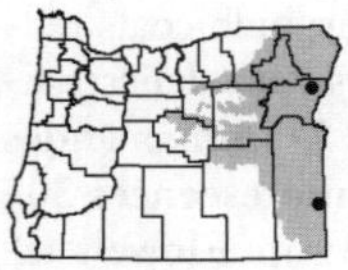

Plants 25–70 cm tall. **Leaves** 20–50 cm × 15–30 mm, glaucous adaxially. **Inflorescences** sterile bracts 0–2. **Flowers** corollas distinctly bilateral; tepals pale blue or blue-violet, 15–35 × 3–6 mm, withering conniventy but later splitting below fruit, veins 3(5); anthers light yellow to dark violet. **Fruits** erect on pedicels and positioned close to stem, 10–20 mm. **Seeds** 5–10 per locule.

Moist areas in montane meadows or along streams near forested areas, pasturelands. Flowering May–Jun. 1200–1300 m. BW, Owy. ID; east to MT and WY, southeast to UT. Native.

Although found in areas near *C. cusickii*, *C. quamash* ssp. *utahensis* differs in having fewer than ten basal leaves, connivent tepals at first withering, and solitary bulbs. Once flowers wither, the tepals separate, and appressed fruits of both taxa are similar, making identification more difficult.

Camassia quamash (Pursh) Greene ssp. *walpolei* (Piper) Gould

Walpole's camas

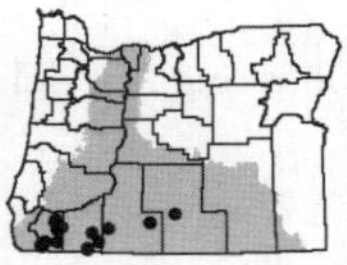

Plants 20–45 cm tall. **Leaves** 12–35 cm × 4–13 mm, not glaucous but uniformly green on both sides. **Inflorescences** sterile bracts 0–2. **Flowers** corollas radial to bilateral; tepals pale blue or blue-violet, 8–20 × 2–5 mm, withering separately and persistent on stem, veins 3(5); anthers mostly bright yellow. **Fruits** erect on incurved pedicels and positioned close to stem, 8–15 mm. **Seeds** 3–6 per locule.

Wet grassy meadows in Ponderosa pine forests, roadside ditches. Flowering Apr–Jun. 50–1400 m. BR, Casc, ECas, Sisk. Native.

The tiny, often radial flowers, combined with tepal venation and fruits held close to the stem typically separate *C. q.* ssp. *walpolei* from other species and subspecies. Although distinctive in southwestern Oregon and often described as the only radially symmetric subspecies of *C. quamash*, some populations of Walpole's camas have plants that vary considerably in this trait.

CHLOROGALUM Amole, soap plant

Frank T. Callahan

Herbs perennial; bulbs ovoid; bulb coats white, brown or reddish brown, with membranous and/or fibrous coats. **Leaves** basal, linear. **Inflorescences** panicles, peduncled, branches ascending; bracts scarious. **Flowers** perianth white, with a yellow-green to green or purple midvein, blooming in the evening; tepals 6, distinct, in 2 whorls, wrapping around ovary post-anthesis; stamens 6, attached to base of perianth; anthers attached at mid-point; ovary superior, 3-chambered; style 3-lobed. **Fruits** capsules, stalked, dehiscence loculicidal. **Seeds** 2 per chamber, ovoid, black, smooth or rough.

Western North America. 5 species; 3 species treated in *Flora*.

The name *Chlorogalum* is Greek (*chloros*, green, and *gala*, milk), referring to the liquid found in the bulbs.

Key to *Chlorogalum* species:

1. Leaves 2–5 mm wide; margins not wavy or barely so . . . . **_C. angustifolium_**

1' Leaves 4–25 mm wide; margins wavy (occasionally not in some *C. pomeridianum*).

  2. Pedicels 2–5 mm, much shorter in length than perianths . . . . **_C. grandiflorum_**

  2' Pedicels 5–35 mm, equal to or greater in length than perianths. . . . . **_C. pomeridianum_**

Chlorogalum angustifolium Kellogg

Narrowleaf amole, narrowleaf soap plant

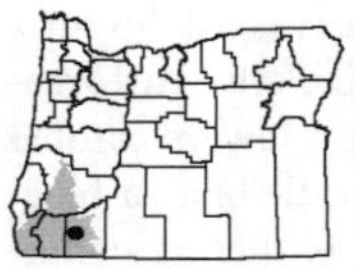

Bulbs 3–5 cm in diameter; bulb coats reddish brown, membranous with delicate fibers. **Leaves** 10–20 cm × 2–5 mm, margins not wavy or barely so. **Inflorescences** 30–75 cm tall; pedicels 1.5–3 mm. **Flowers** tepals recurved or not, oblong, 8–15 mm, midveins green-yellow; stamens 8–12 mm; anthers 1–3 mm, yellow; style 2–8 mm. **Fruits** 1–3 mm. **Seeds** smooth. $2n$=34.

Grasslands in sterile, serpentine, or volcanic soils. Flowering Apr–Jul. 300–400 m. Sisk. CA. Native.

Chlorogalum grandiflorum Hoover

Red Hills soap plant

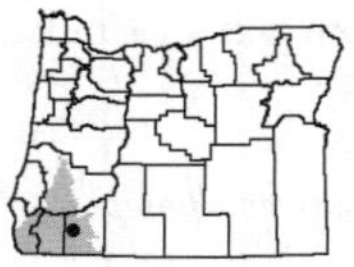

Bulbs 5–7+ cm in diameter; bulb coats reddish brown, membranous with a few delicate fibers. **Leaves** basal, 10–30 cm × 4–12 mm, margins wavy. **Inflorescences** 30–100 cm tall; pedicels 2–5 mm. **Flowers** tepals recurved, linear, 15–30 mm, midveins purple; stamens 12–28 mm; anthers approximately 3 mm, yellow; style 10–25 mm. **Fruits** 4–8 mm. **Seeds** rough.

Rocky soils, sometimes serpentine soils. Flowering Apr–Jul. 400–500 m. Sisk. CA. Native.

Chlorogalum pomeridianum (de Candolle) Kunth

Amole, wavyleaf soap plant

Bulbs 3–10(20) cm in diameter; bulb coats white, brown, or reddish, membranous or not, with many or few coarse fibers. **Leaves** 20–70 × 0.6–3 cm, margins wavy or not. **Inflorescences** 30–260 cm tall, branches symmetrical or asymmetrical, few or many-branched, stout or fragile; pedicels 5–35 mm. **Flowers** tepals spreading, linear, 1.5–2.5 cm, white, with green or purple midveins; anthers cream to yellow; style to 1.5 cm. **Fruits** 5–7 mm. **Seeds** smooth or rough.

California and Oregon. 4 varieties and 1 subspecies; 2 varieties and 1 subspecies treated in *Flora*.

The specific epithet of this species means "post meridiem" or "past mid-day" in Latin, referring to the flowers, which open in the evening.

Key to *Chlorogalum pomeridianum* subtaxa:

1. Bulbs not membranous, covered with many coarse fibers. . . . . ***C. p.*** var. ***pomeridianum***

1' Bulbs membranous, with few course fibers.

   2. Panicles usually not as wide as tall, branching pattern symmetrical; bulb coat white; seeds smooth . . ***C. p.*** ssp. ***austrooreganum***

   2' Panicles nearly as wide as tall, branching pattern asymmetrical; bulb coats reddish; seeds rough . . . . . ***C. p.*** var. ***minus***

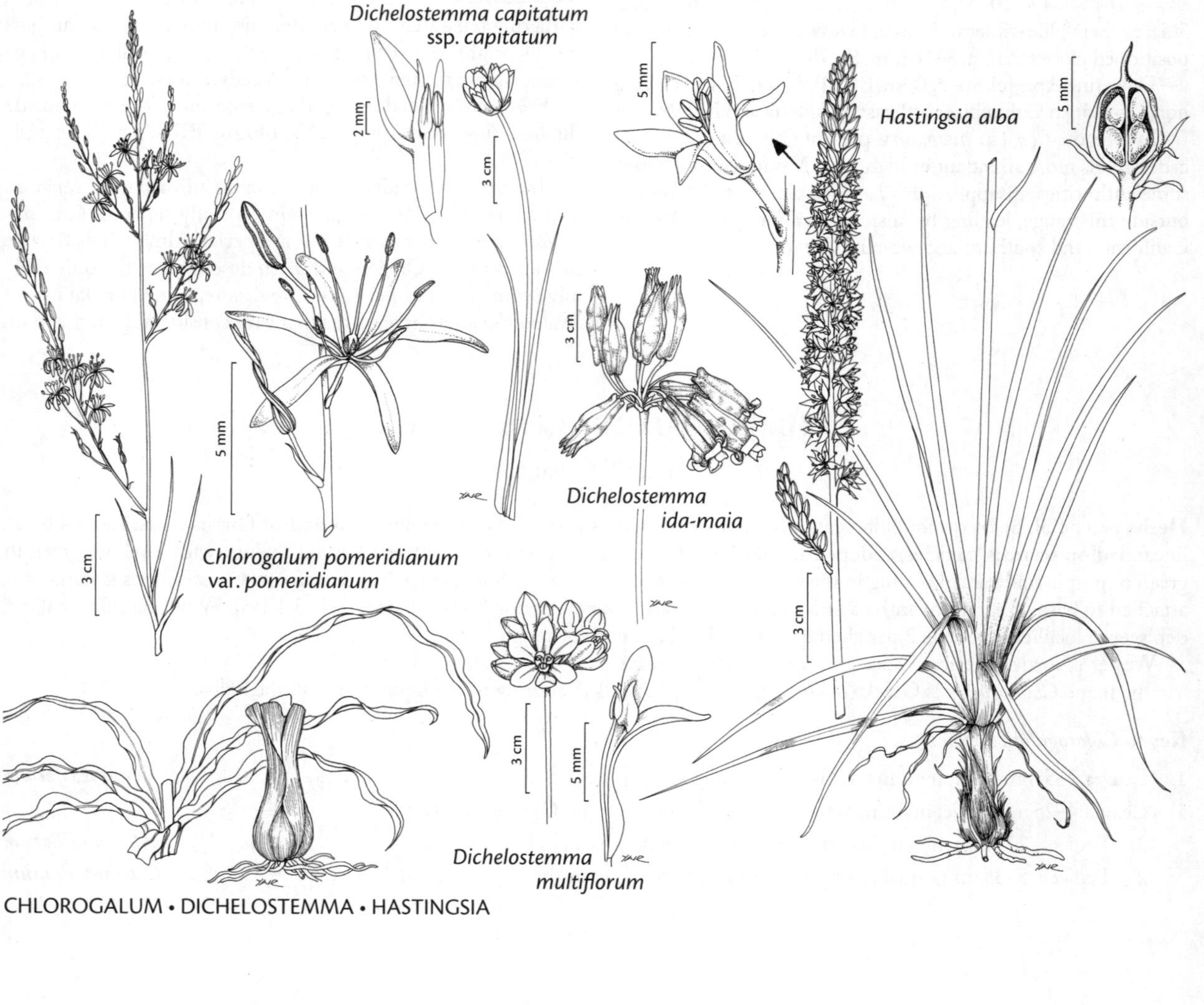

CHLOROGALUM • DICHELOSTEMMA • HASTINGSIA

Chlorogalum pomeridianum (de Candolle) Kunth ssp. ***austrooreganum*** Callahan

Southern Oregon amole

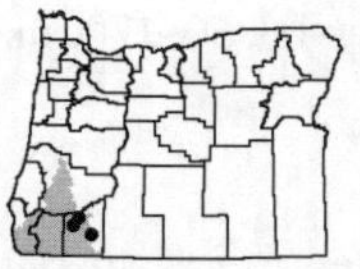

Bulbs to 4 cm in diameter; bulb coats white, membranous, with few coarse fibers. **Leaves** approximately 20–40 × 1.5 cm, margins wavy or not. **Inflorescences** panicles usually not as wide as tall, branching pattern symmetrical, branches few, fragile. **Seeds** smooth. **2*n***=30.

Clay and serpentine soils, talus. 500–900 m. Sisk. CA. Native.

Chlorogalum pomeridianum (de Candolle) Kunth var. ***minus*** Hoover

Small amole

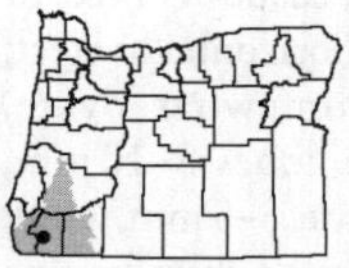

Bulbs to 3.5 cm in diameter; bulb coats reddish brown, with few coarse fibers. **Leaves** 20–35 × 1.5 cm, margins conspicuously wavy. **Inflorescences** panicles nearly as wide as tall, branching pattern asymmetrical, many-branched, stout. **Seeds** rough. **2*n***=30.

Clay and serpentine soils. Flowering May. 300–600 m. Sisk. CA. Native.

Chlorogalum pomeridianum (de Candolle) Kunth var. ***pomeridianum*** (p. 158)

Amole, wavyleaf soap plant

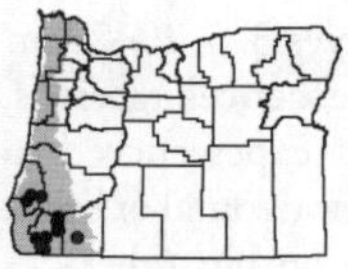

Bulbs to 10(20) cm in diameter; bulb coats brown, not membranous, covered with many coarse fibers. **Leaves** approximately 20–70 × 3 cm, margins wavy. **Inflorescences** panicles erect, many-branched, stout. **Seeds** rough. **2*n***=30, 36.

Clay and serpentine soils. Flowering May–Jul. 0–500 m. CR, Sisk. CA. Native.

DICHELOSTEMMA Snake-lily

Stephen C. Meyers

Herbs perennial; scapose; corms present. **Leaves** basal, 2–5; blades narrowly lanceolate, entire, glabrous, usually glaucous, keeled or not. **Inflorescences** dense or open umbels or racemes, 2–20-flowered; scapes curved to twisted or bent; bracts thin and papery. **Flowers** horizontal, erect or nodding; tepals 6, fused approximately 50% their length, perianth tube cylindric to bell-shaped, perianth appendages protruding from base of perianth lobes and forming a crown, leaning toward or away from anthers; stamens 3 or 6, equal or unequal in length, staminodia 0 or 3; ovary superior; style 1, stigma weakly 3-lobed. **Fruits** capsules, ovoid, 3-angled. **Seeds** ovoid, 3-angled, black.

Western North America. 5 species; 4 species treated in *Flora*.

The taxa within *Dichelostemma* have been included in *Brodiaea* in older texts. The presence of perianth appendages and the lack of staminodia (except *D. volubile*) are the most distinct morphological features separating the two genera.

Dichelostemma volubile has been collected once in Oregon, in 1998, in Jackson County. This species is distinguished from other species within the genus by its conspicuously twisted scapes and the presence of staminodia.

Key to *Dichelostemma* species:

1. Inflorescence open; perianth tube equal to or greater than 20 mm long . . . . ***D. ida-maia***
1' Inflorescence dense; perianth tube less than or equal to 12 mm long.
   2. Stamens 6, unequal in length . . . . ***D. capitatum*** ssp. ***capitatum***
   2' Stamens 3, equal in length.
      3. Perianth appendage tips notched; inflorescences racemes . . . . ***D. congestum***
      3' Perianth appendage tips squared or round; inflorescences umbels . . . . ***D. multiflorum***

Dichelostemma capitatum (Bentham) Alph. Wood ssp. ***capitatum*** (p. 158)

Bluedicks

Brodiaea capitata Benth., *Brodiaea pulchella* (Salisb.) Greene, *Dichelostemma pulchellum* (Salisb.) A. Heller

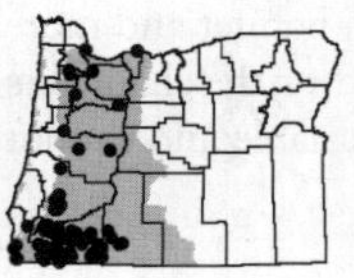

Plants 5–65 cm tall. **Leaves** 2–3, 10–70 cm × 4–10 mm, not or only slightly keeled. **Inflorescences** umbels, dense; bracts ovate-lanceolate; scapes curved to slightly twisted. **Flowers** horizontal or erect; perianth white or pinkish purple to blue, tubes cylindrical to bell-shaped, 5–11 mm, lobes 6–12 mm, usually ascending, perianth appendages lanceolate, 4–6 mm, white, leaning toward anthers, tips notched; stamens 6, outer anthers 2–3 mm, inner anthers 3–4 mm; style 3–7 mm. **2*n***=18, 36, 54, 72, 90.

Meadows, open rocky areas, open forests. Flowering Mar–Jul. 0–1600 m. Casc, ECas, Est, Sisk, WV. CA; south to Mexico. Native.

Dichelostemma capitatum ssp. *capitatum* is the only member of the genus (in Oregon) that has six stamens. All others have three.

Dichelostemma congestum (Smith) Kunth

Ookow

Brodiaea congesta Sm.

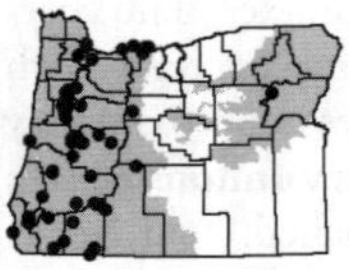

Plants 30–80 cm tall. **Leaves** 3–5, 20–50 cm × 4–8 mm, keeled. **Inflorescences** racemes, dense; bracts lanceolate; scapes curved to slightly twisted. **Flowers** horizontal or erect; perianth purple to bluish purple, tubes cylindrical to ovoid, 8–10 mm, lobes 8–10 mm, usually ascending, perianth appendages lanceolate, 4–6 mm, purple, leaning away from anthers, tips notched; stamens 3, equal in length, anthers 4–5 mm; style 4–6 mm. **2*n***=18, 36.

Meadows, open rocky areas, open forests. Flowering May–Aug. 0–1600 m. BW, Casc, CR, ECas, Est, Sisk, WV. CA, WA. Native.

Rarely, hybrids between *D. congestum* and *D. ida-maia* (*D.* × *venustum*) are found. In Oregon, this is currently only known to occur in Douglas County.

Dichelostemma ida-maia (Alph. Wood) Greene (p. 158)

Firecracker flower

Brodiaea ida-maia (A.W. Wood) Greene

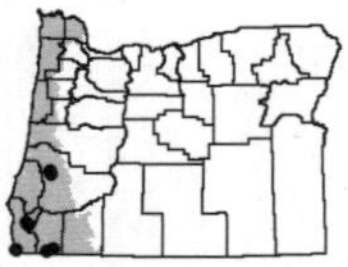

Plants 25–80 cm tall. **Leaves** 3–4, 20–50 cm × 3–8 mm, keeled. **Inflorescences** umbels, open; bracts wide-lanceolate; scapes curved to slightly twisted. **Flowers** nodding; tubes cylindrical, 18–28 mm, red to occasionally purplish red, lobes 7–10 mm, yellowish green, recurved, perianth appendages wide-lanceolate, 3–5 mm, white, leaning away from anthers, tips rounded; stamens 3, equal in length, anthers 6–8 mm; style 7–20 mm. **2*n***=28, 48.

Forest edges, meadows. Flowering May–Jul. 50–1700 m. CR, Est, Sisk. CA. Native.

See also *D. congestum*.

Dichelostemma multiflorum (Bentham) A. Heller (p. 158)

Round tooth snake-lily, round-toothed ookow, wild hyacinth

Brodiaea multiflora Benth.

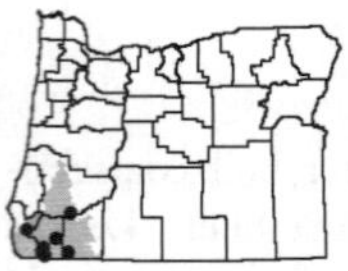

Plants 20–80 cm tall. **Leaves** 3–4, 20–80 cm × 3–6 mm, not keeled. **Inflorescences** umbels, dense; bracts ovate; scapes curved to slightly twisted. **Flowers** horizontal or erect; tubes cylindrical to somewhat ovate, 8–10 mm, pinkish purple to bluish purple, lobes 8–10 mm, spreading, perianth appendages wide-lanceolate, 3–5 mm, white to purple, leaning inward basally and outward distally, tips squared or rounded; stamens 3, equal in length, anthers 4–5 mm; style 5–8 mm. **2*n***=18, 36.

Forest edges, meadows. Flowering May–Jun. 100–1100 m. Sisk. CA. Native.

HASTINGSIA Rush-lily

Kathryn E. Theiss

Roots bulbs. **Leaves** basal, grass-like, simple, strongly V-shaped; cauline leaves 0–4, decreasing in length distally. **Inflorescences** determinate cymes with 0–3(7) branches with subtending bracts. **Flowers** bisexual, each subtended by a bract, perianth closed or open at anthesis; tepals 6, free, narrowly lanceolate to oblong-lanceolate, white, yellowish white, or dark purple; stamens 6, free, 3 longer with dehisced anthers and 3 shorter with undehisced anthers at anthesis, all becoming equal after anthesis; ovary superior, globose, 3-lobed, ovules 2 per locule; style 1; stigma distinctly 3-parted. **Fruits** capsular, ellipsoid to ellipsoid-ovoid, broadly 3-lobed. **Seeds** fusiform, usually flattened adaxially, gray-green to black, reticulate. **2*n***=52.

Southwestern Oregon and California. 2 species.

Hastingsia is virtually always associated with serpentine soil, both in Oregon and in California. The two species are often sympatric in southwestern Oregon in the Illinois River valley but are easily differentiated based on habitat and morphology.

Key to *Hastingsia* species:

1. Tepals 1.5–9 mm long, white or yellowish, spreading and exposing the stamens; primarily dry habitats, sometimes wet seeps . . . . **H. alba**

1' Tepals 7–12 mm long, white or dark purple, fully enclosing the stamens; wet fens . . . . **H. bracteosa**

Hastingsia alba (Durand) S. Watson (p. 158)

White-flowered rush-lily

Hastingsia serpentinicola Becking, *Schoenolirion album* Durand

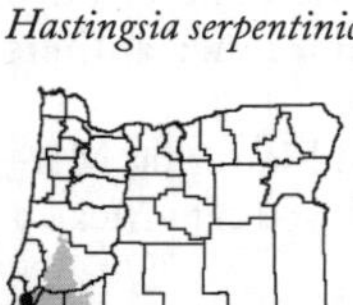

Bulbs 23–56 × 13–31 mm, sometimes with blackish, fibrous tunic. **Scape** 29–85 cm long. **Leaves** 7–78 cm × 3–18 mm. **Inflorescences** 6–90 flowers on a terminal raceme, unbranched or with 1–7 branches. **Flowers** narrow-lanceolate, white or yellowish with yellow, green, or dark purple midvein; tepals 1.5–9 mm long, spreading or sharply reflexed, stamens fully exposed. **Fruits** ellipsoid capsules, 4–10 × 2–6 mm. **Seeds** 3–6 mm, gray-green to black. **2*n***=52.

Dry habitats or wet seeps, especially on serpentine soils, near bogs or meadows. Flowering Apr–Jul. 100–2200 m. Sisk. CA. Native.

Plants on serpentine soils (*Hastingsia serpentinicola*) are sometimes segregated from *H. alba* based on habitat and morphological differences. (Becking 1989). However, the characters used to separate these two taxa are highly variable and cannot be used consistently to separate specimens.

Hastingsia bracteosa S. Watson

Large-flowered rush-lily

Bulbs 28–54 × 18–30 mm, with blackish, fibrous tunic. **Scape** 16–122 cm long. **Leaves** 10–50 cm × 4–13 mm. **Inflorescence** 8–80 flowers on a terminal raceme, unbranched or with 1–6 branches. **Flowers** oblong-lanceolate, creamy white or dark purple, acuminate; tepals 7–12 mm long, fully enclosing the stamens. **Fruit** ellipsoid capsules, 7–11 × 5–8 mm. **Seeds** 4–5 mm, dark grayish green to black.

2 varieties.

Hastingsia bracteosa is almost always associated with *Darlingtonia*. While the two varieties of *H. bracteosa* were previously elevated to species level (Becking 1986), neither recent molecular investigations (Halpin and Fishbein 2013) nor current morphological studies support this. The two varieties can be found sympatrically in fens in the Illinois River valley and are thought to hybridize (Lang and Zika 1997).

Key to *Hastingsia bracteosa* varieties:

1. Flowers dark purple . . . . ***H. b.*** var. ***atropurpurea***

1' Flowers generally creamy white . . . . . ***H. b.*** var. ***bracteosa***

Hastingsia bracteosa S. Watson var. ***atropurpurea*** (Becking) F.A. Lang & Zika

Purple-flowered rush-lily

Hastingsia atropurpurea Becking

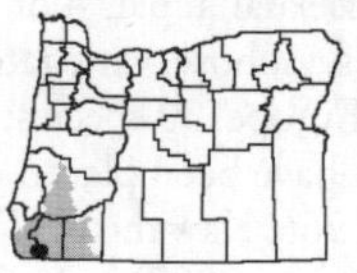

Flowers dark purple.

Bogs and fens on serpentine soils. Flowering Jun–Jul. 300–900 m. Sisk. Native.

Hastingsia bracteosa S. Watson var. ***bracteosa***

Large-flowered rush-lily

Schoenolirion bracteosum (S. Watson) Jeps.

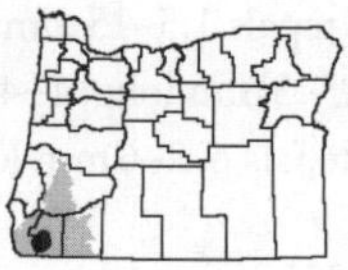

Flowers creamy white, sometimes with green or yellow midvein.

Bogs and fens on serpentine soils. Flowering May–Jul. 300–1000 m. Sisk. Native.

HYACINTHOIDES Bluebell

Stephen C. Meyers

Herbs perennial; bulbs ovoid to globose, completely renewed each year. **Leaves** basal, linear to linear-lanceolate. **Inflorescences** racemes; bracts (1)2, linear-lanceolate, subtending each flower. **Flowers** tepals 6, fused near base, tips slightly to strongly recurved; stamens 6, equal or unequal in length, filaments fused to perianth or inserted at base, anthers cream to yellow or blue; style 1, stigma 1. **Fruits** capsules, subglobose, 3-lobed. **Seeds** 3–30, subglobose to globose, black.

North America, North Africa, and Europe. 4 species; 1 species treated in *Flora*.

Hyacinthoides × *variabilis* (*H. hispanica* × *H. non-scripta*) may occasionally escape from cultivation. Thus far, however, it has not yet become established in Oregon. It is most easily recognized by its whitish to blue anthers of variable length, slightly one-sided racemes and slightly recurved tepals tips.

Hyacinthoides non-scripta (Linnaeus) Chouard (p. 162)

English bluebell

Scilla nonscripta (L.) Hoffmans. & Link

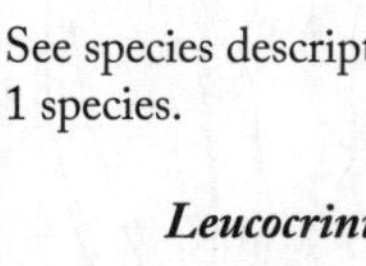

Plants 15–28 cm tall; bulbs approximately 1 mm in diameter. **Leaves** 3–6, 15–40 cm × 5–20 mm. **Inflorescences** racemes 4–16-flowered, obviously 1-sided, nodding toward apex. **Flowers** perianth cylindrical-campanulate; tepals oblong-lanceolate, 1.4–2 cm long, blue to bluish purple, rarely pink or whitish, tips strongly reflexed; stamens unequal in length, outer 3 filaments fused to perianth for 75% of their length, anthers cream to yellow. **2*n***=16, 24.

Shady areas, forests. Flowering Mar–May. 0–400 m. Casc, CR, Est, WV. WA; north to British Columbia; northeastern US; Europe. Exotic.

Easily recognizable, *H. non-scripta* is often one of the first plants to be noticed flowering in the spring.

LEUCOCRINUM Sand-lily

Stephen C. Meyers

See species description.

1 species.

Leucocrinum montanum Nuttall ex A. Gray (p. 162)

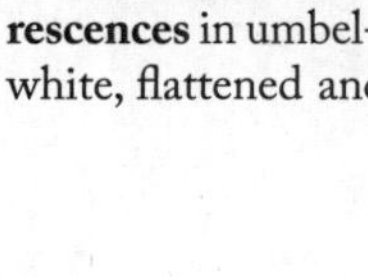

Herbs perennial, 4–10 cm tall; roots short, fleshy. **Stems** 0. **Leaves** basal, grass-like, spreading, tufted, surrounded by a sheath; blades 7–20 cm × 2–7 mm; sheaths thin, dry, membranous, white, up to 7 cm long. **Inflorescences** in umbel-like clusters; pedicels underground. **Flowers** white, flattened and spreading above, connate below; tepals 6, lobes 1.5–2.5 cm × 3–7 mm; perianth tube 4–9 cm long; stamens 6; ovary superior, below ground; style 1; stigmas shortly 3-lobed. **Fruits** capsules, ovoid, 5–8 mm long, 3-angled, underground. **Seeds** black, 3–4 mm long, angled. **2*n***=22, 26, 28.

Dry sandy soil in scrublands, juniper forests, and montane woodlands. Flowering Apr–Jun. 500–1800 m. BR, BW, ECas, Lava, Owy. CA, ID, NV; east to NE, southeast to NM. Native.

The seed capsules of this species are unusual in that as the plant dies, they are pulled down and buried among the roots.

MAIANTHEMUM Solomon's seal, false lily-of-the-valley

Thea Jaster

Herbs perennial, 10–120 cm; rhizomatous, sympodial. **Stems** erect, unbranching. **Leaves** alternate, simple, entire. **Inflorescences** terminal, racemose or paniculate. **Flowers** bisexual; tepals 4 or 6, white; stamens 4 or 6, opposite the tepals; ovary superior; pistil 1, style 1, stigmas 2 or 3. **Fruits** fleshy berries, globose, variegated coloring on immature fruits. **Seeds** globose, brown. **2*n*=36.**

North America, Central America, Asia, Europe. 30 species; 3 species treated in *Flora*.

Formerly, *M. racemosum* and *M. stellatum* have been placed under *Smilacina*. However, molecular and morphological analyses, plus a distinct karyotype (Kawano and Iltis 1966; Kawano et al. 1967; Sen 1975; LaFrankie 1986a; LaFrankie 1986b; Rudall et al. 2000; Yamashita and Tamura 2000; Kim and Lee 2007) have distinguished *Maianthemum*, in the broad sense, as a monophyletic group encompassing all three Oregon species.

Key to *Maianthemum* species:

1. Inflorescences paniculate, 40–500-flowered; tepals 1–2 mm; pedicels 0.5–1.5 mm . . . . ***M. racemosum* ssp. *amplexicaule***
1' Inflorescences racemose, 2–40-flowered; tepals 1.5–15 mm; pedicels 2–14 mm.
   2. Leaves cordate, petiolate; tepals 4, 1.5–3 mm long; 1–4 flowers per node . . . . ***M. dilatatum***
   2' Leaves elliptic-lanceolate, clasping; tepals 6, 2–6 mm long; 1 flower per node . . . . ***M. stellatum***

Maianthemum dilatatum (A.W. Wood) A. Nelson & J.F. Macbride

False lily-of-the-valley, beadruby

Maianthemum bifolium DC. var. *kamtschaticum* (J.F. Gmel.) Jeps.

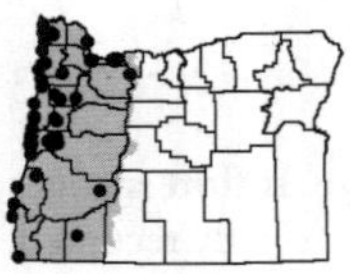

Herbs 10–40 cm; rhizomes 1.3–3(5) mm in diameter. **Leaves** 6–12 × (3)5–10 cm, 2–3 per stalk, long petiolate, the lower 2 cordate, the distal 3rd leaf cordate, lanceolate, ovate, deltoid, or bract-like. **Inflorescences** racemose, 22–40(54)-flowered. **Flowers** tepals 4, 1.5–3 mm long; pedicels (1.5)2–5 mm; stamens 4. **Fruits** globose, 3–7 mm in diameter, red at maturity but green with red speckles approaching maturity.

Near streams and bogs, damp woods under trees. Flowering May–Jul. 0–1300 m. Casc, CR, Est, Sisk, WV. CA, ID, WA; north to AK. Native.

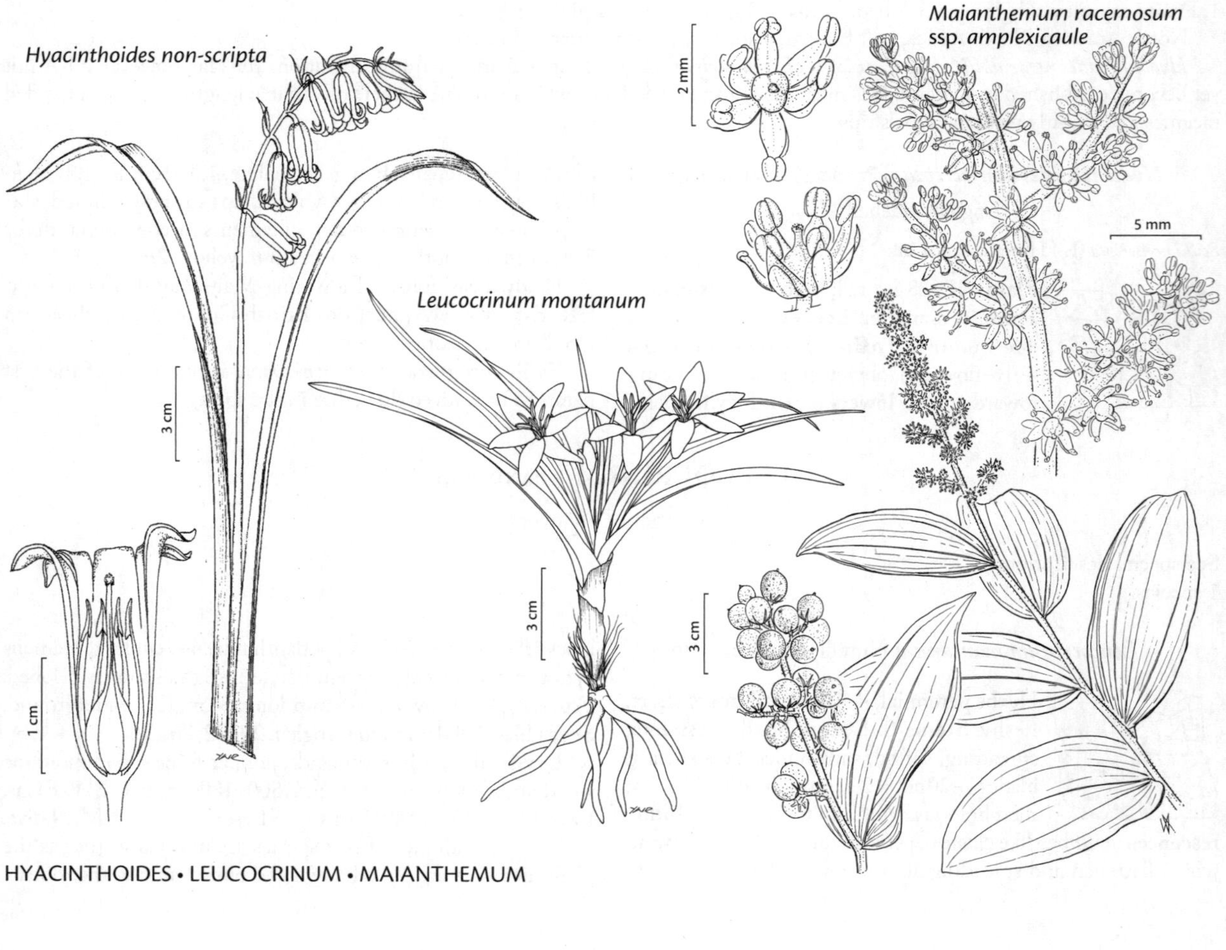

HYACINTHOIDES • LEUCOCRINUM • MAIANTHEMUM

Maianthemum racemosum (Linnaeus) Link (p. 162) ssp. *amplexicaule* (Nuttall) LaFrankie

Large false Solomon's seal, western Solomon plume

Smilacina racemosa (L.) Desf., *Smilacina amplexicaulis* Nutt. var. *amplexicaulis*, *Smilacina amplexicaulis* Nutt. var. *glabra* J.F. Macbr.

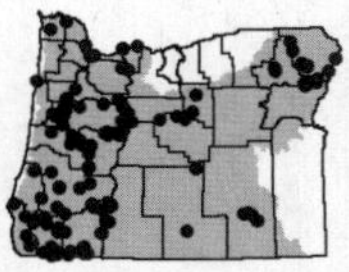

Herbs (15)40–70(120) cm; rhizomes 3–8 mm in diameter. **Leaves** elliptic, lanceolate or sometimes ovate, (4)8–18 × (2)4–8 cm, 5–13 per stalk, clasping the stem. **Inflorescences** paniculate, 40–500-flowered. **Flowers** tepals 6, inconspicuous, 1–2 mm; pedicels 0.5–1.5 mm; stamens 6, flattened, petal-like, protruding past the tepals. **Fruits** globose, 3–7 mm in diameter, speckled with deep red before maturity, turning bright red.

Mesic meadows, creek banks, shaded woods, open rocky summits, exposed hillsides, dry open woods. Flowering Apr–Aug. 0–2700 m. BR, BW, Casc, CR, ECas, Lava, Sisk, WV. CA, ID, NV, WA; northeast to Saskatchewan, southeast to TX. Native.

Maianthemum racemosum ssp. *racemosum* occurs in the eastern United States and Canada. It is principally distinguished from ssp. *amplexicaule* by petiolate leaves that are tapered to the base.

Maianthemum stellatum (Linnaeus) Link

Starry false Solomon's seal, small false Solomon's seal

Smilacina stellata (L.) Desf., *Smilacina liliacea* (Greene) F.L. Wynd, *Smilacina sessilifolia* (Baker) S. Watson

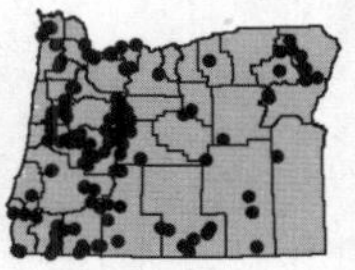

Herbs 20–70(100) cm; rhizomes 1.5–3(5) mm in diameter. **Leaves** (4)7–13 × 2–4 cm, 6–14 per stalk, clasping, lanceolate to elliptic. **Inflorescences** racemose, (2)4–9(13)-flowered. **Flowers** tepals 6, 2–6 mm long; pedicels 3–14(17) mm long; stamens 6. **Fruits** globose, 5–11 mm in diameter, green with dark longitudinal stripes when immature, becoming red.

Creek banks, mesic meadows, woods, swamps. Flowering Apr–Aug. 0–2300 m. Throughout Oregon. CA, ID, NV, WA; throughout Canada and US, south to northwestern Mexico. Native.

MUSCARI Grape hyacinth

Tamra Dickson Prior

Herbs perennial, scapose; bulbs ovoid, brown, without bulblets. **Leaves** basal, 2–7, linear-lanceolate to linear-oblong, glabrous, somewhat succulent. **Scapes** round in cross section. **Inflorescences** terminal, many-flowered racemes; bracts present, small; pedicels descending, slender. **Flowers** fragrant, blue to purple to whitish pink, tubular or urn-shaped, narrower at base; tepals 6, distinct distally, reflexed, nectaries present; stamens 6, anthers dark blue, globose; ovary superior; style 1, short; stigma 3-lobed. **Fruits** capsules, 3-angled, with thin-papery walls. **Seeds** 2 per chamber (6 per fruit), black, round, rough.

Temperate Northern Hemisphere. 30 species. 2 species treated in *Flora*.

The name of this genus comes from the Arabic *muskarimi*, in reference to the flower smelling like musk.

Key to *Muscari* species:

1. Inflorescences 10–20-flowered; scapes equal to or longer than pedicels; pedicels 1–4(5) mm . . . . ***M. botryoides***

1' Inflorescences 40–100-flowered; flowers shorter than pedicels; pedicels 6–25 mm . . . . ***M. comosum***

Muscari botryoides (Linnaeus) Miller (p. 164)

Grape hyacinth, common grape hyacinth

Hyacinthus botryoides L.

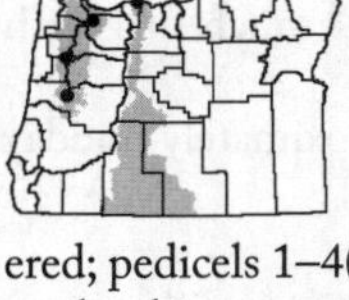

Herbs 5–20(30) cm, erect; bulbs to 2.5 cm. **Leaves** 2–5, flat, linear-oblanceolate, (5)15–35(40) cm × 3–10 mm, ribbed, apex rounded to acute. **Scapes** 20–35 cm, equal to or longer than leaves. **Inflorescences** 10–20-flowered; pedicels 1–4(5) mm. **Flowers** 2–4 mm, light blue to blue; teeth white, acute; sterile flowers lighter in color and smaller; stamens equal to or shorter than corolla. **Fruits** 4–6 mm. **2*n***=18, 36.

Disturbed areas, fields, and roadsides. Flowering Mar–Apr. 0–200 m. ECas, WV. CA, WA; throughout North America; central and southeastern Europe. Exotic.

Muscari comosum (Linnaeus) Miller

Tassel hyacinth, tassel grape hyacinth

Hyacinthus comosus L.

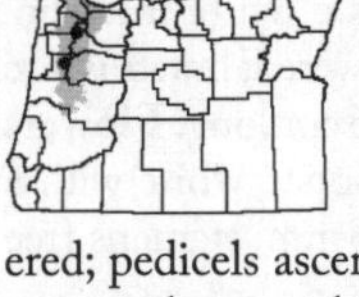

Herbs 5–15(30) cm; bulbs to 3.5 cm. **Leaves** 3–5(7), blade linear-lanceolate, 20–40 cm × 6–20(30) mm, channeled, apex long-attenuate. **Scapes** (15)30–50(80) cm, usually longer than leaves. **Inflorescences** 40–100-flowered; pedicels ascending, those of fertile flowers 4–10(15) mm, more or less equal to the length of the perianth, those of sterile flowers 6–25(40) mm. **Fertile flowers** 5–8 × 2–4 mm, pale to olive-brown, teeth white or lighter brown. **Sterile flowers** 2–6 × 1–2 mm, bright violet. **Fruits** 6–12 × 6–8 mm. **2*n***=18, 27.

Open woods, roadsides, abandoned gardens. Flowering May. 50–200 m. WV. WA; central part of eastern US; northern Africa, Southwest Asia, southern Europe. Exotic.

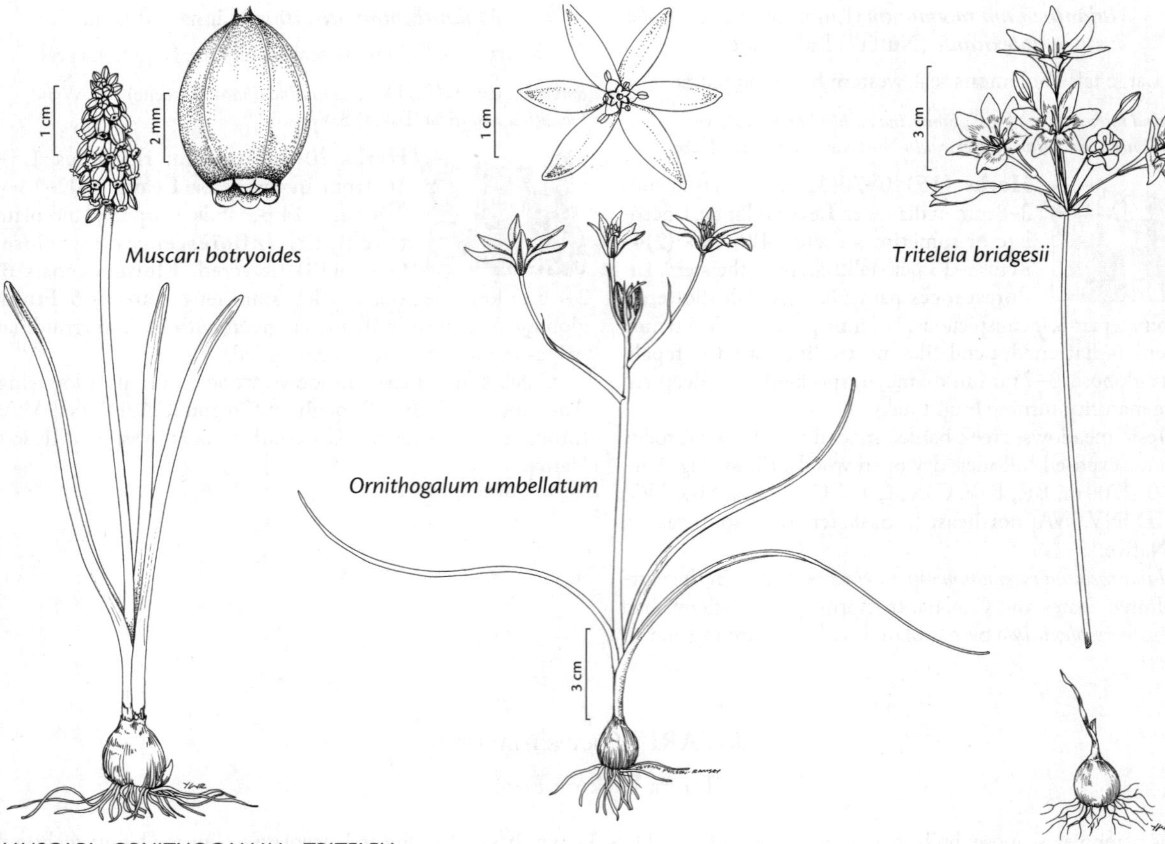

MUSCARI • ORNITHOGALUM • TRITELEIA

ORNITHOGALUM Star of Bethlehem

Stephen C. Meyers

Herbs perennial, from ovoid bulbs, bulblets present. **Leaves** basal, linear. **Inflorescences** corymbs; bracts present. **Flowers** bisexual, erect; tepals 6; stamens 6, distinct, dimorphic; ovary superior, cylindric to round, 6-angled; style 1. **Fruits** capsules, angled, papery, dehiscence loculicidal.

Approximately 150 species; 1 species treated in *Flora*.

Ornithogalum nutans has been reported as escaped in Oregon once, in 1991. It has not persisted nor has it been collected since.

Ornithogalum umbellatum Linnaeus (p. 164)

Star of Bethlehem, nap-at-noon, sleepy dick

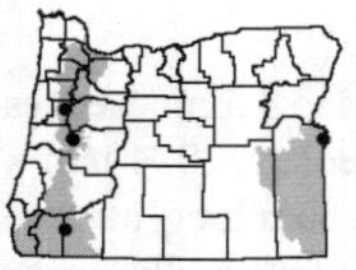

Plants 15–35 cm tall; bulbs papery-scaly. **Leaves** 4–9, 10–30 cm × 2–7 mm; margins entire; usually with a pale white midvein. **Inflorescences** 4–18-flowered; bracts white to brownish white, 1.5–6 cm long. **Flowers** opening with sunlight; tepals spreading, distinct, white with a green stripe, oblong-lanceolate, 15–25 × 5–8 mm; stamens free from perianth, 4–8 mm long; filaments broad and flat, anthers oblong, 2–4 mm; style 3-sided, 3–5 mm, stigma 3-angled. **Fruits** capsules, ovoid, roundish-angular, papery. **Seeds** many, round to ovoid.

Roadsides, disturbed areas. Flowering Apr–May. 100–700 m. Owy, Sisk, WV. CA, ID, WA; throughout much of North America; North Africa, Eurasia. Exotic.

The flowers of this species open at approximately midday and close just prior to sunset.

TRITELEIA Triteleia, triplet-lily

Stephen C. Meyers

Herbs perennial; scapose; corms present. **Leaves** basal, 1–3; blades narrowly lanceolate, entire, glabrous, keeled. **Inflorescences** open umbels, 2–30-flowered; scapes smooth, scabrous or scabrous only at base; bracts lanceolate. **Flowers** tepals 6, fused proximally, often with a darker stripe, lobes ascending to spreading; stamens 6, attached at same or 2 levels on perianth tube, filaments equal in length or of 2 lengths, leaning away from stigma or not; ovary superior, white to green, stalked; style 1, 2–4 mm, stigma 3-lobed. **Fruits** capsules, ovoid. **Seeds** globose or nearly so, ridged, rough.

Western North America. 15 species; 7 species treated in *Flora*.

Triteleia, as a genus, has quite likely been subjected to over-splitting, with local morphological variants described as distinct taxa. Most species are separated by an often confusing combination of flower color and stamen morphology. Here, we conservatively and tentatively present the taxa most often listed as occurring in Oregon.

Triteleia peduncularis has been collected once, in Jackson County. If encountered again, it can be distinguished from other *Triteleia* species by its long ascending pedicels, yellow ovary, and white flowers.

Key to *Triteleia* species:

1. Filaments forked; stamens leaning toward stigma . . . . ***T. ixioides***
1' Filaments not forked (sometimes forked in *T. grandiflora*); stamens leaning away from stigma.
  2. Flowers yellow, sometimes tinged with blue or purple.
    3. Stamens attached at 2 levels; filaments of 2 lengths; flowers bright yellow. . . . . ***T. crocea***
    3' Stamens attached at same level; filaments equal in length or nearly so; flowers pale yellow or bluish to purplish yellow . . . . ***T. hendersonii*** (in part)
  2' Flowers white, pink, purple, or blue.
    4. Filaments of 2 lengths; ovary 2 times the length of stalk . . . . ***T. grandiflora***
    4' Filaments of equal length; ovary 0.25–2 times the length of stalk.
      5. Stamens attached at 2 levels; filaments linear. . . . . ***T. laxa***
      5' Stamens attached at same level; filaments triangular or awl-shaped.
        6. Flowers pink, purple, or blue; tepal lobes shorter than perianth tube . . . . ***T. bridgesii***
        6' Flowers white, sometimes tinged with purple or blue; tepal lobes longer than perianth tube.
          7. Perianth 15–28 mm long; base of perianth tube base tapered. . . . . ***T. hendersonii*** (in part)
          7' Perianth 10–15 mm long; base of perianth tube base bowl-shaped . . . . ***T. hyacinthina***

Triteleia bridgesii (S. Watson) Greene (p. 164)

Bridges' triteleia

Brodiaea bridgesii S. Watson

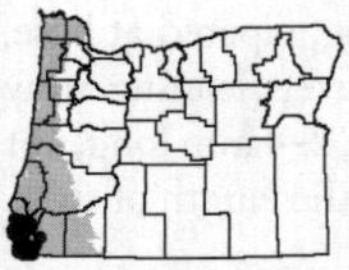

Plants 10–55 cm tall. **Leaves** 20–50 cm × 2–9 mm. **Inflorescences** scapes occasionally scabrous near base. **Flowers** pink, purple, or blue; perianth 25–45 mm long, tube 15–25 mm long, tapered at base, lobes 10–20 mm long; stamens attached at same level, leaning away from stigma; filaments triangular, equal in length, 3–4 mm, not forked; anthers yellow to bluish yellow; ovary 1–1.5 times the length of stalk. **2*n***=16.

Serpentine soils in open areas and coniferous forests. Flowering May–Jul. 0–1400 m. CR, Est, Sisk. CA. Native.

Triteleia bridgesii is nearly identical to *T. laxa*. It differs in having stamens attached at the same level with triangular filaments and by growing in serpentine soils.

Triteleia crocea (Alph. Wood) Greene

Yellow triteleia

Brodiaea crocea (A.W. Wood) S. Watson, *Triteleia crocea* (A.W. Wood) Greene var. *crocea*

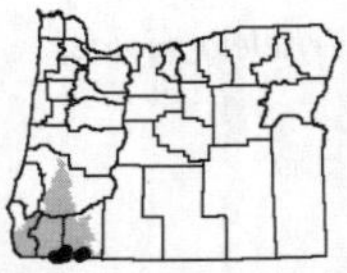

Plants 10–30 cm tall. **Leaves** 5–35 cm × 2–8 mm. **Inflorescences** scapes occasionally scabrous near base. **Flowers** bright yellow; perianth 10–20 mm long, tube 5–10 mm long, tapered at base, lobes 5–10 mm long; stamens attached at 2 levels, leaning away from stigma; filaments linear to slightly widened at base, unequal in length, approximately 1 mm or 3 mm long, not forked; anthers blue; ovary 25–33% the length of stalk. **2*n***=16.

Dry open areas and coniferous forests. Flowering Jun–Jul. 200–2100 m. Sisk. CA. Native.

Triteleia crocea is most easily identifiable by its bright yellow flowers.

Triteleia grandiflora Lindley (p. 166)

Large-flowered triteleia, Howell's triteleia

Brodiaea douglasii S. Watson, *Brodiaea douglasii* S. Watson var. *douglasii*, *Brodiaea douglasii* S. Watson var. *howellii* (S. Watson) M. Peck, *Brodiaea howellii* S. Watson, *Triteleia grandiflora* Lindl. ssp. *grandiflora*, *Triteleia grandiflora* Lindl. ssp. *howellii* (S. Watson) Hoover, *Triteleia howellii* (S. Watson) Greene

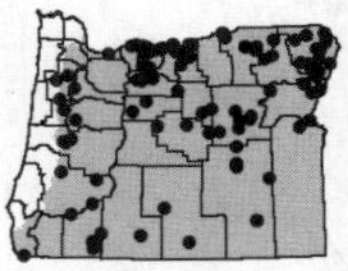

Plants 20–70 cm tall. **Leaves** 15–60 cm × 2–9 mm. **Inflorescences** scapes smooth, not scabrous. **Flowers** white to bluish purple or purple; perianth 15–35 mm long, tube 10–20 mm long, bowl-shaped at base, lobes 10–15 mm long; stamens attached at 2 levels, leaning away from stigma; filaments linear to slightly triangular, unequal in length, 1–2 mm or 3–4 mm long, forked or not; anthers yellow to purple; ovary 2 times the length of the stalk. **2*n***=16, 24, 32, 40, 48, 56.

Dry open areas and forests. Flowering Apr–Jul. 0–3200 m. BR, BW, Casc, Col, ECas, Lava, Owy, Sisk, WV. CA, ID, WA; north to British Columbia, east to WY. Native.

Triteleia hendersonii Greene (p. 166)

Henderson's triteleia

Brodiaea hendersonii S. Watson, *Brodiaea hendersonii* S. Watson var. *galeae* M. Peck, *Brodiaea hendersonii* S. Watson var. *hendersonii*, *Brodiaea leachiae* M. Peck, *Triteleia hendersonii* Greene var. *leachiae* (M. Peck) Hoover

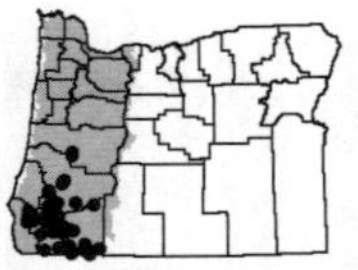

Plants 20–70 cm tall. **Leaves** 15–45 cm × 2–14 mm. **Inflorescences** scapes occasionally scabrous near base. **Flowers** yellow, rarely white; perianth 15–28 mm long, tube 5–10 mm long, tapered at base, lobes 10–18 mm long; stamens attached at same level, leaning away from stigma; filaments awl-shaped, equal in length or nearly so, 3–4 mm, not forked; anthers purple or sometimes white; ovary 50% the length of the stalk. **2*n***=32.

Dry open areas. Flowering Apr–Jul. 0–1800 m. Casc, CR, Sisk, WV. CA. Native.

Plants with white flowers are found in Curry County and are sometimes recognized as a distinct taxon (*T. hendersonii* var. *leachiae*).

Triteleia hyacinthina (Lindley) Greene (p. 166)

Hyacinth triteleia, white triteleia, fool's-onion, hyacinth cluster-lily

Brodiaea dissimulata M. Peck, *Brodiaea hyacinthina* (Lindl.) Baker, *Hesperoscordum hyacinthinum* Lindl.

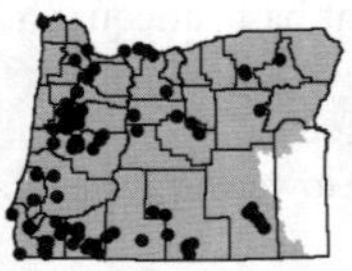

Plants 20–50 cm tall. **Leaves** 10–40 cm × 3–18 mm. **Inflorescences** scapes smooth to scabrous. **Flowers** white, sometimes purple tinged; perianth 10–15 mm long, tube 2–5 mm long, bowl-shaped at base, lobes 6–15 mm long; stamens attached at same level, leaning away from stigma; filaments somewhat triangular, equal in length, 2–4 mm long, not forked; anthers white, yellow, or blue; ovary 2 times the length of stalk. **2*n***=14, 16, 28, 42.

Wet meadows and slopes, rocky seeps. Flowering May–Jul. 0–2500 m. All ecoregions except Owy. CA, ID, NV, WA. Native.

Triteleia ixioides (W.T. Aiton) Greene (p. 166)

Prettyface

Triteleia ixioides (W.T. Aiton) Greene ssp. *anilina* (Greene) L.W. Lenz, *Triteleia ixioides* (W.T. Aiton) Greene ssp. *scabra* (Greene) L.W. Lenz, *Brodiaea scabra* (Greene) Baker var. *anilina* (Greene) Baker, *Calliprora scabra* Greene

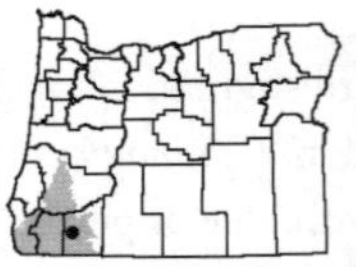

Plants 10–70 cm tall. **Leaves** 10–45 cm × 3–12 mm. **Inflorescences** scapes sometimes scabrous near base. **Flowers** pale yellow, golden yellow, sometimes tinged with blue or purple; perianth 10–30 mm long, tube 4–10 mm long, tapered at base, lobes 5–20 mm long; stamens attached at same level, leaning toward stigma; filaments flattened, unequal in length, 1–2 mm or 3–4 mm long, forked; anthers white, yellow, or blue; ovary 1–1.5 times the length of stalk. **2*n***=10, 16, 24, 32, 40, 42, 50.

Open areas and forests. Flowering Apr–Jul. 400–600 m. Sisk. CA. Native.

As many as five subspecies of *T. ixioides* are recognized by some botanists. These mainly localized variants are separated by slight differences in perianth color and length, or sometimes by anther color.

Triteleia laxa Bentham

Triplet lily, Ithuriel's spear

Brodiaea laxa (Benth.) S. Watson

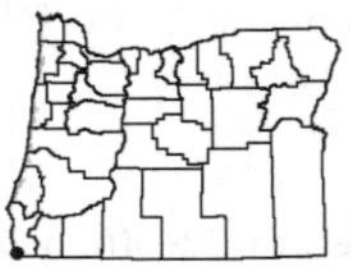

Plants 10–60 cm tall. **Leaves** 15–40 cm × 3–20 mm. **Inflorescences** scapes sometimes scabrous near base. **Flowers** blue, sometimes white or bluish purple; perianth 15–35 mm long, tube 10–55 mm long, tapered at base, lobes 8–20 mm long; stamens attached at 2 levels, leaning away from stigma; filaments linear, equal in length, 2–4 mm long, not forked; anthers blue to purple; ovary 2 times the length of stalk. **2*n***=14, 16, 28, 42.

Open areas and forests. Flowering Apr–Jul. 0–900 m. Est. CA. Native.

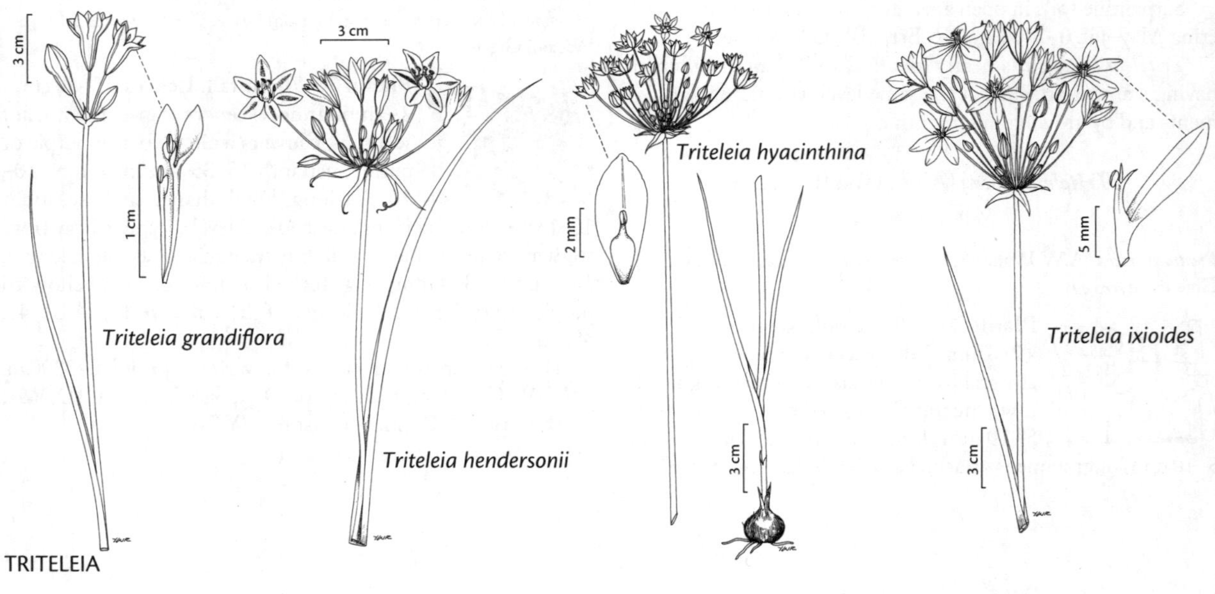

TRITELEIA

CYPERACEAE Sedge family

Barbara L. Wilson, Richard E. Brainerd, and Nick Otting

Herbs annual or perennial; cespitose or not, rhizomatous or not, rarely stoloniferous. **Culms** usually triangular, occasionally terete, rarely compressed or quadrangular, solid or pithy. **Leaves** basal and/or cauline, alternate, usually 3-ranked, bases forming cylindric sheaths enclosing stem, margins fused; ligules often present; blades frequently absent from some basal leaves, rarely from cauline leaves, when present flat, folded, rolled, or terete, linear. **Spikes** with a shortened axis; floral scales 1–many, crowded, spirally arranged, sometimes 2-ranked, usually each subtending a single flower or perigynium, sometimes proximal and/or distal scales empty; perigynia if present closed or open down one side, subtending and enclosing a very short rachilla-bearing 1-pistillate flower (*Carex*), sometimes also (0)3 staminate flowers and empty scales (*Kobresia*); spikes aggregated into secondary inflorescences that are panicles, often modified and sometimes consisting of a single spike; secondary inflorescences usually subtended by a foliaceous or scale-like or bristle-like bract. **Flowers** bisexual (unisexual in *Carex* and *Kobresia*); perianth absent or consisting of (1)3–6(30) bristles and/or scales; stamens usually (1)3, rarely more; ovary with 1–3(4) carpels, fused, locule 1; style undivided or branched, stigmas 2–3(4). **Fruits** achenes, usually trigonous or biconvex.

Cosmopolitan. Approximately 100 genera; 16 genera treated in *Flora*.

Cyperaceae (except *Eleocharis*) is characterized by unusual holocentric chromosomes, in which spindle fibers attach throughout the length of the chromosome, not at discrete centromeres. Also, reduction of chromosome number occurs at the second division of meiosis. These traits are shared with the related family Juncaceae but not by other vascular plants. Chromosome numbers in sedges are often very high, but it is unclear if this represents polyploidy or fragmentation of holocentric chromosomes.

When collecting sedges, get the rhizomes (if present) because some species are distinguished by their cespitose or rhizomatous growth form. If possible, collect extra inflorescences because checking floral traits can be destructive. In some genera, species identification requires mature fruits.

Key to *Cyperaceae* genera:

1. Ovary or achene enclosed in a perigynium as well as subtended by a scale; flowers with either pistil or stamens, not both (plants monoecious).
   2. Perigynium open along one side; habitat alpine . . . . **KOBRESIA**
   2' Perigynium closed (margins fused); habitat various. . . . . **CAREX**

1' Ovary or achene not enclosed in a perigynium, though usually subtended by a scale; flowers usually with both pistil and stamens (perfect).
   3. Ovaries and fruits subtended by 10 or more white to light rusty, smooth, more or less straight bristles usually 2 cm or longer, much exceeding the floral scales that subtend the flowers. . . . . **ERIOPHORUM**
   3' Ovaries and fruits subtended by 0–6(8) scales or bristles of various colors, often scabrous, straight or curly, less than 2 cm long, usually shorter than the floral scales or if much longer, then curly.
      4. Floral scales 2-ranked.
         5. Culms hollow, appearing jointed; inflorescences axillary; perianth bristles present . . . . ***Dulichium arundinaceum*** var. ***arundinaceum***
         5' Culms solid, not appearing jointed; inflorescences terminal; perianth bristles absent . . . . . . **CYPERUS**
      4' Floral scales spirally arranged.
         6. Plants apparently lacking leaf blades; leaves consisting only of sheaths or sometimes with reduced blades much shorter than the sheaths.
            7. Inflorescence consisting of one terminal spike; fruits with a tubercle; inflorescence bract(s), if present, much shorter than spike and resembling floral scales . . . . . . **ELEOCHARIS**
            7' Inflorescence consisting of (1)many spikes; fruits without a tubercle; inflorescence bract(s) present, shorter or longer than spike, not resembling floral scales.
               8. Plants cespitose; culms to 1 mm diameter; spike 1; inflorescences terminal; central and northern Cascades. . . . . **TRICHOPHORUM** (in part)
               8' Plants rhizomatous; the culms sometimes clustered, culms (1)2–10 mm diameter; spikes 1–150+; inflorescences terminal or appearing lateral; widespread . . . . **SCHOENOPLECTUS** (in part)
         6' Plants with leaf blades, these flat, involute, or filiform, sometimes resembling sterile culms.
            9. Leaf blades 0–8 mm long × 0–0.4 mm wide, much shorter than sheath; inflorescence terminal, with 1 spike; inflorescence bracts to 3 mm longer than the spike, blunt; range central and northern Cascades.
               10. Floral scales short-pointed to acute . . . . **TRICHOPHORUM** (in part)
               10' Floral scales long-acuminate to awned . . . . **LIPOCARPHA** (in part)
            9' Leaf blades 2.5–750 mm long × 0.2–22 mm wide, shorter or longer than sheath; inflorescence terminal or apparently lateral, with 1–many spikes; inflorescence bracts various; range widespread.

11. Achenes with a prominent conical or triangular-flattened tubercle 25–100% as long as the achene body . . . . **RHYNCHOSPORA**
11' Achenes lacking a tubercle or tubercle very small.
12. Ovaries and fruits subtended by 1–6 hair-like (rarely strap-like), usually scabrous bristles in addition to the strap-like, smooth filaments beneath each flower.
13. Inflorescence a dense, light brown head about as long as wide, with 5–30+ spikes, spikes apparently sessile; inflorescence bract(s) inconspicuous; southwestern Oregon . . . . **CALLISCIRPUS**
13' Inflorescence differing from the above in some way; range various.
14. Bracts and leaf blades flat (folded if wilted), with prominent midrib forming keel; inflorescences obviously terminal; inflorescence bracts conspicuous, usually leafy.
15. Scales glabrous, never both notched and awned at tip; spikes less than 3.5 mm diameter; achenes 0.5–1.8 mm. . . . . **SCIRPUS** (in part)
15' Scales puberulent on abaxial surface, often glabrescent, notched and awned at tip; spikes mostly 4–12 mm diameter; achenes 2.3–5.5 mm . . . . **BOLBOSCHOENUS**
14' Bracts absent, leaf blades tubular, triangular, V-shaped, or curved to flat in cross section but not prominently keeled; inflorescences terminal or appearing lateral because proximal bract erect and looking like a continuation of the culm.
16. Spikes, or some of them, on obvious pedicels . . . . **SCHOENOPLECTUS** (in part)
16' Spikes all sessile.
17. Plants wiry, with solid culms 0.5–2 mm wide; ligules ciliate; proximal leaf sheaths often disintegrating into fibers . . . . **AMPHISCIRPUS**
17' Plants not wiry, softer, with air spaces, (0.5)1–10 mm wide; ligules not ciliate; proximal leaf sheaths not disintegrating into fibers. . . . . **SCHOENOPLECTUS** (in part)
12' Ovaries and fruits not subtended by bristles, although by strap-like, smooth filaments.
18. Involucral bracts 1–3, the main bract erect or nearly so, stem-like, the inflorescence thus appearing lateral (other bracts, if present, shorter, spreading to reflexed).
19. Stigmas 3; fruits 3-angled . . . . **ISOLEPIS**
19' Stigmas 2; fruits biconvex. . . . . **LIPOCARPHA** (in part)
18' Involucral bracts subtending the inflorescence (1)2–several, variously oriented but more or less leaf-like, the inflorescence thus appearing terminal.
20. Achenes with small tubercles; leaf blades to 0.5 mm wide.. . . . . **BULBOSTYLIS**
20' Achenes lacking tubercles; leaf blades 6-13 mm wide . . . . **SCIRPUS** (in part)

AMPHISCIRPUS Clubrush

Herbs perennial; rhizomatous. **Culms** more or less terete, tough, wiry, internally mostly solid, without evident air spaces, cespitose or not, rhizomatous. **Leaves** all basal; sheaths often disintegrating into fibers; ligules ciliate; blades strongly C-shaped in cross section to subcylindric, tough, wiry. **Inflorescences** terminal, often pseudolateral; spikes 1–6(10), sessile; involucral bracts 1–3, spreading or erect, leaf-like. **Spikes** 5–20 × 3–5 mm; scales 30–60, spirally arranged, each subtending a flower, glabrous. **Flowers** bisexual; perianth of 1–6 bristles, straight, not longer than achene, retrorsely spinulose; stamens 3; styles deciduous, stigmas 2. **Achenes** plano-convex or unequally biconvex.

North and South America. 1 species.

The genus *Amphiscirpus* has been segregated from *Scirpus* mainly on the basis of its solid culms that lack air spaces.

Amphiscirpus nevadensis (S. Watson) Oteng-Yeboah (p. 169)

Nevada clubrush

Scirpus nevadensis S. Watson

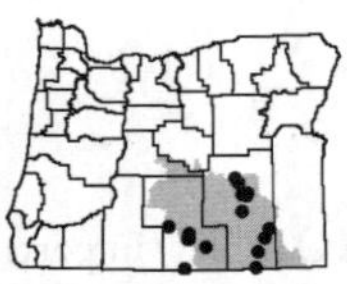

Rhizomes 1–4 mm diameter, tough, hard. **Culms** 10–70 cm × 0.5–2 mm. **Leaves** 5–10; sheaths loose; blades 0.5–1 times as long as culms, hard, without evident internal air spaces; distal blades 3–30 cm × 0.5–2 mm, longer than sheaths. **Inflorescences** proximal; involucral bract 1–15 cm, resembling foliage leaf blade. **Spikes** ovoid to lanceoloid or terete, 5–20 × 3–5 mm; scales pale to dark red-brown, midribs usually stramineous, smooth; proximal 1 or 2 scales often resembling involucral bracts, with awn-like blades, to 15 mm; other scales in proximal part of spike prominently 9-veined, ovate, 4 × 3 mm, apex entire, acute to rounded. **Flowers** perianth bristles unequal. **Achenes** broadly obovoid, 2–2.3 × 1.5–1.7 mm, not beaked.

Alkaline, seasonal wetlands. 1000–1400 m. BR. CA, NV, WA; north to British Columbia and Saskatchewan, east to ND and NE. Native.

Amphiscirpus nevadensis resembles and sometimes grows with similar *Schoenoplectus pungens*, which has distinctly trigonous culms and leaf blades with spongy aerenchyma, glabrous ligules, notched, awned floral scales, and beaked achenes.

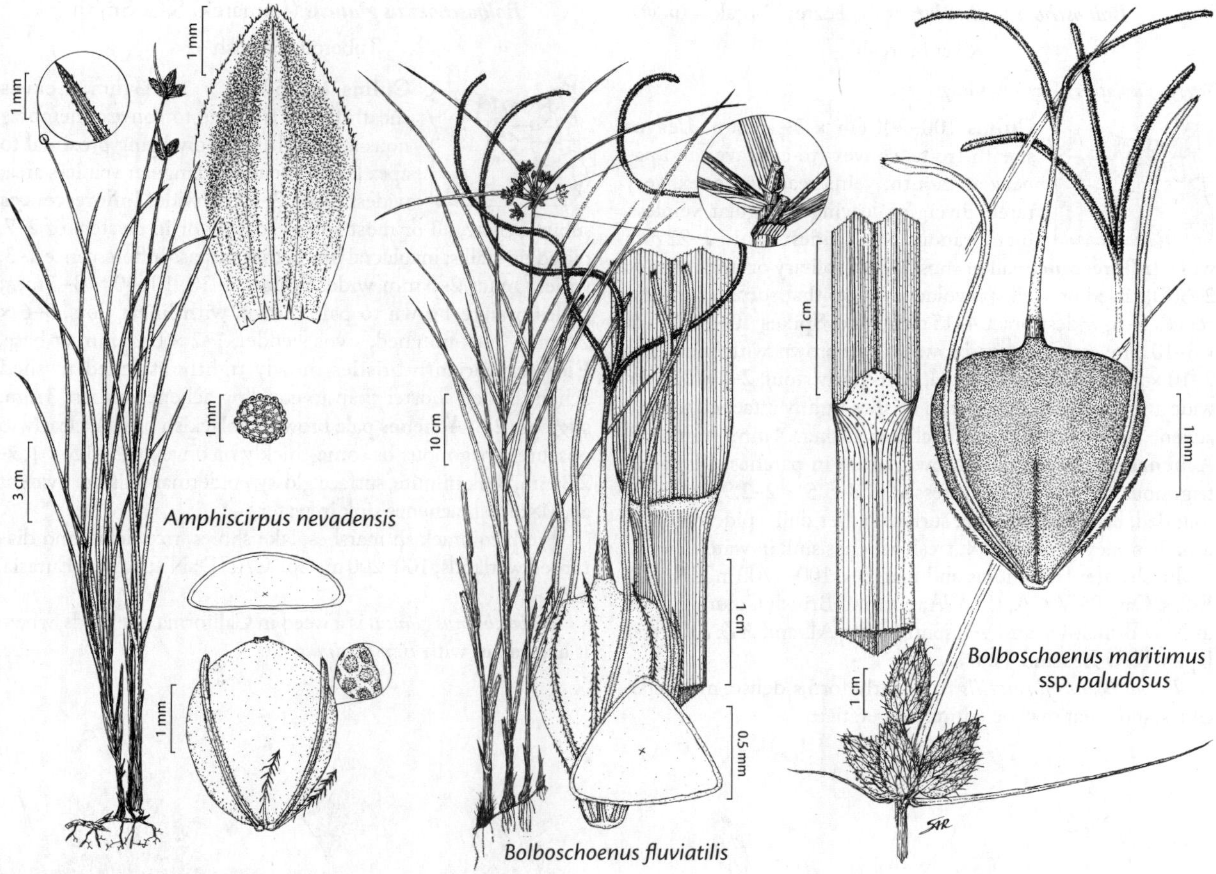

AMPHISCIRPUS • BOLBOSCHOENUS

BOLBOSCHOENUS Tuberous bulrush

Herbs perennial; cespitose or not, rhizomatous. **Culms** cormose at base, sharply trigonous. **Leaves** basal and cauline; sheaths tubular; ligules absent; blades flat, prominently keeled on dorsal surface. **Inflorescences** terminal, with spikes sessile or raised on stalks; spikes 2–40, mostly 4–10 mm wide; involucral bracts surpassing inflorescence, 1–5, spreading or proximal erect, leaf-like. **Spikes** ovoid to lanceolate or cylindric; floral scales 25 or more, spirally arranged, each subtending a flower, dorsally puberulent, often glabrescent, orange-brown to stramineous, often with linear spots, membranous and translucent, apex notched 0.5–1 mm deep, awned. **Flowers** bisexual; perianth of 3–6 bristles; stamens 3, anthers yellow, rarely orange; styles with persistent base, stigmas 2–3. **Achenes** biconvex to trigonous, usually obovate or obpyriform, smooth.

Cosmopolitan. 6–15 species; 3 species treated in *Flora*.

Bolboschoenus is one of the recent segregates from *Scirpus*. The genus has terminal inflorescences, leafy bracts, flat leaves with prominent keels, and hairy floral scales on big, shaggy-looking spikes. *Bolboschoenus* taxonomy is unsettled, particularly in the *B. maritimus* and *B. fluviatilis* complexes. Both *B. fluviatilis* and *B. glaucus* are known to hybridize with *B. maritimus*, but hybrids are rare. *Bolboschoenus* species provide valuable food for wildlife. Some species are significant weeds of rice fields in California.

Measurements of floral scales should exclude awns and be taken without flattening the curved scale. Awn widths can be difficult to measure because the scales may tear, making the awn base appear lower than it really is. Achene lengths include the beak and the persistent remnant of the style base (if present). Culm widths are measured in the middle third of the culm. Inflorescence descriptions are for expanded, flowering or fruiting inflorescences.

Key to *Bolboschoenus* species:

1. Inflorescences with all spikes sessile or not more than 50% on evident branches; perianth bristles usually not persistent on achene; stigmas 2(3); achenes biconvex (or compressed-trigonous); habitats brackish to saline. . . . . **B. maritimus** ssp. ***paludosus***

1' Inflorescences with all or most spikes on evident, elongated branches; perianth bristles usually persistent on achene; stigmas (2)3; achenes equilaterally trigonous to compressed-trigonous (rarely biconvex); habitats fresh to slightly brackish.

2. Widest leaf blade 2–6 mm wide; spikes 3–5 mm wide; achenes 2.5–3.3 mm, beaks minute. . . . . ***B. glaucus***

2' Widest leaf blade 7–22 mm wide; spikes 6–10 mm wide; achenes 3.8–5.5 mm, beaks distinct . . . . ***B. fluviatilis***

Bolboschoenus fluviatilis (J. Torrey) Soják (p. 169)

River bulrush

Scirpus fluviatilis (Torr.) A. Gray

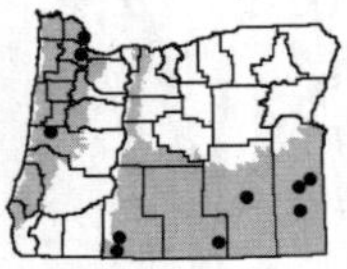

Culms 100–200 cm × 5–15 mm. **Leaves** sheath fronts convex (to concave) and papery at mouth; veins reaching apex, very rarely diverging leaving triangular, veinless, membranous area; widest blade 7–22 mm wide. **Inflorescences** all or most spikes solitary or in clusters of 2–3(8), raised on stalks; involucral bracts that surpass inflorescences 3–6, widest bract 4–15 mm wide. **Spikes** 10–40, 10–25 × 6–10 mm; scales orange-brown to pale brown, with linear dots 7–10 × 3–4 mm; apex notched, awns fairly stout, 2–3 × 0.5 mm wide at base. **Flowers** perianth bristles tightly attached to shed achene, very stout, equaling achene; anthers 4 mm; stigmas 3. **Achenes** grayish or dark brown, often in patches, all clearly trigonous to slightly compressed, 3.8–5.5 × 2–2.9 mm, apex rounded, beak 0.2–0.8 mm, surface rather dull, epidermal cells usually not evident at 20×; fresh achenes sink in water. ***2n***=94.

Freshwater lake shores and marshes. 100–1700 m. BR, CR, ECas, Owy, WV. CA, ID, WA; north to British Columbia, east to New Brunswick and VA, southeast to AL and AZ; Australia, Japan, New Zealand. Native.

Bolboschoenus fluviatilis frequently forms dense, monospecific stands that may be entirely vegetative.

Bolboschoenus glaucus (Lamarck) S.G. Smith

Tuberous bulrush

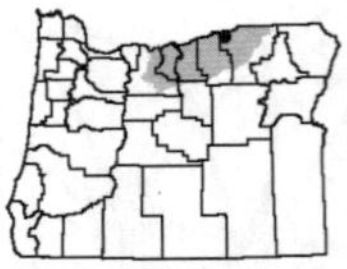

Culms to 150 cm × 1.5–3 mm. **Leaves** sheath fronts truncate to convex, membranous at mouth, veins diverging proximal to apex leaving acutely triangular veinless area; widest blade 2–6 mm wide. **Inflorescences** subumbellate, all or most spikes solitary or in clusters of 2–7, raised on stalks; involucral bracts that surpass inflorescence 1–3, widest bract 2–3 mm wide. **Spikes** 3–30, 10–40 × 3–5 mm; scales orange-brown to pale brown, with linear dots, 5–6 × 2.5 mm; apex notched, awns slender, 1–2 × 0.25 mm at base. **Flowers** perianth bristles mostly tightly attached to shed achene, much shorter than to equaling achene; anthers 3 mm; stigmas (2)3. **Achenes** pale brown to blackish brown, clearly to obscurely trigonous or some thickly biconvex, 2.5–3.3 × 1.3–2.3 mm, beak minute, surface glossy, epidermal cells not evident at 20×; fresh achenes sink in water.

Fresh to brackish marshes, lake shores, rice fields, and disturbed wetlands; 100–200 m. Col. CA, ID; NY; Africa, Eurasia. Exotic.

Bolboschoenus glaucus is a weed in California rice fields, where it hybridizes with *B. maritimus*.

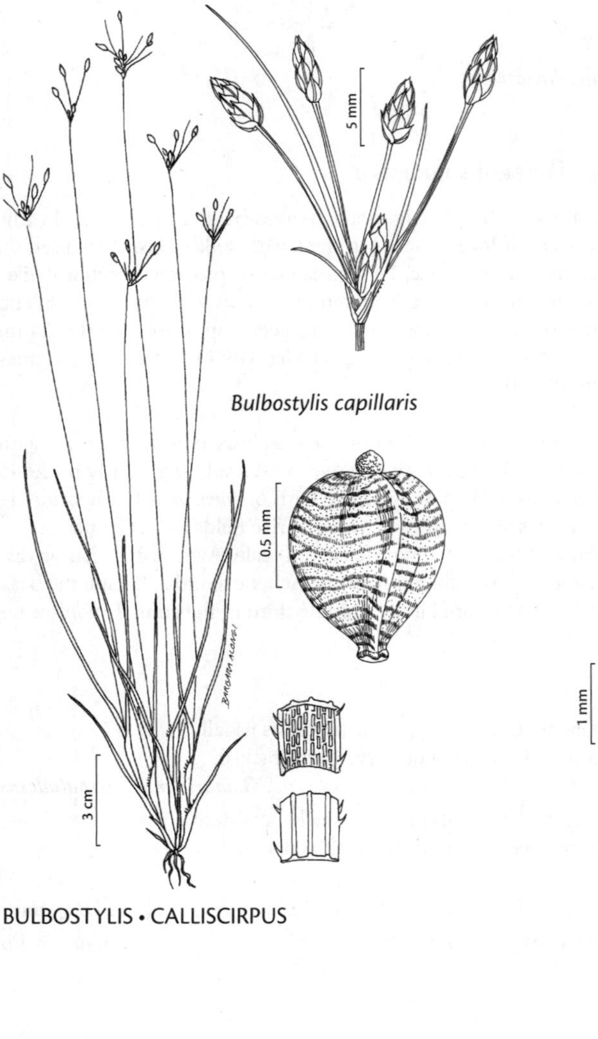

Bulbostylis capillaris

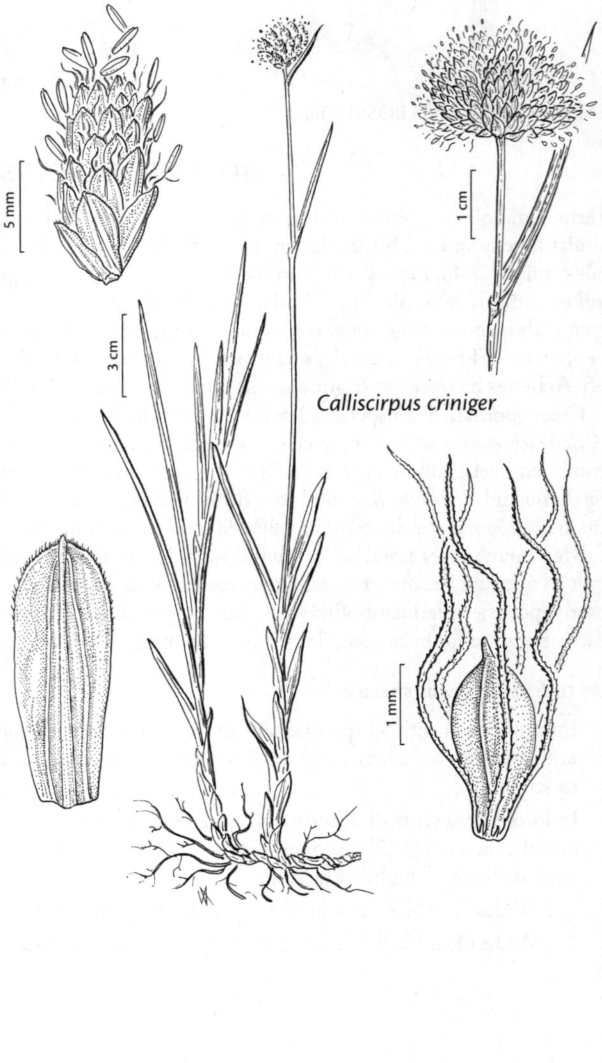

Calliscirpus criniger

BULBOSTYLIS • CALLISCIRPUS

Bolboschoenus maritimus (Linnaeus) Palla ssp. ***paludosus*** (A. Nelson) T. Koyama (p. 169)

Seacoast bulrush

Scirpus maritimus L., *Scirpus maritimus* L. var. *paludosus* (A. Nelson) Kük., *Scirpus pacificus* Britton, *Scirpus paludosus* A. Nelson

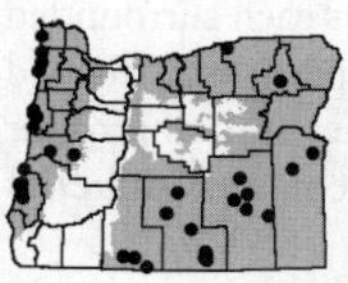

Culms 50–150 cm × (1)3–8 mm. **Leaves** sheath fronts usually membranous at mouth, apex concave to convex, veins diverging proximal to summit leaving triangular, veinless area (often disintegrating); widest blade 2–12 mm wide. **Inflorescences** with all spikes sessile, simply branched with not more than 50% of spikes solitary or in clusters of 2–10 raised on stalks; involucral bracts that surpass inflorescences 1–4, widest bract 1–6 mm wide. **Spikes** (1)2–40, 7–40 × (4)7–10 mm; scales dark brown at the coast, pale brown inland; apex 5–8 × 3–4 mm, acute, notched, awn slender, 1–3 × 0.25 mm at base. **Flowers** perianth bristles not persistent on shed achenes or sometimes 1–few weakly persistent, to 50% achene length; anthers 2–4 mm; stigmas 2(3). **Achenes** usually medium to dark brown, often pale brown east of Cascades, biconvex, 2.3–4.1 × 1.9–2.8 mm, beak 0.1–0.4 mm, surface glossy, epidermal cells usually clearly visible at 10–20×; fresh achenes float on water.

Saline, brackish, and alkaline marshes and shorelines, sometimes in freshwater. 0–1700 m. BR, BW, Col, CR, ECas, Est, Owy, WV. CA, ID, NV, WA; north to AK, east to Prince Edward Island and DE; HI, South America. Native and exotic.

Bolboschoenus maritimus is part of a taxonomically confusing complex. Some North American populations are native, but others may have been introduced from Eurasia. Putative hybrids with *B. fluviatilis* and *B. glaucus* have been found in California.

BULBOSTYLIS Hairsedge

Herbs annual; cespitose, not rhizomatous. **Culms** usually terete, ribbed. **Leaves** basal; sheaths open apically, apex fimbriate-ciliate, rarely entire; ligules absent, lateral tufts of hair at junction of blade and sheath; blades mostly linear or filiform, usually longer than sheaths, 1 mm wide or less, strongly involute, margins scabrid-ciliate. **Inflorescences** terminal on slender scapes; spikes 1–50+; involucral bracts (1)2–8, spreading or erect, leaf-like or scale-like. **Spikes** ovoid to lanceoloid; scales 6–50, spirally arranged, rarely nearly 2-ranked, each subtending a flower or 1–2 proximal scales empty. **Flowers** bisexual; perianth bristles absent; stamens 1(2); styles with enlarged, persistent base, stigmas (2)3. **Achenes** trigonous, obovoid, usually 3-ribbed.

Cosmopolitan in warm-temperate and tropical regions. Approximately 100 species; 1 species treated in *Flora*.

Bulbostylis capillaris (Linnaeus) Kunth ex C.B. Clarke (p. 170)

Densetuft hairsedge

Stenophyllus capillaris (L.) Britton

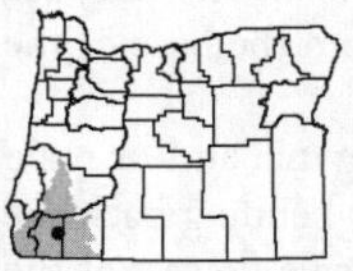

Culms to 30 cm, filiform (rarely with several spikes sessile or subsessile at plant base), prominently ribbed, glabrous. **Leaves** 25–33% length of scapes; sheath borders tan, backs prominently ribbed, glabrous; blades filiform, 0.5 mm wide, surface glabrous. **Inflorescences** solitary or more commonly in simple, open, rarely compact, involucrate groups; proximal involucral bract short-awned or with bristle-like blade, exceeding or exceeded by inflorescence. **Spikes** red-brown, 3–5 mm; fertile scales ovate, 1.5–2 mm; apex acute, glabrous or distally puberulent, short-awned. **Flowers** anthers 0.5–0.7 mm. **Achenes** yellowish to pale brown, 1 mm, faces rugose. **2*n***=84.

Sandy, often disturbed soil. 50–300 m. Sisk. ID, WA; north to AK, east to Greenland and SC, southeast to UT, South America; Eurasia. Native.

CALLISCIRPUS Calliscirpus

Herbs perennial; cespitose, short-rhizomatous. **Culms** trigonous, (10)20–110 cm × 0.6–3 mm. **Leaves** basal and cauline, flat, 5–45 × 1–6 mm, distal leaf longer than sheath; ligule ciliate. **Inflorescences** dense ovoid or globose heads with 1+ usually inconspicuous branches; inflorescence bracts 2–5, scale-like, rarely green and large. **Spikes** 5–30+, oblong-lanceoloid; floral scales brown with pale green or brown markings, sometimes red-spotted, ovate, apex acute. **Flowers** bisexual; perianth bristles 6(12), 3–10 mm long, dull white, antrorsely barbed; stamens 3, anthers 1.2–2 mm long; style deciduous, stigmas 3. **Achenes** 1.3–3 × 0.7–1.1 mm, oblong-obovoid.

Western North America and eastern Asia. 3 species; 1 species treated in *Flora*.

Calliscirpus criniger (A. Gray) C.N. Gilmour, J.R. Starr & Naczi (p. 170)

Fringed cottongrass

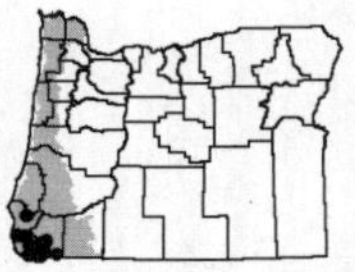

Plants cespitose. **Culms** (10)20–100 cm × 0.6–2 mm. **Leaves** blades flat, to 45 cm × 1–6 mm. **Inflorescences** involucral bracts 2–5, sheathless, scale-like, 0.4–0.8(2.2) cm, apex sometimes mucronate. **Spikes** usually in dense head, subsessile, 3–10 mm long; scales brown, with pale brown or green markings, often red-spotted, 2.5–4 mm. **Flowers** perianth bristles usually 6, 3–9 mm, pale brown, usually antrorsely barbed; anthers 1.2–2 mm. **Achenes** 1.5–3 mm.

Bogs and seepy slopes, usually on serpentine. 300–1900 m. CR, Sisk. CA. Native.

The phylogenetic relationships involving *C. criniger* have been unclear (Ball and Wujek 2002). The species has one close relative in the Sierra Nevada of California and another in northeastern Asia. All three have been placed in both *Scirpus* and *Eriophorum* but do not fit well in either genus. A recent study recommended placing these species in their own genus (Gilmour et al. 2013).

CAREX Sedge

Plants perennial, herbaceous; cespitose or rhizomatous, rarely stoloniferous; culms usually trigonous, sometimes round. **Leaves** sheath fronts closed; ligules membranous; blades linear, usually flat, V-shaped, or M-shaped in cross section, sometimes involute or filiform. **Inflorescences** terminal, consisting of spikes each with few to many perigynia, the spikes solitary or arranged in heads, racemes, or panicles; each spike usually subtended by an inflorescence bract which is leaf-like, bristle-like, or scale-like. **Flowers** unisexual, lacking a perianth; male flowers with 1–3 stamens, each subtended by a staminate scale; female flowers each surrounded by the perigynium, open only at the apex, each perigynium subtended by a pistillate scale; styles deciduous or persistent, divided into 2–3(4) parts each with a stigma; perigynia sometimes stipitate, with a body (wide part around the achene) and a narrower beak (around the style), usually with 2 projections (teeth) on each side of the opening through which the style emerges. **Achenes** biconvex, plano-convex, or trigonous, rarely 4-angled.

Cosmopolitan. Approximately 2000 species; 139 species treated in *Flora*.

Though most diverse in wetlands, *Carex* also occur in dry habitats such as upland forests, alpine ridges, and sagebrush steppes. *Carex* species are important forage plants and excellent soil binders. Foliage and split rhizomes are used for basketry and to make other household items.

The characteristic trait of *Carex* is the perigynium, a modified bract that surrounds the female flower. In *Carex*, this structure is sealed except for a small opening through which the style emerges. Mature perigynia are necessary for identification of most *Carex* species. Inflorescences on late season shoots can have abnormal shapes, bracts, and perigynia. They are often not identifiable.

Many different hybrid combinations have been reported in *Carex*. Most hybrids are completely sterile, with indehiscent anthers and malformed achenes. Some, however, may be partially fertile.

Two species not included in the keys have been documented but apparently are not established in Oregon:

Carex pumila (dwarf sedge) was collected on disturbed, sandy soils in Portland 1906–1916 but has not been documented since. It is strongly rhizomatous, with culms 8–30 cm long and often buried in sand up to the inflorescence. This species is native to Europe.

Carex comans (New Zealand hair sedge) produces numerous seedlings near ornamental plantings and is expected to initiate reproducing populations in the wild. It is densely cespitose, forming a mop of drooping leaves 0.5–1(1.5) mm wide that can be green, gray-green, blue-green, bronze, or yellowish and are often variegated. This species is native to New Zealand.

Carex morphology

Fertile shoots that have leaf blades at the base are phyllopodic. Those that have almost bladeless sheaths at the base are aphyllopodic. The front of the leaf sheath is the side away from the blade. A hyaline sheath front is more or less transparent when fresh but usually whitish when dry. A cross-corrugated sheath front has little horizontal ridges. The sheath front is ladder fibrillose if it disintegrates into a network of fibers with a vertical central fiber and side fibers extending laterally. The mouth is the top of sheath front.

The inflorescence bracts are the bracts at the base of inflorescence spikes. They are often leaf-like, but may be bristle-like, i.e., reduced to little more than a midrib, which does not imply hard or prickly. The inflorescence is measured from the base of the sheath of the lowest inflorescence bract to the tip of the terminal spike.

The spike is the basic inflorescence unit of *Carex*. It is an unbranched section of stem with attached perigynia and/or male flowers. A small bract called a scale subtends each flower. After anthers fall off, male flowers can be found by bending back the scale to observe the persistent filaments. A spike may be staminate (with only male flowers), androgynous (male flowers at the tip and with female flowers at the base), gynaecandrous (with female flowers at the tip and male flowers at the base), or pistillate (with only female flowers).

The perigynium is a modified bract that surrounds the female flower. The perigynium is divided into three main parts. The stipe (sometimes absent) is a short, basal, stalk-like part. The body is the main part around the achene. The beak is a narrowed distal part, often tipped by two teeth. If the perigynium tapers relatively abruptly to the beak, it appears to have shoulders. If it is widest at the base and tapers so gradually you can't tell where the beak begins, it is lance-triangular. Soft, air-filled pithy tissue makes perigynia look swollen and can be identified by sticking a pin easily into it. Wings are expanded, flattened edges of the perigynium body and beak. Two-sided (lenticular) perigynia may be biconvex (rounded on both sides), plano-convex (flat on one side and rounded on the other), or flat except over the achene. The perigynium has two ribs, thick, fibrous, lengthwise vascular bundles. Other vascular bundles, called veins, are usually narrower. Veins are usually raised or rarely may appear sunken into the perigynium tissue (impressed).

Carex have two or three (rarely four) stigmas, producing, respectively, two-sided (lenticular), three-sided (trigonous), or rarely rounded-quadrangular achenes; styles are usually deciduous after flowering. Young deciduous styles can be recognized by an abrupt change in color at the junction of ovary and style. Persistent styles remain attached to the ovary and usually become hard. Compressed between the growing achene and the beak tip, they usually become contorted. Young persistent styles can usually be recognized by the uniform or gradually changing color of the style/achene junction.

Key to *Carex* subkeys:

1. Spike 1 per culm . . . . **Subkey 1**
1' Spikes 2 or more per culm.
  2. All spikes entirely male; no perigynia present . . . . **Subkey 2**
  2' Some or all spikes with female flowers; perigynia present.
    3. Perigynia hairy . . . . **Subkey 3**
    3' Perigynia glabrous (may have cilia on beak margins).
      4. Stigmas 3; achenes 3-sided in cross section.
        5. Styles deciduous; perigynia usually tight against the achenes . . . . **Subkey 4**
        5' Styles persistent; perigynia usually inflated . . . . **Subkey 5**
      4' Stigmas 2; achenes 2-sided in cross section.
        6. Some spikes entirely male; spikes often greater than 1.5 cm long.
          7. Foliage normally orange-brown; female scales white hyaline throughout except midrib; plants of disturbed places, rarely escaping cultivation west of Cascades . . . . **C. buchananii**
          7' Foliage normally green or glaucous; female scales not hyaline throughout; habitat various.
            8. Perigynia winged; habitat disturbed sandy sites in the Portland area . . . . ***C. arenaria*** (in part)
            8' Perigynia not winged; habitat various . . . . **Subkey 6**
        6' Every spike with at least some female flowers; spikes usually less than 1.5 cm long.
          9. Perigynia not winged (may be flat-edged on beak).
            10. Spikes androgynous, female, or with male and female flowers mixed . . . . **Subkey 7**
            10' Spikes gynaecandrous, or occasionally with terminal spike gynaecandrous and lateral spikes female . . . . **Subkey 8**
          9' Perigynia winged (flat-edged) at least on distal part of body.
            11. Plants rhizomatous; spikes staminate, androgynous, or pistillate . . . . ***C. arenaria*** (in part)
            11' Plants cespitose; spikes gynaecandrous or pistillate . . . . **Subkey 9**

Subkey 1: Inflorescence consisting of a single spike.

1. Spike entirely male; no perigynia present . . . . **Subkey 2**
1' Spike entirely or partially female; perigynia present.
  2. Perigynia pubescent or puberulent, at least at base of beak . . . . **Subkey 3**
  2' Perigynia glabrous (sometimes minutely papillose; sometimes serrulate on the edges of the beak).
    3. Stigmas 2; achenes lenticular.
      4. Plants densely cespitose.
        5. Perigynia ovate to orbicular; sessile . . . . ***C. capitata***
        5' Perigynia elliptical; stipitate.
          6. Leaf bases from previous year not persistent; perigynia spreading to reflexed at maturity, with margins and beaks smooth . . . . ***C. micropoda*** (in part)
          6' Leaf bases from previous year persistent; perigynia ascending at maturity, with margins and beaks more or less serrulate . . . . ***C. nardina*** (in part)
      4' Plants rhizomatous or stoloniferous.
        7. Perigynia reflexed when mature; plant rhizomatous . . . . ***C. gynocrates***
        7' Perigynia ascending when mature; plant stoloniferous . . . . ***C. chordorrhiza***
    3' Stigmas 3; achenes trigonous.
      8. Perigynia beakless, blunt, green to tan.
        9. Lower female scales leaf-like, flat, 10+ mm, much longer than the perigynia; leaves 1.5–5.9 mm wide . . . . ***C. cordillerana*** (in part)
        9' All female scales scale-like or involute, not or little longer than the perigynia; leaves 0.4–3.5 mm wide.
          10. Leaves and culms delicate, leaves 0.4–1.3 mm wide; perigynia 2.4–4.9(5.4) mm long; habitat wetlands . . . . ***C. leptalea***
          10' Leaves and culms tough; perigynia (3.6)4–8.5 mm long; habitat uplands.
            11. Culms sharply triangular; leaves greater than or equaling the culms; plants loosely cespitose to short-rhizomatous . . . . ***C. geyeri*** (in part)
            11' Culms more or less cylindric; leaves shorter than culms; plants densely cespitose . . . . ***C. multicaulis***

8' Perigynia beaked, sometimes tapered gradually to the beak, usually tan to brown.
  12. Lowest inflorescence bract leaf-like, much longer than inflorescence, much greater than 2 cm long . . . . . **C. cordillerana** (in part)
  12' Lowest inflorescence bract not leaf-like, usually less than 2 cm long.
    13. Spike female or gynaecandrous . . . . . **C. idahoa**
    13' Spike androgynous.
      14. Perigynia 1–3, 5–7+ mm long; montane . . . . . **C. geyeri** (in part)
      14' Perigynia 5–20, 2–7 mm long; subalpine to alpine.
        15. Plants densely cespitose.
          16. Leaf bases from previous year not persistent; perigynia spreading to reflexed at maturity, with margins and beaks smooth . . . . . **C. micropoda** (in part)
          16' Leaf bases from previous year persistent; perigynia ascending at maturity, with margins and beaks more or less serrulate . . . . . **C. nardina** (in part)
        15' Plants loosely cespitose to long rhizomatous, sometimes with shoots arising in clusters from long rhizomes.
          17. Perigynia 3–4.8 mm wide, oval, ascending . . . . . **C. breweri**
          17' Perigynia 1–1.5 mm wide, lanceolate to elliptic, ascending to spreading or reflexed.
            18. Perigynia (3.0)3.2–4.6(5) mm long, spreading to reflexed at maturity; widespread in mountains (except Coast Range) . . . . . **C. nigricans**
            18' Perigynia 2.5–3.5 mm long, ascending to slightly spreading at maturity; Wallowa Mts. and Steens Mt. . . . . . **C. subnigricans**

Subkey 2: Inflorescences normally entirely male.

1. Terminal spike solitary or much longer than the lateral spikes; lateral spikes usually 0–3 and easy to distinguish.
  2. Leaves 0.3–0.7 mm wide . . . . . **C. gynocrates** (in part)
  2' Leaves 1–5 mm wide.
    3. Some male scales minutely ciliate on edges, at least at tip.
      4. Plants in southwestern Oregon.
        5. Flowering April–early May; ligules wider than long . . . . . **C. serpenticola** (in part)
        5' Flowering June–August; ligules longer than wide.
          6. Leaves 2–3 mm wide; plants more or less cespitose; substrate serpentine . . . . . **C. scabriuscula**
          6' Leaves less than 3 mm wide; plants often rhizomatous; substrate serpentine or not . . . . . **C. scirpoidea** (in part)
      4' Plants elsewhere in Oregon . . . . . **C. scirpoidea** (in part)
    3' Male scales entire to erose, not ciliate.
      7. Mouth of leaf sheath minutely ciliate; culms 10–20 cm long; range southwestern Oregon . . . . . **C. serpenticola** (in part)
      7' Mouth of leaf sheath entire; culms 25–40 cm long; range east of Cascades . . . . . **C. idahoa**

1' Terminal spike similar in length to lateral spikes; lateral spikes many, all or most of them crowded and difficult to distinguish.
  8. Widest leaves 4–8 mm.
    9. Culms sharply angled, serrate on at least one angle; anthers 2.5–5 mm . . . . . **C. macrocephala**
    9' Culms bluntly angled, smooth; anthers 4–6.5 mm . . . . . **C. kobomugi**
  8' Widest leaves 0.3–4 mm.
    10. Widest leaves 0.3–0.7 mm; rhizomes not more than 1 mm wide, mostly without persistent vestigial leaf sheaths; leaf sheaths at base of culms and on rhizomes not or scarcely fibrous; plants delicate; growing in moss mats in bogs and fens . . . . . **C. gynocrates** (in part)
    10' Widest leaves 1+ mm; rhizomes mostly greater than (0.8)1 mm wide, covered with persistent vestigial leaf sheaths; leaf sheaths at base of culms and on rhizomes disintegrating into coarse persistent fibers; plants not delicate; habitats various but usually alkaline or sandy, not growing in moss mats.
      11. Rhizomes less than 2.1 mm thick, brown; leaf blades flat with involute tips or involute throughout.
        12. Inflorescences 1.2–3.5(4.5) cm long; anther awns 0.2–0.4 mm long; common east of Cascades . . . **C. douglasii**
        12' Inflorescences (0.7)1–1.5(2) cm long; anther awns less than 0.1 mm long; rare and local east of Cascades . . . . . **C. duriuscula**
      11' Rhizomes greater than 2 mm thick, brown to blackish; leaf blades flat or V-shaped in cross section.

13. Longest anther awns 0.2–0.4 mm long; filaments generally hidden behind the male scales; habitat coastal sands . . . ***C. pansa***
13' Longest anther awns 0.1–0.2 mm long; filaments generally exserted from the male scales; habitat various but not coastal sands.
  14. Plant bases dark brown to black; anther awns usually hairy (at 20×), the tip tapered or obtuse; habitat meadows, prairies, roadsides, seasonally wet but dry at rhizome depth in summer . . . . ***C. praegracilis***
  14' Plant bases pale to brown; anther awns glabrous (at 20×), the tip obtuse; plant bases gray brown to medium or dark brown; habitat wet meadows, marshes, wet stream banks, where the soil is wet at rhizome depth all year . . . . ***C. simulata***

Subkey 3: Perigynia hairy.

1. Inflorescence entirely female or with male flowers scattered in mainly female spikes.
  2. Leaves involute, 0.3–0.7 mm wide; spike androgynous . . . . ***C. filifolia*** var. ***filifolia***
  2' Leaves flat or V-shaped in cross section, 2.8–4 mm wide; spike female or with some male flowers mixed among the female ones.
    3. Perigynia green to tan at maturity, round in cross section, thick enough to push the female scales away from the axis of the spike, the spike therefore with a jagged outline; perigynium bases tapered; flowering April to May; substrate serpentine . . . . ***C. serpenticola*** (in part)
    3' Perigynia black, brown, or tan at maturity, thinner and not pushing the female scales away from the axis of the spike, the spike therefore with a smooth, cylindrical shape; perigynium bases tapered or truncate; flowering June to August; substrate various, including serpentine.
      4. Female scales narrower than the mature perigynia; perigynia more or less flat . . . . ***C. scabriuscula***
      4' Female scales wider than or equal to the perigynia; perigynia trigonous . . . . ***C. scirpoidea***
1' Inflorescence with terminal spike male (or rarely gynaecandrous).
  5. Lateral female spikes 1–3, less than 1.5 cm long; inflorescence (excluding any basal spike) usually less than 4(6) cm long.
    6. Plants with basal spikes nestled among the plant bases, as well as spikes at the top of the culms; lowest inflorescence bract longer than the inflorescence (except in strictly coastal *C. brevicaulis*).
      7. Perigynia several-veined to at least mid-body; leaves glaucous (occasionally green); lower surface of leaves densely papillose (at 20×); range southwestern Oregon. . . . . ***C. brainerdii***
      7' Perigynia veinless except for 2 ribs (or somewhat veined only at the base); leaves green (occasionally glaucous); lower surface of leaves smooth or sparsely papillose; range various including southwestern Oregon.
        8. Lowest inflorescence bract often shorter than the inflorescence; leaves strongly arching; habitat sandy soils at the immediate coast . . . . ***C. brevicaulis***
        8' Lowest inflorescence bract equaling or longer than the inflorescence; leaves not strongly arching; habitat various, including non-coastal sites and coastal headlands but not coastal sands.
          9. Perigynia from non-basal spikes with beaks 0.4–0.8 mm; culms relatively smooth except near inflorescence; habitat montane . . . . ***C. deflexa*** var. ***boottii***
          9' Perigynia from non-basal spikes with beaks 0.9–1.7 mm; culms definitely scabrous; habitat various, including montane . . . . ***C. rossii***
    6' Plants lacking basal spikes; lowest inflorescence bract shorter than the inflorescence.
      10. Stigmas 4, warty (at 15×); achenes quadrangular at base . . . . ***C. concinnoides***
      10' Stigmas 3, plumose (at 15×); achenes trigonous at base.
        11. Female scales dark purple to black; substrate serpentine . . . . ***C. serpenticola*** (in part)
        11' Female scales brown or reddish brown; substrate non-serpentine.
          12. Male spikes 1–2.5 cm long; lowest inflorescence bract greater than 1 cm long; perigynium beaks 0.4–1.3 mm long . . . . ***C. inops*** ssp. ***inops***
          12' Male spikes 0.3–0.7 cm long; lowest inflorescence bract less than 1 cm long; perigynium beaks to 0.4 mm long . . . . ***C. concinna***
  5' Lateral female spikes 1–many, usually greater than 2 cm long; inflorescence greater than 4 cm long.
    13. Leaf sheath fronts pubescent at least toward top; leaf blades usually hairy.
      14. Perigynium beak teeth less than 0.8 mm; perigynia 2.5–5.3 mm long; style deciduous.
        15. Leaf blades hairy, especially below, 3–12 mm wide; spikes usually overlapping; terminal spike gynaecandrous or staminate . . . . ***C. gynodynama***
        15' Leaf blades glabrous, 2.2–4.5(6) mm wide; spikes usually distant; terminal spike staminate . . . . ***C. pellita*** (in part)
      14' Perigynium beak teeth 0.5–3 mm; perigynia 4.8–12 mm long; style persistent.

16. Perigynium beak teeth (1.2)1.5–3 mm long; perigynia sparsely pubescent . . . . **C. atherodes** (in part)

16' Perigynium beak teeth (0.4)0.6–1.7 mm long; perigynia densely pubescent.

17. Upper male scales with spreading white hairs and with apex short-awned . . . . **C. hirta**

17' Upper male scales glabrous or with appressed hairs near tip, and with apex lacking awns . . . . **C. sheldonii**

13' Leaf sheath fronts glabrous; leaf blades glabrous.

18. Perigynium beak teeth (1.2)1.5–3 mm long . . . . **C. atherodes** (in part)

18' Perigynium beak teeth less than 1 mm long.

19. Perigynia with only a few scattered hairs; plant cespitose to short-rhizomatous . . . . **C. luzulina**

19' Perigynia densely hairy; plant distinctly rhizomatous.

20. Fertile culms phyllopodic; habitat excessively drained soils, usually pumice, in the Cascades. . . . . . . **C. halliana**

20' Fertile culms aphyllopodic; habitat diverse wetlands, widespread.

21. Leaves 2–4.5 mm wide at mid-length, with midvein forming a distinct keel on the lower surface; habitat diverse wetlands including bogs and fens.. . . . . **C. pellita** (in part)

21' Leaves 0.2–2(2.2) mm wide at mid-length (measured as naturally folded), with midvein forming an inconspicuous keel; habitat bogs and fens. . . . . **C. lasiocarpa**

Subkey 4: Stigmas 3; perigynia not pubescent; styles deciduous.

Although the length of the sheath of the lowest inflorescence bract is usually a diagnostic trait, it is occasionally inconsistent. If your plant can't be keyed to a plausible species, try going the other way at lead 8.

1. Lateral spikes each with a single perigynium . . . . **Subkey 1, couplet 11**

1' Lateral spikes each with 2+ perigynia.

2. Leaf blades hairy.

3. Perigynia beakless or nearly so, rounded at tip . . . . **C. hirsutella**

3' Perigynia with a distinct beak.

4. Leaf blades hairy throughout; lowest inflorescence bract with sheath less than 5 mm long. . . . . **C. whitneyi**

4' Leaf blades mostly glabrous, sparsely hairy at base or along margins; lowest inflorescence bract with sheath greater than 5 mm long . . . . **C. mendocinensis** (in part)

2' Leaf blades glabrous.

5. Perigynia 9–15 mm long; inflorescences consisting of a single, dense, much-branched head with many spikes; female inflorescences at least 2 cm wide; habitat sandy soil.

6. Culms sharply angled, serrate on at least one angle; perigynia with 7–9 ventral nerves; perigynium beak at least half as long as perigynium, with sharp teeth; anthers 2.5–5 mm . . . . . **C. macrocephala**

6' Culms bluntly angled, smooth; perigynium with more or less 12 ventral nerves; perigynium beak 33% or less as long as perigynium; anthers 4–6.5 mm . . . . **C. kobomugi**

5' Perigynia to 8 mm long; inflorescences consisting of separate spikes not crowded into a single head, less than 1.5 cm wide; habitats various.

7. Lowest female scales leaf-like and green, similar to the leaf-like lowest inflorescence bract, 10+ mm long. . . . . **C. cordillerana**

7' Lowest female scales not leaf-like, less than 10 mm long.

8. Lowest inflorescence bract with a more or less loose sheath greater than 4 mm long.

9. Leaves broad, 6–18 mm wide.

10. Spikes 3–20+ cm long, drooping; plants 100–200 cm tall . . . . **C. pendula**

10' Spikes 1–3(4) cm long, erect; plants to 60 cm tall.

11. Terminal spike 3–8 mm wide; perigynium beaks straight; perigynia green with reddish or purple spots, or all purplish; habitat moist montane meadows and bogs. . . . . **C. luzulina** (in part)

11' Terminal spike 2–2.7 mm wide; perigynium beaks usually curved toward the dorsal side; perigynia green; habitat lowland to mid-montane mesic forests. . . . **C. hendersonii**

9' Leaves narrower, mostly (0.2)1.5–7 mm wide.

12. One or more of the inflorescence bracts widely spreading, elongate and conspicuously longer than the inflorescence (usually more than twice as long); spikes crowded, stiffly ascending; perigynia widely spreading or reflexed, distinctly beaked; plants densely cespitose . . . . **C. viridula** ssp. **viridula** (in part)

12' Inflorescence bracts not widely spreading, usually not conspicuously longer than the inflorescence (sometimes up to twice as long); other characters various.

13. Perigynia strongly flattened, ovate, and much larger than the achene; terminal spike gynaecandrous.
    14. Spikes dangling on delicate peduncles; plants montane . . . . **C. mertensii** (in part)
    14' Spikes ascending to erect, though the entire culm may nod as the perigynia mature; plants alpine or subalpine . . . . **C. heteroneura** (in part)
13' Perigynia more or less trigonous, not much larger than the achene; terminal spike male or androgynous.
    15. Plants densely to loosely cespitose.
        16. Terminal spike 4–10 mm long, 0.7–1.4 mm wide.
            17. Culms 10–60 cm tall; perigynia 2.3–3.5 mm long; plants of montane to subalpine bogs; Wallowa Mts.. . . . . **C. capillaris**
            17' Culms 1.8–7.5 cm tall; perigynia 1.2–2.1 mm long; plants of alpine wetlands; Steens Mt. . . . **C. tiogana**
        16' Terminal spike 11–66 mm long, 1.2–8 mm wide.
            18. Lateral spikes (4)5–7(8) mm wide. . . . . **C. luzulina** (in part)
            18' Lateral spikes 1.5–4 mm wide.
                19. Largest lateral spikes greater than (2.5)3 cm long . . . . **C. mendocinensis** (in part)
                19' Largest lateral spikes 1–2.5(3) cm long . . . . **C. distans**
    15' Plants rhizomatous.
        20. Perigynium beak 0.5–1.5 mm long.
            21. Perigynia gradually tapered to the beak; leaves green or yellow-green; fertile shoots phyllopodic; plants short-rhizomatous . . . . **C. luzulina** (in part)
            21' Perigynia abruptly contracted to the beak; leaves glaucous; fertile shoots aphyllopodic; plants long-rhizomatous . . . . **C. californica**
        20' Perigynium beak 0–0.4 mm long.
            22. Perigynia spindle-shaped, widest in the middle and tapering to both ends, 3–5 mm long; beak straight; habitat montane and coastal bogs, substrate not serpentine . . . . **C. livida**
            22' Perigynia ovate or obovate, widest above the middle, 1.9–3.6 mm long; beak often curved toward the back; habitat various, meadows and wetlands, including bogs, substrate often serpentine.
                23. Half or more of filled perigynia with 2 stigmas (filled perigynia have well-developed, hard achenes) . . . . **C. hassei**
                23' 90% or more of filled perigynia with 3 stigmas (unfilled perigynia may have 2 stigmas) . . . . **C. klamathensis**

8' Lowest inflorescence bract lacking a sheath or with a sheath less than 4 mm long.
    24. Leaves 8–23 mm wide.
        25. Perigynia strongly flattened. . . . . **C. mertensii** (in part)
        25' Perigynia more or less round or trigonous in cross section, not flat . . . . **C. amplifolia**
    24' Leaves 1–8 mm wide.
        26. Perigynia with conspicuous beak 0.8–2.7 mm long.
            27. Female scales with awns 1–3+ mm long; perigynia ascending . . . . **C. macrochaeta** (in part)
            27' Female scales obtuse to acuminate or with a tiny awn much less than 1 mm long; at least lower mature perigynia spreading to reflexed.
                28. Perigynia veinless or obscurely veined, not exceeding the achenes; styles persistent; habitat bogs . . . . **C. saxatilis** (in part)
                28' Perigynia distinctly veined; styles deciduous; habitat various.
                    29. Female scales brown with pale midrib; terminal spike often gynaecandrous, sometimes male. . . . . **C. serratodens** (in part)
                    29' Female scales pale; terminal spike male or androgynous. . . . . **C. viridula** ssp. **viridula** (in part)
        26' Perigynia beakless or with a short beak up to 0.6 mm long.
            30. Terminal spike male, androgynous, or with male and female flowers mixed.
                31. Terminal spike 1.5–3.5 mm wide, male; perigynia more or less round to trigonous in cross section.
                    32. At least the upper lateral spikes erect to ascending, sessile or on short peduncles; young roots lacking dense yellow hairs.
                        33. Female scales white-hyaline to medium brown.. . . . . **C. viridula** ssp. **viridula** (in part)
                        33' Female scales blackish or dark purplish.
                            34. Perigynia flat and much longer and wider than the achenes . . . . **C. paysonis** (in part)

34' Perigynia not flat, little longer or wider than the achenes.

35. Styles persistent; perigynia dark brown or blackish at least above (rarely green); plants rhizomatous. . . . . **C. saxatilis** (in part)

35' Styles deciduous; perigynia green, gray, yellow-brown, or orange-brown; plants densely to loosely cespitose . . . . **C. raynoldsii**

32' All lateral spikes spreading to dangling on long, thin peduncles; young roots with dense yellow hairs.

36. Female scales blackish, wrapping around perigynia at base; lowest inflorescence bract bristle-like, inconspicuous, shorter than the subtended peduncle . . . . **C. pluriflora** (in part)

36' Female scales medium brown or reddish, not wrapping around perigynia at base; lowest inflorescence bract usually leaf-like and 2+ cm long . . . . **C. limosa**

31' Terminal spike 3.5–6 mm wide, male, androgynous, or with irregularly arranged male and female flowers; perigynia strongly flattened, round, or trigonous in cross section.

37. Achenes nearly filling the perigynia; plants strongly rhizomatous . . . . **C. idahoa** (in part)

37' Achenes much smaller than the perigynia; plants cespitose or short-rhizomatous, forming clumps.

38. Longest female scale awns 1.6–3+ mm long. . . . . **C. macrochaeta** (in part)

38' Longest female scale awns less than 1.5 mm long.

39. Perigynia ovate, well over half as wide as long.

40. Ribs of perigynia at the margins . . . . **C. heteroneura** (in part)

40' Ribs of perigynia displaced from the margins.

41. Spikes 3–5, erect; habitat subalpine and alpine meadows; Wallowas and Elkhorns . . . . **C. paysonis** (in part)

41' Spikes 4–6(9), drooping gracefully; habitat lowland to montane meadows and forest edges; widespread . . . . **C. mertensii** (in part)

39' Perigynia elliptical, to about half as wide as long.

42. Terminal spike male; inflorescence looser, the lower spikes usually stalked and somewhat drooping. . . . . **C. spectabilis**

42' Terminal spike gynaecandrous; inflorescence dense, the lower spikes usually short-stalked and ascending, although the entire culm may nod . . . . **C. heteroneura** (in part)

30' Terminal spike gynaecandrous or female.

43. Perigynia tapering to a beak about 0.5–1 mm long; the beak bidentate, the teeth 0.2–0.4 mm long, minutely bristly; range southwestern Oregon . . . . **C. serratodens** (in part)

43' Perigynia beakless, minutely beaked, or abruptly beaked, the beak 0.2–0.5 mm long; beak teeth absent or short; range various.

44. Plants strongly rhizomatous, often forming large stands of turf.

45. Leaf sheath fronts ladder-fibrillose; perigynia whitish or pale greenish and strongly papillose . . . . **C. buxbaumii**

45' Leaf sheath fronts hyaline, not breaking into fibers; perigynia yellowish to brown or blackish, smooth or minutely papillose.

46. Lateral spikes more or less erect to ascending; peduncles approximately 1.5 mm long . . . **C. idahoa** (in part)

46' Lateral spikes dangling on long, slender peduncles much greater than 1.5 mm long . . **C. pluriflora** (in part)

44' Plants cespitose to short-rhizomatous, forming clumps.

47. Perigynia dark gold with darker beaks . . . . **C. atrosquama**

47' Perigynia brown, dark purplish, or green and yellow.

48. Achene nearly filling the perigynium . . . . **C. media**

48' Achene filling half or less of the perigynium.

49. Spikes usually 1–4 cm long, all drooping; habitat montane but not subalpine, to 1680 m (5500 ft) elevation . . . . **C. mertensii** (in part)

49' Spikes 0.7–2.7 cm long, erect to spreading (though the culm itself may nod); habitat subalpine to alpine, 1740–2740+ m (5700–9000+ ft) elevation.

50. Perigynia dark reddish brown, shiny; spikes 0.7–1 cm long, sessile to short-pedunculate, crowded . . . . **C. pelocarpa**

50' Perigynia yellow-green to dark brown, dull to shiny; spikes 1–2.7 cm long, pedunculate, less crowded. . . . . **C. heteroneura** (in part)

Subkey 5: styles persistent; perigynia usually inflated.

1. Perigynium beaks 0.3–0.8 mm long . . . . *C. saxatilis*
1' Perigynium beaks 1–4.5 mm long.
  2. Lower lateral spikes nodding on drooping stalks.
    3. Perigynium beak teeth spreading or strongly curved, 1.3–3 mm long; plant bases brown. . . . . *C. comosa*
    3' Perigynium beak teeth erect, straight, 0.3–0.9 mm long; plant bases red-purple . . . . *C. hystericina*
  2' Lower lateral spikelets generally erect, stalks sometimes 0.
    4. Spikes crowded, overlapping, except sometimes the lowest . . . . *C. retrorsa*
    4' Spikes well separated.
      5. Leaves hairy at least at top of leaf sheath front . . . . *C. atherodes*
      5' Leaves glabrous.
        6. Perigynia not inflated; plants cespitose; rare introduced plants along the Columbia River. . . . . . . *C. distans*
        6' Perigynia inflated; plants cespitose or rhizomatous; common, native, and widespread.
          7. Dorsal surfaces of basal leaf sheaths with many, regularly spaced crosswalls between the veins, giving the appearance of brickwork; most mature perigynia spreading at right angles from the spike axis. . . . . *C. utriculata*
          7' Dorsal surfaces of basal leaf sheaths with few, irregularly spaced crosswalls between the veins; mature perigynia ascending to somewhat spreading but not at right angles to spike axis.
            8. Longer perigynia 4.0–7.5(8.2) mm long; perigynia contracted to the beak; range mainly east of crest of Cascades . . . . *C. vesicaria*
            8' Longer perigynia 7.5–10.1 mm long; perigynia tapering gradually to the beak; range mainly west of crest of Cascades. . . . . *C. exsiccata*

Subkey 6: Stigmas 2; achenes lenticular; spikes usually greater than 1.5 cm; terminal spike usually entirely male.

In the first lead, when you find one trait that unambiguously sends your specimens to lead 2 or 3, you can be confident in the direction the leads are taking.

1. Perigynia less crowded, (3)4–10(14) per 5 mm on spike, succulent or dry; plants to 50 cm tall; rhizomatous, rhizomes approximately 1 mm wide; plant bases pale or dull medium brown; leaf sheath fronts hyaline, white or transparent; longest lateral spikes to 2.3 cm long; sheath of lowest inflorescence bract (2)4+ mm long; staminate spikes 0–1.
  2. Terminal spike 0.9–2 mm wide; proximal male scales 2–3.5(4) mm long; mature perigynia succulent throughout, divergent; yellow or orange in life (dark brown and waxy-looking when dry); female scales divergent in mature fruit . . . . *C. aurea*
  2' Terminal spike (1.8)2–2.5 mm wide; proximal male scales 3–6(15) mm long; mature perigynia dry or succulent only at base, more or less ascending; green, tan, whitish, or rarely yellowish in life (similar in color or faded when dry); female scales more or less ascending in mature fruit . . . . *C. hassei*
1' Perigynia more crowded, (12)18–25 per 5 mm on spike, dry; plants to 200 cm tall; cespitose or rhizomatous, rhizomes 1.5–6 mm wide; plant bases pale, dull medium brown, reddish, chestnut, or blackish; leaf sheath fronts hyaline or ladder-fibrillose, white, coppery, or densely red-dotted; longest lateral spikes (1)2–15 cm long; sheath of lowest inflorescence bract 0–2(5) mm long; staminate spikes 1–several.
  3. Perigynia very thick-walled and hard, resisting puncture, brown or yellow-brown; achene often with an indentation on one or both margins; range west of the Cascades.
    4. Spikes shorter (the lowest 1.8–5 cm long), brown, blunt at the base, straight but dangling on a peduncle 2–8 cm long; leaf sheath fronts hyaline; leaves dying in winter; habitat coastal salt marshes and tidal wetlands along coastal rivers. . . . . *C. lyngbyei*
    4' Spikes longer (the lowest 2.5–15 cm long), dark brown to blackish, tapering at the base, drooping and sessile or on a straight peduncle to 3 cm (rarely to 9 cm) long; leaf sheath fronts ladder-fibrillose; leaves evergreen; habitat usually freshwater wetlands west of the Cascade crest. . . . . *C. obnupta*
  3' Perigynia membranous or papery to somewhat tough, generally green or tawny, with or without red or purple pigments; achene not indented; range various.
    5. Perigynia distinctly veined on at least one face (in addition to the 2 ribs).
      6. Plants rhizomatous.
        7. Female scales with a prominent, scabrous awn (at least in the lower part of the spike); perigynia often somewhat tough.
          8. Leaf sheaths densely red-dotted and coppery-colored, usually ladder-fibrillose; lowest inflorescence bract much longer than the inflorescence; range southwestern Oregon, California . . . . *C. barbarae*

8' Leaf sheaths usually white-hyaline (occasionally sparsely red-dotted, sometimes coppery), not ladder-fibrillose; lowest inflorescence bract subequal to the inflorescence; widespread east of Cascades, occasionally introduced in and west of the Cascades . . . . ***C. nebrascensis*** (in part)

7' Female scales lacking a scabrous awn, acute, acuminate, or with a short point; perigynium traits various.

9. Perigynia (1)1.5–2.1 mm long, green, widely spaced at base of lowest spike; habitat sandy soil at margins of fast-flowing rivers and streams in and west of the Cascades . . . . ***C. interrupta*** (in part)

9' Perigynia 2.2–5.5 mm long, brownish, straw-colored, or green, not widely spaced at base of lowest spike; habitat other substrates (rarely sand) in wet meadows, ditches, and streams.

10. Mature perigynia with 5–9 veins, thicker and somewhat tough; perigynium beak minutely bidentate; lower leaf sheaths not ladder-fibrillose . . . . ***C. nebrascensis*** (in part)

10' Mature perigynia with 1–3 weak veins, thinner and not tough; perigynium beak not bidentate; lower leaf sheaths usually ladder-fibrillose. . . . . ***C. angustata***

6' Plants cespitose.

11. Plant bases blackish or dark reddish; leaf sheath fronts strongly ladder-fibrillose; perigynia sessile and widest above middle; habitat boulders and cobbles in the scour zone of fast-moving, seasonally flooding rivers, also irrigation ditches . . . . ***C. nudata***

11' Plant bases greenish or brownish; leaf sheath fronts hyaline; perigynia stipitate and widest at or below middle; habitat silts, sands, or gravels at edges of slow-moving or still water, including drawdown zones of reservoirs . . . . ***C. kelloggii***

5' Perigynia veinless or very faintly veined on both faces (but with 2 ribs).

12. Plants densely cespitose . . . . return to **couplet 11**

12' Plants rhizomatous to loosely cespitose.

13. Perigynia inflated toward the top, loosely enveloping the achene.

14. Perigynia (3)3.5–5.5 mm long; usually dark purplish, sometimes greenish, not speckled with red; style persistent . . . . ***C. saxatilis***

14' Perigynia 2.5–2.8 mm long; olive-green, olive-brown, or orangish, with reddish speckles; style deciduous . . . . ***C. aperta***

13' Perigynia flattened and more tightly enveloping the achene.

15. Perigynia (1)1.5–2.1 mm long, green; widely spaced at base of lowest spike; habitat in sandy soil at margins of fast-flowing rivers and streams in and west of the Cascades.. . . . . ***C. interrupta*** (in part)

15' Perigynia 2–4 mm long, brownish, often with red or black spots, not widely spaced at base of lowest spike; habitat and range various.

16. Spikes crowded, with lowest internode 0.3–3.5(4) cm long and lowest two internodes collectively 0.5–4.2(4.7) cm long . . . . ***C. scopulorum*** var. ***bracteosa***

16' Spikes not crowded, with lowest internode 3–20 cm long and lowest two internodes collectively 4.3–26 cm long . . . . ***C. aquatilis***

Subkey 7: Stigmas 2; perigynia unwinged; spikes androgynous, entirely female, or with male flowers mixed irregularly among the female flowers.

1. Leaf sheath fronts strongly cross-corrugated.

2. Leaf blades (4)5–11 mm wide; culm approximately 6 mm wide at mid-length, winged; perigynia 3.6–5.2 mm long . . . . ***C. stipata*** var. ***stipata*** (in part)

2' Leaf blades 1–7 mm wide; culm approximately 1.5–3 mm wide at mid-length, not or only slightly winged; perigynia 2–4 mm long.

3. Inflorescence ovoid, 1.5–2 times as long as wide, dense, greenish to dark brown . . . . ***C. neurophora*** (in part)

3' Inflorescence elongate, greater than 2 times as long as wide, dense or interrupted, gold to light or dark brown.

4. Leaf sheath fronts coppery-tinged near the mouth, also red-dotted; lowest inflorescence node usually with a distinct branch with 5–12 spikes . . . . ***C. cusickii***

4' Leaf sheath fronts white-hyaline or red-dotted, not coppery-tinged near the mouth; inflorescence usually a dense head, if distinctly branched then lowest branch with up to 5 spikes.

5. Leaf sheath fronts either white-hyaline or red-dotted . . . . ***C. diandra***

5' Leaf sheath fronts white-hyaline, often with pale brown dots.

6. Inflorescence usually interrupted; perigynia 2–3.2 mm long; range mainly east of the Cascades.. . . . . ***C. vulpinoidea***

6' Inflorescence dense, not interrupted; perigynia 2.8–4 mm long; range entirely west of the Cascade crest. . . . . ***C. densa***

1' Leaf sheath fronts smooth, or very weakly cross-corrugated.

7. Leaf sheath fronts minutely dotted red, brown, or yellow; plants cespitose. . . . . return to **couplet 4**

7' Leaf sheath fronts white-hyaline to green; plants cespitose or rhizomatous.

8. Plants long-stoloniferous and inconspicuously rhizomatous; introduced in coastal cranberry bogs . . ***C. chordorrhiza***

8' Plants cespitose or rhizomatous but not stoloniferous; habitat various.

9. Inflorescence dark brown, dense, and globose, the individual spikes not distinguishable; habitat alpine. . . . . ***C. vernacula***

9' Inflorescence dark to pale brown, the individual spikes readily distinguishable to somewhat obscure; habitat low elevation to subalpine.

10. Plants densely to loosely cespitose.

11. Perigynia lance-triangular, long-tapered to a poorly defined beak; perigynium base usually swollen with pithy tissue; habitat wetlands, marshes, wet meadows.

12. Inflorescence greater than 3 cm long; widest leaves 5–10 mm wide. . . ***C. stipata*** var. ***stipata*** (in part)

12' Inflorescence less than 2 cm long; widest leaves up to 5 mm wide.

13. Leaf sheath fronts with a thick, white rim at mouth, not cross-corrugated; perigynia 3–4.5 mm; southwestern Oregon . . . . ***C. nervina***

13' Leaf sheath fronts lacking thick rim at mouth, sometimes cross-corrugated; perigynia 2.5–4 mm long; widespread.

14. Leaves crowded near base of culm; leaf sheath front usually not weakly cross-corrugated, mouth usually lacking tongue-like extension; margins of perigynium beaks and upper bodies entire (or very nearly so) . . . . ***C. jonesii***

14' Leaves generally not clustered near culm base; leaf sheath front weakly cross-corrugated, mouth with tongue-like extension; margins of perigynium beaks and upper bodies usually serrate on at least one side . . . . ***C. neurophora*** (in part)

11' Perigynia short-tapered or abruptly narrowed to a distinct beak; perigynium base not swollen with pithy tissue; habitat mesic to dry grasslands.

15. Perigynia bulging dorsally; marginal ribs displaced onto the ventral surface . . . . ***C. vallicola***

15' Perigynia not or slightly bulging; ribs not displaced onto the ventral surface (except sometimes in *C. tumulicola*).

16. Inflorescence a dense, ovoid head, the spikes not easily distinguished; perigynia green with copper-colored center (maturing light brown with dark brown center). . . . . ***C. hoodii***

16' Inflorescence elongate, the lower spikes overlapping but easily distinguished; perigynia brown, sometimes with narrow green margins, not darker over the achene . . . . ***C. tumulicola***

10' Plants rhizomatous.

17. Plants soft, delicate, with rhizomes 1 mm in diameter, dull gray-brown or medium brown; inflorescence linear and much interrupted, the spikes well separated from each other, usually 5 mm long or less with 1–3 perigynia; perigynium beak to 0.25 mm long; habitat usually shaded . . . . ***C. disperma***

17' Plants coarse, tough, with rhizomes greater than 1 mm in diameter, pale brown to blackish; inflorescence dense, though sometimes elongated, the spikes more or less closely aggregated, often greater than 5 mm long, usually with greater than 3 flowers; perigynium beak greater than 0.25 mm long; habitat usually sunny.

18. Perigynia short and squat, 1.8–2.8 mm long, shiny, dark, the beak generally less than 25% of the length of the perigynium body; habitat wetlands with soil submerged or moist at rhizome depth all year long . . . . ***C. simulata***

18' Perigynia longer, 2.6–4.2 mm long, shiny or dull, the beak greater than 33% the length of the perigynium body; habitat uplands to seasonal wetlands that dry out at rhizome depth in summer.

19. Rhizomes less than 2.1 mm thick, brown; leaf blades flat with involute tips, or involute throughout.

20. Perigynia 3.5–4.2 mm long, beaks 0.9–1.9 mm long, more or less equal to the body; stigmas persistent, styles exserted from perigynia; inflorescences on female plants 1.5–3.5 × 0.8–2.7 cm; widespread east of Cascades . . . . ***C. douglasii***

20' Perigynia 2.4–3.9 mm long, beaks 0.3–0.9 mm long, shorter than body; stigmas deciduous, styles not exserted from perigynia; inflorescences on female or bisexual plants less than 2 cm long, less than 1 cm wide; rare and local east of Cascades . . . . **C. duriuscula**

19' Rhizomes greater than 2 mm thick, dark brown to blackish; leaf blades flat or V-shaped in cross section.

21. Female scales usually dark and shiny; female inflorescences usually 1–2 cm wide, ovate; perigynia generally shiny; longest anther awns 0.2–0.4 mm; habitat coastal sands . . . . **C. pansa**

21' Female scales usually tan or dull pale brown; female inflorescences usually less than 1(1.5) cm wide, elliptic or elongate; perigynia more or less dull; longest anther awns 0.1–0.2 mm; habitat inland (but within a few miles of the coast in southwestern Oregon), in more or less alkaline or serpentine soils, including sand . . . . **C. praegracilis**

Subkey 8: Stigmas 2, spikes gynaecandrous; perigynia unwinged.

1. Spikes crowded in a single, dense head, not easily distinguished.

2. Longest inflorescence bracts greater than 3 times as long as the inflorescence; perigynium beak 3–5 mm long, longer than the perigynium body . . . . **C. sychnocephala**

2' Longest inflorescence bracts rarely longer than the inflorescence and never twice as long; perigynium beak 1–1.6 mm long, shorter than the perigynium body.

3. Inflorescence ovoid to elongate, green to straw-colored or pale brown; spikes 5–15; margins of beaks serrulate . . . . **C. arcta** (in part)

3' Inflorescence more or less pyramidal, globose, or ovoid, dark brown or a mix of green and brown and black; spikes 3–9; margins of beaks entire.

4. Inflorescence black and green or black and brown, compact, the base usually more or less truncate; perigynium wings absent . . . . **C. illota**

4' Inflorescence brown or brown and green, elongate, the base usually more or less tapering; perigynium wings very narrow (0.05–0.2 mm wide) . . . . **C. integra** (in part)

1' Spikes remote to overlapping, but individual spikes easily recognized.

5. Plants rhizomatous.

6. Beak (or top of perigynium; it may be virtually beakless) curved to the back; perigynium surface often papillose at least near beak; sheath of lowest inflorescence bract (0)2–5+ mm long. . . . . **C. hassei**

6' Beak not curved to the back; perigynium surface smooth; sheath of lowest inflorescence bract 0–1 mm long . . . . **Subkey 6, Couplet 3**

5' Plants cespitose.

7. Spikes 3+ times as long as wide, sessile or stalked.

8. Perigynium beaks 0.1–0.6 mm long, less than 20% of perigynium length . . . . **Subkey 6, Couplet 3**

8' Perigynium beaks 0.9–2.5 mm long, 28–50% of perigynium length.

9. Beaks 0.9–1.5(1.7) mm, 30–38% of perigynium length; bodies of female scales (2.3)2.7–3.8 mm . . . . **C. leptopoda**

9' Beaks (1.4)1.6–2.7 mm, 38–50% of perigynium length; bodies of female scales 2.1–2.9(3.1) mm.

10. Teeth of perigynium beak 0–0.2(0.4) mm long. . . . . **C. infirminervia**

10' Teeth of perigynium beak (0.2)0.3–1 mm long. . . . . **C. bolanderi**

7' Spikes less than 3 times as long as wide, sessile.

11. Mature perigynia strongly spreading to reflexed.

12. Perigynium beak 0.5–1.2 mm long, 20–44% of perigynium length.

13. Perigynia more or less gradually tapered to the beak; wider leaves 2–4 mm wide. . . . . **C. arcta** (in part)

13' Perigynia abruptly narrowed to the beak; wider leaves 1–2.4(2.7) mm wide. . . . . **C. interior**

12' Perigynium beak 0.9–2 mm long, 33–60% of perigynium length . . . . **C. echinata**

11' Mature perigynia appressed to ascending (sometimes seemingly spreading because the beaks are bent back).

14. Perigynia 3.3–5.3 mm long . . . . return to **couplet 9**

14' Perigynia 1.5–3(3.5) mm long.

15. Spikes crowded, overlapping, the lowest internode no longer than the lowest spike.

16. Perigynium beaks smooth on margins.

17. Beak to 0.5 mm . . . . **C. praeceptorum** (in part)

17' Beak 1–1.6 mm . . . . **C. integra** (in part)

16' Perigynium beaks serrulate on margins.

18. Inflorescence green to tan; perigynia more or less spreading when ripe, exposed by the female scales, 2–3(3.5) × 1.2–1.5 mm . . . . **C. arcta** (in part)
18' Inflorescence brown (with green perigynia when young); perigynia appressed, more or less hidden by female scales, 3.5–4.2 × 1–1.2 mm . . . . **C. leporinella**
15' At least lower spikes remote, the lowest internode longer than the lowest spike.
19. Perigynium beaks 0.4–1.1(1.3) mm long . . . . **C. laeviculmis**
19' Perigynium beaks up to 0.5 mm long.
20. Dorsal suture darker than the surrounding perigynium surface; female scales light brown with lighter center and hyaline edges. . . . . **C. praeceptorum** (in part)
20' Dorsal suture about the same color as the surrounding perigynium surface; female scales white-hyaline (often tinged brownish with age) with green center. . . . . **C. canescens** ssp. **canescens**

Subkey 9: Perigynia winged; stigmas 2.

1. Lowest 1–3 bracts of most inflorescences elongate, at least the lowest as long as or usually much longer than inflorescences; inflorescences dense, head-like.
2. Beak distinctly longer than (sometimes twice as long as) the perigynium body; distance from top of achene to tip of beak 3–5 mm; range southeastern Oregon . . . . **C. sychnocephala**
2' Beak little if at all longer than perigynium body; distance from top of achene to tip of beak 1.1–2.5 mm; range east or west of the Cascades.
3. Inflorescence angled to one side; lowest inflorescence bract more or less leaf-like; range west of the Cascades . . . . **C. unilateralis**
3' Inflorescence erect, located symmetrically at the top of the culm; lowest inflorescence bract usually not leaf-like; range east or west of the Cascades.
4. Beak tip to achene top 1.9–2.5 mm; achenes 0.3–0.4 mm thick; range mainly east of the Cascades, scattered on west of Cascades but not coastal . . . . **C. athrostachya** (in part)
4' Beak tip to achene top 1.2–1.8 mm; achenes 0.5–0.7 mm thick; range coastal, occasionally inland in southwestern Oregon.
5. Perigynia with 3–8 strong veins on the ventral surface, extending beyond the top of the achene; perigynia not leathery . . . . **C. harfordii** (in part)
5' Perigynia with 0–3 veins on the ventral surface, reaching at most the top of the achene; perigynia more or less leathery . . . . **C. subbracteata** (in part)
1' Lowest 1–3 bracts of most inflorescences usually inconspicuous, shorter than the inflorescences; inflorescences dense or elongate.
6. Leaf sheath fronts green and veined nearly to the top, often with a short white-hyaline triangle (less than 6 mm long) at the top.
7. Perigynia lanceolate and wing much narrowed or lacking below middle of body.
8. Widest leaves less than 3 (rarely 4) mm wide; spike apex acute to rounded. . . . **C. scoparia** var. **scoparia** (in part)
8' Widest leaves 3–7 mm wide; spike apex rounded. . . . . **C. tribuloides** var. **tribuloides**
7' Perigynia lanceolate to ovate or obovate, wing usually extending to base of perigynium; if wing much narrowed below middle, perigynium elliptic to ovate.
9. Perigynia obovate, the body widest above middle, narrowed abruptly to the beak; wing 0.5–0.8 mm wide; introduced in coastal wetlands . . . . **C. longii**
9' Perigynia lanceolate to ovate, the body widest near middle, narrowed more gradually to the beak; wing 0.2–0.6 mm wide; native; range various, including coast.
10. Female scales acuminate; perigynia 4.2–5.5(6) mm, elliptic, tapering gradually to the beak . . . . **C. scoparia** var. **scoparia** (in part)
10' Female scales obtuse to acute; perigynia 3.2–4.2 mm, ovate, tapering more abruptly to the beak. . . . **C. feta**
6' Leaf sheath fronts white-hyaline for at least 10 mm at the top (often transparent if plant is fresh).
11. Longer perigynia (6)6.3–8.5 mm long.
12. Perigynia plano-convex or biconvex, 0.5–0.9 mm thick; wing 0.3–0.5 mm wide; habitat lower elevations to subalpine.
13. Inflorescence a round to oval head; spikes all crowded or sometimes the lowest distinct. . . . . **C. pachycarpa** (in part)
13' Inflorescence more or less elongate; spikes overlapping but not crowded.
14. Perigynia rarely as much as 6 mm long; with 0–4(7) ventral veins, these usually shorter than the achene. . . . . **C. praticola** (in part)

14' Perigynia usually greater than 6 mm long; with 4–10 ventral veins, at least 3 of them longer than the achene.
15. Female scales covering more than half the perigynium beak, white-hyaline or with white-hyaline margin 0.2–0.7 mm wide; uncommon but widespread. . . . . **C. petasata**
15' Female scales covering less than half the perigynium beak, with white-hyaline margin 0–0.2 mm wide; rare and local . . . . **C. davyi**
12' Perigynia flat except where distended by the achene, 0.25–0.5 mm thick; wing 0.3–0.9 mm wide; habitat subalpine to alpine.
16. Perigynia much longer than wide; distal 1 mm or more of beak unwinged. . . . . **C. haydeniana** (in part)
16' Perigynia little longer than wide; distal 0–0.7 mm of beak unwinged. . . . . **C. straminiformis** (in part)
11' Longer perigynia 2–5.4(6) mm long.
17. Inflorescence a dense, round to slightly oval head, although occasionally the lowest spike or two may be separated.
18. Heads dark (brown, dark brown, reddish brown, or blackish) or with contrasting dark female scales and pale perigynia (This is an arm's-length trait.)
19. Perigynium wings 0.4–1 mm wide; perigynia 1.8–3.5 mm wide.
20. Perigynia plano-convex, 0.5–0.8 mm thick, wings widest on the beak and much narrowed below middle . . . . **C. pachycarpa** (in part)
20' Perigynia flat except where distended by the achene, 0.2–0.5(0.6) mm thick, wings broad to near the base.
21. Beak usually winged and ciliate-serrulate almost to the tip, sometimes unwinged; brown; parallel-sided for the distal 0.5–0.7 mm; inflorescence with spikes easily distinguished, green or brown, often with contrasting dark female scales and paler perigynia . . . **C. straminiformis** (in part)
21' Beak unwinged; brown; parallel-sided for the distal 1+ mm, entire for 0.3–0.6 mm; inflorescence with spikes crowded and hard to distinguish, dark or contrasting green and dark.
22. Perigynia 4–6.5 mm long, usually at least (2.3)2.6 mm from beak tip to achene; at or above timberline. . . . . **C. haydeniana** (in part)
22' Perigynia 3–4.5(5.2) mm long, usually no more than 2.5 mm from beak tip to achene; montane to near timberline . . . . **C. microptera** (in part)
19' Perigynium wings 0.1–0.5 mm wide; perigynia 1.1–2.4 mm wide.
23. Spikes androgynous; perigynia bodies copper-colored with green margins (maturing dark brown with light brown margins) . . . . **C. hoodii**
23' Spikes gynaecandrous; perigynia usually more or less uniformly colored or with body paler than wings.
24. Perigynium beaks winged to the tip, or with tips less than 0.4 mm long.
25. Perigynia (3.8)4.4–6 mm long; achenes 1.7–2.4 mm long . . . . **C. pachycarpa** (in part)
25' Perigynia 2.4–4.3(4.6) mm long; achenes 1–2 mm long.
26. Lowest two internodes of inflorescence collectively less than 33% of total inflorescence length; habitat moist to wet sites near the coast. . . . . **C. harfordii** (in part)
26' Lowest two internodes of inflorescence collectively greater than 33% of total inflorescence length; habitat mesic to dry sites in mountains . . . . **C. preslii** (in part)
24' Perigynium beaks with tips greater than 0.4 mm long.
27. Perigynia with (8)10–20 veins on the dorsal surface, the veins often sunken; perigynia plano-convex, 0.5–0.8 mm thick, cream-colored to light brown, often with green wings; wings (0.2)0.3–0.4(0.5) mm wide but often narrow at and below middle of perigynium body; habitat at mid to high elevations in mountains. . . . . **C. pachycarpa** (in part)
27' Perigynia with 0–9 veins on the dorsal surface; veins, if present, raised; other traits various.
28. Perigynia with 3–9 strong veins on the ventral surface, extending above the top of the achene.
29. Perigynia (0.5)0.6–0.7 mm thick; habitat at low elevations mainly near the coast. . . . . **C. harfordii** (in part)
29' Perigynia 0.3–0.5(0.6) mm thick; habitat montane to alpine.
30. Distance from achene top to perigynium beak tip (1.6)2–2.3 mm; habitat montane to subalpine . . . . **C. abrupta**
30' Distance from achene top to perigynium beak tip (2.3)2.6–3.8 mm; habitat subalpine to alpine . . . . **C. haydeniana** (in part)

28' Perigynia with 0–8 veins on the ventral surface, the veins if present faint and/or not more than 2 of them extending above the top of the achene.
31. Perigynia 4–6.5 mm long; distance from beak tip to top of achene more than (2.3)2.6–3.8 mm; habitat subalpine to alpine. ***C. haydeniana*** (in part)
31' Perigynia 2.7–4.7(5.7) mm long; distance from beak tip to top of achene 1–2.5(2.8) mm; habitat lowland to subalpine.
32. Perigynium wing entire, 0–0.2 mm wide. . . . . ***C. integra***
32' Perigynium wing minutely ciliate-serrulate, 0.2–0.5 mm wide.
33. Perigynia many, crowded, appressed, and flat except over the relatively small achene, 0.3–0.5 mm thick, thus making the inflorescence more fine-textured; spikes usually green and black . . . . ***C. microptera*** (in part)
33' Perigynia fewer, less crowded, appressed to spreading, plano-convex and more or less filled by the relatively large achene, (0.4)0.5–0.7 mm thick, thus making the inflorescence coarse-textured; spikes usually coppery, brown, or green and black.
34. Perigynium beaks of 2 kinds in the same inflorescence, some more or less flat and winged nearly to the tip, others with tip cylindric, unwinged; habitat montane to subalpine, more or less dry. . . . . ***C. preslii*** (in part)
34' All perigynium beaks with tips cylindric, unwinged; habitat at low to mid elevations, moist to mesic.
35. Perigynia spreading; spikes appearing star-shaped from above; female scales 2.2–3.4(4.2) mm long; perigynia not leathery; range widespread . . . . ***C. pachystachya*** (in part)
35' Perigynia more or less ascending; spikes not appearing star-shaped from above; female scales 3.5–4.5 mm long; perigynia more or less leathery; range coastal or somewhat inland in southwestern Oregon. . . . . ***C. subbracteata*** (in part)

18' Heads pale (whitish, green, straw-colored, or tan) without contrasting dark female scales (If you are unsure, the color is likely not pale).
36. Perigynia 3.5–4.5 times as long as wide.
37. Perigynium beaks winged nearly to the tip; perigynia 0.15–0.35 mm thick. . . . . ***C. crawfordii***
37' Perigynium beak tips unwinged, parallel-sided, and entire for at least 0.4 mm; perigynia 0.35–0.45 mm thick . . . . ***C. athrostachya*** (in part)
36' Perigynia 1.8–3.5 times as long as wide.
38. Inflorescence relatively fine-textured in appearance; perigynia 0.9–1.9 mm wide; habitat lowlands to subalpine.
39. Inflorescence nearly always oval; perigynia tapering to a distinct beak . . . . ***C. subfusca*** (in part)
39' Inflorescence always a dense, roundish head; perigynia tapering gradually to the tip . . . . ***C. athrostachya*** (in part)
38' Inflorescence relatively coarse-textured in appearance; perigynia 1.4–3.4 mm wide; habitat montane to subalpine.
40. Perigynium wings 0.4–1 mm wide; perigynia flat except where distended by the achene, 0.3–0.5 mm thick. . . . . ***C. straminiformis*** (in part)
40' Perigynium wings (0.2)0.3–0.4(0.5) mm wide; perigynia plano-convex, 0.5–0.8 mm thick. . . . . ***C. pachycarpa*** (in part)

17' Inflorescence more or less elongate, varying from a longer oval with overlapping spikes to a line of separated spikes.
41. Spikes green to straw-colored and with a fine texture; female scales acuminate and revealing acuminate perigynium beaks; beaks winged and ciliate-serrulate nearly to the tip; wings usually narrowed or absent below middle of perigynium; range Willamette Valley . . . . ***C. scoparia*** var. ***scoparia*** (in part)
41' Spikes dark or if pale, then differing in some way from above.
42. Inflorescence bent to one side or nodding.
43. Tip of perigynium beak white; ventral surface of perigynia usually lacking veins; female scales with white-hyaline margins 0.1–0.3 mm wide . . . . ***C. praticola*** (in part)
43' Tip of perigynium beak brown; ventral surface of perigynia usually with 3–5+ veins at least as long as the achene; female scales usually lacking white-hyaline margins . . . . ***C. leporina*** (in part)
42' Inflorescence erect.

44. Perigynium wing 0.4–1 mm wide; perigynia 1.8–3.4 mm wide.
   45. Plants of low elevations, usually below 610 m (2000 ft).
      46. Perigynia 2.3–3.2 mm wide, broadly ovate to orbicular. . . . . **C. brevior**
      46' Perigynia to 2.1 mm wide, lanceolate to ovate. . . . . **C. leporina** (in part)
   45' Plants of high elevations, usually above 1525 m (5000 ft) . . . . **C. straminiformis** (in part)
44' Perigynium wing 0.1–0.4(0.6) mm wide; perigynia 0.9–2.2 mm wide.
   47. Inflorescences 3.5–8 cm long; leaf sheath front rounded to acute at summit, prolonged at least 3 mm above attachment point of the leaf blade; plants coarse, usually greater than 60 cm tall. . . . . **C. fracta**
   47' Inflorescences 1–5 cm long; leaf sheath front often U-shaped at summit, usually not prolonged or prolonged less than 2.8 mm above attachment point of leaf blade; plants coarse or delicate, usually less than 60 cm tall.
      48. Lowest two inflorescence internodes collectively at least 10 mm long.
         49. Female scales distinctly shorter than the perigynia; perigynia 2.4–3.5 mm long . . . . **C. subfusca** (in part)
         49' Female scales about as long as the perigynia; perigynia 3.4–6 mm long.
            50. Perigynia widest near middle of total length; habitat subalpine to alpine.
               51. Ventral surface of perigynia with 3–8 strong veins that extend to the top of the achene; perigynia opaque, somewhat leathery, (3.7)4.5–6 mm long . . . . **C. tahoensis** (in part)
               51' Ventral surface of perigynia with 0–4 weak veins; perigynia translucent or brown, not especially tough, 3.8–5.2 mm long . . . . **C. phaeocephala** (in part)
            50' Perigynia widest below middle of total length; habitat in lowlands or montane.
               52. Tip of perigynium beak white; ventral surface of perigynia usually lacking veins; female scales with white-hyaline margins 0.1–0.3 mm wide. . . . . **C. praticola** (in part)
               52' Tip of perigynium beak brown; ventral surface of perigynia usually with 3–5+ veins at least as long as the achene; female scales usually lacking white-hyaline margins . . . **C. leporina** (in part)
      48' Lowest 2 inflorescence internodes collectively 3–10 mm long.
         53. Spikes globose; mature perigynia light to dark brown, 2.5–3.8 mm long; beak winged to the tip on all perigynia; achenes 1–1.3 mm long, 0.6–0.9 mm wide . . . . **C. bebbii**
         53' Inflorescences differing in some way from above.
            54. Inflorescences fine-textured, usually pale, green, whitish, or light brown; achenes 1–1.6 mm long. . . . . **C. subfusca** (in part)
            54' Inflorescences coarse-textured and darker brown, reddish brown, blackish, or sometimes greenish; achenes 1.4–2.4 mm long.
               55. Perigynia 1–1.2 mm wide, boat-shaped; habitat soggy wet in the spring though often drying later . . . . **C. leporinella**
               55' Perigynia 1.3–2.3 mm wide, not boat-shaped; habitat mesic to dry.
                  56. Ventral surface of perigynia with 3–8 strong veins that extend to the top of the achene; habitat dry subalpine to alpine slopes . . . . **C. tahoensis** (in part)
                  56' Ventral surface of perigynia with 0–4 weak veins; habitat mesic lowlands to dry alpine slopes.
                     57. Perigynia widest near middle, gold to brown; habitat dry alpine slopes . . . . **C. phaeocephala** (in part)
                     57' Perigynia widest below middle, brown, coppery, or greenish; habitat mesic lowlands to dry subalpine meadows
                        58. Perigynium beaks of 2 kinds in the same inflorescence, some more or less flat and winged nearly to the tip, others with tip cylindric, unwinged, smooth, brown, and parallel-sided for greater than 0.4 mm; habitat montane to subalpine, dry. . . . . **C. preslii** (in part)
                        58' All perigynium beaks with tips cylindric, unwinged, smooth, brown, and parallel-sided greater than 0.4 mm; habitat low to mid elevations, moist to mesic.
                           59. Perigynia spreading; spikes appearing star-shaped from above; female scales 2.2–3.4(4.2) mm; perigynia not leathery; range widespread . . . . **C. pachystachya** (in part)
                           59' Perigynia more or less ascending; spikes not appearing star-shaped from above; female scales 3.4–4.5(5.7) mm; perigynia more or less leathery; range southwestern Oregon and coastal . . . . **C. subbracteata** (in part)

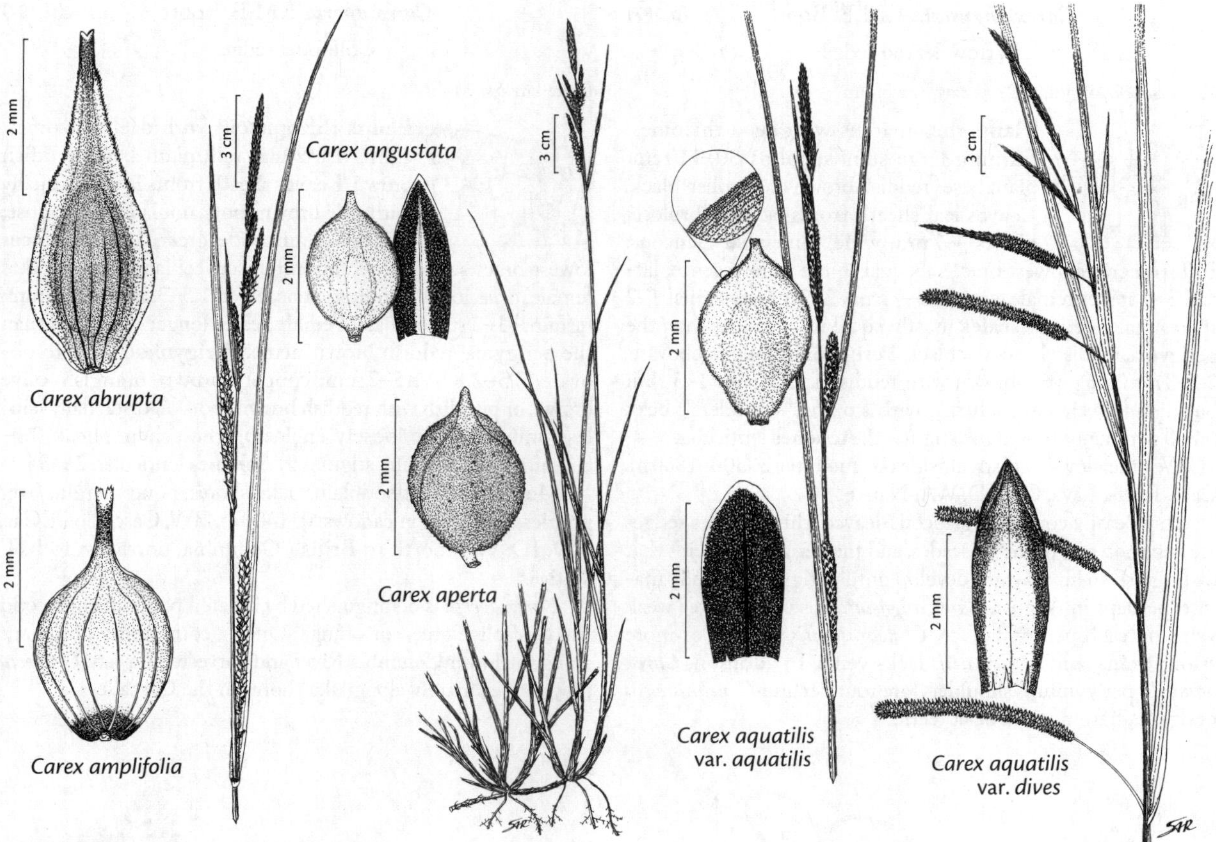

CAREX

Carex abrupta Mackenzie (p. 187)

Abrupt-beaked sedge

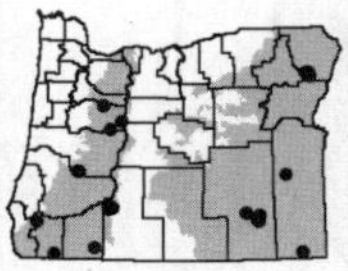

Plants cespitose, 18–70 cm tall. **Leaves** 1.5–3.7(4.9) mm wide. **Inflorescences** dense to slightly elongated, heads 1.2–2.2 cm × (6)9–20 mm; spikes gynaecandrous. **Female scales** 2.4–3.9 mm long, shorter than and as wide as or narrower than the perigynia, reddish brown with lighter or green mid-stripe. **Perigynia** typically elliptic to lance-ovate, sometimes lanceolate and nearly wingless, usually plano-convex, (2.9)3.6–5.4 × 1–2.1 mm, reddish brown or dark brown, lacking metallic sheen, with 3+ conspicuous veins that reach or exceed the top of the achene on each face of the perigynium; wing usually 0.2–0.3 mm wide and sometimes curved toward the ventral surface, making the perigynium boat-shaped; beak tips coppery or reddish brown, usually unwinged, brown, and parallel-sided for the distal 0.5–0.8 mm; (1.6)2–2.3 mm from top of achene to tip of beak; stigmas 2. **Achenes** lenticular, 1.2–1.8 × 0.7–1.1 mm, (0.3)0.4–0.5 mm thick.

Moist meadows and stream banks at moderate to high elevations in mountains. 1200–2800 m. BR, BW, Casc, Owy, Sisk. CA, ID, NV. Native.

Carex abrupta is a "ball-headed" sedge similar to *C. pachystachya* but with more strongly veined perigynia and living at higher elevations. It often grows with *C. microptera* and *C. haydeniana*, which have more translucent, less strongly veined perigynia.

Carex amplifolia F.M.B. Boott (p. 187)

Big-leaf sedge, ample-leaved sedge

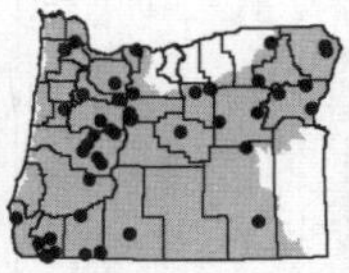

Plants rhizomatous, forming large populations, 50–100(130) cm tall. **Leaves** (3)8–20 mm wide. **Inflorescences** lower spikes female, ascending to spreading, (1.5)3.5–14 cm × (2.5)3.5–6.5 mm; upper spikes male, 5–9.5 cm long. **Female scales** smaller and narrower than the perigynia, often with a short awn. **Perigynia** obovoid, trigonous or somewhat inflated and more or less round, 2.4–3.1 mm long, brownish green, ribbed; beak narrow, 0.7–1.1 mm long; stigmas 3. **Achenes** trigonous.

Streamsides, road ditches, and other wet areas in conifer woodlands with a deciduous overstory component, often in the shade. 100–2300 m. BR, BW, Casc, CR, ECas, Lava, Sisk, WV. CA, ID, NV, WA; north to British Columbia, northeast to MT; NM. Native.

The broad leaves of *C. amplifolia* can be confused with those of native *Scirpus microcarpus* and invasive, cespitose *C. pendula*. *Carex aquatilis* var. *dives* has superficially similar inflorescences but only two stigmas per perigynium.

Carex angustata F.M.B. Boott (p. 187)

Narrow-leaved sedge

Carex eurycarpa Holm, *Carex oxycarpa* Holm

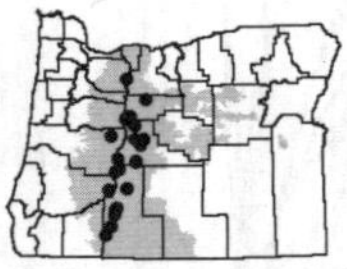

Plants rhizomatous with deep rhizomes; forming large stands; culms 30–110 cm; plant bases reddish brown or purplish black. **Leaves** leaf sheath fronts ladder-fibrillose; blades 4–7 mm wide, somewhat glaucous. **Inflorescences** lowest bract subequal to the inflorescence; lateral 3–4 spikes female, erect, 2.5–7 cm × 3–5 mm; terminal 1–2 spikes male. **Female scales** mostly equal to or longer than the perigynia, reddish brown or black. **Perigynia** elliptic to obovate, 2.2–3 mm long, pale brown with reddish spots; with 1–3 thin but distinct veins at maturity, with sloping "shoulders"; beak 0.2–0.5 mm, not bidentate; stigmas 2. **Achenes** lenticular.

Wet meadows and streamsides in mountains. 300–1800 m. Casc, ECas, Lava. CA, ID, WA. Native.

Three easily confused, glaucous-leaved, rhizomatous sedges are common east of the Cascades, and the perigynium veins that distinguish them may not develop until perigynia are fully mature, perhaps in August. *Carex angustata* has one to three weak veins on each perigynia face, *C. nebrascensis* has five or more strong veins, and *C. aquatilis* lacks veins. In addition, *Carex aquatilis* perigynium "shoulders" are rounder and *C. nebrascensis* perigynia have distinct beak teeth.

Carex aperta F.M.B. Boott (p. 187)

Columbia sedge

Carex accedens Holm

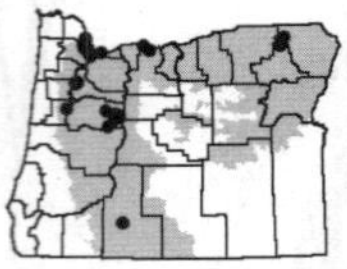

Plants rhizomatous with deep rhizomes; culms 15–90 cm tall; plant bases reddish brown. **Leaves** sheath fronts hyaline, usually with pale brown spots, not ladder-fibrillose; blades 3–6 mm wide, green. **Inflorescences** lowest bract subequal to the inflorescence; lateral 2–3 spikes female, occasionally androgynous, erect, 1.5–3.5 cm × 4–6 mm; terminal 1–2 spikes male. **Female scales** longer or shorter than the perigynia, reddish brown, acute. **Perigynia** elliptic to obovate, 2.5–2.8 × 1.5–2 mm; coppery brown, orangish, olive brown, or purplish with reddish brown spots on distal half; veinless, inflated, only loosely enclosing the achene; beak 0.1–0.3 mm, not bidentate; stigmas 2. **Achenes** lenticular. **2*n***=54.

Montane bogs, floodplains, lake shores, pond margins, wet prairies, and sedge meadows. 0–1400 m. BW, Casc, Col, ECas, WV. ID, WA; north to British Columbia, northeast to MT. Native.

Carex aperta is distinguished by inflated perigynia in an odd shade of olive green or orange. Once a community dominant along the lower Columbia River and harvested for hay, *C. aperta* now thrives mainly along lake shores in the Cascades.

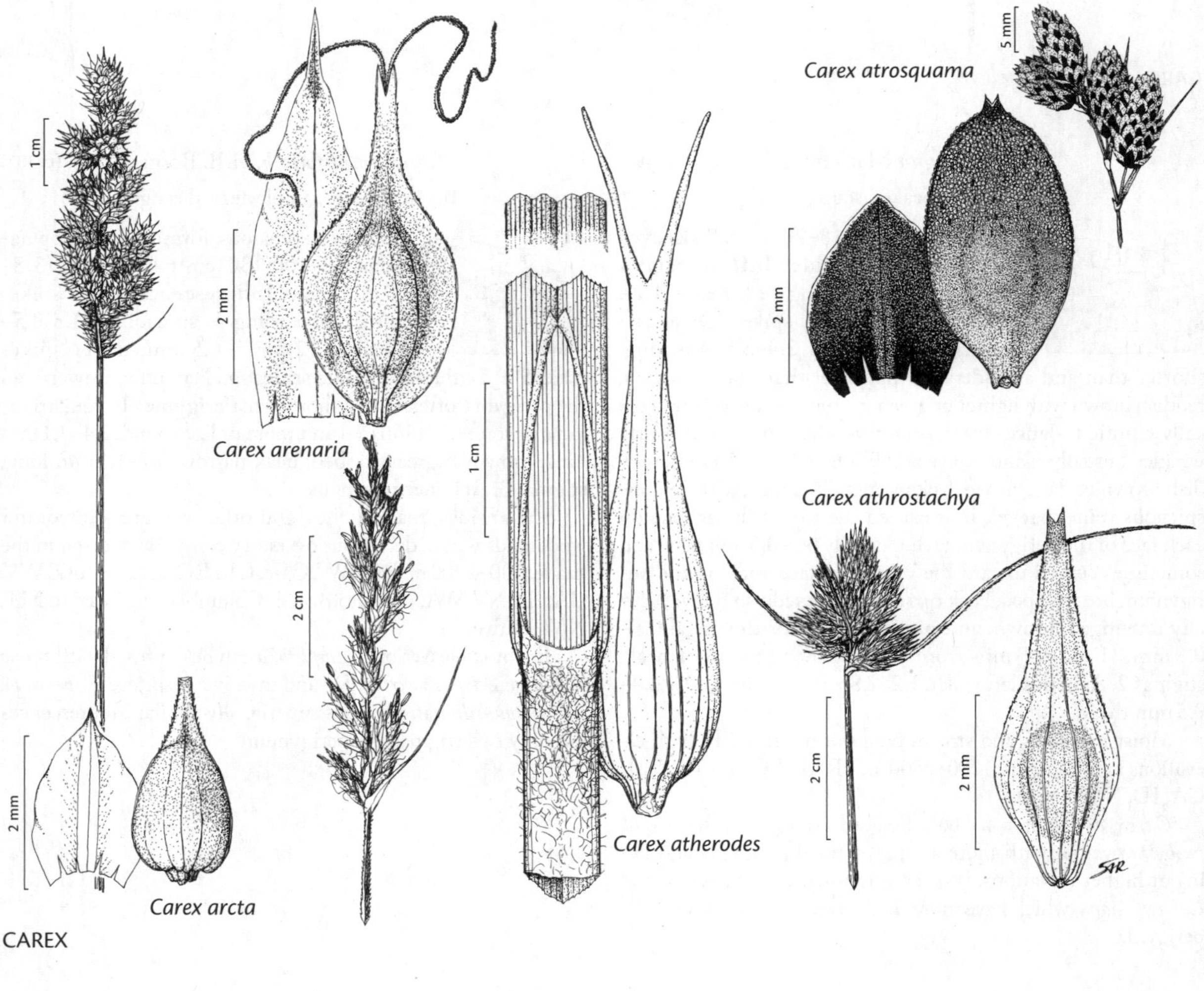

CAREX

Carex aquatilis Wahlenberg

Water sedge

Plants rhizomatous with deep rhizomes; sometimes forming large stands; culms 20–150 cm tall; plant bases reddish brown, chestnut, or brown. **Leaves** sheath fronts not ladder-fibrillose; blades 2.5–18 mm wide, green or glaucous. **Inflorescences** lateral 2–6 spikes female or androgynous, erect or drooping, 1–112 cm × 3–18 mm; bases blunt or tapering; terminal 1–4 spikes male. **Female scales** longer than the perigynia, reddish brown or black. **Perigynia** elliptic to obovate, 1.9–3.6 × 1–2.3 mm; green, tawny, or pale brown with reddish, purplish, or brownish spots; veinless or nearly so, with rounded "shoulders"; beak to 0.4 mm, not bidentate; stigmas 2. **Achenes** lenticular, 1.1–1.8 × 0.7–1.5 mm.

Carex aquatilis is a strongly rhizomatous sedge with two stigmas per perigynium and lenticular achenes. It is distinguished from similar species by its reddish or red-brown plant bases. Many *C. aquatilis* populations, especially in the Cascades, have plants with morphology intermediate between that of the varieties, or even resembling both extremes; these should be reported as *C. aquatilis,* not either variety.

Key to *Carex aquatilis* varieties:

1. Spikes erect, the lower ones usually entirely female, on peduncles up to 4 cm long; perigynia tawny with brown spots toward base; beak of the perigynia tawny, up to 0.2 mm long. . . . . ***C. a.*** var. ***aquatilis***

1' Spikes drooping, the lower ones usually androgynous, on peduncles up to 11 cm long; perigynia green or tawny throughout, often with reddish or purplish brown dots all over; beak of the perigynia generally purplish brown, 0.3–0.4 mm long. ***C. a.*** var. ***dives***

Carex aquatilis Wahlenberg var. *aquatilis* (p. 187)

Water sedge

Carex suksdorfii Kük.

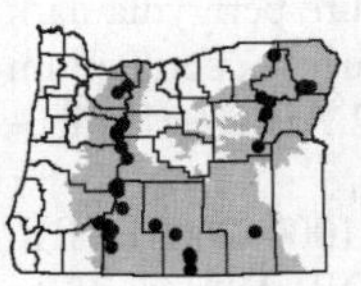

Culms 20–120 cm tall. **Leaves** blades 2.5–8 mm wide. **Inflorescences** lateral 2–6 spikes female, erect, not tapering at the base, 1–10 cm × 3–7 mm. **Perigynia** 2–3.6 × 1.3–2.3 mm, pale brown with reddish or brownish spots; beak to 0.2 mm. **2*n***=72, 74, 76, 80.

Wet sedge meadows, bogs, streamsides, and lake shores, in areas of slow-moving water where soil is inundated in early summer and has moisture in the root zone throughout the year. 500–2600 m. BR, BW, Casc, ECas. CA, ID, NV, WA; north to AK; throughout Canada and all but southeastern US; Greenland; Europe. Native.

This is a common wetland sedge in and east of the Cascades. Leaf traits are insufficient to distinguish this species from *C. angustata* or *C. nebrascensis*; see discussion under *C. angustata.* It has been introduced in a bog in Coos County.

Carex aquatilis Wahlenberg var. *dives* (H.T. Holm) Kükenthal (p. 187)

Sitka sedge, water sedge

Carex sitchensis J.D. Prescott ex Bong.

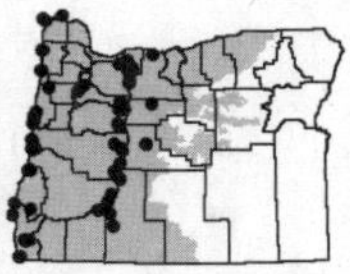

Culms 35–150 cm tall. **Leaves** sheath fronts red-dotted but not densely so, not coppery; blades 5–18 mm wide. **Inflorescences** lateral 2–6 spikes female or androgynous, drooping, often tapering at the base, 4.5–11.5 cm × 4–18 mm. **Perigynia** 1.9–3.5 × 1–1.2 mm, green or tawny with purplish brown spots; beak 0.3–0.4 mm. **2*n***=72, 76, 78, 80.

Wet meadows, bogs, lake shores, and streamsides, sometimes forming large stands. 0–1620 m. Casc, Col, CR, ECas, Est, Lava, Sisk, WV. CA, WA; north to AK, northeast to MT, NC. Native.

This sedge grows mainly west of the Cascades, often with *C. obnupta* and *C. lyngbyei,* both of which have very thick-walled, hard perigynia. Hybrids with *C. obnupta* are known along the Umpqua River.

Carex arcta F.M.B. Boott (p. 189)

Northern clustered sedge

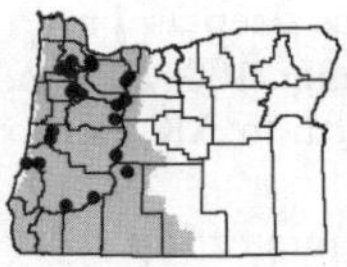

Plants cespitose, 15–80 cm tall. **Leaves** sheath fronts with tiny purplish dots; blades 2–4 mm wide, green, shorter to longer than the culms. **Inflorescences** 1.5–4 cm × 7–12 mm, usually green to light brown; spikes 5–15, crowded, gynaecandrous, oblong, 5–10 mm long, with 10–20 perigynia; perigynia beak tips giving the spikes a somewhat rough outline. **Female scales** shorter than the perigynia, hyaline and often brownish; with green midvein. **Perigynia** ovate, 2–3(3.5) × 1.2–1.5 mm, green to brown, with several veins; beak 0.75–1.25 mm long; dorsal suture conspicuous; stigmas 2. **Achenes** lenticular. **2*n***=60.

Wet places in meadows, woodlands, thickets, and forest opening. 0–1800 m. Casc, CR, ECas, Sisk, WV. CA, ID, WA; north to Yukon, east to Newfoundland and ME. Native.

Carex arcta can appear similar to taxa found in subkey 9, but the perigynia are unwinged. *Carex canescens* has gray-green foliage, fewer spikes, and shorter beaks.

Carex arenaria Linnaeus (p. 189)

Sand sedge

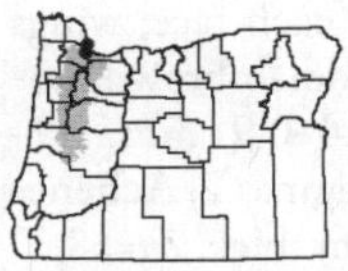

Plants rhizomatous, forming large populations, the rhizomes with a thick pithy cortex; culms 15–60 cm tall, erect. **Leaves** sheath fronts at top yellowish, thickened, concave; blades 1.5–4 mm wide. **Inflorescences** 2–8 cm long, somewhat shaggy looking; spikes 5–15, overlapping, ascending, androgynous or female, or the uppermost male. **Female scales** usually longer than the perigynia, reddish brown with green midrib and hyaline margins. **Perigynia** lanceolate to broadly ovate, plano-convex, (4)4.5–6 × 1.3–3.3 mm, light brown, several-veined, broadly winged above the middle; beak 1.2–2.5 mm long; stigmas 2. **Achenes** lenticular.

In Europe, where it is native, habitat is mainly moving sands near the coast but also inland sandy soils, usually in full sun but persisting in shade, e.g., in pine forests. 0–50 m. WV. DE to NC; Australia, Eurasia. Exotic.

Carex arenaria is strongly rhizomatous, with shaggy inflorescences and large, partly winged perigynia. It was introduced in Portland by 1916, probably in ships' ballast, and rediscovered on Sauvie Island in 2003.

Carex atherodes Sprengel (p. 189)

Awned sedge

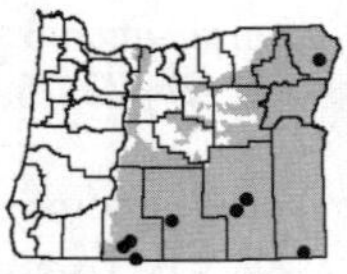

Plants rhizomatous, forming large stands, 35–200 cm tall. **Leaves** sheath front usually pubescent at least near the top, becoming ladder-fibrillose with age; blades 3–10 mm wide, usually sparsely hairy; basal leaf sheaths reddish purple; leaves and sheaths that develop underwater may be glabrous. **Inflorescences** lateral 3–6 spikes female, 2–10 cm, erect or ascending; terminal 2–6 spikes male, 2–6 cm. **Male scales** with scabrous awn, glabrous; rarely sparsely pubescent. **Female scales** narrow, glabrous or scabrous along midvein, with scabrous awn. **Perigynia** (6.5)7–12 × 1.8–3.8 mm, with sparse spreading hairs (occasionally glabrous), and 12–20 veins; beak 2.1–4 mm long, with spreading teeth (1.2)1.5–3 mm long; style persistent, straight; stigmas 3. **Achenes** trigonous.

Often alkaline lake margins, meadows, marshes, streams, and ditches, generally in sites with water as deep as 1 m in spring and remaining wet to moist in summer. 100–2000 m. BR, BW, ECas, Owy. CA, ID, NV, WA; north to AK, east to ME and VA; Eurasia. Native.

Carex atherodes is a tall, rhizomatous, more or less hairy sedge of shallow, usually alkaline wetlands. Similar *C. sheldonii* grows on drier terraces, and has narrower leaves and shorter, more densely pubescent perigynia with shorter teeth. Stands of *C. atherodes* are cut for hay.

Carex athrostachya Olney (p. 189)

Long-bract sedge

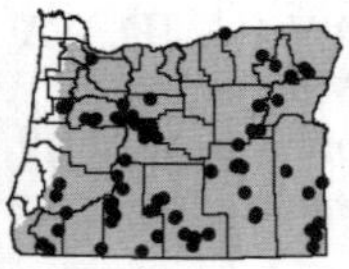

Plants cespitose, (5)20–80 cm tall. **Leaves** (1.5)2–3(5) mm wide. **Inflorescences** lowest 1–3 bracts usually longer than the inflorescence, bristle-like, occasionally leaf-like; inflorescences (0.8)1.5–2.2 cm × 7–20 mm, dense and head-like; spikes gynaecandrous. **Female scales** 2.4–4.3 mm long, shorter and narrower than the perigynia, gold to brown with midrib paler, green, or the same color. **Perigynia** lanceolate, (2.8)3.5–4(4.8) × (0.8)1–1.5(1.8) mm, cream-colored to light brown, with 0–9 veins on each face; wings (0.1)0.2(0.5) mm wide; beak tip gold to reddish brown, unwinged, and parallel-sided for the distal 0.4–0.9+ mm; 1.9–2.5 mm from top of achene to tip of beak; stigmas 2. **Achenes** lenticular, (1)1.2–1.6 × 0.7–1 mm, 0.3–0.4 mm thick. **2*n***=68.

Seasonally wet places, disturbed sites, pond margins, reservoir margins, roadside puddles, and riparian areas. 50–2600 m. BR, BW, Casc, Col, ECas, Lava, Owy, Sisk, WV. CA, ID, NV, WA; north to AK, east to Manitoba and TX; Mexico. Native.

This "ball-headed" sedge usually has conspicuous bracts longer than the fine-textured inflorescences. West of the Cascades, see *C. unilateralis*.

Carex atrosquama Mackenzie (p. 189)

Brass-fruit sedge

Carex atrata L. var. *atrosquama* (Mack.) Cronquist

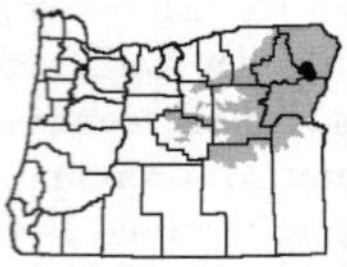

Plants cespitose, 20–50 cm tall. **Leaves** 3–5 mm wide. **Inflorescences** spikes 3–4(6), 0.8–2 cm long; lateral spikes female and erect; terminal spike gynaecandrous. **Female scales** usually shorter than the perigynia, dark purplish black, tip obtuse; the midrib usually dark like the rest of the scale or at least not strongly contrasting with it. **Perigynia** elliptic, more or less flattened except over the achene, 2.5–3.5 × 1.5–1.75 mm, green when young but maturing dark gold with darker tip, papillose (at 10×) distally; beak 0.3–0.5 mm long; stigmas 3. **Achenes** trigonous, nearly filling bodies of perigynia.

Alpine and subalpine grasslands, moist to mesic meadows, and streamsides. 2200–2400 m. BW. ID, WA; north to AK, east to CO. Native.

Carex atrosquama is a high elevation, cespitose plant with dark gold perigynia mostly hidden by blackish female scales. Compare to *C. heteroneura*.

Carex aurea Nuttall (p. 191)

Golden sedge, pumpkin sedge

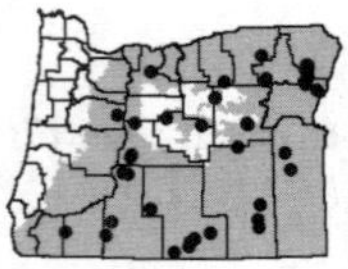

Plants rhizomatous, the rhizomes about 1 mm in diameter; culms 5–40 cm tall; plant bases whitish or medium brown. **Leaves** blades (1.4)2–3.5 mm wide, pale green to slightly glaucous. **Inflorescences** lateral spikes female, 0.4–2 cm × 3–5 mm, with perigynia relatively loosely spaced within the spike, with middle internodes (0.2)0.5–1.5 mm, averaging 0.65 mm; terminal spike usually male (rarely gynaecandrous), when male usually 0.5–1.7 cm long, averaging 0.9 cm × 2 mm. **Female scales** 1.2–2.5 mm, obtuse to acute, sometimes awned, often falling before the perigynia. **Perigynia** globose or obovate, 1.6–3.2 × 1.1–2 mm, nearly veinless at maturity but veined when young; immature perigynia green to tan, often white-powdery; fresh mature perigynia pale greenish becoming orange and more or less succulent when ripe, sometimes white-powdery; dried mature perigynia dark brown and slightly waxy; beak absent or nearly so; perigynium apex sometimes curved to one side; stigmas 2. **Achenes** lenticular, 1.3–1.8(2) × 1–1.6 mm. **2*n***=52.

Springs, riparian areas, or wet meadows. 1000–2600 m. BR, BW, Casc, Col, ECas, Owy, Sisk. CA, ID, NV, WA; north to AK, east to Newfoundland. Native.

When its perigynia are fully ripe, succulent, and orange, *C. aurea* is easy to identify, after which they quickly fall off. Until fully mature, *C. aurea* is easily confused with *C. hassei*; see discussion under that species.

Carex barbarae Dewey (p. 191)

Whiteroot sedge

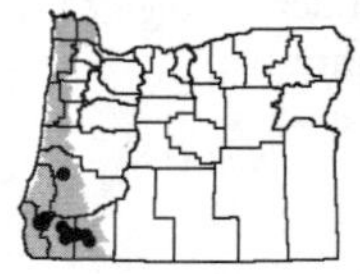

Plants rhizomatous, with deep rhizomes; forming large stands; culms 30–110 cm tall; plant bases reddish brown or purplish black. **Leaves** sheath fronts coppery colored, densely red-dotted, ladder-fibrillose; blades 4–9 mm wide, green to somewhat glaucous. **Inflorescences**

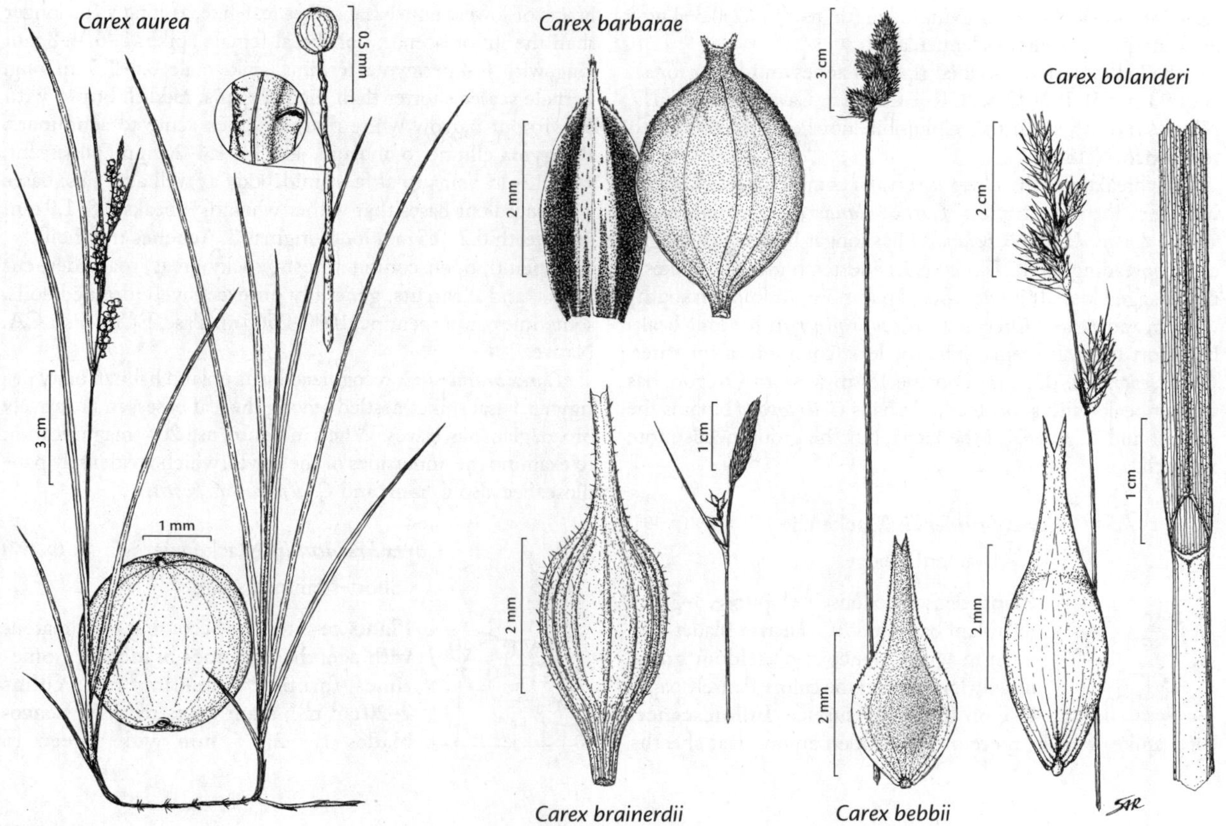

CAREX

lowest bract longer than the inflorescence; lateral 3–6 spikes female, 3–10 cm × 5–6 mm, brown, erect to drooping, tapering at base; upper 2–3 spikes male. **Female scales** reddish brown, acute, usually with a scabrous awn up to 2 mm long. **Perigynia** 3–4 × (1.9)2.2–2.5 mm, brown with reddish spots, faintly veined; beak approximately 0.5 mm, bidentate, scabrous; stigmas 2. **Achenes** lenticular.

Floodplains and terraces, riparian forests and open woodlands, savannas, meadows and ditches. 0–1300 m. CR, Sisk. CA. Native.

Carex barbarae grows with *C. obnupta* and *C. aquatilis* var. *dives*, but its typical habitat is drier, its spikes tend to be shorter and less curved, and its perigynia are not as tough as those of *C. obnupta*. Mostly sterile presumed hybrids of *C. barbarae* and *C. obnupta* occur on some tributaries of the Rogue River.

Carex bebbii (L.H. Bailey) Olney ex Fernald (p. 191)

Bebb's sedge

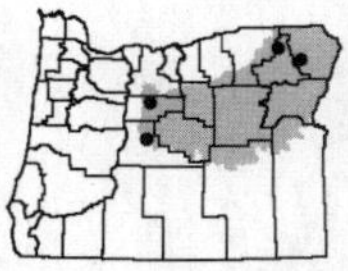

Plants cespitose, 20–90 cm tall. **Leaves** 1.7–4.2 mm wide. **Inflorescences** 1.1–3 cm × 5–14 mm; spikes 3–10, overlapping, more or less round, gynaecandrous, distinct. **Female scales** 2.5–3.5 mm long, shorter and narrower than their perigynia, reddish brown; mid-stripe green, pale, or brown. **Perigynia** ovate or elliptic, plano-convex, 2.5–3.8 × (1)1.2–2 mm, spreading, green when young, becoming reddish brown, with 3+ dorsal veins, the ventral side with 1–3 faint veins or veined only near the base; wings 0.2–0.5 mm wide; beak tips reddish brown, winged, and more or less ciliate-serrulate; 1.2–2.2 mm from top of achene to tip of beak; stigmas 2. **Achenes** lenticular, 1–1.3 × 0.6–0.9 mm, 0.3–0.4 mm thick. **2*n***=68, 70.

Meadows, stream banks, irrigation ponds and roadside ditches. 700–1300 m. BW, Lava. ID, WA; north to AK, east to Newfoundland. Native.

Carex bebbii is difficult to key but easier to recognize by its reddish brown inflorescences with rounded spikes. *Carex subfusca* has similarly small perigynia but its inflorescences are green, straw-colored, or light brown, and proportionately narrower. *Carex pachystachya* has larger, darker inflorescences. Both these species have beak tips unwinged and entire for 0.4–0.7 mm from the tip, unlike *C. bebbii*.

Carex bolanderi Olney (p. 191)

Bolander's sedge

Carex deweyana Schwein. var. *bolanderi* (Olney) W. Boott

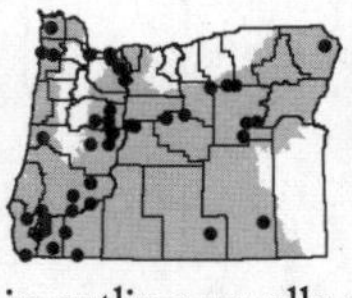

Plants cespitose, 31–115 cm long; culms smooth or minutely antrorsely scabrous (at 20×) at mid-length. **Leaves** 1.8–5.9 mm wide. **Inflorescences** spikes 5–9, light gold to straw-colored (to green), relatively rough in outline; usually gynaecandrous (or female); lowest spikes usually longer than wide, 12–22 (25) mm long, with 14–30+ perigynia; terminal spike 9.6–16 mm long. **Female scales** about as long as or longer than the mature perigynia, gold or straw-colored, sometimes white, usually awned (awn to 1.5 mm), mid-vein green. **Anthers** 1.3–1.8 mm long. **Perigynia** 3.4–5.2 mm long, green, straw-colored, to reddish brown; beak 1.5–2.5 mm long, (38)40–50+% of perigynium length; teeth (0.2)0.3–1 mm long; stigmas 2. **Achenes** lenticular.

Shaded to partially shaded riparian zones and moist forest. 0–2000 m. BR, BW, Casc, CR, ECas, Est, Lava, Sisk. CA, ID, NV, WA; north to British Columbia, northeast to MT, south to Mexico. Native.

The breakup of the *Carex deweyana* complex leaves Oregon with the "identical triplets" *Carex bolanderi, C. infirminervia,* and *C. leptopoda. Carex bolanderi* has longer beaks with longer, often spreading teeth. Those beaks and teeth give the spikes a more jagged look. It is widespread but more common in southwestern and eastern Oregon. *Carex infirminervia* has long beaks but short teeth. It seems to be the least common of the three. *Carex leptopoda,* the common species in western Oregon, has shorter beaks with short teeth. Perhaps *C. leptopoda* blooms the earliest and *C. bolanderi* the latest, but the group needs more research.

Carex brainerdii Mackenzie (p. 191)

Brainerd's sedge

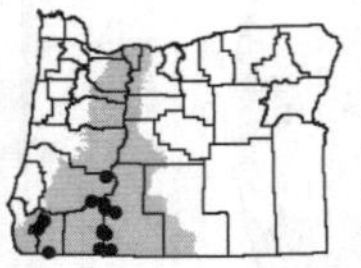

Plants densely to loosely cespitose, 5–35 cm tall; plant bases reddish. **Leaves** blades 1.8–3.8 mm wide, glaucous or seldom green, usually longer than the culms, densely papillose on the lower surface. **Inflorescences** basal spikes female, more or less hidden among leaf sheaths; bracts of lowest non-basal spikes leaf-like, as long as or longer than the inflorescence; non-basal female spikes 2–6, 4–8 mm long, with 1–4 perigynia; terminal spike male, 0.6–1.3 cm long. **Female scales** shorter than the perigynia, reddish brown with or without narrow white margins, apex acute to acuminate. **Perigynia** elliptic to obovoid, 4–5.3 × 1.4–2.2 mm, pubescent, with 12–15 veins to at least mid-body as well as 2 ribs; bases with succulent bases that wither when dry; beaks 0.8–1.9 mm long; teeth 0.2–0.5 mm long; stigmas 3. **Achenes** trigonous.

Upland, open conifer forests, rocky areas, roadsides, cut banks, and clearcuts, generally on excessively drained soils, sometimes on serpentine. 1000–2500 m. Casc, ECas, Sisk. CA. Native.

Carex brainerdii is recognized by its upland habitat, hairy perigynia, basal spikes nestled among the leaf bases, and relatively broad, glaucous leaves. When in doubt, use 20× magnification to examine the undersides of the leaves, which are densely papillose. See also *C. rossii* and *C. deflexa* var. *boottii*.

Carex brevicaulis Mackenzie (p. 192)

Short-stemmed sedge

Plants cespitose or becoming rhizomatous with age, the rhizomes branching, sometimes forming spreading turf; culms 2–20 cm tall; plant bases reddish. **Leaves** blades 1.3–2(3.5) mm wide, green to

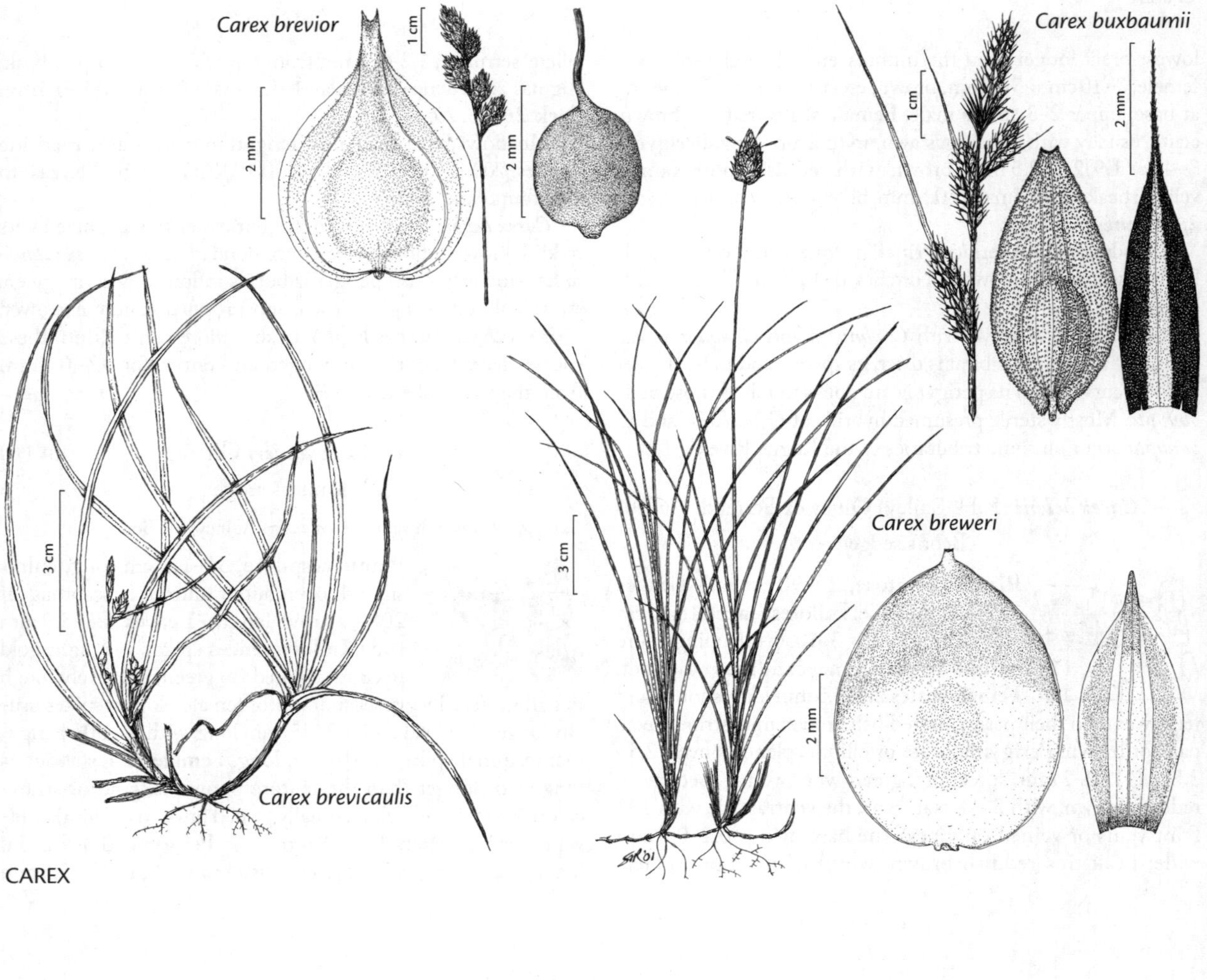

yellowish green, usually longer than the culms, arching and somewhat leathery, not densely papillose on the lower surface. **Inflorescences** basal spikes female, more or less hidden among leaf sheaths; bracts of lowest lateral spikes usually scale-like or bristle-like and shorter than the inflorescence, occasionally leaf-like and longer than the inflorescence; non-basal female spikes 2–5, 0.6–0.8 cm long, with 1–6 perigynia; terminal spike male, 0.5–1.4 cm long. **Female scales** pale to dark reddish brown with narrow white margins, apex acute to acuminate. **Perigynia** from lateral (not basal) spikes, globose, 3.5–4.8 × 1.7–2.1 mm, pubescent, with 2 ribs but otherwise veinless (or finely veined at the base); bases succulent, withering when dry; beaks 0.6–1.6 mm long, with apical teeth 0.2–0.5 mm long; stigmas 3. **Achenes** trigonous. **2*n***=28.

Stabilized coastal sand dunes, usually in dry meadow plant communities, not with European beach grass. 0–400 m. Est. CA, WA; north to British Columbia. Native.

Carex brevicaulis is recognized by its hairy perigynia, basal spikes nestled among the leaves, tough, green, somewhat arching leaves, usually short inflorescence bracts, and its stabilized sand dune habitat. Similar *C. rossii* grows in other habitats including coastal bluffs. Its perigynia average larger, and its inflorescence bracts are usually longer than the inflorescence.

Carex brevior (C. Dewey) Mackenzie (p. 192)

Plains oval sedge

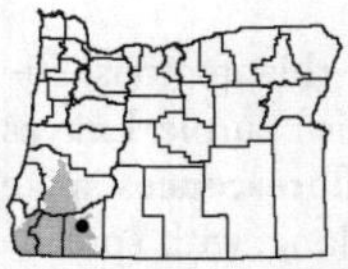

Plants cespitose, 15–120 cm tall. **Leaves** 1.5–3.5 mm wide. **Inflorescences** (1.3)2.5–5(6.5) cm × 5–10(20) mm, spikes gynaecandrous, overlapping but clearly distinguishable. **Female scales** lanceolate to ovate, shorter than the perigynia beaks, narrower than the perigynia, white-hyaline with some brown color, mid-stripe often gold or green, the tips usually acute. **Perigynia** orbicular or broadly ovate, plano-convex, (3)3.4–4.8(5.2) × (2)2.3–3.2 mm, widest near the middle, green or straw-colored, lacking veins or sometimes with 1–5 faint dorsal veins; wing 0.3–0.8 mm wide; beak winged and ciliate-serrulate almost to the tip; 1.5–2.4 mm from top of achene to tip of beak; stigmas 2. **Achenes** lenticular, 1.5–2.2 × 1.3–1.8 mm, 0.5–0.6 mm thick. **2*n***=48, 52, 56, 60, 64, 68.

Mesic grasslands and prairies, sandy stream banks, open woods and road ditches. 400–500 m. Sisk. ID, WA; north to British Columbia, east to ME; Mexico; Japan. Exotic.

With its broad wings that make the perigynium bodies almost round, *C. brevior* is relatively distinct. *Carex straminiformis* also has wide wings, but it grows near timberline. *Carex longii*, introduced in the coast, has perigynium bodies widest above the middle. *Carex molesta*, not yet found in Oregon but introduced in California, is similar to *C. brevior*, but its perigynia have three to six distinct veins over the achene on the ventral surface.

Carex breweri F.M.B. Boott (p. 192)

Brewer's sedge

Carex breweri Boott var. *breweri*

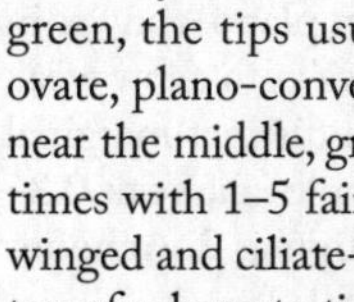

Plants rhizomatous, the rhizomes 2–3 mm thick; culms 15–22(30) cm tall. **Leaves** blades 0.6–1.2 mm wide, more or less cylindrical. **Inflorescences** spike solitary, thick, androgynous, 1.4–2.5 cm × 7–12 mm, lacking inflorescence bract. **Female scales** shorter and narrower than the perigynia, 3–5-veined, white down the center, lower scales sometimes awned. **Perigynia** broadly elliptic, 5–7 × 3–4.8 mm, ascending, with 3–10 short veins on the faces, with rachilla longer than the achene; stigmas 3. **Achenes** trigonous, much smaller than the perigynia.

Dry, open slopes, often on pumice. 700–2900 m. Casc. CA, NV, WA. Native.

Its almost round spikes and perigynia, short stature, and high elevation habitat make *C. breweri* distinctive.

Carex buchananii S. Berggren

Leatherleaf sedge, bronze sedge, red fox sedge, New Zealand copper sedge

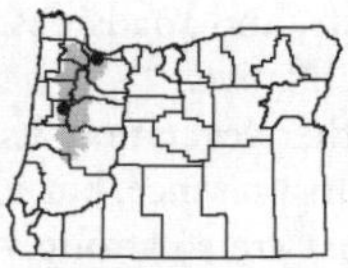

Plants densely cespitose, 10–40(50) cm tall. **Leaves** sheaths dark red-purple to orange; blades 1–1.5 mm wide, orange, red-brown, (olive green in shade); lower surface shiny, upper surface dull; tips often curling. **Inflorescences** spikes 5–6, lateral spikes female, 0.5–3 cm × 4 mm wide; terminal spike 1(2) staminate. **Female scales** shorter than or as long as the perigynia, whitish. **Perigynia** elliptic, plano-convex, 2.5–3 mm long, pale cream with dark brown or purple-black blotches above, occasionally pale throughout, faintly nerved; beak well defined, 33–50% as long as the body, with scabrous margins; stigmas 2. **Achenes** lenticular.

Disturbed urban areas, mesic to wet grasslands. 50–100 m. WV. Australia, New Zealand. Exotic.

Carex buchananii is orange-brown to reddish and commonly planted as an ornamental. It is beginning to escape in the Pacific states. It is considered an invasive weed in southeastern Australia.

Carex buxbaumii Wahlenberg (p. 192)

Brown bog sedge

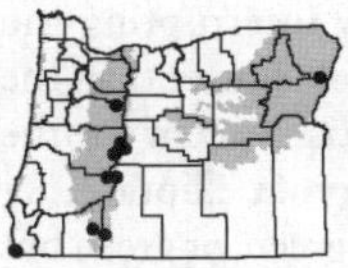

Plants rhizomatous, 25–75 cm tall. **Leaves** sheaths rich reddish brown, ladder-fibrillose; blades 2–3.5 mm wide, glaucous. **Inflorescences** spikes 3–5, 1–2.5 cm long, the lateral spikes female; terminal spike gynaecandrous. **Female scales** light to dark brown with light midvein, mucronate to awned, awn 0–3 mm. **Perigynia** elliptic, 2.5–4 × 1.5–2 mm, whitish, pale green, or gray, strongly papillose, beakless or nearly so; stigmas 3. **Achenes** trigonous, nearly as big as perigynium body.

Bogs and wet meadows. 0–1900 m. BW, Casc, Est. ID, WA; north to AK, east to Greenland, southeast to SC; Africa, Australia, Eurasia. Native.

Its glaucous leaves with rich reddish brown leaf bases, pale spikes with a shaggy look due to the awned, striped female scales, and its pale, strongly papillose perigynia distinguish *Carex buxbaumii* from all similar species. *Carex livida* also has pale, papillose perigynia, but its terminal spike is male, and its female scales are awnless.

Carex californica L.H. Bailey (p. 195)

California sedge

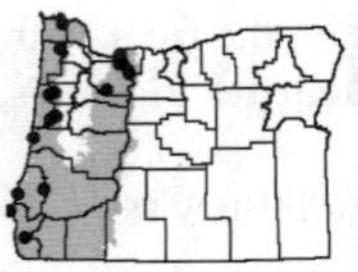

Plants rhizomatous, 15–70 cm tall; plant bases dark reddish brown. **Leaves** lower sheaths lacking blades; blades 1.5–5.5 mm wide, glaucous, flat to folded. **Inflorescences** lateral spikes female or androgynous, 1–3.5(5) cm × 2.5–7 mm; terminal spike usually male or androgynous. **Female scales** as long as the perigynia or shorter, reddish brown to purplish midrib lighter, papillate; apex acute to obtuse, sometimes mucronate or awned. **Perigynia** elliptic, 3.3–4.3(5) mm long, reddish brown or green with reddish brown speckles or blotches, strongly papillose; beak 0.5–1 mm long, distinct; stigmas 3. **Achenes** trigonous.

Meadows, open burned forests, clearcuts, and roadsides. 0–1600 m. Casc, CR, Est, Sisk. CA, ID, WA. Native.

Carex californica often grows mixed in other vegetation, its inflorescences the only obvious evidence of its presence, but it may form clumps on roadsides. Its perigynia are so strongly papillose that they may appear gland-covered. The rich reddish or purplish brown of its plant bases and female scales distinguish it from its paler relatives *Carex livida* and *C. klamathensis.*

Carex canescens Linnaeus ssp. ***canescens*** (p. 195)

Silvery sedge

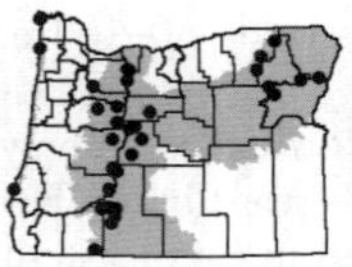

Plants densely cespitose, 15–60 cm tall. **Leaves** sheath fronts white-hyaline, U-shaped at mouth; blades (1.5)2–4 mm wide, usually gray-green (sometimes green), shorter than or equal to the culms; leaf sheath fronts white-hyaline, U-shaped at summit. **Inflorescences** 2–5(7) cm long; usually greenish gray to straw-colored (sometimes dark brown); spikes 4–8, upper spikes crowded but the lower more separated, gynaecandrous, usually oblong but sometimes subglobose, 3–12 mm long with (5)10–20(30) perigynia, the beaks of the perigynia scarcely interrupting the outline of the spikes. **Female scales** equal to or shorter than the perigynia but not hiding them, usually whitish except for the midvein, to green or tinged brownish. **Perigynia** elliptic, 1.8–2.3(3) mm long, ascending, usually pale, gray-green, or green but in some populations yellow-brown or brown; veins many, obscure; beaks 0.3–0.5 mm long, blunt; dorsal suture usually shorter than the beak, inconspicuous; stigmas 2. **Achenes** lenticular. **2*n***=54, 56, 62.

Bogs, wet meadows, wet forests, lake shores, and pond margins. 0–2300 m. BW, Casc, ECas, Est, Lava. CA, ID, NV, WA; most of North America except the Great Plains and southeastern states; South America; Australia, Eurasia. Native.

Typical plants with glaucous foliage and whitish, almost beakless perigynia are distinctive. Dwarf plants with short spikes of dark perigynia can be a problem. *Carex praeceptorum* is similar, but its spikes all overlap, its brown perigynia have a dark brown dorsal suture, darker than the rest of the beak, and it grows at higher elevations. *Carex arcta* is a greener plant with longer perigynia beaks.

Carex capillaris Linnaeus (p. 195)

Hair sedge

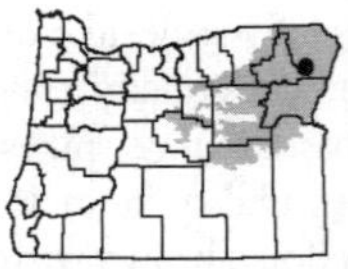

Plants cespitose, (1.8)10–60 cm tall. **Leaves** (0.75)1–4 mm wide, flat. **Inflorescences** sheath of lowest bract well developed; lower spikes usually nodding or drooping; lateral spikes female, with 6–20 perigynia; terminal spike male (rarely gynaecandrous), 4–10 mm long. **Female scales** shorter but wider than the perigynia, light to medium brown at least at the base, obtuse to acute or acuminate, falling early; mid-stripe paler gold to brown or greenish, smooth or slightly scabrous. **Perigynia** oblong-ovate, 2.3–3.5 × 0.8–1.2 mm, light green, gray, straw-colored, or brown, veinless but with 2 marginal ribs; margins smooth or minutely roughened; beak 0.5–1 mm long, indistinct, sometimes minutely roughened; stigmas 3. **Achenes** trigonous. **2*n***=54.

Moss mats in moist areas; bogs, marshes, stream banks, wet meadows, moist woods, usually on limestone substrates. 1600–2800 m. BW. CA, ID, NV, WA; north to AK, east to Greenland and ME, southeast to NM; Eurasia. Native.

Carex capillaris is a delicate sedge with dangling spikes, similar only to *C. tiogana,* a smaller, alpine species found on Steens Mountain.

Carex capitata Linnaeus (p. 195)

Capitate sedge

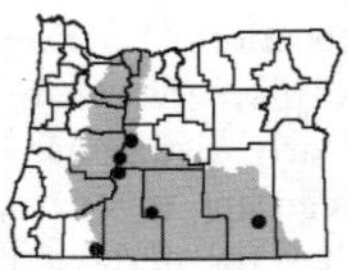

Plants cespitose or short-rhizomatous, 10–30(40) cm tall, bases reddish purple. **Leaves** to about 1 mm wide. **Inflorescences** spike solitary, 0.5–1.1(1.8) cm long, with spreading perigynia. **Female scales** ovate, about as long as the perigynium body, margins narrowly white-hyaline. **Perigynia** narrowly ovate to nearly circular, plano-convex in cross section, 2–4 × 1.5–1.8 mm, dorsal surface with a few weak veins, ventral surface without veins but with distinct marginal ribs; with rachilla, abruptly narrowing to the beak; beak 0.3–0.7 mm long; smooth or slightly serrulate; stigmas 2. **Achenes** lenticular, much smaller than perigynia. **2*n***=50.

Wet or seasonally wet meadows, often alpine but also at lower elevations in cold air drainages or cold springs, often on sandy, acidic soils, usually where snowpack is shallow but the ground remains moist in summer due to snowmelt. 1400–2900 m. BR, Casc, ECas. CA, NV; north to AK, east to Greenland and NH, south to northern Mexico; South America; Europe. Native.

Carex capitata has almost circular, strongly spreading perigynia. Similar *Carex nigricans,* a common subalpine species, is usually turf-forming and has narrowly elliptic perigynia.

Carex chordorrhiza Ehrhart ex Linnaeus f. (p. 195)

Rope-root sedge

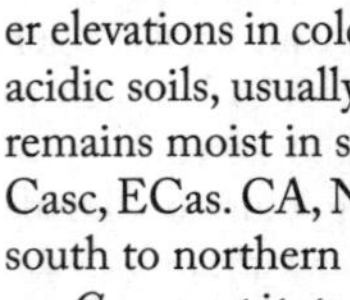

Plants rhizomatous but rhizomes rarely collected; flowering culms 5–35 cm tall; vegetative culms erect when young, becoming horizontal stolons 1+ m long. **Leaves** 0.4–3 mm wide. **Inflorescences** oval, 0.5–1.6 cm long; spikes 2–7, sessile, bractless, erect, androgynous, or lateral spikes female. **Female scales** about as long as the perigynia, brown with greenish center and pale margins. **Perigynia** elliptic,

2–4(4.5) × 1.4–2.2 mm, green or brown, glossy; with many veins as well as 2 marginal ribs; stigmas 2. **Achenes** lenticular. **2*n***=60.

Perennially wet, organic soils of bogs, sedge marshes, and stream banks. 0–100 m. Est. WA; north to AK, east to Greenland; Eurasia. Exotic.

Carex chordorrhiza is recognized by its small oval inflorescences that can be mistaken for single spikes, long stolons, and its bog habitat.

Carex comosa F.M.B. Boott (p. 197)
Bristly sedge

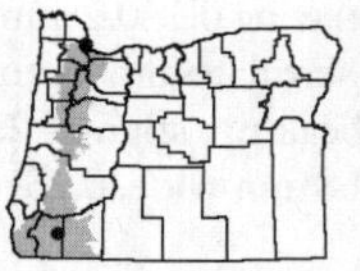

Plants cespitose, 50–120 cm tall; plant bases usually lacking red. **Leaves** 5–18 mm wide. **Inflorescences** lateral 2–5 spikes female, 2–7 cm long, erect or the lower ones nodding; terminal spike to 6 cm long, male, occasionally with some female flowers. **Female scales** lanceolate and tapering to a long, scabrous awn; shorter than the perigynia or lower ones longer. **Perigynia** inflated, (4.8)6.2–8.7 × 1.1–1.8 mm, glabrous, with 14–22 strong veins set close together and separated by less than 2 times their width; gradually tapered to the beak; beak 2–3.8 mm long; teeth widely spreading, 1.3–3 mm long; style persistent, becoming curved; stigmas 3. **Achenes** trigonous.

Marshes, lake margins, wet meadows, bogs, and wet thickets. CA, ID, WA; north to British Columbia; eastern US and Canada from Ontario to ME, SC, TX, south to Mexico. Native.

With its bristly spikes, awned pistillate scales, and long, spreading perigynium teeth, cespitose *C. comosa* resembles only *C. hystericina*, which has shorter, straight perigynium teeth.

Carex concinna R. Brown (p. 197)
Cute sedge

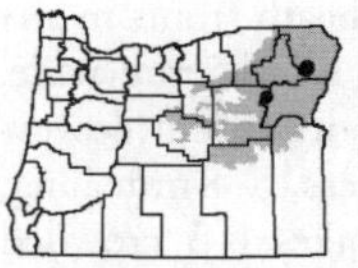

Plants rhizomatous, shoots arising singly or in clumps, 5.5–20 cm tall; plant bases reddish brown. **Leaves** sheath fronts membranous; blades 1–3.1 mm wide. **Inflorescences** basal spikes absent; lowest bract sheath inflated, 2–5 mm long, blade absent or less than 2 cm long, purple-tinged; lateral spikes 2–3, crowded near the terminal spike, 0.4–0.7 × 0.2–0.4 mm; terminal spike male, tiny, 0.3–0.7 cm × 1.1–1.5 mm. **Male scales** dark reddish brown with whitish margins. **Female scales** smaller than the perigynia, reddish brown, ovate, apex obtuse, minutely ciliate. **Anthers** 1.3–1.4 mm long. **Perigynia** elliptic, 2.3–3.3(3.5) × 1.1–1.4 mm, pubescent with crinkled hairs, veinless but with 2 ribs; bases succulent, withering when dry; beak 0.3–0.4 mm long; stigmas 3. **Achenes** Rounded to trigonous. **2*n***=54.

Stream banks, bogs, meadows, and open forests, on stable, usually calcareous substrates that are wet in spring but may dry out later. 1600–2500 m. BW. North to AK, east to Newfoundland. Native.

Carex concinna is recognized by its hairy perigynia, tiny staminate spike, and absent or tiny inflorescence bract. It lacks basal spikes among the leaf bases.

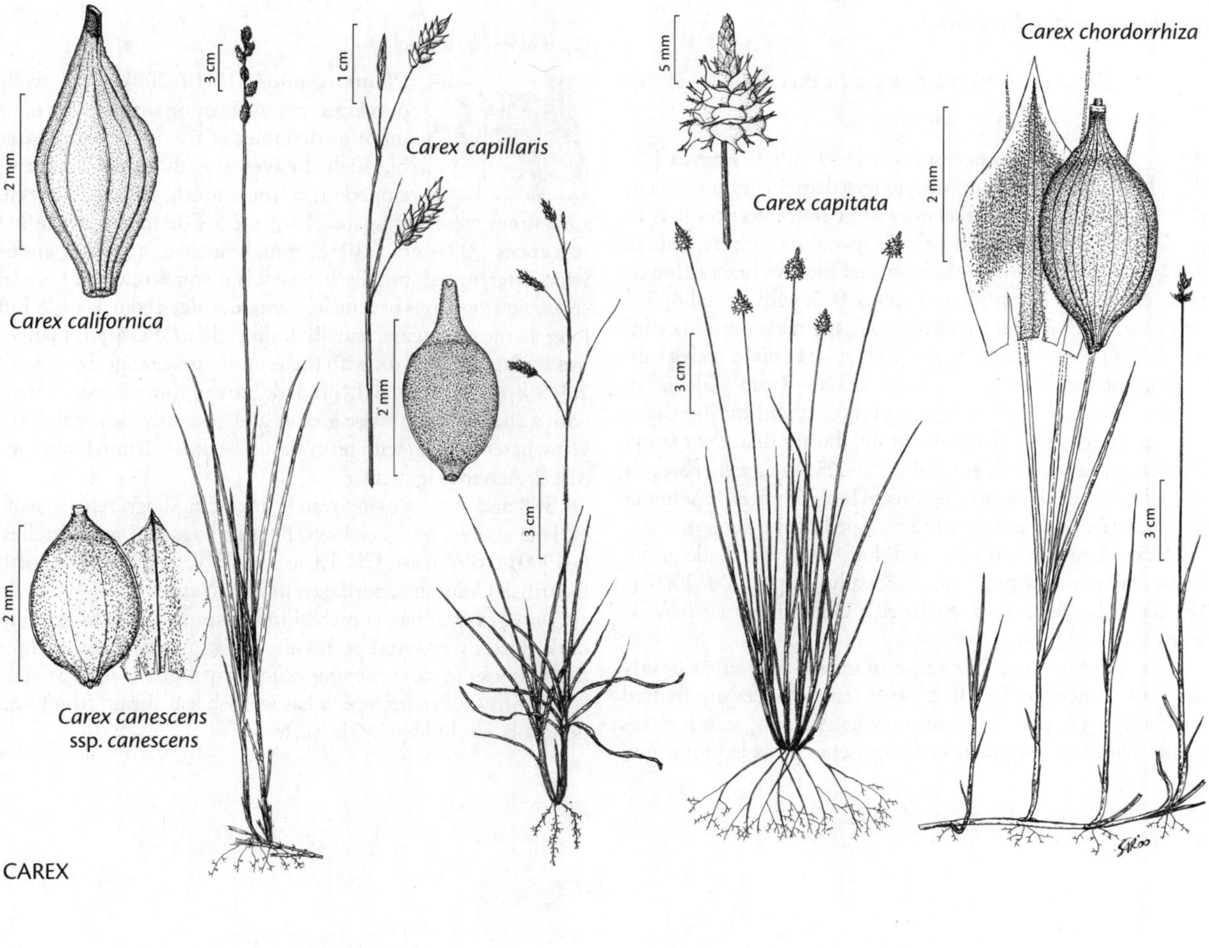

CAREX

Carex concinnoides Mackenzie (p. 197)

Northwestern sedge

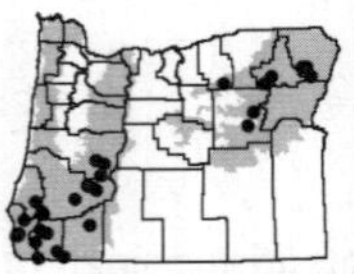

Plants rhizomatous, the shoots usually arising singly and spaced well apart, sometimes a few together, 14–37 cm tall, arching to the ground as the perigynia mature; plant bases reddish brown. **Leaves** sheath fronts membranous, mouth with minute cilia; blades 1.7–4.5 mm wide. **Inflorescences** 1.2–3.3 cm long; basal shoots absent; lowest bract reddish brown, bristle-like, sheathless, 5–8 mm long, sometimes with awn to 1(6) mm; lateral spikes 1–3, crowded near the terminal spike, 0.6–1.3 cm × 3–6 mm; terminal spike male, 0.8–2.2 cm × 1.8–3.1 mm. **Male scales** dark reddish brown or purplish black, with whitish margins. **Female scales** ovate or obovate, smaller than the perigynia, dark reddish brown, minutely ciliate, apex obtuse to acute. **Anthers** 1.9–3.1 mm long. **Perigynia** elliptic to obovoid, 2.5–3 × 1.4–1.7 mm, pubescent with straight hairs, veinless but with 2 ribs; bases succulent, withering when dry; beak 0.5 mm long; stigmas 4. **Achenes** rounded but with somewhat quadrangular bases.

Open conifer forests, sometimes on serpentine substrates. 50–2500 m. BW, Casc, CR, Sisk. CA, ID, WA; north to British Columbia, northeast to MT. Native.

Carex concinnoides is the only North American sedge with 4 stigmas. It blooms early in spring, before nearly all other flowers, and attracts pollinating insects. Its perigynia are hairy, and it lacks basal inflorescences among the leaf bases. Its loosely spaced shoots or tufts with dark green, arching leaves make even sterile plants identifiable. In southwestern Oregon, it often grows with *C. serpenticola*, which has three stigmas per perigynium and is usually dioecious.

Carex cordillerana Saarela & B.A. Ford (p. 197)

Cordilleran sedge

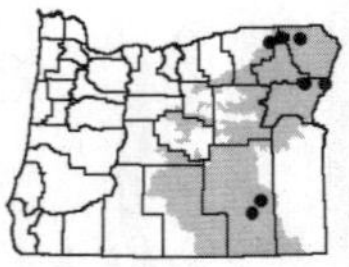

Plants cespitose, 5–40 cm tall. **Leaves** 1.5–5.9 mm wide, longer than the inflorescences. **Inflorescences** some inflorescences may be on relatively long culms, others hidden among the basal leaf sheaths; bract of lowest spike leaf-like or absent; lateral spikes 0–3, with a total of 3–5 perigynia; terminal spike androgynous, the male portion cylindric, 1.7–2.6 mm, hidden by the perigynia. **Female scales** subtending the lower perigynia, leaf-like, 2.5–3 mm wide, often 1–2(7) cm long, longer than the perigynia and hiding them, green; scales of upper perigynia ovate, shorter than the perigynia. **Perigynia** obovoid, 3.9–5.4 × 1.6–2.5 mm, green, abruptly tapered to the beak; beak 0.5–1.6 mm long; stigmas 3. **Achenes** rounded-trigonous and tightly enveloped by the perigynium.

Rocky slopes, in leaf litter and duff, usually in shade of deciduous trees and shrubs or occasionally juniper. 700–2000 m. BR, BW. ID, WA; north to British Columbia, east to WY. Native.

This cespitose grass-like sedge of upland habitats is utterly distinctive, once the leaf-like lower female scales are figured out. Cattle often pull the plants up when grazing, so it survives mainly where woody plants or rocky slopes provide protection.

Carex crawfordii Fernald (p. 197)

Crawford's sedge

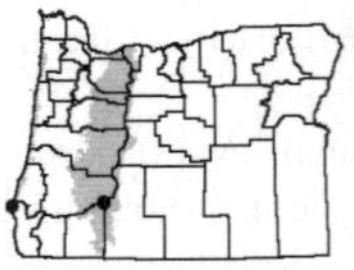

Plants cespitose, 25–60(85) cm tall. **Leaves** 2–4 mm wide. **Inflorescences** 1.8–3 cm × 8–14 mm, dense and head-like; spikes gynaecandrous, crowded or the lower somewhat remote. **Female scales** shorter and narrower than the perigynia, gold to dark brown with mid-stripe lighter or darker, lanceolate, 3–3.8 mm. **Perigynia** flat except over the achene or sometimes plano-convex, 3.4–4.1(4.7) × 0.9–1.3 mm, ascending, whitish to light brown, lanceolate (often narrowly so), veinless or with up to 5 veins on each face; wing 0.1–0.2 mm wide; beak winged and more or less ciliate-serrulate almost to the tip, (1.8)2.1–3 mm from achene top to beak tip; stigmas 2. **Achenes** lenticular, 1.1–1.5 mm long, 0.6–0.8 mm wide, 0.15–0.35 mm thick.

Pond margins that are wet in spring and other seasonally wet spots. 0–1500 m. Casc, Est. ID, WA; north to AK, east to Newfoundland. Native and exotic.

Carex crawfordii is a tufted, yellow-green plant with a tight head. It grows in the "bathtub ring" around a seasonal pond. It resembles *C. athrostachya*, but the inflorescence bracts are usually not prolonged beyond the end of the inflorescence, and the spikes and heads are more slender. Coastal plants were introduced in cranberry bogs, probably from northeastern North America.

Carex cusickii Mackenzie ex Piper & Beattie (p. 197)

Cusick's sedge

Carex obovoidea Cronquist

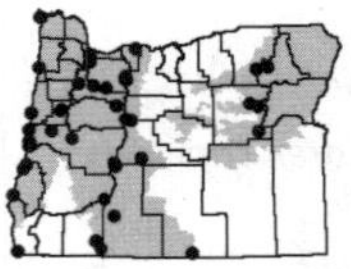

Plants cespitose, (5)30–200 cm tall, well-developed plants forming tussocks 0.3 m or more in diameter at the base; plant bases blackish. **Leaves** sheath fronts coppery-tinged near the mouth and red-dotted, sometimes cross-corrugated; blades 2.5–5(6) mm wide. **Inflorescences** (2)3–8 cm × 10–20 mm, branched, nodding, somewhat interrupted, usually bisexual but sometimes all female; spikes androgynous or female. **Female scales** about as wide and long as the perigynia, usually hiding them. **Perigynia** lance-ovate, plano-convex (or with the ventral surface slightly convex), 2.4–3 × 1.3–1.8 mm, light to dark brown, shiny, dorsal surface with a shallow lengthwise groove and generally also with 7–11 veins; bases swollen with pithy tissue; beak 1–1.3 mm long; stigmas 2. **Achenes** lenticular.

Still and slow-moving waters, including bogs, fens, marshes, lake shores, wet meadows, floating logs and road ditches. 0–1900 m. BW, Casc, CR, ECas, Est, WV. CA, ID, WA; north to British Columbia, northeast to MT. Native.

Carex cusickii has branched inflorescences that hide shiny, dark, dorsally grooved perigynia. Large plants form distinctive tussocks in slow-moving water. Small plants may be confused with *C. diandra*, which has whitish leaf sheath fronts and perigynia not hidden by the scales.

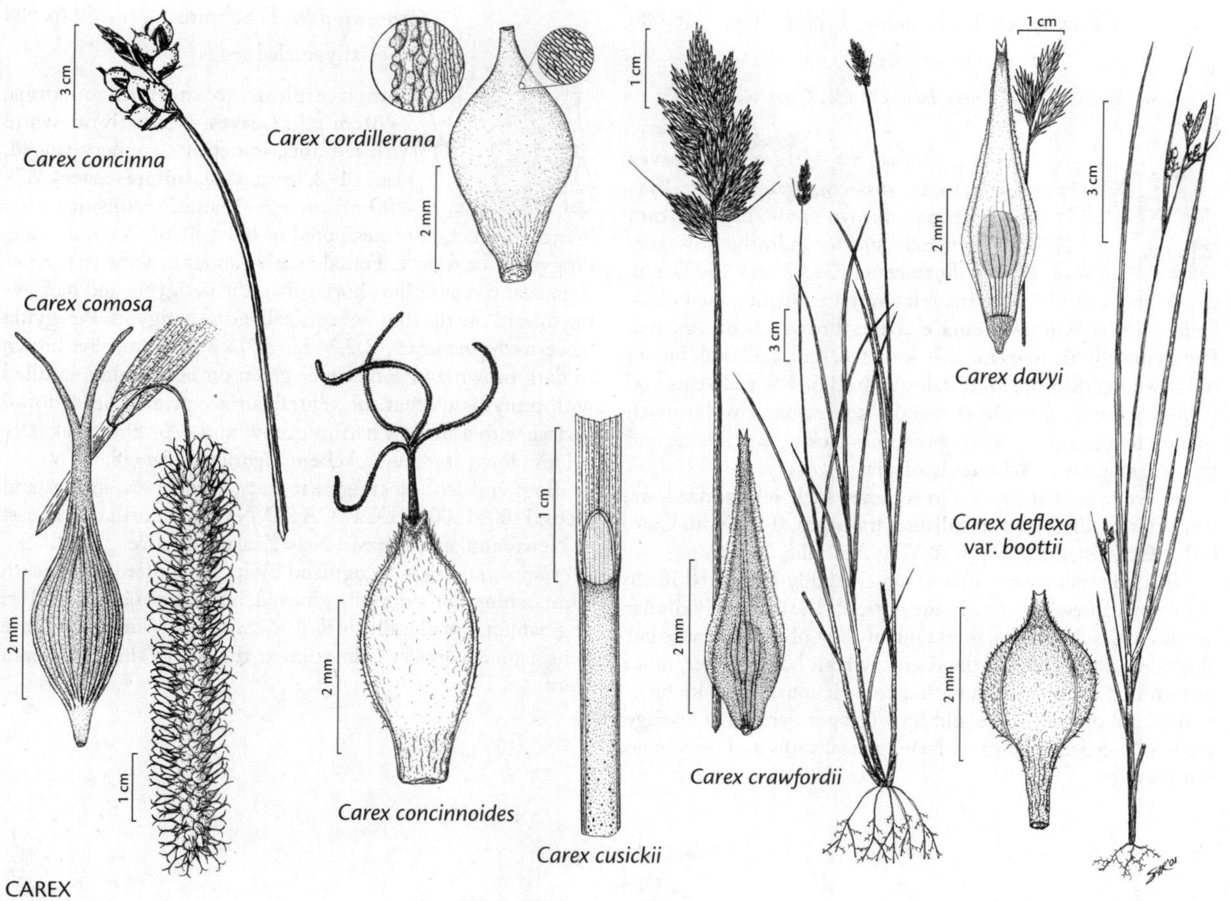

CAREX

Carex davyi Mackenzie (p. 197)

Davy's sedge

Carex constanceana Stacey

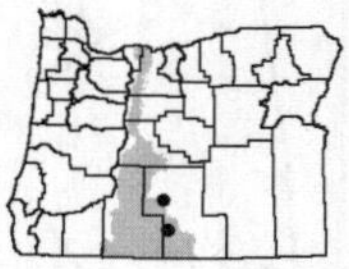

Plants cespitose, 25–45 cm tall. **Leaves** 2–3 mm wide. **Inflorescences** 2.5–4 cm × 13–15 mm, erect; spikes 3(7), gynaecandrous, distinct. **Female scales** 4.5–5.8 mm long, shorter and narrower than the perigynia, revealing most of the beak, reddish brown or coppery with lighter mid-stripe, white margins 0–0.2(0.4) mm wide. **Perigynia** lanceolate to ovate, 6–8.5 × (1.5)2–2.4 mm, appressed to ascending, greenish to straw-colored; 8–16 dorsal veins and 4–7 ventral veins, 3 or more of them extending beyond the achene onto the beak; beak tip unwinged, brown, parallel-sided, and entire for the distal 0.5–0.9 mm; (2.6)3–4.2 mm from top of achene to beak tip, occasionally winged nearly to the tip; dorsal suture inconspicuous; stigmas 2. **Achenes** lenticular, 2–3 × 1.1–1.5 mm, 0.4–0.7 mm thick.

Seasonally moist scablands, open conifer forest, and ephemeral watercourses. 1800–1900 m. ECas. CA, WA. Native.

Carex davyi is a rare sedge with large perigynia and female scales that reveal most of the perigynium beak. *Carex petasata,* a widespread species of sagebrush steppe, has similar large perigynia, but its female scales are nearly as long as the perigynia, concealing most of the beak, and have wide white margins (0.2–0.7 mm wide).

Carex deflexa Hornemann var. ***boottii*** L.H. Bailey (p. 197)

Mountain mat sedge

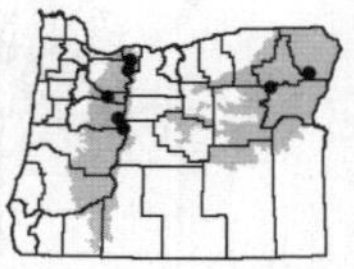

Plants cespitose to mat-forming, 5–31 cm tall; plant bases reddish to purplish or dark brown, often with persistent leaves of previous year. **Leaves** blades 0.9–2.6(3.2) mm wide, green, usually longer than the culms, not densely papillose on the lower surface. **Inflorescences** basal spikes female, 0–3, more or less hidden among leaf sheaths or sometimes raised on stalks much longer than the spike; bract of lowest non-basal spike leaf-like, longer than the inflorescence; non-basal female spikes 1–4, 0.3–0.8 cm long, with 4–15 perigynia; terminal spike male, 0.3–1.1 cm long. **Female scales** shorter than the perigynia, pale to dark reddish brown with white margins, apex acute to acuminate. **Perigynia** of lateral (non-basal) spikes elliptic to obovoid, 2.3–3.1 × 1–1.4 mm, pubescent, veinless but with 2 ribs; bases succulent, withering when dry; beaks 0.4–0.8 mm; teeth 0.1–0.2 mm long; stigmas 3. **Achenes** trigonous.

Establishing in disturbed openings and persisting in open forests in mountains, at elevations where there is snow for much of the winter. 800–2500 m. BW, Casc. CA, ID, WA; north to British Columbia, east to CO. Native.

Carex deflexa var. *boottii* is an upland taxon with hairy perigynia and basal spikes hidden among the leaf sheaths. It is very similar to *C. rossii,* which typically has a longer beak and tougher leaves and grows at lower elevations.

Carex densa (L.H. Bailey) L.H. Bailey (p. 198)

Dense sedge

Carex breviligulata Mack., *Carex dudleyi* Mack., *Carex vicaria* L.H. Bailey

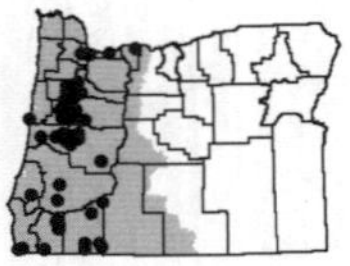

Plants cespitose, 50–110 cm tall. **Leaves** sheath fronts cross-corrugated, white-hyaline with pale brown spots; blades 5 mm wide, generally shorter than the flowering culm. **Inflorescences** 3–5.7 cm × 15–25 mm, dense and cylindrical, with relatively inconspicuous bracts; spikes androgynous. **Female scales** brown, short-awned. **Perigynia** elliptic to ovate, 2.8–4 × 1.5–2 mm, yellowish brown to brown, reddish brown distally, with (3)5–7 dorsal veins and (0)3–5 ventral veins; base usually somewhat swollen with spongy tissue and rounded, sometimes lacking such tissue and tapered; stigmas 2. **Achenes** lenticular.

Moist areas that dry out in summer, such as wet meadows, wet prairies, ditches, and hillside drainages. 0–1500 m. Casc, CR, ECas, Est, Sisk, WV. CA, WA; Australia. Native.

This abundant west-side sedge is readily identified if the cross-corrugated leaf sheaths are noticed. The only truly similar species is *C. vulpinoidea*, found mainly east of the Cascades but also common in the Portland area, which has a longer, more interrupted inflorescence with a conspicuous, hair-like bract at the base of each spike and browner perigynia that average smaller (2–3.5 mm long) with 0–3 dorsal veins and usually no ventral veins.

Carex diandra F. Schrank (p. 198)

Lesser panicled sedge

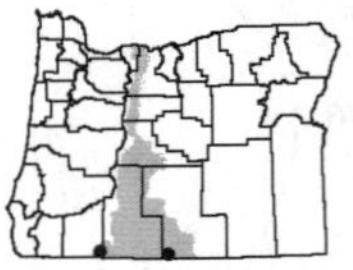

Plants cespitose to short-rhizomatous, 2–90 cm tall. **Leaves** sheath fronts white with red dots, sometimes coppery-tinged; blades 1–2.5 mm wide. **Inflorescences** 1.7–4.0(6.0) cm × 6–15 mm, inconspicuously branched, erect, not interrupted or only slightly so; spikes androgynous or female. **Female scales** about as wide as the perigynia at the base, but shorter than the perigynia and narrowing toward the tip, thus not concealing the perigynia. **Perigynia** lance-ovate, biconvex, (2)2.3–2.5(2.9) × 1–1.4 mm, olive brown to dark brown and sometimes green on beak, shiny, so filled with pithy tissue that the ventral surface often bulges; dorsal surface with a shallow narrow groove and 4–6 veins; beak 0.9–1.1 mm long; stigmas 2. **Achenes** lenticular. ***2n***=48, 50, 60.

Bogs and fens, floating peat mats, lake shores, springs and seeps. 1100–1600 m. ECas. CA, ID, NV, WA; north to AK, east to Newfoundland; Eurasia, New Zealand. Native.

Carex diandra is recognized by its red-dotted leaf sheath fronts, shiny, dark, dorsally grooved, bulging perigynia, and its bog habitat. Small individuals of *C. cusickii* are similar but have conspicuous coppery coloration at the top of the leaf sheath front.

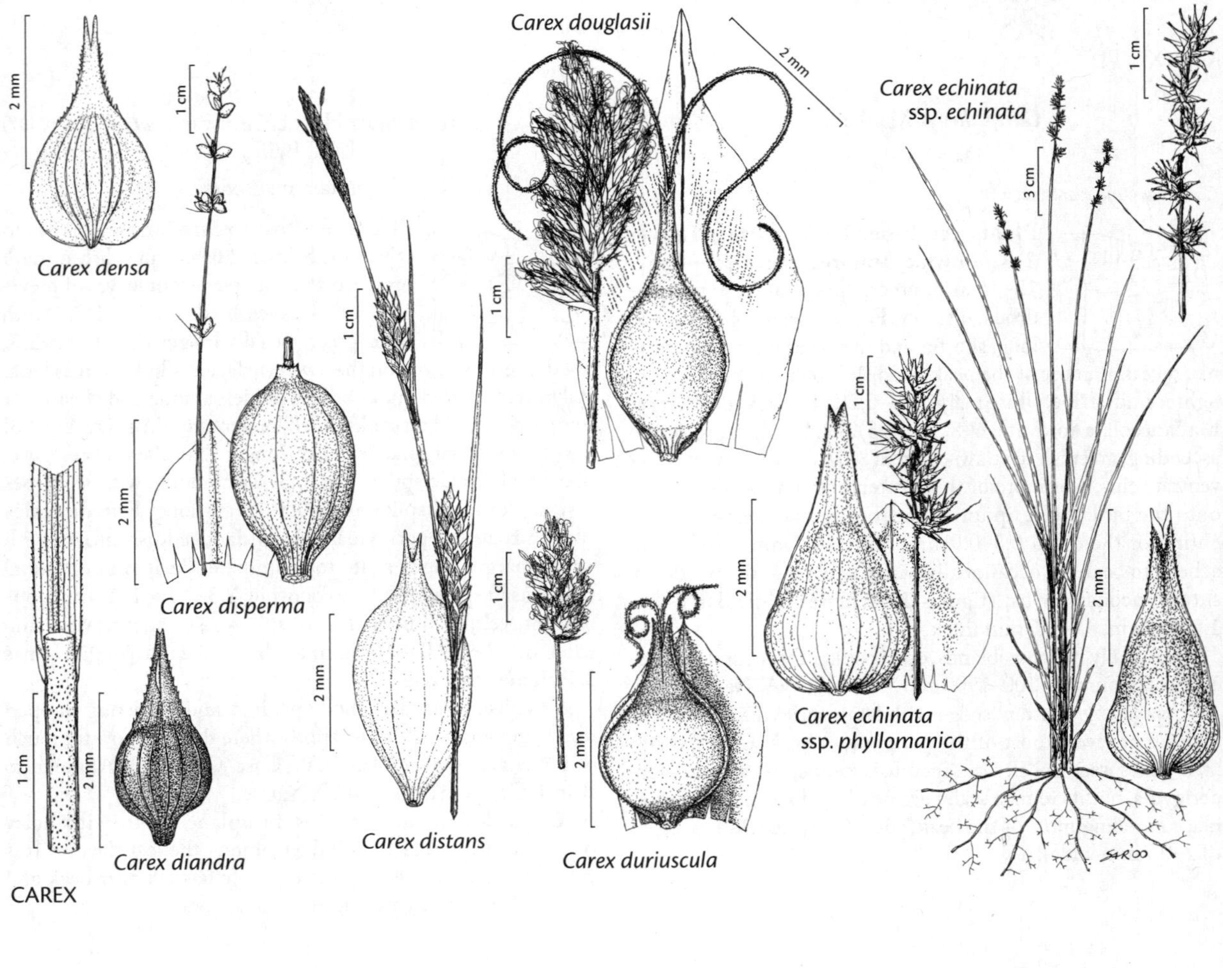

CAREX

Carex disperma C. Dewey (p. 198)

Two-seed sedge

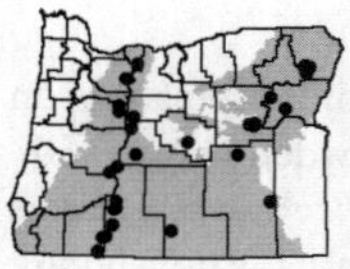

Plants rhizomatous, with slender brown rhizomes; culms 15–60 cm tall, longer than the leaves, very slender, erect when young but usually arching when mature. **Leaves** 0.7–1.5 mm wide, green, flat. **Inflorescences** 1.5–3.5 cm long; lowest bract 5–20 mm long; spikes 1–6, androgynous, each with 1–2 inconspicuous male flowers in addition to 1–3(6) perigynia, the lower spikes separated but the uppermost crowded together. **Female scales** shorter and narrower than the perigynia, white hyaline with green midrib. **Perigynia** ovate, 2.2–3 × 1.3–1.5 mm, green when young, dark brown when mature, shiny, veined; beak about 0.25 mm long; stigmas 2. **Achenes** lenticular. **2*n***=70.

Moist forests, aspen groves, bogs, swamps, often on small hummocks in forested wetlands, often growing in a moss layer, usually in shade. 800–2400 m. BR, BW, Casc, ECas, Sisk. CA, ID, NV, WA; north to AK, east to Greenland, southeast to NM; Eurasia. Native.

The problem with this delicate sedge is seeing it, not identifying it once it's found. The almost thread-like culms with widely separated clusters of egg-shaped, brown perigynia are unique.

Carex distans Linnaeus (p. 198)

Distant sedge

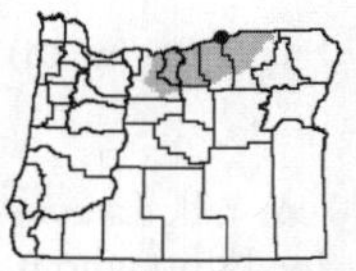

Plants cespitose, 40–80(100) cm tall; plant bases brown to orange-brown. **Leaves** 3–4.6 mm wide, shorter than the culms. **Inflorescences** lowest bract leaf-like, much shorter than inflorescence, with sheath more than 4 mm long; lateral 2–4 spikes female, ascending; terminal spike male, (15)20–32(40) mm long, on stalk 2–5(8.5) cm long. **Female scales** brown, with midrib green or whitish, 3-veined. **Perigynia** elliptical, slightly inflated near the top, 3.5–4.6 × 1.1–1.4 mm, gray green to greenish brown, with 10–16 veins, contracted to the beak; beak scabrous, (0.8)1–1.4 mm long; styles deciduous but leaving a relatively long, often contorted stub on the achene; stigmas 3. **Achenes** trigonous. **2*n***=68, 70–72, 74.

Freshwater and brackish wetlands, wet meadows, on sandy soil. 50–100 m. Col. MA, PA; Africa, Eurasia. Exotic.

Carex distans is a cespitose sedge with a single staminate spike well separated from the pistillate spikes, which are in turn separated from each other.

Carex douglasii F.M.B. Boott (p. 198)

Douglas' sedge

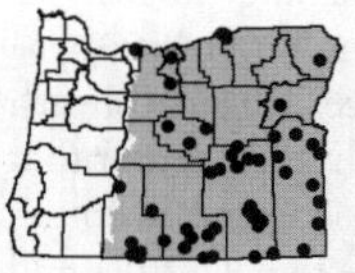

Plants more or less dioecious, rhizomatous, the rhizomes brown, slender, 0.8–1.9 mm in diameter; culms (8)15–40 cm tall; plant bases brown. **Leaves** 1–3(3.5) mm wide. **Inflorescences** 1.2–3.5(4.5) cm long, male inflorescences 7–15 mm wide, female inflorescences 8–27 mm wide, spikes in one plant seemingly all female or all male but usually with a few, nearly undetectable flowers of the opposite sex mixed in the spikes. **Female scales** 4.3–4.7(7.5) mm, pale brown to whitish hyaline, awned or not. **Anthers** tipped with a tiny, bristly awn, often retained long past flowering. **Perigynia** ovate to obovate, (3)3.5–4.2(4.8) × 1.2–2.1 mm, brown; beak (0.9)1.2–1.9 mm long; stigmas 2, long, tangled, often retained at maturity. **Achenes** lenticular.

Mainly upland sites that are moist in spring, sagebrush steppes, short-grass prairies and open forests. 50–2400 m. BR, BW, Col, ECas, Lava, Owy. CA, ID, NV, WA; north to British Columbia, east to SD, southeast to MO, NM, Mexico. Native.

The rhizomatous habit and broad, shaggy, pale, unisexual inflorescences with tangled stigmas make *C. douglasii* easy to identify. It sometimes grows near *C. praegracilis,* which has blackish plant bases, thicker rhizomes, and narrower female inflorescences.

Carex duriuscula C.A. Meyer (p. 198)

Spikerush sedge

Carex eleocharis L.H. Bailey, *Carex stenophylla* Wahlenb.

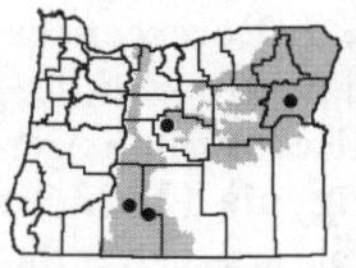

Plants sometimes more or less dioecious, rhizomatous, the rhizomes brown, slender, 0.6–1.8 mm in diameter; culms (6)10–35 cm tall; plant bases brown. **Leaves** 0.6–1.8 mm wide. **Inflorescences** (0.7)1–1.5(2) cm × 5–7 mm; spikes usually androgynous, or sometimes seemingly all female or all male but usually with a few, nearly undetectable flowers of the opposite sex mixed in the spikes. **Female scales** 2.4–4.1 mm, reddish brown with hyaline margins, awned or not. **Anthers** awn tip short, relatively broad, smooth to warty, tipped with a tiny, relatively broad, smooth to warty awn. **Perigynia** broadly elliptical to ovate or obovate, 2.4–3.9 × 1.5–2.1 mm, dark reddish brown; beak 0.3–0.9 mm long; stigmas 2. **Achenes** lenticular. **2*n***=60.

Dry prairies, sagebrush steppes and open forests. 1400–1500 m. BW, ECas. CA, ID; north to AK, east to MI, southeast to AZ; Asia, New Guinea. Native.

Carex duriuscula is an inconspicuous sedge similar to *C. douglasii* but smaller and neater. It has been shifting its range north since the glaciers retreated. Not found in Oregon since 1940, it may be extirpated.

Carex echinata Murray

Star sedge, prickly sedge

Plants cespitose, 10–90(135) cm tall. **Leaves** usually 0.7–3(3.8) mm wide. **Inflorescences** spikes separated or crowded, usually gynaecandrous, with perigynia widely spreading or reflexed at maturity; lateral spikes 3–15.5 mm long, terminal spikes 5–24 mm long, often with a narrow male portion 2–16.5 mm long. **Female scales** inconspicuous, shorter than the mature achene. **Perigynia** tapered gradually to the beak, (2.6)3–4.8 mm long, green to brown, veined or veinless on the ventral surface over the achene, base swollen, pithy when mature; beak 0.9–2 mm long, 33–60% of perigynium length. **Achenes** lenticular.

Cosmopolitan. 2 subspecies.

Narrow-leaved *Carex echinata* is identified by its widely spreading, gradually tapering perigynia in star-like spikes.

Key to *Carex echinata* subspecies:

1. Inflorescences open, lowest internode longer than lowest spike . . . . . *C. e.* ssp. ***echinata*** (in part)

1' Inflorescences dense, lowest internode shorter than lowest spike.

   2. Perigynia 2.9–3.6(4) mm long, ventral surface usually veinless; widest leaves 1–3.3 mm; widespread but not coastal . . . . *C. e.* ssp. ***echinata*** (in part)

   2' Perigynia larger (3.1)3.5–4.75 mm, ventral surface usually with 2–12 veins; widest leaves 1.7–3.8 mm; coastal . . . . *C. e.* ssp. ***phyllomanica***

Carex echinata Murray ssp. ***echinata*** (p. 198)

Star sedge

Carex angustior Mack., *Carex ormantha* (Fernald) Mack.

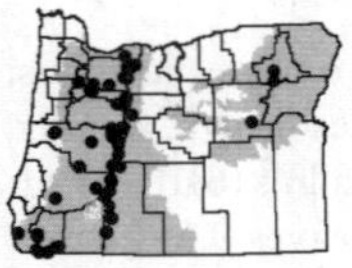

Plants 10–90(135) cm tall. **Leaves** usually 0.7–2.7(3.3) mm wide. **Inflorescences** spikes separated or sometimes crowded, distance between 2 lowest spikes usually longer than lowest spike. **Perigynia** (2.6)2.9–3.6(4) mm long, usually veinless on the ventral surface over the achene; beak (35)38–60% of perigynium length; stigmas 2. **Achenes** lenticular. **2*n***=50, 52, 58.

Bogs, swamps, seeps, streamsides, wet meadows, usually in acidic or serpentine soils, not coastal. 100–2300 m. BW, Casc, ECas, Sisk, WV. CA, ID, NV, WA; north to AK, east to Newfoundland, throughout most of US except Great Plains and southeastern states; HI, Dominican Republic, Mexico, Central America; Australia, Eurasia, New Zealand, Papua New Guinea. Native.

Individuals with well-separated spikes are often confused with *C. interior,* which has proportionately shorter perigynium beaks; because *C. interior* perigynia taper more abruptly to the beak, they have "shoulders." In southwestern Oregon, *C. interior* has beaks at the longer end of that range, and the distinction can be difficult. See *C. echinata* ssp. *phyllomanica.*

Carex echinata Murray ssp. ***phyllomanica*** (W. Boott) Reznicek (p. 198)

Coastal star sedge

Carex phyllomanica W. Boott

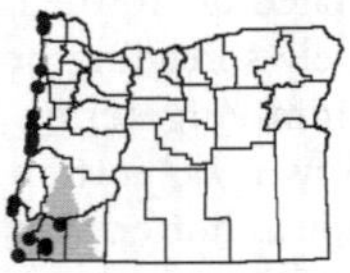

Plants 20–80 cm tall. **Leaves** 1–3.3(3.8) mm wide. **Inflorescences** spikes crowded, distance between 2 lowest spikes shorter than lowest spike. **Perigynia** (3.1)3.5–4.75 mm long, usually with 2–12 veins on the ventral surface over the achene; beak 33–45% of perigynium length. **2*n***=54, 70.

Coastal bogs, deflation plains, swamps, seeps, streamsides and lake shores. 0–1300 m. Est, Sisk. CA, WA; north to AK. Native.

Carex echinata is a densely cespitose sedge with widely spreading, long-beaked perigynia. The similar *C. e.* ssp. *echinata* has narrower leaves, usually less crowded spikes, and shorter perigynia.

Carex exsiccata L.H. Bailey (p. 201)

Western inflated sedge

Carex vesicaria L. var. *major* Boott

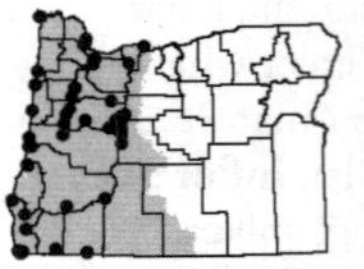

Plants cespitose, or sometimes spreading on short rhizomes with crowded shoots, 30–100 cm tall; plant bases reddish. **Leaves** basal leaf sheath reddish, usually not spongy, nearly lacking crosswalls between the longitudinal veins; blades 2.5–6.2 mm wide, green, flat to W-shaped in cross section. **Inflorescences** lateral spikes 2–5, female, erect or ascending, mostly 2–7 cm long, well separated from the 2–3 male terminal spikes. **Female scales** shorter than the perigynia, apex acute to acuminate, awnless. **Perigynia** tapering gradually to the beak, 7.5–10.1 × 1.5–2.4(2.7) mm, green to straw-colored or reddish brown, with 9–20 veins, somewhat leathery; styles persistent, becoming curved; stigmas 3. **Achenes** trigonous.

Pond margins, lake shores, wet meadows, and seasonal ponds where water pools in winter, generally in shallow water. 0–1800 m. Casc, CR, ECas, Est, Sisk, WV. CA, ID, WA; north to British Columbia, northeast to MT. Native.

Carex exsiccata is the west-side sister taxon to *C. vesicaria,* which grows mostly east of the Cascades. Perigynia of *C. exsiccata* are usually longer and taper more gradually to the beak than do those of *C. vesicaria.* Immature specimens sometimes cannot be distinguished except by geography. Plants in the high Cascades may be intermediate. These two taxa are sometimes treated as varieties of *C. vesicaria.* See also *C. utriculata.*

Carex feta L.H. Bailey (p. 201)

Green-sheath sedge

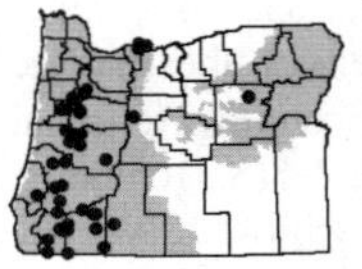

Plants cespitose, 50–100 cm tall. **Leaves** sheath fronts green and veined throughout, except for a white-hyaline triangle to 6 mm long near the top, sometimes prolonged 1.5–6 mm above the base of the blade; blades (2.5)3–4(5) mm wide. **Inflorescences** 3.5–8 cm × 10–13 mm, elongated but with overlapping spikes, green to straw-colored; spikes ovate, gynaecandrous. **Female scales** 2.7–3.5 mm long, shorter and narrower than the perigynia, white hyaline or straw-colored; midrib green, straw-colored, or brown. **Perigynia** ovate to broadly ovate, 3.2–4.2 × 1.7–2.1 mm, green, white, or straw-colored, with 1–8 veins on each face; beak white or straw-colored at tip, winged and ciliate-serrulate; (1.5)1.8–2.4 mm from top of achene to tip of beak; stigmas 2. **Achenes** lenticular, (1.3)1.5–1.8 × 0.9–1.2 mm, 0.4–0.5 mm thick.

Wet meadows and prairies, margins of marshes and road ditches. 100–1500 m. BW, Casc, CR, ECas, Sisk, WV. CA, WA; north to British Columbia, WY. Native.

The neat, tidy appearance, overall green inflorescence, green leaf sheath fronts, and low elevation west-side range identify *C. feta.* Similar *C. fracta* is a coarser, sloppier plant with white-hyaline leaf sheath fronts and silvery inflorescences. The leaf sheath front of *C. scoparia* has such a narrow white-hyaline center that it can appear entirely green, but its usually curving inflorescence has a much finer texture with acuminate female scales and pointed perigynia, and its longer perigynia taper more gradually to the beak. See also discussions of *C. longii* and *C. tribuloides,* which have green sheath fronts.

Carex filifolia Nuttall var. ***filifolia*** (p. 201)

Threadleaf sedge

Carex exserta Mack.

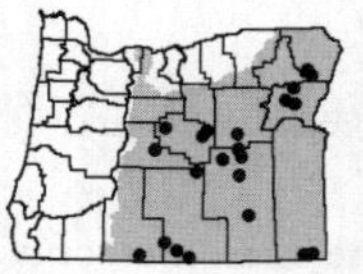

Plants densely cespitose, but plants sometimes growing together in dense stands that might appear rhizomatous, 8–35 cm tall; plant bases brown. **Leaves** sheaths dense, persistent, brown; blades 0.3–0.7 mm wide, very narrow and wiry. **Inflorescences** spike solitary, androgynous, (1)1.2–2.6(3) cm long, with 5–18 perigynia, lacking inflorescence bracts. **Female scales** wider than the perigynia, reddish brown or yellowish with broad brownish or white margins. **Perigynia** obovoid, not compressed and sometimes definitely circular in cross section, abruptly contracted to the beak, (2.8)3–4.8 mm long, pubescent; rachilla well developed; beak 0.2–0.8 mm long, style base often conspicuously exserted from the beak; stigmas 3. **Achenes** trigonous.

Sagebrush steppes on toeslopes, streamside terraces, and in dry meadows with a little extra subsoil moisture so that the habitat is not fully xeric. 800–2600 m. BR, BW, ECas, Lava, Owy. CA, ID, NV, WA; north to AK, east to MN and KS. Native.

With its densely cespitose growth form and very narrow leaves, *C. filifolia* resembles Idaho fescue (*Festuca idahoensis*) or some other fine-leaved bunchgrass. *Carex nardina* is very similar but grows at higher elevations. It has glabrous, flatter, planoconvex perigynia with margins flattened and strongly serrulate distally. *Carex elynoides*, a rhizomatous Great Basin species that may occur in southeastern Oregon, has glabrous perigynia, does not have the style base conspicuously sticking out from the perigynium beak, and lacks a rachilla.

Carex fracta Mackenzie (p. 201)

Fragile-sheath sedge

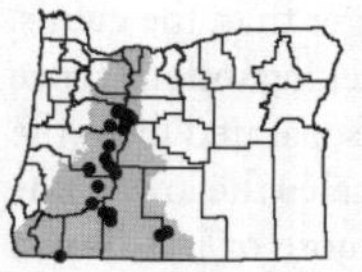

Plants cespitose, 60–115 cm tall. **Leaves** sheath fronts white-hyaline for at least 1 cm from the top, often throughout, extending 3–9 mm above base of blade; blades 4–7 mm wide. **Inflorescences** 3.5–8 cm × 10–16 mm, elongated, silvery to green or straw-colored; spikes elliptical, overlapping, gynaecandrous. **Female scales** (2.8)3.6–4.5 mm long, shorter or longer and narrower than the perigynia, white hyaline or straw-colored; midrib green, straw-colored, or brown. **Perigynia** broadly lanceolate to ovate, 2.9–4(4.8) × 1–1.7 mm, green to straw-colored, with 5–11 veins on each face; beak white or straw-colored at tip, brown, usually unwinged, parallel-sided, and more or less entire for the distal 0.1–0.5 mm, 1.3–2(2.8) mm from top of achene to tip of beak; stigmas 2. **Achenes** lenticular, 1.3–1.6 × 0.8–1.2 mm, (0.3)0.4–0.5 mm thick.

Seasonally damp places, open forests, roadsides, often in partial shade, from foothills to timberline. 200–2100 m. Casc, ECas, Sisk. CA, WA. Native.

Carex fracta is a coarse montane sedge with tall, leaning shoots and whitish or silvery inflorescences. Its loose leaf sheaths are prolonged at the mouth. Compare to *C. feta.*

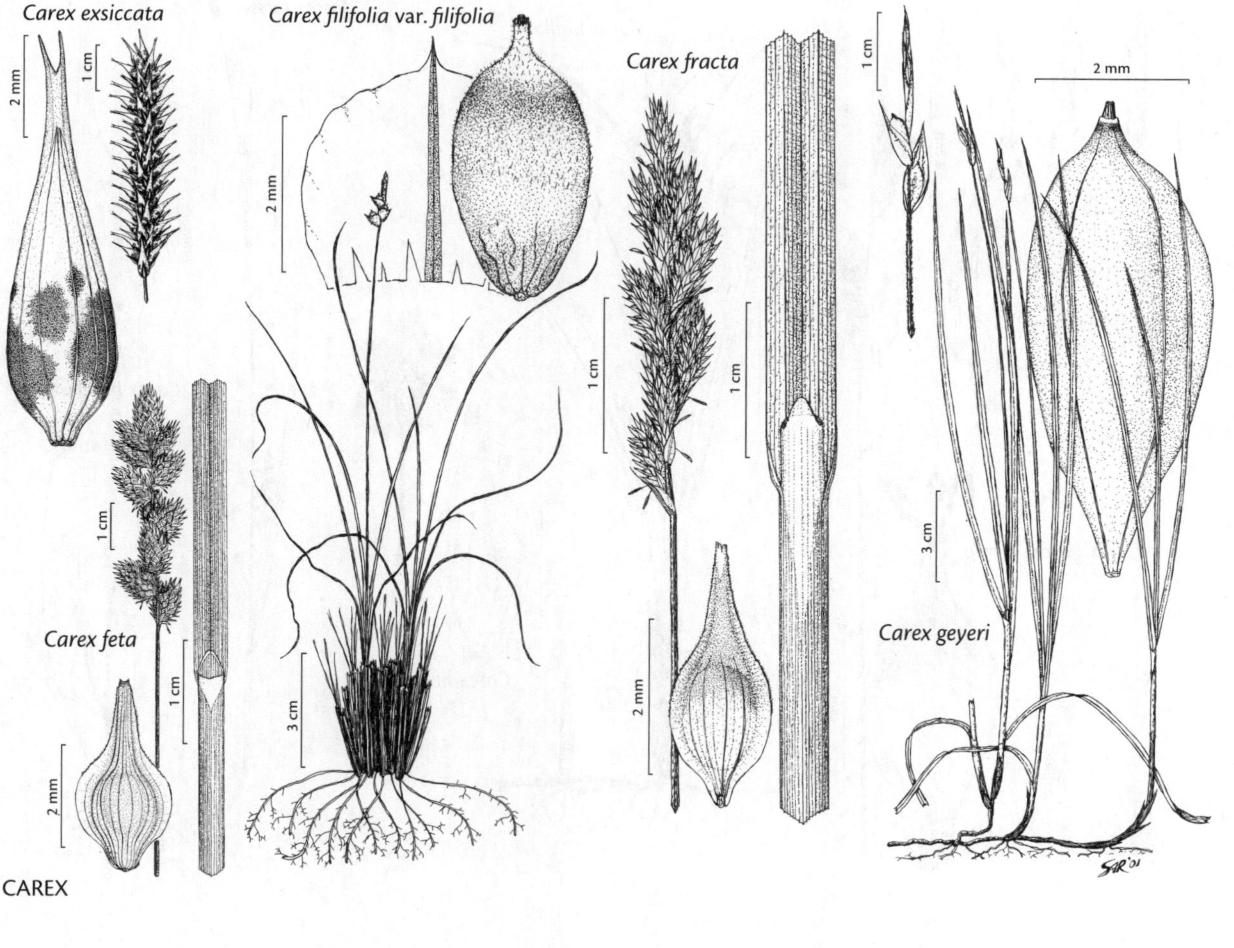

Carex geyeri F.M.B. Boott (p. 201)

Elk sedge

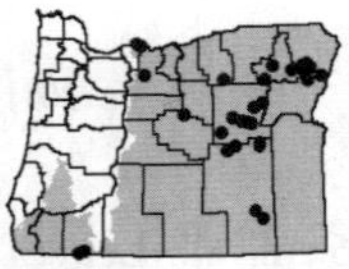

Plants loosely cespitose or short rhizomatous, with thick, dark, shallow rhizomes; culms 12–50 cm, triangular in cross section. **Leaves** 1.1–3.5 mm wide, flat, evergreen, leathery, as long as or longer than the culms. **Inflorescences** spike solitary, androgynous, lacking inflorescence bracts; male terminal portion 1–2.5 cm long, separated from the 1–3 perigynia by a short internode. **Female scales** the lower longer than the perigynia; the upper reduced, more or less brown with paler midrib and margins; pointed to awned. **Perigynia** obovate, rounded at the top, 5–7(8.5) × 1.8–2.8 mm, green, whitish, or brown; beak small; stigmas 3. **Achenes** trigonous.

Well-drained soils in dry, open, conifer forests, aspen stands, and openings, rarely in sagebrush, on both serpentine and nonserpentine substrates. 100–2500 m. BR, BW, Col, ECas, Lava, Owy, Sisk. CA, ID, NV, WA; north to southeastern British Columbia, east to CO, PA. Native.

Carex geyeri is a common ground cover in open forest, especially in northeastern Oregon, recognized by its tough, flat, evergreen leaves from dark, scaly rhizomes. Similar pinegrass, *Calamagrostis rubescens*, usually has hairy internodes and when in flower it is easily distinguished. *Carex geyeri* is important food for wildlife and livestock, especially in winter. In southwestern Oregon, *C. geyeri* can be confused with *C. multicaulis*, which is densely cespitose and has proportionately shorter, usually involute leaves, and longer, green, leaf-like, lower female scales.

Carex gynocrates Wormskjold ex Drejer (p. 202)

Northern bog sedge

Carex dioica L., *Carex dioica* L. var. *gynocrates* (Wormsk. ex Drejer) Ostenf.

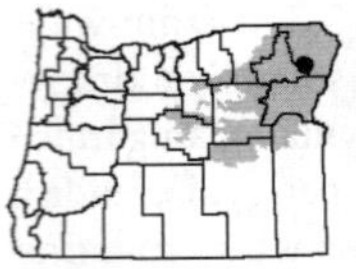

Plants rhizomatous, the rhizomes only 0.3–0.8 mm in diameter; culms slender and round in cross section, 2–30 cm long. **Leaves** 0.3–0.7 mm wide. **Inflorescences** spike solitary, usually female but occasionally androgynous or male; if male, the spike 0.8–1.6 cm × 1–2 mm; if female, the spike more or less oblong, 0.5–1.4 cm × 4–8 mm. **Female scales** mostly shorter but broader than the perigynia, light brown, more or less hyaline. **Perigynia** oblong-ovoid, biconvex, 2.9–3.4 × 1.2–1.7 mm, yellowish or green when young, brown at maturity, with many obscure to conspicuous fine veins; beak 0.5 mm long; stigmas 2. **Achenes** lenticular. **$2n$=46, 48.**

Growing in moss in bogs, usually in openings in conifer forests, sometimes in open bogs or subalpine meadows. 1600–1700 m. BW. ID, NV, WA; north to AK, east to Greenland; Asia. Native.

Carex gynocrates is a delicate little sedge of moss mats in high montane bogs.

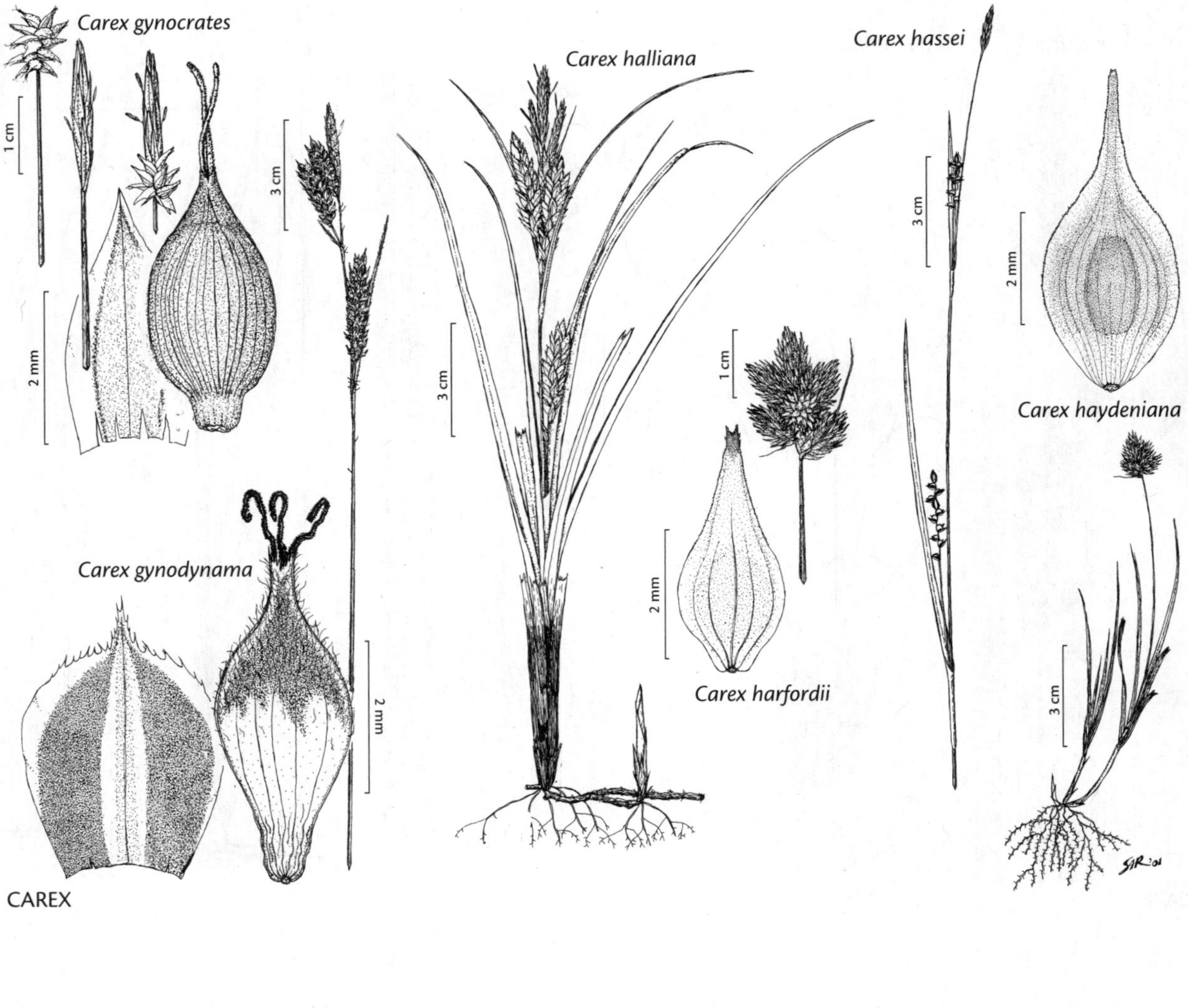

Carex gynodynama Olney (p. 202)

Olney's hairy sedge, wonderwoman sedge

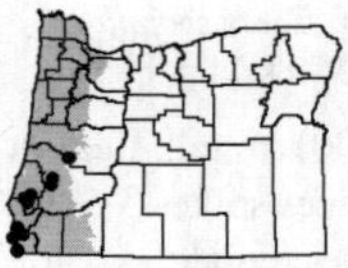

Plants cespitose, 20–100+ cm tall. **Leaves** lower sheaths reddish brown, bladeless, and hairy; upper sheaths dark red to green on back; sheath fronts tan-hyaline, with or without red dots, usually pubescent at the top, sometimes hairy on the margins; blades 3–12 mm wide, flat to W-shaped, sparsely hairy, with ciliate margins. **Inflorescences** bract sheaths well developed, more or less inflated, 5–50 mm long; lateral spikes 2–5, female, cylindrical, 1.2–4 cm × 4–11 mm; terminal spike male (rarely gynaecandrous), sessile or nearly so, 0.8–3 cm long. **Female scales** broadly ovate, shorter than the perigynia, reddish brown with white margins, midrib green, often hairy, awn often hairy. **Perigynia** 3.7–5.3 × 1–2.2 mm, with rather long hairs, light green with reddish blotches at base and near tip, with up to 20 fine veins plus 2 marginal ribs; beak 1 mm long; stigmas 3. **Achenes** trigonous.

Seeps, moist meadows, open woodlands, and ditches, usually in disturbed sites in partial shade, often on serpentine-influenced soils but not in severe serpentine substrates. 0–900 m. CR, Est, Sisk, WV. CA. Native.

Carex gynodynama is recognized by its large, lax culms and hairy leaves and perigynia. The other southwestern Oregon sedges with hairy leaves are *C. whitneyi* and *C. mendocinensis,* both with glabrous perigynia. *Carex gynodynama* and *C. mendocinensis* occasionally hybridize.

Carex halliana L.H. Bailey (p. 202)

Hall's sedge, Oregon sedge

Carex oregonensis Olney

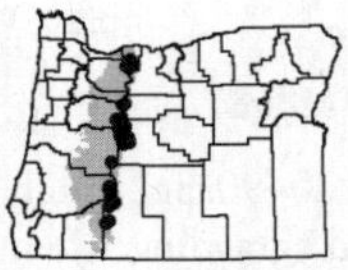

Plants rhizomatous, the shoots mostly arising singly, 10–40(60) cm tall, phyllopodic, with dull brown bases. **Leaves** sheath fronts straw-colored or brown, or the youngest tinged reddish purple; blades 2–5.5 mm wide, green, tough, arching. **Inflorescences** bract of lowest lateral spike leaf-like, about as long as or longer than the inflorescence, with a somewhat inflated sheath (0.4)0.6–1.5 cm long; spikes crowded or the lowest remote; lateral 2–4 spikes female (rarely androgynous), erect to ascending, 1–5 cm long, much longer than wide; terminal 1(3) spikes male, 1–3.5 cm long. **Female scales** typically shorter than the perigynia, green with hyaline or brownish margins, glabrous or with margins scabrous apically. **Perigynia** ovoid, abruptly contracted to the beak, 3.5–5.5 × 1.6–2.2 mm, pubescent; beak 1.2–1.7 mm long; teeth 0.2–0.6 mm long, straight; stigmas 3. **Achenes** trigonous.

Well-drained soils, often pumice, in meadows, open forests and roadsides. 1100–2500 m. Casc. CA, WA. Native.

Carex halliana is a short, upland, rhizomatous sedge with hairy perigynia, tough leaves, and female spikes much longer than wide. Other upland species with hairy perigynia have softer leaves and lateral spikes only slightly longer than wide.

Carex harfordii Mackenzie (p. 202)

Monterey sedge, Harford's sedge

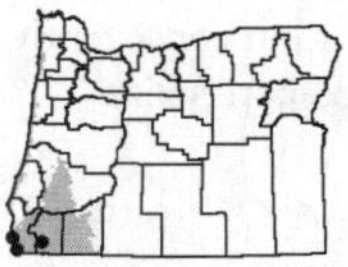

Plants cespitose; culms (30)50–120 cm tall, sometimes falling over, often branched late in the season. **Leaves** sheaths extending 1–4 mm above base of blade; blades usually 2–5(9.9) mm wide. **Inflorescences** 1.5–4 cm × 7–20 mm, dense and head-like, lowest internode less than1.5–2.5 mm; lower bracts usually short and bristle-like, sometimes leaf-like and longer than the inflorescence; spikes gynaecandrous. **Female scales** (2.5)3.5–4.7 mm long, usually shorter and narrower than the perigynia but sometimes longer and covering them, green to dark brown, center often green to gold. **Perigynia** ovate to broadly ovate, (2.6)3.3–4(4.6) × (1.3)1.4–1.6(2) mm, 0.6–0.7 mm thick, green to brown, with metallic sheen; dorsal surface usually green usually with 4–9 veins; ventral surface with 0–9 veins that usually exceed the achene; wing 0.2–0.3(0.4) mm wide; beak tip brown, cylindric, unwinged, and more or less entire for more than 0.4 mm from tip; 1.2–1.8 mm from achene top to beak tip; stigmas 2. **Achenes** lenticular, (1.3)1.5–2.1 × 0.9–1.4 mm, 0.5–0.7 mm thick.

Wet or seasonally wet shores, meadows and open forests near the coast. 0–400 m. Est, Sisk. CA. Native.

This "ball-headed" sedge resembles *C. subbracteata* or an unusually dark *C. pachystachya*. It has less spreading perigynia than *C. pachystachya* and differs from both species in having ventral perigynium veins that exceed the achene, like those of montane *C. abrupta*. Individuals with leaf-like inflorescence bracts might be confused with paler *C. athrostachya,* which has narrower, flatter perigynia.

Carex hassei L.H. Bailey (p. 202)

False golden sedge

Carex garberi Fernald

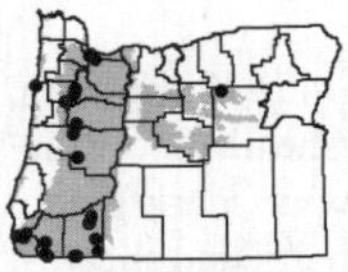

Plants rhizomatous, the rhizomes about 1 mm in diameter. Culms 10–85 cm, averaging 35 cm tall; plant bases whitish or pale to medium brown. **Leaves** blades 2–3(3.7) mm wide, green to glaucous. **Inflorescences** lateral spikes female, (4)10–23 × 3–5 mm, with perigynia densely or loosely spaced, middle internodes 0.2–1.1 mm, averaging 0.5 mm; terminal spike male or less often gynaecandrous, when male 0.6–2.8 cm long, averaging 1.3 cm × (1.8)2–2.5 mm. **Female scales** rounded or obtuse to acute, 2–3.2 mm, brown, sometimes awned, usually persistent. **Perigynia** usually obovate or sometimes elliptical, 1.8–3.2 × 1.1–1.9 mm, pale green, whitish, or tan, with (4)9–15 veins, beakless or nearly so; beak or perigynium tip usually curved to the back; stigmas 2, or a mix of 2 and 3 in the same inflorescence. **Achenes** lenticular, or trigonous if styles 3.

Springs, fens, drier margins of marshes, and mesic or seasonally moist meadows, on various substrates including serpentine. 100–2100 m. Casc, Est, Lava, Sisk, WV. CA, ID, NV, WA; north to British Columbia, southeast to NM, Mexico. Native.

Carex hassei is an inconspicuous grass-like sedge with pale greenish or tan perigynia that have short tips that bend toward the back. It is very similar to *C. aurea,* which has female scales that often fall before the perigynia, and perigynia that become entirely succulent and orange (or white-powdery) when mature. Immature plants cannot be reliably distinguished, but *C. aurea* is usually shorter with generally smaller terminal male spikes. In southwestern Oregon, *C. klamathensis* is similar but consistently has three stigmas per flower (except on aborted perigynia).

Carex haydeniana Olney (p. 202)

Cloud sedge

Carex nubicola Mack.

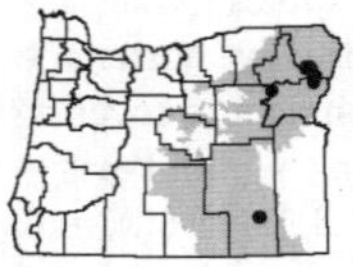

Plants cespitose, 9–40 cm tall. **Leaves** 1.5–4 mm wide. **Inflorescences** 1.1–2.1 cm × 13–18 mm, dense and head-like, with a fine texture because the perigynia are many and thin, not spreading; spikes gynaecandrous. **Female scales** 3–4.8 mm long, shorter and narrower than the perigynia, generally brown or purplish, sometimes with a paler mid-stripe. **Perigynia** flat except over the achene, 4–6.5 × 1.5–2.6 mm, light to dark brown (occasionally greenish) with dark to black tips, with several light dorsal veins and 0–3(8) ventral veins; wings 0.3–0.6(0.8) mm wide; beak tip unwinged, brown, and parallel-sided for at least the distal 1 mm, entire for the distal 0.3–0.6 mm; (2.3)2.6–3.8 mm from achene top to beak tip; stigmas 2. **Achenes** lenticular, (1.2)1.4–1.8 × 0.8–1.1(1.3) mm, 0.3–0.5 mm thick. **2*n***=82.

Moist or mesic, gravelly or rocky, subalpine or alpine slopes and flats, often in snowmelt zones. 1500–3100 m. BR, BW. CA, ID, WA; north to British Columbia, east to CO, southeast to NM. Native.

This "ball-headed" sedge grows near or usually above timberline. Its dense, black or blackish green heads resemble those of *C. microptera*, a close relative with shorter perigynia that grows at lower elevations.

Carex hendersonii L.H. Bailey (p. 205)

Timber sedge

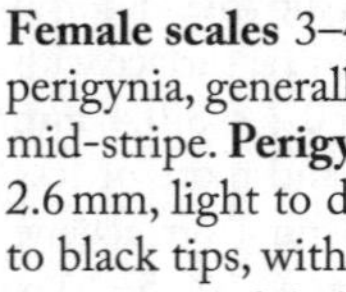

Plants cespitose; culms 45–90 cm long, sprawling, floppy. **Leaves** sheaths loose, inflated; blades 3–16 mm wide, dark green. **Inflorescences** bracts with inflated sheaths; spikes 1.2–4 cm long, widely separated, lateral spikes female, terminal spike male. **Female scales** 2.7–3.1 mm long, acute to awned, whitish with green midrib. **Perigynia** tapering gradually to the beak, 4.6–6 × 1.8–2.3 mm, green, with many veins; beak 0.5–1.8 mm, inflated, bending to the dorsal side; stigmas 3. **Achenes** trigonous.

Shaded riparian zones and moist forest. 0–1500 m. Casc, CR, Est, Sisk, WV. CA, ID, WA; north to British Columbia. Native.

This is a large, sprawling, cespitose sedge of moist forests. The only cespitose sedges that commonly grow with it are *Carex leptopoda* and its relatives, which have narrower, paler green leaves, sheathless inflorescence bracts, shorter spikes, and lenticular perigynia. This plant was named after Louis Henderson, curator of the herbarium at University of Oregon from 1924 to 1939.

Carex heteroneura W. Boott (p. 205)

Different-nerved sedge, smooth-fruit sedge, blackened sedge

Carex atrata L., *Carex atrata* L. var. *erecta* W. Boott, *Carex epapillosa* Mack., *Carex heteroneura* W. Boott var. *epapillosa* (Mack.) F.J. Herm.

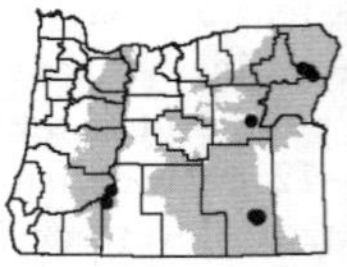

Plants cespitose, 25–60(100) cm tall. **Leaves** 2–6 mm wide. **Inflorescences** spikes 3–7(8), 1–2.7 cm long; lateral spikes female, spreading to erect, the lowest spike sometimes separated from the others and pendent; terminal spike gynaecandrous. **Female scales** shorter than to longer than the perigynia, dark purplish black; margins hyaline, mottled, or pale; midrib pale or dark like the rest of the scale; apex acute or somewhat acuminate. **Perigynia** elliptical to obovate or orbicular, more or less flattened except over the achene, 2.5–4.5 × 1.75–3 mm, green, yellow, or light to dark brown; beak 0.3–0.5 mm long; stigmas 3. **Achenes** trigonous, filling half or less of the perigynium body.

High montane to alpine streamsides, seeps, moist to mesic meadows, and open forests, sometimes in wind-swept sites that have little snow accumulation, but not in fully dry sites. 1300–2800 m. BR, BW, Casc. CA, ID, NV, WA; north to British Columbia, east to CO. Native.

Carex heteroneura is a coarse, cespitose, sedge with erect spikes (except sometimes the lowest) and broad, flat perigynia. Its entire inflorescence tends to nod when fully ripe. The more common *C. spectabilis* is similar, but its spikes lack the contrast between dark female scales and light perigynium margins, its spikes are less crowded, and it is a finer textured plant. Rare *C. atrosquama* has elliptical, golden perigynia.

Two subspecies have been recognized in *C. heteroneura*, but variation is continuous.

Carex hirsutella Mackenzie (p. 205)

Hairy-leaved sedge, fuzzy wuzzy sedge

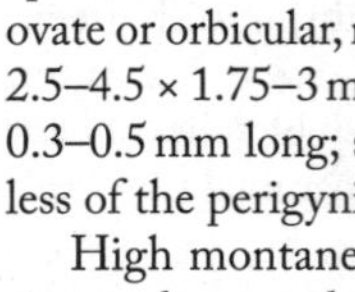

Plants cespitose; culms 20–90 cm, much longer than the leaves and sprawling when mature. **Leaves** sheaths hairy; blades 1.5–4 mm wide, hairy on both sides with hairs usually 0.4–0.6 mm long and bent or wavy. **Inflorescences** lowest bract sheathless or nearly so; lateral spikes female, 8–18 × 3–5.5 mm, often crowded; terminal spike gynaecandrous with at least half the flowers pistillate, 10–20(25) mm. **Female scales** ovate, shorter than the perigynia, white or green to brownish, obtuse, sometimes mucronate. **Anthers** 1.3–2.2 mm. **Perigynia** 2–3 × 1.1–1.6 mm, green, maturing whitish to brown or purplish, glabrous, weakly several-veined; beak absent; stigmas 3. **Achenes** trigonous, 1.6–2.1(2.6) mm. **2*n***=52.

Wet prairie remnants, seasonally moist meadows and open, seasonally wet woodlands. WV. Ontario to ME, south to TX and GA. Exotic.

This eastern North American species was discovered in Oregon in 2013. *Carex hirsutella* is a grass-like sedge with sprawling culms and nearly beakless perigynia. Its distinctive hairy leaves and sheaths are easily overlooked. *Carex pallescens* (known from Vancouver, WA) is similar but has male terminal spikes.

Carex hirta Linnaeus (p. 205)

Hairy sedge

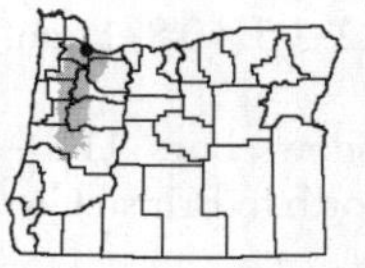

Plants rhizomatous, (10)20–90 cm tall. **Leaves** 2.5–8 mm wide, hairy. **Inflorescences** lateral (1)2–3 spikes female, erect or ascending; terminal 1–3 spikes male. **Male scales** with short, scabrous awns (sometimes lacking on lowest scales), with sparse to dense, spreading, white hairs. **Female scales** usually sparsely pubescent (occasionally glabrous), with scabrous awns. **Perigynia** lanceolate, 4.8–7.8 × 1.7–2.5 mm, more or less densely pubescent with spreading hairs, with 12–20 veins; beak 1.5–2.7 mm long; teeth 0.8–1.7 mm long, spreading; style persistent; stigmas 3. **Achenes** trigonous. **2*n***=112–114.

Disturbed sandy soils, dry to wet meadows, stream banks, roadsides, and open forests. 0–50 m. WV. Northeastern North America; Eurasia. Exotic.

This introduced species resembles *C. sheldonii*, which is native east of the Cascades. *Carex sheldonii* has glabrous, awned female scales, and awnless male scales that are glabrous or have sparse, more or less appressed hairs. Common *C. pellita* has glabrous leaves and smaller, hairy perigynia with shorter teeth. *Carex hirta* was found in Portland in 1916 and not seen in Oregon again until 2010.

Carex hoodii F.M.B. Boott (p. 205)

Hood's sedge

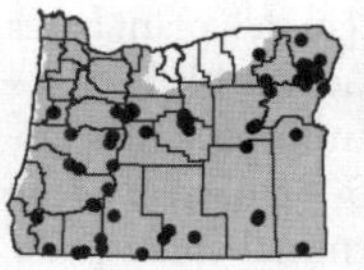

Plants cespitose, 20–80 cm tall. **Leaves** blades to 1–3.5 mm wide, green. **Inflorescences** 0.8–2 cm × 6–15 mm, dense and head-like; spikes androgynous, difficult to distinguish. **Female scales** 3.8–4.3 mm long, brown with a green midvein, acute or short-awned. **Perigynia** 3.2–5 × 1.4–2.5 mm, shiny green with coppery or red-brown center, with age becoming light brown with dark brown center, veinless or with only obscure veins; margins serrulate distally; beak 0.7–1.5 mm long; stigmas 2. **Achenes** lenticular. **2*n***=60.

Mesic to dry montane grasslands, rocky hillsides and openings in forest. 300–3000 m. BR, BW, Casc, CR, ECas, Lava, Owy, Sisk, WV. CA, ID, NV, WA; north to Alaska, east to SD and CO. Native.

This common, cespitose, sedge is one of the most easily distinguished by its shiny, veinless perigynia that have coppery centers contrasting with the green edges. (When old, they have dark brown centers and drab edges.) It is superficially similar to *C. microptera*, *C. pachystachya*, and other "ball-headed" species within subkey 9, but those species have winged perigynia (i.e., perigynia with flat margins) and gynaecandrous spikes.

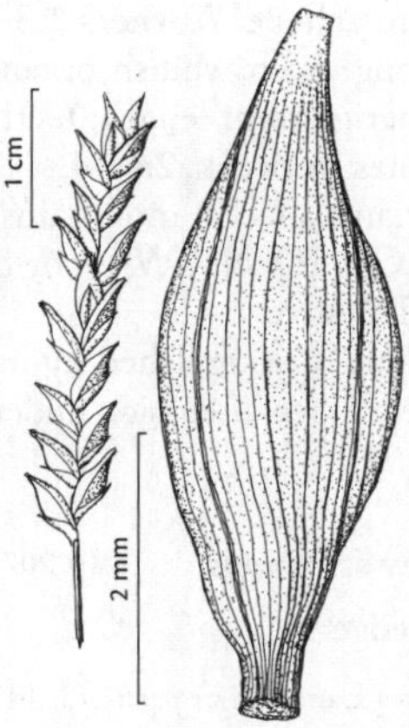

Carex hendersonii

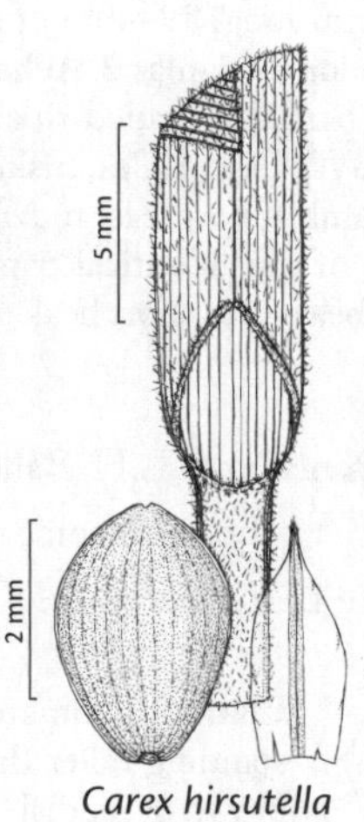

Carex hirsutella

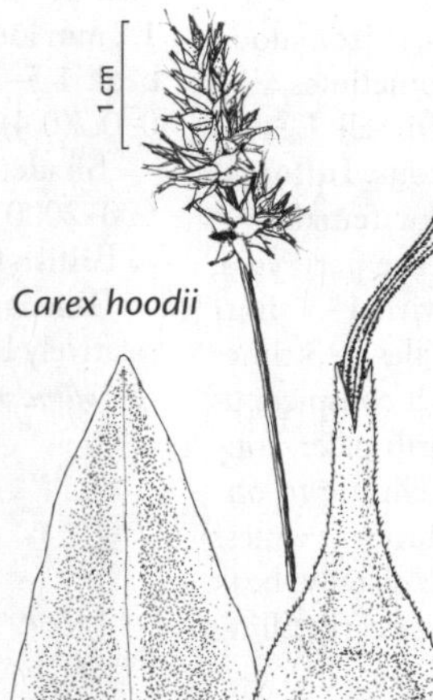

Carex hoodii

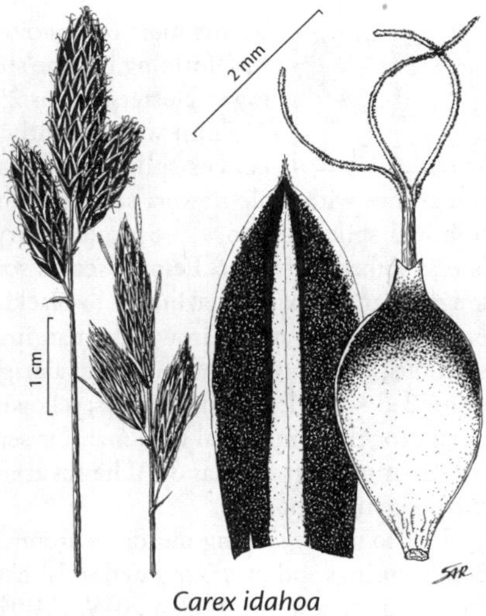

Carex idahoa

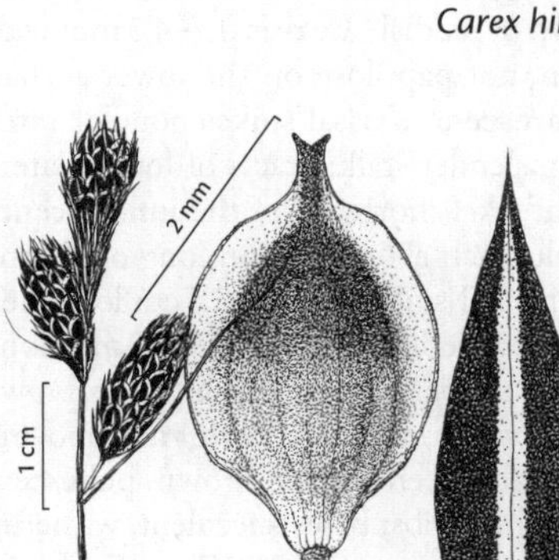

Carex heteroneura

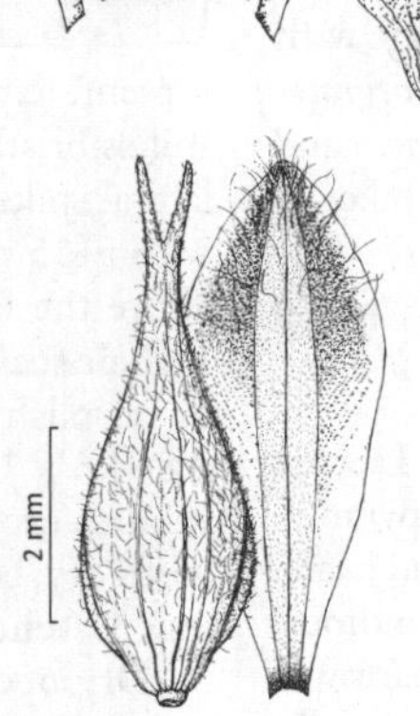

Carex hirta

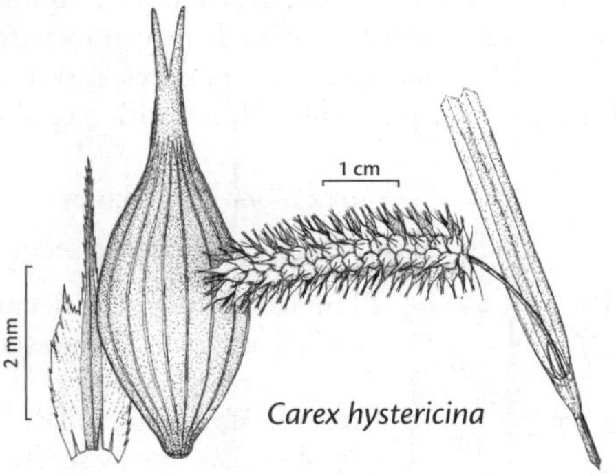

Carex hystericina

Carex hystericina Muhlenberg ex Willdenow (p. 205)

Porcupine sedge, bottlebrush sedge

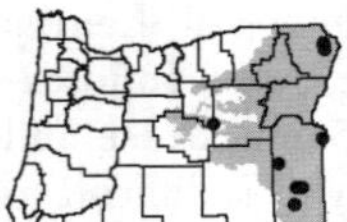

Plants cespitose, 20–100 cm tall; plant bases usually reddish purple. **Leaves** blades 2.5–8.5 mm wide, glabrous. **Inflorescences** lateral 1–4 spikes female, 1.5–4 cm, erect or the lower ones nodding; terminal spike male, 2–4 cm. **Female scales** lanceolate and tapering to a long, scabrous awn, shorter than the perigynia. **Perigynia** inflated, 4.5–7.3 × 1.4–2.1 mm, spreading or the lower ones reflexed when mature, yellow-green, glabrous, with 13–21 strong veins (the space between them usually 3+ times as wide as the veins); beak 1.9–2.8 mm long; teeth 0.3–0.9 mm long, straight; style persistent, becoming curved; stigmas 3. **Achenes** trigonous. **2*n***=58.

Streamsides and mid-slope seeps, usually in partial shade. 500–1200 m. BW, Owy. NV; north to British Columbia, east to Newfoundland, southeast to TX and VA. Native.

With its bristly spikes, awned pistillate scales, and long perigynium teeth, cespitose *C. hystericina* resembles only *C. comosa,* which has longer, outcurved perigynium teeth. More common *C. vesicaria* and *C. utriculata* lack scabrous awns on the female scales and have ascending female spikes.

Carex idahoa L.H. Bailey (p. 205)

Idaho sedge

Carex parryana Dewey ssp. *idahoa* (L.H. Bailey) D.F. Murray

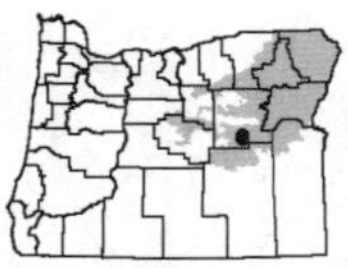

Plants more or less dioecious, rhizomatous, turf-forming but the stems sometimes arising in clusters; culms 25–40 cm tall. **Leaves** 2–5 mm wide, slightly glaucous. **Inflorescences** spikes usually male or female but sometimes with male flowers scattered among the perigynia; terminal spike 2–3 cm × 5–8 mm, solitary or with 1–3 short lateral spikes at its base. **Female scales** some scales 2–3 times longer than the perigynia, brown to blackish with conspicuous paler mid-stripe and narrow pale margins; midrib paler, conspicuous; tip acute to short-awned. **Perigynia** elliptic to obovate, 2–3 × 1.5–1.75 mm, glabrous, yellowish to brown, veinless except for the 2 marginal ribs; margins serrulate distally; beak 0.2–0.3 mm long; stigmas 3. **Achenes** trigonous, nearly filling the perigynium body.

Flat to gently sloping meadows around headwater streams, ponds, springs and in subirrigated soils that dry out on the surface in summer. 1400–1700 m. BW. CA, ID; northeast to MT. Native.

Carex idahoa is a rare, glaucous, turf-forming plant with brown, usually solitary spikes. It may grow with *C. praegracilis,* which has more yellow-green leaves, darker plant bases and rhizomes, and a paler inflorescence with several short spikes.

Carex illota L.H. Bailey (p. 207)

Sheep sedge, small-headed sedge

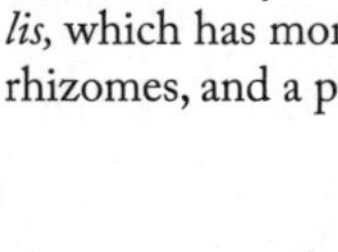

Plants cespitose, 15–40 cm tall. **Leaves** 1.5–2.5 mm wide. **Inflorescences** pyramidal, 1–1.2 cm × 5.5–10 mm, dense and head-like, blackish brown; spikes gynaecandrous. **Female scales** usually shorter and narrower than the perigynia, brown, usually with paler midrib. **Perigynia** ovate, plano-convex, 2.6–3.2 × 0.9–1.3 mm, gold to blackish brown, unwinged but with 2 fairly sharp-edged, entire, lateral ribs; base usually with some pithy tissue; beak unwinged, brown, and entire for the distal 0.4+ mm; 1.2–1.6 mm from achene top to beak tip; stigmas 2. **Achenes** lenticular, 1.2–1.5 × 0.8–1 mm. **2*n***=64.

Montane marshes, streamsides, wet meadows, bogs. 1100–2800 m. BR, BW, Casc. CA, ID, NV, WA; north to British Columbia, east to CO. Native.

This "ball-headed" sedge has a dark, compact, often pyramidal inflorescence and wingless perigynia. It looks like a young or depauperate *C. microptera,* but it lacks perigynium wings and grows in wetter habitats.

Carex infirminervia Naczi (p. 207)

Weak-veined sedge

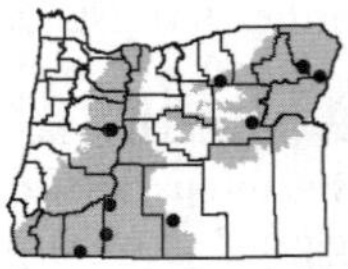

Plants cespitose; culms 10–83 cm long, smooth or more often papillose (at 20×) at mid-length, the fragile papillae like tiny pegs projecting at right angles to the culm on its edges. **Leaves** 1.4–3.8 mm wide. **Inflorescences** spikes (4)5–6(7) usually longer than wide, generally pale to straw-colored, smooth or rough in outline, usually gynaecandrous, sometimes male or female; lowest spike 9.5–18 mm long, with 11–22 perigynia; terminal spike 9.5–16 mm long. **Female scales** 2.9–4.4 mm, silvery white to gold, with green midvein, usually awned; awn to 1.1 mm long; body about as long as or longer than the mature achene. **Anthers** 1.3–1.8 mm long. **Perigynia** 3.7–5.3 mm long, green, whitish, or tan; beak 1.5–2.2 mm long, 39–49% of perigynium length; teeth 0–0.2(0.4) mm long; stigmas 2. **Achenes** lenticular. **2*n***=54.

Shaded to partially shaded riparian zones in mountains. 600–2000 m. BW, Casc, ECas, Sisk. CA, ID, NV, WA; north to British Columbia, northeast to MT. Native.

This is one of the "identical triplets," distinguished by its relatively long beaks and short beak teeth. See discussion under *C. bolanderi*.

Carex inops L.H. Bailey ssp. ***inops*** (p. 207)

Long-rhizome sedge

Carex pensylvanica Lam., *Carex pensylvanica* Lam. var. *vespertina* L.H. Bailey

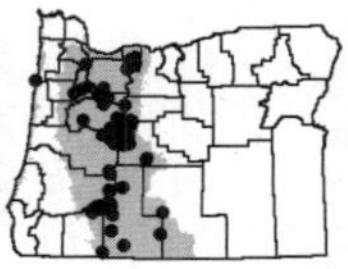

Plants rhizomatous; culms 13–50 cm tall, standing taller than the leaves; plant bases brown or reddish. **Leaves** 0.7–4.5 mm wide, green, not papillose on the lower surface. **Inflorescences** basal spikes none or rarely present, female, on long, slender stalk; bracts of lowest lateral spikes bristle-like or leaf-like, shorter than the inflorescence; lateral spikes 1–4, female, each about 1.5 mm long or shorter, with 5–15 perigynia; terminal spike male, 1–2.5 cm long, held above the female spikes on peduncles (0.8)2.5–20 mm long. **Female scales** subequal to the perigynia, reddish brown, brown, or purplish with narrow white margins. **Perigynia** obovoid, 2.8–4.6 × 1.5–2.2 mm, pale green to pale brown, pubescent, veinless except for 2 marginal ribs; bases succulent, withering when dry; beaks 0.4–1.3 mm long; teeth 0.1–0.7 mm long; stigmas 3. **Achenes** trigonous.

Dry, open, conifer forests, meadows, low elevation prairies and *Quercus garryana* savannas. 100–2300 m. Casc, ECas, Est, WV. CA, WA; north to British Columbia. Native.

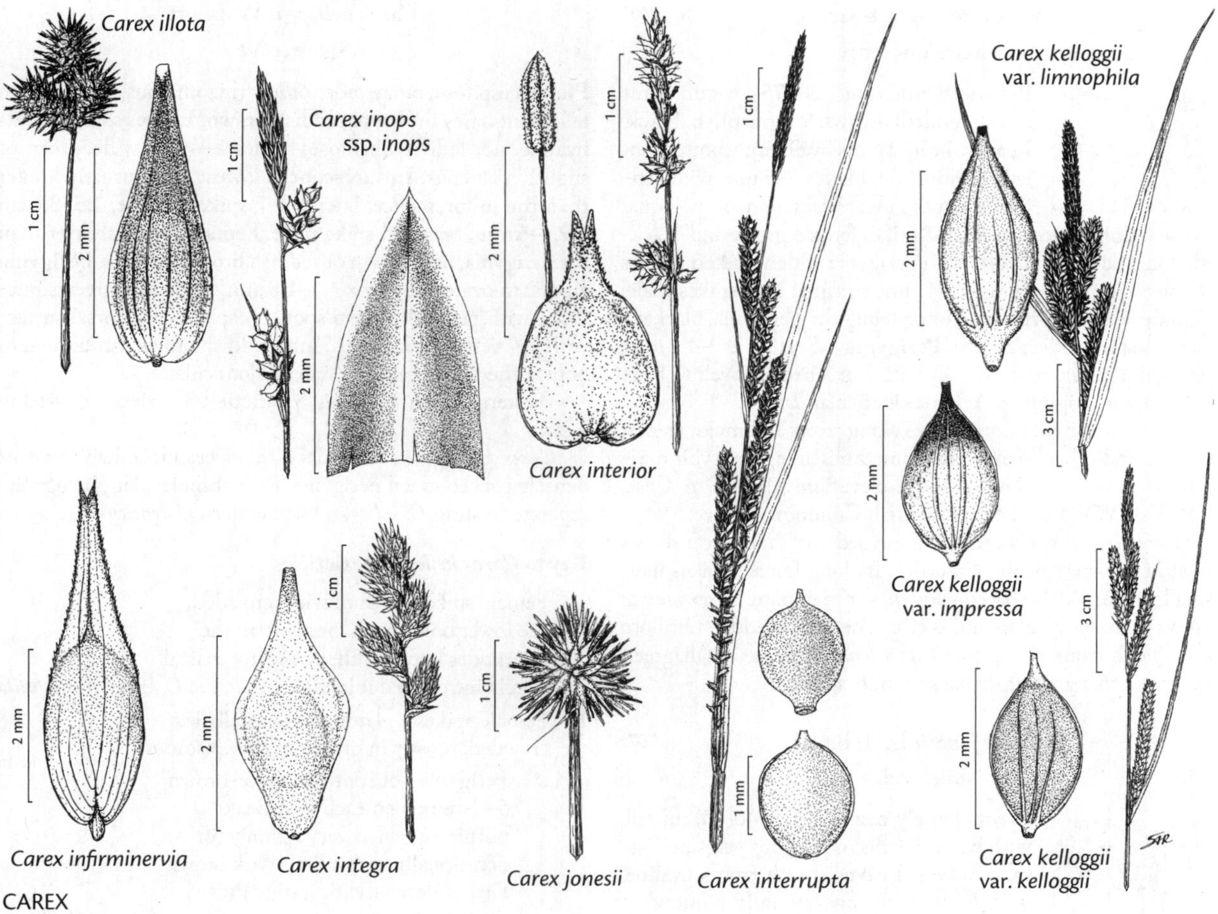

CAREX

Carex inops ssp. *inops* is a common, rhizomatous, upland species with hairy perigynia in short female spikes. *Carex rossii* and *C. deflexa* have similar perigynia but are more or less cespitose and have basal female spikes hidden among the leaf bases and longer inflorescence bracts. Tolerant of trampling, *C. inops* ssp. *inops* is useful for restoration.

Carex integra Mackenzie (p. 207)

Smooth-beak sedge

Plants cespitose, (11)15–55 cm tall. **Leaves** (1.1)1.5–2.6(3.1) mm wide. **Inflorescences** 1.1–2.4 cm × 6.3–14 mm, dense and head-like, often slightly elongate; spikes gynaecandrous. **Female scales** shorter than to slightly longer than the perigynia and about as wide, gold to brown, occasionally dark brown, with paler midrib. **Perigynia** lance-ovate to ovate or obovate, 2.1–3.6 × 0.8–1.4 mm, gold to dark brown, smooth edged or nearly so; wing 0.05–0.2 mm wide; beak tip unwinged, brown, and parallel-sided for at least the distal 0.4 mm; 1–1.6 mm from top of achene to tip of beak; stigmas 2. **Achenes** lenticular, 1.1–1.4 × 0.7–1 mm. ***2n***=82.

Mesic to dry meadows at mid to high elevations in the mountains. 1200–2100 m. BW, Casc, ECas, Sisk. CA. Native.

Carex integra is a cespitose sedge with smallish, pale heads and perigynia. *Carex subfusca* is similar, but its perigynium wings are usually not as narrow. *Carex illota* has no perigynium wings at all, and its heads are darker and more compact.

Carex interior L.H. Bailey (p. 207)

Inland sedge

Plants cespitose or short rhizomatous, 10–95 cm tall. **Leaves** 1–2.4(2.7) mm wide. **Inflorescences** spikes separated, usually gynaecandrous, sometimes female, with perigynia widely spreading or reflexed at maturity; lateral spikes 3–9.5 mm long; terminal spikes 5.2–20 mm long, with a narrow 2.2–14.5 mm long male portion. **Female scales** inconspicuous, shorter than the mature achene. **Perigynia** 1.9–3(3.3) mm long, reddish brown to dark brown, sometimes green, usually veinless on the ventral surface over the achene; base somewhat swollen, pithy when mature; beak 0.4–0.9 mm long, 20–40(43)% of perigynium length, margins usually serrulate; stigmas 2. **Achenes** lenticular.

Springs, wet canyon walls and seeps. 600–2200 m. BW, Casc, ECas, Lava. CA, ID, NV, WA; north to AK, east to Newfoundland and VA, southeast to AR, AZ, and Mexico. Native.

Carex interior is an inconspicuous, slender-leaved, cespitose sedge with an interrupted inflorescence, the star-like spikes with widely spreading perigynia that taper abruptly to the short beak. *Carex echinata* has longer, gradually tapering perigynia. In southwestern Oregon, *C. interior* perigynia tend to have longer beaks than is typical, more similar to those of *C. echinata*.

Carex interrupta Boeckeler (p. 207)

Greenfruit sedge

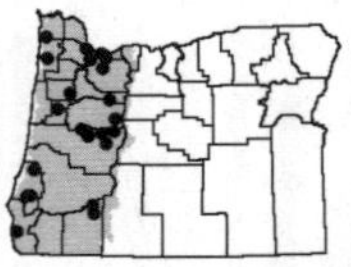

Plants rhizomatous, 20–75 cm tall; plant bases reddish brown or purplish black. **Leaves** sheath fronts hyaline, unspotted, not ladder-fibrillose; blades 3–5 mm wide. **Inflorescences** lowest bract more or less equal to the inflorescence; lateral 4–7 spikes female, green-and-blackish, long-tapering at base, with perigynia widely spaced at base of lowest spike, 4–9 cm × 3–4 mm; terminal 1(2) spikes male. **Female scales** shorter than or equaling the perigynia, blackish, occasionally reddish brown. **Perigynia** 1.5–2.1 × 0.9–1.4 mm, green, not veined (occasionally with 1 or 2 irregular veins); beak 0.1–0.3 mm; stigmas 2. **Achenes** lenticular. **2*n***=68.

Seasonally flooded shorelines along rivers and major creeks, in sand and other coarse soils; sometimes along reservoir margins or in wet meadows at lower elevations. 0–1600 m. Casc, CR, Sisk, WV. WA; north to British Columbia. Native.

Carex interrupta is a rhizomatous sedge of sandy river shores west of the crest of the Cascades. Its long female spikes have small, green, widely spaced perigynia at their base. *Carex nudata* grows on rocky reaches and is cespitose. It has ladder-fibrillose leaf sheath fronts. Cespitose *Carex lenticularis* has small, green perigynia that are tightly packed in its spikes.

Carex jonesii L.H. Bailey (p. 207)

Jones' sedge

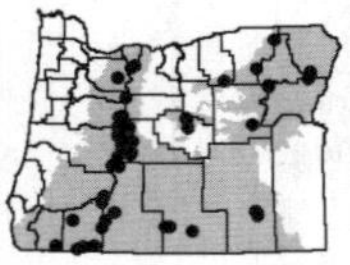

Plants loosely cespitose, 15–40(60) cm tall, with basal sheaths of previous years persisting as fibers. **Leaves** sheath fronts hyaline, usually smooth, occasionally somewhat cross-corrugated, shorter than the blades, with mouth concave; all or most sheaths with well-developed blades; blades 1–4 mm wide, green, all clustered near the base of the shoot, arising at about the same height so that the sheath fronts are mostly hidden. **Inflorescences** 0.8–2(2.5) × 0.6–1.5 cm, dense and head-like; spikes all androgynous, or some female. **Female scales** about as long as the perigynia, dark brown. **Perigynia** 2.5–3.5(4) × 1–1.5 mm, widest near the base (occasionally near middle), brown; strongly veined, with 7–11 veins on the dorsal surface, 5–7 on the ventral surface; base usually pithy and strongly swollen; beak 0.5–1.5(2) mm long, margins smooth or nearly so; stigmas 2. **Achenes** lenticular.

Wet meadows, seeps and streamsides in the mountains. 600–2800 m. BR, BW, Casc, ECas, Sisk. CA, ID, NV, WA; northeast to MT. Native.

Carex jonesii is a common "ball-headed" sedge of montane wetlands, sometimes referred to as the big butt sedge because of its swollen, pithy perigynium bases. *Carex jonesii* often grows with *C. neurophora,* which has more or less cross-corrugated leaf sheath fronts that are usually exposed because the leaves arise at different heights. In southwestern Oregon, see *C. nervina*.

Carex kelloggii W. Boott

Shore sedge

Plants cespitose, rarely more or less rhizomatous, (7)15–90 cm tall; plant bases brown or reddish brown. **Leaves** sheath fronts hyaline, not ladder-fibrillose; blades 2–4 mm wide, green or slightly glaucous. **Inflorescences** lowest bract usually longer than the inflorescence; lateral 3–7 spikes female, 1.5–3.6 cm × 2.5–6 mm; terminal spike male. **Female scales** shorter than the perigynia, pale brown or reddish brown to black. **Perigynia** elliptic to ovate, 1.8–3.5 × 1.1–1.8 mm, whitish or green sometimes with purplish brown spots on apical half, dorsal surface with 1–7 veins; beak 0.1–0.5 mm, reddish brown throughout or only at the tip; stigmas 2. **Achenes** lenticular.

Western North America. 3 varieties; 3 varieties treated in *Flora*.

Carex kelloggii is recognized by its cespitose habit and its densely packed green perigynia. Recent molecular phylogenies separate western *C. kelloggii* from eastern *C. lenticularis*.

Key to *Carex kelloggii* varieties:

1. Female spikes 4–6 mm wide, crowded, the lower ones much longer than the internodes between them; mostly coastal (rarely introduced inland) . . . . ***C. k.*** var. ***limnophila***

1' Female spikes 1–4 mm wide, usually less crowded; mostly in or east of the Cascades.

  2. Perigynia not spotted purple-brown, 5–7-veined on each face; beak purple-brown at very tip only (or occasionally with a little dark brown farther down the tip), stipe greater than 0.2 mm long . . . . ***C. k.*** var. ***kelloggii***

  2' Perigynia spotted purple-brown on apical half, 1–3-veined on dorsal surface; beak entirely purple-brown (or occasionally with some green on the tip), stipe less than 0.2 mm long . . . . ***C. k.*** var. ***impressa***

Carex kelloggii W. Boott var. *impressa* (L.H. Bailey) B.L. Wilson & Otting (p. 207)

Mountain shore sedge

Carex lenticularis Michx. var. *impressa* (L.H. Bailey) L.A. Standl., *Carex paucicostata* Mack.

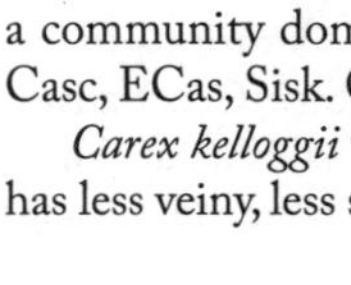

Plants cespitose, rarely more or less rhizomatous, 15–60 cm tall. **Inflorescences** lateral 3–4 spikes female, 3–4 mm wide. **Female scales** pale brown to black. **Perigynia** 1.8–2.5 × 1.1–1.5 mm, green with purplish brown spots on apical half, dorsal surface with 1–3 veins; beak reddish brown throughout or occasionally with some green. **2*n***=92.

Lake shores, reservoir margins, river shorelines, sometimes a community dominant above timberline. 900–2300 m. BW, Casc, ECas, Sisk. CA, ID, WA. Native.

Carex kelloggii var. *impressa* resembles *C. k.* var. *kelloggii* but has less veiny, less strongly stipitate perigynia and brown beaks.

Carex kelloggii W. Boott var. ***kelloggii*** (p. 207)

Kellogg's sedge, lakeshore sedge

Carex lenticularis Michx. var. *lipocarpa* (Holm) L.A. Standl.

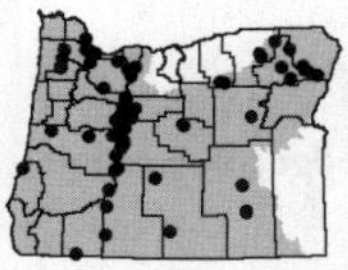

Plants cespitose, 15–90 cm tall. **Inflorescences** lateral 3–7 spikes female, 2.5–4 mm wide. **Female scales** pale brown to black. **Perigynia** 2–3.5 × 1–1.8 mm, whitish or green, with 5–7 veins on each face; beak usually green with reddish brown at the very tip only, or rarely with a little brown extending down the beak.

Lake shores, reservoir margins, pools along rivers and other areas of slow water but often fluctuating water levels. 0–2500 m. BR, BW, Casc, CR, ECas, Est, Lava, Sisk, WV. CA, ID, WA; north to AK, east to CO, southeast to AZ. Native.

Carex kelloggii var. *kelloggii* is the most common and widespread member of this species in the Pacific Northwest. It is recognized by the strongly stipitate, 5–7-veined perigynia with mostly green beaks.

Carex kelloggii W. Boott var. ***limnophila*** (Holm) B.L. Wilson & R.E. Brainerd (p. 207)

Coastal shore sedge

Carex lenticularis Michx. var. *limnophila* (Holm) Cronquist, *Carex hindsii* C.B. Clarke

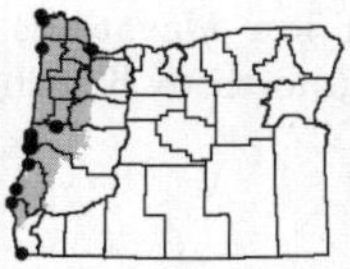

Plants cespitose, (7)17–45(80) cm tall. **Inflorescences** lateral 3–7 spikes female, 4–6 mm wide, crowded, overlapping. **Female scales** reddish brown to black. **Perigynia** 2.5–3.5 × 1.3–1.6 mm, green with purplish brown spots on distal half, with 5–7 veins on each surface; beak green with reddish brown rim at the very tip, occasionally with brown running down the beak. **2*n***=92.

Deflation plains, lake margins, pools, and other shallow wetlands at the coast, occasionally introduced in low elevation, inland sites. 0–300 m. CR, Est, WV. CA, WA; north to AK. Native.

Coastal *C. kelloggii* var. *limnophila* has thick, overlapping spikes that are more than 1.5 cm long. It does not grow with the other *C. kelloggii* varieties, which have longer, narrower leaves and less crowded inflorescences. Inland records are recent introductions.

Carex klamathensis B.L. Wilson & Janeway

Klamath sedge

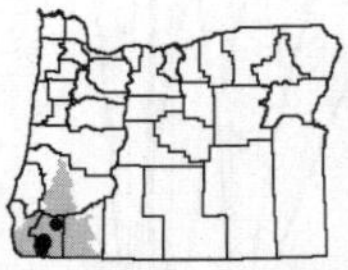

Plants rhizomatous, the rhizomes 1–2 mm in diameter; culms 30–100 cm tall, phyllopodic; plant bases whitish or pale to medium brown. **Leaves** blades flat, the wider blades 2–6 mm wide, glaucous. **Inflorescences** lateral spikes usually 1–3, female, 1–2.5 cm × 4–7 mm; terminal spike usually male, 1.3–2.7 cm × 1–5 mm, but in some populations mostly gynaecandrous, 50–200 flowers. **Female scales** 1.9–2.8 mm long, gold to brown with paler midrib, obtuse, sometimes with an awn up to 1.5 mm long. **Perigynia** obovate or rarely fusiform, 2.1–3.6 × 0.6–1.8 mm, light green, tan, or whitish, sometimes marked with brown distally, with 8–20 faint to strong veins; beak short, often bent toward the dorsal side; stigmas 3. **Achenes** trigonous.

Springs and fens on serpentine substrates, often with *Darlingtonia californica,* usually in a dense thatch of old leaves. 300–1100 m. Sisk. CA. Native.

Carex klamathensis is a rare, glaucous, rhizomatous sedge of serpentine fens, with pale, veined, somewhat papillose perigynia that usually have the apex turned to the back. *Carex livida* has longer perigynia that are consistently spindle-shaped. *Carex hassei* can be very similar and grows in the same habitat but has two stigmas per flower or, in southwestern Oregon, a mix of two-stigma and three-stigma fertile flowers. *Carex klamathensis* may have two stigmas on aborted flowers.

Carex kobomugi Ohwi (p. 210)

Japanese sedge

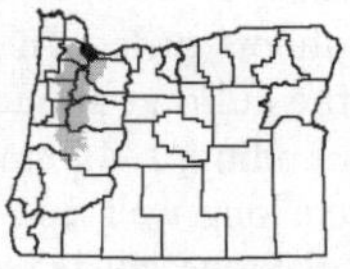

Plants rhizomatous; culms 10–30 cm tall, bluntly 3-angled, smooth near the inflorescence. **Leaves** blades 3–6 mm wide, longer than the culms but curving sideways or down, serrulate on margins. **Inflorescences** dense and head-like, either male or female, growing on separate shoots, the plant thus appearing dioecious, although male and female shoots are connected by rhizomes; male inflorescences 3–4 × 1–2 cm; female inflorescences 3–6 × 2–4 cm, with lower spikes ascending, sometimes spreading, sometimes with inconspicuous male flowers mixed in. **Female scales** long-tapered, sometimes with awn 6–12 mm long. **Anthers** 4–6.5 mm long. **Perigynia** lance-ovate to elliptic, 10–14 mm long, green or yellow to dark brown; wing margins irregular; beak shorter than or about as long as the body; 3–5 mm from the top of the achene to the beak tip; tip notched 0.4–0.6 mm; stigmas 3. **Achenes** rounded-trigonous.

Sandy, often disturbed soils. 0–50 m. WV. WA; MA to NC; Asia. Exotic.

This strongly rhizomatous sedge has wide, tough leaves with strongly serrulate margins and large, dense heads with large perigynia. It was collected in Portland in 1916 in ballast from ships, and rediscovered in 2006. *Carex kobomugi* is invasive and difficult to eradicate where introduced in the Atlantic coast. Compare to native *C. macrocephala.*

Carex laeviculmis Meinshausen (p. 210)

Smoothstem sedge

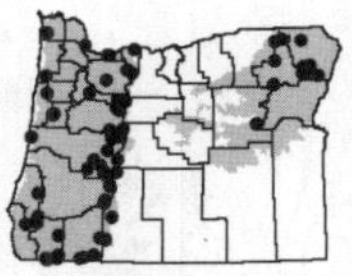

Plants cespitose, 14–66 cm tall. **Leaves** 0.9–2(2.3) mm wide, glaucous to green. **Inflorescences** spikes separated or sometimes upper ones crowded, 4–7, greenish to brown, usually gynaecandrous, with 4–18 female flowers, 5.5–9.8 mm long. **Female scales** shorter than the perigynia, straw-colored to reddish brown. **Perigynia** 2.3–3.7 mm long, ascending (the beaks sometimes spreading, but not the perigynia bodies), green to brown, sometimes reddish brown apically; bases not swollen or pithy; beak 0.4–1.1(1.3) mm long; stigmas 2. **Achenes** lenticular. **2*n***=56.

Forests, wet meadows, bogs, seeps, shady streamsides. 200–2100 m. BW, Casc, CR, Sisk, WV. CA, ID, WA; north to AK, east to WY. Native.

Carex laeviculmis is densely cespitose with several separated, short spikes. Similar *C. leptopoda* and its relatives are coarser plants of more mesic habitats with wider leaves, longer spikes and perigynia, and slightly pithy perigynium bases. Young or small-fruited *C. laeviculmis* are often reported as *C. brunnescens,* which has smaller perigynia with shorter beaks and female scales about as long as the perigynia but not concealing them. *Carex brunnescens* is not currently known from Oregon.

Carex lasiocarpa Ehrhart (p. 210)

Slender woolly sedge, woollyfruit sedge

Carex lasiocarpa Ehrh. var. *americana* Fernald

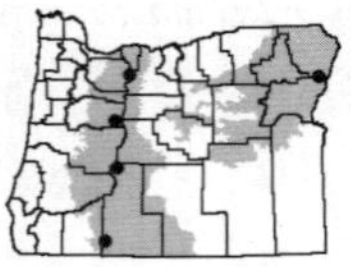

Plants rhizomatous, 40–120 cm tall, aphyllopodic; plant bases reddish brown, shiny. **Leaves** sheaths usually ladder-fibrillose; blades 0.2–2(2.2) mm wide as folded, glaucous, consisting of little more than the midrib, involute to channeled at the base and narrowly triangular near the tip, the tips of leaves on vegetative shoots curled. **Inflorescences** lowest bract diverging from the culm at about 45–90°; lateral 1–3 spikes female, erect or ascending, 1–4.5 cm long, remote; terminal 1–3 spikes male, 1.5–6 cm long, well separated from the female spikes. **Female scales** glabrous, ciliate or with margin scabrous apically, apex acute to awned. **Perigynia** broadly ovoid, abruptly contracted to the beak, 3–4.3 × 1.5–2.2 mm, densely pubescent; beak 0.5–1.1 mm long; teeth 0.2–0.7 mm long, straight; stigmas 3. **Achenes** trigonous. ***2n***=54.

Fens, bogs, lake shores, and wet meadows, sometimes forming floating mats, in deep, organic, acidic soils with low to moderate nutrient levels. 1200–1700 m. BW, Casc, ECas. CA, ID, WA; north to AK, east to Newfoundland, VA; Eurasia. Native.

Carex lasiocarpa is a slender bog sedge. Similar *C. pellita* grows in shallower water and has wider, flatter leaves that do not curl at the tip and more upright lowest inflorescence bracts. The two may grow together in the same bog.

Carex leporina Linnaeus

Oval broom sedge, eggbract sedge, oval sedge, hare sedge

Carex ovalis Gooden., *Carex tracyi* Mack.

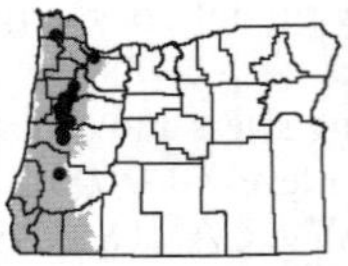

Plants cespitose, (24)35–85 cm tall at flowering; vegetative shoots elongating late in the season and becoming stolons (developing shoots and roots at the nodes), sprawling over shorter vegetation. **Leaves** (1.5)2–3.5(4) mm wide. **Inflorescences** 1.5–4 cm × 10–15 mm, erect to flexuous, not head-like, the lower spikes somewhat separated; spikes gynaecandrous. **Female scales** 3.4–5 mm long, longer or shorter, wider or narrower than the perigynia, straw-colored, brown or green; margins not white-hyaline; the tip often white-hyaline, acute to acuminate. **Perigynia** 3.4–4.7(5.2) × 1.3–2.1 mm, straw-colored or light brown, with 3–9 conspicuous dorsal veins and 2–5 conspicuous ventral veins; wings 0.2–0.5 mm wide; beak brown, parallel-sided, entire for the distal 0.3–0.5 mm (occasionally winged almost to the tip); (1.2)1.5–2 mm from achene top to beak tip; stigmas 2. **Achenes** lenticular, 1.1–1.8 × 0.9–1.2 mm, 0.4–0.5 mm thick. ***2n***=62, 64, 66, 68.

Moist meadows and wet prairies at low elevations. 50–1400 m. CR, Sisk, WV. CA, NV, WA; north to British

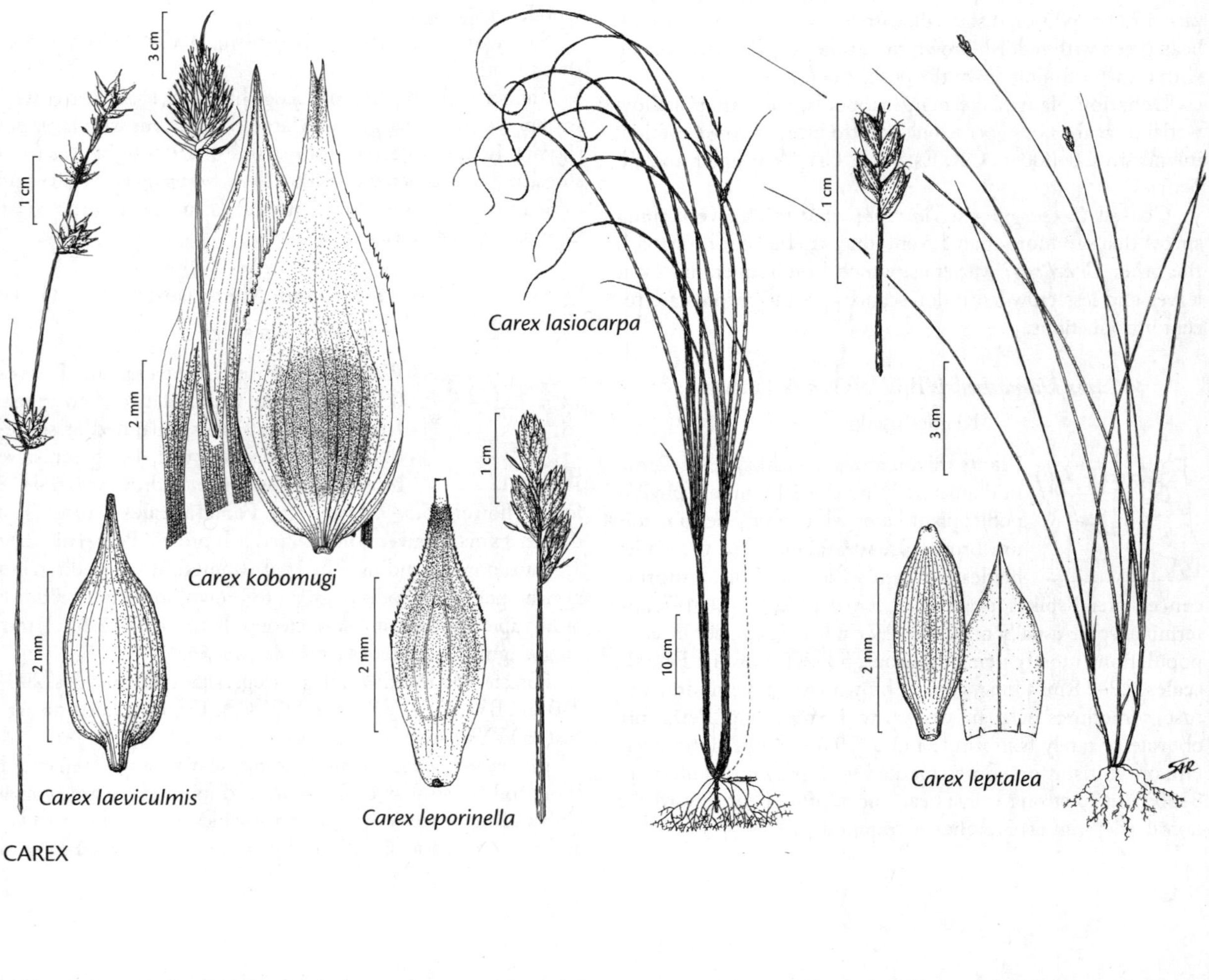

CAREX

Columbia, WI to Prince Edward Island south to KY and NC; Australia, Eurasia, New Zealand. Native.

Carex leporina is a loosely tufted sedge with a light brown, somewhat elongate inflorescence. The stolons form late in the season and are rarely collected. Similar *C. praticola* has female scales with white margins. Its perigynia are veinless or weakly veined on the ventral surface, have longer beaks that are brown, parallel-sided, and wingless for the distal 0.4–1 mm and have the very tip narrowly white-edged.

Carex leporinella Mackenzie (p. 210)

Bog hare sedge

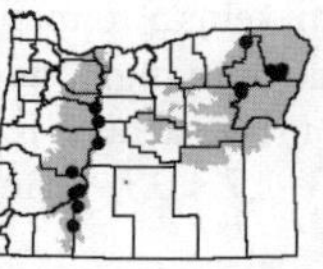

Plants cespitose, 14–30 cm tall. **Leaves** 1–2.5 mm wide. **Inflorescences** 1.4–3 cm × 5–11 mm; spikes ascending to erect, overlapping, gynaecandrous. **Female scales** ovate, (2.5)3.5–6 mm long, usually longer and wider than the perigynium bodies (or the entire perigynia) and more or less hiding them; straw-colored to reddish brown, generally with lighter mid-stripe. **Perigynia** lanceolate, 3.5–4.2 × 1–1.2 mm, straw-colored to whitish or dull brown, opaque, usually with (0)5–10 conspicuous dorsal veins, 0–6 conspicuous ventral veins; wings 0.05–0.2 mm wide, ciliate-serrulate, extending far up the beak and incurved toward the ventral side; beak unwinged, brown, parallel-sided, and entire for the distal 0.3–0.6 mm; 1.5–2.2 mm from achene top to beak tip; dorsal suture usually with a conspicuous white margin; stigmas 2. **Achenes** lenticular, 1.4–1.8 × 0.7–1 mm, 0.3–0.4 mm thick.

Marshes, bogs, lake margins, and puddles, in habitats that are soggy wet early in the season but usually dry out later. 700–2800 m. BW, Casc. CA, ID, NV, WA; east to WY. Native.

Carex leporinella is identified by its short height, ascending spikes, female scales that nearly conceal the perigynia, narrow, boat-shaped perigynia, and shallow-water habitat. *Carex abrupta* can have similar perigynia, but its inflorescence is a dense head, and its female scales are shorter than the perigynia. Compare to *C. phaeocephala* and *C. tahoensis.*

Carex leptalea Wahlenberg (p. 210)

Delicate sedge, jelly bean sedge, bristle-stalked sedge, bristlystalked sedge

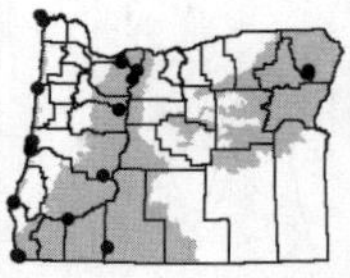

Plants growing in clumps from short, pale rhizomes; culms 10–70 cm, very thin and lax. **Leaves** 0.4–1.3 mm wide, flat, soft. **Inflorescences** spike solitary, androgynous, 0.4–1.5(1.8) cm long; the male portion often short and inconspicuous but occasionally occupying most of the spike; female part with 1–10 more or less overlapping perigynia, lacking inflorescence bracts. **Female scales** usually about half as long as the perigynia, green or brown, sometimes awned; the lowest female scales are sometimes prolonged as thin bristles. **Perigynia** 2.4–4.9(5.4) mm long, green or straw-colored, with 2 marginal ribs and many fine veins; base spongy, stipe-like; apex rounded, virtually beakless; stigmas 3. **Achenes** trigonous. **2*n***=50, 52.

Bogs, wet meadows, pond margins, other wetlands, often in shade, often growing in moss mats, sometimes forming a lawn under willow thickets. 0–1600 m. BW, Casc, ECas, Est, Sisk. CA, ID, WA; throughout North America. Native.

Carex leptalea is distinctive with its solitary spikes and almost beakless perigynia. Its foliage is soft and sometimes forms a grass-like lawn under willows.

Carex leptopoda Mackenzie (p. 212)

Slender-foot sedge, short-scaled sedge

Carex deweyana Schwein. ssp. *leptopoda* (Mack.) Calder & Roy L. Taylor

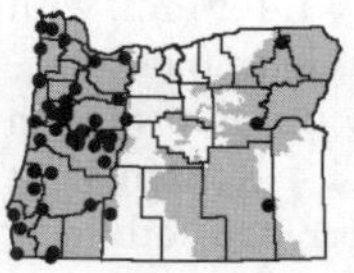

Plants cespitose; culms 21–67 cm long, smooth or papillose (at 20×) at mid-length, the fragile papillae like tiny pegs projecting at right angles to the culm on its edges. **Leaves** 2.4–5.9 mm wide. **Inflorescences** spikes 5–7, usually longer than wide, green and white (maturing light brown), relatively smooth in outline, usually gynaecandrous (or female); lowest spikes 8–15 mm long, usually with 11–23 perigynia; terminal spikes 9–12.7 mm long. **Female scales** white, occasionally straw-colored, with green midvein, usually with awn to 0.7 mm long; body shorter than the mature achene. **Perigynia** 3.3–3.9(4.4) mm long, thin-textured; beak 0.9–1.5(1.7) mm long, 30–38% of perigynium length; teeth to 0.3 mm long; stigmas 2. **Achenes** lenticular. **2*n***=54.

Shaded to partially shaded riparian zones and moist forests. 0–2700 m. BR, BW, Casc, CR, Est, Sisk, WV. CA, ID, WA; north to British Columbia, northeast to MT. Native.

In western Oregon, *Carex leptopoda* is the most common of the "identical triplets" (*Carex bolanderi, C. infirminervia,* and *C. leptopoda*). Its shorter perigynium beaks distinguish it from the other two species. Also see discussion under *C. bolanderi.*

Carex limosa Linnaeus (p. 212)

Mud sedge

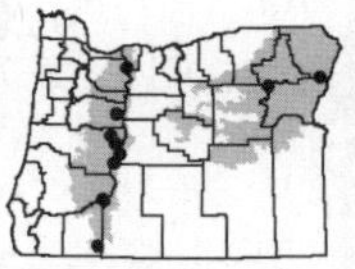

Plants strongly stoloniferous, rhizomes conspicuous, vertical; the shoots arising singly or in small groups; young roots with yellow felty hairiness; culms 20–60 cm tall. **Leaves** 1–2.5 mm wide, few, glaucous, channeled or grooved; margins involute. **Inflorescences** lowest bract 2–7 mm long, leaf-like; lateral spikes female or androgynous, 0.6–2 cm × 4–8 mm; terminal spike male, 0.7–3.5 cm × 1.5–2.5(3) mm. **Female scales** 3–5.5 × 2–3.4 mm, as long as or longer than the perigynia, wider than the perigynia, yellowish brown to dark brown, apex obtuse to more or less acute, sometimes mucronate. **Perigynia** broadly oval, 2.5–4 × 1.8–2.6 mm, pale, green to straw-colored; beak 0.1–0.5 mm long; stigmas 3. **Achenes** trigonous. **2*n***=56, 62, 64.

Nutrient-poor fens and bogs, floating peat mats; wet, peaty meadows and pond margins. 900–2200 m. BW, Casc. CA, NV, WA; north to AK, east to Newfoundland and DE, southeast to UT; Eurasia. Native.

Carex limosa is a spiky little sedge. Each of the shoots bears a couple of short, erect leaves. The bract and the narrow male spikes point upwards. The short, broad female spikes hang downwards. Compare to *C. pluriflora.*

Carex livida (Wahlenberg) Willdenow (p. 212)

Pale sedge, livid sedge

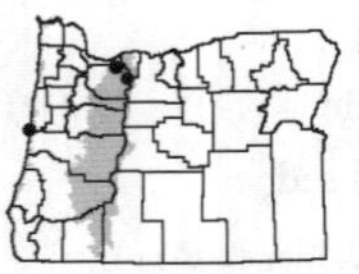

Plants rhizomatous, the rhizomes slender; culms 10–45(55) cm tall. **Leaves** plant bases whitish or pale to medium brown; lower sheaths usually with blades; blades 1–3.5 mm wide, glaucous, channeled. **Inflorescences** lateral spikes usually 1–2, female, 0.7–2.5 cm × 4–5.5 mm; terminal spike male, 1–3.3 cm × 1.4–5 mm, with 16–100 flowers. **Female scales** more or less as long as or somewhat longer than the perigynia, pale to very dark brown with green midrib, obtuse. **Perigynia** 3–4.8(5) × 1.4–2.1 mm, light green or whitish, widest in the middle and tapered to both ends; stigmas 3. **Achenes** trigonous.

Bogs and fens with the water table at or close to the surface, usually with extensive peat deposits but sometimes on mineral substrates next to slow streams. 100–1400 m. Casc, Est. CA, WA; north to AK, east to Newfoundland, IL, NJ; South America; Eurasia, Panama. Native.

From a distance, the grayish color of *C. livida* can give a bog a moldy appearance. This is a glaucous, narrow-leaved, rhizomatous sedge with pale, papillose, spindle-shaped perigynia. Similar *C. klamathensis*, of serpentine fens, has broader leaves and obovate perigynia with the apex usually bent to the dorsal side.

Carex longii Mackenzie (p. 212)

Green-and-white sedge

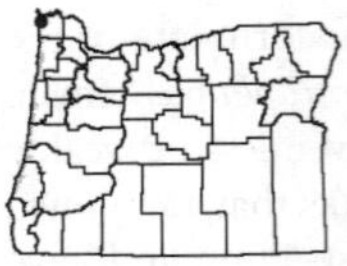

Plants cespitose, 30–120 cm tall, the vegetative culms sometimes rooting at nodes. **Leaves** sheath fronts green and veined except for a small white-hyaline triangle at the top; blades 2–4.5 mm wide. **Inflorescences** 1–4.5(6) cm × 5–14 mm, typically elongate; spikes distinct but overlapping, gynaecandrous, rounded or tapered to tip. **Female scales** 2.2–3.7 mm long, shorter and narrower than the perigynia, white-hyaline, aging silvery brown, with green center, apex obtuse. **Perigynia** obovate, 3–4.6 × 1.6–2.6(2.8) mm, the body widest above the middle, green or light brown; veins 5 to many on each face; wings 0.5–0.8 mm wide; beak tip winged and ciliate-serrulate almost to the tip; 1.4–2.2 mm from achene top to beak tip; stigmas 2. **Achenes** lenticular, 1.3–1.7 × 0.7–1 mm, 0.4–0.5 mm thick. **2n**=58, 60, 62.

Bogs and pond shores, on peaty or sandy soils, often where water levels fluctuate greatly, sometimes a weed in cranberry bogs. 0–50 m. Est. WA; eastern North America, HI, Central America, Caribbean; New Zealand. Exotic.

Carex longii is cespitose and has a silvery-green, elongated inflorescences and green leaf sheath fronts. It strongly resembles native *C. feta*, which has rounder spikes and perigynium bodies widest near the middle.

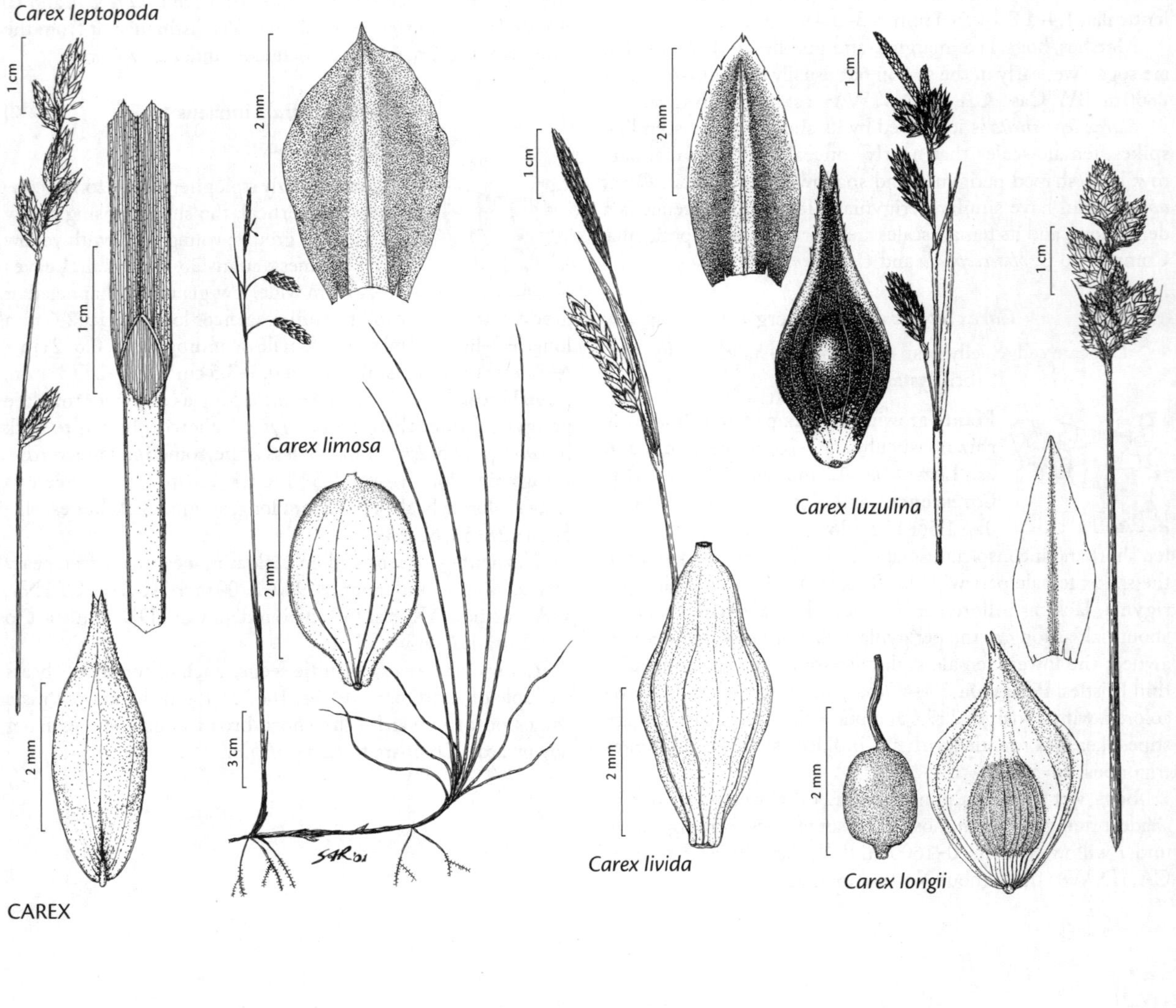

CAREX

Carex luzulina Olney (p. 212)

Woodrush sedge, spring sedge

Carex ablata L.H. Bailey, *Carex luzulina* Olney var. *ablata* (L.H. Bailey) F.J. Herm., *Carex luzulina* Olney var. *luzulina*

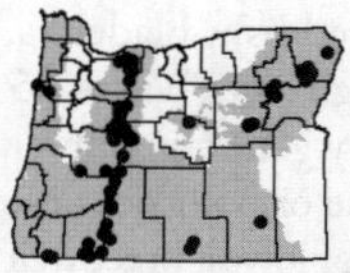

Plants cespitose to short-rhizomatous, 15–90 cm tall; culms much longer than the leaves. **Leaves** sheath fronts whitish hyaline; blades 3–9 mm wide, glabrous. **Inflorescences** lowest bracts with more or less inflated sheaths (5)10+ mm long; lateral spikes crowded around the terminal one or separated, erect and nearly sessile or the lower ones drooping on long, slender peduncles, female or androgynous, 1.2–3.2 cm long; terminal spike male, 1.1–6.6 cm long. **Female scales** shorter than the perigynia, reddish brown to dark purplish with a light mid-stripe; margins sometimes white, glabrous. **Perigynia** (3)3.5–5.5 × 0.9–1.6(1.8) mm, either mostly purple, or green with reddish or purple spots, usually glabrous; sometimes with sparse, spreading to appressed, usually long, soft, thin hairs; with several inconspicuous veins; beak with distance from achene top to beak tip less than 2.5 mm; stigmas 3. **Achenes** trigonous.

Bogs and wet meadows. 600–3000 m. BR, BW, Casc, CR, ECas, Est, Sisk. CA, ID, NV, WA; north to British Columbia, east to WY. Native.

The broad, yellow-green leaves clustered at the base of the culm, together with the inflorescence bracts with more or less inflated sheaths, the pointy perigynia, and the reddish to purplish female scales, make *C. luzulina* unlike any other sedge in Oregon bogs. However, it is hard to point at any one consistent feature as identifying this species. Two varieties (sometimes treated as species) have been recognized in *C. luzulina*, but the traits supposedly distinguishing them vary independently. Variable *C. luzulina* is confused with Californian *C. luzulifolia* and *C. fissuricola*, which usually have perigynia with flat margins more than half as wide as the achene. In *C. fissuricola*, the perigynium faces, at least near the beak, have sparse, spreading-ascending, short, stiff bristles, and the pistillate scales have hairs on the distal margins.

Carex lyngbyei Hornemann (p. 214)

Lyngbye's sedge

Carex lyngbyei Hornem. var. *robusta* (L.H. Bailey) Cronquist

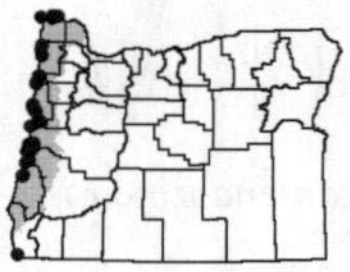

Plants rhizomatous, but the deep rhizomes often not collected; forming dense stands; culms 25–130 cm tall; plant bases reddish brown. **Leaves** sheath fronts hyaline, not spotted, not ladder-fibrillose; blades 3–8 mm wide, green, smooth-edged, dying in winter. **Inflorescences** lowest bract usually longer than the inflorescence; lateral 2–4 spikes female or androgynous, dangling on curved peduncles (but the spike itself does not bend), usually truncate at the base, 1.8–5 cm × 5–7 mm; terminal 2–3 spikes male. **Female scales** longer than the perigynia, reddish brown to dark purplish brown, acuminate. **Perigynia** ovate to obovate, 2.5–3.5 × 1.6–2.5 mm, yellowish brown with pale brown spots, 5–7-veined on each face; beak 0.1–0.3 mm long; stigmas 2. **Achenes** lenticular, often with a dent in the side. **2*n***=68, 70, 72, 76.

Coastal salt marshes and in more or less fresh water along the lower Columbia River. 0–50 m. CR, Est. CA, WA; north to AK, east to Greenland; Europe. Native.

Carex lyngbyei forms extensive stands in coastal salt marshes. Its female spikes are relatively short, with truncate bases, dangling at the ends of curved stalks. Similar *C. obnupta* and *C. aquatilis* var. *dives* grow in fresh to brackish water, sometimes adjacent to *C. lyngbyei* stands. Both have longer, usually sessile female spikes with tapered bases.

Carex macrocephala Willdenow ex Sprengel (p. 214)

Big-head sedge, largehead sedge

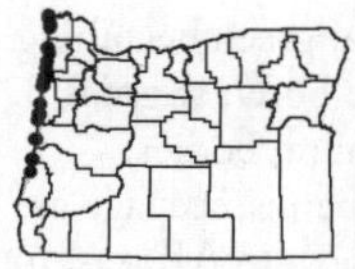

Plants rhizomatous; culms 10–35 cm tall, sharply 3-angled, serrulate near the inflorescence (on at least 1 angle). **Leaves** 4–8 mm wide, longer than the culms but curving sideways or down, thus often not taller; serrulate on margins. **Inflorescences** dense and head-like, either male or female, growing on separate shoots, the plant thus appearing dioecious, although male and female shoots are connected by rhizomes; male inflorescences 3.5–5 × 7–15 mm; female inflorescences 3.5–8 × 2.5–5 cm, with lower spikes more or less spreading. **Female scales** 1.2–4 mm long, with a short awn. **Anthers** 2.5–5 mm long. **Perigynia** 10–15 mm long, dark brown, shiny; wings curved inward over the ventral surface, margins erose; base spongy-thickened; beak about as long as the body; 6.5–9 mm from achene top to beak tip; stigmas 3. **Achenes** rounded-trigonous. **2*n***=74, 78.

Coastal sand dunes. 0–50 m. Est. WA; north to AK; NJ; Asia. Native.

Carex macrocephala is a strongly rhizomatous sedge of coastal sands, with tough, serrulate leaves, big heads, and perigynia a centimeter or more long.

Carex macrochaeta C.A. Meyer (p. 214)

Long-awn sedge, large-awn sedge

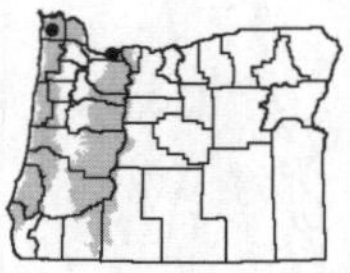

Plants loosely cespitose; young roots with white, cream-colored, or yellowish felty hairiness; culms 20–50(70) cm long. **Leaves** 2–5 mm wide, more or less white papillate (at 15×) on the lower surface. **Inflorescences** spikes 3–5; lateral spikes female, 1–3 cm × 6–8 mm, lower spikes often dangling on moderately long peduncles but upper ones short-pedunculate to sessile; terminal spike male, 1.2–3.2 cm × 3–8 mm. **Female scales** narrower and longer than the perigynia (though the scale body itself may be shorter), dark brown or black, often with a lighter midrib, awned; awns 1–3 mm long in Oregon (to 12 mm long in more northern sites). **Perigynia** narrowly elliptic, 3.8–6.8 × 1.7–2.3 mm, light green or sometimes marked with dark purple, 10–15-veined; beak 0.1–0.3 mm long; stigmas 3. **Achenes** trigonous and filling about 60% of the perigynium body. **2*n***=60.

Spray zones of waterfalls or on seepy, north-facing cliffs. 100–900 m. Casc, CR. WA; north to AK; Asia. Native.

Carex macrochaeta forms dense clumps of leaves that hang down from its cliff-side perch. *Carex scirpoidea* var. *stenochlaena* produces similar tufts in similar habitats, but has only one spike per culm. Oregon's *C. macrochaeta* populations are small remnants left behind as this species retreated north with the glaciers.

Carex media R. Brown (p. 214)

Scandinavian sedge

Carex vahlii Schkuhr

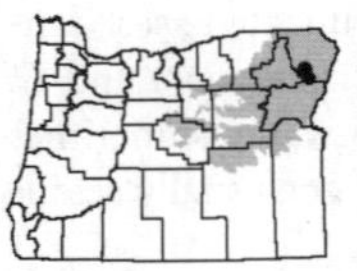

Plants cespitose, sometimes loosely so, 15–50(70) cm tall. **Leaves** 2–4 mm wide. **Inflorescences** spikes 2–5, 5–12 × 3–6 mm, more or less sessile, crowded or sometimes with lowest spike more or less remote and pedunculate; lateral spikes female; terminal spike gynaecandrous. **Female scales** usually much shorter than the perigynia, but sometimes about as long; dark brown or black, including midvein. **Perigynia** elliptic or ovate, 2.5–3.5 × 0.9–1.5 mm, erect to spreading, the lowest sometimes reflexed, light green becoming tan; beak 0.3–0.4 mm long, darker than the body; stigmas 3. **Achenes** trigonous, nearly filling the perigynium bodies.

Montane to subalpine meadows, stream banks, seepage areas, bog margins. 1600–2200 m. BW. WA; north to AK, east to Newfoundland, IA, ME, WY; Eurasia. Native.

Carex media is a montane sedge with small, crowded, mostly sessile spikes with green perigynia that contrast with dark scales. *Carex pelocarpa* is similar, but it has dark inflorescences with female scales about as long as the obovate to circular, relatively long perigynia. It grows in drier habitats above timberline. *Carex atrosquama* has similar crowded inflorescences, but its spikes have short peduncles, and its perigynia are dark golden. *Carex media* is sometimes considered conspecific with Eurasian *C. norvegica*.

Carex mendocinensis Olney ex W. Boott (p. 214)

Mendocino sedge

Carex debiliformis Mack.

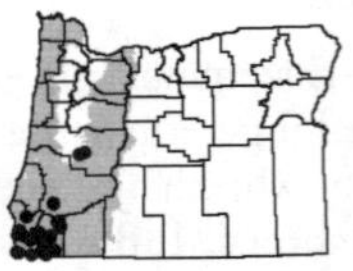

Plants densely cespitose, 25–80 cm tall. **Leaves** lowest sheaths reddish, bladeless, glabrous or with a few rough hairs; blades 2.5–5.5 mm wide, flat, usually glabrous (sometimes sparsely hairy on one or both sides especially near the base). **Inflorescences** sheath of lowest bract well developed, 8–40 mm long; lateral spikes 3–5, female, cylindric, 1–6 cm × 2.5–6 mm; terminal spike male (rarely gynaecandrous), sessile or nearly so, 1.8–3.8 cm long. **Female scales** broadly ovate, shorter than the mature perigynia, more or less brown to gold with white margins and green midrib, with ciliate margins but otherwise glabrous. **Perigynia** oblong-lanceolate, 2.7–5 × 0.8–1.8 mm, green to brown, more or less spotted with red, usually reddish at base, with 12–15 fine veins plus 2 marginal ribs; glabrous or with sparse hairs near the tip; beak 0.3–1 mm long; stigmas 3. **Achenes** trigonous.

Seeps, small streams, and wet meadows, typically on serpentine substrates but sometimes on disturbed non-serpentine sites. 0–1400 m. Casc, CR, Est, Sisk. CA. Native.

Carex mendocinensis is characterized by a dense mound of often yellow-green leaves, slender female spikes, usually ciliate margins of the inflorescence bract sheath and the female scales, and small perigynia. The few hairs on leaves and perigynia might cause confusion with *C. gynodynama*, which has wider, hairier leaves and densely hairy perigynia densely packed in wider spikes.

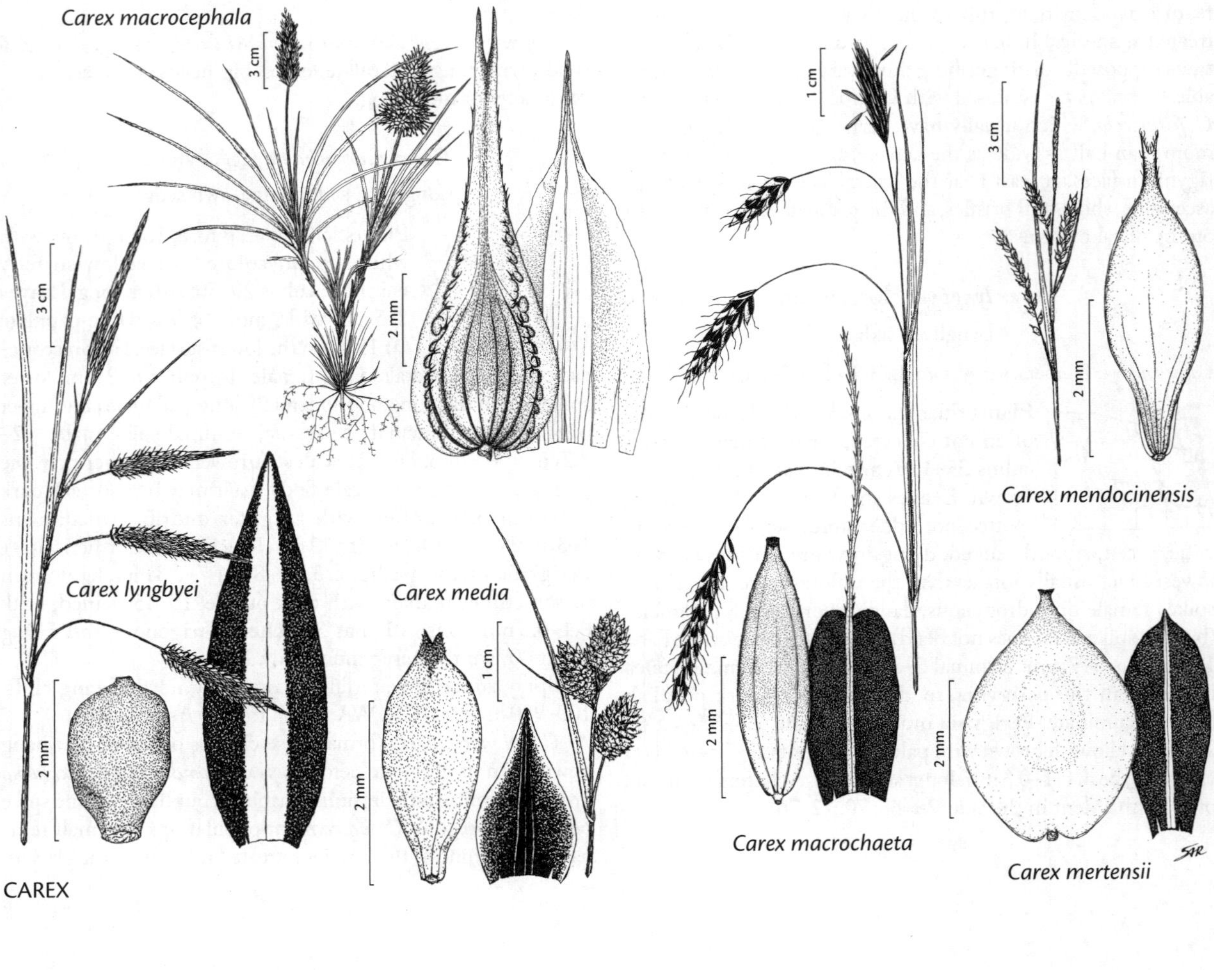

Carex mertensii J.D. Prescott ex Bongard (p. 214)

Mertens' sedge

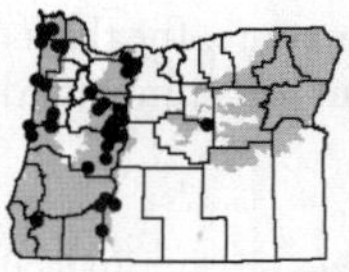

Plants cespitose, 30–100 cm tall. **Leaves** 4–8 mm wide. **Inflorescences** spikes 4–9, 1–4 cm long; lateral spikes female or gynaecandrous, the terminal spike gynaecandrous. **Female scales** brown to dark purplish black, shorter and much narrower than the perigynia, apex obtuse, mucronate. **Perigynia** ovate or orbicular, flattened except over the tiny achene, 4–5 × 2.5–3.5 mm, green to brown; beak 0.2–0.4 mm long; stigmas 3. **Achenes** trigonous, filling the lower half or less of the perigynium body. ***2n***=62.

Mesic to somewhat dry open forests, meadows, roadsides, stream banks, pumice plains and disturbed places at low to moderately high elevations in the mountains, usually in partial shade. 50–2200 m. BW, Casc, CR, Est, Sisk. CA, ID, WA; north to AK, northeast to MT; Asia. Native.

Distinctive *C. mertensii* has thick, dangling, sausage-shaped spikes and is common on roadsides and other disturbed sites in mesic montane forest.

Carex micropoda C.A. Meyer (p. 217)

Timberline sedge

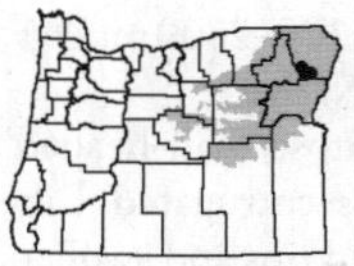

Plants cespitose, 5–30(40) cm tall. **Leaves** blades 0.25–1.5 mm wide, involute; occasionally flat and to 2 mm wide, but always with involute leaves on the same plant. **Inflorescences** spike solitary, androgynous, 0.7–2 cm × 4–7 mm, lacking inflorescence bract. **Female scales** lanceolate, dark reddish brown to black, deciduous. **Perigynia** lanceolate to elliptic, 3–4(5) × 1–1.3 mm, ascending when young, later spreading or sometimes reflexed, dark brown or blackish distally, paler in lower half, glossy, veinless but with 2 marginal ribs; beak dark brown or black, short, poorly defined; stigmas 2 or 3. **Achenes** lenticular or trigonous. ***2n***=62.

Moist alpine meadows and snowmelt basins. 2200–3100 m. BW. CA, ID, NV, WA; north to AK, east to CO; Asia. Native.

Carex micropoda is a cespitose, narrow-leaved, single-spiked sedge of high elevations. With age, its perigynia spread and may even point downwards. Inflorescences of *C. nigricans* are very similar, but *C. nigricans* is rhizomatous, and its leaves are flatter and wider. *Carex micropoda* is sometimes considered conspecific with European *C. pyrenaica*.

Carex microptera Mackenzie (p. 217)

Small-wing sedge

Carex festivella Mack., *Carex limnophila* F.J. Herm.

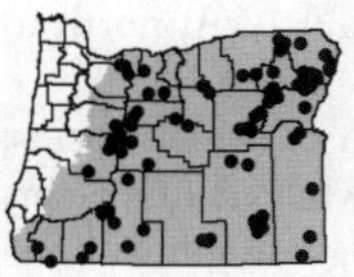

Plants cespitose, 20–110 cm tall. **Leaves** blades 2–5 mm wide. **Inflorescences** (0.8)1–2.6 cm × 9.5–17.5 mm, dense and head-like, usually 2-toned, with a fine texture due to the numerous thin, appressed perigynia; spikes gynaecandrous (rarely female). **Female scales** 2.4–3.5 mm long, shorter and narrower than the perigynia, usually brown or purplish with a paler mid-stripe. **Perigynia** flat except over the achene, (2.8)3.4–4.5(5.2) × 1.1–2.4 mm, green or straw-colored but with dark beak tips; 3–8+ dorsal veins, 0–8 ventral veins; wing 0.2–0.5 mm wide; beak tip unwinged, brown, and parallel-sided for the distal 0.6–1.3 mm; 1.5–2.5(2.8) mm from achene top to beak tip; stigmas 2. **Achenes** lenticular, 1.1–1.6 × 0.7–1.2 mm. ***2n***=80, 82.

Common in mesic forest openings and grasslands at moderate to high elevations in the mountains. 600–2800 m. BR, BW, Casc, Col, ECas, Lava, Owy, Sisk. CA, ID, NV, WA; north to Yukon, northeast to Manitoba and SD, southeast to NM and Mexico. Native.

Carex microptera is one of the two most common "ball-headed" sedges, with *C. pachystachya*. *Carex microptera* has fine-textured inflorescences and perigynia that are flat except over the small achene. In *C. pachystachya,* the fewer, thicker, more spreading, plano-convex perigynia produce a coarser-textured inflorescence. In the alpine zone, see *C. haydeniana*. *Carex microptera* is variable, and small-headed extremes have sometimes been segregated as different species.

Carex multicaulis L.H. Bailey (p. 217)

Many-stemmed sedge, stick sedge

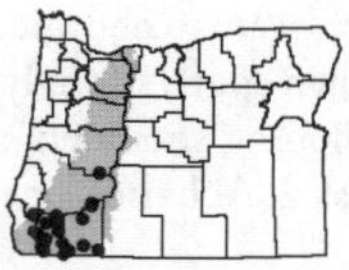

Plants densely cespitose; culms 15–72 cm tall, round in cross section, more or less erect to arching. **Leaves** blades 0.7–2 mm wide, involute, usually much shorter than the culms. **Inflorescences** spike solitary, lacking inflorescence bracts; terminal part male, 0.4–3.5(5.2) cm × 0.8–2.3 mm. **Female scales** the lowest leaf-like, awned, 1.1–1.3 cm long and much longer than the perigynia; the upper reduced, green to brown. **Perigynia** obovate or elliptic, 1–5(10), (3.5)4–7.7 × 1.1–3(3.5) mm, light green to light brown, widest near or above middle; almost beakless; stigmas 3. **Achenes** trigonous.

Dry slopes or seasonally moist areas in open conifer, oak, or mixed forests or in chaparral, often on serpentine substrates. 300–2400 m. Casc, Sisk. CA. Native.

At a glance, an observer could easily mistake this plant as a rush with its stiff, densely clumped culms and short, inconspicuous leaves. *Carex geyeri* has similar perigynia, but is loosely cespitose to short-rhizomatous, with many flat leaves, triangular culms and lower female scales that are short and not leaf-like.

Carex nardina Fries (p. 217)

Spikenard sedge, spike sedge

Carex hepburnii Boott, *Carex nardina* Fr. var. *hepburnii* (Boott) Kük.

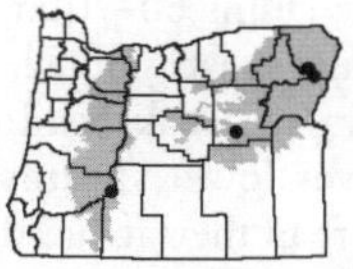

Plants densely cespitose, to 15 cm tall; plant bases yellow-brown. **Leaves** sheaths dense, persistent, yellow-brown; blades to 0.5 mm wide, very narrow and wiry. **Inflorescences** spike solitary, androgynous, to about 1.5 cm long, with 5–15 perigynia, lacking inflorescence bracts. **Female scales** about as wide as the perigynia, reddish brown to yellowish with broad brownish or white margins. **Perigynia** elliptic or lanceolate to oblong-obovate, 3.5–4.5 × 1.2–2 mm, green or light brown at maturity, glabrous with serrulate distal margins; margins flattened distally; beak brown or hyaline, 0–0.8 mm long; stigmas 2 or 3. **Achenes** lenticular or trigonous. ***2n***=68, 70.

Dry alpine ridges and slopes, abrasion plateaus, heaths and fellfields. 2300–3000 m. BW, Casc. ID, NV, WA; north to AK; Eurasia. Native.

Carex filifolia, C. nardina, and *Kobresia myosuroides* are densely cespitose, fine-leaved "bunchgrass" sedges with yellow-brown, persistent leaf sheaths. The first two always grow in the alpine zone, and *C. filifolia* may join them there. Normal perigynia of *C. filifolia* are pubescent, at least in the upper part, and more 3-dimensional than those of *C. nardina,* but aborted *C. filifolia* perigynia can be confusing.

Carex nebrascensis C. Dewey (p. 217)

Nebraska sedge

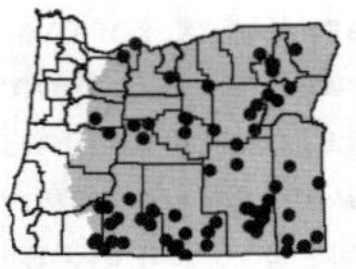

Plants rhizomatous, but the deep rhizomes not often collected; culms 20–90 cm tall; plant bases brown or reddish brown. **Leaves** sheath fronts hyaline, not ladder-fibrillose, usually not spotted; blades (3)4–12 mm wide, usually glaucous. **Inflorescences** lowest bract more or less equal to the inflorescence; lateral 2–4 spikes female (or androgynous), erect, blunt or only slightly tapering at base, 3–5.5 cm × 5–8 mm; terminal 1–3 spikes male. **Female scales** reddish brown to dark purplish brown, acute to awned; awn 0–0.5 mm long, sometimes scabrous. **Perigynia** elliptic to ovate to obovate, 2.6–4 × 1.6–2.5 mm, brown with reddish brown spots, strongly 5–9-veined on each face at maturity, with "sloping shoulders"; beak 0.3–0.6 mm, minutely bidentate; stigmas 2. **Achenes** lenticular. **2*n***=66, 68.

Common in riparian areas, wet meadows and other wetlands, often in alkaline sites. 100–2500 m. BR, BW, Casc, Col, ECas, Lava, Owy. CA, ID, NV, WA; north to Alberta, east to NE, southeast to AZ, MO, IL; Japan. Native.

This important forage and soil-binding species is a common, rhizomatous, broad-leaved, usually glaucous, lower-elevation sedge with strong perigynium veins. Some traits traditionally used to distinguish it from relatives (leaves W-shaped vs. V-shaped in cross section or blue vs. green) do not work. *Carex aquatilis,* which grows in more or less acidic wetlands, has veinless perigynia with rounded shoulders. *Carex angustata* has ladder-fibrillose leaf sheath fronts and perigynia with one to three weak veins on each face. *Carex nebrascensis* populations west of the Cascades are probably recent introductions.

Carex nervina L.H. Bailey (p. 217)

Sierra sedge

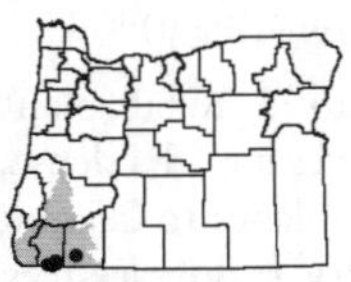

Plants cespitose, forming a thick tuft of gracefully arching leaves; culms 60–70 cm tall, without persisting basal sheaths of previous years, somewhat winged, weak, bending down in fruit. **Leaves** lower sheaths herbaceous and veined, not splitting, longer than their reduced blades; distal sheaths hyaline, faintly veined, not cross-corrugated, with pale, thickened, concave mouths; blades 5 mm wide, green, arising at different heights so that the sheaths are mostly visible. **Inflorescences** 1.5–2.2 cm × (6)9–12 mm, dense and head-like; spikes androgynous or some female. **Female scales** about as long as the perigynia, brown. **Perigynia** lance-triangular, to 4.5 × 2 mm, widest near the base, brown, strongly veined, with about 15 veins on the dorsal surface and 7–12 veins on the ventral surface; base only somewhat pithy and swollen, cordate, with stipe to 0.2 mm long; beak to 2 mm long, poorly defined; stigmas 2. **Achenes** lenticular.

Wet meadows, seeps, and streamsides. 1500–2100 m. Sisk. CA, NV. Native.

Carex nervina produces graceful mounds of leaves with tall, weak culms that bend and look broken when mature. The thickened rim of the leaf sheath front is unique but sometimes hard to see. Compare to *C. neurophora.*

Carex neurophora Mackenzie (p. 219)

Veined sedge, alpine nerved sedge

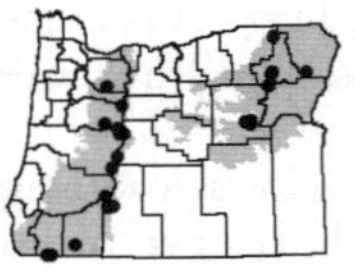

Plants loosely cespitose and often forming small mounds. **Culms** 35–60 cm tall, lacking persistent basal sheaths of previous years. **Leaves** sheaths of lower leaves herbaceous and veined, not splitting, longer than their reduced blades; upper sheaths hyaline, faintly veined, usually cross-corrugated, with the mouth convex; blades 5 mm wide, green, arising at different heights so that the sheaths are mostly visible. **Inflorescences** 1–2.5 cm × 6–15 mm, dense and head-like; spikes androgynous or some female. **Female scales** shorter than or about equal to the perigynia, brown. **Perigynia** lance-triangular, to 2.9–3.8 × 1–1.5 mm, widest near the base, brown, strongly veined, with 9–11 veins on the dorsal surface, 5–7 on the ventral surface; base pithy and somewhat swollen; beak to 2 mm long, gradually tapered; stigmas 2. **Achenes** lenticular.

Wet meadows, seeps, and streamsides. 1100–2500 m. BW, Casc, Sisk. CA, ID, WA; northeast to MT. Native.

Carex neurophora is a relatively tall, cespitose "ball-headed" sedge. Its leaf sheath fronts are usually cross-corrugated. Like *C. jonesii,* it has veined perigynia with tapered tips and swollen, pithy bases. *Carex jonesii* has the leaves all clustered at the base so that the usually smooth leaf sheath fronts are hard to see. In southwestern Oregon, compare to *C. nervina.*

Carex nigricans C.A. Meyer (p. 219)

Black alpine sedge

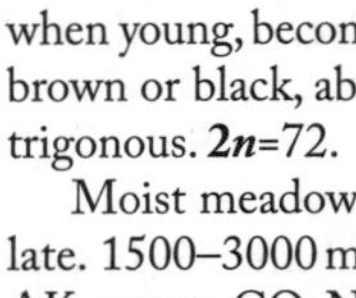

Plants rhizomatous and turf-forming, or appearing loosely cespitose if growing among rocks, 5–30 cm tall. **Leaves** (1.5)2–4 mm wide, flat. **Inflorescences** spike solitary, (0.7)1–2 cm × 6–10 mm, androgynous, lacking inflorescence bract. **Female scales** lanceolate, dark reddish brown to black, deciduous before the perigynia mature. **Perigynia** lanceolate to elliptic, (3.0)3.2–4.6(5) × 1–1.2 mm, dark brown or blackish distally, paler in lower half, glossy; ascending when young, becoming spreading to reflexed; veinless; beak dark brown or black, about as long as the body; stigmas 3. **Achenes** trigonous. **2*n***=72.

Moist meadows, streamsides, and basins where snow melts late. 1500–3000 m. BR, BW, Casc, ECas. CA, WA; north to AK, east to CO. Native.

Common *Carex nigricans* forms a turf in snowmelt basins; spikes of young plants are dark, dense, ovals, while mature plants look like a different species because the perigynia spread horizontally or even reflex. Young plants resemble *C. subnigricans* or cespitose *C. micropoda,* but those species have narrow, involute leaves.

Carex nudata W. Boott (p. 219)

Torrent sedge, naked sedge

Carex suborbiculata Mack.

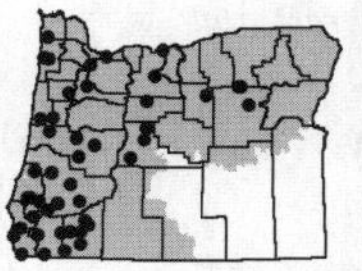

Plants densely cespitose, with blackish bases that usually cannot be wrenched from their rocky strongholds; culms 35–70 cm tall; plant bases blackish. **Leaves** sheath fronts coppery with reddish brown dots, conspicuously ladder-fibrillose; blades 2–4 mm wide, green or somewhat glaucous; plant bases reddish brown or blackish. **Inflorescences** lowest bract shorter than the inflorescence; lateral 2–4 spikes female, 2–4.5 cm × (4)5–6 mm, erect; terminal 1–2 spikes male. **Female scales** much shorter than the perigynia, dark reddish brown to black, awnless. **Perigynia** elliptical to ovate to obovate, 2.2–4 × 1.2–1.8 mm, pale brown with red spots and often a big black blotch near the tip, with 3–9 veins on each face; beak 0.1–0.3 mm; stigmas 2. **Achenes** lenticular, about half as long as the perigynia. **2*n***=70, 72.

Cracks in rocks or among cobbles within the scour zones of large, fast-moving rivers, including small rocky islands; occasionally along irrigation canals. 0–1300 m. BW, Casc, Col, CR, ECas, Est, Lava, Sisk, WV. CA, WA. Native.

Carex nudata forms dense tussocks with narrow, gracefully drooping leaves among rocks in fast-flowing rivers. It can be identified from a passing car. Rhizomatous *C. interrupta* may grow adjacent to *C. nudata* tussocks on stream banks in sandy or gravelly soils; its perigynia are smaller, and its lowest spike bases have perigynia widely spaced.

Carex obnupta L.H. Bailey (p. 219)

Slough sedge

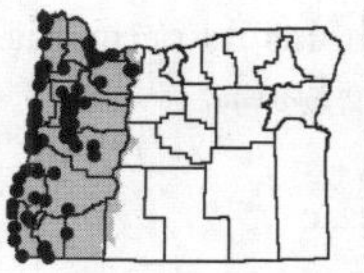

Plants rhizomatous but deep rhizomes often not collected; culms 20–120(200) cm tall; plant bases reddish brown or blackish. **Leaves** sheath fronts coppery, densely red-dotted, ladder-fibrillose; blades 3–7 mm wide, evergreen, tough, green when young, aging yellowish green with sharp, antrorse marginal prickles; leaf sheath fronts coppery, densely red-dotted, ladder-fibrillose. **Inflorescences** lowest inflorescence bract longer than the inflorescence; lateral 2–5 spikes female, dark brown, erect to drooping, tapering at base, 2.5–15 cm × 4.5–10 mm; upper 2–3(5) spikes male. **Female scales** longer than the perigynia, reddish brown, acute, awnless. **Perigynia** 2.2–3.8 × 1.4–2.2 mm, dark brown, hard, veinless; beak 0.1–0.3 mm; stigmas 2. **Achenes** lenticular, often indented on one or both sides. **2*n***=70, 72, 74, 76.

Marshes, wet meadows, streamsides, pond margins, low spots in riparian woodlands—often with *Fraxinus latifolia* (Oregon ash), stabilized coastal dunes, brackish upper reaches of salt marshes and road ditches. 0–1300 m. Casc, CR, Est, Sisk, WV. CA, WA; north to British Columbia. Native.

Only this species and *C. lyngbyei* have such hard perigynia. *Carex lyngbyei* grows in salt marshes and has smooth-edged leaves that die in winter, yellow-brown perigynia, and straight, usually truncate-based spikes on arching peduncles. *Carex aquatilis* var. *dives* has softer leaves and perigynia, and leaf sheath fronts that lack coppery color and are not ladder-fibrillose.

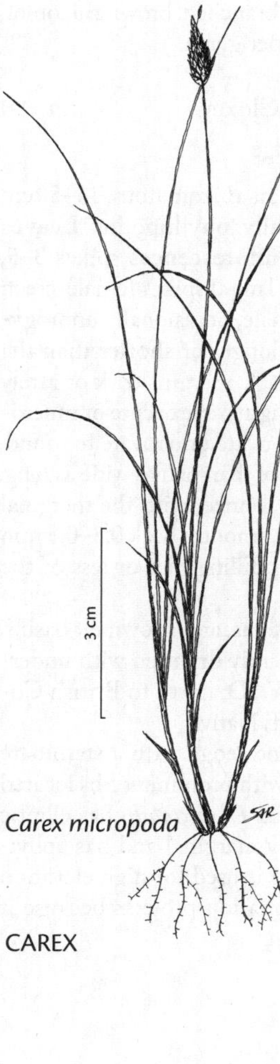

Carex micropoda

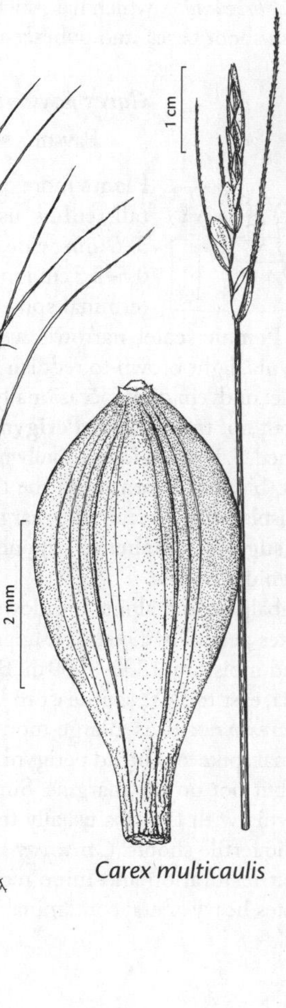

Carex multicaulis

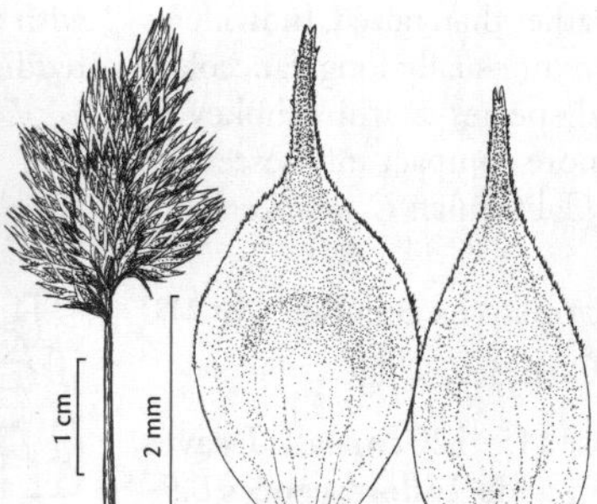

Carex microptera

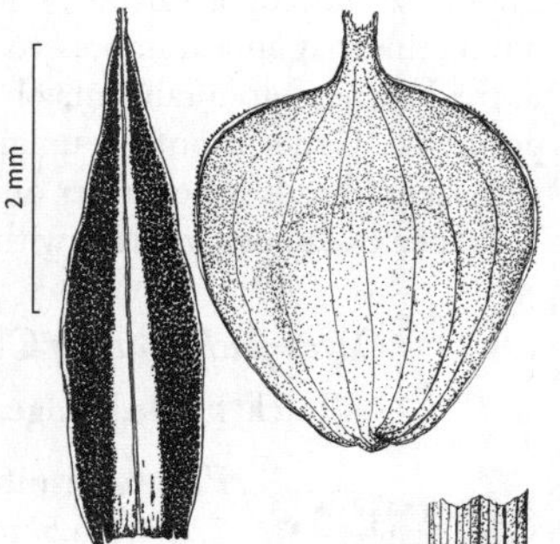

Carex nebrascensis

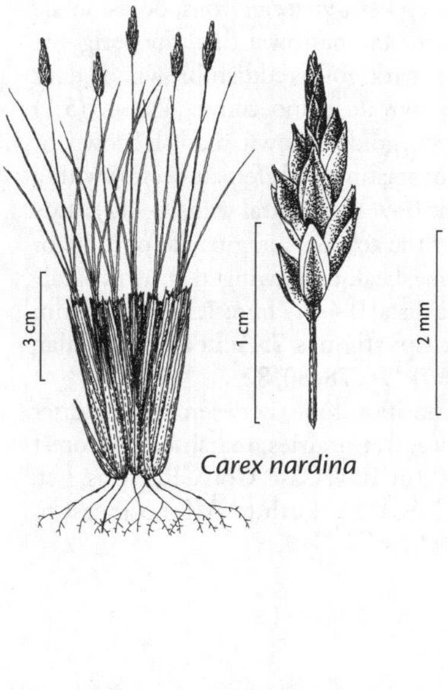

Carex nardina

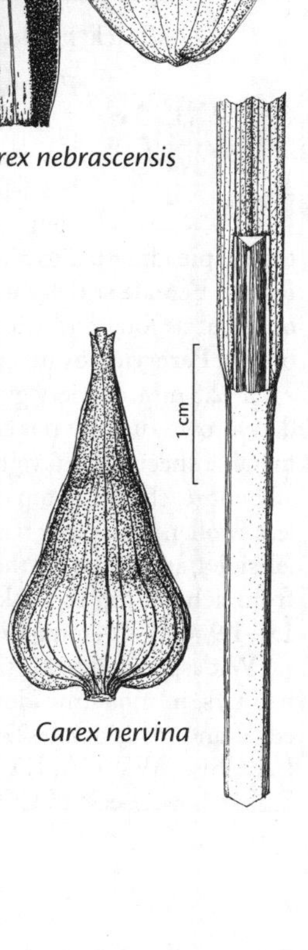

Carex nervina

CAREX

Carex barbarae of southwestern Oregon has brown, usually erect spikes and scabrous awns at the tips of the female scales. Presumed hybrids with *C. aquatilis* var. *dives* and *C. barbarae* have been collected in Oregon. *Carex obnupta* holds its perigynia through the winter, releasing them on spring floods.

Carex pachycarpa Mackenzie

Many-rib sedge

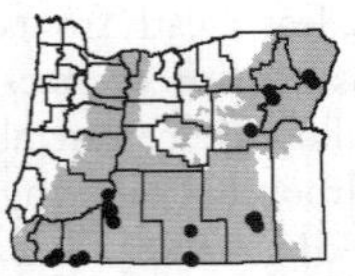

Plants cespitose, 14–110 cm tall. **Leaves** (2)3–5 mm wide. **Inflorescences** 1.6–3 cm × 12–20 mm, dense and head-like; spikes gynaecandrous, occasionally female. **Female scales** (3.2)3.9–4.8 mm, gold or light brown with paler midrib and margins, longer and narrower than the perigynia. **Perigynia** smoothly oval to obovate, plano-convex, (3.8)4.4–5.8(6.5) × (1.4)1.8–2.8(3.3) mm; cream-colored or tan with green "shoulders," becoming uniformly straw-colored with age; dorsal veins (8)10–17 and on dried specimens usually appearing sunken; ventral veins (0)4–12; wing (0.2)0.3–0.4(0.5) mm wide, broad near the beak but narrow lower on the body; beak winged nearly to the tip; 2–2.8(3.3) mm from achene top to beak tip; stigmas 2. **Achenes** lenticular, 1.7–2.4 × 1.1–1.7 mm.

Dry grasslands, mesic meadows and open forests. 600–2800 m. BR, BW, Casc, ECas, Sisk. CA, ID, NV, WA; northeast to MT. Native.

Carex pachycarpa is a "ball-headed" bunchgrass sedge with large, pale, shaggy heads. Its pale perigynia have smooth oval bodies and wings that are broad near the beak but taper smoothly to become narrow beside the achene. Dorsal perigynium veins may appear impressed rather than raised. Individuals in the Warner Mountains may have unusually long, lanceolate perigynia. Other montane upland species (within subkey 9) with dense heads have darker or more compact inflorescences. This species can be confused with Californian *C. multicostata*.

Carex pachystachya Chamisso ex Steudel (p. 219)

Thick-headed sedge, Chamisso sedge

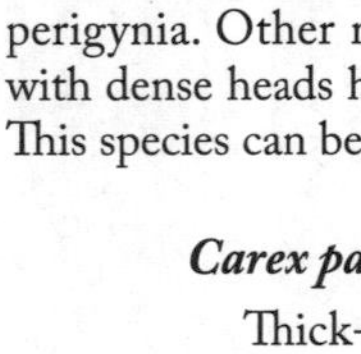

Plants cespitose, 15–120 cm tall. **Leaves** 1.2–4.2(6.5) mm wide. **Inflorescences** 0.9–2.8 cm × 6.7–16.3 mm, usually dense and head-like but occasionally slightly elongate, rough in texture because of relatively few, thick, spreading perigynia; spikes gynaecandrous, occasionally female. **Female scales** shorter and narrower than the perigynia or about as long and wide, dark gold, reddish brown, or dark brown. **Perigynia** ovate to obovate, plano-convex, 2.8–4.7(5.1) × 1.1–2.3 mm, coppery green, golden brown, reddish brown, or dark brown (usually not contrasting with female scales) with a metallic sheen; dorsal veins 0–9(16), ventral veins 0–4(7), usually not reaching the top of the achene; margins flat or more or less swollen with pithy tissue; beak tip unwinged, brown, parallel-sided, and entire for the distal 0.4–0.7 mm; 1.5–2.2(2.8) mm from achene top to beak tip; stigmas 2. **Achenes** lenticular, 1.4–1.9 × 0.7–1.4 mm. ***2n***=74, 76, 78, 80, 82.

Widespread in mesic transition zones between wet and drier habitats, in moist meadows, wet prairies, marsh edges, forest edges and roadsides. 0–2700 m. BW, Casc, Col, CR, ECas, Est, Lava, Sisk, WV. CA, ID, NV, WA; north to AK, northeast to Saskatchewan, east to CO; Siberia. Native.

Carex pachystachya is a common sedge of the transition zones at the edges of wetlands. It has short, dense heads usually with much green color. The spikes often have a star-burst look because the perigynia are thick and spreading. Compare to *C. harfordii*, *C. microptera*, *C. preslii*, and *C. subbracteata*.

Carex pansa L.H. Bailey (p. 219)

Sanddune sedge

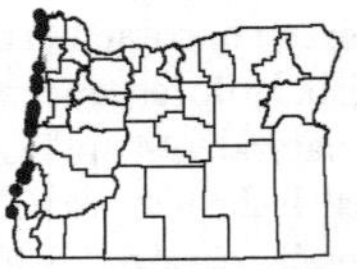

Plants more or less dioecious; rhizomatous, the rhizomes little-branched, blackish, stout, 1.8–2.6 mm in diameter; culms 8–40 cm tall, arising singly or in small clusters; plant bases dark brown to blackish. **Leaves** 1.7–3.8 mm wide, mainly clustered at the base of the shoot. **Inflorescences** 1.2–2.5 cm long, male inflorescences 5–9(11) mm wide, female inflorescences (6)8–16 mm wide; spikes in one plant seemingly all female or all male but usually with a few, nearly undetectable flowers of the opposite sex mixed in the spikes. **Female scales** dark reddish brown to purplish black with hyaline margins, similar in color to or darker than the perigynia, awned or not, hiding the perigynia. **Anthers** with a tiny bristly awn. **Perigynia** ovate to obovate, 3.1–4.2 × 1.3–1.8 mm, dark brown to nearly black; beak 0.7–1.5 mm long; stigmas 2. **Achenes** lenticular.

Coastal sand dunes and sandy meadows. 0–50 m. Est. CA, WA; north to British Columbia. Native.

Carex pansa forms a green or yellow-green turf on stabilized sand dunes. Its female inflorescences are large and dark. *Carex praegracilis*, widespread east of the Cascades and in southwestern Oregon, is extremely similar, but has narrower, lighter-colored, narrower female inflorescences. *Carex pansa* may grow with *C. brevicaulis*, which has much-branched, brown rhizomes, reddish shoot bases, and pubescent perigynia.

Carex paysonis Clokey (p. 219)

Payson's sedge

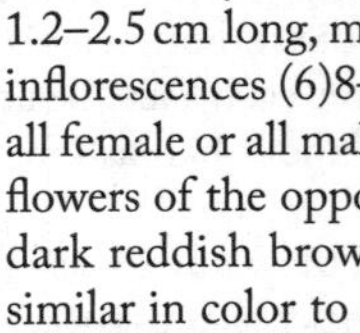

Plants more or less rhizomatous, 15–50 cm tall; culms usually phyllopodic. **Leaves** 3–6 mm wide. **Inflorescences** spikes 3–5, 0.5–2.5 cm long; lateral spikes female, erect; terminal spike male, occasionally androgynous. **Female scales** narrower and longer or shorter than the perigynia, light brown to reddish black, margins dark or rarely hyaline; midvein dark (occasionally light); apex acute or mucronate but not truly awned. **Perigynia** ovate or obovate to round, flattened, 3.5–4 mm long, usually more than half as wide as long, purple-black at least toward the tip, veined, with the marginal ribs displaced away from the margin, smooth; beak 0.3–0.5 mm long; stigmas 3. **Achenes** trigonous, filling half or less of the perigynium bodies.

Subalpine and alpine meadows, talus, and snowmelt basins, the sites generally moist or if apparently dry then with underground moisture. 1800–3000 m. BW. ID; north to British Columbia, east to WY, southeast to UT. Native.

Carex paysonis is a large montane sedge with a staminate terminal spike and broad perigynia with prominent ribs located near, but not on the margins. Similar *C. spectabilis* has elliptic perigynia with the ribs usually truly marginal and has aphyllopodic fertile shoots. *Carex paysonis* is used for high elevation habitat restoration and mine reclamation projects because it tolerates heavy-metal contamination.

Carex pellita Muhlenberg ex Willdenow (p. 220)

Woolly sedge

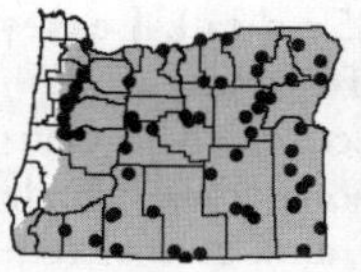

Plants rhizomatous, 15–70 cm tall, aphyllopodic; plant bases reddish, shiny. **Leaves** sheaths glabrous, rarely very short-pubescent, lower ones ladder-fibrillose; blades (2)2.2–4.5(6) mm, yellow-green, flat or M-shaped in cross section, the tips not curled. **Inflorescences** lowest bract erect or nearly so, longer than the inflorescence; lateral 1–3 spikes female, erect or ascending, well separated from the 1–3 terminal male spikes. **Female scales** glabrous or with scabrous margin apically; acute to awned. **Perigynia** broadly ovoid, abruptly contracted to the beak, 2.4–5.2 × 1.7–2.8 mm, densely pubescent, ascending; beak (0.6)0.8–1.6 mm long; teeth 0.4–0.8 mm long, straight; stigmas 3. **Achenes** trigonous.

Marshes, stream banks, ditches, pond margins, and wet meadows, often in places that are dry at the surface in summer, often in habitats with a history of disturbance. 0–2300 m. BR, BW, Casc, Col, ECas, Lava, Owy, Sisk, WV. CA, ID, NV, WA; north to AK, east to Newfoundland, south to Mexico. Native.

Carex pellita is a common, slender, rhizomatous wetland sedge, sometimes forming dense stands but often mixing inconspicuously with other vegetation. Its densely hairy perigynia taper abruptly to short beaks. It often fails to flower; sterile plants can be identified by the slender, aphyllopodic, glabrous, shiny, varnished-looking, red-brown shoot bases with ladder-fibrillose leaf sheath fronts. Compare to *C. lasiocarpa*.

Carex pelocarpa F.J. Hermann (p. 220)

Duskyseed sedge

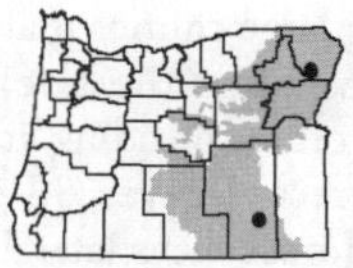

Plants cespitose, 10–40 cm tall. **Leaves** 3–4 mm wide. **Inflorescences** spikes 2–4, sessile, crowded or rarely the lowest one remote; lateral spikes female, 0.7–1 cm × 5–7 mm; terminal spike gynaecandrous. **Female scales** lanceolate, about as long as the perigynia; dark brown or black, including the margins and midvein; apex acute or mucronate. **Perigynia** obovate or circular, nearly flat, abruptly narrowed to the beak, 3.5–4.5 × 2.5–3 mm, glossy dark reddish brown, often with lighter, yellowish or green margins; beak 0.5–0.8 mm long; stigmas 3. **Achenes** trigonous, filling less than half of the perigynium body.

Alpine slopes, ridges, boulder fields, talus slopes, and rocky streamsides. 2500–3100 m. BR, BW. ID, NV; northeast to MT, east to UT and CO. Native.

Carex pelocarpa is an alpine sedge with dark, sessile spikes. It is similar to *C. media*, which also has sessile spikes, but *C. media* grows in lower elevation, often wetter habitats and has smaller, green perigynia that are nearly filled by the achene. Its spikes usually have more contrast of dark scales with light perigynia, in part because the scales are usually less than half as long as the perigynia. *Carex pelocarpa* and *C. nova* have sometimes been treated as the same species.

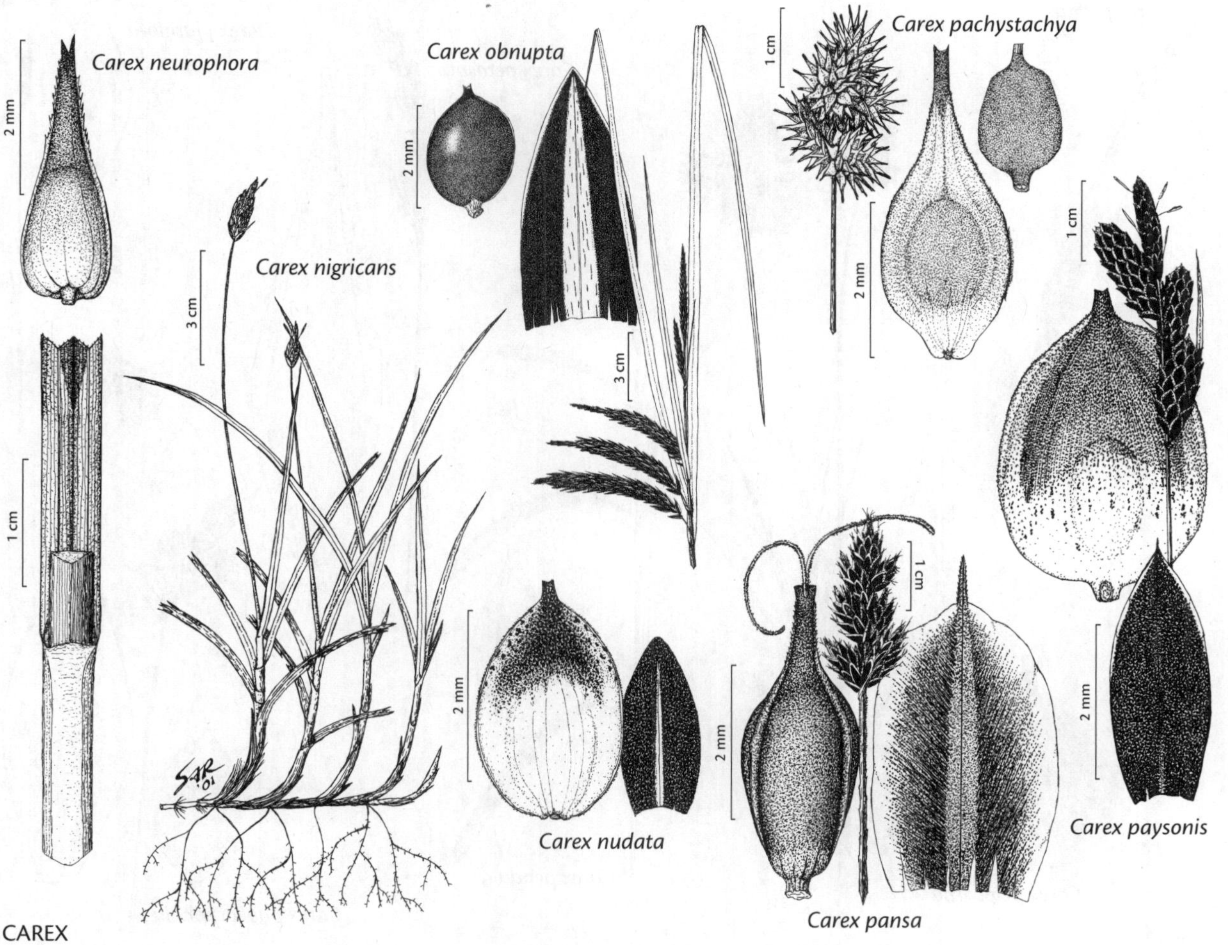

CAREX

Carex pendula W. Hudson (p. 220)

Pendulous sedge

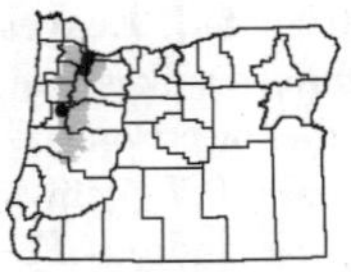

Plants cespitose, forming large clumps that may be a foot or more across at the base; culms 100–200 cm long, often spreading at an angle rather than erect. **Leaves** 80–130 cm × 8–20+ mm. **Inflorescences** lateral spikes female, drooping, 10–30+ cm × 5–8 mm; terminal spike male, arching. **Female scales** as long as or a little longer than the perigynia, brown with pale midrib. **Perigynia** elliptic, 2.6–4 mm long, green to brown; beak 0.5 mm long; stigmas 3. **Achenes** trigonous.

Shaded streamsides and ravines in riparian forests. 0–200 m. WV. CA, WA; IL, VA; Africa, Eurasia, New Zealand. Exotic.

Carex pendula is a large, clump-forming sedge with long, drooping spikes. Sterile shoots resemble those of *Scirpus microcarpus* and *C. amplifolia*, but those species are rhizomatous. *Carex pendula* is planted as an ornamental, but this abundant seed producer is escaping along riparian zones. It has great potential to be invasive along Oregon streams.

Carex petasata C. Dewey (p. 220)

Liddon's sedge

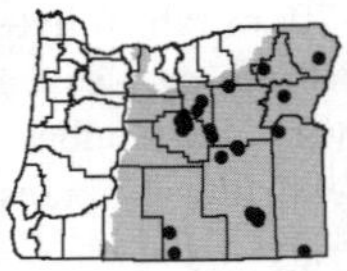

Plants cespitose, 30–85 cm tall. **Leaves** 2–4(5) mm wide. **Inflorescences** (2)2.5–4.5(6) cm × 9–16 mm, erect, somewhat elongate; spikes distinct, gynaecandrous. **Female scales** 5.8–7.6 mm long, dark gold or red-brown, with white margins 0.2–0.7 mm wide, acute, approximately as long and wide as the perigynia, concealing the bodies and more than half the beak. **Perigynia** lanceolate to ovate, 6–8 × 1.7–2.4 mm, whitish green to brown, with 10+ dorsal veins and 4–10 ventral veins; beak tip unwinged, brown, parallel-sided, and entire for up to 1 mm, or sometimes winged and ciliate-serrulate almost to the tip, (2.8)3–4.6 mm from achene top to beak tip; dorsal suture white-edged and relatively conspicuous; stigmas 2. **Achenes** lenticular, 2.2–3 × (1.1)1.3–1.8 mm.

Sagebrush steppes, dry to moist meadows, and open forests. 500–2700 m. BR, BW, ECas, Lava, Owy. CA, ID, NV, WA; north to AK, east to CO. Native.

Carex petasata has ascending spikes in an erect inflorescence and large perigynia that are mostly concealed by the female scales. *Carex praticola* has smaller perigynia and a usually nodding inflorescence. *Carex davyi* has shorter female scales that reveal most of the beak and have no white margins or very narrow ones.

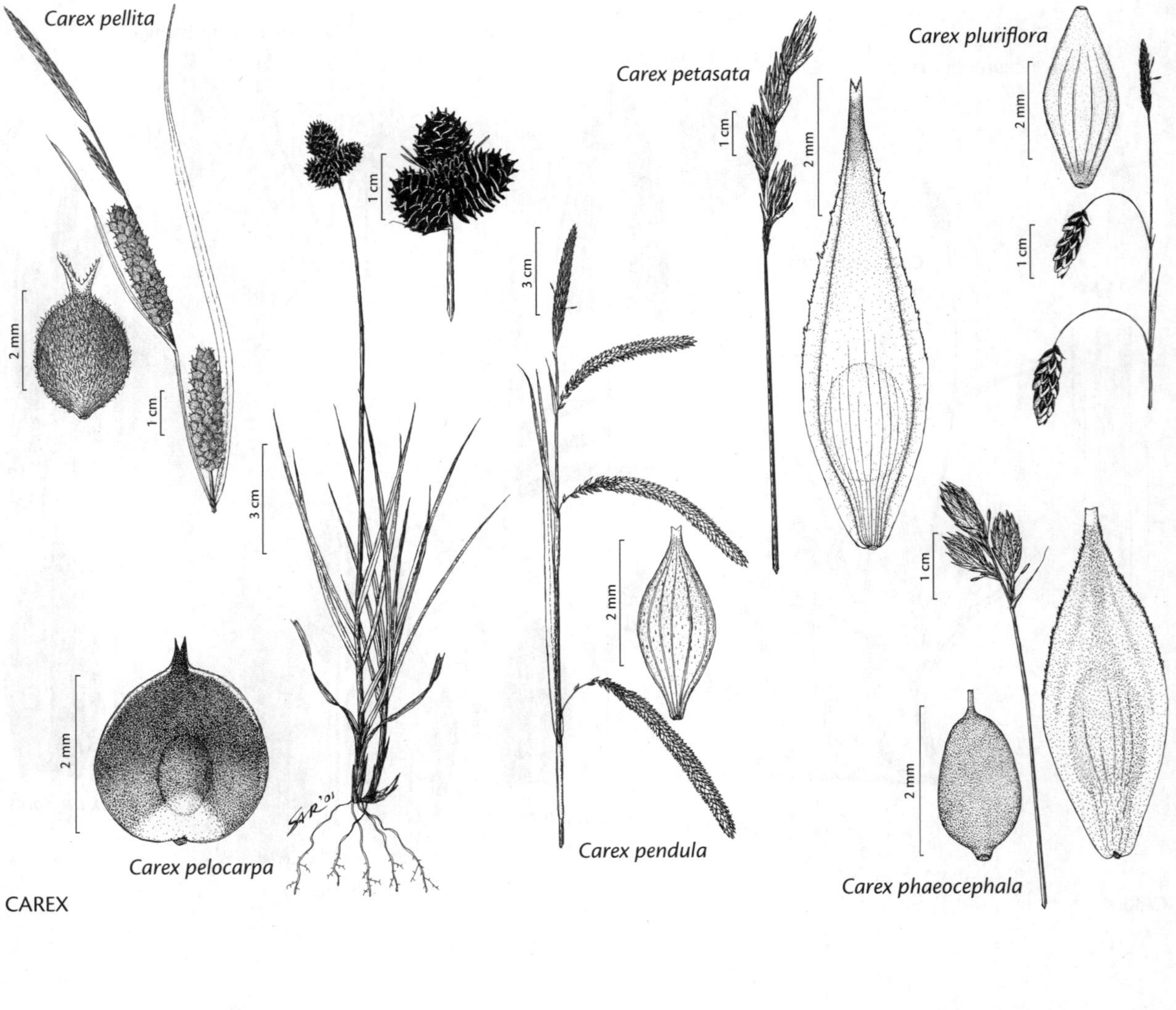

Carex phaeocephala Piper (p. 220)

Dunhead sedge, alpine hare sedge

Carex eastwoodiana Stacey

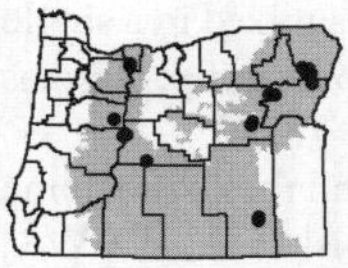

Plants cespitose, (5)15–45 cm tall. **Leaves** 1–2.5(3) mm wide, folded, somewhat tough, the edges sometimes revolute. **Inflorescences** 1.5–3.5(4) cm × 10–15 mm, somewhat elongated, erect or somewhat angled to the side; spikes gynaecandrous, narrow, overlapping, ascending to erect. **Female scales** 3.7–5.1 mm long, gold to brown with paler midrib, more or less covering the perigynia. **Perigynia** ovate, 3.5–4.5(5.2) × 1.5–2.3 mm, gold to dark brown, translucent, veinless or with up to 5 faint ventral veins and up to 9 dorsal veins; beak tip unwinged, brown, and parallel-sided for the distal 0.6 mm, with the perigynium wings extending up to the base of the cylindric tip; 1.8–2.2 mm from beak tip to achene top; stigmas 2. **Achenes** lenticular, 1.5–2 × 0.8–1.1(1.2) mm. ***2n***=84.

Dry, rocky alpine or subalpine slopes. 1500–3000 m. BR, BW, Casc, ECas. CA, ID, NV, WA; north to AK, east to CO. Native.

Carex phaeocephala is a tough, cespitose sedge of rocky alpine slopes, with overlapping, ascending spikes, and female scales that hide the perigynia. Its somewhat translucent, ovate perigynia are usually veinless or have faint veins on the ventral surface. It is very similar to *C. tahoensis,* which grows at slightly lower elevations and has larger perigynia that are tough, strongly veined, opaque, and brown.

Carex pluriflora Hultén (p. 220)

Black bog sedge, many-flower sedge

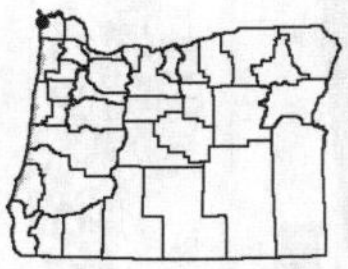

Plants rhizomatous, the shoots arising singly or in small groups; young roots with yellow, dense, felty hairs; culms 20–60 cm tall. **Leaves** 1.5–4 mm wide, glaucous, flat, margins sometimes revolute. **Inflorescences** lowest bract 0.5–2.5 mm long, bristle-like; lateral spikes female, 1.2–2 cm × 6–8 mm; terminal spike male, 1–2.5 cm × 2.2–3 mm, rarely gynaecandrous. **Female scales** 3.5–4.5 × 2.1–3.8 mm, slightly wider and shorter than the perigynia, black; apex acute to acuminate, mucronate or with awn to 1.5 mm long. **Perigynia** 3.2–4.2 × 1.5–2.6 mm, pale green to dark brown or blackish at maturity; beak 0; stigmas 3. **Achenes** trigonous.

Bogs, marshes, fens, wet meadows, streamsides and lake shores. 0–50 m. Est. WA; north to AK; northeastern Asia. Native.

Carex pluriflora is a spiky, rhizomatous sedge with few, ascending leaves and dangling, black female spikes. Similar *C. limosa* has brown female scales. It is known from one coastal bog in Oregon where it is threatened by succession to woody vegetation.

Carex praeceptorum Mackenzie (p. 222)

Teacher's sedge, early sedge

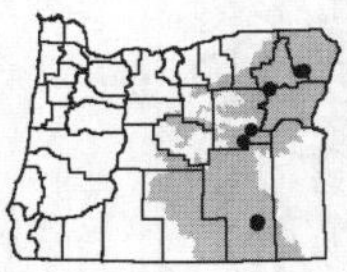

Plants loosely or densely cespitose, 10–20(30) cm tall. **Leaves** 1.5–2.5 mm wide, green or gray-green, shorter than the culms. **Inflorescences** usually brownish, (1)1.5–2 cm long; spikes 3–5, crowded or sometimes the lowest one remote, gynaecandrous, oblong, 0.4–0.7 cm long with 8–15(20) perigynia. **Female scales** brown with whitish centers and margins, about as long as the perigynia. **Perigynia** elliptic, (1.9)2.0–2.3(2.4) mm long, ascending, light brown (sometimes dark brown with age), with several distinct veins; beaks blunt, to 0.5 mm long; dorsal suture about as long as the beak, dark brown, darker than the surrounding perigynium surface; stigmas 2. **Achenes** lenticular.

Bogs, wet meadows, wet edges of lakes and streams. 1400–2800 m. BR, BW. CA, ID, NV, WA; north to British Columbia, east to CO. Native.

Carex praeceptorum is a small, cespitose plant with brownish inflorescences with overlapping spikes. The dorsal suture of the perigynium is about as long as the beak, dark brown, and darker than the rest of the perigynium. Compare to *C. canescens.*

Carex praegracilis W. Boott (p. 222)

Black creeper sedge, clustered field sedge

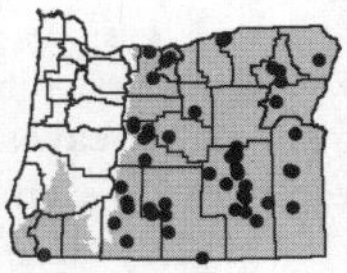

Plants more or less dioecious, rhizomatous, the rhizomes blackish, stout, 1.8–3 mm in diameter; culms (10)25–80 cm tall, arising singly or in small clusters; plant bases dark brown to blackish. **Leaves** blades 1–3(3.5) mm wide, mainly clustered at the base of the shoots. **Inflorescences** 0.9–4 cm long, male inflorescences 4–9 mm wide, female inflorescences (4.5)6–12 mm wide; spikes in one plant seemingly all female or all male but usually with a few, nearly undetectable flowers of the opposite sex mixed in the spikes. **Female scales** pale, straw-colored to pale reddish brown with hyaline margins, more or less hiding the perigynia, similar in color or paler than the perigynia, awned or not. **Anthers** with a tiny bristly awn. **Perigynia** ovate to obovate, (2.2)2.6–3.7 × 1.2–1.9 mm, brown to nearly black, sometimes paler, the base often filled with pithy tissue; beak 0.7–1.2 mm long; stigmas 2. **Achenes** lenticular.

Usually found in alkaline or serpentine soils, in seasonally moist prairies and plains, also roadsides salted in winter. 50–2300 m. BR, BW, Col, ECas, Lava, Owy, Sisk. CA, ID, NV, WA; north to Yukon, east to ME and VA, south to Mexico. Native.

Carex praegracilis is a common, rhizomatous, more or less dioecious sedge, usually found in alkaline areas of eastern Oregon. It also grows in low-elevation serpentine meadows of southwestern Oregon. Its non-descript female inflorescences hide the perigynia. Rhizomes and shoot bases are blackish; shoots arise singly. *Carex douglasii,* which grows with *C. praegracilis,* has slender brown rhizomes and broader, ovoid female inflorescences. It tolerates more alkaline conditions.

Carex praticola Rydberg (p. 222)

Meadow sedge

Carex piperi Mack. ex Piper & Beattie

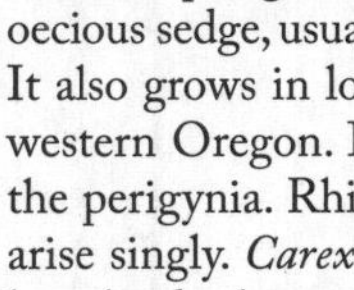

Plants cespitose, (16)35–70(100) cm tall. **Leaves** (1.5)2–3(4) mm wide. **Inflorescences** (1.7)2.5–5 cm × 10–15(17) mm, bent to the side, somewhat elongate, the spikes distinct and the lowest one remote; spikes gynaecandrous. **Female scales** (3.4)4.2–5.8 mm long, as long as or longer than but narrower than the perigynium body, straw-colored, reddish brown, or brown with similar or paler midstripe; white margin 0.1–0.3 mm wide. **Perigynia** (3.7)4.5–6 × 1.2–2 mm, with margins brown distally but often whitish over the achene, usually with 4–11 dorsal veins and 0–4 ventral veins; wings (0.1)0.2–0.4(0.5) mm wide; beak brown, parallel-sided, and entire for the distal 0.4–1 mm, with the very tip white, (1.6)1.9–3 mm from achene top to beak tip; stigmas 2. **Achenes** lenticular, 1.4–2.1(2.7) × (0.8)1–1.5 mm. ***2n***=76, 78.

Moist to mesic meadows and forest openings, often in somewhat disturbed spots. 50–2100 m. BW, Casc, CR, ECas, WV. CA, ID, NV, WA; western and northern North America. Native.

Carex praticola has a greenish to pale brown, nodding or bent inflorescence with white-margined female scales. Similar *C. petasata* has longer perigynia. West of the Cascades, similar *C. leporina* has female scales that are not white-margined, though they may have white-hyaline tips or bases. Its perigynium beaks are brown at the very tip. Its perigynia average shorter with shorter beaks and are usually veined on the ventral surface.

Carex preslii E.G. Steudel (p. 222)
Presl's sedge

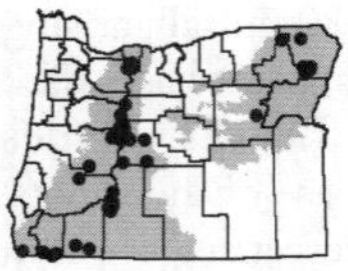

Plants cespitose, 23–55 cm tall. **Leaves** 1.7–3.6 mm wide. **Inflorescences** 1.4–3 cm × (5)8–15 mm, head-like but slightly elongated; lowest internode 3–7 mm; spikes gynaecandrous, occasionally female. **Female scales** 2.8–3.7 mm, shorter or longer and narrower than the perigynia, gold to red-brown, occasionally white; mid-stripe green, off-white, or dark; white margin 0.1–0.2(0.5) mm wide. **Perigynia** ovate to obovate, (3.3)3.5–4.3 × (1.3)1.5–2 mm, green, straw-colored, or golden-brown with green margins that contrast in color with the female scales, with 0–5 ventral veins; beak winged nearly to the tip, or unwinged, brown, parallel-sided, and entire for up to 0.6 mm (both types mixed in a single inflorescence), 1–2.4 mm from achene top to beak tip; stigmas 2. **Achenes** lenticular, 1.5–2 × 1–1.5 mm. ***2n***=78, 80.

Roadsides, dry open slopes, forest edges and mesic subalpine meadows. 900–2800 m. BW, Casc, ECas, Sisk. CA, ID, WA; north to AK, northeast to MT. Native.

Carex preslii perigynium beaks are of two forms within the same inflorescence—some are winged nearly to the tip, and others are unwinged, brown and parallel-sided for more than 0.4 mm. *Carex preslii* seems to have differentiated from *C. pachystachya* more physiologically (in its ability to withstand dryness) than morphologically. The diversity of perigynium tips and a drier habitat are the main features distinguishing *C. preslii* from runty *C. pachystachya*. All *C. pachystachya* perigynia have the beak tips unwinged, brown, and parallel-sided for 0.4–0.7 mm. *Carex pachystachya* tends to have more and larger spikes in a more compact inflorescence, its darker perigynia (golden brown or coppery to dark brown) have a metallic sheen that *C. preslii* lacks, and it grows in a more mesic meadow or meadow edge habitat.

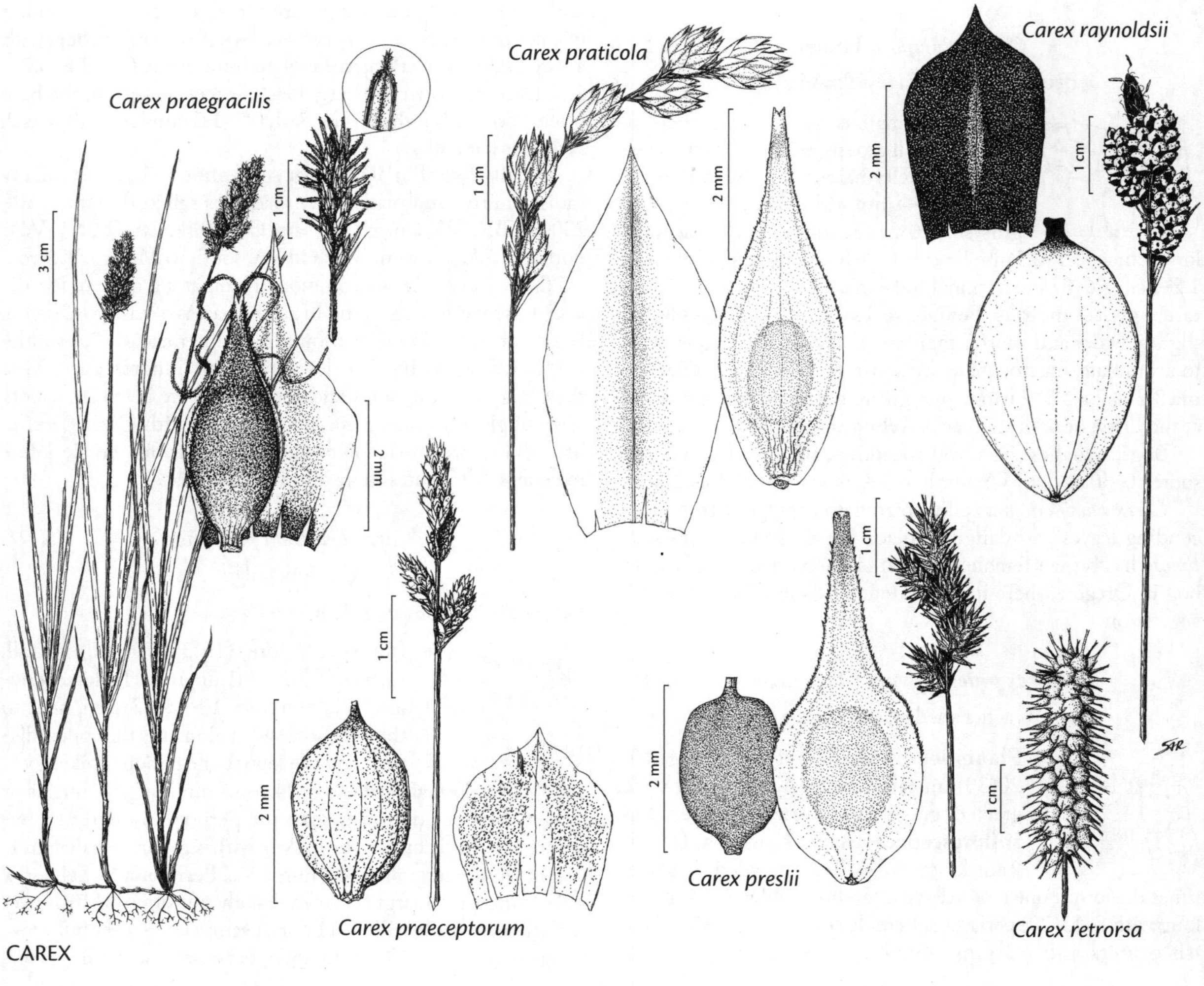

CAREX

Carex raynoldsii C. Dewey (p. 222)

Raynolds' sedge

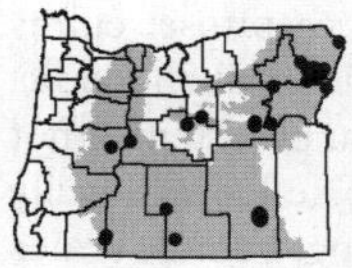

Plants cespitose, often loosely so, 20–75 cm tall. **Leaves** 3–7 mm wide. **Inflorescences** spikes 3–6, erect to ascending, usually overlapping and clustered near the tip of the culm, but sometimes the lowest remote; lateral spikes on short stalks, female, 1–2 cm × 3–5 mm; terminal spike male. **Female scales** shorter than or as long as the perigynia, narrower than the perigynia at least above, dark brown or blackish, the margin and sometimes the midvein lighter, mostly acute (but varying from obtuse to acuminate). **Perigynia** elliptic or obovate, unusually thick and swollen, (3)3.3–4.5 × 1.75–2 mm, greenish yellow to yellow-brown, sometimes gray, often orange-brown above, contrasting with the darker female scales; beak 0.3–0.5 mm long; stigmas 3. **Achenes** trigonous, 1.8–2.4 mm long, nearly filling the perigynium bodies. ***2n***=58.

Moist to dry meadows, often associated with seeps, in mountains to near timberline. 1300–2800 m. BR, BW, Casc, ECas. CA, ID, NV, WA; north to British Columbia, east to CO. Native.

Carex raynoldsii is a cespitose sedge with short spikes that contain plump, often orangish perigynia and contrasting dark scales. The achenes are large, nearly filling the perigynium bodies.

Carex retrorsa Schweinitz (p. 222)

Retrorse sedge, knotsheath sedge

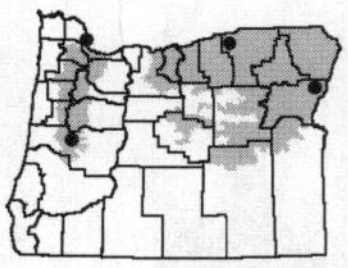

Plants densely cespitose, 10–100 cm tall; plant bases often brown or tinged with red. **Leaves** widest blades 3–10 mm wide, green, flat to W-shaped in cross section. **Inflorescences** bracts 3–9 times longer than inflorescence; lateral spikes ascending to spreading, female or androgynous, (1.5)2–4.6 cm, crowded or with the lowest somewhat separated; terminal 1(3) spike(s) male, not or barely elevated above the lateral spikes. **Female scales** shorter than the perigynia, apex acute to acuminate, awnless. **Perigynia** 6–10 × (1.6)2.1–3.4 mm, green or straw-colored, with 6–13 veins; apex contracted to the beak; beak 2.1–4.5 mm long; stigmas 3 style persistent, becoming curved. **Achenes** trigonous.

Floodplain forests, lake shores, swamps, streamsides, wet thickets, wet meadows, in sites that are soggy wet in spring but may dry out on the surface later in the season. 0–1000 m. BW, Col, WV. ID, WA; north to British Columbia, east to New Brunswick. Native.

Carex retrorsa is a robust, cespitose sedge with its spikes all nestled near the bases of its long inflorescence bracts. Its male spikes are located among the clustered female spikes. One's first impression might be that this is a deformed *C. utriculata* or *C. vesicaria*; both of these species have spikes more separated, with the male spikes held distinctly above the female ones, and with inflorescence bracts not more than 2.5 times as long as the inflorescence. In addition, *C. utriculata* is strongly rhizomatous.

Carex rossii F.M.B. Boott (p. 224)

Ross' sedge

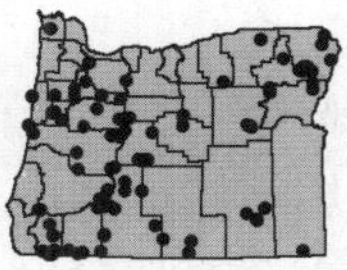

Plants cespitose, often loosely so, sometimes short-rhizomatous; culms 7–30(40) cm tall, usually arising in small clusters; plant bases reddish. **Leaves** sheaths ladder-fibrillose; blades (0.5)0.7–2.5(3.5) mm wide, green, rarely glaucous in southwestern Oregon, not densely papillose on the lower surface. **Inflorescences** basal spikes female, more or less hidden among leaf sheaths, with stalks usually shorter than or as long as the spikes; bracts of lowest non-basal spikes leaf-like, as long as or longer than the inflorescence; non-basal female spikes 1–4, 0.5–0.9(1.1) cm long, with 3–10 perigynia; terminal spike male, (0.5)0.6–1.3 cm long. **Female scales** shorter than the perigynia, pale to dark reddish brown with white margins, apex acute to acuminate. **Perigynia** from lateral (not basal) spikes, elliptic to obovoid, 3.1–4.5 × 1.4–1.7 mm, pubescent, veinless except for the 2 ribs, bases succulent, withering when dry; beaks 0.9–1.7 mm long; teeth 0.2–0.4 mm long; stigmas 3. **Achenes** trigonous.

Uplands, including open forests, sagebrush steppes, meadows, grassy coastal headlands, roadsides, cut banks and clearcuts. 0–2900 m. Throughout Oregon. CA, ID, NV, WA; north to AK, east to OH, southeast to NM. Native.

Carex rossii is a common more or less cespitose upland sedge. It has very short basal female spikes nestled among the shoot bases. Remnants of the basal spikes can be found long after its early spring flowering.

Carex saxatilis Linnaeus (p. 224)

Russet sedge, rock sedge

Carex saxatilis L. var. *major* Olney

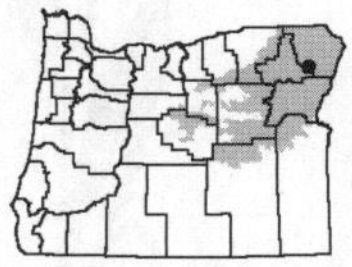

Plants rhizomatous, 8–90 cm tall; plant bases reddish brown. **Leaves** 0.9–5 mm wide, green, flat to V-shaped in cross section. **Inflorescences** lateral 1–3 female spikes, 1–3 cm long, erect or ascending or the lower one often nodding on its slender peduncle; terminal 1–3 spikes male, 1–4 cm long, well separated from the female spikes. **Female scales** as long as or shorter than the perigynia, apex acute to acuminate, awnless. **Perigynia** 2.2–5.5 × 1.1–2.9 mm, green to straw-colored or often reddish brown to blackish, shiny, with 2 ribs and otherwise veinless or with obscure dorsal veins, apex abruptly narrowed at the beak; beak 0.2–0.8 mm long; teeth 0.3 mm long, straight; stigmas 2(3); styles persistent, usually becoming curved. **Achenes** lenticular or trigonous. ***2n***=78, 80.

Fens, bogs, and lake shores at high elevations. 1800–2400 m. BW. WA; north to AK, east to Greenland, southeast to UT; Eurasia. Native.

Carex saxatilis is a slender, rhizomatous sedge, with shiny perigynia and a narrow terminal spike well separated from the female spike(s). Vegetatively it resembles *C. pellita* and *C. lasiocarpa*, which have hairy perigynia. The persistent styles differentiate *C. saxatilis* from *C. aquatilis* and its relatives. *Carex saxatilis* hybridizes with *C. vesicaria* and *C. utriculata*.

Carex scabriuscula Mackenzie (p. 224)

Siskiyou sedge

Carex gigas (Holm) Mack.

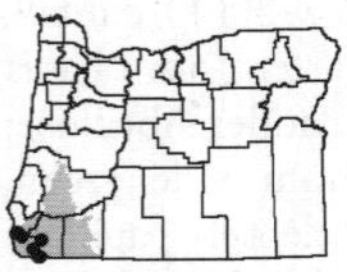

Plants usually dioecious, loosely cespitose, with short rhizomes; culms 30–65 cm tall, about twice as tall as leaves, phyllopodic. **Leaves** blades 2–3 mm wide, blades of previous year persistent at base of flowering shoots. **Inflorescences** spike 1, occasionally 2 or 3, unisexual, with 20–40 flowers. **Female scales** lanceolate, shorter than and about half as wide as the perigynia, reddish brown to purple. **Perigynia** broadly elliptic, 2.5–4 × 1–2.5 mm, purplish to blackish at maturity, slightly hairy on distal 33%; beak 0.5 mm long; stigmas 3 (occasionally 4). **Achenes** trigonous (or quadrangular).

Moist to mesic serpentine substrates in meadows, lake margins, fens and seeps in open forest. 600–1400 m. Sisk. CA. Native.

Carex scabriuscula forms dense mounds of leaves in moist, serpentine habitats. *Carex scirpoidea* ssp. *pseudoscirpoidea* is dioecious like *C. scabriuscula*, but is shorter, does not have persistent leaf bases, and has narrower perigynia that are nearly filled by mature achenes. *Carex mendocinensis* has a terminal male spike and lateral female spikes, the lowest of them well down the stem.

Carex scirpoidea A. Michaux

Northern singlespike sedge

Plants dioecious, rhizomatous or loosely cespitose; culms 5–31(40) cm long, erect or lax, aphyllopodic or phyllopodic. **Leaves** to 3.5 mm wide, blades of previous year persistent or not at base of flowering shoots. **Inflorescences** spike 1, occasionally 2 or 3, unisexual; female spikes 1.5–4 cm long. **Female scales** ovate to lanceolate, about 2.5–3.5 mm long, shorter than, as long as, or longer than the perigynia, reddish brown to purple, with hyaline margin that may be fringed with hairs; sometimes hairy on the back. **Perigynia** ovate, (1.5)2–4(5) × 0.9–1.5(1.6) mm, red-brown or purplish, short-hairy especially toward the beak; stigmas usually 3 (occasionally 2). **Achenes** trigonous (occasionally lenticular), 1.2–2 × 0.8–1.2 mm.

Circumboreal. 4 subspecies; 3 subspecies treated in *Flora*.

Carex scirpoidea forms leafy, more or less spreading clumps or turf and produces cylindrical, unisexual, usually solitary spikes. The perigynia and often the female scales are covered with hairs. Mature plants are easily distinguished from *C. scabriuscula,* which has perigynia that are black, broad, and loose around the achene. Immature plants can be hard to differentiate. Rare *C. idahoa* may have superficially similar spikes but its perigynia and female scales are glabrous and plants are more strongly rhizomatous.

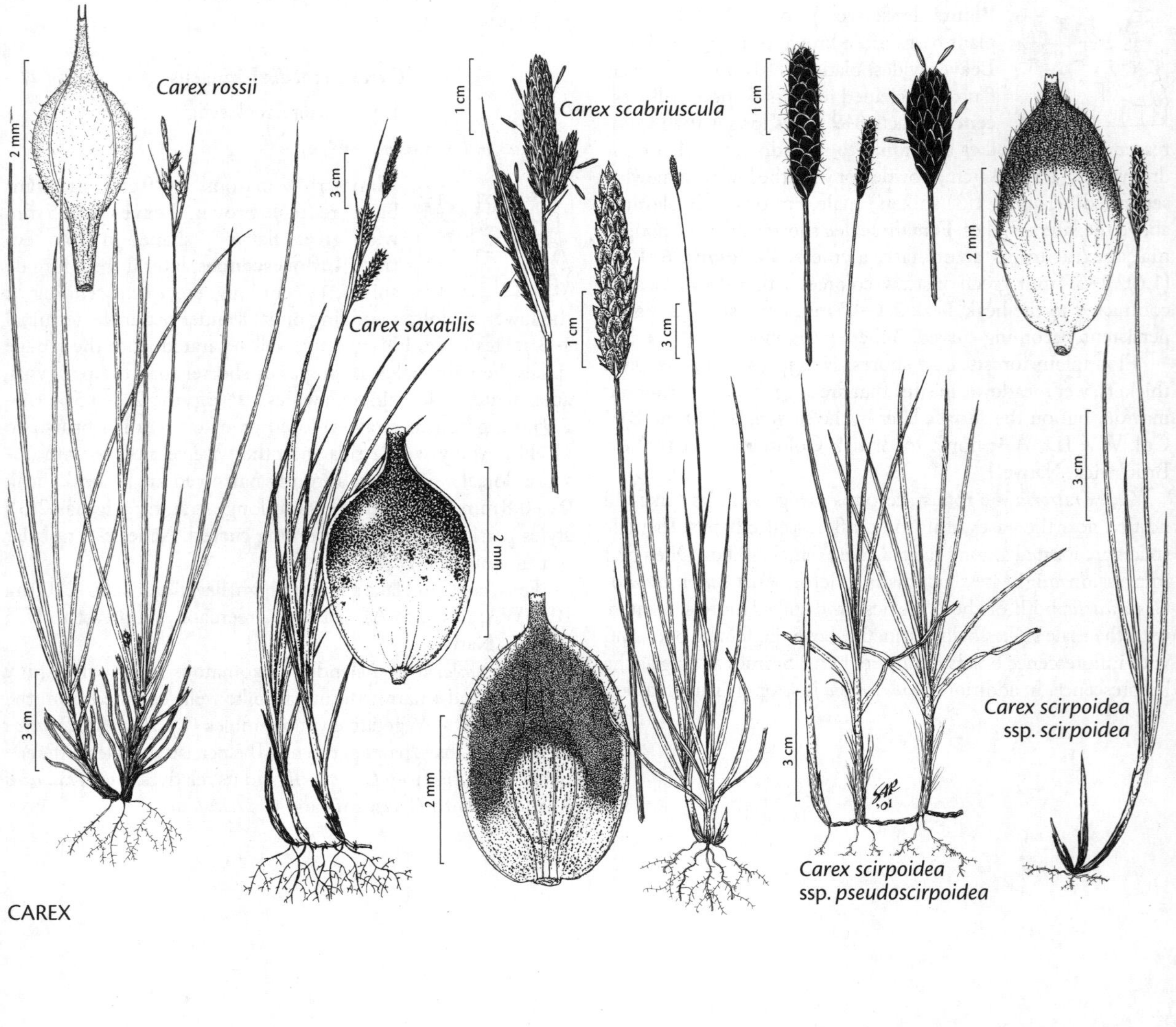

CAREX

The three subspecies of *C. scirpoidea* have a continuum of ecological requirements. *Carex s.* ssp. *pseudoscirpoidea* grows at the highest, mostly alpine elevations, on the driest sites and diverse substrates. *Carex s.* ssp. *scirpoidea* grows on alpine to montane sites with intermediate moisture, on limestone or other non-acidic substrates. *Carex s.* ssp. *stenochlaena* grows at low to moderate elevation, wet, weakly acidic sites, not limestone. All three have limited ranges in Oregon.

Key to *Carex scirpoidea* subspecies:

1. Plants staminate.
   2. Fertile culms phyllopodic; plant more or less rhizomatous. . . ***C. s.*** ssp. ***pseudoscirpoidea*** (in part)
   2' Fertile culms aphyllopodic; plant more or less cespitose.
      3. Habitat wet . . . . . . ***C. s.*** ssp. ***stenochlaena*** (in part)
      3' Habitat mesic to dry . . . . ***C. s.*** ssp. ***scirpoidea*** (in part)

1' Plants pistillate.
   4. Perigynia lanceolate to elliptical, (2.8)3–4+ mm long, mostly greater than 2.5 times as long as wide; female scales lanceolate, 3.5 mm long . . . . ***C. s.*** ssp. ***stenochlaena*** (in part)
   4' Perigynia ovate, (1.3)2–2.8(3) mm long, less than 2.5 times as long as wide; female scales ovate, 2.5–3 mm long.
      5. Fertile culms phyllopodic; plant more or less rhizomatous . . . . ***C. s.*** ssp. ***pseudoscirpoidea*** (in part)
      5' Fertile culms aphyllopodic; plant more or less cespitose . . . . . ***C. s.*** ssp. ***scirpoidea*** (in part)

Carex scirpoidea A. Michaux ssp. ***pseudoscirpoidea*** (Rydberg) D.A. Dunlop (p. 224)

Western singlespike sedge

Carex pseudoscirpoidea Rydb., *Carex scirpoidea* Michx. var. *pseudoscirpoidea* (Rydb.) Cronquist

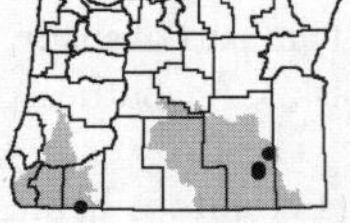

Plants rhizomatous; culms 5–31 cm long, erect, phyllopodic. **Leaves** to 3.5 mm wide (narrower in southwestern Oregon), blades of previous year persistent at base of flowering shoots. **Inflorescences** spikes cylindrical; female spikes 1.5–4 cm long. **Female scales** ovate, 2.8 mm long, usually shorter than or as long as the perigynia. **Perigynia** ovate, (1.5)2–2.8(3) × 1.5 mm, tightly enveloping the achenes. **Achenes** 1.5–1.8 × 0.9–1.2 mm, filling the perigynia.

Mesic to dry ridges and fellfields, generally in areas that have subsoil moisture often due to snowmelt, often among rocks and on a variety of substrates including calcareous soils and rarely serpentine. 2000–2900 m. BR, Sisk. CA, ID, WA; north to British Columbia, east to CO. Native.

In *C. scirpoidea* ssp. *pseudoscirpoidea,* the fertile culm begins growth the year prior to flowering, and dead leaves of the previous year can be found at its base (phyllopodic). In the other *C. scirpoidea* subspecies, the flowering culm completes its growth during the same season that it flowers. Therefore, all leaf blades on the flowering culm are typically alive at flowering and the lowest leaves are represented by nearly bladeless sheaths that protected the young shoot (aphyllopodic).

Carex scirpoidea A. Michaux ssp. ***scirpoidea*** (p. 224)

Northern singlespike sedge

Carex scirpoidea Michx. var. *scirpoidea*

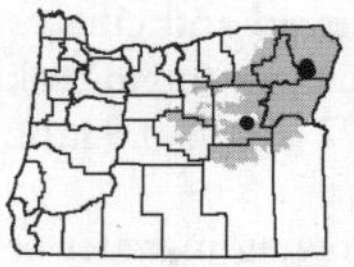

Plants cespitose, with short rhizomes; culms (5)10–35(40) cm tall, erect, aphyllopodic. **Leaves** lowest leaves reduced to bladeless sheaths; blades 2.5 mm wide, blades of previous year not persistent at base of flowering shoots. **Inflorescences** spikes cylindrical. **Female scales** ovate, 2.5 mm long, usually shorter than the perigynia. **Perigynia** ovate, (1.8)2–2.5(3) × 1–1.2 (1.5) mm, tightly enveloping the achenes. **Achenes** 1.5–1.8 × 0.8–1.2 mm, filling the perigynia.

Mesic to dry meadows, ridges and fellfields, generally in areas that have subsoil moisture due to snowmelt, usually on calcareous soils but also on granodiorite and serpentine. 1600–2800 m. BW. ID, NV, WA; north to AK, east to Greenland; Eurasia. Native.

Carex scirpoidea ssp. *scirpoidea* is distinguished from other subspecies by its smallish perigynia and the bladeless sheaths at the base of its flowering shoots.

Carex scirpoidea A. Michaux ssp. ***stenochlaena*** (H.T. Holm) Á. Löve & D. Löve

Alaska singlespike sedge

Carex scirpoidea Michx. var. *stenochlaena* H.T. Holm

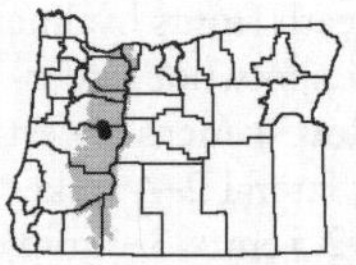

Plants cespitose, with short rhizomes; culms 24–34 cm long, lax, those of female inflorescences tending to droop, aphyllopodic. **Leaves** lowest leaves reduced to bladeless sheaths; blades 2.5 mm wide, blades of previous year not persistent at base of flowering shoots. **Inflorescences** spikes somewhat shaggy-looking, widest near the top, loosely flowered below. **Female scales** lanceolate, 3.5 mm long, often longer than the perigynia, reddish brown to purple. **Perigynia** lanceolate, (2.8)3–4(5) × 0.9–1.4(1.6) mm, dark purplish, tightly enveloping only the lower 75% of the achene. **Achenes** 1.2–2 × 0.8–1.2 mm, not filling the distal portion of the perigynium.

Moist meadows, stream banks, and rocky slopes, the more southerly populations on waterfalls and seepy cliffs on somewhat acidic substrates. 500–2000 m. Casc. WA; north to AK, northeast to MT. Native.

Carex scirpoidea ssp. *stenochlaena* can be distinguished from the other subspecies by its longer, narrower perigynia and its preferences for wet, seepy, somewhat acidic habitats.

Carex scoparia Schkuhr ex Willdenow var. ***scoparia*** (p. 227)

Pointed broom sedge

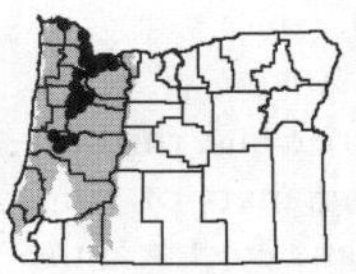

Plants cespitose, 20–100 cm tall, the vegetative culms occasionally producing shoots and roots at the nodes and functioning as stolons. **Leaves** sheath fronts narrowly white-hyaline, often hidden in dried specimens and thus seeming green; blades 1.4–3(3.5) mm wide. **Inflorescences** 1.5–6 cm × 5–20 mm, erect to angled or arching; spikes separated or overlapping but always distinguishable, football-shaped to round, gynaecandrous. **Female scales**

3.4–4 mm long, shorter and much narrower than the perigynia, hyaline brown, often with a green mid-stripe, apex acuminate. **Perigynia** lanceolate, 4.2–5.5(6) × 1.2–2 mm, green to light brown, with about 5 veins on each face, or fewer on the ventral side; wings 0.2–0.6 mm wide; beak tip winged and ciliate-serrulate almost to the tip, 2.2–4.8 mm from achene top to beak tip; stigmas 2. **Achenes** lenticular, 1.3–1.7 × 0.7–0.9 mm. **2*n***=56–70.

Marshes and wet meadows, generally growing in water or saturated soils. 0–1100 m. Casc, CR, WV. CA, ID, WA; north to British Columbia, east to Labrador, southeast to GA; Australia, Europe, New Zealand. Native.

Carex scoparia is a greenish to straw-colored sedge with very fine-textured spikes due to the relatively flat, appressed perigynia and the narrowly pointed tips of both the female scales and the perigynium beaks. It is often confused with *C. leporina* and *C. feta,* which have coarser-textured spikes and smaller perigynia.

Carex scopulorum H.T. Holm var. *bracteosa* (L.H. Bailey) J.F. Hermann (p. 227)

Mountain sedge

Carex campylocarpa Holm, *Carex gymnoclada* Holm

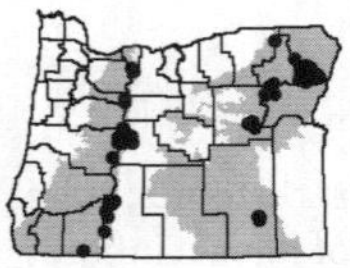

Plants rhizomatous, but rhizomes short and plants sometimes clump-forming; culms 11–65 cm tall; plant bases reddish brown to purplish black. **Leaves** sheath fronts hyaline with light brown or purple spots, not ladder-fibrillose; blades 3–6 mm wide, glaucous. **Inflorescences** lowest bract usually shorter than the inflorescence; lateral 2–4 spikes female, erect, usually not tapered at base, 1–2.5 cm × 3–5 mm; terminal 1–2 spikes male. **Female scales** about as along as the perigynia, dark purplish or blackish. **Perigynia** obovate, 2–4 × 1.2–2.3 mm, green or brown with reddish brown spots, usually extensively blotched with purplish black on the upper half, veinless, tip obtuse to rounded; beak 0.2–0.3 mm long; stigmas 2(3). **Achenes** lenticular.

Seasonally wet habitats including streamsides and wet meadows. 1300–2800 m. BR, BW, Casc, Sisk. CA, ID, WA; north to Yukon, east to CO. Native.

Despite great variability in perigynium color and spike length, *C. scopulorum* is recognizable because of its glaucous leaves, usually blackish inflorescence, and montane wetland habitat. *Carex spectabilis,* in dryish to mesic sites near and above timberline, is similar but always has three stigmas per perigynium, and its spikes are often less crowded, the lower sometimes dangling. Vegetatively, *C. scopulorum* resembles lower-elevation *C. aquatilis,* which usually has broader leaves, more separated spikes, and the inflorescence bract longer than the inflorescence.

Carex serpenticola Zika (p. 227)

Serpentine sedge

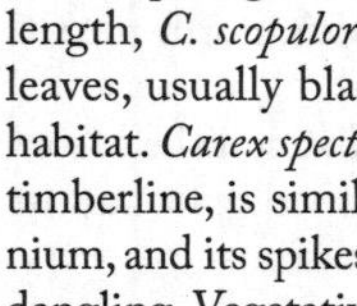

Plants dioecious or monoecious, rhizomatous, sometimes forming mats or rings; culms 8–40 cm tall, short and erect at anthesis, female flowering culms elongated in fruit and arching to the ground; plant bases purplish black. **Leaves** (1.5)2.2–3.5(5) mm wide, green, usually longer than the culms, not papillose on the lower surface. **Inflorescences** lowest bract usually bristle-like, usually shorter than the inflorescence; basal spikes none, or, if present, female and on elongated stalks; non-basal inflorescences male, female, or with lateral female spikes and one terminal male spike; solitary male spikes 1.1–2.4 cm long; sometimes with a small, infertile female spike at the base; lateral spikes female, 0–3, 0.6–1.1 cm long; solitary female spikes 1–1.5 cm long. **Female scales** longer and narrower than the perigynia, purplish black, white-margined, with pale midrib. **Perigynia** obovate, 3.1–3.6 × 1.5–1.8 mm, green, ripening tan or dark purple, pubescent, veinless except for the 2 marginal ribs (rarely with up to 5 veins to mid-length); beaks 0.5–1 mm long; teeth 0.2 mm long; stigmas 3. **Achenes** globose-trigonous.

Serpentine savannas, in seasonally moist uplands, road ditches, abandoned roadbeds, dry creek terraces and edges of seeps. 50–1300 m. Est, Sisk. CA. Native.

Carex serpenticola is a short, upland species with pubescent perigynia and purplish black leaf sheaths and male and female scales, growing in serpentine savanna. It flowers in April or early May, and perigynia fall off by June. No similar species is normally dioecious. Compare to *C. rossii* and especially to *C. concinnoides,* which may grow with *C. serpenticola* and also has dark purplish black male and female scales, though its shoot bases are reddish brown.

Carex serratodens W. Boott (p. 227)

Two-tooth sedge, saw-toothed sedge

Carex bifida Boott

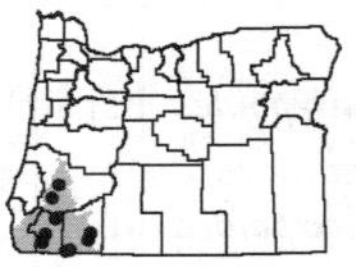

Plants cespitose; culms 30–120 cm long, often sprawling. **Leaves** 2–4 mm wide. **Inflorescences** spikes 3–5, 0.8–3 cm long; lateral spikes female; terminal spike usually gynaecandrous, sometimes male. **Female scales** light or dark brown, hyaline-margined; midvein and broad center area conspicuous, lighter in color than the body, the midvein sometimes raised and/or with minute spines; apex acute to mucronate, sometimes with prominent hairs. **Perigynia** broadly elliptical or ovate, (2)3–5 × 1.2–2 mm, green or light brown, green or light brown, veined; beak 0.5–1 mm long, with 2 small teeth; stigmas 3. **Achenes** trigonous, nearly filling the body of the perigynia.

Moist meadows, hillsides, and seeps, usually in partial shade, often on serpentine substrates. 200–1100 m. Sisk. CA; southeast to AZ. Native.

Carex serratodens can be recognized by its cespitose habit, short, thick spikes, gynaecandrous terminal spikes, striped female scales, and toothed perigynia. *Carex gynodynama* is somewhat similar but has much wider, hairy leaves and longer, narrower, usually darker spikes.

Carex sheldonii Mackenzie (p. 227)

Sheldon's sedge

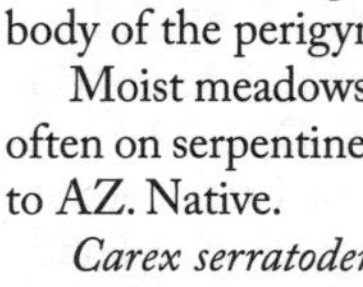

Plants rhizomatous, forming large populations; fertile culms 40–80 cm tall; plant bases reddish purple or drab. **Leaves** sheath fronts pubescent at least near the top, becoming ladder-fibrillose with age; blades 2.5–6 mm wide, usually spreading hairy (occasionally glabrous). **Inflorescences** lateral (1)2–3 spikes female, 2–6 cm long, erect or ascending; terminal 1–4 spikes male, 1.5–4.5 cm long. **Male scales** awnless, glabrous or sparsely appressed-pubescent near

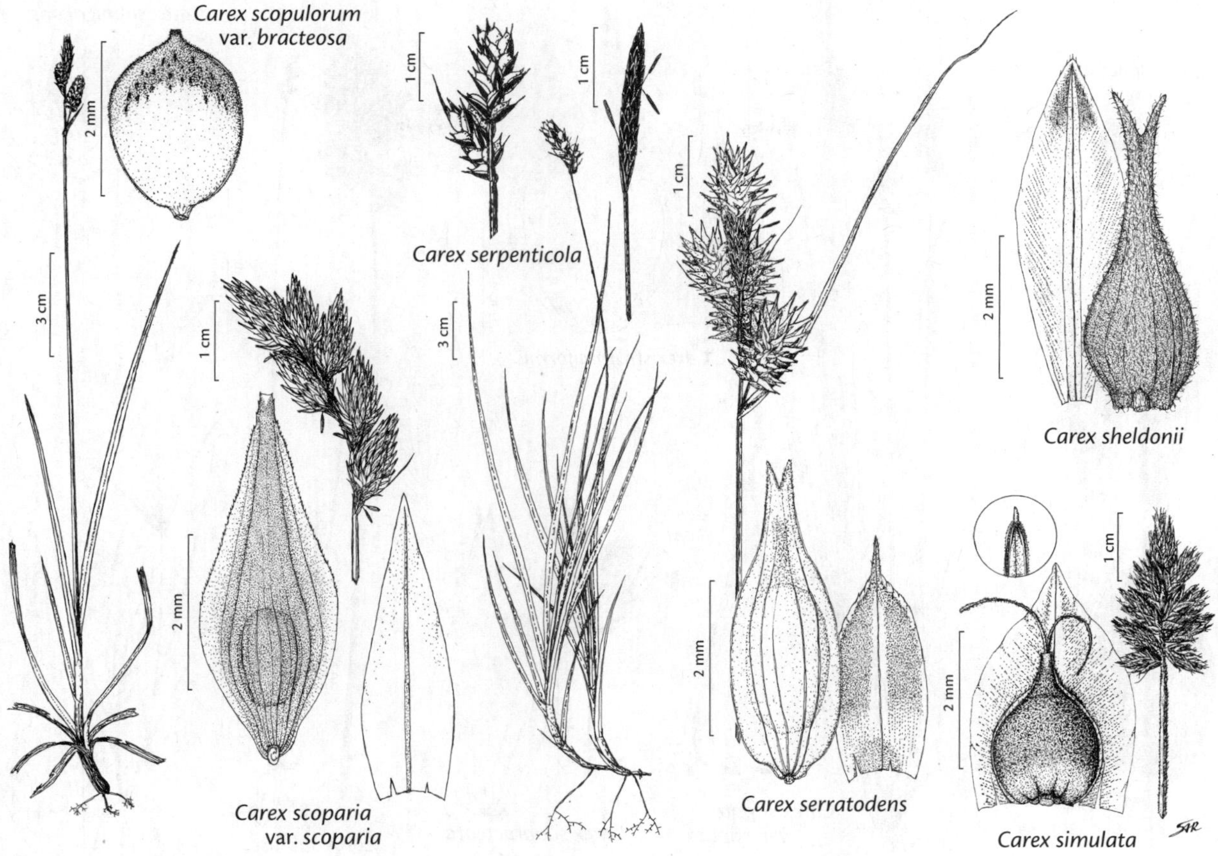

CAREX

the tip. **Female scales** scabrous awn; narrow, scabrous along midvein but not hairy. **Perigynia** lanceolate or lance-ovate, 4.8–6.5 × 1.4–2.4 mm, sparsely to densely pubescent, with 12–18 veins; beak 1.4–2.6 mm long; with spreading teeth (0.4)0.6–1.4 mm long; style persistent; stigmas 3. **Achenes** trigonous.

Wet meadows, ditch banks, riparian meadows, openings in riparian forests, on stable fluvial surfaces above the annual flood line, often in somewhat alkaline sites. 700–1890 m. BR, BW, ECas, Owy. CA, ID, NV. Native.

Carex sheldonii is a rhizomatous sedge with hairy, strongly beaked perigynia and more or less hairy leaf blades and leaf sheath fronts. It is similar to *C. atherodes*, which grows in deeper water and has wider leaves, longer perigynia with longer teeth, and more sparsely hairy perigynia. *Carex hirta*, introduced in Portland, resembles *C. sheldonii* but has male scales with dense, spreading, white hairs.

Carex simulata Mackenzie (p. 227)

Short-beaked sedge, analogue sedge

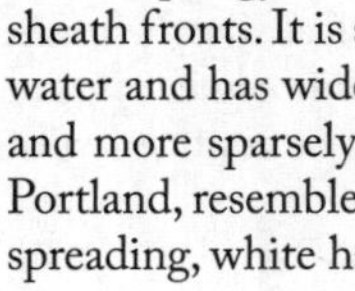

Plants more or less dioecious, rhizomatous, forming large, dense stands; rhizomes pale to brown, 1.5–2.8 mm in diameter; culms (10)20–50(100) cm tall; plant bases brownish. **Leaves** blades 1–3.7 mm wide. **Inflorescences** 1–3 cm long, male inflorescences (3)3.8–6.5(9) mm wide, female inflorescences (4)5–14 mm wide; inflorescences in one plant seemingly all female or all male but usually with a few, nearly undetectable flowers of the opposite sex mixed in the spikes. **Female scales** reddish brown with hyaline margins. **Anthers** with a tiny smooth to warty awn. **Perigynia** 1.8–2.8 × 1.1–1.7 mm, dark brown, glossy; beak 0.25–0.5 mm long; stigmas 2. **Achenes** lenticular.

Wet meadows and springs, on stable stream terraces that are not scoured, where the water table is at or above the soil surface until late summer, generally on calcareous or slightly alkaline substrates; in and east of the Cascades. 700–2500 m. BR, BW, Casc, ECas, Lava, Owy. CA, ID, WA; northeast to Alberta, east to CO, southeast to NM. Native.

Carex simulata has short, dark, plump perigynia. Male plants can be identified based on the narrow leaves, rhizomatous habit, dull plant bases, clonal habit, and wet habitat. *Carex praegracilis* and *C. douglasii* grow in drier habitats and have longer perigynium beaks. *Carex praegracilis* has blackish rhizomes.

Carex spectabilis C. Dewey (p. 228)

Showy sedge

Carex tolmiei Boott

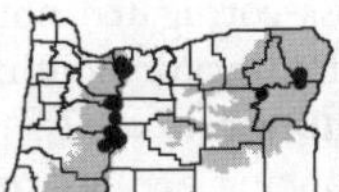

Plants loosely cespitose to short rhizomatous, (10)25–50 cm tall, aphyllopodic. **Leaves** 2–5 mm wide, slightly glaucous. **Inflorescences** spikes (2)3–5(9), 0.8–2 cm long; lateral spikes female, the upper ones erect and the lower ones drooping, often on long peduncles; terminal spike male. **Female scales** narrower and shorter than to sometimes as long as the perigynia, brown or black with dark margins, midvein light and conspicuous (rarely black); awn

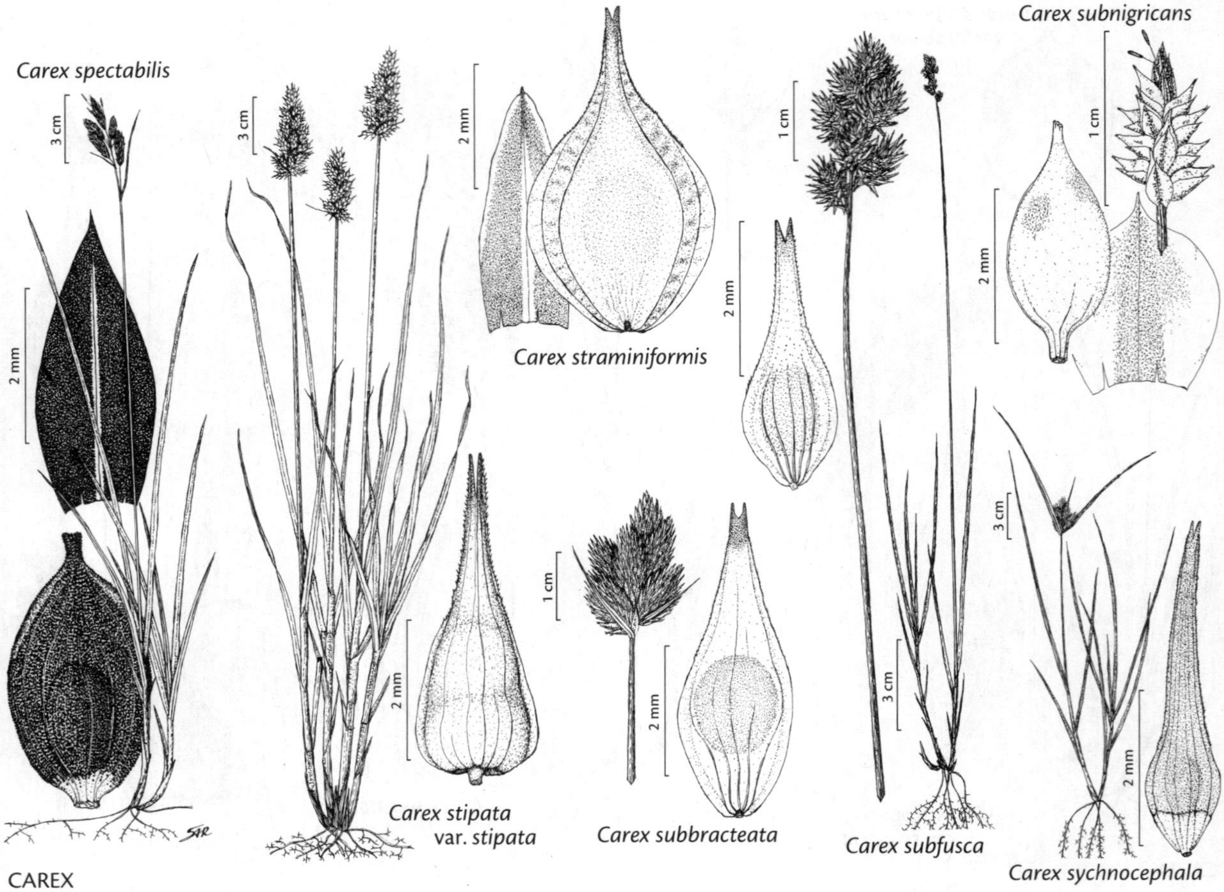

CAREX

0–1 mm long. **Perigynia** ovate, flattened, 3.5–5 × 1.75–2 mm, greenish or purplish black, veined, with the marginal ribs at the edges or sometimes displaced away from the margin; beak 0.4–0.5 mm long; stigmas 3. **Achenes** trigonous, filling half or less of the perigynium bodies. **2*n***=±42.

Subalpine and alpine meadows and roadsides. 1200–2700 m. BW, Casc, Sisk. CA, ID, WA; north to AK, east to WY. Native.

Very similar *C. paysonis* has blades on the lower leaves of its flowering culms and broader, rounder perigynia with the marginal veins consistently displaced away from the margins. Other similar species have gynaecandrous terminal spikes; see *C. heteroneura* and *C. raynoldsii*.

Carex stipata Muhlenberg ex Willdenow var. ***stipata*** (p. 228)

Awl-fruit sedge, sawbeak sedge

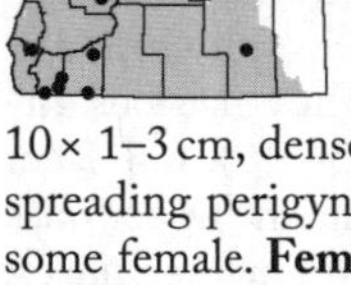

Plants densely cespitose; culms 30–120 cm tall, thick, winged, spongy. **Leaves** sheaths with blades, the fronts cross-corrugated, not veined, fragile, hyaline, with concave mouth; blades 5–11 mm wide. **Inflorescences** (2)3–10 × 1–3 cm, dense and head-like, prickly-looking because of spreading perigynium beaks; spikes androgynous, sometimes some female. **Female scales** hyaline to brownish, shorter than the perigynia. **Perigynia** lance-triangular, (3.6)4–5.2 × 2 mm, widest near the base, green, maturing yellowish to pale brown, strongly veined, with 15 reddish brown veins on the dorsal surface, 7 on the ventral surface; base pithy, swollen, cordate; apex tapering gradually to the beak; beak about 2.5 mm long, poorly defined; stigmas 2. **Achenes** lenticular. **2*n***=52.

Marshes, edges of thickets, and other wet spots with still water, usually in full sun. 0–1600 m. BR, BW, Casc, CR, ECas, Est, Sisk, WV. CA, ID, NV, WA; north to AK, east to Newfoundland, southeast to SC; Asia. Native.

Carex stipata is densely cespitose with a large, spiky inflorescence, large, green to straw-colored perigynia, and cross-corrugated leaf sheath fronts. *Carex densa* and *C. vulpinoidea* grow in the same habitat and have cross-corrugated leaf sheath fronts, but they have smaller perigynia in narrower inflorescences, and they lack winged culms. Runty *C. stipata* can be confused with *C. neurophora,* which has smaller perigynia.

Carex straminiformis L.H. Bailey (p. 228)

Shasta sedge

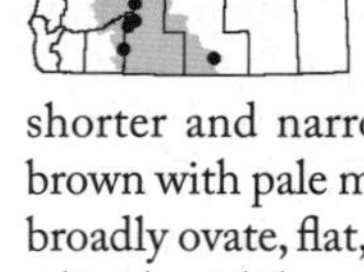

Plants cespitose, 20–50 cm tall. **Leaves** 3–4 mm wide, flat or folded. **Inflorescences** 1.3–3 cm × 12–20 mm; spikes 3–8, crowded but distinguishable, gynaecandrous. **Female scales** lanceolate to ovate, 3.2–4.8 mm long, shorter and narrower than the perigynia, white to reddish brown with pale mid-stripe; apex acute to acuminate. **Perigynia** broadly ovate, flat, 4–5.8 × 1.8–3.4 mm, straw-colored, green, or whitish with brown tip, widest below middle of total perigynium length, with 10–20 dorsal veins and 1–12 ventral veins;

wings 0.4–1 mm wide, the edges often crinkled; beak winged and ciliate-serrulate to near the tip, or sometimes unwinged, brown, parallel-sided, and entire for the distal 0.5–0.7 mm; 2.2–3 mm from achene top to beak tip; stigmas 2. **Achenes** lenticular, located below the center of the perigynium, (1.4)1.7–2.4 × 1–1.6 mm.

In rocky or gravelly soils, dry open slopes, sometimes in open forests, at or near timberline, often near persistent snowbanks. 1500–2800 m. Casc, ECas. CA, ID, NV, WA; northeast to MT, southeast to UT. Native.

Carex straminiformis is a coarse, tough sedge with spikes that show a lot of contrast between pale perigynia and dark scales. The broad, flat, "potato chip" perigynia clinch the identification. *Carex brevior* has equally wide but plano-convex perigynia, and it grows only at low elevations.

Carex subbracteata Mackenzie (p. 228)

Smallbract sedge, small-bracted sedge

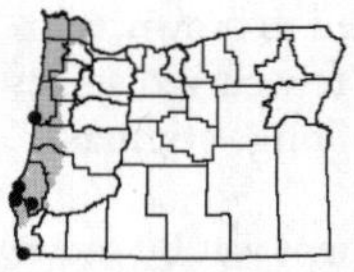

Plants cespitose; culms 25–105 cm, sometimes lax and leaning. **Leaves** sheaths sometimes extending above base of blade; blades 1.3–4.6 mm wide. **Inflorescences** 1.3–3.5 cm × 7–23 mm, dense and head-like, often elongate on late season shoots, lowest internode 2–3(4.5) mm, generally shorter than lowest spike; bracts usually short and bristle-like, sometimes leaf-like and longer than the inflorescence; spikes gynaecandrous. **Female scales** 3.5–4.5 mm long, usually shorter and narrower than the perigynia, sometimes longer and covering them; brown, red-brown, or coppery, with whitish, green, or pale brown mid-stripe; margins white, 0–0.4 mm wide. **Perigynia** ovate to broadly ovate, plano-convex, (2.9)3.5–4.7(5.7) × 1.3–1.7(2.2) mm, usually coppery with a metallic sheen and brown margin, leathery, with (0)5–9 conspicuous veins dorsally, 0–7 conspicuous veins ventrally, the ventral veins shorter than or as long as the achenes; wing 0.2–0.3(0.4) mm wide; base rounded to bluntly angled; beak tip brown or white, cylindric, unwinged, and more or less entire for 0.4–0.8 mm; 1.5–2.5 mm from achene top to beak tip; stigmas 2. **Achenes** lenticular, (1.3)1.5–2.1 × 0.9–1.7 mm.

Seasonally moist soil, coastal grasslands to open forests. 0–100 m. CR, Est. CA, WA. Native and Exotic.

Carex subbracteata looks like a particularly dark *C. pachystachya*, but in that species the mature spikes are star-shaped in outline because of the spreading perigynia, and the female scales do not cover the perigynia and are dark throughout or have white margins less than 0.1 mm wide. Very similar *C. harfordii* has ventral perigynium veins usually reaching the base of the beak. Late-season *C. subbracteata* shoots with leaf-like inflorescence bracts might be confused with paler *C. athrostachya*.

Carex subfusca W. Boott (p. 228)

Rusty sedge, pale broom sedge

Carex teneriformis Mack.

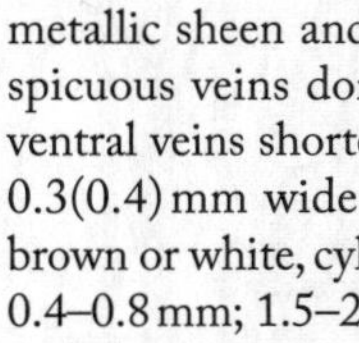

Plants cespitose, 20–105 cm tall. **Leaves** sheath fronts white-hyaline; blades 1.2–3(3.7) mm wide. **Inflorescences** 1.1–3 cm × 5.4–16.3 mm, usually green to straw-colored (sometimes brownish in northern plants); spikes gynaecandrous, overlapping but more or less distinguishable, often crowded, fine-textured because the perigynia are small. **Female scales** 2.1–3.5(3.9) mm long, shorter than or more or less covering the perigynia, white-hyaline to gold, occasionally brown. **Perigynia** lance-ovate to ovate, 2.4–3.5 × 0.9–1.9 mm, green, whitish, straw-colored, or light brown, with (0)2–11 dorsal veins and 0–6 ventral veins; wings 0.15–0.3(0.4) mm wide; beak tip winged and ciliate-serrulate nearly to the tip, or unwinged, parallel-sided, and more or less entire for the distal 0.4–0.7 mm; 1.2–2 mm from achene top to tip of beak; stigmas 2. **Achenes** lenticular, 1–1.6 × 0.7–1.25 mm. $2n$=84.

Mesic to moist meadows, seeps, roadsides, and lake shores, sometimes in moderately disturbed sites, usually in the sun but sometimes in partial shade. 50–2200 m. BR, BW, Casc, Col, ECas, Est, Sisk, WV. CA, ID, NV, WA; southeast to NM and Mexico; HI. Native.

Carex subfusca has pale to greenish, fine-textured, erect, somewhat elongated inflorescences and small achenes. It can be confused with other species in subkey 9, particularly with individuals that have small perigynia. *Carex bebbii* has brown female scales and perigynia, and rounder spikes.

Carex subnigricans Stacey (p. 228)

Nearlyblack sedge, dark mountain sedge

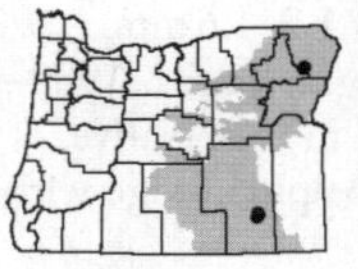

Plants rhizomatous, the rhizomes 1 mm thick; culms 5–20 cm tall. **Leaves** sheath fronts uniformly pale brown or colorless; blades 0.4–1 mm wide, round to semicircular in cross section. **Inflorescences** spike solitary, androgynous, 0.7–2 cm × 4–6 mm, lacking an inflorescence bract. **Female scales** about as wide as the perigynia but shorter, 1-veined, yellowish brown; apex obtuse to acute, flat, not awned. **Perigynia** elliptic, 2.5–3.5 × 1.5 mm, ascending; brown above, darker brown through the middle, and pale below; veinless, with rachilla shorter than the achene; stigmas 3. **Achenes** trigonous, much smaller than the perigynium.

Moist, rocky, alpine to subalpine slopes and meadows. 2300–2800 m. BR, BW. CA, ID, NV; east to WY. Native.

Carex subnigricans is a short, turf-forming, high elevation, wetland species with narrow, involute leaves. Its single, oval spikes resemble those of young *C. nigricans*, which has wider, flat leaves. *Carex breweri* has broader heads and veined perigynia and lives in drier habitats.

Carex sychnocephala J. Carey (p. 228)

Many-headed sedge

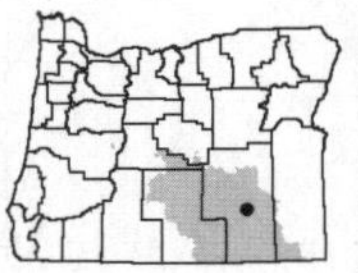

Plants cespitose, 8–40 cm tall. **Leaves** sheath fronts white-hyaline; blades 1.2–3 mm wide. **Inflorescences** 1.6–3 cm × 7–15 mm, dense and head-like, greenish to golden; lower bracts leaf-like and usually at least 5 times as long as the inflorescence; spikes gynaecandrous. **Female scales** 3.5–4.5 mm long, shorter or longer than the perigynia, white- or gold-hyaline. **Perigynia** narrowly lanceolate, (4.6)5.5–7.3 × 0.7–1.2 mm, green, straw-colored, or light brown, with 3–12 dorsal veins, ventrally veinless or with up to 9 veins; wing 0.1 mm wide on the body; beak winged and ciliate-serrulate to near the tip; 3–5 mm from achene top to beak tip; stigmas 2. **Achenes** lenticular, 1–1.8 × 0.6–0.8 mm. $2n$=64.

Open, seasonally wet shores of ponds and rivers. BR. WA; north to AK, east to NY. Native.

Carex sychnocephala is unmistakable with its extremely long, leaf-like inflorescence bracts and its long, narrow perigynia that are more beak than body. *Carex athrostachya* has elongated inflorescence bracts and may occur in similar habitats, but its bracts and lanceolate perigynia are much shorter.

Carex tahoensis F.J. Smiley (p. 230)

Tahoe hare sedge, Tahoe sedge

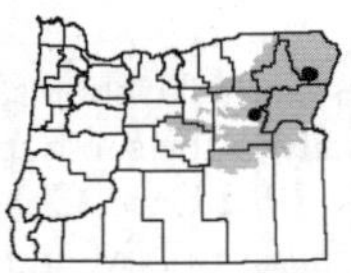

Plants cespitose, 15–45 cm tall. **Leaves** 1.5–2.5(3) mm wide, folded, somewhat tough, the edges sometimes revolute. **Inflorescences** (1.5)2–3(3.7) cm × 6–12 mm, somewhat elongated, erect or somewhat angled to one side; spikes gynaecandrous, narrow, ascending, overlapping. **Female scales** 4–5 mm long, gold to brown with paler midrib, more or less covering the perigynia. **Perigynia** ovate, (3.7)4.5–6 × 1.5–2.6 mm, dark reddish brown with green or pale edges, opaque, and somewhat tough when mature, strongly 3–8-veined on the ventral surface, with 7–15 strong veins on the dorsal surface; wings 0.2–0.4(0.5) mm wide; beak tip unwinged, brown, parallel-sided, and entire for the distal 0.2–0.7 mm, with the perigynium wings extending up to the base of the cylindric tip; (0.6)1–2 mm from beak tip to achene top; stigmas 2. **Achenes** lenticular, 1.9–2.4 × 1.2–1.6 mm.

Dry subalpine and alpine ridges and grasslands. 1900–2900 m. BW. CA, WA; north to Yukon, east to CO. Native.

Carex tahoensis is a tough, cespitose, subalpine sedge with ascending to appressed, opaque perigynia that are mostly hidden by the female scales. *Carex phaeocephala* is superficially very similar, but its perigynia are more or less translucent, usually paler, veinless (or with few faint veins), and shorter, narrower, and particularly less thick. Plants with immature perigynia cannot be distinguished.

Carex tiogana D.W. Taylor & J. Mastrogiuseppe

Tioga Pass sedge

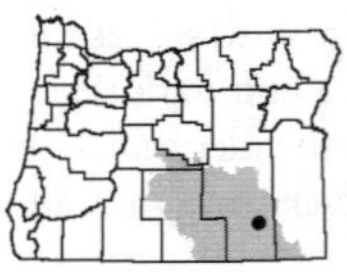

Plants cespitose, 1.8–7.5 cm tall. **Leaves** 1.5–3 mm wide, falcate, with serrulate margins and the midrib serrulate at least in the distal 33%. **Inflorescences** bracts with well-developed sheaths; lateral spikes female, with 3–9 perigynia, the lower erect to nodding; terminal spike male, 0.35–0.5 cm long. **Female scales** obtuse to acuminate, shorter but wider than the perigynia, mostly white with scabrous, gold midrib; falling before the perigynia. **Perigynia** oblong-ovate, 1.2–2.1 × 0.4–1 mm, straw-colored to brown, veinless but with 2 marginal ribs, usually with a few curved spines on the distal margins; beak 0.2–0.5 mm long; stigmas 3. **Achenes** trigonous.

Rocky benches in cirques with thin soil kept wet by snowmelt. 2700–2800 m. BR. CA. Native.

Carex tiogana is a tiny, alpine sedge with arching leaves and small brown perigynia subtended by largely white female scales that fall early. At first glance, it might be dismissed as a small

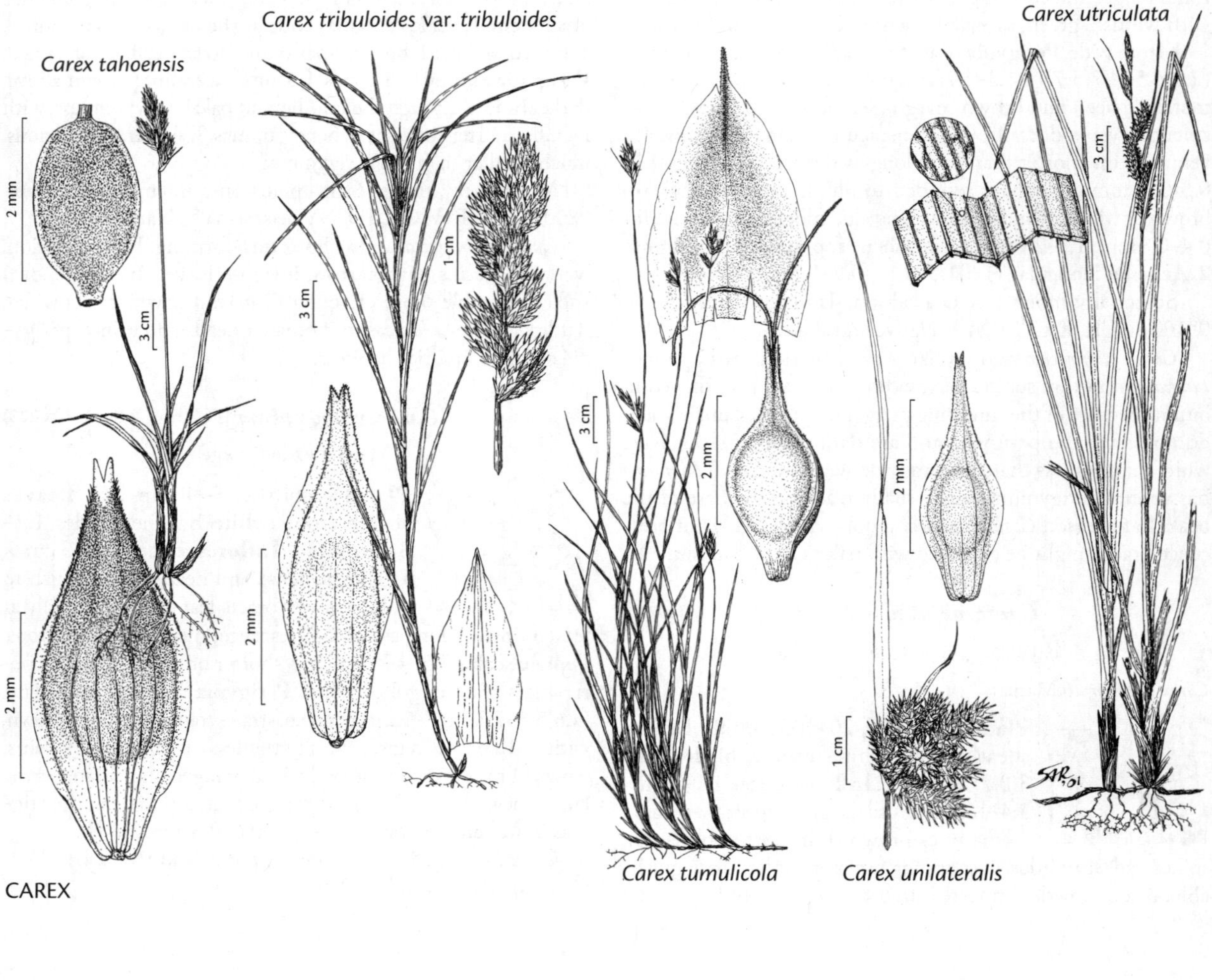

CAREX

alpine grass, perhaps an *Agrostis*. *Carex tiogana* is an alpine relative of *C. capillaris*, a taller plant with flat leaves and larger perigynia. In *C. capillaris*, the midrib of the female scales and the margins and midribs of its leaves are usually smooth.

Carex tribuloides Wahlenberg var. ***tribuloides*** (p. 230)

Tribulation sedge, blunt broom sedge

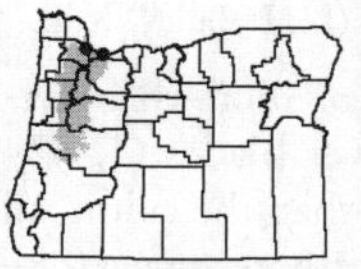

Plants cespitose, 50–110 cm tall, the vegetative culms sometimes rooting at the nodes and functioning as stolons. **Leaves** sheath fronts green and veined; blades 3.2–7 mm wide. **Inflorescences** 2–5(8) cm × 10–20 mm, brown, erect, often dense and head-like, sometimes elongated but dense distally; spikes gynaecandrous. **Female scales** (1.9)2.5–3 mm long, narrower than and half as long as the perigynia, white-hyaline or silvery brown with green midstripe; apex acute to acuminate. **Perigynia** ovate-lanceolate to diamond-shaped, (3.3)3.6–5.4 × 1.1–1.5 mm, green to light brown, with 3–6 dorsal veins and 2–4 ventral veins (which may be faint); wing 0.1–0.3 mm wide, often asymmetric, abruptly narrowed and often lacking below the middle; beak winged and ciliate-serrulate almost to the tip; 1.4–2 mm from achene top to beak tip; stigmas 2. **Achenes** lenticular, 1–1.8 × 0.6–0.9 mm. **2*n***=70.

Floodplain forests, wet thickets, ditches, wet meadows. 0–400 m. WV. WA; MN to Nova Scotia, south to KS and FL. Exotic.

The aptly named *C. tribuloides* looks like many other species of subkey 9, until the green leaf sheath fronts are noticed. In the native *C. feta*, the most common species with green sheaths, the perigynia are ovate and mostly smaller, and the perigynium wings extend to the perigynium base. *Carex scoparia* has narrower leaves and usually longer perigynia.

Carex tumulicola Mackenzie (p. 230)

Foothill sedge

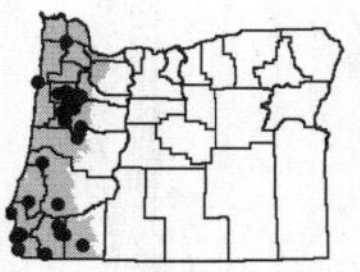

Plants rhizomatous, but the stout rhizomes short, so that the plant forms patches; culms 20–80 cm tall. **Leaves** widest blades 1.5–2.5 mm wide, green, tough. **Inflorescences** 1.5–5 cm × 5–8 mm, usually bent somewhat to the side; lowest inflorescence bract usually longer than the lowest spike, sometimes longer than the inflorescence; spikes androgynous, relatively narrow. **Female scales** 3.3–5.2 mm long, about as long as the perigynia and somewhat wider, tan or brown with green midrib; awn to 1 mm long. **Perigynia** 3.5–5 × 1.5–2 mm, light green or light brown, the margins not flat (but with ribs); beak 1–3 mm long; stigmas 2. **Achenes** lenticular.

Grasslands, oak savanna, dry slopes, and openings in forest. 0–800 m. CR, Est, Sisk, WV. CA, WA; north to British Columbia. Native.

Carex tumulicola forms patches of tough green leaves in mesic to dry grasslands. It produces brown inflorescences that usually bend to the side, often with the lowest bract diverging. Its perigynia are mostly hidden by the female scales. *Carex hoodii* has a denser, relatively shorter inflorescence, and its perigynia are greenish with contrasting chestnut brown center (or light brown with contrasting dark brown center, when old). Sedges sold as *C. tumulicola* for native lawns are often introduced *C. divulsa*.

Carex unilateralis Mackenzie (p. 230)

One-sided sedge, lateral sedge

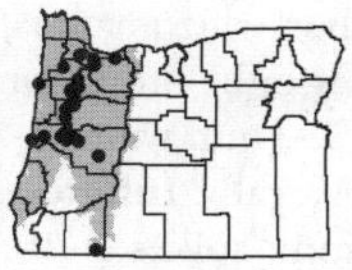

Plants cespitose, 35–75 cm tall. **Leaves** 2–3(4) mm wide. **Inflorescences** (1)1.4–2.5 cm × 9–17 mm, dense and head-like, usually tilted to one side; bracts leaf-like, longer than the inflorescence; spikes gynaecandrous. **Female scales** 3.3–4.8 mm long, shorter and narrower than the perigynia, gold to brown with midrib green or gold. **Perigynia** ovate to lanceolate, 3.5–5 × 1.3–1.75 mm, cream-colored to light brown, with 0–11 dorsal veins; wings 0.2–0.3 mm wide; beak tip usually winged and ciliate-serrulate to the tip, but occasionally unwinged, brown, parallel-sided, and entire for the distal 0.5–0.6 mm; (1.4)1.7–2.5 mm from achene top to beak tip; stigmas 2. **Achenes** lenticular, (1.2)1.5–1.9 × 0.75–1(1.2) mm.

Wet prairies, vernal pools, pond margins, marshes, road ditches and other seasonally wet places. 0–1300 m. Casc, CR, Est, WV. CA, WA; north to British Columbia; Japan. Native.

With its long, leaf-like bracts and inflorescence angled to one side of the culm, *C. unilateralis* is usually easy to identify. Some plants are intermediate between this and *C. athrostachya*, which lives mainly east of the Cascades but occasionally on the west side. *Carex athrostachya* inflorescences are erect, perched directly atop the culm, and its lower inflorescence bracts are more bristle-like. Its beaks are unwinged, brown, and entire, but in *C. unilateralis* the majority of beaks are flattened and ciliate-serrulate almost to the tip.

Carex utriculata F.M.B. Boott (p. 230)

Southern beaked sedge, Northwest Territory sedge, inflated sedge

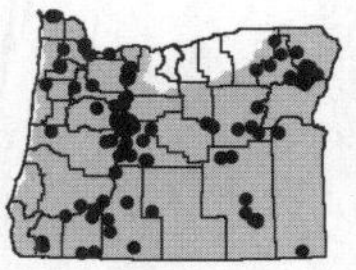

Plants rhizomatous, forming large stands, 25–120 cm tall; plant bases often brown to tinged reddish. **Leaves** leaf sheaths somewhat spongy, with crosswalls common between the vertical veins, making a pattern like brickwork; blades to 4.5–12(15) mm wide, green, flat to broadly W-shaped in cross section, smooth, not papillose. **Inflorescences** lateral 2–5 spikes female, well separated from terminal spikes, erect to ascending, mostly 2–10 cm long, with 8 spiraling columns of perigynia; terminal 2–5 spikes male, usually 2–7 cm long. **Female scales** mostly shorter than the perigynia, apex acute to acuminate, awnless. **Perigynia** inflated, (3.2)4–8.6 × 1.7–3 mm, green, straw-colored, or reddish brown, with 9–15 veins; apex abruptly narrowed to the beak; beak (1)1.2–2.7 mm long; teeth 0.2–1.3 mm long, straight; styles persistent, becoming curved; stigmas 3. **Achenes** trigonous. **2*n***=72, 76, 82.

Lake shores, marshes, streamsides, bogs, fens, areas with high water table, usually on organic soils. 100–2800 m. BR, BW, Casc, CR, ECas, Lava, Owy, Sisk, WV. CA, ID, NV, WA; throughout Canada and most of US; Mexico; Eurasia. Native.

Carex utriculata is a common, robust species, forming dense stands in shallow water. It has large, inflated perigynia with short, straight beak teeth. The backs and sides of its leaf sheaths have many crosswalls between the vertical veins. *Carex vesicaria* and *C. exsiccata* are similar, but they lack this "brickwork" of crosswalls and have perigynia that taper less abruptly to the beak.

Carex vallicola C. Dewey (p. 232)

Valley sedge

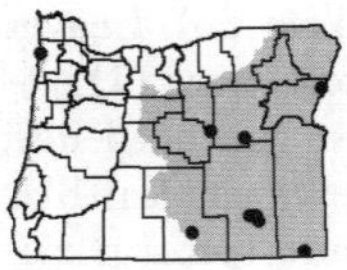

Plants cespitose or very short-rhizomatous, 12–60 cm tall; plant bases light brown or warm tan. **Leaves** blades 1–3 mm wide, inconspicuously papillose ventrally. **Inflorescences** 0.5–3 cm × 4–8 mm; spikes 5–10, short, crowded, androgynous. **Female scales** slightly shorter than or as long as the perigynia, whitish or pale brown hyaline with green center, apex acute or short-awned. **Perigynia** oval, 3.3–4 × 1.8–2.3 mm, light brown, shiny; broadly rounded on the dorsal side so that the 2 marginal ribs appear to be on the flat ventral surface, with several (7–15) more or less obscure dorsal veins and no ventral veins; base more or less spongy, 0.7–1 mm long, drying wrinkled; tip abruptly contracted to the beak; beak 0.5–1 mm long, smooth-margined or serrulate; stigmas 2. **Achenes** lenticular.

Mesic depressions and slopes in dry grasslands or sage steppes, often in thickets or aspen stands, generally in gravelly loam, often with big sage (*Artemisia tridentata*). 1300–2300 m. BR, BW, Est, Lava, Owy. CA, ID, NV, WA; north to British Columbia, east to SD, south to NM and Mexico. Native.

Carex vallicola is a cespitose sedge with a fairly narrow inflorescence composed of short spikes. Its shiny, plump perigynia have the marginal veins displaced to the ventral side. It often grows with *C. hoodii,* which has darker, more compact inflorescences and perigynia that have ribs at the margins where they belong and are green with copper centers (light brown with dark brown centers when fully mature).

Carex vernacula L.H. Bailey (p. 232)

Native sedge, foetid sedge

Carex foetida All., *Carex foetida* All. var. *vernacula* (L.H. Bailey) Kük.

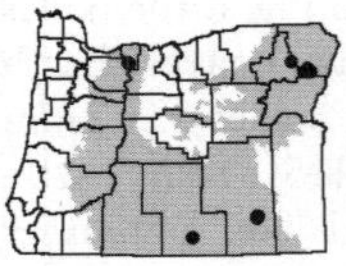

Plants loosely cespitose or short-rhizomatous, 8–40 cm tall. **Leaves** blades (1.5)2–3.5(4) mm wide, shorter than the culms, all arising near the base. **Inflorescences** 0.8–1.6 cm long, dense and head-like; spikes 5–15, crowded, indistinguishable, androgynous. **Female scales** about as long and wide as the perigynia, reddish brown to blackish, often with paler midrib, more or less translucent, shiny. **Perigynia** narrowly to broadly elliptic, flattened, 3.3–4.6 × 1.1–1.9 mm, pale yellowish brown at the base but reddish brown to blackish above, shiny; beak 0.9–1.5 mm long; stigmas 2. **Achenes** lenticular.

Alpine and subalpine wet meadows, rocky slopes that receive snowmelt, edges of headwater streams, and lake shores. 2300–2800 m. BR, BW, Casc, ECas. CA, ID, WA; east to WY and CO. Native.

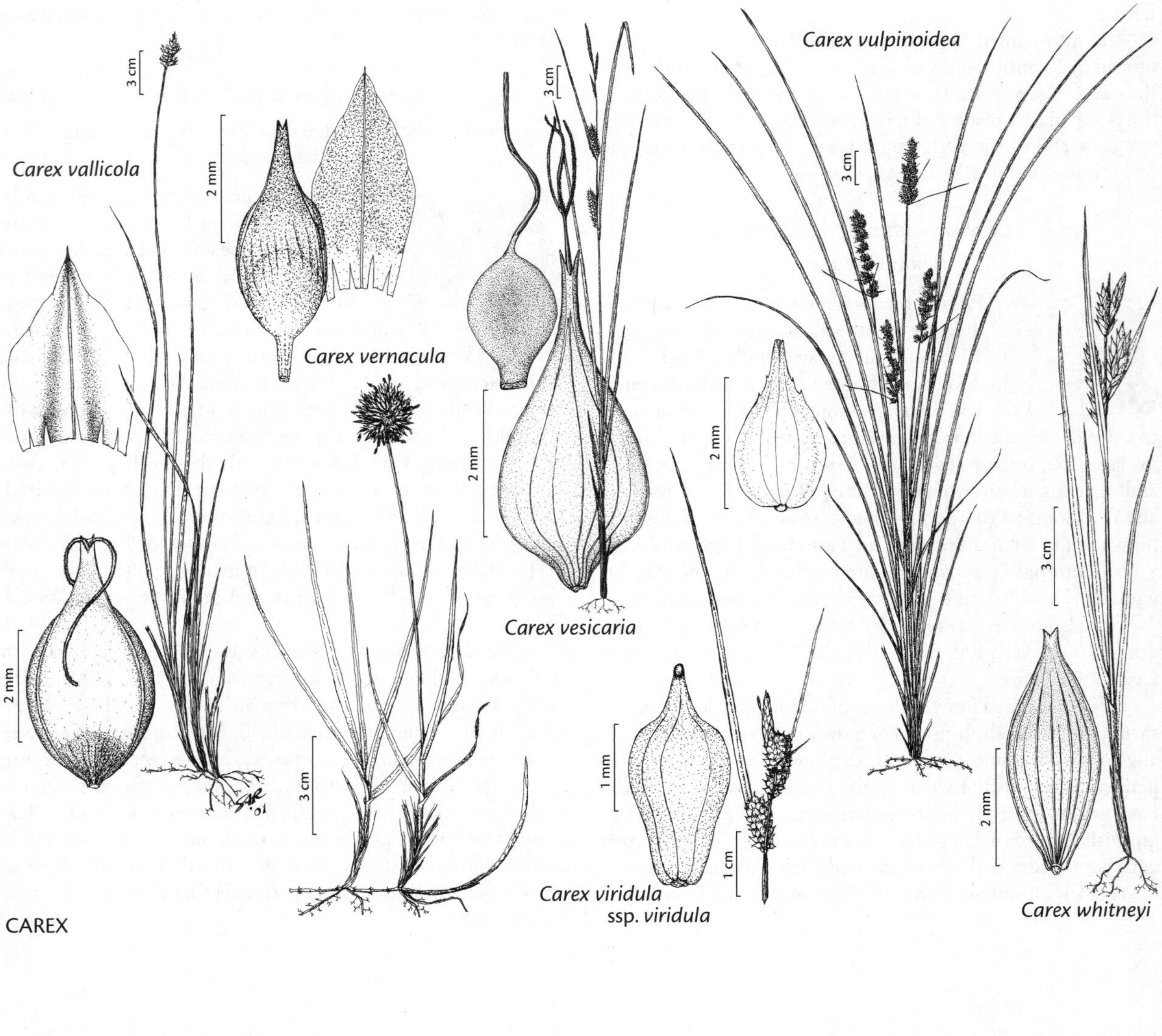

This alpine "ball-headed" sedge has dense, dark heads, stipitate perigynia, and persistent anthers. Similar *C. illota* has smaller inflorescences and uniformly dull dark brown perigynia that lack stipes. *Carex jonesii* has the perigynium base swollen with pithy tissue.

Carex vesicaria Linnaeus (p. 232)

Inflated sedge, blister sedge

Carex vesicaria L. var. *vesicaria*

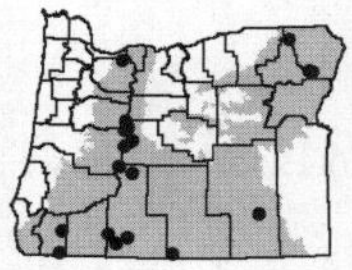

Plants cespitose, or less often short-rhizomatous, 15–100 cm tall; plant bases reddish. **Leaves** basal leaf sheath not spongy, with few crosswalls between the vertical veins; blades to 1.8–6.5 mm wide, green, V- or W-shaped in cross section. **Inflorescences** lateral 1–3 spikes female, erect or ascending, usually 1.5–5 cm long; terminal 1–3 spikes male, 2–7 cm, well separated from the female spikes. **Female scales** shorter than the perigynia, apex acute to acuminate, awnless. **Perigynia** (3.5)4–7.5(8.2) × (1.4)1.7–3.3 mm, green to straw-colored or reddish brown, with 7–12 veins; beak 1.1–3 mm long; teeth 0.3–0.7 mm long, straight; styles persistent, becoming curved; stigmas 3. **Achenes** trigonous. ***2n***=74, 82, 86.

Pond margins, lake shores, wet spots in meadows, generally in shallow water. 900–2300 m. BR, BW, Casc, ECas, Sisk. CA, WA; north to British Columbia, east to Newfoundland, southeast to VA; Eurasia. Native.

Broadly cespitose *C. vesicaria* can be confused with *C. utriculata,* which grows in slightly deeper water. *Carex utriculata* has leaf sheaths with "brickwork" of crosswalls. In and west of the Cascades, see *C. exsiccata.*

Carex viridula A. Michaux ssp. *viridula* (p. 232)

Little green sedge, green sedge

Carex oederi Retz., *Carex oederi* Retz. var. *viridula* (Michx.) Kük. *Carex viridula* Michx. var. *viridula*

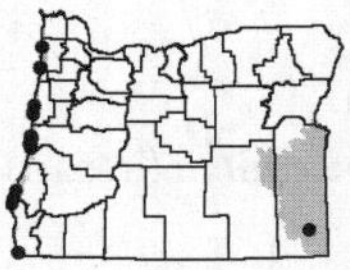

Plants cespitose, 2–40 cm tall. **Leaves** 1–3.1(4.5) mm wide. **Inflorescences** lowest bract diverging at 90° or more from the culm, usually lacking a sheath, but if lowest spike is remote, then its bract may have a sheath greater than 4 mm long; lateral spikes 1–8, female or androgynous, 0.4–1.4 cm × 3–11 mm, crowded or sometimes the lowest remote; terminal spike male or androgynous, sessile or pedunculate, 0.8–2.5 cm × 1–3.3 mm. **Female scales** shorter than perigynia. **Perigynia** (1.8)2.2–3.3(3.9) × 0.8–1.6 mm, yellowish green to green, strongly veined; beak 0.3–1.3 mm long, straight or slightly bent toward the dorsal side; stigmas 3. **Achenes** trigonous. ***2n***=70, 72.

Deflation plains among coastal sand dunes; also margins of hot springs. 0–1300 m. Est, Owy. CA, ID, NV, WA; north to AK, east to Greenland, southeast to NM; Eurasia. Native.

Carex viridula is a strictly cespitose sedge with crowded green spikes, widely spreading perigynia, and a divergent inflorescence bract.

Carex vulpinoidea A. Michaux (p. 232)

Fox sedge

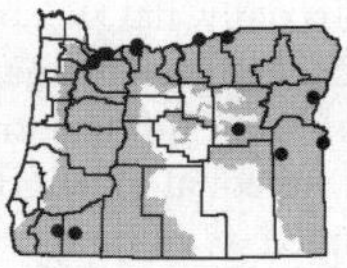

Plants cespitose, 30–100 cm tall. **Leaves** sheath fronts cross-corrugated, white-hyaline with brown spots; blades approximately 5 mm wide, generally longer than the flowering culms. **Inflorescences** elongate, (3)7–10 cm × 15 mm, somewhat interrupted; bracts hair-like, usually conspicuous; spikes many, short, androgynous; lowest node usually with a short branch with at least 1 side spike, thus the lowest node often seeming to produce 2 spikes. **Female scales** brown, short-awned. **Perigynia** elliptic to ovate, 2–3.2 × 1.3–1.8 mm, dull grayish brown, with 0–3 dorsal veins, veinless ventrally; base obtuse or rounded, depending on the amount of pithy tissue within; stigmas 2. **Achenes** lenticular. ***2n***=52, 54.

Moist to wet meadows, marshes and ditches. 0–1300 m. BW, Casc, Col, ECas, Owy, Sisk, WV. CA, ID, WA; throughout most of Canada and US, south to Mexico; Eurasia, New Zealand. Native.

Carex vulpinoidea is a cespitose sedge with cross-corrugated leaf sheaths and long, interrupted inflorescences with conspicuous though almost hair-thin bracts. *Carex densa,* its equivalent west of the Cascades, has a shorter, uninterrupted inflorescence with less conspicuous bracts and yellower perigynia that average larger, with more dorsal and ventral veins. Its leaves are shorter than the flowering culms. *Carex vulpinoidea* is native in eastern Oregon and along the Columbia, but southwestern Oregon records are recent introductions.

Carex whitneyi Olney (p. 232)

Whitney's sedge

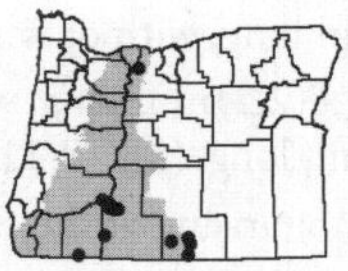

Plants cespitose; culms 25–100 cm tall, much longer than the leaves. **Leaves** 2–6 mm wide, pubescent. **Inflorescences** bracts leaf-like; lateral spikes 2–3, female, 0.7–3 cm long, stalked; terminal spike male, 0.5–3 cm long. **Female scales** shorter than the perigynia, white-hyaline with green mid-stripe, usually with a short awn. **Perigynia** ovate, 3.7–5.5 mm long, glabrous, green, veined; beak 0.2–1 mm long; stigmas 3. **Achenes** trigonous.

Montane dry meadows and openings in conifer forests, on sandy to gravelly soils. 1500–2500 m. Casc, E. Casc, Sisk. CA, NV. Native.

Its hairy leaf blades, glabrous perigynia, and upland, montane habitat make *C. whitneyi* easy to identify. Introduced *C. hirsutella* shares these traits but has almost beakless, weakly veined perigynia and grows at lower elevations.

CYPERUS Umbrella-sedge, flatsedge

Herbs perennial or annual, cespitose or not, rhizomatous, rarely tuberous. **Culms** solitary or clustered, trigonous or round. **Leaves** usually basal; ligules absent; blades keeled abaxially, flat, V-, or M-shaped in cross section. **Inflorescences** terminal, the spikes either all arising close to one another in head-like clusters or arising separately in cylindrical clusters; some or all clusters usually raised on stalks (rays) (0)2–12(30) cm, if rays absent inflorescence a dense often lobed head of overlapping spikes, secondary rays sometimes present; clusters 1–80; rays usually unequal in length, produced singly from the axils of inflorescence bracts; involucral bracts 2–12(25), 1–50 cm long, spirally arranged at culm apex, spreading to erect, leaf-like. **Spikes** scales to 76, 2-ranked, each subtending a flower, cylindric to compressed, borne spicately or digitately at ends of rays. **Flowers** bisexual, in axils of floral scales, scale bases often decurrent onto rachilla as more or less hyaline wings; perianth bristles absent; stamens 1–3; style base deciduous or persistent; stigmas 2–3. **Achenes** biconvex, flattened, or trigonous.

Cosmopolitan in tropical and temperate areas. Approximately 600 species; 10 species treated in *Flora*.

Cyperus is characterized by terminal inflorescences and usually flat spikes with 2-ranked scales.

Cyperus aggregatus was collected in Oregon in 1916 and *C. odoratus* in 1915. Both were found on ballast in Linnton (Multnomah County).

Key to *Cyperus* species:

1. Spikes in head-like clusters, all arising at about the same place on the inflorescence axis.
    2. Stigmas 2; achenes biconvex to flattened, subcylindric, or cylindric.
        3. Plants perennial, tufted on slender rhizomes; achenes ellipsoid; heads usually compact; plants 5–40(95) cm tall . . . . . **C. niger**
        3' Plants annual, cespitose with fibrous roots; achenes obovoid to ovoid; heads usually more open; plants 3–25(30) cm tall . . . . . **C. bipartitus**
    2' Stigmas 3; achenes usually trigonous.
        4. Floral scales 2-keeled in proximal 30–60%.
            5. Plants annual; longest inflorescence bract erect to ascending; anthers 0.5 mm long . . . . . **C. acuminatus**
            5' Plants perennial; longest inflorescence bract horizontal to ascending at 30°; anthers 1–1.2 mm long . . . . . **C. eragrostis**
        4' Floral scales 1-keeled or broadly rounded on the back.
            6. Plants perennial, more or less rhizomatous, with corm-like bases . . . . . **C. schweinitzii** (in part)
            6' Plants annual, tufted, corm-like bases lacking.
                7. Floral scales 0.6–0.8 mm long, with or without a short apical point to 0.3 mm long, 3-veined. . **C. difformis**
                7' Floral scales (1.2)1.3–1.8(2.2) mm long, with strongly outcurved awn-like tip an additional 0.5–1(1.3) mm long, (5)7–9(11)-ribbed . . . . . **C. squarrosus** (in part)

1' Spikes in more open, cylindric clusters, not arising at the same place on the rachis.
    8. Floral scales with strongly outcurved awn 0.5–1(1.3) mm long; plants annual . . . . . **C. squarrosus** (in part)
    8' Floral scales lacking an awn-tip or with a short awn that is not strongly outcurving; plants annual or perennial.
        9. Spikes (2.8)3.2–4(4.5) mm wide . . . . . **C. schweinitzii** (in part)
        9' Spikes 1–2 mm wide.
            10. Floral scales 1.3–1.5 mm long . . . . . **C. erythrorhizos**
            10' Floral scales 1.8–4.5(6) mm long.
                11. Floral scales 3.2–4.5(6) mm; plants not producing tubers . . . . . **C. strigosus**
                11' Floral scales 1.8–2.7(3.4) mm; plants producing tubers. . . . . **C. esculentus** var. **leptostachyus**

Cyperus acuminatus J. Torrey & W.J. Hooker (p. 236)

Short flatsedge, sharp-pointed flatsedge

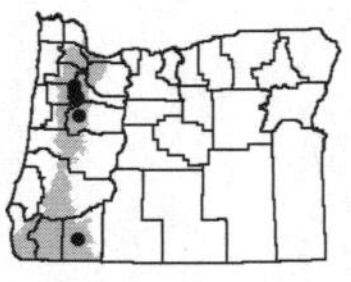

Plants annual, cespitose. **Culms** roundly trigonous, (10)20–30(45) cm × 0.4–1.2 mm. **Leaves** 8–12(16) cm × 1–2(4) mm. **Inflorescences** clusters head-like, hemispheric, 7–12(15) mm wide; rays 1–3(5), 0.5–2 cm long; longest inflorescence bracts erect to ascending. **Spikes** (15)25–50(75), ovoid, compressed, 4–7 × 2–3 mm; floral scales 8–20(35), ovate-lanceolate, 1–2 × 1–1.4 mm, yellowish, yellowish green, or light reddish brown, with light brown or greenish brown mid-stripe, 2-keeled, subtly 1-ribbed; apex acute, mucronate, straight to curved outward, glabrous or sparsely scabridulous. **Flowers** anthers 0.5 mm; styles 0.8–1.1 mm; stigmas 0.5 mm. **Achenes** broadly ellipsoid, abruptly contracted to base, 0.8–1.1 × 0.3–0.4 mm long, stipe 0.1 mm long.

Disturbed seasonal wetlands, usually in sandy soils. 0–500 m. Sisk, WV. CA, WA; east to NH, south to Mexico. Native.

Cyperus acuminatus is most often confused with *C. squarrosus,* which has longer, more outcurved awns and more veins on the floral scales.

Cyperus bipartitus J. Torrey

Shining flatsedge

Cyperus niger Ruíz & Pav. var. *rivularis* (Kunth) V.E. Grant, *Cyperus rivularis* Kunth

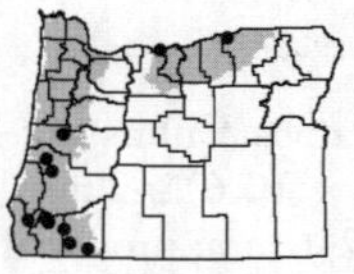

Plants annual, densely cespitose, with fibrous roots. **Culms** roundly trigonous, 3–25(30) cm × 0.3–1.4 mm. **Leaves** 1–8 cm × 1–2 mm. **Inflorescences** cluster head-like, 1, loosely ovoid, compressed, 7–14 × 9–14 mm; rays 1–4, to 2(3) cm; inflorescence bracts 2–3, horizontal to ascending at 45(75)°. **Spikes** 3–5(8), flattened, oblong to oblong-lanceoloid, 8–18 × 2–2.6(3) mm; floral scales (6)10–26(32), closely overlapping, 1.9–2.7 × 1.8–2.3 mm, oblong to ovate, light to dark brown, with light brown mid-stripe; lateral ribs 0, medial ribs 2–3; 2-keeled; apex obtuse. **Flowers** anthers 0.4–0.5 mm; styles 0.6–1 mm; stigmas 1–1.5 mm. **Achenes** with network of ridges forming isodiametric or square cells, obovoid to ovoid, 1–1.3(1.5) × (0.6)0.8 mm.

Shorelines exposed by seasonal low water levels, ditches, disturbed wetlands. 0–800 m. Col, CR, Sisk, WV. CA, ID, NV, WA; northeastern Canada, most of US, south to South America. Native.

Cyperus bipartitus closely resembles *C. niger*, a taller rhizomatous perennial with clustered shoots that are often collected without the rhizomes.

Cyperus difformis Linnaeus

Asian flatsedge

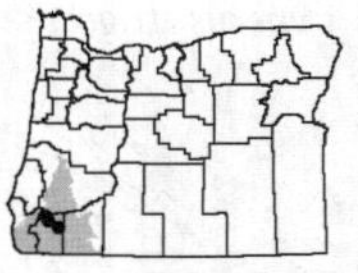

Plants annual, cespitose. **Culms** 1–15, trigonous, 7–30 cm × 1.2–2.5 mm, soft (flattened in pressing). **Leaves** (2)7–22 cm × 2.2–4 mm. **Inflorescences** clusters head-like, dense, 7–17(35) mm wide; rays (0)1–5, 2–32 mm; longest inflorescence bract erect or nearly so, appearing as continuation of culm, other bracts horizontal to ascending. **Spikes** 30–120, oblong-ellipsoid, compressed, (2)3–5(6) × 0.8–1.2 mm, greenish brown to purplish brown; floral scales (6)12–20(30), obovate to orbiculate, 0.6–0.8 × 0.6–0.8 mm, stramineous to deep purple with clear margins; mid-stripe greenish, stramineous, or purplish; lateral ribs 0, medial ribs 3; apex mucronulate. **Flowers** anthers 0.1 mm; styles 0.1 mm; stigmas 0.1–0.3 mm. **Achenes** obovoid-ellipsoid, 0.6–0.8 × 0.3–0.4 mm (as long as subtending scale), base cuneate.

Disturbed, sandy or muddy soils, sand bars, roadsides, sometimes on serpentine. 100–400 m. Sisk. CA, ID; east to NJ, south to Mexico; South America; Africa, Australia, Eurasia, islands of the Pacific and Indian Oceans; Exotic.

Cyperus difformis is recognized by its tight heads of small spikes with unusually small scales. The first Oregon collection was made in 2004.

Cyperus eragrostis Lamarck (p. 236)

Tall flatsedge

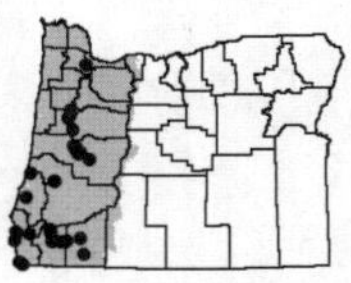

Plants perennial, cespitose, rhizomatous. **Culms** trigonous to terete, (25)40–60(90) cm. **Leaves** (10)25–50 cm × 5–8(12) mm. **Inflorescences** clusters head-like, globose, (10)20–40 mm diameter; rays (0)3–10, 2.5–5(12) cm; inflorescence bracts 4–8, horizontal to ascending at 15–30°, 3–30(50) cm × 1.5–8(12) mm. **Spikes** (20)30–50(70), oblong, compressed, 5–20 × 2.2–3 mm; floral scales (12)20–30(50), ovate-deltate, 2–2.3 × 1–1.4 mm, off-white to golden brown, basally 2-keeled, lateral ribs 0, declined 30–40° from rachilla. **Flowers** anthers 1–1.2 mm; styles 1–1.2 mm; stigmas 0.5 mm. **Achenes** broadly ellipsoid, 1.2–1.4 × 0.5–0.6 mm, stipe to 0.1 mm, beak 0.2–0.3 mm.

Disturbed wetlands, roadside ditches. 0–800 m. Casc, CR, Est, Sisk, WV. CA, WA; north to British Columbia, eastern US; South America; Europe. Native?

Cyperus eragrostis is a southern species that has been spreading northwards. Southwestern Oregon populations may be native. Further research is warranted.

Cyperus erythrorhizos Muhlenberg (p. 236)

Red-rooted flatsedge

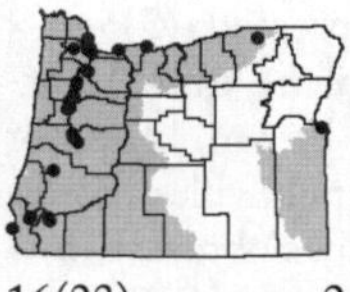

Plants annual, cespitose. **Culms** trigonous to roundly trigonous, (0.5)5–25(105) cm × 1–2.5(7.5) mm. **Leaves** 5–25(90) cm × 2–5(11) mm. **Inflorescences** clusters cylindric, dense, 1–3(6), (4)10–30(45) × (6)10–16(23) mm; rays 2–6(12), 1–8(28) cm long; inflorescence bracts horizontal to ascending at 30°. **Spikes** (20)40–80, linear, quadrangular in cross section, 3–8(11) × 1–1.5 mm; rachilla persistent, at maturity becoming laterally free, remaining firmly attached proximally, wings 0.3(0.4) mm wide; floral scales deciduous, 6–16(30), appressed, oblong-ovate to obovate, 1.3–1.5 × 0.8–1.2 mm, light brown with reddish speckles and greenish midstripe; lateral ribs 0, medial ribs 3; apex obtuse, mucronulate. **Flowers** anthers 0.2–0.3 mm, styles 0.7–0.8 mm; stigmas 0.4 mm. **Achenes** ovoid, (0.4)0.7–1 × 0.4–0.6 mm.

Shorelines exposed by low water levels. 0–700 m. Casc, Col, CR, ECas, Est, Owy, Sisk, WV. CA, NV, WA; throughout most of US and southern Canada, south to Mexico. Native.

Cyperus esculentus Linnaeus var. ***leptostachyus*** Boeckeler (p. 236)

Yellow nutsedge, yellow nutgrass

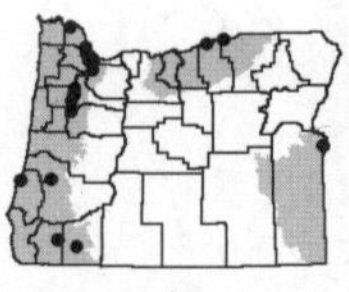

Plants perennial, stoloniferous or rhizomatous; stolons or rhizomes bearing tubers (3)6–11 mm diameter. **Culms** trigonous, 15–60(100) cm × 0.6–3.4 mm. **Leaves** (6)20–40(80) cm × 2–4(6.5) mm. **Inflorescences** clusters usually cylindric, sometimes head-like, (12)18–30 × (12)18–35 mm; rays 4–10, (0.3)2–12 cm; inflorescence bracts more or less horizontal to ascending at 45(75)°. **Spikes** (3)10–20(28), divaricate, forming 75–90° angle with rachis; linear to linear-lanceoloid, compressed-quadrangular, (10)15–20 × (1.2)1.5–2 mm; yellowish brown to dark brown; rachilla persistent, wings hyaline, 0.3–0.5 mm wide; floral scales persistent, 6–34, spreading or appressed, ovate-lanceolate, 1.8–2.7(3.4) × (1)1.5–1.8(2) mm, widest at mid-length, with 7–9-ribs, yellowish brown to dark brown with brownish, reddish, or greenish mid-stripe, lateral ribs 7–9, medial ribs 3, apex acute or subacute. **Flowers** anthers (1)1.3(1.6) mm; styles (0.7)1–1.2 mm; stigmas (1.2)1.8(2.5) mm. **Achenes** seldom maturing, ellipsoid, (1.1)1.3–1.5(1.6) × 0.3–0.6(0.8) mm.

Disturbed soils, cultivated fields. 0–1100 m. Col, CR, Est, Owy, Sisk, WV. CA, ID, NV, WA; throughout most of US and much of southern Canada; South America; Europe. Exotic.

The tubers of this taxon are unique in our *Cyperus* flora. Frustratingly, however, they usually detach from the roots if the plant is pulled from the ground. To collect them, reach below the level of the tubers and remove the soil as well as the intact rhizomes. As a species, *Cyperus esculentus* is widespread and variable. As many as seven varieties have been named, though not all are widely accepted. Only one is known from Oregon.

Cyperus niger H. Ruiz & Pavon

Black flatsedge

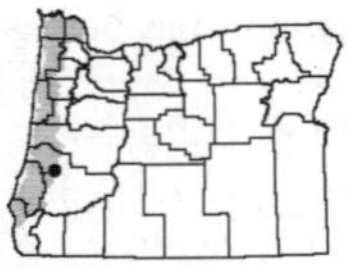

Plants perennial, rhizomatous. **Culms** 5–40(95) cm. **Leaves** 3–20(30) cm × 1.5–3 mm. **Inflorescences** clusters densely head-like, ovoid to hemispheric, 7–16(20) mm diameter; rays (0)1–2, 0.3–4.5 cm; inflorescence bracts 2–3, horizontal to reflexed downward. **Spikes** (3)5–25(60), linear to oblong-linear, compressed, (3)5–9 × 1.8–2.3 mm; floral scales 4–18, closely imbricate, ovate to orbiculate, 1.5–2.1 × 1.4–2.2 mm, chestnut brown, black, or brown, medially green or greenish brown, laterally ribless, medially 2(3)-ribbed, distinctly 2-keeled basally, apex obtuse. **Flowers** anthers 0.6–0.8 mm, styles 0.4–0.6 mm; stigmas 0.7–2.1 mm. **Achenes** ellipsoid, 1.2–1.4 × 0.6–0.8 mm, with network of ridges forming isodiametric or square cells.

Swamps, ditches, wet meadows. 50–100 m. CR. CA; east to Oklahoma; South America. Exotic.

Cyperus niger is commonly collected without rhizomes and therefore confused with the similar but shorter *C. bipartitus*. *Cyperus niger* is variable, and as many as six varieties have been recognized, but the pattern of variation seems too complex to fit into simple taxonomic categories.

Cyperus schweinitzii J. Torrey

Schweinitz's flatsedge

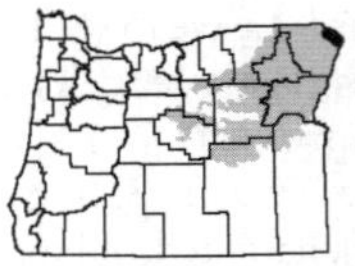

Plants perennial, rhizomatous. **Culms** trigonous, (10)20–50 cm × (0.6)1–2 mm. **Leaves** (2)20–35 cm × 2–6 mm; proximal leaves bladeless. **Inflorescences** clusters head-like to cylindric, 10–25 × 7–16 mm; rays 3–5, 1.5–8(15) cm; inflorescence bracts erect (or declined at most 25–30° from vertical). **Spikes** (1)5–10(20), oblong to linear-oblong, quadrangular in cross section, 7–10(18) × (2.8)3.2–4(4.5) mm; rachilla more or less deciduous, wingless; floral scales deciduous, (1)5–10(14), spreading, broadly ovate to oblong-ovate, (2.3)2.8–3.2 × (2)2.4–3.2 mm, stramineous or dull whitish, sometimes also red-spotted, with green mid-stripe, lateral ribs 2–3, medial ribs 3; apex broadly rounded, often clear-erose, cusp 0.1–0.4 mm in proximal scales, 0.3–1 mm in distal scales. **Flowers** anthers 0.8–1.4 mm; styles 0.1–0.3 mm; stigmas 1.5–3 mm. **Achenes** broadly ellipsoid, infrequently obovoid, 2–2.4 × 0.9–1.4 mm, base cuneate to more or less stipe-like. **2*n***=166.

Sand bars in rivers, sandy lake shores, and sandy disturbed meadows. 200–400 m. BW. WA; east to MA, south to AZ, TX, Mexico. Exotic.

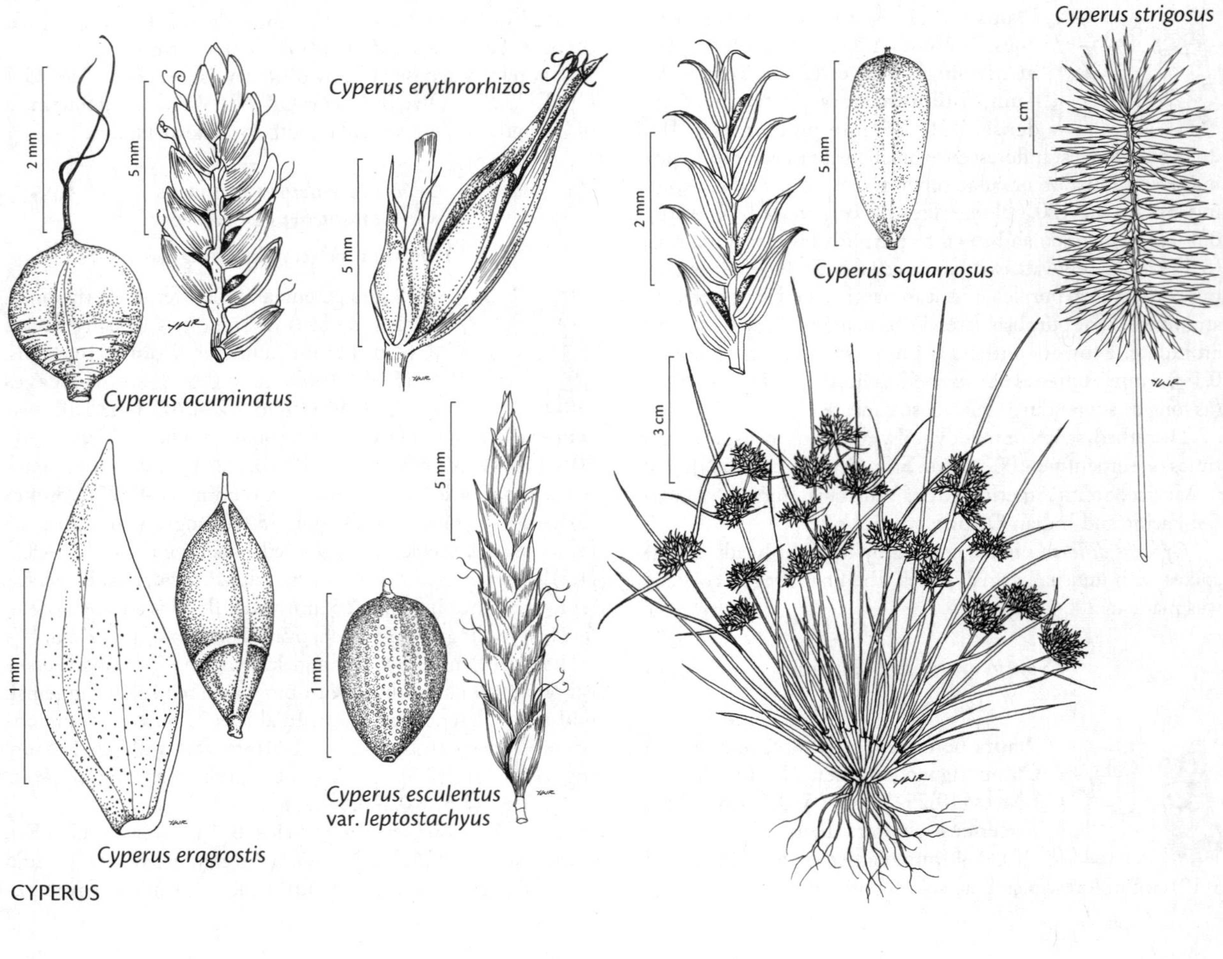

CYPERUS

The first Oregon specimens of this species were collected in 1994 along the Snake River and the lower reaches of its major tributaries. These specimens have been variously identified as *C. schweinitzii*, *C. lupulinus* ssp. *lupulinus*, and their hybrid *C.* × *mesochorus*. *Cyperus lupulinus* differs from *C. schweinitzii* in having more (15–60) spikes per inflorescence, forming a denser, nearly globose cluster that forces the inflorescence bracts to spread widely or reflex. Its anthers are 0.3–0.6 mm long, its styles are about 1 mm long, and its stigmas are about 1–1.5 mm long. The hybrids differ from *C. lupulinus* in having obconic spikes and ascending leafy bracts. They differ from *C. schweinitzii* in having smoother, not scabrous culms, a denser cluster of spikes, and spikes that are more flattened along the keels.

Cyperus squarrosus Linnaeus (p. 236)

Awned flatsedge

Cyperus aristatus Rottb., *Cyperus inflexus* Muhl.

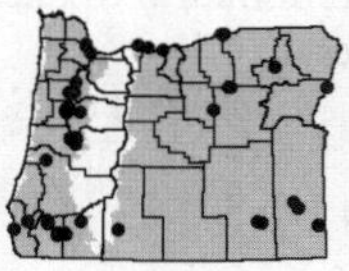

Plants annual, cespitose, with fibrous roots. **Culms** 1–20, trigonous, 2–16 cm × 0.5–0.8 mm. **Leaves** (1)5–10(15) × 0.5–2.5 mm. **Inflorescences** clusters usually densely head-like to cylindric, 1, loosely to densely ovoid to oblong, 6–20(40) × 9–15(20) mm; rays absent or 1–3(6), 0.4–4 cm; longest inflorescence bracts more or less erect. **Spikes** (2)6–20(40), ovoid-lanceoloid to oblong, somewhat compressed, 2.5–10(20) × 1.3–2.2 mm (excluding awns), greenish to reddish brown; rachilla more or less deciduous, wingless; floral scales deciduous, (4)10–20(34), oblong-lanceolate, (1.2)1.3–1.8(2.2) × (0.5)0.7–0.8(1) mm, greenish to stramineous or brownish red with greenish mid-stripe, (5)7–9(11)-ribbed nearly to margins, apex awned, the strongly outcurved awn an additional 0.5–1(1.3) mm. **Flowers** anthers 0.2–0.3 mm; styles 0.3–0.5 mm; stigmas 0.4–0.7 mm. **Achenes** obovoid (occasionally linear-spatulate or linear-oblong), 0.7–0.8(1.1) × (0.2)0.3–0.4(0.5) mm, stipe 0.05–0.1 × 0.1 mm.

Lake shores and river margins exposed as water falls, puddles, disturbed shallow wetlands. 0–1300 m. BR, BW, Col, CR, ECas, Lava, Owy, Sisk, WV. CA, ID, NV, WA; throughout US and adjacent Canada, south to South America; nearly cosmopolitan. Native.

Cyperus squarrosus is easily recognized by its small size, strongly outcurved scale awns, and many-veined scales. It is sometimes confused with *C. acuminatus,* which has 3-veined floral scales.

Cyperus strigosus Linnaeus (p. 236)

False nutsedge

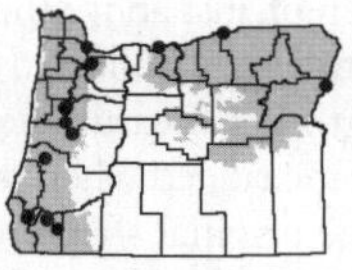

Plants perennial, often flowering first year, rhizomatous. **Culms** basally swollen and corm-like, trigonous, (1)20–40(90) cm × 1–6 mm. **Leaves** (10)20–40 cm × 1–4(8) mm. **Inflorescences** clusters loosely cylindric, 1–4, (6)10–28(50) × 10–40 mm; rays (0)3–6(8), 0.5–7(25) cm; inflorescence bracts ascending at 30–45°. **Spikes** (5)12–50, linear-lanceoloid, compressed, 5–30 × 0.6–0.9 mm; rachilla more or less deciduous, wings 0.4–0.5 mm wide; floral scales more or less deciduous, 3–11, appressed, oblong-obovate, 3.2–4.5(6) × 1–1.8 mm, stramineous to pale brown with green mid-stripe, lateral ribs 3–4, medial ribs (1)3; apex more or less acute, overlapping lower 25–50% of next scale. **Flowers** anthers 0.3–0.5 mm; styles 1–2 mm; stigmas 3–4 mm. **Achenes** more or less stipitate, narrowly oblong, (1.5)1.8–2.4 × 0.5–0.6 mm.

Lake shores and river margins exposed as water level falls, ditches, other disturbed wet soils, sometimes a weed in croplands. 0–600 m. BW, Col, CR, Sisk, WV. CA, ID, NV, WA; north to Saskatchewan, east to ME and FL. Native.

Cyperus strigosus is common in Oregon. Individuals that flower the first year are easily confused with *C. erythrorhizos*, *C. esculentus*, and *C. odoratus. Cyperus erythrorhizos* has shorter floral scales; *C. esculentus* has persistent floral scales; *C. odoratus* has subcylindric spikes.

DULICHIUM Three-way sedge

Herbs perennial, rhizomatous. **Culms** terete. **Leaves** cauline; ligules present; 3-ranked, proximal leaves bladeless, distal with well-developed blades, flat. **Inflorescences** axillary with 3–10 spikes; primary bracts leaf-like, each subtending an inflorescence. **Spikes** scales 4–8, distichously arranged, each subtending a flower. **Flowers** bisexual; perianth of 6–9 retrorsely barbed bristles, slightly longer than achene; stamens 3; style bases persistent, linear; stigmas 2. **Achenes** biconvex. **2*n***=32.

North America. 1 species.

Dulichium arundinaceum (Linnaeus) N.L. Britton var. ***arundinaceum*** (p. 239)

Three-way sedge

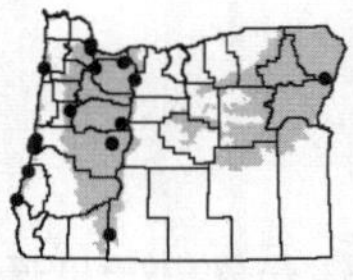

Rhizomes 3–5 mm thick; internodes 2–5 cm. **Culms** 2–14, erect, unbranched, jointed, aphyllopodic, terete, 20–100 cm × 2–5 mm; internodes hollow. **Leaves** sheaths green, often with white dots between the veins, with brown-hyaline 1–2 mm margin at mouth, firm; distal leaf blades 4–15 cm × 3.5–8 mm, more or less leathery, glabrous. **Inflorescences** 1–17, in sheath axils, 2–5 cm. **Spikes** 2-ranked, appressed-ascending to ascending-spreading, linear-lanceolate, 1.5–3 cm × 1.2–2.5 mm; rachilla winged, wings hyaline; floral scales decurrent, spongy-based, 5–9.5 × 1.4–2.4 mm, each subtending a flower. **Flowers** 2-ranked, sessile; scales green with gold-hyaline margins; perianth bristles 6–9, 4–7.5 mm, stiff; stamens 3, longer than perianth bristles, anthers 3–4 mm; styles 3.8–6 mm, base persistent on achene stigmas 2, 1–3.8 mm. **Achenes** stipitate, 2–4 × 0.8–1.1 mm (excluding style base), shorter than perianth bristles.

Bogs, lake margins, marshes. 0–2200 m. BW, Casc, Est, WV. CA, ID, WA; north to AK, east to SD; from Manitoba, south to TX, east to Atlantic Ocean. Native.

Dulichium arundinaceum has stiff, short leaf blades diverging in three directions from the jointed stems. Inflorescences arise in the axes of the leaves.

ELEOCHARIS Spikerush

Herbs annual or perennial, usually cespitose, often rhizomatous, sometimes stoloniferous; rhizomes rarely with terminal tubers or bulbs, horizontal and long or ascending and caudex-like, scales conspicuous to absent. **Culms** sometimes solitary, terete or 4-angled, smooth or ridged when dry. **Leaves** basal, 2 per culm; ligules absent; blades absent or a mucro or awn (tooth) at apex of sheath. **Inflorescences** terminal; spike 1; involucral bracts absent, rarely a proximal scale of spike resembling short bract. **Spikes** ovoid to cylindric, scales 4–500 or more, spirally arranged, each subtending a flower or the proximal 1–2(3) empty. **Flowers** bisexual; perianth of (0)3–6(10) bristles, straight or curved, shorter than to 2 times longer than achene, margins spinulose or smooth; stamens 1–3; style, bases forming a tubercle that is usually persistent, usually enlarged, usually different in appearance from achene; stigmas 2–3. **Achenes** ellipsoid, obovoid, or obpyriform, biconvex, plano-convex, or trigonous to subterete. **Tubercles** sharply distinct from or merging with the achene.

Cosmopolitan. Approximately 200 species; 16 species treated in *Flora.*

When collecting *Eleocharis,* get the underground parts and a lot of extra spikes! Identification usually requires ripping up spikes to view the rachilla and the flower, or lack thereof, just above the lower two floral scales. Most of these plants produce mature fruits in late summer or fall; they may not be identifiable if collected early in the season.

For the purposes of this key, achene length is a measurement of the achene only, without the tubercle (persistent style base). Culms that are smooth when fresh may become ridged when dry. Culms of pressed specimens become flattened even if the culms were quadrangular in life, so determining the original shape requires rehydrating a section of culm. Culm widths are for culms pressed flat. Widths of floral scales are for flattened scales and can be determined by measuring the distance from midrib to one margin and doubling that. The *proximal* scale is the lowest in the spike; the *subproximal* scale is the second lowest.

Reports of *E. coloradoensis* in Oregon remain unsubstantiated. If found, this species would key to either *E. quinqueflora* or *E. suksdorfiana*, however, unlike these species, the achenes of *E. coloradoensis* are less than 1.3 mm long.

Key to *Eleocharis* species:

1. Stems acutely rectangular in cross section; floral scales with 15+ prominent to obscure longitudinal veins; spikes cylindric, pale, (15)20–76 mm long . . . . . ***E. quadrangulata***

1' Stems round to flattened in cross section; floral scales with 1(10) widely spaced longitudinal veins; spikelets ovoid to cylindric, dark, 1.5–25(40) mm long.

  2. Culms 0.2–0.5(0.7) mm wide; achenes with 9–13 rows of fine horizontal ridges between much more prominent longitudinal ridges.

    3. Plants annual (rarely perennial), tufted (rarely rhizomatous), 1–7 cm tall; floral scales 1–1.5 mm long; anthers 0.3–0.5 mm; achenes (0.55)0.65–0.75 mm, with 20–30 fine horizontal ridges . . . . . ***E. bella***

    3' Plants perennial, rhizomatous, 1–60 cm tall; floral scales 1.5–2.5(3.5) mm; anthers 0.7–1.5 mm; achenes 0.7–1.1 mm, with 30–60 fine horizontal ridges . . . . . ***E. acicularis***

  2' Culms 0.3–5 mm wide; achenes without fine horizontal ridges.

    4. Plants annual, tufted.

      5. Tubercles not more than 25% as high as achene, 10–40% as high as wide. . . . . ***E. engelmannii***

      5' Tubercles 25% or more as high as achene, 33% as high as wide.

        6. Tubercles 0.3–0.5 mm wide; spikes ovoid; Willamette Valley & Columbia Plateau . . . . . ***E. ovata***

        6' Tubercles 0.5–0.8 mm wide; spikes broadly ovoid to lanceoloid (or ellipsoid or subcylindric); widespread. . . . . ***E. obtusa***

    4' Plants perennial, cespitose, rhizomatous, or stoloniferous.

      7. Tubercles continuous with the achene (though perhaps marked off by color changes).

        8. Culms to 9 cm tall; floral scales 1.4–2.7 mm long; achenes 0.9–1.2 mm long; anthers 0.7–1.2 mm long . . . . . ***E. parvula***

        8' Culms 5–100 cm tall; floral scales 2.5–6 mm long; achenes 1.5–2.5 mm long; anthers 1.5–3.5 mm long.

          9. Culms often arching and rooting at the tips; rhizomes absent; rachilla internodes at base and middle of spike similar in height and thickness . . . . . ***E. rostellata***

          9' Culms erect, not rooting at the tips; rhizomes present; basal rachilla internodes shorter and thicker than middle rachilla internodes.

            10. Proximal floral scale usually with a flower; perianth bristles unequal, shorter than or reaching the top of the tubercle; culms 0.2–0.5(1.2) mm thick; achenes 1.6–2.3 mm long. . . . . ***E. quinqueflora***

            10' Proximal floral scale usually without a flower; perianth bristles equal, usually much exceeding the tubercle; culms 0.5–1.2 mm wide; achenes 1.8–2.7 mm. . . . . ***E. suksdorfiana***

      7' Tubercles set off from achene by a well-marked constriction or rim.

11. Stigmas mostly 2 (3); achenes biconvex (trigonous), smooth or finely rugulose at 10–20×; plants with evident rhizomes.
    12. Proximal scale of spikes clasping 67(75)% of culm; subproximal scale always empty; culms 0.5–5 mm wide . . . . ***E. palustris***
    12' Proximal scale of spikes clasping 75–100% of culm; subproximal scale subtending a flower or empty; culms 0.3–2.5(3.5) mm wide.
        13. Proximal scale of all spikes clasping 95–100% of the culm; subproximal scale always subtending a flower; culms 0.3–1.4 mm wide; range southeastern Oregon . . . . ***E. erythropoda***
        13' Proximal scale of some or all spikes clasping less than 95% of culm; subproximal scales of at least some spikes empty; culms 0.5–2.5(3.5) mm wide; range throughout Oregon . . . . ***E. macrostachya***

11' Stigmas mostly (2)3; achenes trigonous or subcircular in cross section, or sometimes biconvex, smooth to prominently variously sculptured at 10×; habit various.
    14. Distal leaf sheath often with an abrupt tooth to about 1 mm long; rhizomes generally evident with longer internodes 5–30 mm; spikes 3–20 mm . . . . ***E. parishii***
    14' Distal leaf sheath lacking tooth; rhizomes more or less hidden by culms and roots, with longer internodes 5 mm; spikes 3–8 mm.
        15. Culms 0.3–2 mm wide; rhizome scales 20–25 mm, papery, disintegrating to fibers; floral scales 3–3.5 mm; perianth bristles mostly greater than or equal to tubercles; tubercles well developed, often about as high as wide; habitat fens in southwestern Oregon . . . . ***E. decumbens***
        15' Culms 0.3–0.5 mm wide; rhizome scales not evident; floral scales 2–3 mm; perianth bristles to about 50% as long as achenes; tubercles much lower than wide, often nearly flat; habitat seasonally dry stream channels in eastern Oregon . . . . ***E. bolanderi***

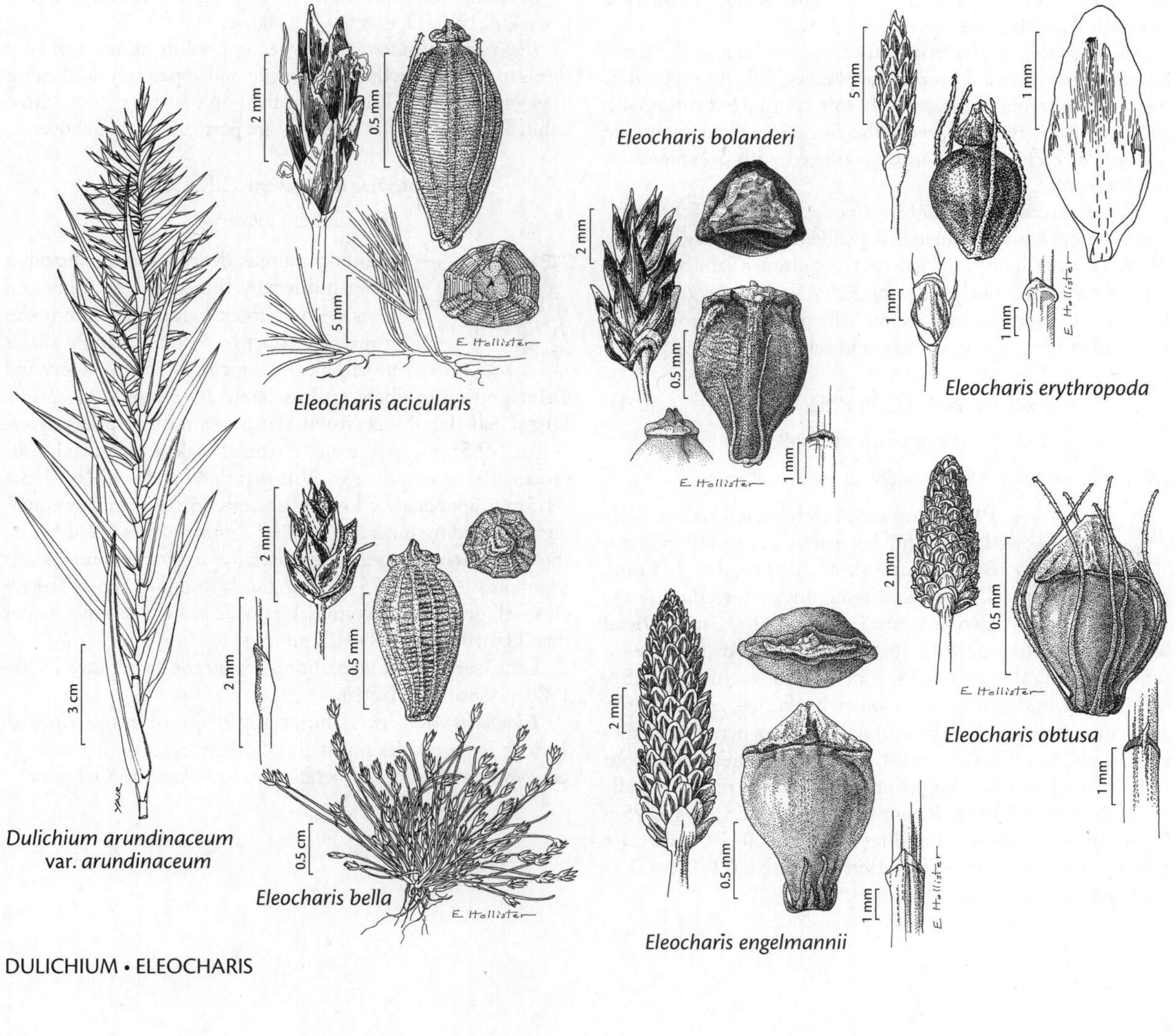

Dulichium arundinaceum var. *arundinaceum*
Eleocharis acicularis
Eleocharis bella
Eleocharis bolanderi
Eleocharis engelmannii
Eleocharis erythropoda
Eleocharis obtusa

Eleocharis acicularis (Linnaeus) J.J. Roemer & J.A. Schultes (p. 239)

Needle spikerush

Eleocharis acicularis (L.) Roem. & Schult. var. *acicularis*, *Eleocharis acicularis* (L.) Roem. & Schult. var. *occidentalis* Svenson

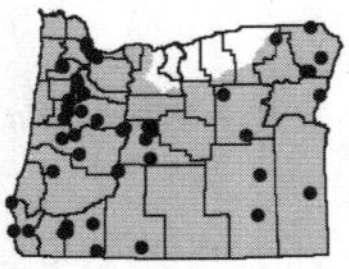

Plants perennial; rhizomes 0.25–0.5 mm thick, internodes 5–15 mm, sometimes with resting buds. **Culms** sometimes arching, terete to sometimes distinctly compressed, 1–60 cm × 0.2–0.5(0.7) mm. **Distal leaf sheaths** often splitting, apex rounded (to acute). **Spikes** 2–8 × 1–2 mm; floral scales 4–25, 4–6 per mm of rachilla, 1.5–2.5(3.5) × 1–1.5 mm, apex blunt to acute. **Flowers** perianth bristles mostly absent, uncommonly 2–4, shorter than to equaling achene; anthers 0.7–1.5 mm; stigmas 3. **Achenes** with angles plus about 8–12 obscure to prominent longitudinal ridges, narrowly to broadly obovoid to obpyriform, 0.7–1.1 × 0.35–0.6 mm, fine horizontal ridges 30–60, clearly evident to crowded and obscure, spaces between trabeculae sometimes translucent. **Tubercles** pyramidal to much depressed, (0.05)0.1–0.2 × 0.15–0.25 mm. **2*n*=20.**

Ponds, shallow wetlands, and exposed shorelines and mudflats, vernal pools, and other, often disturbed wetlands. 0–2300 m. All ecoregions except Col. CA, ID, NV, WA; north to AK, east to Greenland and GA, south to South America; Australia, Eurasia. Native.

Plants of this species frequently grow as submerged or floating mats that do not flower. These plants differ from typical *E. acicularis* in having terete, smooth, soft, translucent culms with conspicuous partitions between the air cavities. Such plants are difficult to identify and can be confused with *Schoenoplectus subterminalis*.

Eleocharis acicularis is highly variable, but much of the variation is probably environmental, dependent particularly on water depth. As a result of this plasticity, recognition of any of the described varieties is likely unwarranted. *Eleocharis bella* is similar but annual, cespitose, only occasionally produces rhizomes and has smaller floral scales, anthers, and achenes.

Eleocharis bella (C.V. Piper) Svenson (p. 239)

Delicate spikerush, pretty spikerush

Eleocharis acicularis (L.) Roem. & Schult. var. *bella* Piper

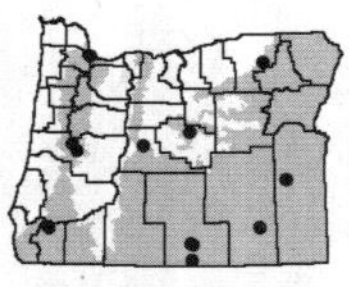

Plants annual, rarely perennial, usually densely tufted; rhizomes rarely evident, 0.2–0.3 mm thick, internodes 1–5 mm. **Culms** often ascending or spreading, 4-angled or terete 1–7 cm × 0.2–0.3 mm. **Distal leaf sheaths** often splitting, apex acute. **Spikes** ovoid, 1.5–4 × 0.8–2 mm; floral scales 4–15, 8 per mm of rachilla, 1–1.5 × 0.5–0.7 mm, midrib obscure to somewhat keeled; apex narrowly acute to acuminate, slightly recurved. **Flowers** perianth bristles absent; anthers 0.3–0.5 mm; stigmas 3. **Achenes** with about 6–10 rather prominent angles and longitudinal ridges, broadly ovoid, less than 2 times longer than wide, (0.55)0.65–0.75 × 0.3–0.4 mm, fine horizontal ridges distinct, 20–30. **Tubercles** mostly appressed, pyramidal, often depressed, 0.1–0.2 × 0.1–0.25 mm.

Exposed lake margins and streamsides, wet meadows. 0–2100 m. BR, BW, ECas, Owy, Sisk, WV. CA, ID, NV, WA; northeast to MT, southeast to NM, south to Mexico. Native.

Eleocharis bella has sometimes been considered a variety of *E. acicularis*, which is a rhizomatous perennial with larger floral scales, anthers, and achenes. Hybrids between the two are unknown.

Eleocharis bolanderi A. Gray (p. 239)

Bolander's spikerush

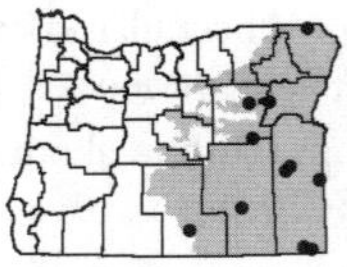

Plants perennial, densely cespitose; rhizomes caudex-like, mostly hidden by culms and roots, short, 1.5–3 mm thick, internodes very short. **Culms** subterete, 10–30 cm × 0.3–0.5 mm. **Distal leaf sheaths** persistent, not splitting, apex obtuse, rarely hardened, lacking a tooth. **Spikes** ovoid, 3–8 × 2–3 mm; proximal scale clasping, entire; subproximal scale with flower; floral scales 8–30, 4–5 per mm of rachilla, 2–3 × 1.5 mm, apex entire, acute, often keeled in distal part of spike. **Flowers** perianth bristles 3–6, often unequal, from rudimentary to 50% of achene length; anthers 0.9–1.4 mm; stigmas 3. **Achenes** slightly to greatly compressed-trigonous, rarely thickly lenticular, 0.9–1.2 × 0.65–0.8 mm. **Tubercles** flat to shallowly pyramidal, lower than wide, 0.1–0.3 × 0.4–0.65 mm.

Shallow, rocky, ephemeral streams. 1100–2100 m. BR, BW, Owy. CA, NV, ID; east to CO. Native.

Eleocharis bolanderi is easily recognized in summer when it forms masses of dry, brown, dormant and apparently dead culms in dry stream beds. It is easily pulled up to expose its cespitose habit. The virtually flat tubercles are particularly distinctive.

Eleocharis decumbens C.B. Clarke

Decumbent spikerush

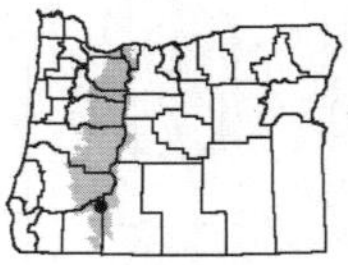

Plants perennial, densely tufted; rhizomes often hidden by the crowded culms and roots, 3–4 mm thick, hard, longer internodes from very short to 5 mm; rhizome scales usually clearly evident, 2–2.5 cm, papery and disintegrating to fibers. **Culms** terete 10–50 cm × 0.3–2 mm. **Distal leaf sheaths** persistent, lacking a tooth. **Spikes** ovoid, 3–8 × 2–2.5 mm, apex acute; proximal scale amplexicaul; subproximal scale empty or with flower; floral scales 10–20, 3–3.5 × 1.5 mm, apex acute. **Flowers** perianth bristles 6, mostly equaling or exceeding tubercle, (0.5)1–2.2 mm; anthers 1.2–1.5 mm; stigmas 3. **Achenes** nearly equilaterally- to greatly compressed-trigonous, 1–1.3 × 0.75–0.9 mm, neck absent or short. **Tubercles** well developed, pyramidal, as wide as high to much wider than high, 0.2–0.6 × 0.4–0.7 mm.

Fens, seeps, and lake shores, montane to alpine. 1500–1700 m. Casc. CA. Native.

Eleocharis decumbens is most easily identified by the crowded, 2 cm or longer scales on its thick rhizomes. It occurs with *E. suksdorfiana* in montane wetlands in southwestern Oregon.

Eleocharis engelmannii E.G. Steudel (p. 239)

Engelmann's spikerush

Eleocharis engelmannii Steud. var. *monticola* (Fernald) Svenson, *Eleocharis obtusa* (Willd.) Schult. var. *engelmannii* (Steud.) Gilly

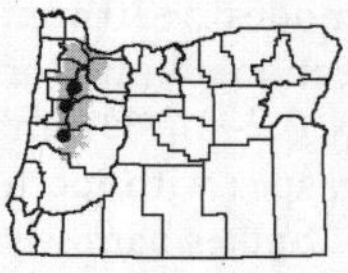

Plants annual, tufted. **Culms** terete, 2–40 cm × 0.5–1.5(2) mm. **Distal leaf sheaths** persistent, apex of distal leaf sheath obtuse to acute, with tooth to 0.3 mm. **Spikes** lanceoloid to subcylindric, 5–10(20) × 2–3(4) mm; proximal scale empty, encircling approximately 67% of culm; floral scales 25–100(200), 8–12 per mm of rachilla, 2(2.5) × 1–1.3 mm, apex narrowly rounded to subacute. **Flowers** perianth bristles present or often absent, (0)5–8, about as long as the achenes; anthers 0.3–0.7(1) mm; stigmas 2–3. **Achenes** biconvex or to 33% greatly compressed trigonous, 0.9–1.1(1.5) × 0.7–1.1 mm. **Tubercles** depressed, subdeltoid, 0.1–0.3(0.4) × 0.6–0.9(1) mm, 10–40% as high as wide, 25% or less as high as achene, 90% as wide as achene.

Freshwater shores exposed by seasonal low water levels, marshes, disturbed wetlands. 50–500 m. WV. CA, ID, WA; north to British Columbia, east to MA, southeast to AL. Native.

Eleocharis engelmannii is similar to *E. ovata* and the much more common *E. obtusa* but has markedly shorter tubercles and usually more cylindric spikes.

Eleocharis erythropoda E.G. Steudel (p. 239)

Bald spikerush

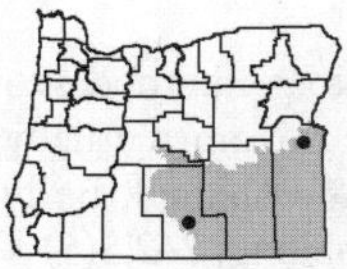

Plants perennial, mat-forming; rhizomes evident, long, 0.5–1.5 mm thick, soft to firm, cortex often breaking loose, longer internodes 5–30 mm. **Culms** terete, 8–80 cm × 0.3–1.4 mm. **Distal leaf sheaths** persistent, apex broadly obtuse to subacute, tooth sometimes present, to 0.1 mm. **Spikes** ovoid to lanceoloid or nearly cylindric, 3–18 × 2–3(4) mm; proximal scale clasping all or nearly all of culm; subproximal scale with flower; floral scales 15–50, 4–5 per mm of rachilla, 2–3.5 × 1.5–1.7 mm, in proximal part of spike apex rounded, in distal part apex entire, acute. **Flowers** perianth bristles 4 or absent, usually equal, equaling achene to slightly exceeding tubercle; anthers 1–1.8 mm; stigmas 2. **Achenes** biconvex, 0.9–1.6 × 0.7–1.2 mm **Tubercles** pyramidal, much higher than wide to lower than wide, 0.35–0.65 × 0.2–0.6 mm. **2*n***=16, 18, 19, 20.

Marshes, wet meadows, fens, disturbed wetlands. 500–1500 m. BR, Owy. ID, WA; north to AK, east to New Brunswick and NC, southeast to AZ and MS. Native.

Eleocharis erythropoda is the most delicate member of the *E. palustris* complex. Its proximal scales consistently clasp the entire culm, and its subproximal scales consistently subtend flowers. Intermediates between *E. erythropoda* and *E. macrostachya* may be more common in Oregon than pure *E. erythropoda.*

Eleocharis macrostachya N.L. Britton

Confusing spikerush, pale spikerush

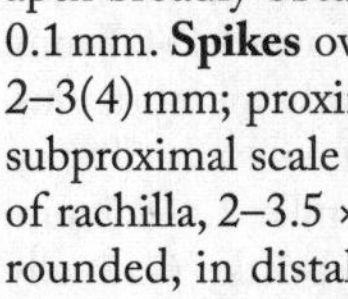

Plants perennial; rhizomes evident, long, 1–2 mm thick, firm, cortex persistent, longer internodes 10–55 mm. **Culms** terete to markedly compressed, to 3 times wider than thick, 10–100 cm × 0.5–2.5(3.5) mm. **Distal leaf sheaths** persistent, apex truncate to obtuse, tooth sometimes present on some or all culms, 0.1–0.6(1) mm. **Spikes** narrowly lanceoloid to ovoid, 5–40 × 2–5 mm; proximal scale clasping (6)75(100)% of culm usually variably in same plant; subproximal scale usually empty in some spikes and subtending a flower in other spikes in the same clone; floral scales 30–80, 3–5 per mm of rachilla, 2.5–5.5 × 1.5–2.5 mm. **Flowers** perianth bristles 4(5), sometimes rudimentary or absent, much shorter than achene to equaling tubercle; stamens 2(3), anthers 1.3–2.7 mm; stigmas 2. **Achenes** biconvex to plano-convex, 1.1–1.9 × 0.8–1.5 mm. **Tubercles** pyramidal, as high as or sometimes much higher than wide, 0.35–0.7 × 0.25–0.7 mm. **2*n***=18, 19, 38.

Fresh to slightly brackish or alkaline marshes, lake shores, stream beds, vernal pools, deflation plains, and wet meadows. 0–1800 m. All ecoregions except Casc. CA, ID, NV, WA; north to AK, east to Quebec and MS, south to Mexico; South America. Native.

Eleocharis macrostachya is a diploid-polyploid complex consisting (at least in part) of populations that originated as hybrids of *E. palustris* with *E. erythropoda* and *E. uniglumis.* As a result, *E. macrostachya* is one of the most variable species in the genus. Plants vary from short, narrow-culmed plants resembling *E. erythropoda* to fairly robust individuals resembling *E. palustris.* The extent to which the proximal scale clasps the culm is usually greater than in *E. palustris* and often varies within a clone, as does the present or absence of a flower in the subproximal scale.

Strictly coastal individuals have shiny dark chestnut brown or blackish achenes with brownish midribs. The proximal scales usually clasp 95–100% of the culm, and the subproximal scales usually subtend flowers. The culms are tough and sometimes arching. These plants can be referred to as *E. macrostachya* variant C (coastal form) or *E. macrostachya* tending toward *E. uniglumis,* according to Smith et al. (2002). Most inland Oregon plants are referred to as *E. m.* variant B.

Eleocharis obtusa (Willdenow) J.A. Schultes (p. 239)

Blunt spikerush

Eleocharis obtusa (Willd.) Schult. var. *gigantea* (C.B. Clarke) Fernald, *Eleocharis obtusa* (Willd.) Schult. var. *obtusa*

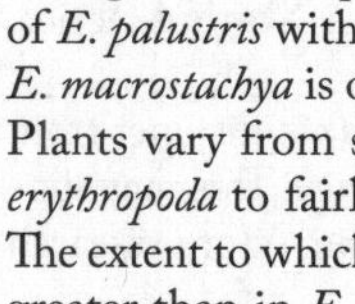

Plants annual, tufted. **Culms** terete, 3–50(90) cm × 0.2–2 mm. **Distal leaf sheaths** persistent, apex of distal leaf sheath obtuse to acute, tooth to 0.3 mm. **Spikes** usually broadly ovoid, (2)5–13 × (2)3–4 mm; floral scales 15–150+, 8–20 per mm of rachilla, 1.5–2.5 × 1–1.5 mm, apex broadly rounded. **Flowers** perianth bristles (5)6–7, rarely 0, slightly to usually greatly exceeding tubercle; stamens (2)3, anthers 0.3–0.6 mm; stigmas usually 3 and 2 in the same spike. **Achenes** biconvex or to 33% greatly compressed trigonous, 0.9–1.2(1.3) × 0.7–0.9 mm. **Tubercles** deltoid 0.35–0.5 × (0.4)0.5–0.8 mm, 33–50% as high as wide, 33–50% as high and 67–90% as wide as achene. **2*n***=10.

Freshwater shores exposed by seasonal low water levels, marshes, and disturbed wetlands. 0–1300 m. BW, Casc, Col, CR, ECas, Est, Sisk, WV. CA, ID, WA; north to British Columbia, east to Nova Scotia, southeast to FL, HI. Native.

Eleocharis obtusa is our most common tough, annual, cespitose spikerush. It is sometimes considered to be the same species as *E. ovata,* which generally has two stamens and two stigmas and has narrower tubercles. See also *E. engelmannii.*

Eleocharis ovata (A.W. Roth) J.J. Roemer & J.A. Schultes (p. 243)

Ovoid spikerush

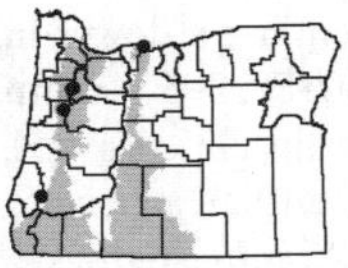

Plants annual, tufted. **Culms** terete, 2–35 cm × 0.3–1 mm. **Distal leaf sheaths** persistent, apex of distal leaf sheath obtuse to acute, tooth to 0.2 mm. **Spikes** ovoid, 2–8 × 2–4 mm; floral scales 25–100+, approximately 10 per mm of rachilla, 1.5–2 × 1 mm, midribs often keeled in distal part of spike, apex rounded to subacute. **Flowers** perianth bristles present, rarely absent, (5)6–7, exceeding tubercle; stamens 2(3), anthers 0.3 mm; stigmas 2 or some 3. **Achenes** biconvex or to 33% greatly compressed trigonous, 0.75–1 × 0.6–0.85 mm. **Tubercles** deltoid, 0.3–0.5 × 0.3–0.5 mm, 60% to as high as wide, 33–67% as high and 50–75% as wide as achene. **2*n***=10.

Freshwater shores exposed by seasonal low water levels, marshes, and disturbed wetlands. 100–1300 m. ECas, Sisk, WV. WA; north to British Columbia; northeastern North America. Native?

Pacific Northwest occurrences of *E. ovata* are disjunct from the species northeastern North American range. This may suggest that the species is introduced in Oregon. More research is needed.

See also *E. obtusa.*

Eleocharis palustris (Linnaeus) J.J. Roemer & J.A. Schultes (p. 243)

Common spikerush, marsh spikerush, creeping spikerush

Eleocharis palustris (L.) Roem. & Schult. var. *major* Sond., *Eleocharis palustris* (L.) Roem. & Schult. var. *palustris*

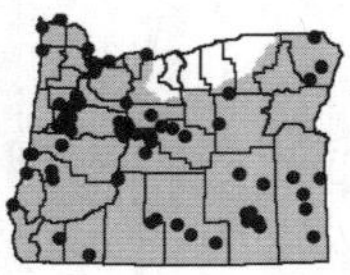

Plants perennial, mat-forming; rhizomes evident, long, 1.5–4.5 mm thick, cortex persistent, longer internodes 10–35 mm. **Culms** terete or slightly compressed, 30–115 cm × 0.5–5 mm. **Distal leaf sheaths** persistent or sometimes disintegrating, often splitting adaxially, apex lacking tooth. **Spikes** ovoid to lanceoloid, 5–25 × 3–7 mm; proximal scale clasping 67(75)% of culm; subproximal scales 1–2, empty; floral scales 30–100, 4–8 per mm of rachilla, 3–5 × 1.5–2.5 mm, apex entire, acute or subacute, often keeled in distal part of spike. **Flowers** perianth bristles 4(5), sometimes absent, much shorter than achene to equaling tubercle, rarely to 2 times as long as achene; anthers 1.5–2.2 mm; stigmas 2, very rarely some 3. **Achenes** biconvex, 1.1–2 × 1–1.5 mm. **Tubercles** pyramidal to mammillate, as high as wide to 2 times higher, 0.3–0.7 × 0.35–0.7 mm. **2*n***=16, 27, 36.

Freshwater marshes, lake shores, streamsides, and ponds. 0–2300 m. All ecoregions except Col. CA, ID, NV, WA; north to AK, east to Newfoundland, south to Mexico; Eurasia, New Zealand. Native.

Eleocharis palustris is the most common species in a taxonomically difficult, circumboreal complex of the genus. Its pure, rhizomatous stands can dominate plant communities in shallow wetlands. It can usually be recognized by its subequal proximal and subproximal scales, both lacking flowers. Culm widths overlap those of *E. macrostachya,* but the very widest spikerush culms in Oregon belong to *E. palustris.*

Eleocharis parishii N.L. Britton (p. 243)

Parish's spikerush

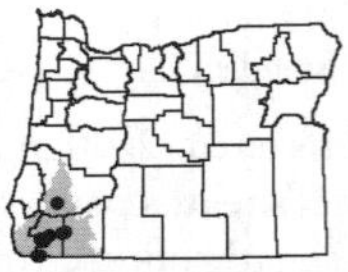

Plants perennial, mat-forming; rhizomes evident, long, 0.5–1(2) mm thick, cortex persistent, longer internodes 5–30 mm. **Culms** terete or cross section elliptic (or rectangular), 10–50 cm × 0.2–0.7(1) mm. **Distal leaf sheaths** persistent, not splitting, apex with tooth often present on most or all culms, to 1 mm. **Spikes** narrowly lanceoloid, 3–20 × 1.5–2.5 mm; proximal scale amplexicaul, entire; subproximal scale empty or with flower; floral scales 15–40, 3–4 per mm of rachilla, 2–3 × 1 mm; apex entire, rounded to obtuse in proximal part of spike, acute in distal part, keeled, at least in distal part of spike. **Flowers** perianth bristles (0)3–7, often unequal, rudimentary to slightly exceeding tubercle; anthers 1.1–2 mm; stigmas 3. **Achenes** compressed-trigonous, 0.8–1.4 × 0.5–0.7 mm. **Tubercles** pyramidal, often higher than wide, 0.3–0.4 × 0.25–0.35 mm. **2*n***=10.

Fresh to brackish wetlands, often on drying lake shores, pond margins, streamsides and springs. 200–500 m. Sisk. CA, NV; east to Kansas, south to Mexico. Native.

Eleocharis parvula (J.J. Roemer & J.A. Schultes) Link ex Bluff, Nees & J.C. Schauer (p. 243)

Small spikerush

Eleocharis parvula (Roem. & Schult.) Link ex Bluff, Nees & Schauer var. *parvula*, *Scirpus nanus* Spreng.

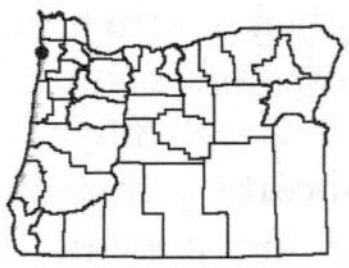

Plants perennial, rhizomatous, internodes to 4 cm; tubers terminating rhizomes usually markedly J- or horseshoe-shaped, body (apart from apical bud) oblong, 2–2.5(5) × 0.5–1 mm; tubers among culm bases straight, narrowly fusiform, 4–5 mm. **Culms** terete, 2–9 cm × 0.2–0.5 mm. **Distal leaf sheaths** often disintegrating, thinly membranous, apex rounded. **Spikes** 2–4 × 1–2 mm; proximal scale 50% or more of spike length; floral scales 6–10 per spike, 1.4–2.7 mm, apex rounded to subacute. **Flowers** perianth bristles 6, usually equaling achene to slightly exceeding tubercle, sometimes unequal and some 50% of achene, very rarely rudimentary, anthers 0.7–1.2 mm; stigmas (2)3. **Achenes** thickly trigonous, 0.9–1.2 × 0.55–0.75 mm. **Tubercles** 0.1–0.2 × 0.15 mm.

Brackish or saline tidal marshes and other coastal wetlands. 0–50 m. Est. CA, WA; throughout much of North America; Eurasia. Native.

When growing submerged in deeper water, the plants may not produce spikes and can be confused with *E. acicularis* and *Schoenoplectus subterminalis.*

Eleocharis quadrangulata (A. Michaux) J.J. Roemer & J.A. Schultes

Square-stemmed spikerush, four-angled spikerush

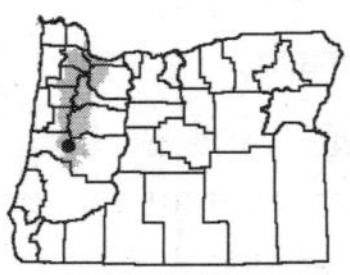

Plants perennial; rhizomes 1.5–4 mm thick, longer internodes 3–8 cm, tubers absent. **Culms** acutely quadrangular, (30)45–105 cm × (1)2–5.4 mm. **Distal leaf sheaths** persistent, apex narrowly acute to acuminate, sometimes prolonged into a blade-like portion to 8 cm. **Spikes** (15)20–76 × 3–5(6) mm; proximal scale empty, amplexicaul,

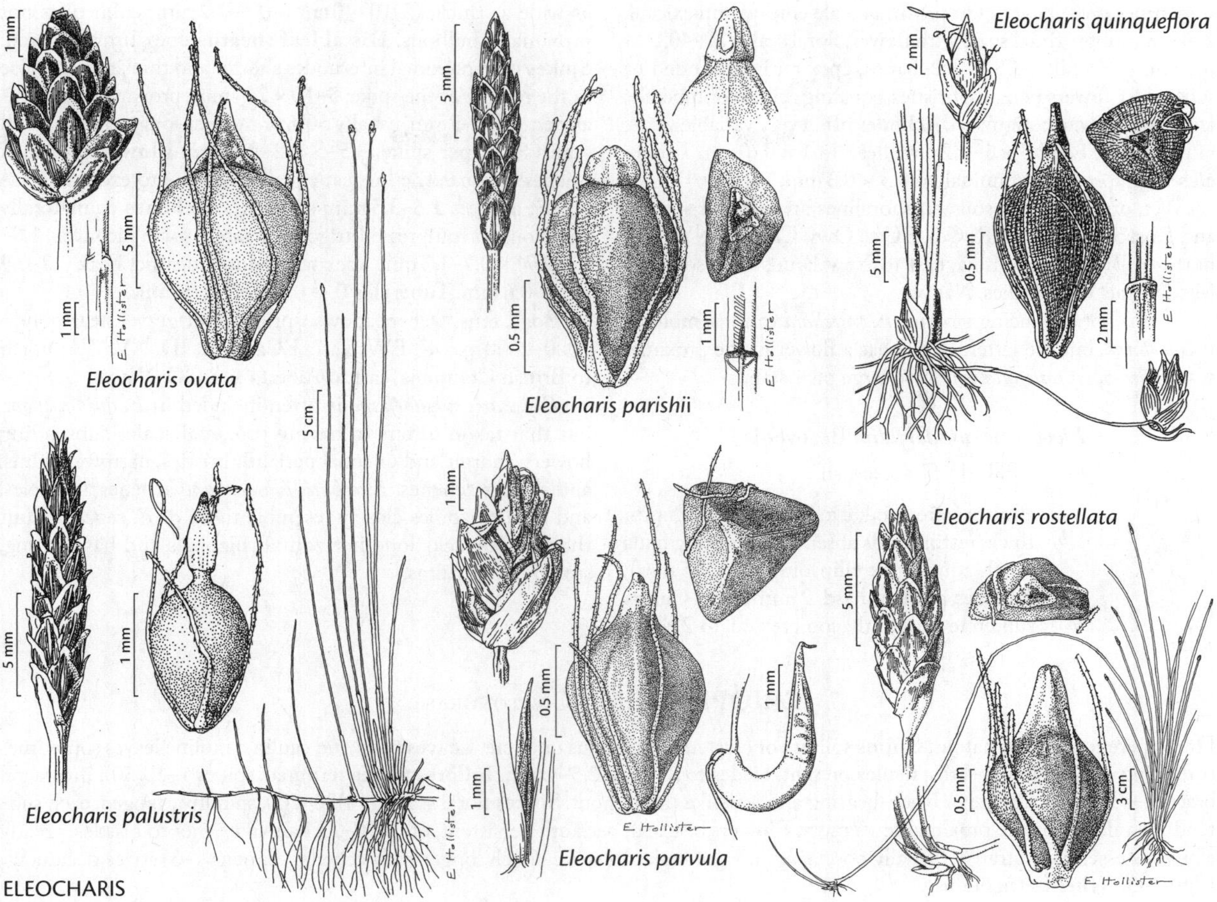

ELEOCHARIS

(1)2.2–5.4 mm; floral scales (28)45–135, 2–3 per mm of rachilla, (4)4.5–6.2 × 2.8–5 mm, apex rounded to obtuse. **Flowers** perianth bristles 6–7, often markedly unequal, shorter than achene or some equaling tubercle; anthers 2.3–2.9 mm; stigmas 3, sometimes 2. **Achenes** biconvex, 1.8–3 × 1.3–2 mm, almost smooth to markedly sculptured at 10–15×, each face with 19–38 rows of almost linear, transversely elongated cells, which are sometimes isodiametric at achene base. **Tubercles** deltoid to high-pyramidal or lanceoloid, 0.7–1.5 × 0.4–1 mm, often spongy.

Shallow water of freshwater marshes, ditches, wet meadows, and lake shores. 100–200 m. WV. CA; south to Mexico; eastern North America. Exotic.

Eleocharis quadrangulata is a robust spikerush with long, pale spikes and rectangular culms.

Eleocharis quinqueflora (F.X. Hartmann) O. Swartz (p. 243)

Few-flowered spikerush

Eleocharis pauciflora (Lightf.) Link, *Eleocharis pauciflora* (Lightf.) Link var. *pauciflora*, *Eleocharis pauciflora* (Lightf.) Link var. *suksdorfiana* (Beauverd) Svenson, *Scirpus pauciflorus* Lightf.

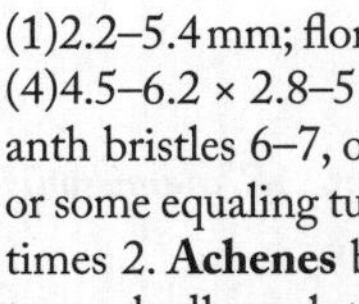

Plants perennial, rhizomatous; rhizomes 0.2–1 mm thick; resting buds often present on rhizomes or among culm bases, broadly to narrowly ovoid, 3–6(10) × 2–5 mm; caudices absent, rarely present, soft or rarely hard, 0.5 mm thick. **Culms** subterete to slightly compressed, to 2 times wider than thick, 5–15 cm × 0.2–0.5(1.2) mm; culm tufts often proximally bulbous (if bulbous then tunicated by papery-fibrous scales). **Distal leaf sheaths** apex often reddish, subtruncate to acute. **Spikes** with proximal internodes shorter and thicker than those in the middle of the spike; 3–8 × 1.5–4 mm; proximal scale subtending a flower, seldom empty, 2–5 mm, 50% or more as long as spike; floral scales 3–10 per spike, 2.5–6 × 1.5–2.5 mm. **Flowers** perianth bristles (0)3–6, often unequal, rudimentary to equaling tubercle; anthers 1.5–2.7(3.5) mm; stigmas 3. **Achenes** equilaterally trigonous to compressed-trigonous, rarely some biconvex, 1.6–2.3 × 0.7–1.3 mm. **Tubercles** rarely absent, 0.3–0.4 × 0.2–0.3 mm.

Fens, wet meadows, seeps, springs, hot springs. 500–2900 m. BR, BW, Casc, ECas, Owy, Sisk. CA, ID, NV, WA; north to AK, east to Greenland and DE, southeast to NE and NM; Eurasia. Native.

See also *E. suksdorfiana*.

Eleocharis rostellata (J. Torrey) J. Torrey (p. 243)

Walking spikerush, walking sedge

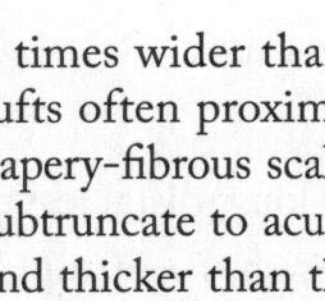

Plants densely tufted, stoloniferous, mat-forming by means of rooting culm tips. **Culms** 1.5–3 times as wide as thick, 20–100 cm × 0.35–2 mm, wiry; some culms arching or decumbent and rooting at tips. **Distal leaf sheaths** not splitting, apex usually reddish. **Spikes** 5–17 × 2.5–5 mm; spikes on stolons rudimentary and sometimes producing bulbs; proximal scale empty, amplexicaul, 2–4 mm; subproximal scale with flower; floral scales 20–40, 2–3 per mm of rachilla, 3.5–6 × 2–3 mm, apex entire, rounded to subacute. **Flowers** perianth bristles, equaling achene or tubercle; anthers 2–2.4 mm; stigmas 3. **Achenes** often very variable within one plant, 1.5–2.5 × 1–1.2 mm, beak to 1 × 0.6 mm. **Tubercles** when present pyramidal, to 0.5 × 0.3 mm.

Wet, often alkaline soils on shorelines and around springs and fens. 50–1500 m. BR, Col, ECas, Owy. CA, ID, NV, WA; north to British Columbia, east to New Brunswick, south to Mexico and West Indies. Native.

When not producing stolons, *E. rostellata* can resemble *E. suksdorfiana*, but the latter species has a flower in the proximal scale. *Eleocharis rostellata* can form large pure stands.

Eleocharis suksdorfiana Beauverd

Suksdorf's spikerush

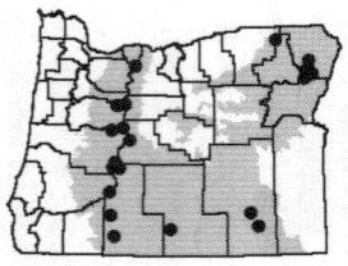

Plants perennial; rhizomes 0.5–1.5 mm thick; resting buds absent (non-resting buds on rhizome apex ellipsoid, 10 × 2–5 mm); caudices present, hard, 2 mm thick. **Culms** subterete to slightly compressed, to 2 times as wide as thick, (5)10–40 cm × 0.5–1.2 mm; culm tufts not proximally bulbous. **Distal leaf sheaths** apex brown to red. **Spikes** with proximal internodes shorter and thicker than those in the middle of the spike; 5–10 × 2–4 mm; proximal scale usually empty, 3–5 mm, usually 50% or more as long as spike; floral scales 8–12 per spike, 3.5–5 × 2–2.5 mm. **Flowers** perianth bristles 6, equal, the longest equaling achene to exceeding tubercle; anthers 1.6–3.5 mm; stigmas 3. **Achenes** equilaterally trigonous to compressed trigonous, rarely some biconvex, 1.8–2.2(2.7) × 0.7–1.3 mm, apex tapered to a distinct beak 0.2–0.9 × 0.3–0.6 mm. **Tubercles** 0.4–0.5 × 0.3–0.5 mm.

Bogs, fens, wet meadows, springs, wet gravel near ponds. 1000–2600 m. BR, BW, Casc, ECas. CA, ID, NV, WA; north to British Columbia, east to Alberta and CO. Native.

Eleocharis suksdorfiana is often included in *E. quinqueflora*, but that taxon differs in having proximal scales subtending flowers, shorter and unequal perianth bristles, narrower culms, and shorter achenes. *Eleocharis suksdorfiana* achenes, tubercles, and hard rhizomes closely resemble those of *E. rostellata*, but that species lacks long horizontal rhizomes and has arching, stoloniferous culms.

ERIOPHORUM Cottongrass

Herbs perennial, rhizomatous. **Culms** solitary or clustered, trigonous or terete. **Leaves** basal and cauline; cauline leaves sometimes consisting of bladeless sheaths; ligules present; blades to 25 cm × 2.5–4 mm. **Inflorescences** terminal, spikes 1–10(30); involucral bracts 1–several, scale-like or blade-bearing and leaf-like throughout. **Spikes** floral scales (10)20–200, spirally arranged, each subtending a flower or the proximal few empty. **Flowers** bisexual; perianth persistent, of (8)10–25 hair-like, smooth bristles, greatly elongate, essentially straight, usually obscuring most of scales in spike, much longer than achene; stamens 1–3; styles deciduous, stigmas 3. **Achenes** trigonous.

Mostly in cool temperate, alpine, and arctic regions of the northern hemisphere. Approximately 25 species; 3 species treated in *Flora*.

Key to *Eriophorum* species:

1. Spikes solitary; bristles rusty, 20–40 mm long; distal leaves bladeless or with blades to 1 cm long . . . . ***E. chamissonis***
1' Spikes usually 2+; bristles white to yellowish, 10–30 mm long; distal leaves with blades more than 1 cm long.
   2. Anthers 2–5 mm long; longest inflorescence bract 2–12 cm long; longer floral scales 5–10 mm long . . . . ***E. angustifolium*** ssp. ***angustifolium***
   2' Anthers 1–2(2.5) mm long; longest inflorescence bract 1–2(2.5) cm long; longer floral scales 3–4.5 mm long . . ***E. gracile***

Eriophorum angustifolium Honckeny ssp. *angustifolium*

Tall cottongrass, many-spike cottongrass

Eriophorum polystachion L.

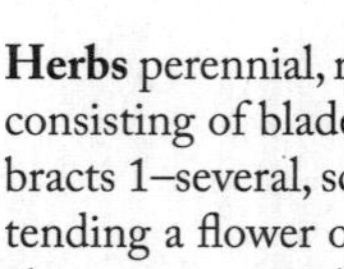

Culms 20–100 cm × (0.8)1–1.2 mm distally. **Leaves** blades flat, tip trigonous, channeled in cross section, to 40 cm × 1.5–6(8) mm; distal leaf blade much longer than sheath. **Inflorescences** blade-bearing involucral bracts 1–3, blade and often sheath black proximally, leaf-like, longest (2)3–7(12) cm. **Spikes** (1)2–10; floral scales lanceolate or ovate, 5–10 mm, brown to dark gray proximately, paler distally, usually with broad white margins; with prominent midrib fading proximal to tip, apex more or less acute. **Flowers** perianth bristles 15–30 mm, white or pale yellow brown; anthers 2–5 mm. **Achenes** oblanceoloid, 2–5 mm.

Marshes, bogs, fens, meadows, shores. 100–1900 m. Casc, ECas. ID, WA; north to AK, east across Canada and northern US to Greenland; CO, NM; Eurasia. Native.

Eriophorum chamissonis C.A. Meyer (p. 245)

Russet cottongrass

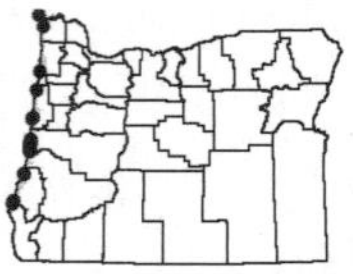

Culms somewhat trigonous, (20)30–70(80) cm, stout, to 1.5(4) mm thick basally. **Leaves** blades filiform, trigonous-channeled, (2)3–10 cm × 1–2 mm; distal 1–2 leaf sheaths bladeless. **Inflorescences** involucral bracts absent. **Spikes** 1; floral scales blackish to purplish brown; proximal 1–7 scales empty, obovate or trigonous-obovate, 4–20 mm, margins white or paler, as wide as dark portion, apex blunt; distal scales obovate to ovate-lanceolate, margins hyaline, 1 mm wide or wider. **Flowers** perianth bristles 20–40 mm, red-brown to white; anthers (0.6)1.5–3 mm. **Achenes** oblong-obovoid, 2–2.7 mm, as wide as long. **2*n***=58.

Coastal bogs and marshes. 0–50 m. Est. ID, WA; north to AK, east to WI, southeast to CO; Eurasia. Native.

Eriophorum chamissonis is a lovely plant with large, shiny, dense heads of long, red-brown or bronze hairs.

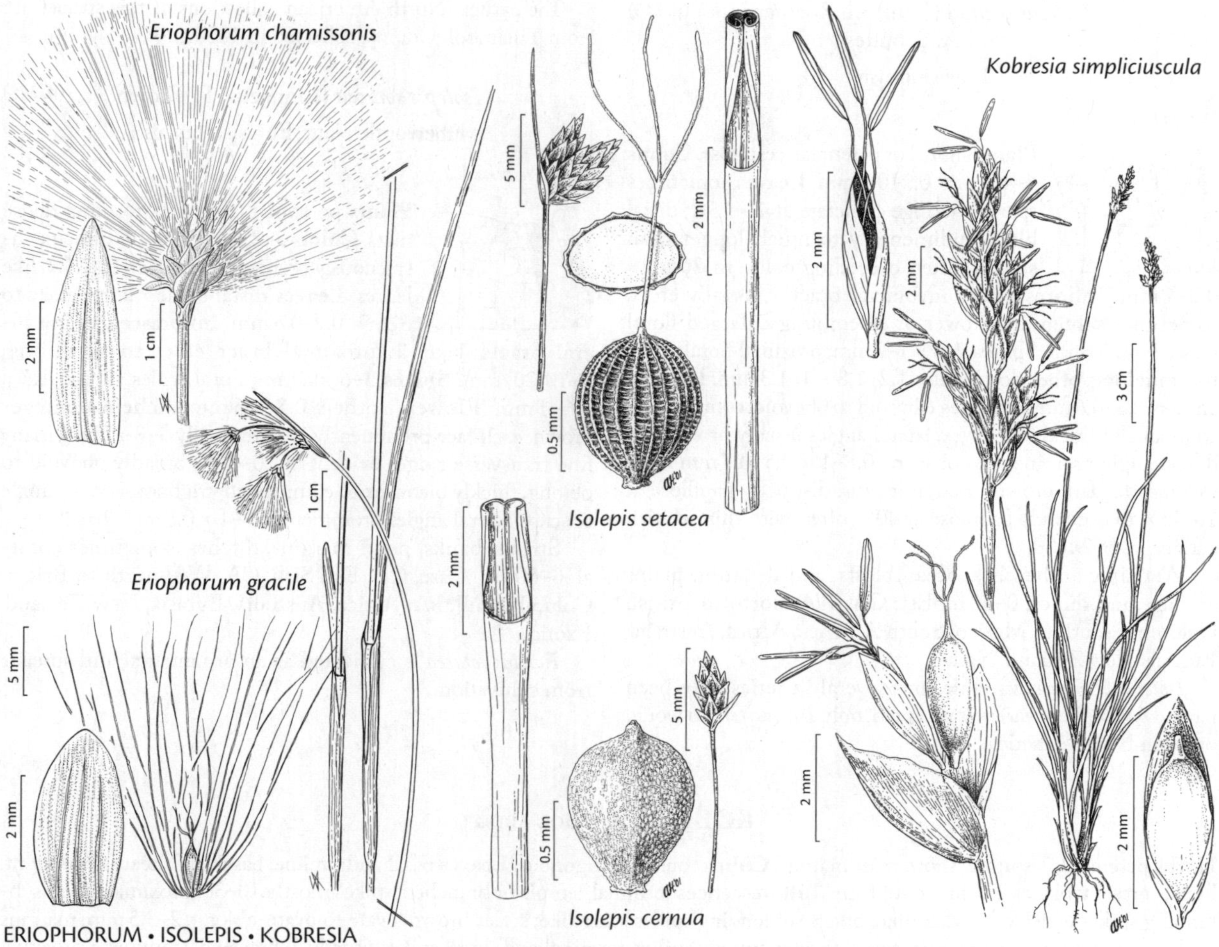

ERIOPHORUM • ISOLEPIS • KOBRESIA

Eriophorum gracile **W.D.J. Koch** (p. 245)

Slender cottongrass

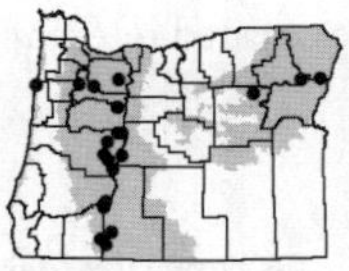

Culms 20–60 cm × 0.5–0.8 mm. **Leaves** blades trigonous-channeled in cross section, to 30 cm × 1–2 mm; distal leaf blade 1–4 cm × 1–1.5 mm, shorter than its sheath. **Inflorescences** blade-bearing involucral bracts solitary, similar to distal leaf, blade usually gray or black proximally, 0.6–2 cm. **Spikes** (1)2–5, floral scales broadly ovate, 3–4.5 mm, black or dark gray with black tip, margins absent or to 0.1 mm wide, 5–9-ribbed, midrib reaching tip, apex obtuse to more or less acute. **Flowers** perianth bristles 10–15 mm, white; anthers 1–2.5 mm. **Achenes** narrowly obovoid, 1.5–3 mm.

Meadows, bogs, shores, usually peaty, acidic substrates. 900–1800 m. BW, Casc, ECas, Est, WV. CA, ID, NV, WA; north to Alaska, east to Newfoundland and DE, southeast to CO; Eurasia. Native.

ISOLEPIS Bulrush

Herbs annual or perennial, cespitose, rhizomes present but short and often obscured by culm bases. **Culms** terete. **Leaves** all basal; sheaths usually reddish proximally; ligules absent; blades rudimentary to exceeding culms. **Inflorescences** terminal but usually apparently lateral because involucral bract is erect; spikes 1–3(15); involucral bracts 1(2), more or less erect, leaf-like. **Spikes** floral scales 8–25, spirally arranged, each subtending a flower. **Flowers** bisexual; perianth absent; stamens 1–3; styles linear, base persistent, sometimes slightly enlarged; stigmas 3 or mixed 2 and 3. **Achenes** falling separately from the scales, biconvex or both biconvex and trigonous; papillose or longitudinally ribbed.

Cosmopolitan in cool-tropical and temperate areas. 69 species; 2 species treated in *Flora*.

Isolepis is a recent segregate from *Scirpus*. The plants are small, more or less cespitose, with pseudolateral, sessile inflorescences and narrowly linear leaves that resemble the culms.

Key to *Isolepis* species:

1. Achenes not ribbed, smooth to minutely textured; spike 1 . . . . *I. cernua*

1' Achenes prominently longitudinally ribbed at 10× and minutely transversely ridged at 20–30×; spikes 1–3. . . . . . . *I. setacea*

Isolepis cernua (Vahl) J.J. Roemer & J.A. Schultes (p. 245)

Low clubrush

Scirpus cernuus Vahl

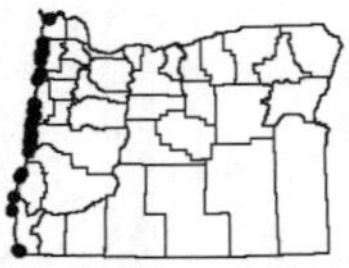

Plants annual or perennial; cespitose. **Culms** 4–40 cm × 0.2–0.5 mm. **Leaves** sometimes sparsely orange-punctate at 10–15×; distal blade rudimentary to much longer than sheath, often exceeding culm, to 20 cm × 0.2–1 mm. **Inflorescences** involucral bract 1, usually erect, sometimes subtending flower or resembling enlarged floral scale, 2–6(23) mm. **Spikes** 2–5 × 1–2 mm; proximal floral scales to 2 mm long, other floral scales 1.2–1.8 × 1–1.3 mm. **Flowers** anthers 0.3–0.6 mm. **Achenes** ellipsoid to obovoid, compressed-trigonous to thickly biconvex, lateral angles usually prominent, dorsal angle prominent to obscure, 0.8–1 × 0.5–0.7 mm long, medium to dark brown or stramineous, distinctly papillose at 10–15× to obscurely papillose at 40×, often with thin whitish surface layer. **2*n***=30.

Wet sites on beaches, ocean bluffs, and deflation plains among sand dunes. 0–50 m. Est. CA, WA; north to British Columbia, south to Mexico; South America; Africa, Australia, Eurasia, New Zealand. Native.

Isolepis cernua is variable, and several varieties have been named. Were these varieties accepted, only *I. c.* var. *cernua* would occur in North America.

The earliest North American collections of this species are from Tillamook County, collected in 1888.

Isolepis setacea (Linnaeus) R. Brown (p. 245)

Fiber-optic plant, Eurasian bulrush

Scirpus setaceus L.

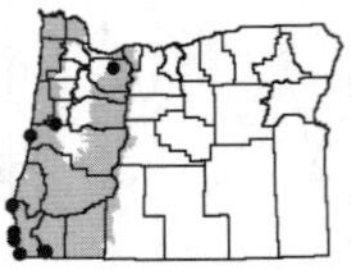

Plants perennial (annual?), forming dense mats. **Culms** 3–25 cm × 0.2–0.3 mm, inconspicuously orange-punctate at 10–15×, like leaves. **Leaves** distal blade rudimentary to 6 cm × 0.2–0.5 mm. **Inflorescences** involucral bracts 1 or 2; proximal bract erect to spreading, 3–10(20) mm. **Spikes** 3–6 × 2 mm; floral scales all 1.2–1.6 × 0.6–1 mm. **Flowers** anthers 0.3–0.5 mm. **Achenes** orange-brown, each face prominently longitudinally 5–8-ribbed, many fine transverse ridges evident at 20–30×, broadly obovoid to oblong, thickly biconvex to compressed-trigonous, dorsal angle obscure, lateral angles prominent, 0.8–1 × 0.5 mm. **2*n***=28.

Stream banks, pond margins, ditches, sometimes coastal. 0–600 m. Casc, CR, Est, Sisk. CA, WA; north to British Columbia; NJ, PA; Africa, Australia, Eurasia, New Zealand. Exotic.

Isolepis setacea is cultivated as an ornamental and spreads from cultivation.

KOBRESIA Bog sedge

Herbs perennial, cespitose, short-rhizomatous. **Culms** rounded-trigonous. **Leaves** basal and cauline; basal leaf sheaths persistent; ligules present; blades involute to filiform. **Inflorescences** terminal, simple or branched, spikes mostly 10–30; proximal bracts subtending spikes leaf-like or scale-like; bracts subtending spikes scale-like; scales brown, ovate to ovate-oblong, 2–3.5 mm, margins hyaline, tip obtuse to acute, sometimes short-pointed. **Spikes** terminal and distal spikes usually 1-flowered, staminate; proximal spikes 1-flowered and pistillate, or 2–4-flowered and bisexual with 1 pistillate flower proximally and 1–3 staminate flowers distally, all enclosed by a scale-like bract open to one side (perigynium), perigynium sometimes enclosing additional sterile scales. **Flowers** unisexual; perianth absent; stamens 3; anthers 1–2 mm; perigynium brown, smooth, margins free to base; style bases persistent, stigmas usually 3. **Achenes** usually trigonous, included in perigynium.

High mountains, arctic, and subarctic areas of the Northern Hemisphere. Approximately 30 species; 2 species treated in *Flora*.

Kobresia apparently evolved within the *Carex* clade and may be merged into that genus in the future.

Key to *Kobresia* species:

1. Inflorescences simple, unbranched . . . . ***K. myosuroides***

1' Inflorescences branched . . . . ***K. simpliciuscula***

Kobresia myosuroides (Villars) A. Fiori

Bellard's bog sedge

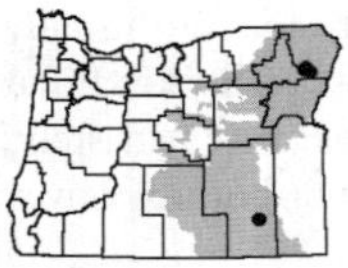

Culms 5–20(35) cm. **Leaves** basal sheaths somewhat glossy, bladeless; blades filiform, 2–20 cm × 0.2–0.5 mm. **Inflorescences** simple, 10–30 × 2–3 mm; proximal spikes 2-flowered and androgynous, rarely 1-flowered and pistillate; distal spikes 1-flowered, staminate. **Flowers** perigynia 2–3.5 mm. **Achenes** 2–2.8 mm. **2*n***=52, 56–58.

Alpine grasslands on dry to wet ground. 2400–2900 m. BR, BW. CA, ID; north to Alaska, east to Newfoundland, southeast to NM; Eurasia. Native.

This sedge has shiny, persistent leaf sheaths similar to those of *Carex filifolia*.

Kobresia simpliciuscula (Wahlenberg) Mackenzie (p. 245)

Simple bog sedge

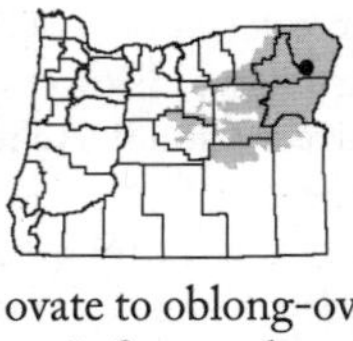

Culms 5–35 cm. **Leaves** basal sheaths dull, base of blade usually persistent; blades more or less channeled, 2–20 cm × 0.2–1.5(2) mm. **Inflorescences** usually compound, (8)10–35 × (2)3–8 mm; bracts brown, margins hyaline, ovate to oblong-ovate; proximal spikes 1-flowered and pistillate or 2-flowered and androgynous; terminal and distal spikes 1-flowered, staminate. **Flowers** perigynia 2.5–3.2 mm. **Achenes** 2–3 mm. **2*n***=76.

Alpine and subalpine fens and wet rocky slopes, usually on calcareous substrates. 1700–2500 m. BW. ID; north to AK, east to Greenland and MT, southeast to NM; Eurasia. Native.

LIPOCARPHA Lipocarpha

Herbs annual, usually densely cespitose, not rhizomatous. **Culms** terete, 0.2–0.5 mm thick. **Leaves** basal, proximal bladeless, distal with distinct blade, glabrous; ligules absent; blades setaceous to involute or flat. **Inflorescences** terminal, appearing lateral because the inflorescence bract is erect and resembles a continuation of the culm, spikes 1–4; spikes (20)50–150 per spike; involucral bracts 1–4, spreading or erect, leaf-like. **Spikes** floral (outer, 1st) scales spirally arranged; flowers bisexual; inner (2nd) scales 0–1, hyaline, transparent, subtending achene on the ventral side; stamen 1; stigmas 2. **Achenes** trigonous or terete, minutely papillose.

North America, worldwide in tropical and wet, warm-temperate regions. Approximately 35 species; 3 species treated in *Flora*.

The evolutionary origin of the structures we term outer (1st) and inner (2nd) scales is uncertain. It seems likely that the outer is a floral scale like those in other sedges, and perhaps the inner scales are highly modified petals, like the bristles in many *Scirpus* and *Schoenoplectus*. However, it is possible that each flower plus its scales is the remnant of a whole spike like that in *Cyperus*. If so, the outer scale might be an inflorescence bract, and the inner scale would like be a modified floral scale. Fortunately, specimens can be identified without resolving this issue.

All the Oregon species appear rare, perhaps because they are overlooked due to their small size and phenology. They flower in late summer to fall.

Key to *Lipocarpha* species:

1. Outer scale widest in lower half of total length (including awn-like tip); culms 0.5–5 cm. . . . . ***L. occidentalis***

1' Outer scale widest at or above mid length; culms 2–20 cm.

  2. Inner scale 0.1–0.2 mm or absent. . . . . ***L. micrantha***

  2' Inner scale 0.5–0.8 mm . . . . ***L. aristulata***

Lipocarpha aristulata (F.V. Coville) G.C. Tucker (p. 248)

Awned lipocarpha

Hemicarpha aristulata (Coville) Smyth, *Hemicarpha micrantha* (Vahl) Britton var. *aristulata* Coville

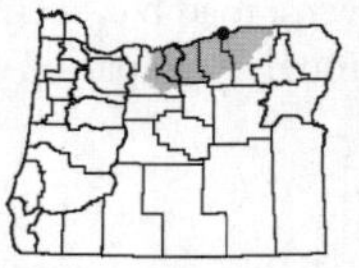

Culms 2–15 cm × 0.3–0.5 mm. **Leaves** 0.4–0.6 mm wide. **Inflorescences** spikes 1–2, 1–5(8) × 1–2.5 mm; bracts 1–2, longest erect, 1–3.5 cm. **Spikes** floral scale 0.9–1.6 × 0.3–0.6 mm, widest at mid length, long-acuminate, curving outward; inner scale oblong, veinless (occasionally with 2–4 reddish veins), 0.5–0.8 × 0.2 mm. **Achenes** obovoid to terete, 0.5–0.8 × 0.25–0.35 mm.

Fruiting late summer to early fall. Typically on sandy soils, emergent shorelines, stream banks, ponds, ditches. 400–700 m. Col. CA, WA; east to MN, southeast to TX. Native.

Lipocarpha micrantha (Vahl) G.C. Tucker (p. 248)

Small-flowered lipocarpha

Hemicarpha micrantha (Vahl) Britton

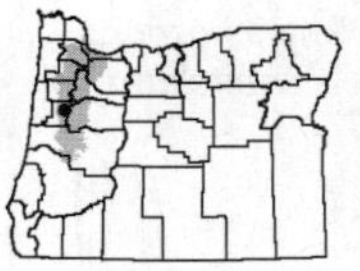

Culms 2–20 cm × 0.3–0.5 mm. **Leaves** 0.3–0.5 mm wide. **Inflorescences** spikes 1–2(3), 1–3(5) × 1–2 mm; bracts 1–2(3), longest erect, others spreading, 2–5 cm. **Spikes** floral scale widest at or distal to mid length, 0.5–1 × 0.3–0.4 mm; inner scale notched, sometimes absent, 0.1–0.2 × 0.1 mm. **Achenes** obovoid to broadly ellipsoid, 0.4–0.6(0.7) × 0.2–0.3 mm.

Emergent shorelines, rarely freshwater tidal shores. 100–200 m. WV. CA, WA; north to British Columbia, east from ND, TX, and Ontario to Atlantic, south to South America; Africa. Native.

Lipocarpha occidentalis (A. Gray) G.C. Tucker (p. 248)

Western halfchaff sedge

Hemicarpha occidentalis A. Gray

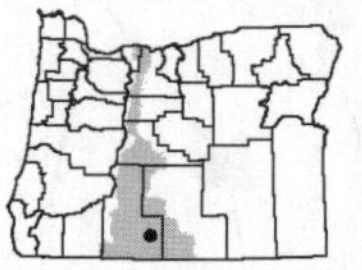

Culms 0.5–5 cm × 0.2–0.3 mm. **Leaves** 0.5–1 mm wide. **Inflorescences** spikes 1(2), 2–3 × 1.5–3 mm; bracts 2–3, 0.4–2 cm. **Spikes** floral scale 1.8–2.8 × 0.2–0.4 mm, widest proximal to mid length, apex long-acuminate, to 0.6–1.4 mm; inner scale oblong, 0.6–0.8 × 0.2–0.3 mm. **Achenes** obovoid, 0.5–0.8 × 0.3 mm.

Sandbars and shorelines exposed by seasonal low water levels. 600–1700 m. ECas. CA, WA. Native.

With its relatively long outer scales, *L. occidentalis* is easy to distinguish from other members of the genus.

RHYNCHOSPORA Beak-rush

Herbs perennial, cespitose, rhizomes absent. **Culms** usually trigonous, principal leaves mostly overtopped by the culms. **Leaves** basal and cauline, mostly 3-ranked; sheaths open apically, glabrous; ligules present or absent; blades flat, V-shaped in cross section or terete, typically keeled abaxially, margins involute or revolute, usually scabrid or scabridulous. **Inflorescences** terminal, paniculate, corymbose, or capitate; spikes 3–100+; involucral bracts 1–6, spreading or ascending, leaf-like. **Spikes** scales spirally arranged; 1 or more proximal scales empty. **Flowers** all bisexual or sometimes distal-most staminate; perianth absent or of 2–12(20) bristles, usually persistent in fruit; stamens 2–3; styles deeply cleft into 2(3) linear stigmatic branches, style base persistent as tubercle on fruit, distinct, enlarged. **Achenes** borne on pedicellar joint, directly distal to compact, dilated receptacle; body various shades of brown, stipitate, flattened, lenticular, obovate, transversely striate to smooth; tubercle triangulate-subulate.

Cosmopolitan, mostly in areas with acidic soils. Approximately 250 species; 2 species treated in *Flora*.

For the purposes of the descriptions in this genus, the fruit is the achene plus attached structures that fall with it. Fruit length is the distance from the pedicellar joint (which is distal to the receptacle and can almost always be identified because it is very contracted) to the distal tip of the tubercle at the top of the achene. Tubercle length is measured from the middle of the top of the achene to the tip of the tubercle. Measurements should be taken from mature fruit, after the distal portion of the style falls.

Key to *Rhynchospora* species:

1. Spikes pale, whitish or straw-colored to pale dull brown; perianth bristles 10–12; flower clusters 1–3; leafy inflorescence bracts not or scarcely exceeding the inflorescence . . . . ***R. alba***

1' Spikes darker, red-brown to dark brown; perianth bristles 6; flower clusters 1–5+; leafy inflorescence bracts mostly exceeding the inflorescence . . . . ***R. capitellata***

Rhynchospora alba (Linnaeus) Vahl (p. 248)

White beakrush

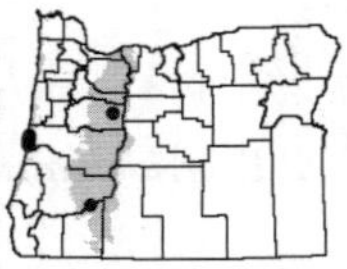

Plants 6–75 cm. **Leaves** narrowly linear to filiform, proximally flat, 0.5–1.5 mm wide. **Inflorescences** clusters 1 or 2–3 (then widely spaced), narrowly turbinate to hemispheric, 1.5–2.5 cm wide; subtending leafy bracts often exceeded by distal cluster. **Spikes** ellipsoid, 3.5–5.5 mm, pale brown to nearly white, apex acute; fertile scales elliptic, 3–3.5(4) mm, apex acute or acuminate, midrib excurrent as mucro. **Flowers** perianth bristles 10–12, slightly overtopping tubercle. **Achenes** 1(2) per spike, (2.3)2.5–3 mm; body 1.5–1.8(2) × 0.9–1.2 mm, pale brown with paler center; surfaces transversely striate, relatively smooth, rim narrow, confluent with tubercle base; tubercle 0.5–1.2 mm.

Acidic bogs and fens, often with *Sphagnum*, often on floating mats. 0–1500 m. Casc, Est. CA, ID, WA; north to Alaska, east to Newfoundland, south from Minnesota to Georgia; Puerto Rico; Eurasia. Native.

Rhynchospora capitellata (A. Michaux) Vahl (p. 248)

Brownish beakrush

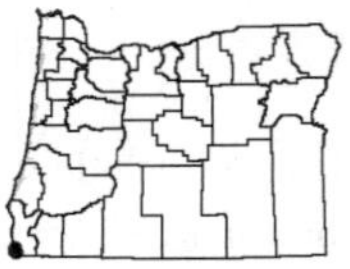

Plants 20–100 cm. **Leaves** flat, to 3 mm wide. **Inflorescences** clusters 1–5+, turbinate or hemispheric, 1–1.5 cm wide; subtending leafy bracts mostly exceeding subtended clusters. **Spikes** lance-ellipsoid, 3.5–4(5) mm, rich deep brown, rarely pale brown; fertile scale elliptic, 2.7–3 mm, apex acute or rounded, midrib excurrent as short mucro or not. **Flowers** perianth bristles 6. **Achenes** (1)2–3(5) per spike, (2)2.5–3 mm; body 1.2–1.5 × 0.7–1(1.2) mm, red-brown with pale central disc, surfaces smooth; tubercle (0.8)1–1.2(1.6) mm.

Moist to wet meadows, seeps, fens, and bogs. 0–100 m. Est. CA; eastern Canada south to TX and Mexico. Native.

This species has been collected repeatedly from one location, in Curry County, since 1919. Recently, however, a road has been built near the population and a power line through it, jeopardizing its long term existence.

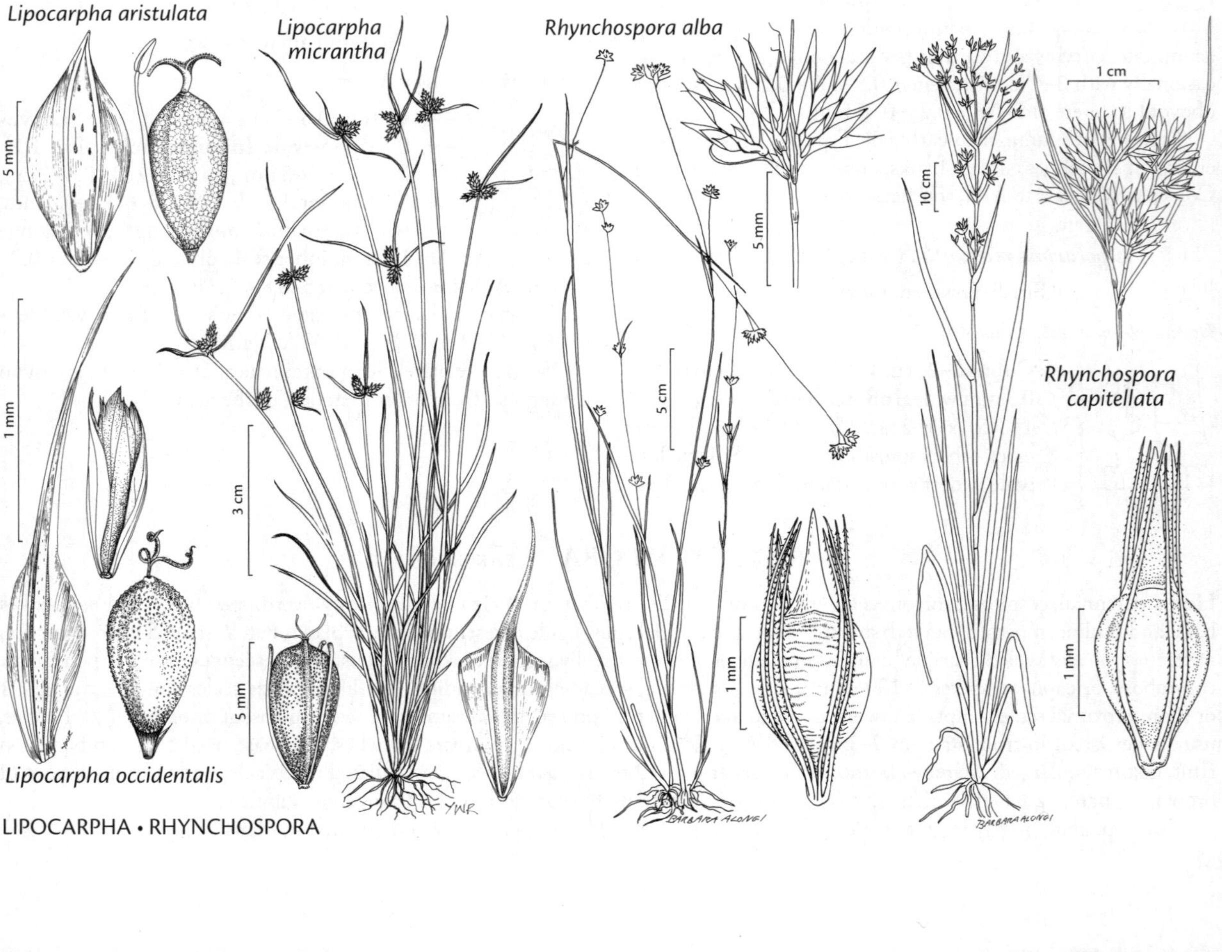

LIPOCARPHA • RHYNCHOSPORA

SCHOENOPLECTUS Naked-stemmed bulrush

Herbs perennial or annual, cespitose or not, rhizomatous or not. **Culms** cylindric to strongly trigonous, smooth, glabrous, spongy with internal air cavities. **Leaves** basal; sheaths tubular; ligules membranous, glabrous; blades well developed to rudimentary, cross section flat to C-shaped or laterally compressed, soft, smooth or margins sometimes distally scabrous or spinulose. **Inflorescences** terminal, capitate to openly paniculate; spikes 1–100+; involucral bracts 1–5, leaf-like, proximal bract erect to spreading. **Spikes** terete, 3–25 × 2–5 mm; floral scales 8+, spirally arranged, each subtending flower or proximal scale empty, margins ciliate, smooth or abaxial surface scabrous; apex entire or notched, usually mucronate or short-awned. **Flowers** bisexual; perianth of 0–6(8) bristles, straight or curved, thread-like or strap-like, margins usually fringed with hairs, shorter than to somewhat exceeding achene; stamens 3, anthers 0.8–3 mm; styles linear, stigmas 2–3, base not or scarcely enlarged, deciduous in fruit. **Achenes** ovoid to obovoid or obpyriform, biconvex to trigonous, with apical beak, 2.5–3.5 mm including 0.1–2 mm beak; gray, brown, or blackish; faintly to prominently rugose or with transverse wavy ridges.

Cosmopolitan. Approximately 77 species; 9 species treated in *Flora*.

Schoenoplectus is a segregate from *Scirpus*. These bulrushes are rhizomatous and usually large, with pseudolateral, sessile or branched inflorescences, triangular to C-shaped leaves in cross section, and triangular or cylindrical culms that are spongy inside. They are often community dominants that provide habitat and food for birds and other animals.

The leaf blade cross sectional shape should be assessed at mid-length. The lengths of achenes include the beak, and the lengths of floral scales include the awn. Culm thickness is measured in the middle third of the culm. Spikes elongate during flowering. Achenes reach full size and become firm while still pale and not fully mature.

Key to *Schoenoplectus* species:

1. Plants usually aquatic; leaves lax, leaf tips floating; culms sometimes floating as well; spikes strictly solitary, sessile . . . . ***S. subterminalis***

1' Plants often growing in shallow water; leaves (if present) and culms stiff and emergent for most of their lengths; spikes usually 2 or more, occasionally solitary, sessile or stalked.

- 2. Culms triangular in cross section; spikes 1–150+, sessile or on branches, 1–150+.
  - 3. Inflorescence bract strongly divergent to reflexed, rarely erect; spikes 4–20; floral scale tips entire . . . ***S. mucronatus***
  - 3' Inflorescence bract more or less erect, rarely divergent to reflexed; spikes 1–150; floral scale tips clearly emarginate.
    - 4. Inflorescences branched, or rarely some with all spikes sessile.
      - 5. Inflorescences to 2 times branched, with 1–35 spikes; culms 2–4 mm in diameter; perianth bristle thread-like, with narrow, sharply pointed, retrorse hairs . . . . ***S. triqueter***
      - 5' Inflorescences 3 or more times branched, with 25–150+ spikes; culms 4–10 mm in diameter; perianth bristles strap-like, with blunt, spreading hairs . . . . ***S. californicus*** (in part)
    - 4' Inflorescences unbranched; spikes sessile.
      - 6. Floral scale apex notch 0.1–0.4 mm, awns 0.2–0.6 mm; proximal inflorescence bract 1–6 cm, other bracts without blades; sides of mid culm deeply concave, rarely nearly flat; distal leaf blade much shorter than to nearly equaling, rarely 1.5 times longer than sheath . . ***S. americanus***
      - 6' Floral scale apex notch (0.3)0.5–1 mm, awns 0.5–1.5(2.5) mm; proximal inflorescence bract (1)3–20 cm, other bracts with blades; sides of mid culm shallowly concave to nearly flat; distal leaf blade much longer than to rarely equaling sheath . . . . ***S. pungens***
- 2' Culms cylindric; spikes 20–150+, on branches.
  - 7. Perianth bristles 2–4, strap-like, fringed with blunt, spreading hairs; leaf blades 0–20 mm long; culms usually bluntly trigonous near inflorescence . . . . ***S. californicus*** (in part)
  - 7' Perianth bristles 4–6(8), filiform to narrowly flattened, sharply pointed, retrorse hairs; leaf blades 2–220 mm long; culms cylindric throughout.
    - 8. Achenes trigonous; stigmas 3; perianth bristles 4(5), 2 usually much shorter than the others; floral scales scabrous only on awn; spikes all solitary (or some in pairs); one collection. . . . . ***S. heterochaetus***
    - 8' Achenes usually more or less lenticular, rarely trigonous; stigmas 2, mixed, or all 3; perianth bristles (4)6(8), usually nearly equal; floral scales scabrous on awn and often body, especially in proximal parts of spikes; spikes all clustered to all solitary.
      - 9. Floral scale awns straight to bent, 0.2–0.8 mm; scales (except often midribs) orangish or pale and prominently spotted at 10×; spikes all solitary . . . . ***S. tabernaemontani***
      - 9' Floral scale awns mostly strongly contorted, rarely all straight, often broken, 0.5–2 mm; scales usually pale and prominently spotted at 10×; most spikes clustered . . . . . . ***S. acutus*** var. ***occidentalis***

Schoenoplectus acutus (Muhlenberg ex J.M. Bigelow) Á. Löve & D. Löve var. *occidentalis* (S. Watson) S.G. Smith (p. 250)

Tule, hardstem bulrush

Scirpus acutus Muhl. ex J.M. Bigelow var. *occidentalis* (S. Watson) Beetle, *Scirpus rubiginosus* Beetle

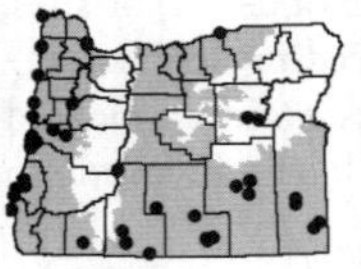

Rhizomes 5–15 mm thick; tubers absent. **Culms** cylindric, 100–400 cm × 4–12 mm, with air cavities in the distal quarter 1–2.5 mm wide. **Leaf blades** C-shaped to dorsiventrally flat in cross section, usually much shorter than sheath, distal blade 8–120 × 3–7 mm, margins often scabridulous. **Inflorescences** 2(3) times branched, open or compact, branches 6(18) cm, sometimes drooping; proximal bract usually erect, thickly C-shaped in cross section, 1–9 cm. **Spikes** 4–190, solitary or in clusters of 2–8, never all solitary, 6–24 × 4 mm; floral scales reddish to orange-brown to straw-colored, often variable on same spike, usually wholly or partly straw-colored and with prominent spotted lines at 10×; scale or midrib ovate, 3–4 × 2–3 mm, often green when young, sparsely to often densely covered with reddish or straw-colored papillae or scabrous distally or on most of surface, margins ciliate, hairs long, contorted, apex acute to obtuse, notch 0.3–0.5 mm deep; awn on at least some scales in spike usually strongly contorted, 0.5–2 mm, their tips usually black, often broken off. **Flowers** perianth members (4)6(8), brown, bristle-like, equaling achene body or sometimes much shorter, rarely rudimentary, spinulose; stigmas 2 in most of the spike with a few flowers with 3 near the tip, sometimes most or all flowers with 3 stigmas. **Achenes** plano-convex or compressed trigonous, (1.5)2–3 × 1.2–1.7 mm; beak 0.2–0.4 mm.

Freshwater to brackish marshes, lakes, and slow streams, often in shallow water. 0–1600 m. BR, Col, CR, ECas, Est, Lava, Owy, Sisk, WV. CA, ID, NV, WA; north to British Columbia, northeast to MT, south to Mexico. Native.

Schoenoplectus acutus has been considered a strictly New World species, but recent specimens from Eurasia and a Pacific island may belong to this species. *Schoenoplectus acutus* hybridizes with *S. californicus*, *S. heterochaetus*, and *S. tabernaemontani*. *Schoenoplectus acutus* var. *acutus* occurs in Washington. All of its flowers have two stigmas, and its achenes are all plano-convex. It also has ascending inflorescence branches and firmer culms with smaller air cavities in the distal quarter.

Schoenoplectus americanus (C.H. Persoon) Volkart ex H. Schinz & R. Keller (p. 250)

Olney's three-square bulrush, chairmaker's bulrush

Scirpus chilensis Nees & Meyen ex Kunth, *Scirpus olneyi* A. Gray

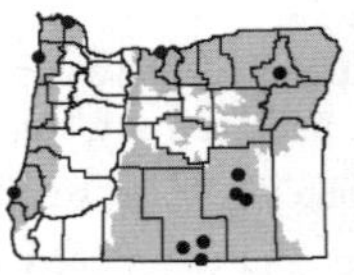

Rhizomes 2–5 mm thick. **Culms** sharply trigonous, sides deeply concave throughout to rarely nearly flat, 40–250 cm × 3–10 mm. **Leaf blades** V-shaped near base, otherwise laterally flattened-trigonous in cross section; distal blade 0.2–1.5 times as long as sheath, 25–200 × 2–8 mm. **Inflorescences** proximal bract usually erect, resembling leaf blade, 1–6 cm. **Spikes** 2–20, sessile or on branches to 5 mm long, 5–15 × 3–5 mm; floral scales broadly ovate, 2.7–4 × 2–3 mm, bright

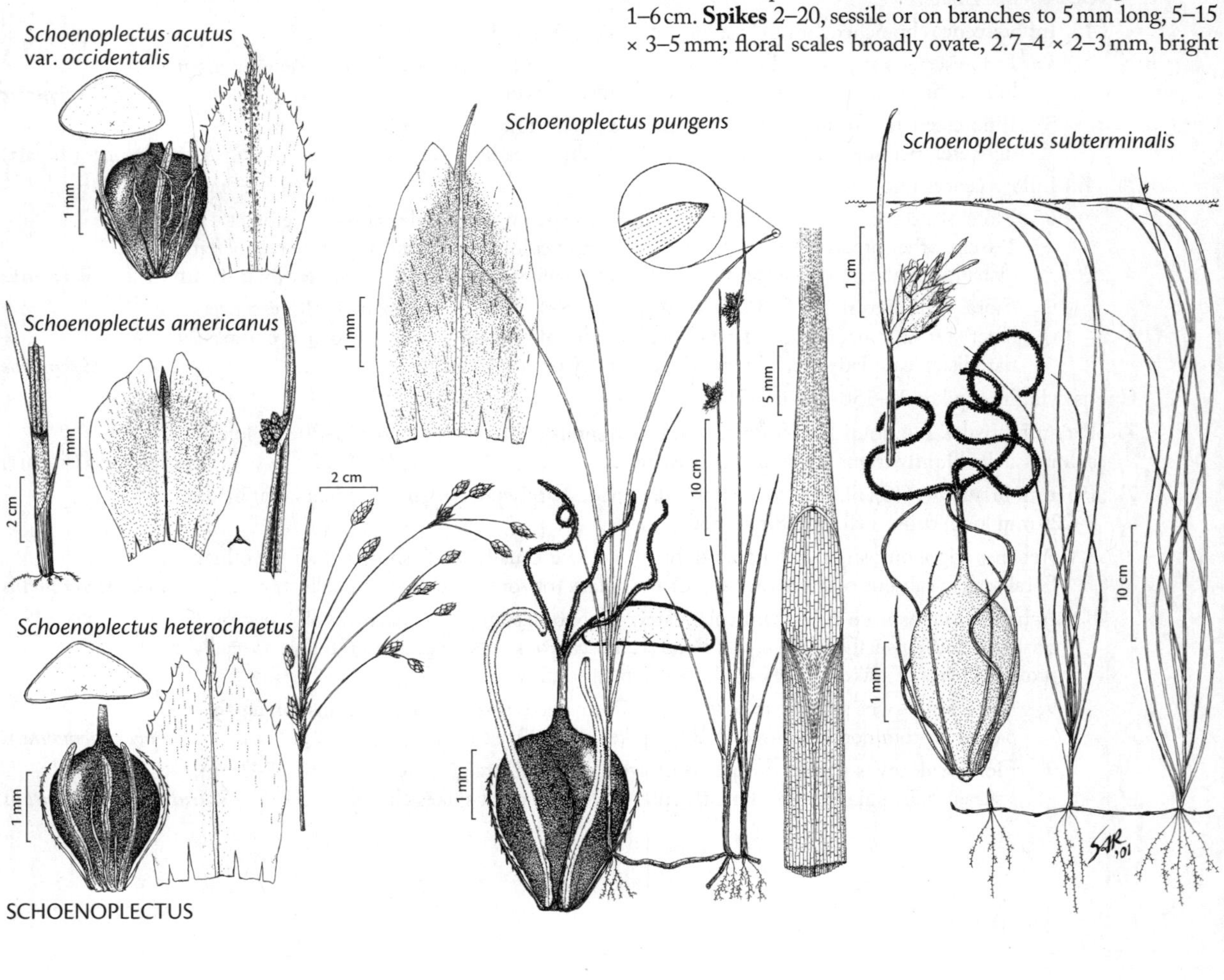

SCHOENOPLECTUS

orange- to red- or purple-brown to straw-colored, usually with linear dots, smooth or awn sparsely spinulose, margins deciduously ciliolate, apex rounded to acute, notch 0.1–0.4 mm deep, awn not contorted, 0.2–0.6 mm. **Flowers** perianth members (2)5–6(7), bristle-like, often unequal, to 50% as long as achene body, retrorsely spinulose, yellow-brown; stigmas 2, or 2 and 3. **Achenes** thickly plano-convex or unequally biconvex or compressed obtusely trigonous, 1.8–2.8 × 1.3–2 mm; beak 0.1–0.3 mm. ***2n***=78.

Brackish or alkaline marshes, lake shores, fens. 0–1500 m. BR, BW, Col, CR, ECas, Est. CA, ID, NV, WA; north to British Columbia, east to Nova Scotia, south to South America. Native.

Schoenoplectus americanus is similar to and confused with *S. pungens,* which has more deeply notched floral scales with longer awns, and distal leaf blades mostly longer than their sheaths. Eastern North American populations have seriously declined in recent years.

Schoenoplectus californicus (C.A. Meyer) Soják

California bulrush, tule

Scirpus californicus (C.A. Mey.) Steud.

Rhizomes 10–15 mm thick, tubers absent. **Culms** bluntly trigonous with nearly flat sides throughout to cylindric proximally, obscurely trigonous distally, 100–400 cm × 4–10 mm. **Leaf blades** dorsiventrally flat, not more than 50% of sheath length, distal blade to 20 × 2 mm. **Inflorescences** proximal bract usually erect, obtusely trigonous to subcylindric-channeled, 1–8 cm. **Spikes** 25–150+, solitary and in clusters of 2 or more, on branches to 21 cm long, 5–11 × 3 mm; floral scales broadly oval, 2.5 × 2 mm, orange-brown, with linear dots, margins ciliolate, apex rounded, notch 0.1–0.2 mm deep, awn not contorted, usually bent outward, 0.3 mm, sparsely scabrous. **Flowers** perianth members 2–4, thickly strap-shaped or trigonous, equaling achenes, dark red-brown, fringed with brownish, soft, blunt hairs; stigmas 2. **Achenes** unequally biconvex or plano-convex, 1.8–2.2 × 1.3 mm, dark gray-brown when ripe; beak 0.2 mm.

Brackish to freshwater marshes and lake shores, often emergent in water. 0–50 m. Est. CA, NV; east to NC, south to South America; New Zealand, Pacific Islands. Exotic?

Schoenoplectus californicus resembles common *S. acutus* and *S. tabernaemontani,* but the unusual bristles permit confident identification. Hybrids of *S. californica* and *S. acutus* var. *occidentalis* are known in California.

Schoenoplectus heterochaetus (M.A. Chase) Soják (p. 250)

Pale great bulrush, slender bulrush

Scirpus heterochaetus Chase

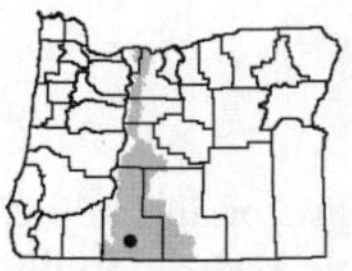

Rhizomes 5–8 mm thick; tubers absent. **Culms** cylindric, 150–250 cm × 4–8 mm. **Leaves** thin, C-shaped to dorsiventrally flat in cross section, much shorter to longer than sheath, distal blade to 220 × 5 mm. **Inflorescences** proximal bract usually erect, thickly C-shaped to subterete in cross section, 1–15 cm. **Spikes** 5–30, solitary or rarely 2 together, on stalks to 9 cm, 5–15 × 3–4 mm; floral scales light orange-brown to nearly colorless, usually with faint linear dots; midribs ovate-oblong, 3–4 × 1.5 mm, often greenish, sparse reddish or straw-colored scabrous on awn, sometimes distal midrib, and rarely adjacent parts of sides, margins ciliate, hairs contorted, apex obtuse, notch 0.5 mm deep, awn straight to slightly contorted, 0.3–1 mm. **Flowers** perianth members 4(5), bristle-like, equaling achene body or 2 of them much shorter, sparsely retrorsely spinulose, brown; stigmas 3. **Achenes** compressed-trigonous, 2.2–3.2 × 1.4–1.9 mm; beak 0.3–1 mm. ***2n***=38.

Freshwater marshes and lakes, often emergent in shallow water. 1000–1500 m. ECas. CA, ID; northeast to Alberta, east to MA, south to WY, TX, KY, and NY. Native.

This species hybridizes with *S. acutus* and *S. tabernaemontani.*

Schoenoplectus mucronatus (Linnaeus) E. Palla

Bog bulrush

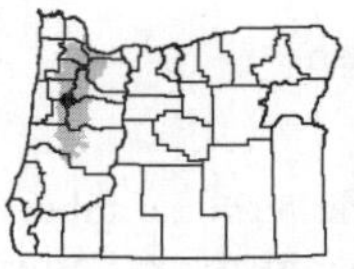

Rhizomes short. **Culms** sharply trigonous, 40–100 cm × 2–3 mm. **Leaf blades** absent. **Inflorescences** proximal bract divergent to reflexed or rarely erect, trigonous, adaxially channeled, 1–10 cm. **Spikes** 4–20, sessile, 7–12 × 4 mm; floral scales broadly obovate, 3–3.5 × 2–2.5 mm, orange-brown to straw-colored, central region often greenish, margins ciliolate, apex obtuse to broadly acute, entire, mucronate. **Flowers** perianth members 6, bristle-like, equaling achene, retrorsely spinulose, brown; stigmas 3(2 in same spike). **Achenes** thickly plano-convex to obtusely trigonous 1.7–2.2(2.5) × 1.2–1.7 mm; beak 0.2 mm.

Freshwater ponds, ditches, and rice fields, disturbed wet soil. 50–100 m. WV. CA, WA; scattered in eastern North America; Africa, Australia, Eurasia. Exotic.

The dense heads, bladeless leaves, and strongly divergent, blunt inflorescence bracts make *S. mucronatus* unmistakable. This species is an important weed of rice fields in California. It has been cultivated as food for wildlife in Clark County, Washington.

Schoenoplectus pungens (Vahl) E. Palla (p. 250)

Common three-square, western bulrush, common three-square bulrush

Schoenoplectus pungens (Vahl) Palla var. *badius* (C. Presl) S.G. Sm., *Schoenoplectus pungens* (Vahl) Palla var. *longispicatus* (Britton) S.G. Sm., *Scirpus americanus* Pers. var. *monophyllos* (J. Presl & C. Presl) T. Koyama, *Scirpus pungens* Vahl, *Scirpus pungens* Vahl var. *longispicatus* (Britton) Cronquist

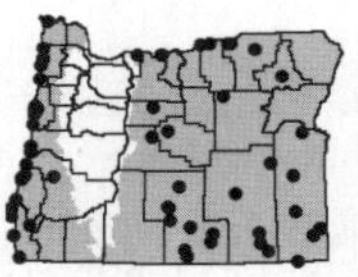

Rhizomes often vertical, 1–6 mm thick. **Culms** sharply trigonous, sides convex to concave proximally, deeply concave to flat distally, 10–200 cm × 1–6 mm. **Leaf blades** proximally V-shaped, distally trigonous to asymmetrically laterally flattened in cross section; distal blade (1)2–5 times as long as sheath, 50–750 × 2–9 mm. **Inflorescences** proximal bract usually erect, resembling leaf blade but trigonous proximally, (1)3–20 cm. **Spikes** 1–5(10), sessile, 5–23 × 3–5(7) mm; floral scales bright (to very dark) orange-, red-brown, or purplish brown to straw-colored, often with prominent linear dots; midrib mostly paler, ovate, 3.5–6 × 2–3 mm, smooth or awn sparsely spinulose; margins deciduously ciliolate, apex acute (to obtuse), notch (0.3)0.5–1 mm deep, awn mostly irregularly bent, 0.5–1.5(2.5) mm. **Flowers** perianth members 4–8, sometimes fewer, bristle-like, equal or unequal, all equaling achene body to all rudimentary, retrorsely spinulose, brown; stigmas 2–3. **Achenes** biconvex to compressed bluntly trigonous, (2)2.5–3.5 × 1.3–2.3 mm; beak 0.1–0.5 mm. ***2n***=74, 78.

Freshwater to brackish marshes, lake shores, and fens, often emergent in shallow water. 0–2000 m. BR, BW, Col, CR, ECas, Est, Lava, Owy, Sisk. CA, ID, NV, WA; north to AK, throughout US and southern Canada, south to Mexico; South America; Australia, Europe, New Zealand. Native.

Schoenoplectus pungens and *S. americanus* share the oxymoronic and yet oddly descriptive common name "Three-square." Their culms are triangular with prominent angles and the inflorescences are sessile. *Schoenoplectus pungens* shows considerable variation and has been divided into controversially delimited varieties. The group needs more research.

Schoenoplectus subterminalis (J. Torrey) Soják (p. 250)

Water bulrush, water clubrush

Scirpus subterminalis Torr.

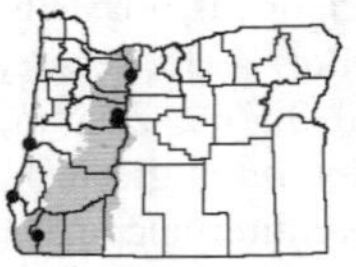

Rhizomes to 1 mm thick; terminal tubers sometimes present, to 3 mm thick, fleshy. **Culms** subcylindric, 20–150 cm × 0.5–1 mm. **Leaf blades** when erect C-shaped in cross section proximally, when submerged usually flaccid and filiform to narrowly ribbon-like, probably longer than sheath, 0.2–1 mm wide. **Inflorescences** proximal bract erect, 7–60 mm. **Spikes** solitary, sessile, 5–15 × 3–7 mm; floral scales ovate-lanceolate, 4–6 × 1.5 mm, pale brown, or central region green when young, smooth, margins sometimes ciliolate apically at 40×, apex acute, entire, minutely mucronate. **Flowers** perianth members 6, bristle-like, from 50% of to equaling achene beak, spinulose, brown; stigmas 3. **Achenes** strongly trigonous, (2.5)3–3.5 × 1.5–1.7 mm; beak 0.3–0.8 mm. **2*n***=72.

Freshwater lakes and bogs, submerged to emergent in shallow water, occasionally terrestrial if water levels fall. 0–1100 m. Casc, Est, Sisk. CA, ID, WA; north to AK, east to Alberta and WY, eastern Canada south to MO and FL. Native.

Schoenoplectus subterminalis is an aquatic plant with lax, floating leaves and firmer, usually emergent flowering culms. The narrow leaved, sterile plants can form extensive underwater mats and are easy to confuse with sterile aquatic individuals of *Eleocharis acicularis* and *E. parvula*.

Schoenoplectus tabernaemontani (C.C. Gmelin) E. Palla

Tule, soft-stem bulrush

Scirpus tabernaemontani C.C. Gmel., *Scirpus validus* Vahl, *Scirpus validus* Vahl var. *creber* Fernald, *Scirpus validus* Vahl var. *validus*

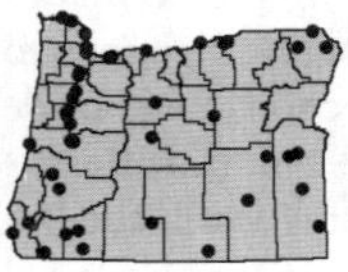

Rhizomes 3–10 mm thick. **Culms** cylindric, 50–300 cm × 2–10 mm. **Leaf blades** C-shaped to dorsiventrally flat in cross section, usually much shorter than sheath, distal blade 2–200 × 1–4 mm, margins often scabridulous. **Inflorescences** proximal bract usually erect, thickly C-shaped to subterete, 1–8 cm. **Spikes** 15–200, solitary or less often in clusters of 2–4(7), clusters on branches to 15(25) cm long; spikes 3–17 × 2.5–4 mm; floral scales ovate, 2–3.5 × 1.5–2 mm, uniformly dark to pale orange-brown, sometimes straw-colored, sometimes with prominent linear dots, midrib often pale or green, sparsely (rarely densely) reddish or straw-colored scabrous on awn and distal parts of midrib and sometimes flanks, margins ciliate, hairs contorted; apex obtuse to rounded, notch 0.2–0.3 mm deep, awn straight or bent, 0.2–0.8 mm. **Flowers** perianth bristles 6, more or less equaling achene, densely retrorsely spinulose, brown; stigmas 2, sometimes 3 near spike apex. **Achenes** plano-convex, 1.5–2.8 × 1.2–1.7 mm; beak 0.2–0.4 mm. **2*n***=42.

Freshwater to brackish marshes, bogs, lakes, stream banks, and disturbed wetlands, often emergent in shallow water. 0–1600 m. Throughout Oregon. CA, ID, NV, WA; throughout North America; southern South America; Africa, Australia, Eurasia, New Zealand, Pacific Islands. Native.

Schoenoplectus tabernaemontani is a cosmopolitan, variable taxon. North American populations have sometimes been treated as *S. validus*. *Schoenoplectus acutus,* also common in Oregon, is very similar. *Schoenoplectus tabernaemontani* hybridizes with *S. acutus, S. californicus S. glaucus, S. heterochaetus,* and S. *triqueter*. Hybrids tend to be fertile. Yellow-striped forms of *S. tabernaemontani* are cultivated as ornamentals.

Schoenoplectus triqueter (Linnaeus) E. Palla

Streambank bulrush, triangular clubrush

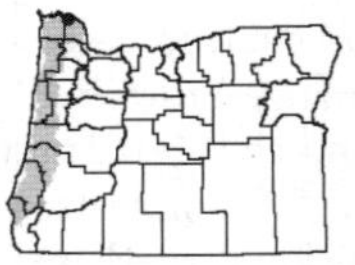

Rhizomes 1.5–5 mm thick. **Culms** sharply trigonous, 50–150 cm × 2–4 mm. **Leaf blades** thickly V-shaped in cross section, shorter than to equaling sheath length. **Inflorescences** proximal bract usually erect, trigonous, 2–7 cm. **Spikes** 1–35, in clusters or solitary, clusters sessile or on branches to 4 cm long; spikes 5–12 × 3–4 mm; floral scales ovate, 3–3.5 × 2.5 mm, straw-colored to orange-brown or midrib greenish, usually with linear spots, midrib proximally long-scabrous, apex acute to rounded, notch 0.3 mm deep, awn 0.3 mm, sparsely long-scabrous. **Flowers** perianth members 4–6, bristle-like, equaling to less than 50% of achene body, retrorsely spinulose, brown; stigmas 2. **Achenes** biconvex, 2–2.5 × 1.5–2 mm; beak 0.1–0.2 mm. **2*n***=42.

Freshwater tidal shores, marshes. 0–50 m. CR. WA; Eurasia. Exotic.

Although it has triangular culms, *S. triqueter* is most closely related to *S. tabernaemontani*. The two produce fertile hybrids, (*S.* × *kuekenthalianus*) where they occur sympatrically along the lower Columbia River.

SCIRPUS Bulrush

Herbs perennial, cespitose or not, rhizomatous or not. **Culms** more or less trigonous. **Leaves** basal and cauline or all cauline; sheaths not fibrous; ligules present, rarely absent, glabrous; blades flat or V-shaped in cross section, prominently keeled dorsally. **Inflorescences** terminal, sometimes also axillary in 1–3 distal leaves, subumbellate or more or less paniculate; spikes 50–500; involucral bracts usually 3, leaf-like. **Spikes** less than 3.5(5) mm diameter; scales 10–50, spirally arranged, each scale subtending a flower, glabrous. **Flowers** bisexual; perianth of (0)3–6 bristles to 8+ mm long; bristles straight or strongly curled, smooth or variously toothed or barbed, shorter to much longer than achene; stamens 1–3; styles with base persistent; stigmas (2)3. **Achenes** brown to whitish, trigonous, biconvex or plano-convex, 0.6–1.8 mm, minutely papillose.

Worldwide except Africa. Approximately 35 species; 7 species treated in *Flora*.

As traditionally delimited, the genus *Scirpus* included a great diversity of taxa. Recently, the genus has been dismembered, and the segregate genera are much more morphologically uniform. *Scirpus* in the narrow, modern sense consists of plants with terminal inflorescences having stalked to sessile spikes; leafy bracts; flat, keeled leaves; and glabrous floral bracts. See also its segregates *Amphiscirpus, Bolboschoenus, Calliscirpus, Isolepis, Schoenoplectus,* and *Trichophorum*.

Key to *Scirpus* species:

1. Most spikes more or less stalked, solitary; perianth bristles contorted in fruit.
    2. Perianth bristles exceeding floral scales; achenes 0.6–0.9 mm . . . . . **S. atrocinctus**
    2' Perianth bristles not exceeding floral scales; achenes 1–1.8 mm . . . . . ***S. pendulus***

1' Most spikes sessile in head-like clusters; perianth bristles contorted or curved to straight in fruit.
    3. Stigmas 2(3); achenes 2(3)-angled; leaves 5–15(20) mm wide. . . . . ***S. microcarpus***
    3' Stigmas (2)3; achenes (2)3-angled; leaves 3–10(12) mm wide.
        4. Perianth bristles absent to as long as the achenes, straight or curved, not contorted.
            5. Perianth bristles 0–3, much shorter than the achenes, smooth or with teeth only near the tips . . . ***S. georgianus***
            5' Perianth bristles 6, as long as the achenes, with retrorse bristles in the distal 30–50%. . . . . ***S. pallidus***
        4' Perianth bristles longer than the achenes, contorted.
            6. Perianth bristles greatly exceeding floral scales at maturity, therefore the inflorescences looking woolly when achenes mature; plants forming dense tussocks; habitat low to moderate elevation wetlands. . . . . ***S. cyperinus***
            6' Perianth bristles slightly exceeding floral scales at maturity; plants spreading on well-developed rhizomes; habitat high elevation bogs and fens. . . . . ***S. congdonii***

Scirpus atrocinctus Fernald

Blackgirdle bulrush

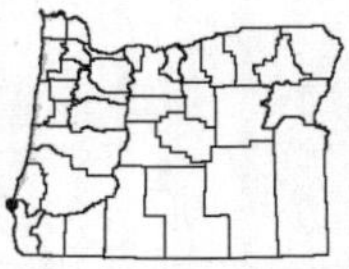

Plants forming dense tussocks; rhizomes short. **Leaves** blades 25–63 cm × 3–6 mm. **Inflorescences** terminal, open. **Spikes** in small stalked clusters, the central spike of each cluster sessile, others usually stalked, spikes ovoid or ellipsoid, 4–7 × 2–2.7 mm; floral scales 1.3–1.8 mm, usually blackish, at least distally, apex rounded or with very small apical point 0–0.05 mm. **Flowers** perianth bristles persistent, 6, contorted, much longer than achene, smooth, projecting beyond floral scales, mature inflorescence appearing woolly; stigmas 3. **Achenes** plumply trigonous or plano-convex, 0.7–0.9 × 0.4–0.5 mm. ***2n***=68.

Marshes, cranberry bogs, disturbed wetlands. 0–50 m. Est. WA; northeast to Nunavut, east to Newfoundland, souh to WY, IL, and DE. Exotic.

Scirpus atrocinctus resembles *S. cyperinus,* with which it hybridizes.

Scirpus congdonii N.L. Britton

Congdon's bulrush

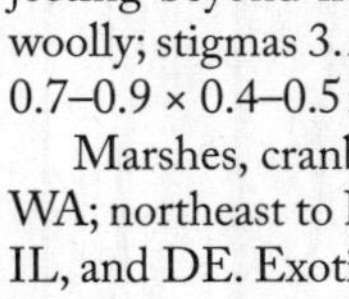

Plants spreading, rhizomatous. **Leaves** 13–26 cm × 3–7 mm. **Inflorescences** terminal. **Spikes** in dense clusters of 2–20, spikes sessile, ovoid or narrowly ovoid, 2.5–6 × 1–3 mm; floral scales 1.5–2.7 mm, black with pale midribs, apex short-awned, awn 0.2–0.4 mm. **Flowers** perianth bristles persistent, 6, slender, contorted, much longer than achene and projecting beyond it, with scattered, often inconspicuous, antrorse teeth in distal 50%, enclosed within floral scales or some bristles projecting; stigmas 3. **Achenes** plumply trigonous or plano-convex, 0.9–1.3 × 0.6–0.8 mm.

Lake margins and wet meadows. 900–1900 m. Casc, Sisk. CA, NV. Native.

Strongly rhizomatous, *S. congdonii* looks much like a dwarfed *S. microcarpus* with very dense inflorescences and blackish floral scales. It is common in montane marshes within its limited range.

Scirpus cyperinus (Linnaeus) Kunth

Woolgrass

Scirpus cyperinus (L.) Kunth var. *brachypodus* (Fernald) Gilly

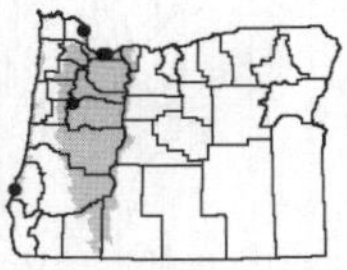

Plants forming dense tussocks; rhizomes branching, short. **Leaves** 22–80 cm × 3–10 mm. **Inflorescences** terminal, open; branches ascending or spreading. **Spikes** in dense cymes of 2–15, central spike of each cyme sessile, others sessile or pedicellate, spikes broadly ovoid, ovoid, or sometimes cylindric, 3.5–8 × 2.5–3.5 mm; floral scales 1.1–2.2 mm, reddish brown, brownish, or blackish, apex apiculate or mucronate, apiculus or mucro to 0.1 mm. **Flowers** perianth bristles persistent, 6, contorted, much longer than achene, smooth, projecting beyond floral scales, mature inflorescence appearing woolly; stigmas 3. **Achenes** plumply trigonous or plano-convex, 0.6–0.9 × 0.3–0.5 mm. ***2n***=66.

Marshes, edges of ponds, ditches, wet meadows, disturbed wetlands. 0–100 m. Casc, Est, WV. WA; south to Mexico; eastern North America. Exotic.

Scirpus cyperinus is identified by its big, open inflorescence with small spikes that look woolly when achenes are mature because of the contorted perianth bristles that are much longer than the floral scales.

Scirpus georgianus R.M. Harper

Georgia bulrush

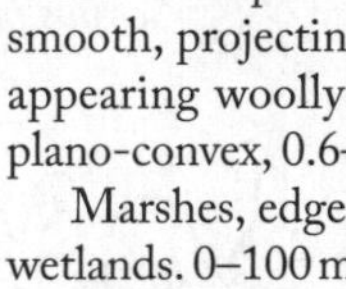

Plants cespitose; rhizomes short. **Leaves** blades 19–50 cm × 6–13 mm. **Inflorescences** terminal. **Spikes** in clusters of 4–35, spikes sessile, ovoid, 2–4 × 1–2 mm; floral scales 1–1.8 mm, brown or blackish brown

with pale midribs, apex mucronate, mucro 0.1–0.3 mm. **Flowers** perianth bristles persistent when present, 0–3, slender, straight or curved, much shorter than the achene, smooth or rarely with retrorse teeth near apex, enclosed within floral scales; stigmas 3. **Achenes** plumply trigonous or plano-convex, 0.6–1.2 × 0.4–0.5 mm. **2*n***=50, 52, 54.

Ditches, marshes, disturbed wetlands. 0–200 m. WV. From MN and New Brunswick, south TX to GA; New Zealand. Exotic.

Native to southeastern North America, *S. georgianus* has established disjunct populations north to Canada as a result of accidental seed transport by humans. The Oregon populations were originally identified as *S. pallidus*, which has perianth bristles as long as the achenes.

Scirpus microcarpus J. Presl & C. Presl
Small-fruited bulrush

Scirpus microcarpus J. Presl & C. Presl var. *longispicatus* M. Peck, *Scirpus microcarpus* J. Presl & C. Presl var. *microcarpus*

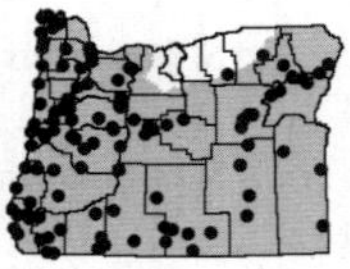

Plants spreading; rhizomes long, with conspicuous nodes and internodes. **Leaves** 23–60(75) cm × 5–15(20) mm. **Inflorescences** terminal; branches spreading and ascending. **Spikes** in dense clusters of (1)3–18 spikes, sessile, ovoid or narrowly ovoid, 2–8 × 1–3.5 mm; floral scales 1.1–3.4 mm, green or black, apex rounded or acute to mucronate, mucro 0–0.2 mm. **Flowers** perianth bristles persistent, (3)4(6) per flower, straight or curved, shorter than to 1.5 times as long as achene, retrorsely spinulose almost to base, enclosed within (occasionally weakly projecting from) floral scales; stigmas 2(3). **Achenes** biconvex to plano-convex, 0.7–1.6 × 0.8–1 mm. **2*n***=64, 66.

Marshes, ditches, wet meadows. 0–2100 m. All ecoregions except Col. CA, ID, NV, WA; north to AK, east to Newfoundland, KY, and DE, southeast to NM, south to Mexico; eastern Asia. Native.

Scirpus microcarpus is a common, widespread species. The broad leaves can be confused with *Carex amplifolia* if inflorescences are not found. Some botanists prefer to treat the plants from eastern North America as a distinct species, *S. rubrotinctus*, but the differences are small.

Scirpus pallidus (N.L. Britton) Fernald (p. 254)
Pale bulrush

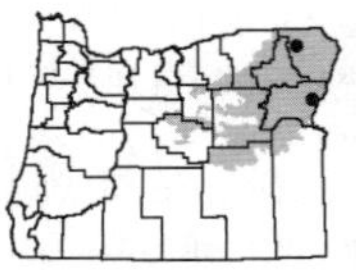

Plants cespitose; rhizomes short. **Leaves** 20–55 cm × 8–16 mm. **Inflorescences** terminal; branches ascending and drooping. **Spikes** aggregated in a few dense clusters of 12–130, spikes sessile, narrowly ovoid, 4–5 × 1.8–2.3 mm; floral scales 1.6–2.8 mm, black or brownish black with pale midribs, apex with awn 0.4–0.6(1.2) mm. **Flowers** perianth bristles persistent, 6, rather stout, straight or curved,

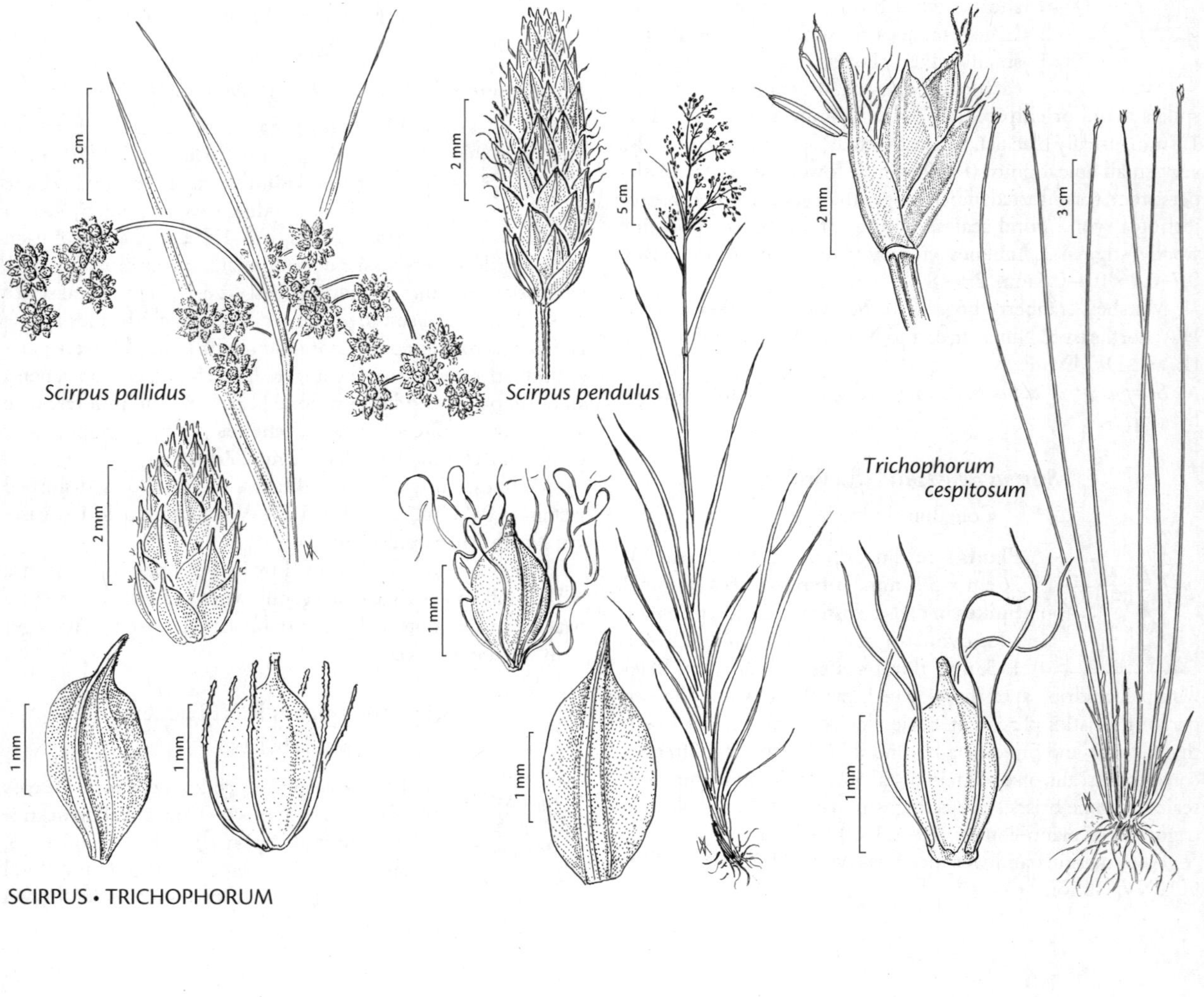

SCIRPUS • TRICHOPHORUM

longest bristles equaling achene, with retrorse teeth in distal 0.3–0.5, enclosed within floral scales; stigmas 3. **Achenes** plumply trigonous or plano-convex, 0.8–1.2 × 0.4–0.6 mm. **2*n***=56.

Marshes, streamsides, ditches. 0–700 m. BW. ID, WA; north to British Columbia, east to PA, southeast to TX. Native.

Scirpus pallidus is similar to introduced *S. georgianus,* which has reduced perianth bristles.

Scirpus pendulus Muhlenberg (p. 254)
Drooping bulrush

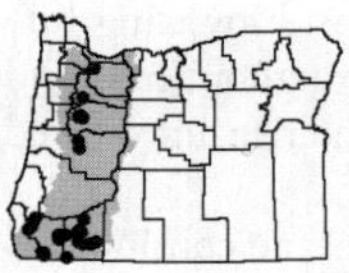

Plants cespitose; rhizomes short. **Leaves** 15–40 cm × 4–8(12) mm. **Inflorescences** terminal; branches ascending and drooping. **Spikes** in open cymes, central spike of each cyme sessile, others long-pedicellate, spikes ovoid to lance-ovoid or subcylindric, 5–10(12) × 2–3 mm; scales ovate, 2 mm, brown to red-brown with green midribs, apex mucronate, mucro 0.1–0.3 mm. **Flowers** perianth bristles persistent or breaking off near base, 6, slender, strongly contorted, much longer than achene and often projecting far beyond it despite being strongly contorted, smooth, enclosed within floral scales or scarcely projecting beyond them; stigmas 3. **Achenes** plumply trigonous or plano-convex, 1–1.2 × 0.6–0.8 mm. **2*n***=40.

Marshes, moist meadows, ditches. 50–1500 m. Casc, Sisk, WV. CA; east from MT to New Brunswick, south to Mexico; Australia. Native.

Scirpus pendulus is recognized by its cylindrical spikes on drooping branches, with prominent green midribs on the brown floral scales.

TRICHOPHORUM Clubrush

Herbs perennial, cespitose, rhizomatous or not. **Culms** terete. **Leaves** basal or sub-basal; sheaths bladeless or distal sheaths with blade to 5 mm; ligules present; blades rudimentary, elongate, linear, less than 1 cm × 1 mm. **Inflorescences** terminal; spike 1; involucral bract 1, suberect, scale-like, apex mucronate or awned. **Spikes** scales 3–9, spirally arranged, each subtending a flower. **Flowers** bisexual; perianth of 0–6 bristles, straight, shorter than to about 20 times as long as achene, smooth or scabrous; stamens 3; styles with base not forming a tubercle; stigmas 3. **Achenes** trigonous or plano-convex.

Circumpolar to circumboreal North America, Andean, and tropical Southeast Asian mountains. 9 species; 1 species treated in *Flora*.

This odd genus superficially resembles an *Eleocharis* but has short leaf blades and short inflorescence bracts. It has traditionally been placed in *Scirpus*.

Trichophorum cespitosum (Linnaeus) C.J. Hartman (p. 254)
Tufted clubrush

Scirpus cespitosus L.

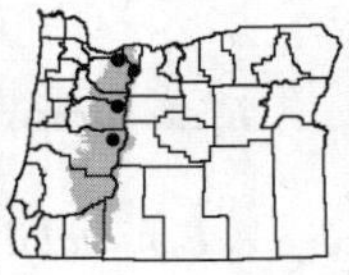

Plants densely cespitose; rhizomes absent. **Culms** grooved, terete, 5–45 cm, smooth. **Leaves** blades absent or 1.5–8 × 0.3–0.4 mm, much shorter than culms. **Inflorescences** spikes 3–9-flowered, 3.3–7 × 1.2–3 mm; bracts equaling spikes or to 3 mm longer, 3.3–4.5 mm. **Spikes** scales reddish brown to dark brown, apex mucronate to acute. **Flowers** perianth bristles 6, terete, equaling or exceeding achenes, brown, smooth to scabrous; anthers 1.5–2.6 mm. **Achenes** compressed trigonous, 1.4–1.7 × 0.6–0.9 mm. **2*n***=104.

Sphagnum bogs and fens in the Cascades. 800–1700 m. Casc. ID, WA; east to Greenland, southeast to CO, southeastern US; Eurasia. Native.

HYDROCHARITACEAE Tape-grass or frog-bit family

Stephen C. Meyers (except where noted)

Herbs monoecious or dioecious, annual or perennial, aquatic, freshwater or marine. **Stems** rhizomatous, creeping, leafy. **Leaves** basal or cauline, opposite or whorled, usually sheathing at base, glabrous. **Inflorescences** 1-flowered or a cyme; flowers 1–few, subtended by sheathing, free-floating. **Flowers** perianth present or not, radial; perianth tube 0 or elongated, peduncle-like in flower; sepals 0 or 3, green; petals 0 or 3(4), translucent or white; stamens (1)3–many; ovary inferior; style lobes 2–4, lobed or notched. **Fruits** achene- or berry-like, dehiscence irregular, ovoid to lance-ellipsoid to cylindric or fusiform, submersed. **Seeds** many.

Cosmopolitan. 17 genera; 4 genera treated in *Flora*.

Key to Hydrocharitaceae genera:

1. Leaves basal . . . . **VALLISNERIA**
1' Leaves cauline.
    2. Perianth 0; fruits enclosed by a thin, translucent membrane . . . . **NAJAS**
    2' Perianth present; fruits not enclosed by a thin, translucent membrane.
        3. Whorls with 5 or more leaves per node; petals greater than 6 mm long . . . . **EGERIA**
        3' Whorls with 2–3 leaves per node (at proximal nodes, distal nodes with up to 7 leaves) or leaves opposite; petals less than 6 mm long . . . . **ELODEA**

EGERIA Brazilian waterweed

Lareina Shipler

Plants dioecious, perennial; rooted in mud, submersed or floating in fresh waters; rhizomes and stolons absent. **Stems** branched or not, elongate. **Leaves** cauline, whorled, 5 or more leaves per node; blades linear, 1-nerved, sessile, apex rounded. **Inflorescences** 1-flowered, sessile; bracts slender. **Flowers** floating on surface, projected from an elongated floral tube; petals white. **Staminate flowers** sepals 3; petals 3(4); stamens 3–9. **Pistillate flowers** sepals 3; petals 3; stigmas 3, 3–4 lobed. **Fruits** ovoid, smooth. **Seeds** fusiform, 1–5, mucilaginous.

Fresh water habitats nearly worldwide. 2 species; 1 species treated in *Flora*.

Egeria densa Planchon (p. 257)

Brazilian waterweed, South American waterweed

Elodea densa (Planch.) Casp.

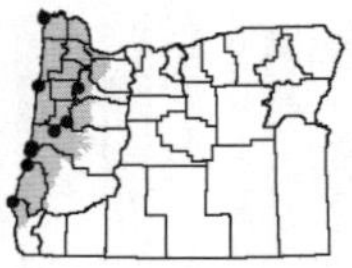

Stems densely leafy; 2–3 mm thick. **Leaves** narrowly lanceolate, 1.5–4 cm × 2–5 mm, margin finely serrate. **Staminate flowers** petals 8–10 mm, white, round; pedicel to 8 cm long. **2*n***=46.

Shallow parts of lakes, ponds and streams. Flowering Jul–Oct. 0–300 m. CR, Est, WV. CA, WA; throughout mid and southern North America; native to South America, introduced throughout the world. Exotic.

A popular aquarium plant, this species has and continues to escape cultivation when individuals carelessly empty their fish tanks. Interestingly, only staminate plants are known to exist anywhere outside the native range of the species. These plants are obviously reproducing vegetatively.

ELODEA Waterweed

Plants perennial, submersed in fresh waters, dioecious or rarely bisexual; rooting at nodes; rhizomes and stolons absent. **Stems** erect, branched or unbranched, elongate. **Leaves** cauline, whorled, 2–3 leaves per node proximally, usually 2–3(4) per node distally or leaves opposite, sessile; blade linear to linear-lanceolate, apex acute, margins minutely serrate. **Inflorescences** solitary, axillary, sessile; bracts present. **Flowers** floating on surface projected by an elongated perianth tube, sessile; petals white. **Staminate flowers** stamens 7–9, anthers oval. **Pistillate flowers** styles 3, lobed or not. **Fruits** cylindric to ovoid, smooth, irregular dehiscence. **Seeds** with or without basal hairs.

Western Hemisphere and Europe. 5–12 species; 2 species treated in *Flora*.

Elodea bifoliata has been collected once, near Bend, by Morton Peck (date not recorded on specimen). This species is distinguished from all *Elodea* species in Oregon by having some nodes with two leaves (as opposed to always three or four) and its small seeds (less than 4 mm long).

Key to *Elodea* species:

1. Leaves (1)1.5–3.5(5) mm wide, often crowded at stem tips; styles more than 2 mm long; sepals 2–5 mm long; seeds without basal hairs present . . . . **_E. canadensis_**

1' Leaves 0.9–1.5(2) mm wide, usually not crowded at stem tips; style 2 mm long or less; sepals 1–2 mm long; seeds with basal hairs present . . . . **_E. nuttallii_**

Elodea canadensis Michaux

Canadian waterweed

Anacharis canadensis (Michx.) Planch., *Elodea planchonii* Casp.

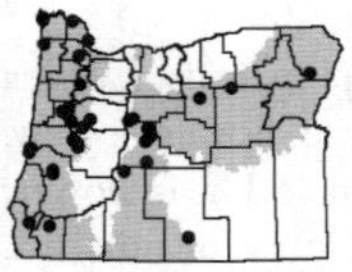

Leaves nodes mostly with 3(4) leaves, often crowded at stem tips, linear to narrow-elliptic, 5–15 mm long, (1)1.5–3.5(5) mm wide. **Inflorescences** staminate and pistillate bracts cylindric, 8–17 mm long. **Staminate flowers** sepals 3–5 m long; stamens 7–9, anthers 1.5–3 mm long. **Pistillate flowers** sepals 2–3 mm long; styles 2.5–4 mm long. **Seeds** fusiform, 4–6 mm long, without basal hairs present. **2*n***=24.

Lakes, ponds, rivers. Flowering Jul–Aug. 0–2300 m. BW, CR, ECas, Est, Lava, Sisk, WV. CA, ID, NV, WA; throughout most of North America; Asia, Australia, Europe. Native.

Elodea nuttallii (Planchon) H. St. John (p. 257)

Nuttall's waterweed, western waterweed

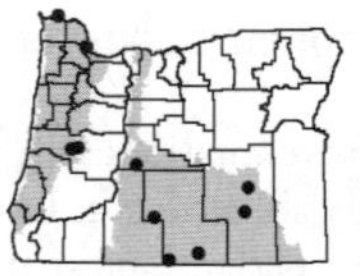

Leaves nodes mostly with 3(4) leaves, usually not crowded at stem tips, linear to lanceolate, 5–14 mm long, 0.9–1.5(2) mm wide. **Inflorescences** staminate bracts 2–4 mm long; pistillate bracts 8.5–15 mm long. **Flowers** sepals 1–2 mm long. **Staminate flowers** sepals approximately 1 mm long; stamens 9, anthers 1–1.5 mm long. **Pistillate flowers** sepals approximately 1 mm long; styles mostly 1–2 mm long. **Seeds** fusiform, 4–5 mm long, with basal hairs present. **2*n***=48.

Lakes, ponds, rivers. Flowering Jul–Aug. 0–1600 m. BR, CR, ECas, WV. CA, ID, WA; throughout much of US, north just into southern Canada; Asia, Europe. Native.

NAJAS Water nymph

Lareina Shipler

Plants monoecious, annual, submersed in fresh waters. **Stems** branched, slender. **Leaves** simple, cauline, more or less opposite or whorled; forming sheaths; blades generally linear, sessile; margins usually toothed, teeth similar to blades. **Inflorescences** 1(few), axillary, usually inconspicuous. **Flowers** calyx and corolla absent; floating on surface projected from an elongated floral tube. **Staminate flowers** stamen 1, subsessile, subtended by 0–2 minute involucres. **Pistillate flowers** style 1, short, stigma 2–4 lobed; generally not subtended by involucres. **Fruits** achene-like, fusiform, enclosed in a thin, translucent membrane. **Seeds** fusiform to ovoid, hard, brittle.

Cosmopolitan. Approximately 50 species; 1 species treated in *Flora*.

Some authorities place *Najas* within the monogeneric family Najadaceae.

Najas guadalupensis has been collected twice in the Portland area, in the 1880s. It has also been collected once recently (2006) along the along Owyhee River in Malheur County. This species is distinguished from the more common *N. flexilis* by leaves that are not crowded on the branchlets and dull, pitted seeds.

Najas flexilis (Willdenow) Rostkovius & W.L.E. Schmidt (p. 257)

Wavy water nymph, slender water nymph, slender najas

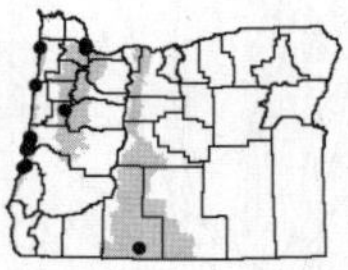

Plants slender, mat-like or not, often floating. **Leaves** numerous, crowded on branchlets, narrow and linear except at base, 1.5–2.5 cm long, 1-nerved, margins minutely serrulate. **Staminate flowers** 1–3 mm long, stamen enclosed in a spathe. **Pistillate flowers** 2–5 mm long. **Fruits** achene-like, fusiform; enclosed in a scarious sac-like coat. **Seeds** pale, narrowly to widely obovate, approximately 2.5 mm long and less than 1.5 mm wide. $2n$=12, 24.

Ponds, slow streams, shallow bodies of brackish waters. Flowering Jun–Sep. 0–1300 m. ECas, Est, WV. CA, ID, WA; north to AK, east to Newfoundland, southeast to AZ; north central and northeastern states; northern Europe. Native.

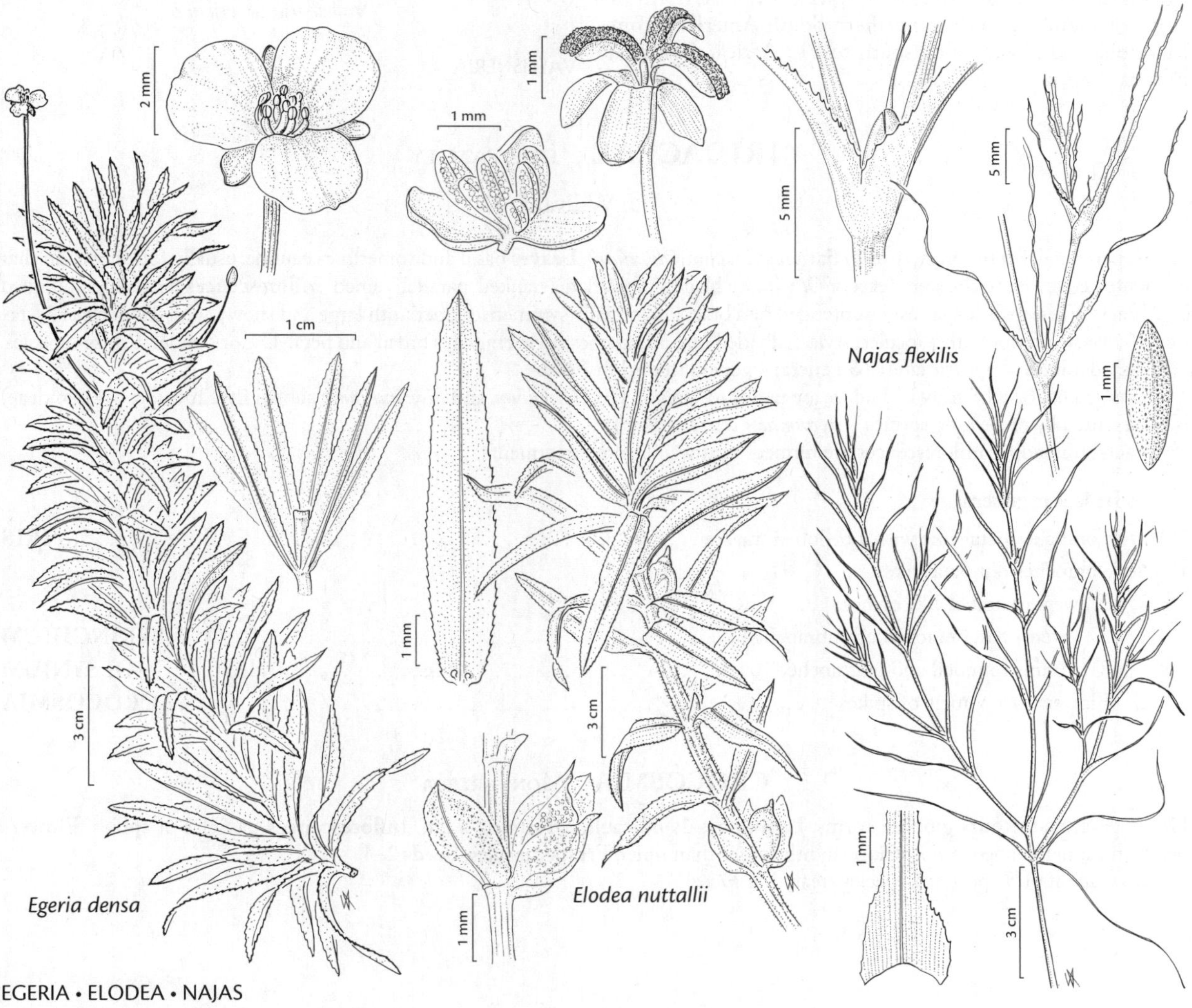

EGERIA • ELODEA • NAJAS

VALLISNERIA Tape-grass, eelgrass

Plants dioecious, perennial, submersed or floating in fresh or brackish waters. **Leaves** basal, sessile. **Inflorescences** cymes. **Flowers** enclosed in spathes. **Staminate flowers** sessile. **Pistillate flowers** solitary, pedicellate, floating. **Fruits** retracted under water by pistillate coils after anthesis. **Seeds** few.

Cosmopolitan. 2 species; 1 species treated in *Flora*.

Vallisneria americana Michaux (p. 258)

Tape-grass, American eelgrass, American water celery

Vallisneria neotropicalis L.

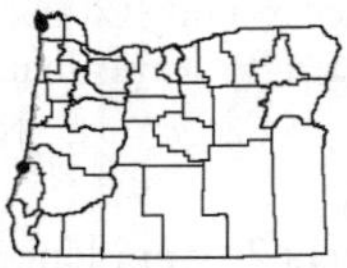

Plants herbs; rhizomes and stolons present. **Stems** unbranched. **Leaves** 10–75(150) × (0.1)0.3–1.5 cm; blades linear, thin, flat, ribbon-like, lighter colored in middle, apex obtuse to slender pointed, midveins of 8–10 rows. **Inflorescences** long-pedunculate; bracts present. **Flowers** sepals 3, unequal, 1 smaller than others; petals 3, linear, white to transparent. **Staminate flowers** stamens (1)2(3), filaments distinct, floating to surface, anthers round. **Pistillate flowers** stigmas 2–3-lobed, style 1. **Fruits** cylindric to elliptical, ridged, 5–10 cm long, dehiscence irregular. **Seeds** elliptical, glabrous. **2*n***=20.

Fresh to brackish rivers, streams, lakes, ponds, and bays. Flowering Jul–Sep. 0–50 m. Est. CA, ID, NV, WA; north to British Columbia, south to northern South America; From Manitoba and Nova Scotia south to FL; Africa, Australia, Eurasia. Native.

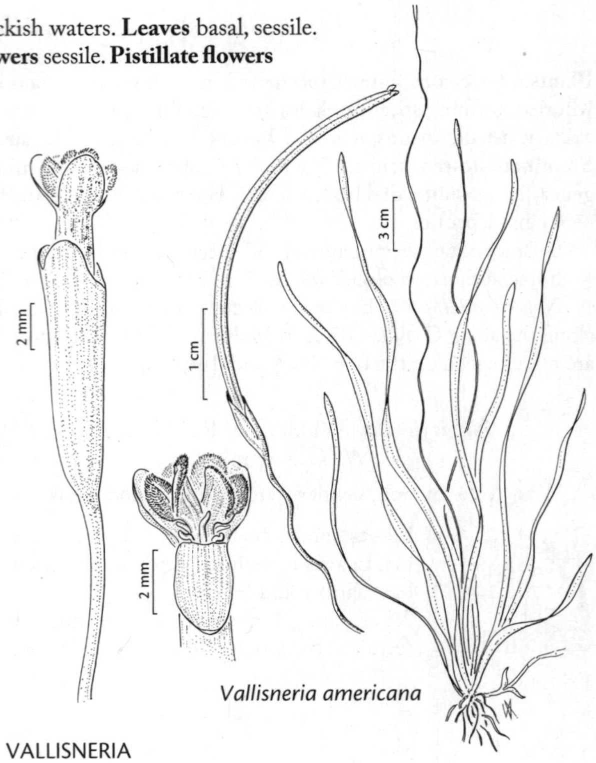

VALLISNERIA

IRIDACEAE Iris family

Ann Willyard

Herbs perennial. **Stems** erect; usually flattened (except *Olsynium*). **Leaves** basal and sometimes cauline, usually forming a fan that is oriented edgewise to the stem (except *Olsynium*); blade simple, flat, 2-ranked, parallel-veined. **Inflorescences** terminal, subtended by 2 bracts. **Flowers** bisexual, each subtended by a bract and radially symmetrical; perianth large and showy, with 2 whorls of 3; stamens 3; ovary inferior with 3 locules; style 1, divided into 3 branches, style branches broad and petal-like or slender. **Fruits** capsules.

Cosmopolitan. Approximately 65 genera; 4 genera treated in *Flora*.

Three genera of subfamily Iridoideae are native to Oregon (*Iris*, *Olsynium*, and *Sisyrinchium*); subfamily Crocoideae (=Ixioideae) is represented by one exotic species (*Crocosmia* × *crocosmiiflora*).

Bracts subtending inflorescences are termed "spathes" in some treatments.

Key to Iridaceae genera:

1. Style branches petal-like with a bi-lobed tip . . . . **IRIS**

1' Style branches not petal-like.

  2. Flowers in umbel-like cymes.

    3. Stems flat, branched or unbranched . . . . **SISYRINCHIUM**

    3' Stems rounded and unbranched . . . . **OLSYNIUM**

  2' Flowers in a panicle of spikes. . . . . **CROCOSMIA**

CROCOSMIA Montbretia

Herbs perennial. **Roots** globose corms. **Leaves** basal and cauline, forming a fan. **Inflorescences** panicles of spikes. **Flowers** perianth funnel-shaped, outer whorl slightly larger than inner. **Fruits** capsules. **Seeds** 2–4.

Cosmopolitan. 9 species; 1 species treated in *Flora*.

Crocosmia* × *crocosmiiflora (Lemoine) N.E. Brown (p. 260)

Montbretia

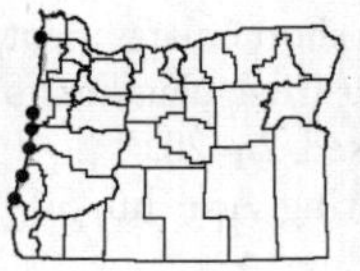

Herbs 50–100 cm tall; deciduous. **Roots** 15–25 mm in diameter; often with stolons. **Stems** usually 2–4 branched. **Leaves** 5–8, mostly basal; basal leaves much larger than cauline, blades lanceolate, 8–20 mm wide. **Inflorescences** 4–8-flowered, 2-ranked; outer bract usually exceeding inner, apex acute, inner bract apex forked. **Flowers** tepals 15–25 × 10–15 mm, orange to red; style branches entire. **Fruits** globose. **Seeds** 2–3 mm long, black.

Coastal habitats. Flowering Jul–Aug. 0–100 m. Est. CA; scattered throughout North America; South Africa. Exotic.

This species is an escaped ornamental. Some plants are partially or completely sterile.

IRIS Iris

Herbs 20–150 cm tall. **Roots** present or not; rhizomes present; bulbs present or not. **Leaves** cauline 0–4, usually similar to basal, subtending each branch, decreasing in length distally, sometimes bract-like and sheathing stem. **Inflorescences** cymes or flowers solitary. **Flowers** perianth united in tube; sepals clawed; signal a ridge, with a line of contrasting colors or a band of multicellular hairs along midline of sepal; filaments free, flattened and appressed to the under surface of style; styles divided into 3 branches, each branch further lobed into 2 petal-like tips; stigma a lip of tissue on abaxial surface of petal-like style. **Fruits** capsules, loculicidal. **Seeds** few to many.

North America, Eurasia. 280 species; 9 species treated in *Flora*.

Oregon has one exotic member of subgenus *Iris* (bearded iris; *I. germanica*). There are seven native species and one exotic member of subgenus *Limniris* (beardless iris). The Pacific Coast irises (Section *Limniris* ser. *Californicae*) is the most species-rich group of *Iris* in Oregon, represented by *I. bracteata*, *I. chrysophylla*, *I. douglasiana*, *I. innominata*, and *I. tenax*. The geographic range of these five species extends into California, and there are an additional six species of Pacific Coast iris that are endemic to California. Hybrids are reported among many pairs of species in section *Californicae*, making identification of individual plants in areas of sympatry challenging. The other series in section *Limniris* each have one species in the Oregon flora: series *Laevigatae* (*I. pseudacorus*, an exotic species) and series *Longipetalae* (*I. missouriensis*, a native species). We have one member of section *Lophiris* (crested iris), *I. tenuis*.

Several specialized terms have been applied to Iris flowers: "falls" (sepals); "standards" (petals). The signal along the midline of the sepal claw is called a "beard" if it is a band of hairs and a "crest" if ridged. The bilobed, petal-like style tips are termed "style crests" by some. We avoid this term because of possible confusion with signal crests. The shape of the style tips and the length of the perianth tube are useful characters to distinguish among species, especially in series *Californicae*. Great care must be taken to preserve these characters in pressed specimens.

Iris × *thompsonii* (hypothesized to be a hybrid between *I. douglasiana* and *I. innominata*) has been consistently collected in Curry County for decades where the ranges of the putative parental species overlap.

Key to *Iris* species:

1. Signal bearded with multicellular hairs . . . . ***I. germanica***

1' Signal not bearded.

 2. Signal an inconspicuous crest with low, yellow ridges; rhizomes heterogeneous with stem innovations appearing at some distance from parent rhizome . . . . ***I. tenuis***

 2' Signal of lines or colored patches; rhizomes of one kind.

  3. Rhizomes stout, 1–4 cm diameter; roots fleshy.

   4. Leaves 20–30 mm wide; stem branched; flowers many, bright yellow or cream, often growing in water; escaped from cultivation . . . . ***I. pseudacorus***

   4' Leaves 3–12 mm wide, stem unbranched; flowers generally 2, lilac or whitish; not growing in water; native . . . . ***I. missouriensis***

  3' Rhizome slender, 3–10 mm diameter; roots fibrous.

   5. Perianth tube shorter than 10 mm.

    6. Leaves deciduous and less than or equal to 5 mm wide, light green; inflorescence usually 1(2)-flowered . . . . ***I. tenax***

    6' Leaves evergreen and equal to or greater than 6 mm wide, dark green on one side, pale yellow on the other; inflorescence usually 2-flowered . . . . ***I. bracteata***

   5' Perianth tube longer than 10 mm.

    7. Perianth tube longer than 40 mm and slender; style tips very long and narrow; inflorescence bracts broad and closed . . . . ***I. chrysophylla***

    7' Perianth tube shorter than 40 mm and not slender; style tips rounded and reflexed; inflorescence bracts not as above.

     8. Stem unbranched, flowers 1–2; leaves 2–4 mm wide; woodland habitats . . . . ***I. innominata***

     8' Stem sometimes branched; flowers 2–3; leaves to 22 mm wide; coastal habitats . . . . ***I. douglasiana***

Iris bracteata S. Watson

Siskiyou iris

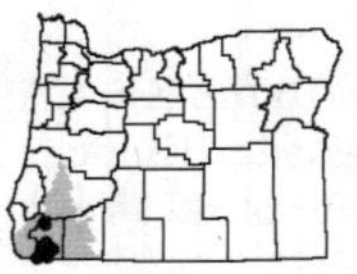

Herbs to 60 cm tall. **Roots** fibrous; rhizomes sheathed with old leaf bases. **Stems** unbranched. **Leaves** basal leaves evergreen, thick and rigid, 6–10 mm wide, glossy green on one side and light yellow-green on other, often red or pink basally; cauline leaves 3–6, some bract-like, overlapping and sheathing the stem. **Inflorescences** usually 2-flowered; bracts opposite and closed. **Flowers** perianth cream to buff-yellow; sepals with deeper yellow signal, veined with purple or brown; floral tube less than 10 mm, funnel-shaped; style tips oblong, reflexed and toothed; stigmas triangular. **Fruit** capsules barrel-shaped, tapering into pedicels. **Seeds** dark brown, irregular in shape, wrinkled. **2*n***=40.

Shady forests, particularly ponderosa pine. Flowering Apr–Jun. 200–600 m. Sisk. CA. Native.

Iris bracteata hybridizes with *I. chrysophylla*, *I. douglasiana*, *I. innominata*, and *I. tenax*.

Iris chrysophylla Howell (p. 260)

Yellow-leaf iris, slender-tubed iris

Iris californica Leichtlin, *Iris macrosiphon* Torr.

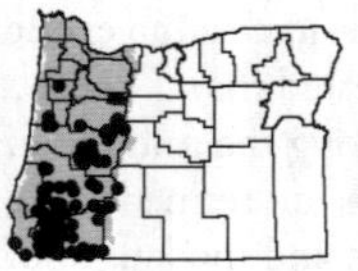

Herbs to 20 cm tall. **Roots** fibrous; rhizomes compact, dark brown, slender, covered with old leaf bases. **Stems** unbranched or stemless. **Leaves** evergreen, narrow, 3–5 mm wide, light green, often pink or red basally. **Inflorescences** usually 2-flowered; bracts broad, opposite, closed, outer much shorter and narrower than inner. **Flowers** perianth cream or very pale yellow, segments membranous and narrow, with deeper yellow or lavender veins; sepals with bluish tinge and veins; floral tube long and slender, 45–120 mm long, with short enlargement at top; style tips very long and narrow; stigmas triangular. **Fruits** capsules, ovoid. **Seeds** buff, slightly compressed. **2*n***=20.

Open conifer forests, roadsides. Flowering Apr–Jul. 50–2000 m. Casc, CR, Sisk, WV. CA. Native.

Iris chrysophylla hybridizes with *I. bracteata*, *I. douglasiana*, *I. innominata*, and *I. tenax*.

Iris douglasiana Herbert (p. 260)

Douglas iris

Iris douglasiana Herb. var. *oregonensis* R.C. Foster

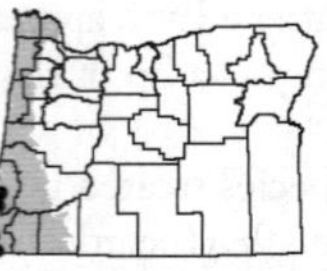

Herbs 20–80 cm tall. **Roots** fibrous; rhizomes slender, freely branching, covered with old leaf bases, forming large colonies. **Stems** sometimes branched. **Leaves** evergreen, to 22 mm wide, shiny yellow-green above, duller green below, often red or purple basally; cauline leaves reduced. **Inflorescences** 2–3-flowered; bracts opposite or separate, open or closed, 5–10 cm long. **Flowers** red-purple, lavender, gray-blue, cream, or white, with gold signal and blue or purple veins; floral tube 10–28 mm long, usually widening to bowl shape; style tips rounded to oblong, reflexed, overlapping, and coarsely toothed; stigmas triangular. **Fruits** sharply triangular in cross section, angles ridged, with nipple-like projection above. **Seeds** pear-shaped, dark brown, wrinkled. **2*n***=40.

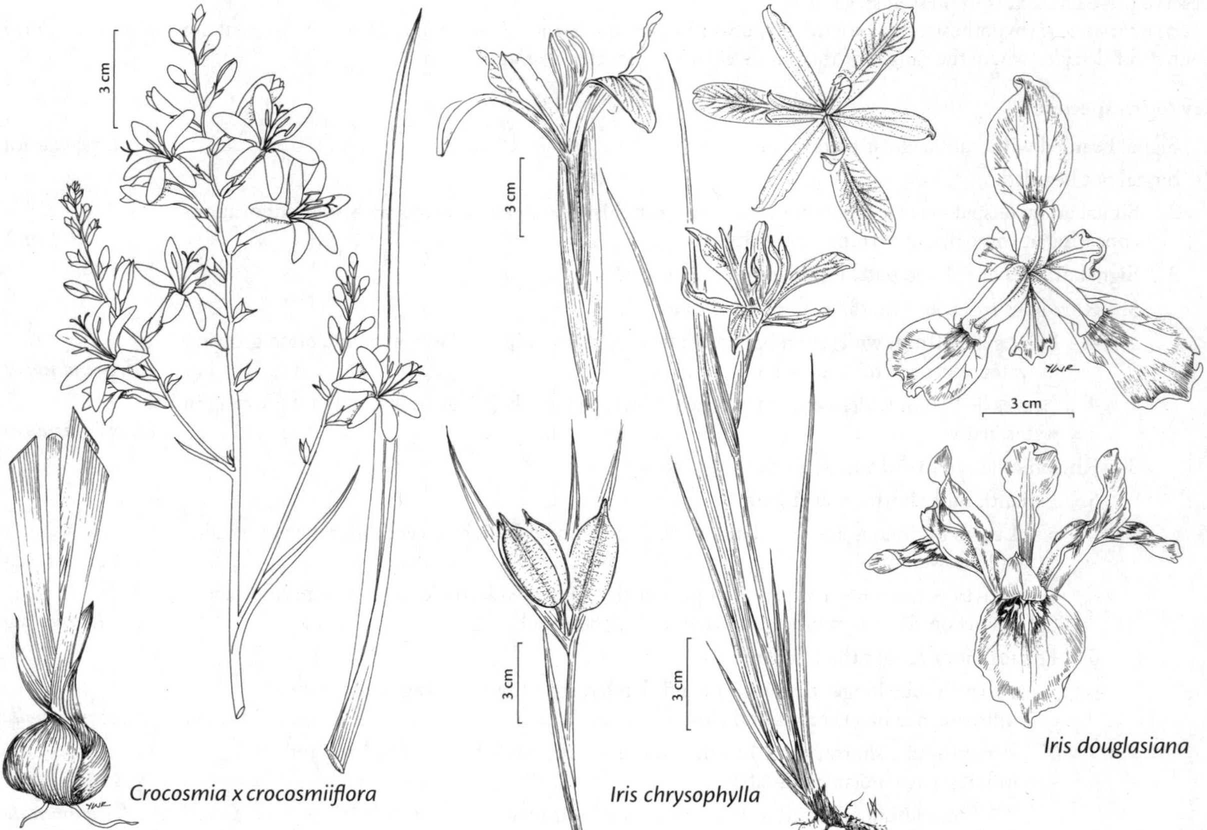

Crocosmia x crocosmiiflora *Iris chrysophylla* *Iris douglasiana*

CROCOSMIA • IRIS

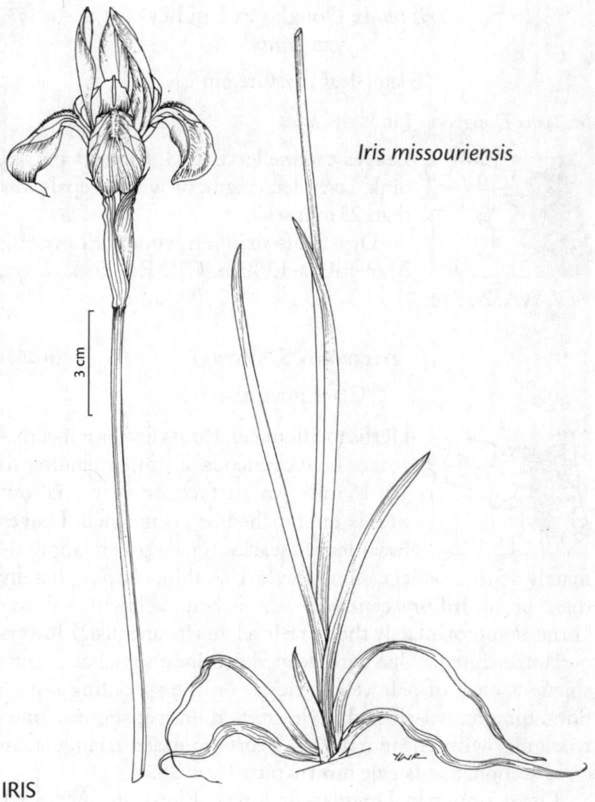

IRIS

Coastal habitats, open woods, and fields. Flowering Mar–Jul. 0–400 m. CR, Est, Sisk. CA. Native.

The leaves of *I. douglasiana* are unpalatable to livestock, and the species is considered a noxious weed in pastures. In Oregon, *I. douglasiana* hybridizes with *I. bracteata*, *I. chrysophylla*, *I. innominata*, and *I. tenax*. In California, it hybridizes with *I. tenuissima*, *I. fernaldii*, *I. hartwegii*, *I. macrosiphon*, *I. munzii*, and *I. purdyi*. Cultivars have been developed for the nursery trade from some of these hybrids.

Iris germanica Linnaeus

German iris, bearded iris, garden iris

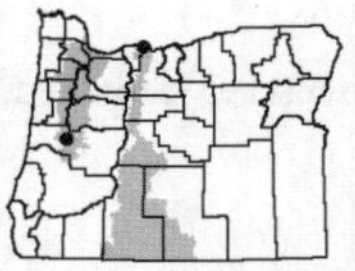

Herbs to 120 cm tall. **Roots** fleshy; rhizomes homogeneous, many-branched. **Stems** branched. **Leaves** unifacial, sword-shaped, 3.5 cm wide, forming an obvious fan. **Inflorescences** 1–3-flowered; bracts green, sometimes with a purple base. **Flowers** perianth blue-violet, yellow, brown, or white with various patterns of color; sepals with beard along midrib of claw and lower part of limb; petals erect. **Fruits** roundly 3-lobed. **Seeds** red-brown, wrinkled.

Persisting after cultivation near old homesteads. Flowering Apr–May. 100–200 m. ECas, WV. CA, WA; throughout North America; Europe. Exotic.

Iris germanica is presumably a cultivar resulting from hybridization between *I. pallida* and *I. variegata*. Some populations may also contain genetic material from other species, in which case other proposed names may be more appropriate (i.e., *I.* × *conglomerata*). The genetic makeup of Oregon plants is currently unknown.

Iris innominata L.F. Henderson

Golden iris, Del Norte County iris

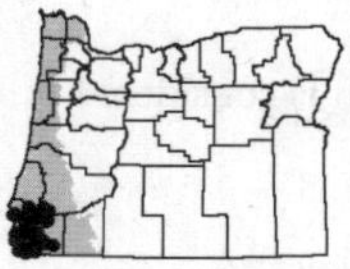

Herbs to 20 cm tall. **Roots** fibrous; rhizomes slender, freely branching, covered with leaf bases that form dense clumps. **Stems** unbranched. **Leaves** evergreen; basal leaves narrow, 2–4 mm wide, grass-like, longer than flowering stem, dark green and glossy, pink or purple basally; cauline leaves 2–4, spreading. **Inflorescences** 1(2)-flowered; bracts opposite, closed, broad. **Flowers** usually dark golden yellow, sometimes cream, pink, lavender, or purple, usually heavily veined purple or brown; floral tube slender, 15–30 mm; style tips rounded, reflexed, overlapping; stigmas triangular. **Fruits** round in cross section. **Seeds** dark brown, sharply angled, wrinkled and pitted. **2*n***=40.

Dry, open woods. Flowering Mar–Jul. 0–1500 m. CR, Est, Sisk. CA. Native.

Iris innominata hybridizes with *I. bracteata*, *I. chrysophylla*, *I. douglasiana*, and *I. tenax*.

Iris missouriensis Nuttall (p. 261)

Western blue flag, Rocky Mountain iris

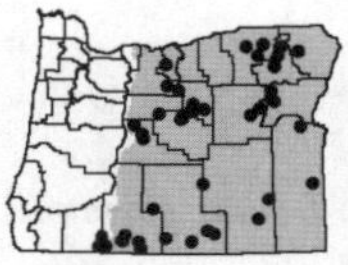

Herbs to 60 cm tall. **Roots** fleshy; rhizomes freely branching, stout, covered with old leaf bases. **Stems** unbranched or up to 2-branched. **Leaves** basal leaves light green, usually white basally or sometimes purplish, 3–12 mm wide, glaucous; cauline leaves similar to basal. **Inflorescences** 2–3-flowered; bracts keeled, outer usually shorter than inner. **Flowers** perianth blue, lavender, or white, veined deeper violet; sepals deeply veined lilac-purple, with yellow-white signal at base of limb; claws yellowish white, veined and dotted with purple; floral tube less than 12 mm; style tips rounded, margins irregularly toothed; stigmas 2-lobed. **Fruits** round in cross section, with 6 ridges. **Seeds** globular to pear-shaped, light brown, wrinkled. **2*n***=38.

Wet meadows, roadside ditches, margins of streams. Flowering May–Jul. 300–2000 m. BR, BW, Col, ECas, Lava, Owy. CA, ID, NV, WA; north to British Columbia, northeast to MN and southeast to NM, south to Mexico. Native.

The leaves of *I. missouriensis* are unpalatable to livestock, and the species is considered a noxious weed in pastures. This species is morphologically variable across its large geographic range.

Iris pseudacorus Linnaeus

Yellow flag, yellow water iris

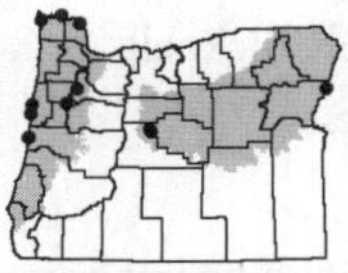

Herbs to 150 cm tall. **Roots** fleshy; rhizomes pink, freely branching, producing extensive clumps with remains of old leaves. **Stems** usually 1-branched. **Leaves** deciduous, stiff, erect at first then recurved; blades dark green with prominent median thickening, 20–30 mm wide; cauline leaves equal in height to inflorescence. **Inflorescences** 4–12-flowered; bracts subequal, green with brown margins, outer bract strongly keeled. **Flowers** perianth bright yellow or cream; signal a darker yellow basal patch delimited by short, brown lines; petals without veining; tube 12–13 mm; stigmas rounded with prominent tongue. **Fruits** angled, with obvious groove at each angle. **Seeds** D-shaped, flattened, corky, lustrous. **2*n***=24, 30, 32, 34.

Swamps, wet shores of rivers and lakes, irrigation ditches. Flowering May–Jun. 0–1000 m. BW, CR, Est, Lava, WV. CA, ID, NV, WA; throughout North America; northern Africa, Eurasia. Exotic.

A native to Eurasia and Africa, this taxon is considered invasive in Oregon.

Iris tenax Douglas ex Lindley

Tough-leaf iris, Oregon iris

Herbs to 30 cm tall. **Roots** fibrous; rhizomes slender, branched, forming dense clumps. **Stems** slender, unbranched. **Leaves** deciduous; basal leaves slender, approximately 5 mm wide, longer than stem, light green, pinkish basally; cauline leaves reduced, clasping basally then spreading. **Inflorescences** 1(2)-flowered; bracts narrow, open, attached up to 3 cm apart, outer longer than inner. **Flowers** purple, pink, lavender, cream, yellow, or white; floral tube less than 10 mm long; style tips rounded, reflexed, toothed; stigmas triangular. **Fruits** triangular in cross section, prominently ribbed. **Seeds** brown, wrinkled.

2 varieties.

Iris tenax hybridizes with *I. bracteata*, *I. chrysophylla*, *I. douglasiana*, and *I. innominata*.

Key to *Iris tenax* varieties:

1. Flowers yellow; cauline leaves 2; sepals more than 25 mm wide; endemic to Columbia and Washington counties . . . ***I. t.*** var. ***gormanii***

1' Flowers not yellow; cauline leaves 2–3; sepals less than 25 mm wide; widespread throughout western Oregon . . . . ***I. t.*** var. ***tenax***

Iris tenax Douglas ex Lindley var. *gormanii* (Piper) R.C. Foster

Gorman's iris

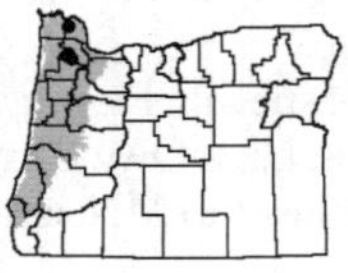

Leaves cauline leaves 2. **Flowers** bright yellow; sepals more than 25 mm wide.

Dry fields or open woods. Flowering May–Jun. 50–200 m. CR, WV. Native.

Iris tenax Douglas ex Lindley var. *tenax* (p. 263)

Tough-leaf iris, Oregon iris

Iris tenax Douglas ex Lindl. ssp. *tenax*

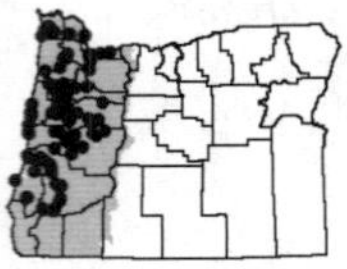

Leaves cauline leaves 2–3. **Flowers** purple, pink, lavender, cream, or white; sepals less than 25 mm wide.

Dry fields or open woods. Flowering Mar–Jul. 0–1300 m. CR, Est, Sisk, Casc, WV. WA. Native.

Iris tenuis S. Watson (p. 263)

Clackamas iris

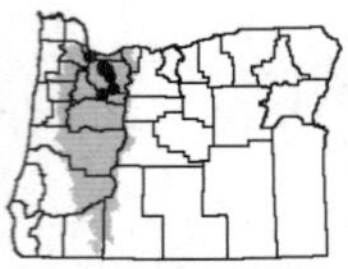

Herbs to 30 cm tall. **Roots** few or none; rhizomes heterogeneous, 2 mm, expanding to 10–15 mm, on surface or very shallow. **Stems** unbranched or 1-branched. **Leaves** basal leaves scarious, pale green, approximately 15 mm wide; cauline leaves 1–2, thin, clasping basally then spread. **Inflorescences** 2–3-branched, each with 1 flower borne at approximately the same level; bracts unequal. **Flowers** perianth white or blue with deep violet lines; signal an inconspicuous crest of pale yellow ridges on the spreading sepals; floral tube funnel-shaped; style crests rounded; stigmas long-triangular with entire margins. **Fruits** rounded triangular in cross section. **Seeds** pale brown, pitted. **2*n***=28.

Open slopes in Douglas-fir forest. Flowering May–Jul. 100–1000 m. Casc, WV. Native.

Iris tenuis is the only member of subgenus *Limniris*, section *Lophiris* (crested iris) in western North America. This species is a narrow endemic, currently known only from three counties in Oregon.

OLSYNIUM Grass widows, purple-eyed grass

Herbs to over 40 cm tall. **Rhizomes** indistinct. **Stems** simple, rounded. **Leaves** basal, alternate, 2–4. **Inflorescences** 1–5-flowered. **Flowers** perianth segments distinct, not clawed. **Fruits** capsules. **Seeds** many.

North America, South America. Approximately 12 species; 1 species treated in *Flora*.

Olsynium douglasii (A. Dietrich) E.P. Bicknell

Grass widows, purple-eyed grass

Sisyrinchium douglasii A. Dietr.

Stems entirely to partly concealed by long sheathing leaf bases. **Leaves** basal, narrow, grass-like, not forming a fan. **Inflorescences** scorpioid cymes; bracts unequal in length. **Flowers** perianth united in short tube, 12–30 mm long, reddish purple; lower 50% of filaments fused, filament column enlarged, glabrous, exuding nectar; styles 3, slender, extending between anthers. **Fruits** globose. **Seeds** brown, angular. **2*n***=64.

2 varieties.

Olsynium has been included in *Sisyrinchium* by some. In addition to rounded, unbranched stems, the genus is distinguished by the glabrous filament column which produces nectar, the only member of Iridaceae to secrete nectar from filaments.

Key to *Olsynium douglasii* varieties:

1. Filament column abruptly and broadly flared at base . . . . ***O. d.*** var. ***inflatum***

1' Filament column slightly or not flared . . ***O. d.*** var. ***douglasii***

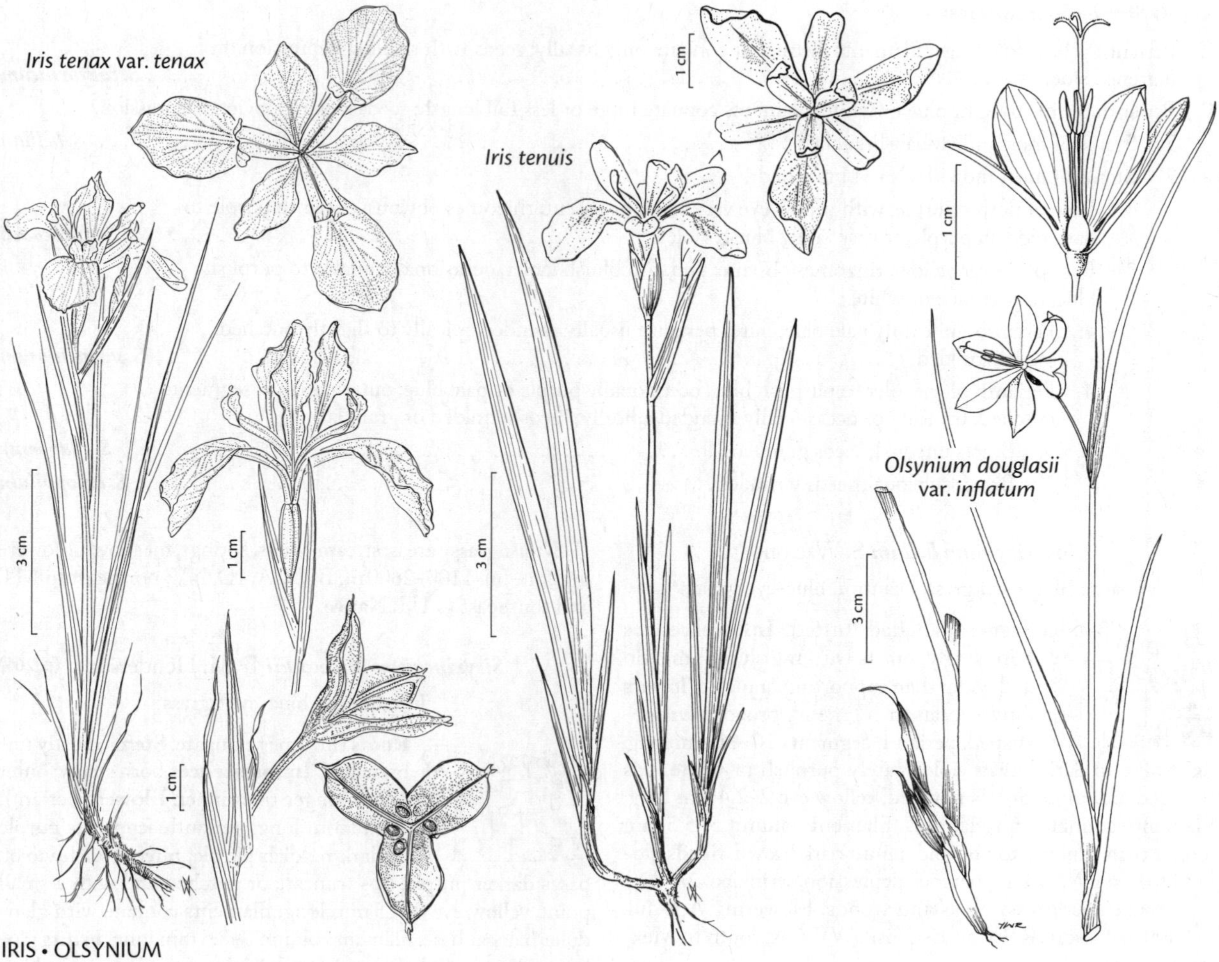

IRIS • OLSYNIUM

Olsynium douglasii (A. Dietrich) E.P. Bicknell var. ***douglasii***

Douglas grass widows, purple-eyed grass

Sisyrinchium douglasii A. Dietr. var. *douglasii*

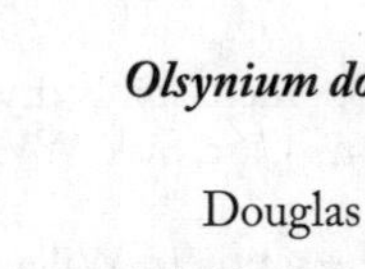

Flowers filament columns tapering evenly to base or slightly flared.

Vernally moist open areas, oak woodlands, coniferous forests, sagebrush, lake margins. Flowering Feb–May. 0–1500 m. BW, Casc, Col, CR, ECas, Lava, Sisk, WV. CA, WA; north to British Columbia. Native.

Olsynium douglasii (A. Dietrich) E.P. Bicknell var. ***inflatum*** (Suksdorf) Cholewa & D.M. Henderson (p. 263)

Inflated grass widows, inflated purple-eyed grass

Sisyrinchium douglasii A. Dietr. var. *inflatum* (Suksd.) P.K. Holmgren, *Sisyrinchium inflatum* (Suksd.) H. St. John

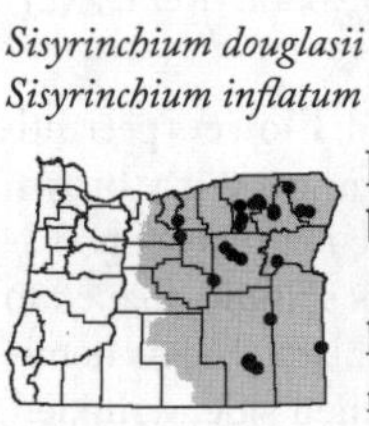

Flowers filament columns abruptly and broadly flared at base.

Vernally moist open areas, oak woodlands, coniferous forests, sagebrush, lake margins. Flowering Apr–Jun. 300–2000 m. BR, BW, Col, Lava, Owy. NV, ID, WA; north to British Columbia, southeast to UT. Native.

SISYRINCHIUM Blue-eyed grass

Herbs perennials, tufted. **Roots** obscurely rhizomatous. **Stems** flat, to 60 cm, winged or not. **Leaves** narrow, grass-like, glabrous; bases not persisting. **Inflorescences** scorpioid cymes, often condensed and umbel-like; bracts glabrous, green. **Flowers** ephemeral; perianths free, tips usually with a small point; filaments free to fused; styles 3, slender (not petal-like), connate at least basally, usually extending beyond anthers. **Fruits** capsules. **Seeds** more or less ovoid, wrinkled.

North America, South America. 7 species; 6 species treated in *Flora*.

Because of variability within each species, identification often requires examination of more than one individual. Floral characters (e.g., fusion of filament column and shape of perianth) and bract shape are critical for species identification but are only preserved in pressed specimens with the utmost of care. In addition, perianth color and shape should be recorded before collecting.

Key to *Sisyrinchium* species:

1. Perianth yellow to orange; filaments distinct or connate only basally; seeds with shallow depression on flattened side. . . . . ***S. californicum***

1' Perianth purple to light blue or white; filaments connate more or less full length; seeds lacking obvious depression.

  2. Stems of most individuals branched . . . . ***S. bellum***

  2' Stems of most individuals not branched.

    3. Base of flower purple, with yellow eye very small or absent; rhizomes obvious; perianth purple to dark reddish purple or very rarely bluish violet . . . . ***S. hitchcockii***

    3' Base of flower yellow; rhizomes obscure; perianth bluish violet, occasionally purple to purplish blue, to pale blue or white.

      4. Perianth uniformly pale blue; outer perianth usually rounded apically to slightly notched; seeds wrinkled . . . . ***S. sarmentosum***

      4' Perianth blue-violet to purplish blue, occasionally purple or pale blue; outer perianth segments notched, truncate or occasionally rounded apically; seeds wrinkled or granular.

        5. Bracts unequal; seeds granular . . . . ***S. idahoense***

        5' Bracts subequal; seeds wrinkled . . . . ***S. halophilum***

Sisyrinchium bellum S. Watson

Western blue-eyed grass, beautiful blue-eyed grass

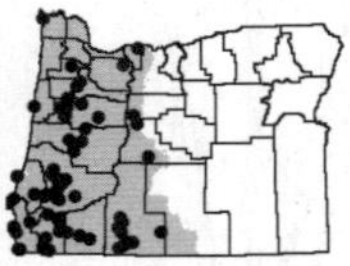

Stems branched, tufted. **Inflorescences** born singly; bracts with translucent margin and wider than supporting branch. **Flowers** outer perianth segments broadly wedge-shaped; perianth segments 10–11 mm long, light blue to dark bluish violet, rarely purplish or white, tips rounded and more or less notched, yellow eye 2–2.4 mm long; filaments connate and glabrous, filament column 5–5.5 mm long. **Fruits** globose to obovoid, tan to dark brown. **Seeds** globose to obconic, lacking obvious depression, wrinkled. **2*n***=32.

Open moist grassy areas and woods. Flowering Apr–Jul. 0–1400 m. Casc, CR, ECas, Est, Sisk, WV. CA; south to Mexico. Native.

Sisyrinchium californicum (Ker Gawłowska) W.T. Aiton (p. 265)

Golden-eyed grass

Hydastylus brachypus E.P. Bicknell, *Hydastylus californicus* (Ker Gawl.) Salisb.

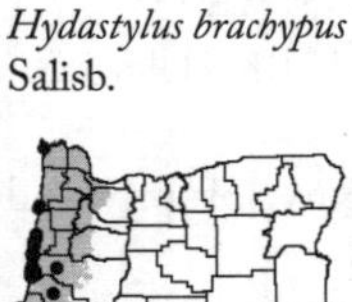

Stems unbranched, tufted. **Flowers** perianth segments yellow to orange with brown veins, tips rounded or acute; filaments distinct and glabrous. **Fruits** spindle-shaped to obconic, dark brown to black. **Seeds** hemispherical, with shallow depression on flattened side, wrinkled. **2*n***=32, 34, 38.

Moist coastal areas. Flowering May–Aug. 0–150 m. CR, Est, Sisk, WV. CA, WA; north to British Columbia. Native.

Sisyrinchium halophilum Greene (p. 265)

Nevada blue-eyed grass

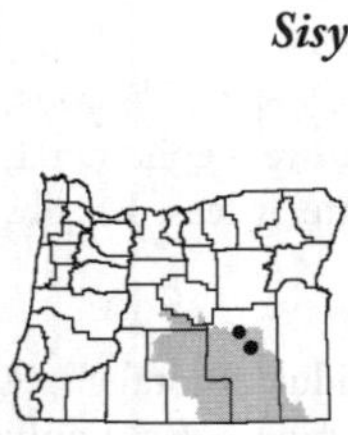

Stems unbranched, tufted. **Inflorescences** born singly; bracts subequal. **Flowers** perianth segments bluish violet with yellow bases, tips truncate with a small point; filaments connate with glandular hairs at base. **Fruits** globose, beige. **Seeds** globose to obconic, lacking obvious depression, wrinkled. **2*n***=32.

Moist grassy areas, stream banks, springs, meadows. Flowering Jun–Jul. 1100–2600 m. BR. CA, ID, NV; northeast to MT and southeast to UT. Native.

Sisyrinchium hitchcockii D.M. Henderson (p. 265)

Hitchcock's blue-eyed grass

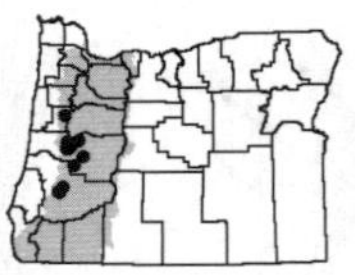

Roots rhizomes elongate. **Stems** mostly unbranched. **Inflorescences** born singly; outer bracts longer than inner. **Flowers** perianth 14–17 mm long; perianth segments purple to dark reddish purple, rarely bluish violet, bases darker purple, tips truncate or notched and with a small point, yellow eye 0–1.5 mm long; filaments connate with glandular hairs at base, filament column 8–8.5 mm long. **Fruits** globose, beige to light brown. **Seeds** globose to obconic, lacking obvious depression, wrinkled. **2*n***=64.

Grassy areas, openings in woods, mostly where soil is dry late in season. Flowering May–Jun. 50–300 m. Casc, Sisk, WV. Native.

Sisyrinchium hitchcockii is endemic to the southern Willamette Valley and the Roseburg area. It is often in mixed populations with *S. bellum* but is genetically distinct. The lack of obvious yellow perianth bases, longer perianth, and longer filament column distinguish *S. hitchcockii* from *S. bellum*.

Sisyrinchium idahoense E.P. Bicknell

Idaho blue-eyed grass

Roots rhizomes obscure. **Stems** to 45 cm tall, tufted, usually unbranched, obviously winged. **Inflorescences** born singly; outer bracts longer than inner, keels entire to toothed, inner bracts with narrow translucent margin. **Flowers** perianth segments 8–17 mm, dark to pale blue to blue-violet with yellow base, tips rounded to deeply notched and with a small point; filaments bluish violet, connate, with glandular hairs at base. **Fruits** globose to pear-shaped, beige to light or dark brown, sometimes with purple blotches. **Seeds** globose to obconic, lacking obvious depression, usually granular.

Western North America. 4 varieties; 2 varieties treated in *Flora*.

Sisyrinchium idahoense is a highly variable species throughout its range; its varieties intergrade.

Key to *Sisyrinchium idahoense* varieties:

1. Upper stem margins entire; outer perianth segments less than 14 mm long; inner bract less than 21 mm long; stems less than 2 mm wide . . . . *S. i.* var. ***occidentale***

1' Upper stem margins with a few teeth; outer perianth segments greater than 13 mm long; inner bract greater than 20 mm long; stems greater than 2 mm wide . . . . *S. i.* var. ***idahoense***

Sisyrinchium idahoense E.P. Bicknell var. ***idahoense***

Idaho blue-eyed grass

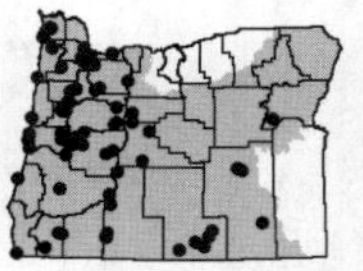

Stems 1.5–2 mm wide; margins of upper stem wings with a few small teeth. **Inflorescences** inner bracts 20–30 mm long, keels toothed. **Flowers** outer perianth segments 13–17(20) mm long. **2*n***=64, 96.

Moist meadows, stream banks, edges of springs. Flowering May–Jul. 0–1900 m. BR, BW, CR, ECas, Est, Lava, Sisk, Casc, WV. CA, ID, WA; north to British Columbia, northeast to MT, southeast to NM. Native.

Sisyrinchium idahoense E.P. Bicknell (p. 265)
var. ***occidentale*** (E.P. Bicknell) D.M. Henderson

Western blue-eyed grass

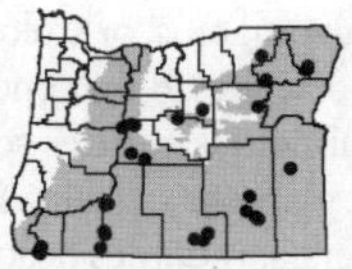

Stems 2–2.5 mm wide; margins of upper stem wings entire. **Inflorescences** inner bracts 12–21 mm long; keel entire. **Flowers** outer perianth segments 12–14 mm long. **2*n***=64.

Moist meadows, stream banks, edges of springs. Flowering Jun–Aug. 400–2200 m. BR, BW, Casc, ECas, Owy, Sisk. CA, ID, NV WA; northeast to MT, southeast to NM. Native.

Sisyrinchium sarmentosum Suksdorf ex Greene

Pale blue-eyed grass, Suksdorf's blue-eyed grass

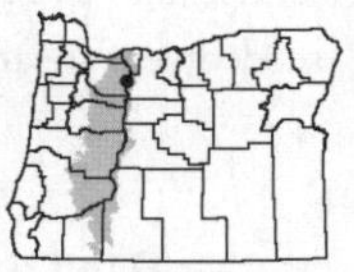

Stems unbranched, tufted, usually not winged. **Inflorescences** born singly; outer bract longer than inner. **Flowers** perianth segments pale blue with a yellow base, tips rounded with a small point; filaments connate with some glandular hairs at base. **Fruits** globose, tan to brown. **Seeds** globose to obconic, lacking obvious depression, wrinkled. **2*n***=96.

Moist, grassy areas. Flowering May–Aug. 500–1000 m. Casc. WA. Native.

The pale blue flowers with rounded apices on the outer perianth segments, generally unwinged stems, and wrinkled seeds set this species apart from *S. idahoense*, the most morphologically similar *Sisyrinchium* species.

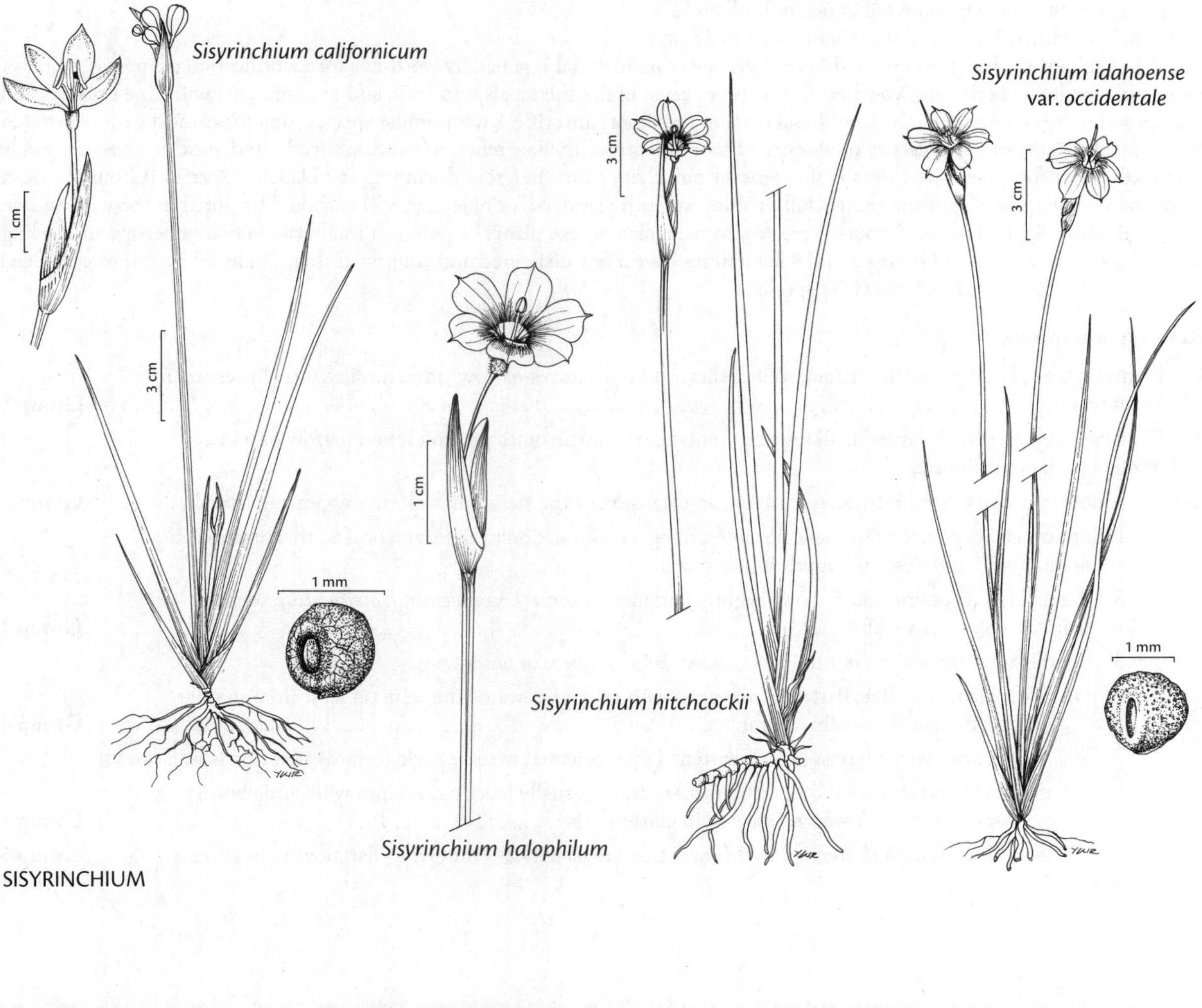

SISYRINCHIUM

JUNCACEAE Rush family

Peter Zika

Plants annual or perennial herbs, cespitose, rhizomatous, or stoloniferous. **Stems** (culms) herbaceous, usually erect. **Leaves** basal and/or alternate, spirally arranged, each consisting of a sheath that encircles the stem, and, if present, a simple blade, sometimes reduced to a short awn; leaf sheaths open or closed; membranous auricles sometimes present at junction of sheath and blade; leaf blades simple, usually linear to lanceolate with parallel veins, glabrous or hairy. **Inflorescences** terminal, sometimes appearing lateral under an erect bract, flowers solitary or in loose to dense clusters, sometimes head-like, or in open cymes, umbels, or racemes. **Flowers** usually bisexual, rarely unisexual; perianth of (4)6 inconspicuous tepals in 2 alternating whorls, free; stamens (0)2–6, anthers 2-chambered; ovary superior; style 1, carpels 3, primarily wind-pollinated. **Fruit** a loculicidal capsule. **Seeds** 3–200, apiculate or with an elongate white tail or aril.

Cosmopolitan. 8 genera; 2 genera treated in *Flora*.

Key to Juncaceae genera:

1. Leaf blades glabrous or absent; capsules (4)10–200-seeded; leaf sheaths open; seeds usually narrower, 0.2–0.5(0.6) mm wide, apiculate or with slender tail . . . . *Juncus*

1' Leaf blades hairy on margins; capsules 3-seeded; leaf sheaths closed; seeds plump, 0.5–1.1 mm wide, often with aril . . . . *Luzula*

JUNCUS Rush

Plants annual with delicate fibrous roots, or perennial herbs with coarse roots, cespitose, rhizomatous, or uncommonly stoloniferous with lax stems rooting at nodes, usually erect, glabrous. **Stems** round or flattened. **Leaves** sometimes absent, mostly basal or on lower 33% of stem; sheaths open, awned; blades linear to filiform, hollow and with chambered crosswalls, or solid and very thin to inrolled, or flattened laterally and incompletely septate (iris-like), or flattened adaxially/abaxially, glabrous, base of blade sometimes with auricles, veins parallel. **Bracts** leafy or membranous. **Inflorescences** racemose to cymose with flowers solitary or in small clusters, or in dense clusters, often head-like; pedicels with small bract at base, flowers with (0)1–2 small bractlets at base. **Flowers** perfect, tepals (4)6(8), often lanceolate; stamens (2)3 or 6, anthers shorter or longer than linear filaments; stigmas 3. **Capsules** dry, 1- or 3-chambered, usually ovoid-trigonous, sometimes crested apically, beaked or not. **Seeds** (4)10–200, 0.2–0.5(0.6) mm wide, apiculate or with a slender white tail at one or both ends.

Cosmopolitan. 315 species; 56 species treated in *Flora*.

Although anther characters are used in the key, flowering material is generally worthless for identification purposes and never makes an acceptable herbarium voucher. The anthers persist under the tepals into fruit, and only material with ripe capsules and mature seeds should be used in the keys. Basal parts of the shoots are critical, with similar species sometimes most easily separated by proximal sheaths or the presence or absence of long rhizomes. The key relies on delicate sheaths and auricles; these are easily damaged unless specimens are carefully dug—never pull them from the ground. Among the bladeless species (Group 2), fresh plants have varied and diagnostic stems: dull or shiny, smooth or ridged, or blue-green. This should be noted in the field and on specimen labels. Septate leaves (Group 3) are easy to recognize on live plants by pulling a leaf between two fingertips and feeling for "bumps." Populations of Groups 3 and 4 sometimes have a few disfigured and congested stems, induced by gall insects. Seed lengths include the tails unless otherwise specified.

Key to *Juncus* species:

1. Plants annuals, usually less than 10 cm; roots delicate, fibrous; leaves narrow, often inrolled, usually less than 1 mm wide . . . . **Group 1**

1' Plants perennials, usually larger in all respects; roots coarse or rhizomes present; leaves narrow or wide, inrolled to flat or cylindric.

  2. Inflorescence bracts cylindrical, resembling an extension of the stem; inflorescence appearing lateral . . . . **Group 2**

  2' Inflorescence bracts not cylindrical, not resembling a continuation of the stem, or if so then channeled along inner side; inflorescence appearing terminal.

    3. Leaf blades cylindric and hollow (slightly flattened in some *J. nevadensis*, *J. articulatus*); septate, with complete crosswalls . . . . **Group 3**

    3' Leaf blades flat or wiry, not hollow; crosswalls incomplete or absent.

      4. Leaf blades iris-like, flattened, arranged with an edge towards the stem (at least from middle to tip); crosswalls usually incomplete . . . . **Group 4**

      4' Leaf blades wiry or grass-like, if flattened then oriented with flat side towards stem; crosswalls absent.

        5. Flowers subtended by 2 or 0 bractlets; leaves usually less than 1.3 mm wide, usually wiry or channeled; flowers solitary or in clusters . . . . **Group 5**

        5' Flowers each subtended by 1 bractlet; leaves usually 2–7 mm wide, flat; flowers in clusters . . . . . . **Group 6**

Group 1, annuals:

1. Stems usually branched and leafy; flowers single or in clusters at nodes or congested in terminal cluster; stamens 3 or 6 per flower.
    2. Outer tepals 2–4 mm.
        3. Flowers scattered along upper stem, solitary at the nodes; tepals green or pale brown to reddish . . ***J. bufonius*** (in part)
        3' Flowers 3–10 in a terminal cluster; tepals dark brown to blackish. . . . . ***J. planifolius*** (in part)
    2' Outer tepals usually 4–10 mm.
        4. Capsule apex usually truncate, more or less equaling inner tepals; inner tepals more or less blunt. . . . . . . ***J. ranarius***
        4' Capsule apex usually blunt or acute, shorter than inner tepals; inner tepals acute to acuminate. . ***J. bufonius*** (in part)

1' Stems unbranched, leafless, leaves all basal; flowers or inflorescences terminal; stamens 2 or 3 per flower.
    5. Inflorescence bracts leafy and longer than inflorescence; outer tepals longer and more acuminate than inner tepals . . . . ***J. capitatus***
    5' Inflorescence bracts inconspicuous, membranous, shorter than inflorescence; outer and inner tepals similar in size and shape, sometimes outer tepals shorter than inner.
        6. Tepals dark brown to blackish. . . . . ***J. planifolius*** (in part)
        6' Tepals green to pink or red.
            7. Seeds not longitudinally striate at 30×; flowers 1 per stem; bractlets 0 or 1–2, tips truncate to acute.
                8. Inflorescence bractlets 0; stem thickened below flower . . . . ***J. hemiendytus*** (in part)
                8' Inflorescence bractlets 1–2; stem not thickened below flower.
                    9. Inflorescence bractlet 1, tips truncate, bases encircling stem; stamens 2–3 per flower . . . . ***J. uncialis***
                    9' Inflorescence bractlets 1–2, tips acute to blunt, bases not encircling stem; stamens usually 2–4 per flower.
                        10. Tepals usually 6, shiny, incurved. . . . . ***J. bryoides***
                        10' Tepals usually 4, dull, erect or outcurved . . . . ***J. hemiendytus*** (in part)
            7' Seeds longitudinally striate at 30×; flowers 1–4(7) per stem; bractlets 2, tips acute to acuminate.
                11. Tepals usually 4 per flower.
                    12. Capsules much paler than tepals; seeds 0.5–0.8 mm, fewer than 4 seeds per chamber, fewer than 10 seeds per capsule; capsules subglobose to obovoid, usually shorter than the tepals. . . . . ***J. capillaris*** (in part)
                    12' Capsules and tepals both pale green or pink; seeds 0.3–0.5 mm, more than 4 seeds per chamber, more than 10 seeds per capsule; some capsules ellipsoid to oblong, equaling or longer than the tepals . . . . ***J. tiehmii***
                11' Tepals usually 6 per flower.
                    13. Tepals usually distinctly longer than and darker than capsule; seeds fewer than 10 per capsule, seeds 0.5–0.8 mm, with striations usually faint at 30× . . . . ***J. capillaris*** (in part)
                    13' Tepals usually shorter than and same color as capsule; seeds more than 40 per capsule, seeds 0.35–0.55 mm, with striations obvious at 30×. . . . . ***J. kelloggii***

Group 2, inflorescence appears lateral, leaf blades often reduced to vestigial awns (or terete and stem-like):

1. Rhizomes long and creeping, stems solitary, scattered or in loose colonies.
    2. Leaf blades well developed on the distal sheaths of some stems, blades terete and stem-like . . . . ***J. mexicanus***
    2' Leaf blades absent, or reduced to vestigial awns.
        3. Fruiting tepals usually 6–8 mm long; coastal sandy habitats.
            4. Heads dense, capitate, branches obscured; older sheaths shiny, dark brown; stems usually somewhat flattened, often arching, twisted . . . . ***J. breweri***
            4' Heads somewhat loose or sparse, at least some branches obvious; older sheaths dull to slightly shiny, green to dark brown or blackish distally; stems usually round and erect, usually not twisted. . . . ***J. lescurii***
        3' Fruiting tepals 2.5–5.5(6) mm long; various habitats including coastal sand.
            5. Tepals green or pale brown; inflorescence usually attached near or below middle of stems (stems here including erect inflorescence bracts above inflorescences). . . . . ***J. filiformis***
            5' Tepals with dark brown to black stripes; inflorescence attached well above middle of stem (including erect inflorescence bracts) . . . . ***J. balticus*** ssp. ***ater***

1' Rhizomes short, stems cespitose, usually in dense tufts.
  6. Stamens 6 per capsule.
    7. Inflorescences with 1–7 flowers; seeds including tails 0.7–2.7 mm, tails 0.3–0.9 mm; fresh stems green to dark green, ridges inconspicuous.
      8. Leaf blades absent or reduced to vestigial awns; capsule apices notched . . . . ***J. drummondii***
      8' Leaf blades well developed on distal sheaths, resembling stems; capsule apices acute to acuminate . . . . ***J. parryi***
    7' Inflorescences with 10–100+ flowers; seeds 0.4–0.6 mm, blunt or apiculate; fresh stems blue-green, ridges prominent (roll fresh stems between fingers to feel ridges).
      9. Capsules ovate to elliptic, often dark or blackish distally, apex pointed; styles persistent, forming prominent beaks; inner tepals usually shorter than capsules; pith of fertile stems chambered; rare introduction in eastern Oregon . . . . ***J. inflexus*** ssp. ***inflexus***
      9' Capsules subglobose, pale or reddish, apex rounded to subacute; styles deciduous or inconspicuous; inner tepals usually longer than capsules; pith of fertile stems solid; native in western Oregon . . . . ***J. patens***
  6' Stamens 3 per capsule.
    10. Fresh upper stems with prominent ridges, dull, not shiny; dried stems coarsely ridged, ridges visible in high relief at 10×, dried ridges capped with shiny bulging cells; inflorescence bracts swollen at base of inflorescence, often somewhat reflexed in fruit. . . . . ***J. conglomeratus***
    10' Fresh upper stems with relatively inconspicuous ridges, shiny or slightly so; dried stems with lower ridges, fine or coarse, ridges visible in low relief at 10×, dried ridges capped with low dull cells; inflorescence bracts not swollen, erect in fruit.
      11. Upper sheath apices usually strongly asymmetrical on fruiting stems.
        12. Sheath apices thickened, with raised (convex) rims; sheaths usually dark brown to black; fruiting stems stout, usually 2–3.5 mm diameter above sheath . . . . ***J. effusus*** (in part)
        12' Sheath apices thin with broad membranous wings, flattened and lacking raised rims; sheaths green (fresh) to pale or medium brown (dried); fruiting stems slender, usually 0.8–1.9 mm diameter above sheath . . . . ***J. hesperius***
      11' Upper sheath apices usually symmetrical on fruiting stems.
        13. Visible stem ridges 6–16 per side (at 10×), low and relatively coarse or wide when dried; proximal sheaths smooth (at 10×); fruiting stems slender, 0.6–2.6 mm diameter above sheath; tepals usually with medium to dark brown stripes.
          14. Distal half of distal sheaths green to pale brown, thin, dull, nerves prominent, apices thin, slightly inrolled towards stem. . . . . ***J. exiguus***
          14' Distal half of distal sheaths medium brown, dark brown or black, thick and glossy, nerves obscure, apices thickened, not inrolled. . . . . ***J. laccatus***
        13' Visible stem ridges usually 18–26 per side (at 10×), slender and relatively inconspicuous when dried; proximal sheaths papillose (at 10×); fruiting stems stout, 2.2–4.9 mm diameter above sheath; tepals usually pale brown. . . . . ***J. effusus*** (in part)

Group 3, septate hollow leaves:

1. Seeds tailed, 0.8–2.8 mm including tails, seeds narrowly ellipsoid to linear.
  2. Inflorescence a single terminal cluster of 1–3 flowers partially hidden by sheathing base of inflorescence bracts; leaves 0.3–0.6 mm wide; native in montane limy wetlands, Wallowa Mountains. . . . . ***J. triglumis*** var. ***albescens***
  2' Inflorescence branched, of 2–50 clusters with 2–50 flowers per cluster, not obscured by inflorescence bract; leaves 1–2 mm wide; introductions in coastal acidic wetlands, in or near cranberry farms.
    3. Clusters mostly with 5 or more flowers, often subglobose or hemispheric; tepals usually 0–1 mm shorter than capsules; seeds including tails 1.25–1.9 mm . . . . ***J. canadensis***
    3' Clusters mostly with 2–5 flowers, ellipsoid to narrowly obconic; tepals 1–1.5 mm shorter than capsules; seeds including tails 0.8–1.1 mm . . . . ***J. brevicaudatus***
1' Seeds apiculate, without elongate tails, less than 0.8 mm, seeds ellipsoid to ovoid.
  4. Flowers and fruits 1–2(3) per node . . . . ***J. pelocarpus***
  4' Flowers and fruits in clusters of 3–80 per node.
    5. Inflorescence bearing leafy bulblets among fruits; plants often stoloniferous with stems trailing and rooting at nodes.

6. Tepals 3.4–5.4 mm; inner tepal tips acuminate . . . . *J. supiniformis* (in part)
6' Tepals 1.8–3.4 mm; inner tepal tips usually acute to blunt, hooded.
7. Capsules usually 0.1–0.4 mm longer than the tepals, capsule apex usually blunt below beak . . . . . . *J. bulbosus*
7' Capsules usually 0.5–1.4 mm longer than the tepals, capsule apex tapered to beak . . . . . . *J. articulatus* ssp. *articulatus* (in part)

5' Inflorescence not bearing leafy bulblets among fruits; plants cespitose or rhizomatous, stems erect and not trailing and rooting at nodes.
8. Most or all flowers with 3 stamens.
9. Plants rhizomatous; inflorescences dense, spherical clusters 1–3(5), 40–70-flowered; tepals dark brown or blackish . . . . . . *J. bolanderi*
9' Plants cespitose; inflorescences open, clusters more than 5, usually fewer than 15-flowered; tepals usually green, red or medium-brown.
10. Capsules usually shorter than or equaling tepals . . . . . . *J. acuminatus*
10' Capsules longer than the tepals.
11. Tepals 2–3.2 mm; capsules pale brown or reddish, often with bronzy or metallic gloss. . . . . . *J. diffusissimus*
11' Tepals 3.4–5.4 mm; capsules brown to dark brown, without metallic gloss . . . . . . *J. supiniformis* (in part)
8' Most or all flowers with 6 stamens.
12. Plants rhizomatous, rhizomes swollen or tuberous at nodes; capsules gradually tapered to acuminate tip. . . *J. torreyi*
12' Plants cespitose or rhizomatous, rhizomes never swollen and tuberous at nodes; capsules blunt or notched or acute, never gradually tapered to acuminate tip.
13. Anthers longer than filaments.
14. Inflorescence clusters 1–2; capsule apices usually blunt or notched. . . . . . *J. mertensianus* (in part)
14' Inflorescence clusters usually more than 5; capsule apices usually acute.
15. Capsules gradually tapered to beaks; capsules usually longer than the tepals. . . . . . *J. articulatus* ssp. *articulatus* (in part)
15' Capsules abruptly tapered to beaks; capsules usually shorter than or equaling tepals . . . . . . *J. nevadensis* (in part)
13' Anthers shorter than or equaling filaments.
16. Inflorescence clusters more than 5; capsules longer than the tepals. . . . . . Go to lead 6
16' Inflorescence clusters usually 1–3; capsules usually shorter than or equaling tepals.
17. Capsules abruptly tapered to distinct beaks; leaves slightly to strongly flattened though still inconspicuously hollow, blue-green; coastal. . . . . . *J. nevadensis* (in part)
17' Capsules blunt or notched at apex, usually beakless; leaves tubular and obviously hollow, green; montane . . . . . . *J. mertensianus* (in part)

Group 4, leaves iris-like, with incomplete septa:

1. Capsules gradually tapered to long beaks . . . . . . *J. oxymeris*
1' Capsules abruptly tapered to short beaks.
2. Stamens 3 per flower; tepals often dark brown to black; inflorescences often with few, large head-like clusters . . . . . . *J. ensifolius*
2' Stamens 6 per flower; tepals often green to medium brown; inflorescences often with many small clusters . . . . . . *J. saximontanus*

Group 5, narrow wiry leaves:

1. Seeds tailed, 1.3–2.8 mm, including tails; flowers without bractlets; inflorescence a small terminal cluster of 1–3(5) flowers partially hidden by swollen base of inflorescence bract; tepals white to pale brown; high elevation limestone wetlands; Wallowa Mountains . . . . . . *J. triglumis* var. *albescens*
1' Seeds apiculate, without elongate tails, 0.3–0.7 mm; flowers subtended by 2 bractlets; inflorescences open to condensed cymes of 10–100 flowers, never hidden by inflorescence bract, bract not swollen at base; tepals green to red, or brown to blackish; low to moderate elevations, not restricted to limestone; widespread including Wallowa Mountains.
2. Tepal tips blunt, hooded and incurved, often black-striped; tepals shorter than capsules; often in saline habitats.
3. Filaments subequal to anthers; anther/filament ratio usually 0.67–2; northern Oregon. . . . . . *J. compressus*
3' Filaments much shorter than anthers; anther/filament ratio usually 2.6–4; coastal salt marshes . . . . . . *J. gerardi* ssp. *gerardi*

2' Tepal tips acute or acuminate, straight or gently outcurved, never hooded and incurved, green, brown or red, never black-striped; tepals often longer than capsules; often in fresh water habitats.
- 4. Auricles hardened and thickened distally (plastic-like), shiny, rounded, yellowish . . . . *J. dudleyi*
- 4' Auricles soft and thin distally (membranous or scarious), dull, rounded or acute to acuminate, dirty white to translucent.
  - 5. Capsules 3-chambered, intruding partitions straight, united except at apex.
    - 6. Style 0.05–0.2 mm; tepals equal or subequal, with heavy medium-brown stripes flanking a green midvein; inflorescences usually dense and small, 0.5–2.5 cm long, not including inflorescence bracts . . . . *J. confusus*
    - 6' Style 0.3–1 mm; tepals unequal, the outer usually much longer than the inner, more nearly concolorous, greenish to pale brown; inflorescence generally open, 2–20 cm . . . . *J. trilocularis*
  - 5' Capsules 1-chambered, intruding partitions curved, separated except at base.
    - 7. Capsules strongly ridged or crested distally, apex usually truncate or slightly notched . . . . *J. occidentalis*
    - 7' Capsules weakly or not ridged distally, usually acute to blunt at apex.
      - 8. Auricles less than 1 mm; rounded, two-textured, base usually thicker and more opaque than thin apical margin; bractlets subtending flowers acuminate to aristate; base of plant usually pink; 4–6 strong longitudinal stem ridges visible on one side . . . . *J. interior*
      - 8' Auricles usually 1–5(8) mm long (seek long auricles on young shoots, before mid-summer; auricles produced late in the season are usually short), often acute to acuminate (auricles on late season shoots are often blunt), entirely thin and scarious to translucent; bractlets subtending flowers usually acute; base of plant usually green to brown; 0–6 strong longitudinal stem ridges visible on one side.
        - 9. Capsules usually more than 2.5 mm, more than 75% of the length of the tepals; plants usually delicate, less than 50 cm; inflorescence usually less than 10 cm; individual flowers often longer than adjacent internodes; stems with 0–1(2) strong ridges visible on a side; anthers usually 0.4–0.6(0.8) mm; common widespread native in moist disturbed sites . . . . *J. tenuis*
        - 9' Capsules less than 3 mm, less than 75% of the length of the tepals; plants usually coarse, 60–90 cm tall; inflorescence 6–21 cm; individual flowers shorter than adjacent internodes; stems with 2–6 strong ridges visible on a side; anthers usually 0.6–0.8(1) mm; rare introduction west of Cascades . . . . *J. anthelatus*

Group 6, grass-like leaves:

1. Stamens 3 per flower.
   - 2. Stem base with knobby tubers; anthers purplish red; plants cespitose . . . . *J. marginatus*
   - 2' Stem base without tubers; anthers yellow; plants rhizomatous or cespitose . . . . *J. planifolius*

1' Stamens 6 per flower.
- 3. Seeds slender, with 1 or 2 tails that are at least one half the length of the seed body.
  - 4. Tails shorter than seed body; inner tepals usually longer than the outer; inflorescence clusters usually 5–20, clusters usually 2–8-flowered; leaves up to 4.5 mm wide . . . . *J. howellii*
  - 4' One or both tails longer than seed body; inner tepals usually shorter than or as long as the outer; inflorescence clusters 1–5, clusters usually 10–30-flowered; leaves less than 3 mm wide . . . . *J. regelii*
- 3' Seeds more or less plump, apiculate, tails absent or short points, much shorter than seed body.
  - 5. Tepals usually 2–3.6 mm; capsules longer than or equaling tepals . . . . *J. covillei*
  - 5' Tepals usually 4–6 mm; capsules usually shorter than or equaling tepals.
    - 6. Auricles prominent, usually rounded, more than 1 mm . . . . *J. longistylis*
    - 6' Auricles absent or inconspicuous, usually acute, less than 1 mm.
      - 7. Tepals less than 1.5 mm wide; capsule brown; inflorescence with 3–12 clusters; interior meadows and wetlands . . . . *J. orthophyllus*
      - 7' Tepals more than 1.5 mm wide; capsule blackish; inflorescence with 1–3 clusters; coastal sand . . . . *J. falcatus* ssp. *sitchensis*

Juncus acuminatus Michaux

Tapered rush, knotty-leaf rush, sharp-fruited rush

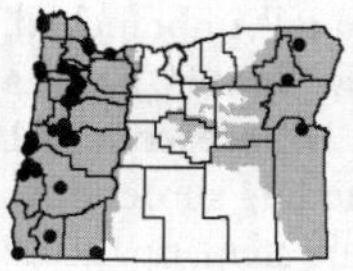

Plants perennial, 14–100 cm tall, cespitose. **Leaves** round, hollow, septate; auricles 1–1.5 mm. **Inflorescences** open, cymose, (2)5–50 clusters or head-like clusters, clusters usually 5–15(50)-flowered. **Flowers** tepals 6, green, red or light brown; stamens 3, filaments 0.5–0.6 mm, anthers 0.5–0.9 mm; styles 0.2–0.3 mm. **Capsules** usually shorter than or equaling tepals, light brown to red, 1-chambered. **Seeds** ellipsoid to ovoid, 0.3–0.5 × 0.15–0.2 mm, apiculate. **2*n***=40.

Shores, springs, wet meadows, peatlands, ditches. 0–1300 m. BW, Casc, CR, Est, Owy, Sisk, WV. CA, ID, WA; throughout North America, South America; New Zealand. Native.

Juncus anthelatus (Wiegand) R.E. Brooks

Kentucky rush, poverty rush

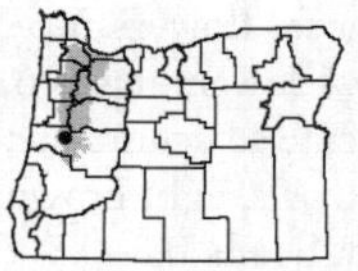

Plants perennial, (30)60–90 cm tall, usually coarse, cespitose, base usually green to brown, stems with 2–6 strong longitudinal ridges visible on one side. **Leaves** thin and wiry, blade flat and slightly inrolled; auricles soft and thin, 1.3–4 mm (longest auricles on early season shoots, those produced late in season are usually short), scarious or dirty white to translucent dull, acute or acuminate (those on late season shoots often blunt). **Inflorescences** large, cymose, 6–21 cm, often reddish, bractlets subtending flowers usually acute, flowers solitary at the nodes, internodes longer than flowers. **Flowers** tepals 6, green to light brown or reddish, tepal tips acuminate, erect; stamens 3, filaments 0.8–1.1 mm, anthers usually 0.6–0.8(1) mm; styles 0.2–0.4 mm. **Capsules** 2–3 mm, usually less than 75% of the length of the tepals, light brown to red, apex rounded to acute, not crested, 1-chambered. **Seeds** 0.3–0.6 × 0.2–0.25 mm, apiculate. **2*n***=80.

Shores, wet prairie remnants, seasonally wet ground, ditches. 100–500 m. WV. CA, WA; south to Mexico; central and eastern North America; New Zealand. Exotic.

Juncus anthelatus is a rare introduction, often with a much larger inflorescence than *J. tenuis*.

Juncus articulatus Linnaeus ssp. ***articulatus*** (p. 272)

Joint-leaved rush, jointed rush

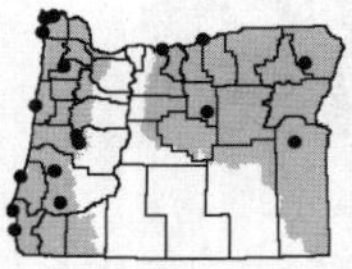

Plants perennial, 10–60 cm tall, cespitose to loosely cespitose or short-rhizomatous, rhizomes not tuberous at nodes, stems sometimes prostrate, trailing, or stoloniferous, rooting at lower nodes. **Leaves** round or somewhat flattened, hollow, septate; auricles 0.5–1 mm. **Inflorescences** panicles of (3)5–50 heads, clusters 3–12-flowered, flowers sometimes forming leafy bulblets. **Flowers** tepals 6, 1.8–2.6 mm, brown to dark brown, outer tepal tips acute to acuminate, hyaline; inner tepal margins wide, flat, tips generally acute to blunt, hooded; stamens 6, filaments 0.5–1 mm, shorter to longer than anthers, anthers 0.4–0.9 mm; styles 0.2–0.3 mm. **Capsules** usually (0.5)0.7–1.4 mm, longer than the tepals, light brown to dark brown, gradually tapered to beaks, apices acute, 1-chambered. **Seeds** ellipsoid to ovoid, 0.45–0.6 × 0.2–0.25 mm, apiculate. **2*n***=80.

Shores, seeps, ditches. 0–1100 m. BW, Col, CR, Est, Lava, Owy, Sisk, WV. CA, NV, ID, WA; north to AK, east to NC, southeast to NM; North Africa, Eurasia, probably introduced in South Africa, Australia, New Zealand, Peru. Native.

Juncus balticus Willdenow ssp. ***ater*** (Rydberg) Snogerup

Baltic rush, dark Baltic rush

Juncus balticus Willd. ssp. *balticus*, *Juncus balticus* Willd. var. *balticus*, *Juncus balticus* Willd. var. *montanus* Engelm.

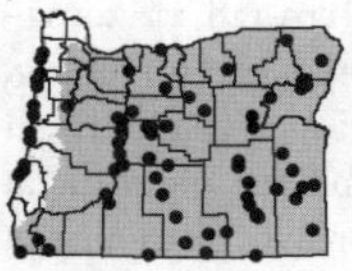

Plants perennial, 25–150 cm tall, rhizomatous, stems without strong ridges visible fresh or dried. **Leaves** lacking blades, distal sheaths 4–16 cm, pale to dark brown, shiny or dull, apex usually symmetrical, thin. **Inflorescences** more or less diffuse, with 5–45 flowers, branches obvious, attached well above middle of stem (stems here including erect inflorescence bracts above inflorescence); inflorescence bracts not swollen at base. **Flowers** tepals 6, 4–5.5(6) mm, dark brown- or black-striped; stamens 6, filaments 0.4–0.7 mm, anthers 1.2–2.2 mm; styles 0.7–1.3 mm. **Capsules** 3.5–4.5 mm, usually shorter than or equaling the tepals, brown to dark brown, apex acute to acuminate, 1-chambered. **Seeds** 0.7–0.9 × 0.35–0.45 mm, apiculate.

Shores, dunes, saline marshes, meadows, swales, ditches. 0–3000 m. All ecoregions except CR. CA, NV, ID, WA; east to SD. Native.

A widespread and common taxon across the state, *J. balticus* ssp. *ater* never shows the leaf blades characteristic of the otherwise similar *J. mexicanus*. The division of North American plants into two subspecies is questionable.

Juncus bolanderi Engelmann

Bolander's rush

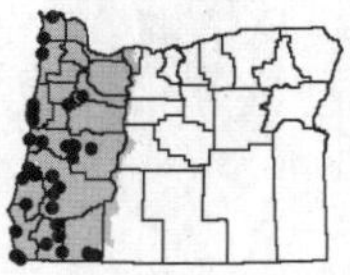

Plants perennial, 25–85 cm tall, rhizomatous. **Leaves** round, hollow, septate; auricles 2–4 mm. **Inflorescences** dense, clusters 1–3(5), 40–70-flowered. **Flowers** tepals 6, brown to dark brown or blackish; stamens 3, filaments 1–1.5 mm, anthers 0.4–0.7 mm; styles 0.1–0.15 mm. **Capsules** shorter than or equaling tepals, dark brown, 1-chambered. **Seeds** ellipsoid to ovoid, 0.4–0.5 × 0.2–0.25 mm, apiculate.

Marshes, wet meadows, shores. 0–1300 m. Casc, CR, Est, Sisk, WV. CA, WA; north to British Columbia. Native.

Juncus brevicaudatus (Engelmann) Fernald

Short-tailed rush

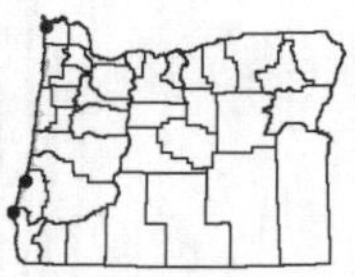

Plants perennial, 14–55 cm tall, cespitose. **Leaves** basal and cauline, round, hollow, septate, 1–2 mm wide; auricles 0.5–3 mm. **Inflorescences** 2–35 clusters, clusters mostly 2–5-flowered, elliptic to narrowly obconic, lacking bulblets. **Flowers** tepals 6, green, red, pale or medium brown, tips acuminate; stamens 3, filaments 0.6–0.8 mm, anthers 0.3–0.5 mm; styles 0.1–0.3 mm. **Capsules** 1–1.5 mm longer than the tepals, dark brown, 1-chambered. **Seeds** narrowly ellipsoid to linear, 0.8–1.1 × 0.2–0.3 mm, body 0.5–0.6 mm, tails 0.2–0.4 mm, 0.5–1 times as long as seed body. **2*n***=80.

Wet acidic sandy banks, shores, mossy ditches, cranberry farms. 0–50 m. Est. ID, WA; northeastern US, Great Lakes area, southern Canada; AZ, CO, WY. Exotic.

This species is native to northeastern Washington, northern Idaho, Montana, and eastern North America, but it is introduced in Oregon.

Juncus breweri Engelmann (p. 272)

Brewer's rush, salt rush

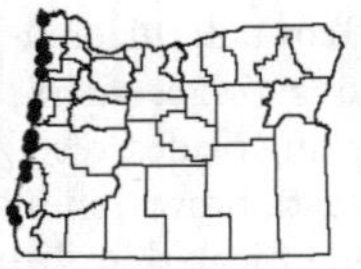

Plants perennial, 20–140 cm tall, rhizomatous, stems usually flattened, often arching, twisted, without strong ridges visible fresh or dried. **Leaves** lacking blades, older distal sheaths 2–13 cm, dark brown, shiny; apex usually symmetrical, thin. **Inflorescences** dense head-like clusters of 10–20 flowers, branches obscured. **Flowers** tepals 6, usually 6–8 mm, dark brown- or black-striped; stamens 6, filaments 0.4–0.7 mm, anthers 1.6–2.8 mm; styles 0.7–1.4 mm. **Capsules** 3–5 mm, shorter than or equaling the tepals, dark brown, apex obtuse to acute, 3-chambered. **Seeds** 0.7–1.1 × 0.4–0.6 mm, not apiculate.

Dunes, sandy swales, sandy ditches. 0–50 m. Est. CA, WA, north to British Columbia. Native.

The little-known *Juncus lescurii* may represent a series of fertile hybrids between *J. balticus* and *J. breweri* and is intermediate in inflorescence characters. *Juncus breweri* has a denser and more nearly globose inflorescence than the other two taxa, with its inflorescence branches obscured by the compact arrangement of flowers or capsules.

Juncus bryoides F.J. Hermann

Mosslike dwarf rush, moss rush

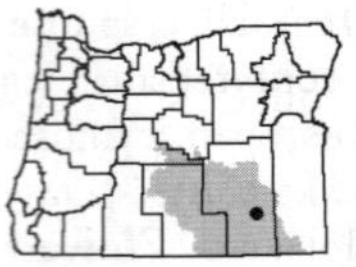

Plants annual, 0.3–2.5 cm tall, unbranched, stems not thickened below flowers. **Leaves** 0.1–0.2 mm thick, basal. **Inflorescences** 1 terminal flower, 1–2 bractlets subtending flowers, bases not sheathing stem, tips acute to blunt. **Flowers** tepals usually 6, brown to dark brown, shiny, incurving; usually 3(2–4) stamens, filaments 0.4–0.6 mm, anthers 0.15–0.25 mm; styles 0–0.1 mm. **Capsules** red to purple, usually shorter than the tepals. **Seeds** 0.3–0.5 × 0.2–0.3 mm, not striate (at 10×).

Moist depressions, sagebrush flats. 2200–2300 m. BR. CA, NV, ID; southeast to CO. Native.

Juncus bufonius Linnaeus

Toad rush

Plants annual, branched. **Leaves** usually less than 1 mm wide, usually 1–5 on stem. **Inflorescences** cymose, flowers usually solitary, clustered apically in one variety. **Flowers** tepals 6, green to red or black-striped, inner tepals acute to acuminate; stamens (3)6. **Capsules** shorter than inner tepals, light brown to red, usually blunt to acute, 1-chambered. **Seeds** 0.3–0.5 × 0.2–0.35 mm, apiculate, smooth or faintly reticulate.

6 varieties; 3 varieties treated in *Flora*.

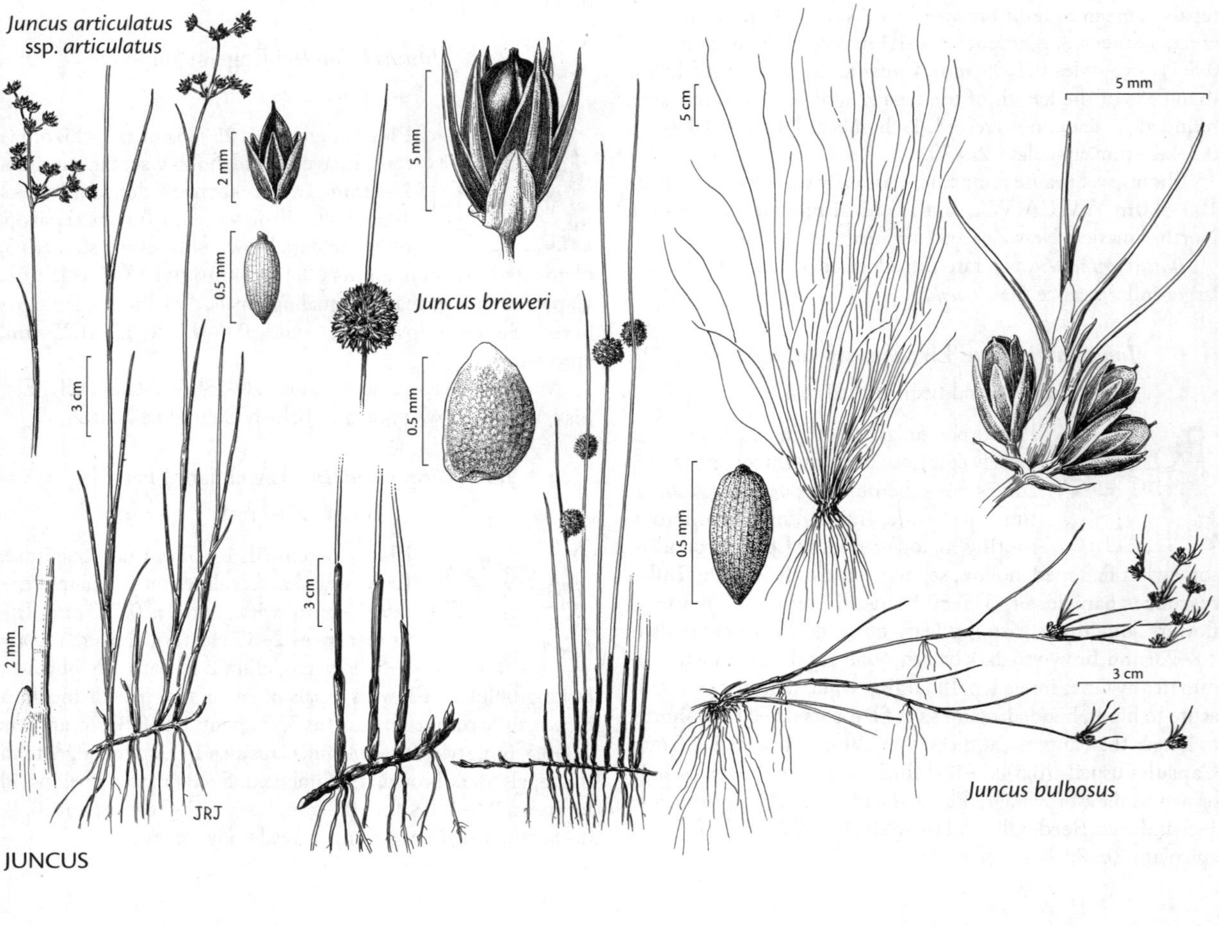

JUNCUS

Key to *Juncus bufonius* varieties:

1. Outer tepals 2–4 mm . . . . ***J. b.* var. *occidentalis***
1' Outer tepals usually 4–10 mm.
   2. Flowers scattered, not clustered apically; usually non-saline soil . . . . ***J. b.* var. *bufonius***
   2' Flowers clustered apically; often on saline soil . . . . ***J. b.* var. *congestus***

Juncus bufonius Linnaeus var. ***bufonius***

Toad rush

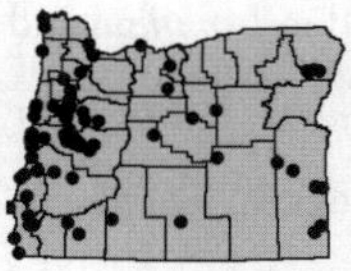

Plants 0.5–40 cm tall. **Inflorescence** open, flowers solitary at nodes, rarely paired. **Flowers** outer tepals usually 4–10 mm; filaments 0.7–0.9 mm, anthers 0.3–0.9 mm; styles 0.4–0.5 mm. **2*n***=108.

Shores, swales, wet prairie, wet balds, seeps, vernal pools, damp disturbed ground, ditches; usually non-saline soil. 0–2200 m. Throughout Oregon. CA, NV, ID, WA; most of North America; nearly worldwide. Native.

The native or introduced status of many southern hemisphere records of this taxon is debated. The elongate outer tepals and well-spaced flowers mark this variety. Extremely dwarfed plants with only one or two flowers have stem leaves and thus are distinguished from our other native annual species, which have only basal leaves.

Juncus bufonius Linnaeus var. ***congestus*** Wahlenberg

Swedish toad rush

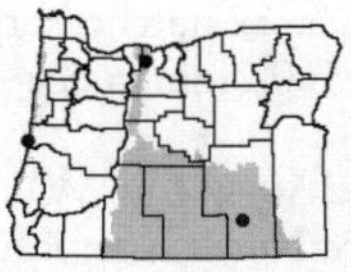

Plants 2–12(36) cm tall. **Inflorescence** congested, dense; flowers solitary at nodes proximally, in clusters distally. **Flowers** outer tepals 4–10 mm; filaments 0.5–0.9(1.5) mm, anthers 0.4–0.7 mm; styles 0.05–0.2 mm.

Depressions in dunes, swales, pools, saline marshes, ditches, often on saline soil. 700–1400 m. BR, ECas, Est. CA, WA; scattered across North America; Europe. Exotic.

The glomerules of flowers of this variety are distinctive; the other varieties have solitary flowers, rarely with flowers in pairs.

Juncus bufonius Linnaeus var. ***occidentalis*** F.J. Hermann

Western toad rush

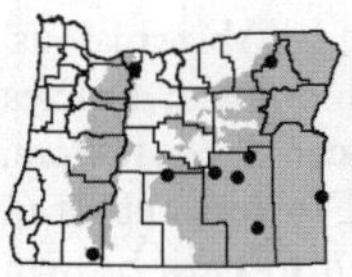

Plants 3–10 cm tall. **Inflorescences** flowers scattered along upper stem, solitary at the nodes. **Flowers** tepals green to reddish, outer tepals 2–4 mm; filaments 0.5–0.8 mm, anthers 0.2–0.4 mm; styles 0–0.2 mm.

Swales, seeps, moist meadows, mud flats. 100–1800 m. BR, BW; Casc. Owy, CA, WA; north to British Columbia, southeast to CO and AZ. Native.

The small flowers, especially the short outer tepals of this species are distinctive. All other varieties have some or all outer tepals longer than 4 mm.

Juncus bulbosus Linnaeus (p. 272)

Bulbous rush, spreading rush

Juncus supinus Moench

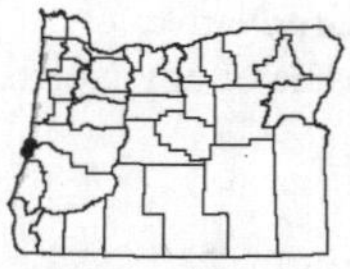

Plants perennial, sometimes flowering in the first year, 1–30 cm tall, cespitose, sometimes tuberous at nodes, stems sometimes prostrate, trailing, stoloniferous, rooting at nodes, upper nodes sometimes forming leafy bulblets. **Leaves** round, hollow or several-chambered, septate at least in part; auricles 1–2 mm. **Inflorescences** open, (2)6–16 clusters, clusters 2–12-flowered, flowers often forming leafy bulblets. **Flowers** tepals 6, 1.8–3.4 mm, green, red, or brown (dark brown), inner tepal margins flat, tips usually blunt, hooded; stamens (3)6, filaments 0.6–1 mm, longer than or equaling anthers, anthers 0.2–0.8 mm; styles 0.2–0.5 mm. **Capsules** usually 0.1–0.4 mm longer than or equaling tepals, brown to dark brown, abruptly tapered to apices, blunt or acute, 1-chambered. **Seeds** ellipsoid to ovoid, 0.4–0.6 × 0.2–0.25 mm, apiculate. **2*n***=40.

Damp grassy disturbed ground, lakeshores. 0–50 m. Est. CA, WA; north to British Columbia, northeastern Canada and NY; Australia, Europe, Mauritius, New Zealand. Exotic.

This species is uncommon but can invade natural wetland systems. The capsule is rounded at the tip, differing from the acute to acuminate capsules of *J. supiniformis*. *Juncus supiniformis* capsules are more than 1 mm longer than the tepals. *Juncus articulatus* is coarser and has capsules more exserted than *J. bulbosus*.

Juncus canadensis J. Gay ex Laharpe

Canada rush, Canadian rush

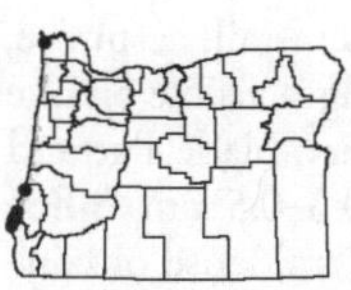

Plants perennial, 30–100 cm tall, cespitose. **Leaves** basal and cauline, round, hollow, septate, 1–2 mm wide; auricles 1–1.2 mm. **Inflorescences** panicles of 3–50 clusters, clusters 5–50-flowered, not obscured by inflorescence bract. **Flowers** tepals 6, green, reddish, or pale brown, tips acuminate; stamens 3, filaments 0.8–1.5 mm, anthers 0.3–0.5 mm; styles 0.2–0.3 mm. **Capsules** usually 0–1 mm longer than the tepals, dark brown, 1-chambered. **Seeds** narrowly ellipsoid to linear, 1.25–1.9 × 0.2–0.25 mm, bodies 0.5–0.8 mm, tails 0.3–0.5 mm, 0.5–1 times as long as seed body. **2*n***=80.

Shores, ponds, peatlands, disturbed sandy acidic wet ground, cranberry farms, ditches. 0–100 m. Est. WA; north to British Columbia; eastern North America; Europe, New Zealand. Exotic.

This species was introduced by cranberry agriculture and is spreading into natural wetlands. Some authors suggest rare flowers have 6 stamens.

Juncus capillaris F.J. Hermann

Hairstem rush

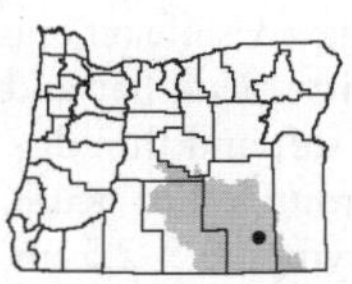

Plants annual, 0.9–6 cm tall, unbranched. **Leaves** capillary, basal. **Inflorescences** 1–2(3) flowers in a terminal cluster, flowers subtended by 2 bractlets, tips acute. **Flowers** tepals usually 4(6), red, brown or dark brown; stamens usually 2(3), filaments 0.6–1.1 mm, anthers 0.3–0.4 mm; styles 0.1–0.3 mm. **Capsules** subglobose to obovoid, usually shorter than the tepals, pale green (red), apex sometimes red or blackish, acuminate to attenuate, capsule much paler than tepals. **Seeds** 4–9 per capsule, 2–3 per chamber, 0.5–0.8 × 0.3–0.45 mm, longitudinally striate, though sometimes faintly so (at 30×). **2*n*=36.**

Stream banks, seeps. 1800–1900 m. BR. CA. Native.

This species has the fewest seeds per capsule of any North American *Juncus*.

Juncus capitatus Weigel

Capitate rush, leafybract dwarf rush

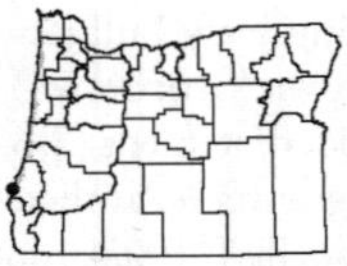

Plants annual, 2–13.5 cm tall, unbranched. **Leaves** capillary, basal. **Inflorescences** 1(3) terminal head-like clusters, 2–15(26) flowers per cluster; inflorescence bracts leafy and longer than inflorescence. **Flowers** tepals 6, brown, outer longer than and more acuminate than the inner; stamens 3, filaments 0.6–1.3 mm, anthers 0.2–0.6 mm; styles 0.3–0.5(1) mm. **Capsules** shorter than the tepals, brown, reddish brown or dark brown. **Seeds** 0.25–0.35 × 0.15–0.2 mm, slightly reticulate (at 10×). **2*n*=18.**

Moist bare ground. 50–1000 m. Est. CA; southeastern US, South America; Africa, Australia, Eurasia, New Zealand. Exotic.

This species was discovered in Oregon in 2010 on the Oregon Islands National Wildlife Refuge by Dave and Diane Bilderback. The large involucral bract is distinctive among our annuals.

Juncus compressus Jacquin

Round-fruited rush

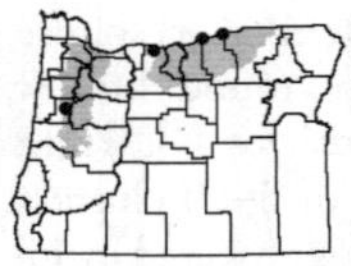

Plants perennial, 10–40 cm tall, cespitose, stems with 0–1 strong ridges visible on one side. **Leaves** thin and wiry, blade flat and slightly inrolled, auricles 0.3–0.5 mm. **Inflorescences** cymose, 1.5–8 cm, loose or congested, never hidden by inflorescence bract, bract not swollen at base; flowers solitary at the nodes, subtended by 2 bractlets. **Flowers** tepals 6, often dark brown- or black-striped, tips blunt, hooded and incurved; stamens 6, filaments 0.4–0.7 mm, subequal to anthers 0.4–1 mm, anther/filament ratio usually 0.67–2; styles 0.2–0.4 mm. **Capsules** 2.5–3.5 mm, longer than the tepals, brown to dark brown, apex rounded to truncate, not crested, 1-chambered. **Seeds** 0.35–0.6 × 0.2–0.25 mm, apiculate, reticulate. **2*n*=44.**

Shores, damp pastures, often saline soil. 0–100 m. Col, WV. Northern North America; Eurasia. Exotic.

Both *J. compressus* and *J. gerardi* are easily separated from other species in Group 5 by the hooded, dark-striped tepals.

Juncus confusus Coville

Colorado rush

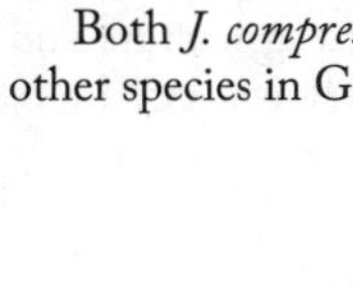

Plants perennial, 20–50 cm tall, cespitose, stems with 0–5 strong ridges visible on one side. **Leaves** thin and wiry, blade flat and slightly inrolled, auricles soft and thin distally, 0.3–0.8 mm, dull, rounded or acute, dirty white to translucent. **Inflorescences** cymes, 0.5–2.5 cm, usually dense and small. **Flowers** tepals 6, with broad medium brown stripes flanking a green midvein, tepal tips acuminate, erect, equal or subequal; stamens 6, filaments 0.7–1.1 mm, anthers 0.5–0.7 mm; styles 0.05–0.2 mm. **Capsules** 2.5–3.5 mm, shorter than or equaling tepals, brown to dark brown, apex notched, crested, 3-chambered. **Seeds** 0.4–0.5 × 0.2–0.3 mm, apiculate, longitudinally ridged. **2*n*=80.**

Seasonally moist or wet meadows, springs, swales, shores, ditches. 800–2700 m. BR, BW, ECas, Owy, Sisk. CA, NV, ID, WA; north to British Columbia, east to SD, southeast to NM. Native.

This species is often confused with *Juncus occidentalis,* which shares the brown-striped tepals and crested capsules but does not grow east of the Cascades. Where they overlap in southwestern Oregon and California, they are separable by the internal structure of the capsules: three-chambered in *J. confusus* and one-chambered in *J. occidentalis.*

Juncus conglomeratus Linnaeus

Compact rush

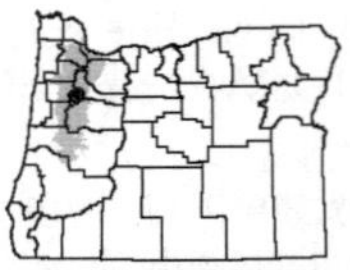

Plants perennial, 40–100 cm tall, cespitose, fresh upper stems dull, not shiny, with (10)14–20 prominent strong ridges visible fresh or dried, dried ridges capped with shiny bulging cells (at 10×). **Leaves** lacking blades, distal sheaths 5–20(23) cm, yellow-brown to red-brown proximally; dull to shiny, papillose, green to light brown distally; apex symmetrical, thin. **Inflorescences** a dense head-like cluster, sometimes lobed, branches usually obscured, inflorescence bract base swollen, bract often somewhat reflexed in fruit. **Flowers** tepals 6, usually green to light brown (dark brown-striped); stamens 3, filaments 0.4–0.8 mm, anthers 0.5–0.8 mm; styles 0–0.15 mm. **Capsules** 2–2.5 mm, shorter than or equaling the length of the tepals, light to dark brown, apex notched or obtuse, 1-chambered. **Seeds** 0.45–0.55 × 0.15–0.2 mm, reticulate, apiculate. **2*n*=84.**

Damp disturbed ground, shores, ditches. 50–200 m. WV. WA; north to British Columbia; northeastern North America, South America; North Africa, Eurasia, New Zealand. Exotic.

Last collected in Oregon in 1921, *J. conglomeratus* is a rare weed in the western United States but is more common in southwestern British Columbia.

Juncus covillei Piper (p. 275)

Coville's rush

Juncus covillei Piper var. *covillei*, *Juncus covillei* Piper var. *obtusatus* (Engelm.) C.L. Hitchc., *Juncus obtusatus* Engelm.

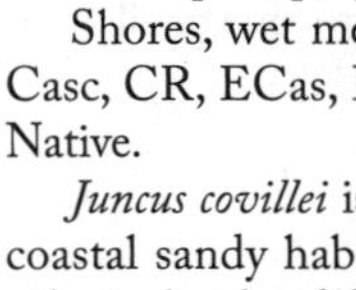

Plants perennial, 2–25 cm tall, rhizomatous. **Leaves** flat, grass-like, blue-green; auricles 0–1.5 mm. **Inflorescences** 2–6 clusters, clusters 3–7-flowered. **Flowers** tepals 6, usually 2–3.6 mm, brown to dark brown; stamens 6, filaments 0.5–1 mm, anthers 0.8–1.4 mm; styles 0.8–1 mm. **Capsules** notched, longer than or equaling tepals, light to dark brown, 3-chambered. **Seeds** 0.3–0.4 × 0.2–0.25 mm, plump, apiculate, reticulate. **2*n*=36, 38.**

Shores, wet meadows, peatlands, ditches. 0–1800 m. BW, Casc, CR, ECas, Est, Sisk, WV. CA, ID, WA; north to AK. Native.

Juncus covillei is similar to *J. falcatus*, which is restricted to coastal sandy habitats and has larger clusters that are more spherical or head-like. Reports of this species from northwestern Asia are *J. prominens.*

Juncus diffusissimus Buckley
Slim pod rush

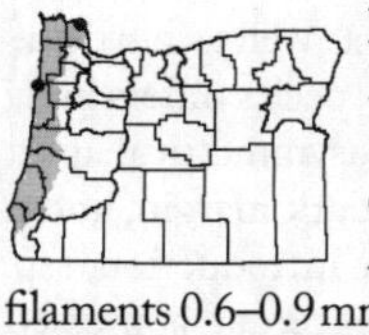

Plants perennial, 25–65 cm tall, cespitose. **Leaves** round, hollow, septate; auricles 1–3 mm. **Inflorescences** open, clusters 20–70, clusters 2–5(10)-flowered. **Flowers** tepals 6, 2–3.2 mm, green or pale brown; stamens 3, filaments 0.6–0.9 mm, anthers 0.4–0.5 mm; styles 0–0.1 mm. **Capsules** 4–5.2 mm, 0.5–2 mm longer than the tepals, pale brown or reddish, often with metallic or bronzy gloss, 1-chambered. **Seeds** ellipsoid to ovoid, 0.3–0.4 × 0.2–0.3 mm, apiculate, reticulate.

Moist river banks, wet disturbed gravel. 0–1000 m. Est. CA, WA; southeastern US; South America. Exotic.

Juncus drummondii E. Meyer (p. 275)
Drummond's rush, threeflower rush

Juncus drummondii E. Mey. var. *subtriflorus* (E. Mey.) C.L. Hitchc.

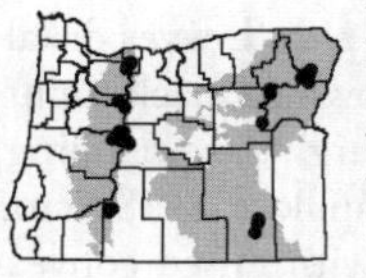

Plants perennial, 10–40 cm tall, cespitose. **Leaves** lacking blades, the blades reduced to vestigial awns, distal sheaths 2–6 cm, light brown. **Inflorescences** a loose cluster of 1–3(5) flowers, inflorescence bracts erect and stem-like, sometimes longer than inflorescence. **Flowers** tepals 6, brown to dark brown; stamens 6, filaments 0.5–1.3 mm, anthers 1–1.6 mm; styles 0.1–0.25 mm. **Capsules** 4.5–8 mm, longer than or equaling the tepals, brown to dark brown, apices notched, 3-chambered. **Seeds** (0.7)1.2–2.3 × 0.2–0.3 mm, bodies 0.5–0.6 mm, striate, tails prominent, 0.4–0.9 mm.

Montane, subalpine, and alpine meadows, damp slopes, swales, gravel bars, springs, pumice. 700–2800 m. BR, BW, Casc. CA, NV, ID, WA; north to AK, southeast to NM. Native.

This species sometimes grows with the similar *J. parryi*, which has elongate leaf blades resembling stems, and often favors drier habitats.

Juncus dudleyi Wiegand (p. 275)
Dudley's rush

Juncus tenuis Willd. var. *dudleyi* (Wiegand) F.J. Herm.

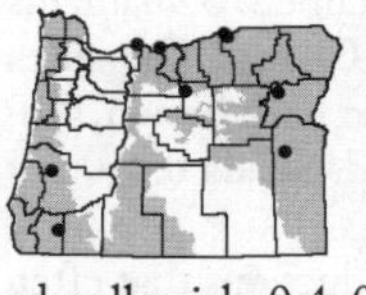

Plants perennial, 20–100 cm tall, cespitose, stems with 0–6 strong ridges visible on one side. **Leaves** thin and wiry, blade flat and slightly inrolled; auricles hardened and thickened distally, plastic-like, shiny, rounded, yellowish, 0.4–0.6 mm. **Inflorescences** cymes, 1.5–5(9) cm, never hidden by inflorescence bract, bract not swollen at base. **Flowers** tepals 6, green to light brown, tips acuminate, erect; stamens 6, filaments 0.7–1 mm, anthers 0.5–0.7(0.9) mm; styles 0.1–0.3 mm. **Capsules** 2–3.6 mm, shorter than the tepals, light brown, apex obtuse, not or weakly crested, 1-chambered. **Seeds** 0.4–0.5(0.65) × 0.2–0.3 mm, apiculate, reticulate.

Fresh water shores, moist banks, springs, ditches. 0–1700 m. BW, Col, CR, Owy, Sisk. CA, NV, ID, WA; most of North America; Europe. Native.

The sturdy yellowed auricles easily distinguish this species from others in Group 5.

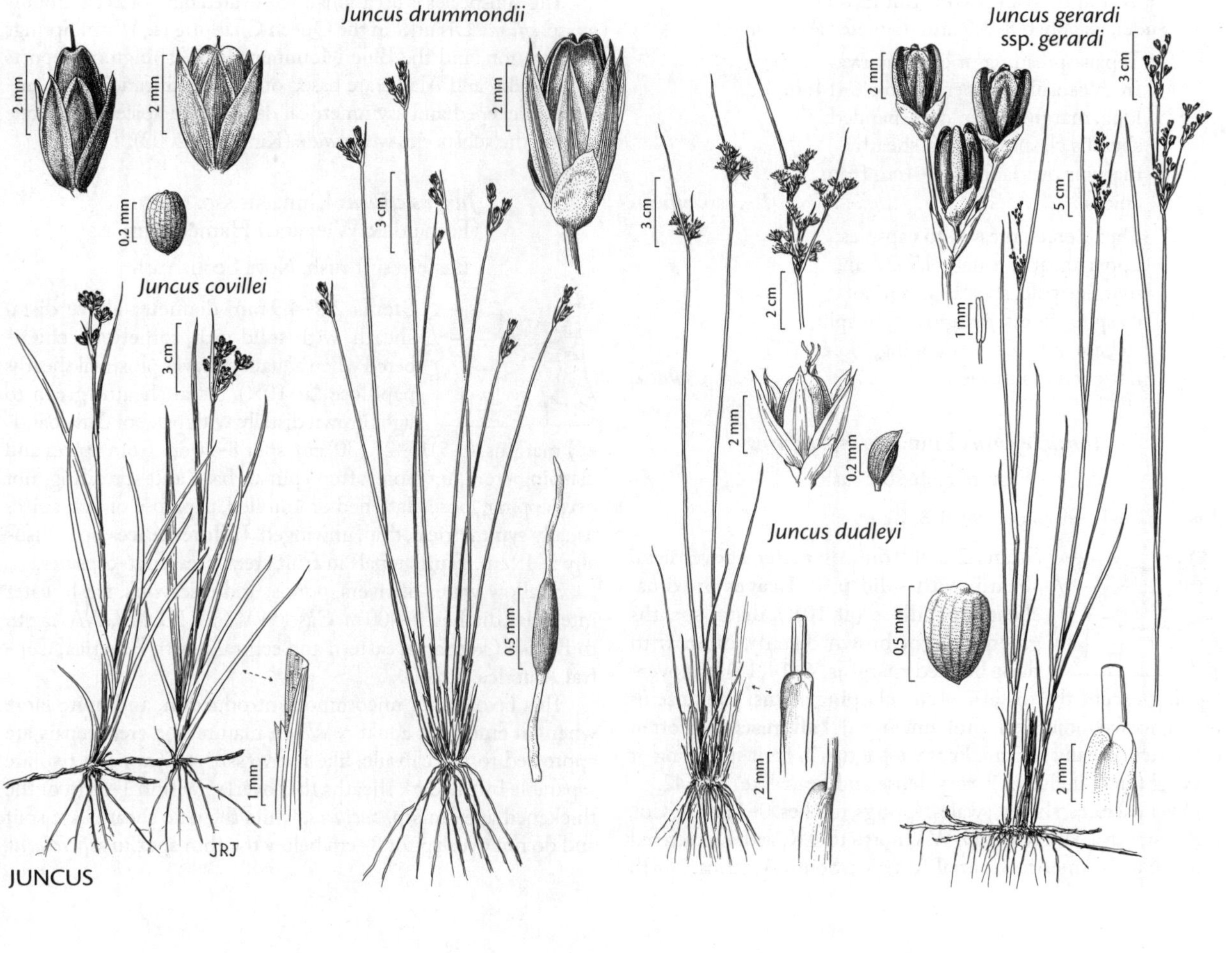

JUNCUS

Juncus effusus Linnaeus

Pasture rush, soft rush

Plants perennial, 50–150 cm tall, cespitose; fruiting stems stout, usually 2–4.9 mm diameter above distal sheath; fresh upper stems shiny, smooth, dried stems with (12)18–26(32) weak low ridges, slender and relatively inconspicuous, visible on one side when dried, dried ridges capped with low dull cells (at 10×). **Leaves** lacking blades, distal sheaths dull or shiny, smooth or papillose, apex symmetrical or asymmetrical, thin or thickened, winged or not. **Inflorescences** dense or open, inflorescence bracts not swollen, erect in fruit. **Flowers** tepals 6, usually green to pale brown (dark brown-striped); stamens 3, filaments 0.4–0.7 mm, anthers 0.4–0.7 mm; styles 0–0.4 mm. **Capsules** 1.6–2.5 mm, shorter than the tepals, light to dark brown, apices obtuse to acute (notched), 1-chambered. **Seeds** 0.4–0.5 × 0.2 mm, reticulate, apiculate.

The various subspecies include two introductions that often grow mingled with our native plants; the species can dominate wet pastures with heavy grazing.

9 subspecies; 3 subspecies in Oregon.

Key to *Juncus effusus* subspecies:

1. Sheath apices thickened, with raised (convex) rims; sheaths usually dark brown to black; fruiting stems stout, usually 2–3.5 mm diameter above sheath . . . . . . ***J. e.*** ssp. ***pacificus***

1' Sheath apices thin with broad membranous wings, flattened and lacking raised rims; sheaths green (fresh) to pale or medium brown (dried); fruiting stems slender, usually 0.8–1.9 mm diameter above sheath.

  2. Tepals spreading or curving away from capsules; upper sheaths 6–14 cm long, margins often dark-banded; sheaths clasping stems, sheath margins overlapping 2–4 cm from apices . . . . ***J. e.*** ssp. ***effusus***

  2' Tepals erect, pressed to capsules; upper sheaths usually 15–27 cm, margins pale; sheaths often not clasping stems, margins often split to base and not overlapping, loose, flattened or unrolled. . . . . ***J. e.*** ssp. ***solutus***

Juncus effusus Linnaeus ssp. *effusus*

Pasture rush, soft rush

Juncus effusus L. var. *compactus* Lej. & Court.

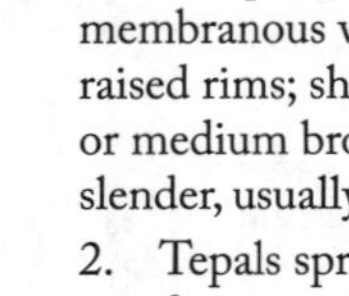

Stems 2.2–4.9 mm diameter above distal sheath, with solid pith. **Leaves** proximal sheaths papillose (at 10×), distal sheaths green to light brown distally, often with dark-banded margins, 6–14(15) cm, overlapping except the distal 2–4 cm, clasping stems, papillose or not, apices symmetrical, thin, unwinged. **Inflorescences** often congested, usually 1–5 cm. **Fruits** tepal tips in fruit spreading or curving (erect if cluster is very dense and head-like). **2*n***=42.

Wet pastures, shores, swales, springs, ditches. 0–1100 m. Col, CR, Est, Lava, WV. CA, ID, WA; north to AK, widely scattered across North America, Central America, South America; North Africa, East Asia, southwestern Asia, Australia, Europe, New Zealand. Exotic.

This taxon is a common introduction often mistaken for native *Juncus effusus* ssp. *pacificus* in pastures as well as horticultural and restoration settings. Our native subspecies differs in its dark sheaths with a winged, thickened and asymmetrical apex; ssp. *effusus* has green sheaths often with a dark margin, and a symmetrical thin unwinged apex. Another introduction, ssp. *solutus*, has longer sheaths, a larger inflorescence, and appressed tepals in fruit; ssp. *effusus* has sheaths less than 15 cm, a variable but often small and dense inflorescence, and the tepals tend to spread or curve away from the ripe capsules.

Juncus effusus Linnaeus ssp. *pacificus* (Fernald & Wiegand) Lint

Common rush, Pacific rush

Juncus effusus L. var. *pacificus* Fernald & Wiegand

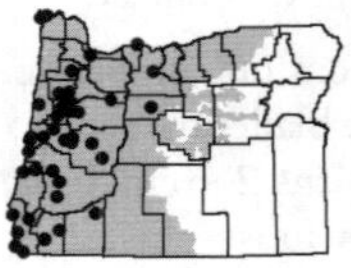

Stems usually 2–3.5 mm diameter above distal sheath, with solid pith. **Leaves** distal sheaths usually dark brown to black (light brown) distally, with margins overlapping except distal 1–3 mm, papillose, 8–18.5 cm, apices strongly asymmetrical, thickened, with raised convex rims, widely winged. **Inflorescences** often open, usually 3–15 cm. **Fruits** tepals in fruit erect.

Wet meadows, wet prairie, shores, swales, springs, wet pastures, ditches. 0–700 m. Casc, CR, Est, Sisk, WV. CA, NV, ID, WA; north to British Columbia, northeast to MT, southeast to AZ. Native.

This subspecies is occasionally cultivated but not as commonly as ssp. *effusus*. Disjunct in the Queen Charlotte Is., Warm Springs Reservation, and the Blue Mountains in Washington. Reports from Idaho and Alaska are based on misidentifications. The unusual thickened and asymmetrical dark sheath apices are unique among the subspecies worldwide (Kirschner 2002).

Juncus effusus Linnaeus ssp. *solutus* (Fernald & Wiegand) Hämet-Ahti

Eastern soft rush, Nova Scotia rush

Stems 2.2–4.9 mm diameter above distal sheath, with solid pith, sometimes chambered when aquatic. **Leaves** proximal sheaths papillose (at 10×), distal sheaths green to light brown distally, with pale (or dark-banded) margins, (9.5)15–27 (30) cm, split 8–20 cm from apices and clasping stem, or more often split to base and spreading, not overlapping, loose, flattened or unrolled, papillose or not, apices usually symmetrical, thin, unwinged. **Inflorescences** open, usually 5–19 cm. **Fruits** tepals in fruit erect, pressed to capsules.

Shallow water of rivers, ponds, and reservoirs, fresh water intertidal, ditches. 0–400 m. Casc, WV. CA, ID, NV, WA; north to British Columbia, eastern and central North America, Central America. Exotic.

This taxon is an uncommon introduction, and quite large when an emergent aquatic. When mature, the erect tepals are appressed to the capsule, like native ssp. *pacificus*. The two are separable by the dark sheaths that overlap within 1–3 cm of the thickened apex in ssp. *pacificus*, while the pale sheaths are split and do not overlap for 8+ cm below the thin apex in ssp. *solutus*.

Juncus ensifolius Wikström

Dagger-leaved rush, swordleaf rush

Juncus ensifolius Wickstr. var. *ensifolius*, *Juncus tracyi* Rydb.

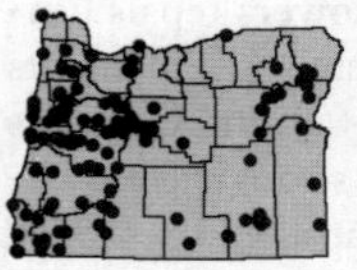

Plants perennial, 15–65 cm tall, rhizomatous. **Leaves** flat, iris-like, blue-green, partially septate, auricles usually absent (0.1–0.4 mm). **Inflorescences** 1–12(22) clusters, clusters often head-like, 3–70-flowered. **Flowers** tepals 6, brown to often dark brown or black; stamens 3, filaments 0.6–1.1 mm, anthers 0.4–0.7 mm; styles 0.2–0.8(1) mm. **Capsules** longer than or equaling tepals, dark brown to blackish, apices obtuse or acute and abruptly tapered to short beak, 1-chambered. **Seeds** 0.4–0.5(0.6) × 0.15–0.2 mm, apiculate, reticulate.

Shores, swales, springs, wet meadows, wet prairie, damp dunes, peatlands, ditches. 0–2500 m. Throughout Oregon. CA, ID, NV, WA; north to Alaska, east to Saskatchewan, SD and CO, south to AZ and Mexico; Ontario and Quebec south to WI and NY; HI; northeastern Asia, Europe, New Zealand. Native.

With three stamens, this species is reliably separated from *J. saximontanus*, which has six stamens. Rare plants with variable stamen numbers in an inflorescence may represent hybrids and need investigation.

Juncus exiguus (Fernald & Wiegand) Lint

Klamath rush, weak rush

Juncus effusus L. var. *exiguus* Fernald & Wiegand

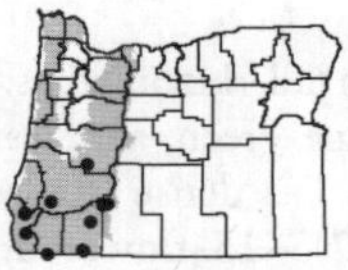

Plants perennial, 40–90 cm tall, cespitose, fresh stems shiny (dull), green, smooth (with relatively inconspicuous ridges); dried fruiting stems slender, 0.7–1.5 mm diameter above distal sheath, often turning blue-green with 6–11(13) strong low ridges per side becoming visible; dried ridges coarse or wide, capped with low dull cells. **Leaves** lacking blades, proximal sheaths smooth (at 10×), distal sheaths 3–14 cm, distal half green to pale brown, thin, dull, not papillose, nerves prominent; apices symmetrical, unwinged, thin, dark and often slightly inrolled towards stem. **Inflorescences** loose clusters, branches visible, inflorescence bracts not swollen, erect in fruit. **Flowers** tepals 6, usually mid to dark brown-striped; stamens 3, filaments 0.3–0.8 mm, anthers (0.3)0.4–0.8(0.9) mm; styles 0.2–0.3 mm. **Capsules** 2.2–3 mm, shorter than or equaling the tepals, light to dark brown, 1-chambered. **Seeds** 0.35–0.5 × 0.2–0.25 mm, reticulate, apiculate.

Shores, wet meadows, marshes, stream banks, springs, peatlands. 900–2300 m. CR, Sisk, Casc. CA; southeast to AZ. Native.

The fresh upper stems of this species are variable, but are often smooth, green, and shiny. Upon drying they become blue-green and clearly ridged, so the plants look quite different in the field and in a herbarium. Similar *Juncus hesperius* is found at lower elevations and has papillose proximal sheaths; *J. exiguus* is montane and is never papillose.

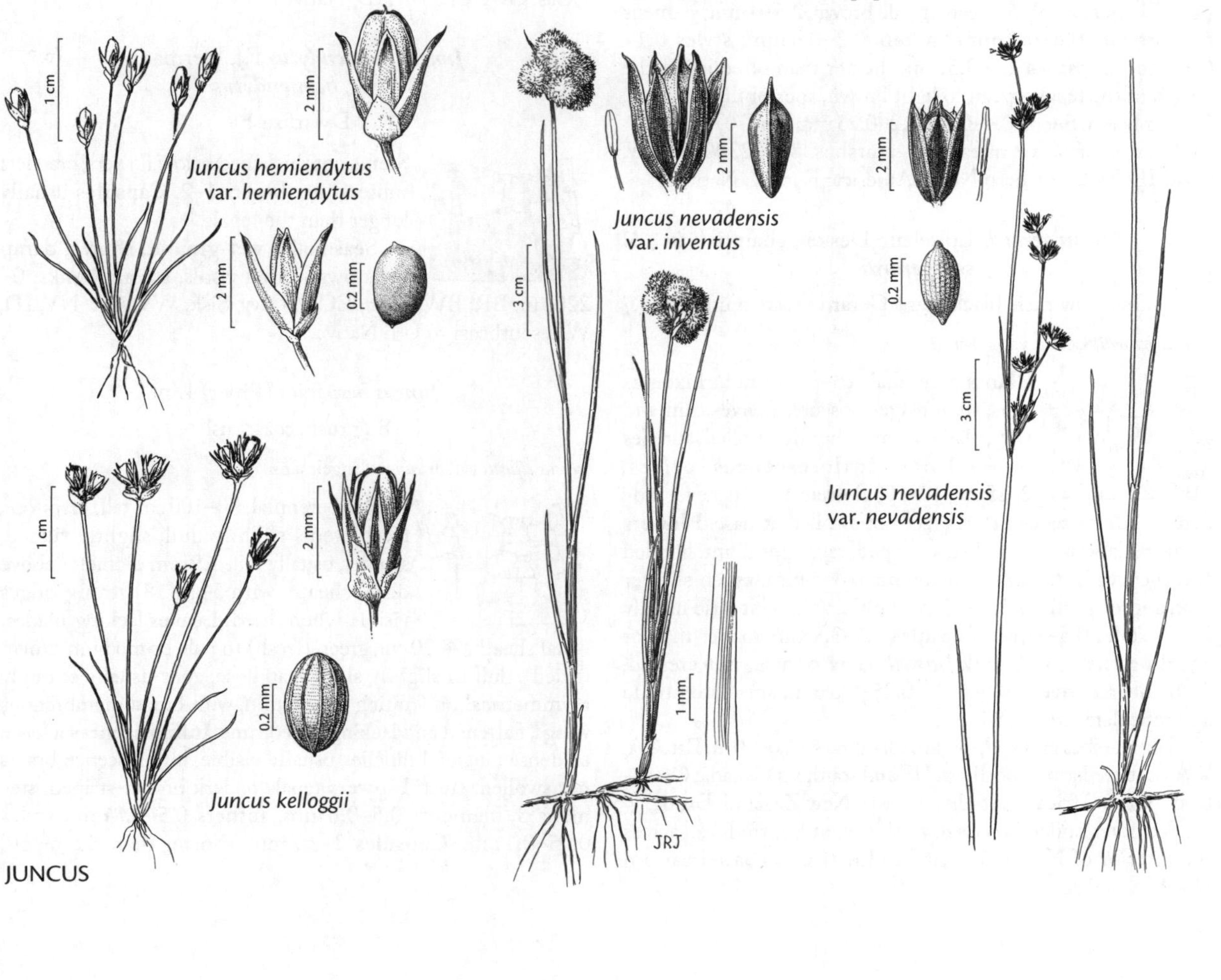

JUNCUS

Juncus falcatus E. Meyer ssp. ***sitchensis*** (Buchenau) Hultén

Sickle-leaved rush

Juncus falcatus E. Mey. var. *sitchensis* Buchenau

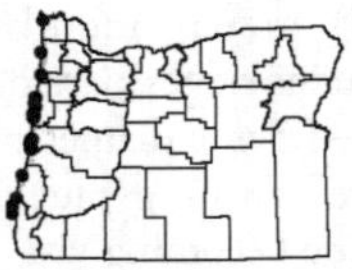

Plants perennial, 5–30 cm tall, rhizomatous. **Leaves** flat, grass-like, blue-green; auricles 0–1 mm. **Inflorescences** 1–3 dense head-like clusters, clusters (2)5–15-flowered. **Flowers** tepals 6, dark brown, 4–6 × 1.5–2 mm; stamens 6, filaments 0.5–1.3 mm, anthers 0.8–1.4 mm; styles 0.3–0.8 mm. **Capsules** subglobose or obovoid, notched, shorter than the tepals, blackish, 3-chambered. **Seeds** 0.6–0.8 × 0.3–0.5 mm, reticulate. **2*n***=38.

Damp dunes, shores, peatlands, sandy ditches, cranberry farms. 0–50 m. Est. CA, WA; north to AK; Asia, Australia. Native.

This species includes native populations in Asia and Australia; their taxonomy needs study.

Juncus filiformis Linnaeus

Thread rush

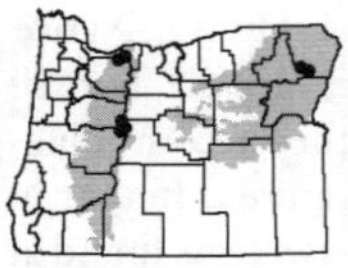

Plants perennial, 10–80 cm tall, rhizomatous, stem ridges obscure. **Leaves** lacking blades, distal sheaths 3–10 cm, light brown. **Inflorescences** a more or less loose cluster of 4–15 flowers, usually attached near or below middle of stems (stems here including erect inflorescence bracts above inflorescences); inflorescence bracts not swollen at base. **Flowers** tepals 6, green or pale brown, 2.5–4 mm; stamens 6, filaments 0.4–0.9 mm, anthers 0.3–0.6 mm; styles 0.1–0.15 mm. **Capsules** 2.5–3.5 mm, shorter than or equaling the length of the tepals, green to light brown, apex obtuse to acute, 3-chambered. **Seeds** 0.4–0.5 × 0.2–0.25 mm. **2*n***=70, 80, 84.

Pond shores, wet meadows, marshes. 800–2300 m. BW, Casc. ID, WA; northern North America; Eurasia. Native.

Juncus gerardi Loiseleur-Deslongchamps ssp. ***gerardi*** (p. 275)

Saltmeadow rush, black grass, Gerard's rush, mud rush

Juncus gerardii Loisel. var. *gerardii*

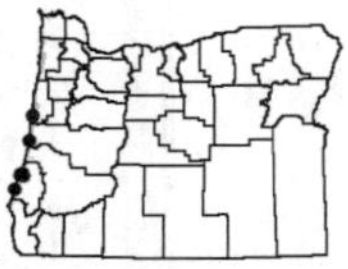

Plants perennial, 10–80 cm tall, rhizomatous, stem ridges obscure. **Leaves** thin and wiry, blade flat and slightly inrolled, auricles 0.4–0.8 mm. **Inflorescences** cymes, 2–15 cm, 10–100 flowers, flowers never hidden by inflorescence bract, bract not swollen at base. **Flowers** tepals 6, dark brown- to black-striped, tepal tips blunt, hooded and incurved; stamens 6, filaments 0.4–5 mm, much shorter than anthers, anthers 1.3–1.7 mm, anther/filament ratio usually 2.6–4; styles 0.5–1 mm. **Capsules** 2.5–3.8 mm, longer than or equaling the tepals, dark brown, apex obtuse, not crested, 3-chambered. **Seeds** 0.5–0.7 × 0.25–0.3 mm, apiculate, striate and reticulate. **2*n***=80, 84.

Salt marshes and wet, disturbed, saline soil. 0–50 m. Est. CA, WA; scattered across northern US and southern Canada; Greenland; North Africa, Australia, Eurasia, New Zealand. Exotic.

See notes under *J. compressus*. This plant is introduced across the interior of North America and native to coastal eastern North America.

Juncus hemiendytus F.J. Hermann

Dwarf rush

Plants annual, 0.2–3.6 cm tall, unbranched. **Leaves** capillary, basal. **Inflorescences** 1 terminal flower. **Flowers** tepals usually 4(6), reddish, erect or outcurved; stamens 2(3), filaments 0.5–1.4 mm, anthers 0.2–0.7 mm; styles 0.1–0.4 mm. **Capsules** usually longer than the tepals, reddish, not strongly contrasting with the tepals. **Seeds** about 70 per capsule, 0.3–0.55 × 0.2–0.35 mm, apiculate, smooth or faintly reticulate (at 10×), not striate.

2 varieties.

Key to *Juncus hemiendytus* varieties:

1. Inflorescence bractlets 0; stem thickened below flower . . . . ***J. h.*** var. ***abjectus***

1' Inflorescence bractlets 1–2; stem not thickened below flower . . . . ***J. h.*** var. ***hemiendytus***

Juncus hemiendytus F.J. Hermann var. ***abjectus*** (F.J. Hermann) Ertter

Center Basin rush, least rush

Juncus abjectus F.J. Herm.

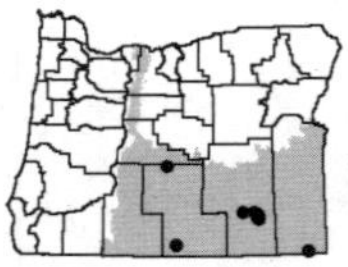

Stems thickened below flowers, bractlets subtending flowers 0. **Capsules** usually equaling tepals.

Springs, shores, creekbeds, wet meadows, seasonally wet ground. 1400–2700 m. BR, ECas, Owy. CA, NV, ID. Native.

Juncus hemiendytus F.J. Hermann var. ***hemiendytus*** (p. 277)

Dwarf rush

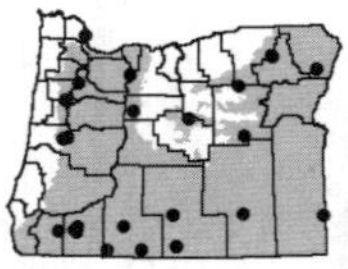

Stems not thickened below flowers, bractlets subtending flowers 1–2. **Capsules** usually longer than the tepals.

Seasonally wet ground, shores, damp meadows, vernal pools, stream banks. 0–2200 m. BR, BW, Casc, ECas, Owy, Sisk, WV. CA, NV, ID, WA, southeast to UT. Native.

Juncus hesperius (Piper) Lint

Bog rush, coast rush

Juncus effusus var. *brunneus* Engelmann

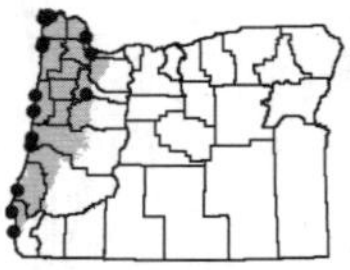

Plants perennial, 30–100 cm tall, cespitose, fresh stems slightly dull, slightly ridged, slender, usually 0.8–1.9 mm diameter above distal sheath, with 5–14(18) strong ridges visible when dried. **Leaves** lacking blades, distal sheaths 4–20 cm, green (fresh) to pale or medium brown (dried), dull to slightly shiny, papillose, apex usually strongly asymmetrical on fruiting stems, thin, with broad membranous wings, flattened and lacking raised rims. **Inflorescences** a loose to dense cluster, branches usually visible, inflorescence bracts not swollen, erect. **Flowers** tepals 6, dark brown-striped; stamens 3, filaments 0.5–0.6 mm, anthers 0.5–0.8 mm; styles 0.05–0.1 mm. **Capsules** 2–2.7 mm, shorter than the tepals, brown to dark brown, apex acute to notched or obtuse, 1-chambered. **Seeds** 0.4–0.5 × 0.2–0.25 mm, reticulate, apiculate.

Shores, springs, swales, peatlands, wet pastures, cranberry farms, ditches. 0–800 m. CR, Est, WV. CA, WA; north to British Columbia. Native.

This species rarely hybridizes with *J. laccatus*. Montane *J. exiguus* has a symmetrical sheath apex and smooth sheath, differing from the more coastal *J. hesperius*, which has an asymmetrical sheath apex and papillose sheaths.

Juncus howellii F.J. Hermann
Howell's rush

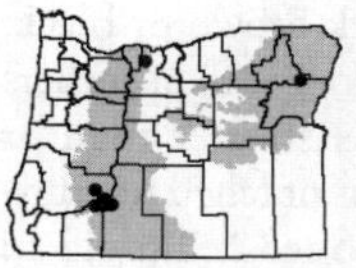

Plants perennial, 20–60 cm tall, rhizomatous. **Leaves** flat, grass-like, blue-green, up to 4.5 mm wide; auricles absent or 0.2–2 mm. **Inflorescences** (2)5–20 clusters, clusters usually 2–8-flowered. **Flowers** tepals 6, dark brown, papillose, inner usually longer than the outer; stamens 6, filaments 0.6–1.1 mm, anthers 1.2–2.6 mm; styles 1.2–3.2 mm. **Capsules** obovoid, shorter than the tepals, brown, acute, 3-chambered. **Seeds** slender, 0.6–1 × 0.25–0.3 mm, body 0.4–0.6 mm, tail 0.1–0.3 mm, 1 or 2 tails at least half the length of the seed body.

Springs, gravel bars, wet slopes, peatlands. 600–2200 m. BW, Casc. CA, ID, WA. Native.

Auricle characters are occasionally unreliable, so this species is best separated from *J. orthophyllus* and *J. regelii* by the intermediate length of the tail on the seed, with one or both seed tails about half as long as the seed body. *Juncus orthophyllus* has apiculate seeds without a tail, and *J. regelii* has seed tails approximately as long as the seed body.

Juncus inflexus Linnaeus ssp. *inflexus*
European blue rush

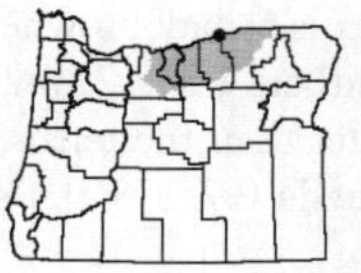

Plants perennial, 50–120 cm tall, cespitose, fresh stems dull, blue-green, with 5–11 prominent strong longitudinal ridges visible fresh or dried; pith of fertile stems chambered. **Leaves** lacking blades, distal sheaths 5–15 cm, dark brown to dark purple distally, shiny, smooth, apex symmetrical, thin, not winged. **Inflorescences** a well-developed, loose cluster of 10–100+ flowers, branches usually visible, inflorescence bracts not swollen. **Flowers** tepals 6, dark brown-striped; stamens 6, filaments 0.5–0.7 mm, anthers 0.6–0.9 mm; styles 0.5–0.55 mm. **Capsules** ovate to elliptic, 2.5–4 mm, brown, often dark brown or blackish distally, apex pointed; styles forming a prominent beak, capsule shorter or longer than the tepals, apex acute, 3-chambered. **Seeds** 0.4–0.6 × 0.2–0.3 mm, reticulate, apiculate. **2*n***=40.

Damp pastures, swales, ditches. 0–100 m. Col. Eastern North America; Africa, Australia, Eurasia, New Zealand. Exotic.

Juncus interior Wiegand
Interior rush

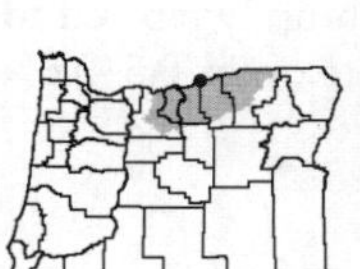

Plants perennial, (11)20–60 cm tall, cespitose, base usually pink, 4–6 strong stem ridges visible per side. **Leaves** thin and wiry, blade flat and slightly inrolled; auricles soft, 2-textured, usually thicker proximally and more opaque than thin distal margin, 0.4–0.8(1.2) mm, dirty white to translucent, dull, rounded. **Inflorescences** cymes, 1.5–7 cm; bractlets subtending flowers acuminate to aristate. **Flowers** tepals 6, light brown, tepal tips acuminate; stamens 6, filaments 0.8–1.1 mm, anthers 0.3–0.7(0.9) mm; styles 0.1–0.4(0.5) mm. **Capsules** 3.3–4.7 mm, longer than the tepals, light brown, apex obtuse to truncate, not ridged or crested, 1-chambered. **Seeds** 0.3–0.45 × 0.15–0.25 mm, apiculate. **2*n***=80.

River shores. 50–200 m. Col. CA, ID, WA; throughout North America. Native.

Juncus kelloggii Engelmann (p. 277)
Kellogg's rush

Juncus brachystylus (Engelm.) Piper

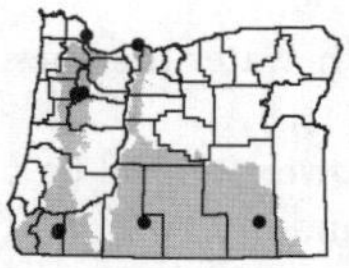

Plants annual, 1–5.5 cm tall, unbranched. **Leaves** capillary, basal. **Inflorescences** 1 terminal cluster, with 1–4 flowers per cluster; inflorescence bracts inconspicuous, flowers subtended by 2 bractlets, bractlet tips usually acute (acuminate). **Flowers** tepals (4)6, midveins dark reddish, tips acute to acuminate; stamens (2)3, filaments 0.7–1.3 mm, anthers 0.3–0.5 mm; styles 0.2–0.5 mm. **Capsules** equaling or slightly longer than the tepals, dark red, apex acute to acuminate. **Seeds** approximately 50 per capsule, 0.35–0.55 × 0.25–0.35 mm, longitudinally striate (obvious at 30×), apiculate. **2*n***=34.

Creek banks, floodplains in wet prairie, swales, pool margins. 100–1900 m. BR, Sisk, WV. CA, NV, WA; north to southern British Columbia. Native.

Juncus laccatus Zika
Shiny rush

Juncus effusus L. var. *gracilis* Hook.

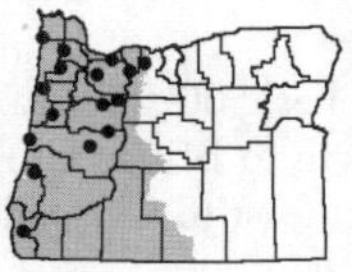

Plants perennial, 25–108 cm tall, cespitose, fresh stems shiny, smooth, dried fruiting stems slender, 0.6–2.6 mm diameter above distal sheath, remaining green, with 6–16 strong ridges becoming visible per side, dried ridges coarse or wide, capped with low dull cells. **Leaves** lacking blades; proximal sheaths smooth, not papillose (at 10×); distal sheaths 5–18 cm, distal half dark brown or black, shiny, thick and glossy, nerves obscure, apices symmetrical, unwinged, uniformly thick, dark, not inrolled. **Inflorescences** loose clusters, branches usually visible, inflorescence bracts not swollen. **Flowers** tepals 6, medium to dark brown-striped; stamens 3, filaments 0.4–0.9 mm, anthers 0.4–0.9 mm; styles 0.05–0.3 mm. **Capsules** 1.7–2.2 mm, shorter than the tepals, brown to dark brown, apex obtuse to acute, 1-chambered. **Seeds** 0.4–0.6 × 0.2–0.25 mm, reticulate, apiculate. **2*n***=40, 80.

Shores, swales, springs, seasonally wet ground, peatlands, and ditches. 0–1600 m. Casc, CR, Est, WV. CA, WA; north to British Columbia, southeast to Arizona. Native.

Occasionally cultivated, the thick shiny dark and smooth sheath of this species is distinctive; *J. exiguus* has a thinner, paler, and more textured sheath; *J. hesperius* has a less shiny, paler, and papillose sheath.

Juncus lescurii Bolander

San Francisco rush

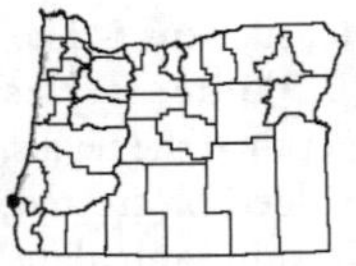

Plants perennial, 40–140 cm tall, rhizomatous, fresh stems usually round and erect, usually not twisted, smooth, with 12–30 weak ridges visible per side when dried. **Leaves** lacking blades, distal sheaths 5–12 cm, green to dark brown or blackish distally, dull or slightly shiny, smooth, apex symmetrical, thin, not winged. **Inflorescences** variable, a loose or sparse cluster, often lobed, at least some branches usually visible, inflorescence bracts not swollen. **Flowers** tepals 6, usually 6–8 mm, dark brown-striped; stamens 6, filaments 0.3–0.9 mm, anthers 0.9–2.3 mm; styles 1–1.8 mm. **Capsules** 4–5 mm, shorter than the tepals, dark brown, apex acute, 1-chambered. **Seeds** 0.6–1.1 × 0.4–0.6 mm, often silvery, apiculate.

Damp dunes, sandy shores, coastal fresh-water marshes. 0–50 m. Est. CA. Native.

This species is identified by the large flowers typical of *J. breweri* and the more open inflorescence found in *J. balticus*. It is possibly derived from hybrids between those two species and often grows with them. It was long thought to be endemic to the San Francisco Bay area (Lint 1977).

Juncus longistylis Torrey

Long-styled rush

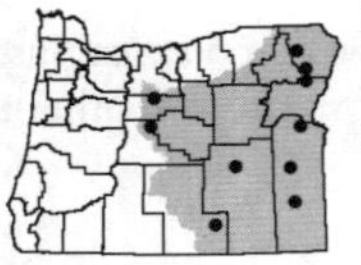

Plants perennial, 15–55 cm tall, rhizomatous. **Leaves** flat, grass-like, blue-green; auricles prominent, usually rounded, (0.7)1–1.8(2.5) mm. **Inflorescences** 1–4(5) clusters, clusters 3–8(12)-flowered. **Flowers** tepals 6, usually 4–6 mm, dark brown, sometimes papillose; stamens 6, filaments (0.5)0.7–1 mm, anthers (1.2)1.5–1.7(2) mm; styles 0.6–1.3 mm. **Capsules** ovoid, obtuse or truncate, shorter than the tepals, dark brown, 3-chambered. **Seeds** 0.4–0.6 × 0.2–0.25 mm, faintly reticulate, apiculate. ***2n***=40.

Wet meadows, springs, floodplains, ditches. 500–2000 m. BR, BW, Lava, Owy. CA, NV, ID, WA; north to British Columbia, east to Newfoundland and MI, southeast to CO and NM. Native.

This species is similar to *J. regelii* but with apiculate seeds. The well-developed auricles help separate it from *J. orthophyllus*.

Juncus marginatus Rostkovius

Grass rush

Juncus marginatus Rostk. var. *marginatus*

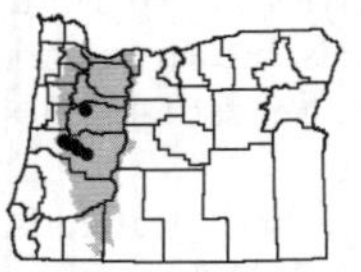

Plants perennial, 30–100 cm tall, cespitose, base with knobby tubers. **Leaves** flat, grass-like, green; auricles 0.5–1.7 mm. **Inflorescences** 5–40 clusters, clusters 3–10-flowered. **Flowers** tepals 6, brown to dark brown, red-spotted or not, papillose; stamens 3, filaments 1.1–1.8 mm, anthers 0.3–1.3 mm, purplish red; styles 0–0.3 mm. **Capsules** subglobose to obovoid, more or less obtuse, shorter or longer than the tepals, brown to dark brown, red-spotted or not, 3-chambered. **Seeds** 0.4–0.6 × 0.15–0.2 mm, striate, apiculate. ***2n***=38, 40.

Wet prairies, pastures, shores, ditches. 100–300 m. Casc, WV. CA; scattered across North America; South America. Exotic.

Juncus marginatus is considered adventive in California as well as British Columbia and Oregon. The red-speckling on the tepals or capsules (when present) and knobby stem base are distinctive.

Juncus mertensianus Bongard

Mertens' rush

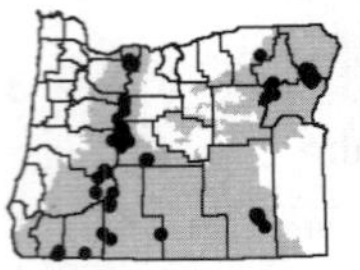

Plants perennial 10–40 cm tall, loosely cespitose to short-rhizomatous, rhizomes not tuberous at nodes. **Leaves** green, tubular, obviously hollow, septate; auricles 1–2 mm. **Inflorescences** 1(2) head-like clusters, clusters 12–60-flowered. **Flowers** tepals 6, dark brown or black; stamens 6, filaments 0.6–1.5 mm, shorter to longer than anthers 0.4–0.8 mm; styles 0.5–0.7 mm. **Capsules** shorter than or equaling the tepals, dark brown, apices often notched or truncate, more or less beakless, 1-chambered. **Seeds** ellipsoid to ovoid, 0.4–0.6 × 0.2–0.25 mm, apiculate, reticulate. ***2n***=40, 80.

Wet meadows, shores, peatlands, springs, wet ledges, ditches. 50–2900 m. BR, BW, ECas, Sisk, Casc. CA, ID, NV, WA; north to AK, east to SD, southeast to NM. Native.

This species is similar to *J. nevadensis*, which regularly has more than one cluster per inflorescence.

Juncus mexicanus Willdenow

Mexican rush

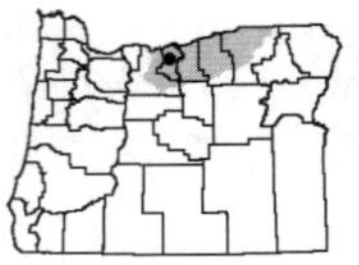

Plants perennial, 15–65 cm tall, rhizomatous, fresh stems smooth, slightly shiny, dark green to blue-green, dried stems with 0–14 weak ridges visible per side. **Leaves** with some well-developed blades on distal sheaths 4–10 cm, pale brown to dark brown distally, dull or shiny, smooth, apex symmetrical, thin, not winged. **Inflorescences** loose clusters, at least some branches usually visible, inflorescence bracts not swollen. **Flowers** tepals 6, dark brown-striped; stamens 6, filaments 0.2–0.4 mm, anthers 1.2–2.2 mm; styles 0.8–0.9 mm. **Capsules** 4–5 mm, shorter than the tepals, dark brown, apex acute, 1-chambered. **Seeds** 0.7–1 × 0.4–0.45 mm, apiculate.

Floodplains, canyon bottoms. 500–600 m. Col. CA, NV; southeast to CO and TX; south to South America. Native.

This species looks much like *J. balticus* but has distinctive, stem-like leaf blades on some stems.

Juncus nevadensis S. Watson

Sierra rush

Juncus badius Suksd., *Juncus columbianus* Coville, *Juncus nevadensis* S. Watson var. *badius* (Suksd.) C.L. Hitchc., *Juncus nevadensis* S. Watson var. *columbianus* (Coville) H. St. John

Plants perennial, 10–70 cm tall, rhizomatous, rhizomes elongate, not tuberous at nodes. **Leaves** round to flattened, hollow, septate; auricles 1–3.2 mm. **Flowers** tepals 6, brown to dark brown; stamens 6; styles 0.7–1.8 mm. **Capsules** usually shorter than or equaling the tepals, dark brown, abruptly tapered to distinct beak, 1-chambered. **Seeds** ellipsoid to ovoid, 0.3–0.6 × 0.2–0.3 mm, apiculate, reticulate.

4 varieties. 2 varieties treated in *Flora*.

Key to *Juncus nevadensis* varieties:

1. Anthers longer than filaments; not usually coastal . . . . *J. n.* var. ***nevadensis***
1' Anthers shorter than or equaling filaments; coastal. . . . . *J. n.* var. ***inventus***

Juncus nevadensis S. Watson var. ***inventus*** (p. 277)
(L.F. Henderson) C.L. Hitchcock

Oregon rush

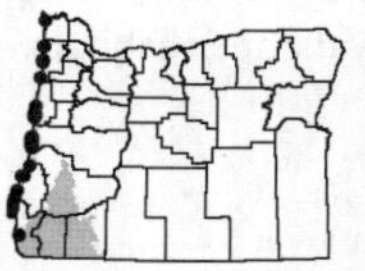

Leaves slightly to strongly flattened, hollow, slightly to strongly blue-green, septate. **Inflorescences** usually 1–3(14) clusters or head-like clusters, clusters 20–45-flowered. **Flowers** filaments 0.7–1.4 mm, anthers 0.6–1.2 mm, usually equaling the filaments.

Coastal dunes, shores, sandy swales, springs, cranberry farms, ditches. 0–50 m. Est, Sisk. CA, WA; north to British Columbia. Native.

The peculiar leaves of this taxon are usually flattened but hollow and completely septate, which helps separate it from *J. phaeocephalus*, a California species with iris-like leaves that are solid and incompletely septate. *Juncus phaeocephalus* has been reported in error from Oregon several times.

Juncus nevadensis S. Watson var. ***nevadensis*** (p. 277)

Sierra rush

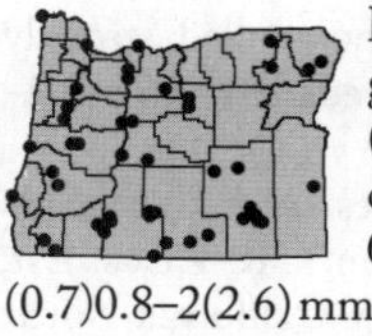

Leaves round or slightly flattened, hollow, green or blue-green, septate. **Inflorescences** (1)6–30 clusters or head-like clusters, clusters 3–15-flowered. **Flowers** filaments 0.3–0.7(0.8) mm, shorter than anthers (0.7)0.8–2(2.6) mm.

Shores, seasonally wet meadows, marshes, swales, ditches. 0–2500 m. Throughout Oregon. CA, NV, ID, WA; northeast to Saskatchewan, southeast to CO. Native.

The color, thickness, and flattening of the leaf blades is highly variable.

Juncus occidentalis (Coville) Wiegand

Prairie rush, western rush

Juncus tenuis Willd. var. *congestus* Engelm.

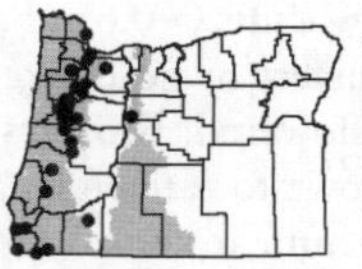

Plants perennial, 30–60 cm tall, cespitose, 0–3(5) strong stem ridges visible per side. **Leaves** thin and wiry, blade flat and slightly inrolled; auricles soft and thin distally, 0.1–1.1 mm, dull, rounded or acute, dirty white. **Inflorescences** cymes, 1–7 cm; bractlets subtending flowers

JUNCUS

blunt to acute. **Flowers** tepals 6, brown-striped, tepal tips acuminate; stamens 6, filaments 0.6–1.1 mm, anthers 0.4–0.9 mm; styles 0.1–0.3 mm. **Capsules** 2.7–4.5 mm, shorter than the tepals, brown, apex usually truncate to slightly notched, strongly ridged or crested distally, 1-chambered. **Seeds** 0.4–0.6 × 0.2–0.25 mm, apiculate. **2*n***=80.

Wet prairies, swales, shores, riparian zones, seeps, wet meadows, balds, vernal pools, ditches. 0–2100 m. CR, ECas, Est, Sisk, WV. CA, WA; British Columbia, Baja California Norte, Mexico. Native.

Juncus occidentalis is a western Oregon sister species to *J. confusus*, separable by the construction of the capsules (three-chambered in *J. confusus*, one-chambered in *J. occidentalis*).

Juncus orthophyllus Coville

Straight-leaved rush

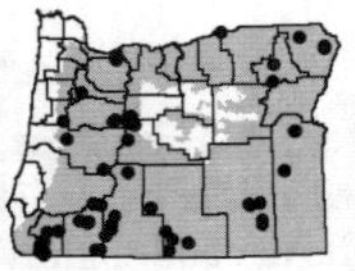

Plants perennial, 20–40 cm tall, rhizomatous. **Leaves** flat, grass-like; auricles absent or inconspicuous, usually acute, 0–0.6 mm. **Inflorescences** 3–12 clusters or head-like clusters, clusters 4–10-flowered. **Flowers** tepals 6, 5–6 × 1–1.4 mm, greenish brown, brown to dark brown, papillose; stamens 6, filaments 0.4–0.8(1) mm, anthers 1.6–2.8 mm; styles 1.3–2.5 mm. **Capsules** oblong-ovoid, obtuse or truncate, shorter than the tepals, brown, 3-chambered. **Seeds** 0.5–0.7 × 0.2–0.25 mm, reticulate, apiculate, tails less than 0.2 mm. **2*n***=40.

Floodplains, wet meadows, marshes, wet prairie, springs, ditches, serpentine wetlands. 300–1900 m. BR, BW, Casc, Col, ECas, Owy, Sisk, WV. CA, NV, ID, WA; east to MT. Native.

This species is often confused with *J. longistylis*, which has prominent long auricles; *J. orthophyllus* auricles are inconspicuous or absent. Many reports from the southern Cascades are misidentifications of *J. howellii*, which grows at higher elevations and has more elongated tails on the seeds.

Juncus oxymeris Engelmann (p. 281)

Pointed rush

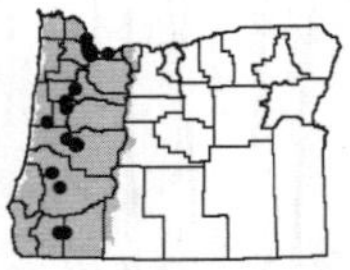

Plants perennial, 30–90 cm tall, rhizomatous. **Leaves** flat, iris-like, blue-green, partially septate, auricles absent. **Inflorescences** 10–70 clusters, clusters (2)3–7(11)-flowered. **Flowers** tepals 6, green to pale brown, usually red-tipped; stamens 6, filaments 0.5–0.8 mm, anthers 0.7–1.6 mm; styles 0.4–1 mm. **Capsules** longer than the tepals, brown to dark red, usually gradually tapered to long acuminate tip, 1-chambered. **Seeds** 0.4–0.5 × 0.2–0.25 mm, apiculate, reticulate.

Shores, swales, marshes, wet meadows, fresh water intertidal. 0–700 m. Casc, CR, Sisk, WV. CA, WA; north to British Columbia. Native.

Juncus parryi Engelmann

Parry's rush

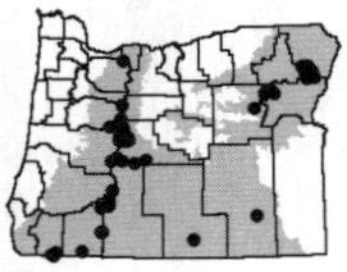

Plants perennial, 5–30 cm tall, cespitose. **Leaves** with some terete, stem-like blades, distal sheaths 1–5 cm, light brown or red-brown. **Inflorescences** a loose cluster of 1–3 flowers, inflorescence bracts erect and stem-like, often longer than inflorescence. **Flowers** tepals 6, brown to dark brown; stamens 6, filaments 0.6–1 mm, anthers 1–1.6 mm; styles 0.3–0.6 mm. **Capsules** 6–9 mm, longer than or equaling the tepals, brown, apices acute to acuminate, 3-chambered. **Seeds** (1.1)1.7–2.7 × 0.2–0.3 mm, bodies 0.6–0.9 mm, striate, tails prominent, (0.3)0.4–0.9 mm.

Rocky slopes, scree, pumice, dry meadows, shores, ridgelines, open conifer forest, peatlands. 1100–2800 m. BR, BW, ECas, Sisk, Casc. CA, NV, ID, WA; northeast to Alberta, southeast to NM. Native.

This species is often confused with *J. drummondii*, which lacks leaf blades and sometimes is in slightly damper habitats.

Juncus patens E. Meyer

Gray rush, spreading rush

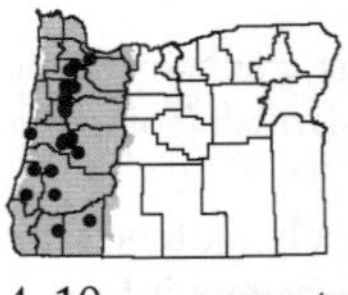

Plants perennial, 30–90 cm tall, cespitose, fresh stems ridged, dull, blue-green, with (8)10–15 prominent strong ridges visible per side when dried, pith of fertile stems solid. **Leaves** without blades, distal sheaths 4–10 cm, green to pale brown distally, dull or shiny, smooth, apex symmetrical, thin, not winged. **Inflorescences** a well-developed, loose cluster of 10–100+ flowers, branches usually visible, inflorescence bracts not swollen. **Flowers** tepals 6, green to pale brown, tips acute; stamens 6, filaments 0.5–1 mm, anthers 0.4–0.65 mm; styles 0.1 mm. **Capsules** subglobose, 2–2.7 mm, pale brown, apex rounded or subacute; styles deciduous or inconspicuous, capsule shorter than the tepals, 1-chambered. **Seeds** 0.4–0.5 × 0.2–0.3 mm, apiculate.

Shores, floodplains, wet prairies, wet forest, swales, marshes, ditches. 0–1000 m. Casc, CR, Sisk, WV. CA, WA; south to Mexico. Native.

The blue-green stems separate *J. patens* from all other cespitose species in Group 2, except the rare exotic *J. inflexus*, which has a chambered pith and dark tepals. However, the Australian introduction *J. usitatus* L.A.S. Johnson has become an invasive pest in northern California, where it has been mistakenly used in native plant restoration work and landscaping because it closely resembles *J. patens*. They are separable by the blunt not acute petals of *J. usitatus*, its three stamens per flower, and its tepals much shorter than the capsules. *Juncus patens* has six stamens per flower, and its tepals are longer than the capsules.

Juncus pelocarpus E. Meyer

Brown-fruited rush

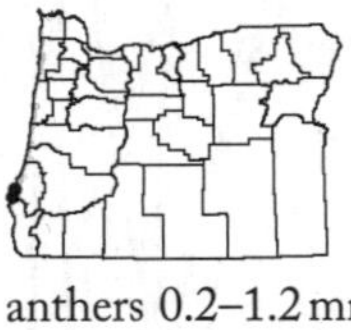

Plants perennial, 3–70 cm tall, rhizomatous. **Leaves** round, hollow, septate; auricles 0.3–1 mm. **Inflorescences** loose cymes, 1–2(3) flowers per node. **Flowers** tepals 6, brown or reddish; stamens 6, filaments 0.2–0.8 mm, anthers 0.2–1.2 mm; style 0.3–0.5 mm. **Capsules** shorter than or longer than the tepals, pale brown to dark brown, 1-chambered. **Seeds** ellipsoid to ovoid, 0.3–0.5 × 0.2–0.3 mm, apiculate, faintly reticulate. **2*n***=40.

Sandy shores, ditches, cranberry farms. 0–50 m. Est. WA; eastern North America; north to British Columbia. Exotic.

The diffuse inflorescence is distinct. It is native in eastern North America.

Juncus planifolius R. Brown

New Zealand rush, flat-leaved rush

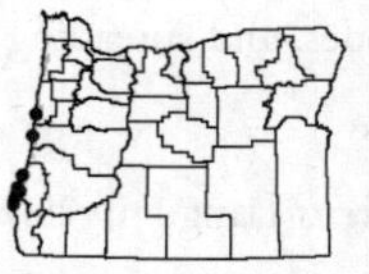

Plants perennial, rarely annual, 1–50 cm tall, cespitose, branching, base without tubers. **Leaves** flat, grass-like; auricles absent. **Inflorescences** 1–35 clusters, clusters 3–10-flowered, sometimes bearing leafy bulblets, inflorescence bracts inconspicuous, membranous, shorter than or equaling inflorescence. **Flowers** tepals 6, 1.8–3.2 mm, dark brown to black, smooth; stamens 3, filaments 0.8–1.1 mm, anthers 0.4–0.6 mm, yellow; style 0.2–0.3 mm. **Capsules** elliptic to ovoid, shorter than or equaling the tepals, brown to dark brown, 1-chambered. **Seeds** 0.3–0.45 × 0.2–0.3 mm, reticulate, apiculate.

Muddy or sandy shores, sometimes in shallow water, ditches, cranberry farms, peatlands, damp dune hollows. 0–50 m. Est. CA; HI, South America; Australia, Ireland, New Zealand. Exotic.

Juncus planifolius was first collected in Oregon in 1988 and is rapidly spreading. It is native south of the equator.

Juncus ranarius Songeon & E. P. Perrier

Frog rush

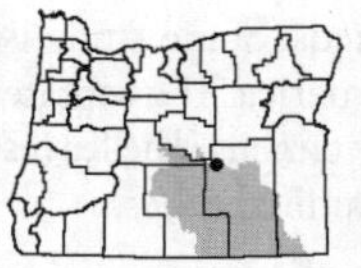

Plants annual, 3–10(17) cm tall, branched. **Leaves** usually less than 1 mm wide, stem leaves usually 1–2. **Inflorescences** cymose, flowers solitary at nodes. **Flowers** tepals 6, green to light brown, outer tepals usually 4–5 mm, inner more or less blunt; stamens 6, filaments 0.7–1.5 mm, anthers 0.4–0.8 mm; styles 0.3–0.4 mm. **Capsules** usually truncate (blunt to subacute), (shorter than) more or less equaling inner tepals, brown, 1-chambered. **Seeds** 0.35–0.5 × 0.25–0.35 mm, apiculate. **2*n*=**34.

Moist clay, stream banks, disturbed wet ground. 0–2000 m. BR. CA; scattered across North America; North Africa, Eurasia. Exotic.

Juncus ranarius is similar to *J. bufonius*, as well as Eurasian *J. ambiguus*, a misapplied name in our flora. The taxonomy of these species is controversial (Balslev 1996), and the complex needs worldwide revision. It is presumably exotic in North America and easily overlooked.

Juncus regelii Buchenau (p. 281)

Regel's rush

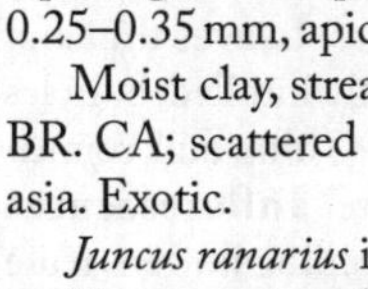

Plants perennial, 15–60 cm tall, rhizomatous. **Leaves** flat, grass-like, less than 3 mm wide; auricles 0–1.6 mm. **Inflorescences** 1–5 clusters, usually head-like, clusters usually 10–30-flowered. **Flowers** tepals 6, 4–6 × 1.1–1.6 mm, dark brown, papillose, inner usually shorter than or as long as the outer; stamens 6, filaments 0.8–1.3 mm, anthers 1–1.5 mm; style 1.1–2.8 mm. **Capsules** oblong-ovoid, shorter than the tepals, brown, obtuse or truncate, 3-chambered. **Seeds** 0.8–1.4(1.7) × 0.15–0.2 mm including tails, seed body 0.3–0.5 mm, seed tails (0.2)0.3–0.6 mm, 1 or both tails longer than the seed body.

Wet meadows, swales, shores, seeps. 100–2600 m. BR, BW, ECas, Lava, Casc. CA, NV, ID, WA; north to British Columbia, northeast to MT, southeast to WY. Native.

This species is often confused with *Juncus howellii*, *J. orthophyllus*, and *J. longistylis*; the complex is sorted most reliably using seed characters; see the key.

Juncus saximontanus A. Nelson

Rocky Mountain rush

Juncus ensifolius Wikstr. var. *brunnescens* (Rydb.) Cronquist, *Juncus ensifolius* Wikstr. var. *montanus* (Engelm.) C.L. Hitchc., *Juncus saximontanus* A. Nelson var. *robustior* M. Peck, *Juncus saximontanus* A. Nelson var. *saximontanus*

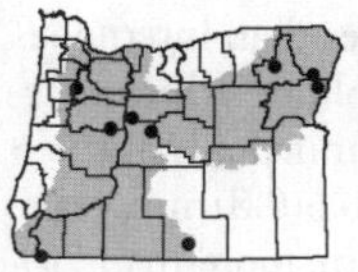

Plants perennial, 20–60 cm tall, rhizomatous. **Leaves** flat, iris-like, blue-green, partially septate; auricles 0–1.2 mm. **Inflorescences** (2)10–20 clusters, clusters 4–25-flowered. **Flowers** tepals 6, often green (red) to medium brown (dark brown); stamens usually 6, filaments 0.6–1.1 mm, anthers 0.4–0.9 mm; style 0.4–0.9 mm. **Capsules** shorter than to longer than the tepals, brown to dark brown or blackish, obtuse or acute, 1-chambered. **Seeds** 0.4–0.5 × 0.25–0.3 mm, apiculate, reticulate.

Shores, creek banks, springs, swales, peatlands, ditches. 100–2500 m. BW, Casc, ECas, Lava, Sisk, WV. CA, ID, NV, WA; north to AK, northeast to Manitoba, southeast to TX, south to Mexico. Native.

Juncus saximontanus has six stamens per flower and is usually found east of the Cascade Range. It is weakly separated from *J. ensifolius*, which has three stamens per flower. Tepal color, inflorescence shape, and style length are unreliable means to separate them. A few plants have three to six stamens and are of uncertain status, perhaps representing hybrids.

Juncus supiniformis Engelmann

Hair-leaved rush, spreading rush

Juncus oreganus S. Watson

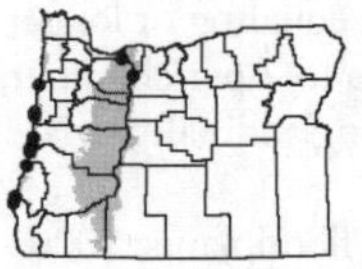

Plants perennial, 3–50 cm tall, cespitose, forming mats if submerged, sometimes stoloniferous with stems trailing and rooting at nodes. **Leaves** numerous, capillary, 0.2–0.3 mm wide if aquatic, round, hollow, septate; auricles 0.8–2.1 mm. **Inflorescences** open, cymes of 2–9 clusters, clusters 2–5(12)-flowered, often bearing leafy bulblets. **Flowers** tepals 6, 3.4–5.4 mm, brown, greenish or reddish, inner tepal margins usually inrolled, tips acuminate; stamens 3 or 6, filaments 0.8–1.5 mm, anthers 0.5–0.8 mm; styles 0.1–0.4 mm. **Capsules** usually 1–1.5 mm longer than the tepals, brown to dark brown, without metallic gloss, 1-chambered. **Seeds** ellipsoid to ovoid, 0.6–1 × 0.2–0.3 mm, apiculate, striate.

Shores, peatlands, marshes, wet dune hollows, often in shallow water. 0–1100 m. Est, Casc. CA, WA, north to AK. Native.

Like introduced *Juncus bulbosus*, this species can bear bulblets in the inflorescence, and both are variable in habit depending on water levels. They differ in the shape of the tepal tips and length of the capsules. *Juncus articulatus* is similar but has blunter tepal tips and is common on disturbed wet ground.

Juncus tenuis Willdenow

Path rush, poverty rush, slender rush

Juncus tenuis Willd. var. *tenuis*

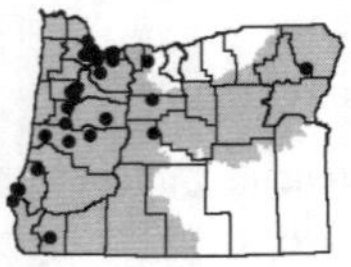

Plants perennial, 15–50 cm tall, cespitose, usually delicate, base usually green to brown, with 0–1(2) strong longitudinal stem ridges visible on a side. **Leaves** thin and wiry, blade flat and slightly inrolled, usually 1–8 mm on early season shoots, dirty white or translucent, scarious, acute or acuminate, auricles soft and thin. **Inflorescences** cymes, usually small, 1–6 cm; individual flowers often longer than internodes, bractlets subtending flowers usually acute (blunt). **Flowers** tepals 6, 3–4.1 mm, green to reddish, tips acuminate; stamens 6, filaments 0.6–1.2 mm, anthers usually 0.4–0.6(0.8) mm; styles 0.1–0.3(0.5) mm. **Capsules** usually 2.5–3 mm, more than 75% the length of, and shorter than the tepals, pale brown, apex usually blunt (acute), not crested, 1-chambered. **Seeds** 0.4–0.5 × 0.2–0.25 mm, apiculate. **2*n*=**40, 80.

Shores, swales, springs, dune hollows, wet woods, marshes, damp paths, ditches, cranberry farms, moist disturbed sites. 0–1800 m. BW, Casc, CR, Est, Lava, Sisk, WV. CA, NV, ID, WA; throughout most of North America. Native.

Juncus tenuis is usually much smaller than *Juncus anthelatus*. The blunt unridged capsules separate *J. tenuis* from *J. confusus*, *J. occidentalis*, and *J. trilocularis*.

Juncus tiehmii Ertter

Tiehm's dwarf rush

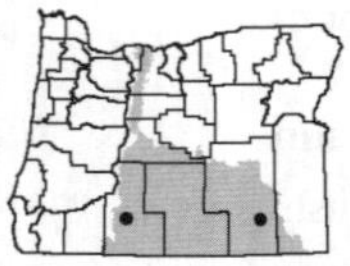

Plants annual, 0.5–6 cm tall, unbranched. **Leaves** capillary, basal. **Inflorescences** 1–4(7) flowers in a terminal cluster, flowers subtended by 2 bractlets, tips acute. **Flowers** tepals usually 4, pale green or pink; stamens 2(3), filaments 0.6–0.9 mm, anthers 0.3–0.4 mm; styles 0.1–0.3 mm. **Capsules** usually elliptic to oblong, equaling or longer than the tepals, pale green or pink. **Seeds** up to 7 per chamber, 12–30 per capsule, 0.3–0.5 × 0.2–0.3 mm, longitudinally striate (at 10×). **2*n*=**34.

Damp depressions, stream banks, sunny floodplains. 1000–2000 m. BR, ECas. CA, NV, ID, WA; south to Mexico. Native.

Plants with few seeds per capsule have larger seeds, and thus some specimens might suggest *J. capillaris*, but there is no overlap. *Juncus capillaris* always has fewer than 10 seeds per capsule.

Juncus torreyi Coville

Torrey's rush

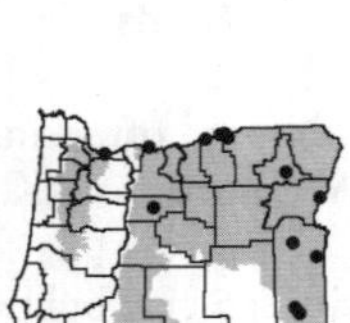

Plants perennial, 30–100 cm tall, rhizomatous, rhizomes elongate, usually swollen or tuberous, at least at some nodes, perennial. **Leaves** round, hollow, septate, often stiff and divaricate; auricles 1–3.5 mm. **Inflorescences** 1–23 dense head-like clusters, clusters globose with 26–80 flowers. **Flowers** tepals 6, brown, greenish or reddish, tips acuminate; stamens 6, filaments 0.6–1.2 mm, anthers 0.4–0.8 mm; styles 0.1–0.2 mm. **Capsules** equaling or longer than the tepals, pale brown to dark brown, gradually tapered to acuminate tips, 1-chambered. **Seeds** ellipsoid to ovoid, 0.4–0.5 × 0.2–0.3 mm, apiculate. **2*n*=**40.

Shores, swales, springs, ditches. 50–1300 m. BW, Col, Lava, Owy. CA, NV, ID, WA; across the US and southern Canada, south to Mexico. Native.

This is primarily an eastern Oregon species, with large globose clusters, peculiar swollen rhizome nodes, and spreading stiff foliage.

Juncus triglumis Linnaeus var. ***albescens*** Lange (p. 281)

Three-flowered rush

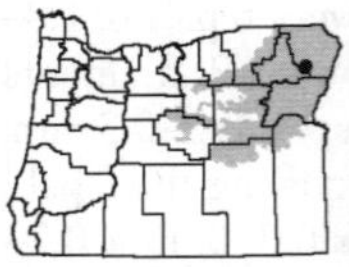

Plants perennial, 5–30 cm tall, cespitose. **Leaves** mostly basal, round, hollow, more or less septate, very thin, 0.3–0.6 mm wide; auricles 0–1 mm. **Inflorescences** 1 small terminal cluster of 1–3(5) flowers, often partially hidden by sheathing base of inflorescence bracts; flowers without subtending bractlets. **Flowers** tepals 6, white to pale brown, tips acute to obtuse, hooded, blunt; stamens 6, filaments 2–3 mm, anthers 0.5–0.8 mm; styles 0.3–0.4 mm. **Capsules** longer than the tepals, pale brown to dark brown, 1-chambered. **Seeds** narrowly ellipsoid to linear, 1.3–2.8 × 0.2–0.3 mm, tails 0.4–0.8 mm, 0.5–1 times as long as seed body. **2*n*=**132.

Limy shorelines and marshes in the high peaks of the Wallowa Mountains. 2300–2400 m. BW. ID; across North America; eastern Russia. Native.

Juncus triglumis is a well-marked species with a small terminal flower cluster, pale flowers, and tailed seeds. Some study is needed to clarify the subspecies in North America. The septate leaf character is difficult to detect in some specimens. The leaves are minutely hollow, usually with two longitudinal tubes.

Juncus trilocularis Zika

Foothill rush

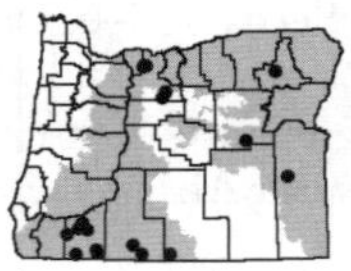

Plants perennial, 10–78 cm tall, cespitose, lacking strong stem ridges. **Leaves** thin and wiry, blade flat and slightly inrolled; auricles soft and thin distally, 0.3–3 mm, dull, rounded or acute, dirty white. **Inflorescences** cymes, 2–20 cm, usually open; bractlets subtending flowers acute to acuminate, often aristate. **Flowers** tepals 6, 3.3–5.1 mm, greenish to pale brown, nearly concolorous, tips acuminate, outer tepals distinctly longer than inner tepals; stamens 6, filaments 0.5–0.8 mm, anthers (0.6)0.8–1.7 mm; styles 0.3–1 mm. **Capsules** 2.8–4.1 mm, shorter than the tepals, pale brown to dark brown, apex truncate to notched (acute), crested, 3-chambered. **Seeds** 0.4–0.6 × 0.2–0.3 mm, apiculate, reticulate.

Seasonally damp meadows, springs, shores, creek banks, moist open forest, ditches. 300–1900 m. BW, Col, ECas, Owy, Sisk, Casc. CA, ID, NV, WA. Native.

This is an uncommon species found east of the Cascade Range and in the Siskiyou Mountains that has long been confused with *Juncus brachyphyllus*, a plant of grasslands and glades in the Midwest, which occurs no further west than Nebraska.

Juncus uncialis Greene

Inch-high rush, twelfth rush

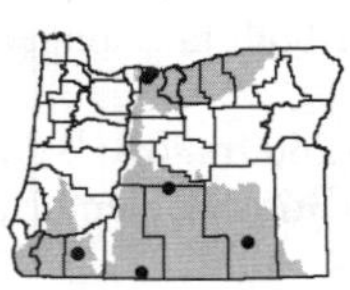

Plants annual, 0.3–3.5 cm tall, unbranched, stems not thickened below flowers. **Leaves** capillary, basal. **Inflorescences** 1 terminal flower, subtended by solitary bractlet, base encircling stem, tip truncate. **Flowers** tepals usually 6 (4 or 8), green or reddish; stamens (2)3, filaments 0.9–1.6 mm, anthers 0.3–0.5 mm; styles 0.1–0.3 mm. **Capsules** usually shorter than or equaling the tepals, (green) orange, reddish or dark purple and often darker than tepals. **Seeds** approximately 100–140 per capsule, 0.3–0.4 × 0.15–0.25 mm, sometimes faintly reticulate (at 10×), not striate. **2*n***=32.

Vernal pools, shores exposed by receding water, stony swales. 50–2700 m. BR, Col, ECas, Sisk. CA, NV, WA. Native.

LUZULA Woodrush

Plants perennial herbs, erect; cespitose, rhizomatous or stoloniferous. **Leaves** sheaths closed, auricles absent, blades hairy, especially along margins, leaf tips obtuse and slightly swollen, or narrow and acute to acuminate. **Inflorescences** racemes, spikes, cymes, or paniculate with flowers solitary or in small clusters, or in dense clusters, often head-like or lobed; pedicels with small bract at base, flowers with 1–2 small bracts at base. **Flowers** perfect; tepals 6, usually lanceolate; stamens 6, anthers shorter or longer than filaments, filaments linear; style 1, stigmas 3. **Capsules** 1-chambered, ovoid-trigonous, dehiscing into 3 segments. **Seeds** 3, plump, 0.5–1.1 mm wide, usually with a fleshy basal aril.

Cosmopolitan. 115 species; 12 species treated in *Flora*.

Measurements of seed length include the arils (broad, fleshy, white appendages or caruncles), when present, but not the few crinkled hairs often attached to the usually basal arils. Hybrids poorly documented but seemingly occasional in mixed populations in the *L. comosa* group.

Key to *Luzula* species:

1. Flowers solitary (occasionally in pairs) at ends of panicle branches.
    2. Plants long-rhizomatous, stems scattered; anthers 0.9–1.6 mm; styles usually 0.7–1.1 mm . . . . ***L. hitchcockii***
    2' Plants cespitose or loosely cespitose, stems clustered; anthers 0.5–0.9 mm; styles 0.1–0.8 mm
        3. Inflorescence of 4–10 flowers in a corymb or compound corymb, with little secondary branching; styles usually 0.7–0.8 mm; seeds with fleshy aril 0.3–1.3 mm . . . . ***L. forsteri*** ssp. ***forsteri***
        3' Inflorescence of more than 10 flowers in an uneven panicle, secondary and tertiary branching common; styles 0.1–0.5 mm; seeds without fleshy aril, occasionally with short dry tail 0.05–0.2 mm.
            4. Stem leaves blue-green, glaucous, dull; leaf tips acuminate, not blunt or swollen; tepals dark brown or dark purple; stem bases pale brown. . . . . ***L. piperi***
            4' Stem leaves bright yellow-green, shiny, never glaucous; leaf tips acuminate or blunt and swollen; tepals often pale; stem bases often red-brown . . . . ***L. parviflora***

1' Flowers clustered in dense spikes, heads, or umbels of spikes.
    5. Mature tepals (in fruit) pale or medium brown.
        6. Plant base rhizomatous, not bulbous; wet soil; seeds, including aril, 1.1–1.5 × 0.5–0.9 mm . . . ***L. cascadensis*** (in part)
        6' Plant base bulbous, not rhizomatous; usually dry or mesic soil; seeds, including aril, 1–2 × 0.7–1.1 mm . . . ***L. comosa***
    5' Mature tepals (in fruit) dark brown or blackish, at least on proximal half.
        7. Inflorescence usually terminated with a solitary spike or head, not umbellate.
            8. Peduncle or spike nodding . . . . ***L. spicata*** ssp. ***spicata***
            8' Peduncle and spike erect.
                9. Inflorescence mostly 1.5–2.5 cm; tepals 2.5–3.5 mm; seeds including aril 1.2–2.1 × 0.7–1.1 mm, aril 0.2–0.6 mm; prairies, ledges, and dunes, low elevations in western Oregon . . . . . . ***L. subsessilis***
                9' Inflorescence 0.6–0.9 cm; tepals 2.3–2.9 mm; seeds including aril 0.9–1 × 0.5–0.6 mm, aril 0.1–0.25 mm; subalpine meadows and shores, northeastern Oregon. . . . . ***L. orestera***
        7' Inflorescence composed of multiple discrete peduncled spikes, usually in an umbel, sometimes slightly congested.
            10. Plants with horizontal rhizomes or stolons, loosely cespitose or stems scattered; stem base never bulbous.
                11. Seed arils 0.4–0.8 mm; anthers 1–1.8 mm; styles 0.5–1.5 mm; introduction on dry disturbed soil below 300 m . . . . ***L. campestris*** ssp. ***campestris***
                11' Seed arils 0.2–0.4 mm; anthers 0.35–1.2 mm; styles 0.15–0.7 mm; native on wet soil of peatlands and springs, often above 900 m in the mountains . . . . ***L. cascadensis*** (in part)
            10' Plants without horizontal rhizomes or stolons, usually densely cespitose (rarely with a short vertical rhizome, perhaps a response to burial); stem base often bulbous.
                12. Styles usually 0.6–2.2 mm; anthers 1.1–3 mm; capsules at base of individual spikes usually slightly scattered or loosely arranged . . . . ***L. macrantha***
                12' Styles usually 0.3–0.8 mm; anthers 0.4–1.6 mm; capsules at base of individual spikes usually dense . . . . ***L. multiflora***

Luzula campestris (Linnaeus) de Candolle ssp. ***campestris***

Field woodrush

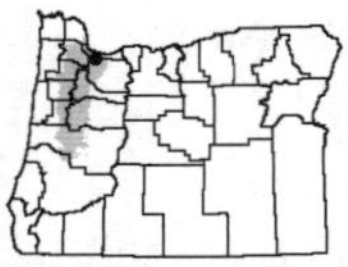

Plants 15–40 cm, rhizomes horizontal, elongate, or stoloniferous, loosely cespitose or stems scattered, base not bulbous. **Leaves** tips obtuse and slightly swollen. **Inflorescences** umbels of 2–7 discrete peduncled spikes, sometimes slightly congested. **Flowers** tepals 2.8–4.2 mm, usually dark brown or blackish; filaments 0.3–0.6 mm, anthers 1–1.8 mm; styles 0.5–1.5 mm. **Capsules** 2.1–3 mm, apex acute. **Seeds** including arils 1.6–2 × 0.7–1 mm, arils (0.4)0.6–0.8 mm. ***2n***=12.

Dry to mesic sunny sites on disturbed soil. 0–300 m. WV. WA; across northern North America; North Africa, Australia, Europe, Falkland Islands, Macaronesia, New Zealand. Exotic.

Luzula cascadensis Zika

Cascades woodrush

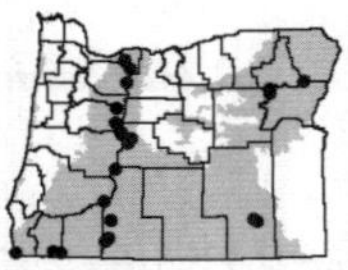

Plants 22–68 cm, with horizontal rhizomes, loosely cespitose or stems scattered, base not bulbous. **Leaves** tips obtuse and slightly swollen. **Inflorescences** umbels of (2)3–10(19) dense spikes, sometimes slightly congested in loose clusters; rays 0.5–8 cm, straight or some arching. **Flowers** tepals 1.9–3.3 mm, medium brown to dark brown; filaments 0.4–0.6 mm, anthers 0.35–1.2 mm; styles 0.15–0.7 mm. **Capsules** 1.9–2.4 mm, apex obtuse to acute. **Seeds** including arils 1.1–1.5 × 0.5–0.9 mm, arils 0.2–0.4 mm.

Wet soil of peatlands, springs, often above 900 m in the mountains. 50–2200 m. BR, BW, ECas, Est, Sisk, Casc. CA, NV, ID, WA; northeast to MT. Native.

Luzula cascadensis is an overlooked species with distinctive fragile horizontal rhizomes that are difficult to extract from the peaty substrate. Growth in saturated soil is unusual among *Luzula* species; similar species are usually on damp margins of wetlands, or in mesic to dry soils, and have a bulbous, non-rhizomatous base. Rarely, a bulbous-based species (like *L. comosa* or *L. subsessilis*) may produce an elongate vertical rhizome, possibly as a response to slumping soil or burial by fossorial rodents. It is similar to *L. arcuata* ssp. *unalaschkensis,* a much smaller boreal plant with acuminate unswollen leaf tips. A couple of collections from Clackamas and Jackson counties suggest a hybrid between *L. cascadensis* and *L. comosa* var. *laxa.* They are morphologically intermediate and found with the putative parents.

Luzula comosa E. Meyer

Pacific woodrush

Plants (12)15–60 cm, cespitose, base bulbous. **Leaves** tips obtuse and slightly swollen. **Flowers** tepals 2.5–4 mm, usually light or medium brown. **Capsules** 2.3–3 mm, apex acute to obtuse.

2 varieties.

Key to *Luzula comosa* varieties:

1. Inflorescence open, umbels of spikes; coastal and inland . . . . ***L. c.*** var. ***laxa***

1' Inflorescence dense, capitate or nearly so; coastal . . . . ***L. c.*** var. ***comosa***

Luzula comosa E. Meyer var. ***comosa***

Pale woodrush

Luzula multiflora (Ehrh.) Lej. var. *comosa* (E. Mey.) Fernald & Wiegand

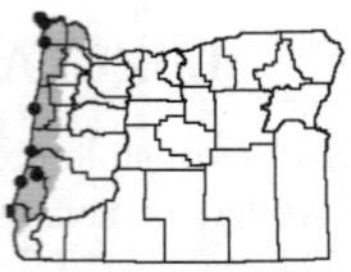

Inflorescences dense clusters, capitate or nearly so, sometimes lobed, rarely with 1–2 short lateral branches. **Flowers** filaments 0.5–0.6 mm; anthers 0.6–1 mm; styles 0.4–0.5 mm. **Seeds** including arils 1–1.2 × 0.8–0.9 mm, arils 0.4–0.55 mm. ***2n***=24.

Open forest, stabilized dunes, lakeshores, damp to mesic ground, sun or shade, usually near the ocean. 0–900 m. CR, Est. CA; north to AK. Native.

Occasional plants of this variety have atypically dark inflorescences, but the mature clusters are often wider than long, and are separable from *L. subsessilis,* which usually have fruiting clusters longer than wide

Luzula comosa E. Meyer var. ***laxa*** Buchenau

Pacific woodrush

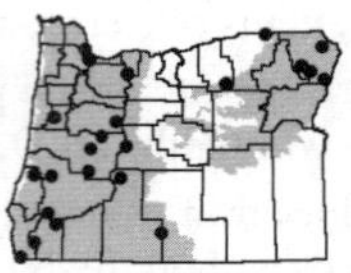

Inflorescences open umbels of (1)2–6 loose spikes. **Flowers** filaments 0.3–0.8 mm; anthers 0.5–1.5 mm; styles 0.3–1.1 mm. **Seeds** including arils 1.1–2 × 0.7–1.1 mm, arils 0.2–0.7 mm. ***2n***=12.

Meadows, ledges, dry forest, mesic spring margins and shores. 100–2000 m. BW, Casc, CR, ECas, Sisk, WV. CA, ID, WA; north to British Columbia, northeast to MT. Native.

Luzula forsteri (Smith) de Candolle ssp. ***forsteri***

Southern woodrush

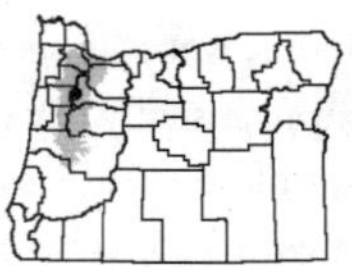

Plants 10–41 cm, cespitose, stems clustered. **Leaves** tips obtuse and slightly swollen. **Inflorescences** corymbs (compound corymbs) of 4–8(10) solitary flowers, with little secondary branching. **Flowers** tepals 3–3.6 mm, red-brown; filaments 0.7–0.9 mm, anthers 0.7–0.8(0.9) mm; styles usually 0.7–0.8 mm. **Capsules** 2.8–3.3 mm, apex acute to acuminate. **Seeds** including arils 1.4–2.4 × 1–1.4 mm, arils 0.3–1.3 mm. ***2n***=24.

Lawns; disturbed forest. 50–200 m. WV. WA; North Africa, Eurasia. Exotic.

Luzula hitchcockii Hämet-Ahti

Hitchcock's woodrush, smooth woodrush

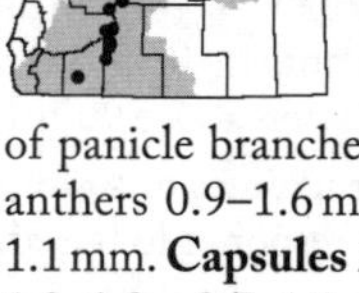

Plants 15–50 cm, long rhizomatous, stems scattered. **Leaves** tips acute to acuminate, thickened but not swollen. **Inflorescences** cymes, branches often spreading or nodding, flowers solitary, occasionally paired at ends of panicle branches. **Flowers** tepals 2.5–3.5 mm, dark brown; anthers 0.9–1.6 mm, filaments 0.3–0.6 mm; styles (0.5)0.7–1.1 mm. **Capsules** 2.6–3.8 mm, apex acute to acuminate. **Seeds** 1.3–1.8 × 0.7–1.1 mm, arils lacking. ***2n***=24.

Dry subalpine and alpine meadows, rocky slopes, open forest. 1300–2600 m. BW, Casc, ECas, Sisk, WV. ID, WA, north to British Columbia, east to WY. Native.

Luzula macrantha Zika & B.L. Wilson

Prairie woodrush, large-anthered woodrush

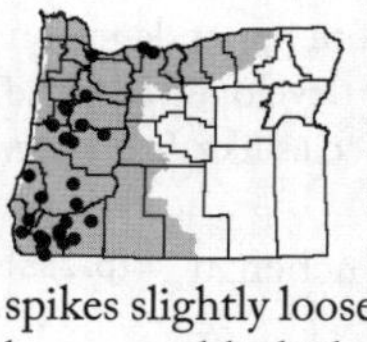

Plants 12–41 cm, usually densely cespitose (short vertical rhizome), base often bulbous. **Leaves** tips obtuse and slightly swollen. **Inflorescences** umbels of (1)2–6 discrete peduncled spikes, flowers at base of individual spikes slightly loosely arranged. **Flowers** tepals 2.9–5 mm, dark brown or blackish; filaments 0.2–0.7 mm, anthers 1.1–3 mm; styles 0.6–2.2 mm. **Capsules** 2.4–3 mm, apex acute. **Seeds** including arils 1.2–2 × 0.7–1.2 mm, arils (0.1)0.2–0.5 mm.

Dry forest, rocky slopes, meadows, sometimes on serpentine. 0–1100 m. Casc, Col, CR, ECas, Sisk, WV. CA, WA; north to British Columbia. Native.

This species is often mistaken for *L. multiflora* but has much longer anthers and longer styles. The lectotypes chosen in Kirschner (2002) suggest this is a synonym of *L. subsessilis*, but *L. macrantha* has an open branching, more or less umbellate, terminal inflorescence, while *L. subsessilis* has a dense unbranched terminal inflorescence with an occasional peduncled cluster arising from a lower node. Where the two grow together in dunes and prairies, a few intermediate plants have been seen that suggest hybrids.

Luzula multiflora (Ehrhart) Lejeune

Common woodrush

Luzula campestris (L.) DC. var. *multiflora* (Ehrh.) Célak., *Luzula multiflora* (Ehrh.) Lej. var. *multiflora*

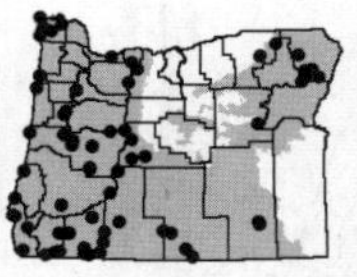

Plants 15–38 cm, usually densely cespitose, without horizontal rhizomes, base often fibrous and bulbous. **Leaves** tips obtuse and slightly swollen. **Inflorescences** umbels of 4–8 discrete peduncled spikes, bases of individual spikes dense. **Flowers** tepals 2.5–3.9 mm, dark brown or blackish (medium brown); filaments 0.4–0.6 mm, anthers 0.4–1.6 mm; styles usually 0.3–0.7(0.8) mm. **Capsules** 2.4–2.6 mm, apex obtuse to acute. **Seeds** including arils 1–1.6 × 0.6–0.9 mm, arils 0.1–0.5 mm. ***2n***=48, 72.

Meadows, open woods, ditches and clearings. 0–2400 m. BR, BW, Casc, CR, ECas, Est, Sisk, WV. CA, WA; scattered across northern North America, South America; North Africa, Australia, Eurasia, New Zealand. Native and Exotic.

Low elevation records on disturbed ground, east and west of the Cascades, are most likely introductions of the common Eurasian weed, sometimes called ssp. *multiflora*. However, records from subalpine meadows, indistinguishable morphologically but perhaps belonging to ssp. *frigida*, may be boreal natives just reaching as far south as Oregon. Seemingly native populations are found in similar montane habitats in the Olympic and Cascade Ranges in Washington. The complex needs study and a worldwide revision.

Luzula orestera Sharsmith

Sierra woodrush, alpine woodrush

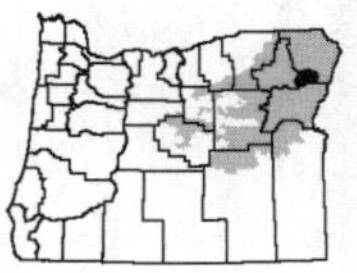

Plants 2.5–14(21) cm, cespitose, often reddish. **Leaves** tips obtuse and slightly swollen. **Inflorescences** solitary, erect, dense, head-like clusters, 0.6–0.9 cm. **Flowers** tepals 2.3–2.9 mm, dark brown or blackish; filaments 0.4–0.6 mm, anthers 0.4–0.6 mm; styles 0.1–0.3 mm. **Capsules** 1.8–2 mm, apex obtuse. **Seeds** including arils 0.9–1 × 0.5–0.6 mm, arils 0.1–0.25 mm. ***2n***=20, 22.

Moist to wet subalpine meadows, shores. 2000–2500 m. BW. CA. Native.

Luzula orestera is often considered endemic to the central Sierra Nevada mountains of California, but Oregon populations, all from high elevation meadows in the Wallowa Mountains, are indistinguishable ecologically or morphologically. Its closest relative appears to be *L. groenlandica* of the American arctic. Oregon plants release pollen within the closed tepals, apparently self-pollinating.

Luzula parviflora (Ehrhart) Desvaux (p. 288)

Small-flowered woodrush

Luzula parviflora (Ehrh.) Desv. var. *melanocarpa* (Michx.) Buchenau, *Luzula parviflora* (Ehrh.) Desv. var. *parviflora*

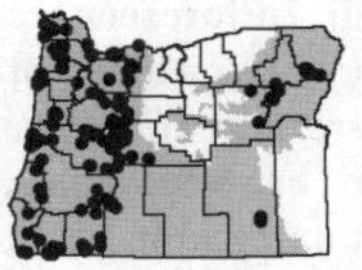

Plants 15–100 cm, cespitose or loosely cespitose, stems often red-brown (pale brown) at base, clustered. **Leaves** stem leaves bright yellow-green, shiny, never glaucous; leaf tips acuminate, or blunt and swollen. **Inflorescences** uneven panicles or cymes of more than 10 flowers, secondary and tertiary branches common, variously arched, stiffly spreading, and/or nodding within a population, flowers solitary, occasionally paired at ends of panicle branches. **Flowers** tepals variable, often pale to medium brown, 1.8–2.5 mm; filaments 0.5–0.6 mm, anthers 0.3–0.5 mm; styles 0.1–0.5 mm. **Capsules** 1.8–2.4 mm, apex acute to acuminate. **Seeds** including tail 1.1–1.5 × 0.6–0.85 mm, lacking aril, occasionally with a short dry tail 0.05–0.2 mm. ***2n***=22, 24.

Forest, rocky slopes, meadows, shores, springs. 0–2800 m. BR, BW, Casc, CR, ECas, Est, Sisk, WV. CA, NV, ID, WA; across North America; Eurasia. Native.

Within this species, the proximal end of the seed bears short hairs. Eastern North American populations, with dark tepals and blue-green leaves, have been called var. *melanocarpa*, but the trend is inconsistent. Plants with primary inflorescence branches stiffly divaricate have been misidentified as the Sierra Nevada endemic *L. divaricata* or called *L. parviflora* ssp. *fastigiata*. They are merely one extreme of a long continuum of variation in branching patterns in trans-Pacific plants. A species with both acuminate and blunt swollen leaf tips is unusual for *Luzula*, but leaf tip shape in *L. parviflora* does not appear to be correlated with branching patterns and needs study.

Luzula piperi (Coville) M.E. Jones

Piper's woodrush

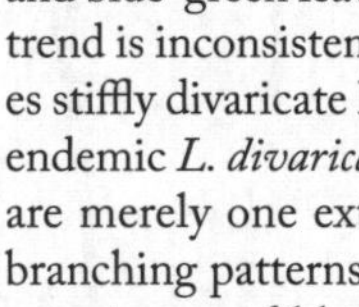

Plants 10–50 cm, cespitose or loosely cespitose, stems pale brown at base, usually clustered. **Leaves** stem leaves dull, blue-green, glaucous, leaf tips acuminate, sometimes thickened but not blunt or swollen. **Inflorescences** uneven panicles or cymes, secondary and tertiary branches common, often arched or nodding; flowers more than 10, solitary, occasionally paired at ends of panicle branches. **Flowers** tepals 1.8–2.5 mm, dark brown or dark purple; filaments 0.5–0.6 mm, anthers 0.3–0.8 mm; styles 0.2–0.4 mm. **Capsules** 1.7–2.4 mm, apex acute to acuminate. **Seeds** 1–1.4 ×

0.45–0.6 mm, arils lacking, occasionally with short dry tail 0.05–0.2 mm. **2*n***=24.

Subalpine and alpine meadows, rocky slopes, snowbeds, open forest. 1500–2800 m. BW, ECas, Casc. CA, NV, ID, WA; north to AK, northeast to Alberta, east to WY; northeastern Asia. Native.

Luzula piperi has dull blue-green leaves in the field, while *L. parviflora* has shiny bright green leaf blades. The seeds of *L. piperi* do not bear a developed fleshy aril, although there are a few crinkled hairs at the base and sometimes a small dry flap of basal or apical tissue, somewhat like the apiculate, abbreviated tails of many *Juncus* seeds.

Luzula spicata (Linnaeus) de Candolle ssp. ***spicata*** (p. 288)

Spiked woodrush

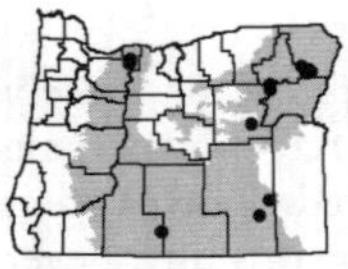

Plants 5–30 cm, cespitose. **Leaves** tips more or less acute, not swollen. **Inflorescences** solitary, dense, often head-like or lobed or slightly interrupted clusters, peduncle or spike nodding, 0.6–3 cm. **Flowers** tepals, 2.2–2.8 mm, dark brown or blackish at least on proximal half; filaments 0.4–0.5 mm, anthers 0.4–0.7 mm; styles 0.1–0.4 mm. **Capsules** 1.6–2.3 mm, apex acute. **Seeds** 0.9–1.1 × 0.5–0.7 mm, apiculate, arils lacking. **2*n***=24.

Subalpine and alpine dry to damp meadows, rocky windy slopes, cliffs, scree, moraines. 1000–3100 m. BR, BW, ECas, Casc. CA, NV, ID, WA; across North America; Eurasia. Native.

The seeds of this species have short white appendages on either end, about 0.1 mm. They are not well developed like the fleshy arils of low elevation, ant-dispersed species like *L. comosa*.

Luzula subsessilis (S. Watson) Buchenau (p. 288)

Prairie woodrush

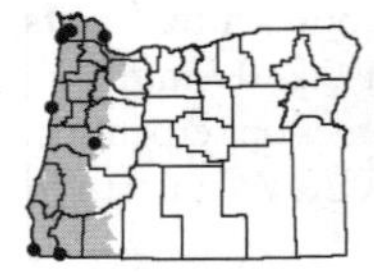

Plants 8–60 cm, cespitose, base bulbous. **Leaves** tips obtuse and slightly swollen. **Inflorescences** usually solitary, erect, dense, head-like or lobed cluster, mostly 1.5–2.5(3.6) cm, occasionally with smaller secondary cluster on elongate erect peduncle from lower axil. **Flowers** tepals 2.5–3.5 mm, dark brown or blackish at least on proximal half; filaments 0.4–0.7 mm, anthers 0.4–1.1 mm; styles (0.1)0.2–0.6 mm. **Capsules** 1.9–2.4 mm, apex truncate, blunt or acute. **Seeds** including arils 1.2–2.1 × 0.7–1.1 mm, arils 0.2–0.6 mm.

Dry to wet prairie, ledges, dunes, balds, rocky slopes, headlands, low elevations in western Oregon. 0–1400 m. CR, Est, Sisk, WV. CA, WA; north to British Columbia. Native.

The dense, dark inflorescence of this species is distinctive in its low elevation habitat. Where it mixes with *L. macrantha*, occasional intermediate plants are found, presumably hybrids.

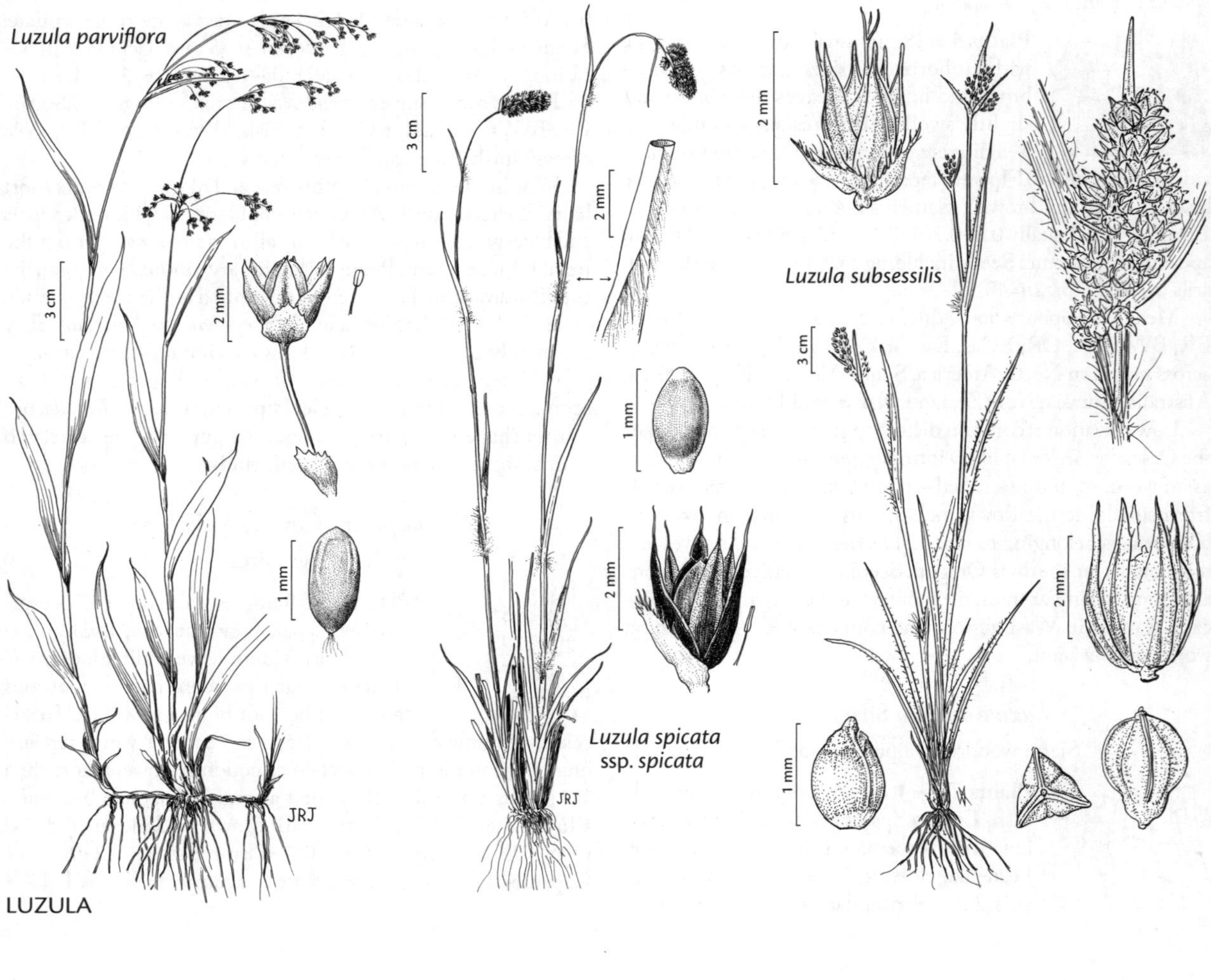

LUZULA

JUNCAGINACEAE Arrow-grass family

Rachel E. Newton

Herbs terrestrial and aquatic; acaulescent, rhizomatous. **Leaves** basal, erect, emersed; sheaths clasping the stem, ligulate; blades simple, somewhat succulent. **Inflorescences** axillary, solitary flowers or terminal scapose spikes or racemes. **Flowers** unisexual or bisexual, pistillate and staminate on the same plant; tepals lacking or 6 in 2 series (3 in 1 series), free, greenish, occasionally purple-tinged; stamens 1, 3, or 6, free, epipetalous; pistils 1, 3, or 6; ovaries superior. **Fruits** dry.

Cosmopolitan in coastal and wetland areas. 3 genera; 1 genus treated in *Flora*.

Juncaginaceae is a member of the order Alismatales, an early diverging branch of monocotyledons composed of mainly aquatic and wetland plants. The center of specific diversity for the family is Australia, where up to 20 species are found (von Mering and Kadereit 2010).

TRIGLOCHIN Arrow-grass

Herbs annual or perennial. **Leaves** elliptical to round in cross section; ligules entire to deeply bilobed. **Inflorescences** shorter than to greatly exceeding leaves. **Flowers** anthers more or less sessile; styles lacking to greatly elongating in fruit, stigmas capitate to more or less plumose.

Temperate circumboreal areas and higher elevations in the tropics. Approximately 12 species; 4 species treated in *Flora*.

Key to *Triglochin* species:

1. Plants annual; inflorescences spikes with bisexual and unisexual flower. . . . . *T. scilloides*

1' Plants perennial; inflorescences spike-like racemes with bisexual flowers.

  2. Fertile carpels 3, fruits roughly as long as wide . . . . *T. striata*

  2' Fertile carpels 6, fruits longer than wide.

    3. Ligules entire to notched; leaves 2–5 mm wide. . . . . *T. maritima*

    3' Ligules deeply bilobed; leaves 1–1.5 mm wide . . . . *T. concinna*

Triglochin concinna Burtt Davy

Graceful arrow-grass, slender arrow-grass

Triglochin concinna Burtt Davy var. *concinna*, *Triglochin concinna* Burtt Davy var. *debilis* (M.E. Jones) J.T. Howell

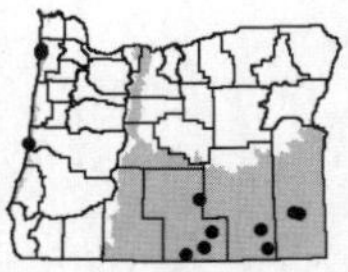

Herbs perennial. **Leaves** round in cross section, 7–18.5 cm × 1–1.5 mm; ligules deeply bilobed, about 1 mm. **Inflorescences** racemes mostly exceeding leaves, (8.5)13–60 cm. **Pedicels** stiffly ascending, 1.5–4 mm. **Flowers** tepals 1.5–2 mm; fertile carpels 6. **Fruits** schizocarps, linear-clavate, 3–4 mm. **2*n***=48, 96.

Coastal salt and inland freshwater marshes and moist meadows, especially brackish, saline, or alkaline areas. Flowering Apr–Jul. 0–1100 m. BR, ECas, Est, Owy. Throughout North America, South America; Asia, Europe. Native.

Triglochin maritima Linnaeus (p. 290)

Seaside arrow-grass

Triglochin maritimum L.

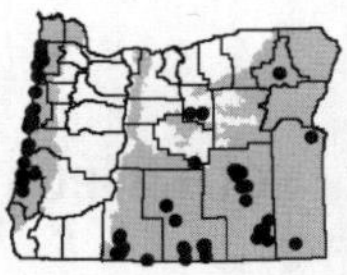

Herbs perennial. **Leaves** elliptical in cross section, 22–60 cm × 2–5 mm; ligules entire to notched, 1–2 mm. **Inflorescences** racemes exceeding leaves, 45–130 cm. **Pedicels** ascending, 2–6 mm. **Flowers** tepals 1.5–2 mm; fertile carpels 6. **Fruits** schizocarps, oblong, 3–5 mm. **2*n***=12, 24, 30, 36, 48, 60, 120, 144.

Coastal salt and inland freshwater marshes and moist meadows, especially brackish, saline, or alkaline areas. Flowering Apr–Aug. 0–1700 m. BR, BW, CR, ECas, Est, Owy. CA, ID, NV, WA; throughout North America; South America; Asia, Europe. Native.

Triglochin concinna and *T. maritima* are frequently considered the same species; the treatment in *Flora of North America* synonymizes *T. concinna* with *T. maritima* (Haynes and Hellquist 2000). Both species produce a cyanide-containing sugar called triglochinin, which can poison livestock when eaten in large amounts. Production is highest in the spring (Majak et al. 1980), and because the compound remains potent in a dried state, care should be taken to prevent inclusion of young *Triglochin* in hay bales.

Triglochin scilloides (Poiret) Mering & Kadereit

Flowering quillwort

Lilaea scilloides (Poir.) Hauman, *Lilaea subulata* Bonpl.

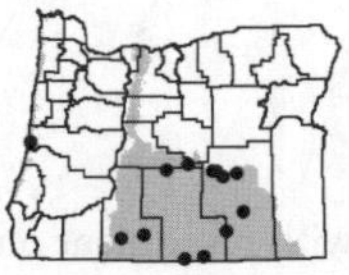

Herbs annual. **Leaves** round in cross section, 9–35 cm × 1–9 mm; ligules acute, 1–2 mm. **Inflorescences** terminal spikes of unisexual and bisexual flowers shorter or longer than leaves in length, 3.5–25 cm; solitary pistillate flowers sessile in leaf axils. **Flowers** tepals lacking. **Bisexual flowers** with caducous bracts; stamen 1; pistil 1. **Staminate flowers** generally above pistillate and bisexual flowers on racemes, with caducous bracts; stamen 1. **Pistillate flowers** solitary and sessile in leaf axils or mixed in with bisexual flowers on racemes, bractless; pistil 1; style elongating in fruit, 4–25 cm. **Fruits** nutlets, 2–3 mm. **2*n***=12.

Vernal pools and other areas of shallow water and adjacent mud flats along lakes, ponds, and streams. Flowering Mar–Oct. 0–1900 m. BR, ECas, Est. CA, NV, WA; north to Alberta, British Columbia, and Saskatchewan, east to MT, south to Mexico and South America. Native.

Triglochin scilloides is uncommon within Oregon's borders but not considered immediately imperiled (ORBIC 2010). This taxon has faced taxonomic uncertainty throughout the years, most commonly considered a monotypic genus (*Lilaea*) within the Juncaginaceae, although occasionally placed in its own family (Lilaeaceae). Only recently has molecular work placed it firmly within the genus *Triglochin* (von Mering and Kadereit 2010).

Triglochin striata Ruiz & Pavon

Three-ribbed arrow-grass

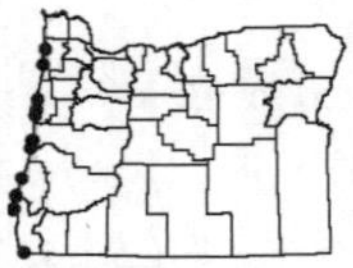

Herbs perennial. **Leaves** 5–40 cm × 1–2 mm, elliptical in cross section; ligules entire, 1–2 mm. **Inflorescences** racemes conspicuously shorter than or equal to leaves in length, 5–40 cm. **Pedicels** spreading to ascending, 1–2 mm. **Flowers** tepals 1 mm; fertile carpels 3. **Fruits** schizocarps, globose, up to 2 × 2 mm. **2*n***=18, 24.

Coastal brackish, alkaline, and freshwater marshes. Flowering Jun–Aug. 0–50 m. Est. CA, WA; south to Mexico, West Indies, and South America; disjunct populations in southeastern US. Native.

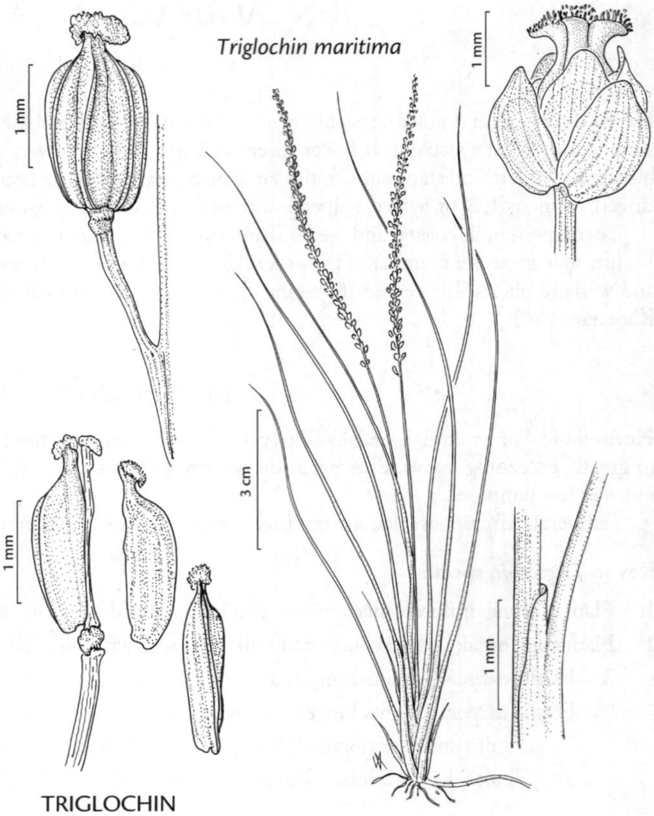

TRIGLOCHIN

LEMNACEAE

(see Araceae)

LILIACEAE Lily family

(also see Amaryllidaceae, Asparagaceae, Melanthiaceae, Nartheciaceae, Smilacaceae, Tecophilaeaceae, Tofieldiaceae)

Herbs perennial; from bulbs or rhizomes. **Stems** underground or erect, scapose or leafy. **Leaves** 1–many, basal, cauline, or both, alternate, subopposite, or whorled, linear to ovate or square-shaped; petioles short or absent. **Inflorescences** solitary, panicles, racemes, or umbels. **Flowers** radially symmetrical, erect or pendent; perianth segments 6, distinct, in 2 petal-like whorls; stamens 3 or 6; ovary superior; style 1, stigmas entire or 3-lobed. **Fruits** capsules or berries. **Seeds** flat or angled.

Cosmopolitan. 15 genera; 9 genera treated in *Flora*.

Due to taxonomic changes proposed by Stevens (2012), most genera formerly contained within Liliaceae have been placed in other families, most notably Amaryllidaceae and Asparagaceae.

Key to Liliaceae genera:

1. Plants from rhizomes.
    2. Leaves basal.
        3. Flowers yellow-green to green, with prominent purplish brown lines; inner and outer tepals dissimilar .... **SCOLIOPUS**
        3' Flowers white, yellow, pink, purple, or red, without lines; inner and outer tepals identical.
            4. Flowers not nodding; fruits berries.......... **CLINTONIA**
            4' Flowers usually nodding; fruits capsules.......... **ERYTHRONIUM** (in part)
    2' Leaves cauline.
        5. Inflorescences 1(2)-flowered, axillary.......... **STREPTOPUS**
        5' Inflorescences 2–7-flowered, terminal .... **PROSARTES**

1' Plants from bulbs.
 6. Inner and outer perianth segments dissimilar (having obviously differentiated sepals and petals) . . **CALOCHORTUS**
 6' Inner and outer perianth segments (tepals) identical.
  7. Leaves basal and cauline, not or only slightly reduced upwards.
   8. Tepals 1–4 cm long, oblanceolate to ovate, round or square-shaped. **FRITILLARIA**
   8' Tepals (3)4–10.5 cm long, generally lanceolate. . . . . **LILIUM** (in part)
  7' Leaves basal or greatly reduced upwards.
   9. Tepals less than 2 cm long, not or slightly recurved . . . . **GAGEA**
   9' Tepals 2 cm long or greater, often strongly recurved.
    10. Leaves primarily cauline . . . . **LILIUM** (in part)
    10' Leaves basal . . . . **ERYTHRONIUM** (in part)

CALOCHORTUS Mariposa lily, cat's ear, star-tulip

Frank T. Callahan

Herbs perennial; bulbs with membranous tunics, often with offsetting bulblets. **Leaves** basal or cauline, linear to lanceolate, V-shaped to flat in cross section, glabrous or glaucous; basal leaves solitary, rarely with hairs; bulblets in leaf axils or not. **Inflorescences** solitary to many flowers, erect or nodding. **Flowers** perianth spherical to campanulate, nearly closed to open; sepals 3, generally lanceolate; petals 3, broadly ovate, few to many hairs adaxially, nectary adaxially situated near base of petal; stamens 6, filaments tapering from a dilated base, anthers fixed at base; ovary superior, broadly ovate to nearly linear; style 1, stigmas 3-lobed, often strongly recurved. **Fruits** capsules, ovate to oblong to linear, 3-sided or strongly winged, chambers 3, dehiscence septicidal. **Seeds** purple or reddish to white or cream.

Western North America and Central America. Approximately 80 species; 18 species treated in *Flora*.

Calochortus amabilis was collected once, near the coast in Curry County, in 2010. If seen again and/or persisting, it is easily distinguished from all other Oregon taxa by its nodding yellow flowers. *Calochortus luteus* was seen once and last collected in 1939, in Jackson County by Margaret Knowles Small. It is best recognized by its red-streaked, deep yellow petals. Very near the Oregon-California border in Jackson County, *C. persistens* was once collected in 1998. The most obvious diagnostic characters of this species are its fringed, pink to purple petals and its nodding capsules with the perianth segments still attached (hence the name "persistens").

Bulbs of all *Calochortus* species are edible. Many species were regularly harvested by Native Americans.

Above the nectary glands of many species a darker marking, referred to as a blotch, crescent, or chevron is present. As the terms imply, a blotch resembles an ink blot, a crescent is shaped like a waning moon or croissant, and a chevron is a thin, bent, horizontal line similar to those on military uniforms.

Key to *Calochortus* species:

1. Petals yellow . . . . ***C. monophyllus***
1' Petals white to purple (base of inner petals may be yellow).
 2. Petals smoky blue, light blue, light pink, or light purple to purple.
  3. Inner petal surface hairy throughout.
   4. Color of inner surface markedly lighter in color than outer surface (edges often same color as outer surface) . . . . ***C. greenei***
   4' Color of upper 67% of inner surface similar to outer surface.
    5. Petals 8–12 mm long . . . . ***C. coeruleus***
    5' Petals 12–25 mm long . . . . ***C. tolmiei*** (in part)
  3' Inner petal surface sparsely hairy and/or hairy only at base near nectary.
   6. Petals uniformly colored.
    7. Fruits nodding; petals 14–28 mm long . . . . ***C. uniflorus*** (in part)
    7' Fruits erect; petals 12–16 mm long . . . . ***C. nudus*** (in part)
   6' Basal inner petal surface differing in color.
    8. Fruits nodding; inner petal surface without a marking above basal nectary . . . . ***C. indecorus***
    8' Fruits erect; inner petal surface with a blotch, chevron or crescent above basal nectary.
     9. Basal leaves withering by flowering.
      10. Petals 40–60 mm long. . . . . ***C. macrocarpus*** (in part)
      10' Petals 10–40 mm long. . . . . ***C. leichtlinii*** (in part)

9' Basal leaves persistent at flowering.
11. Blotch present above nectary gland . . . . *C. eurycarpus* (in part)
11' Chevron present above nectary gland.
12. Bulblets present in leaf axils; chevron above basal nectary reddish purple . . . . . . *C. longebarbatus*
12' Bulblets not present in leaf axils; chevron above basal nectary light to deep purple.
13. Petals 20–35 mm long . . . . *C. nitidus*
13' Petals 12–16 mm long . . . . *C. nudus* (in part)
2' Petals white to cream or mostly so (lower 33% may be yellow or purple).
14. Inner petal surface sparsely hairy and/or hairy only at base near nectary.
15. Petals uniformly colored.
16. Fruits nodding; petals 14–28 mm long . . . . *C. uniflorus* (in part)
16' Fruits erect; petals 12–16 mm long . . . . *C. nudus* (in part)
15' Basal inner petal surface differing in color (including color of crescent or chevron).
17. Basal nectary similar in color to petal (excluding color of chevron) . . . . *C. nudus* (in part)
17' Basal nectary yellow (or appearing so due to yellow hairs).
18. Outer petal surface with distinct green band.
19. Petals 20–40 mm long . . . . *C. bruneaunis*
19' Petals 40–60 mm long . . . . *C. macrocarpus* (in part)
18' Outer petal surface without distinct green band.
20. Green band present between yellow basal nectary and reddish to purple crescent . . . . *C. eurycarpus* (in part)
20' Green band not present between yellow basal nectary and dark red to dark purple blotch or crescent (green band may be present above crescent).
21. Fruits nodding; nectary oblong-lunate; plants of Douglas County . . . . . . *C. umpquaensis* (in part)
21' Fruits erect; nectary ovate; plants of Jackson and Lake counties . . . . *C. leichtlinii* (in part)
14' Inner petal surface hairy throughout.
22. Inner base of petals yellow (or appearing so due to yellow hairs).
23. Sepals with a purple blotch near base . . . . *C. subalpinus*
23' Sepals without a purple blotch near base . . . . *C. coxii*
22' Inner base of petals purple to dark purple (or appearing so due to purple hairs).
24. Fruits erect . . . . *C. howellii*
24' Fruits nodding.
25. Stems usually with 2 branches; prominent (bract-like) cauline leaves present . . . . *C. tolmiei* (in part)
25' Stems not usually branching; cauline leaves not present or not prominent.
26. Stems 5–20 cm long; upper surface of leaves glabrous . . . . *C. elegans*
26' Stems 20–30 cm long; upper surface of leaves pubescent . . . . *C. umpquaensis* (in part)

Calochortus bruneaunis A. Nelson & J.F. Macbride

Bruneau mariposa lily

Calochortus nuttallii Torr. var. *bruneaunis* (A. Nelson & J.F. Macbr.) Ownbey

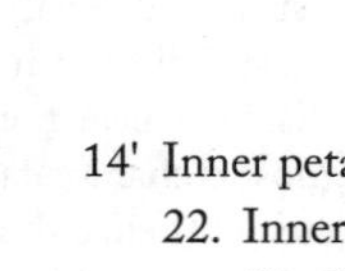

Stems 10–40 cm long. **Leaves** basal leaves very narrow, 10–20 cm long, withering at anthesis. **Inflorescences** 1–4-flowered, erect; bracts several, 2–8 cm long. **Flowers** sepals 10–40 mm long; petals oblanceolate, 20–40 mm long, with median longitudinal green stripe on outer surface, yellow zone surrounding nectary with a dark red to dark purple blotch or crescent above, tips acuminate; nectary round, depressed, surrounded by scant hairs, hairs in a tight oblong grouping clustering to a center; filaments 5–6 mm long, anthers oblong, 5–9 mm long, reddish. **Fruits** upright, lance-linear, triangular in cross section, 3–7 cm long. **Seeds** flat, light tan. **2*n***=14.

Dry sagebrush zones. Flowering May–Jul. 700–2200 m. BR, BW, Owy. CA, ID, NV; east to MT and UT. Native.

Calochortus coeruleus (Kellogg) S. Watson

Beavertail-grass, blue cat's ear

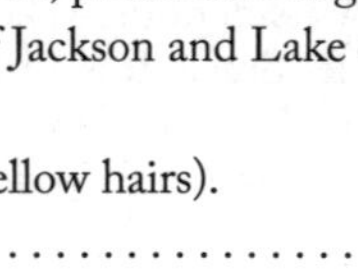

Stems 3–20 cm long, flexuous. **Leaves** basal leaves 10–20 cm long, persistent; cauline leaves rare. **Inflorescences** 1–12-flowered; bracts lance-linear, 2–7 cm, acuminate. **Flowers** sepals lance-oblong, to 10 mm long, glabrous; petals obovate, 8–12 mm long, light blue, clawed, with fringed margins, inner surface very hairy; nectary with a fringed scale; filaments approximately 4 mm long; anthers to 4 mm long. **Fruits** nodding, 10–15 mm long, nearly as broad. **Seeds** irregular-shaped, light lavender.

Mineral soils to open grasslands. Flowering May–Jun. 1000–1500 m. Casc, ECas. CA. Native.

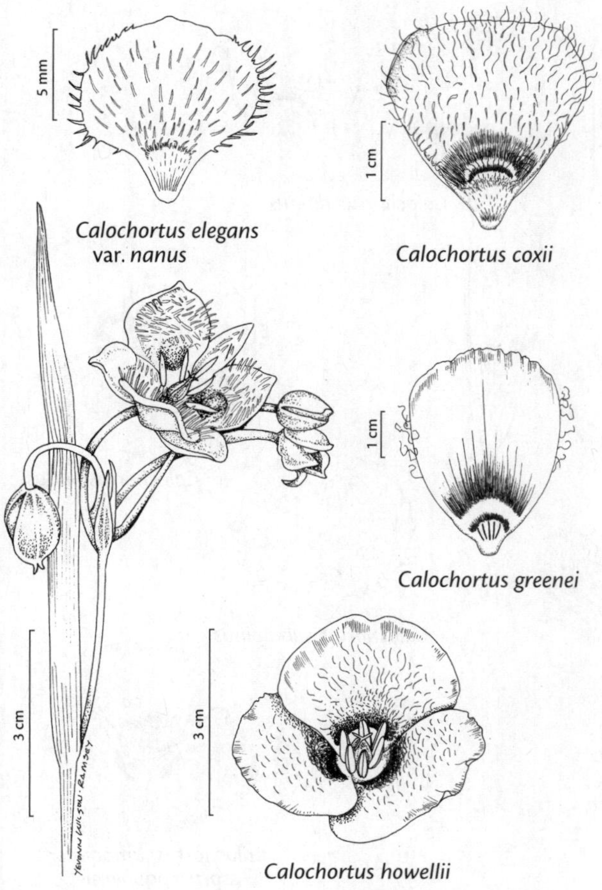

CALOCHORTUS

Calochortus coxii Godfrey & Callahan (p. 293)

Cox's mariposa lily, Cox's cat's ear

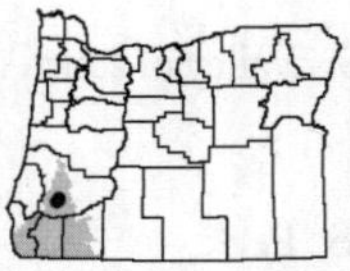

Stems 15–20 cm long, flexuous. **Leaves** basal leaves to 30 cm long, with short hairs in lines on adaxial surface, persistent. **Inflorescences** 1–5-flowered. **Flowers** sepals to 25 mm long; petals to 25 mm long, densely covered with long simple white hairs approximately 50% way to nectary where hairs become yellow, margins somewhat fringed; petal claw with fine penciling/spots (nectary guides); nectary lunate, crested with branching hairs; filaments to 7 mm long, anthers 3–7 mm long, reddish brown, apiculate. **Fruits** nodding, 3–4 × approximately 1.5 cm, with a long recurved tip approximately 4 mm, 3 winged. **Seeds** light straw-yellow.

Serpentine grasslands and summit balds. Flowering Jun–Jul. 200–800 m. Sisk. Native.

Calochortus elegans Pursh

Elegant cat's ear, elegant mariposa lily

Stems 5–20 cm long, flexuous, usually not branching. **Leaves** basal leaves broadly linear, 10–20 cm long, glabrous, persistent. **Inflorescences** 1–7-flowered; bracts lanceolate, to approximately 3 mm long. **Flowers** sepals lanceolate, ovate-lanceolate, or elliptic, 12–16 mm long, inner surface somewhat papillate; petals oblanceolate, 10–20 mm long, upper portion of inner surface white, lower portion purple, entire inner surface densely hairy; filaments to 4 mm long, anthers to 5 mm long, white or purple, long acuminate. **Fruits** nodding, 2 × 2 cm, winged. **Seeds** irregular-shaped, light purple. ***2n***=20.

Northwestern US. 3 varieties; 2 varieties treated in *Flora*.

Calochortus elegans is reported to hybridize with *C. subalpinus* in a small geographic area (Lane-Douglas County border) where their ranges overlap.

Key to *Calochortus elegans* varieties:

1. Nectary membrane fringed; sepals not tinged purple at base. . . . . ***C. e.*** var. ***elegans***

1' Nectary membrane with sharp or rounded teeth; sepals tinged purple at base . . . . ***C. e.*** var. ***nanus***

Calochortus elegans Pursh var. *elegans* Pursh

Elegant cat's ear

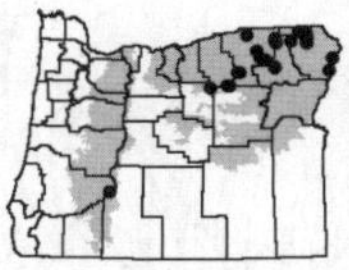

Leaves basal leaves 5–15 cm long. **Inflorescences** 1–3-flowered. **Flowers** sepals ovate-lanceolate, not tinged purple at base; nectary membrane fringed.

Open sparse grasslands. Flowering May–Jun. 500–1700 m. BW, Casc, Col. ID, WA; east to MT and UT. Native

Calochortus elegans Pursh var. *nanus* A. Wood (p. 293)

Dwarf cat's ear

Calochortus nanus (A.W. Wood) Piper

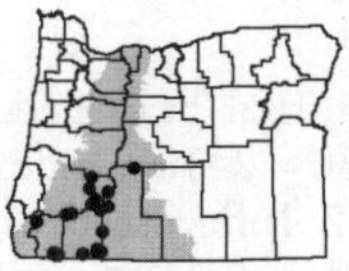

Leaves basal leaves 5–20 cm long. **Inflorescences** 1–7-flowered. **Flowers** sepals lanceolate to elliptic, tinged purple at base; nectary membrane with sharp or rounded teeth.

Open grasslands and summits, scree habitats. Flowering May–Jul. 1100–2300 m. Casc, ECas, Sisk. CA. Native.

Calochortus eurycarpus S. Watson

Bigpod mariposa lily

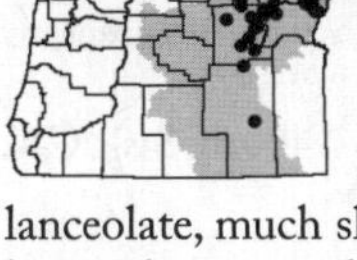

Stems to 50 cm long. **Leaves** basal leaves flat, to 30 cm long; bract-like cauline leaf midway up stem. **Inflorescences** 1–5-flowered; bracts narrow-lanceolate to linear, to 5 cm long, unequal. **Flowers** sepals ovate to lanceolate, much shorter than petals; petals obovate, to 45 mm long, white to pink, usually rounded at apex, with a reddish to purple blotch above nectary; nectary crested with long branching yellow hairs, transversely oval; filaments to 10 mm long, anthers to 8 mm long. **Fruits** erect, to 25 mm long, rounded at apex, with a short beak, winged. **Seeds** light yellow. ***2n***=20.

Grasslands, scablands, open coniferous forests. Flowering Jun–Aug. 200–2500 m. BR, BW, Lava. NV, WA. Native.

A disjunct population of this species exists in the Humboldt Mountains of Nevada.

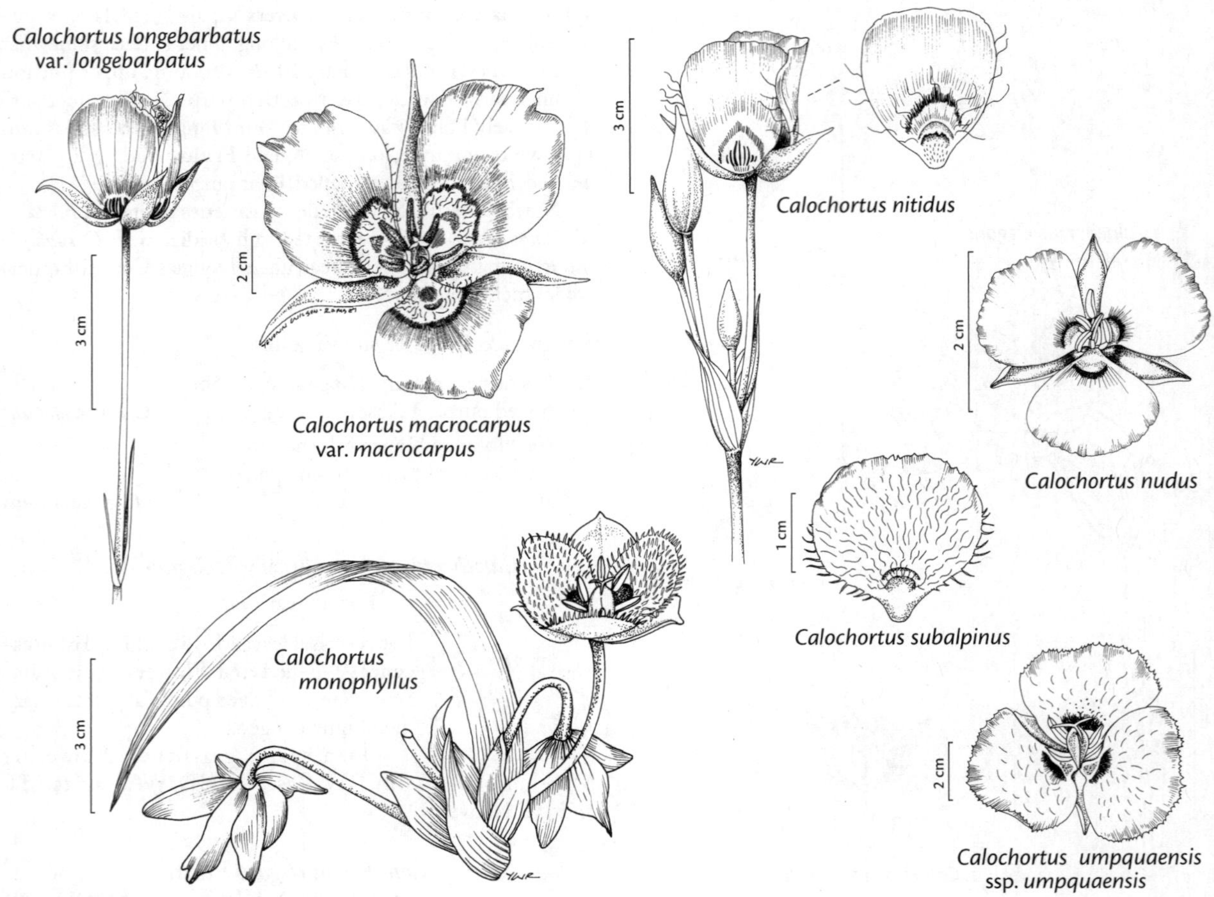

CALOCHORTUS

Calochortus greenei S. Watson (p. 293)

Greene's mariposa lily

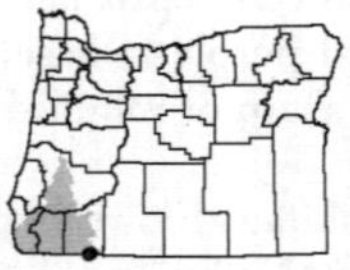

Stems 10–40 cm long, often branched. **Leaves** basal leaves 10–30 × 2–3 cm wide; cauline leaves generally 2. **Inflorescences** 1–6-flowered; bracts 2–4, lance-linear, to 4 cm long. **Flowers** not fully opening; sepals 25–30 mm long; petals ovate, to 30 mm long, pink, inner surface markedly lighter in color than outer surface (edges often same color as outer surface), inner surface base sometimes with a dark red chevron or spots and a yellow band with some penciling, inner surface hairs white above and yellow near base; nectary lunate, deeply sunken, crested with branching hairs; filaments to 14 mm long, anthers obtuse, to 12 mm long. **Fruits** erect, to 25 × 15 mm, with a short beak.

Open grasslands. Flowering Jun–Jul. 900–1400 m. Sisk. CA. Native.

Calochortus howellii S. Watson (p. 293)

Howell's mariposa lily

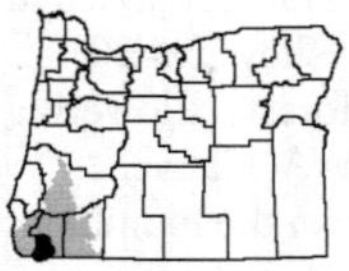

Stems 20–40 cm long. **Leaves** basal leaves 20–40 cm long, upper surface with rows of short hairs; cauline leaves few, lanceolate, to 5 cm long. **Inflorescences** 1–3-flowered; bracts to 4 cm. **Flowers** sepals ovate, to 20 mm long, acuminate; petals to 25 mm long, white, rounded at apex, inner surface with short dense white hairs above and purple hairs on a green base near nectary (may appear to be brown hairs); petal claw papillate; nectary crested with branching hairs; filaments to 7 mm long, anthers oblong, to 7 mm long, acuminate. **Fruits** erect, elliptic, to 20 mm long, acute.

Rocky serpentine soils. Flowering Jun–Jul. 300–800 m. Sisk. Native.

This species is known only from the Illinois Valley of southern Oregon. See also *C. umpquaensis.*

Calochortus indecorus Ownbey & Peck

Sexton Mountain mariposa lily

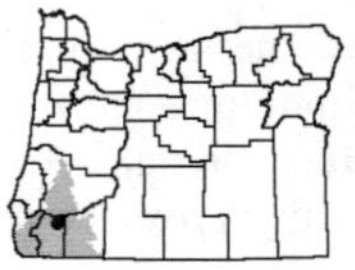

Stems 16–22 cm long. **Leaves** basal leaves 20–30 cm long; cauline leaves to 5 cm. **Inflorescences** 1–4-flowered. **Flowers** petals broadly ovate, to 25 mm long, purple, with few hairs near nectary; nectary triangular, centered well in from the petal edges, crested with simple hairs; filaments to 8 mm long, anthers oblong, to 4 mm long, lilac. **Fruits** nodding, 15–20 mm long. **Seeds** irregular-shaped, pale purple.

Open serpentine vernal grasslands. Flowering May. 400–500 m. Sisk. Native.

This species is very similar to *C. uniflorus* except for its upright stature and usual lack of bulblets. This species is known only from the base of Sexton Mountain in Josephine County.

Calochortus leichtlinii Hooker f.

Leichtlin's mariposa lily

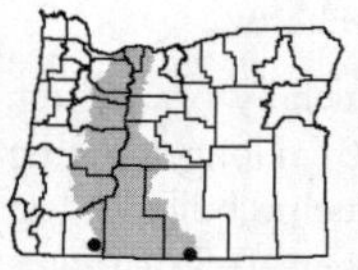

Stems 20–60 cm long, bearing a spur-like bulb. **Leaves** basal leaves flat, linear, 10–15 cm long, margins revolute, withering; cauline leaves to 9 cm. **Inflorescences** 1–5-flowered; bracts 1–5 cm long. **Flowers** sepals 10–20 mm long; petals 10–40 mm long, smoky blue, with a dark purple blotch above nectary, sparsely hairy around nectary; nectary oblong, lacking membrane, clustered with simple hairs; filaments to 5 mm long, anthers 5–7 mm long, white to yellow. **Fruits** erect, lanceolate, 3–5 cm long. **Seeds** flat, straw yellow.

Sagebrush, open montane conifer forest. Flowering Jul. 1800–2500 m. Casc, ECas. CA. Native.

Calochortus longebarbatus Watson

Long-haired star-tulip, long-bearded mariposa lily

Stems 10–20 cm long. **Leaves** with bulblets present in axils; basal leaves 20–30 cm long; cauline leaves 1–2. **Inflorescences** 1–5-flowered; bracts to 6 cm long. **Flowers** sepals ovate-lanceolate, 15–20 mm long, acuminate; petals obovate, pink to purple, 20–30 mm long, irregularly toothed at tip, dark red chevron above nectary gland, with sparse long hairs near nectary; nectary centered well away from the petal edges; filaments to 10 mm long, anthers approximately 4 mm long, short pointed. **Fruits** erect, broadly elliptic, 20–30 mm long, sharply pointed at tip, winged. **Seeds** irregular-shaped, light brown.

2 varieties.

Key to *Calochortus longebarbatus* varieties:

1. Petals not sharply bent at nectary. . ***C. l.*** var. ***longebarbatus***

1' Petals sharply bent at nectary . . . . ***C. l.*** var. ***peckii***

Calochortus longebarbatus Watson var. *longebarbatus* (p. 294)

Long-haired star-tulip, long-bearded mariposa lily

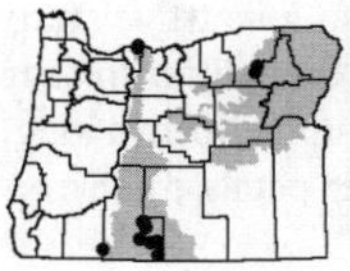

Inflorescences 1–4-flowered. **Flowers** petals not sharply bent at nectary. $2n$=20.

Grasslands and prairies. Flowering Jun–Aug. 1100–1700 m. BW, ECas. CA, WA Native.

Calochortus longebarbatus Watson var. *peckii* Ownbey

Peck's mariposa lily

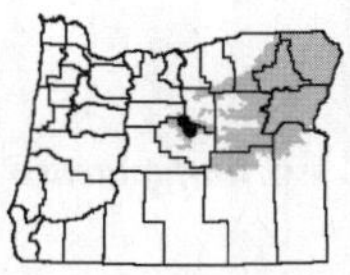

Inflorescences 1–5-flowered. **Flowers** petals sharply bent at nectary. $2n$=30.

Grasslands and prairies. Flowering Jun–Jul. 1400–1600 m. BW. Native.

Calochortus longebarbatus var. *peckii* is a sterile triploid that reproduces only through bulb division.

Calochortus macrocarpus Douglas

Sagebrush mariposa lily, green-banded star-tulip

Stems 20–55 cm long. **Leaves** basal leaves linear, 5–10 cm long, glaucous, margins revolute, withering. **Inflorescences** 1–2-flowered. **Flowers** sepals lanceolate, 40–50 mm long; petals obovate, 40–60 mm long, white to pink or purple, with a chevron or crescent or not, lower 33% of inner petal with yellow scattered hairs, outer petal surface with distinct green band; nectary depressed, triangular; filaments 8–9 mm long, anthers to 10 mm long, pink to light purple. **Fruits** erect, lance-linear, 5–10 mm wide. **Seeds** wafer-like, straw-colored.

2 varieties.

Key to *Calochortus macrocarpus* varieties:

1. Petals purple, with or without a dark purple chevron or crescent above nectary; fruits 8–10 mm wide . . . . ***C. m.*** var. ***macrocarpus***

1' Petals white, with reddish purple chevron or crescent above nectary; fruits 5–7 mm wide . . . . ***C. m.*** var. ***maculosus***

Calochortus macrocarpus Douglas var. *macrocarpus* (p. 294)

Sagebrush mariposa lily

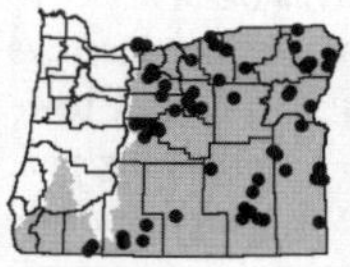

Flowers petals purple, with or without a dark purple chevron or crescent above nectary. **Fruits** 8–10 mm wide.

Sagebrush, dry plains, and montane slopes. Flowering Jun–Aug. 50–2300 m. BR, BW, Col, ECas, Lava, Owy, Sisk. CA, ID, WA; north to British Columbia, northeast to MT. Native.

Calochortus macrocarpus Douglas var. *maculosus* (A. Nelson & J.F. Macbride) A. Nelson & J.F. Macbride

Lewiston mariposa lily

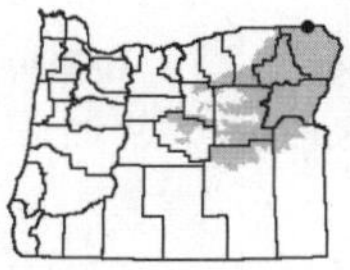

Flowers petals white with reddish purple chevron or crescent above nectary. **Fruits** 5–7 mm wide.

Canyons. Flowering Jun–Aug. 700–1300 m. BW. ID, WA. Native.

Calochortus monophyllus (Lindley) Lemaire (p. 294)

Yellow cat's ear

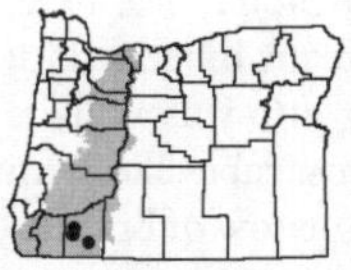

Stems 8–20 cm long, flexuous. **Leaves** basal leaf 1, 10–30 cm long, persistent; cauline leaves 0–3, lanceolate to linear, to 7 cm long. **Inflorescences** 1–6-flowered. **Flowers** sepals broadly lanceolate, 16–20 mm long, greenish yellow; petals 12–20 mm long, deep yellow, hairy throughout, margins sparsely ciliate, claw with chestnut-colored base; nectary lunate, with branching hairs; filaments to 5 mm long; anthers to 4 mm long, yellow. **Fruits** nodding, widely elliptic, 10–20 mm long, winged. **Seeds** irregular-shaped, brown. $2n$=20.

Conifer forests in mineral soils, sometimes serpentine. Flowering Apr–May. 500–1800 m. Casc, Sisk. CA. Native.

Calochortus nitidus Douglas

(p. 294)

Broadfruit mariposa lily

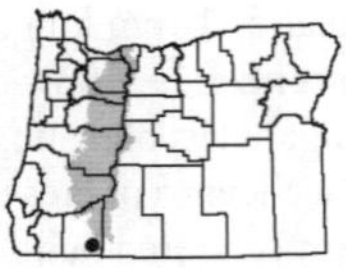

Stems 20–40 cm long. **Leaves** basal leaves flat, tapering both ends, 10–30 cm × 10–25 mm, persistent; cauline leaf 1. **Inflorescences** 1–4-flowered; bracts 2, linear-lanceolate, 2–10 cm long. **Flowers** sepals ovate to lanceolate, shorter than the petals; petals obovate, to 20–35 mm long, purple above, lower 33% white, dark purple chevron above nectary; nectary yellow-green with a series of vertical dark brown stripes, membrane present but much reduced, crested with long branching clear to yellow hairs; filaments to 10 mm long, anthers, oblong, 6–10 mm long. **Fruits** erect, broadly ovate, to 20 mm long, with a trifid beak, winged. **Seeds** irregular-shaped, light yellow. **2*n***=40.

Open grasslands and volcanic scablands. Flowering Jun–Aug. 1700–1900 m. Casc. ID, WA. Native.

The main range of this species is in the Palouse Hills of Idaho and Washington. There, the species is now rare due to the influence of farming. A disjunct population in Jackson County was extirpated.

Calochortus nudus S. Watson

(p. 294)

Naked star-tulip, Shasta star-tulip

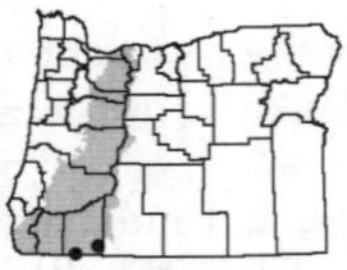

Stems 10–30 cm long. **Leaves** basal leaves to 5–20 cm long, persistent. **Inflorescences** 1–6-flowered; bracts 2, to 5 cm long. **Flowers** sepals lance ovate, 10–12 mm long; petals 12–16 mm long, white to purple, purple chevron above nectary, with sparse simple hairs near nectary; nectary with a membrane, crested with short simple hairs; filaments to 5 mm long, anthers to 4 mm long, purple. **Fruits** erect, elliptic, 10–20 mm long, acutish at both ends. **Seeds** irregular-shaped, light brown.

Grasslands. Flowering May–Jul. 1800–2000 m. Casc, Sisk. CA. Native.

Calochortus subalpinus Piper

(p. 294)

Cascade mariposa lily, mountain cat's ear, subalpine mariposa lily

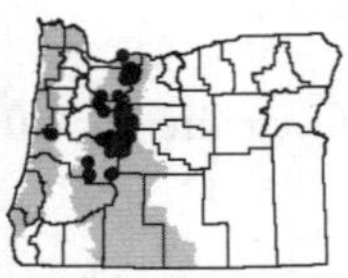

Stems 5–20 cm long, flexuous. **Leaves** basal leaves 10–30 cm × 2–15 mm. **Inflorescences** 1–5-flowered; bracts 2 or more, lanceolate-linear, 1–5 cm long, acuminate. **Flowers** sepals to 25 mm, with a dark blotch near base; petals obovate to 30 mm, white to yellowish cream, inner surface hairy throughout, white hairs above grading into yellow hairs near nectary; nectary strongly lunate and almost tube-like, with a shallow membrane, front crested with a single row of branching hairs, back crested with sparse branching hairs; filaments to 7 mm long, anthers to 8 mm long, yellow. **Fruits** nodding, oblong-obtuse, to 3 cm long. **Seeds** irregular-shaped, light yellow, aging to brown. **2*n***=20.

Rocky grasslands and rocky summit balds. Flowering May–Aug. 800–2500 m. Casc, CR, ECas. WA. Native.

Calochortus tolmiei Hooker & Arnott

Tolmie's cat's ear

Calochortus maweanus Leichtlin, *Calochortus purdyi* Eastw.

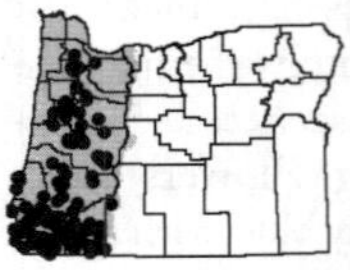

Stems 10–40 cm long, usually branching. **Leaves** basal leaves 10–40 cm long, tapering at both ends, rarely producing bulbs in leaf axils. **Inflorescences** flowers 1–5 or more-flowered; bracts 2 or more, linear-lanceolate, to 7 cm long, **Flowers** sepals 10–15 mm long, acuminate; petals obovate or wedge-shaped, 12–25 mm long, white above and purple near nectary, inner surface with dense simple hairs throughout; nectary lunate, membrane crested with 2 rows of short hairs, margins well away from petal edges; filaments to 8 mm long, anthers lanceolate to 5 mm long, white, pink, or purple. **Fruits** nodding, elliptic-oblong, 20–30 mm long, winged. **Seeds** irregular-shaped, purple, aging to brown. **2*n***=20.

Grasslands, forests, gravelly slopes. Flowering Mar–Jul. 0–2300 m. Casc, CR, Est, Sisk, WV. CA. Native.

Calochortus tolmiei is a highly variable species, which has resulted in many populations being described as distinct taxa (and subsequently "lumped" or synonymized back into the overall species concept). *Calochortus tolmiei* is known to hybridize with *C. uniflorus*.

Calochortus umpquaensis Fredricks

Umpqua mariposa lily

Stems 20–30 cm long, usually not branching. **Leaves** basal leaves narrowly lanceolate, to 40 cm long, upper surface pubescent; cauline leaf 1, not prominent. **Inflorescences** 1–5-flowered; bracts 2, subopposite, narrowly lanceolate. **Flowers** sepals lanceolate-acuminate, to 20 mm long; petals to 40 mm long, white to cream-colored, with dark purple crescent above nectary, inner surface hairy throughout or sparsely hairy, upper inner surface minutely papillose or not; filaments 7 mm long, anthers to 7 mm long. **Fruits** nodding, 3–5.5 cm long. **Seeds** irregular-shaped, yellow. **2*n***=20.

Southwestern Oregon. 3 subspecies treated in *Flora*.

Similar to *C. howellii*, this species is also a southwestern Oregon endemic found on serpentine soils. They differ in that *C. umpquaensis* has nodding fruits and petals up to 40 mm long, whereas *C. howellii* has erect fruits and shorter petals, to 25 mm long.

Key to *Calochortus umpquaensis* subspecies:

1. Nectary membrane absent .... ***C. u.*** ssp. ***confertus***

1' Nectary membrane present.

  2. Upper inner petal surface minutely papillose; without band of long yellow hairs above nectary .... ***C. u.*** ssp. ***umpquaensis***

  2' Upper inner petal surface not minutely papillose; with band of long yellow hairs above nectary . . ***C. u.*** ssp. ***flavicomus***

Calochortus umpquaensis Fredricks ssp. *confertus* Callahan

Callahan Meadows mariposa lily

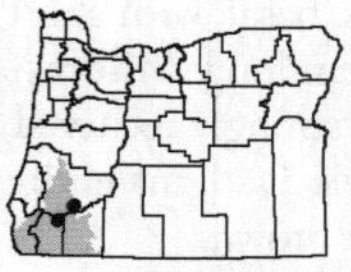

Flowers upper inner petal surface not minutely papillose (basal inner petal surface minutely papillose); hairs above nectary purple; nectary membrane absent, dense cluster of hairs present instead.

Serpentine meadows. 700–900 m. Sisk. Native.

The name *confertus* means "dense or compact" in Latin, referring to the dense cluster of hairs present instead of the nectary membrane.

Calochortus umpquaensis Fredricks ssp. *flavicomus* Callahan

Umpqua mariposa lily

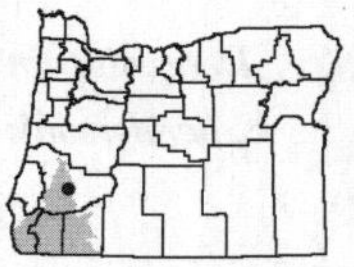

Flowers upper inner petal surface not minutely papillose (basal inner petal surface minutely papillose); transverse band of yellow hairs above nectary; nectary membrane present.

The name *flavicomus* means "yellow hair" referring to the yellow hairs above the nectary.

Serpentine ridgelines. 300–700 m. Sisk. Native.

Calochortus umpquaensis Fredricks ssp. *umpquaensis* (p. 294)

Umpqua mariposa lily

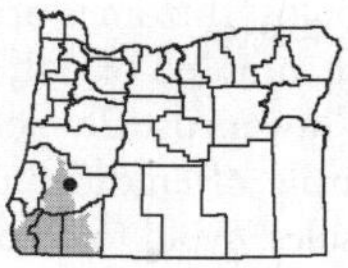

Flowers upper inner petal surface minutely papillose; hairs above nectary purple; nectary membrane present.

Serpentine meadows and rocky soils. 200–800 m. Sisk. Native.

Calochortus uniflorus Hooker & Arnott

Large-flowered star-tulip, Monterey mariposa lily

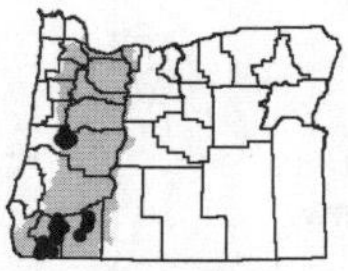

Stems to 10 cm long; bulblets present. **Leaves** basal leaves 10–40 cm long, clasping stem at base, persistent; cauline leaves 1–3, linear. **Inflorescences** 1–5-flowered; bracts 2, linear, to 4 cm long. **Flowers** sepals lance elliptic, 12–16 mm long, acuminate; petals widely wedge-shaped to ovate, 14–28 mm long, white to pink or purple; nectary with pocket-like triangular membrane, crested with simple hairs; filaments to 8 mm long; anthers to 4 mm long, pink to light purple. **Fruits** nodding, elliptical, 10–30 mm long. **Seeds** irregular-shaped, purple, aging to brown. **2*n***=20, 40.

Vernal meadows in heavy clays or serpentine soils. Flowering Apr–Jun. 100–800 m. Casc, Sisk, WV. CA. Native.

Calochortus uniflorus is known to hybridize with *C. tolmiei*.

CLINTONIA Bluebead

Lareina Shipler

Herbs perennial, 10–80 cm; from underground spreading rhizomes bearing thin, fibrous roots. **Stems** lacking. **Leaves** basal, 2–6, obovate, oblanceolate or elliptic, arising from rhizome crown, sessile; venation parallel, central vein prominent basally; petioles sheathing, chaffy in age; base attenuate to cuneate; margins entire, apex acute to abruptly short acuminate, often mucronate. **Inflorescences** elongated and narrowly 1 or an umbel, short racemes or umbel-like; bracteate; pedicels unequal. **Flowers** tepals 6, narrowly oblong to oblanceolate; stamens 6, distinct, hairy (or not), filaments dilated, basally pubescent, anthers oblong to obovate to oblong-linear; ovary superior; style slender, columnar and deciduous, stigma weakly 2–3 lobed. **Fruits** ellipsoid to ovoid, blue to black, glabrous. **Seeds** 4–30, abaxially round, shiny brown to black, angled with 2–several faces.

Temperate North America, eastern Asia. 6 species; 2 species treated in *Flora*.

Native Americans used *Clintonia* as an eye and heart medicine, as well as a dermatological and gynecological aid (Moerman 1986).

Key to *Clintonia* species:

1. Flowers many, umbellate, pink to red; leaves 4–6, tomentose . . . . *C. andrewsiana*

1' Flowers usually solitary, white; leaves 2–3, glabrous except ciliate margins . . . . *C. uniflora*

Clintonia andrewsiana Torrey (p. 299)

Red clintonia, redwood clintonia, Andrew's clintonia

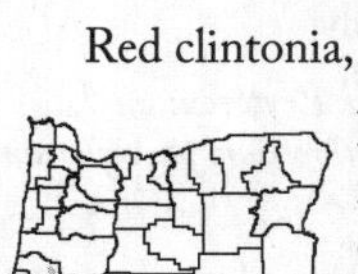

Plants 30–80 cm tall from slender rootstock; rhizomes short. **Leaves** 4–6, tomentose, ovoid, short-acuminate, 15–30 × 5–12 cm. **Inflorescences** in 1 or more umbel-like clusters, many-flowered; peduncle 20–50 cm long, hairy; bract 1. **Flowers** perianth campanulate; tepals narrowly oblong to oblanceolate, 1–2 cm long, pink to reddish (purple). **Fruits** berries, globose, 8–15 mm long, bright blue to bluish black. **2*n***=28.

Shaded, moist, coastal, redwood forests. Flowering May–Jun. 100–200 m. Sisk. CA. Native.

Clintonia uniflora (Schultes) Kunth (p. 299)

Queen's cup, beadlily, bride's bonnet, bluebead

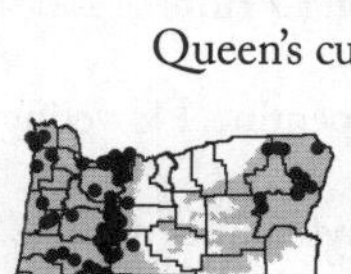

Plants 10–30 cm; rhizomes slender and creeping. **Leaves** 2–3, glabrous except the ciliate margins; oblanceolate to oblong, mostly short-acuminate, 7–15 × 2–6 cm wide. **Inflorescence** solitary (rarely 2); peduncle 4–8 cm long, hairy; bract 1. **Flowers** perianth campanulate; tepals oblanceolate, 15–25 mm long, creamy white. **Fruits** berries, globose to pear-shaped, 6–12 mm long, bright blue to deep blue. **2*n***=28.

Moist, shaded understory of coniferous forests from coastline to timberline. Flowering May–Aug. 0–2300 m. BW, Casc, CR, ECas, Est, Sisk, WV. CA, ID, WA; north to AK, east to Alberta and MT. Native.

ERYTHRONIUM Fawn-lily, trout-lily

Stephen C. Meyers

Herbs perennial, scape-like; from elongate bulbs, rhizomes present. **Leaves** 2(3), or 1 in non-flowering plants, basal, with short petioles; blades lanceolate to elliptic, 5–25 cm, mottled or not, glabrous, margins entire or wavy. **Scapes** 5–35 cm long. **Inflorescences** peduncled racemes, 1–10-flowered. **Flowers** usually nodding; tepals 6, recurved, distinct, similar, often strongly recurved, lanceolate to ovate, white, yellow, pink, or purple, often different colored at base; stamens 6; ovary superior; style 1, stigma entire or 3-lobed, recurved when lobed. **Fruits** capsules, ovoid to oblong, 3-sided, dehiscence loculicidal. **Seeds** ovoid, brown.

Northern Hemisphere. 27 species; 8 species treated in *Flora*.

A small population of *E. multiscapideum* was discovered in Jackson County in 2010. The collector (Frank Callahan), however, notes that the number of plants was not abundant and that the population was suffering from browsing deer. Whether this species is currently present in Oregon is unknown at this time.

Key to *Erythronium* species:

1. Tepals completely yellow . . . . ***E. grandiflorum*** var. ***grandiflorum***
1' Tepals not yellow or not completely yellow.
  2. Tepals purple to pink.
    3. Filaments flattened, more than 2 mm wide . . . . ***E. revolutum***
    3' Filaments linear and not flattened, less than 1 mm wide . . . . ***E. hendersonii***
  2' Tepals white, cream-colored, white tinged with pink or white with a yellow base (these taxa may age to light pink).
    4. Leaves mottled (first year leaves of some *E. oregonum* plants not mottled).
      5. Stigma entire or lobed less than 1 mm . . . . ***E. citrinum*** var. ***citrinum***
      5' Stigma lobed more than 1 mm.
        6. Filaments 2–3 mm wide . . . . ***E. oregonum***
        6' Filaments 0.5–2 mm wide . . . . ***E. elegans*** (in part)
    4' Leaves not mottled.
      7. Stigma entire . . . . ***E. klamathense***
      7' Stigma lobed.
        8. Filaments flattened, more than 0.8 mm wide . . . . ***E. elegans*** (in part)
        8' Filaments linear, less than 0.8 mm wide . . . . ***E. montanum***

Erythronium citrinum S. Watson var. *citrinum*

Pale fawn-lily

Erythronium howellii S. Watson

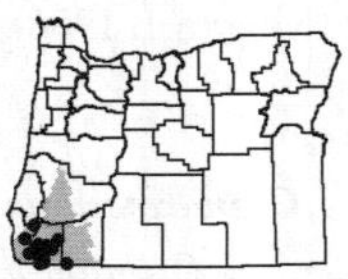

Leaves 8–15 cm; blades mottled brown or white, margins wavy. **Inflorescences** 1–3-flowered. **Flowers** tepals 20–45 mm, white, sometimes aging to light pink, usually yellow at base, inner tepals with or without auricles at base; stamens 10–15 mm, filaments linear, less than 1 mm wide, white to pink, anthers white to red; style white to pink, stigma entire or lobes less than 1 mm long. **Fruits** capsules 2–5 cm. **2*n***=24.

Open, dry woods and slopes, often on serpentine. Flowering Mar–Jun. 300–1800 m. Sisk. CA. Native.

Erythronium citrinum is known to hybridize with *E. hendersonii*.

Erythronium elegans P.C. Hammond & K.L. Chambers

Elegant fawn-lily

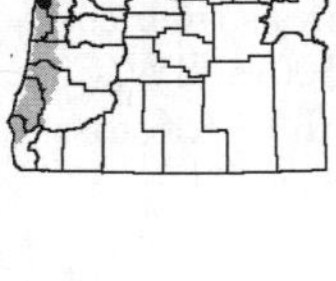

Leaves 6–15 cm; blades ovate-lanceolate, not mottled or slightly mottled with brown or white, margins usually wavy. **Inflorescences** 1–2(4)-flowered. **Flowers** tepals 20–40(50) mm, white or white tinged with pink, usually more pinkish with age, yellow band at base, inner tepals with auricles at base; stamens 12–20 mm, filaments flattened, 0.8–2 mm wide, white, anthers yellow; style white, stigma with recurved lobes 2–4 mm long. **Fruits** capsules 2–5 cm. **2*n***=48.

Open sites on rocky slopes and cliffs. Flowering May–Jun. 400–1100 m. CR. Native.

Erythronium elegans is a rare species, known only from the Coast Range in northern Oregon.

Erythronium grandiflorum Pursh var. *grandiflorum* (p. 299)

Glacier-lily, yellow fawn-lily

Erythronium grandiflorum Pursh ssp. *grandiflorum*, *Erythronium grandiflorum* Pursh var. *pallidum* H. St. John, *Erythronium parviflorum* (S. Watson) Goodd.

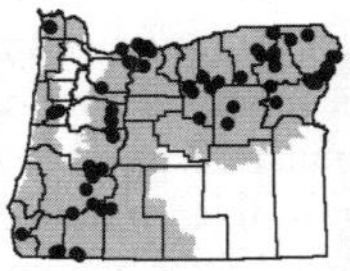

Leaves 5–20 cm; blades lanceolate, not mottled, margins wavy. **Inflorescences** 1(3)-flowered. **Flowers** tepals 20–35 mm, yellow, pale at base, inner tepals with auricles at base; stamens 10–18 mm, filaments, linear, less than 1 mm wide, white, anthers cream to yellow to purple-red; style white, stigma with recurved lobes (1)2–4 mm long. **Fruits** capsules 2–5 cm. **2*n***=24.

Subalpine meadows, woods, rock outcrops. Flowering Mar–Jul. 0–2300 m. BW, Casc, Col, CR, ECas, Lava, Sisk. CA, ID, WA;

north to British Columbia and Alberta, southeast to NM. Native.

As the common name implies, this lily is often found emerging from melting snow in subalpine meadows.

Erythronium hendersonii S. Watson (p. 299)

Henderson's fawn-lily

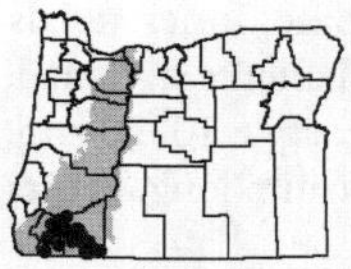

Leaves 10–25 cm; blades oblong to lanceovate, mottled brown or white, margins entire to wavy. **Inflorescences** 1–4-flowered. **Flowers** tepals 20–38 mm, purple to pink, dark purple at base, inner tepals with auricles at base; stamens 10–15 mm, filaments linear, less than 1 mm wide, purple, anthers brown to purple; style purple, stigma entire or with lobes shorter than 1 mm. **Fruits** capsules 2–4 cm. ***2n***=24.

Openings in dry woods, occasionally fields or meadows. Flowering Mar–May. 200–1800 m. Casc, Sisk. CA. Native.

Erythronium klamathense Applegate

Klamath fawn-lily

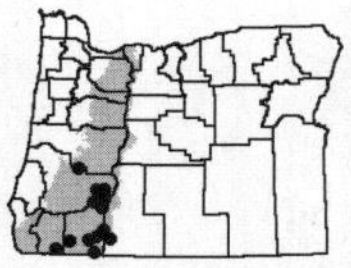

Leaves 6–17 cm; blades lanceolate to elliptic, not mottled, margins entire to wavy. **Inflorescences** 1–3-flowered. **Flowers** tepals 20–35 mm, white, with large zone of yellow at base, inner tepals with auricles at base; stamens 8–15 mm, filaments linear, less than 1 mm wide, white, anthers yellow; style 4–10 mm, white, stigma entire. **Fruits** capsules 2–5 cm. ***2n***=24.

Montane meadows, forest openings, rock outcrops. Flowering May–Jul. 400–2100 m. Casc, Sisk. CA. Native.

Erythronium montanum S. Watson (p. 299)

Avalanche-lily

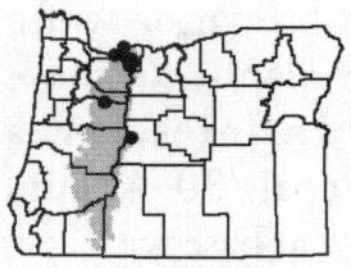

Leaves 10–20 cm; blades ovate to broadlanceolate, not mottled, margins wavy. **Inflorescences** 1–3-flowered. **Flowers** tepals 25–45 mm, white to cream-colored, yellow at base, inner tepals with auricles at base; stamens 10–25 mm, filaments, linear, less than 0.8 mm wide, white, anthers yellow; style 12–25 mm, white, stigma with recurved lobes 1–5 mm long. **Fruits** capsules 3–6 cm. ***2n***=24.

Forest openings and meadows at montane and subalpine elevations. Flowering Jun–Aug. 800–1800 m. Casc. WA; north to British Columbia. Native.

Similar to its presumed close relative *E. elegans*, this species is best identified by its linear, very narrow (less than 0.8 mm) filaments. Its flowers also tend to be more outward-facing than *E. elegans* and our other species, which are usually nodding. Like *E. grandiflorum*, this lily is also found emerging from melting snow in montane and subalpine meadows.

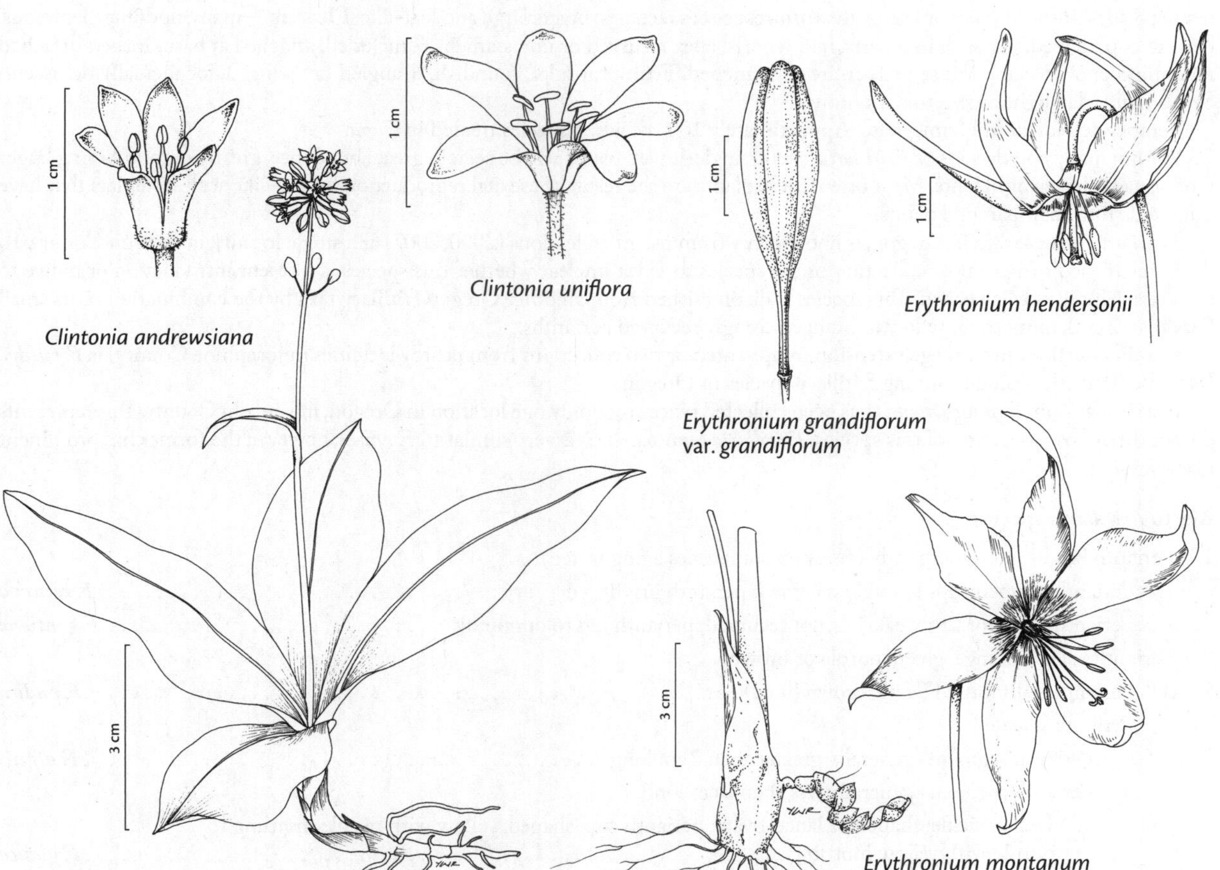

Clintonia andrewsiana

Clintonia uniflora

Erythronium hendersonii

Erythronium grandiflorum var. *grandiflorum*

Erythronium montanum

Erythronium oregonum Applegate

Oregon fawn-lily

Erythronium giganteum Lindl., *Erythronium oregonum* Applegate ssp. *leucandrum* (Applegate) Applegate, *Erythronium oregonum* Applegate var. *leucandrum* (Applegate) M. Peck, *Erythronium oregonum* Applegate var. *oregonum*

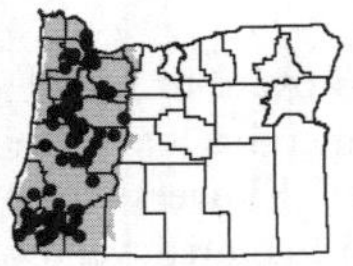

Leaves 12–25 cm; blades ovate to broad-lanceolate, mottled with brown or white (first year leaves of some plants not mottled), margins wavy. **Inflorescences** 1–3-flowered. **Flowers** tepals 30–45 mm, white to cream-colored, becoming more yellow at base with age, inner tepals with auricles at base; stamens 10–25 mm, filaments flattened, 2–3 mm wide, white, anthers cream-colored to yellow; style white, stigma with recurved lobes 3–6 mm long. **Fruits** capsules oblong to narrowly obovoid, 3–5 cm. **2*n***=24.

Openings in woodlands and meadows. Flowering Mar–Jun. 0–1800 m. Casc, CR, Sisk, WV. CA, WA; north to British Columbia. Native.

The logo of the Oregon Flora Project is a silhouette of *E. oregonum*.

Erythronium revolutum Smith (p. 301)

Pink fawn-lily

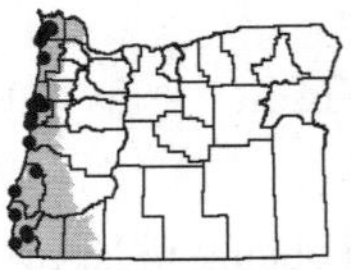

Leaves 10–25 mm; blades broad-lanceolate to ovate, distinctly mottled brown or white, margins entire to wavy. **Inflorescences** 1–3-flowered. **Flowers** tepals 25–40 mm, purple-pink, yellow at base, inner tepals with auricles at base; stamens 10–25 mm, filaments flattened, 2–3 mm wide, white to pink, anthers yellow; style 10–20 mm, white to pink, stigma with recurved lobes 4–6 mm long. **Fruits** capsules 3–6 cm.

Wet woods, riparian areas. Flowering Mar–May. 0–1200 m. CR, Est, Sisk. CA, WA; north to British Columbia. Native.

While generally considered a coastal species, populations of *E. revolutum* can also be found up to 50 km inland.

FRITILLARIA Fritillary

Stephen C. Meyers

Herbs perennial; from bulbs with fleshy scales, usually with many small rice-like scales or bulblets. **Stems** solitary, erect (0 in non-flowering plants). **Leaves** 1 in immature plants to many in flowering plants, basal and cauline, alternate, subopposite or whorled, sessile; blades linear to more or less ovate. **Inflorescences** racemes; bracts large and leaf-like. **Flowers** 1–many, nodding, 3-merous, bell- to cup-shaped, perianth in 6 parts, in 2 whorls; apex recurved or not; stamens 6, included, attached at base, anthers attached near middle; ovary superior; style 1, entire or branched. **Fruits** capsules, roundish, 6-angled or -winged, loculicidally dehiscent. **Seeds** many, flat, light brown to dark brown.

Temperate Northern Hemisphere. Approximately 100 species; 6 species treated in *Flora*.

Within many populations of *Fritillaria*, small, single-leaved plants will be seen in great abundance, greatly outnumbering larger (and usually flowering) plants. Most often, these plants are the result of asexual reproduction by rice-like scales/bulblets that have migrated from their parental plants.

Fritillaria eastwoodiae, in Oregon, is known only from recent collections (2000, 2007) at a single locality, in Jackson County. To date, this is the northernmost collection of the species. It is yet unclear whether this species is a recent introduction or native to our state. More study is needed. This species is distinguished from all other Oregon fritillary taxa by the combination of its small (less than 2 cm), unmottled, yellow to orange, strongly recurved perianths.

Another northernmost range extension (represented by two collections from nearby localities in Josephine County) is *F. purdyi*. Its white perianth is unique among fritillary species in Oregon.

In a similar vein, *F. camschatcensis* has been collected twice from only one location in Oregon, in Lincoln County. This represents the southernmost collection of this species. *Fritillaria camschatcensis* is very similar to *F. affinis*, however, the former has prominent nectaries.

Key to *Fritillaria* species:

1. Perianth bright red to burgundy (*F. pudica* sometimes aging to red).
    2. Perianth segment tips strongly recurved; perianth bright red . . . . ***F. recurva***
    2' Perianth segment tips weakly or not recurved; perianth red to burgundy . . . . ***F. gentneri***

1' Perianth yellow, orange, green, purple or brown.
    3. Perianth not mottled (some brown lined) . . . . ***F. pudica***
    3' Perianth mottled.
        4. Perianth segments generally greater than 2 cm long . . . . ***F. affinis***
        4' Perianth segments generally less than 2 cm long.
            5. Leaves sickle-shaped or lance-ovate; perianth cup-shaped, yellow with purple mottling to purple with yellow mottling . . . . ***F. glauca***
            5' Leaves mostly straight, linear to linear-lanceolate; perianth widely spreading, brownish purple with yellow mottling . . . . ***F. atropurpurea***

Fritillaria affinis (Schultes & Schultes f.) Sealy (p. 301)

Checker-lily, mission bells, chocolate lily

Fritillaria affinis (Schult.) Sealy var. *affinis*, *Fritillaria lanceolata* Pursh

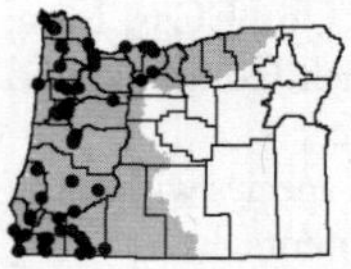

Stems 10–70 cm. **Leaves** in whorls of 2–8 per node and alternate, 4–14 cm long; blades linear-lanceolate to ovate. **Flowers** cup-shaped; perianth segments oblong to ovate, (1.5)2–4 cm, brown-purple with yellow mottling to yellow-green with purple mottling, rarely unmottled, apex not recurved; nectaries yellow with purple dots, equal to or over 50% of perianth length; style branched approximately 50% its length. **Fruits** capsules winged. ***2n***=24, 36.

Meadows, edges of forests, open oak and pine forests, occasionally on somewhat rocky slopes and outcrops. Flowering Mar–Jun. 0–2100 m. Casc, Col, CR, ECas, Est, Sisk, WV. CA, ID, WA; north to British Columbia. Native.

Fritillaria affinis is a highly variable species. Some morphologically distinct populations may deserve taxonomic recognition.

Fritillaria atropurpurea Nuttall

Spotted fritillary

Fritillaria adamantina M. Peck

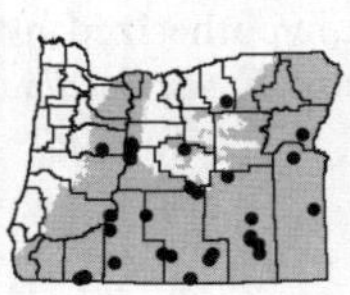

Stems 10–60 cm. **Leaves** in whorls of 2–3 leaves per node, 4–13 cm long; blades linear to lanceolate. **Flowers** widely spreading; perianth oblong to more or less square-shaped, 1–2(2.5) cm long, segments brownish purple with yellow or rarely white mottling, apex not recurved; nectaries obscure, covering most of perianth segments; style branched for more than 50% its length. **Fruits** capsules angled.

Dry or grassy slopes, often under trees or bushes. Flowering May–Jul. 100–2500 m. BR, BW, Casc, ECas, Owy, Sisk. CA, ID, NV; east to ND and NB, southeast to NM. Native.

Fritillaria gentneri Gilkey (p. 301)

Gentner's fritillary

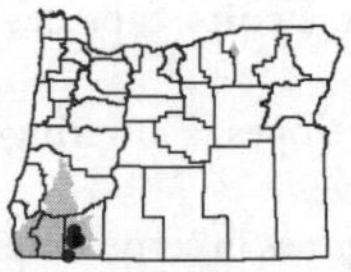

Stem 10–70 cm. **Leaves** in 1–3 whorls of 3–5 leaves per node or alternate, 5–15 cm long; blades broadly linear to oblanceolate. **Flowers** perianth segments oblong to oblanceolate, 3–4 cm long, red to purple with yellow mottling, apex weakly recurved or not; nectaries linear, approximately 50% perianth segment length; style branched for 33–50% its length. **Fruits** capsules winged. ***2n***=36.

Dry woodlands. Flowering Apr–May. 100–1600 m. Sisk. Native.

In addition to morphological traits, DNA evidence implies that *F. gentneri* is a hybrid between *F. affinis* and *F. recurva* (Meyers et al. 2006). Whether this proposed hybridization event was recent or older warrants more study.

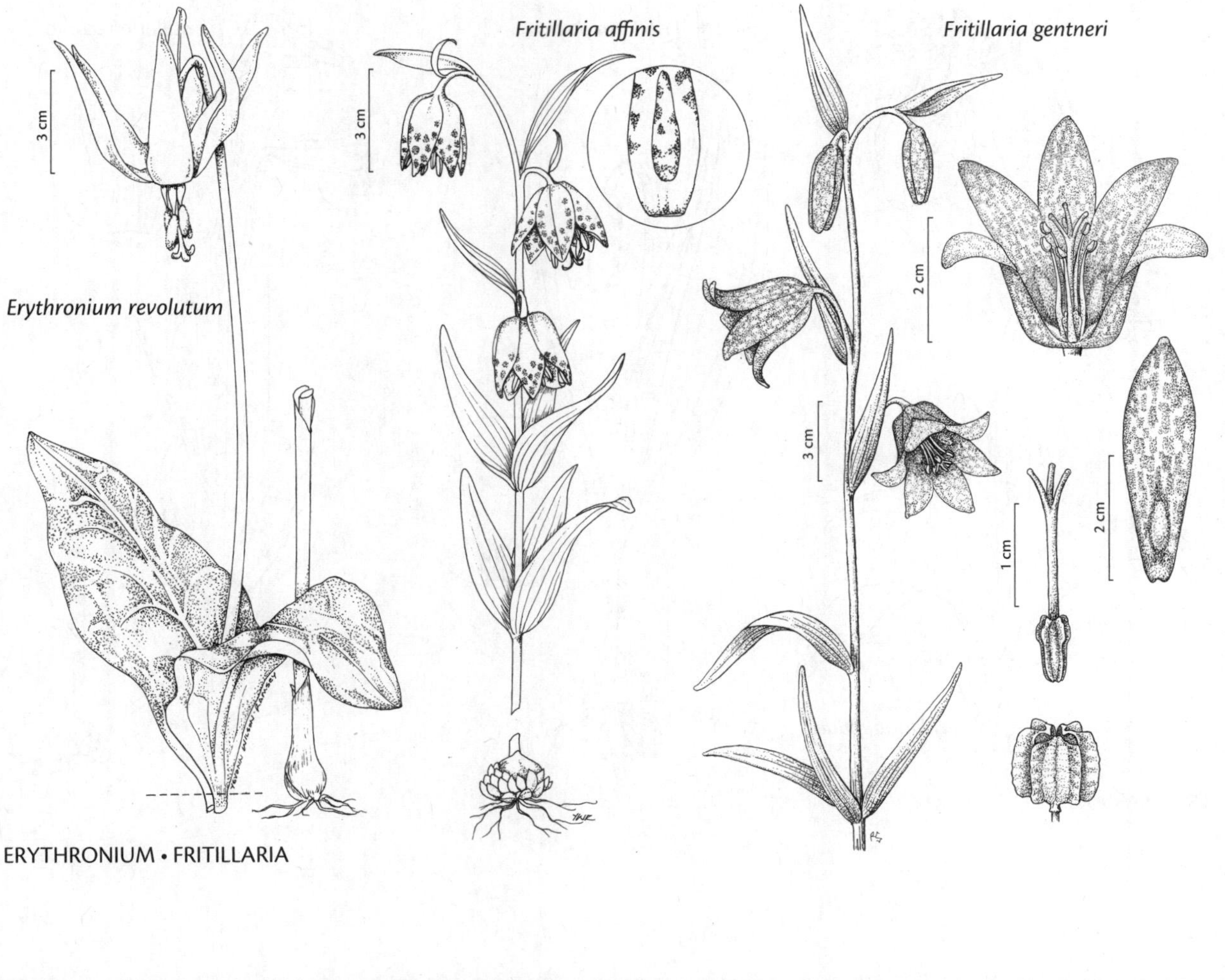

ERYTHRONIUM • FRITILLARIA

Fritillaria glauca Greene (p. 302)

Siskiyou fritillary

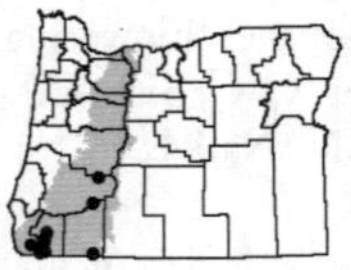

Stems 5–20 cm. **Leaves** 1–4(10), alternate, 3–11 cm long; blades lance-ovate, sickle-shaped, strongly glaucous. **Flowers** cup-shaped; perianth segments oblanceolate to oblong, 1.5–2(2.5) cm long, yellow with purple mottling to purple with yellow mottling, apex not recurved; nectaries broad-lanceolate, 25-33% perianth segment length; style branched for 50% its length. **Fruits** capsules winged. $2n$=24.

Serpentine and non-serpentine gravelly slopes. Flowering Mar–Jun. 400–2100 m. Casc, Sisk. CA. Native.

Fritillaria glauca is our only species that grows in serpentine soils. This species also grows in non-serpentine habitats. This begs the question: Was *F. glauca* historically a serpentine or non-serpentine endemic? The sickle-shaped and strongly glaucous leaves of *F. glauca* are distinct among Oregon fritillaries.

Fritillaria pudica (Pursh) Sprengel (p. 302)

Yellow bells, yellow fritillary

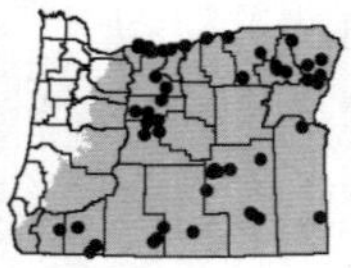

Stems 5–25 cm. **Leaves** 2–8, subopposite or alternate, 3–15(18) cm long; blades linear to lanceolate. **Flowers** perianth segments 1–2.5 cm long, yellow to orange, sometimes aging to red, some lined brown and/or brown at base; nectaries, shallow and at base of perianth segments, oval to round; style unbranched. **Fruits** capsules angled. $2n$=24.

Grassy or shrubby areas, open woods, often on slopes. Flowering Mar–May. 0–2600 m. BR, BW, Casc, Col, ECas, Lava, Owy, Sisk. CA, ID, NV, WA; north to British Columbia and Alberta, east to NB, southeast to NM. Native.

Fritillaria pudica is unique among Oregon species within the genus because of its unmottled perianth segments.

Fritillaria recurva Bentham (p. 302)

Scarlet fritillary

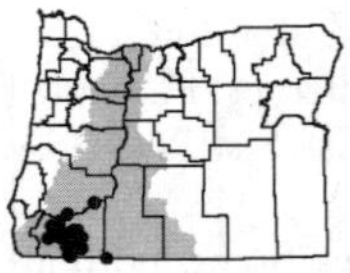

Stems 20–70 cm. **Leaves** in 1–3 whorls of 2–5 leaves per node or alternate, 3–10(15) cm long; blades linear to narrowly lanceolate. **Flowers** perianth segments oblong to oblanceolate, 2–4 cm long, red to burgundy with checkered yellow mottling within and purple mottling without, apex strongly recurved; nectaries narrowly lanceolate, up to 25% of perianth segment length; style branched for 25–50% its length. **Fruits** capsules winged. $2n$=24, 36.

Dry open woods and scrubland. Flowering Mar–May. 200–1500 m. Casc, ECas, Sisk. CA, NV. Native.

Fritillaria recurva and *F. micrantha* are hypothesized by some authorities to be the parental species of *F. eastwoodiae* (McFarlane 1978). More study is needed.

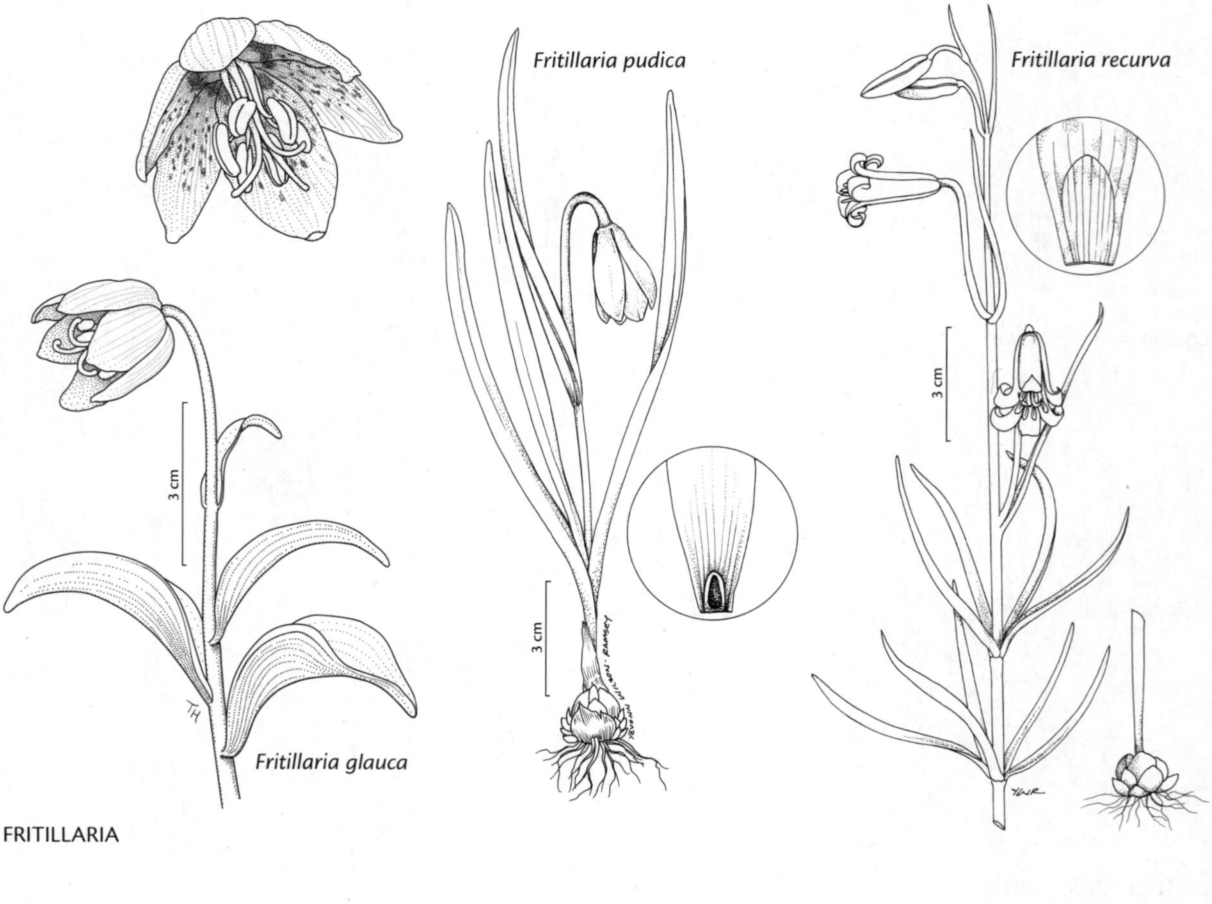

FRITILLARIA

GAGEA Gagea

Stephen C. Meyers

Herbs perennial, erect, caulescent; rhizomes present. **Leaves** linear to linear-lanceolate, reduced upwards. **Inflorescences** terminal, loosely flowered. **Flowers** fragrant; tepals 6, with a small gland at base; stamens 6; ovary superior; style 1, stigma shortly 3-lobed. **Fruits** capsules, dehiscence loculicidal, 3-loculed.

Western North America, Eurasia. Approximately 200 species; 1 species treated in *Flora.*

Gagea serotina (Linnaeus) Ker-Gawler (p. 303)

Alp lily, common alp lily

Lloydia serotina (L.) Salisb. ex Rchb. ssp. *serotina*, *Lloydia serotina* (L.) Salisb. ex Rchb. var. *serotina*

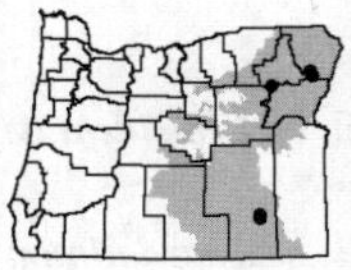

Plants glabrous, 7–20 cm tall, surrounded by old leaf sheaths below; from small, oblong, papery, bulb-like rhizomes. **Leaves** basal leaves 2, 3–20 cm × approximately 1 mm; cauline leaves alternate, linear-lanceolate, 0.5–4 cm × 1–2 mm. **Inflorescences** terminal racemes, 1–2-flowered. **Flowers** yellow-white to cream, rose-tinged on back, with green to purple veins; tepals distinct, spreading, sometimes slightly recurved, oblong, 0.8–1.3 cm long. **Fruits** ovoid, triangular in cross section, 6–8 mm long. **Seeds** angled, black. **2*n***=24, 36, 48.

Rocky slopes. Flowering Jul–Aug. 1900–2900 m. BR, BW. ID, NV, WA; north to AK, southeast to NM; Eurasia. Native.

In Oregon, *G. serotina* is known only from a few mountainous, rocky locations.

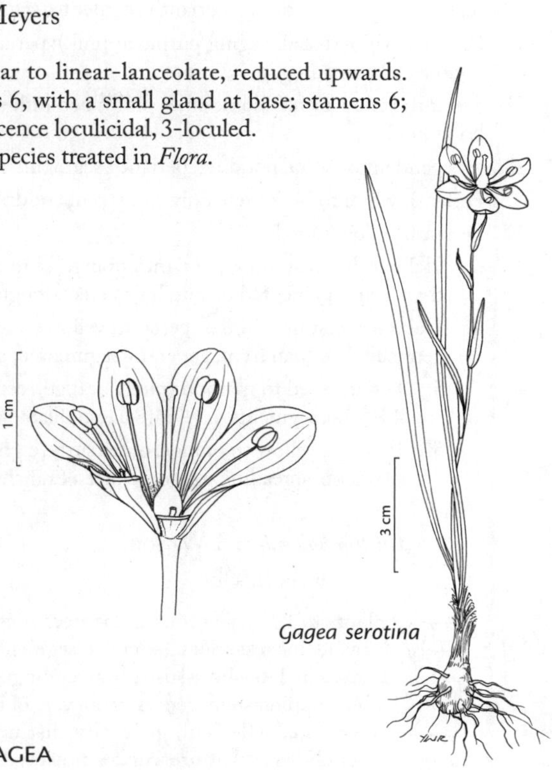

GAGEA

LILIUM Lily

Mark W. Skinner

Herbs perennial, generally glabrous; from scaly rhizomes (bulbs), most not clonal; stems erect. **Leaves** primarily cauline, mostly whorled (often some scattered), sessile, horizontal, and drooping at tips to ascending, generally more or less elliptic; veins generally 3; stipules 0. **Inflorescences** flowers axillary, 1–45 or more; bracts generally 2 per flower. **Flowers** generally radial, generally bell- or trumpet-shaped; perianth segments 6, in 2 petal-like whorls, commonly lanceolate and narrowed at base, generally red-purple-spotted on inner base; stamens 6, anthers attached near middle; style 1; stigma 3-lobed. **Fruits** capsules, erect, generally smooth, loculicidal. **Seeds** many, flat, triangular, in 6 stacks.

Worldwide in north temperate zone, and tropical mountains of Asia (India, Philippines). Approximately 100 species; 6 species treated in *Flora*.

Several *Lilium* taxa have been mistakenly reported to occur in Oregon. Various subspecies of *Lilium pardalinum* have been misidentified as *Lilium parvum*, Sierra tiger lily, but the latter drops out north of Lake Tahoe in California, about 200 miles from Oregon. *L. rubescens*, redwood lily, makes a fairly close approach to Oregon, though the nearest I have seen it is near Forks of Salmon in Siskiyou County, CA, about 50 miles from the state line. A number of specimens at the Oregon State University Herbarium—most from the Dead Indian Plateau in Jackson County—were previously annotated as *L. rubescens*; all proved to be *L. washingtonianum* ssp. *purpurascens*. Cascade lily is distinguished from *L. rubescens* by its more robust stature and larger flowers [(5.6)6.2–9.4(10.5) vs. 4.2–6.4 cm] which are held horizontally rather than ascending to erect, and by nocturnal nectar and heavy scent production associated with sphingid (hawkmoth) pollination.

Within-species variation in North American *Lilium* is significant, and specific and subspecific boundaries are loose; hybridization is common, particularly where humans have altered natural barriers to gene flow (Skinner 2002). Many *Lilium* species are declining from habitat destruction and collecting; only a few thrive in gardens. In Oregon, most flower between June and August, with a strong peak in July. Deer crave most liliaceous plants including *Lilium*, and cattle can eliminate them.

This treatment is primarily field-based, and all measurements are from live material. In bulb scale descriptions, 1-segmented = unsegmented, i.e., consisting of a single segment. Anther measurements are after dehiscence and are highly variable due to anther shrinkage.

Key to *Lilium* species:

1. Flowers nodding to ascending, perianth segments recurved but not reflexed.
   2. Perianth white (usually aging purple or pink); perianth segments generally 6.2–9.5 cm; flowers nodding to ascending . . . . *L. washingtonianum* ssp. *purpurascens*
   2' Perianth yellow to red (sometimes pink); perianth segments less or equal to 5 cm; flowers nodding to horizontal . . . . *L. bolanderi*

1' Flowers pendent or some nodding, perianth segments reflexed.
   3. Perianth white to pink (generally aging pink, midribs yellow) . . . . *L. kelloggii*
   3' Perianth yellow to red.
      4. Habitat dry in summer; perianth orange, segments 3.4–6.1(7.1) cm; longest bulb scales greater than 3 cm, segmented or not; bulbs erect, roughly ovoid . . . . *L. columbianum*
      4' Habitat moist in summer; perianth yellow to reddish, segments 3.4–9.2 cm; longest bulb scales generally less than 3 cm, generally segmented; bulbs horizontal, rhizome-like.
         5. Perianth red to maroon (rarely orange), center green to yellow, outer surface green on basal 40–50%, segments 4.3–7.6(8.1) cm; filaments parallel or nearly so; rhizome unbranched; coastal . . *L. occidentale*
         5' Perianth yellow to orange, outer surface green only on basal 20%, segments 3.4–9.2 cm; filaments spreading; rhizome often branched; mostly inland . . . . *L. pardalinum*

Lilium bolanderi S. Watson (p. 305)

Bolander's lily

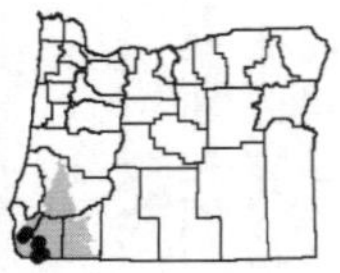

Plants to 1.1 m, glaucous; bulbs erect, roughly ovoid; longest scales 3–6 cm, unsegmented. **Leaves** in 2–6 whorls, usually ascending and often cupping stem, generally obovate or oblanceolate, 1.8–7 cm, generally distinctly glaucous; margin generally wavy. **Inflorescences** flowers 1–9, nodding to horizontal. **Flowers** narrowly bell-shaped, not fragrant; perianth segments 3–4.7 cm, red or magenta (rarely salmon or pale yellow), inner basal 30–50% portion usually yellowish, distal 20–40% portion somewhat recurved; stamens less than or equal to perianth, filaments parallel or nearly so, anthers 3–8 mm, reddish or magenta, pollen red-brown, orange, or yellow; pistil 2–3.5 cm. **Fruits** 2–4 cm. **2*n***=24.

Serpentine soil in chaparral, conifer forests. Flowering Jun–Aug. 400–1400 m. Sisk. CA. Native.

Lilium bolanderi can be fairly common in its preferred scrubby serpentine savanna habitat. Hybrids with *L. pardalinum* and *L. washingtonianum* ssp. *purpurascens* are possible. The range of corolla color variation from decidedly yellowish through red to maroon is extreme for the genus in North America—only *L. canadense* is similarly variable—but most are pale red to magenta. In Oregon, Bolander's lily is primarily pollinated by rufous hummingbirds.

Lilium columbianum Leichtlin (p. 305)

Columbia lily, Oregon lily, tiger lily

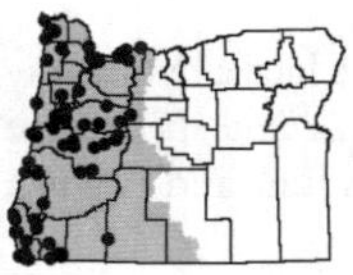

Plants to 1.7 m, glaucous or not; bulbs erect to ascending, oblique or roughly ovoid; longest scales 3–7.3 cm, (1)2–3(5)-segmented. **Leaves** in 1–9 whorls, generally ascending, weakly oblanceolate to obovate, 1.5–16 cm; margin generally wavy. **Inflorescences** flowers 1–25(45), pendent to nodding. **Flowers** widely bell-shaped, not fragrant; perianth segments 3.4–6.1(7.1) cm, orange (sometimes reddish outside), distal 50–60% portion reflexed; stamens longer than perianth, filaments spreading, anthers 5–13 mm, yellowish to very pale magenta, pollen orange or yellow; pistil 2.4–3.7 cm. **Fruits** 2–5.4 cm. **2*n***=24.

Dry scrub, coastal prairie, gaps and roadsides in conifer forest especially along coast. Flowering May–Aug. 0–2000 m. Casc, CR, ECas, Est, Sisk, WV. CA, ID, WA; north to British Columbia. Native.

This widespread lily is rather variable. Oregon plants tend to have exserted stamens and the "Turk's-cap" flower form associated with large butterfly pollination. Patches of flowering plants can be thick with western tiger swallowtails and pale swallowtails. *Lilium columbianum* hybridizes with *L. pardalinum* ssp. *wigginsii* and *vollmeri*, and extensively with *L. occidentale* in southern coastal Oregon. *Lilium columbianum* may intergrade with *L. kelloggii* along Highway 199 at the border between California and Oregon; these plants are slightly fragrant and the bulb scales unsegmented, all characteristics of Kellogg's lily.

Native Americans used *Lilium columbianum* bulbs as a food or peppery condiment. For many, it was a staple food (Pojar and MacKinnon 1994).

Lilium kelloggii Purdy (p. 307)

Kellogg's lily

Plants to 2.1 m; bulbs erect, more or less ovoid; longest scales 3–6.5 cm, unsegmented. **Leaves** in 2–7 whorls, generally narrowly elliptic, 6–16.5 cm; margin generally not wavy. **Inflorescences** flowers 1–27, pendent. **Flowers** widely bell-shaped, fragrant; perianth segments (3.2)3.7–6.1(7.2) cm, pink or white (generally aging pink), midrib yellow, distal 67% portion reflexed; stamens longer than perianth, filaments spreading, anthers 5–14 mm, pale red-orange or magenta, pollen orange; pistil 2.9–4.2 cm. **Fruits** 3–6 cm. **2*n***=24.

Gaps and roadsides in conifer forests or chaparral. Flowering Jun–Aug. 300–500 m. Sisk. CA. Native.

Kellogg's lily occurs from Humboldt County, California to very near or just over the Oregon border in Josephine County, near Oregon Mountain. Like *Lilium bolanderi* and *L. pardalinum* spp. *vollmeri* and ssp. *wigginsii*, it is endemic to the Klamath Mountains. The southernmost populations near Eureka, CA occur on soils derived from sedimentary rock, but others thrive on serpentine soils.

The current existence of *Lilium kelloggii* in Oregon needs documentation. A sheet from "southern Josephine County" (OSC 49769) is the only hard evidence for a presence in Oregon. Other specimens at the Oregon State University Herbarium, assumed to be *L. kelloggii,* are either planted or are other taxa, e.g., *L. columbianum.* Confirmed specimens (CAS513171, CAS376602) at Hazelview Summit on Highway 199 in Del Norte County, CA, are less than two miles from the border with Josephine County, and other specimens in nearby areas are also within a few miles of Oregon. Beane and Vollmer (1955) reported *L. kelloggii* from one mile south of the Oregon border along the Patrick Creek road in Del North County, CA. All these collections or records for extreme northern Del Norte County are at least 35 years old.

With its white or pink "Turk's-cap" flowers and overpowering diurnal fragrance, *Lilium kelloggii* is as unmistakable as it is elusive. It is pollinated by pale and western tiger swallowtail butterflies.

Lilium occidentale Purdy (p. 307)

Western lily

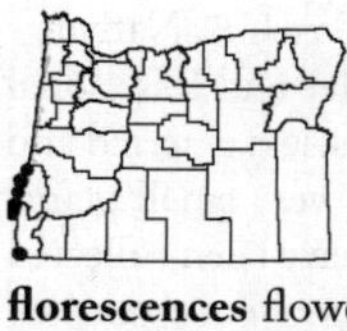

Plants to 2.2 m; bulbs horizontal, rhizome-like, unbranched; longest scales 0.9–2.5 cm, 1–2(3)-segmented. **Leaves** in 1–9 whorls (rarely scattered), generally elliptic, sometimes linear, 4–27 cm; margin not wavy. **Inflorescences** flowers 1–13(35), pendent. **Flowers** widely bell-shaped, not fragrant; perianth segments 4.3–7.6(8.1) cm, distal 60% portion reflexed, 2-toned, inner surface red to maroon (rarely orange) on distal 50–60% portion, base yellow to green, sometimes with an intervening band of orange or yellow, outer surface strongly green on basal 40–50% portion; stamens longer than perianth, filaments more or less parallel, anthers 5–14 mm, dull red or magenta, pollen generally red-brown (to orangey); pistil 3.0–5.5 cm. **Fruits** 2.1–5.4 cm. ***2n***=24.

Coastal scrub or prairie, sphagnum bogs and marshes, gaps in coniferous forests. Flowering Jun–Aug. 0–100 m. Est. CA. Native.

In Oregon, *Lilium occidentale* occurs in scattered disjunct colonies from Coos Bay south to Brookings within six kilometers of the ocean. Always rare, western lily is threatened by habitat loss, browsing and grazing, competition, collecting, and hybridization with *L. columbianum.*

Plants from marshes and bogs tend to have narrower leaves often scattered along the stem, especially in smaller plants. *Lilium occidentale* is distinguished from similar species such as *L. pardalinum* by the closely parallel stamens and inner green star formed by the conspicuously green sepal and petal bases, a characteristic also discernible in the outer surface of the buds. This green star is present in nearly all flowers when first opened but changes to yellow or yellow-orange as the flowers age. Frequent hybrids with *Lilium columbianum* in Oregon occur primarily where human disturbance has introduced that species into boggy areas via drier open pathways such as powerline rights-of-way and roads.

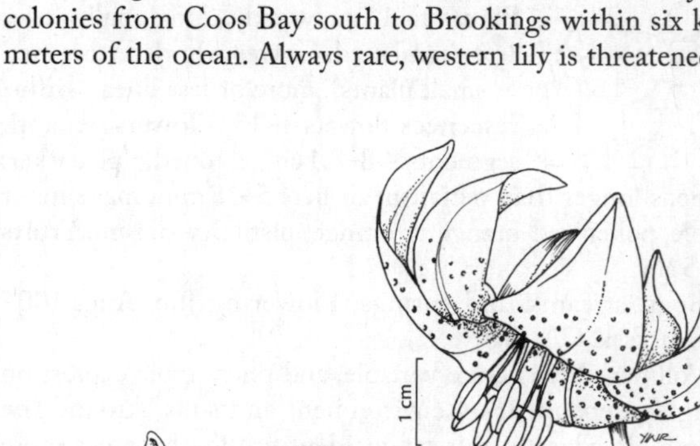

LILIUM

Lilium pardalinum Kellogg

Leopard lily

Plants to 2.8 m, somewhat to very clonal; bulbs horizontal, rhizome-like, often branched; scales (1)2–4-segmented, longest 1–3.3 cm. **Leaves** in 1–8 whorls or sometimes scattered, 4–27 cm, generally elliptic, less often linear; margin generally not wavy. **Inflorescences** flowers 1–28(35), pendent. **Flowers** more or less widely bell-shaped, generally not fragrant; perianth segments 3.4–6.1(7.1) cm, distal 67–75% potion reflexed, generally 2-toned, inner surface generally pale orange to red on distal 25–60% portion, lighter near base, maroon spots (if near tip) margined yellow or orange, outer surface paler and green on basal 20% portion; stamens longer than perianth, filaments spreading more or less widely, anthers 5–22 mm, magenta to orange or yellow, becoming darker, pollen red-brown to yellow, becoming lighter; pistil 3–7.6 cm. **Fruits** 2.3–6 cm. ***2n***=24.

California and southwestern Oregon. 5 subspecies; 4 subspecies in Oregon.

The subspecies of *Lilium pardalinum* display a classic pattern of discrete geographical ranges with intervening zones of introgression, and no two occur sympatrically without intermixing. Plants in the hybrid zones are confusing in appearance and cannot readily be assigned to subspecies; however, each is fairly well marked within its core distribution. *L. pardalinum* subspecies in Oregon can be fairly common plants in the proper habitats within their rather narrow distributions, but since all are restricted to wetlands of one type or another, distribution is never continuous across the landscape. In introgression zones, plants often show indicators of genetic instability that include flowers lacking stamens or with shrunken and deformed anthers, filaments, or even petals.

Leaf size and shape are quite variable in *Lilium pardalinum* subspecies and often clearly dependent on environment. In species or populations that typically have narrow, ascending leaves, for example *L. pardalinum* ssp. *vollmeri*, shaded plants often have wide, horizontal leaves. This hampers taxonomic separation as well as identification, especially of herbarium specimens. In addition to flower size, color, and habitat, anther size and the degree of stamen exsertion and spread are helpful characters for subspecies identification.

Lilium pardalinum is primarily pollinated by western tiger swallowtails and pale swallowtails; in Oregon, Anna's and rufous hummingbirds are also important visitors, especially when butterflies are rare. Native Americans reportedly ate *Lilium pardalinum* bulbs steamed or baked in an earth oven (Moerman 1986).

Key to *Lilium pardalinum* subspecies:

1. Perianth generally uniformly yellow or yellow-orange (green at base) . . . . ***L. p.*** ssp. ***wigginsii***

1' Perianth essentially 2-toned, with green, yellow, or orange centers, darker orange or red tips.

  2. On serpentine soils, leaves generally linear (length 7–34 times the width), ascending, often scattered and concentrated basally; pollen red-orange or orange; anthers magenta or purple. . . . . ***L. p.*** ssp. ***vollmeri***

  2' Not on serpentine soils, leaves generally elliptic but sometimes linear (length 3–17 times the width) and horizontal, whorled and evenly distributed along stem; pollen yellow to brown-orange; anthers orange to magenta.

    3. Pistil 4.5–7.6 cm; perianth segments 5.6–9.3 cm; anthers 9–22 mm, pale magenta or magenta; pollen red- to brown-orange; strongly clonal. . . . . ***L. p.*** ssp. ***pardalinum***

    3' Pistil 3.3–4.4 cm; perianth segments (3.7)4.0–7.6 cm; anthers 5–14 mm, orange to magenta; pollen yellow to bright orange (occasionally red-orange); weakly clonal . . . . ***L. p.*** ssp. ***shastense***

Lilium pardalinum Kellogg ssp. *pardalinum*

Leopard lily

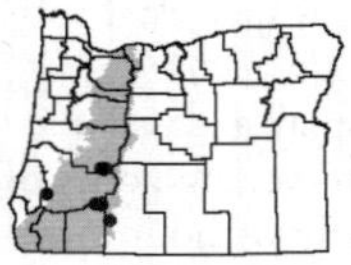

Plants to 2.8 m, strongly clonal; bulb scales (1)2(3)-segmented. **Leaves** whorled. **Inflorescences** flowers 1–28. **Flowers** rarely fragrant; perianth segments 5.6–9.3 cm, 2-toned, tips darker; stamens much longer than perianth, anthers 9–22 mm, pale magenta or magenta, becoming purple or rarely yellowish, pollen red- to brown-orange; pistil 4.5–7.6 cm. **Fruits** 3–6 cm. **2*n***=24.

Stream banks, seeps, springs, fens, and roadside ditches. Flowering Jul–Aug. 300–1600 m. Casc, Sisk. CA. Native.

Quite variable in flower size, color, and leaf size and shape, ssp. *pardalinum* is the easiest of the western lilies to cultivate, although the strongly clonal material of the best California stock is preferred for its vigor and strongly marked flowers. Oregon plants often have narrowly elliptic leaves and slightly smaller perianth segments and paler flowers than their California counterparts. Plants north of Pelican Bay in Klamath County are near the northern boundary of ssp. *shastense* and show the floral deformities characteristic of introgression zones, including missing stamens and shrunken, malformed petals.

There are some differences between the Oregon and California spp. *pardalinum* plants, and some might advocate description of a sixth subspecies to encompass the Cascadian material. Cascade plants are generally less clonal and less robust with small-spotted, pale orange flowers, broad petals, and pale reddish tepal tips, in contrast to California ssp. *pardalinum* with its tightly clumped stems, blazing reddish tepal tips and dramatic leopard spots. However, the appearance of plants that recapitulate the essence of California ssp. *pardalinum,* on non-serpentine basaltic soils, to the north of well-marked ssp. *vollmeri* complicates the picture; these plants, north of Agnes in Curry County, are robust, quite clonal, and have the broadly elliptic leaves and large flowers typical of spp. *pardalinum.* Similar populations appear various places in the Oregon Cascades, especially north of Prospect around Kiter and Mill Creeks. Therefore it seems that the current California circumscription of ssp. *pardalinum* can and should be slightly enlarged to embrace the Oregon plants here included in it.

Lilium pardalinum Kellogg ssp. *shastense* (Eastwood) M.W. Skinner

Shasta lily

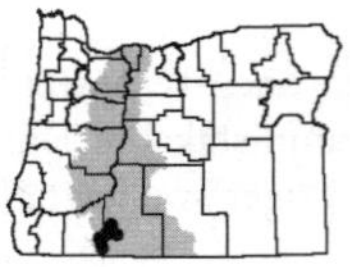

Plants to 2.1 m, weakly clonal; bulb scales (1)2–4-segmented. **Leaves** whorled (scattered in young plants). **Inflorescences** flowers 1–25(35). **Flowers** perianth segments (3.7)4.0–7.6 cm, 2-toned, tips darker; stamens longer than perianth, anthers 5–14 mm, orange to magenta, pollen yellow to bright orange, occasionally red-orange; pistil 3.3–4.4 cm. **Fruits** 2.2–4.3 cm.

Wet meadows, lakeshores, streamsides in coniferous forests. Flowering Jun–Aug. 1200–1800 m. Casc, ECas. CA. Native.

Subspecies *shastense* is in essence a smaller and less clonal version of ssp. *pardalinum* with smaller flowers, less exserted and spreading stamens, and brighter pollen. The very small plants from Lake of the Woods in Jackson County have been collected for decades. The northern limits of ssp. *shastense* are reached around Daley Creek in Jackson County and Lake of the Woods and Pelican Bay in Klamath County; some of these populations are variable and display the malformed flowers characteristic of *Lilium pardalinum* introgression zones. To the north, ssp. *pardalinum* eventually prevails around the town of Prospect and near Agency Lakes respectively, but in the intervening area, identification to subspecies may not be possible.

Lilium pardalinum Kellogg ssp. *vollmeri* (Eastwood) M.W. Skinner

Vollmer's lily

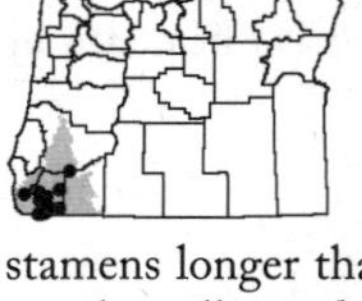

Plants to 1.7 m, weakly clonal; bulb scales 1–2-segmented. **Leaves** whorled (scattered in small plants), more or less linear. **Inflorescences** flowers 1–13. **Flowers** perianth segments 4.8–8.1 cm, 2-toned, tips darker; stamens longer than perianth, anthers 5–18 mm, magenta or purple, pollen red-orange or orange; pistil 3.5–5.3 cm. **Fruits** 2.5–5 cm.

Bogs, streams, and springs. Flowering Jun–Aug. 100–1500 m. Sisk. CA. Native.

Vollmer's lily is quite variable, and phenotypic expression is highly dependent on environment and soils. Around the geographic edges of this range, identification becomes more difficult, and ssp. *vollmeri* intergrades with other subspecies of *pardalinum* along all its boundaries to produce morphologically jumbled intermediates. Even within the core range of the subspecies, plants in deep shade generally have scattered, wider leaves that much resemble ssp. *pardalinum*.

Like most *L. pardalinum*, Vollmer's lily is pollinated in Oregon by pale and western tiger swallowtails and also visited by rufous hummingbirds.

Lilium pardalinum Kellogg ssp. ***wigginsii*** (p. 307)
(Beane & Vollmer) M.W. Skinner

Wiggins' lily

Lilium wigginsii Beane & Vollmer

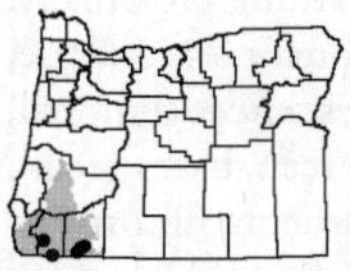

Plants to 1.7 m, weakly clonal; bulb scales 2–4-segmented. **Leaves** whorled (scattered in small plants). **Inflorescences** flowers 1–15. **Flowers** perianth segments (3.4)4.4–6.8(7.1) cm, not 2-toned, generally uniformly orange or yellow-orange; stamens often malformed or shrunken, stamens longer than perianth, anthers 5–13 mm, pale yellow, pollen yellow or orange; pistil 3–4.3 cm. **Fruits** 2.3–4.2 cm.

Wet thickets, meadows, streams among conifers. Flowering Jul–Aug. 1000–2000 m. Sisk. CA. Native.

Subspecies *wigginsii* is a Klamath Mountains endemic that occurs widely along the county boundary between Del Norte and Siskiyou counties, California, and east through southeastern Josephine County, Oregon, to Mount Ashland in Jackson County (Ballantyne 1983). It intergrades with ssp. *vollmeri* near Grayback Mt. in Josephine County, and reaches its purest, most consistent expression high on Mt. Ashland in Jackson County. The solid orange flowers, broad perianth segments, yellow anthers, and wet-ground habitat are diagnostic. Genetic instability in this subspecies is frequently expressed as malformed flowers with shrunken or missing reproductive structures.

Lilium washingtonianum Kellogg ssp. ***purpurascens***
(Stearn) M.W. Skinner

Cascade lily, Washington lily

Lilium washingtonianum Kellogg var. *purpurascens* (Stearn) M.W. Skinner

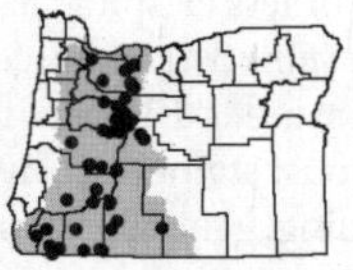

Plants to 2.6 m, often glaucous; bulbs ascending to erect most scales indistinctly 2(3)-segmented, longest 3.3–12 cm. **Leaves** in 1–9(14) whorls, horizontal to nearly clasping stem, generally oblanceolate, 3–13 cm; margin wavy or not. **Inflorescences** flowers 1–15(33), nodding to ascending. **Flowers** slightly bilateral, roughly trumpet-shaped, strongly fragrant; perianth segments (5.6)6.2–9.5(10.5) cm, white, becoming deep pink or lavender, magenta spots minute, inner segments wider and strongly oblanceolate, distal 33% portion recurved, outer surface generally purplish (often faintly so); stamens slightly longer in length than perianth, filaments roughly parallel, anthers 8–15 mm, cream, becoming yellow, pollen pale (bright) yellow; pistil 7.5–10.4 cm. **Fruits** 2.7–5.8 cm, generally ribbed. **2*n***=24.

Douglas-fir forests, roadsides, burned clearcuts. Flowering Jun–Aug. 50–2000 m. Casc, ECas, Sisk, WV. CA. Native.

The Cascade lily is pollinated by sphinx and doubtless other moths in the family Sphingidae. Nectar secretion and scent production are timed to their arrival at dusk. The open nature of the flower and copious nectar secretion attract many diurnal visitors including a variety of butterflies, bees, and flies, but these insects are not significant pollinators.

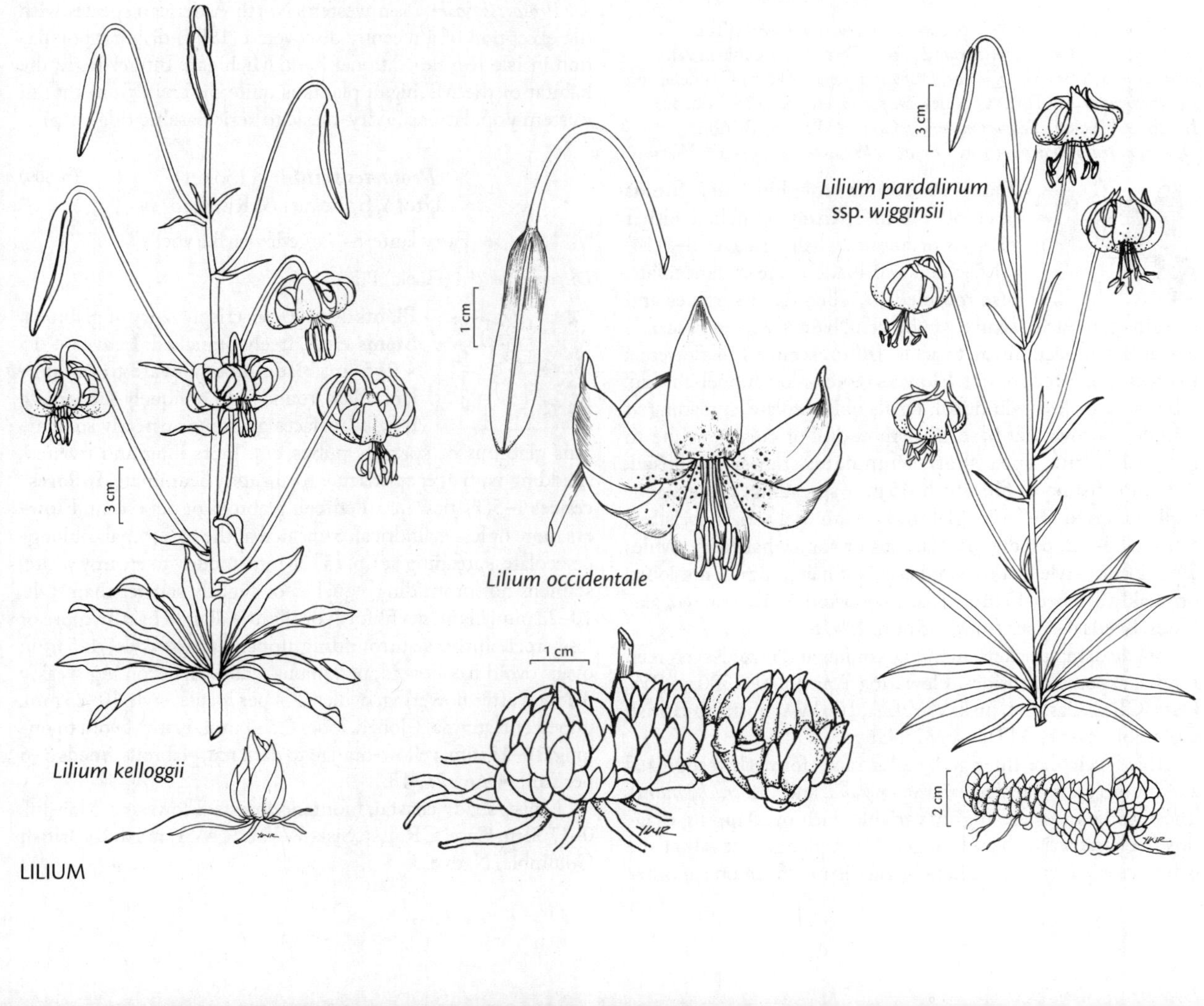

LILIUM

PROSARTES Fairybells

Tamra Dickson Prior

Herbs generally pubescent throughout; roots fibrous, rhizomes slender, knotty. **Stems** erect; hairs absent or sharp-branched, generally not glandular, some or all shedding in age; branched distally, scaly below, leafy above; 2–5 bracts, papery, sheathing proximally. **Leaves** alternate, sessile or subsessile to more or less clasping; leaf blades ovate to elliptic or oblanceolate, base more or less oblique, cordate to acute, tip acute to acuminate, main veins 3 or more, converging, minor veins forming loose net. **Inflorescences** terminal, umbel-like; flowers 1–7, pendulous; pedicels slender. **Flowers** tepals 6, in 2 whorls, free, white to more or less green, bases green, generally weakly gibbous proximally, deciduous; stamens 6, hypogynous, free but adnate to tepals at base, filaments filiform to basally dilated, anthers linear-oblong, protruding, generally shorter than filaments, generally glabrous; ovary superior, (1)3-locular, narrowly ellipsoid to obovoid, sessile, ovules 2–6 per locule, pendulous or horizontal; style slender, protruding or not, stigma either entire or weakly 3-lobed. **Fruits** berries, straw-colored to orange or red, more or less fleshy. **Seeds** ellipsoid to oblong, white or pale yellow to orange-brown, smooth.

North America. 5 species; 3 species treated in *Flora.*

In the past, *Prosartes* and *Disporum* have been sometimes linked together as one genus. *Prosartes* species, however, are strictly North American, and *Disporum* species are of Asian origin.

Key to *Prosartes* species:

1. Ovary broadly ovoid to obovoid or inverted cone to pear shape, becoming 3-lobed with age; leaf margins moderately hairy; ovules horizontal; 6 or more seeds per fruit . . . . *P. trachycarpa*

1' Ovary ellipsoid or ovoid to obovoid; leaf margins glabrous or short-haired; ovules pendulous; less than six seeds per fruit.

  2. Stamens not protruding; stems freely branched; leaf margins glabrous or hairs sparse and long; tepals spreading at tips; berries 3–6-seeded . . . . *P. smithii*

  2' Stamens exposed; stems sparingly branched; leaf margins with many short hairs; tepals spreading from middle; berries 4–6-seeded . . . . *P. hookeri*

Prosartes hookeri Torrey (p. 309)

Hooker's fairybells, Oregon fairybells, drops-of-gold

Disporum hookeri (Torr.) G. Nicholson, *Disporum hookeri* (Torr.) G. Nicholson var. *hookeri*, *Disporum hookeri* (Torr.) G. Nicholson var. *oreganum* (S. Watson) Q. Jones, *Disporum hookeri* (Torr.) G. Nicholson var. *trachyandrum* (Torr.) Q. Jones, *Disporum oreganum* (S. Watson) Benth. & Hook. f., *Disporum parvifolium* (S. Watson) Britton, *Disporum trachyandrum* (Torr.) Britton, *Prosartes parvifolia* S. Watson

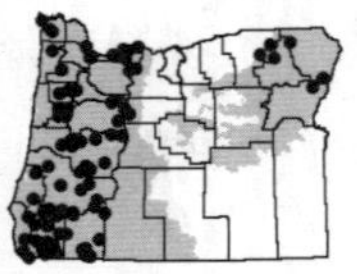

Plants 30–100 cm, roughly hairy. **Stems** erect, 30–80 cm, sparsely branched, either glabrous or minutely hairy. **Leaves** 3–9(14) × 1.5–5 cm; leaf blade ovate to lanceolate, base rounded to subcordate; surfaces and margins glabrate to sparsely pubescent, hairs sharp and ascending; apex long-acuminate to acute. **Inflorescence** 1–4-flowered. **Pedicels** pubescent or not. **Flowers** more or less funnel-shaped, tube more or less cylindrical; tepals oblanceolate, spreading at middle, narrowed at base, 8–18 mm, dull or creamy white to greenish white, apex short-acuminate; stamens exposed, 8–15 mm, filaments filiform, 6–15 mm, spreading, anthers generally exserted, 3–5 mm, glabrous or minutely sharp-bristled; ovary ellipsoid, pendulous, glabrous or somewhat hairy, ovules 2 per locule; style glabrous or basal half hairy, stigma not lobed or weakly 3-lobed. **Fruits** ovoid to obovoid, 7–12 mm, red, glabrous. **Seeds** 4–6 per fruit, 5–6 mm. $2n$=18.

Shady damp thickets, montane conifer and mixed-evergreen forests, exposed roadsides. Flowering Apr–Jul. 0–2300 m. BW, Casc, CR, ECas, Est, Sisk, WV. CA, ID, WA; north to British Columbia, east to MT and MI. Native.

This species has three geographic races, formerly recognized as varieties (*P. h.* var. *hookeri*, var. *oregana*, and var. *trachyandra*). These three races are highly variable with overlapping diagnostic characteristics and ranges. Only plants that fall at the extremes can be recognized as being distinct from one another.

Glandular hairy forms of *P. hookeri* are referred to as distinct species (*P. parvifolia*) by some authorities.

Prosartes hookeri is a western North American species with the exception of a recently discovered (1997) disjunct population in Isle Royale National Park, Michigan. Interestingly, the habitat of the Michigan plants is quite different from that of western populations (a dry-mesic to xeric basaltic ridge top).

Prosartes smithii (Hooker) Utech, Shinwari & Kawano (p. 309)

Fairy lanterns, largeflower fairybells

Disporum smithii (Hook.) Piper

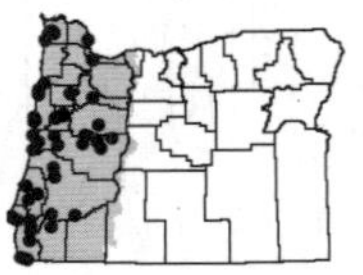

Plants 30–100 cm, crisply hairy or glabrous. **Stems** erect, freely branched. **Leaves** 4–15 × 2–5 cm; blade mostly ovate to oblong-lanceolate, rounded to obliquely subcordate basally, surfaces glabrous or nearly so, margins glabrous or sparsely pubescent, hairs long and twisted, shedding with age; apex acute to abruptly acuminate. **Inflorescences** 1–5(7)-flowered. **Pedicels** glabrous or pubescent. **Flowers** more or less cylindrical, truncated at the base; tepals oblong-lanceolate, spreading at tip, 15–28 mm, muddy to creamy white; stamens not protruding, equal to or slightly shorter than style, 10–23 mm, filaments filiform or slightly dilated basally, more or less erect, anthers surrounding upper part of style, 4–5 mm; ovary ovoid to obovoid, pendulous glabrous, becoming weakly 3-lobed after flowering, ovules 2–4 per locule; style 10–23 mm, pubescent, stigma 3-lobed, lobes 0.5–1 mm. **Fruits** ovoid to oblong, 10–15 mm, yellow-orange to dark red, glabrous. **Seeds** 3–6 per fruit, 4 mm. $2n$=16.

Moist, shady coastal, montane forests. Flowering Mar–Jul. 0–1700 m. Casc, CR, Est, Sisk, WV. CA, WA; north to British Columbia. Native.

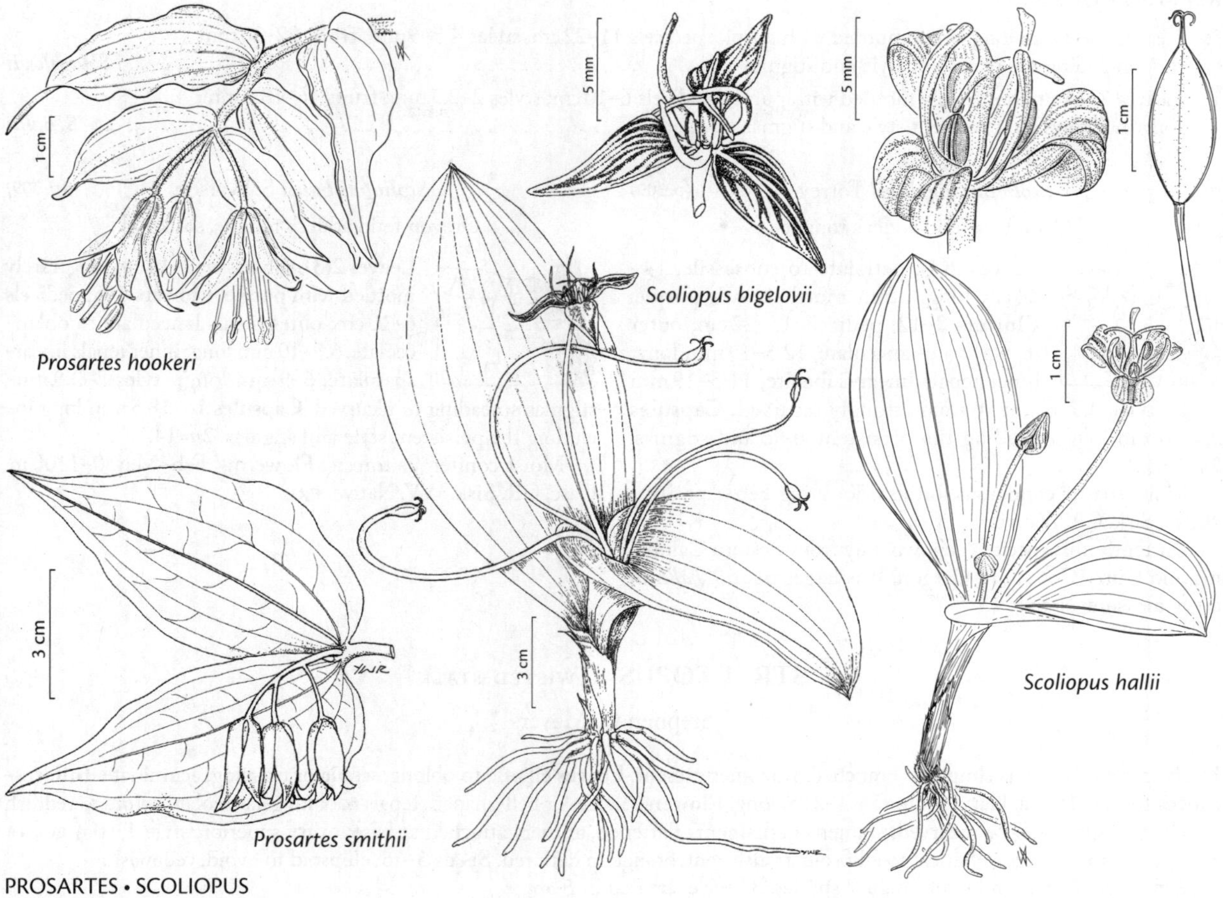

PROSARTES • SCOLIOPUS

Prosartes trachycarpa S. Watson

Roughfruit fairybells, wartberry, rough-fruited mandarin

Disporum trachycarpum (S. Watson) Benth. & Hook. f.

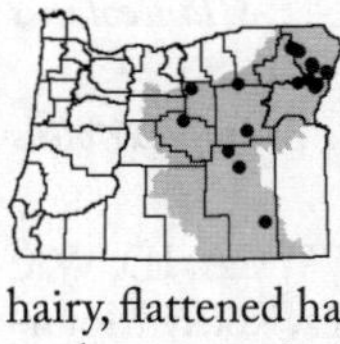

Plants 30–80 cm, glabrous to minutely hairy. **Stems** erect, few-branched, pubescent. **Leaves** 4–12 × 2–5 cm; blade ovate to oblong-lanceolate, base subcordate to oblique, abaxial surface and margins moderately hairy, flattened hairs scattered, 7–9 prominent veins, apex acute to short-acuminate. **Inflorescences** 1–2(3)-flowered. **Flowers** narrowly bell-shaped; tepals narrowly oblanceolate, 8–15 mm, creamy to greenish white; stamens generally protruding, filaments filiform, 10–15 mm, anthers 3–4(5) mm; ovary broadly ovoid to obovoid, becoming 3-lobed and obpyriform after flowering, finely papillose, ovules 2–6 per locule, horizontal; style 0.9–1.2 cm, glabrous or pubescent, stigma unlobed or 3-lobed. **Fruits** depressed-globose, 12–18 mm, red-orange to bright red, surface strongly papillose. **Seeds** 6–12(18) per fruit, 4–5 mm. **2*n***=22.

Shady deciduous groves to open coniferous forests. Flowering Apr–Jun. 700–1700 m. BR, BW, Lava. NV, ID, WA; north to British Columbia, east to Ontario, south to NM. Native.

SCOLIOPUS Fetid adder's tongue

Katie E. Mitchell

Herbs glabrous; short rhizomatous. **Stems** subterranean, short. **Leaves** 2(4), basal, green, sometimes mottled with purple, oblong to elliptic, sheathing at the base; petiolate to subsessile; parallel-veined. **Inflorescences** umbellate, nearly sessile; peduncles short, (mostly) subterranean. **Pedicels** sharply 3-angled, twisted, appearing scapose, recurved in fruit. **Flowers** pale yellow-green to green, with prominent purplish brown lines, malodorous; outer tepals 3, spreading or recurving, ovate to lanceolate to oblanceolate; inner tepals 3, erect, linear; stamens 3, anthers attached near the middle; ovary superior, 3-angled; stigmas 3, persistent. **Fruits** capsules, elliptic, 3-angled, brownish purple, with persistent style and stigmas forming a beak. **Seeds** oblong, elaiosome present.

Western California and Oregon; 2 species.

Species of this genus are adapted to seed dispersal by ants. The seeds have elaiosomes, and in fruit the pedicels recurve, allowing the capsules to touch the ground and eventually dehisce in the duff.

Key to *Scoliopus* species:

1. Leaves 14–24 cm long, often mottled with purple; pedicels 11–22 cm; styles 4.5–6 mm; fruits 22–30 mm long including the persistent style and stigmas . . . . *S. bigelovii*

1' Leaves 8–14 cm long, rarely mottled with purple; pedicels 6–10 cm; styles 2–2.5 mm; fruits 16–18.5 mm long including the persistent style and stigmas . . . . *S. hallii*

Scoliopus bigelovii Torrey (p. 309)

California fetid adder's tongue

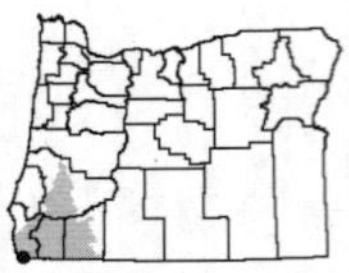

Leaves 2(4), petiolate to subsessile, 14–24 cm long, often mottled with purple. **Flowers** 3–12; pedicels 11–22 cm; outer tepals ovate-lanceolate, 12.5–19 mm long; inner tepals linear-subulate, 11.5–19 mm long; styles 4.5–6 mm stigmas strongly recurved. **Capsules** 22–30 mm long including the persistent style and stigmas. **2*n***=16.

Moist coastal coniferous forests. Flowering Feb–Apr. 300–400 m. Sisk. CA. Native.

Currently, this species is known only from western California and Curry County in Oregon. It is larger than *S. hallii* in every measure. Native.

Scoliopus hallii S. Watson (p. 309)

Oregon fetid adder's tongue, slinkpod

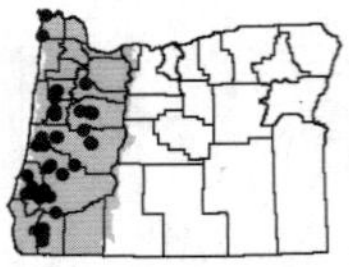

Leaves 2(3), subsessile, 8–14 cm long, rarely mottled with purple. **Flowers** 1–5; pedicels 6–10 cm; outer tepals lanceolate to oblanceolate, 6.5–10 mm long; inner tepals linear-spatulate, 6–9 mm long; styles 2–2.5 mm, stigmas spreading to recurved. **Capsules** 16–18.5 mm long including the persistent style and stigmas. **2*n***=14.

Moist coniferous forests. Flowering Feb–May. 0–1100 m. Casc, CR, Sisk, WV. Native.

STREPTOPUS Twisted-stalk

Stephen C. Meyers

Herbs perennial; stems simple or branched; rhizomes present. **Leaves** elliptic to oblong, sessile or clasping, acuminate. **Inflorescences** flowers 1(2), axillary; peduncles 1–2 cm long. **Flowers** saucer- or bell-shaped; tepals 6, oblong-lanceolate, white to reddish purple, tips spreading or recurved; stamens 6, filaments flattened, anthers attached at base; ovary superior; style 1, stigma 1 or 3-lobed. **Fruits** berries, spherical to ellipsoid, translucent, orange to dark red. **Seeds** 3–15, ellipsoid to ovoid, yellowish.

Temperate Northern Hemisphere. 7 species; 3 species treated in *Flora*.

Key to *Streptopus* species:

1. Flowers saucer-shaped; plants 10–20 cm tall; tepals 3–5.5 mm long; style approximately 0.5 mm long . . . . *S. streptopoides* var. *brevipes*

1' Flowers bell-shaped; plants 15–120 cm tall; tepals 6–15 mm long; style 3–5 mm long.
   2. Plants 15–40(70) cm tall; pedicels not or rarely twisted; tepals with pink to reddish spots or streaks, spreading but not recurved at tips; leaves not or rarely clasping . . . . *S. lanceolatus*
   2' Plants 50–120 cm tall; pedicels twisted; tepals not spotted or streaked, spreading and recurved at tips; leaves clasping . . . . *S. amplexifolius*

Streptopus amplexifolius (Linnaeus) de Candolle (p. 311)

Claspleaf twisted-stalk, clasping twisted-stalk

Streptopus amplexifolius (L.) DC. var. *americanus* Schult., *Streptopus amplexifolius* (L.) DC. var. *chalazatus* Fassett

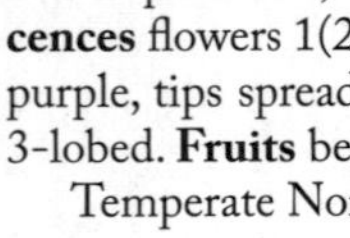

Plants 50–120 cm tall; stems branched, glabrous or pubescent basally. **Leaves** 5–12(15) × 2–5 cm, clasping, margins without hairs. **Flowers** bell-shaped; tepals narrowly oblong-lanceolate, 9–15 mm, white to greenish yellow, inner tepals clasping stamens, tips spreading to somewhat recurved; pedicels twisted; stamens unequal, inner series 2–3 mm, outer series approximately 1 mm, anthers 3–4 mm; style 4–5 mm, stigma unlobed or minutely 3-lobed. **Fruits** ellipsoid, 5–9 mm, whitish green when young, maturing to orange or red. **Seeds** approximately 3 mm long. **2*n***=16, 32.

Moist forests, shady stream banks. Flowering May–Jul. 0–2600 m. BW, Casc, CR, ECas, Est, Sisk, WV. CA, ID, WA; northwestern and northeastern North America, Rocky Mountains; Eurasia. Native.

Hybrids between *S. amplexifolius* and *S. lanceolatus* have been encountered where their ranges overlap in eastern North America. Thus far, hybrids have not been documented in western North America.

Streptopus lanceolatus (Aiton) Reveal

Rosy twisted-stalk, simplestem twisted-stalk

Streptopus curvipes Vail, *Streptopus lanceolatus* (Aiton) Reveal var. *curvipes* (Vail) Reveal, *Streptopus roseus* Michx. var. *curvipes* (Vail) Fassett

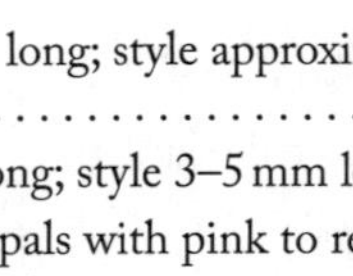

Plants 15–40(70) cm tall; stems simple or occasionally branched, nodes sparsely pubescent. **Leaves** 4–9(10) × 2–5 cm, generally sessile, rarely clasping, margins with ciliate hairs. **Flowers** bell-shaped; tepals narrowly

oblong-lanceolate, 6–10 mm, white to greenish yellow to nearly pink, with pinkish to reddish spots or streaks, tips slightly spreading, not recurved; pedicels not or rarely twisted; stamens more or less equal, 1–2 mm, anthers approximately 2 mm; style 3–4 mm, stigma 3-lobed. **Fruits** somewhat spherical and slightly angled, 5–10 mm, reddish purple to red. **Seeds** approximately 3 mm long. ***2n***=16, 48.

Moist forests. Flowering May–Jul. 900–2100 m. Casc, CR. WA; northwestern and northeastern North America. Native.

See also *S. amplexifolius.*

Streptopus streptopoides (Ledebour) Frye & Rigg var. ***brevipes*** (p. 311)

Small twisted-stalk

Kruhsea streptopoides (Ledeb.) Kearney

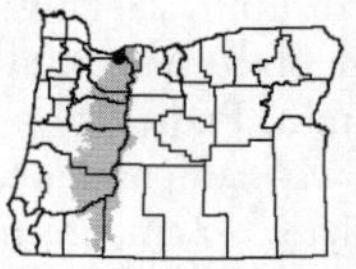

Plants 10–20 cm tall; stems simple, nodes ciliate. **Leaves** 3–5 × 1–2 cm, sessile, margins without hairs. **Flowers** saucer-shaped; tepals oblong-lanceolate, 3–5.5 mm, light purple to reddish purple or greenish with purple tinges, tips yellow to yellowish green, recurved; pedicels not twisted; stamens more or less equal, 1–2 mm, anthers approximately 1 mm; style approximately 0.5 mm, stigma unlobed. **Fruits** spherical, 5–6 mm, orangish red. **Seeds** 2–3 mm long. ***2n***=16, 54.

Moist forests. Flowering Jun–Aug. 600–1200 m. Casc. WA; north to AK, northeast to Alberta. Native.

Oregon plants represent the southernmost (and slightly disjunct) extent of this taxon.

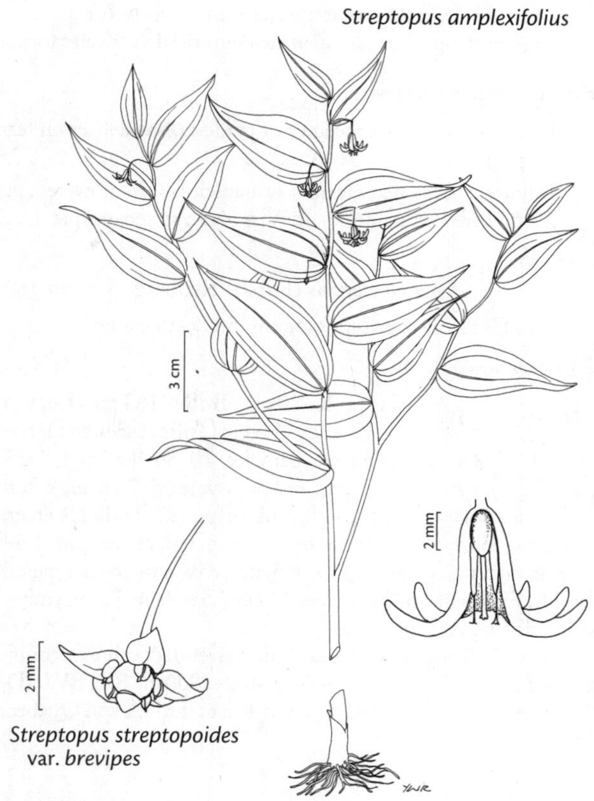

STREPTOPUS

MELANTHIACEAE False-Hellebore family

Tamra Dickson Prior

Herbs perennial, from bulbs or rhizomes. **Leaves** alternate, whorled, or spirally arranged, basal or cauline, simple; persistent after 1 year or not. **Scape** present or not. **Inflorescences** racemes, panicles or solitary flowers. **Flowers** perianth parts 6, either 6 tepals in 2 whorls or distinct sepals and petals, free or fused basally, spreading; stamens 6, arising from perianth, anthers attached basally or near middle; ovary superior or partly inferior, 3-chambered; styles 3, persistent. **Fruits** capsules, loculicidal or septicidal dehiscence.

Northern hemisphere. 10 genera; 5 genera treated in *Flora.*

Key to Melanthiaceae genera:

1. Leaves linear.
   2. Cauline leaves many, tough, persistent after 1 year . . . . **XEROPHYLLUM**
   2' Cauline leaves 0–2, not tough, deciduous.
      3. Tepals narrow, not clawed, bearing 1 bilobed gland . . . . **ANTICLEA**
      3' Tepals clawed, bearing 1 rounded gland . . . . **TOXICOSCORDION**

1' Leaves lanceolate to widely ovate.
   4. Leaves 3, in a whorl; plant less than 0.7 m tall . . . . **TRILLIUM**
   4' Leaves more than 3, alternate; plant 1–2 m tall . . . . **VERATRUM**

ANTICLEA Death camas

Bulbs narrowly ovoid. **Leaves** linear to oblanceolate, not tough, deciduous; cauline leaves 1–2. **Inflorescences** racemes or panicles. **Flowers** bell-shaped; tepals cream, yellow, green, purple, red, brown or a combination thereof, narrow but not clawed; gland 1, bilobed; ovary partly inferior.

North America. 11 species; 2 species treated in *Flora*.

This genus and *Toxicoscordion* are comprised of species formally assigned to the polyphyletic genera *Stenanthium* and *Zigadenus*.

Key to *Anticlea* species:

1. Inflorescences loose racemes or panicles, but denser on terminal end, flowers 10–50, upright; flowers cream to green . . . . *A. elegans*

1' Inflorescences open racemes or panicles, 3-25-flowered, pendent (up to 6-flowered on racemes and up to 25-flowered on panicles); flowers green to red-purple . . . . *A. occidentalis*

Anticlea elegans (Pursh) Rydberg (p. 312)

Glaucous zigadenus, mountain death camas

Zigadenus elegans Pursh

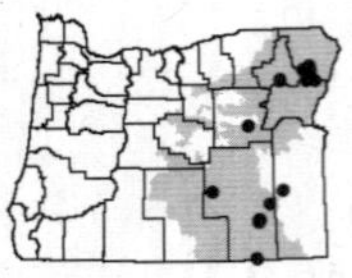

Plants 20–80 cm. **Bulbs** 1–3 cm. **Leaves** 10–30 × 0.3–1.5 cm. **Inflorescences** loose racemes or panicles, 10–50-flowered, 9–25 × 3–6 cm; bracts, ovate, 5–20 mm, often edged with pink or purple. **Pedicels** erect, 1–2.3 cm. **Flowers** rotate to rotate-campanulate, upright, 1.5–2 cm long; tepals ovate, 7–12 × 4–5 mm, cream to green, tapered basally. **Fruits** narrowly conical, 1–2 cm. **Seeds** 4–5 mm, straw-colored. **2*n***=32.

Moist and boggy areas, grasslands, rivers, lake shores, coniferous forests. Flowering Jul–Aug. 400–2900 m. BR, BW. ID, NV, WA; north to AK, southeast to TX, east to IA and Quebec. Native.

Anticlea occidentalis (A. Gray) Zomlefer & Judd

Western stenanthium, bronze bells

Stenanthium occidentale A. Gray

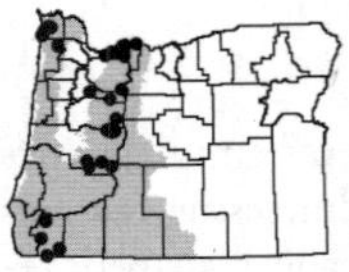

Plants 10–50(60) cm. **Bulbs** 2–4 cm. **Leaves** 10–30 × 0.3–2.5 cm. **Inflorescences** open 3–6-flowered (racemes) or up to 25-flowered (panicles), 10–20 cm; bracts lance-linear to lanceolate, 6–10 mm. **Pedicels** ascending to erect, 0.5–3 cm. **Flowers** narrowly campanulate, pendent, 1–2 cm long; tepals oblong-lanceolate, 1–2 cm, green to reddish purple, recurved distally. **Fruits** lanceolate, 1.5–2 cm. **Seeds** 3–4 mm, brown to black. **2*n***=16.

Moist mountain meadows, wet cliffs, rocky crevices, and mossy scree. Flowering May–Jul. 50–1800 m. Casc, CR, ECas, Sisk. CA, ID, WA; north to British Columbia, northeast to MT. Native.

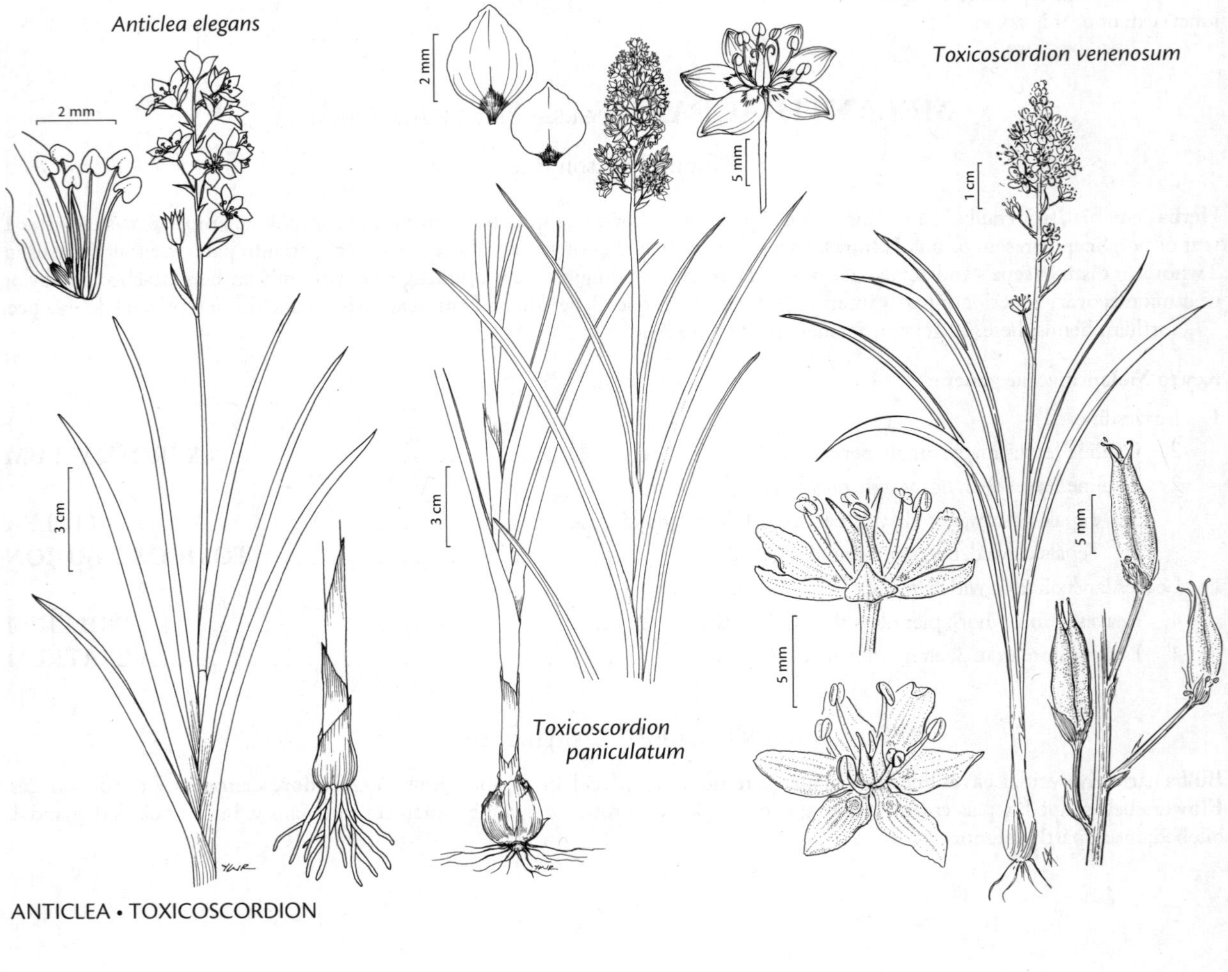

ANTICLEA • TOXICOSCORDION

TOXICOSCORDION Death camas

Stems scapose. **Leaves** deciduous, basal, linear, curved and folded, margins entire. **Inflorescences** racemes or panicles; bracts 1(2). **Flowers** mostly bisexual, occasionally staminate or sterile; tepals 6, white to cream-colored, clawed, free or fused to ovary base; (0)1(2) rounded yellow to green glands near base; stamens 6, free or attached to tepals; ovary superior; styles 3, stigmas 3. **Fruits** oblong or cylindrical, with septicidal dehiscence. **Seeds** to 8 mm, brown, 2–20 per capsule.
Western and central North America. 8 species; 5 species in Oregon.

Key to *Toxicoscordion* species:

1. Stamens equal in length to perianth; tepals 2–8 mm.
 2. Proximal inflorescence branches 60–90° from main axis when in fruit . . . . ***T. micranthum***
 2' Proximal inflorescence branches 10–60° from main axis when in fruit.
  3. Inflorescences panicles; outer tepals usually not clawed . . . . ***T. paniculatum***
  3' Inflorescences racemes; outer tepals usually clawed . . . . ***T. venenosum***

1' Stamens usually longer than perianth; tepals 4–15 mm.
 4. Bulbs clumped; outer tepals reflexed below mature fruit . . . . ***T. exaltatum***
 4' Bulbs not clumped; outer tepals not reflexed below mature fruit . . . . ***T. fremontii***

Toxicoscordion exaltatum (Eastwood) A. Heller
Giant zigadenus

Zigadenus exaltatus Eastw.

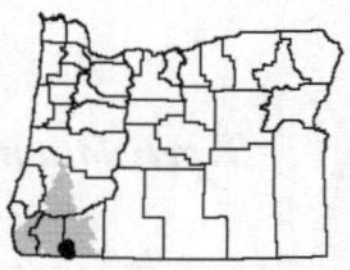

Plants 30–100 cm tall. **Bulbs** clumped, ovoid, 2–6 × 2–5 cm, outer layer dark brown. **Stems** glabrous. **Leaves** 20–60 cm × 5–30 mm. **Inflorescences** panicles, open, 10–80-flowered, 15–40 cm; bracts 3–25 mm long, green to white-membranous; pedicels 10–40 mm. **Flowers** mostly bisexual, those on lower branches staminate or with stunted ovary, 10-20 mm in diameter; tepals ovate to elliptic or obtuse, 5–12 × 2–5 mm, outer set not clawed, often persistently reflexed when fruiting, inner set clawed, 1–2 mm; gland yellow to greenish yellow. **Fruits** oblong, (9)10–20(25) × 4–9 mm. **2*n***=22.

Meadows, wooded slopes. Flowering Jun. 900–1400 m. Sisk. CA, NV. Native.

Toxicoscordion fremontii (Torrey) Rydberg
Fremont's star lily, Fremont's death camas

Zigadenus fremontii (Torr.) Torr. ex S. Watson

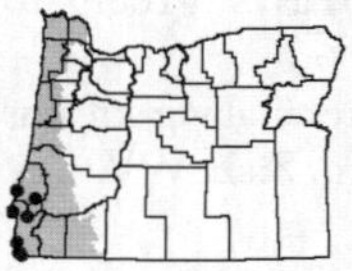

Plants (10)40–90 cm tall. **Bulbs** not clumped, ovoid to spherical, 1–4 × 0.5–4 cm, outer layer black. **Stems** glabrous. **Leaves** (5)20–50 cm × (2)5–30 mm. **Inflorescences** panicles or racemes, open, 2–80-flowered, 5–40 cm; proximal branches 33–50% length of inflorescence, spreading at 60–90° from main axis; bracts 5–50 mm long, green; pedicels 10–50 mm. **Flowers** bisexual, 15–25 mm in diameter; tepals widely ovate to obtuse, 5–15 × 2–6 mm, outer and inner tepals clawed, persistent in fruit; gland yellow. **Fruits** cylindrical, (8)10–35 × 4–11 mm. **2*n***=22.

Grassy or wooded slopes, outcrops. Flowering Apr–Jul. 0–500 m. CR, Est, Sisk. CA; south to northern Baja California. Native.

Toxicoscordion micranthum (Eastwood) A. Heller
Small-flowered zigadenus

Zigadenus micranthus Eastw., *Zigadenus micranthus* Eastw. var. *micranthus*

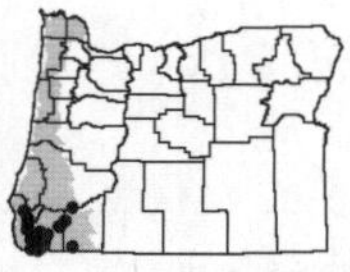

Plants 10–70 cm tall. **Bulbs** not clumped, ovoid, 10–25(35) × 8–30 mm, outer layer dark brown to black. **Stems** scabrous on lower part. **Leaves** 10–45 cm × 2–10 mm. **Inflorescences** narrow panicles, occasionally racemes, 10–60-flowered, 3–40 × 3–7 cm; proximal branches less than 50% the length of inflorescence, spreading at 60–90° from main axis in fruit; bracts 5–40 mm long, green to membranous white; pedicels 12–40 mm. **Flowers** bisexual, 5–12 mm in diameter; tepals ovate to elliptic, 3–8 × 1–5 mm, outer and inner tepals clawed, persistent in fruit; gland yellow. **Fruits** cylindrical, 6–18 × 2–7 mm. **2*n***=22.

Dry slopes, flats. Flowering Mar–Jun. 50–1600 m. CR, Sisk. CA. Native.

Toxicoscordion paniculatum (Nuttall) Rydberg (p. 312)
Panicled death camas, panicled zigadenus, foothill death camas

Zigadenus paniculatus (Nutt.) S. Watson

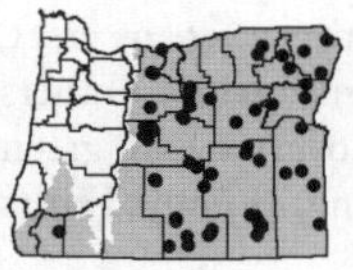

Plants 20–70 cm tall. **Bulbs** not clumped, ovoid, tunicate, (18)30–50 × 8–30(35) mm, outer layer dark brown to black. **Stems** glabrous. **Leaves** (15)20–45 cm × (3)6–15 mm. **Inflorescences** narrow panicles, 10–80-flowered, 10–30 cm; proximal branches less than half the length of entire inflorescence, spreading at 10–60° from main axis when fruiting; bracts 5–15 mm long, green; pedicels 3–25 mm. **Flowers** either bisexual or whole branches of sterile or staminate flowers, 5–10 mm in diameter; tepals ovate, 2–6 × 1–5 mm, outer tepals not clawed, inner tepals clawed, persistent in fruit; gland greenish yellow. **Fruits** cylindrical, 5–20 × 3–8 mm. **2*n***=22.

Dry sagebrush scrubs, conifer forests. Flowering Apr–Jul. 400–2500 m. BR, BW, Col, ECas, Lava, Owy, Sisk. CA, ID, NV, WA; northeast to MT, southeast to NM. Native.

Toxicoscordion venenosum (S. Watson) Rydberg (p. 312)

Meadow death camas, poison camas

Zigadenus venenosus S. Watson var. *gramineus* (Rydb.) O.S. Walsh ex M. Peck, *Zigadenus venenosus* S. Watson var. *venenosus*

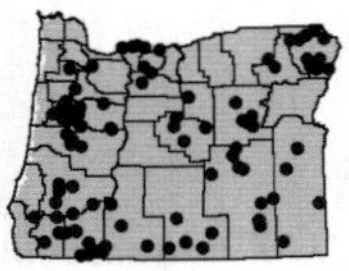

Plants 20–70 cm tall. **Bulbs** not clumped, tunicate, widely ovate, 10–30 × 8–20 mm, outer layer brown. **Stems** glabrous. **Leaves** 12–50 cm × 2–10 mm. **Inflorescences** generally racemes, occasionally panicles, 10–50-flowered, 2–20 × 2–5 cm; branches less than 33% length of entire inflorescence, at 10–60° from main axis when fruiting; bracts 5–25 mm long, white to green; pedicels 3–25 mm. **Flowers** bisexual, campanulate 5–10 mm in diameter; tepals ovate to obtuse, 2–6 × 1–3 mm, outer and inner tepals clawed, persistent in fruit; gland greenish yellow. **Fruits** cylindrical 8–14 × 4–5 mm. **2*n***=22.

Moist meadows to dry rocky hillsides. Flowering Apr–Jul. 0–2600 m. All ecoregions except Est. CA, ID, NV, WA; north to British Columbia, east to UT. Native.

TRILLIUM Trillium, wakerobin

Herbs perennial. **Rhizomes** short, thick, tapered to a point. **Roots** adventitious, contractile, ringed. **Leaves** 3, in a whorl subtending flower, sessile or petiolate, ovate to cordate; major veins 3–5, venation palmate-reticulate. **Scapes** 1-several, erect, straight, round, usually aerial (below ground in *T. petiolatum*). **Inflorescences** terminal, 1 per stem. **Flowers** bisexual; sepals 3, ovate or obovate to elliptical, green (sometimes with maroon markings), free, alternating with leaves, leaf-like, persistent; petals 3, ovate to linear, white, pink, yellow, maroon, or purple, free, spreading or recurved, sometimes clawed, withering with age; stamens 6, alternating in 2 whorls, erect, recurved or divergent, filaments wider at base, anthers white or purple; ovary superior; stigmas 3, linear to awl-shaped, spreading, twisted, or erect, usually persistent. **Fruits** berry-like capsules, 0.5–3 cm long, fleshy. **Seeds** many, ovoid to elliptical, 2–4 mm, with a large white or yellowish oil body (elaiosome), ant dispersed. **2*n***=10.

North America, Asia. 43 species; 5 species in Oregon.

Key to *Trillium* species:

1. Scapes mostly subterranean . . . . *T. petiolatum*

1' Scapes above ground.

  2. Leaf petioles 1–3 cm . . . . *T. rivale*

  2' Leaf petioles less than 0.5 cm or leaves sessile.

    3. Flowers pedicellate . . . . *T. ovatum* ssp. *ovatum*

    3' Flowers sessile.

      4. Flowers maroon . . . . *T. kurabayashii*

      4' Flowers white to cream-colored (base sometimes rose pink to purple, rarely entire petal purple) . . . . *T. albidum*

Trillium albidum J.D. Freeman

Giant trillium, sessile trillium

Leaves ovate to broadly ovate, 6–20 × 5–15 cm, with weak mottling, spots often fading, not glossy, apex rounded to obtuse; sessile. **Scapes** 1–3 (rarely more), 20–55 cm long. **Flowers** sessile; sepals lanceolate, 15–65 × 4–15 mm, green, margins entire, apex acute; petals erect or slightly spreading, linear to obovate, 2–8 × 0.5–3 cm, white to cream-colored, base sometimes rose pink to purple, rarely entire petal purple, base often cuneate, apex rounded to acute, margins entire; stamens erect, 10–25 mm, filaments 1–4 mm, white to greenish white, anthers erect, 10–20 mm, yellow or greenish white; ovary ovoid, green or sometimes purple; stigmas erect to spreading, 3–7 mm. **Fruits** ovoid to globose, green to maroon.

2 subspecies.

The two subspecies of *T. albidum* are known to hybridize where their ranges overlap. This hybrid is sometimes referred to as *T.* × *oregonum*.

Key to *Trillium albidum* subspecies:

1. Petals usually more than 5 cm long, obovate (lanceolate when young) . . . . . . *T. a.* ssp. *albidum*

1' Petals equal to or less than 5 cm long, linear to linear-lanceolate . . . . *T. a.* ssp. *parviflorum*

Trillium albidum J.D. Freeman ssp. *albidum*

Giant trillium, sessile trillium

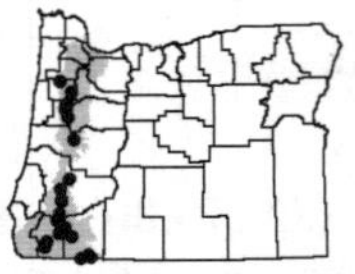

Scapes 25–55 cm. **Leaves** 10–20 × 5–15 cm. **Flowers** petals obovate (lanceolate when young), 5–8 × 2–3 cm; stamens 15–25 mm, filaments 3–4 mm. **Fruits** green to purple-green.

Moist areas, fields, pastures, coniferous forests, along streams and rivers. Flowering Mar–Jun. 50–1900 m. Sisk, WV. CA. Native.

Trillium albidum J.D. Freeman ssp. *parviflorum* (V.G. Soukup) K.L. Chambers & S.C. Meyers

Small-flowered trillium

Trillium parviflorum V.G. Soukup

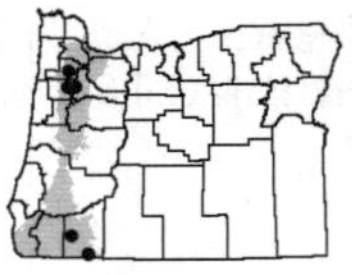

Scapes 20–30 cm. **Leaves** 6–16 × 5–9 cm. **Flowers** petals linear to linear-lanceolate, 2–5 × 0.5–1 cm; stamens 10–15 mm, filaments 1–3 mm. **Fruits** purple to maroon.

Mixed forests, generally with thick shrubby undergrowth. Flowering Mar–May. 50–500 m. Sisk, WV. WA. Native.

Trillium kurabayashii J.D. Freeman

Giant purple wakerobin

Trillium angustipetalum (Torr.) J.D. Freeman

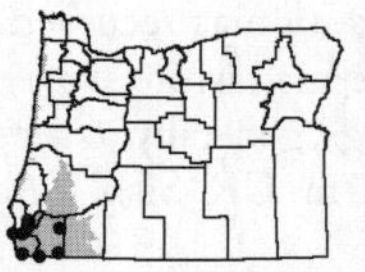

Leaves ovate to broadly ovate, (9)11–18(25) × (8)12–17 cm, bright green, usually with prominent light and dark green spots, occasionally lightly mottled with mottling fading with age, somewhat glossy, apex acuminate; sessile. **Scapes** 2, 25–55 cm, glabrous. **Flowers** sessile; sepals oblong- to lance-linear, 3.5–6(7.5) × 1–1.5 cm, green with purple streaks, margins entire, apex acute to round-acute; petals maroon, spreading-erect to erect, linear to oblanceolate, 5.5–11 × 2–3.5 cm, maroon, base cuneate, apex round-acute, margins entire; stamens erect, (11)15–26 mm, filaments 2–4 mm, dark purple, anthers 13–24 mm, dark maroon; ovary ovoid, dark purple, rounded or with 6 angles; stigmas erect, 6–8 mm. **Fruits** ovoid to ellipsoid, dark reddish purple.

Along streams, meadows, forests, found often with oak trees. Flowering Mar–May. 0–1500 m. Est, Sisk. CA. Native.

Trillium ovatum Pursh ssp. ***ovatum*** (p. 315)

Western trillium, western wakerobin

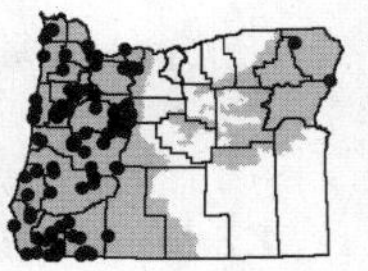

Leaves ovate-rhombic, 4–20 × 5–20 cm, green, not mottled or rarely so, base rounded, apex acute; petioles absent or less than 0.5 cm. **Scapes** 1–2, 20–50 cm, glabrous. **Flowers** pedicels 2–6 cm; sepals lanceolate, 1–6 × 0.2–2.5 cm wide, green, margins entire, apex acute; petals ascending, lanceolate to ovate, 1.5–7 × 1–4(4.5) cm, white, turning pink with age, margins flat to undulate, apex acuminate; stamens 10–18 mm, prominent, slightly recurved and spreading or straight, filaments white, anthers 4–16 mm, yellow; ovary ovoid, 5–12 mm, green or white, 6-angled; stigmas recurved, 6–10 mm. **Fruits** broadly ovoid, green or white.

Coniferous and mixed forests, meadows. Flowering Mar–Jul. 0–2300 m. BW, Casc, CR, ECas, Est, Sisk, WV. CA, ID, WA; north to British Columbia, east to CO and MT. Native.

Trillium petiolatum Pursh (p. 315)

Round-leaved trillium, petioled wakerobin

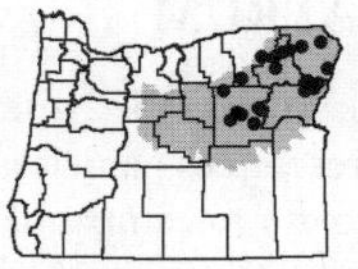

Leaves round or ovate to elliptic, 7–14 × 5.5–10.2 cm, green, not mottled, not glossy, apex round or obtuse; petioles 5–17 cm. **Scapes** usually 1, 4–20 cm, glabrous, generally subterranean (occasionally vertical). **Flowers** sessile; sepals oblong-elliptic to oblanceolate, 20–45 × 7–10 mm, green, margins entire, apex acute; petals erect to incurved, linear-lanceolate, 3–5.5 × 0.4–1 cm, light maroon-red or purple, sometimes yellow to greenish, margins entire, apex acute; stamens erect, (15)22–30 mm, filaments (3)5–7 mm, dark green, anthers 16–20 mm, purple-brown to dark green; ovary ovoid, 4–9 mm, white to greenish, purplish distally; stigmas erect, 7–20 mm. **Fruits** ovoid.

Seasonally wet places, conifer and mixed forests, near streams and rivers. Flowering Apr–Jun. 400–1900 m. BW, Lava. ID, WA. Native.

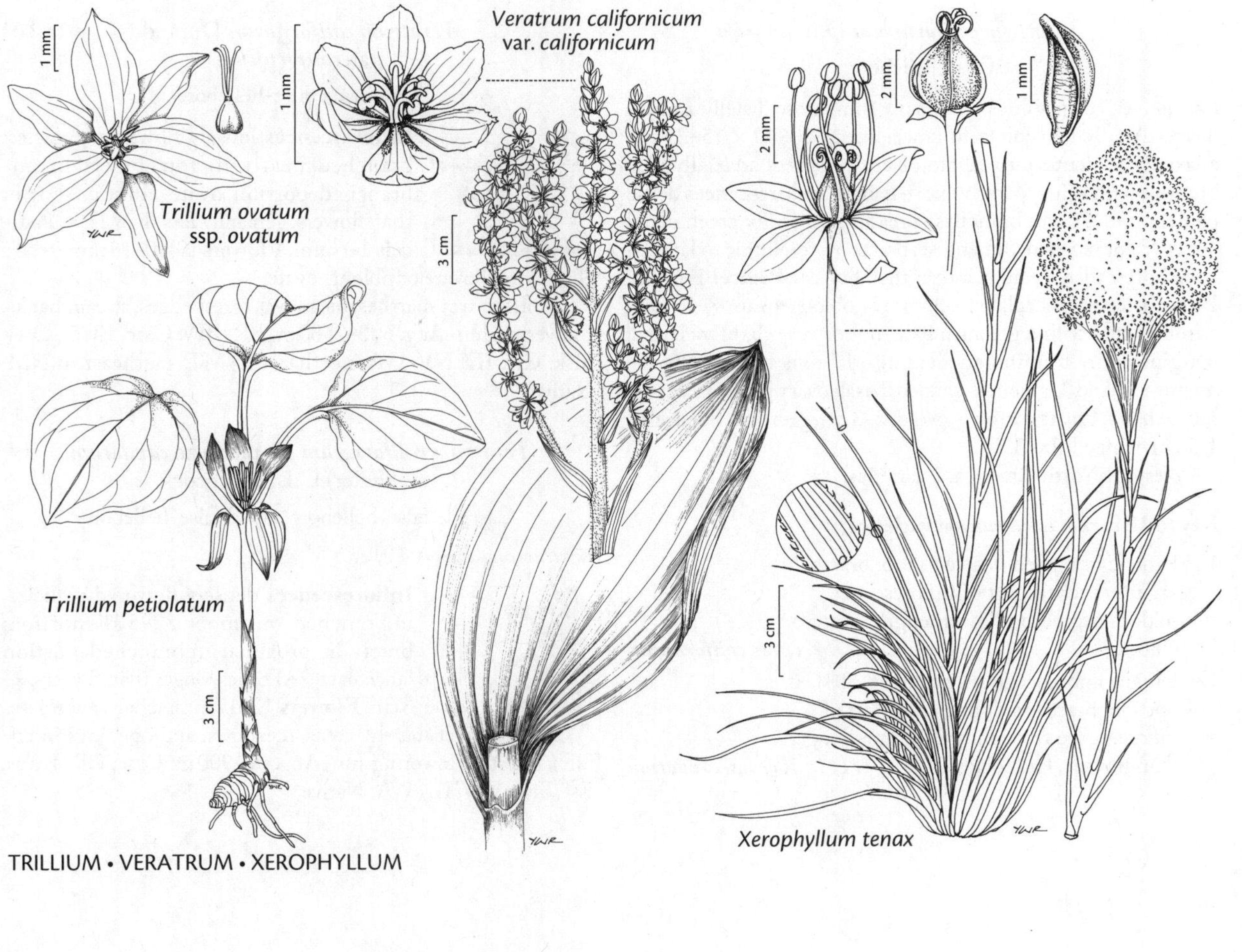

TRILLIUM • VERATRUM • XEROPHYLLUM

Trillium rivale S. Watson

Brook trillium, brook wakerobin

Pseudotrillium rivale (S. Watson) S.B. Farmer

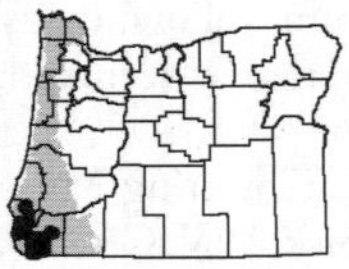

Leaves 1.7–11 cm × 0.8–6 cm, blue-green, glossy, leathery, apex acute to acuminate, base lanceolate to cordate; petioles (0.5)1–3(6) cm. **Scapes** 1–2, 4–20 cm, glabrous. **Flowers** pedicel 2-10 cm; sepals 3, green, lanceolate or ovate to oblong, 10–23 × 4–8(11) mm, margins entire, apex rounded to apiculate; petals erect to spreading, ovate-cordate to orbicular, 1.3–2.8(3) × (0.8)1–2.4 cm, white, usually with dark purple spots, base rounded to cuneate, margins entire, apex apiculate to acuminate, veins inconspicuous or sometimes with pale venation; stamens 6–15 mm, erect, filaments white, anthers straight, 3–6 mm, yellow; ovary ovoid, 3–9 × 4 mm, white to cream, weakly 6-angled; stigmas recurved, 2–4 mm, white to cream. **Fruits** globose, greenish white.

Stream banks, talus slopes, forest clearings, generally in serpentine soils. Flowering Mar–Jul. 50–1300 m. CR, Sisk. CA. Native.

This species is sometimes placed in its own monotypic genus, *Pseudotrillium.*

VERATRUM False-hellebore, corn lily

Roots fleshy; rhizomes thick, short, vertical. **Stems** 0.5–2 m erect, hollow. **Leaves** alternate, lanceolate to widely ovate, reduced upwards, apex acute, strongly veined. **Inflorescences** panicles, flowers many, pubescent, bracts present. **Flowers** bisexual or staminate; tepals 6, white or green to yellow, rotate to campanulate, free, spreading, equal to subequal, not clawed, margins mostly entire, persistent; nectary glands 1–2, basal; stamens 6, attached to perianth; filaments distinct, filiform; anthers 1-locular, cordate-reniform, persistent, dehiscence apical; ovary partly inferior, 3-chambered; styles 3, distinct, erect to divergent, short; stigmas long, extending down inner surfaces of style. **Fruits** 3-lobed, beaks persistent, dehiscence septicidal. **Seeds** winged.

Northern temperate zones. Approximately 25 species; 3 species treated in *Flora*.

Key to *Veratrum* species:

1. Tepals creamy white to yellow; ovary and fruit woolly; outer perianth parts with shallow irregular fringe . . . . . . ***V. insolitum***

1' Tepals dark green to greenish yellow or creamy white with a greenish base; ovary and fruit glabrous or minutely hairy, not woolly; outer perianth parts entire to denticulate.

2. Tepals creamy white with a greenish base; inflorescence dense, branches spreading to ascending. . . . . ***V. californicum***

2' Tepals dark green to greenish yellow; inflorescence open, lower branches drooping. . . . . . . ***V. viride*** var. ***eschscholzianum***

Veratrum californicum Durand

False-hellebore

Stems 1–2.5 m, glabrous proximally, tomentose distally. **Leaves** ovate, distally lanceolate to lance-linear, 20–40 × 15–25 cm, abaxially tomentose to ciliate or curly-haired, adaxially glabrous or with veins sparsely short-haired. **Inflorescences** dense panicles, 30–70 cm; branches spreading to stiffly erect, distal 33–50% unbranched, tomentose; bracts ovate-elliptic to lanceolate, shorter than to much longer than flowers. **Pedicel** 1–6 mm. **Flowers** tepals lanceolate to elliptic or oblong-ovate, 8–17 mm, creamy white with a greenish base, not or very slightly clawed, margins entire to minutely dentate, glabrous to abaxially tomentose; gland 1, green, V-shaped, basal; ovary glabrous or minutely hairy. **Fruits** narrowly ovoid, 2–3 cm, glabrous. **Seeds** flat, 1–2 cm, winged. **2*n***=32.

Western North America. 2 varieties.

Key to *Veratrum californicum* varieties:

1. Panicle branched nearly to top; bracts on unbranched part of inflorescence seldom exceeding flower length, ovate-elliptic. . . . . *V. c.* var. ***californicum***

1' Panicle unbranched on upper 33–50% portion; bracts on unbranched part of inflorescence frequently 2–3 times longer than flowers, lanceolate. . . . . *V. c.* var. ***caudatum***

Veratrum californicum Durand var. ***californicum*** (p. 315)

California false-hellebore

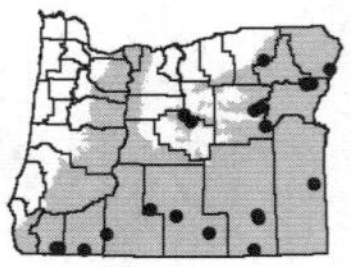

Inflorescences loosely flowered panicles; branched nearly to top; bracts in unbranched portion ovate-elliptic, shorter than flowers, generally less than 1 cm. **Pedicels** 1–6 mm. **Flowers** (8)10–15 mm; tepals elliptic to obovate or oblong-ovate.

Moist to wet marshes, meadows, forest edges, stream banks. Flowering Jun–Aug. 600–2500 m. BR, BW, Casc, ECas, Owy, Sisk. CA, ID, NV, WA; northeast to MT, southeast to NM. Native.

Veratrum californicum Durand var. ***caudatum*** (A. Heller) C.L. Hitchcock

Cascade false-hellebore, tailed false-hellebore

Veratrum caudatum A. Heller

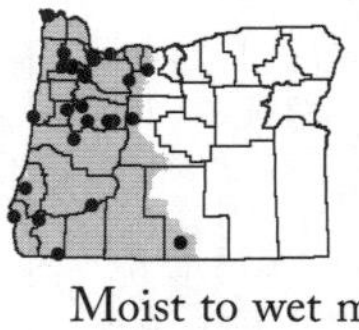

Inflorescences densely flowered panicles; unbranched on upper 33–50% portion; bracts in proximal unbranched portion lanceolate, 2–3 times longer than flowers, 2–3 cm. **Flowers** 12–17 mm; tepals lanceolate.

Moist to wet marshy areas, meadows, and openings in conifer forests. Flowering Jun–Aug. 0–1900 m. Casc, CR, ECas, Est, Sisk, WV. ID, WA. Native.

Veratrum insolitum Jepson

Siskiyou false-hellebore

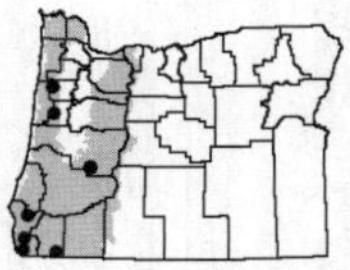

Stems 1–1.5 m, distally hairy. **Leaves** elliptic, 10–35 × 5–20 cm, distal leaves lanceolate, reduced, gray woolly hairs below and on sheathes, at least when young. **Inflorescences** 20–50 cm; branches ascending, gray woolly-haired; bracts ovate to lanceolate, shorter than flowers. **Pedicel** 6–15 mm. **Flowers** 6–9(15) mm; tepals widely ovate, creamy white to yellow, not clawed; gland 1, green, elliptic, V-shaped, basal; stamens equal to the length of the tepals; ovary densely woolly. **Fruits** oblong-ovoid, 2–3 cm, woolly. **Seeds** flat, 10–15 mm, wings broad. ***2n***=32.

Moist meadows and stream banks, mixed-evergreen forest margins, open areas in thickets. Flowering Jun–Aug. 700–1900 m. Casc, CR, Sisk. CA, WA. Native.

Veratrum viride Aiton var. ***eschscholzianum*** (Roemer & Schultes) Breitung

Indian poke, green false-hellebore

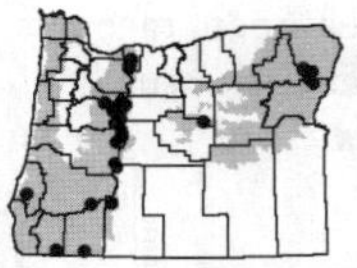

Stems 0.5–2 m, with a gradient from nearly glabrous proximally to densely hairy distally. **Leaves** elliptic to widely ovate, 15–30 × 10–18 cm, minutely curly-ciliate, abaxially hairy, adaxially veins minutely short-haired. **Inflorescences** open, 30–70 cm, woolly; usually with branches spreading, or sometimes drooping on lower part. **Pedicel** 2–5 mm. **Flowers** (5)6–10(12) mm, erect; tepals oblanceolate to oblong, dark green to greenish yellow, irregularly toothed, adaxially hairy; glands 2, dark green or yellow, elliptic; stamens approximately half the length of the tepals; ovary glabrous. **Fruits** 2–3 cm, oblong-ovoid. **Seeds** 7–10 mm, broadly winged. ***2n***=32.

Moist subalpine meadows, openings in coniferous forests. Flowering Jul–Sep. 900–2500 m. BW, Casc, CR, Sisk. CA, ID, WA; north to AK, east to WY, eastern North America. Native.

XEROPHYLLUM Beargrass, basket-grass

Plants not flowering for several years, dying after fruiting; rhizomes present, tuber-like. **Leaves** spirally arranged, basal and cauline, linear, wiry, tough, persistent after 1 year, margins serrulate, apex pointed; basal leaves tufted. **Inflorescences** dense racemes; bracts present. **Flowers** tepals 6, oblong, white to cream, distinct, 5–7-veined, persistent; stamens 6; ovary superior; styles 3, approximately 1 mm, recurved, distinct, persistent, stigmas slender. **Fruits** capsules, round to ovoid, dehiscence loculicidal. **Seeds** oblong, angled, 3–4 mm, brown to black.

North America. 2 species; 1 species treated in *Flora.*

Xerophyllum tenax (Pursh) Nuttall (p. 315)

Beargrass, bear-grass, western turkeybeard

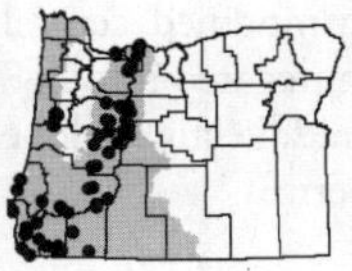

Plants up to 1.5 m tall. **Leaves** up to 80 cm × 2–5 mm; basal leaves in tufted and dense clumps; cauline leaves shorter and reduced upwards. **Inflorescences** to 60 cm tall; pedicels 2–5 cm. **Flowers** tepals 5–9 × 2–3 mm. **Fruits** 5–7 mm. **Seeds** 4 per locule. ***2n***=30.

Dry slopes and ridges, montane conifer forests. Flowering May–Jul. 0–2200 m. Casc, CR, ECas, Est, Sisk. CA, ID, WA; north to British Columbia, northeast to MT. Native.

Populations of this species have irregular cycles of abundant and sparse flowering. The cause of this is currently unknown but worthy of investigation.

NAJADACEAE

(see Hydrocharitaceae)

NARTHECIACEAE Bog-asphodel family

Tamra Dickson Prior

Herbs perennial. **Stems** simple, glabrous. **Leaves** basal and cauline, along rhizome; blades linear, grass-like. **Inflorescences** racemes. **Flowers** radially symmetrical, perianth segments 6, in 2 whorls; ovary superior; style 1. **Seeds** many, bristle-tipped on both ends.

Temperate Northern Hemisphere. 5 or 6 genera; 1 genus treated in *Flora.*

NARTHECIUM Bog-asphodel

Herbs from slender creeping rhizomes. **Roots** fibrous. **Stems** erect, more or less scapose. **Leaves** mostly basal, base folded and enclosing adjacent leaf, overlapping, 3–6 upper leaves, cauline leaves reduced; blades greenish yellow to yellow, glabrous, apex acute. **Inflorescences** terminal, simple. **Pedicels** with bracts at base, usually with bracteole near middle. **Flowers** perianth segments free, linear-lanceolate, yellow to green-yellow, persistent, erect with age, glandless, weakly 3-veined, margins scarious; septal glands absent; stamens 6, filaments densely woolly, anthers linear, curved, 2-locular, attached at the back, flexible, turned inwards; style very short or none, stigma minutely 3-lobed. **Fruits** capsules, weakly 3-lobed, oblong-lanceolate, thin walled, dehiscence loculicidal. **Seeds** filiform-ellipsoid, brown or pale yellow.

Temperate Northern Hemisphere. 4–8 species; 1 species treated in *Flora.*

Narthecium californicum Baker (p. 318)

Bog-asphodel, California bog-asphodel

Stems 20–60 cm. **Leaves** 10–30 cm × 3–5 mm. **Inflorescences** open racemes, 8–15 cm; bract 1, bractlet (0)1, usually above middle. **Pedicels** overlapping or not, 5–12(15) mm. **Flowers** 5–10 × 1.5–2.5 mm; stamens 3–4 mm, anthers 3–4 mm. **Fruits** capsules, lanceolate, 8–12 mm including beaks, approximately twice the length of the perianth. **Seeds** 10–15 in each cell, 1.5–2 × 6–8 mm including bristle-tails. **2*n*=**26.

Wet mountain meadows, springs, seeps, bogs, and stream banks. Flowering Jun–Aug. 200–1400 m. CR, Sisk. CA. Native.

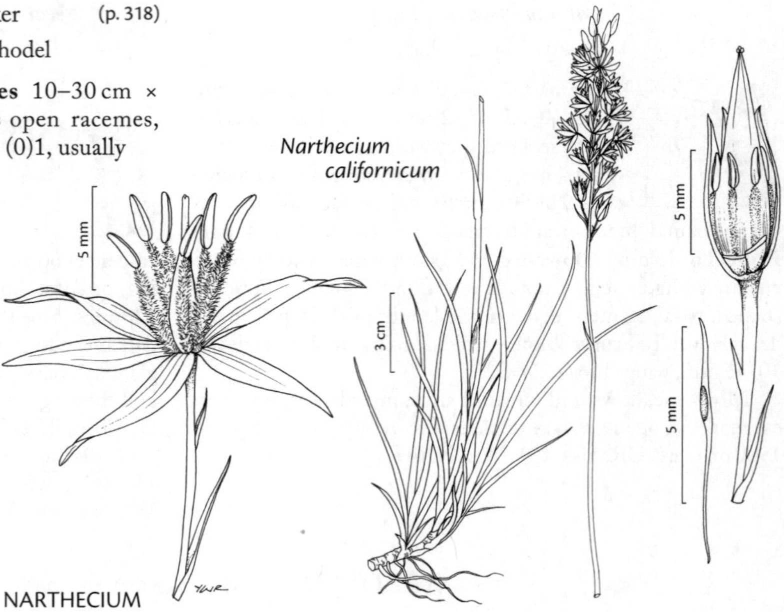

NARTHECIUM

ORCHIDACEAE Orchid family

James P. Riser II

Herbs usually perennial, strongly mycotrophic, usually photosynthetic, sometimes non-photosynthetic and heterotrophic. **Leaves** simple, basal, or cauline; alternate, opposite, or whorled; plicate, convolute, conduplicate, cylindric, or flattened; margins entire; petioled or not; stipules 0. **Inflorescences** racemes, panicles, spikes, or rarely cymose; flowers 1 to many, terminal or lateral; generally with bracts. **Flowers** bisexual, usually strongly bilaterally symmetric; sepals 3, often petaloid; petals 3, highly modified, central petal modified into lip; stamens 1 or 2 (rarely 3), opposite lip, fused to style to form column; pollen generally aggregated into 2–8 pollinia; ovary inferior, 3-carpellate with 1 or 3 locules; style fused to anther filaments; stigma 3-lobed with middle stigma lobe modified into a small beak (rostellum). **Fruits** capsules. **Seeds** numerous, minute and dust-like, lacking endosperm.

Cosmopolitan. Approximately 880 genera; 9 genera treated in *Flora*.

Achlorophyllous plants within this family are known to obtain nutrition from forest litter through fungal intermediates. Fungal association is crucial to seed germination and plant establishment and often persists for the life of the plant.

The "lip" within flower descriptions refers to a central petal that is distinguished by its irregular, often larger, shape. Its purpose is to attract and serve as a guide to pollinators.

Key to Orchidaceae genera:

1. Plants non-photosynthetic and lacking leaves.
    2. Plants pure white with yellow spot on lip .... **CEPHALANTHERA**
    2' Plants variously colored, but not pure white .... **CORALLORHIZA**
1' Plants photosynthetic with obvious leaves.
    3. Lip inflated or pouch-like.
        4. Lip distinctly inflated and slipper-shaped; stamens 2 .... **CYPRIPEDIUM**
        4' Lip pouch-like or boat-shaped; stamen 1.
            5. Leaves cauline .... **EPIPACTIS**
            5' Leaves basal.
                6. Single basal leaf and single flower per stem .... **CALYPSO**
                6' Several basal leaves and several flowers per stem .... **GOODYERA**
    3' Lip not inflated or pouch-like.
        7. Spur on lip present .... **PLATANTHERA**
        7' Spur on lip absent.
            8. Leaves cauline, wide, and opposite .... **NEOTTIA**
            8' Leaves basal, narrow, and whorled .... **SPIRANTHES**

CALYPSO Fairy slipper, calypso orchid

Herbs perennial; from an underground corm. **Stems** simple, fleshy, erect. **Leaves** single, ribbed, often prostrate on ground. **Inflorescences** terminal scapes. **Flowers** sepals 3, distinct; petals 3, distinct, 1 modified into large saccate lip; stamen 1.

Circumboreal. 1 species.

Calypso bulbosa (Linnaeus) Oakes var. ***occidentalis*** (Holzinger) B. Boivin (p. 320)

Fairy slipper, calypso orchid

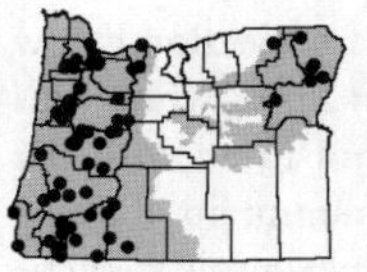

Herbs 4–20 cm tall. **Leaves** ovate to cordate, 4–8 × 2–3 cm long. **Flowers** solitary; sepals and petals rose-purple (rarely pure white) except lip; lip large, saccate with 2 distal spurs, white with lavender mottling. **Fruits** erect. $2n=28$.

Mixed or coniferous forests in humus. Flowering Mar–Jun. 50–2300 m. BW, Casc, CR, ECas, Est, Sisk, WV. CA, ID, WA; north to British Columbia. Native.

One of Oregon's more showy native orchids, flowering fairy slippers are often locally common in the spring, in coniferous forests. Following fruiting, this species is summer dormant, the leaves reemerging with the fall rains. A species with four varieties, ours can be distinguished from the only other North American variety by the white beard hairs on the lip (versus the bright yellow hairs in var. *americana*).

CEPHALANTHERA Phantom orchid, snow orchid

Herbs perennial, non-photosynthetic, 20–60(100) cm tall; from rhizomes. **Stems** simple, erect. **Leaves** alternate, cauline, often keeled or pleated. **Inflorescences** solitary terminal spikes; often with foliaceous floral bracts. **Flowers** sepals 3, distinct; petals 3, distinct, 1 modified into 3-lobed lip; stamen 1.

Western North America, North Africa, Eurasia. Approximately 15 species; 1 species treated in *Flora*.

This genus is widely distributed in Eurasia and North Africa where about 15 species are known. The Old World species are green and at least partially capable of photosynthesis, although all members probably maintain mycorrhizal associations.

Cephalanthera austiniae (A. Gray) A. Heller (p. 320)

Phantom orchid, snow orchid

Eburophyton austiniae (A. Gray) A. Heller

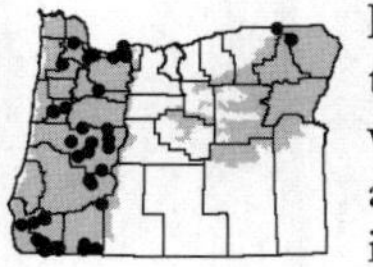

Herbs 20–50 cm tall, white. **Leaves** reduced to sheathing bracts. **Flowers** 5–20 or more, white, with a yellow spot on lip; dorsal sepal and both petals arch over the column making a loose hood; lip divided into 2 distinct parts, inner portion with upright triangular lobes encircling column, outer portion curved downward with bright yellow spot. **Fruits** ellipsoid, white, erect.

Coniferous forests in humus. Flowering May–Aug. 50–2300 m. BW, Casc, CR, Sisk, WV. CA, ID, WA; north to British Columbia. Native.

The phantom orchid is unmistakable in that it is the only orchid in our flora that is always pure white. This achlorophyllous orchid is a mycoheterotroph, meaning it derives all its nutrients from a symbiotic fungus.

CORALLORHIZA Coralroot

Herbs perennial, non-photosynthetic, to 60 cm tall. **Stems** erect, leafless, often tan to purplish. **Leaves** reduced to sheathing bracts. **Inflorescences** racemes, with few (2–3) to many (10 or more) flowers. **Flowers** perianth segments spreading to converging but not united; sepals and petals lanceolate to oblanceolate; lip broadly obovate, entire, often basally toothed or lobed; stamen 1; ovary occasionally with a short rounded projection (mentum).

North America, Europe, and Asia. 10–11 species; 5 species treated in *Flora*.

Individuals within populations of most *Corallorhiza* species may have reduced amounts of anthocyanins, or lack them entirely, with coloration ranging from greenish yellow to yellow to white. Only *C. trifida* is consistently greenish yellow. In addition, the lips of all species may or may not have markings. Several individuals in a population should be examined when encountering light-colored plants. All species have the namesake coral-like, warty, multi-branched rhizomes and are mycoheterotrophic.

Key to *Corallorhiza* species:

1. Edges of lip upturned; sepals and petals strongly striped . . . . ***C. striata***

1' Edges of lip not upturned; sepals and petals not strongly striped.

  2. Stems, sepals, and petals consistently yellow-green . . . . ***C. trifida***

  2' Stems, sepals, and petals generally reddish, purplish, or pinkish.

    3. Lip without basal lobes, teeth, or auricles . . . . ***C. wisteriana***

    3' Lip with basal lobes, teeth, or auricles.

      4. Lateral sepals wide-spreading; distinct projection present; lip generally unspotted or merely splotched, small lobes or teeth on lip margin . . . . ***C. mertensiana***

      4' Lateral sepals not wide-spreading; projection lacking or obscure; lip distinctly spotted, distinct forward-directed lobes at base of lip . . . . ***C. maculata***

Corallorhiza maculata (Rafinesque) Rafinesque (p. 320)

Spotted coralroot

Corallorhiza maculata (Raf.) Raf. var. *immaculata* M. Peck, *Corallorhiza maculata* (Raf.) Raf. var. *maculata*, *Corallorhiza maculata* (Raf.) Raf. var. *occidentalis* (Lindl.) Ames

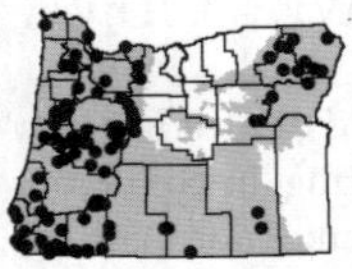

Herbs 15–50 cm tall. **Stems** pinkish red to purple. **Inflorescences** loosely 5–30-flowered. **Flowers** sepals oblanceolate, approximately 8 × 2 mm, yellowish brown to bronze with reddish purple margins and tips, occasionally spotted, somewhat closed to widely spreading, 3-nerved, obscure short rounded projection present or missing below the lip; petals oblanceolate, approximately 7 × 2 mm, similar in color to sepals or somewhat lighter, often purple spotted; lip approximately 6 × 5 mm, ovate with 3 distinct lobes; lateral lobes small and tooth-like, directed forward; central lobe wider with undulate margin, white and spotted with purple. **2*n***=42.

Decaying leaf and needle litter of mixed conifer forests. Flowering Apr–Aug. 0–2300 m. BR, BW, Casc, CR, ECas, Est, Sisk, WV. CA, ID, NV, WA; throughout North America. Native.

Our *C. maculata* is sometimes subdivided into two varieties, with overlapping ranges. Variety *maculata* is distinguished by the middle lobes of the lips not or only slightly expanded with parallel sides. Within variety *occidentalis*, the middle lobes of the lips are distinctly expanded with clearly diverging sides.

Corallorhiza mertensiana Bongard (p. 320)

Western coralroot, Mertens' coralroot

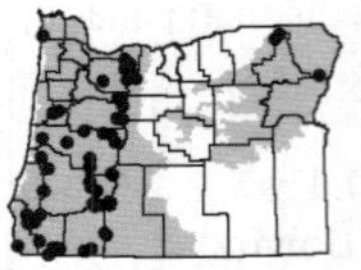

Herbs 15–50 cm tall. **Stems** reddish to purplish. **Inflorescences** loosely to relatively densely 10–30-flowered. **Flowers** perianth segments widely spreading; dorsal sepal oblanceolate, 8–9 × approximately 2 mm, pale pink to yellowish brown, converging to but not fused with the petals; lateral sepals oblanceolate, 8–9 × approximately 2 mm, forming a small but distinct, short, rounded projection below the lip; petals oblanceolate, approximately 8 × 2 mm, white with pink-purple splotches, converging to but not fused with the dorsal sepal and forming a hood over the column; lip obovate, 7–8 × 3–4 mm, bright purple or white with pink-purple splotches, shallowly 3-lobed with 2 small lateral teeth and an erose-margined middle lobe. **2*n***=40.

Decaying leaf and needle litter of dense mixed conifer forests. Flowering May–Aug. 0–2300 m. BW, Casc, CR, ECas, Sisk, WV. CA, ID, WA. Native.

Mertens' coralroot is recognized by its narrow petals and sepals, wide-open flowers, and distinct projection below the lip. Individuals may be colored a rich purple throughout (including the lip) or be lighter with yellow or even bronze highlights on the petal and sepal tips, these having pink-purple splotched lips.

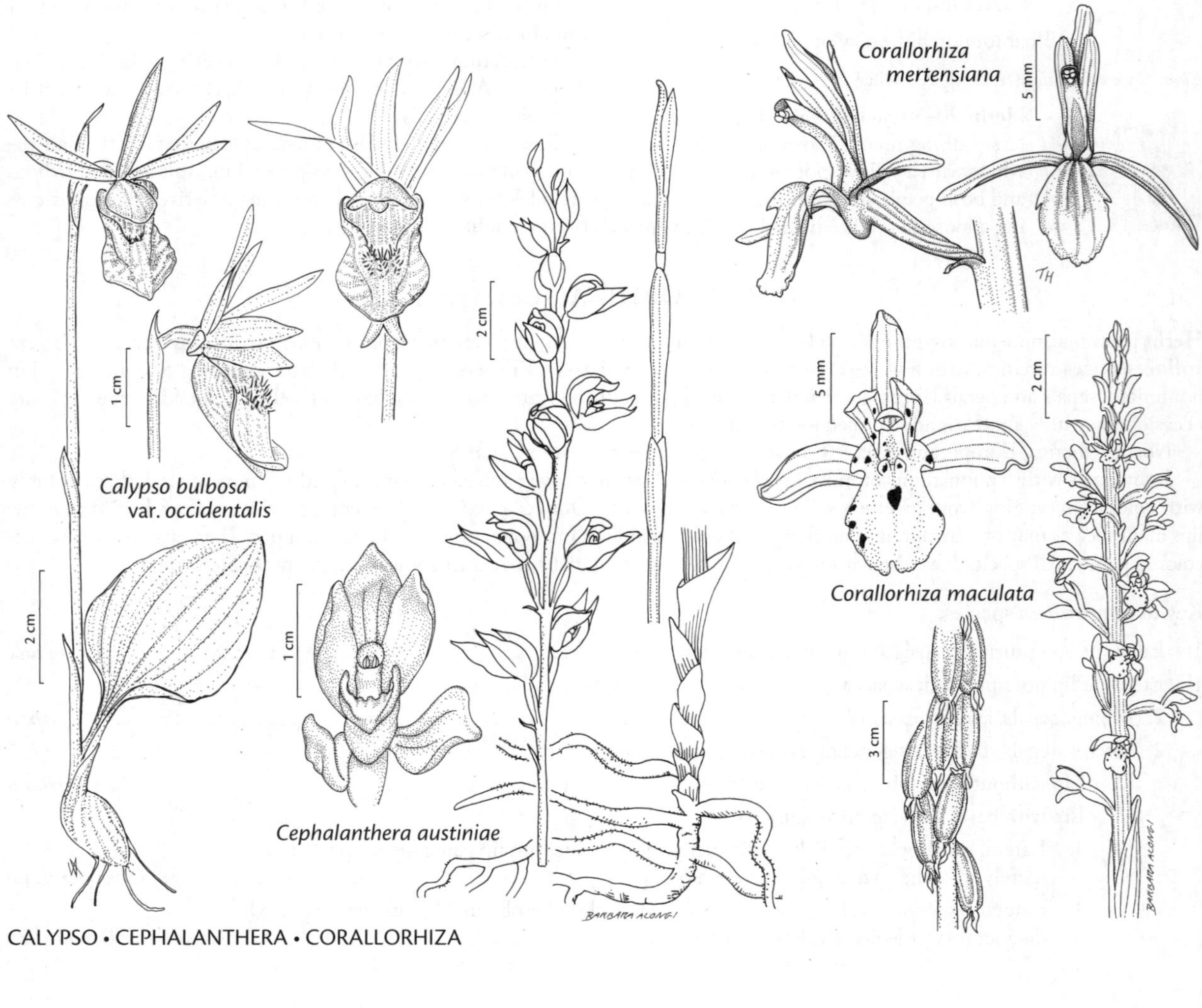

CALYPSO • CEPHALANTHERA • CORALLORHIZA

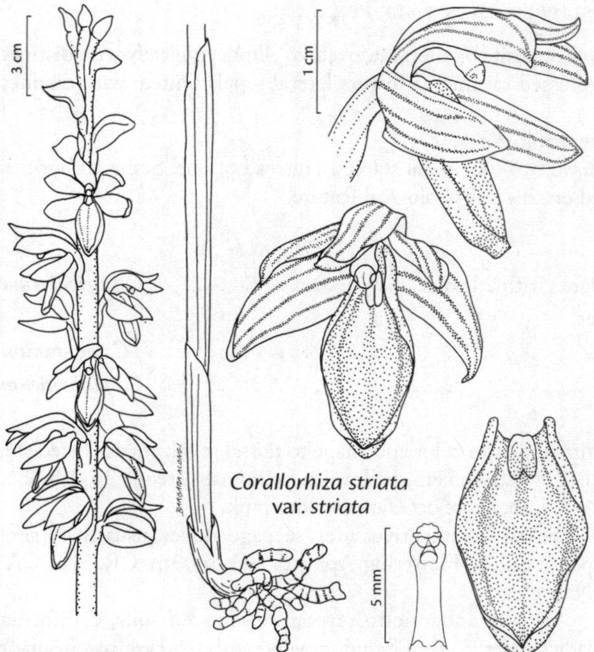

CORALLORHIZA

Corallorhiza striata Lindley

Striped coralroot

Herbs 15–60 cm tall. **Stems** robust, reddish to purplish. **Inflorescences** loosely 5–30-flowered. **Flowers** perianth segments spreading to converging, often drooping; dorsal sepal oblong to oblanceolate, 6–18 × 2–5 mm, cream to yellowish pink with 3 purple veins; lateral sepals, similar in size, shape, and color to dorsal sepal, oblique, projection absent; petals elliptic-oblanceolate, 6–16 × 2–5 mm, cream to yellowish pink, with 5 purple veins; lip obovate to oblanceolate, 4–12 × 2–8 mm, cream to yellowish white, with 5 purple veins becoming confluent towards a solid purple tip, margins entire and upturned. ***2n*=42**.

2 varieties.

Corallorhiza striata is the only striped coralroot. It is also the largest and showiest of the coralroots.

Key to *Corallorhiza striata* varieties:

1. Flowers vivid, veins reddish purple to purple; sepals and petals 10–18 mm ..... ***C. s.*** var. ***striata***

1' Flowers dull, veins reddish brown to pinkish; sepals and petals 6–8 mm .... ***C. s.*** var. ***vreelandii***

Corallorhiza striata Lindley var. ***striata*** (p. 321)

Striped coralroot

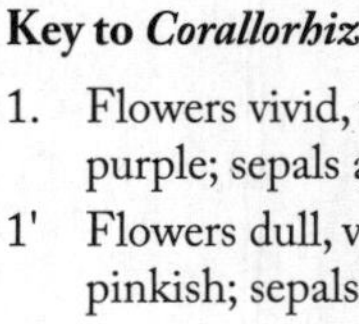

Herbs to 60 cm tall. **Inflorescences** loosely 5–30-flowered. **Flowers** perianth segments spreading to converging; sepals and petals 10–18 × 3–5 mm, cream to yellowish pink with reddish purple to purple veins; lip 8–12 × 6–8 mm, cream to yellowish white with reddish purple to purple veins.

Decaying leaf and needle litter of mixed conifer forests. Flowering Mar–Jun. 50–1900 m. BW, Casc, CR, Sisk, WV. CA, ID, WA; Rocky Mountains and northern North America. Native.

Corallorhiza striata Lindley var. ***vreelandii*** (Rydberg) L.O. Williams

Vreeland's striped coralroot

Corallorhiza bigelovii S. Watson

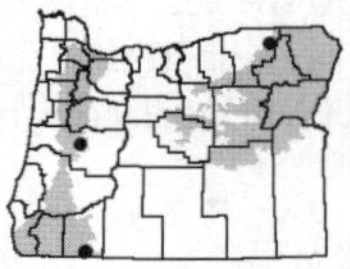

Herbs to 36 cm tall. **Inflorescences** loosely 5–20-flowered. **Flowers** perianth segments converging; sepals and petals 6–8 × 2–3 mm, cream to yellowish with pink to dull purple veins; lip 4–7 × 2–4 mm, cream to yellowish with pink to dull purple veins.

Decaying leaf and needle litter of mixed conifer forests. Flowering May–Jul. 200–1500 m. BW, Sisk, WV. CA, ID, NV; western North America, Quebec. Native.

Vreeland's striped coralroot is a diminutive and less showy version of the nominate variety. It is widespread but less frequently encountered.

Corallorhiza trifida Châtelain

Early coralroot, yellow coralroot, northern coralroot

Corallorhiza corallorhiza (L.) H. Karst.

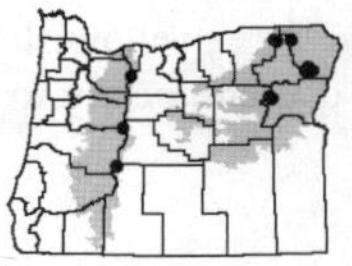

Herbs 10–30 cm tall. **Stems** yellow to light green. **Inflorescences** loosely 3–20-flowered. **Flowers** sepals and petals linear-oblanceolate to oblanceolate, 4–6 × 1–1.5 mm, yellow to yellowish green, the tips sometimes red-brownish, occasionally forming an obscure short rounded projection; lip 3-lobed, with 2 small tooth-like lateral lobes and a much larger rounded middle lobe, 2–5 × 1–3 mm, white, occasionally with purple spots. ***2n*=42**.

Decaying leaf and needle litter of moist mixed conifer forests. Flowering Jun–Jul. 1000–2200 m. BW, Casc. CA, ID, NV, WA; northern North America; Eurasia. Native.

Early coralroot is the only species of *Corallorhiza* that occurs outside of North America. It is the only *Corallorhiza* species that is consistently pale yellowish green, although the flowers of some individuals may have a bronze hue. Its drooping lateral sepals help to identify this species from yellow forms of other coralroots.

Corallorhiza wisteriana Conrad

Wister's coralroot, spring coralroot

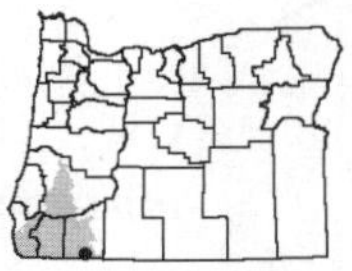

Herbs to 30 cm tall. **Stems** yellow-brown to purplish. **Inflorescences** loosely 5–25-flowered. **Flowers** sepals and petals lanceolate to oblanceolate, 5–7 × 1–2 mm, yellow to green suffused with red-brown or purple, forming a hood over the column and converging over lip, projection present; lip ovate, 4–6 × 24 mm, white spotted with pink or purple, margin undulate.

Decaying leaf and needle litter of mixed conifer forests. Flowering Jun–Jul. 300–1000 m. Sisk. ID; Rocky Mountains, southeastern North America. Native.

Wister's coralroot reaches its westernmost distribution in Oregon. It is much more common in the Rocky Mountains and the southeastern portion of North America.

CYPRIPEDIUM Lady slipper, lady's-slipper

Herbs perennial, to 130 cm tall. **Stems** erect, often pubescent. **Leaves** opposite or alternate, ovate to elliptic, typically with distinct pleating, with sheathing bases. **Inflorescences** elongated or compressed racemes. **Flowers** lateral sepals united; with distinct inflated lip; stamens 2.

North America, Eurasia. Approximately 50 species; 3 species treated in *Flora*.

Lady slippers are unique among Oregon orchids in that they have two functional stamens, instead of one. Some authorities have treated *Cypripedium* as a distinct family, Cypripediaceae, based on this morphological feature.

Key to *Cypripedium* species:

1. Lip green with purple or brown mottling; flowers clustered; plants with 2 leaves.. . . . . ***C. fasciculatum***

1' Lip white; flowers not clustered; plants with more than 2 leaves.

  2. Petals purple, twisted . . . . ***C. montanum***

  2' Petals yellow to yellow-green, flat . . . . ***C. californicum***

Cypripedium californicum A. Gray

California lady slipper

Herbs to 80(100 or more) cm tall. **Leaves** alternate, 5–10, 5–15 × 2–6 cm, grading into bracts in the inflorescence. **Inflorescences** elongated, 3–20-flowered. **Flowers** dorsal sepal elliptic, 1.5–2 × 1–1.5 cm, yellowish green to pale brownish yellow; lateral sepals connate nearly to tip, similar in color and shape to dorsal sepal; petals lanceolate, 1–2 × 0.3–0.5 cm, yellow to yellowish green; lip 1.5–2 × 1–1.5 cm, white occasionally with pink.

Small streams, riversides, seepage slopes, usually on serpentine soils. Flowering Apr–Jul. 200–1600 m. CR, Sisk. CA. Native.

Usually endemic to serpentine-derived soils, California lady slipper is often found growing with the carnivorous plant *Darlingtonia californica* in seepage slopes and small streams.

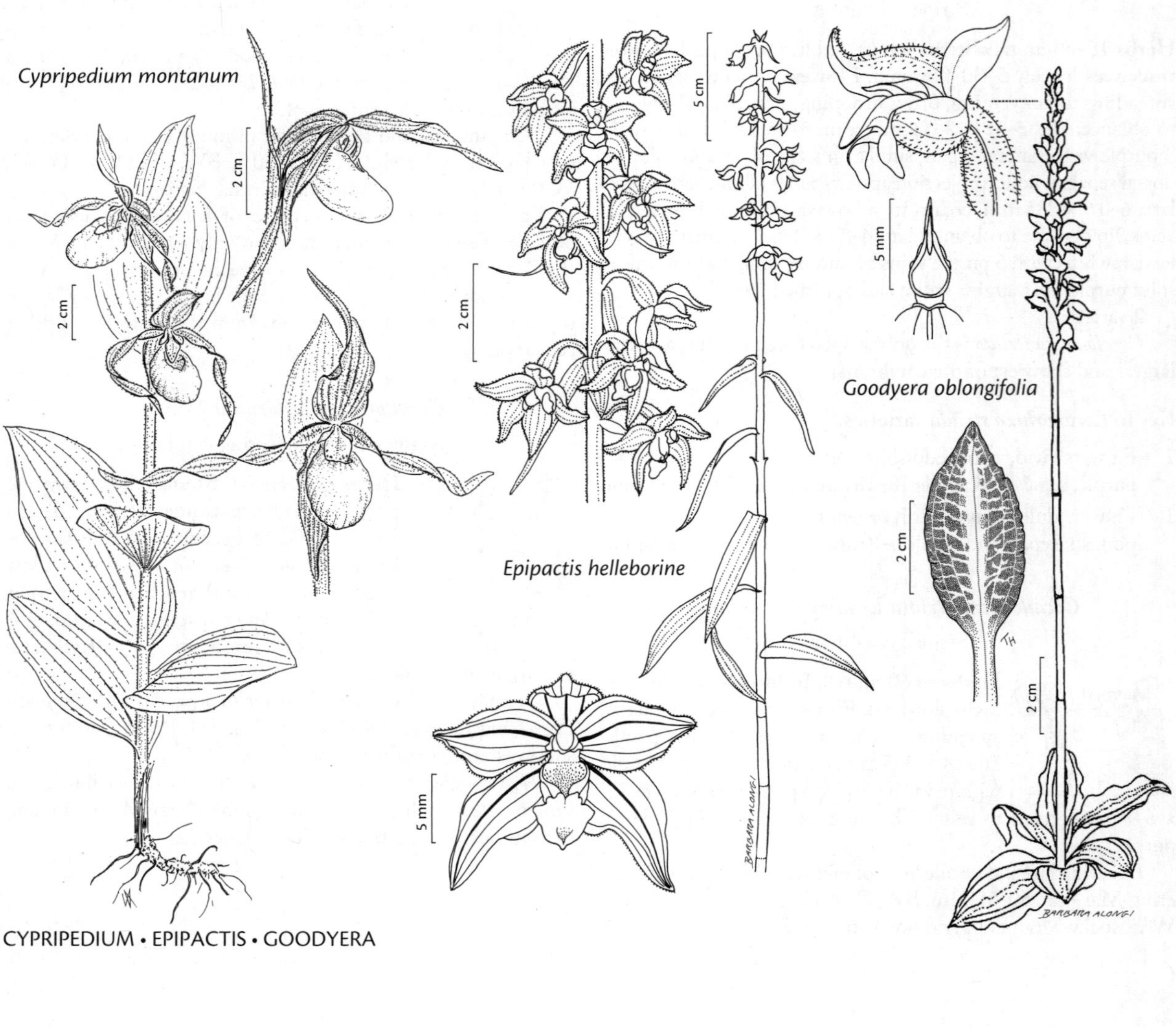

CYPRIPEDIUM • EPIPACTIS • GOODYERA

Cypripedium fasciculatum Kellogg ex S. Watson
Clustered lady slipper

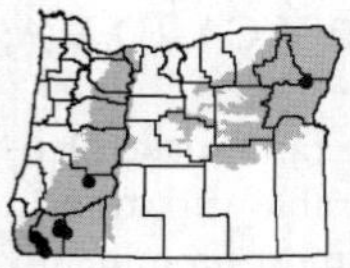

Herbs to 20 cm tall. **Leaves** opposite, 2, 5–11 × 4–7.5 cm. **Inflorescences** compressed with 2–5-clustered flowers, nodding. **Flowers** dull yellow-green with purplish and/or brownish mottling throughout; dorsal sepal lanceolate to ovate, 1.5–2 × 0.3–0.9 cm; lateral sepals connate nearly to tip, similar in color and shape to dorsal sepal; petals lanceolate to ovate, 1.5–2 × 0.6–1.5 cm; lip 0.8–1.3 × 0.5–1 cm. **2*n***=20.

Mixed conifer forests. Flowering Apr–Jul. 50–1900 m. BW, Casc, Sisk. CA, ID, WA; east to CO. Native.

The pedicel of this species elongates dramatically when in fruit, increasing the height of the plant from two to three times the flowering height.

Cypripedium montanum Douglas ex Lindley (p. 322)
Mountain lady slipper

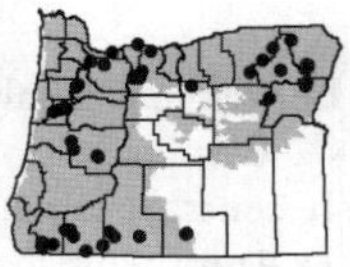

Herbs to 70 cm tall. **Leaves** alternate, 4–6, 8–17 × 4–10 cm. **Inflorescences** elongated, 1–3-flowered. **Flowers** dorsal sepal lanceolate to ovate, 3–6 × 1–1.5 cm, green suffused with purple; lateral sepals connate almost to tip, similar in color and shape to dorsal sepal; petals linear-lanceolate and twisted, 4–7 × 0.3–0.6 cm, purple; lip 2–3 × 1.2–1.7 cm, white with purple spots inside.

Mixed conifer forests, open shrubby woodlands, roadsides. Flowering Apr–Jul. 50–2200 m. BW, Casc, Col, CR, ECas, Sisk, WV. CA, ID, WA; north to AK. Native.

EPIPACTIS Helleborine

Herbs perennial, 4–100 cm or more tall. **Stems** erect, leafy. **Leaves** alternate, lanceolate, 2–10 or more, grading into bracts within inflorescence. **Inflorescences** terminal racemes, 5–60-flowered. **Flowers** more or less spreading; sepals and petals lanceolate, variously colored; lip divided into 2 distinct jointed segments, basal portion of lip (hypochile) deeply concave, terminal portion of lip (epichile) triangular or cordate and often mobile, spur lacking; stamen 1.

North America and Eurasia. 61–64 species; 2 species treated in *Flora*.

Key to *Epipactis* species:

1. Lip three-lobed, distal segment oblanceolate and elongated . . . . *E. gigantea*
1' Lip not three-lobed, distal segment triangular and recurved . . . . *E. helleborine*

Epipactis gigantea Douglas ex Hook.
Chatterbox, stream orchid, giant helleborine

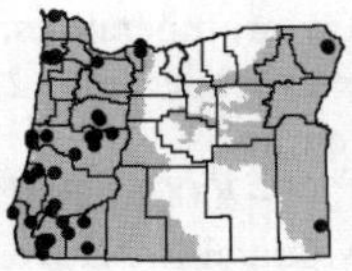

Herbs 30–100 cm tall; roots fibrous. **Stems** green. **Inflorescences** loose racemes, 5–30-flowered; with prominent floral bracts. **Flowers** sepals ovate-lanceolate, 15–20 × 6–8 mm, greenish yellow with purple stripes, lateral sepals spreading, dorsal sepal arched over column with petals; petals ovate-lanceolate, 13–15 × 6–8 mm, greenish yellow grading into pinkish towards the tips, with purple stripes; lip 14–18 mm, clearly 3-lobed, basal portion of lip deeply concave with blunt triangular lobes, yellowish or pinkish with purple veins, terminal portion of lip oblanceolate, yellowish with 2 prominent orange ridges. **2*n***=40.

Streamsides, springs, lakeshores. Flowering May–Sep. 0–1700 m. BW, Casc, CR, ECas, Est, Owy, Sisk, WV. CA, ID, WA; north to British Columbia, east to MT, SD, TX. Native.

The terminal portion of the lip is hinged to the basal portion and is quite mobile, giving rise to the common name "chatterbox."

Epipactis helleborine (Linnaeus) Crantz (p. 322)
Broad-leaved helleborine, helleborine

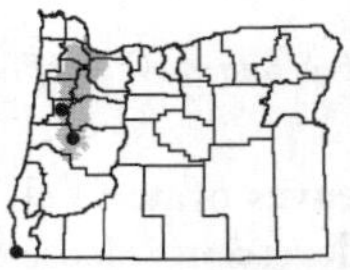

Herbs 10–80 cm tall; roots fibrous. **Stems** green. **Inflorescences** loose racemes, often 1-sided, 15–50-flowered; with prominent floral bracts. **Flowers** sepals ovate-lanceolate, 10–13 × 5–6 mm, greenish often suffused with purple, lateral sepals oblique, somewhat closed to widely spreading and directed downward; petals ovate-lanceolate, 9–11 × 4–6 mm, green, yellow, pink, or purplish, often with a prominent central vein; lip 10–11 mm overall, basal portion of lip deeply concave, purplish to brownish, terminal portion of lip triangular, strongly recurved, pink with 2 prominent brownish ridges. **2*n***=36, 38, 40, 44.

Variable habitat, often calcareous. Flowering Jul–Oct. 100–200 m. WV. CA, WA; scattered across North America. Exotic.

GOODYERA Rattlesnake plantain

Herbs perennial, from a creeping rhizome. **Leaves** several in a basal rosette, sometimes with reticulate markings. **Inflorescences** a loosely flowered terminal spike, spiraled or one-sided. **Flowers** sepals 3, distinct; petals 3, distinct, 1 petal modified into a larger saccate lip; stamen 1.

Cosmopolitan. 40–100 species; 1 species treated in *Flora*.

Goodyera is distributed nearly worldwide (except in Africa) in shady montane forests. The genus occurs in temperate to tropical climates and is usually terrestrial, but epiphytic and lithophytic species are known. It is unclear how many species there are in the genus, but the number is probably close to 100. *Goodyera* is often cultivated for its distinctively patterned leaves.

Goodyera oblongifolia Rafinesque (p. 322)

Giant rattlesnake plantain, western rattlesnake plantain, Menzies' rattlesnake plantain

Peramium decipiens (Hook.) Piper

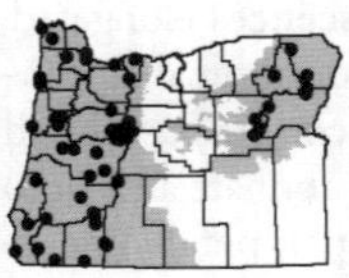

Herbs up to 40 cm tall. **Leaves** in a basal rosette, oblong, 3–7, 3–7 × 2–3 cm, green to gray-green, with whitish central stripe. **Flowers** approximately 10 × 4 mm, whitish; lateral sepals spreading; dorsal sepal converging with but not fused with petals forming a hood over the lip; lip deeply concave, saccate or boat-shaped. **Fruits** erect. **2*n***=30.

Dry or moist coniferous forests. Flowering Jul–Sep. 0–2300 m. BW, Casc, CR, ECas, Est, Sisk, WV. CA, ID, WA; throughout US and Canada. Native.

Goodyera oblongifolia is the largest of the rattlesnake plantains in North America. There is considerable variation in the whitish reticulations on the leaves. The most prominently patterned individuals have been described as variety or forma *reticulata*, but recognition of this seems superfluous.

NEOTTIA Twayblade

Herbs perennial, to 30 cm tall. **Stems** erect, green. **Leaves** cauline, 2(3), opposite, ovate to elliptic to cordate. **Inflorescences** loosely to densely flowered terminal racemes, 5–40 or more-flowered. **Flowers** widely spreading; sepals and petals ovate, elliptic, lanceolate to linear, spreading to acutely reflexed; lip basally constricted; margin entire or minutely ciliolate, basally toothed or lobed; stamen 1. **Fruits** ellipsoidal to subglobose.

Temperate and boreal North America and Eurasia. 50–60 species; 4 species treated in *Flora*.

The opposite leaves are an easily distinguishable hallmark of this genus.

Key to *Neottia* species:

1\. Lip deeply cleft into 2 linear-lanceolate lobes . . . . . ***N. cordata***

1' Lip entire or shallowly cleft into 2 lobes.

  2\. Distal portion of lip more or less the same width as the basal portion . . . . . ***N. banksiana***

  2' Distal portion of lip much wider than basal portion.

    3\. Base of lip tapered to a narrow claw with 2 small lateral lobes lacking associated glands. . . . . ***N. convallarioides***

    3' Base of lip narrowed with 2 distinct forward-thrust horns arising from a pair of glands; 2 distinct stripes on lip originating from basal glands. . . . . ***N. borealis***

Neottia banksiana Reichenbach f.

Northwestern twayblade

Listera banksiana Lindl., *Listera caurina* Piper, *Ophrys caurina* (Piper) Rydb.

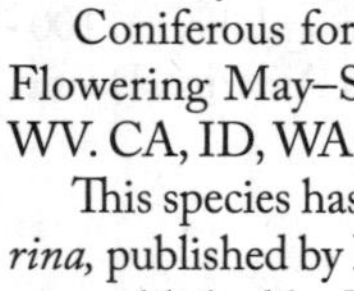

Herbs 10–30 cm tall. **Leaves** ovate to elliptic, 2.5–5 × 2–4 cm. **Inflorescences** loosely 5–20-flowered. **Flowers** sepals lanceolate, approximately 4 × 1 mm, yellowish green, widely spreading to reflexed; petals linear, 3 × 0.5 mm, yellowish green, widely spreading to reflexed; lip ovate, 6 × 4 mm, yellowish green, horizontal, basal discs or glands 2, each with a small forward-thrust horn, apex rounded or shallowly cleft, margin entire.

Coniferous forests, sheltered edges of meadows, roadsides. Flowering May–Sep. 100–2200 m. BW, Casc, CR, Est, Sisk, WV. CA, ID, WA; north to AK, east to WY. Native.

This species has long been known by the specific epithet *caurina*, published by Piper in 1898; however, the epithet *banksiana* was published by Lindley in 1840 and therefore has priority.

Neottia borealis (Morong) Szlachetko

Northern twayblade

Listera borealis Morong

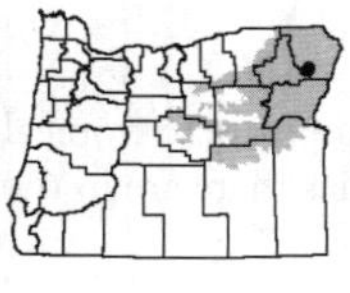

Herbs 7–15 cm tall. **Leaves** ovate to lanceolate, 2–3.5 × 1–2.5 cm. **Inflorescences** loosely 5–20-flowered. **Flowers** sepals elliptic to lanceolate, approximately 5 × 2 mm, pale green, strongly reflexed; petals linear, approximately 3 × 0.5 mm, pale green, strongly reflexed; lip oblong, slightly tapered in middle, approximately 10 × 6 mm, pale green, horizontal, darker central leading to a pair of low protuberances, base broadly lobed and bluntly angled, apex expanded into 2 lobes with a small tooth in between, margin entire.

Coniferous forests. Flowering Jun–Aug. 1300–2000 m. BW. ID, WA; boreal North America and Rocky Mountains. Native.

Neottia borealis reaches its southwestern most extent in Oregon.

Neottia convallarioides (Swartz) Richard (p. 325)

Broad-lipped twayblade

Listera convallarioides (Sw.) Nutt. ex Elliott, *Ophrys convallarioides* (Sw.) W. Wight

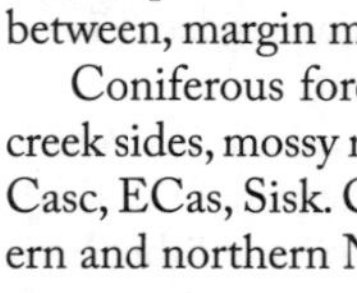

Herbs 10–20 cm tall. **Leaves** ovate to elliptic, 2–5.5 × 2–4 cm. **Inflorescences** loosely 5–20-flowered. **Flowers** sepals lanceolate, 4–5 × 1–2 mm, pale green, strongly reflexed; petals linear to lanceolate, approximately 4 × 1 mm, pale green, strongly reflexed; lip cuneate, approximately 10 × 5 mm, translucent to pale green, horizontal to descending, base constricted with a narrow claw and pair of blunt lateral teeth, apex shallowly cleft into 2 lobes with a small tooth in between, margin minutely ciliolate. **2*n***=36.

Coniferous forests, damp sheltered roadsides, wet ditches, creek sides, mossy rocks. Flowering Jun–Aug. 300–2300 m. BW, Casc, ECas, Sisk. CA, ID, NV, WA; throughout much of western and northern North America. Native.

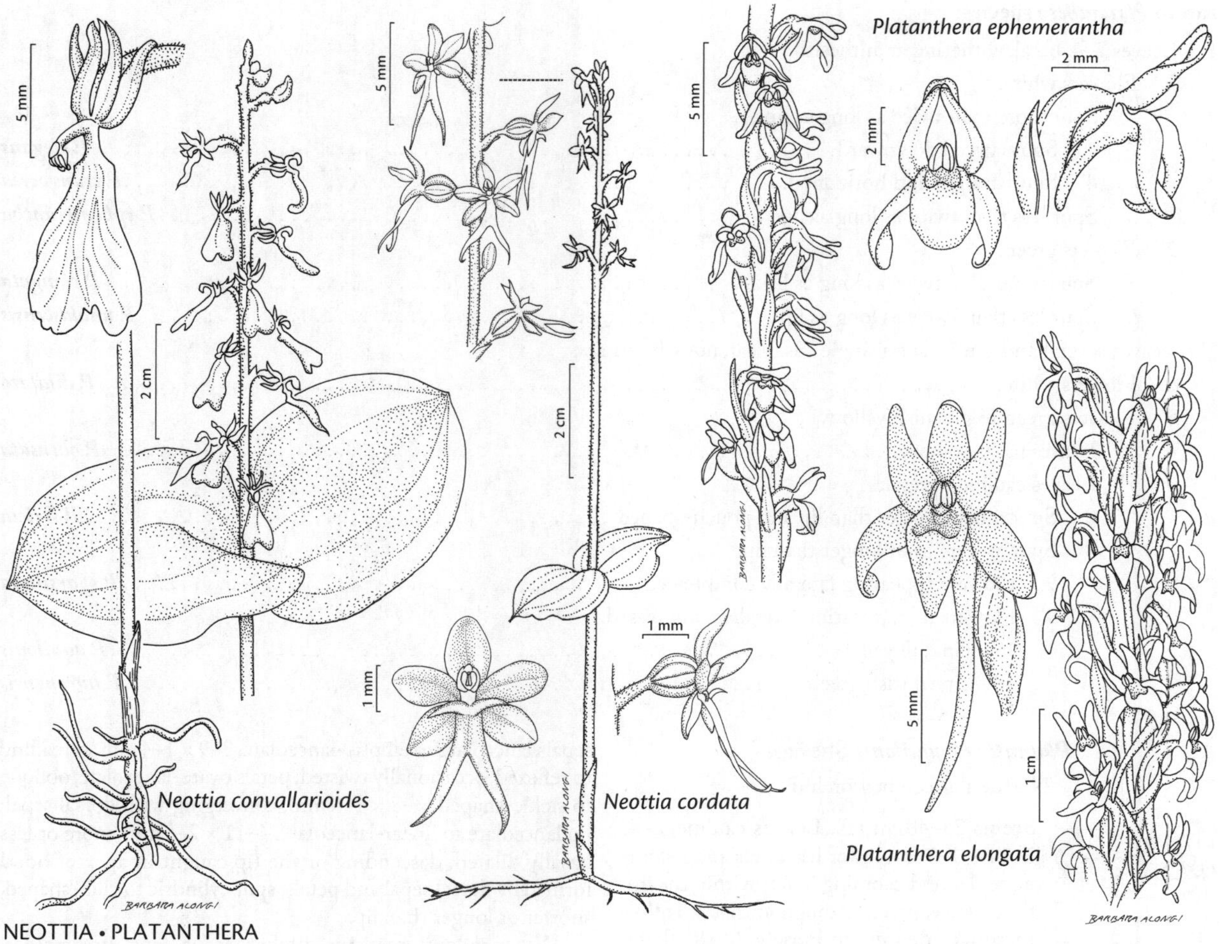

NEOTTIA • PLATANTHERA

Neottia cordata (Linnaeus) Richard (p. 325)

Western heart-leaved twayblade

Listera cordata (L.) R. Br., *Listera cordata* (L.) R. Br. var. *cordata*, *Listera cordata* (L.) R. Br. var. *nephrophylla* (Rydb.) Hultén, *Ophrys cordata* L.

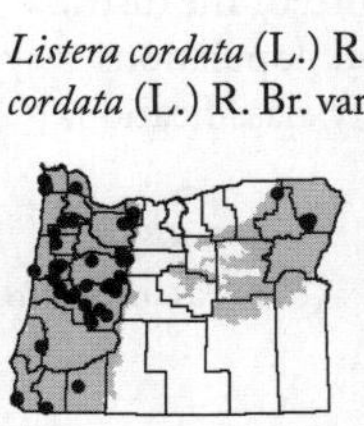

Herbs 6–25 cm tall. **Leaves** ovate to cordate, 1.5–2.5 × 1–3 cm. **Inflorescences** loosely to densely 5–20-flowered. **Flowers** sepals and petals similar, ovate, 2–3 × 1 mm, yellowish green sometimes tinged with purple to reddish purple, widely spreading to slightly reflexed; lip linear, 3–4 × 1–1.5 mm, yellowish green sometimes tinged with purple to reddish purple, descending, basal disc or gland with pair of spreading lobes. **2*n***=36, 38, 40, 42.

Coniferous forests, stream sides, seeps, *Sphagnum* bogs. Flowering Mar–Aug. 0–2200 m. BW, Casc, CR, Est, Sisk, WV. CA, ID, NV, WA; throughout North America; Eurasia. Native.

Not unexpectedly, given its large range, *Neottia cordata* is a highly variable species, both in terms of overall size and flower color. This variation has been recognized at the varietal rank, under *Listera*, as *L. cordata* var. *nephrophylla*. The appropriate combination within *N. cordata* has not been made and proposed, however. Variety *nephrophylla* is characterized by more robust plants with exclusively green flowers. Yet, size characters overlap within varieties, and western plants can have a mixture of flower colors, thus undermining the argument for separate varieties. At this time, recognition of varieties in *N. cordata* seems unwarranted.

PLATANTHERA Bog orchid, rein orchid, fringed orchid

Herbs perennial, to 100 cm tall. **Stems** erect, delicate to robust, green. **Leaves** basal to cauline, 1–several, elliptic, lanceolate, oblanceolate or ovate, becoming reduced towards inflorescence. **Inflorescences** lax to dense terminal racemes with 2–100 or more flowers. **Flowers** upside down due to twisting of pedicel, more or less spreading; dorsal sepal and lateral petals often converging and forming a hood over the short column; sepals ovate, elliptic, to lanceolate; petals ovate to lanceolate; lip entire; spur present and distinct, linear-elliptic to lanceolate; stamen 1. **Fruits** elliptic to ellipsoidal.

Circumboreal in predominantly temperate woodlands, meadows, fens, bogs, coastal bluffs, montane forests, some tropical forests. 100–200 species; 11 species treated in *Flora*.

Although separated by some authorities, recent molecular studies have consistently shown the segregate genus *Piperia* as being nested within an expanded *Platanthera*.

Key to *Platanthera* species:

1. Leaves 2–4, basal, withering at anthesis.
    2. Flowers white.
        3. Spur more than twice as long as lip.
            4. Spur curved downward . . . . . ***P. elegans***
            4' Spur straight and horizontal . . . . . ***P. transversa***
        3' Spur less than twice as long as lip . . . . . ***P. ephemerantha***
    2' Flowers green.
        5. Spur more than twice as long as lip . . . . . ***P. elongata***
        5' Spur less than twice as long as lip . . . . . ***P. unalascensis***
1' Leaves several and cauline, or a single basal leaf, not withering.
    6. Flowers white . . . . . ***P. dilatata***
    6' Flowers green to greenish yellow.
        7. Leaves basal, solitary . . . . . ***P. obtusata***
        7' Leaves cauline, several.
            8. Spur much shorter than lip and pouch-shaped . . . . . ***P. stricta***
            8' Spur as long as or longer than lip.
                9. Flowers appearing laterally compressed . . . . . ***P. sparsiflora***
                9' Flowers not appearing laterally compressed.
                    10. Lip dull yellow . . . . . ***P. aquilonis***
                    10' Lip whitish green . . . . . ***P. huronensis***

Platanthera aquilonis Sheviak

Northern green bog orchid

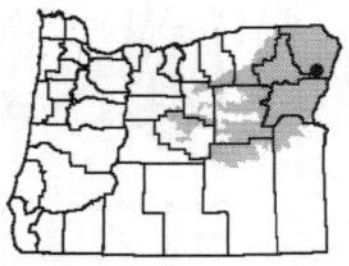

Stems 25–40 cm tall. **Leaves** cauline, 2–4, 5–14 × 1–3 cm; lower leaves clasping stem; upper leaves becoming bracts within the inflorescence, green through anthesis. **Inflorescences** densely to loosely 20–40-flowered. **Flowers** yellowish green, more or less spreading; dorsal sepal ovate, 3–7 × 2–4 mm; lateral sepals elliptic-lanceolate and oblique, 3–6 × 1–3 mm, spreading to reflexed; petals ovate-lanceolate, sickle-shaped, 3–6 × 1–3 mm; lip rhombic-lanceolate, basally dilated, dull yellowish, 2.5–6 × 2–3 mm, descending or the apex caught in the tip of the converging dorsal sepal and petals; spur more or less equal in length to lip, cylindrical and tapering, 4–6 mm.

Wet meadows, roadside ditches, streamsides, fens, seeps, springs. Flowering May–Aug. 1500–1900 m. BW. CA, ID, NV, WA; most of Canada and the northern US. Native.

Flowers of *P. aquilonis* commonly self-pollinate: the pollen masses often obviously contact the stigma in mature flowers. *Platanthera aquilonis* and *P. huronensis* can generally be distinguished from one another, in the field, by the dull yellow lip in the former and the whitish green lip in the latter. Unfortunately, this character is often indiscernible on pressed specimens.

Platanthera dilatata (Pursh) Lindley ex Beck

White bog orchid, bog candles, rein orchid

Stems 15–100 or more cm tall. **Leaves** cauline, 4–12, linear, linear-lanceolate, oblong or oblanceolate, 6–27 × 1–4 cm; lower leaves clasping stem and upper leaves becoming bracts within the inflorescence, green through anthesis. **Inflorescences** densely to loosely 20–100 or more-flowered. **Flowers** more or less spreading, white; dorsal sepal ovate-elliptic, 3–7 × 1.5–4 mm, converging with petals forming a hood over the column; lateral sepals lanceolate to elliptic-lanceolate, 3–9 × 1–4 mm, spreading to reflexed, occasionally twisted; petals ovate-lanceolate, oblique to sickle-shaped, 3–8 × 1–4 mm, converging with dorsal sepal; lip lanceolate to linear-lanceolate, 4–11 × 2–5 mm, more or less basally dilated, descending or the tip caught in apex of hood formed by dorsal sepal and petals; spur cylindric to club-shaped, shorter or longer than lip.

Wet meadows, roadside ditches, seeps, fens, streamsides, bogs.

3 varieties.

The highly variable spur length in *P. dilatata* has resulted in three weakly separated varieties being recognized. The distributions of all three varieties overlap, and mixed populations may be found, including individuals that may defy classification.

Key to *Platanthera dilatata* varieties:

1. Spur length approximately equal to lip . . ***P. d.*** var. ***dilatata***
1' Spur either shorter or longer than lip.
    2. Spur much longer than the lip, cylindric . . . . . ***P. d.*** var. ***leucostachys***
    2' Spur much shorter than the lip, club-shaped . . . . . ***P. d.*** var. ***albiflora***

Platanthera dilatata (Pursh) Lindley ex Beck var. ***albiflora*** (Chamisso) Ledebour

Short-spurred bog orchid

Habenaria dilatata (Pursh) Hook. var. *albiflora* (Cham.) Correll

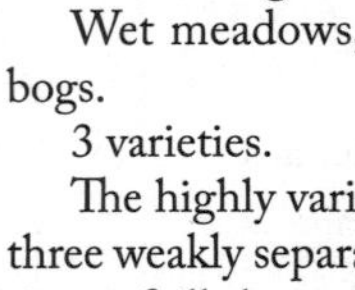

Spur generally clavate to clavellate, much shorter than lip.

Flowering Jun–Aug. 1000–2400 m. BW. ID, NV, WA; north to AK, east to CO. Native.

Platanthera dilatata (Pursh) Lindley ex Beck var. ***dilatata***

White bog orchid, bog candles, rein orchid

Habenaria dilatata (Pursh) Hook. var. *dilatata*, *Limnorchis dilatata* (Pursh) Rydb.

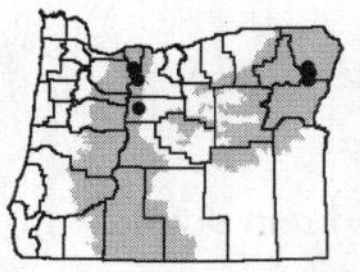

Spur cylindric, length approximately equal to lip.

Flowering Jun–Jul. 200–2400 m. BW, Casc, ECas. ID, NV, WA; throughout northern North America. Native.

The nominate variety is widespread but generally less common in Oregon than var. *leucostachys*.

Platanthera dilatata (Pursh) Lindley ex Beck var. ***leucostachys*** (Lindley) Luer

Long-spurred bog orchid, Sierra rein orchid

Habenaria dilatata (Pursh) Hook. var. *leucostachys* (Lindl.) Ames, *Habenaria leucostachys* (Lindl.) S. Watson, *Limnorchis leucostachys* (Lindl.) Rydb. var. *leucostachys*, *Limnorchis leucostachys* (Lindl.) Rydb. var. *robusta* Rydb., *Platanthera leucostachys* Lindl.

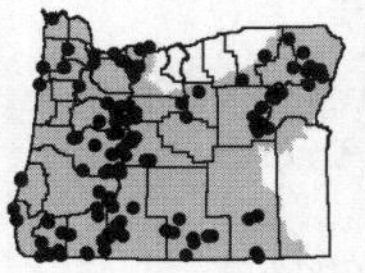

Spur cylindric, distinctly longer than lip.

Flowering Jun–Aug. 0–2800 m. BR, BW, Casc, CR, ECas, Est, Lava, Sisk, WV. CA, ID, NV, WA; north to AK, east to CO. Native.

This is the most common variety of bog candle in Oregon. The lip may be only moderately dilated compared to the other varieties.

Platanthera elegans Lindley

Elegant rein orchid, hillside rein orchid

Habenaria greenei Jeps., *Piperia elegans* (Lindl.) Rydb. ssp. *elegans*, *Piperia maritima* (Greene) Rydb., *Piperia multiflora* Rydb.

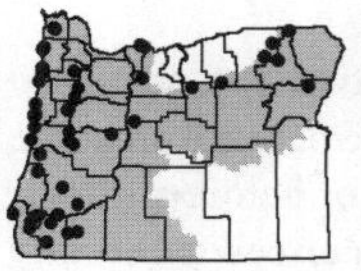

Stems 15–65 cm tall. **Leaves** 2–4, oblong to oblanceolate, 10–15 × 1–4 cm; apex obtuse; basal leaves prostrate and flaccid, withering at or before anthesis. **Inflorescences** densely to loosely 20–60 or more-flowered. **Flowers** more or less spreading, white to whitish green; dorsal sepal ovate-lanceolate, 3–5 × 1–2 mm, often with a greenish central stripe; lateral sepals oblong-lanceolate, slightly oblique, 3–5 × 1–2 mm, spreading to recurved; petals ovate-lanceolate, slightly oblique, 2–5 × 1–2 mm, spreading, often with a greenish central stripe; lip triangular-lanceolate, dilated at base and constricted below the middle, 2–6 × 2–3 mm, descending to deflexed; spur curved and descending, much longer than lip, 15–18 mm, often white or greenish white fading to green at tip. **2*n***=42.

Shady to open forests and woodlands, meadows, shrubby hillsides, roadsides, coastal bluffs. Flowering Jun–Aug. 0–1400 m. BW, Casc, CR, ECas, Est, Lava, Sisk, WV. CA, ID, WA; north to British Columbia, northeast to MT. Native.

Plants in open sites or on coastal bluffs are often compact with densely flowered stems.

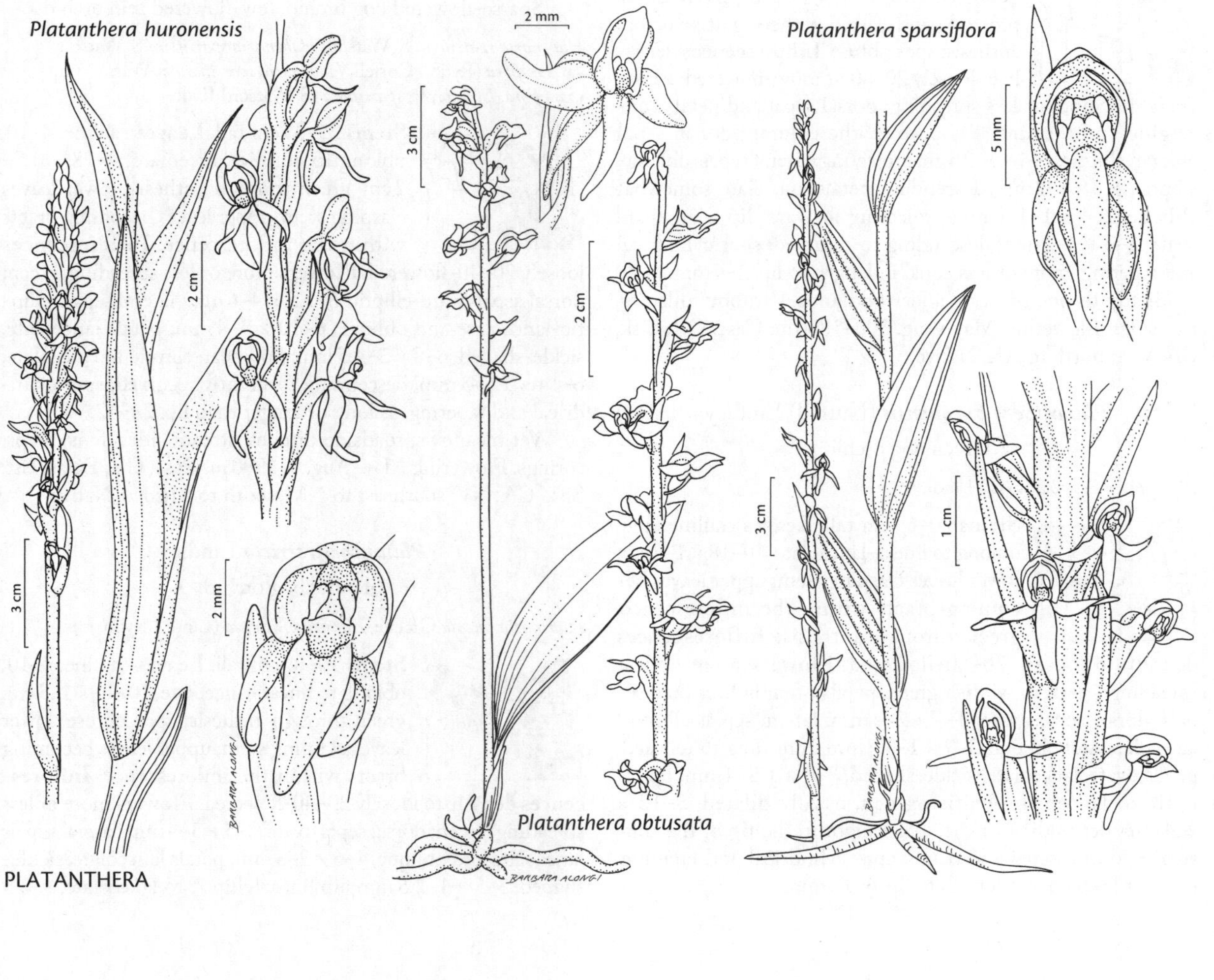

PLATANTHERA

Platanthera elongata (Rydberg) R.M. Bateman (p. 325)

Long-spurred piperia, chaparral orchid, wood rein orchid

Habenaria unalascensis (Spreng.) S. Watson var. *elata* (Jeps.) Correll, *Piperia elongata* Rydb.

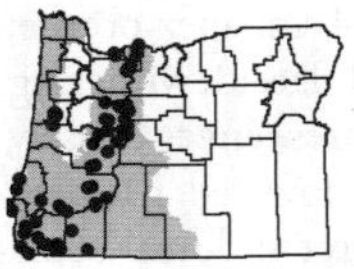

Stems 20–40 cm tall. **Leaves** 2–4, oblong to oblanceolate, 6–30 × 2–8 cm; basal, prostrate and flaccid, withering at or before anthesis; apex obtuse. **Inflorescences** generally densely 20–60 or more-flowered. **Flowers** green, more or less spreading; dorsal sepal ovate-lanceolate, 2.5–6 × 1–3 mm; lateral sepals oblong-lanceolate, slightly oblique, 2.5–6 × 1–3 mm, spreading to recurved; petals ovate-lanceolate, oblique to sickle-shaped, 3–6 × 1.5–2.5 mm, spreading; lip triangular-lanceolate, dilated at base, 2–5 × 2–3 mm, strongly deflexed; spur curved and descending, much longer than lip, 9–15 mm, often white or greenish white fading to green at tip. **2*n***=42.

Shady to open forests and woodlands, meadows, shrubby hillsides, roadsides. Flowering Jul–Aug. 300–1400 m. BW, Casc, Sisk. CA, ID, WA; north to British Columbia, east to MT. Native.

This species in known to have a sweet nocturnal scent.

Platanthera ephemerantha R.M. Bateman (p. 325)

White-flowered piperia

Piperia candida Morgan & Ackerman

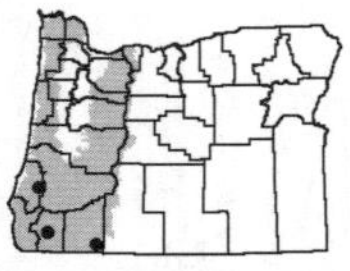

Stems 25–45 cm tall. **Leaves** 2–4, oblong to oblanceolate, 5–18 × 1–4 cm; basal leaves prostrate and flaccid withering at or before anthesis; apex obtuse. **Inflorescences** densely to loosely 20–60 or more-flowered. **Flowers** white, more or less spreading; dorsal sepal and petals converging and forming a hood over the column; dorsal sepal oblong-elliptic, 2–4 × 1–2 mm, projecting; lateral sepals oblong-elliptic, 2–4 × 1–2 mm, descending; petals lanceolate, somewhat oblique, 2–4 × 1–1.5 mm, projecting forward; lip triangular-ovate, 2–3 × 1–3 mm, descending to recurved; spur cylindrical and tapering, more or less equal in length to lip, 2–4 mm.

Shady to open forests and woodlands, shrubby hillsides, roadsides. Flowering May–Aug. 300–1800 m. Casc, CR, Sisk. CA, WA; north to AK. Native.

Platanthera huronensis (Nuttall) Lindley (p. 327)

Green bog orchid

Platanthera hyperborea (L.) Lindl.

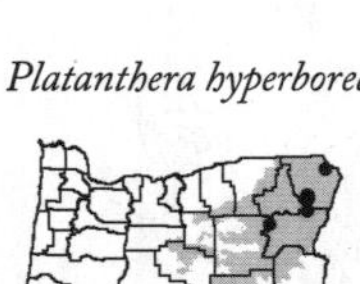

Stems 20–65 cm tall. **Leaves** cauline, 2–4, oblong to linear-lanceolate, 10–18 × 1–2 cm; lower leaves clasping stem; upper leaves becoming bracts within the inflorescence, green through anthesis. **Inflorescences** densely to loosely 20–60-flowered. **Flowers** more or less spreading; green to whitish green, petals often lighter than sepals; dorsal sepal ovate, 3–7 × 2–4 mm; lateral sepals elliptic-lanceolate and oblique, 3–9 × 1–3.5 mm, spreading to reflexed; petals ovate-lanceolate, sickle-shaped, 3–8 × 1.5–4 mm; lip elliptic to lanceolate whitish green, basally dilated, 5–12 × 2–4 mm, descending or the apex caught in the tip of the converging dorsal sepal and petals; spur cylindrical and tapering, more or less equal in length to lip, 4–12 mm.

Wet meadows, roadside ditches, stream sides, fens, seeps, springs. Flowering Jun–Aug. 100–2000 m. BW. Throughout northern North America. Native.

Platanthera huronensis can be difficult to differentiate from *P. aquilonis*, especially in dried, pressed examples (see *P. aquilonis*).

Platanthera obtusata (Banks ex Pursh) Lindley (p. 327)

Small northern bog orchid, blunt-leafed rein orchid

Habenaria obtusata (Banks ex Pursh) Richardson

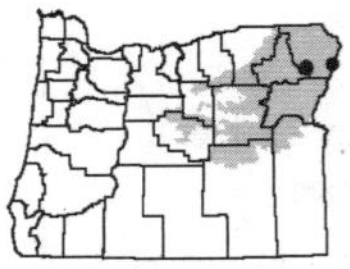

Stems 10–25 cm tall. **Leaves** basal, 1, clasping base of stem, obovate to oblanceolate, 5–8 × 1.5–3 cm, green through anthesis. **Inflorescences** loosely 10–15-flowered. **Flowers** more or less spreading, green to greenish white; dorsal sepal broadly ovate with blunt apex, 2–5 × 2–4 mm; lateral sepals lanceolate and oblique, 3–5 × 1.5–2 mm, strongly reflexed; petals lanceolate, oblique, tapering abruptly at middle, 2–4 × 1–2 mm; lip narrowly lanceolate, 3–6 × 1–2 mm, descending and reflexed; spur cylindrical and tapering, as long or longer than lip, to 10 mm.

Mixed conifer forests. Flowering Jun–Aug. 1500–1800 m. BW. ID, WA; throughout northern North America; northern Eurasia. Native.

The single basal leaf is diagnostic for *P. obtusata.*

Platanthera sparsiflora (S. Watson) Schlechter (p. 327)

Sparse-flowered bog orchid, few-flowered rein orchid

Habenaria sparsiflora S. Watson, *Habenaria sparsiflora* S. Watson var. *laxiflora* (Rydb.) Correll, *Habenaria sparsiflora* S. Watson var. *sparsiflora*, *Limnorchis sparsiflora* (S. Watson) Rydb.

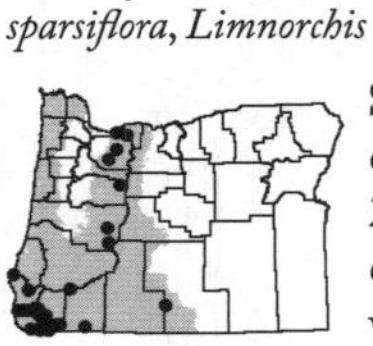

Stems 25–65 cm tall. **Leaves** cauline, 4–10, oblong to elliptic-lanceolate, 9–18 × 0.5–2 cm, green through anthesis; lower leaves clasping stem; upper leaves becoming bracts within the inflorescence. **Inflorescences** loosely 20–40-flowered. **Flowers** more or less spreading, green; dorsal sepal ovate-elliptic, 5–7.5 × 4–6 mm; lateral sepals elliptic-lanceolate and oblique, 6–10 × 3–4 mm; petals lanceolate, sickle-shaped, 6–8 × 3–4.5 mm; lip linear-elliptic to lanceolate, 6–14 × 1.5–3 mm, descending to slightly recurved; spur cylindrical and tapering, as long or longer than lip. **2*n***=42.

Wet meadows, roadside ditches, stream sides, fens, seeps, springs. Flowering May–Aug. 0–1900 m. Casc, CR, ECas, Est, Sisk. CA, NV; southeast to NM, south to Mexico. Native.

Platanthera stricta Lindley

Slender bog orchid

Habenaria saccata Greene, *Limnorchis stricta* (Lindl.) Rydb.

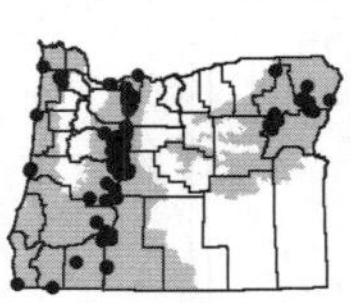

Stems 10–65 cm tall. **Leaves** cauline, 4–10; oblong to linear-lanceolate, 10–11 × 1–3 cm, green through anthesis; apex obtuse; lower leaves clasping stem; upper leaves becoming bracts within the inflorescence. **Inflorescences** densely to loosely 20–40-flowered. **Flowers** more or less spreading, green; dorsal sepal ovate, 3–5 × 3–4 mm; lateral sepals lanceolate and oblique, 4–6 × 2–3 mm; petals lanceolate, sickle-shaped, 3–5 × 1–2.5 mm; lip linear-elliptic to lanceolate, 4–8 ×

1–2 mm, descending; spur shorter than lip and distinctly pouch-shaped. **2*n***=42.

Wet meadows, roadside ditches, stream sides, fens, seeps, springs. Flowering Jun–Aug. 0–2300 m. BW, Casc, CR, ECas, Est, Sisk. CA, ID, WA; north to AK, east to WY. Native.

Widespread, *P. stricta* is easily identified by its distinctive pouch-like spurs.

Platanthera transversa (Suksdorf) R.M. Bateman

Flat-spurred piperia, Suksdorf's piperia

Piperia transversa Suksd.

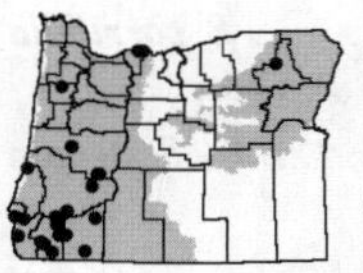

Stems 20–40 cm tall. **Leaves** basal, oblong to oblanceolate, 2–4; 6–20 × 1–5 cm, withering at or before anthesis; apex obtuse. **Inflorescences** generally densely 20–60-flowered. **Flowers** more or less spreading, white with green or yellow midveins; dorsal sepal ovate-lanceolate, 3–5 × 1–2 mm; lateral sepals oblong-lanceolate, slightly oblique, 3–5 × 1–2 mm, spreading to recurved; petals ovate-lanceolate, slightly oblique, 2–5 × 1–2 mm, spreading; lip oblong to ovate-elliptic, 2–5 × 2–3 mm, descending to deflexed; spur straight, much longer than lip and horizontal, 7–12 mm. **2*n***=42.

Shady to open forests and woodlands, meadows, shrubby hillsides, roadsides. Flowering Jun–Aug. 50–1300 m. BW, Casc, CR, ECas, Sisk, WV. CA, WA; north to British Columbia. Native.

This species is known to have a strong, clove-like nocturnal scent.

Platanthera unalascensis (Sprengel) Kurtz (p. 329)

Alaska rein orchid, short-spurred piperia

Habenaria unalascensis (Spreng.) S. Watson, *Piperia unalascensis* (Spreng.) Rydb.

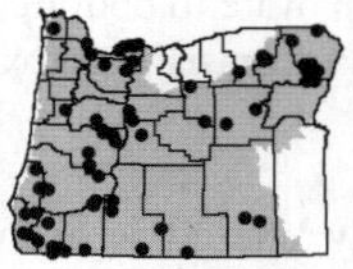

Stems 20–60 cm tall. **Leaves** basal, 2–4, oblanceolate, 10–12 × 1–3 cm, prostrate and flaccid; withering at or before anthesis; apex obtuse. **Inflorescences** densely to loosely 20–60 or more-flowered. **Flowers** more or less spreading, green to greenish yellow; dorsal sepal ovate, 2–4 × 1–2 mm, converging with petals; lateral sepals ovate, 2–4 × 1–2 mm, spreading to recurved; petals ovate-lanceolate, sickle-shaped, 2–4 × 1–1.5 mm, converging with dorsal sepal and forming a hood over the column; lip ovate-elliptic, 2–4 × 1–3 mm, often descending; spur more or less equal in length to lip, cylindrical and tapering, 3–4 mm. **2*n***=42.

Shady forests, meadows, rocky hillsides. Flowering Jun–Aug. 0–2200 m. BR, BW, Casc, CR, ECas, Lava, Sisk, WV. CA, ID, NV, WA; throughout western and northern North America. Native.

Coastal individuals may be distinctly shorter and more densely flowered than those from inland populations.

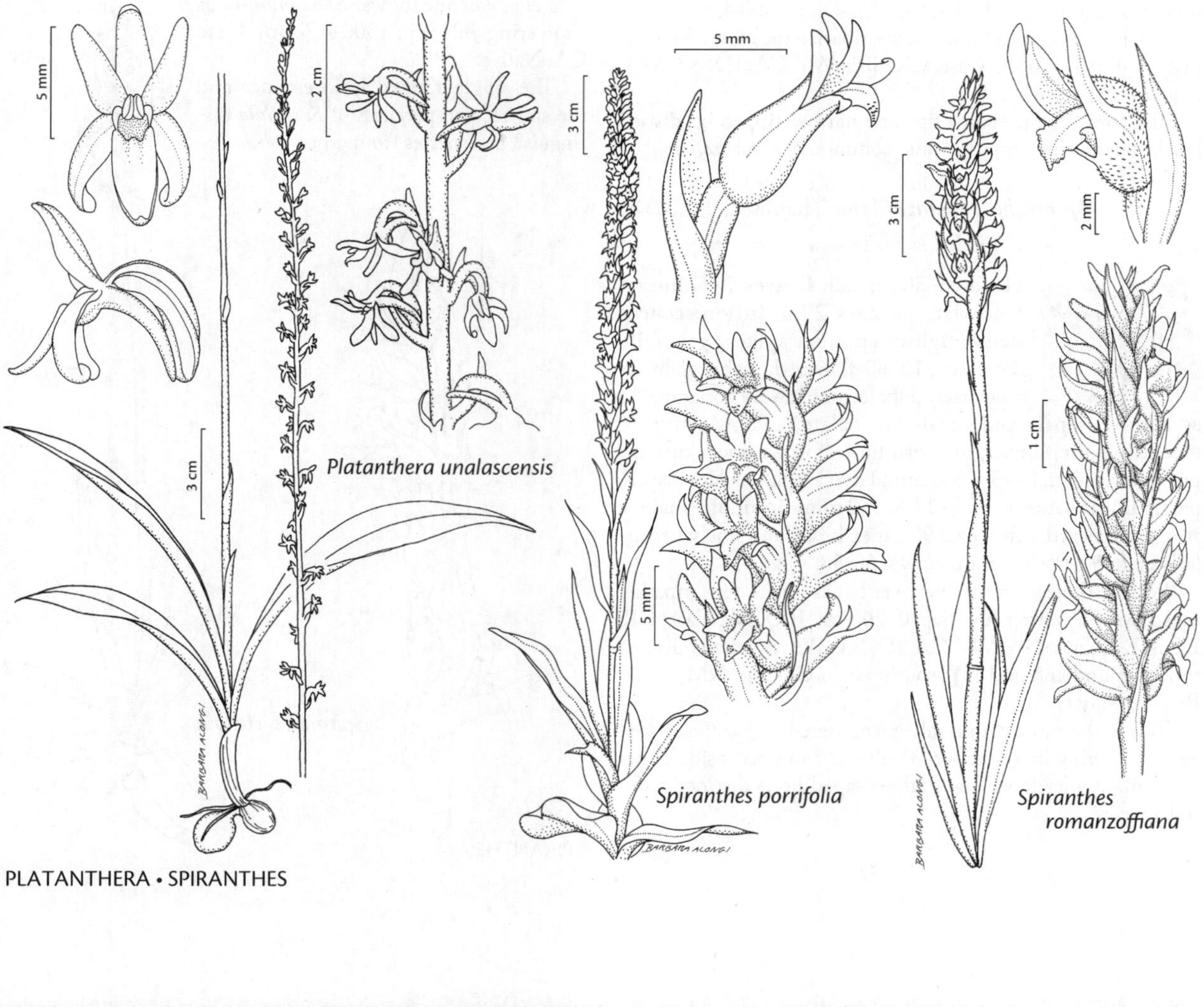

Platanthera unalascensis

Spiranthes porrifolia

Spiranthes romanzoffiana

PLATANTHERA • SPIRANTHES

SPIRANTHES Ladies tresses, Ladies'-tresses

Herbs perennial, to 55 cm tall. **Stems** erect with foliaceous sheaths. **Leaves** basal, 3–7, lanceolate to elliptic, grading into sheathing bracts. **Inflorescences** loose to dense spikes, single or more than 3-ranked, usually with many (10 or more) flowers; rachis glabrous to sparsely pubescent. **Flowers** sepals and petals similar, horizontal or ascending, lanceolate to oblanceolate, yellowish to creamy-yellow to white, often converging with lip; lip ovate to obovate, more or less reflexed; stamen 1.

North America, South America, Australia, Eurasia. Approximately 30 species; 3 species treated in *Flora*.

Key to *Spiranthes* species:

1. Lip fiddle-shaped, broad-tipped, strongly reflexed; sepal and petal tips appressed to form a hood over lip. . ***S. romanzoffiana***

1' Lip not fiddle-shaped, narrow-tipped, gently reflexed or merely spreading; sepal and petal tips more or less spreading.

 2. Inflorescences usually 3-ranked; lip concave, not strongly folded; sepals and petals converging for most of their length, only the tips spreading; flowers yellowish to creamy-yellow. . . . . ***S. porrifolia***

 2' Inflorescences usually single-ranked; lip strongly folded; sepals and petals converging for only about half of their length, the distal third widely spreading; flowers pale cream to white . . . . ***S. stellata***

Spiranthes porrifolia Lindley (p. 329)

Western ladies tresses

Spiranthes romanzoffiana Cham. var. *porrifolia* (Lindl.) Ames & Correll

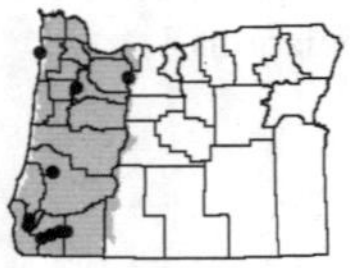

Herbs 25–55 cm tall. **Leaves** 3–4, elliptic-lanceolate, to 35 × 3 cm. **Inflorescences** dense tightly spiraled spikes, generally 3-ranked, 10–60-flowered; rachis glabrous to sparsely pubescent. **Flowers** ascending, sepals and petals lanceolate, creamy-yellow to yellowish, converging with lip to form a tube, tips spreading, 7–8 × 1–2.5 mm; lateral sepals oblique; lip ovate to lanceolate, concave to loosely folded with a narrowed apex, 6–8 × 3–4 mm. ***2n***=44.

Stream banks, river banks, seeps, riparian meadows. Flowering Jun–Jul. 0–2700 m. Casc, CR, Sisk, WV. CA, ID, NV, WA. Native.

The spreading perianth tips and narrow-tipped lip distinguish this species from the more common *S. romanzoffiana.*

Spiranthes romanzoffiana Chamisso (p. 329)

Hooded ladies tresses

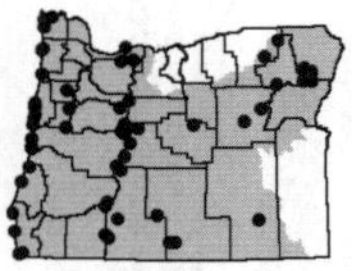

Herbs 6–30 cm tall. **Leaves** 3–6, linear-lanceolate, to 25 × 2 cm. **Inflorescences** dense tightly spiraled spikes, generally 3-ranked, 10–60-flowered; rachis glabrous to sparsely pubescent. **Flowers** horizontal to ascending, sepals and petals lanceolate, creamy-white to greenish white, converging with lip to form a hood, tips appressed together, 8–13 × 3–4 mm; lateral sepals sickle-shaped; petals linear-lanceolate, 8–11 × 1–2 mm; lip fiddle-shaped, strongly reflexed, often past 90°, apex broad and finely irregularly cleft, 5–10 × 2–6 mm. ***2n***=44, 66, 88.

Seaside dunes, stream and river banks, seeps, fens, riparian meadows. Flowering Jun–Sep. 0–2600 m. BR, BW, Casc, CR, ECas, Est, Lava, Sisk, WV. CA, ID, NV, WA; north to AK, east to Newfoundland and NY, southeast to CO and NM; Great Britain. Native.

This is the most commonly encountered *Spiranthes* in Oregon, occurring in a number of habitats from oceanside dunes to montane meadows. The fiddle-shaped lip is distinctive for this species.

Spiranthes stellata P.M. Brown, L.A. Dueck & K.M. Cameron (p. 330)

Starry ladies tresses

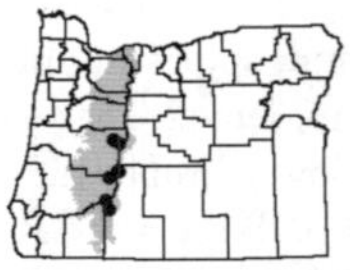

Herbs 15–45 cm tall. **Leaves** 4–6, lanceolate to oblanceolate, 16 × 2 cm. **Inflorescences** loosely spiraled spikes, usually single-ranked, 10–30-flowered; rachis glabrous. **Flowers** ascending, sepals and petals linear to oblanceolate, white to cream, converging with lip to form a short tube, distal 33% widely spreading, 5–12 × 1–3 mm; lip ovate to lanceolate, reflexed approximately 45°, strongly folded with a pointed apex, 4.5–10 × 1.7–6 mm.

Fens, wet meadows, *Sphagnum* bogs. Flowering Jul–Sep. 1300–1700 m. Casc. CA. Native.

The single-ranked inflorescence and the star-shaped perianth of *S. stellata* distinguish this species from other *Spiranthes*.

Spiranthes stellata

SPIRANTHES

POACEAE Grass family

Barbara L. Wilson, Richard Brainerd, and Nick Otting (except where noted)

Plants annual or perennial; cespitose, rhizomatous or stoloniferous. **Culms** usually herbaceous, erect to sprawling; internodes hollow or solid; nodes prominent and usually swollen. **Leaves** alternate, 2-ranked, each consisting of a sheath that encircles the culm and a blade; sheaths open or closed with margins fused for much or all of their lengths; auricles present or absent; ligules usually present at the sheath-blade junction, on the side toward the culm, membranous or of hairs, rarely absent; blades usually linear to lanceolate with parallel veins. **Inflorescences** usually complex aggregations of spikelets into panicles, racemes, or spikes, usually with a main axis; spikelets usually supported on pedicels; disarticulation above or below the glumes, usually beneath each floret, sometimes in or below the inflorescence axis. **Spikelets** consisting of 2 glumes subtending 1–many florets arranged on either side of the rachilla, laterally or dorsiventrally compressed, occasionally round in cross section. **Glumes** (0)2, with an odd number of veins, sometimes awned. **Florets** usually bisexual, sometimes pistillate, staminate or sterile, usually consisting of a lemma and palea with the flower between them, the flower itself reduced to stamens, a pistil, and 2 lodicules. **Lemmas** usually with an odd number of veins, awned or awnless, with calluses; lemma awns, if present, arising at the tip, along the back or near the base, sometimes curved, bent, or twisted. **Paleas** usually 2-keeled, often with additional veins. **Stamens** (0)3(6). **Fruits** usually a caryopsis.

Cosmopolitan. Approximately 700 genera; 99 genera treated in *Flora.*

With approximately 11,000 species, *Poaceae* is the fourth largest plant family in the world and the most economically and ecologically important. Grasses grow in most terrestrial habitats and are dominant or co-dominant in grasslands and savannas. The world's major grain crops are grasses, including corn, rice, wheat, barley, sorghum, millet, oats, and rye. Some grass species are serious weeds, threatening natural ecosystems, harming livestock, and competing with cultivated crops.

Arundo donax, giant reed, has spread very locally from ornamental plantings in Jackson County. Plants are strongly rhizomatous; culms (2)3–10 m tall, 2+ cm wide near the base; leaves 30–100 × 2–7(9) cm; inflorescence a fluffy panicle; lemmas 8–12 mm with hairs 4–9 mm. Efforts are being made to eradicate this population. It is an invasive weed in California.

The following identification hints may be helpful when keying within this family: If the inflorescence is a dense spike, break the inflorescence axis to isolate a node and count the spikelets attached there. Spikelet and lemma measurements exclude the awns. To determine spikelet compression, break off a single spikelet and lay it down. If both glumes are visible at the same time, it is laterally compressed. If only one glume is visible, it is dorsiventrally compressed.

Key to Poaceae groups:

1. Spikelets enclosed in a spiny, bur-like fascicle .... **CENCHRUS**
1' Spikelets not enclosed in a bur-like fascicle.
  2. Florets normally producing tiny bulbs instead of seeds .... ***Poa bulbosa***
  2' Florets normally producing seeds.
    3. Inflorescence a single spike, spike-like panicle, dense head, raceme, or single spikelet; panicle branches not easily seen at arm's length .... **Subkey 1**
    3' Inflorescence a panicle, not spike- or head-like; panicle branches easily seen at arm's length.
      4. Panicle branches spikes or spike-like, the inflorescence sometimes digitate; spikelets often paired .... **Subkey 2**
      4' Panicle branches not spike-like, sometimes further branched 2 or more times; spikelets usually not paired.
        5. Florets 1 per spikelet .... **Subkey 3**
        5' Florets 2 or more per spikelet (1 or more florets sometimes minute).
          6. Leaf sheaths closed for at least 25% of their length .... **Subkey 4**
          6' Leaf sheaths open to the base or very nearly so .... **Subkey 5**

Subkey 1: Inflorescence a spike, spike-like, or a head; branches none or obscure.

1. Spikelets 1 per inflorescence.
  2. Plants perennial .... ***Danthonia unispicata***
  2' Plants annual .... **BROMUS** (in part)
1' Spikelets 2 or more per inflorescence.
  3. Plant less than 20 cm tall and spikelets mostly hidden in short clusters of leaves that have stiff, sharp tips; plants of the Great Basin .... **MUNROA**
  3' Plants in one or more aspects not as above.
    4. Leaf sheaths closed more than half their length.
      5. Disarticulation above the glumes, below the florets; leaf sheaths usually with spreading or shaggy hairs; plants tufted to rhizomatous, annual or perennial. .... **BROMUS** (in part)
      5' Disarticulation below the glumes; leaf sheaths minutely scabrous; plants cespitose, perennial .... ***Melica stricta*** (in part)

4' Leaf sheaths open, or closed less than 25% of their length.
- 6. Inflorescence 1-sided or with 1-sided clusters of spikelets.
  - 7. Individual spikelets 20–40(50) mm long, spreading to reflexed. . . . . **PLEUROPOGON**
  - 7' Individual spikelets 3–25 mm long, if 20–25 mm long, then spikelets appressed.
    - 8. Glumes absent or vestigial; leaves 0.5–1 mm wide, stiff and sharp pointed . . . . **NARDUS**
    - 8' Glumes present; leaves 0.5–14 mm wide, not stiff, not sharp pointed.
      - 9. Glumes rounded at tip; base of mature inflorescence usually hidden in leaves . . . . ***Sclerochloa dura***
      - 9' Glumes pointed to awned at tip; base of mature inflorescence well exserted from leaves.
        - 10. Inflorescence open and raceme-like; spikelets on curved pedicels; leaf sheaths closed to near the top. . . . . ***Melica stricta*** (in part)
        - 10' Inflorescence a densely condensed panicle or with densely condensed panicle branches; leaf sheaths open at least 75% of its length.
          - 11. Spikelets paired, 1 fertile and the other sterile and appearing to be a set of empty bracts . . . . **CYNOSURUS**
          - 11' Spikelets not paired, all normally fertile.
            - 12. Spikelets extremely flattened, parallel, arranged like teeth of a comb . . . . **SPARTINA**
            - 12' Spikelets moderately flattened, variously oriented, and not looking like the teeth of a comb . . . . ***Poa unilateralis*** (in part)
- 6' Inflorescence 2-sided or more or less round in cross section; clusters of spikelets not 1-sided.
  - 13. Spikelets clearly subtended by bristles longer than the spikelets. . . . . **SETARIA**
  - 13' Spikelets not subtended by bristles longer than the spikelets.
    - 14. Inflorescence axis thickened; all spikelets (or at least the lower spikelets) strongly appressed or sunken into the axis.
      - 15. Glumes with awns 6–9 mm long (at least on the upper spikelets). . . . . **AEGILOPS**
      - 15' Glumes acute to acuminate, awnless.
        - 16. Glumes 1 per spikelet (except on the terminal spikelet); spikelets oriented with 1 edge against the inflorescence axis.
          - 17. Spikelets with 1 functional floret (a 2nd, reduced floret sometimes present). . . . . . . **HAINARDIA**
          - 17' Spikelets with 2–20+ functional florets . . . . **LOLIUM** (in part)
        - 16' Glumes 2 per spikelet on all spikelets; spikelets oriented with the flat side against the inflorescence axis.
          - 18. Lemmas awnless. . . . . **PARAPHOLIS**
          - 18' Lemmas awned, the awns 2–4 mm long. . . . . **SCRIBNERIA**
    - 14' Inflorescence axis not clearly thickened; spikelets neither strongly appressed nor sunken into inflorescence axis.
      - 19. Spikelets 2–4 per node, often sessile (1 spikelet elevated on a stalk to 15 mm long in *Leymus flavescens*); inflorescences spikes or condensed racemes (branched in *L. condensatus*) . . . . . **Subkey 1a**
      - 19' Spikelets 1 per node, often on a short stalk; inflorescences spikes or condensed panicles or racemes.
        - 20. Florets 1 per spikelet.
          - 21. Each lemma with 3 awns . . . . **ARISTIDA**
          - 21' Each lemma with 0–1 awn.
            - 22. Spikelets 10–13 mm long.
              - 23. Leaves 4–8 mm wide when flat but usually involute and 0.5–2.5 mm in diameter; perennial plants of coastal sands . . . . **AMMOPHILA**
              - 23' Leaves 6–15+ mm wide, flat; annual plants of disturbed, usually inland habitats . . . . **TRITICUM**
            - 22' Spikelets 0.8–10 mm long.
              - 24. Glume keels with coarse hairs.
                - 25. Glumes awned.
                  - 26. Spikelets with a stipe and disarticulating below the glumes; panicle tending to be lobed . . . . **POLYPOGON** (in part)
                  - 26' Spikelet lacking a stipe, initially disarticulating above glumes (later also below the glumes); panicle an even-sided cylinder . . . . **PHLEUM**
                - 25' Glumes not awned.
                  - 27. Ligules membranous; lemmas awned . . . . **ALOPECURUS**
                  - 27' Ligules of hairs; lemmas awnless. . . . . **CRYPSIS**

24' Glume keels lacking coarse hairs.

28. Glumes swollen at base where distended by the floret, long-tapered; upper glumes 3–5 times as long as lemma body . . . . **GASTRIDIUM**

28' Glumes not swollen at base, tip obtuse to acuminate; upper glumes (excluding awn) to 2.5 times as long as lemma body.

29. Glumes clearly awned.

30. Spikelet disarticulating above glumes, floret falling alone; lemma 3-veined; lemma awn usually shorter than lemma body, sometimes lacking . . . . **MUHLENBERGIA** (in part)

30' Spikelet disarticulating below glumes and stipe, falling as 1 unit; lemma 5-veined; lemma awn lacking or usually longer than lemma body. . . . . **POLYPOGON** (in part)

29' Glumes truncate to long-tapered or mucronate, not clearly awned.

31. Lemma awned from the middle or below.

32. Glumes fused to each other in the lower half; calluses glabrous . . . . . . ***Alopecurus myosuroides***

32' Glumes free to the base; calluses with hairs, which may be small.

33. Paleas less than half as long as lemmas; callus hairs inconspicuous, much less than half as long as lemmas; glumes 1.5–5(7) mm . . . . **AGROSTIS**

33' Paleas subequal to lemmas; callus hairs conspicuous, more than half as long as lemmas; glumes (4)5–7(9) mm . . . . **CALAMAGROSTIS**

31' Lemma awned from the tip or awnless.

34. Glumes 2–8 mm long, often winged; lemmas 2–6.8 mm long; culm (10)20–230 cm tall; leaves (0.5)2–20 mm wide; inflorescence 0.8–4 cm wide . . . **PHALARIS** (in part)

34' Glumes 1.5–3.2 mm long, not winged; lemmas 1.5–2.6(3.2) mm long; culm (3)5–30 cm tall; leaves 0.5–6 mm wide; inflorescence 0.1–1.7 cm wide . . . . **MUHLENBERGIA** (in part)

20' Florets 2–20+ per spikelet.

35. Fertile florets 1, usually subtended by 2 sterile florets (may be reduced to tiny soft bristles in *Phalaris*).

36. Sterile florets longer than the fertile floret, awned . . . . **ANTHOXANTHUM**

36' Sterile florets usually shorter than the fertile floret, awnless . . . . **PHALARIS** (in part)

35' Fertile florets (1)2 or more; sterile florets, if any, above fertile floret(s).

37. Glumes 1, on outside edge of spikelet (terminal spikelet with 2 glumes) . . . . **LOLIUM** (in part)

37' Glumes 2 on all spikelets.

38. Spikelets sessile or subsessile.

39. Spikelets (4.8)5–6 times as long as wide; inflorescences nodding . . . . **BRACHYPODIUM** (in part)

39' Spikelets 1–4 times as long as wide; inflorescences nodding or erect . . . . **Subkey 2a**

38' Spikelets on short stalks.

40. Spikelets (15)20–40(50) mm long . . . . **BRACHYPODIUM** (in part)

40' Spikelets 2.5–12 mm long.

41. Upper and lower glumes of different shapes (widest at different points along their total lengths).

42. Upper glumes 1.5–2.5 mm long, rounded to truncate; lemmas 1.9–2.8 mm long. . . . . **SPHENOPHOLIS**

42' Upper glumes 2.5–5 mm long, acute; lemmas 2.5–6.5 mm long . . . . **KOELERIA** (in part)

41' Glumes nearly alike in shape, though often differing in length.

43. Lemmas not awned.

44. Plants of sea cliffs; panicle branches papillate with very sparse scabers; panicles dull . . . . ***Poa unilateralis*** (in part)

44' Plants usually inland, if on sea cliffs then panicle branches covered with soft hairs; panicles usually shiny.

45. Panicle branches antrorsely scabrous, not covered with soft hairs. . . . . ***Trisetum wolfii***

45' Panicle branches covered with soft hairs. . . . . **KOELERIA** (in part)

43' Lemma awned from tip or back.

46. Lemma awned from the tip. . . . . **VULPIA**

46' Lemmas awned from the back.

47. Spikelets 2.8–3.8 mm long . . . . ***Aira praecox***

47' Spikelets 4–7 mm long . . . . **TRISETUM**

Subkey 1a: *Triticeae* Tribe–Wheatgrasses.

Most perennial wheatgrasses traditionally classified in the genus *Agropyron* have been transferred to *Elymus* or placed in the smaller segregate genera *Leymus, Pascopyrum, Pseudoroegneria*, and *Thinopyrum*. Except for thick-glumed *Thinopyrum*, these segregate genera are hard to distinguish concisely from *Elymus* in its current circumscription. Therefore, *Elymus* taxa appear repeatedly in this key.

When keying within this group, the number of spikelets per node should be assessed near the middle of the spike. If the number of spikelets per node is not obvious, break the spike to isolate a node. Glumes and lemmas may be reduced and awn-like. Curvature of lemma awns increases with maturity; if the plant appears to have any strongly curved lemma awns, it should be taken through the "strongly curved" side of the key.

1. Inflorescence axis thickened; lower spikelets strongly appressed or sunken into axis. . . . . **AEGILOPS**
1' Inflorescence axis not clearly thickened; spikelets neither strongly appressed nor sunken into axis.
  2. Spikelets 1 per node at most nodes.
    3. Lemma marginal veins keeled and strongly ciliate, lemmas otherwise smooth and glabrous. . . . . **SECALE**
    3' Lemma marginal veins not ciliate (but sometimes scabrous or hairy if the whole lemma surface is hairy), generally not keeled.
      4. Plant annual.
        5. Lemmas 5–7.5 mm long; spikes 1.3–2.4 cm long (excluding awns). . . . . **EREMOPYRUM**
        5' Lemmas 10–15 mm long; spikes (3.5)6–18 cm long (excluding awns) . . . . **TRITICUM**
      4' Plant perennial.
        6. Inflorescence axis internodes usually less than 3 mm long; spikelets more than 3 times as long as the internodes . . . . **AGROPYRON**
        6' Inflorescence axis internodes 4–15+ mm long; spikelets 0.5–3 times as long as the internodes.
          7. Glumes very thick, stiff and hard, truncate to acute, lacking awns; lemmas usually awnless . . . . **THINOPYRUM**
          7' Glumes thin and flexible, obtuse to acuminate, awned or not; lemmas awned or awnless.
            8. Lemmas with awns outcurving at maturity.
              9. Inflorescence disarticulating when mature; anthers 1–3 mm long.
                10. Inflorescence axis internodes 2.5–5(7) mm long; anthers 1–2 mm long, dehiscent or not. . . . . **ELYMUS** (in part)
                10' Inflorescence axis internodes (5)7–9 mm long; anthers about 2.5 mm long, indehiscent . . . . ×**PSEUDELYMUS**
              9' Inflorescence not disarticulating; anthers 2–8 mm long.
                11. Glume veins 1–3, glumes narrowly lanceolate . . . . ***Elymus wawawaiensis***
                11' Glume veins (3)4–5(7), glumes lanceolate to oblanceolate . . . . **PSEUDOROEGNERIA** (in part)
            8' Lemmas awnless or with straight awns.
              12. Plants cespitose, sometimes loosely so; lemmas glabrous to sparsely hairy.
                13. Anthers 4–8 mm long; spikes loose and open; spikelets usually 1.1–1.5 times as long as the internodes; inflorescence axis usually plainly visible. . . . . **PSEUDOROEGNERIA** (in part)
                13' Anthers 0.7–2.5(3.5) mm long; spikes compact; spikelets usually 1.5–3 times as long as the internodes of the inflorescence axis; inflorescence axis partly concealed to plainly visible. . . . . **ELYMUS** (in part)
              12' Plants strongly rhizomatous; lemmas glabrous to densely long-hairy.
                14. Spikelets 2 per node at most or all nodes in the middle of the spike, 1 per node near the base and tip of the spike . . . . **LEYMUS** (in part)
                14' Spikelets 1 per node at most or all nodes in all parts of the spike.
                  15. Blades 5–10 mm wide, usually flat; lower sheaths usually hirsute-pilose; lemma awns, if present, straight . . . . **ELYMUS** (in part)
                  15' Blades normally strongly involute or considerably less than 6 mm broad; lower sheaths usually glabrous; lemma awns, if present, often divergent.
                    16. Glumes oblong-lanceolate, broadest at or above mid-length, 3–9-veined; glume midvein straight; lemmas glabrous to copiously hairy. . . . . **ELYMUS** (in part)
                    16' Glumes lanceolate, tapered from near the base, 3–5-veined; glume midvein leaning slightly to one side; lemmas glabrous to moderately pubescent . . . . **PASCOPYRUM**

2' Spikelets 2–4(8) per node.

17. Plants sterile or nearly sterile; anthers usually aborted or narrow (less than 0.2 mm wide in smaller species and less than 0.5 mm wide in the larger) and indehiscent; caryopses usually not formed.

18. Rachis not disarticulating; awns 0–7 mm long, straight; plants rhizomatous; glumes 10–25 mm long (excluding awns) with 3–5 veins; glumes and lemmas generally partly or entirely hairy . . . ***Leymus* × *vancouverensis***

18' Rachis disarticulating or not; awns 0–50+ mm long, straight or curving; plants cespitose or rhizomatous; glumes 5–12(25) mm long, with 1–5 veins; glumes and lemmas usually glabrous or scabrous.

19. Longer awns more than 10 mm long, outcurving to straight.

20. Glume veins 2–3(5); internodes generally 4–10 mm long . . . . **ELYMUS** hybrids

20' Glume veins 1(5); internodes generally 2.5–7 mm long. . . . . **×ELYHORDEUM**

19' Longer awns 0–10 mm long, straight.

21. Lemmas with dense, long, spreading hairs; glumes (excluding awns) 8.5–18 mm long . . . . ***Leymus flavescens* × *L. triticoides***

21' Lemmas usually glabrous or scabrous; glumes (excluding awns) 5–7.5 mm long; spikelets sessile to pedicellate.

22. Spikes (excluding awns) 3–5 mm wide . . . . **×*Elyleymus aristatus***

22' Spikes (excluding awns) (8)14–30 mm wide. . . . . **×*Elyhordeum stebbinsianum***

17' Plants fertile; anthers not narrow and dehiscent at maturity; caryopses usually formed at maturity.

23. Spikelets consistently 3 per node.

24. Plants annual or perennial; central spikelet with 1 floret; lateral spikelets usually on pedicels. . . . **HORDEUM**

24' Plants perennial; central spikelet usually with 2 florets; lateral spikelets sessile and sterile . . . . ***Elymus elymoides*** ssp. ***hordeoides***

23' Spikelets 2–4(8) per node at most nodes, but not consistently 3 per node.

25. Some awns longer than the axis of the inflorescence; plants annual. . . . . **TAENIATHERUM**

25' Awns, if present, shorter than the axis of the inflorescence; plants perennial.

26. Lemma awns more than 3 mm long; inflorescence axis disarticulating or not at maturity.

27. Inflorescence axis disarticulating at maturity; awns outcurving at maturity . . . . . . **ELYMUS** (in part)

27' Inflorescence not disarticulating; awns straight or outcurving at maturity.

28. Glumes 1.5–4 mm wide; lemmas densely hairy with hairs 0.5–1 mm long . . . . ***Leymus mollis***

28' Glumes 0.6–2 mm wide; lemmas usually glabrous or scabrous . . . . **ELYMUS** (in part)

26' Lemma awns 0–3 mm long; inflorescence axis not disarticulating.

29. Glumes tapering very gradually from the base or from a point well below the middle . . . . **LEYMUS** (in part)

29' Glumes parallel-sided to above the middle, then tapering more abruptly.

30. Long-rhizomatous plants of unstable, sandy soils near the coast or Columbia River; inflorescences 1–3 cm wide . . . . ***Leymus flavescens***

30' Short-rhizomatous to loosely cespitose plants of more stable, better developed substrates, widely distributed; inflorescences 0.5–1 cm wide . . . . **ELYMUS** (in part)

Subkey 2: Inflorescence umbel-like or branches spike-like.

1. Inflorescence axis much shorter than the spike-like branches, the branches usually attached at more or less the same height.

2. Spikelets dorsiventrally compressed, awnless. . . . . **Subkey 2a**

2' Spikelets laterally compressed.

3. Lemmas conspicuously awned, the awns as long as or longer than the lemmas . . . . **CHLORIS**

3' Lemmas awnless or with inconspicuous awns shorter than the lemmas.

4. Plants perennial, creeping with shoots scattered . . . . **CYNODON**

4' Plants annual, tufted, all shoots arising from a concentrated area . . . . **ELEUSINE**

1' Inflorescence axis as long as or longer than most of the branches, the branches attached at markedly different heights.

5. Glumes semicircular, wrinkled, with inflated wings. . . . . **BECKMANNIA**

5' Glumes not semicircular, not winged or wings flat.

6. Spikelets dorsiventrally compressed; lower glume much reduced; lowest lemma sterile, glume-like . . . . . **Subkey 2a**

6' Spikelets laterally compressed; lower glume reduced or not; lowest lemma fertile, not glume-like.

7. Spikelets very flat and very tightly crowded on 1-sided branchlets, like teeth of a comb. . . . . **SPARTINA**

7' Spikelets not crowded, branchlets not 1-sided.

8. Lemmas (1)3-veined .... **DIPLACHNE**
8' Lemmas (3)5–9-veined.
 9. Lemmas awnless .... **DESMAZERIA**
 9' Lemma awns usually 1–22 mm long .... **VULPIA**

Subkey 2a: Spikelets dorsiventrally flattened; *Paniceae* Tribe.

1. Inflorescence a condensed, cylindrical panicle; each spikelet subtended by 1–many bristles .... **SETARIA**
1' Inflorescence not as above; spikelets not subtended by bristles though pustulose-based hairs may occur on the pedicel.
 2. Panicle branches more or less branched; panicles sometimes diffuse.
  3. Ligules lacking; upper glumes acuminate to long-awned .... **ECHINOCHLOA** (in part)
  3' Ligules of hairs; upper glumes obtuse to acute or acuminate, not awned.
   4. Basal rosettes well developed, usually persistent; leaves of 2 forms, the basal leaves shorter and usually wider than the cauline leaves; plants perennial .... **DICHANTHELIUM**
   4' Basal rosettes not well developed, not persistent; leaves all similar in form; plants annual or perennial .... **PANICUM**
 2' Panicle branches unbranched, spike-like; panicles not diffuse
  5. Ligules lacking; glumes and lemmas usually awned .... **ECHINOCHLOA** (in part)
  5' Ligules present; glumes and lemmas awnless.
   6. Plants annual or rarely perennial; fertile lemma flexible, thinner than sterile lemma and upper glumes; back of floret facing away from inflorescence axis .... **DIGITARIA**
   6' Plants perennial; fertile lemma firm to hard, thick, similar to sterile lemma and upper glume; back of fertile lemma facing inflorescence axis or orientation difficult to determine .... **PASPALUM**

Subkey 3: Inflorescence a panicle, branches not spike-like; floret 1 per spikelet.

1. Glumes lacking.
 2. Culms 1–5(10) cm tall; panicles with umbel-like clusters of branches .... **COLEANTHUS**
 2' Culms 20–300 cm tall; inflorescences more or less normal panicles.
  3. Leaf margins and surfaces sharply scabrous, apt to cut skin and tending to stick to clothing; plants to 150 cm tall; all parts of the panicle similar; spikelets bisexual .... **LEERSIA**
  3' Leaf margins and surfaces smooth or scabrous but not sharp and not sticking to clothing; plants 20–900 cm tall; lower part of panicle with staminate spikelets (which are narrower and fall early), upper part of panicle with wider pistillate spikelets that persist longer .... **ZIZANIA**
1' Glumes 2, well developed.
 4. Spikelets dorsiventrally flattened; spikelet consisting of 2 florets, the sterile lower floret glume-like; spikelet with a small lower glume (sometimes reduced to a shallow rim), a larger upper glume, a glume-like sterile lemma and a harder fertile lemma that often clasps the palea .... **Subkey 5, couplet 4**
 4' Spikelets more or less laterally flattened, sometimes round in cross section; spikelet consisting of 2 more or less similar glumes and 1 floret that may or may not be harder than the glume.
  5. Glumes winged with the keel expanded as thin, membranous tissue.
   6. Inflorescence open, branches easily discernible; glumes semicircular; wings wrinkled and more or less inflated .... **BECKMANNIA** (in part)
   6' Inflorescence dense, branches obscure; glumes not semicircular; wings flat, not wrinkled or inflated .... **PHALARIS** (in part)
  5' Glumes wingless.
   7. Glumes and lemmas awnless.
    8. Lemma stiffly membranous to hard, tightly enfolding the palea and fruit, usually round in cross section, not keeled .... **ACHNATHERUM**
    8' Lemma thin and flexible, not tightly enfolding palea and fruit, usually laterally flattened, often keeled.
     9. Lemma base clearly hairy-tufted, if lemmas evenly hairy, the hairs at least 1 mm long.
      10. Spikelets 10–13 mm, subsessile; inflorescence dense .... **AMMOPHILA**
      10' Spikelets 3–10 mm, usually clearly stalked; inflorescence dense to open.
       11. Lemma longer than the glumes; culms generally solid; paleas well developed, mostly subequal to the lemmas .... **MUHLENBERGIA** (in part)
       11' Lemma shorter than the glumes; culms never solid; paleas often reduced or lacking .... **AGROSTIS** (in part)
     9' Lemma base not hairy-tufted, if lemmas evenly hairy, the hairs less than 1 mm long.

12. Lower glume (or both glumes) shorter than the lemma.
13. Leaf sheath closed; leaves 2–13 mm wide; inflorescences open panicles . . . . **CATABROSA** (in part)
13' Leaf sheath open; leaves 0.6–6 mm wide; inflorescences condensed or open panicles.
14. Ligule of hairs; leaf sheath with a tuft of hairs at the top . . . . **SPOROBOLUS**
14' Ligule membranous (sometimes also minutely ciliolate); leaf sheath without a tuft of hairs at the top . . . . **MUHLENBERGIA** (in part)
12' Both glumes equaling or longer than lemma.
15. Spikelets disarticulating below the glumes; leaves 7–20 mm wide and panicle open.
16. Leaves 7–20 mm wide; panicle open . . . . **CINNA** (in part)
16' Leaves 1–6 mm wide; panicle condensed, somewhat lobed . . . . ***Polypogon viridis***
15' Spikelets disarticulating above the glumes; leaves 0.3–20 mm wide, if leaves more than 6 mm wide then panicle condensed.
17. Glumes (4)4.5–8.1 mm long; wider leaf blades more than 10 mm wide . . . . ***Phalaris arundinacea***
17' Glumes 1.5–4.3(5) mm long; leaf blades 2–8(10) mm wide.
18. Rachilla prolonged 0.1–1.9 mm beyond the base of the floret; palea nearly as long as the lemma. . . . . **PODAGROSTIS** (in part)
18' Rachilla not prolonged beyond the base of the floret; palea 0–75% as long as the lemma.
19. Anthers 0.6–0.7 mm long and palea 67–75% as long as the lemma; glumes 1.6–2 mm long; inflorescence rarely as much as 5 mm wide . . . . ***Podagrostis humilis***
19' Anthers 0.2–2.5 mm long; paleas missing to 67% as long as the lemma; glumes 1.5–5(7) mm long; inflorescence 3–150 mm wide . . . . **AGROSTIS** (in part)

7' Glumes or lemmas awned.
20. Lemmas with 3 awns . . . . **ARISTIDA**
20' Lemmas with 1 awn.
21. Lemmas stiffly membranous to hard, tightly enfolding the palea and caryopsis, generally round in cross section; awn 0.9–225 mm long . . . . **Subkey 3a**
21' Lemmas thin and flexible, not tightly enfolding the palea and caryopsis, usually laterally flattened; awn to 16 mm long.
22. Upper or both glumes abruptly short-pointed or awned from tip or from between 2 short teeth.
23. Glume awn as long as or longer than the glume body. . . . . **POLYPOGON**
23' Glume awn much shorter than the glume body.
24. Palea less than 67% as long as lemma; lemma 5-veined or veins too faint to see . .**AGROSTIS** (in part)
24' Palea generally as long as lemma; lemma 3-veined. . . . . **MUHLENBERGIA** (in part)
22' Glumes obtuse, acute, or tapered to tip, not awned.
25. Lemma awns more than 2 times as long as the spikelets.
26. Lemma awns subterminal, generally straight . . . . **APERA**
26' Lemma awns terminal, bent twice . . . . **Subkey 3a**
25' Lemma awns lacking or less than 2 times as long as the spikelets.
27. Lemmas awned below middle.
28. Lemmas with conspicuous callus hairs, easily visible with the naked eye or a 10× hand lens. . . . . **CALAMAGROSTIS** (in part)
28' Lemmas glabrous or with inconspicuous callus hairs.
29. Palea less than 60% as long as the lemma; rachilla not prolonged beyond the base of the floret as a bristle. . . . . **AGROSTIS** (in part)
29' Palea 67% to as long as the lemma; rachilla prolonged beyond the base of the floret as a bristle, though sometimes for only 0.1 mm.
30. Callus hairy, the hairs 0.5–4.5 mm long; inflorescences dense, branches spikelet-bearing throughout; culms (10)35–100+ cm long . .**CALAMAGROSTIS** (in part)
30' Callus hairless or with hairs to 0.5 mm long; inflorescence open, branches spikelet-bearing only in the outer half; culms 5–50 cm long. . **PODAGROSTIS** (in part)
27' Lemmas abruptly short-pointed or awned at or near the tip.
31. Inflorescence branches spreading; leaves 1–20 mm wide; spikelet disarticulating below the glumes . . . . **CINNA** (in part)
31' Inflorescence branches appressed; leaves 0.5–1.6(4.2) mm wide; spikelet disarticulating above the glumes. . . . . **MUHLENBERGIA** (in part)

Subkey 3a: *Stipeae* Tribe–Needlegrasses.

1. Lemma awns 3–11 mm long or early deciduous and absent.
    2. Lemma margins separate their entire length; awns usually persistent . . . . **PIPTATHEROPSIS**
    2' Lemma margins overlapping; awns deciduous . . . . **ACHNATHERUM** (in part)
1' Lemma awns 12–225 mm long.
    3. Hairs on basal segment of awn 3–8 mm long . . . . **PAPPOSTIPA**
    3' Hairs on basal segment of awn 0–2 mm long.
        4. Lemma surface minutely warty and glabrous; lemma apex with a membranous crown 0.1–0.2 mm long, edged with hairs less than 0.5 mm long . . . . **NASSELLA**
        4' Lemma surface smooth to hairy and not minutely warty; lemma apex with hairs or lobes or swellings but lacking a crown.
            5. Awns 65–225 mm long; calluses 1.5–6 mm long, sharply pointed . . . . **HESPEROSTIPA**
            5' Awns 1–70 mm long; calluses 0.1–2 mm long, blunt or sharply pointed . . . . . . **ACHNATHERUM** (in part)

Subkey 4: Inflorescence a panicle; florets 2+ per spikelet; leaf sheaths closed.

1. Spikelets subsessile in dense 1-sided clusters at the tips of comparatively few stiff panicle branches . . . . **DACTYLIS**
1' Spikelets not borne in dense 1-sided clusters.
    2. Culms bulbous-based. . . . . **MELICA** (in part)
    2' Culms not bulbous-based or culm bases not available.
        3. Upper 2–4 florets sterile, enclosing one another and lacking paleas; leaf sheaths usually closed their full length; auricles lacking; plants perennial. . . . . **MELICA** (in part)
        3' Upper florets normal though usually reduced and often staminate, a palea present in each; leaf sheaths with at least a V-shaped opening near the top; auricles sometimes present; plants perennial or annual.
            4. Lemmas awned.
                5. Plants annual . . . . **BROMUS** (in part)
                5' Plants perennial.
                    6. Spikelets 15–80 mm; lemma awns often arising from between lemma teeth that are visible at 10× magnification; leaf sheath of fertile culms closed (with a V-shaped opening at top) . . . . **BROMUS** (in part)
                    6' Spikelet 7–17 mm; lemma awns terminal or very nearly so; leaf sheaths of fertile culms usually split by growing culm and therefore open. . . . . ***Festuca rubra***
            4' Lemmas awnless.
                7. Lemma veins faint to conspicuous, not equally spaced their whole length, usually curved and converging but not actually meeting near the tip, or too faint to see; leaf tips prow-shaped or not.
                    8. Spikelets 16–40(45) mm long, tips of ovary and seed densely short-hairy; callus never with cobwebby hairs . . . . **BROMUS** (in part)
                    8' Spikelets 2–17 mm long; tip of ovary and seed glabrous; callus sometimes with cobwebby hairs.
                        9. Leaf tips not prow-shaped; leaf blades without translucent lines on each side of the midvein; glumes equaling or shorter than the lowest lemmas; callus lacking hairs; hila (attachment scars on seeds) linear. . . . . **FESTUCA** (in part)
                        9' Leaf tips prow-shaped, usually conspicuously so; leaf blades with translucent lines on each side of the midvein; glumes often equaling or longer than the lowest lemmas; callus often with hairs, the hairs sometimes cobwebby; hila round to oval . . **POA** (in part)
                7' Lemma veins conspicuous, equally spaced their whole length, not converging; leaf tips prow-shaped.
                    10. Spikelets sessile or subsessile on one side of the inflorescence axis . . . . ***Sclerochloa dura*** (in part)
                    10' Spikelets pedicellate in open or contracted panicles.
                        11. Spikelets with (1)2(3) florets; lower glume veinless; rachilla internode of lower floret absent and of upper one relatively long; spikelet open . . . . **CATABROSA** (in part)
                        11' Spikelets with 2–16 florets; lower glume with 1 vein; rachilla internodes relatively short so the florets strongly overlap, concealing the rachilla; spikelet more compact . . **GLYCERIA**

Subkey 5: Inflorescence a panicle; florets 2+ per spikelet; leaf sheaths open.

1. Spikelets dorsiventrally flattened; lemmas of fertile florets firmer or thicker than glumes.
  2. Spikelets with 3–4 florets; spikelets not or only slightly dorsiventrally flattened. . . . . *Scolochloa festucacea*
  2' Spikelets with 2 florets, 1 floret so modified it may not be recognized as a floret; spikelets dorsiventrally flattened.
    3. Both glumes well developed, hiding the small lemmas; spikelets in 3s (though the 2 staminate spikelets may fall early) . . . . **SORGHUM** (in part)
    3' One glume much reduced, lowest lemma glume-like; spikelets solitary or in pairs . . . . **Subkey 2a**
1' Spikelets laterally flattened or terete; lemmas of fertile florets and glumes alike in texture or sometimes the glumes or lemmas thicker.
  4. Mature glumes leathery, shiny, thicker than fertile lemma. . . . . **SORGHUM** (in part)
  4' Mature glumes and lemmas alike in texture.
    5. Inflorescence 1-sided and dense.
      6. Spikelets paired, 1 spikelet fertile and the other sterile and resembling an empty set of bracts; plants (5)9–70(180) cm tall. . . . . **CYNOSURUS**
      6' Spikelets not paired, similar in appearance, all normally fertile; plants 2–15(30) cm tall . . . . *Sclerochloa dura* (in part)
    5' Inflorescence not 1-sided and dense.
      7. Plants usually more than 2 m tall; culms usually more than 10 mm in diameter; spikelets with long silky hairs; inflorescences large, fluffy panicles.
        8. Plants densely cespitose, usually growing in uplands; lemmas long-hairy, especially below middle . . . . **CORTADERIA**
        8' Plants strongly rhizomatous, growing in water at least part of the year; lemmas glabrous, though spikelet axis long-soft-hairy . . . . **PHRAGMITES**
      7' Plants differing from the above in some way; plants usually less than 2 m tall; culms usually less than 10 mm in diameter; spikelets lacking silky hairs; inflorescences various.
        9. Top of leaf sheath with a distinct tuft of long hairs, hairs on remainder of sheaths shorter or lacking.
          10. Plants strongly rhizomatous, unisexual; leaves stiffly 2-ranked. **DISTICHLIS**
          10' Plants cespitose, bisexual; leaves neither stiff nor strongly 2-ranked.
            11. Paleas with conspicuously densely hairy keels, the tufts of hairs visible with naked eye; culms disarticulating at maturity; leaf sheaths somewhat inflated and hiding axillary inflorescences . . . . **TRIPLASIS**
            11' Paleas not conspicuously hairy; culms not disarticulating at maturity; leaf sheaths not inflated and not hiding axillary inflorescences (except occasionally in *Danthonia*).
              12. Lemmas awnless.
                13. Glumes longer than the adjacent floret. . . . . *Danthonia decumbens*
                13' Glumes shorter than the adjacent floret.
                  14. Plants annual or perennial, often with glands; spikelets with (1)2–60 florets; leaf blades not disarticulating . . . . **ERAGROSTIS** (in part)
                  14' Plants perennial, lacking glands; spikelets with (1)2–5 florets; leaf blades eventually disarticulating from sheaths . . . . **MOLINIA**
              12' Lemmas awned.
                15. Lemmas with distinct tufts of hairs near the callus base and elsewhere on the lemma, hairs often in 1 or more transverse row(s) above the callus and/or at mid-length. . . . . **RYTIDOSPERMA**
                15' Lemmas with tufts of hairs only at the base of the callus, otherwise glabrous or with hairs evenly distributed or mostly marginal . . . . **DANTHONIA**
        9' Top of leaf sheath lacking a distinct tuft of long hairs, or sheaths hairy with all hairs more or less the same length.
          16. Plants annual; lemma awns easily visible without dissecting the spikelets, bent, arising below the middle of the lemma; panicles open at maturity and much-branched; leaves evenly distributed.
            17. Lemma awns 15–45 mm long; glumes 14–33 mm long; spikelets pendent at maturity . . . **AVENA**
            17' Lemma awns 2–9 mm long; glumes 1–12 mm long; spikelets spreading to erect at maturity.

18. Glume veins 5–9, conspicuous; lemmas 5–10 mm long; 1 lemma in each spikelet with awn arising from below mid-length, the other with awn arising from the tip .... **VENTENATA**
18' Glume veins 1–3, inconspicuous; lemmas 1.3–3 mm long; both lemmas in each spikelet with awn arising from below mid-length.
19. Spikelets 4–9 mm long .... ***Deschampsia danthonioides*** (in part)
19' Spikelets 1.5–3.5 mm long .... **AIRA** (in part)

16' Plants not as above in one or more aspects.
20. Spikelets each with 1 fertile floret above 2 sterile, modified, often greatly reduced florets.
21. Lower florets vestigial and minute, to about 75% as long as the uppermost fertile floret; plants without a sweet odor when crushed .... **PHALARIS** (in part)
21' Lower florets longer than the uppermost fertile floret, 3–5 mm long, staminate; plants with a sweet odor when crushed .... **HIEROCHLOË** (in part)
20' Spikelets with (1)2–many fertile florets, the sterile florets, if present, above the fertile florets.
22. Spikelets with 2(3) florets, the awn of the lower lemma distinctly different from the awn of the upper floret or the upper floret awnless.
23. Plants annual, delicate; leaves and nodes glabrous .... ***Aira caryophyllea***
23' Plants perennial, not delicate; leaves and nodes glabrous or pubescent.
24. Lower lemma awnless; disarticulation below the glumes .... **HOLCUS**
24' Lower lemma awned; disarticulation above the glumes .... **ARRHENATHERUM**
22' Spikelets with 2–many florets, awnless or with awns that are similar (but may vary in length).
25. Lemmas with 3 conspicuous veins.
26. Florets 1–2 per spikelet.
27. Leaf sheath open to the base; leaves 0.6–6 mm wide, if more than 2 mm wide then inflorescence a condensed panicle; glumes and lemmas generally short-pointed to awned .... **MUHLENBERGIA**
27' Leaf sheath open partially, closed on the lower half; leaves 2–13 mm wide, inflorescence an open panicle; glumes and lemmas rounded to truncate .... **CATABROSA** (in part)
26' Florets (3)5–40 per spikelet; glumes and lemmas obtuse to acuminate.
28. Spikelets attached to secondary or tertiary branches .... **ERAGROSTIS** (in part)
28' Spikelets attached directly to the primary branches .... **DIPLACHNE**
25' Lemmas with (1–3)5–many faint to distinct veins.
29. Lemmas 7–20 mm long; inflorescences dense, stiff, main axis usually hidden by spikelets .... **LEYMUS**
29' Lemmas 2–6(7) mm long; inflorescences dense to open, flexible, main axis usually visible.
30. Some lemmas with awns more than 1 mm long.
31. Lemma awns arising at or below middle.
32. Lemma bodies 9–12 mm long, apical teeth 1.5–7 mm long; lemma awns 5.5–13 mm long, bent and twisted; leaf sheaths usually with dense, long, spreading hairs .... ***Danthonia unispicata***
32' Lemmas 1–6 mm long, entire, erose, or with short teeth; lemma awns 0.5–9 mm long, straight or bent; leaf sheaths glabrous, scabrous or retrorsely hairy.
33. Plants delicate annuals.
34. Glumes 1.5–3.5 mm; lemma tips with 2 slender teeth .... **AIRA** (in part)
34' Glumes 3.5–9 mm; lemma tips acute to acuminate, not toothed .... ***Deschampsia danthonioides*** (in part)
33' Plants moderately robust perennials.
35. Ligules acute, 2–13 mm, always some more than 3.5 mm .... **DESCHAMPSIA**
35' Ligules truncate to obtuse, 1.5–3.5 mm long .... ***Vahlodea atropurpurea***
31' Lemma awns arising from above middle or from the tip.
36. Lower glume longer than lowest floret .... **TRISETUM** (in part)
36' Lower glume shorter than or as long as lowest floret.
37. Panicle branches soft-hairy; leaf tips strongly prow-like .... **KOELERIA** (in part)
37' Panicle branches glabrous to scabrous; leaf tips not prow-like.
38. Upper glume as long as or longer than lowest floret; lemma awn arising more than 1 mm below the lemma tip .... **TRISETUM** (in part)

38' Upper glume shorter than lowest floret; lemma awn arising at or within 1 mm of the tip.
39. Plants annual; stamens usually 1 . . . . **VULPIA**
39' Plants perennial; stamens 3.
40. Auricles broadly flaring and rounded or comma- or claw-like and/or leaf blade bases flaring . . . . **SCHEDONORUS** (in part)
40' Auricles lacking, represented by inconspicuous swellings or small upright flaps; leaf blade bases not expanded or rounded . . . . **FESTUCA** (in part)
30' Lemmas awnless or awns less than 1 mm long.
41. Lemmas wider than long . . . . **BRIZA**
41' Lemmas longer than wide.
42. Lemma veins more or less equally spaced, parallel, not converging toward lemma tip, conspicuous.
43. Lower glume at least half as long as the lowest lemma; lemma veins (3)5 . . . . **POA** (in part)
43' Lower glume usually less than 33% as long as the lowest lemma; lemma veins 5–9.
44. Lemma veins prominent; leaf blades flat, 4–15 mm wide; ligules erose or lacerate, 3–9 mm long . . . . **TORREYOCHLOA** (in part)
44' Lemma veins not prominent; leaf blades usually involute, usually less than 3 mm wide (rarely more than 4 mm wide); ligules entire or nearly so, rarely more than 3 mm long . . **PUCCINELLIA**
42' Lemma veins not equally spaced, generally curved and converging but not actually meeting at tip or too faint to see.
45. Plants emergent in shallow water, rhizomatous.
46. Lemmas 4–9 mm; calluses densely hairy, the hairs 1–1.5 mm; culms 6–8 mm thick . . . . **SCOLOCHLOA**
46' Lemmas 2–3.6 mm; calluses glabrous; culms 0.6–4.8 mm thick . . . . **TORREYOCHLOA** (in part)
45' Plants of moist meadows or uplands, not emergent, rhizomatous or cespitose.
47. Florets 3, the 2 lowest sterile, upper 1 fertile; leaves with a sweet odor when crushed . . . . **HIEROCHLOË** (in part)
47' Florets 2–12, the lower 1-few fertile, the upper 1-few sometimes sterile; leaves without a sweet odor when crushed.
48. Upper glumes rounded to truncate, more or less hooded, length 2–3(3.5) times width . . . . **SPHENOPHOLIS** (in part)
48' Upper glumes acute to acuminate, rarely rounded, not hooded, length usually more than 3 times width.
49. Panicle branches covered with soft hairs . . . . **KOELERIA** (in part)
49' Panicles branches glabrous, scabrous, or strigose, not covered with soft hairs.
50. Leaves with claw-like or clasping auricles or with expanded flaring area at base of leaf blade . . . . **SCHEDONORUS** (in part)
50' Leaves without auricles, without expanded flaring rounded area at base of leaf blades.
51. Leaf tips usually prow-shaped; leaves with a translucent stripe on each side of the midrib, otherwise opaque; calluses sometimes with cobwebby hairs; sheaths of upper culm leaves closed part of their length . . . . **POA** (in part)
51' Leaf tips not prow-shaped or only minutely so; leaves not as above; calluses lacking cobwebby hairs; sheaths of upper culm leaves open to near the base (sometimes torn open by growing culms).
52. Leaves 0.5–1.5 mm wide, generally folded or rolled . . . . **FESTUCA** (in part)
52' Leaves 1.6–10+ mm wide, flat, sometimes rolling lengthwise when dry.
53. Plants dioecious, rhizomatous . . . . **LEUCOPOA**
53' Plants bisexual, rhizomatous or cespitose.
54. Lower glume shorter than lowest lemma; florets 3–10 (or as few as 1–2 in dwarf alpine plants) . . . . **FESTUCA** (in part)
54' Lower glume generally exceeding the lowest lemma; florets 2(3) . . . . ***Trisetum wolfii*** (in part)

ACHNATHERUM Needlegrass

Plants perennial; tightly to loosely cespitose. **Leaves** sheaths open; ligules hyaline to membranous, sometimes ciliate; blades flat, convolute, or involute; tips acute but not sharp. **Inflorescences** panicles, usually narrow; disarticulation above the glumes. **Spikelets** with 1 floret. **Glumes** longer than the floret, usually lanceolate, 1–7-veined, acute to acuminate, occasionally obtuse. **Calluses** 0.1–4 mm, blunt or sharp, usually hairy. **Lemmas** fusiform, usually round in cross section or sometimes more or less laterally compressed, stiffly membranous to leathery, usually hairy; lemma margins not or slightly overlapping (much overlapping in some species with glabrous lemmas); terminal lobe 0.05–3 mm; lobes usually thin and flexible, sometimes thickened and stiff; lemma awns 3–80 mm, usually scabrous, sometimes hairy, sometimes deciduous, bent once, twice, or not at all, the lower segment(s) usually twisted; terminal segment straight or flexuous. **Paleas** from 33% as long as to slightly longer than the lemmas, usually pubescent, not keeled. **Anthers** 3. **Caryopses** fusiform; style bases persistent.

North America and Eurasia. Approximately 30 species; 11 species treated in *Flora*.

Achnatherum is Oregon's largest fragment of the subdivided genus *Stipa*. It also includes some species once considered members of *Oryzopsis*. *Achnatherum* is probably polyphyletic and may be split further. Other Oregon genera that used to be considered part of *Stipa* include *Amelichloa*, *Hesperostipa*, *Nassella*, and *Pappostipa*.

Identification notes: Glume width is the distance between the midvein and the margin. Floret lengths include the callus but not the apical lobes. Floret thickness is measured at the thickest part of the floret. Callus traits should be examined on the dorsal side, the side away from the overlapping lemma margins. The border between the callus and the lemma body is indicated by a change in texture and usually color, but the border may be concealed by hairs. If the border is not apparent, callus length can be determined by measuring the distance from the base of the palea to the tip of the callus.

Key to *Achnatherum* species:

1. Basal awn segments glabrous, scabrous, or with hairs shorter than 0.5 mm; awns persistent or deciduous (often absent).
    2. Lemmas glabrous, shiny, dark brown, elliptic.
        3. Panicles drooping, branches flexuous and often diverging . . . . . ***A. wallowaense***
        3' Panicles erect, branches straight, ascending to appressed . . . . . ***A. hendersonii***
    2' Lemmas more or less hairy, dull to shiny, pale or dark, oval or cylindrical.
        4. Panicle branches ascending or spreading to strongly divergent.
            5. Awns 15–25 mm, persistent, bent twice; spikelets drooping. . . . . ***A. richardsonii***
            5' Awns 3–20 mm, deciduous, straight or bent once; spikelets not drooping . . . . . ***A. hymenoides***
        4' Panicle branches appressed to strongly ascending.
            6. Lemmas densely long-hairy, longest hairs more than 2 mm.
                7. Awns persistent, 13–15 mm, bent twice. . . . . ***A. pinetorum***
                7' Awns rapidly deciduous, 4–11 mm, straight or bent once . . . . . ***A. webberi***
            6' Lemmas sparsely hairy, longest hairs up to 2 mm.
                8. Awn appearing off center on the floret; floret laterally compressed; lemma apical lobe thick, stiff, about 0.1 mm; palea 75% to as long as the lemma . . . . . ***A. lemmonii*** ssp. ***lemmonii***
                8' Awn centered on the floret; floret round in cross section; lemma apical lobe thin, membranous, 0.1–0.4 mm; palea 50% as long as the lemma.
                    9. Calluses blunt, the glabrous area on the dorsal side about as long as wide, with a straight to rounded boundary between the glabrous tip and the hairy portion. . . . . ***A. nelsonii*** ssp. ***dorei***
                    9' Calluses sharp, the glabrous area on the dorsal side longer than wide, with an acute boundary between the glabrous tip and the hairy portion.
                        10. Lower awn segment scabrous; calluses not quite as sharp, with a shorter extension of the glabrous area into the hairy part of the callus; widest leaves to 5 mm wide . . . . . ***A. nelsonii*** ssp. ***nelsonii***
                        10' Lower awn segment scabrous to short-hairy; calluses sharper, with a longer extension of the glabrous area into the hairy part of the callus; widest leaves to 2 mm wide . . . . . ***A. occidentale*** ssp. ***californicum*** (in part)

1' Basal awn segments with spreading hairs, at least some of the hairs 0.5–8 mm (check several awns); awns persistent.
    11. Uppermost culm leaves with ligules 3–8 mm; lemmas with 1 apical lobe, the lobe to 0.1 mm, thick, leathery . . . . . ***A. thurberianum***
    11' Uppermost culm leaves with ligules 0.3–3 mm; lemmas usually with 2 apical lobes, sometimes not lobed, lobes to 1 mm, thin, membranous.
        12. Apical lemma hairs shorter than or as long as the lower awn hairs; hairs of lower awn segment all more or less alike in length and angle to the awn, thus looking tidy, gradually and evenly becoming shorter toward the first bend. . . . . ***A. occidentale*** ssp. ***pubescens***

12' Apical lemma hairs longer than most of the lower awn hairs; hairs of lower awn segment varying in length and in angle to the awn, thus looking untidy.

13. Calluses 0.5–0.7 mm, dorsal side with a rounded to acute boundary between the glabrous tip and the hairy portion . . . . ***A. nevadense***

13' Calluses 0.8–1.2 mm long, dorsal side with a narrowly acute boundary between the glabrous tip and the hairy portion . . . . ***A. occidentale*** ssp. ***californicum*** (in part)

Achnatherum hendersonii (Vasey) Barkworth

Henderson's ricegrass

Oryzopsis hendersonii Vasey

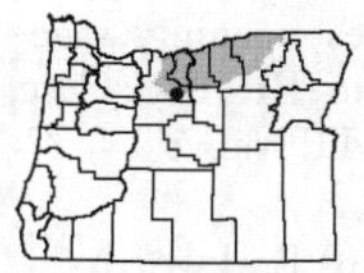

Plants 10–35 cm tall; cespitose. **Culms** 0.3–0.9 mm thick, pubescent below the nodes, glabrous or pubescent elsewhere. **Leaves** basal sheaths glabrous or nearly so; margins sometimes ciliate distally; collars glabrous; ligules 0.4–1 mm; blades convolute, to 1 mm in diameter, outer surface scabrous; inner surface pubescent. **Inflorescences** 4–12 × 2–5 cm; branches straight, appressed to ascending, longest branches 2–7 cm. **Glumes** 3.5–5.5 mm subequal, 1–1.5 mm wide, 5-veined, rounded to obtuse or acute; upper glumes to 1 mm, shorter than lower glumes. **Florets** 3.5–4.5 × 1–1.5 mm, laterally compressed. **Calluses** 0.3–0.6 mm, blunt. **Lemmas** leathery, glabrous, shiny, dark brown; apical lobes thick, about 0.2 mm; lemma awns 6–10 mm, deciduous, not bent, scabrous. **Paleas** 75% to as long as the lemmas, hard, flat, glabrous. **Anthers** 1.4–2(2.5) mm, with a tuft of short hairs at the end.

Shallow, rocky soil in scablands, sagebrush-steppe, or ponderosa pine forest, often in areas with frost heaving. 900–1500 m. Col. WA. Native.

This species has dark brown, shiny, more or less elliptic lemmas that are nearly visible through the translucent glumes. The lemma awns fall off early in development.

Achnatherum hymenoides (Roemer & Schultes) Barkworth

Indian ricegrass

Oryzopsis hymenoides (Roem. & Schult.) Ricker, *Stipa hymenoides* Roem. & Schult.

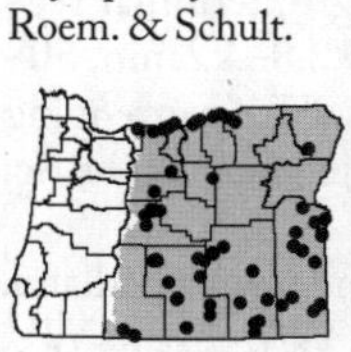

Plants 25–70 cm tall; cespitose. **Culms** 0.7–1.3 mm thick, glabrous or minutely scabrous in part. **Leaves** sheaths glabrous or minutely scabrous, sometimes puberulent on the distal margins; hairs to 0.8 mm; ligules 1.5–4 mm; blades usually tightly folded, 0.1–1 mm in diameter; outer surface smooth or scabrous; inner surface pubescent. **Inflorescences** 9–20 × 8–14 cm; branches usually strongly divergent (sometimes ascending), longest branches 3–15 cm. **Glumes** 5–9 mm, subequal, 0.8–2 mm wide, sac-like near the base, tapering above mid-length to an acuminate tip, hairs to approximately 0.1 mm; lower glumes 5-veined at the base, 3-veined at mid-length; upper glumes 5–7-veined at the base. **Florets** 3–4.6 mm long, 1–2 mm thick, obovoid. **Calluses** 0.4–1 mm, sharp. **Lemmas** hard, light to dark brown; dense white hairs 2.5–6 mm, easily rubbed off tips not lobed; lemma awns 3–20 mm, deciduous, more or less straight, scabrous. **Paleas** about the same length and texture as the lemmas, with few hairs. **Anthers** 1.5–2 mm, with a brush-like tuft of short hairs at the tip. **2*n***=46, 48.

Dry, well-drained, usually sandy soils. 50–1800 m. BR, BW, Col, ECas, Lava, Owy. CA, ID, NV, WA; north to Yukon, east to Manitoba and MO, south to Mexico. Native.

Achnatherum hymenoides is distinguished by its pale, very open inflorescence with strongly diverging branches. It hybridizes with many other species of *Achnatherum* (including *A. occidentale* and *A. thurberianum*) and with *Pappostipa speciosa, Nassella viridula*, and species of *Piptatheropsis*. In general, the hybrids differ from *A. hymenoides* in having more cylindrical florets, longer awns, and indehiscent anthers that contain few poorly formed pollen grains.

Achnatherum lemmonii (Vasey) Barkworth ssp. ***lemmonii***

Lemmon's needlegrass

Stipa columbiana Macoun var. *columbiana*, *Stipa lemmonii* (Vasey) Scribn. var. *jonesii* Scribn., *Stipa lemmonii* (Vasey) Scribn. var. *lemmonii*

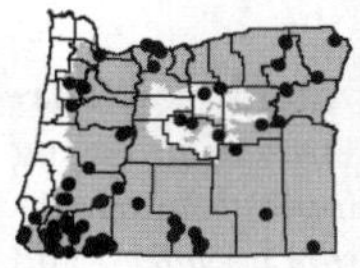

Plants 15–90 cm tall; cespitose; glaucous. **Culms** 0.7–1 mm thick, glabrous or pubescent. **Leaves** basal sheaths glabrous or pubescent; collars glabrous or sparsely pubescent, hairs shorter than 0.5 mm; ligules of basal leaves 0.5–1.2 mm, hyaline; those of upper leaves to 2.5 mm; blades folded to conduplicate, those of vegetative shoots 0.5–1.5 mm wide; the outer surface glabrous; inner surface hairs 0.3–0.5 mm; culm leaves to 3 mm wide, sometimes glabrous. **Inflorescences** 7–21 × 1 cm; branches appressed, longest branches 4–5 cm. **Glumes** 7–11.5 mm, subequal; lower glumes 0.9–1.1 mm wide, 4–5-veined; upper glumes 3-veined. **Florets** 5.5–7 mm long, 0.8–1.3 mm thick, somewhat laterally flattened. **Calluses** 0.4–1.2 mm, blunt. **Lemmas** leathery, evenly pubescent, hairs 0.4–1 mm; tip with hairs 0.4–0.8 mm, with 1 thick, stiff, inconspicuous lobe 0.1 mm; lemma awns 16–30 mm, persistent, appearing off center on the lemma tip, bent twice (occasionally once), all segments scabrous, terminal segment straight. **Paleas** 75% to as long as the lemmas; pubescent, hairs not exceeding the tips. **Anthers** 3.2–3.5 mm, lacking a tuft of short hairs at the tip. **2*n***=34.

Grassy balds, upland prairie, sagebrush steppe and ponderosa pine forest. 50–2700 m. BR, BW, Casc, Col, ECas, Owy, Sisk, WV. CA, ID, NV, WA; north to British Columbia, southeast to UT, south to Mexico. Native.

Achnatherum lemmonii is a common, widespread, fine-leaved bunchgrass that lacks spreading hairs on its long, bent awns. Its lemmas are slightly laterally compressed and have a slight, hard bulge at the apex, forcing the awn off center. Vegetatively, it can be confused with fine-leaved fescues, but *A. lemmonii* has rougher, stiffer leaves. *Achnatherum occidentale* has narrow leaves like *A. lemmonii*, but its lemmas are round in cross section and it has spreading hairs on the lower segments of its awns. Awns of *A. nelsonii* are similar to those of *A. lemmonii*, but *A. nelsonii* has wider leaves and its lemmas are round in cross section, with centered awns.

Achnatherum nelsonii (Scribner) Barkworth
Nelson's needlegrass

Stipa nelsonii Scribn.

Plants 40–175 cm tall; cespitose. **Culms** 0.7–2.4 mm thick, glabrous, sometimes slightly pubescent below the lower nodes. **Leaves** basal sheaths glabrous to sparsely or densely pubescent; margins sometimes ciliate; collars glabrous or somewhat pubescent, lacking tufts of hairs; ligules of basal leaves 0.2–0.7 mm; those of upper leaves 1–1.5 mm; blades flat in life but folded or rolled when dry, those of vegetative shoots 1.2–2.5 mm wide, culm leaves to 5 mm wide; outer surface glabrous but with scabrous margins, inner surface scabrous to pubescent, rarely glabrous. **Inflorescences** 9–36 × 0.8–2 cm; branches appressed, straight. **Glumes** 6–12.5 × 0.7–1.1 mm. **Florets** 4.5–7 × 0.6–0.9 mm, usually round in cross section when mature. **Calluses** 0.2–1 mm, blunt to sharp, the dorsal side with a straight to acute boundary between the glabrous tip and the hairy portion. **Lemmas** evenly pubescent, hairs at mid-length 0.5–1 mm; tips with hairs to 2 mm and erect or ascending, with flexible, membranous lobes 0.1–0.4 mm; lemma awns 19–45 mm, persistent, bent twice, the first 2 segments scabrous or hairs shorter than 0.5 mm, terminal segment glabrous, straight. **Paleas** 33–67% as long as the lemmas, pubescent, hairs usually not exceeding the tips. **Anthers** 2–3.5 mm, lacking hairs at the tip. **2*n***=36, 44.

2 subspecies.

Achnatherum nelsonii is often confused with its close relatives. *Achnatherum lemmonii* differs from *A. nelsonii* in having narrower leaves and relatively longer paleas. In addition, *A. lemmonii* has thick, stiff lemma lobes. *Achnatherum nevadense* and *A. occidentale* differ from *A. nelsonii* in having spreading hairs on the lower two awn segments, but some *A. occidentale* individuals have short awn hairs and can be hard to distinguish from *A. nelsonii* var. *nelsonii*. Individuals intermediate between *A. nelsonii* and *A. occidentale* are occasionally encountered.

Key to *Achnatherum nelsonii* subspecies:

1. Calluses blunt, the glabrous area on the dorsal side about as long as wide, with a straight to rounded boundary between the glabrous tip and the hairy portion .... ***A. n.*** ssp. ***dorei***

1' Calluses sharp, the glabrous area on the dorsal side longer than wide, with an acute boundary between the glabrous tip and the hairy portion .... ***A. n.*** ssp. ***nelsonii***

Achnatherum nelsonii (Scribner) Barkworth ssp. *dorei* (Barkworth & J.R. Maze) Barkworth
Dore's needlegrass

Stipa nelsonii Scribn. var. *dorei* (Barkworth & J.R. Maze) Dorn

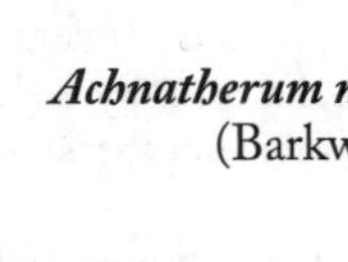

Calluses blunt, glabrous tips 0.02–0.06 mm on the dorsal side, about as long as wide, the dorsal side with a straight to rounded boundary between the glabrous tip and the hairy portion. **Lemma awns** 19–31 mm.

Montane to subalpine meadows and forest. 300–3000 m. BR, BW, Casc, ECas, Owy, Sisk. CA, ID, NV, WA; north to Yukon, east to Saskatchewan and SD, southeast to NM. Native.

Achnatherum nelsonii (Scribner) Barkworth ssp. *nelsonii* (p. 346)
Nelson's needlegrass

Stipa columbiana Macoun var. *nelsonii* (Scribn.) Hitchc., *Stipa occidentalis* Thurb. ex S. Watson var. *nelsonii* (Scribn.) C.L. Hitchc., *Stipa williamsii* Scribn.

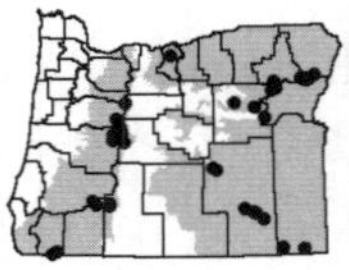

Calluses sharp, glabrous tips 0.05–0.15 mm on the dorsal side, longer than wide, the dorsal side with an acute boundary between the glabrous tip and the hairy portion. **Lemma awns** 19–45 mm.

Mesic to moist areas in meadows and forest openings, sagebrush steppe, juniper woodland. 400–2600 m. BR, BW, Casc, Col, Owy, Sisk. ID, NV, WA; northeast to MT. Native.

Achnatherum nevadense (B.L. Johnson) Barkworth
Nevada needlegrass

Stipa nevadensis B.L. Johnson

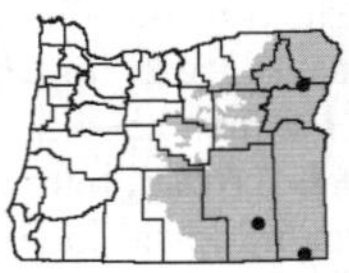

Plants 20–85 cm tall; cespitose. **Culms** 0.8–1.2 mm thick, usually with downward-pointing hairs below the lower nodes and otherwise glabrous, but sometimes completely glabrous or completely pubescent. **Leaves** sheaths glabrous or upper sheaths sometimes pubescent or minutely scabrous, usually glabrous at the throat; collars glabrous; ligules often wider than the blades; basal ligules 0.2–0.7 mm; upper ligules 0.3–1 mm; blades usually 10–25 cm, usually involute, 1–3 mm wide, outer surface glabrous, inner surface more or less puberulent. **Inflorescences** 6–25 × 0.4–1.5 cm; branches appressed, longest branches 2–7.5 cm. **Glumes** 7–14 mm; lower glumes 0.6–0.9 mm wide. **Florets** 5–6.5 mm long, 0.6–1 mm thick, round in cross section. **Calluses** 0.5–0.7 mm, sharp, dorsal boundary between the glabrous tip and the region with callus hairs rounded to acute. **Lemmas** evenly pubescent, hairs 0.5–1 mm at mid-length; tip with hairs 0.8–2 mm, longer than those at mid-length and those at the base of the awn and with membranous lobes about 0.4 mm; lemma awns 20–30 mm, persistent, bent twice, the first 2 segments with spreading hairs 0.5–1.5 mm, varying in length; terminal segment minutely scabrous to smooth. **Paleas** 2.8–4.2 mm, 50–75% as long as the lemmas, pubescent, the distal hairs extending beyond the tips. **Anthers** about 2.5 mm, lacking hairs at the tip. **2*n***=68.

Montane to subalpine sagebrush steppe and open woodland. 1800–2200 m. BR, BW, Owy. CA, ID, NV, WA; east to WY. Native.

Achnatherum nevadense is difficult to distinguish from *A. occidentale*, especially *A. o.* ssp. *californicum* (which may be a parent of allopolyploid *A. nevadense*). In *A. nevadense*, the boundary between the glabrous and hairy parts of the callus is rounded; in *A. o.* ssp. *californicum*, the boundary is narrowly acute. Culms of *A. nevadense* are usually glabrous except pubescent below the nodes; those of *A. o.* ssp. *californicum* are usually glabrous or pubescent throughout. Paleas of *A. nevadense* are 50–75% as long as the lemmas; those of *A. o.* ssp. *californicum* are shorter, 40–60% as long as the lemmas.

Achnatherum occidentale (Thurber) Barkworth

Western needlegrass

Plants 30–120(180) cm tall; cespitose. **Culms** 0.3–2 mm thick, glabrous or with sparse to dense short hairs. **Leaves** basal sheaths glabrous or with sparse to dense short hairs; collars often with tufts of hairs at the sides; ligules 0.2–1.5 mm; blades 10–30 cm, flat or convolute, 0.5–0.8 mm in diameter, outer surface smooth or slightly scabrous, inner surface pubescent to pilose. **Inflorescences** 5–30 cm, 0.5–1.5 cm wide; branches usually appressed, straight, longest branches 1–7 cm. **Glumes** 9–15 × 0.6–0.9 mm, 3–5(7)-veined, more or less acuminate. **Florets** 5.5–7.5 × 0.5–0.9 mm, round in cross section. **Calluses** 0.8–1.2 mm, sharp, boundary between the glabrous tip and hairy portion narrowly acute. **Lemmas** hard, evenly pubescent, hairs 0.5–1.5 mm at mid-length; tips with membranous lobes 0.3–0.5 mm; lemma awns 15–55 mm, bent twice, the first 2 segments evidently hairy or occasionally scabrous, terminal segment glabrous, scabrous, or pilose. **Paleas** 40–60% as long as the lemmas, pubescent, hairs at the tip often exceeding the tips. **Anthers** 2.5–3.5 mm, lacking hairs at the tip.

3 subspecies; 2 subspecies treated in *Flora*.

The spreading hairs on the lower awn segments distinguish *A. occidentale* from the other needlegrasses except for *A. thurberianum*, which has longer ligules.

Key to *Achnatherum occidentale* subspecies:

1. First 2 awn segments scabrous or with hairs of mixed lengths; hairs at the tip of the lemma longer than the hairs on the lower part of the awn . . . . *A. o.* ssp. ***californicum***

1' First 2 awn segments with hairs that gradually and evenly become shorter toward the first bend; hairs at the tip of the lemma about as long as the hairs on the lower part of the awn . . . . *A. o.* ssp. ***pubescens***

Achnatherum occidentale (Thurber) Barkworth ssp. *californicum* (Merrill & Burtt Davy) Barkworth (p. 346)

California needlegrass

Stipa californica Merr. & Burtt Davy, *Stipa occidentalis* Thurb. ex S. Watson var. *californica* (Merr. & Burtt Davy) C.L. Hitchc.

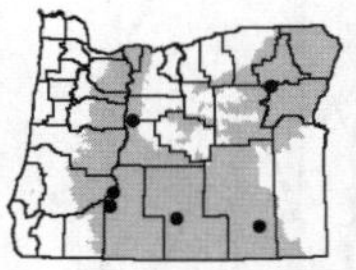

Plants 30–100(180) cm tall. **Culms** 0.5–2 mm thick; internodes glabrous or pubescent, sometimes densely pubescent. **Leaves** sheaths glabrous, pubescent, or pilose; blades 0.8–2 mm wide, inner surfaces with spreading hairs. **Inflorescences** 8–30 cm. **Lemmas** apical hairs to 1.8 mm, longer than the basal awn hairs; lemma awns 18–55 mm, scabrous or pilose on the first 2 segments, hairs of mixed lengths, some at least 0.5 mm; terminal segment usually scabrous, occasionally pilose at the base. **2*n***=36.

Sagebrush scrub, coniferous forest. 1100–2500 m. BR, BW, Casc, ECas. CA, ID, NV, WA; east to WY. Native.

This widespread subspecies is the most confusing component of the *A. occidentale* complex. It differs from the other *A. occidentale* subspecies because its apical lemma hairs are longer than the lemma body hairs and the lower awn hairs, and its awn hairs look untidy because they vary in length and in angle to the awn surface. In this trait, it resembles *A. nevadense,* which differs in having the boundary between the glabrous and hairy parts of the callus rounded to more or less acute and in being polyploid. Occasional individuals of *A. o.* ssp. *californicum* that have short-hairy or even scabrous lower awn segments can be confused with *A. nelsonii* ssp. *nelsonii*, which has wider leaves and less sharp calluses.

Achnatherum occidentale (Thurber) Barkworth ssp. *pubescens* (Vasey) Barkworth (p. 346)

Common western needlegrass

Stipa elmeri Piper & Brodie ex Scribn.

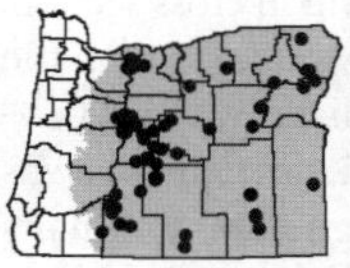

Plant 30–120 cm tall. **Culms** 0.8–1.3(2) mm thick; basal internodes puberulent to pubescent. **Leaves** basal sheaths usually pubescent, occasionally glabrous; blades 1–3 mm wide, inner surfaces with very short to longer spreading hairs. **Inflorescences** 10–30 cm. **Lemmas** apical hairs shorter than to as long as the basal awn hairs; lemma awns 24–50 mm, pilose on the first 2 segments, hairs gradually becoming shorter distally, terminal segment scabrous or glabrous. **2*n***=36.

Sagebrush scrub, coniferous forest. 100–2800 m. BR, BW, Casc, Col, ECas, Lava, Owy. CA, ID, NV, WA; east to WY. Native.

Achnatherum occidentale ssp. *pubescens* intergrades with *A. o.* ssp. *californicum* and *A. nelsonii. A. nelsonii* differs in having longer paleas and scabrous to short-hairy lower awn segments.

Achnatherum pinetorum (M.E. Jones) Barkworth

Pinewoods needlegrass

Stipa pinetorum M.E. Jones

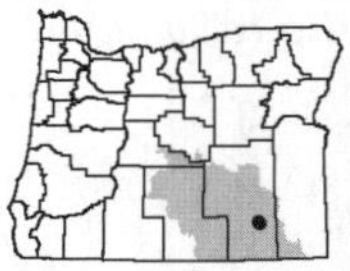

Plants 14–50(80) cm tall; cespitose; older plants spreading while dying in the center, forming rings. **Culms** 0.4–0.9 mm thick, mostly glabrous but lower internodes often puberulent or pubescent, especially below the nodes. **Leaves** basal sheaths usually glabrous, throat sometimes with a few hairs about 0.2 mm; collars glabrous; ligules of basal leaves 0.2–0.5(0.8) mm; those on upper culm leaves to 2 mm; blades 3–12 cm, usually involute, 0.2–0.4 mm in diameter, sometimes flat and 0.5–1 mm wide, often curved distally, outer surface minutely scabrous, inner surface hairs about 0.1 mm. **Inflorescences** 4.5–11(20) × 0.5–1 cm; branches appressed, longest branches 1–5 cm. **Glumes** 7–11 × 0.6–0.9 mm, lanceolate, not sac-like toward the base. **Florets** 3.5–5.5 mm long, 0.6–0.8 mm thick, round in cross section. **Calluses** 0.4–0.6 mm, sharp. **Lemmas** densely and evenly covered with long white hairs, hairs 1.5–3.5 mm at mid-length; tips with hairs to 5 mm, with thin, inconspicuous terminal lobes 0.3–2 mm; lemma awns 13–15 mm, persistent, bent twice, first 2 segments scabrous. **Paleas** 33% to as long as the lemmas, pubescent. **Anthers** 1.8–2.6 mm, lacking hairs at the tip. **2*n***=32.

Dry subalpine meadows. 2500–2900 m. BR. CA, ID, NV; northeast to MT, southeast to CO and NM. Native.

Achnatherum pinetorum has lemmas with long hairs, though the hairs on the lower half may be somewhat shorter than the apical hairs. *Achnatherum webberi* has similar long-pubescent lemmas but shorter, deciduous awns. In Oregon, *A. pinetorum* is known only from Steens Mountain.

Achnatherum richardsonii (Link) Barkworth
Richardson's needlegrass

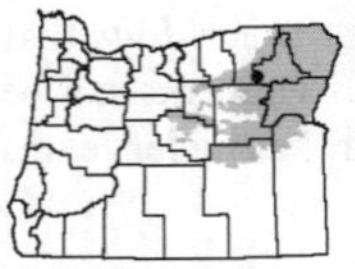

Plants 30–100 cm tall; cespitose. **Culms** 1–1.5 mm thick, glabrous. **Leaves** basal sheaths glabrous; margins minutely ciliate; collars glabrous, lacking tufts of hair on the sides; ligules 0.1–0.5 mm; blades convolute when dry, 0.8–3 mm wide, outer surface minutely scabrous, inner surface glabrous. **Inflorescences** 7–25 × 7–15 cm, often purplish; branches divergent, flexuous, with spikelets few and near the tips; longest branches 7–10 cm. **Glumes** lower glumes 7.5–11 × 0.9–1.2 mm, 1.5–3 mm longer than the upper glumes. **Florets** 5–6 mm long, 0.6–0.9 mm thick, round in cross section. **Calluses** 0.4–0.7 mm, blunt. **Lemmas** evenly pubescent on the proximal part, often becoming glabrous distally, hairs 0.2–0.5 mm; tips lacking lobes or with lobes to 0.1 mm; lemma awns 15–25 mm, persistent, bent twice, first 2 segments with hairs about 0.1 mm, terminal segment straight. **Paleas** 50–67% as long as the lemmas, pubescent, hairs not exceeding the tips. **Anthers** 2.5–3 mm, with a brush-like tuft of hairs at the tip.

Grasslands, sagebrush steppe, open woodlands, often on sandy or gravelly substrates. 1100–1500 m. BW. WA; north to Yukon, northeast to Manitoba, east to SD. Native.

Achnatherum richardsonii is unique among our needlegrasses in its combination of wavy, spreading inflorescence branches and persistent awns.

Achnatherum thurberianum (Piper) Barkworth
Thurber's needlegrass

Stipa thurberiana Piper

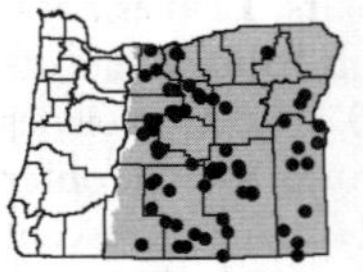

Plants 30–75 cm tall; cespitose. **Culms** 0.5–1.7 mm thick, glabrous or pubescent, more likely to be pubescent on lower internodes; nodes usually pubescent. **Leaves** basal sheaths usually glabrous; collars glabrous, lacking tufts of hair on the sides; ligules of basal leaves 1.5–6 mm; those of upper leaves to 8 mm; blades 10–30 cm, convolute, 0.5–2 mm wide, outer surface scabrous, inner surface scabrous or hairs about 0.3 mm. **Inflorescences** 7–15 × 0.5–2.5 cm; branches appressed or strongly ascending, longest branches 1.5–6 cm. **Glumes** often purplish, 3(5)-veined; lower glumes 10–15 × 1.2–2 mm, to 2 mm longer than the upper glumes. **Florets** 6–9 mm long, 0.7–1.2 mm thick, round in cross section. **Calluses** 0.9–1.5 mm, sharp. **Lemmas** leathery, evenly pubescent or the back becoming glabrous distally, all hairs 0.5–0.8 mm; tips with a thick lobe 0.1 mm on 1 margin; lemma awns 32–56 mm, bent twice, first 2 segments with spreading hairs 0.8–2 mm, terminal segment glabrous, often scabrous. **Paleas** 75% to nearly as long as the lemmas, sparsely pubescent toward the base. **Anthers** 2.5–3.5 mm, lacking hairs at the tip. **2*n***=34.

Sagebrush steppe and ponderosa pine woodland. 100–2300 m. BR, BW, Col, ECas, Lava, Owy. CA, ID, NV, WA; northeast to MT, southeast to CO. Native.

Achnatherum thurberianum is a common bunchgrass of arid regions east of the Cascades. It is distinguished by long ligules and the spreading hairs on the first two lemma awn segments.

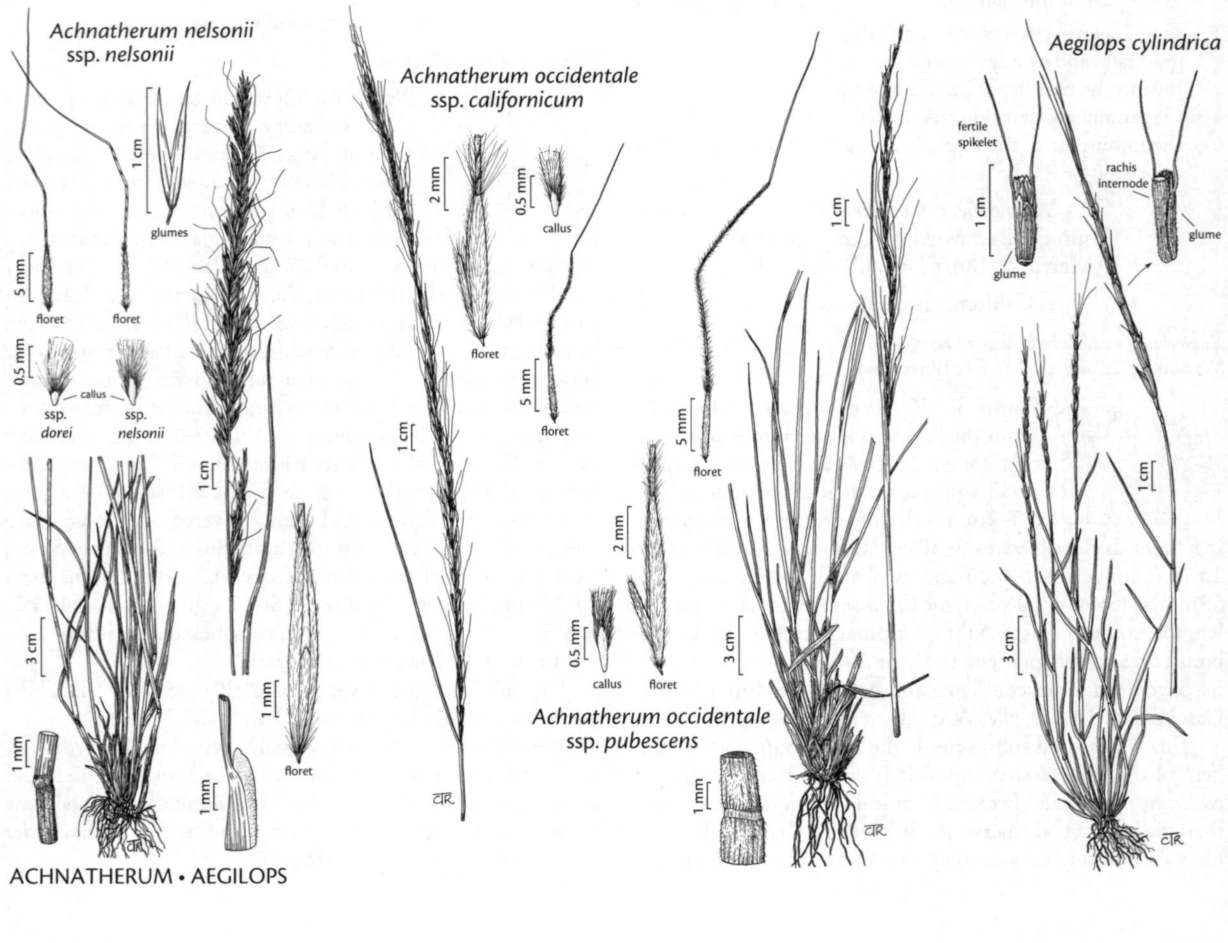

ACHNATHERUM • AEGILOPS

Achnatherum wallowaense J.R. Maze & K.A. Robson

Wallowa ricegrass

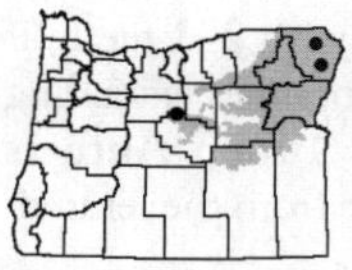

Plants (10)15–40(45) cm tall; cespitose. **Culms** 0.5–0.7 mm thick, glabrous. **Leaves** basal sheaths becoming flat with age, glabrous; collars glabrous; ligules of basal leaves 0.8–1.3 mm, those of upper leaves to 3–8 mm; blades tightly folded lengthwise, 0.5–0.8 mm in diameter, outer surface minutely scabrous, inner surface hairs less than 0.05 mm. **Inflorescences** (4)5–13(15) cm, to 10 cm wide; branches divergent, flexuous, sometimes drooping; longest branches 2–10 cm. **Glumes** obtuse to acute; lower glumes 3.5–7 × 0.8–1.3 mm, 5(7)-veined; upper glumes 3–6.5 mm, 3-veined. **Florets** 3–5.5 mm long, 1–1.5 mm thick, more or less round in cross section. **Calluses** (0.1)0.3–0.5 mm, blunt. **Lemmas** 3.5–5 mm, leathery, shiny, glabrous, dark brown to black at maturity; margins overlapping at maturity; tips thickened on one side; lemma awns 5–11 mm, deciduous. **Paleas** 2.8–4.5 mm, leathery, glabrous. **Anthers** 1.4–2.4 mm, with a tuft of short hairs at the tip.

Shallow, dry, rocky soil. 1200–1600 m. BW. Native.

Achnatherum wallowaense has shiny, tough, dark brown lemmas in a drooping panicle with divergent branches. It looks nothing like a typical needlegrass. The only similar species is *A. hendersonii*, which has appressed inflorescence branches. *Achnatherum wallowaense* is endemic to Oregon and is only known from a few sites in Crook and Wallowa counties.

Achnatherum webberi (Thurber) Barkworth

Webber's needlegrass

Oryzopsis webberi (Thurb.) Benth. ex Vasey, *Stipa webberi* (Thurb.) B.L. Johnson

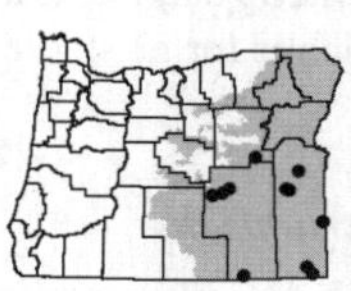

Plants 12–35 cm tall; densely cespitose. **Culms** 0.4–0.7 mm thick, smooth or minutely scabrous. **Leaves** basal sheaths glabrous, smooth or minutely scabrous; collars glabrous; ligules of basal leaves 0.1–1 mm; those of upper leaves 1–2 mm; blades 3–9 cm, usually folded to involute and about 0.5 mm in diameter, 0.5–1.5 mm wide when flat, stiff, outer surface smooth or scabrous, inner surface scabrous. **Inflorescences** 2.5–7 × 0.5–2 cm; branches appressed, longest branches 1–2 cm. **Glumes** 6–10 × 0.6–0.9 mm, lanceolate, not sac-like at the base. **Florets** 4.5–6 mm long, 0.7–1 mm thick, more or less round in cross section. **Calluses** 0.3–0.8 mm, blunt. **Lemmas** 6–7 mm, densely and evenly covered with long, soft, straight hairs 2.5–3.5 mm; tips with membranous lobes 0.6–1.9 mm; lemma awns 4–11 mm, deciduous, straight or bent once, scabrous. **Paleas** as long as or slightly longer than the lemmas. **Anthers** 1.5–2 mm, lacking hairs at the tip. **2*n***=32.

Dry, open flats and rocky slopes. 700–1900 m. BR, BW, Owy. CA, ID, NV. Native.

Achnatherum webberi has long, abundant lemma hairs. The only other needlegrass with similar lemma hairs is *A. pinetorum*, which has longer, persistent awns.

AEGILOPS Goatgrass

Plants annual. **Culms** erect or bent at the lowest node. **Leaves** mainly cauline; sheaths open; ligules membranous, truncate; blades flat. **Inflorescences** terminal spikes with 1 spikelet per node, occasionally hidden in leaf sheaths; spikelets usually embedded in the inflorescence axis, with 2–13 spikelets and usually with 1–3 additional rudimentary spikelets at the base; internodes 6–12 mm; disarticulation at the base of the spikes or in the inflorescence axis. **Spikelets** 0.5–2(3) times the length of the internodes, tangential to the axis, appressed or ascending; the upper spikelet(s) sometimes sterile; fertile spikelets with 2–7 florets. **Glumes** ovate to rectangular, rounded on the back, with several prominent veins, sometimes hard at maturity; tips truncate, toothed, or bearing 1–3 awns. **Lemmas** rounded on the back; tips toothed, frequently awned; lemma awns terminal, often between teeth. **Paleas** tough, 2-keeled, the keels ciliate. **Anthers** 3. **Caryopses** lanceolate to lance-ovate.

Temperate regions worldwide. Approximately 23 species; 2 species treated in *Flora*.

Aegilops comosa and *A. neglecta* had persisted as weeds at the Oregon State University Experimental Farm near Corvallis in the past but have since been eradicated.

In the following descriptions, it should be noted that spike and spikelet lengths exclude the rudimentary spikelets at the base and the awns.

Key to *Aegilops* species:

1. Glumes of terminal spikelet 1-awned, glumes of lower spikelets 1-awned or with a long tooth . . . . ***A. cylindrica***

1' Glumes of terminal spikelet 3-awned, glumes of lower spikelets 2–5-awned . . . . ***A. triuncialis*** var. ***triuncialis***

Aegilops cylindrica Host (p. 346)

Jointed goatgrass

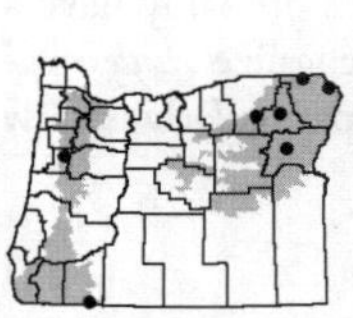

Plants 14–75 cm tall. **Leaves** blades 3–15 cm × 2–5 mm. **Inflorescences** spikes 2.2–12 cm, up to 0.5 cm wide, uniformly narrowly cylindrical, 10–45 times longer than wide, with (2)3–8(12) fertile spikelets and 0–2 rudimentary spikelets; disarticulation first at the base of the spikes and later in the inflorescence axis; spikelets remaining attached to the adjacent internodes. **Spikelets** 9–12 mm, narrowly cylindrical, scabrous or pubescent, with 3–5 florets, the lower (1)2–3 florets fertile. **Glumes** of flower spikelets 8–9.1 mm, 1-awned, awns 2–5 mm; glumes of apical spikelets 7–9 mm, scabrous; tips usually with 2 teeth, always awned, awn 50–80 mm. **Lemmas** of lower spikelets 9–10 mm, ventral surfaces velvety distally; tips mucronate to awned; lemmas of apical spikelets with 2 teeth and 1 awn; lemma awns of lower spikelets 10–50 mm, those of apical spikelets 40–80 mm. **Anthers** 3–3.5 mm. **Caryopses** 6–7 mm, adhering to the lemmas and paleas. **2*n***=28.

Disturbed roadsides, railroads, and agricultural fields. 100–1500 m. BW, Sisk, WV. CA, ID, NV, WA; widespread in North America; worldwide. Exotic.

Jointed goatgrass is an odd-looking annual weed with the florets sunken into the inflorescence axis. At maturity, the axis breaks into barrel-shaped units, each with an embedded seed. Similar *A. triuncialis* differs in having shorter spikes with 3-awned glumes on the terminal spikelet. Jointed goatgrass is a troublesome weed in winter wheat with which it forms sterile hybrids.

Aegilops triuncialis Linnaeus var. *triuncialis*

Barbed goatgrass

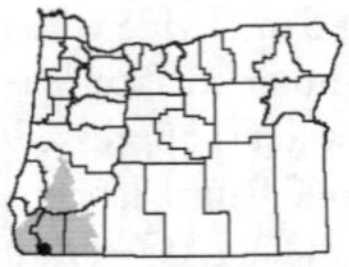

Plants 17–60 cm tall. **Leaves** blades 1.5–7 cm × 2–3 mm. **Inflorescences** spikes 2.2–6 × 0.4–0.6 cm, somewhat broadened at the base, more or less cylindric distally, 2–7 fertile spikelets and (2)3 rudimentary spikelets; disarticulation at the base of the spikes. **Lower fertile spikelets** 7–13 mm, lanceolate to ovate, with 3–5 florets, the lower 1–2 florets fertile; upper spikelets 7–9 mm. **Glumes** of lower spikelets 6–10 mm, 2–3-awned; awns 25–60 mm, glabrous, scabrous, or velvety, if 3-awned the central awn often shorter than the lateral awns, sometimes reduced to a tooth; glumes of apical spikelets 6–8 mm, 3-awned, central awns 25–80 mm, longer than the lateral awns. **Lemmas** 7–11 mm, with 2–3 teeth, if 3-toothed the central tooth the longest, sometimes extending into an awn; lemma awns, if present, up to 10 mm. **Anthers** 4.0–4.5 mm. **Caryopses** 5–8 mm, falling free from the lemmas and paleas. **2*n***=28.

Disturbed sites, wheat fields, rangelands and relatively undisturbed grasslands and savanna on serpentine substrates. 300–600 m. Sisk. CA, NV; Asia, Mediterranean. Exotic.

Aegilops triuncialis resembles *A. cylindrica,* which has longer, narrowly cylindrical spikes, but *A. triuncialis* is more abundantly awned, particularly on the uppermost spikelet, which has three awns on each glume. *Aegilops triuncialis* is capable of invading grasslands and savanna on strongly serpentine substrates. It was found in southwestern Oregon in 2003 but has since been eradicated. The species is expanding its range in northern California and may appear in Oregon again.

AGROPYRON Wheatgrass

Plants perennial; usually cespitose. **Culms** erect or geniculate; internodes hollow. **Leaves** sheaths open; auricles usually present; ligules membranous, often erose; blades flat to involute. **Inflorescences** terminal spikes, usually with crowded spikelets; disarticulation above the glumes and beneath the florets. **Spikelets** usually more than 3 times as long as the internodes, usually spreading from the rachis, with 3–16 florets. **Glumes** shorter than the adjacent lemmas, lance-ovate to lanceolate, 1–5-veined, asymmetrically keeled, sometimes with a secondary keel on the wider side; keels glabrous or with hairs; tips acute and entire, sometimes awned. **Lemmas** 5–7-veined, asymmetrically keeled; tips awned to acute; lemma awns, if present, to 4.5 mm. **Anthers** 3, yellow. **Caryopses** usually falling with the lemma and palea attached.

Temperate regions of the world. Approximately 15 species; 1 species treated in *Flora*.

Classifying *Agropyron*, as well as other members of the wheatgrass tribe (*Triticeae*), is difficult because their evolution has been characterized by hybridization. As a result, no matter which traits are used, many species are morphologically intermediate. Starting in the 1980s, grass taxonomists began to settle on a classification for the wheatgrass tribe that is largely based on chromosome arrangements. This approach has resulted in groups that have some morphological coherence, although the distinguishing morphological features may be subtle. In this process, the traditional genus *Agropyron* has been divided into a number of genera. Most of the wheatgrasses with one spikelet per node have been distributed among *Elymus* and the smaller genera *Eremopyrum, Leymus, Pseudoroegneria, Pascopyrum,* and *Thinopyrum.* Currently, the genus *Agropyron* is restricted to the "crested wheatgrasses." *Agropyron fragile*, Siberian wheatgrass, was collected once in northeastern Oregon. It is awnless, with linear to narrowly lanceolate spikes and spikelets more appressed to the inflorescence axis (diverging less than 30° from it).

Agropyron cristatum (Linnaeus) Gaertner (p. 353)

Crested wheatgrass

Agropyron desertorum (Fisch. ex Link) Schult.

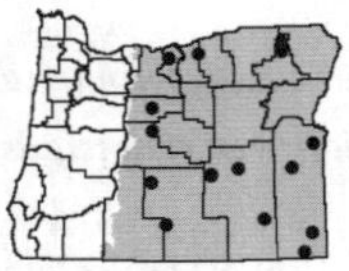

Plants 25–110 cm tall; cespitose, rarely more or less rhizomatous. **Leaves** ligules to 1.5 mm; blades 1.5–6 mm wide, flat. **Inflorescences** terminal spikes 1.3–10.5(15) × 0.5–2.5 cm, lanceolate, rectangular, or ovate in outline; internodes (0.2)0.7–5(8) mm. **Spikelets** 7–16 mm, with 3–6(8) florets, closely imbricate, spreading-ascending, diverging 30–95° from the inflorescence axis at maturity. **Glumes** 3–6 mm, (1)3–9-veined, glabrous or with coarse hairs on the keels, strongly keeled or winged, entire, usually awned, awns (0)1.5–3(6) mm and usually somewhat divergent. **Lemmas** 5–9 mm, with 5–7 veins, chartaceous to leathery, glabrous or pubescent; tips acute, awned; lemma awns (0.1)0.4–3(3.8) mm. **Anthers** 3–5 mm. **2*n***=14, 28, 42.

Disturbed shrub-steppe, roadsides. 50–2300 m. BR, BW, Col, ECas, Lava, Owy. CA, ID, NV, WA; widespread in North America; worldwide. Exotic.

Three species of *Agropyron* have been crossed to produce Oregon's cultivated and naturalized crested wheatgrasses. As a result, these three grasses have lost their integrity as species within North America. Most Oregon plants probably have a hybrid background. One can sometimes recognize *A. desertorum* ancestry within awned *A. cristatum* with particularly narrow spikes.

AGROSTIS Bentgrass

Plants perennial or annual; usually cespitose, sometimes rhizomatous or stoloniferous. **Culms** (3)5–120 cm tall, usually erect. **Leaves** sheaths open, usually glabrous or scabrous; ligules membranous, truncate to acute, erose or lacerate, rarely entire; blades flat, folded, or involute, usually smooth and glabrous, sometimes minutely scabrous. **Inflorescences** panicles, narrowly cylindrical to very open and diffuse; disarticulation above the glumes and beneath the florets, sometimes also at the panicle base. **Spikelets** laterally compressed, lanceolate to narrowly oblong or ovate, with 1 floret (rarely 2); rachilla not prolonged beyond the floret. **Glumes** 2, longer than the lemmas, 1(3)-veined, glabrous; veins often scabrous; tips acute to acuminate or awn-tipped; lower glumes usually slightly longer than the upper glumes. **Calluses** blunt, glabrous or hairy. **Lemmas** membranous or hyaline, usually smooth and glabrous, rarely pubescent or tuberculate, 3–5-veined; veins extremely faint, not convergent, sometimes excurrent as 2–5 tiny teeth; tips acute or obtuse to truncate, sometimes erose, awned or awnless; lemma awns arising from near the lemma bases to near the tip, usually bent. **Paleas** absent, minute, or as long as the lemmas, hyaline, often veinless. **Anthers** (1)3. **Caryopses** with hard or liquid endosperm.

Widely distributed in temperate regions and tropical mountains. Approximately 150–200 species; 14 species treated in *Flora.*

As traditionally defined, *Agrostis* is a polyphyletic grouping of superficially similar species. In recent treatments, some of the more distantly related taxa that have the rachilla prolonged beyond the base of the distal floret have been moved to the genera *Apera* and *Podagrostis*. In the following key and descriptions, awn development should be assessed on several spikelets at different locations along the branches.

Key to *Agrostis* species:

1. Paleas at least 40% as long as the lemmas.
   2. Lemma awns 4–16 mm, the longer awns more than twice as long as the glumes . . . . **APERA**
   2' Lemma awns 0–3.5 mm, less than twice as long as the glumes.
      3. Anthers 0.3–0.8 mm; rachilla sometimes prolonged as a bristle beyond the base of the floret.
         4. Palea 0.9–3 mm; rachilla prolonged 0.1–1.9 mm beyond the base of the floret, sometimes with hairs near the tip; ligules 0.2–4(5) mm, if greater than 3 mm then inflorescence open. . . . . **PODAGROSTIS**
         4' Palea 0–0.5(0.7) mm; rachilla not prolonged; ligules (1.5)3–8(13) mm long; inflorescence condensed.
            5. Back of glume fine-scabrous throughout; palea 0.5–0.7 mm and about 33% as long as the lemma . . . . ***A. densiflora*** (in part)
            5' Back of glume glabrous or more or less scabrous; palea 0–0.5 mm . . . . ***A. exarata*** (in part)
      3' Anthers (0.7)0.9–1.4 mm; rachilla never prolonged.
         6. Ligules of the upper leaves longer than wide, 2–7.5 mm; usually some lower panicle branches spikelet-bearing to the base.
            7. Inflorescences usually open after anthesis, usually reddish (except in shade-grown plants), 8–30 cm; longest lower panicle branches 4–9 cm; rhizomes present, stolons absent; culms 20–120 cm tall . . . . ***A. gigantea***
            7' Inflorescences usually contracted after anthesis, usually light green to tan, 3–20 cm; longest lower panicle branches 2–6 cm; rhizomes absent, stolons present; culms 8–60 cm tall . . . . ***A. stolonifera***
         6' Ligules of upper leaves usually shorter than wide, 0.3–3 mm; lower panicle branches with spikelets confined to the distal 33–50%.
            8. Calluses glabrous or with a few hairs to 0.1 mm; panicles erect, 3–20 cm; awns rarely present, to 2 mm; lemmas glabrous . . . . ***A. capillaris***
            8' Calluses abundantly hairy, hairs to 0.6 mm; panicles somewhat lax, 10–30 cm; awns often present on the terminal spikelet of a branch, to 5 mm; lemmas occasionally pubescent on the lower half. . . . . ***A. castellana***

1' Paleas absent or less than 40% as long as the lemmas.
   9. Longest lemma awns 3.5–10 mm; plants perennial or annual.
      10. Inflorescences open; plants perennial.
         11. Lemmas (2.2)2.5–3 mm long; lemma awns 4–6 mm long; plant cespitose. . . . . ***A. howellii***
         11' Lemmas 1–2(2.3) mm long; lemma awns to 4(5) mm long; plant often short-stoloniferous . . . . ***A. canina***
      10' Inflorescences dense, spike-like; plants annual.
         12. Lemmas 2.5–4 mm, teeth to 1.5 mm; awns (5)8–10 mm; blades 1–4.5 cm. . . . . ***A. hendersonii***
         12' Lemmas 1.5–2.3 mm, teeth to 0.5 mm; awns 2.3–5(8) mm; blades 3–15 cm . . . . ***A. microphylla***
   9' Longest lemma awns 0–3.5 mm; plants perennial.
      13. Floret callus with abundant hairs (0.8)1–2 mm. . . . . ***A. hallii***
      13' Floret callus with sparse to abundant hairs to 0.3(1) mm.

14. Anthers more than 1 mm.
    15. Plant more or less rhizomatous; primary branches of the inflorescence branched from below mid-length, spikelet-bearing to the base . . . . ***A. pallens*** (in part)
    15' Plant cespitose; primary branches of the inflorescence branched mainly above mid-length, spikelet-bearing on the distal half only . . . . ***A. oregonensis*** (in part)
14' Anthers less than 1 mm, or absent.
    16. Panicles dense, often spike-like; lower branches usually shorter than 3(5) cm, appressed to ascending, usually hidden by the spikelets.
        17. Leaf blade 0.5–2 mm wide, usually inrolled or folded . . . . ***A. variabilis***
        17' Leaf blade usually 2–7 mm wide, usually flat in life (may fold or roll if wilted before pressing).
            18. Lower inflorescence branches 2–5 cm; plants rhizomatous; anthers 0.7–1.8 mm . . . . ***A. pallens*** (in part)
            18' Lower inflorescence branches 0.3–2(4) cm; plants cespitose, sometimes short-rhizomatous when buried in sand; anthers 0.3–0.6 mm.
                19. Back of glumes fine-scabrous throughout; paleas 0.5–0.7 mm and approximately 33% as long as the lemmas . . . . ***A. densiflora*** (in part)
                19' Back of glumes glabrous or more or less scabrous; paleas 0–0.5 mm, approximately 20% as long as the lemma . . . . ***A. exarata*** (in part)
    16' Panicles open to diffuse, or somewhat contracted but not spike-like, lower branches 0.5–12 cm, erect to spreading, readily visible.
        20. Plant rhizomatous . . . . ***A. pallens*** (in part)
        20' Plant cespitose or habit difficult to determine.
            21. Wider leaf blades to 2 mm wide, 1–14 cm long, flat or involute; leaves mostly basal.
                22. Lower panicle branches 4–12 cm; whole panicle often detaching at the base at maturity forming a tumbleweed; leaf blades 4–14 cm . . . . ***A. scabra***
                22' Lower panicle branches 1–4 cm; panicle not detaching at maturity, not forming a tumbleweed; leaf blades 1–7 cm . . . . ***A. idahoensis***
            21' Wider leaf blades usually 2–7 mm wide, 5–30 cm long, flat; leaves mostly cauline to mostly basal.
                23. Primary branches of the inflorescence branched from below mid-length, spikelet-bearing to the base; plants rhizomatous . . . . ***A. pallens*** (in part)
                23' Primary branches of the inflorescence branched above mid-length, spikelet-bearing on outer half only; plants cespitose . . . . ***A. oregonensis*** (in part)

Agrostis canina Linnaeus

Velvet bentgrass

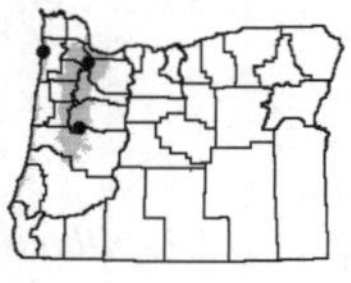

Plants perennial, 15–75 cm tall; loosely cespitose to stoloniferous. **Leaves** ligules 1–4 mm, truncate to obtuse to acute; blades 1–10 cm × 1–3 mm, flat. **Inflorescences** 3–10 × 1–7 cm, open, lanceolate to broadly ovate, lowest node with 3–7 branches; branches erect to spreading, spikelet-bearing in the distal 33%; lower branches 3–5 cm long. **Glumes** 1.7–3 mm long, subequal, 1-veined, acute. **Calluses** with hairs to 0.1 mm long. **Lemmas** 1–2(2.3) mm long, glabrous but minutely pubescent at base, 5-veined; veins obscure or prominent; tips acute to obtuse; usually awned. **Lemma awns** 4(5) mm long, bent, arising near the base, rarely absent or to 0.2 mm. **Paleas** 0–0.2 mm long, less than 40% as long as the lemmas. **Anthers** 1–1.5 mm long.

Grasslands, disturbed sites, roadsides, river margins, in areas cool in summer. 0–500 m. Est, WV. western North America; KY, VT; Eurasia. Exotic.

Agrostis canina differs from the more common introduced species in having consistently awned lemmas, the awns more than 3 mm long, as well as short paleas. *Agrostis howellii* has longer awns and more open panicles and is restricted to the Columbia Gorge area.

Agrostis capillaris Linnaeus

Colonial bentgrass, browntop

Agrostis tenuis Sibth., *Agrostis tenuis* Sibth. var. *aristata* (Parnell) Druce

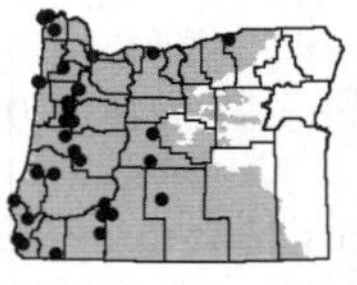

Plants perennial, 10–75 cm tall; rhizomatous or stoloniferous, rhizomes or stolons to 5 cm. **Leaves** ligules shorter than wide, 0.3–2 mm; blades 3–10 cm × 1–5 mm, flat. **Inflorescences** 3–20 cm, less than half the length of the culm, (1)2–12 cm wide, erect, widely ovate, open, lowest node with (2)3–9(13) branches; branches spreading during and after anthesis, with spikelets usually confined to the outer 50%, pedicels divergent so the spikelets are not clustered; lower branches 1.5–7 cm. **Glumes** 1.7–3 mm, 1-veined, acute. **Calluses** glabrous or with a few hairs to 0.1 mm. **Lemmas** 1.2–2.5 mm, glabrous, 3(5)-veined; veins usually excurrent at lemma tips; tips obtuse to acute, usually entire, usually awnless; lemma awns, if present, 0–2 mm, originating at mid-length, straight or bent. **Paleas** 0.6–1.2(1.4) mm, typically at least 50% as long as the lemmas. **Anthers** 3, 0.8–1.3 mm. **2*n***=28.

Roadsides, meadows, prairies, sand dunes, disturbed areas, lawns. 0–2300 m. BR, Casc, Col, CR, ECas, Est, Lava, Sisk, WV. CA, ID, NV, WA; western and eastern North America; Eurasia. Exotic.

Agrostis capillaris, one of four similar introduced bentgrasses, is often a community dominant in disturbed sites and degraded grasslands. It spreads by rhizomes or stolons and has an open panicle and short ligules. Similar *A. stolonifera* and *A. gigantea* have longer ligules, and many of their panicle branches are spikelet-bearing to the base. The latter also tend to grow in wetter habitats. See *A. castellana.*

Agrostis castellana Boissier & Reuter

Dryland bentgrass, highland bentgrass, dryland browntop

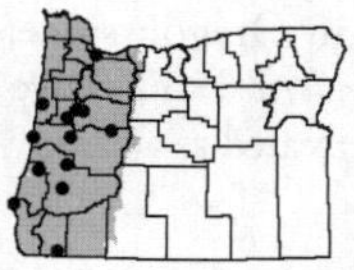

Plants perennial, 3–80 cm tall; rhizomatous, rhizomes to 10(40) cm. **Leaves** ligules 0.5–3 mm, those of culm leaves shorter than wide, those of sterile shoots as long as or slightly longer than wide; blades 4–10 cm × 1–3 mm, flat. **Inflorescences** 10–30 × 3–8 cm, loosely ovate, somewhat contracted after anthesis, lowest nodes with (1)2–7 branches; branches spreading to ascending, spikelets usually confined to the distal 33%; lower branches 3–9 cm; pedicels not divergent, usually shorter than the spikelets, so that the spikelets seem to arise in clusters. **Glumes** 2–3 mm, lanceolate, 1-veined, acute or acuminate. **Calluses** with many hairs 0.3(0.6) mm, but sometimes lacking hairs on awnless lemmas. **Lemmas** 1.3–1.9 mm, translucent (3)5-veined; veins excurrent at lemma tips, sometimes pubescent on the back; tips truncate to obtuse (occasionally acute), usually awned at least in terminal spikelets; lemma awns 0–5 mm, varying from none to well developed on a single plant, arising from the lower 33% of the lemma (or occasionally represented by a minute bristle arising above the middle), usually bent. **Paleas** 0.6–1.1 mm, 50–67% as long as the lemmas. **Anthers** 3, 1–1.5 mm.

Disturbed sites, lawns, golf courses. 50–1300 m. Casc, CR, Est, Sisk, WV. WA; Eurasia. Exotic.

Agrostis castellana is the least common of the introduced bentgrasses. It usually has bent awns arising from near the base of the lemmas, at least in spikelets near the tips of the panicle branches. Lemmas of spikelets lower on the panicle branches may be awned or unawned, and the awns may be very short and attached in the upper half of the lemma. *Agrostis castellana* is widely planted for lawns and golf courses because of its drought tolerance. For this reason, it likely is more widespread than is currently documented.

Agrostis densiflora Vasey

California bentgrass

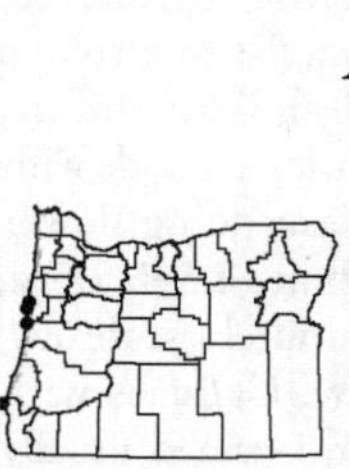

Plants perennial, 9–85 cm tall; cespitose, occasionally rhizomatous when buried by moving sand. **Leaves** ligules 1–3(7.5) mm, dorsal surfaces scabrous, membranous; blades 2–12 cm × 2–10 mm, flat; upper blades wider than lower blades. **Inflorescences** 2–10 × 0.5–2 cm, narrow, dense, ellow-green, sometimes tinged with purple, bases often enclosed in the upper sheaths at maturity; branches appressed, to 1.5 cm, mostly hidden by the spikelets. **Glumes** 2–3.3 mm, 1-veined, densely scabrous over the back, strongly scabrous on the veins; tips narrowly acute to acuminate or mucronate. **Calluses** hairs to 0.3 mm, the hairs usually dense. **Lemmas** 1.5–2.1 mm, (3)5-veined; veins extended as teeth to 0.3 mm; tips acute to obtuse, sometimes toothed, awnless or awned; lemma awns 0–3.5 mm, arising from above mid-length, straight, readily falling off. **Paleas** (0.3)0.5–0.7 mm. **Anthers** 0.3–0.7 mm. **2*n***=42.

Sandy soils on coastal dunes and bluffs. 0–50 m. Est. CA. Native.

With its dense inflorescence, *A. densiflora* can be confused with *A. exarata* and coastal forms of *A. pallens*. *Agrostis exarata* usually has shorter paleas and longer ligules, although there is considerable overlap in these characters. *Agrostis pallens* is usually strongly rhizomatous, its anthers are more than 1 mm, its inflorescence is exserted from the upper leaf sheaths, and it usually has narrower leaves.

Agrostis exarata Trinius (p. 353)

Spike bentgrass

Agrostis ampla Hitchc., *Agrostis exarata* Trin. ssp. *exarata*, *Agrostis exarata* Trin. ssp. *minor* (Hook.) C.L. Hitchc., *Agrostis exarata* Trin. var. *exarata*, *Agrostis exarata* Trin. var. *minor* Hook., *Agrostis exarata* Trin. var. *monolepis* (Torr.) Hitchc., *Agrostis longiligula* Hitchc.

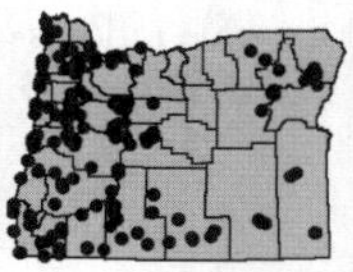

Plants perennial, sometimes flowering in the first year, 8–100(150) cm tall, slightly glaucous; usually cespitose, rarely rhizomatous. **Leaves** ligules (1)1.7–8(11.2) mm, truncate to obtuse; blades 4–15(25) cm × 2–7 mm, flat. **Inflorescences** (3)5–30 × 0.5–4 cm, contracted, spike-like, usually dense, sometimes interrupted near the base, bases usually exserted from the upper sheaths at maturity, lowest node with 1–5 branches; branches ascending to appressed, spikelet-bearing to or near the base, usually hidden by the spikelets; lower branches 1–2(4) cm. **Spikelets** crowded. **Glumes** 1.5–3.5 mm, scabrous on the midvein and sometimes on the back, 1(3)-veined; tips acute or long-acuminate, with an awn-like tip to 1 mm. **Calluses** with sparse to abundant hairs up to 0.3 mm. **Lemmas** 1.2–2.2 mm, smooth, 5-veined; veins sometimes extended as teeth to 0.1 mm; tips acute, entire to toothed, awnless or awned; lemma awns 0–3.5 mm, arising from above mid-length, straight or bent. **Paleas** absent or to 0.5 mm. **Lodicules** 0.3–0.5 mm. **Anthers** 3, 0.3–0.6 mm. **2*n***=28, 42, 56.

Moist ground, streamsides, ditches, sand bars, lake margins, also seasonally moist disturbed spots in grasslands, shrublands, and forest openings, including coastal sites. 0–2700 m. Throughout Oregon. CA, ID, NV, WA; throughout North America except midwest; Eurasia. Native.

Agrostis exarata is a common plant of wet, often disturbed habitats. It is variable in size, glume shape, and other details but can be recognized by its dense, narrow inflorescences and tiny anthers. This species can be difficult to separate from two closely related taxa, *A. densiflora* and *A. microphylla.* For both of these, habitat is an important clue for identification. *Agrostis densiflora* is a short plant of coastal sands and bluffs, while *Agrostis microphylla* is a vernal pool annual.

Agrostis gigantea Roth

Redtop

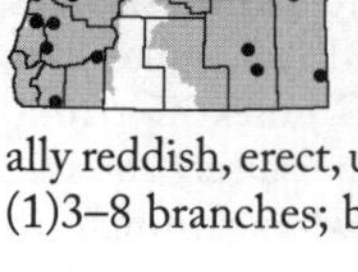

Plants perennial, 20–120 cm tall; rhizomes to 26 cm. **Leaves** ligules longer than wide; basal ligules 1–4.5 mm; upper ligules 2–7 mm; blades 4–10 cm × 3–8 mm, flat. **Inflorescences** 8–25(30) × (1.5)3–15 cm, usually reddish, erect, usually open, broadly ovate, lowest node with (1)3–8 branches; branches usually spreading during and after anthesis, usually some branches spikelet-bearing to the base, many short branches generally present so the inflorescence looks crowded; lower branches 4–9 cm. **Glumes** 1.7–3.2 mm, 1-veined; tips acute to acuminate; scabrous on the distal half of the midvein. **Calluses** with sparse hairs to 0.5 mm, or glabrous. **Lemmas** 1.5–2.2 mm, 3–5-veined; tips usually acute, sometimes obtuse or truncate, often awnless; lemma awns, if present, 0.4–1.5(3) mm, straight, arising from near the tip to near the base. **Paleas** 0.7–1.4 mm, 40–60% as long as the lemmas. **Anthers** 3, 1–1.4 mm. **2*n*=42.**

Fields, meadows, ditches, and disturbed, open, mesic sites. 0–1800 m. BR, BW, Casc, Col, CR, Est, Lava, Owy, Sisk, WV. CA, ID, NV, WA; throughout North America; worldwide. Exotic.

Agrostis gigantea is a rhizomatous, turf-forming grass with a large, open, dark reddish purple inflorescence. It is confused with *A. stolonifera* because both have long ligules on their upper culm leaves. *Agrostis stolonifera* is stoloniferous and has somewhat contracted, green to drab, usually shorter panicles. Individuals of *A. gigantea* growing in shade may have pale inflorescences with ascending branches, superficially like *A. stolonifera.*

Agrostis hallii Vasey

Hall's bentgrass

Agrostis hallii Vasey var. *hallii*, *Agrostis hallii* Vasey var. *pringlei* (Scribn.) Hitchc.

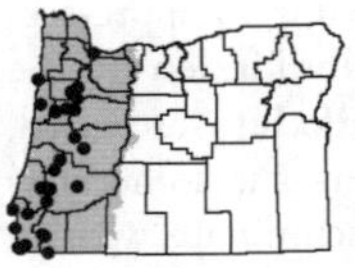

Plants perennial, 17–100 cm tall, slightly glaucous; rhizomatous. **Leaves** ligules 2.3–7 mm; tips acute, usually lacerate; blades 6–20 cm × 2–5 mm, flat. **Inflorescences** 7–22 × 1.5–5.5(7) cm, narrow to widely open, lowest node with 2–15 branches; branches ascending to more or less appressed, mostly branching at or above mid-length, some branching near base. **Spikelets** lanceolate, yellow-green, often tinged with purple. **Glumes** 2.5–4 mm, scabrous or minutely so on the margin at least distally, occasionally over the back; tips acute, sometimes toothed; teeth to 0.2 mm. **Calluses** with conspicuous hairs 0.8–2 mm. **Lemmas** 2–3 mm, 5-veined; tips acute, entire to erose, sometimes toothed; teeth to about 0.2 mm, awnless. **Paleas** absent or minute. **Anthers** 1.5–2.3 mm. **2*n*=42.**

Open oak or mixed forest, oak savanna, coastal and inland meadows. 0–1300 m. Casc, CR, Est, Sisk, WV. CA. Native.

Agrostis hallii is a rhizomatous grass with conspicuous callus hairs. Its inflorescence branches may be widely spreading or appressed. In the field, its slightly glaucous herbage helps distinguish it from other grasses. This variable native grass has a superficial resemblance to the invasive introduced *Agrostis* species, but it lacks a palea. *Agrostis hallii* closely resembles native *A. pallens,* which has sparsely short-hairy calluses.

Agrostis hendersonii Hitchcock

Henderson's bentgrass

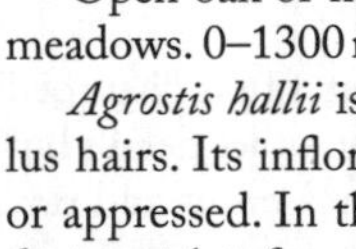

Plants annual, 6–70 cm tall. **Ligules** 0.5–5 mm, acute to obtuse, lacerate, sometimes ciliolate; blades 1–4.5 cm × 0.5–1(2) mm, flat or somewhat involute. **Inflorescences** 1–5 × 0.5–1.5 cm, cylindrical, spike-like, dense, sometimes interrupted near the base. **Glumes** 3.5–5 mm, 1(3)-veined, scabrous over the midvein, often sparsely and minutely scabrous over the body; tips narrowly acuminate to awn-tipped, awns, if present, to 2 mm. **Calluses** with abundant hairs to 0.7 mm. **Lemmas** 2.5–4 mm, 5-veined, minutely scabrous on the veins and occasionally over the back; tips acute, with 2 teeth 0.2–1.5 mm; awned; lemma awns arising at about mid-length, (5)8–10 mm, more or less bent. **Paleas** absent. **Anthers** 0.4–0.7 mm. **2*n*=42.**

Clay or rocky soils around vernal pools. 200–500 m. Sisk. CA. Native.

Agrostis hendersonii is an annual of vernal pool habitats in southwestern Oregon. It has very long lemma awns for an *Agrostis* and long, awned glumes. It was collected in Jackson County in 1930 and has not been found in Oregon since, though it still exists in California. *Agrostis hendersonii* may have evolved from *A. microphylla* and is sometimes treated as a variety of that species.

Agrostis howellii Scribner

Howell's bentgrass

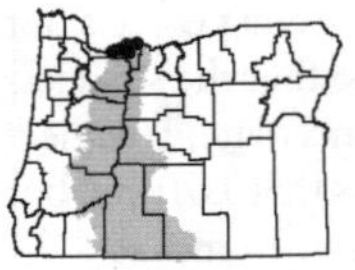

Plants perennial, 40–80 cm tall; cespitose. **Ligules** (0.9)2.7–5 mm, truncate to acute, erose to lacerate; blades 15–20 cm × 3–5 mm, flat. **Inflorescences** 10–25 × 3.5–15 cm, open and diffuse, lowest node with (1)2–6 branches; branches wavy, spreading, usually spikelet-bearing only on the outer 50%; lower branches 3–10 cm. **Glumes** 2.3–3.5 mm, 1(3)-veined; veins minutely scabrous; tips acute to acuminate. **Calluses** with sparse to abundant hairs to 0.3 mm. **Lemmas** (2.2)2.5–3 mm, 5-veined; veins extending as teeth to 0.5 mm; tips acute to obtuse, 4-toothed, awned; lemma awns 4–6 mm, bent, arising from the lower 33% of the lemma. **Paleas** absent or minute. **Anthers** 1–1.3 mm.

Moist rocks and cliffs, waterfall spray zones, and moist shady woodlands. 50–500 m. Casc, ECas. WA. Native.

Agrostis howellii is recognized by its long, exserted, bent awns that arise from the lower third of the lemma and an open panicle.

Agrostis idahoensis Nash

Idaho redtop

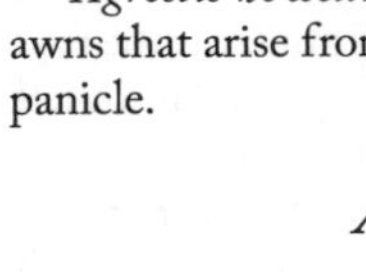

Plants perennial, 8–40 cm tall; cespitose. **Leaves** ligules 0.7–3.8 mm, rounded to truncate, rarely acute, erose to lacerate; blades 1–7 cm × 0.5–2 mm, flat to involute. **Inflorescences** 3–13 × 1–6(8) cm, diffuse, usually reddish to purple, occasionally dull tan, lowest node with 1–6(10) branches; branches more or less ascending; lower branches 1–4 cm. **Spikelets** not crowded. **Glumes** 1.5–2.5 mm, subequal, 1-veined, usually scabrous or minutely scabrous, sometimes smooth; tips acute to acuminate. **Calluses** with sparse hairs to 0.3 mm. **Lemmas** 1.2–2.2 mm, 5-veined, usually smooth; tips acute to obtuse, entire, awnless. **Paleas** absent or to 0.2 mm and thin. **Anthers** 0.3–0.6 mm.

Montane to subalpine bogs, fens, seeps, and wet meadows. 50–2700 m. BR, BW, Casc, ECas, Sisk. CA, ID, NV, WA; north to AK, northeast to MT, southeast to NM; South America. Native.

Agrostis idahoensis is a delicate cespitose grass with narrow, mainly basal leaves and an open, dark reddish to purple inflorescence. It is sometimes confused with dwarf forms of *A. scabra,* which grows in drier habitats and has longer lower panicle branches and longer leaves. *Agrostis variabilis* is also a delicate, narrow-leaved bentgrass, but its inflorescence is a narrow, condensed panicle with shorter panicle branches.

Agrostis microphylla Steudel

Small-leaf bentgrass

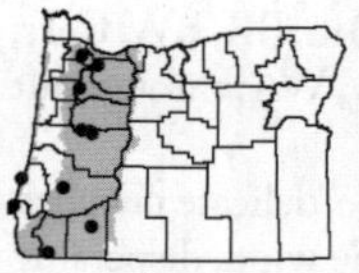

Plants winter annuals, 8–65 cm tall. **Leaves** upper sheaths usually inflated; ligules 1.4–4.5 mm, truncate to acute, lacerate to erose; blades 3–15 cm × 0.7–2.5(3.6) mm, flat, becoming involute. **Inflorescences** 2–12 × 0.4–2 cm, cylindrical, dense, spike-like, occasionally lobed or interrupted near the base, appearing bristly with awns. **Glumes** 2.5–5 mm, scabrous over the midveins and often minutely scabridulous on the body, 1-veined; tips narrowly acuminate to awn-tipped, awns, if present, to 1.5 mm. **Calluses** usually densely hairy, the hairs to 0.5(1) mm. **Lemmas** 1.5–2.3 mm, scabrous, 5-veined; veins extended into 2(4) teeth 0.1–0.5 mm; tips acute, awned; lemma awns 2.3–5(8) mm, the longer awns generally more than 3.5 mm, bent, arising from about mid-length or above. **Paleas** absent or to 0.2 mm. **Lodicules** up to 0.2 mm. **Anthers** 0.4–0.5 mm, often retained at the tips of the caryopses.

Vernally moist rocky soils, vernal pools, cliffs, occasionally serpentine substrates. 0–1300 m. Casc, Est, Sisk, WV. CA, WA; north to British Columbia, south to Mexico. Native.

Agrostis microphylla is an annual grass similar to the perennial *A. exarata*. *Agrostis microphylla* tends to have narrower leaves, longer and more tapered glumes and longer glume awns, but there is much overlap between the two. Lodicules are consistently shorter in *A. microphylla* than *A. exarata*, but this feature can be difficult to assess. *Agrostis hendersonii*, a small annual collected once in southwestern Oregon, has longer glumes and longer awns.

Agrostis oregonensis Vasey

Oregon redtop

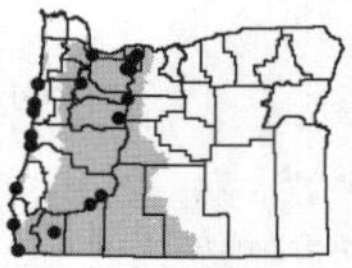

Plants perennial, 12–75 cm tall; cespitose. **Leaves** ligules 1.2–6.3 mm, truncate to rounded, erose-lacerate; blades 10–30 cm × (1)2–5 mm, flat. **Inflorescences** 8–35(70) × (1.5)2.5–20 cm, usually 3 times longer than wide, open, lowest node with 1–15 branches, usually dark red or purple; branches ascending, spikelets somewhat clustered toward the branch tips; lower branches 2–10 cm, primary branches branched above mid-length. **Glumes** 2–3.6 mm, 1(3)-veined, scabrous on the midvein, occasionally minutely scabrous over the body, acute to acuminate. **Calluses** with sparse hairs to 0.2 mm. **Lemmas** 1.5–2.5 mm, usually smooth, sometimes minutely scabrous or pubescent, 5-veined; veins sometimes extended into teeth to 0.4 mm; tips acute to obtuse, usually entire to toothed,

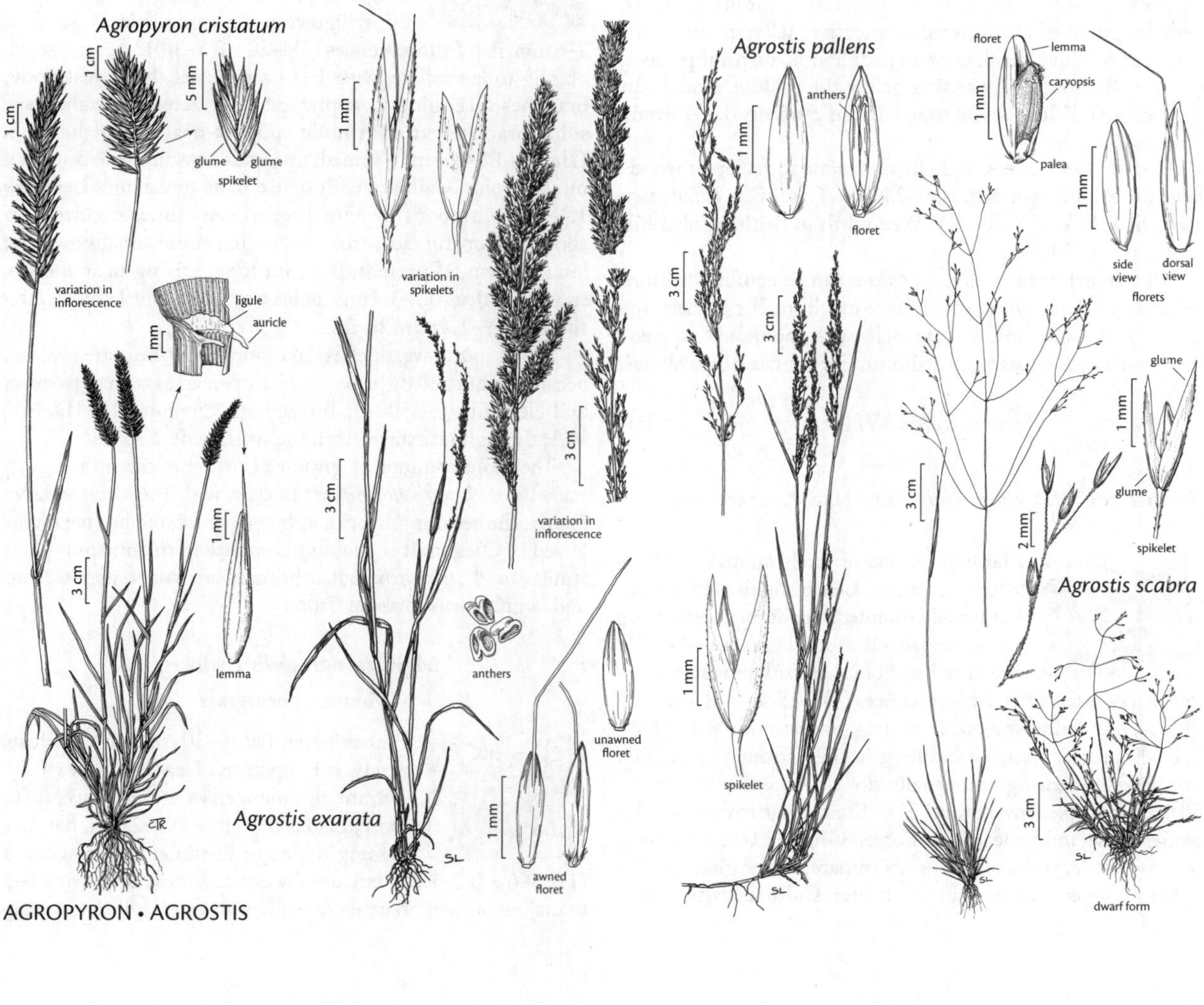

AGROPYRON • AGROSTIS

usually awnless but sometimes with awned and awnless spikelets on the same plant; lemma awns 0–2 mm, arising from about mid-length, straight (rarely bent), not exserted. **Paleas** absent or to 0.2 mm and thin. **Anthers** 0.5–1.2 mm.

Lakeshores, streamsides and wet meadows, sometimes with sphagnum. 0–2300 m. Casc, ECas, Est, Sisk, WV. CA, ID, NV, WA; north to British Columbia, east to WY. Native.

Agrostis oregonensis has an open, dark reddish panicle with primary branches not branched or spikelet-bearing on the lower half. *Agrostis scabra* has narrower leaves and an even more open, diffuse panicle that breaks off at the base and rolls away like a tumbleweed.

Agrostis pallens Trinius (p. 353)

Thin bentgrass, seashore bentgrass

Agrostis diegoensis Vasey, *Agrostis foliosa* Vasey, *Agrostis lepida* Hitchc.

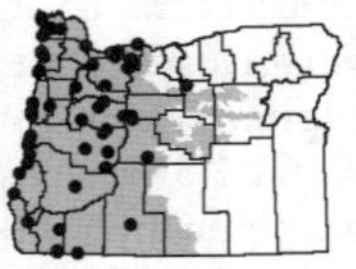

Plants perennial, 10–70 cm tall; rhizomes to 10 cm. **Leaves** ligules 1–4(6) mm; tips truncate to rounded or acute, often lacerate to erose; blades 1.5–11.5 cm × 1–4(6) mm, flat, becoming involute. **Inflorescences** 5–20 × 0.4–6(8) cm, condensed to open, purple or pale, lowest node with 1–8 branches; branches usually ascending, spikelet-bearing to the base; lower branches 2–5 cm, primary branches branched from below mid-length. **Glumes** 2–3.5 mm, scabrous on midvein, sometimes sparsely scabrous on the body, 1(3)-veined; tips acute to acuminate. **Calluses** with sparse hairs to 0.3(1) mm. **Lemmas** 1.5–2.5 mm, smooth, minutely scabrous or warty, 5-veined; veins often extended into teeth to 0.2 mm; tips acute, entire or toothed, awnless or awned; lemma awns, if present, 0.5–2.5(4) mm, usually arising below the middle, straight to slightly bent. **Paleas** absent or to 0.2 mm. **Anthers** 0.7–1.8 mm. **2*n***=42, 56.

Coastal sand dunes and cliffs, meadows, dry open woodlands, subalpine woodlands. 0–2300 m. Casc, CR, ECas, Est, Lava, Sisk, WV. CA, ID, NV, WA; north to British Columbia, northeast to MT. Native.

The coastal form of *Agrostis pallens* can be confused with *A. exarata* and *A. densiflora*, which have much smaller anthers. Inland *A. pallens* has more open panicles with appressed to spreading branches. *Agrostis hallii* is also similar but has hairy calluses.

Agrostis scabra Willdenow (p. 353)

Ticklegrass

Agrostis scabra Willd. var. *geminata* (Trin.) Hitchc., *Agrostis scabra* Willd. var. *scabra*

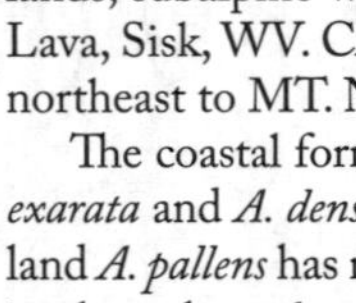

Plants perennial or rarely annual, (7.5)15–90 cm; cespitose. **Leaves** ligules 0.7–5 mm; tips usually rounded, sometimes truncate or acute, erose-ciliate; blades 4–14 cm × 1–2 mm; basal blades mostly involute; cauline blades usually flat. **Inflorescences** (4)8–25(50) × 0.5–20 cm, diffuse, often almost as wide as long, lowest node with (1)2–7(12) branches; branches capillary, flexible, strongly spreading, not spikelet-bearing and usually not branched on the lower 50%, dark purple; lower branches 4–12 cm; disarticulation at the panicle base and beneath the florets. **Glumes** 1-veined; keels scabrous at least distally; tips acuminate; lower glumes 1.8–3.4 mm; upper glumes slightly shorter. **Calluses** with sparse hairs to 0.2 mm. **Lemmas** 1.4–2 mm, scabrous to smooth, 5-veined; veins extended into minute teeth; tips acute to obtuse, usually entire, awnless or awned; lemma awns, if present, 0.2–3 mm, arising below mid-length, bent or straight. **Paleas** absent or to 0.3 mm. **Anthers** 0.4–0.8 mm, usually not retained after anthesis. **2*n***=42.

Moist to dry grasslands, pastures, shrub-steppe, forests and roadsides, especially in disturbed sites. 0–3000 m. BR, BW, Casc, CR, ECas, Est, Lava, Owy, WV. CA, ID, NV, WA; throughout North America; Asia. Native.

Agrostis scabra has a large, extremely diffuse, delicate purplish inflorescence which breaks off and rolls in the wind, dispersing seeds like a tumbleweed. The inflorescence is often more than half as tall as the entire plant. Similar *A. idahoensis* is a short-leaved plant of subalpine and alpine wetlands. Its lower panicle branches rarely exceed 4 cm, and the inflorescences do not break off at maturity.

Agrostis stolonifera Linnaeus

Creeping bentgrass

Agrostis alba L. var. *maritima* (Lam.) G. Mey., *Agrostis alba* L. var. *palustris* (Huds.) Pers., *Agrostis alba* L. var. *stolonifera* (L.) Sm., *Agrostis maritima* Lam., *Agrostis palustris* Huds., *Agrostis stolonifera* L. var. *palustris* (Huds.) Farw., *Agrostis stolonifera* L. var. *stolonifera*

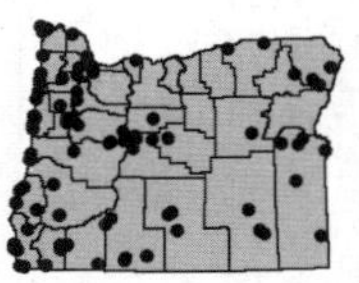

Plants perennial, (8)15–60 cm tall; stolons rooting at the nodes and sometimes forming a dense mat. **Leaves** ligules of sterile shoots longer than wide; basal ligules 0.7–4 mm; upper ligules 3–7.5 mm; blades 2–10 cm × 2–6 mm, flat. **Inflorescences** (3)4–20 × 0.5–3(6) cm, contracted, oblong to lanceolate, with 1–7 branches at the lowest node; branches ascending to appressed, contracted after anthesis, some branches at each node spikelet-bearing to the base. **Glumes** 1.6–3 mm, 1-veined; tips acute to acuminate or with a minute point. **Calluses** with sparse hairs to 0.5 mm. **Lemmas** 1.4–2 mm, smooth, 5-veined; veins sometimes excurrent to about 0.1 mm; tips acute to obtuse, often entire, usually awnless; lemma awns, if present, 0–1 mm long, arising near the tip, straight. **Paleas** 0.7–1.4 mm, at least as long as the lemmas. **Anthers** 3, 0.9–1.4 mm. **2*n***=28.

Moist meadows, ditches, lakeshores, marshes, streamsides, ocean beaches and disturbed, moist, open areas such as pastures and clearcuts. 0–2500 m. Throughout Oregon. CA, ID, NV, WA; throughout North America; worldwide. Exotic.

The more common *A. gigantea* is often misidentified as *A. stolonifera*. *Agrostis stolonifera* can cross with *Polypogon monspeliensis*. The resulting hybrid, ×*Agropogon lutosus*, has been collected in Oregon. It is a loosely cespitose to rhizomatous plant similar to *A. stolonifera*, but it has glume awns to about 2 mm and lemma awns to about 3 mm.

Agrostis variabilis Rydberg

Mountain bentgrass

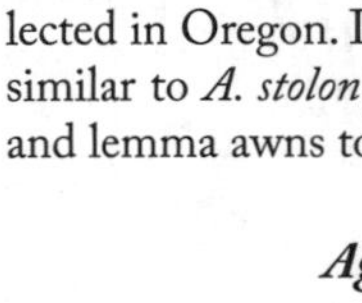

Plants perennial, 5–30 cm tall; cespitose, rarely rhizomatous. **Leaves** ligules 0.7–2.8 mm; tips rounded to truncate, lacerate to erose; blades 3–7 cm × 0.5–2 mm, flat, becoming folded or involute. **Inflorescences** (1)2.5–6 × 0.3–1.2(2) cm, usually dense, lowest node with 1–5 branches, purple; branches ascending to erect, spikelet bearing to the base; lower branches 0.5–1.5 cm. **Glumes** 1.8–2.5 mm, 1-veined, smooth, or slightly, acute to acuminate. **Calluses** with sparse to abundant hairs to 0.2 mm. **Lemmas** 1.5–2 mm, smooth, (3)5-veined; tips acute, entire, often awnless; lemma awns, if present, 0–1(2.8) mm, arising above the midpoint, usually not exceeding the lemma. **Paleas** 0.2–0.4 mm. **Anthers** 0.4–1 mm.

Alpine and subalpine meadows, forests and talus slopes. 1200–3000 m. BR, BW, Casc, ECas. CA, ID, NV, WA; north to British Columbia, east to WY, southeast to NM. Native.

Agrostis variabilis is a tufted, fine-leaved grass with a narrow, dark purple inflorescence. Its foliage might be mistaken for a fine-leaved fescue.

AIRA Hairgrass

Plants annual. **Culms** 1–55 cm tall, erect or decumbent. **Leaves** cauline; sheaths open, glabrous; ligules membranous; blades glabrous; tips prow-shaped. **Inflorescences** open or contracted, sometimes spike-like panicles. **Spikelets** 1.5–3.8 mm long, laterally compressed, with 2 bisexual florets, one or both awned; disarticulation above glumes and below florets. **Glumes** 2, equal or subequal, longer than the florets, 1–3-veined, awnless. **Calluses** with short hairs. **Lemmas** 5-veined; tips bifid, awned or awnless, awns attached below mid-length. **Anthers** 3. **Caryopses** shorter than the lemmas, remaining attached to the lemmas and/or paleas.

North America, North Africa, western Asia, and Europe. 8 species; 3 species treated in *Flora*.

Key to *Aira* species:

1. Inflorescence condensed, narrow, branches appressed; pedicels shorter than spikelets. . . . . ***A. praecox***

1' Inflorescence open, branches spreading; pedicels longer than spikelets.

  2. Lower lemma lacking an awn; pedicels 1–8 times as long as the spikelets; spikelets 1.7–2.5 mm. . . . . . . ***A. elegantissima***

  2' Both lemmas awned; pedicels 1–2 times as long as the spikelets; spikelets 2–3.5 mm . . ***A. caryophyllea* var. *caryophyllea***

Aira caryophyllea Linnaeus var. *caryophyllea* (p. 358)

Silver hairgrass

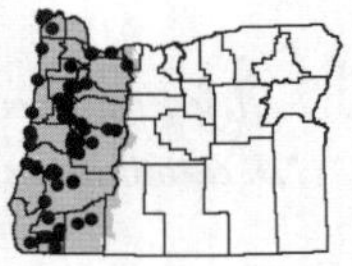

Plants 4.5–55 cm tall, tufted. **Leaves** blades 0.3–13.5 cm × 0.3–2.5 mm, involute, glabrous. **Inflorescences** open panicles, 1.2–13.5 × 1.5–6 cm; branches spreading, pedicels appressed to the secondary branches, 0.9–7 mm, 1–2 times as long as the spikelets. **Spikelets** 2–3.5 mm. **Glumes** 1-veined or the upper sometimes 3-veined, acute. **Lemmas** 2–2.6 mm, usually with both lemmas awned; lemma awns 2.1–3 mm, straight or bent. **Paleas** 1.4–1.7 mm. **Anthers** 0.2–0.5 mm, yellow to orange. **Caryopses** 1.4–1.5 mm. **2*n***=24.

Roadsides and in other open areas with little or no canopy cover, including disturbed grasslands, rocks, cliffs and talus slopes, occasional on edges of wetlands, often in sites with very shallow soil over rock or asphalt. 0–1900 m. Casc, CR, Est, Sisk, WV. CA, ID, WA; scattered throughout North America; Mediterranean. Exotic.

This abundant but easily overlooked annual is recognized by its delicate, open panicle. Its spikelets are more clustered toward the ends of branches than are those of *A. elegans*, and it usually has two awns sticking out of the spikelet.

Aira elegantissima Schur

Delicate hairgrass

Aira caryophyllea L. var. *capillaris* (Mert. & W.D.J. Koch) Bluff & Fingerh., *Aira elegans* Willd. ex Roem. & Schult.

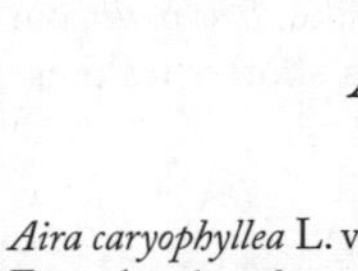

Plants 4.5–55 cm tall, tufted. **Leaves** blades 0.3–13.5 cm × 0.3–2.5 mm, involute, glabrous or scabrous. **Inflorescences** open panicles, 1.2–13.5 × 1.5–13 cm; branches spreading, pedicels divergent, 2–11.3 mm, 1–8 times as long as the spikelets. **Spikelets** 1.7–2.5 mm. **Glumes** 1-veined, acute. **Lemmas** 1.3–2.1 mm, usually only the upper lemmas awned; lemma awns 2.1–3.9 mm, straight or bent. **Paleas** 1–1.3 mm. **Anthers** 0.2–0.4 mm, yellow-orange or purple. **Caryopses** 0.9–1.2 mm. **2*n***=14.

Disturbed areas, especially on shallow soil over rock or pavement. 0–600 m. Est, Sisk, WV. CA, WA; NM east to NJ and FL; northern Africa, southern Europe. Exotic.

This grass is recognized by its panicles that form delicate clouds of isolated spikelets. This species is sometimes treated as a subspecies of *A. caryophyllea.*

Aira praecox Linnaeus

Spike hairgrass

Aspris praecox (L.) Nash

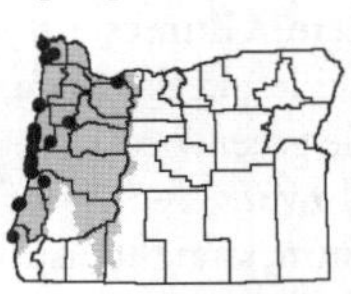

Plants 1–36 cm tall, tufted. **Leaves** blades 0.1–5 cm × 0.3–2 mm, involute, minutely antrorsely scabridulous. **Inflorescences** spike-like panicles, 0.5–4.1 × 0.3–0.7 cm; branches appressed to the rachis, pedicels 0.7–3.2 mm. **Spikelets** 2.7–3.8 mm. **Glumes** 1–3-veined, acute. **Lemmas** 2.4–3.3 mm, both lemmas awned; lemma awns 3–4.5 mm. **Paleas** 1.7–2.1 mm, membranous. **Anthers** 0.2–0.4 mm, yellow. **Caryopses** 1.3–1.7 mm. **2*n***=10, 14.

Coastal sands in deflation plains and disturbed areas, occasionally inland in disturbed sites, roadsides. 0–1000 m. Casc, CR, Est, WV. CA, WA; north to British Columbia, Nova Scotia south to VA; Europe. Exotic.

Aira praecox is a delicate tufted grass with a spike-like inflorescence. Unusually robust individuals can resemble depauperate *Anthoxanthum odoratum,* but that species rarely has blades less than 3 mm wide. Delicate *Anthoxanthum aristatum* may resemble *Aira praecox* but has more complex spikelets with two sterile lemmas and one fertile lemma.

ALOPECURUS Meadow foxtail

Plants annual or perennial; cespitose or short-rhizomatous. **Leaves** sheaths open; upper sheaths sometimes inflated; ligules membranous, entire to lacerate; blades flat or involute, glabrous to scabrous. **Inflorescences** terminal, dense, spike-like, cylindrical panicles; disarticulation below the glumes. **Spikelets** strongly laterally compressed, oval in outline, with 1 floret. **Glumes** equaling or exceeding the florets, free or fused in at least the lower half, narrowing from above mid-length, 3-veined, keeled, the keels ciliate at least basally; tips obtuse, acute, or short-awned. **Calluses** blunt, glabrous. **Lemmas** 3-veined; lower half of margins often fused; tips truncate to acute, awned; lemma awns arising on the lemma back from just above the base to mid-length, bent or straight. **Paleas** lacking. **Anthers** 3.

Temperate regions of the world. Approximately 36 species; 6 species treated in *Flora*.

Some *Alopecurus* have a superficial resemblance to *Phleum*, which has truncate, abruptly awned glumes and a harsh feel to the inflorescence. Superficially, *Crypsis* looks like a small, annual *Alopecurus*, but differs in having ligules of hairs, 1-veined glumes and 1-veined, awnless lemmas. *Alopecurus* shares the common name "foxtail" with the distantly related genus *Setaria*.

Key to *Alopecurus* species:

1. Spikelets (excluding awns) more than 4 mm.
   2. Plants perennial . . . . *A. pratensis*

   2' Plants annual.
      3. Glumes glabrous (keels ciliate); upper sheaths inconspicuously inflated . . . . *A. myosuroides*

      3' Glumes pubescent; upper sheaths conspicuously inflated . . . . *A. saccatus* (in part)

1' Spikelets (excluding awns) less than 4 mm.
   4. Spikes neat and trim because the inconspicuous awns are hidden in the glumes or exceed the glumes by less than 1 mm . . . . *A. aequalis* var. *aequalis*

   4' Spikes shaggy with awns that exceed the glumes by 1.5–5 mm.
      5. Spikelets more than 3 mm; upper leaf sheath conspicuously inflated; awn exceeding glume by 3–5 mm . . . . *A. saccatus* (in part)

      5' Spikelets usually less than 3 mm; upper leaf sheath not or slightly inflated; awn exceeding glume by 1.5–4 mm.
         6. Plants perennial, rooting at lower nodes; anthers (0.9)1.4–2.2 mm . . . . *A. geniculatus*

         6' Plants annual, not rooting at the nodes; anthers 0.3–1 mm . . . . *A. carolinianus*

Alopecurus aequalis Sobol var. *aequalis*

Short-awn foxtail

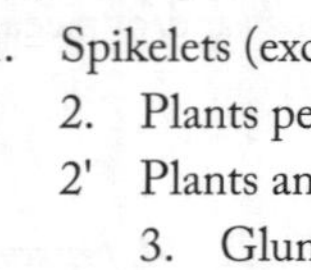

Plants perennial, 9–75 cm tall; cespitose. **Culms** often decumbent. **Leaves** upper sheaths not inflated; blades 2–10 cm × 1–5(8) mm, flat. **Inflorescences** dense, spike-like, 1–9 × 0.3–0.6 cm. **Glumes** 1.8–3 mm, fused near the base, pubescent on the sides; keels ciliate, not winged; tips obtuse, sometimes erose, pale green, occasionally purplish. **Lemmas** 1.5–2.5(3.5) mm; lower 33–50% of margins fused, glabrous; lemma awns 0.7–3 mm, straight, usually not exceeding the lemmas or exceeding by less than 1 mm. **Anthers** 0.5–0.9 mm.

Wet sites including vernal pools, springs, meadows, forest openings and disturbed areas. 0–2800 m. BR, BW, Casc, Col, ECas, Lava, Owy, Sisk, WV. CA, ID, NV, WA; most of Canada and US; Eurasia. Native.

Alopecurus aequalis var. *aequalis* is a variable grass of vernally wet habitats. It has inconspicuous awns that barely extend out of the glumes. *Alopecurus aequalis* produces sterile hybrids with *A. geniculatus*.

Alopecurus carolinianus Walter

Tufted foxtail

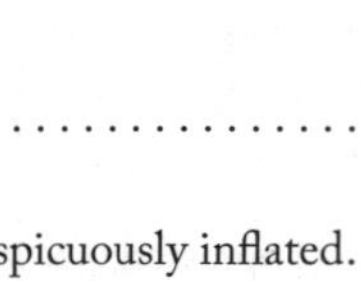

Plants annual, 5–50 cm tall. **Culms** erect or decumbent. **Leaves** upper sheaths not or only slightly inflated; blades 3–15 cm × 0.9–3 mm, flat. **Inflorescences** dense, spike-like, 1–7 × 0.3–0.6 cm. **Glumes** 2.1–3.3 mm, fused at the base, sparsely pubescent; keels ciliate, not winged; tips obtuse, pale green to pale yellow. **Lemmas** 1.9–2.7 mm; lower 50% of margins fused, glabrous; lemma awns 3–6.5 mm, bent, exceeding the lemmas by 1.5–4 mm. **Anthers** 0.3–1 mm, yellow or orange. **2*n***=14.

Wet meadows and disturbed wet areas. 800–1700 m. BR, Col, Owy. CA, ID, WA; throughout US north into southwestern Canada. Exotic.

Morphologically, *A. carolinianus* is much like *A. aequalis*, but it is annual and more delicate, usually with shorter inflorescences and with conspicuous exserted awns.

Alopecurus geniculatus Linnaeus
Water foxtail

Alopecurus pallescens Piper

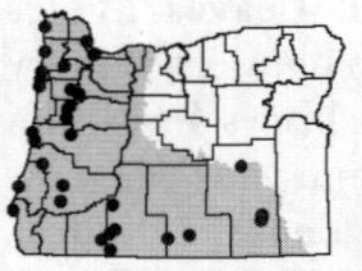

Plants perennial, (5)10–60 cm tall; cespitose. **Culms** erect or decumbent, rooting at the nodes. **Leaves** upper sheaths somewhat inflated; blades 2–12 cm × 1–4(7) mm, flat. **Inflorescences** dense, spike-like, 1.5–7 × 0.4–0.8 cm. **Glumes** 1.9–3.5 mm, fused at the base, membranous, pubescent; keels ciliate and not winged; tips obtuse, parallel, often purplish. **Lemmas** 2.5–3 mm; lower 50% of margins fused, glabrous or with a few hairs at the tips; lemma awns 3.5–6 mm, bent, exceeding the lemmas by (1.2)2–4 mm. **Anthers** (0.9)1.4–2.2 mm, yellow. **2*n***=28.

Vernal pools, wet prairies, disturbed places with shallow water. 0–2800 m. BR, Casc, CR, ECas, Est, Sisk, WV. CA, ID, NV, WA; most of Canada and US; Eurasia. Native?

Alopecurus geniculatus is similar to *A. aequalis* and lives in similar habitats but has longer, fairly conspicuous awns and longer anthers. There is uncertainty whether populations in western North America are native or exotic.

Alopecurus myosuroides Hudson
Blackgrass

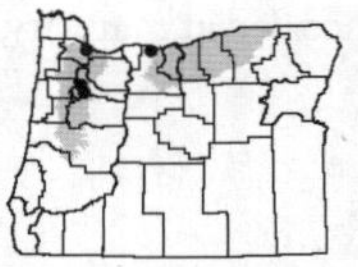

Plants annual, (10)40–85 cm tall. **Culms** erect. **Leaves** upper sheaths inconspicuously inflated; blades 5–20 cm × (2)3.5–6 mm, flat. **Inflorescences** dense, spike-like, 4–12 × 0.3–0.7 cm. **Glumes** 4.5–7.5 mm, lower 50% fused, leathery, dark purple to green, sides glabrous; keels short-ciliate, winged, blue or purple, scabrous toward the top; lateral veins ciliate basally or glabrous; tips acute, convergent to parallel. **Lemmas** 4–7 mm; lower 33–50% of margins fused, glabrous; lemma awns to 12 mm, bent, exceeding the lemmas by 3–6 mm. **Anthers** 2.4–4.1 mm, yellow. **2*n***=14, 28.

Grain and grass seed fields. 0–200 m. Col, WV. CA, WA; sparsely scattered across North America; Eurasia. Exotic.

Alopecurus myosuroides is a relatively tall, erect annual weed of grain crops, especially winter wheat. It is unique among Oregon *Alopecurus* as an annual with relatively large spikelets.

Alopecurus pratensis Linnaeus (p. 358)
Meadow foxtail

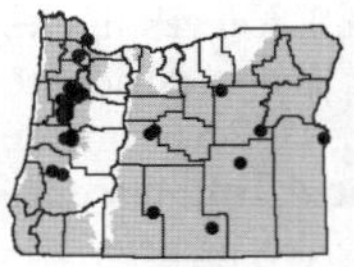

Plants perennial, 30–110 cm tall; rhizomatous. **Culms** erect. **Leaves** upper sheaths not inflated; blades 6–40 cm × 1.9–8 mm, flat. **Inflorescences** dense, spike-like, 3.5–9 × 0.6–1 cm. **Glumes** 4–6 mm; lower 25% fused, membranous, sides pubescent; keels ciliate, not winged; tips acute, parallel or convergent. **Lemmas** 4–6 mm; lower 33% of margins, glabrous or the keels ciliate toward the tip; lemma awns 5–10.5 mm, bent, exceeding the lemmas by (1)2.2–5.5 mm. **Anthers** 2–4 mm, yellow, orange, reddish, or purplish. **2*n***=28, 42.

Wet or mesic meadows, pastures, ditches, streamsides, roadsides. 0–2700 m. BR, BW, CR, ECas, Lava, Owy, Sisk, WV. CA, ID, NV, WA; most of Canada and US; North Africa, Eurasia. Exotic.

Alopecurus pratensis is a common, turf-forming perennial that blooms early in spring before the superficially similar *Phleum pratense.* The latter generally grows in drier habitats and has truncate glumes that are abruptly awned or mucronate, giving each spikelet a 2-horned look.

Alopecurus saccatus Vasey
Pacific meadow foxtail

Alopecurus howellii Vasey

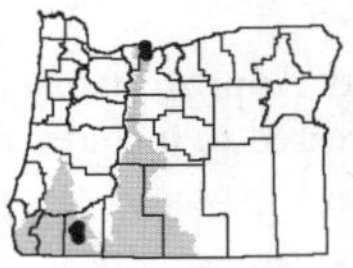

Plants annual, 12–45 cm tall. **Culms** erect or decumbent. **Leaves** upper sheaths conspicuously inflated; blades 4–12 cm × 1.2–4 mm, flat. **Inflorescences** dense, spike-like, 1.5–6.5 × 0.5–1 cm. **Glumes** 3–5 mm, fused at the base, pubescent, ciliate; keels not winged; tips obtuse. **Lemmas** 3–5 mm; lower 33–50% of margins fused, glabrous; lemma awns 6–10 mm, bent, exceeding the lemmas by 3–6 mm. **Anthers** 0.7–1.8 mm, yellow to rusty brown.

Vernal pools, streamsides, and meadows. 100–700 m. ECas, Sisk. CA, WA. Native.

Alopecurus saccatus grows in vernal pools and puddles, similar to *A. aequalis* and *A. geniculatus.* The latter, however, have shorter glumes and awns.

AMMOPHILA Beachgrass

Plants perennial, strongly rhizomatous. **Culms** 20–130 cm, erect, glabrous. **Leaves** mostly basal; sheaths open; ligules membranous, sometimes with minutely ciliate margins; blades 0.5–8 mm wide, involute or convolute. **Inflorescences** dense, spike-like panicles. **Spikelets** laterally compressed with 1 floret; disarticulation above the glumes, below the florets. **Glumes** equaling or exceeding the florets, subequal, linear-lanceolate, keeled, acute to acuminate, the lower 1-veined, the upper 3-veined. **Calluses** pubescent. **Lemmas** papery in texture, linear-lanceolate, obscurely 3–5-veined; tips entire or minutely bifid; lemma awns lacking or 0.2–0.5 mm. **Anthers** 3.

Western and eastern North America, North Africa, and Europe. 2 species.

Both of our *Ammophila* species were introduced to Oregon to stabilize coastal sands. *Ammophila arenaria*, in particular, has been so effective that it has created a landform new to Oregon: the high foredune. The foredunes trap sand blowing off the beach, thus preventing the formation of dunes that move inland. This stabilization benefits roads and homes near the ocean but prevents formation of moving dunes and the associated deflation plains, unique habitats needed by native plants and wildlife. Both species of *Ammophila* hybridize with species of *Calamagrostis*.

Key to *Ammophila* species:

1. Ligules 10–35 mm, acute and bifid or lacerate . . . . ***A. arenaria*** ssp. ***arenaria***
1' Ligules 1–4.6 mm, truncate to obtuse. . . . . ***A. breviligulata*** ssp. ***breviligulata***

Ammophila arenaria Linnaeus ssp. *arenaria* (p. 358)

European beachgrass, marram grass

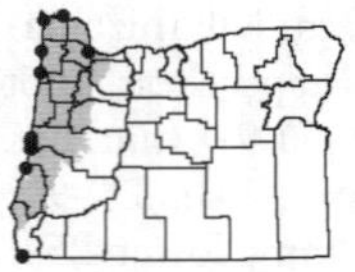

Plants (50)60–120 cm tall. **Leaves** ligules 10–35 mm, acute, bifid or lacerate; blades to 6 mm wide when flat, (0.5)1–2.5 mm wide when rolled, sharply pointed. **Inflorescences** dense, cylindric, spike-like, panicles 12–35 × 1.5–2.5 cm. **Spikelets** 12–14 mm. **Glumes** 10–14 mm, glabrous. **Calluses** hairs 2–6 mm. **Lemmas** 8–12 mm. **Anthers** 4–7 mm. **2*n***=28.

Coastal sand dunes, beaches, open sandy ground. 0–50 m. CR, Est, WV. CA, WA; north to British Columbia, south to Mexico, scattered in eastern North America; worldwide. Exotic.

The dense, cylindrical inflorescences of *A. arenaria* superficially resemble those of *Leymus mollis,* which grows in the same coastal habitat but has inflorescences that are true spikes, wider leaves, and a more open growth form.

Ammophila breviligulata Fernald ssp. *breviligulata* (p. 358)

American beachgrass

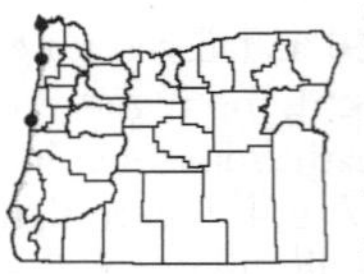

Plants 50–130 cm tall. **Leaves** ligules 1–3(4.6) mm, truncate to obtuse, with very short hairs on the margin; blades 4–8(9) mm wide when flat, 0.5–2.5 mm wide when involute. **Inflorescences** dense, cylindric, spike-like, 9.5–40 × 1–2.5 cm. **Spikelets** 8–15 mm. **Glumes** 8–15 mm, glabrous, sometimes sparsely hairy on keels. **Calluses** hairs 1–3 mm. **Lemmas** 8–11.5(14) mm. **Anthers** 3.5–7 mm.

Coastal sand dunes and dry, sandy shores. 0–50 m. Est. CA, WA; north to British Columbia; the Great Lakes region and east coast of North America. Exotic.

Ammophila breviligulata has been sparingly introduced to the Oregon coast. It closely resembles *A. arenaria*, which has much longer ligules.

ANTHOXANTHUM Vernalgrass

Plants annual or perennial, densely to loosely cespitose. **Leaves** sheaths open; auricles present or absent; ligules membranous, sometimes shortly ciliate; blades flat or rolled. **Inflorescences** dense, spike-like panicles; disarticulation above the glumes, the florets falling together. **Spikelets** laterally compressed, with 3 florets, the 2 lowest florets sterile and awned, hiding the distal floret; distal florets bisexual, awnless. **Lower 2 florets** lemmas strongly compressed, 3-veined, strigose, hairs brown; tips bilobed, awned from the back; lemma awns arising dorsally. **Glumes** 2, unequal, the larger 1 exceeding the florets, lanceolate, keeled. **Calluses** blunt, hairy. **Distal lemmas** somewhat hard, shiny, awnless. **Anthers** 2. **Caryopses** shorter than the lemmas, concealed at maturity, tightly enclosed by the florets.

Temperate and arctic regions of the world. Approximately 20 species; 2 species treated in *Flora*.

Sometimes the genus *Anthoxanthum* is treated to include the species in the genus *Hierochloë*.

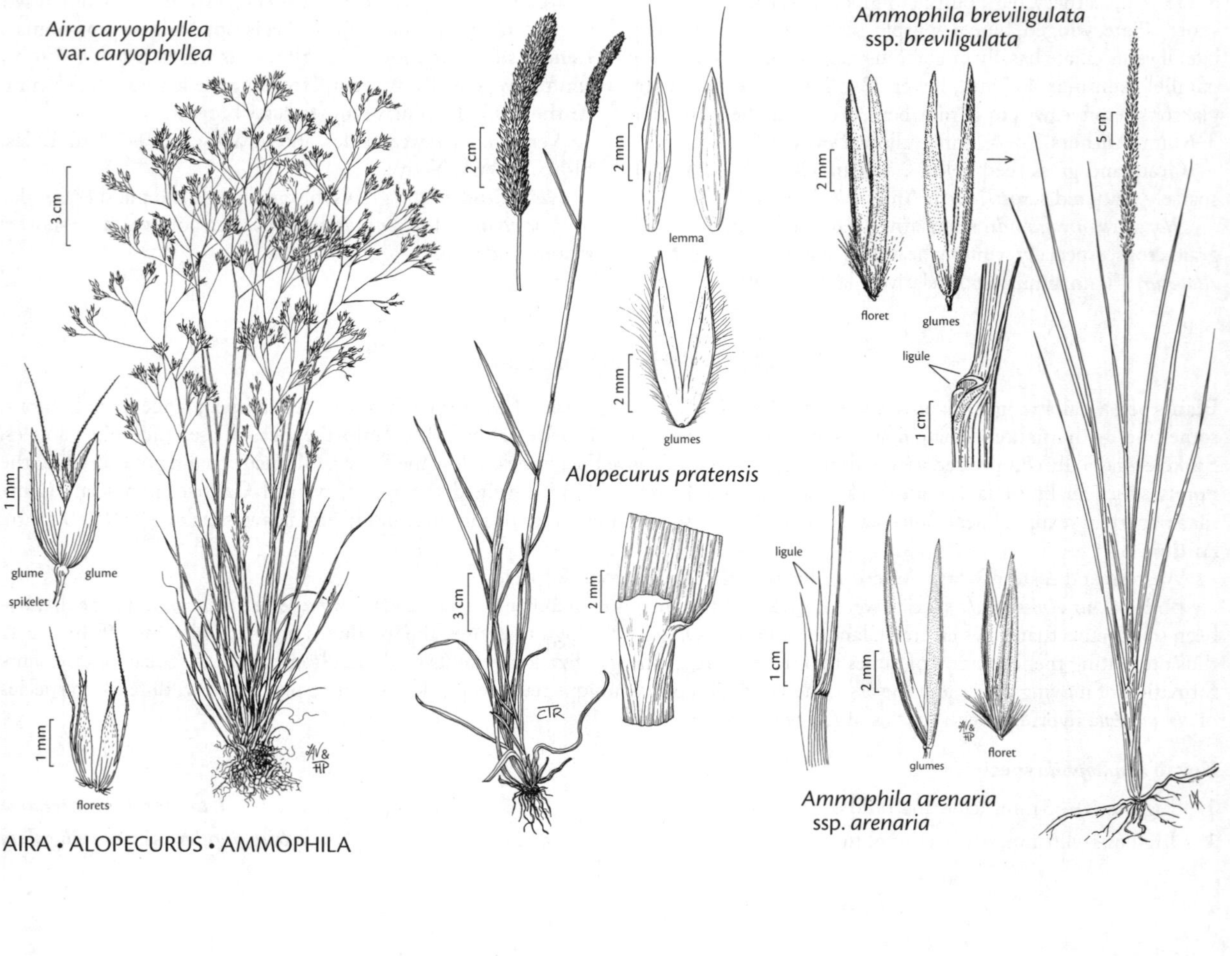

AIRA • ALOPECURUS • AMMOPHILA

Key to *Anthoxanthum* species:

1. Annual; leaf blade 1–2.7(5) mm wide; inflorescences 1–4 cm; most longer lemma awns exserted 2–3 mm beyond the tip of the upper glume; upper glume usually less than 7 mm . . . . ***A. aristatum*** ssp. ***aristatum***

1' Perennial; leaf blade 3–10 mm wide; inflorescences 3–14 cm; most longer lemma awns exserted 0–2 mm beyond the tip of the upper glume; upper glume usually more than 7 mm . . . . ***A. odoratum***

Anthoxanthum aristatum Boissier ssp. ***aristatum***

Annual vernalgrass

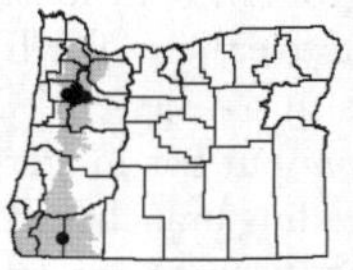

Plants annual, delicate, 5–60 cm tall. **Culms** often bent at base. **Leaves** auricles 0–0.5 mm, sometimes with a few hairs; ligules 1–2(3) mm, obtuse to acute; blades 0.8–6 cm × 1–2.7(5) mm. **Inflorescences** dense, spike-like, 1–4 × 0.2–1 cm; lowest branches 0.8–1.5 cm. **Spikelets** 4–9 mm. **Glumes** unequal, exceeding the florets; lower glumes 3–5 mm; upper glumes 5–7.5 mm. **Lemmas** lower 2 staminate, about 3 mm, pubescent; tips awned; lemma awns arising below mid-length, straight or bent, awn of lowest lemma 3.5–5 mm; awn of second lemma 6–10 mm, exceeding the upper glume by 2–3 mm, straight or bent; upper lemmas bisexual, about 2 mm, hard, glabrous or pubescent; tips acute, awnless. **Anthers** 2.8–4.1 mm. ***2n***=10, 20.

Margins of vernal pools, disturbed seasonally wet habitats. 50–400 m. Sisk, WV. CA, WA; sparsely scattered in eastern North America; Europe. Exotic.

Anthoxanthum aristatum closely resembles *A. odoratum* but is an annual and therefore lacks the remains of the previous year's leaves at the base. *Anthoxanthum aristatum* also has narrower leaves and longer awns.

Anthoxanthum odoratum Linnaeus (p. 360)

Sweet vernalgrass

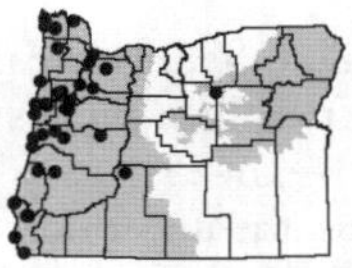

Plants perennial, (10)25–60(100) cm tall; cespitose. **Culms** erect or ascending. **Leaves** auricles 0.5–1 mm, pubescent; ligules 2–7 mm, truncate; blades 1–31 cm × 3–10 mm. **Inflorescences** dense, spike-like, (3)4–14 × 0.7–1.8 cm, often widest near the base, lowest branches 1–2.5 cm. **Spikelets** 6–10 mm. **Glumes** unequal, exceeding the florets; lower glumes 3–4 mm, 1(3)-veined; upper glumes 7–10 mm, 3-veined. **Lemmas** lower 2 staminate, 3–4 mm, pubescent; tips awned; lemma awns of lowest lemma about 3 mm, arising near or above mid-length, straight or bent; those of the second lemma 7–10 mm, arising near the base, bent and twisted, equaling or slightly longer than upper glumes; upper lemmas bisexual, 1–2.5 mm, hard, glabrous, brown; tips acute, awnless. **Anthers** (2.9)3.5–4.8(5.5) mm. ***2n***=10, 20.

Meadows, roadsides, waste places, openings in coniferous forests, occasionally in dense shade. 0–1700 m. BW, Casc, CR, ECas, Est, Sisk, WV. CA, ID, WA; scattered throughout North America; worldwide. Exotic.

This abundant weed of grasslands blooms early in the spring and is recognized by its dense, pointed spikes. Similar *A. aristatum* is an annual with proportionately longer awns.

APERA Windgrass

Plants annual. **Culms** solitary or tufted, erect or bent at the base. **Leaves** sheaths open; ligules membranous, often lacerate; blades flat or slightly involute. **Inflorescences** panicles, the branches strongly ascending to divergent. **Spikelets** slightly laterally compressed, with 1 floret (occasionally with additional vestigial florets); rachilla prolonged as a bristle 0.2–0.6 mm beyond the base of the floret; disarticulation above the glumes. **Glumes** unequal, shorter than to slightly longer than the florets, 3-veined. **Calluses** blunt, sparsely hairy. **Lemmas** firmer than the glumes, folded to nearly terete, faintly 5-veined; tips entire, awned; lemma awns subterminal, more than 2 times as long as the spikelets. **Paleas** 2-veined or 2-keeled, 75% to as long as the lemmas. **Anthers** 3. **Caryopses** shorter than the lemmas, elliptic, slightly curved.

Temperate regions of the world. 3 species; 2 species treated in *Flora*.

The genus *Apera* has been segregated from *Agrostis*. It differs from *Agrostis* by having lemmas that are somewhat firmer than the glumes, prolonged rachillas, and awns that are at least twice as long as the lemmas.

Key to *Apera* species:

1. Anthers 0.3–0.5 mm; panicle branches appressed, short, spikelet bearing to within 2 mm of the base . . . . ***A. interrupta***

1' Anthers 1–2 mm; panicle branches spreading, naked for the basal 5 mm or more, spikelets usually borne near the tips of the branches . . . . ***A. spica-venti***

Apera interrupta (Linnaeus) P. Beauvois (p. 360)

Interrupted windgrass

Agrostis interrupta L.

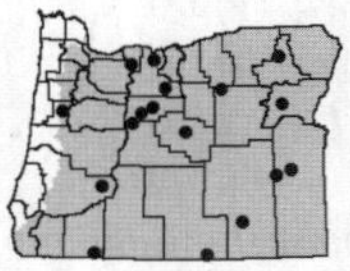

Plants (5)10–50(75) cm tall, tufted. **Leaves** sheaths glabrous, often purplish; ligules 1.5–5 mm; blades 4–12 cm × 0.3–4 mm, flat or convolute, glabrous. **Inflorescences** 3–15(20) × 0.4–1.5(3) cm; branches erect to ascending, spikelet-bearing to within 2 mm of the base; pedicels 0.5–2 mm. **Spikelets** 2–2.8 mm. **Glumes** scabrous; lower glumes 1–2.2 mm; upper glumes 2–2.8 mm, approximately equal in length to the floret. **Lemmas** 1.5–2.5 mm, 5-veined, mostly glabrous, the distal 33% scabrous to short pubescent; lemma awns 4–10(16) mm, arising above mid-length or just below the lemma apex, straight or curved near the base. **Anthers** 0.3–0.5 mm. ***2n***=14, 28.

Disturbed areas, roadsides, lawns, grain fields, sandy open ground. 50–1700 m. BR, BW, Casc, Col, ECas, Lava, Owy, Sisk, WV. CA, ID, NV, WA; north to British Columbia, east to MT and NM, scattered in eastern North America; temperate regions worldwide. Exotic.

Apera interrupta is a delicate annual with a very narrow, usually interrupted inflorescence. It is similar to some *Agrostis* species but has awns at least twice as long as the spikelets.

Apera spica-venti (Linnaeus) P. Beauvois
Silky windgrass

Agrostis spica-venti L.

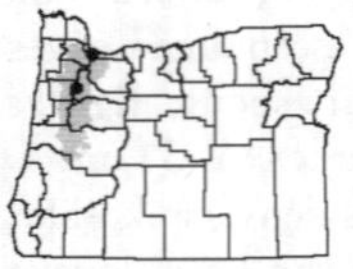

Plants 20–80(120) cm tall, tufted. **Leaves** sheaths glabrous, often purplish; ligules 3–5(12) mm; blades 6–16(25) cm × 2–5(10) mm, flat, glabrous, the uppermost occasionally exceeding the inflorescence. **Inflorescences** (5)10–35 × 2–15 cm, sometimes partially hidden in upper sheaths; branches spreading, naked at the base for at least 5 mm; spikelets usually borne toward the distal ends; pedicels 1–3 mm. **Spikelets** 2.4–3.2 mm. **Glumes** scabrous, 1–3-veined; lower glumes 1.5–2.5 mm; upper glumes 2.4–3.2 mm, slightly shorter than to slightly longer than the florets. **Lemmas** 1.6–3 mm, 5-veined, scabrous, more densely so in the distal 33%; lemma awns 5–12 mm, arising above mid-length or sometimes just below apex, straight or curved to base. **Anthers** 1–2 mm. **2*n***=14.

Waste places, grass seed fields, sandy ground, roadsides. 800–1700 m. WV. CA, WA; sparsely scattered to eastern North America; Europe. Exotic.

Apera spica-venti is similar to *A. interrupta* but has longer anthers and a broader inflorescence with spreading branches. It was first collected in Portland in 1909 and has become a weed in grass seed fields in the Willamette Valley.

ARISTIDA Three-awn

Plants annual or perennial; cespitose. **Culms** erect or bent at base, usually pith-filled, sometimes hollow, nodes glabrous or with straight hairs less than 0.5 mm. **Leaves** basal and cauline; sheaths open; ligules of hairs or membranous with marginal cilia; blades flat to involute. **Inflorescences** panicles, sometimes racemes or spikes. **Spikelets** with 1 floret, florets terete or weakly laterally compressed; disarticulation above the glumes. **Glumes** 2, often longer than the florets, thin, acute to acuminate; glume awns, if present, arising between 2 slender inconspicuous hyaline teeth. **Calluses** hairy, often sharp. **Lemmas** 3-veined, hardened, terete, enclosing the palea at maturity. **Lemma awns** 3, arising at the tip of the lemma, ascending, spreading, sometimes coiled or contorted, persistent or disarticulating at maturity, subequal or sometimes the 2 lateral awns reduced to short teeth. **Anthers** 1 or 3. **Caryopses** fusiform.

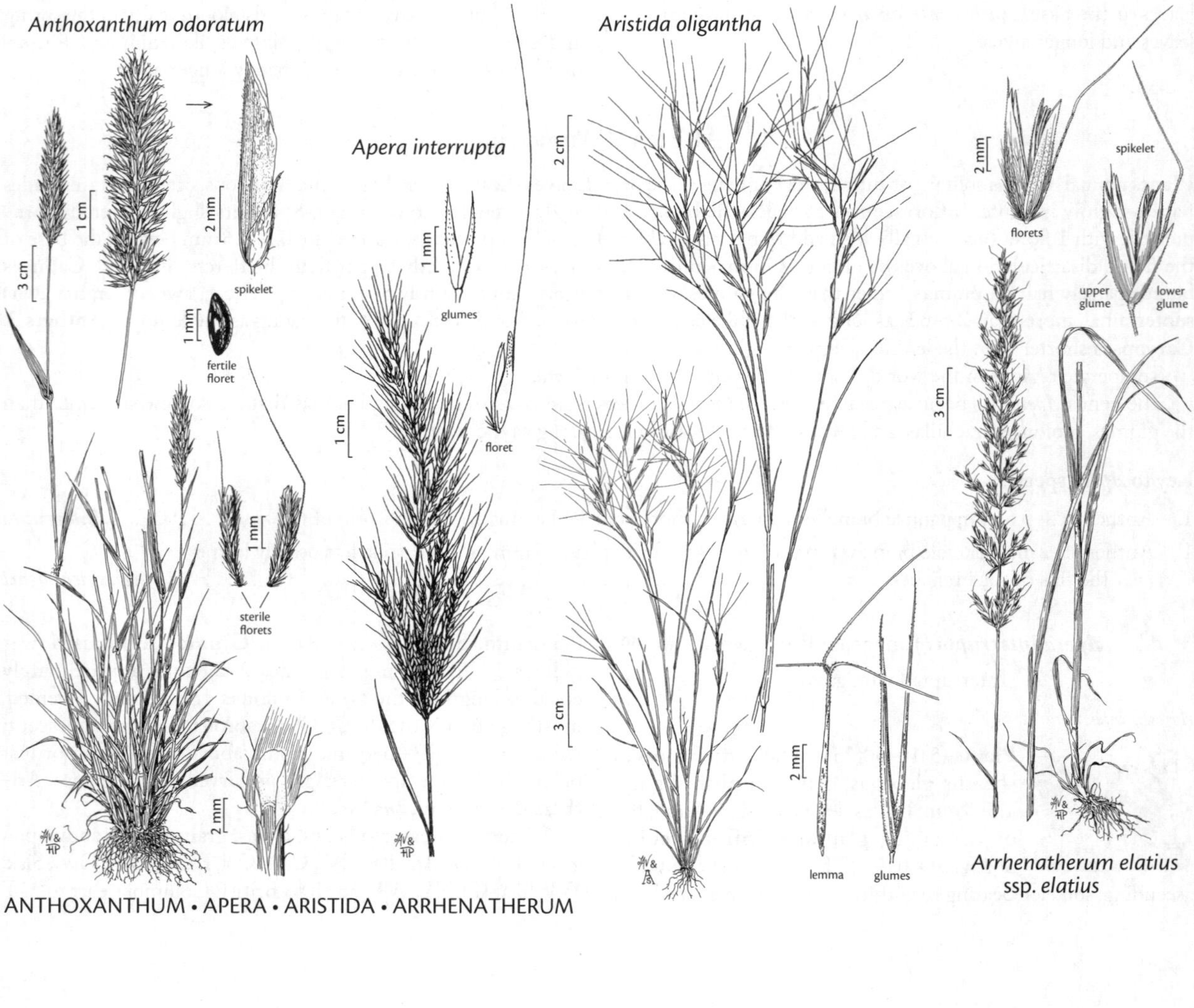

ANTHOXANTHUM • APERA • ARISTIDA • ARRHENATHERUM

Tropical to warm-temperate regions of the world. Approximately 250–300 species; 3 species treated in *Flora*.

Aristida populations often increase in size when associated with heavy grazing. The awns aid in dispersal on animal fur and by wind and also help with seedling establishment by driving the seeds into the soil as the awns twist with changes in humidity. The awns and the sharp callus can harm grazing animals.

Key to *Aristida* species:

1. Plants annual; lower glumes (9)12–22(28) mm with 3–7 veins and with awns 1–13 mm; anthers usually 1 and less than 0.5 mm (if 3, then 3–4 mm) . . . . ***A. oligantha***

1' Plants perennial or annual; lower glumes 4–13 mm with 1(2) veins, acuminate or with awns to 2 mm; anthers 3, 0.3–2 mm.

  2. Plants annual; central lemma awns 7–15(20) mm; inflorescences dense panicles; upper glumes 6–11 mm; ligules 0.4–1 mm . . . . ***A. adscensionis***

  2' Plants perennial; central lemma awns 40–100(140) mm; inflorescences open racemes or sparsely branched panicles; upper glumes (14)16–25 mm; ligules to 0.5 mm long . . . . ***A. purpurea* var. *longiseta***

Aristida adscensionis Linnaeus

Sixweeks three-awn

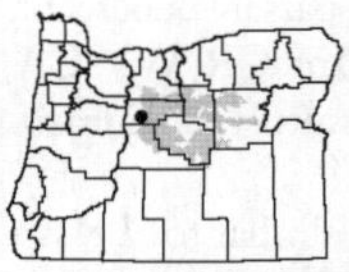

Plants annual, (3)10–50(80) cm tall, tufted. **Culms** much branched above the base, with solid internodes. **Leaves** sheaths glabrous; ligules about 0.5 mm, with a fringe of dense hairs; blades 2–14 cm × 1–2.5 mm, flat to involute. **Inflorescences** panicles, 5–15(20) × 0.5–3 cm; branches erect to ascending. **Spikelets** 5–8 mm. **Glumes** subequal to lemmas, 1-veined, glabrous, acuminate to short-awned; lower glumes 4–8 mm; upper glumes 6–11 mm. **Calluses** 0.5–0.8 mm, hairy. **Lemmas** 6–9 mm, 3-veined, leathery, midveins scabrous, glabrous elsewhere; lemma awns 3, the central awn 7–15(20) mm, the lateral awns somewhat shorter, occasionally only 1–2 mm, flattened near the base, straight to somewhat curved. **Anthers** 3, 0.3–0.7 mm. **2*n*=**22, 33, 44.

Waste ground, roadsides, degraded rangelands, and dry hillsides, often in sandy soils. 800–1200 m. Lava. CA, ID, NV, WA; southwestern states and scattered elsewhere in US; South America. Exotic.

Aristida oligantha Michaux (p. 360)

Oldfield three-awn

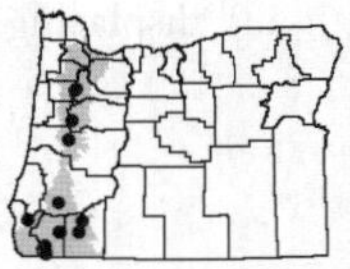

Plants annual, 25–55 cm tall, tufted. **Culms** much branched above the base, with solid internodes. **Leaves** sheaths slightly scabrous, sometimes hairy at the throat; ligules less than 0.5 mm, with a fringe of hairs shorter than the membranous portion; blades 4–12 cm × 0.5–1.5 mm, flat to loosely involute, glabrous or scabridulous. **Inflorescences** usually racemes, (5)7–20 × 2–4 cm; primary branches erect to ascending. **Spikelets** 9–28 mm. **Glumes** unequal, 3–7-veined; lower glumes (9)12–22(28) mm, midvein extended into a 1–13 mm awn; upper glumes (7)11–20(24) mm. **Calluses** 0.5–2 mm, hairy, sharp. **Lemmas** (9)12–23 mm, 3-veined, leathery, glabrous, light colored and often mottled; lemma awns 3, (8)12–65(7–0) mm, subequal, curved at the base, spreading. **Anthers** usually 1 and less than 0.5 mm; rarely 3 and 3–4 mm. **2*n*=**22.

Waste places, dry fields, roadsides, along railroads, and burned areas, usually in sandy soils, sometimes on serpentine. 50–600 m. Sisk, WV. CA, ID; most of US, British Columbia, Quebec. Native.

Aristida oligantha has abundant awns and long, slender florets that resemble the panicle branches. It is distinguished from the other *Aristida* species by its long glumes with many veins.

Aristida purpurea Nuttall var. ***longiseta*** (Steudel) Vasey

Red three-awn

Aristida longiseta Steud. var. *robusta* Merr., *Aristida purpurea* Nutt. var. *fendleriana* (Steud.) Vasey, *Aristida purpurea* Nutt. var. *robusta* (Merr.) A.H. Holmgren & N.H. Holmgren

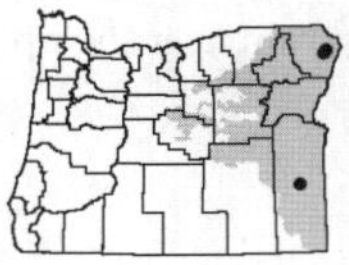

Plants perennial, 10–40(50) cm tall; cespitose. **Culms** usually unbranched above the base, with hollow internodes. **Leaves** sheaths scabrous or glabrous, hairy at the throat; collars glabrous or with sparse, straight hairs at the sides; ligules less than 0.5 mm, with a fringe of hairs longer than the membranous portion; blades 4–16 cm × 1–1.5 mm, involute, glabrous or sometimes scabridulous on the outer surface. **Inflorescences** panicles or racemes, 5–15 × 2–12 cm; branches few, sometimes absent, appressed or ascending basally, stout to slender, straight to drooping toward their tips. **Spikelets** 14–25 mm. **Glumes** unequal, 1(2)-veined, smooth or scabridulous, awnless or with awns to 2 mm; lower glumes 8–13 mm; upper glumes (14)16–25 mm. **Calluses** 0.5–1.8 mm, hairy, sharp. **Lemmas** 12–16 mm, 3-veined, leathery, minutely scabrous or bumpy; tips 0.3–0.8 mm wide; lemma awns 3, 40–100(140) mm, subequal, 0.2–0.5 mm wide at the base, straight, ascending to spreading; central awn thicker than the others; all awns approximately equal in length, occasionally the lateral awns shorter. **Anther** 1, 0.7–2 mm. **2*n*=**22, 44, 66.

Open slopes, hills, and sandy or rocky flats, slopes and plains and in disturbed soils. 600–1000 m. BW, Owy. CA, ID, NV, WA; north to British Columbia, east to Manitoba and LA, south to northern Mexico. Native.

Very long awns distinguish *Aristida purpurea* var. *longiseta* from our other *Aristida* species.

ARRHENATHERUM Oatgrass

Plants perennial; usually cespitose. **Leaves** sheaths open; ligules membranous; blades flat or convolute. **Inflorescences** panicles, the branches spreading through anthesis, then loosely appressed; disarticulation above the glumes, the florets usually falling together. **Spikelets** laterally compressed, with 2 florets; lower floret staminate; upper floret pistillate or bisexual; occasionally with a third distal, vestigial floret. **Glumes** unequal, the lower 1- or 3-veined, the upper 3-veined and about as long as the lower floret. **Calluses** short, blunt, pubescent. **Lower lemmas** 3–7-veined, awned from below mid-length; upper lemmas 7-veined, sometimes tougher in texture, awnless or awned from near the tip; lemma awns of lower floret twisted and bent, those of upper floret usually straight. **Anthers** 3. **Caryopses** dorsally compressed to terete, pubescent.

Temperate regions of the world. Approximately 6 species; 1 species treated in *Flora.*

Arrhenatherum elatius (Linnaeus) P. Beauvois ex J. Presl & C. Presl

Tall oatgrass

Plants 50–140(180) cm tall; cespitose, sometimes with rhizomes. **Culms** sometimes swollen and tuberous or bulbous at the base, nodes glabrous or pubescent. **Leaves** blades 5–32 cm × (1)3–8(10) mm, flat, usually glabrous, rarely with short spreading hairs, sometimes scabrous. **Inflorescences** 7–30(36) × 1–6(10) cm; branches ascending to divergent, usually spikelet-bearing to the base. **Spikelets** 7–11 mm. **Glumes** 2, glabrous, entire, 3-veined; lower glumes 4–7 mm; upper glumes 7–10 mm, approximately equal to the adjacent lemma. **Lemmas** (4)7–10 mm, 7-veined, membranous or tough, glabrous or pubescent; tips bifid; lemma awns of lower lemma 10–20 mm, twisted and bent, arising below mid-length, awns of upper lemma usually absent, if present, usually straight and arising from near the tip. **Anthers** 3, 3.6–5(6) mm. **2*n***=14, 28, 42.

2 subspecies.

Key to *Arrhenatherum elatius* subspecies:

1. Lower internodes of culm 2–4 mm thick, not swollen, not bulb-like . . . . ***A. e.* ssp. *elatius***
1'. Lower internodes of culm 5–10 mm thick, swollen, bulb-like . . . . ***A. e.* ssp. *bulbosum***

Arrhenatherum elatius (Linnaeus) P. Beauvois ex J. Presl & C. Presl ssp. *bulbosum*

Tall oatgrass

Arrhenatherum elatius (L.) P. Beauv. ex J. Presl & C. Presl var. *bulbosum* (Willd.) Spenn.

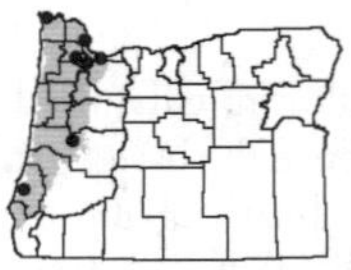

Culms lower internodes 5–10 mm thick, swollen; nodes densely to sparsely pubescent.

Cultivated fields. 0–50 m. CR, WV. CA, WA; midwestern and eastern US; Eurasia. Exotic.

Arrhenatherum elatius ssp. *bulbosum* is very similar to *A. e.* ssp. *elatius,* differing in its bulbous culm bases. It is less common, but it can be a serious weed in cultivated fields where tilling spreads the bulb-like basal internodes.

Arrhenatherum elatius (Linnaeus) P. Beauvois ex J. Presl & C. Presl ssp. *elatius* (p. 360)

Tall oatgrass

Arrhenatherum elatius (L.) P. Beauv. ex J. Presl & C. Presl var. *elatius*

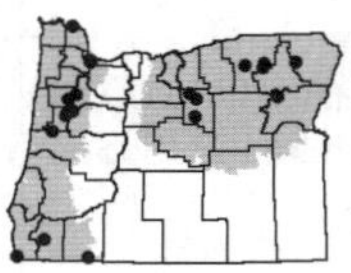

Culms lower internodes 2–4 mm thick, not swollen; nodes usually glabrous.

Grasslands and roadsides. 0–1600 m. BW, Col, CR, Est, Lava, Sisk, WV. CA, ID, NV, WA; north to British Columbia, east to Nova Scotia and VA; Eurasia. Exotic.

Arrhenatherum elatius ssp. *elatius* is abundant in meadows and roadsides. It is a serious weed of upland prairies where it forms an upper canopy layer, competing with and displacing shorter native prairie species.

AVENA Oats

Plants annual. **Culms** 8–200 cm tall, often prostrate when young but becoming erect. **Leaves** cauline; sheaths open; ligules membranous; blades flat. **Inflorescences** open panicles. **Spikelets** 15–50 mm, pedicellate, laterally compressed, with 1–6(8) florets; disarticulation above the glumes, usually also between the florets (cultivated species not disarticulating). **Glumes** usually exceeding the florets, membranous, glabrous, 3–11-veined, awnless. **Calluses** glabrous to hairy. **Lemmas** more or less hard, enclosing the caryopses at maturity, 5–9-veined, pubescent below mid-length, sometimes glabrous; tips bifid with the teeth sometimes long and bristle-like; lemma awns usually present, arising on the lemma back, usually bent and twisted. **Anthers** 3. **Caryopses** shorter than the lemmas, concealed at maturity, terete, pubescent.

Temperate and cold regions worldwide. Approximately 29 species worldwide; 3 species treated in *Flora.*

Wild oats are widely distributed weeds of disturbed grasslands, becoming community dominants in areas with Mediterranean climates such as California and western Oregon. *Avena sterilis* has been reported from Oregon, but the reports cannot be confirmed. It is much like *A. fatua,* but its florets remain together, falling from the glumes as one unit. Its awns are 30–90 mm; the longer awns are twice the length of awns of our other wild oats. Its awns may be hairy below the midpoint; those of our other species are glabrous to minutely scabrous.

Key to *Avena* species:

1. Lemma tips with 2 elongated bristle-like teeth 2–4 mm . . . . ***A. barbata***

1' Lemma tips not with 2 elongated bristle-like teeth, bifid tips less than 1 mm, or lacking.

  2. Callus glabrous; second lemma awnless or with awn much shorter than the awn of lowest lemma and usually straight to slightly twisted; florets not disarticulating from the glumes, remaining attached to the plant even when mature . . . . ***A. sativa***

  2' Callus hairy; second lemma with awn usually subequal to the awn of the lowest lemma, twisted below the bend; florets disarticulating at maturity . . . . ***A. fatua***

Avena barbata Pott ex Link

Slender wild oats

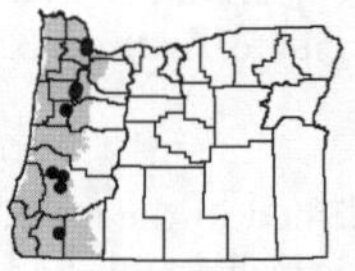

Plants 60–80(150) cm tall. **Leaves** blades 6–30 cm × 2–20 mm, flat to involute, glabrous or pilose. **Inflorescences** 15–35(50) × 6–12 cm; branches spreading; disarticulation beneath each floret. **Spikelets** 21–30 mm with 2–3 florets. **Glumes** 15–30 mm with 7–9 veins, awnless. **Calluses** hairs 2–3 mm. **Lemmas** 15–26 mm with 5–9 veins, leathery, with dense, appressed hairs below mid-length; tips acute, with 2 elongated, bristle-like teeth 2–4 mm; lemma awns 30–45 mm, arising from about mid-length, twisted, bent. **Anthers** 2.5–4 mm. **2*n***=28.

Disturbed areas, roadsides, and upland grasslands. 0–600 m. CR, Sisk, WV. CA, NV, WA; northeast to MT, southeast to NM; central Asia, Mediterranean region. Exotic.

Avena barbata is characterized by long, bristle-like lemma teeth. This aggressive annual displaces native species in moderately dry grasslands.

Avena fatua Linnaeus (p. 364)

Wild oats

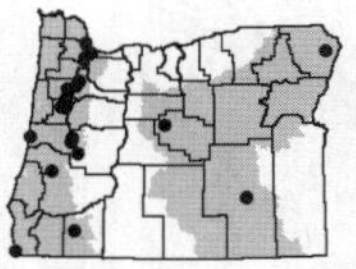

Plant 8–160 cm tall. **Leaves** blades 10–45 cm × 3–15(25) mm, flat or occasionally involute, scabrous. **Inflorescences** 7–40 × 6–12 cm; branches spreading; disarticulation beneath each floret. **Spikelets** 18–32 mm with 2–3(5) florets. **Glumes** 18–32 mm with 9–11 veins, awnless. **Calluses** hairs to 25% as long as the lemmas. **Lemmas** 14–22 mm with 5–9 veins, leathery, thin and membranous at the tip, densely pubescent below mid-length, sometimes sparsely pubescent or glabrous; tips bifid; teeth 0.3–1.5 mm but lacking long bristle-like tips; lemma awns 23–42 mm, arising in the middle 33% of the lemma backs. **Anthers** about 3 mm. **2*n***=42.

Disturbed areas, roadsides, grain fields, upland grasslands. 0–1100 m. BR, BW, CR, Est, Lava, Sisk, WV. CA, ID, NV, WA; throughout North America; worldwide. Exotic.

Avena fatua invades upland grasslands, displacing native prairie species. It also can be a serious weed in grain crops. It may be the ancestor of cultivated *A. sativa*. *Avena fatua* × *A. sativa* hybrids resemble *A. sativa* but lack lobes on the wings of the lodicules and may have an awn on the lowest lemma.

Avena sativa Linnaeus

Cultivated oats

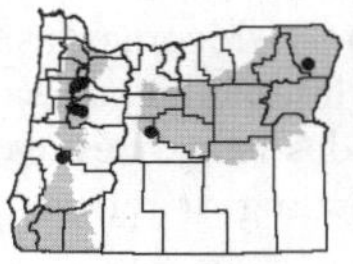

Plants 35–180 cm tall. **Leaves** sheaths glabrous; blades 8–45 cm × 3–14(25) mm, flat or occasionally involute, minutely scabrous. **Inflorescences** (6)15–40 × 5–15 cm; branches spreading; not disarticulating or disarticulating tardily. **Spikelets** (18)25–32 mm with 1–2(7) florets. **Glumes** (18)20–32 mm, 9–11-veined, awnless. **Calluses** glabrous. **Lemmas** 14–18 mm with 5–9 veins, (membranous), glabrous or occasionally with sparse appressed hairs; tips erose to toothed, longest teeth 0.2–0.5 mm; lemma awns usually absent, if present, 15–30 mm. **Anthers** (1.7)3–4.3 mm. **2*n***=42.

Along railways and roadsides and in fallow farm fields. 0–1000 m. BW, Lava, Sisk, WV. CA, ID, NV, WA; throughout North America; worldwide. Exotic.

Avena sativa is usually shorter, sturdier, and has thicker spikelets than other *Avena* species. It is grown for grain and is planted for erosion control. It is generally found as a waif along roads, railroads, and equestrian facilities.

BECKMANNIA Sloughgrass

Plants annual or perennial, tufted. **Culms** erect. **Leaves** sheaths open; ligules membranous, acute; blades flat. **Inflorescences** dense, narrow panicles, the branches 1-sided with closely arranged spikelets; disarticulation below the glumes. **Spikelets** laterally compressed, circular or ovate in side view, nearly sessile, with 1 fertile floret. **Glumes** subequal, semi-circular, inflated, keeled, awnless. **Calluses** blunt, glabrous. **Lemmas** lanceolate, inconspicuously 5-veined, awnless. **Anthers** 3. **Caryopses** concealed at maturity.

North America and Eurasia. 2 species; 1 species treated in *Flora*.

Beckmannia syzigachne (Steudel) Fernald (p. 364)

American sloughgrass

Beckmannia eruciformis (L.) Host

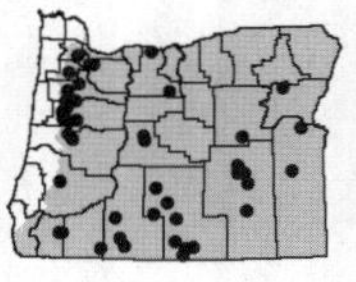

Plants annual, 20–120 cm tall, cespitose. **Culms** erect or ascending. **Leaves** ligules pubescent, entire or lacerate, usually folded back; blades to 25 cm × 4–10(20) mm, flat, scabrous. **Inflorescences** panicles 7–30 cm × 0.5–2.5 mm, occasionally partially hidden in the sheath of the uppermost leaf, green or straw-colored; branches more or less erect. **Spikelets** 2–3 mm with 1–2 florets, the second, when present, poorly developed. **Glumes** about 3 mm, 3-veined, semi-circular, rugose with transverse wrinkles, somewhat inflated, scabrous, awnless but very short-pointed. **Lemmas** 2.4–3.5 mm, 5-veined, glabrous; tip acuminate, sometimes mucronate, awnless. **Anthers** 0.5–1.5 mm. **Caryopses** less than 2 mm. **2*n***=14.

Freshwater wetlands including marshes, floodplains, the

edges of ponds, lakes, streams, and ditches, often in standing water. 0–1800 m. BR, BW, Casc, Col, ECas, Lava, Owy, Sisk, WV. CA, ID, NV, WA; north to AK, east to Newfoundland and VA; Asia. Native.

Beckmannia syzigachne is distinctive in having rows of crowded, almost circular spikelets.

BRACHYPODIUM Falsebrome

Plants annual or perennial; cespitose or rhizomatous. **Culms** erect or decumbent, often rooting at lower nodes, sometimes branched above the base; nodes often pubescent. **Leaves** sheaths open; ligules membranous, sometimes ciliate; blades flat or convolute. **Inflorescences** spike-like racemes, most nodes with 1 spikelet, most or all spikelets appressed or strongly ascending; disarticulation above the glumes and beneath the florets. **Spikelets** 14–80 mm, terete to laterally compressed, with (3)5–24 florets. **Glumes** unequal, 50% to as long as the lowest lemma, tips obtuse to acuminate, lower glumes 3–7-veined, upper glumes 5–9-veined. **Lemmas** membranous to leathery at maturity, rounded on the back, (5)7–9-veined, tips obtuse or acute, awnless or awned; lemma awns if present, terminal. **Paleas** with 2 strongly keeled veins, sometimes with fainter veins between, the keels strongly ciliate. **Anthers** 3. **Caryopses** oblong, flattened, tips pubescent.

North America and Eurasia. Approximately 18 species worldwide; 1 species treated in *Flora*.

Three *Brachypodium* species have been recorded in Oregon, but only one, *B. sylvaticum*, is currently established in the state. *Brachypodium distachyon* was collected in Portland in 1902. Though this population did not persist, the species is well established in California. *Brachypodium pinnatum* escaped from an experimental planting in Jackson County in 1949 but has not persisted.

Brachypodium sylvaticum (p. 364)
(Hudson) P. Beauvois
Falsebrome

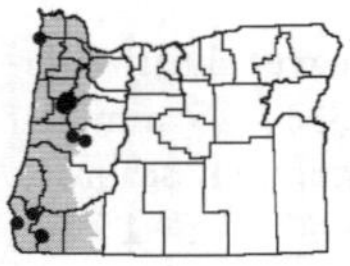

Plants perennial, 30–120(200) cm tall, cespitose, sometimes short-rhizomatous. **Culms** glabrous or lower internodes pubescent near the nodes; nodes hairy. **Leaves** sheaths open, with spreading or reflexed hairs; blades 8–35 cm × 4–12(15) mm, flat, scabrous, with sparse spreading hairs, lower surfaces often with a tuft of hairs near the base, yellow-green. **Inflorescences** spike-like racemes, (2)4–20 × 3–12 cm, nodding to nearly erect. **Spikelets** 17–30(55) mm, with 6–16(22) florets, more or less round in cross section, nearly sessile or on short pedicels to 2 mm. **Glumes** usually pubescent; lower glumes 6–9 mm, 5–7-veined, upper glumes 8–11 mm, 7–9-veined. **Lemmas** 6–12(13.5) mm, 7-veined, hairy, tips

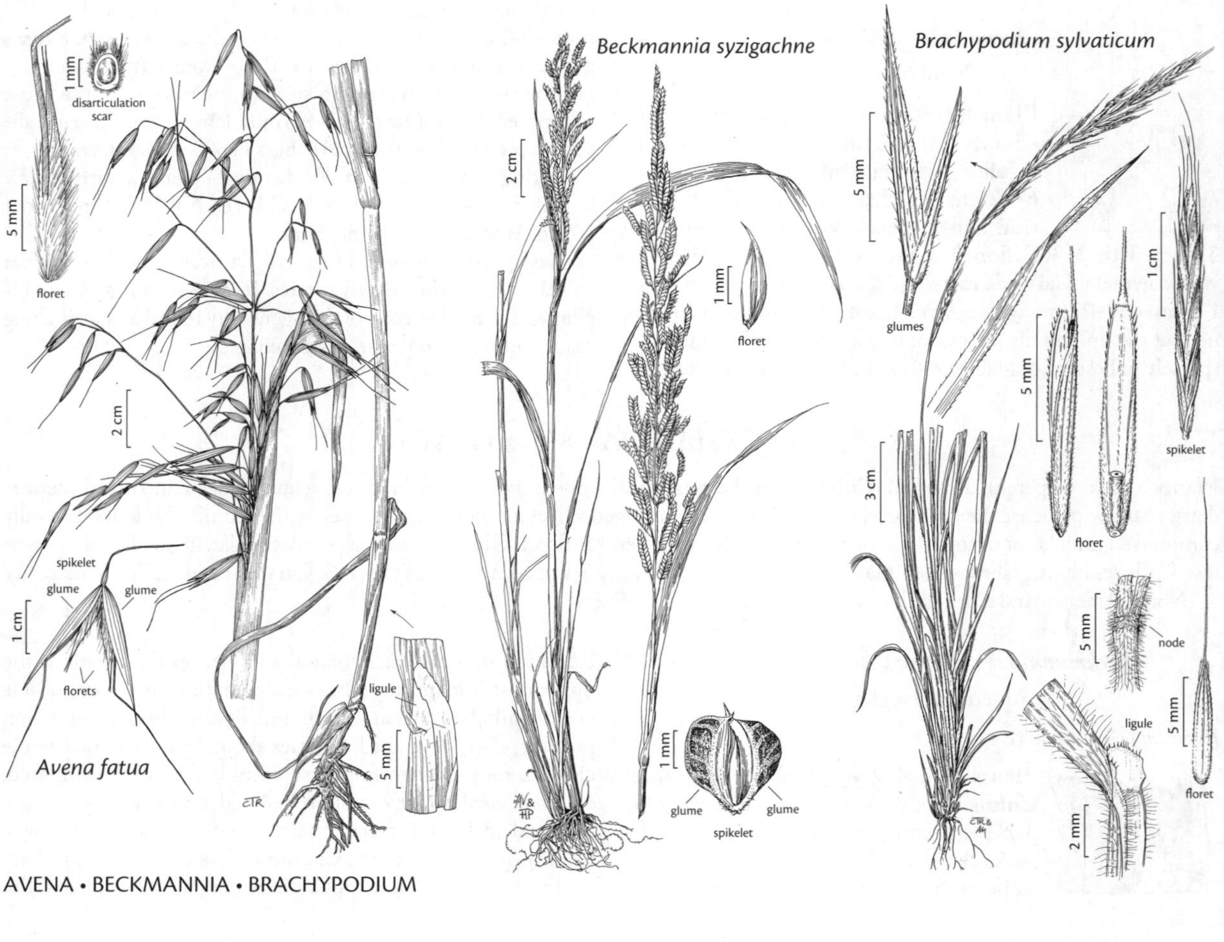

AVENA • BECKMANNIA • BRACHYPODIUM

obtuse, awned; lemma awns 7–15 mm, arising from the tip, straight or curved to base. **Anthers** 3–4(5.5) mm. **Caryopses** 5–8.7 mm.

Usually in full to partial shade, occasionally in full sun, usually in woodland but invading savanna, prairie, seasonal wetlands, roadsides, and disturbed areas. 0–700 m. CR, Est, Sisk, WV. CA, WA; NY, VA; North Africa, Eurasia. Exotic.

Brachypodium sylvaticum has clusters of arching, slightly yellow-green leaves and nodding, unbranched inflorescences. The combination of long-hairy leaf sheaths and glabrous culms help identify it. It is most often confused with *Bromus vulgaris*, which has darker green leaves, closed sheaths and a very similar inflorescence except that some of the spikelets are on long pedicels or branches. This species has become a problematic weed in Oregon.

BRIZA Quaking grass

Plants annual or perennial, cespitose. **Culms** usually erect, unbranched. **Leaves** sheaths open; ligules hyaline; blades flat. **Inflorescences** open panicles, spikelets usually hanging down. **Spikelets** laterally flattened but glumes and lemmas with broadly rounded backs and strongly divergent from the rachilla, 4–12(15) florets, distal florets much reduced; disarticulation above the glumes and beneath the florets. **Glumes** subequal, shorter or longer than the adjacent lemmas, faintly 3–7-veined, rounded over the back, unawned. **Calluses** short, glabrous. **Lemmas** inflated, about as wide as long, with broadly rounded backs, similar in shape to glumes but somewhat heart-shaped, awnless, margins hyaline. **Anthers** 3. **Caryopses** shorter than the lemmas, concealed at maturity, usually falling with the lemma and palea.

Cosmopolitan. Approximately 20 species; 3 species treated in *Flora*.

These small grasses readily establish in disturbed sites. The inflorescences with attractive pendent spikelets are often used in ornamental arrangements.

Key to *Briza* species:

1\. Plants perennial; rhizomatous; spikelets 4–5.5 mm . . . . ***B. media***

1' Plants annual; tufted or culms solitary; spikelets 2–20 mm.

  2\. Spikelets 2–4(7) mm . . . . ***B. minor***

  2' Spikelets 10–19(27) mm . . . . ***B. maxima***

Briza maxima Linnaeus

Big quaking grass, rattlesnake grass

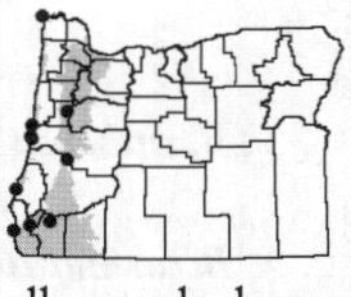

Plants annual, 20–80 cm tall. **Leaves** blades 2.5–20 cm × 2–8 mm, flat, margins strigose to glabrous. **Inflorescences** 3.5–10 × mostly 1–5 cm. **Spikelets** 10–19(27) mm, oval to elliptic, with 4–12(15) florets. **Glumes** usually purple; lower glumes 5–5.5 mm; upper glumes 6–6.5 mm, 7-veined. **Lemmas** usually glabrous proximally and with appressed hairs distally, tips obtuse, awnless; lowermost lemmas 7–9 mm, 7–9-veined, upper lemmas reduced. **Anthers** 1.2–1.5 mm. **Caryopses** 2–3 mm. **2*n***=10, 14.

Sand dunes, rocky areas, open woodlands, roadsides and pastures, often in sandy soils usually near the coast, but spreading to inland habitats. 0–300 m. Est, Sisk, WV. CA, WA; north to British Columbia; scattered in the eastern US and Canada; the Mediterranean region; worldwide. Exotic.

The large, inflated, dangly spikelets make this species almost unmistakable. The only similar species is *Bromus briziformis*, which has closed leaf sheaths and more pointed spikelet bases. Both species are grown as ornamentals. *Briza maxima* is mainly coastal in Oregon at this time but is expanding its range to inland sites.

Briza media Linnaeus

Perennial quaking grass

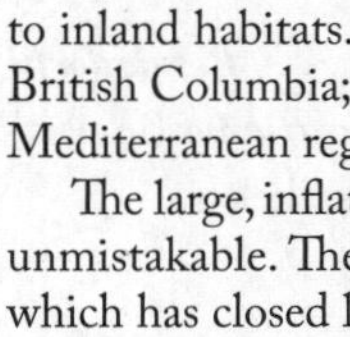

Plants perennial, 15–75 cm tall, short-rhizomatous. **Leaves** blades 4–151–12 cm × 1.9–3.2 mm, flat, glabrous or scabridulous. **Inflorescences** 8–20 cm long, almost equally wide. **Spikelets** 4–5.5 mm, usually oval, with 3–6(10) florets. **Glumes** usually purple; lower glumes 2.5–3.2 mm; upper glumes 2.5–4 mm. **Lemmas** 3–4 mm, often irregular in shape, obtuse, glabrous or minutely scurfy, with 9–10 indistinct veins, awnless. **Anthers** 1.3–2 mm. **Caryopses** 1.2–1.5 mm.

Disturbed areas. 1000–1400 m. BW. Northeastern North America; Europe. Exotic.

Briza media is grown as an ornamental and occasionally escapes.

Briza minor Linnaeus (p. 369)

Little quaking grass

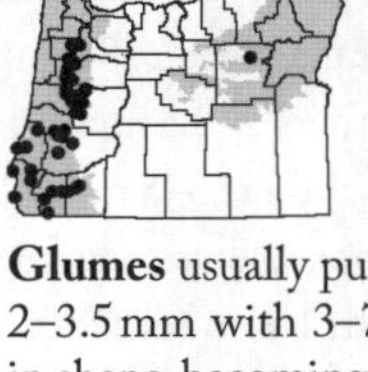

Plants annual, delicate, 7.5–80 cm tall. **Leaves** blades 1–12 cm × 1–8(10) mm, flat, slightly scabrous. **Inflorescences** (2)4–14(18) × 3–11 cm wide. **Spikelets** 2–4(7) mm, triangular, with 4–7(13), florets. **Glumes** usually purple; lower glumes 2–2.5 mm; upper glumes 2–3.5 mm with 3–7 veins. **Lemmas** 1.6–2 mm, often irregular in shape, becoming hyaline distally, glabrous or minutely scurfy, with indistinct veins, awnless. **Anthers** 0.4–0.5 mm. **Caryopses** 0.8–1 mm. **2*n***=10, 14.

Disturbed dry meadows, seasonal wetlands around vernal pools, open woodlands, sand hills, roadsides, and pastures. 0–1300 m. BW, CR, Est, Sisk, WV. CA, WA; east to NC, south to Mexico; worldwide. Exotic.

Briza minor is a delicate grass with tiny, pendent, triangular spikelets. *Aira caryophyllea* is equally delicate, but its spikelets are simpler, and its lemmas are not inflated.

BROMUS Brome

Plants annual, biennial, or perennial, 5–190 cm tall; usually cespitose, sometimes rhizomatous. **Leaves** sheaths closed nearly to the top, usually with spreading hairs; ligules membranous, usually erose or lacerate; blades usually flat, rarely involute. **Inflorescences** panicles, sometimes racemes or with a single spikelet in depauperate specimens; disarticulation above the glumes, below the florets. **Spikelets** terete to laterally compressed with 3–30 florets. **Glumes** unequal, usually shorter than the adjacent lemmas, always shorter than the spikelets, usually acute, rarely mucronate; lower glumes 1–7(9)-veined; upper glumes 3–9(11)-veined. **Lemmas** 5–13-veined, keeled to rounded on the back; tips entire to toothed, usually awned; lemma awns (0)1(3), terminal or subterminal. **Paleas** ciliate on the keels, adnate to the caryopses. **Anthers** (2)3. **Ovaries** and caryopses with densely hairy tips.

Temperate and cool regions worldwide. 100–400 species; 22 species treated in *Flora*.

Bromes are distinguished by closed leaf sheaths, usually hairy foliage, and relatively large spikelets with several florets. Most have awned lemmas. *Bromus* is the second most diverse grass genus in Oregon, after *Poa*. Some of our native species interbreed, and intermediate plants are found fairly often. Within this genus, it is helpful to remember that culm diameters are measured near the middle of one of the lowest elongated internodes. The distance from the base of the lemma awn to the lemma apex should be measured on the distal florets of a spikelet. Lemma lengths are measured on the lowest florets in a spikelet. Lengths of lemmas and spikelets do not include awns. The lower panicle branches are measured from the node on the inflorescence axis to the end of the farthest floret (excluding awns).

Key to *Bromus* species:

1. Plants perennial or biennial.
   2. Lemmas keeled on the back, V-shaped in cross section, at least near the tip.
      3. Lemmas lacking awns, or with awns to 4 mm. . . . . ***B. catharticus*** var. ***catharticus*** (in part)
      3' Lemmas awned, the awns (2)4–17 mm.
         4. Panicles dense; spikelets crowded and overlapping, usually longer than the pedicels and panicle branches; culms 20–70 cm tall . . . . ***B. maritimus***
         4' Panicles more open; spikelets not crowded, often not overlapping, shorter than some of the pedicels and panicle branches; culms 30–180 cm tall.
            5. Lemmas hairy throughout, with long, soft, hairs, especially toward the base; glumes with conspicuous hairs; lemmas only slightly keeled . . . . ***B. pacificus*** (in part)
            5' Lemmas glabrous or hairy but without long, soft hairs; glumes glabrous to hairy but less conspicuously so than in *B. pacificus*; lemmas strongly keeled distally, at least when mature.
               6. Longest panicle branches usually more than 10 cm, spreading to drooping; culms 3–7 mm thick; some leaves more than 10 mm wide . . . . ***B. sitchensis***
               6' Longest panicle branches usually less than 10 cm, spreading to erect or reflexed; culms less than 4 mm thick; leaves less than 10 mm wide . . . . ***B. carinatus***
   2' Lemmas rounded on back, C-shaped in cross section.
      7. Plants strongly rhizomatous; lemmas awnless or awns 1–2(4) mm. . . . . ***B. inermis***
      7' Plants not rhizomatous; lemmas usually with awns more than 4 mm.
         8. Inflorescence narrow and dense, branches appressed to erect, usually shorter than the spikelets.
            9. Inflorescence nodding; longer lemma awns (4)6–12 mm; ligules 2–6 mm. . . . . ***B. vulgaris*** (in part)
            9' Inflorescence erect; lemma awns 2–8 mm; ligules up to 3 mm.
               10. Awns 2–5 mm; anthers 2–3.5 mm. . . . . ***B. suksdorfii***
               10' Awns (4)5–8 mm; anthers 3–5 mm . . . . ***B. orcuttianus*** (in part)
         8' Inflorescence wider, more open, lower branches reflexed to ascending, usually longer than the spikelets.
            11. Most lower glumes with 1 vein (occasionally some with 3 veins).
               12. Glumes glabrous.
                  13. Lemma awns 2–4(5) mm; anthers 1.5–2 mm . . . . ***B. ciliatus***
                  13' Lemma awns 4–12 mm; anthers 2–5 mm.
                     14. Inflorescence nodding; ligules 2–6 mm . . . . ***B. vulgaris*** (in part)
                     14' Inflorescence erect; ligules 1–3 mm . . . . ***B. orcuttianus*** (in part)
               12' Glumes hairy.
                  15. Inflorescence nodding; panicle branches appressed to ascending. . . . . ***B. vulgaris*** (in part)
                  15' Inflorescence upright; panicle branches widely spreading when mature, the lower branches sometimes reflexed.
                     16. Culms with 2–4 nodes; panicles 7–13.5 cm; inland. . . . . ***B. orcuttianus*** (in part)
                     16' Culms with (5)6–8 nodes; panicles 10–25 cm; coastal. . . . . ***B. pacificus*** (in part)

11' Most lower glumes with 3 veins (occasionally some with 1 vein).

17. Upper glumes 5-veined; panicle branches more or less flexuous. . . . . **B. laevipes**

17' Most upper glumes 3-veined (some may be 5-veined); panicle branches more or less stiff . . . . **B. orcuttianus** (in part)

1' Plants annual.

18. Lemmas strongly keeled, V-shaped in cross section . . . . **B. catharticus** var. **catharticus** (in part)

18' Lemmas more or less rounded over the back, C-shaped in cross section.

19. Lemma tips narrowly attenuate, the teeth acuminate to acuminate-aristate; lemma awns usually at least 12 mm; first glume 1-veined, the second glume usually 3-veined.

20. Panicle branches stout, mostly erect, most much shorter than the spikelets; panicles strongly contracted. . . . . **B. rubens**

20' Panicle branches slender, mostly spreading to reflexed, often longer than the spikelets; panicles rather open and loose.

21. Spikelets 10–20 mm; upper glumes 7–13 mm long . . . . **B. tectorum**

21' Spikelets more than 20 mm; upper glume usually more than 13 mm long.

22. Lower glumes 15–25 mm; upper glumes 25–35 mm; lemmas 22–30 mm. . . . . **B. diandrus**

22' Lower glumes 8–10 mm; upper glumes 12–15 mm; lemmas 14–20 mm. . . . . **B. sterilis**

19' Lemma tips more or less rounded to acute, usually shallowly bidentate; lemma awns often much less than 12(16) mm; first glume 3(5)-veined, the second 5(9)-veined.

23. Lemmas usually awnless or with awn-like tips scarcely 1 mm, strongly inflated; spikelets with 9–15 florets, 15–30 mm . . . . **B. briziformis**

23' Lemmas usually with awns at least 2 mm, not strongly inflated, occasionally awnless but then otherwise not as above.

24. Inflorescence consisting of 2 or more spikelets.

25. Panicle more or less dense; pedicels mostly shorter than their spikelets, sometimes a minority of pedicels longer than their spikelets; mature lemma veins thickened and raised when dry; lemma texture papery; lemmas usually hairy.

26. Lemma veins thickened and raised when dry; lemma usually hairy (occasionally glabrous); lemma texture papery; anthers 0.5–1.5(2) mm. . . . . **B. hordeaceus** (in part)

26. Lemma veins not or only slightly thickened and raised; lemmas usually glabrous over the back; lemma texture leathery or firm; anthers 0.7–2(3) mm . . . . . . **B. commutatus** (in part)

25' Panicle more open, most of the pedicels longer than their spikelets; mature lemma veins obscure to distinct, rarely thickened and raised; lemma texture more or less leathery; lemmas glabrous or hairy.

27. Lower glumes 7–10 mm; upper glumes 8–12 mm; panicle branches conspicuously sinuous; awns erect to weakly spreading, 10–16 mm; seldom collected . . . . **B. arenarius**

27' Lower glumes 4–7 mm; upper glumes 5–8 mm; panicle branches sometimes sinuous; awns erect to strongly divergent, (0)3–13 mm; widespread.

28. Awns originating 1.5 mm or more below the lemma tips (check several middle lemmas of spikelets), 8–13(16) mm; spikelets often 20–40 mm.

29. Lemmas with hyaline margins 0.3–0.6 mm wide, slightly angled above the middle; spikelets with 6–12 florets . . . . **B. japonicus**

29' Lemmas with hyaline margins 0.6–0.9 mm wide, strongly angled above the middle; spikelets with 8–30 florets . . . . **B. squarrosus**

28' Awns originating less than 1.5 mm below lemma tips, (0)3–10 mm; spikelets 10–20(30) mm.

30. Lemma margins smoothly curved; anthers 1–3 mm; leaf blades 1–12 mm wide. . . . . **B. secalinus**

30' Most lemma margins broadly angled; anthers 0.5–2(2.3) mm; leaf blades 1–4 mm wide.

31. Lemma veins thickened and raised when dry; lemma usually hairy (occasionally glabrous), papery in texture; anthers 0.6–1.5(2) mm . . . . **B. hordeaceus** (in part)

31' Lemma veins not or only slightly thickened and raised; lemmas usually glabrous over the back, leathery in texture; anthers 0.7–2(3) mm . . . . **B. commutatus** (in part)

24' Inflorescence consisting of a single spikelet.

32. Lemma veins thickened and raised when dry; lemma usually hairy, papery in texture; anthers 0.6–1.5(2) mm . . . . **B. hordeaceus** (in part)

32' Lemma veins not or only slightly thickened and raised; lemmas often glabrous over the back, leathery in texture; anthers 0.7–2(3) mm . . . . **B. commutatus** (in part)

Bromus arenarius Labillardière

Australian brome

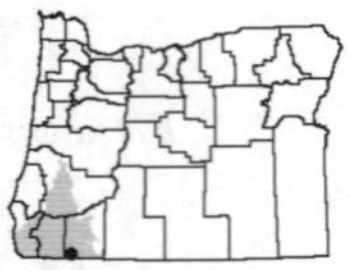

Plants annual, 20–40 cm tall. **Leaves** sheaths densely retrorsely pilose; blades 7–8 cm × 1–5 mm, pilose on both surfaces. **Inflorescences** (4)10–15 × 4–7 cm, open, nodding; at least some pedicels longer than the spikelets, spreading or ascending, sinuous. **Spikelets** 10–20 mm, lanceolate, terete to moderately laterally compressed, with 5–9(11) florets, floret bases usually concealed at maturity; rachilla internodes concealed at maturity. **Glumes** densely pilose; lower glumes 7–10 mm, 3-veined; upper glumes 8–12 mm, (5)7-veined. **Lemmas** 9–11(13) × 1–1.8 mm, lanceolate, densely pilose, distinctly 7-veined; margins rounded, not inrolled at maturity; tips acute, bifid; teeth shorter than 1 mm; lemma awns 10–16 mm, straight to slightly spreading, arising 1.5 mm or more below the lemma tips. **Anthers** 0.7–1 mm. **Caryopses** as long as or shorter than the paleas, thin, weakly inrolled.

Dry disturbed soil. 1000–1300 m. Sisk. CA, NV; southeast to TX; PA; Australia, New Zealand. Exotic.

Bromus arenarius differs from the other annual bromes with blunt lemmas by its long glumes, usually longer awns, and sinuous panicle branches.

Bromus briziformis Fischer & C.A. Meyer

Rattlesnake brome

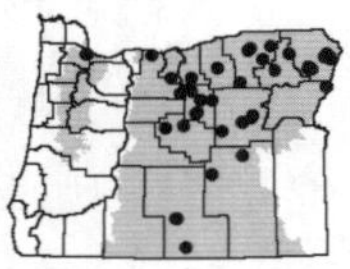

Plants annual, 20–65 cm tall. **Leaves** sheaths densely pilose; blades 3–13 cm × 2–4 mm, pilose to pubescent on both surfaces. **Inflorescences** 5–15 × 3–7 cm, open, nodding; pedicels sometimes longer than the spikelets, curved to reflexed. **Spikelets** 15–30 × 8–12 mm, ovate, laterally compressed, usually dangling, with 9–15 florets, floret bases usually concealed at maturity; rachilla internodes concealed at maturity. **Glumes** smooth or minutely scabrous; lower glumes 5–6 mm, 3–5-veined; upper glumes 6–8 mm, 7–9-veined. **Lemmas** 9–10 mm long, 6–8 mm wide, inflated, obovate or rhombic, leathery, smooth or scabridulous, obscurely 9-veined; veins usually not raised and thickened; margins hyaline, 1–1.3 mm wide, abruptly angled, not inrolled at maturity; tips acute to obtuse, bifid; teeth shorter than 1 mm, usually awnless; lemma awns, if present, to 1 mm, arising less than 1.5 mm below the lemma tips. **Anthers** 0.7–1 mm. **Caryopses** about as long as the paleas. **2*n***=14.

Disturbed places, roadsides, forest openings, sand dunes, overgrazed rangeland. 0–1800 m. BR, BW, Col, ECas, Lava, WV. CA, ID, NV, WA; throughout North America; southwestern Asia, Europe. Exotic.

Unusual inflated florets make *Bromus briziformis* unique among Oregon bromes, but it can be confused with *Briza maxima*. In the latter species, the florets spread more widely, so that the angle between the glumes is almost 180°. Records of *B. briziformis* west of the Cascades have often been misidentified *B. maxima*. The dry, mature inflorescences of this species look somewhat like the tails of rattlesnakes and also make a disturbing rattling sound as one walks through a population of ripe plants.

Bromus carinatus Hooker & Arnott

California brome

Plants perennial, often flowering the first year, 45–120(180) cm tall; loosely cespitose; culms to 3 mm thick. **Leaves** sheaths mostly glabrous or retrorsely soft pilose, throats usually hairy; blades 8–30 cm × 1–10 mm, flat or becoming involute, glabrous or sparsely pilose to pubescent on 1 or both surfaces. **Inflorescences** 5–40 cm, more or less open; branches erect to spreading, lower branches usually shorter than 10 cm, 1–4 per node, ascending to strongly divergent or reflexed, with 1–4 spikelets variously distributed. **Spikelets** 20–40(53) mm, shorter than some pedicels and branches, elliptic to lanceolate, strongly laterally compressed, not crowded or overlapping, 4–11 florets. **Glumes** glabrous or pubescent; lower glumes 7–11 mm, 3–7(9)-veined; upper glumes 9–13 mm, 5–9(11)-veined. **Lemmas** 10–17(22) mm, lanceolate, laterally compressed, strongly keeled distally, uniformly pubescent or pubescent on the margins only, sometimes glabrous or scabrous, 7–9-veined; tips entire or with acute teeth less than 1 mm; lemma awns 4–17 mm, straight or occasionally slightly bent. **Anthers** 1–6 mm. **2*n***=56.

Grasslands and savannas.

2 varieties.

West of the Cascades, *Bromus carinatus* intergrades with *B. sitchensis*, which is taller with wider leaves and longer panicles. Most individuals are obviously distinct, but some are frustratingly intermediate and can be assigned to species only arbitrarily. It is possible that *B. carinatus* and *B. sitchensis* should be considered subtaxa of one species.

Key to *Bromus carinatus* varieties:

1. Awns 8–17 mm; panicle branches mostly spreading to reflexed . . . . **B. c.** var. ***carinatus***

1' Awns 4–7 mm; panicle branches mostly erect to ascending . . . . **B. c.** var. ***marginatus***

Bromus carinatus Hooker & Arnott var. *carinatus* (p. 369)

California brome

Bromus carinatus Hook. & Arn. var. *hookerianus* (Thurb.) Shear, *Bromus carinatus* Hook. & Arn. var. *linearis* Shear

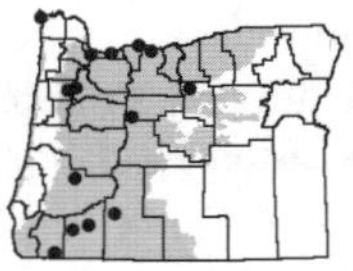

Plants 50–100 cm tall. **Leaves** sheaths mostly glabrous or retrorsely soft pilose, throats usually hairy; blades 10–30 cm × 3–5 mm, flat, usually sparsely pilose on both surfaces, sometimes glabrous. **Inflorescences** 15–40 cm, lax, open; lower panicle branches usually shorter than 10 cm, 2–4 per node, ascending to strongly divergent or reflexed. **Spikelets** with 6–11 florets. **Glumes** lower glumes 8–10 mm, 3(5)-veined; upper glumes 9.5–12 mm, 5(7)-veined. **Lemmas** 12–16 mm, usually uniformly pubescent, sometimes scabrous, 7-veined; lemma awns 8–17 mm. **Anthers** 1–5 mm.

Grasslands, savannas, shrublands and forest openings. 0–2300 m. Casc, Col, ECas, Est, Lava, Sisk, WV. CA, ID, NV, WA; north to British Columbia, southeast to NM and Baja California. Native.

Bromus carinatus var. *carinatus* is distinguished from *B. c.* var. *marginatus* by its longer awns and usually lax panicle branches. It is more common west of the Cascades. In western Oregon, it can be confused with *B. sitchensis,* which has broader leaves and more open panicles. Intermediate forms cannot be assigned to either taxon reliably.

Bromus carinatus Hooker & Arnott var. *marginatus* (Nees) Barkworth & Anderton

Mountain brome

Bromus breviaristatus Buckley, *Bromus marginatus* Nees ex Steud. var. *latior* Shear, *Bromus marginatus* Nees ex Steud. var. *marginatus*, *Bromus marginatus* Nees ex Steud. var. *seminudus* Shear, *Bromus subvelutinus* Shear

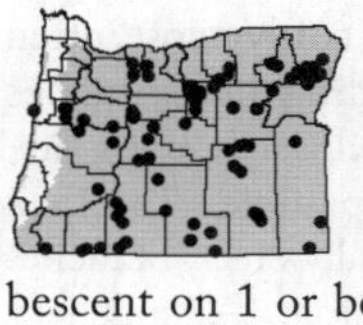

Plants 45–120(180) cm tall. **Leaves** sheaths generally sparsely retrorsely pilose but varying from densely pilose to glabrous, throat pilose; blades 8–25 cm × 1–12 mm, flat or involute, glabrous or sparsely pilose to pubescent on 1 or both surfaces. **Inflorescences** 5–20(30) cm, erect; lower panicle branches shorter than 10 cm, 1–4 per node, erect or ascending. **Spikelets** with 4–9 florets. **Glumes** lower glumes 7–11 mm, 3–7(9)-veined; upper glumes 9–13 mm, 5–9(11)-veined. **Lemmas** 10–14(17) mm, leathery, pubescent on the backs and/or margins or glabrous, 7–9-veined; lemma awns 4–7 mm. **Anthers** 1–6 mm.

Grasslands, shrublands, open forest, mainly montane. 0–2600 m. All ecoregions except CR. CA, ID, NV, WA; north to British Columbia, east to Saskatchewan, southeast to NM and northern Mexico. Native.

Bromus carinatus var. *marginatus* is distinguished from *B. carinatus* var. *carinatus* by its shorter awns and usually erect panicle branches.

Bromus catharticus Vahl var. *catharticus*

Rescue grass

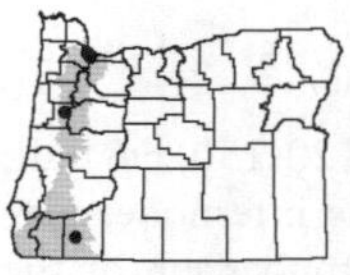

Plants annual, biennial, or perennial, 30–120 cm tall; loosely cespitose. **Culms** 2–4 mm thick. **Leaves** sheaths densely, often retrorsely, hairy, hairs sometimes confined to the throat; blades 4–26 cm × 3–10 mm, flat, glabrous or pilose on both surfaces. **Inflorescences** 9–28 cm, open, erect or nodding. lower nodes with 1–4 branches, lower branches less than 10 cm, spreading or ascending, with 1–5 spikelets. **Spikelets** 17–30 mm, shorter than some pedicels and branches, elliptic to lanceolate, strongly laterally compressed, not crowded, with 6–12 florets. **Glumes** glabrous, scabrous, or pubescent; lower glumes 7–12 mm, 5–7-veined; upper glumes

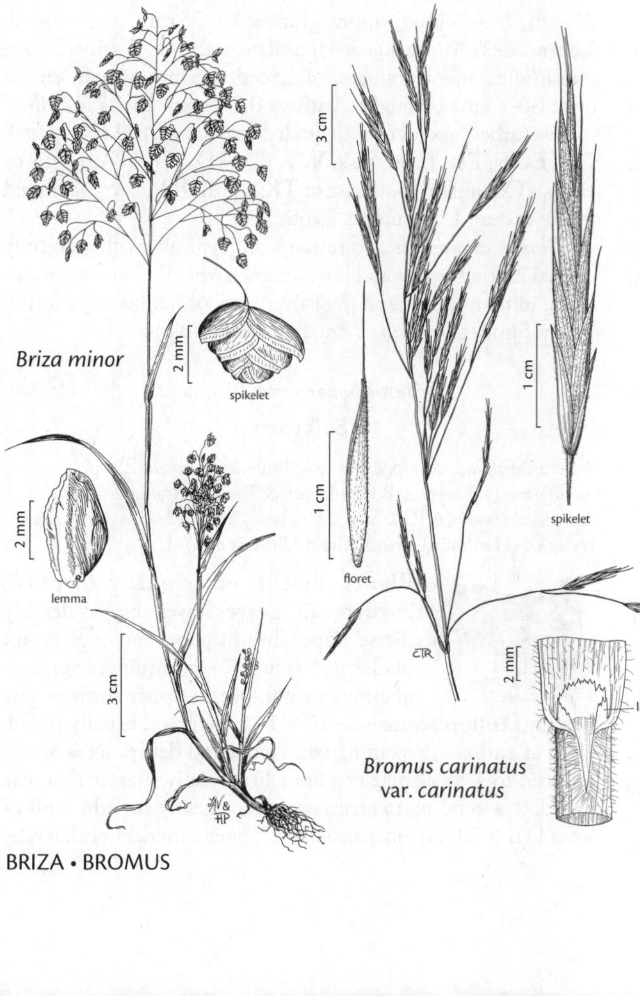

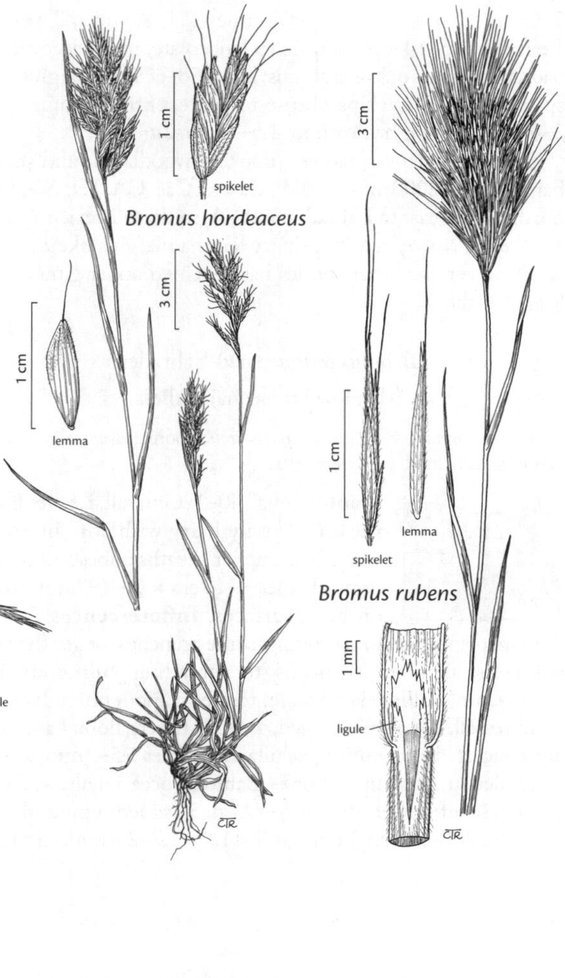

9–13 mm, (7)9(11)-veined. **Lemmas** 11–20 mm, lanceolate, laterally compressed, strongly keeled, glabrous or scabrous, sometimes pubescent distally, (9)11–13-veined; veins often raised; margins hyaline, whitish or partly purplish; tips entire or toothed; teeth acute, shorter than 1 mm, awnless or short-awned; lemma awns, if present, up to 4 mm. **Anthers** 2–4 mm in cross-pollinating florets, approximately 0.5 mm in self-pollinating florets. ***2n***=42.

Disturbed open areas. 0–500 m. Sisk, WV. CA, NV; southern US, scattered in northern US and Canada; South America. Exotic.

Bromus catharticus has strongly flattened spikelets and keeled lemmas that are awnless or with very short awns. Similar *B. carinatus* has longer lemma awns.

Bromus ciliatus Linnaeus

Fringed brome

Bromus richardsoni Link var. *pallidus* (Hook.) Shear

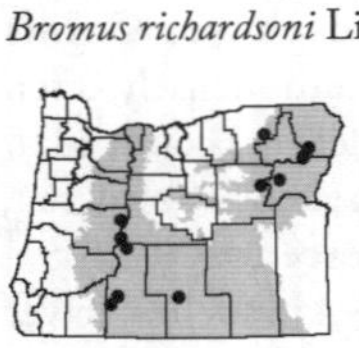

Plants perennial, 45–120(150) cm tall; loosely cespitose. **Culms** internodes glabrous; nodes (3)4–7(8), pubescent or the lower culms glabrous. **Leaves** basal sheaths retrorsely pilose, sometimes glabrous; upper sheaths glabrous, throats glabrous or pilose; ligules 0.4–1.4 mm; blades 13–25 cm × 4–10 mm, flat; lower surface glabrous, sometimes pilose; upper surface usually pilose, sometimes glabrous. **Inflorescences** 10–20 × approximately 12 cm, open, nodding; branches ascending, spreading, or drooping. **Spikelets** 15–25 mm, elliptic to lanceolate, terete to moderately laterally compressed, with 4–9 florets. **Glumes** glabrous; lower glumes 5.5–7.5 mm, 1(3)-veined; upper glumes 7.1–8.5 mm, 3-veined. **Lemmas** 9.5–14 mm, elliptic to lanceolate, rounded over the back, backs glabrous or scabrous; lower 50–67% of margins conspicuously hirsute; tips obtuse to acute, entire; lemma awns 2–4(5) mm, straight. **Anthers** 1.5–2 mm. ***2n***=14.

Moist montane meadows, thickets, woodlands, and stream banks. 1000–2000 m. BR, BW, Casc, ECas. CA, ID, NV, WA; north to AK, east to Labrador and VA, south to Mexico. Native.

Bromus ciliatus has long-hairy lemma margins like *B. laevipes*, however, the latter species has 5-veined upper glumes and longer anthers.

Bromus commutatus Schrader

Meadow brome, hairy chess

Bromus commutatus Schrad. var. *apricorum* Simonk., *Bromus commutatus* Schrad. var. *commutatus*

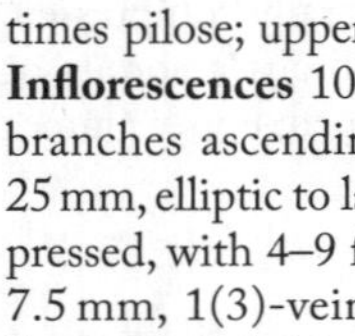

Plants annual, 40–120 cm tall. **Leaves** lower sheaths densely hairy with stiff, often retrorse hairs; upper sheaths pubescent or glabrous; blades 9–13 cm × 2–4(9) mm, pilose on both surfaces. **Inflorescences** 7–16 × 3–6 cm, open, erect to ascending; some branches longer than the spikelets, slender, ascending to spreading. **Spikelets** 14–18(30) mm, oblong-lanceolate, terete to moderately laterally compressed, not purple-tinged, 4–9(13) florets, floret bases usually concealed at maturity; rachilla internodes 1.5–2 mm, usually concealed at maturity. **Glumes** glabrous, occasionally scabrous or pubescent; lower glumes 5–7 mm, 5-veined; upper glumes 6–9 mm, 7(9)-veined. **Lemmas** 8–11.5 × 1.7–2.6 mm, elliptic to lanceolate, leathery, backs usually glabrous, rounded over the back, distinctly 7(9)-veined; veins not raised and thickened; margins scabrous or pubescent, bluntly angled, inrolled or not at maturity; tips acute to obtuse, bifid; teeth shorter than 1 mm; tips awned; lemma awns 3–10 mm, straight, arising less than 1.5 mm below the lemma tips, awn of the lowest lemma shorter than the others. **Paleas** 1–2 mm shorter than lemmas. **Anthers** 0.7–2(3) mm. **Caryopses** equal to or shorter than paleas, weakly to strongly inrolled. ***2n***=14, 28, 56.

Dry, disturbed areas. 0–1900 m. All ecoregions except Est. CA, ID, NV, WA; north to Yukon, east to Newfoundland and GA; Europe and the Baltic region. Exotic.

Bromus commutatus is a common, weedy, annual brome. It is similar to *B. racemosus*, which has shorter lemmas and longer anthers. It is often confused with *B. secalinus*, which has paleas subequal to the lemmas and shorter lemmas.

Bromus diandrus Roth

Ripgut brome

Bromus rigidus Roth, *Bromus rigidus* Roth var. *gussonei* (Parl.) Coss. & Durieu, *Bromus rigidus* Roth var. *rigidus*

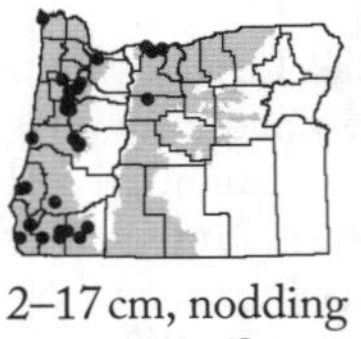

Plants annual, 20–90 cm tall. **Culms** puberulent. **Leaves** sheaths softly pilose, often with retrorse or spreading hairs; blades 3.5–27 cm × 1–9 mm, pilose on both surfaces. **Inflorescences** panicles 13–25 × 2–17 cm, nodding at maturity but occasionally erect; branches 1–7 cm, stiffly erect to ascending or spreading, 1–2 spikelets. **Spikelets** 25–70 mm, moderately laterally compressed, with 4–11 florets. **Glumes** smooth or scabrous; lower glumes 15–25 mm, 1–3-veined; upper glumes 25–35 mm, 3–5-veined. **Lemmas** 22–30 mm, linear-lanceolate, scabrous, 7-veined; margins hyaline; tips acuminate, bifid; teeth 3–5 mm, awned; lemma awns 30–65 mm, straight. **Anthers** 0.5–1 mm. ***2n***=28, 42, 56.

Disturbed open areas, degraded grasslands. 0–1300 m. Col, CR, ECas, Est, Lava, Sisk, WV. CA, ID, NV, WA; north to British Columbia, southeast to TX, south to Mexico, scattered in the eastern US; Europe. Exotic.

Bromus diandrus is a common weed, typically with gracefully curved inflorescences and long lemma awns. The long awns can lodge in the mouths and digestive tracts of herbivores, causing injury. Similar *B. sterilis* is smaller in all its parts.

Bromus hordeaceus Linnaeus (p. 369)

Soft chess

Bromus hordeaceus L. ssp. *hordeaceus*, *Bromus hordeaceus* L. ssp. *molliformis* (J. Lloyd ex Billot) Maire & Weiller, *Bromus hordeaceus* L. ssp. *pseudothominei* (P.M. Sm.) H. Scholz, *Bromus hordeaceus* L. ssp. *thominei* (Hardouin) Braun-Blanq., *Bromus mollis* L.

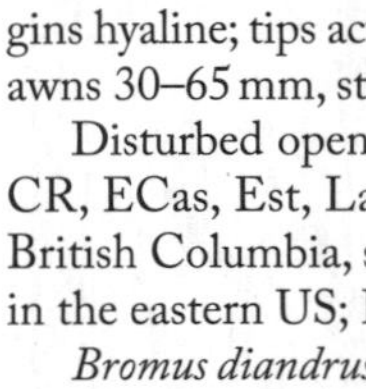

Plants annual, occasionally biennial, 2–70 cm tall. **Leaves** lower sheaths densely pilose; upper sheaths pubescent or glabrous; blades 2–19 cm × 1–4 mm; lower surfaces glabrous or pubescent; upper surfaces pubescent. **Inflorescences** 1–13 × 1–4 cm, erect, usually ovoid, open at anthesis, becoming contracted and dense, occasionally reduced to 1 or 2 spikelets; branches usually shorter than the spikelets, ascending to erect, straight or nearly straight. **Spikelets** (11)14–20(23) mm, lanceolate, terete to moderately laterally compressed, 5–10 florets, floret bases usually concealed at maturity; rachilla internodes concealed at maturity. **Glumes** pilose to glabrous; lower glumes 5–7 mm, 3–5-veined; upper glumes 6.5–8 mm, 5–7-veined. **Lemmas** 6.5–11 × 3–5 mm, lanceolate, chartaceous, antrorsely pilose to pubescent, occasionally glabrous near the base or throughout, 7–9-veined; hyaline margins abruptly or bluntly angled, not inrolled at maturity; lateral veins prominent, thickened and raised; tips rounded to acute, bifid; teeth shorter than 1 mm; lemma awns 6–8 mm, usually arising less than 1.5 mm below the lemma tips, straight to recurved at maturity. **Anthers** 0.6–1.5(2) mm. **Caryopses** equal to or shorter than paleas, thin, weakly inrolled to flat. **2*n*=28.**

Disturbed areas, degraded grasslands. 0–1900 m. All ecoregions except BR. CA, ID, NV, WA; north to AK, south to Mexico, east to Newfoundland and SC; North Africa, southern Europe; scattered elsewhere worldwide. Exotic.

Bromus hordeaceus has been divided into subspecies, two of which may occur in Oregon (*B. h.* ssp. *hordeaceus* and *B. h.* ssp. *pseudothominei*). Most Oregon specimens, however, cannot be identified confidently to either subspecies.

Bromus inermis Leysser

Smooth brome

Bromus inermis Leyss. ssp. *inermis*

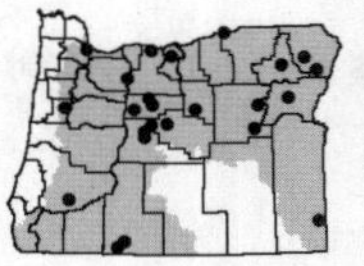

Plants perennial, 50–130 cm; rhizomatous. **Culms** internodes usually glabrous, rarely pubescent; nodes (2)3–5(6), usually glabrous, rarely pubescent. **Leaves** sheaths glabrous, rarely pubescent or pilose; ligules to 3 mm; blades 11–35(42) cm × 5–15 mm, flat, glabrous, rarely pubescent or pilose. **Inflorescences** 10–20 × 1–4(7) cm, open, erect; branches ascending or spreading. **Spikelets** 20–40 mm, elliptic to lanceolate, terete to slightly laterally compressed, (5)8–10 florets. **Glumes** glabrous; lower glumes (4)6–9 mm, 1(3)-veined; upper glumes (5)7–10 mm, 3-veined. **Lemmas** 9–13 mm, elliptic to lanceolate, rounded over the back, glabrous and smooth, occasionally scabrous, the basal part of the backs rarely puberulent; margins sometimes sparsely puberulent; tips acute to obtuse, entire, awnless or awned; lemma awns if present 1–2(4) mm, straight. **Anthers** 3.5–6 mm. **2*n*=28, 42, 56.**

Mesic uplands, stream terraces, valley bottoms, open forests, roadsides, and ditches. 0–2400 m. BW, Casc, Col, ECas, Lava, Owy, Sisk, WV. CA, ID, NV, WA; north to AK, east to Newfoundland and NC, south to Mexico; Eurasia. Exotic.

Rhizomatous *B. inermis* is most likely to be confused with two bromes that have been reported from Oregon but have not been confirmed. *Bromus riparius* has longer lemma awns, and its basal sheaths decay into a reticulate, thatch-like mass. It is an Asian species introduced for forage and recently collected from a roadside seeding in southwestern Oregon. *Bromus pumpellianus* is similar but has pubescent lemmas, nodes, and leaf blades.

Bromus japonicus Thunberg

Japanese brome

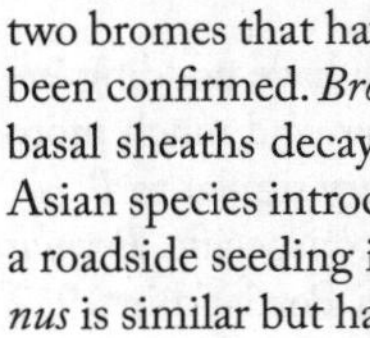

Plants annual, (22)30–70 cm tall. **Leaves** sheaths densely pilose; upper sheaths sometimes pubescent to glabrous; blades 10–20 cm × 2–4 mm, generally pilose on both surfaces. **Inflorescences** 10–22 × 4–13 cm, open, nodding; branches usually longer than the spikelets, spreading to ascending, slender, flexuous, somewhat drooping, sometimes sinuous. **Spikelets** 20–40 mm, lanceolate, terete to moderately laterally compressed, with 6–12 florets, floret bases usually concealed at maturity; rachilla internodes concealed at maturity. **Glumes** smooth or scabrous; lower glumes 4.5–7 mm, (3)5-veined; upper glumes 5–8 mm, 7-veined. **Lemmas** 7–9 × 1.2–2.2 mm, lanceolate, leathery, smooth proximally, scabrous on the distal half, obscurely (7)9-veined; veins usually not raised and thickened; margins hyaline, 0.3–0.6 mm wide, obtusely angled above the middle, not inrolled at maturity; tips rounded to acute, bifid; teeth shorter than 1 mm; lemma awns 8–13 mm, strongly divergent at maturity, occasionally straight, twisted, flattened at the base, arising 1.5 mm or more below the lemma tips. **Anthers** 0.7–1.5 mm. **Caryopses** equal to or shorter than the paleas, thin, weakly inrolled or flat. **2*n*=14.**

Dry, disturbed areas. 0–2200 m. BW, Casc, ECas, Lava, Owy, WV. CA, ID, NV, WA; throughout North America; Eurasia. Exotic.

Bromus japonicus is an annual brome with relatively long spikelets and curved awns. *Bromus commutatus*, *B. racemosus* and *B. secalinus* have shorter spikelets and have lemma awns arising less than 1.5 mm below the lemma apex. *Bromus squarrosus* differs in having more strongly angled lemma margins with broader hyaline margins, more florets per spikelet, and usually only one spikelet per panicle branch.

Bromus laevipes Shear

Chinook brome

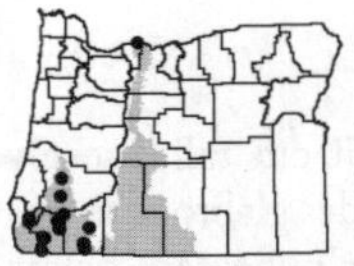

Plants perennial, 50–150 cm tall; loosely cespitose. **Culms** often rooting at the lower nodes; internodes usually glabrous, often puberulent-pubescent below the nodes, rarely puberulent throughout; nodes 3–5(6), pubescent. **Leaves** sheaths glabrous, sometimes slightly pubescent near the throat; ligules 2–4.2 mm, glabrous, obtuse, lacerate; blades 13–26 cm × 4–10 mm, light green or glaucous, flat, glabrous, sometimes scabrous on both surfaces. **Inflorescences** 10–30 × 2–5 cm, open, nodding; branches ascending to spreading, sometimes drooping. **Spikelets** 23–35 cm, elliptic to lanceolate, terete to moderately laterally compressed, with 5–11 florets. **Glumes** glabrous, sometimes scabrous; margins often bronze-tinged; lower glumes 6–9 mm, 3-veined; upper glumes 8–12 mm, 5-veined. **Lemmas** 12–16 mm, elliptic to lanceolate, rounded over the back, backs sparsely pilose, pubescent, or scabrous; margins densely pilose at least on the lower half, often bronze-tinged; tips acute to obtuse, entire, rarely slightly emarginate with lobes less than 1 mm; lemma awns 4–6 mm, straight. **Anthers** 3.5–5.5 mm. **2*n*=14.**

Open slopes and woodlands. 100–1400 m. ECas, Sisk. CA. Native.

Bromus laevipes has a more or less open inflorescence with ascending to spreading branches. The lemma margins have dense, long, spreading hairs, contrasting with the scabrous to sparsely hairy lemma backs. The upper glumes have 5 veins, which is unusual among the perennial bromes that have rounded lemma backs.

Bromus maritimus (Piper) Hitchcock

Maritime brome

Bromus carinatus Hook. & Arn. var. *maritimus* (Piper) C.L. Hitchc.

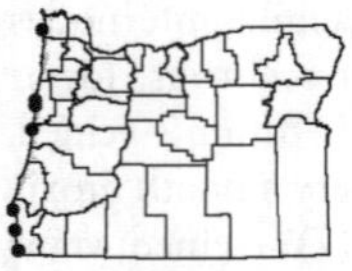

Plants perennial, 20–70 cm tall; loosely cespitose to rhizomatous. **Culms** to 3 mm thick. **Leaves** sheaths smooth or minutely scabridulous, sometimes slightly pubescent distally, not pilose at the throat; blades 6–13 cm × 6–8 mm, flat, both surfaces smooth or scabrous. **Inflorescences** 9–20 × 2–2.5 cm, dense, erect. **Spikelets** 20–40 mm, usually longer than the branches and pedicels, elliptic to lanceolate, strongly laterally compressed, crowded, overlapping, with 3–7 florets. **Glumes** pubescent; lower glumes 8–12 mm, (3)5(7)-veined; upper glumes 10–13 mm, 7(9)-veined. **Lemmas** 12–14 mm, lanceolate, distinctly 9–11-veined, strongly keeled at least toward the tip, more or less uniformly hairy; margins hyaline, often bronze-colored; tips entire or with acute teeth less than 1 mm; lemma awns (2)4–7 mm. **Anthers** 2–4 mm. ***2n***=84.

Coastal sands. 0–50 m. Est. CA. Native.

Bromus maritimus is distinguished by its very dense inflorescences, laterally compressed spikelets, and coastal habitat on sand dunes and bluffs. Wind and salt spray in coastal habitats can induce other bromes to have abnormally dense inflorescences superficially resembling *B. maritimus.* Like many native sand dune species, *B. maritimus* is becoming rare due to habitat loss from sand dune stabilization and coastal development.

Bromus orcuttianus Vasey

Orcutt's brome

Bromus orcuttianus Vasey var. *orcuttianus*

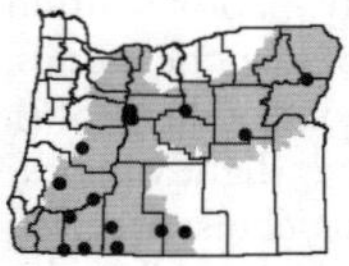

Plants perennial, 90–150 cm tall; loosely cespitose. **Culms** internodes glabrous to pubescent, pilose to densely pubescent below the nodes; nodes 2–4, pubescent or puberulent. **Leaves** sheaths sparsely to densely pilose; basal sheaths with hairs 2–4 mm, occasionally glabrous; upper sheaths with hairs to 1 mm; collars pilose with hairs to 4 mm or glabrous; ligules 1–3 mm; blades 7–24 cm × 3–12 mm, flat, glabrous, sometimes hairy. **Inflorescences** 7–13.5 × 2–12 cm; branches erect or ascending, appressed to slightly spreading. **Spikelets** 20–40 mm, elliptic to lanceolate, terete to moderately laterally compressed, with 3–9(11) florets. **Glumes** glabrous, occasionally scabrous or pubescent; lower glumes 5–9 mm, 1(3)-veined; upper glumes 7–11 mm, 3(5)-veined, sometimes short-pointed. **Lemmas** 9–16 mm, elliptic, rounded over the back, backs pubescent, sometimes glabrous or scabrous; margins pubescent or scabrous, the hairs similar to those on the lemma backs; tips obtuse, entire, awned; lemma awns (4)5–8 mm, straight. **Anthers** 3–5 mm. ***2n***=14.

Dry montane conifer forests, especially ponderosa pine; chaparral, openings, sometimes in moister sites. 700–2200 m. BW, Casc, ECas, Lava, Sisk. CA, WA; southeast to AZ. Native.

Bromus orcuttianus is a grass of open, upland forests. The inflorescence of *B. orcuttianus* is upright and open with stiff, spreading branches. *Bromus suksdorfii* has a denser inflorescence, and *B. laevipes* has more flexuous branches and 5-veined upper glumes. *Bromus orcuttianus* is an uncommon native grass that seems to have declined in numbers, probably due to grazing.

Bromus pacificus Shear

Pacific brome

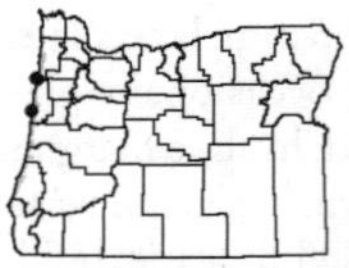

Plants perennial, 60–170 cm tall; cespitose. **Culms** internodes glabrous, sometimes pubescent near the nodes; nodes (5)6–8, pubescent. **Leaves** sheaths pilose; ligules 2–4 mm; blades 20–35(37) cm × 6–16 mm wide, flat; lower surfaces glabrous; upper surfaces pilose. **Inflorescences** 10–25 cm, open, more or less erect; branches spreading to drooping. **Spikelets** 20–30 mm, elliptic to lanceolate, terete to moderately laterally compressed, with (4)6–10 florets. **Glumes** pubescent; lower glumes 6–8.5 mm, 1(3)-veined; upper glumes 8–11.5 mm, 3-veined. **Lemmas** 10–12 mm, lanceolate, rounded over the back, backs pubescent; margins more densely pubescent; tips acute, entire; lemma awns 3.5–7 mm, straight. **Anthers** 2–4 mm.

Moist thickets and forest edges along the coast. 0–1700 m. Est. North to AK. Native.

Bromus pacificus is a tall brome of coastal thickets, with widely spreading inflorescence branches and broad leaves. It most closely resembles *B. sitchensis,* which has glabrous or scabrous glumes. *Bromus pacificus* is often reported from inland sites west of the Cascades, but none of these reports have been confirmed and they are likely *B. sitchensis.*

Bromus rubens Linnaeus (p. 369)

Red brome, foxtail chess

Bromus madritensis L. ssp. *rubens* (L.) Husn.

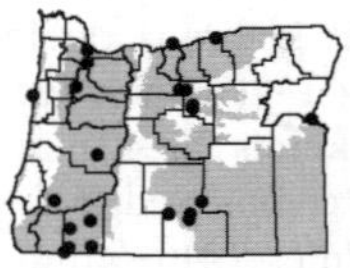

Plants annual, 10–40 cm tall. **Culms** often puberulent. **Leaves** sheaths softly pubescent to pilose; blades 1–15 cm × 1–5 mm, pubescent on both surfaces. **Inflorescences** 2–10 × 2–5 cm, erect, dense, often reddish brown; branches 0.1–1 cm, ascending, never drooping, not readily visible, 1–2 spikelets. **Spikelets** 18–25 mm, much longer than the panicle branches, densely crowded, subsessile, moderately laterally compressed, with 4–8 reddish florets. **Glumes** pilose; lower glumes 5–8 mm, 1(3)-veined; upper glumes 8–12 mm, 3–5-veined. **Lemmas** 10–15 mm, linear-lanceolate, pubescent to pilose, 7-veined; margins hyaline; tips acuminate, with teeth 1–4 mm, awned; lemma awns 8–20 mm, straight. **Anthers** 0.5–1 mm. ***2n***=14, 28.

Disturbed places, bare droughty soils. 0–1400 m. BR, Casc, Col, Est, Lava, Owy, Sisk, WV. CA, ID, NV, WA; east to CO and TX; MA, NY; Europe. Exotic.

Similar *B. diandrus* and *B. sterilis* usually have nodding inflorescences with the spikelets angling down and much more open inflorescences. *Bromus madritensis* is very similar to *B. rubens* with which it is sometimes considered conspecific. *Bromus madritensis* was collected once on ballast in Portland in 1912 and apparently did not persist. It has a somewhat looser inflorescence with longer, exposed, erect panicle branches.

Bromus secalinus Linnaeus (p. 377)

Rye brome

Bromus secalinus L. var. *secalinus*, *Bromus secalinus* L. var. *velutinus* (Schrad.) Schübeler & G. Martens

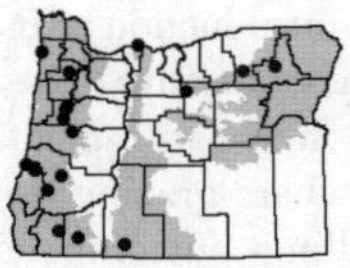

Plants annual, 20–80(120) cm tall. **Leaves** sheaths glabrous or loosely pubescent and becoming glabrous; blades 15–30 cm × 1–12 mm; lower surfaces pilose or glabrous; upper surfaces pilose. **Inflorescences** 5–23 × 2.5–12 cm, open, nodding; branches spreading to ascending; lower branches slightly drooping; some pedicels longer than their spikelets. **Spikelets** 10–20 mm, ovoid-lanceolate or ovate, laterally compressed, not purple-tinged, with 4–9(10) florets, floret bases visible at maturity; rachilla internodes visible at maturity. **Glumes** scabrous or glabrous; lower glumes 4–6 mm, 3–5-veined; upper glumes 6–7 mm, 7-veined. **Lemmas** 6.5–8.5(10) × 1.7–2.5 mm, elliptic, leathery, rounded over the back, backs glabrous, sometimes pubescent, scabrous to puberulent on the margins and near the tips; obscurely 7-veined; veins usually not raised and thickened; margins smoothly curved, inrolled at maturity; tips acute to obtuse, bifid, with teeth less than 1 mm, awned; lemma awns (0)3–6(9.5) mm, straight or flexuous, arising less than 1.5 mm below the lemma tips. **Anthers** 1–3 mm. **2*n***=14, 28.

Disturbed areas, ditches, moist meadows, shorelines. 0–1500 m. BW, CR, ECas, Est, Sisk, WV. CA, ID, NV, WA; throughout most of North America; Europe. Exotic.

Mature *B. secalinus* is identified by the strongly inrolled lemmas enfolding the inrolled caryopses. The spreading florets expose the florets and rachilla joints.

Bromus sitchensis Trinius

Sitka brome, Alaska brome

Bromus sitchensis Trin. var. *sitchensis*

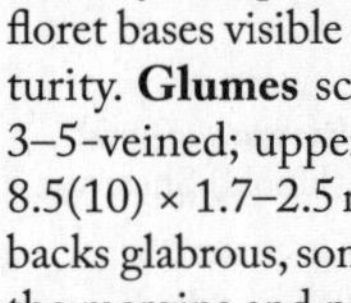

Plants perennial, 120–180 cm tall; loosely cespitose. **Culms** 3–7 mm thick. **Leaves** sheaths glabrous or sparsely pilose; blades 20–40 cm × (2)5–20 mm, flat, sparsely pilose on the upper surface or both surfaces. **Inflorescences** 25–35 × 30(40) cm, drooping, open; lower nodes with 2–4(6) branches; lower branches to 20 cm, spreading, often drooping, with 1–3 spikelets on the outer half. **Spikelets** 18–38 mm, shorter than some branches and pedicels, elliptic to lanceolate, strongly laterally compressed, (5)6–9 florets. **Glumes** glabrous, sometimes scabrous; lower glumes 6–10 mm, 3–5-veined; upper glumes 8–11 mm, 5–7-veined. **Lemmas** 12–15 mm, lanceolate, laterally compressed, strongly keeled at least toward the tip, 7–11-veined, glabrous but sometimes with very small coarse hairs; margins sometimes sparsely pilose; tips entire or with acute teeth, teeth less than 1 mm; lemma awns 5–10 mm. **Anthers** to 6 mm. **2*n***=42, 56.

Moist forest edges, riparian corridors, oak woodlands and savanna, disturbed areas. 0–1600 m. Casc, CR, ECas, Est, Sisk, WV. CA, ID, WA; north to AK, east to WY. Exotic.

Bromus sitchensis has large, open panicles and broad leaves. Similar *B. carinatus* usually has narrower panicles with shorter lower panicle branches and narrower leaves and culms. In the Willamette Valley and northwestern Oregon, many plants are intermediate between *B. sitchensis* and *B. carinatus,* suggesting hybridization.

Bromus squarrosus Linnaeus

Squarrose brome

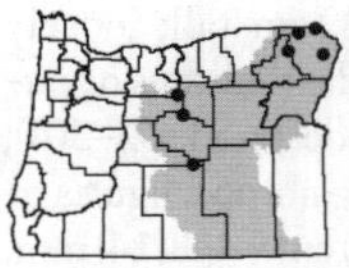

Plants annual, 20–60 cm tall. **Leaves** lower sheaths densely pilose; upper sheaths glabrous; blades 5–15 cm × 4–6 mm, densely pilose on both surfaces. **Inflorescences** 7–20 × 4–8 cm, open, nodding, often with few spikelets, usually 1-sided; branches sometimes longer than the spikelets, ascending-spreading, flexuous, slightly curved. **Spikelets** 15–30 mm, lanceolate to elliptic, terete to moderately laterally compressed, 8–30 florets, floret bases concealed at maturity; rachilla internodes concealed at maturity. **Glumes** smooth, scabrous, or densely hairy, hairs to 0.8 mm; lower glumes 4.5–7 mm, 3–5(7)-veined; upper glumes 6–8 mm, 7-veined. **Lemmas** 8–11 × 2–2.4 mm, lanceolate, chartaceous, smooth, scabridulous, or densely pubescent with hairs to 0.8 mm, 7–9-veined; margins hyaline, 0.6–0.9 mm wide, strongly angled above the middle, not inrolled at maturity; veins not raised and thickened; tips rounded to acute, bifid; teeth shorter than 1 mm; lemma awns 8–10 mm, flattened and sometimes twisted at the base, spreading at maturity, arising 1.5 mm or more below the lemma tips. **Anthers** 0.8–1.3 mm. **Caryopses** as long as the paleas, thin, weakly inrolled or flat. **2*n***=14.

Dry, disturbed areas, roadsides. 500–1900 m. BR, BW, Lava. CA, ID, NV, WA; scattered throughout Canada and US; southern Europe, central Russia. Exotic.

Bromus squarrosus is an uncommon annual with few long, dangling spikelets. Its lemma margins are relatively strongly angled above the middle, with broad hyaline edges. It is most likely to be confused with *B. japonicus,* which has narrower hyaline lemma margins and shorter spikelets with fewer florets.

Bromus sterilis Linnaeus

Sterile brome, poverty brome

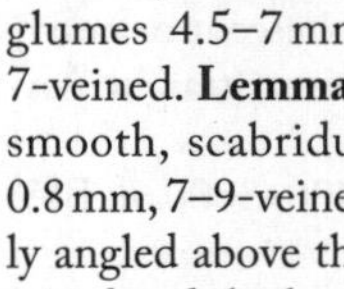

Plants annual, 35–100 cm tall. **Culms** glabrous. **Leaves** sheaths densely pubescent; blades 4–20 cm × 1–6 mm, hairy on both surfaces. **Inflorescences** 10–20 × 5–12 cm, open, nodding; branches spreading, usually longer than the spikelets; with 1–2 spikelets. **Spikelets** 20–35 mm, moderately laterally compressed, with 5–9 florets. **Glumes** smooth or scabrous; lower glumes 8–10 mm, 1(3)-veined; upper glumes 12–15 mm, 3(5)-veined. **Lemmas** 14–20 mm, narrowly lanceolate, pubescent or puberulent, 7(9)-veined; margins hyaline; tips acuminate, bifid, with teeth 0.2–3 mm, awned; lemma awns 15–30 mm, straight. **Anthers** 1–1.4 mm. **2*n***=42, 56.

Dry disturbed areas, overgrazed rangeland, disturbed grasslands. 0–1700 m. Casc, Col, CR, Sisk, WV. CA, ID, NV, WA; north to British Columbia, east to CO and TX, eastern North America; Europe. Exotic.

This weedy brome with its long, dangling awns is very similar to *B. diandrus* except that it is more delicate with its smaller glumes, lemmas, and awns.

Bromus suksdorfii Vasey
Suksdorf's brome

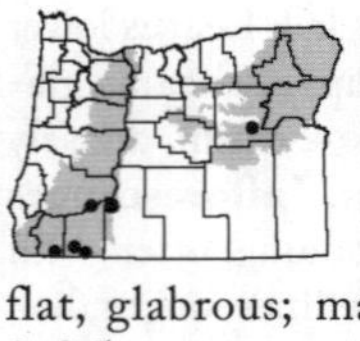

Plants perennial, 50–100 cm tall; loosely cespitose. **Culms** internodes glabrous or pubescent just below the nodes; nodes 2–3(4), glabrous. **Leaves** sheaths glabrous; ligules to 1 mm; blades (8)12–19(24) cm × 4–8(14) mm, flat, glabrous; margins scabrous. **Inflorescences** 6–14 × 1–2.5 cm, erect, contracted; branches erect or ascending. **Spikelets** 15–30 mm. elliptic to lanceolate, terete to moderately laterally compressed, with (3)5–7 florets. **Glumes** glabrous or sparsely pubescent; lower glumes 7–11 mm, 1(3)-veined; upper glumes 9–12 mm, 3-veined. **Lemmas** 12–15 mm, elliptic, rounded over the back, backs and margins pubescent or nearly glabrous; tips obtuse, entire; lemma awns 2–5 mm, straight. **Anthers** 2–3.5 mm. **2*n***=14.

Open slopes and forests in mountains. 400–3000 m. BW, Casc, Sisk. CA, NV, WA. Native.

Bromus suksdorfii is identified by its dense inflorescence and strictly erect panicle branches. Both *B. orcuttianus* and *B. laevipes* have more open inflorescences. *Bromus inermis* is strongly rhizomatous, and at flowering its inflorescence is more open.

Bromus tectorum Linnaeus (p. 377)
Cheatgrass, downy chess, downy brome

Bromus tectorum L. var. *glabratus* Spenn., *Bromus tectorum* L. var. *nudus* Klett & Richt., *Bromus tectorum* L. var. *tectorum*

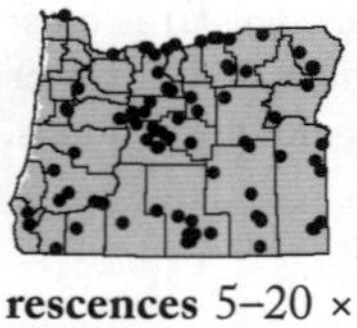

Plants annual, 5–90 cm tall. **Culms** puberulent. **Leaves** sheaths densely and softly retrorsely pubescent to pilose; upper sheaths sometimes glabrous; blades 1–16 cm × 1–6 mm, softly hairy on both surfaces. **Inflorescences** 5–20 × 3–8 cm, open, nodding; branches 1–4 cm, drooping, 1-sided and longer than the spikelets, usually at least 1 branch with 4–8 spikelets. **Spikelets** 10–20 mm, moderately laterally compressed, not densely crowded, 4–8 florets. **Glumes** villous, pubescent, or glabrous; lower glumes 4–9 mm, 1-veined; upper glumes 7–13 mm, 3–5-veined. **Lemmas** 9–12 mm, lanceolate, glabrous or pubescent to pilose, 5–7-veined; tips acuminate, hyaline, bifid, with teeth 0.8–2(3) mm, awned; lemma awns 10–18 mm, straight. **Anthers** 0.5–1 mm. **2*n***=14.

Disturbed areas, sagebrush steppe, degraded grasslands, roadsides. 0–2400 m. BR, BW, Casc, Col, CR, ECas, Lava, Owy, Sisk, WV. CA, ID, NV, WA; throughout North America; worldwide. Exotic.

Bromus tectorum is a relatively short grass with drooping inflorescences. Similar *B. sterilis* and *B. diandrus* have longer glumes, lemmas, and awns, and spikelets that hang down at a shallower angle than those of *B. tectorum*. The introduction of *B. tectorum* to shrub steppe habitats during a time of massive overgrazing in the late 1800s has made restoration of native plant communities difficult or impossible, even where grazing no longer occurs. Fast-growing *B. tectorum* seedlings outcompete slower growing native grass seedlings for water in drying soils. At maturity, the awns make *B. tectorum* unpalatable to livestock.

Bromus vulgaris (Hooker) Shear
Columbia brome

Bromus vulgaris (Hook.) Shear var. *eximius* Shear, *Bromus vulgaris* (Hook.) Shear var. *robustus* Shear, *Bromus vulgaris* (Hook.) Shear var. *vulgaris*

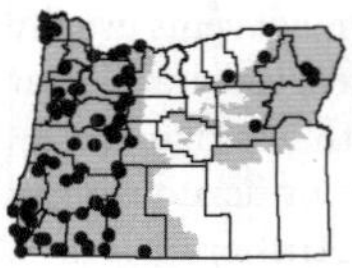

Plants perennial, 60–120 cm tall; loosely cespitose. **Culms** internodes glabrous; nodes (3)4–6(7), usually pilose. **Leaves** sheaths pilose or glabrous; ligules 2–6 mm; blades 13–25(33) cm × up to 14 mm, flat; upper and lower surfaces usually glabrous, sometimes pilose. **Inflorescences** 10–15 cm, usually narrow, nodding, the branches ascending to appressed. **Spikelets** 15–30 mm, elliptic to lanceolate, terete to moderately laterally compressed, with (3)4–9 florets. **Glumes** glabrous or pilose; lower glumes 5–8 mm, 1(3)-veined; upper glumes 8–12 mm, 3-veined. **Lemmas** 8–15 mm, lanceolate, rounded over the back, backs sparsely hairy or glabrous; margins coarsely pubescent, sometimes glabrous; tips subulate to acute, entire, awned; lemma awns (4)6–12 mm, straight. **Anthers** 2–4 mm. **2*n***=14.

Forests, in shade or partial shade. 0–2400 m. BW, Casc, CR, ECas, Est, Sisk, WV. CA, ID, WA; north to British Columbia, northeast to Alberta, east to WY. Native.

Bromus vulgaris is a common forest grass of western Oregon and the Blue and Wallowa mountains. Its closed sheaths with spreading hairs and its long, narrow spikelets distinguish it from most other woodland grasses. Invasive *Brachypodium sylvaticum* has similar nodding inflorescences with long spikelets, but all the spikelets are connected directly to the inflorescence axis by very short, unbranched stalks, and the leaf sheath margins are overlapping.

CALAMAGROSTIS Reedgrass

Plants perennial; cespitose or rhizomatous. **Leaves** sheaths open; ligules membranous, usually truncate to obtuse, sometimes acute, entire or lacerate; blades flat to involute, smooth to scabrous, rarely pubescent. **Inflorescences** terminal open or contracted to spike-like panicles; disarticulation above the glumes. **Spikelets** weakly laterally compressed with 1(2) floret; rachillas prolonged beyond the base of the distal floret, usually pubescent. **Glumes** subequal, equal to, or longer than the lemmas, rounded on the back or keeled, smooth or scabrous, rarely long-scabrous with bent projections, the 4 lateral veins often extending into minute teeth. **Calluses** with hairs 0.2–6.5 mm, sparse to abundant. **Lemmas** 3(5)-veined, smooth or scabrous; lemma awns arising from near the tip to near the base of the midvein, straight or bent. **Anthers** 3. **Caryopses** oblong, usually glabrous.

Cool temperate regions of the world. Approximately 100 species; 10 species treated in *Flora*.

Some species of *Calamagrostis* can be confused with *Agrostis*. In general, *Calamagrostis* plants are larger and coarser than *Agrostis*, with firmer glumes and proportionately longer paleas. Very few *Agrostis* have hairy calluses like *Calamagrostis*. *Calamagrostis* species have prolonged rachillas, which *Agrostis* lack.

Calamagrostis is taxonomically difficult because of interspecific hybridization, polyploidy, and asexual seed production. Select for identification only individuals that seem typical of the population. When using the key, read each lead in its entirety to determine if the plant matches all of the traits. Inflorescence colors are provided for many species. Occasional albino plants lacking purple or brown coloration occur in all species. When keying within this genus, it is helpful to remember that awns are considered exserted if they extend beyond the glume tips. If the awn is bent, consider it exserted if it would be significantly longer than the glumes if it were straight. Measure the longest hairs present when determining the length of callus or rachilla hairs. When measuring callus hairs, be careful not to measure hairs originating on the prolonged rachilla.

Key to *Calamagrostis* species:

1. Lemma awns geniculate, at least some exserted beyond the glume tips by (1.5)2–11 mm.
  2. Lemma awns 10–16 mm; panicle open, branches spreading . . . . *C. howellii*
  2' Lemma awns 5–8.5(10) mm; panicle open or congested, branches spreading to erect.
    3. Wider leaf blades 6–13 mm wide; plants (47)60–120(150) cm tall. . . . . *C. tweedyi*
    3' All leaf blades (1.5)2–2.5(4) mm wide; plants (20)30–80(95) cm tall.
      4. Panicles open; leaf blades 0.4–1.7 mm wide . . . . *C. breweri*
      4' Panicles congested, narrow; leaf blades (1.5)2–5(8) mm wide . . . . *C. tacomensis* (in part)

1' Lemma awns straight or sometimes geniculate but not exserted beyond the glume tips by more than 1(1.5) mm.
  5. Plants coastal; collars thickened, veins obscure or lacking; leaf blades thick and tough, (2)4–10(20) mm wide. . . . . *C. nutkaensis*
  5' Plants inland or occasionally near the coast, collars not thickened, veins evident; leaf blades not particularly thickened, 2–5(8) mm wide.
    6. Callus hairs usually 20–50(60)% as long as the lemma.
      7. At least some collars hairy, the hairs longer than those of the adjacent blade or sheath. . . . *C. rubescens* (in part)
      7' Collars glabrous or only puberulent, the hairs, if any, no longer than those of the sheath or blade.
        8. Plants cespitose; foliage glaucous, glabrous, not scabrous; range Steens Mt. . . . . *C. tacomensis* (in part)
        8' Plants more or less rhizomatous, foliage green, glabrous to short-hairy, sometimes scabrous; range various.
          9. Leaves densely short-hairy on ventral surface, scabrous on upper surface . . . . *C. purpurascens*
          9' Leaves sparsely short-hairy or sparsely scabrous to glabrous on either or both surfaces.
            10. Anthers (1)1.3–2(2.6) mm long; collars often hairy. . . . . *C. rubescens* (in part)
            10' Anthers 2–3.5 mm; collars glabrous . . . . *C. koelerioides*
    6' Callus hairs usually 67% as long to longer than the lemma.
      11. Panicle relatively congested, rarely over 2 cm broad, branches erect or ascending, not readily distinguished; awns stout . . . . *C. stricta*
      11' Panicle usually open, mostly more than 2 cm broad, branches tending to spread, rarely obscured; awns delicate. . . . . *C. canadensis*

Calamagrostis breweri Thurber

Brewer's reedgrass, shorthair reedgrass

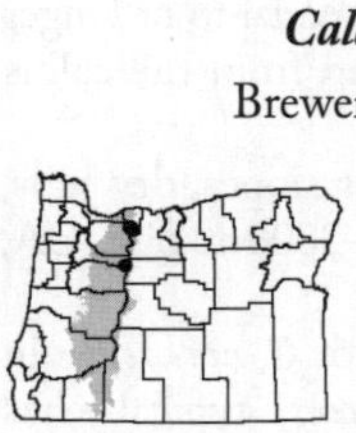

Plants (12)15–30(50) cm tall, more or less cespitose, sometimes loosely so; rhizomes to 5 cm long, 1–2 mm thick. **Leaves** sheaths and collars smooth or slightly scabrous; blades (2)10–15 cm × 0.4–1.7 mm when flat, 0.4–0.6 mm in diameter when dry and involute; outer surfaces scabrous, inner surfaces sparsely pubescent. **Inflorescences** (4)5.7–8.5 × 0.7–5.2 cm, more or less erect, dark purple; branches (1)2–3(3.5) cm, smooth; spikelets usually confined to the ends of the branches. **Glumes** 3–5 mm, rounded over the back. **Calluses** with hairs 0.3–1.2 mm, 20–50% as long as the lemmas, sparse. **Lemmas** 2.5–4 mm, 0.5–1 mm shorter than the glumes; lemma awns 3.5–5.5 mm, arising from the lower portion of the lemmas, exserted from the spikelets, bent. **Anthers** 1.3–2.6 mm. **2*n***=42.

Moist subalpine and alpine meadows, lake margins, and stream banks. 1400–2800 m. Casc. CA. Native.

This little subalpine grass looks like a small *Agrostis* but differs in its conspicuous lemma hairs, its palea almost as long as the lemma, and its awn arising near the base of the lemma. *Podagrostis* species differ in having poorly developed callus hairs and shorter awns. *Calamagrostis breweri* is a globally rare grass with an interrupted distribution high on mountains of California's central Sierra Nevada, the Klamath Range and Oregon's northern Cascades.

Calamagrostis canadensis (Michaux) P. Beauvois

Bluejoint reedgrass

Plants (32)65–112(180) cm tall; rhizomatous and often forming large stands; shoots sometimes clumped; rhizomes 2–15+ cm long, 1–3 mm thick. **Leaves** sheaths smooth or scabrous; collars scabrous, rarely smooth or pubescent; blades (10)16–31(50) cm × 2–8(11) mm, flat, lax; lower surfaces scabrous; upper surfaces strongly scabrous, rarely smooth or with scattered hairs, often glaucous. **Inflorescences** (6)9–17(25) × (1)2–4(8) cm, often contracted when young, more or less open at maturity, nodding, usually purplish; branches 2.7–6(12) cm, scabrous spikelets concentrated on the distal 67%. **Glumes** 2–4.5(5.2) mm, rounded or keeled on the backs. **Calluses** with hairs (1.5)2–3.5(4.5) mm, 0.5–1.5 times the length of the lemmas, abundant. **Lemmas** 2–3.1(4) mm, 0–2.1 mm shorter than the glumes; lemma awns 0.9–3.1 mm, arising from the lower 20–67% of the lemma, usually not exserted, delicate, often difficult to distinguish from the callus hairs, usually straight. **Anthers** (0.8)1.2–1.6(2.6) mm. **2*n***=42, 66.

Wet stream and lake margins, meadows, and forest openings. 3 varieties; 2 varieties treated in *Flora*.

This more or less glaucous grass forms large stands on montane lake shores or narrow, straggling populations along shaded creeks. The brown or purplish inflorescences nod and have relatively open, lobed inflorescences with distinct panicle branches. The callus hairs are so long that they can be seen without separating the glumes. *Calamagrostis canadensis* can produce seed either sexually or asexually. As a result, large populations may be clones, creating a kind of uniformity that highlights any variation between populations and encourages botanists to excesses of nomenclature. Despite the many reported differences between them, the two varieties of *C. canadensis* in Oregon intergrade morphologically and overlap geographically and may not be worth recognizing.

Key to *Calamagrostis canadensis* varieties:

1. Spikelets (3.5)4–4.5(5.2) mm; glumes usually scabrous over the entire surface, the prickles on the keels often curved as well as angled forward at the base; glume apices distinctly acuminate . . . . ***C. c.* var. *langsdorffii***

1' Spikelets 2–4 mm; glumes smooth or scabrous on only part of the surface, the prickles on the keels angled forward at the base but not curved; glume apices acute, rarely acuminate . . . . ***C. c.* var. *canadensis***

Calamagrostis canadensis (Michaux) P. Beauvois var. ***canadensis*** (p. 377)

Calamagrostis canadensis (Michx.) P. Beauv. var. *acuminata* Vasey ex Scribn., *Calamagrostis canadensis* (Michx.) P. Beauv. var. *pallida* (Vasey & Scribn. ex Vasey) Stebbins, *Calamagrostis canadensis* (Michx.) P. Beauv. var. *robusta* Vasey, *Calamagrostis scribneri* Beal

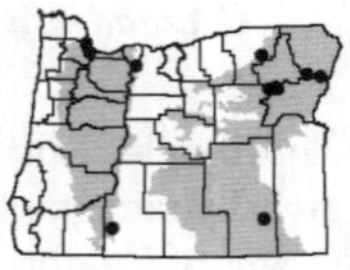

Ligules (1)4–6(12) mm; blades 2–5(8) mm wide. **Inflorescences** with branches 2.9–4.5(5.7) cm. **Spikelets** 2–4 mm. **Glumes** rounded to broadly keeled, smooth or scabrous; prickles on keel straight; tips usually acute, rarely acuminate. **Calluses** with hairs (1.7)2.5–2.9(3.1) mm. **Lemma awns** 0.9–2.6 mm.

400–2600 m. BR, BW, Casc, WV. CA, ID, NV, WA; throughout North America. Native.

Calamagrostis canadensis (Michaux) P. Beauvois var. ***langsdorffii*** (Link) Inman (p. 377)

Calamagrostis canadensis (Michx.) P. Beauv. var. *lactea* (Suksd. ex Beal) C.L. Hitchc., *Calamagrostis canadensis* (Michx.) P. Beauv. var. *scabra* (J. Presl) Hitchc.

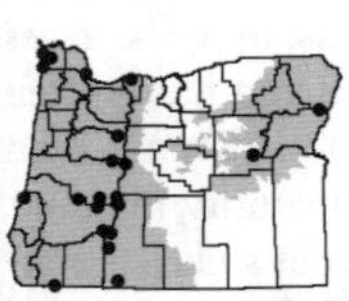

Ligules (3)5–8(12) mm; blades (2)3–7(11) mm wide. **Inflorescences** with branches (2.7)3.5–6(12) cm. **Spikelets** (3.5)4–4.5(5.2) mm. **Glumes** keeled scabrous throughout, rarely smooth, prickles on keels hair-like, often bent; tips acuminate. **Calluses** with hairs (1.5)3–3.3(4.5) mm. **Lemma awns** 1.2–3.1 mm.

0–2500 m. BW, Casc, CR, ECas, Est, Sisk, WV. CA, ID, WA; north to AK; east through Canada; Greenland. Native.

Calamagrostis howellii Vasey

Howell's reedgrass

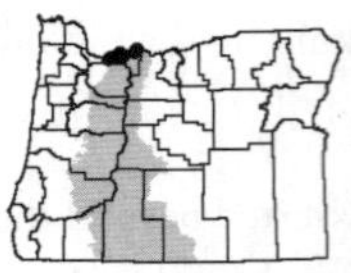

Plants (25)35–45(60) cm tall; densely cespitose. **Leaves** sheaths and collars smooth; blades (9)12–20(25) cm × 1–3 mm, flat to involute; lower surfaces smooth; upper surfaces finely scabrous, glabrous or sparsely pubescent. **Inflorescences** (5)7–12(15) × (2)3.5–6.5(8) cm, loose, open, straw-colored or green, rarely purplish; branches (2)3.5–5(7) cm, smooth or sparsely scabrous, spikelets usually confined to the distal half. **Glumes** (5.5)6–8 mm, rounded on the back to slightly keeled, acuminate. **Calluses** with hairs 2–3(4.5) mm, 40–60(67%) as long as the lemmas, abundant. **Lemmas** 4.5–5 mm, approximately 2 mm shorter than the glumes; lemma awns 10–16 mm, arising from the lower 20–40% of the lemmas, exserted, easily distinguished from the callus hairs, strongly bent. **Anthers** 2–3(4) mm.

Cliffs, bluffs, dry rocky slopes. 100–500 m. Casc, ECas. WA. Native.

Calamagrostis howellii has long awns, long tapered glumes, and an open inflorescence.

Calamagrostis koelerioides Vasey

Dense-pine reedgrass

Calamagrostis densa Vasey

Plants (26)60–85(120) cm tall, often densely cespitose; rhizomes 2–6 cm long, 2–4 mm thick. **Leaves** sheaths and collars usually scabrous, rarely smooth, glabrous; blades (2)9–20(30) cm × 2–4.5(8) mm, flat, slightly scabrous; upper surfaces glabrous or sparsely pubescent. **Inflorescences** (4)10–13(16) × 1–2(3) cm, contracted, erect to slightly nodding, straw-colored or pale green to pale purple; branches (1.1)2.8–4(6) cm, scabrous, spikelet-bearing to the base. **Glumes** 4–6(7) mm, slightly keeled. **Calluses** hairs 1.5–2 mm, 33–40% as long as the lemmas, sparse. **Lemmas** 3.5–5(6) mm, 0.5–1.5 mm shorter than the glumes; lemma awns 4–5.5 mm, attached to the lower portion of the lemma, equal to or longer than the glumes, stout, easily distinguished from the callus hairs, bent. **Anthers** 2–3.5 mm. **2*n***=28.

Montane meadows, conifer forest, talus slopes, dry hills, ridges, occasionally on serpentine soils. 400–2800 m. Sisk. CA, ID, WA; east to MT and WY. Native.

Calamagrostis koelerioides intergrades with *C. nutkaensis* in southwestern Oregon. *Calamagrostis nutkaensis* typically has broader leaves, longer callus hairs, shorter lemma awns, and shorter anthers.

Calamagrostis nutkaensis (J. Presl) Steudel

Pacific reedgrass

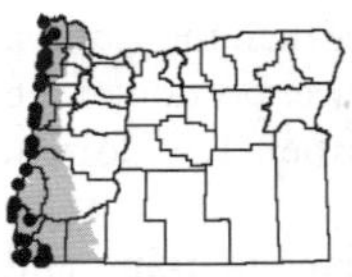

Plants (42)55–105(150) cm tall; densely cespitose; rhizomes shorter than 3(6) cm long, 1.5–3 mm thick. **Leaves** sheaths and collars smooth; blades (4)10–41(56) cm × (2)4–10(20) mm, flat, tough; lower surfaces smooth; upper surfaces smooth or slightly scabrous, glabrous or sparsely pubescent. **Inflorescences** (8)12–23(33) × (1.1)2–4.5(9) cm, contracted to somewhat loose, erect to slightly nodding, purple-tinged or sometimes green to straw-colored in shade-grown plants; branches 2.7–7(10.5) cm, sparsely scabrous, spikelets usually confined to the distal half. **Glumes** 4.5–6.5(8) mm, keeled. **Calluses** with hairs (1)2–3 mm, (20)50–67% as long as the lemmas, sparse. **Lemmas** (3)4–4.5(5) mm, (0.4)0.8–1.2(1.9) mm shorter than the glumes; lemma awns 1–3 mm, attached on the lower 33–50% of the lemma, not exserted, easily distinguished from the callus hairs, straight or slightly bent. **Anthers** (1)2.4–2.5(3.3) mm. **2*n***=28.

Wetlands, openings in conifer forests, beaches and dunes, cliffs, usually close to the ocean. 0–1100 m. CR, Est, Sisk. CA, WA; north to AK. Native.

Calamagrostis nutkaensis blends with *C. koelerioides* in more inland habitats in southwestern Oregon. In general, *C. nutkaensis* has wider leaves and shorter awns, but many southwestern Oregon plants are intermediate and cannot be identified clearly to either taxon.

Calamagrostis purpurascens R. Brown

Purple reedgrass

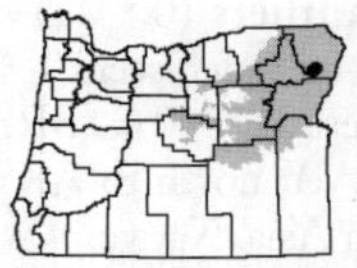

Plants 15–50 cm tall; densely cespitose; rhizomes 1–4 cm long, 1–2 mm thick. **Leaves** sheaths scabrous; collars usually scabrous or pubescent, rarely smooth; blades (4)5–17(30) cm × 2–5(6) mm, flat or involute, stiff; lower surfaces scabrous; upper surfaces usually densely short-pubescent, rarely sparsely pubescent. **Inflorescences** 4–13(15) × 0.9–2(2.8) cm, erect, condensed, often red- or purple-tinged; branches 1.3–3.5 cm, scabrous, with long almost hair-like prickles, spikelet-bearing to the base. **Glumes** (4.5)5.5–6.5(8) mm long, keeled. **Calluses** with hairs 0.9–1.5(2.4) mm, 20–40(67)% as long as the lemmas, sparse. **Lemmas** (3.5)4–4.5(5) mm, usually 1–2.5 mm shorter than (rarely equal to) the glumes; lemma awns (4.5)6–7(9) mm, attached to the lower portion of the lemma, usually exserted, stout, easily distinguished from the callus hairs, bent. **Anthers** (1.3)1.7–2.9 mm. **2*n***=28, 40, 58, 84.

Rocky alpine ridges and summits. 1200–3000 m. BW. CA, ID, NV, WA; throughout Canada and western US; eastern Asia. Native.

Calamagrostis purpurascens is a relatively short, neat plant with a dense, purplish inflorescence that grows at high elevations in the Wallowa Mountains. Its leaves, densely but inconspicuously pubescent on the upper surface, distinguish it from all other *Calamagrostis* species in Oregon.

Calamagrostis rubescens Buckley

Pinegrass

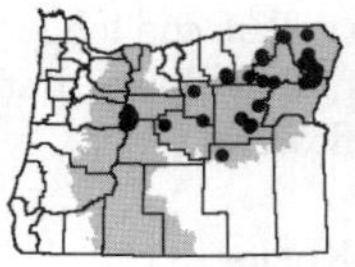

Plants (50)60–100(105) cm tall; rhizomatous, sometimes appearing loosely cespitose; rhizomes to 15 cm long or longer, 1.5–2 mm thick. **Leaves** sheaths smooth or slightly scabrous, rarely pubescent; collars usually

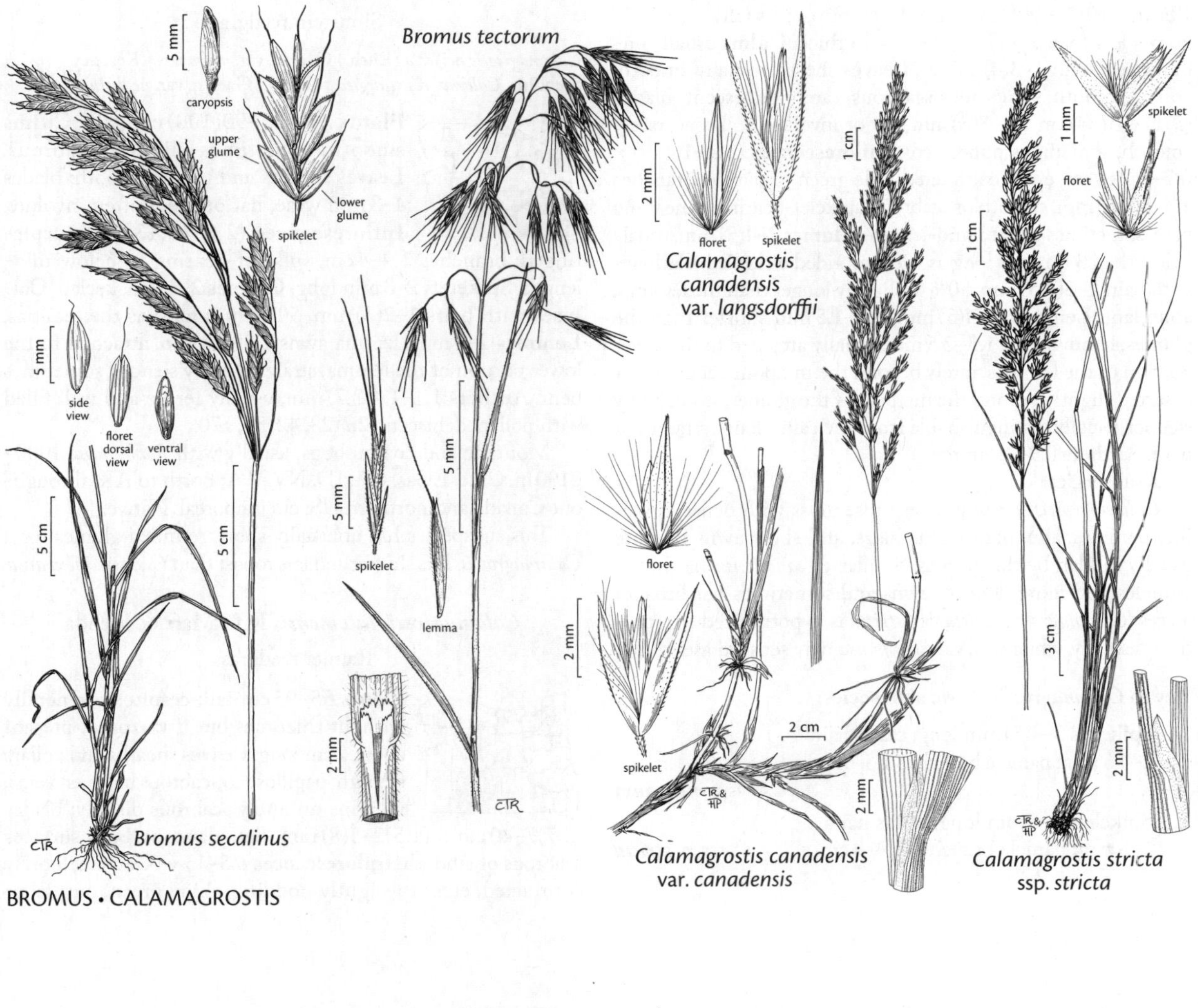

BROMUS • CALAMAGROSTIS

pubescent, rarely glabrous; blades (6)8–40(42) cm × (1)2–5(8) mm, usually flat; lower surfaces smooth or slightly scabrous; upper surfaces smooth, scabrous or sparsely pubescent. **Inflorescences** (5)6–15(25) × 1–2.5 cm, contracted to somewhat open, erect, usually greenish to straw-colored, uncommonly purplish; branches (1.2)2–3(6) cm, slightly scabrous, rarely densely long-scabrous, spikelet-bearing to the base. **Glumes** (3)4–4.5(5.5) mm, rounded to slightly keeled. **Calluses** with hairs 0.5–1.5(2.5) mm, 20–50(60)% as long as the lemmas, sparse. **Lemmas** 2.5–4 mm, 0.5–2 mm shorter than the glumes; lemma awns 2.8–3.5(4.5) mm, usually attached to the lower 20% of the lemma, rarely higher, exserted from the spikelet, stout, readily distinguished from the callus hairs, strongly bent. **Anthers** 1.2–2.6 mm. **2*n***=28, 42, 56.

Dry montane conifer forests. 800–2500 m. BW, Casc, ECas, Lava. CA, ID, NV, WA; north to British Columbia, northeast to Saskatchewan, southeast to CO. Native.

Calamagrostis rubescens is a common community dominant with *Carex geyeri* in dry montane conifer forests east of the Cascades. Though it is usually identified by the curly, spreading hairs on its leaf collars, some collars may be glabrous. Examining several leaves will usually reveal pubescent collars somewhere on the plant. Similar *C. koelerioides* has glabrous collars and longer lemmas and awns. Established populations rarely flower, but low-intensity forest fires stimulate vigorous flowering.

Calamagrostis stricta (Timm) Koeler

Slimstem reedgrass

Plants (10)35–90(120) cm tall; cespitose; usually with rhizomes less than 5 cm long, 1–1.5 mm thick. **Culms** usually unbranched, with 1–3(4) nodes. **Leaves** sheaths usually smooth; collars smooth, sometimes scabrous, rarely pubescent; blades (5)11–25(34) cm × 1–5(6) mm, flat or involute, scabrous, rarely smooth, sometimes puberulent. **Inflorescences** (2)4–18(29) × 0.7–2(2.8) cm, erect, contracted, pale green to purple; branches 1.4–5(9.5) mm, smooth or scabrous, spikelet-bearing to near the base, sometimes only to mid-length. **Glumes** 2–4(5) mm, usually less than 3 times as long as wide, rounded or keeled. **Calluses** with hairs 1–3(4.5) mm, 50% to slightly longer than the lemmas, abundant. **Lemmas** 2–4(5) mm, 0.1–1.5 mm shorter than the glumes; lemma awns 1.5–2.5 mm, usually attached to the lower portion of the lemmas, rarely beyond the midpoint, equaling or exserted slightly beyond the margins of the glumes, stout, rarely slender, usually distinguishable from the callus hairs, straight or bent. **Anthers** 0.9–2.4 mm.

2 subspecies.

Calamagrostis stricta is a cespitose grass with dense, spike-like panicles, abundant callus hairs, and short awns that are usually hidden by the glumes. Similar *C. canadensis* has looser panicles and more delicate awns and sometimes has broader leaves. *Calamagrostis stricta* var. *stricta* is hypothesized to reproduce sexually, while *C. s.* var. *inexpansa* may set seed asexually.

Key to *Calamagrostis stricta* subspecies:

1. Spikelets 3–4(5) mm long; callus hairs 2–4.5 mm; panicle branches 1.5–9.5 cm . . . . ***C. s.*** ssp. ***inexpansa***

1' Spikelets 2–3 mm long; callus hairs 1–3 mm; panicle branches 1.4–4 cm. . . . . . ***C. s.*** ssp. ***stricta***

Calamagrostis stricta (Timm) Koeler ssp. *inexpansa* (A. Gray) C.W. Greene

Northern reedgrass

Calamagrostis californica Kearney, *Calamagrostis crassiglumis* Thurb., *Calamagrostis inexpansa* A. Gray var. *barbulata* Kearney, *Calamagrostis inexpansa* A. Gray var. *brevior* (Vasey) Stebbins, *Calamagrostis inexpansa* A. Gray var. *inexpansa*, *Calamagrostis inexpansa* A. Gray var. *robusta* (Vasey) Stebbins

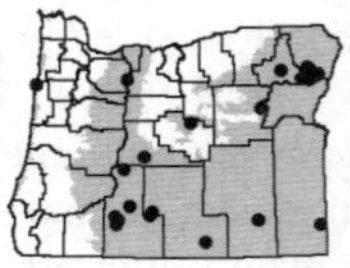

Plants (29)35–75(120) cm tall. **Culms** scabrous, sometimes smooth. **Leaves** sheaths usually smooth; collars smooth, sometimes scabrous, rarely pubescent; blades 1.5–5(6) mm wide, flat. **Inflorescences** (6)8–11(29) cm, pale green, sometimes purple-tinged; branches 1.5–5(9.5) cm, spikelet-bearing to the base. **Spikelets** 3–4(5) mm long. **Glumes** 3–4(5) mm, broadly keeled or rounded over the back. **Calluses** with hairs 2–3(4.5) mm, 50% to slightly longer than the lemmas. **Lemmas** 2.5–3.5(5) mm; lemma awns 2–2.5 mm, attached to the lower portion of the lemma, rarely above the midpoint, straight or bent. **Anthers** (0.9)1.5–1.8(2.4) mm, often poorly developed, sterile and indehiscent.

Diverse wetlands including alkaline sites. 0–2200 m. BR, BW, Casc, ECas, Est, Owy. CA, ID, NV, WA; north to AK, east to Newfoundland and NY; northeastern Asia. Native.

This subspecies is coarser and larger than *C. stricta* ssp. *stricta*.

Calamagrostis stricta (Timm) Koeler ssp. *stricta* (p. 377)

Slimstem reedgrass

Calamagrostis neglecta (Ehrh.) Gaertn. var. *micrantha* (Kearney) Stebbins, *Calamagrostis neglecta* (Ehrh.) Gaertn. var. *neglecta*

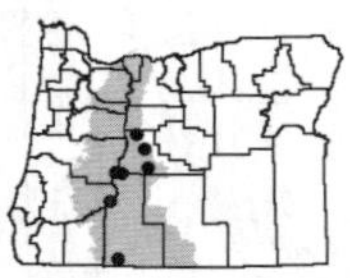

Plants (10)35–90(100) cm tall. **Culms** smooth, sometimes slightly scabrous. **Leaves** sheaths and collars smooth; blades 1–3 mm wide, flat or more often involute. **Inflorescences** (2)8–10(13) cm, purple-tinged; branches 1.4–4 cm, spikelet-bearing to below mid-length. **Spikelets** 2–3 mm long. **Glumes** 2–3 mm, keeled. **Calluses** with hairs 1–2(3) mm, 50% to as long as the lemmas. **Lemmas** 2–3 mm; lemma awns 1.5–2.5 mm, attached to the lower portion of the lemma, straight, rarely slender, sometimes bent. **Anthers** 1.1–1.4(2.7) mm, usually fertile and well filled with pollen, dehiscent. **2*n***=28, 42, 56, ±70.

Montane and coastal bogs, usually with *Sphagnum*. 1000–2100 m. Casc, ECas. CA, ID, NV, WA; north to AK; throughout Canada and northern US; circumboreal. Native.

This subspecies has unusually short, rounded glumes for a *Calamagrostis*. It is shorter and less robust than *C. s.* ssp. *inexpansa*.

Calamagrostis tacomensis K.L. Marr & Hebda

Rainier reedgrass

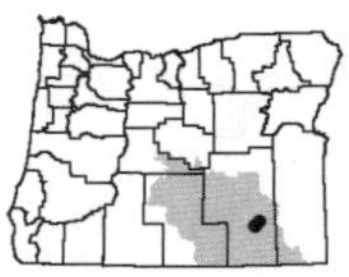

Plants 55–95 cm tall; cespitose; generally without rhizomes but if rhizomes present about 2 cm long. **Leaves** sheaths and collars smooth, papillose to scabrous between veins; margins minutely scabrous distally; blades (3.7)7–40 cm × (1.5)2–5(8) mm, flat, glaucous, both surfaces scabrous or smooth. **Inflorescences** 6.3–15 × 0.7–5 cm, loosely contracted, erect to slightly nodding, shiny green, purple, or brown; branches 1.5–6 cm, scabrous, spikelet-bearing to the base or nearly so. **Glumes** 3.2–5.8 mm, keeled, scabrous on keel, scabridulous elsewhere. **Calluses** with hairs 1.7–3 mm, about 50% as long as the lemma, abundant. **Lemmas** 3.1–3.7(4) mm, 0.5–1.5 mm shorter than the glumes; lemma awns 3.4–5.5 mm, attached to the lower portion of the lemma, exserted 0.5–2.3 mm beyond the glumes, easily distinguished from the callus hairs, strongly bent. **Anthers** (1)2–3.5 mm.

Montane to alpine slopes, along rocky streamsides. 1900–2700 m. BR. WA. Native.

Oregon plants treated here as *C. tacomensis* differ from typical members of the species in habit, foliage color, and spikelet traits. These plants will likely be described as a distinct species.

Calamagrostis tweedyi (Scribner) Scribner
Cascade reedgrass, Tweedy's reedgrass

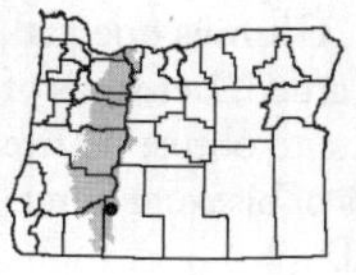

Plants (47)60–120(150) cm tall, lacking sterile culms; loosely cespitose; rhizomes 1–10 cm long, 2–4 mm thick. **Leaves** sheaths and collars smooth; blades (3)4–20(38) cm × (2)3–13 mm, flat; lower surfaces smooth; upper surfaces smooth or slightly scabrous, glabrous to sparsely to moderately pubescent on the veins. **Inflorescences** 7–16(19) × (1)1.5–2 cm, erect, usually contracted, pale to dark purple; branches (0.2)2.4–6.7(7.7) cm, smooth, sometimes sparsely scabrous distally, spikelet-bearing to the base. **Glumes** (4.5)5.5–8(9) mm, keeled. **Calluses** with hairs 0.8–1 mm, 20–33% as long as the lemmas, sparse. **Lemmas** 4–6.5(7.5) mm, 0–1.5 mm shorter than the glumes, minutely scabrous; lemma awns 6–11 mm, attached to the lower portion of the lemma, exserted more than 2 mm from the spikelet, stout, bent. **Anthers** 2–3.5 mm.

Montane to subalpine meadows and conifer forests. 1700–2100 m. Casc. WA. Native.

This uncommon *Calamagrostis* is recognized by its combination of broad leaves and long awns. Similar *C. howellii*, confined to the Columbia Gorge, has longer awns.

CATABROSA Whorlgrass

Plants perennial, often stoloniferous. **Leaves** sheaths closed; ligules membranous; blades flat. **Inflorescences** open panicles; disarticulation above the glumes and beneath the florets. **Spikelets** pedicellate, laterally compressed to terete, with (1)2(3) florets, the lower spikelet sessile and the upper a more or less elongated rachilla internode. **Glumes** unequal, much shorter than the lemmas, veinless or the upper with 1 vein at the base, rounded to truncate and often erose, awnless. **Calluses** short, blunt, glabrous. **Lemmas** glabrous, conspicuously 3-veined, rounded over the midvein; tips rounded to truncate, erose to scarious, awnless. **Caryopses** concealed at maturity.

North America, South America, and Eurasia. 2 species; 1 species treated in *Flora*.

Catabrosa aquatica (Linnaeus) P. Beauvois (p. 381)
Brookgrass, water whorlgrass

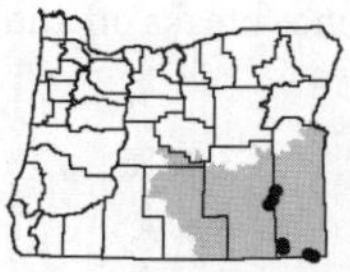

Plants 10–60 cm tall, often stoloniferous. **Leaves** sheaths closed half their length, glabrous; ligules 1–8 mm; blades (1)3–15(20) cm × 2–22 mm, flat, glabrous. **Inflorescences** 3–35 cm × (1)2–10(12) mm, often partially hidden in uppermost leaf sheath, with 3 to many very unequal branches per node. **Spikelets** 1.5–3.5(4) mm, with (1)2(3) florets, lowest floret sessile, next floret on an elongated rachilla internode. **Glumes** much shorter than the lemmas; lower glumes 0.7–1.3 mm, veinless; upper glumes 1.2–2.2 mm, 0–1-veined, truncate to notched. **Lemmas** 2–3 mm, prominently 3-veined, glabrous, rounded to truncate. **Paleas** similar in color and texture to the lemmas, often angled away from the lemma so that the floret gapes open. **Anthers** 2–3 mm.

In and adjacent to creeks and springs. 1400–2200 m. BR, Owy. ID, NV, WA; north to AK, northeast to Greenland, southeast to NM; South America; Eurasia. Native.

From a distance, this grass looks like a *Glyceria*, but a closer look reveals truncate, spreading glumes much shorter than the lemmas. The lower floret is nearly sessile, but the upper is raised on a long rachilla internode. The paleas look like additional lemmas and are usually gaping wide open.

CENCHRUS Sandbur

Plants annual or perennial. **Leaves** sheaths open; ligules membranous, ciliate; blades flat or folded; margins cartilaginous. **Inflorescences** spike-like panicles of highly reduced branches called fascicles or burs which consist of sharp, retrorsely scabridulous bristles surrounding and sometimes almost concealing 1–4 spikelets; outer bristles terete or flattened; inner bristles usually strongly flattened, fused at least at the base. **Spikelets** sessile, dorsiventrally compressed, with 2 florets, the lower floret sterile and the upper bisexual; disarticulation at base of the fascicles. **Glumes** 2; lower glumes ovate, scarious and 1-veined, acute to acuminate; upper glumes ovate, 3–9-veined. **Lemmas** like upper glumes, awnless. **Paleas** as long as lemmas. **Anthers** 3.

Temperate and tropical regions of the world. 16 species; 1 species treated in *Flora*.

Sandburs are mostly weedy plants of grasslands, open forests, and disturbed areas, usually on sandy soil. The spines of sandburs are modified inflorescence branches, homologous to the bristles below the spikelets of *Setaria* and *Pennisetum*. *Cenchrus spinifex* was reported from Portland in the early 1900s, probably due to misidentifications, but the relevant specimens have not been found. It has shorter spikelets than *C. longispinus* with fewer bristles, wider inner bristles, and flattened (rather than terete) outer bristles.

Cenchrus longispinus (Hackel) Fernald (p. 381)

Longspine sandbur

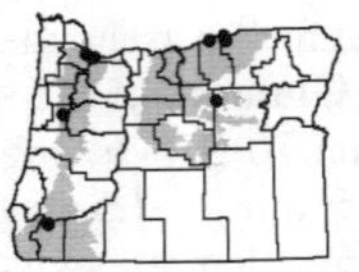

Plants annual, 20–90 cm tall. **Leaves** sheaths keeled and laterally compressed; blades 4–27 cm × 1.5–5(7.5) mm; upper surface scabrous or with sparse hairs. **Inflorescences** spike-like panicles, 1.5–8(10) cm × 10–20 mm, with spikelets enclosed in bur-like fascicles of hard, often spine-tipped branches. **Fascicles** 8.3–11.9 × 3.5–6 mm, more or less globose, pubescent, with 45–75 bristles, the longer bristles 3.5–7 mm. **Spikelets** 2–3(4) per fascicle, (4)5.8–7.8 mm. **Glumes** lower glumes 0.8–3 mm; upper glumes 4–6 mm; lower lemmas 4–6.5 mm; upper lemmas 4–7.6 mm. **Anthers** of lower florets 1.5–2 mm; anthers of upper florets 0.7–1 mm and poorly developed. **2*n***=34.

Disturbed sandy soil. 0–600 m. Col, Lava, Sisk, WV. CA, ID, NV, WA; north to British Columbia, east to New Brunswick and FL, south to Venezuela. Native.

The round, painfully spiny burs make this grass unmistakable. These plants flower and set seed late in the growing season.

CHLORIS Windmill grass

Plants annual or perennial; cespitose. **Culms** with pith-filled internodes. **Leaves** sheaths open, strongly keeled; ligules membranous, erose to lacerate or ciliate; blades often with long, coarse hairs near the base of the upper surface and margins. **Inflorescences** terminal panicles with (1)5–30 spike-like branches borne in 1–3 whorls, sometimes with an isolated branch below. **Spikelets** laterally compressed, with 1–3(5) florets, with a sessile fertile floret and a staminate or sterile floret on a short stalk. **Glumes** unequal, exceeded by the florets, lanceolate, acute to acuminate, usually awnless, awns, if present, to 0.3 mm. **Calluses** bearded. **Lemmas** of bisexual florets 3-veined; marginal veins pubescent; midveins usually glabrous, sometimes scabrous; tips truncate to obtuse, entire to bilobed; midveins and sometimes the lemma lobes extending into an awn, sometimes mucronate; lemma awns of bisexual florets to 37 mm (central awns) or to 0.6 mm (awns on lemma lobes). **Anthers** 3. **Caryopses** ovoid, elliptic or obovoid.

Mainly tropical to subtropical regions. Approximately 60 species; 1 species treated in *Flora.*

Chloris species resemble *Cynodon, Eleusine*, and, to a lesser extent, *Digitaria. Chloris radiata* was collected in Portland in 1915. This plant has wider leaves (to 10 mm wide) and smaller floral parts (second florets 0.4–1.7 mm; glumes 0.7–2.7 mm; lower lemmas 0.4–0.6 mm wide) than *C. verticillata.*

Chloris verticillata Nuttall (p. 381)

Tumble windmill grass

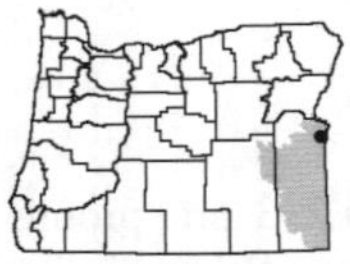

Plants perennial, 14–40 cm tall; cespitose. **Leaves** sheaths mostly glabrous, with hairs to 3 mm adjacent to the ligule; blades to 15 cm × 2–3 mm, with basal hairs, otherwise glabrous. **Inflorescences** with 10–16 branches in several well-separated whorls plus a solitary vertical branch at the tip; branches 5–15 cm; disarticulation at the upper cauline node, panicles falling intact. **Glumes** lower glumes 2–3 mm; upper glumes 2.8–3.5 mm. **Lemmas** lower lemmas 2–3.5 × 1.5–1.9 mm, elliptic to lanceolate; keels glabrous or appressed pubescent, sides not conspicuously grooved; margins glabrous or appressed pubescent; tips acute to obtuse; lower lemma awns 4.8–9 mm; upper lemmas 1.1–2.3 mm, oblong, somewhat inflated, truncate to inconspicuously bilobed, sterile; upper lemma awns 3.2–7 mm. **Caryopses** 1.3–1.5 × 0.5 mm. **2*n***=28, 40, 63.

Disturbed areas, lawns, sandy or gravelly soils. 500–900 m. Owy. CA, NV; throughout most of US. Exotic.

In 1994, a population of *C. verticillata* was found in Vale, Oregon. It has been expanding. The inflorescence breaks off and blows in the wind like a tumbleweed.

CINNA Woodreed

Plants perennial; cespitose, sometimes rhizomatous. **Culms** often rooting at the lower nodes, usually glabrous. **Leaves** cauline; sheaths open, glabrous; ligules membranous; blades flat, smooth or scabrous. **Inflorescences** panicles with spreading to ascending branches; disarticulation below the glumes. **Spikelets** laterally compressed, with 1 floret, rarely with a second rudimentary or fertile floret, rachilla usually prolonged beyond the uppermost floret. **Glumes** 2, slightly shorter to slightly longer than the floret, 1- or 3-veined; margins hyaline, keeled, the keels scabrous; tips acute, sometimes minutely awn-tipped. **Calluses** short, glabrous. **Lemmas** 3- or 5-veined, sometimes faintly so; tips acute, minutely bifid, usually awned; lemma awns subterminal. **Anthers** 1 or 2.

Circumboreal and South America. 4 species; 1 species treated in *Flora.*

Cinna latifolia (Treviranus ex Göppert) Grisebach (p. 381)

Slender woodreed

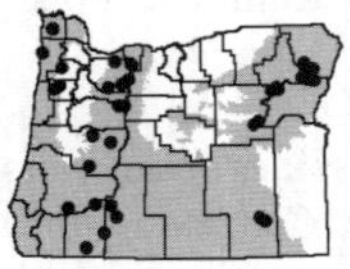

Plants 20–190 cm tall; loosely cespitose; sometimes rhizomatous. **Leaves** sheaths glabrous; ligules 2–8 mm, pubescent; blades to 28 cm × 1–20 mm, flat, scabrous. **Inflorescences** 3–46 × 0.5–20 cm, nodding, sometimes partially hidden in the uppermost leaf sheath; branches spreading, sometimes ascending. **Spikelets** 2–4.7 mm, with 1(2) floret; rachillas prolonged 0–1.3 mm beyond the floret. **Glumes** with scabrous keels and sometimes backs; tips narrowly acute; lower glumes (1.8)2.5–4(4.7) mm, 1-veined; upper glumes 1.9–4(5) mm, 3-veined. **Lemmas** 1.8–3.8 mm, 3-veined, midveins scabrous, glabrous elsewhere; tips acute, minutely bifid, usually awned; lemma awns 0.1–2.5 mm, arising from sinus of lemma apex, straight or curved. **Anthers** 0.4–1 mm. **Caryopses** 1.8–2.8 mm. **2*n***=28.

Moist woods, wooded wetlands, and shady streamsides. 100–2500 m. BR, BW, Casc, CR, ECas, Est, Sisk. Throughout Canada and western and northern US; circumboreal. Native.

Cinna latifolia resembles a woodland *Agrostis* except that its spikelets disarticulate below the glumes, leaving bare branch tips.

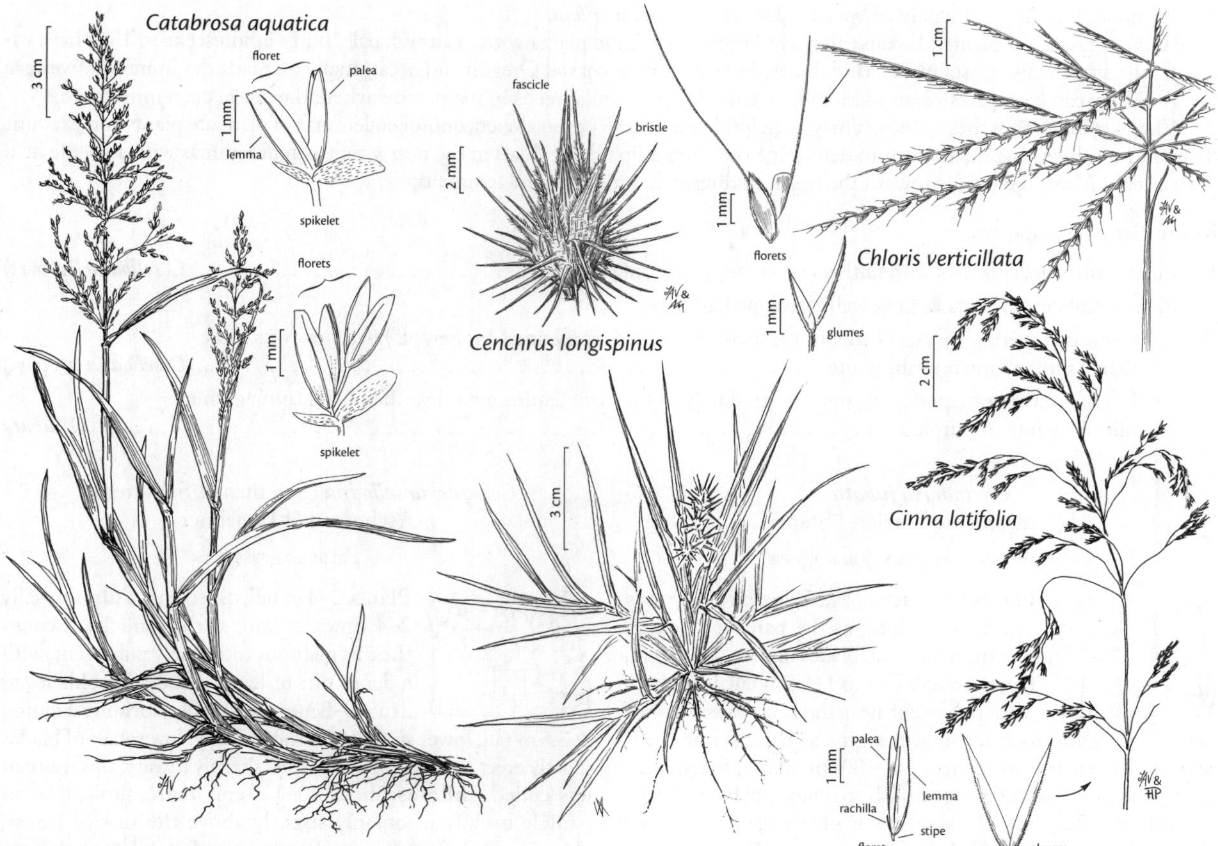

CATABROSA • CENCHRUS • CHLORIS • CINNA

COLEANTHUS Mud grass

Plants annual, tufted. **Culms** ascending or decumbent, up to 10 cm tall. **Leaves** cauline; sheaths closed, inflated; ligules membranous; blades folded. **Inflorescences** clusters of umbel-like panicles; disarticulation below the floret. **Spikelets** slightly laterally compressed, with 1 floret, pedicellate, pedicels pubescent. **Glumes** absent. **Calluses** short, glabrous. **Lemmas** 1-veined, narrowed to awn-like tips. **Anthers** 2. **Caryopses** elliptic, exceeding the lemma and palea, falling free of them.

North America and Eurasia. 1 species.

Coleanthus subtilis (Trattinnick) Seidl (p. 383)
Mud grass

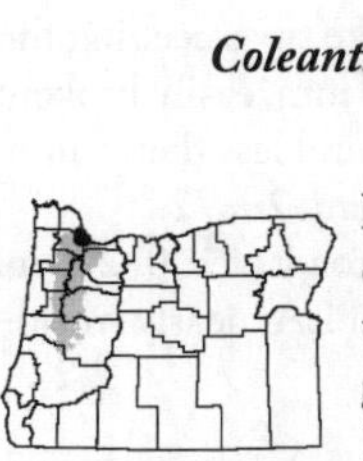

Plants 2–7(10) cm tall. **Culms** decumbent or ascending. **Leaves** sheaths inflated; ligules 1–1.5 mm; blades 1–2 cm × 0.5–1.5 mm, glabrous. **Inflorescences** 1–5 cm × 0.3–1 mm panicles with umbel-like clusters of branches. **Spikelets** 0.75–1.5 mm with 1 floret. **Glumes** absent. **Lemmas** 0.75–1.5 mm, 1-veined; awn-like tips about equaling the lower portions. **Paleas** about 0.5 mm; keels slightly prolonged. **Anthers** about 0.3 mm. **Caryopses** about 1 mm, longer than the lemma when mature.

Disturbed, wet, muddy to sandy sites, usually on receding shores of lakes and rivers. 0–50 m. WV. WA; north to British Columbia; Eurasia. Native.

This tiny pioneer species is more likely to be confused with an annual *Juncus* than with any other grass species. It is easy to overlook not only because it is small but also because it flowers in September through November and sometimes again in April or May.

CORTADERIA Pampas grass

Plants perennial, often dioecious or monoecious; densely cespitose. **Culms** 2–7 m tall, erect. **Leaves** mainly basal; sheaths open, glabrous or pubescent; margins overlapping; ligules of hairs; blades flat to folded, arching, edges usually sharply serrate. **Inflorescences** plumose panicles, 30–130 cm, subtended by a long, ciliate bract; disarticulation above the glumes and below the florets. **Spikelets** somewhat laterally compressed, usually unisexual but sometimes bisexual, 2–9 florets. **Glumes** unequal, nearly as long as the spikelets, hyaline, 1-veined. **Calluses** pilose. **Lemmas** 3–5(7)-veined, long-acuminate, bifid and awned or entire and mucronate; lemmas of pistillate and bisexual florets with long, silky hairs, lemmas of staminate florets less hairy or glabrous; lemma awns straight. **Paleas** about half as long as the lemmas, 3-veined. **Anthers** of staminate or bisexual florets 3, 1.5–6 mm, those of pistillate florets smaller or absent. **Caryopses** 1.5–3 mm.

Cosmopolitan. Approximately 25 species; 2 species treated in *Flora*.

Pampas grasses are planted because they are impressively large plants with beautiful, pale, fluffy inflorescences. They have escaped into disturbed sites, stabilized sand dunes, and estuaries in coastal Oregon and occasionally on roadsides in interior western Oregon. They can be aggressive invaders and are considered noxious weeds in many areas where they have been introduced.

When keying within these genus, it may be helpful to use fresh or moistened, unbroken lemmas of pistillate plants and examine them under a dissecting microscope to determine the lemma tip's shape. Flatten the nearly transparent lemmas with a needle held horizontally. Move the needle toward the tip, unfolding it. Examine several lemma tips.

Key to *Cortaderia* species:

1. Plant staminate; spikelets with anthers unless they have fallen out . . . . **C. selloana** (in part)

1' Plant pistillate; spikelets lacking well-developed anthers.

  2. Lemmas tapering very gradually to the awn (lemma tips sometimes broken off); lemma awns more than 1 mm; glume midribs white . . . . **C. selloana** (in part)

  2' Lemmas acute or tapering abruptly or gradually to the awn; lemma awns less than 0.5(1) mm; glume midribs white or purple . . . . ***C. jubata***

Cortaderia jubata (Lemoine ex Carrière) Stapf (p. 383)

Purple pampas grass, jubatagrass

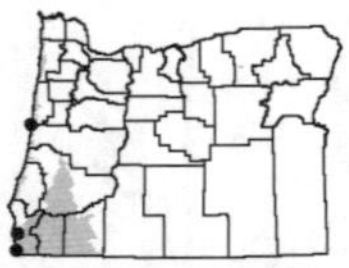

Plants 1.5–7 m tall; densely cespitose. **Culms** 4–7 times as long as the panicles. **Leaves** sheaths pubescent; blades to 2 m × 2–10 mm, flat, forming a clump 1–1.5 m tall; lower surfaces pubescent near the base; lower half of blades ascending to spreading; margins sharply serrate; tips scarcely coiled. **Inflorescences** 30–100 cm, dense, fluffy, deep violet when young, becoming pinkish to dingy tan, base of panicle usually 1–2 m above the main clump of leaves; branches usually purplish or dark. **Spikelets** 14–16 mm, pistillate, with 3–5 readily disarticulating florets. **Glumes** usually with white or purple midribs. **Calluses** about 0.6 mm. **Lemmas** 10–13 mm including the awn, with hairs 6–10 mm; tip usually acute or with a definite short taper to the awn, occasionally tapered gradually to the awn, tapered tip equal to or up to 5 mm longer than the lemma hairs; lemma awns less than 0.5(1) mm. **Stigmas** usually not exserted. **Anthers** absent. **Caryopses** to 2.5 mm. **2*n*=108.**

Grasslands, sand dunes, clearcuts and roadsides. 0–600 m. Est, Sisk. CA, WA; South America, Australia. Exotic.

The two species of pampas grass introduced to Oregon can be difficult to distinguish. Often, *C. selloana* can be recognized by its long panicles that usually have their lower edge in or just above the mass of leaves. *Cortaderia jubata* panicles are held well above the mass of leaves. However, this difference in stature can be obscured in depauperate plants. The purple midveins of *Cortaderia jubata* glumes, if present, permit confident identification. In *C. jubata*, the lemma tip is usually acute or tapers abruptly to an awn. The awn, if present, is short. In *C. selloana*, the lemma tapers so gradually to the awn that it is hard to tell where the awn begins. The awn is parallel-sided, scabrous on both sides, and longer than the awn of *C. jubata*.

Cortaderia selloana (Schultes & Schultes f.) Ascherson & Graebner

Pampas grass

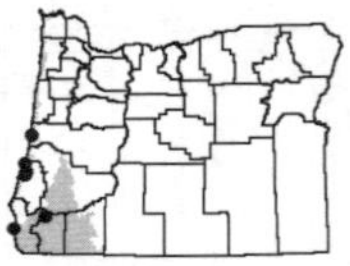

Plants 2–4 m tall, dioecious. **Culms** usually 2–4 times as long as the panicles. **Leaves** sheaths glabrous to sparsely pubescent, with a dense tuft of hairs at the collar; blades to 2 m × 3–8 mm, mostly flat, forming a clump 2–3.5 m tall; lower surface glabrous basally; lower half of blades mostly erect to ascending; margins sharply serrate; tips more or less coiled. **Inflorescences** 30–130 cm, dense, fluffy, base of panicle usually in or only slightly above the tuft of leaves, gleaming white to cream-colored in pistillate plants, white to violet when young, becoming white to pinkish tan in staminate plants, base of panicle 0–1 m above the main tuft of leaves in pistillate plants or 0–2 m above the main tuft of leaves in staminate plants; branches greenish or straw-colored. **Spikelets** 15–17 mm, with 5–7 florets in pistillate plants or with 3–5 florets in staminate plants, florets not easily separated from the rachilla. **Glumes** white, those of staminate plants sometimes purplish at the base with white midribs. **Calluses** to 1 mm, with hairs to 2 mm. **Lemmas** pistillate plants 15–20 mm, with hairs 6–10 mm, tapering gradually to a scabrous awn, awn-like tip exceeding the lemma hairs by 5–12 mm; lemma awns 2–5 mm, easily broken; staminate plants 10–13 mm, with hairs few and less than 5 mm. **Stigmas** exserted. **Anthers** sometimes present. **2*n*=72.**

Grasslands, sand dunes, clearcuts, and roadsides. 0–200 m. Est, Sisk. CA, NV, WA, southeast to SC and LA; nearly worldwide. Exotic.

See comments for *C. jubata*.

CRYPSIS Pricklegrass

Plants annual. **Culms** sometimes branching above the base, nodes usually exposed. **Leaves** sheaths open, often inflated; ligules of hairs; blades often disarticulating. **Inflorescences** terminal and sometimes axillary, dense, spike-like or head-like panicles often partially enclosed in 1 or more of the uppermost leaf sheaths; disarticulation above or below the glumes. **Spikelets** strongly laterally compressed, oval in outline, 1 floret. **Glumes** 1-veined, strongly keeled. **Lemmas** 1-veined, glabrous, strongly keeled, awnless but sometimes mucronate. **Anthers** 3. **Caryopses** oblong, pericarp loosely enclosing the seed and easily removed when wet.

North America, Eurasia, and Africa. 8 species; 2 species treated in *Flora*.

Crypsis species vary greatly in height, leaf length and color depending on the available moisture. *Alopecurus* species have superficially similar congested panicles and live in similar wet habitats but differ in having membranous ligules, 3-veined glumes and 3-veined, awned lemmas.

Key to *Crypsis* species:

1. Spikelets 1.5–2.8 mm; panicles 7–8 times longer than wide, usually completely exserted from the uppermost sheath at maturity. . . . . *C. alopecuroides*

1' Spikelets 2.5–3.5 mm; panicles 1–5 times longer than wide, the bases usually enclosed in the uppermost sheath at maturity. . . . . *C. schoenoides*

Crypsis alopecuroides (Piller & Mitterpacher) Schrader

(p. 383)

Foxtail pricklegrass

Heleochloa alopecuroides (Piller & Mitterp.) Host

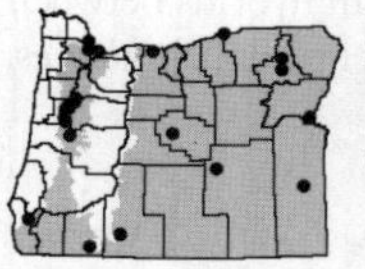

Plants (3)5–75 cm tall. **Culms** rarely branched above the base. **Leaves** sheaths glabrous; collars glabrous; blades 5–12 cm × 1.2–2.5 mm. **Inflorescences** spike-like, 1.5–6.5 × 0.3–0.6 cm, 7–8 times longer than wide, dense, often purplish, completely exserted from the uppermost leaf sheath on peduncles at least 1 cm; disarticulation below the glumes. **Spikelets** 1.5–2.8 mm; lower glumes 1.2–2 mm; upper glumes 1.4–2.4 mm. **Lemmas** 1.7–2.5 mm. **Anthers** 0.5–0.6 mm. **Caryopses** 0.9–1.1 mm. **2*n*=16.**

Sandy, drying lakeshores and streamsides. 0–1400 m. BR, BW, Col, ECas, Lava, Owy, Sisk, WV. CA, ID, NV, WA; east to MT and CO; North Africa, western Asia, Europe. Exotic.

Larger, greener *C. alopecuroides* individuals resemble an *Alopecurus,* as the name suggests. *Alopecurus* species differ in having membranous ligules and awned lemmas. *Crypsis schoenoides* resembles *C. alopecuroides* but has shorter, thicker inflorescences that are usually not fully exserted from the leaf sheaths, and its spikelets are larger. The very dark pigment in the inflorescences helps the plants in two ways. It absorbs heat, speeding the maturation of the flower and seed, and it protects against ultraviolet light. *Crypsis alopecuroides* plants develop the dark color when growing on open lake margins, while plants growing along shady streamsides have green and white inflorescences.

Crypsis schoenoides (Linnaeus) Lamarck

Swamp pricklegrass

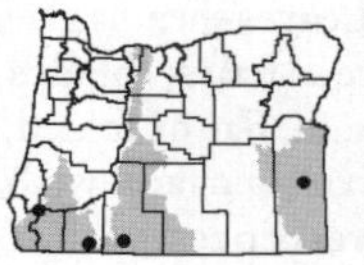

Plants 2–75 cm tall. **Culms** usually not branching above the base, but sometimes profusely branched. **Leaves** sheaths glabrous or with hairs at the throat, often inflated; blades 2–10 cm × 2–6 mm. **Inflorescences**

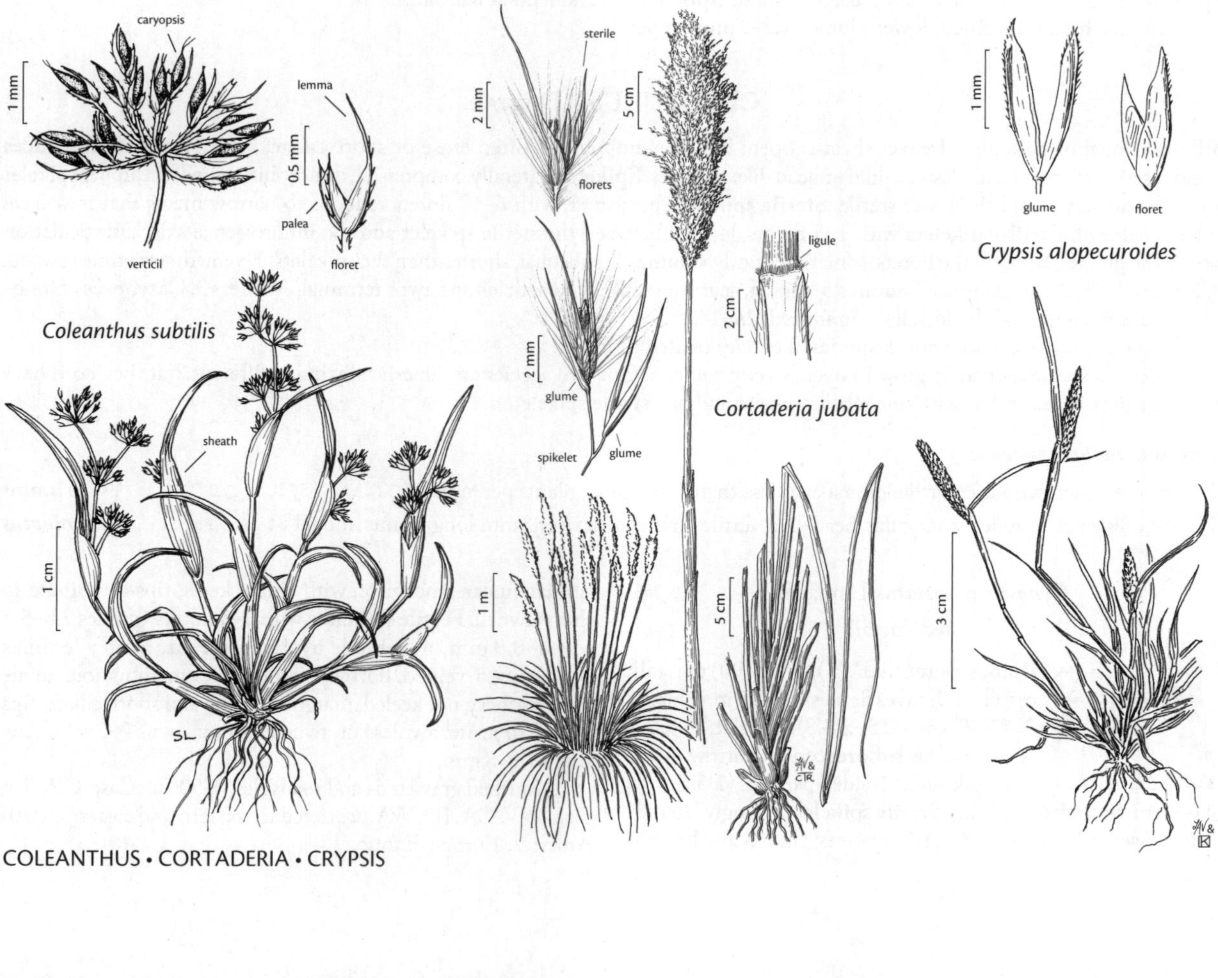

COLEANTHUS • CORTADERIA • CRYPSIS

spike-like, 0.3–4(7.5) × 0.5–0.7(1.5) cm, 1–5 times longer than wide, dense, bases usually enclosed in the uppermost sheaths at maturity. **Spikelets** 2.5–3.5 mm; lower glumes 1.8–2.3 mm; upper glumes 2.2–2.7 mm. **Lemmas** 2.4–3 mm. **Anthers** 0.7–1.1 mm. **Caryopses** about 1.3 mm. **2*n***=32.

Clay or sandy soils of drying lake margins and vernal pools. 100–1300 m. ECas, Owy, Sisk. CA, NV; southeast to UT and AZ, scattered in eastern US; northern Africa and southern Europe to India. Exotic.

See comments for *C. alopecuroides*.

CYNODON Bermudagrass

Plants perennial; stoloniferous and often also rhizomatous, often turf-forming. **Leaves** sheaths open; ligules of hairs or membranous; blades flat, folded, or involute. **Inflorescences** panicles with spike-like branches attached at approximately the same point like spokes of an umbrella; branches 1-sided, with 2 rows of solitary, appressed, overlapping spikelets. **Spikelets** laterally compressed, with 1(3) florets (if more than one, only the lowest fertile); disarticulation above the glumes. **Glumes** shorter than the lemmas, keeled, 1-veined or the upper sometimes 3-veined, sometimes short-awned. **Lemmas** keeled, sometimes winged; tip pointed. **Anthers** 3.

Warmer regions worldwide. 9 species; 1 species treated in *Flora*.

Cynodon species are widely planted as lawn and pasture grasses. Some of the cultivated forms are derived from hybrids between *C. dactylon* and *C. transvaalensis*. These hybrids are smaller plants than *C. dactylon* with narrower leaf blades, fewer panicle branches, and anthers that are indehiscent at maturity.

Cynodon dactylon (Linnaeus) Persoon (p. 385)

Bermudagrass

Cynodon dactylon (L.) Pers. var. *aridus* J.R. Harlan & de Wet, *Cynodon dactylon* (L.) Pers. var. *dactylon*

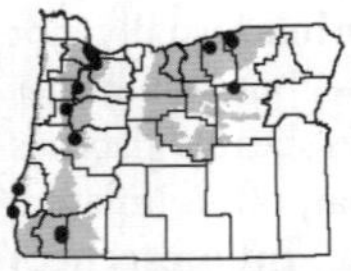

Plants 5–40(50) cm tall. **Leaves** sheaths glabrous or with scattered hairs; blades 1–6(16) cm × (1)2–4(5) mm, flat or folded, glabrous or upper adaxial surface with spreading hairs. **Inflorescences** with 1 whorl of (3)4–7(9) spike-like branches 1.4–6 cm × 1–2 mm; spikelets in 2 rows on one side of each branch. **Spikelets** 2–3.2 mm. **Glumes** 1-veined; lower glumes 1.2–2 mm; upper glumes 1.4–2.3 mm. **Lemmas** 1.9–3.1 mm, 3-veined, usually densely pubescent on the keel, sometimes pubescent on the margins, awnless. **Anthers** about 1 mm, dehiscent. **2*n***=36.

Lawns, waste areas in urban areas and sandbars on rivers. 0–1200 m. Col, Est, Lava, Sisk, WV. CA, ID, NV, WA; throughout most of North America; worldwide. Exotic.

This grass is commonly planted for forage, lawns, and erosion control in warm climates. It readily escapes, usually in disturbed urban settings and along transportation corridors but sometimes into more natural habitats. *Cynodon dactylon* is highly diverse in inflorescence branch length, branch number, and lemma hairiness.

CYNOSURUS Dogtail

Plants annual or perennial. **Leaves** sheaths open; ligules membranous, often erose or short-ciliate; blades flat. **Inflorescences** condensed, terminal, 1-sided, spike-like or head-like panicles. **Spikelets** laterally compressed, dimorphic, paired, the upper spikelet in each pair fertile and the lower sterile. **Sterile spikelets** persistent, with 6–18 florets reduced to narrow bracts that may have a terminal awn. **Fertile spikelets** with 1–5 florets, located between the sterile spikelet and the inflorescence axis; disarticulation above the glumes, beneath the florets, sometimes tardy. **Glumes** 2, subequal, shorter than the spikelets, 1-veined, sometimes awned. **Calluses** short, blunt, glabrous. **Lemmas** 5-veined, acute, sometimes awned; lemma awns terminal. **Anthers** 3. **Caryopses** oblong-ellipsoid, subterete, slightly dorsally compressed. **2*n***=14.

Temperate regions worldwide. 8 species; 2 species treated in *Flora*.

Cynosurus species generally grow in open, weedy habitats. Our two species are superficially quite different, but they both have spikelets in pairs, each pair with one fertile spikelet and one sterile spikelet.

Key to *Cynosurus* species:

1. Inflorescence cylindric; fertile lemma awn less than 3 mm long; plants perennial . . . . *C. cristatus*

1' Inflorescence rounded-triangular, head-like; fertile lemma awn over 3 mm long; plants annual . . . . *C. echinatus*

Cynosurus cristatus Linnaeus (p. 385)

Crested dogtail

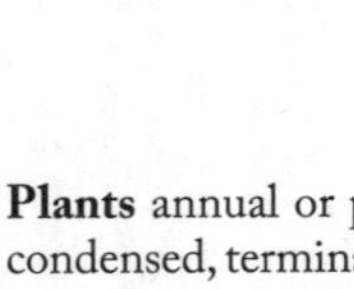

Plants perennial, (5)15–75(90) cm tall; cespitose. **Leaves** ligules 0.5–2.5 mm; blades 3–15(19) cm × 0.5–2(4.3) mm, flat, glabrous or pubescent. **Inflorescences** narrow, cylindric, spike-like, 1-sided panicles (1)3.5–14 × 0.4–1 cm. **Spikelets** 3–7 mm. **Sterile spikelets** strongly laterally compressed, fan-like, with 6–11(18) florets; glumes and lemmas similar, linear-lanceolate, with ciliate keels; tips acuminate to short-awned. **Fertile spikelets** with 2–5 florets; glumes 2.8–5.1 × 0.6–0.9 mm, 1-veined, hyaline, keeled, acute; lemmas 3–4.5 mm, 5-veined, dorsally compressed, membranous to almost leathery, not keeled; margins hyaline and short-ciliate; tips obtuse to acute, awnless or awned; lemma awns 0–3 mm. **Anthers** 1.8–3 mm.

Disturbed grasslands and roadsides. 0–900 m. Casc, CR, Est, Sisk, WV. CA, ID, WA; scattered in western and eastern North America; Europe. Exotic.

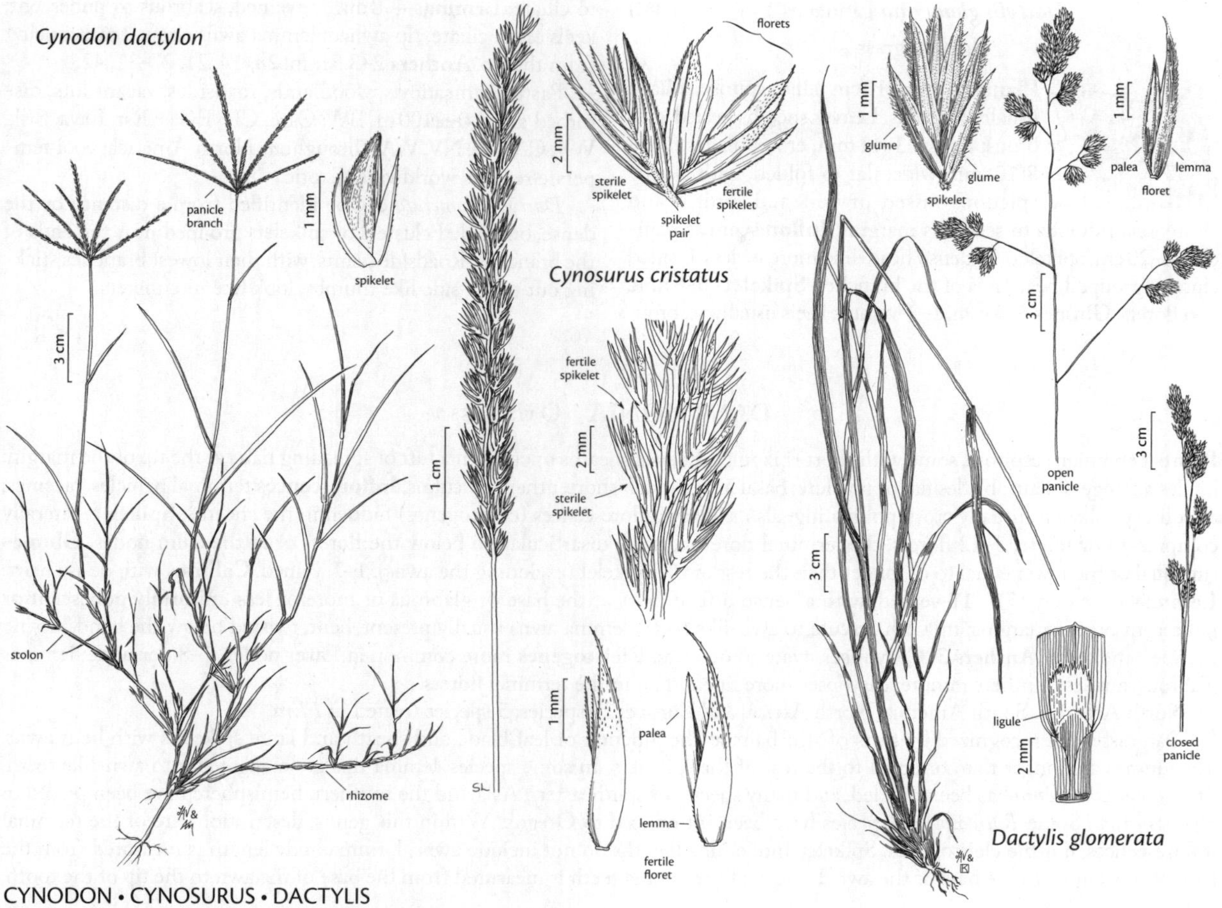

CYNODON • CYNOSURUS • DACTYLIS

This species is distinguished from Oregon's other grasses with spike-like panicles by the one-sided inflorescence with a wavy outline and the fan-like sterile spikelets closed protectively around the fertile spikelets.

Cynosurus echinatus Linnaeus

Bristly dogtail

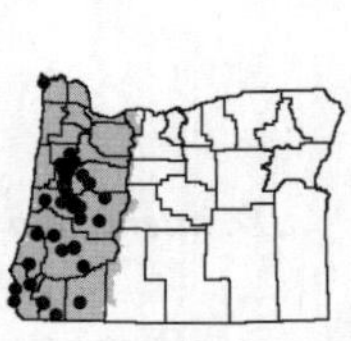

Plants annual, 9–70(180) cm tall, tufted. **Leaves** ligules 2.5–5(10) mm; blades 3–13.5(23) cm × 2.5–14 mm, flat, scabrous. **Inflorescences** dense, round-triangular, head-like, 1-sided panicles 1–4(8) × 0.7–2 cm. **Spikelets** 7–14 mm. **Sterile spikelets** strongly laterally compressed, 6–18 florets; glumes and lemmas similar, narrowly subulate to linear-lanceolate, almost awn-like throughout; awns to 8 mm. **Fertile spikelets** with 1–5 florets; glumes 5.5–12 × 0.4–0.9 mm, laterally compressed, white-hyaline, glabrous, 1-veined, keeled, awned; glume awns 0.5–2.2 mm; lemmas 4–7 mm, slightly dorsally compressed, chartaceous or leathery, not keeled; tips hyaline, short-ciliate, entire or bidentate, awned; lemma awns 5–18(25) mm. **Anthers** 1–4 mm.

Open, disturbed habitats, including grasslands, roadsides, and forest margins. 0–1600 m. Casc, CR, Est, Sisk, WV. CA, ID, WA; north to British Columbia; western, eastern and southern US; temperate regions worldwide. Exotic.

Cynosurus echinatus is recognized by its dense, one-sided, head-like, round-triangular inflorescences that are widest near the base. It is an aggressive invasive in dry grasslands of western Oregon, displacing native upland prairie species.

DACTYLIS Orchard grass

Plants cespitose perennials. **Culm** bases laterally compressed. **Leaves** mostly basal; sheaths closed for at least half their length; ligules membranous; blades flat to folded. **Inflorescences** panicles, the primary branches with more or less 1-sided clusters of subsessile spikelets distally; disarticulation above the glumes, beneath the florets. **Spikelets** laterally compressed, oval to elliptic in outline, 2–6 florets. **Glumes** shorter than the florets, 1–3-veined, ciliate-keeled. **Calluses** short, blunt. **Lemmas** 5-veined; keels scabrous to ciliate, tapering to a short awn; lemma awns terminal, short. **Anthers** 3. **Caryopses** falling free or adhering to the lemma and/or palea.

Cool temperate regions worldwide. 1 species.

Dactylis is a variable, monotypic genus that is sometimes split into as many as five species.

Dactylis glomerata Linnaeus (p. 385)

Orchard grass

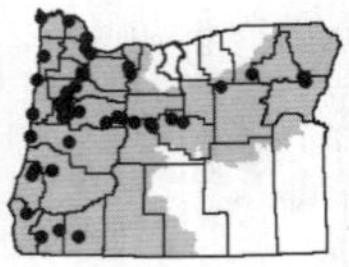

Plants to over 210 cm tall; cespitose, foliage usually glaucous. **Leaves** sheaths keeled, glabrous; ligules 3–11 mm, erose; blades (2)4–8(10) mm wide, flat to folded, with a conspicuous raised or widened midrib and white, scabridulous to scabrous margins. **Inflorescences** panicles, 4–20 cm; spikelets in dense head-like more or less 1-sided clusters grouped near ends of the branches. **Spikelets** 5–8 mm, 2–6 florets. **Glumes** 3–5 mm, 1–3-veined; keels usually scabrous to ciliate. **Lemmas** 4–8 mm, 5-veined, scabrous to pubescent; keels often ciliate; tip awned; lemma awns about 1 mm, arising from the tip. **Anthers** 2–3.5 mm. **2*n***=14, 21, 27–31, 42.

Pastures, meadows, woodlands, roadsides, vacant lots, disturbed areas. 0–2100 m. BW, Casc, CR, ECas, Est, Lava, Sisk, WV. CA, ID, NV, WA; throughout North America; cool temperate regions worldwide. Exotic.

Dactylis glomerata can be identified from a distance by the dense, one-sided clusters of spikelets grouped near the ends of the branches. Roadside plants, with their lowest branches sticking out to the side like thumbs, look like hitchhikers.

DANTHONIA Oatgrass

Plants perennial; cespitose, some with short rhizomes. **Leaves** sheaths open, with a tuft of spreading hairs at the tip of the margin; ligules a fringe of hairs; blades flat or involute; basal leaves much shorter than the culms. **Inflorescences** terminal panicles, racemes, or solitary spikelets, usually cross-pollinating; also axillary inflorescences (cleistogenes) hidden in the sheaths. **Spikelets** laterally compressed or terete, 3–12 florets, the terminal floret reduced; disarticulation below the florets or at the culm nodes. **Glumes** subequal or the lower equal to or longer than the rest of the spikelet (excluding the awns), 1–7-veined. **Calluses** with dense hairs. **Lemmas** obscurely (5)7–11-veined, with a dense tuft of hairs at the base or glabrous or more or less uniformly pubescent or pubescent on the margins; tip with 2 acute to awn-like teeth; lemma awns usually present, bent, twisted below the bend, arising between the teeth. **Anthers** 3. **Caryopses** ovate to obovate. **Cleistogenes** more common at lower nodes, 1–8 per node; lemmas glabrous, awnless; anthers minute; caryopses more linear than in the terminal florets.

North America, South America, North Africa, and Europe. 27 species; 5 species treated in *Flora*.

Oatgrasses are recognized by tufts of stiff hairs at the junction of leaf blade and sheath and large spikelets with bent awns. The glumes are longer than or equal to the rest of the spikelets. In some species, lemma tips are elongated into awn-like teeth. The genus *Danthonia* has been divided, and many species of southeastern Asia and the southern hemisphere have been placed in *Rytidosperma*. Some *Rytidosperma* species have been introduced to Oregon. Within this genus, descriptions are of the terminal inflorescences, not the cleistogenes. Spikelet and lemma lengths do not include awns. Lemma body length is measured from the base of the lemma to the base of the awn. Length of the lemma teeth is measured from the base of the awn to the tip of the tooth.

Key to *Danthonia* species:

1. Pedicels of lowest inflorescence branch generally longer than the spikelets; mature branches usually spreading. . ***D. californica***

1' Pedicels of lowest inflorescence branch generally shorter than or as long as the spikelets; mature branches (if present) ascending to erect.

 2. Lemmas not awned (may have a short, straight point between the short apical teeth) . . . . ***D. decumbens***

 2' Lemmas with a twisted, geniculate awn.

  3. Spikelets 1(4), if 2–4, the inflorescence a raceme with each spikelet on a pedicel attached to the inflorescence axis; lemma bodies 5.5–11 mm . . . . ***D. unispicata***

  3' Spikelets (4)5–13(18), inflorescence usually a panicle with some spikelets on pedicels attached to inflorescence branches; lemma bodies 2.5–6 mm.

   4. Calluses of middle florets longer than wide, concave dorsally; caryopses (2)2.3–3 mm long; lemma bodies glabrous on back . . . . ***D. intermedia***

   4' Calluses of middle florets about as long as wide, convex dorsally; caryopses 1.5–2(2.3) mm long; lemma bodies usually pilose (sometimes glabrous) on back . . . . ***D. spicata***

Danthonia californica Bolander

California oatgrass

Danthonia americana Scribn., *Danthonia californica* Bol. var. *americana* (Scribn.) Hitchc., *Danthonia californica* Bol. var. *californica*

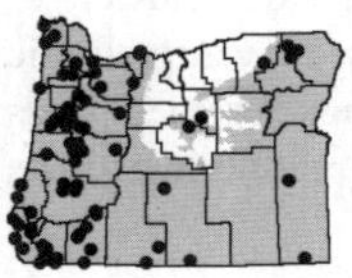

Plants (10)30–130 cm tall; cespitose, erect or ascending, with short rhizomes. **Leaves** sheaths glabrous or pilose; upper sheaths usually glabrous or unevenly pilose; hairs of collar and throat spreading and usually 1–2 mm; blades (1)2–5(6) mm wide; uppermost culm leaf blade spreading to reflexed. **Inflorescences** 2–4.5(7) × 1–8 cm, with (1)2–4(6+) spikelets; branches strongly divergent to reflexed at maturity; pedicels equal to or longer than the spikelets. **Spikelets** (10)14–26(30) mm, with 3–6(12) florets. **Glumes** usually 14–18 mm, with 4–7 veins, glabrous or scabrous. **Calluses** of middle florets longer than wide, dorsally concave. **Lemmas** bodies 5–10 mm, glabrous or sparsely pilose over the back; margins pubescent, rarely glabrous, with 9–11 veins; apical teeth (2)4–6(7) mm, awn-like; lemma awns (7)8–12 mm, usually bent and twisted below the bend. **Paleas** subequal to lemma. **Anthers** 2.5–4 mm. **Caryopses** 2.5–4.2 × 1.3–1.6 mm. **2*n***=36.

Prairies, meadows, fields, and savannas. 0–2300 m. BR, BW, Casc, CR, ECas, Est, Owy, Sisk, WV. CA, ID, NV, WA; north to British Columbia, east to Saskatchewan and NM, MA; Chile. Native.

In *D. californica,* the culm angles sideways from the base; the blades spread stiffly sideways or down, and the panicle branches hold the few large spikelets out at stiff angles. *Danthonia unispicata* is similar but usually has fewer spikelets that are more appressed and racemose inflorescences. It is a hairier plant with pedicels shorter than the spikelets.

Danthonia decumbens (Linnaeus) de Candolle

Mountain heathgrass

Sieglingia decumbens (L.) Bernh.

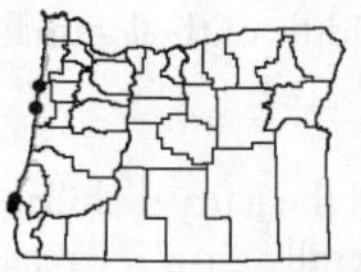

Plants 8–60 cm tall; cespitose. **Leaves** sheaths glabrous or pilose; blades 0.5–4 mm wide; uppermost culm blades ascending. **Inflorescences** 4–10 × 0.8–1.6 cm, with up to 15 spikelets; branches erect, pedicels shorter than the spikelets, except for the terminal spikelet on each branch. **Spikelets** 6–15 mm, with 3–5 florets. **Glumes** 7–10 mm, with 4–7 veins, glabrous. **Calluses** of middle florets equal to or slightly longer than wide, dorsally convex. **Lemmas** bodies 5–6 mm, glabrous or margins pilose for most of their length, with 7 faint veins, awnless, sometimes with short, inconspicuous points between the teeth, apical teeth short, acute. **Paleas** subequal to lemma. **Anthers** 0.2–0.4 mm, or about 2 mm in rare cross-pollinating florets. **Caryopses** 2.1–2.5 × 1.1–1.8 mm. **2*n***=24, 36, 124.

Sandy meadows, stabilized dunes, and roadsides near the coast. 0–100 m. Est. WA; east coast of North America from Labrador to ME; western Asia, Europe. Exotic.

Danthonia decumbens lacks awns and has smaller, more rounded spikelets than the other *Danthonia* species. The shape of its inflorescence and spikelets suggests *D. spicata,* but that species has awns and occurs at moderately high elevations away from the coast.

Danthonia intermedia Vasey

Timber oatgrass

Danthonia intermedia Vasey var. *cusickii* T.A. Williams, *Danthonia intermedia* Vasey var. *intermedia*

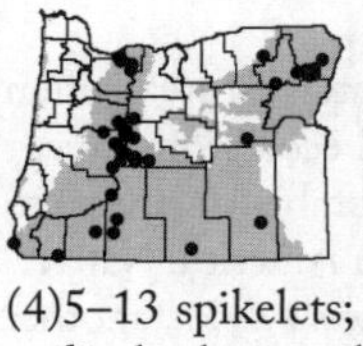

Plants 10–50(70) cm tall; cespitose, with short rhizomes. **Leaves** sheaths glabrous or pilose; blades 1–3.5 mm wide; uppermost culm blades strongly divergent or reflexed. **Inflorescences** 2–5 × 0.5–2.5 cm, with (4)5–13 spikelets; branches appressed or strongly ascending, pedicels shorter than or as long as spikelets. **Spikelets** 11–15(19) mm, with 3–12 florets. **Glumes** (11)13–17 mm, with 4–7 veins, glabrous. **Calluses** of middle florets longer than wide, dorsally concave. **Lemmas** bodies 3–6 mm, nearly glabrous on back, pilose on the margins, with (5)7–11 veins, apical teeth acute to awn-like, 1.5–2.5 mm; lemma awns 6.5–8 mm, usually bent and twisted below the bend. **Paleas** subequal to lemma. **Anthers** usually tiny, sometimes to 4 mm in rare cross-pollinating florets. **Caryopses** (2)2.3–3 × 0.7–1.1 mm. **2*n***=18, 36.

Open woods, meadows, rocky outcrops, in montane and alpine zones. 400–2900 m. BR, BW, Casc, ECas, Est, Sisk. CA, ID, NV, WA; north to AK, throughout Canada and western US; eastern Russia. Native.

Danthonia intermedia produces its large spikelets in a dense head, unlike any other Oregon *Danthonia. Danthonia spicata* usually has narrower, spike-like inflorescences, shorter lemmas with much shorter teeth, and sparse, long, white hairs over the lemma back as well as along the margins.

Danthonia spicata (Linnaeus) P. Beauvois ex Roemer & Schultes

Poverty oatgrass

Danthonia thermale Scribn.

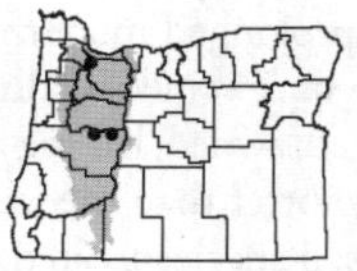

Plants (7)10–70(100) cm tall; cespitose, without rhizomes. **Leaves** sheaths glabrous or pilose; blades 0.8–3(4) mm wide, usually curled or twisted (straight in shade-grown plants); uppermost cauline blades erect to ascending. **Inflorescences** 2.5–6 × 0.4–1.1 cm, 5–10(18) spikelets; branches appressed to ascending, pedicels on the lowest branch shorter than or equal to the spikelets. **Spikelets** 7–15 mm, with 3–12 florets. **Glumes** (8)9–12 mm, with (3)4–5(7) very faint veins, glabrous. **Calluses** of middle florets about as long as wide, dorsally convex. **Lemmas** bodies 2.5–5 mm, usually sparsely pilose or sometimes glabrous on back, the longest hairs 0.5–2 mm, with (5)7–11 veins; apical teeth acute to acuminate or awn-like, 0.5–2 mm; lemma awns 5–8 mm, usually bent and twisted below the bend. **Paleas** shorter than lemma. **Anthers** to 2.5 mm. **Caryopses** 1.5–2(2.3) × 0.7–1 mm.

Meadows, rocky slopes, usually in the sun, on sandy or rocky substrates. 50–1700 m. Casc, WV. ID, NV, WA; throughout North America. Native.

The inflorescence of *D. spicata* is narrow, almost spike-like, and the lemmas are usually short with short apical teeth and long, white hairs over the back and on the margins. See discussion of *Danthonia intermedia.* The florets in the terminal inflorescence of *D. spicata* are usually cleistogamous like those hidden in the leaf sheaths. Occasional plants with outcrossing flowers tend to have more spreading inflorescence branches.

Danthonia unispicata (Thurber) Munro ex Vasey (p. 390)

One-spike oatgrass

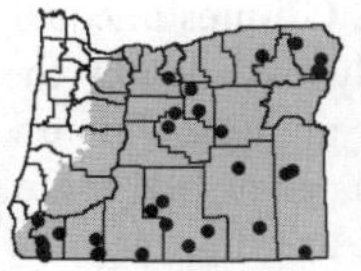

Plants (10)15–30(42) cm tall; cespitose, without rhizomes. **Leaves** sheaths usually densely pilose with spreading, often papillose-based hairs, rarely glabrous; hairs of collar and throat spreading, usually 3–4 mm; blades 1–3 mm wide; upper cauline blades usually more or less ascending. **Inflorescences** 8–21 × 0.4–1.3 cm, with 1(4) spikelets, unbranched, pedicels shorter than or as long as the spikelets. **Spikelets** (8)12–26 mm, 3–12 florets. **Glumes** lower glume 15–23 mm; upper glume similar or slightly shorter, all glumes 3–5-veined, glabrous. **Calluses** of middle florets longer than wide, dorsally concave. **Lemmas** bodies 5.5–11 mm, acute to awn-like, glabrous over the back, rarely with a few scattered hairs; margins pilose, rarely glabrous, with (5)7–11 veins; apical teeth 2, acute to awn-like, 1.5–7 mm; lemma awns 5.5–13 mm, usually bent and twisted below the bend. **Paleas** shorter than lemma. **Anthers** less than 3.5 mm. **Caryopses** approximately 2.2–4 × 1 mm. **2*n***=36.

Dry grasslands, shrub steppe, and open forest, up to timberline. 300–2400 m. BR, BW, Casc, Col, ECas, Lava, Owy, Sisk. CA, NV, ID, WA; north to British Columbia, east to Saskatchewan and CO. Native.

Danthonia unispicata is usually easy to identify by its single spikelet per culm and densely hairy leaf blades and sheaths. Occasional populations, however, have glabrous leaves. Occasional individuals with more than one spikelet per culm resemble *D. californica*, but the panicle branches are more or less erect, and the hairs of the collar and throat are usually longer.

DESCHAMPSIA Hairgrass

Plants annual or perennial; cespitose. **Culms** hollow, erect. **Leaves** mainly basal, or mainly cauline in some annuals; sheaths open, glabrous; auricles none; ligules membranous, decurrent, long, obtuse or acute to acuminate; blades rolled, folded, or flat. **Inflorescences** panicles, open with spreading branches to very narrow with appressed branches; disarticulation above the glumes. **Spikelets** with 2(3) florets; rachillas hairy and usually prolonged more than 0.5 mm beyond the base of the distal floret. **Glumes** exceeding the lower and sometimes the upper floret, 1- or 3-veined. **Calluses** with hairs up to half as long as the lemmas. **Lemmas** smooth, shiny, obscurely (3)5–7-veined, erose to toothed, awned; lemma awns attached from near the base to about the middle of the lemma, straight to bent, sometimes twisted. **Paleas** shorter than the lemmas. **Anthers** 3.

Cosmopolitan. Approximately 30 species; 3 species treated in *Flora*.

Deschampsia species are recognized by the basal tuft of leaves, the two small florets hidden by the glumes, and the hairy rachillas prolonged beyond the base of the distal floret. Their ligules are typically long and acute to acuminate. They generally grow in areas that are wet at least in early spring. *Deschampsia* has been divided, and one Oregon species is now placed in *Vahlodea*. Within this genus, lemma traits should be examined on the lowest floret of each spikelet.

Key to *Deschampsia* species:

1. Plants annual; leaf blades mainly cauline, filiform, rarely as much as 1.5 mm wide; glumes 5–8.5 mm ...... ***D. danthonioides***
1' Plants perennial; leaf blades mainly basal, often more than 1.5 mm wide; glumes often less than 5 mm.
    2. Basal leaf blades approximately 1 mm wide .... ***D. elongata***
    2' Basal leaf blades 1–6 mm wide........ ***D. cespitosa***

Deschampsia cespitosa (Linnaeus) P. Beauvois (p. 390)

Tufted hairgrass

Aira caespitosa L., *Aira holciformis* (J. Presl) Steud., *Deschampsia caespitosa* (L.) P. Beauv. ssp. *holciformis* (J. Presl) W.E. Lawr., *Deschampsia cespitosa* (L.) P. Beauv. ssp. *beringensis* (Hultén) W.E. Lawr., *Deschampsia cespitosa* (L.) P. Beauv. ssp. *cespitosa*, *Deschampsia cespitosa* (L.) P. Beauv. var. *cespitosa*, *Deschampsia cespitosa* (L.) P. Beauv. var. *longiflora* Beal

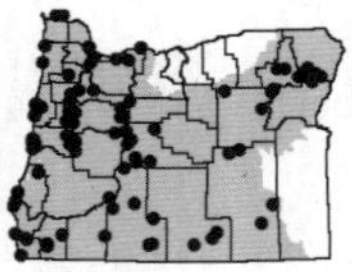

Plants perennial, (7)35–150 cm tall; cespitose. **Leaves** mostly basal; ligules membranous, 2–13 mm, acute (sometimes obtuse on some leaves); blades flat, folded, or rolled, 1–6 mm wide. **Inflorescences** 8–30(40) cm; branches spreading. **Spikelets** 2.5–7.6 mm. **Glumes** unequal, acute to long-tapered, exceeding or slightly exceeded by the distal floret; lower glume 2.7–7 mm; upper glume 2–7.5 mm. **Lemmas** 2–5(7) mm, smooth, shiny; lemma awns 0.5–8 mm, arising from near lemma base to about mid-length, straight or slightly bent. **Anthers** 1.5–3 mm. ***2n***=26–28.

Wet prairie, wet meadows, and salt marshes. 0–2900 m. BR, BW, Casc, CR, ECas, Est, Lava, Sisk, WV. CA, ID, NV, WA; Canada and all but southeastern US; circumboreal, Australia, New Zealand. Native.

From a distance a *D. cespitosa*-dominated area is recognized by the light, open cloud of tan inflorescences floating a meter or so above the leaves, which form dense tussocks in wet prairie or salt marshes. The combination of dense basal tufts of leaves, delicate inflorescences, and long, acuminate ligules is unique among our common grass species. Three subspecies have been recognized, but the variation among them seems continuous, and many Oregon plants have combinations of traits that do not correspond to any named subspecies.

Deschampsia danthonioides (Trinius) Munro

Annual hairgrass

Aira danthonioides Trin.

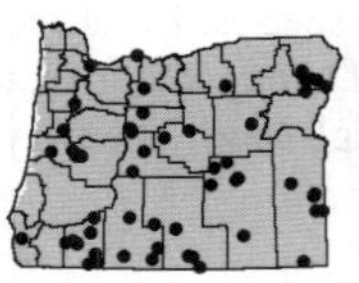

Plants annual, 10–40(70) cm tall. **Leaves** mostly cauline, filiform; ligules (0.5)1.5–3(4.7) mm, acute to acuminate; blades 0.3–1.5 mm wide, involute to flat. **Inflorescences** 5–15(25) cm; branches appressed when young, spreading when mature. **Spikelets** 4–9 mm. **Glumes** subequal, 5–8.5 mm, acute, exceeding the distal floret, with 0–3 veins, glabrous to scabridulous. **Lemmas** 2–3(3.5) mm, smooth, shiny; lemma awns (4)5–9(10.5) mm, arising near base to about middle of lemma, strongly bent, twisted at base. **Anthers** 0.3–0.5 mm. ***2n***=26.

Vernal pools, scablands, stream banks, disturbed areas, often sites that are wet in spring. 50–2100 m. All ecoregions except Est. CA, ID, NV, WA; north to British Columbia, northeast to MT, south to Mexico; ME, NY, OH; South America. Native.

Annual hairgrass is very similar to *Ventenata dubia*, another delicate annual with open panicles. *Ventenata dubia* glumes have more veins and longer lemmas. Additionally, one lemma of each spikelet has an awn that is terminal and straight, whereas the awn of the other lemma is dorsal and bent. Young *D. danthonioides* with unexpanded inflorescences can be confused with *D. elongata*, which has a dense tuft of basal leaves.

Deschampsia elongata (Hooker) Munro
Slender hairgrass

Aira elongata Hook.

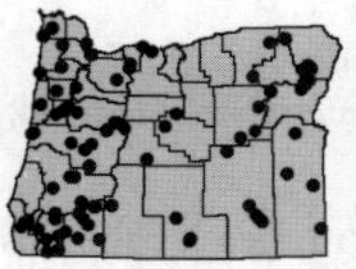

Plants perennial, (10)30–120 cm tall; cespitose. **Leaves** mostly basal; ligules 2.5–8(9) mm, acute to acuminate; blades 0.2–2 mm wide, usually involute; basal blades approximately 1 mm wide. **Inflorescences** 5–30(35) cm, nodding; branches erect to ascending. **Spikelets** 3–6.7 mm. **Glumes** subequal, 3–5.5(6.5) mm, acute, exceeding the distal floret. **Lemmas** 1.7–3.3 mm, smooth, shiny; lemma awns (1.5)3–6 mm, arising near middle of lemma, straight or slightly bent. **Anthers** 0.3–0.7 mm. **2*n***=26.

Roadsides, disturbed meadows, shores of lakes, ponds, and streams. 0–2700 m. Throughout Oregon. CA, ID, NV, WA; north to the Yukon, northeast to MT, southeast to NM and northern Mexico; Chile. Native.

Deschampsia elongata produces a dense tuft of narrow leaves and slender, nodding, often silvery panicles. Immature *D. danthonioides* may look similar, but it has few basal leaves and usually longer glumes. *Deschampsia elongata* commonly grows in moist disturbed habitats and is sometimes cultivated for habitat restoration projects including road decommissioning.

DESMAZERIA Ferngrass

Plants annual. **Leaves** basal and cauline; sheaths open; ligules membranous, longer than wide; blades flat, sometimes convolute when dry. **Inflorescences** panicles or racemes. **Spikelets** laterally compressed, subsessile, 4–25 florets, the distal florets reduced; disarticulation above the glumes, beneath the florets. **Glumes** shorter than or subequal to the adjacent lemmas, 1–5-veined, awnless. **Calluses** blunt. **Lemmas** narrowly elliptic, leathery at maturity, inconspicuously 5-veined, glabrous; apices acute to obtuse, sometimes bifid, awnless. **Anthers** 3. **Caryopses** ellipsoid-oblong, dorsally flattened, falling with the paleas.

North America and the Mediterranean region. 6 or 7 species; 1 species treated in *Flora.*

Desmazeria rigida (Linnaeus) Tutin
Ferngrass

Scleropoa rigida (L.) Griseb.

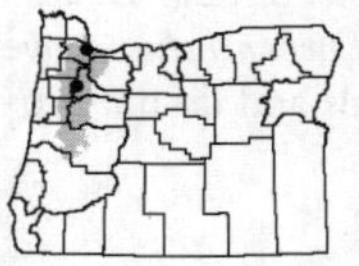

Plants to 60 cm tall. **Culms** decumbent or ascending. **Leaves** sheaths glabrous; upper margins hyaline and continuous with the sides of the ligules; ligules 1.5–4 mm, hyaline, pointed, lacerate; blades 2–8(12) cm × 1–3(4) mm, flat to folded, glabrous. **Inflorescences** panicles or occasionally racemes, 1–12(18) × 1.2–3 cm; branches stiff and somewhat spreading at maturity. **Spikelets** 4–10 mm, with 5–12 florets. **Glumes** 1.5–2.3 mm, 1–3-veined, scabrous on keel. **Lemmas** 2–3 mm, 5-veined, rounded on the back, glabrous; tips acute to obtuse, often with a short point, awnless. **Anthers** 0.4–0.6 mm. **2*n***=14.

Disturbed sites. 0–200 m. WV. CA, NV, WA; southeastern US from TX to FL; Europe. Exotic.

Desmazeria rigida is a "generic grass," resembling a small *Festuca* or a *Poa* with narrow lemmas. It is distinguished by its hyaline sheath margins that are continuous with the sides of hyaline pointed ligules.

DICHANTHELIUM Panic grass, rosette grass

Plants annual or perennial; cespitose. **Culms** 15–210 cm tall, hollow, erect to sprawling, usually branched from the middle or lower nodes. **Leaves** cauline and basal, producing winter rosettes of short broad leaves that are well differentiated from the mature longer narrower cauline leaves; ligules membranous with or without cilia, of hairs only or absent; blades flat, glabrous to spreading-pubescent. **Inflorescences** of 2 types; primary panicles cross-pollinating, produced in early spring and sometimes also in late fall; secondary panicles usually cleistogamous with closed self-pollinating flowers produced in late spring to fall; disarticulation below the glumes. **Spikelets** 0.8–5.2 mm, dorsiventrally compressed, sometimes almost round in cross section, 2 florets, the lower sterile or staminate, the upper bisexual. **Glumes** unequal, glabrous to scabrous; lower glumes shorter than the spikelets, 1–5-veined; upper glumes approximately as long as the spikelets, 5–11-veined. **Lower lemmas** similar to the upper glume. **Upper lemmas** more or less hard and leathery, shiny, glabrous; margins involute and usually clasping the palea. **Upper paleas** striate. **Anthers** 3. **Caryopses** smooth.

North America, South America, and the Caribbean. Approximately 72 species; 2 species treated in *Flora.*

The species of *Dichanthelium* have often been placed in *Panicum*. However, in *Panicum* the leaves do not form a winter rosette. *Panicum* have C4 photosynthesis and are warm-season grasses; *Dichanthelium* have C3 photosynthesis and are cool-season grasses.

Key to *Dichanthelium* species:

1. Spikelets 1–2.5 mm, slightly less than 1 mm wide; upper glumes 7-9-veined .... ***D. acuminatum*** ssp. ***fasciculatum***
1' Spikelets 2.7–3.5 × 1–2 mm; upper glumes 9-veined .... ***D. oligosanthes*** ssp. ***scribnerianum***

Dichanthelium acuminatum (Swatrz) Gould & C.A. Clark ssp. *fasciculatum* (Torrey) Freckmann & Lelong

Hairy panic grass, tapered rosette grass

Dichanthelium acuminatum (Sw.) Gould & C.A. Clark var. *fasciculatum* (Torr.) Freckmann & LeLong, *Dichanthelium lanuginosum* (Elliott) Gould var. *fasciculatum* (Torr.) Spellenberg, *Panicum lassenianum* Schmoll, *Panicum occidentale* Scribn., *Panicum pacificum* Hitchc. & Chase

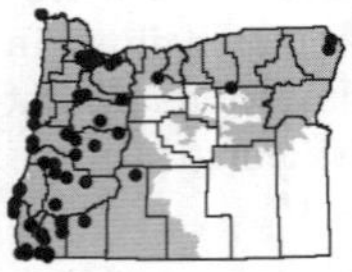

Plants perennial, 15–75 cm. **Leaves** sheaths with dense ascending or spreading papillose-based hairs, occasionally with inconspicuous shorter hairs; margins ciliate or glabrous; ligules of hairs, 1–5 mm; blades 5–12 cm × 6–12 mm, flat, spreading to ascending; lower surfaces usually pubescent; upper surfaces pilose or glabrous, bases rounded to subcordate and usually ciliate. **Primary inflorescences** 3–12 × 0.75–9 cm, well exserted; branches ascending. **Spikelet** 1.5–2.5 mm, slightly less than 1 mm wide. **Glumes** lower glumes usually 25–50% as long as the spikelets, acute; upper glumes 1.5–1.8 mm, with 7–9 indistinct veins. **Lemmas** lower lemmas similar to upper glumes; upper lemmas 1.2–1.5 mm, tough or hard. **Anthers** 0.5–0.8 mm. **2*n***=18.

Disturbed areas, grasslands, open woodlands, shores of streams and lakes. 0–1500 m. BW, Casc, Col, CR, ECas, Est, Sisk, WV. CA, ID, NV, WA; throughout most of North America. Native.

Dichanthelium acuminatum ssp. *fasciculatum* has broad fuzzy leaves with disproportionately tiny, round spikelets. Similar *D. oligosanthes* has larger spikelets.

Dichanthelium oligosanthes ssp. *scribnerianum* (Nash) Freckmann & Lelong (p. 390)

Scribner's panic grass, Heller's rosette grass

Dichanthelium oligosanthes (Schult.) Gould var. *scribnerianum* (Nash) Gould, *Panicum oligosanthes* Schult. var. *scribnerianum* (Nash) Fernald, *Panicum scribnerianum* Nash

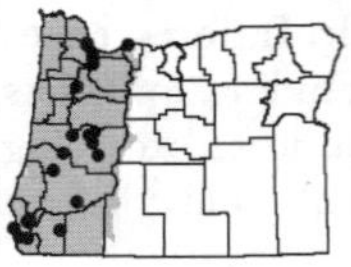

Plants perennial, 20–50 cm. **Leaves** sheaths glabrous or sparsely papillose-hispid, rarely puberulent, with ciliate margins; ligules of hairs, 1–1.5 mm; blades of cauline leaves 5–12 cm × 6–15 mm, flat, glabrous or sparsely pubescent abaxially, bases ciliate, rounded to truncate; margins cartilaginous. **Primary inflorescences** 5–9 × 3–6 cm, sometimes partially hidden in leaf sheaths; branches ascending. **Spikelets** 2.7–3.5 × 1–2 mm. **Glumes** lower glumes 1–1.6 mm, acute, similar in texture to upper glumes; upper glumes 2.7–3.5 mm, strongly veined, with orange or purplish spot at base, with 9 veins. **Lemmas** lower lemmas more or less pubescent; upper lemmas 2.5–3 mm, tough or hard. **Anthers** about 1 mm. **2*n***=18.

Meadows, prairies, sandy or clayey shores of rivers and lakes, disturbed areas. 0–400 m. Casc, CR, Est, Sisk, WV. CA, ID, WA; throughout most of North America. Native.

Dichanthelium oligosanthes has short broad leaves and stiffly spreading panicles with small round spikelets. Similar *D. acuminatum* has smaller spikelets. *Panicum and Dichanthelium* have similar spikelets, but our *Panicum* are annuals and do not produce a well-developed basal rosette of leaves.

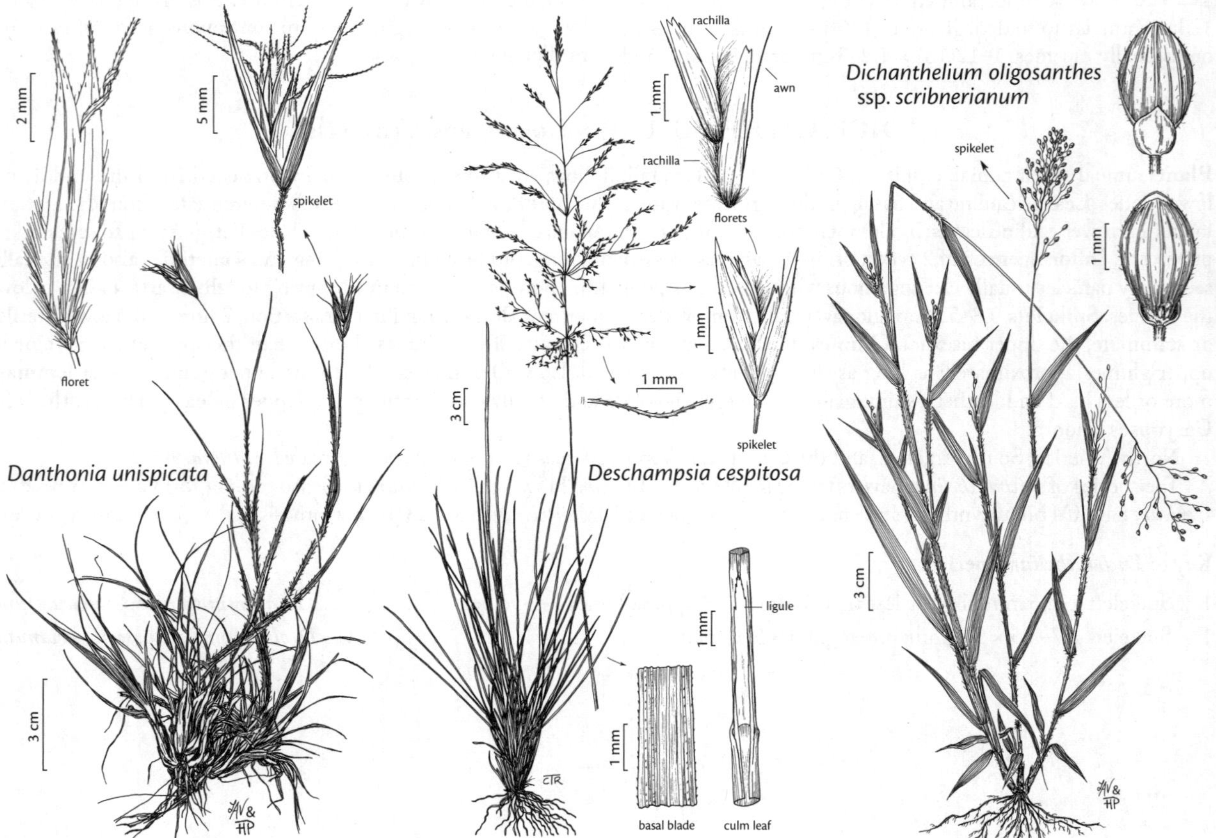

DIGITARIA Crabgrass

Plants annual or perennial. **Culms** branched at the base or at higher nodes, often rooting at the lower nodes. **Leaves** sheaths open; ligules membranous, sometimes ciliate; blades usually flat. **Inflorescences** terminal, sometimes also axillary, panicles of 1-sided spike-like branches, branches arising close together, wing-margined, bearing spikelets on one side, in 2 rows, usually in groups of 2–5 with unequal pedicels. **Spikelets** lanceolate to elliptic, dorsiventrally compressed; tips obtuse to acuminate, awnless, 2 florets, the lower one sterile; disarticulation beneath the glumes. **Glumes** lower glumes absent to 25% as long as the spikelet; upper glumes from 12% to equal the length of the spikelet, 0–5-veined. **Lemmas** lower lemmas usually as long as the upper lemmas, usually pubescent, (3)5–7(13)-veined; lower paleas reduced or absent; upper lemmas stiffly chartaceous to cartilaginous, obscurely veined with 0.5–1 mm hyaline margins that embrace the upper palea; upper paleas similar to upper lemmas. **Anthers** 3. **Caryopses** plano-convex.

Cosmopolitan. Approximately 200 species; 2 species treated in *Flora*.

Interpreting the *Digitaria* spikelet can be a challenge. In our species, the lower glume is absent or present as a tiny triangle up to 0.4 mm long. The upper glume and lower lemma may be relatively similar, but in our species the upper glume has three veins, while the lower lemmas have seven. This difference can be hard to detect; the midvein and the next lateral veins are obvious, on the middle and near the edges of the nearly flat lemma back. The other lemma veins are crowded together on the rounded lemma sides. In *D. ischaemum,* the upper glume is easy to find because it is about half to two-thirds as long as the spikelet. In both species, the upper, fertile lemma is tough and shiny and easy to mistake for the seed.

Key to *Digitaria* species:

1. Leaf sheaths, blades with many long, papillose-based hairs; upper glume about 67% as long as the spikelet . . ***D. sanguinalis***

1' Leaf sheaths, blades nearly glabrous though with some hairs on upper margins of leaf sheath and lower margins of leaf blades; upper glume about as long as the spikelet . . . . ***D. ischaemum***

Digitaria ischaemum (Schreber) Schreber ex Muhlenberg
Smooth crabgrass

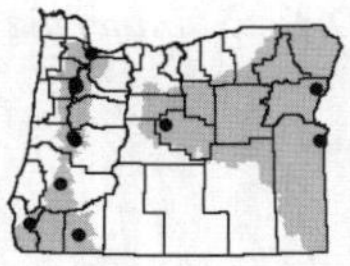

Plants annual or perennial, 20–55(70) cm tall, tufted, mat-forming, or rhizomatous. **Leaves** sheaths glabrous or sparsely pubescent, often pubescent near the top; blades 1.5–9 cm × 3–5 mm, flat, glabrous, sometimes with a few papillose-based hairs on margins near base. **Inflorescences** often purple with 2–7 spike-like branches, 6–15.5 cm × 0.7–2.5 mm, ascending to spreading. **Spikelets** 1.7–2.3 mm. **Glumes** lower glumes absent or reduced to a membranous rim; upper glumes 1.3–2.3 mm, 3-veined, 75% to as long as the upper lemmas with appressed hairs that may form whitish lines between the veins or cover the entire surface. **Lemmas** lower lemmas 1.7–2.3 mm, pubescent, 7-veined, 3 middle veins usually widely spaced, outer veins tightly spaced and near the margins; upper lemmas stiff, pubescent, dark brown at maturity; tips acute, awnless. **Anthers** 0.4–0.6 mm. **2*n***=36.

Disturbed areas, lawns, roadsides. 50–900 m. BW, Lava, Owy, Sisk, WV. CA, ID, NV, WA; throughout most of North America; warm-temperate regions worldwide. Exotic.

Digitaria ischaemum is recognized by its mostly glabrous foliage that is pubescent on the upper margins of the leaf sheaths and lower margins of the leaf blades. The matted hairs on the upper glumes and lower lemmas may form pale lines between the veins, but they may be longer, extensive, and gland-tipped, forming what can look like a coating of mold. *Digitaria sanguinalis* is similar and more common. Its foliage is sparsely covered with papillate-based hairs, and its glumes and sterile lemmas never appear moldy.

Digitaria sanguinalis (Linnaeus) Scopoli (p. 392)
Hairy crabgrass

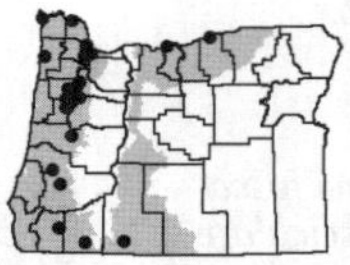

Plants annual, 20–70(112) cm tall, tufted or sometimes mat-forming. **Leaves** sheaths with sparse papillose-based hairs; blades 2–11(14) cm × 3–8(12) mm, flat, usually with papillose-based hairs on both surfaces, sometimes glabrous. **Inflorescences** often purple, with 4–13 spike-like primary branches 3–30 cm × 0.7–1.5 mm, ascending to spreading. **Spikelets** 1.7–3.4 mm. **Glumes** lower glumes 0.2–0.4 mm; upper glumes 0.9–2 mm, 33–50% as long as the spikelets, 3-veined, pubescent on the margins. **Lemmas** lower lemmas less than to 0.2 mm longer than the upper floret, glabrous, 7-veined; lateral or all veins scabrous throughout or smooth on the lower 33–50% and scabrous distally, 3 middle veins usually widely spaced, outer veins tightly spaced and near the margins; upper lemmas 1.7–3 mm, yellow or gray, often purple-tinged when young and becoming brown at maturity. **Anthers** 0.5–0.9 mm. **2*n***=36, 28, 34, 54.

Disturbed areas, lawns, roadsides. 0–1400 m. Col, CR, ECas, Est, Sisk, WV. CA, ID, NV, WA; throughout most of North America; worldwide. Exotic.

Digitaria sanguinalis is a common, late-season weed characterized by its spike-like branches that spread like the spokes of an umbrella. The very similar *D. ischaemum* lacks hairs on most of its foliage. *Cynodon* has a similar though more delicate inflorescence, but it is a strongly stoloniferous perennial.

DIPLACHNE Sprangletop

Plants annual or short-lived perennial; cespitose. **Culms** usually erect, sometimes prostrate. **Leaves** cauline, sometimes in basal rosettes; sheaths open; ligules membranous; blades flat, involute when dry; tips attenuate. **Inflorescences** panicles with 1-sided, spike-like branches that are spikelet-bearing throughout their length and have spikelets in 2 rows; lower branches occasionally secondarily branched; axillary panicles, if present, concealed in the lower leaf sheaths. **Spikelets** laterally compressed; disarticulation beneath the florets in primary panicles; axillary panicles not disarticulating. **Glumes** 2, unequal, exceeded by the florets, awnless, occasionally mucronate. **Calluses** distinct or poorly developed, pubescent or glabrous. **Lemmas** 3(5)-veined, usually pubescent at least in the proximal part, awned or not. **Anthers** 1–3. **Caryopses** elliptic to ovate or obovate, falling free of lemmas and paleas.

Tropical and warm-temperate areas of the world. 32 species; 1 species treated in *Flora*.

Diplachne fusca (Linnaeus) P. Beauvois ex Roemer & Schultes

Sprangletop

Leptochloa fusca (L.) Kunth.

Plants 5–110 cm tall. **Ligules** 2–8 mm, membranous, attenuate, lacerate when mature; blades 3–50 cm × 2–7 mm, flat to involute, glabrous or scabrous. **Inflorescences** panicles 1.5–72 cm × 4–22 mm; branches 3–35, 1.5–20(22) cm, spreading to reflexed. **Spikelets** 5–12(14) mm with (2)3–12(20) florets. **Glumes** 1-veined, glabrous, scabrous, or hairy; lower glumes 1–3(4.9) mm; upper glumes 1.8–5 mm. **Lemmas** 2–6 mm, glabrous to hairy, sometimes with a dark spot at the base; tips truncate to acute, often shortly bilobed, sometimes mucronate or awned; lemma awns 0–3.5 mm. **Anthers** 0.2–1 mm. **Caryopses** 0.8–2 mm. **2*n***=20.

3 subspecies; 2 subspecies treated in *Flora*.

Diplachne fusca can be a challenge to identify when first encountered. Its ephemeral wetland habitat and several florets per spikelet might suggest *Puccinellia*, and individuals with dark gray spikelets recall some species of *Muhlenbergia*, but the spike-like branches and the long ligules distinguish it from both these genera.

Key to *Diplachne fusca* subspecies:

1. Uppermost leaf blades exceeding the panicles; panicles usually partially enclosed in the uppermost leaf sheaths; mature lemmas smoky white with a dark smudge in the basal half . . . . ***D. f.* ssp. *fascicularis***

1' Uppermost leaf blades exceeded by the panicles; panicles usually completely exserted; mature lemmas dark green to dark gray, lacking a dark spot . . . . ***D. f.* ssp. *uninervia***

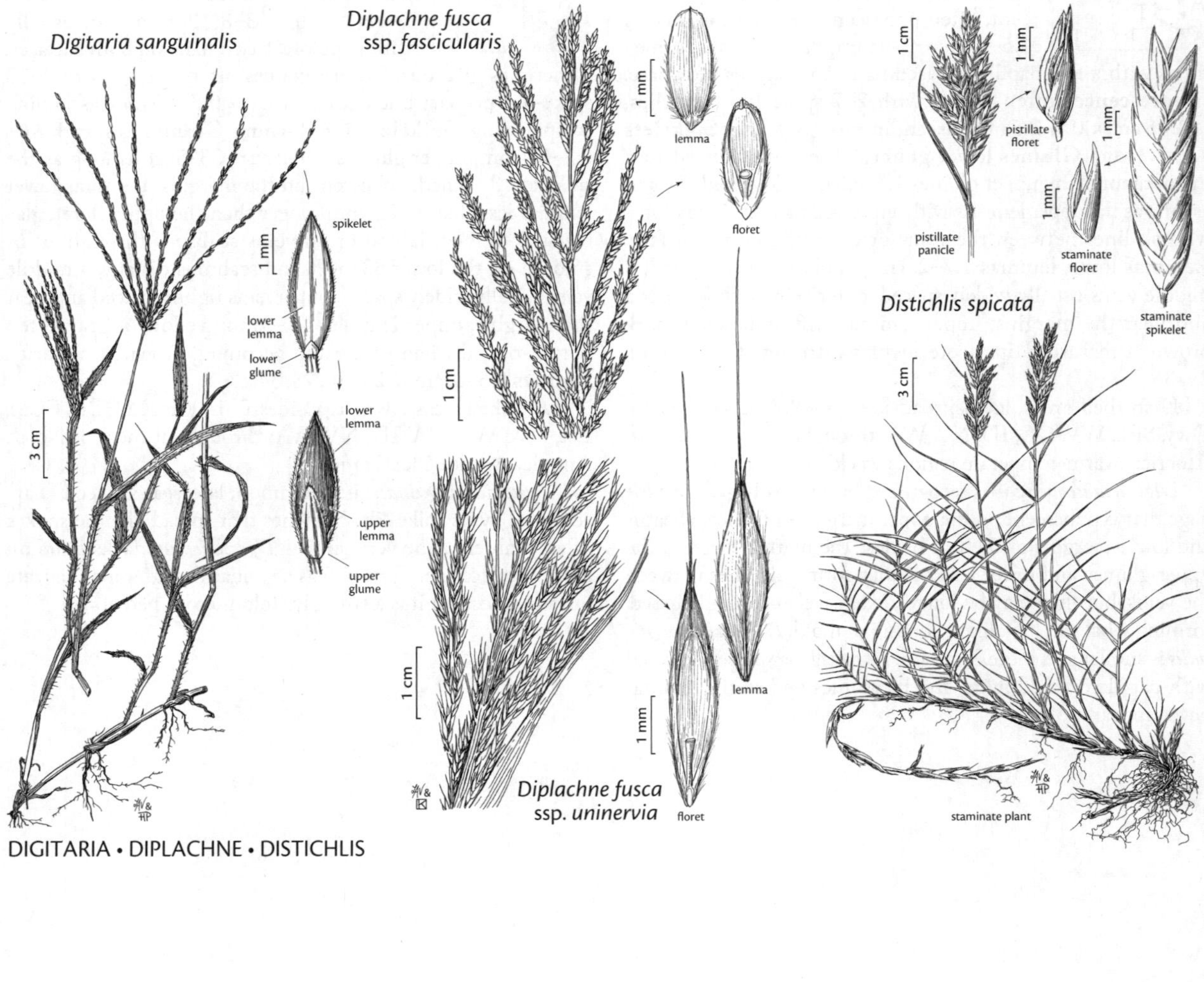

DIGITARIA • DIPLACHNE • DISTICHLIS

Diplachne fusca (Linnaeus) P. Beauvois ex Roemer & Schultes ssp. ***fascicularis*** (Lamarck) P.M. Peterson & N. Snow (p. 392)

Bearded sprangletop

Leptochloa fascicularis (Lam.) A. Gray, *Leptochloa fusca* (L.) Kunth ssp. *fascicularis* (Lam.) N. Snow

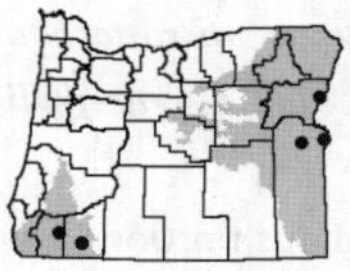

Culms prostrate to erect. **Leaves** blades glabrous or scabrous; uppermost blades often exceeding the inflorescence. **Inflorescences** 10–72 cm, usually partially enclosed in the uppermost leaf sheaths; branches 3–12(22) cm, often spreading. **Spikelets** 5–12 mm. **Glumes** lower glumes 2–3 mm, lanceolate, sometimes asymmetric; upper glumes 2.5–5 mm, elliptic to ovate. **Lemmas** 3.5–4(5) mm, smoky white, often with a dark spot on the basal half, sometimes entirely dark; tips acute, mucronate, or awned; lemma awns 0–3.5 mm. **Anthers** 1–3, 0.2–0.5 mm. **Caryopses** 0.8–2 mm.

Ephemeral wetlands and river banks. 0–900 m. BW, Owy, Sisk. CA, ID, NV, WA; throughout most of North America; South America. Native.

Diplachne fusca var. *fascicularis* is native to ephemeral wetlands in southwestern Oregon and along the Snake River and its tributaries.

Diplachne fusca (Linnaeus) P. Beauvois ex Roemer & Schultes ssp. ***uninervia*** (J. Presl) P.M. Peterson & N. Snow (p. 392)

Mexican sprangletop

Leptochloa fusca (L.) Kunth ssp. *uninervia* (J. Presl) N. Snow, *Leptochloa uninervia* (J. Presl) Hitchc. & Chase

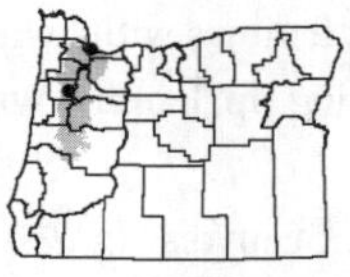

Culms more or less erect, often branching from the lower nodes. **Leaves** blades usually densely scabrous on both surfaces; uppermost leaves not exceeding the panicle. **Inflorescences** 10–57 cm, usually completely exserted from the uppermost leaf sheaths; branches 2–11 cm, mostly ascending. **Spikelets** 5–10 mm. **Glumes** lower glumes 1–2.6 mm, narrowly triangular to ovate; upper glumes 1.8–2.8 mm, obovate to widely obovate. **Lemmas** 2–3.6 mm, light brown, dark green, or lead-colored, without a basal dark spot; tips truncate or obtuse, rarely broadly acute or mucronate, awnless. **Anthers** 3, 0.2–0.6(1) mm. **Caryopses** 1–1.5 mm.

Disturbed areas. 50–200 m. WV. CA, NV; throughout most of North America; South America. Exotic.

DISTICHLIS Saltgrass

Plants perennial, usually dioecious, strongly rhizomatous and/or stoloniferous. **Culms** erect, solid. **Leaves** conspicuously 2-ranked, the lower leaves reduced to bladeless, scale-like sheaths; upper sheaths strongly overlapping, often with a tuft of hairs near the top; ligules membranous, less than 1 mm, more or less ciliolate; blades stiff, ascending to spreading. **Inflorescences** dense panicles or racemes. **Spikelets** laterally compressed, 2–20 florets; disarticulation of pistillate spikelets above the glumes and below the florets; staminate spikelets not disarticulating. **Glumes** 3–7-veined. **Lemmas** 9–11-veined, awnless; lemmas of pistillate florets leathery, staminate florets thinner. **Paleas** with keels narrowly or broadly winged, serrate to toothed. **Anthers** 3. **Caryopses** free from the palea at maturity.

North America, South America, and Australia. 5 species; 1 species treated in *Flora*.

Distichlis spicata (Linnaeus) Greene (p. 392)

Saltgrass

Distichlis spicata (L.) Greene var. *borealis* (J. Presl) Beetle, *Distichlis spicata* (L.) Greene var. *stricta* (Torr.) Scribn., *Distichlis stricta* (Torr.) Rydb., *Distichlis stricta* (Torr.) Rydb. var. *dentata* (Rydb.) C.L. Hitchc., *Distichlis stricta* (Torr.) Rydb. var. *stricta*

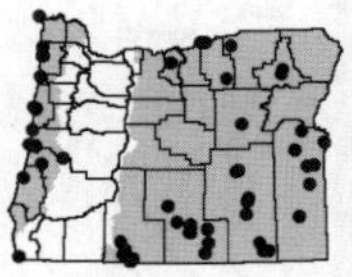

Plants 10–60 cm tall; rhizomatous, sometimes stoloniferous. **Leaves** sheaths glabrous; collars glabrous or with hairs 0.5–3 mm; ligules about 0.5 mm, erose, front often with hairs 0.5–3 mm; blades 1–8(20) cm × 2–4 mm, involute except at the base, often stiffly spreading, glabrous or sometimes with sparse, spreading hairs on the upper surface. **Inflorescences** 1–7 cm × 5–20 mm. **Spikelets** 5–20 × 4–7 mm, with 5–20 florets, unisexual. **Glumes** lower 2–3 mm; upper 3–4 mm. **Lemmas** 3.5–6 mm, firm, smooth, staminate lemmas narrower and thinner in texture; paleas on the pistillate spikelets bowed out. **Anthers** 3–4 mm. **Caryopses** 2–5 mm. ***2n***=40.

Coastal salt marshes and interior alkali flats, desert seeps, and sandy lakeshores. 0–1900 m. BR, BW, Col, CR, ECas, Est, Lava, Owy. CA, ID, NV, WA; throughout most of North America; South America; Australia. Native.

This grass exhibits wide intraspecific variation, and, as a result, many taxa have been described at the species or infraspecific rank.

ECHINOCHLOA Barnyard grass

Plants annual or perennial, with or without rhizomes. **Culms** sometimes rooting at nodes; internodes hollow or solid. **Leaves** sheaths open, laterally compressed; ligules usually absent; blades flat with a prominent midrib. **Inflorescences** panicles with spike-like branches that may be further branched. **Spikelets** subsessile, crowded, dorsiventrally compressed, more or less flat on one side, rounded on the other, with 2(3) florets; disarticulation below the glumes (cultivated taxa not disarticulating). **Glumes** lower glumes 20(50+)% as long as the spikelets, awnless or minutely awn-tipped; upper glumes as long as the spikelet, awnless or short-awned. **Lemmas** lower lemmas similar to the upper glumes in length and texture, awnless or awned; lower paleas absent or present as thin membranes; upper lemmas tough, dorsally rounded, usually smooth and shiny; tips short or elongate, awnless; upper paleas similar in texture to the upper lemmas, partially clasped by the upper lemmas but free at the tips. **Anthers** 3. **Caryopses** ellipsoid, broadly ovoid, or spheroid.

Warm temperate regions worldwide. Approximately 45 species; 2 species treated in *Flora*.

Barnyard grasses are common weeds that mature later in summer than most of our grasses. Two species are widespread in Oregon: the abundant introduced *E. crus-galli* and the less common native *E. muricata* var. *microstachya.* In general, plants with long awns are *E. crus-galli*, while awnless or short-awned plants should be looked at more closely because they might belong to either species.

Key to *Echinochloa* species:

1. Lemmas gradually tapering to a stiff tip; lemmas with awns 0–6(10) mm . . . . ***E. muricata*** var. ***microstachya***
1' Lemmas narrowing abruptly to a withering tip; lemmas with awns 0–50 mm . . . . ***E. crus-galli***

Echinochloa crus-galli (Linnaeus) P. Beauvois (p. 394)

Barnyard grass

Echinochloa crus-galli (L.) P. Beauv. var. *crus-galli*, *Echinochloa crus-galli* (L.) P. Beauv. var. *mitis* (Pursh) Peterm.

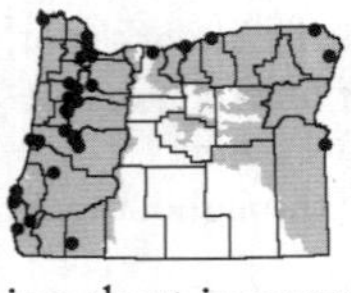

Plants annual, 30–200 cm tall. **Culms** spreading to erect. **Leaves** blades to 65 cm × 5–30 mm, glabrous but occasionally sparsely hirsute. **Inflorescences** panicles, 5–25 cm, the primary branches 1.5–10 cm and bearing short inconspicuous side branches; papillose-based hairs sparse to common, to 5 mm and sometimes longer than the spikelets. **Spikelets** 2.5–4 × 1.1–2.3 mm. **Glumes** lower glumes about 50% as long as the spikelets; upper glumes about as long as the spikelets, the longer hairs of the lateral veins 0.5–0.7 mm. **Lemmas** lower lemmas awnless to awned, sometimes within the same inflorescence; lemma awns 0–50 mm; upper lemmas about as long as the spikelet, broadly ovate to elliptical, the tough portion rounded distally with an abrupt transition to the early-withering acuminate membranous tip; tip marked off from the leathery portion by a line of tiny hairs visible at 25×. **Paleas** lower paleas about as long as the lemmas. **Anthers** 0.5–1 mm. **Caryopses** 1.3–2.2 × 1–1.8 mm, ovoid or oblong, brownish.

Disturbed, often moist areas. 0–1600 m. BW, Casc, Col, CR, Est, Owy, Sisk, WV. CA, ID, NV, WA; throughout most of North America; warm temperate regions worldwide. Exotic.

Echinochloa crus-galli is similar to the less common *E. muricata,* which is distinguished by lemmas that have tough, non-withering tips, without a line of minute hairs. *Echinochloa crus-galli* is variable in size, color, inflorescence branching, and awn development. Herbicide-resistant forms are known to have evolved.

Echinochloa muricata (P. Beauvois) Fernald var. ***microstachya*** Wiegand

American barnyard grass

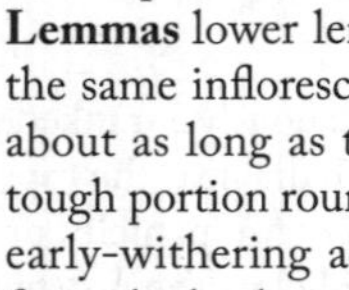

Plants annual, 80–160 cm tall. **Culms** erect or spreading, sometimes rooting at the lowest nodes, often developing short axillary fertile shoots at the upper nodes when mature. **Leaves** blades 1–27 cm × 0.8–30 mm. **Inflorescences** panicles 7–35 cm, the primary branches 2–8 cm, often bearing short inconspicuous side branches; papillose-based hairs absent or present to 3 mm. **Spikelets** 2.5–3.8 mm, purple or streaked with purple, usually hispid, hairs papillose-based. **Glumes** lower glumes 0.9–1.6 mm; upper glumes 2.8–3.8 mm, longer hairs of lateral veins 0.5–0.7 mm. **Lemmas** lower lemmas awnless or awned; lemma awns 0–6(10) mm; upper lemmas broadly obovoid or orbicular, narrowing to an acute or acuminate leathery portion that extends into the stiff membranous tip, boundary between the leathery and membranous portions not marked by tiny hairs. **Anthers** 0.4–0.7 mm. **Caryopses** 1.2–2.5 mm, broadly obovoid or spheroid, yellowish.

Disturbed, often moist areas. 50–1400 m. BR, BW, Col, CR, Est, Owy, Sisk, WV. CA, ID, NV, WA; north to British Columbia, southeast to TX, south to northern Mexico. Native.

In Oregon, *E. muricata* is awnless or nearly so; most *E. crus-galli* have long awns. In *E. crus-galli,* the smooth, shiny, fertile lemma changes abruptly to the green, withering tip, and the line separating the two parts is lined with tiny hairs, visible at 25×. By contrast, in *E. muricata*, the fertile lemma's main body and tip are of the same tough material, and no line of hairs separates the two.

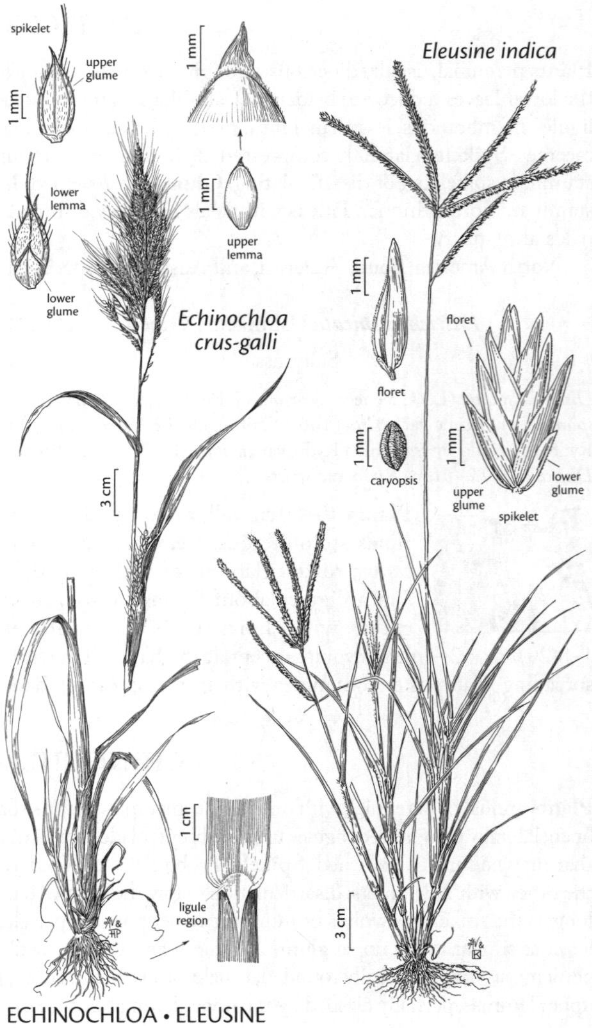

ECHINOCHLOA • ELEUSINE

ELEUSINE Goosegrass

Plants annual or perennial; cespitose. **Culms** 10–150 cm tall, glabrous, branching at and above the base. **Leaves** sheaths open; ligules membranous, ciliate. **Inflorescences** panicles with spike-like branches. **Spikelets** laterally compressed, 2–15 florets; disarticulation above the glumes and beneath the florets (cultivated forms not disarticulating). **Glumes** 2, unequal, shorter than the lower lemmas; lower glumes 1–3-veined; upper glumes 3–5(7)-veined. **Lemmas** 3-veined, glabrous, keeled; tips entire, awnless. **Anthers** 3. **Caryopses** modified; when moist, the seeds separate easily from the thin pericarp; usually trigonous, surfaces ornamented.

Warmer regions of the world. 9 species; 1 species treated in *Flora*.

One species of *Eleusine* is cultivated for grain, and others have been used as food in times of starvation. Asian *E. tristachya* was collected in Portland in 1915 but apparently has not persisted. It has 1–3 inflorescence branches that are 1–3(8) cm × 5–14 mm and attached at nearly the same point on the inflorescence axis.

Eleusine indica (Linnaeus) Gaertner (p. 394)

Goosegrass

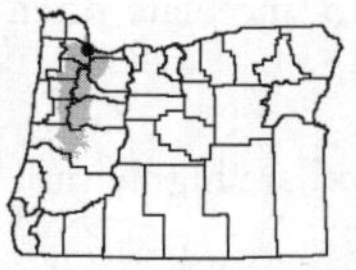

Plants annual, 30–90 cm tall. **Culms** erect or ascending, somewhat compressed; longer internodes 1.5–2 mm thick. **Leaves** sheaths conspicuously keeled; margins often with long, papillose-based hairs, particularly near the throat; ligules 0.2–1 mm, truncate, erose; blades 15–40 cm × 3–7 mm, with prominent white midveins; margins and/or upper surfaces often with basal papillose-based hairs. **Inflorescences** panicles with 4–10(17) spike-like branches, most in a terminal cluster, often with 1 or more branches attached as much as 3 cm lower; branches (3.5)7–16 cm × 3–5.5 mm, linear. **Spikelets** 4–7 × 2–3 mm, 5–7 florets. **Glumes** lower 1.1–2.3 mm, 1-veined; upper 2–2.9 mm; lateral nerves crowded near the keel. **Lemmas** 2.4–4 mm, awnless. **Paleas** with narrowly winged keels. **Anthers** 0.3–0.7 mm. **2*n***=18.

Disturbed places, often in trampled, compacted soils in full sun. 0–50 m. WV. CA, ID, NV, WA; eastern Canada, most of US; warmer regions worldwide. Exotic.

Oregon's only record of *Eleusine indica* is a collection from Portland in 1915, but the species has been collected many times in Idaho and eastern Washington, suggesting that it may be overlooked in eastern Oregon.

×ELYHORDEUM Hybrid wheatgrass

Plants perennial; usually cespitose, occasionally shortly rhizomatous. **Inflorescences** terminal, spikes or spike-like, with 1–3(7) spikelets per node; lateral spikelets usually on short pedicels; central spikelets sessile or nearly so; disarticulation at the rachis nodes and beneath the florets, tardy. **Spikelets** 1–4 florets. **Glumes** subulate to narrowly lanceolate, usually awned. **Lemmas** usually awned; anthers sterile. **Caryopses** rarely formed.

Cosmopolitan, wherever the parent taxa coexist. Approximately 10 species; 2 species treated in *Flora*.

×*Elyhordeum* refers to hybrids between species of *Elymus* and *Hordeum*. Some hybrids are fairly common, and most appear to be sterile. ×*Elyhordeum* hybrids resemble hybrids between *Elymus elymoides* or *E. multisetus* and other *Elymus* species because they also have tardily disarticulating inflorescences and awned glumes and lemmas; however, ×*Elyhordeum* taxa are more likely to have two to three spikelets at each node. All species are highly variable and may differ from the descriptions given below. They are best identified by evaluating the *Elymus* and *Hordeum* species present at the collection site.

Key to ×*Elyhordeum* species:

1. Spikelets 1–2 per node . . . . ×***E. macounii***
1' Spikelets (1)3 per node . . . . ×***E. stebbinsianum***

×*Elyhordeum macounii* (Vasey) Barkworth & D.R. Dewey

Macoun's wheatgrass

×*Agrohordeum macounii* (Vasey) Lepage, *Elymus* ×*macounii* Vasey

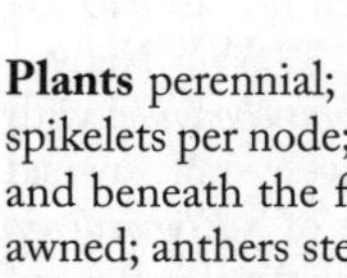

Culms 50–100 cm tall. **Leaves** blades 9–16 cm × (0.5)2–2.5(5) mm, smooth, scabrous or very sparsely hairy. **Inflorescences** (4)7.5–15 cm × 3–5 mm excluding the awns, 10–15 mm wide including the awns, erect; lower nodes with 1–2 spikelets; upper nodes with 1 spikelet; internodes 2.5–4.5 mm, much shorter than the spikelets; disarticulation tardy, in inflorescence axis. **Spikelets** about 10 mm, those at the lower nodes frequently with 3 glumes. **Glumes** (5)6–11 mm excluding the awns, 15–25 mm long including the awns, 0.6–0.8 mm wide, linear to very narrowly lanceolate, 1(3)-veined, not hard at the base, awns 8–13 mm, as long as or longer than the glume bodies. **Lemmas** 6–11 mm, oblong-lanceolate, glabrous or sometimes scabrous distally; lemma awns (2.5)9–20 mm. **Anthers** 1–1.6 mm, usually indehiscent.

Disturbed mesic grasslands and open forest. 600–1400 m. BR, BW, ECas, Sisk. CA, ID, NV, WA; north to AK, northeast to Quebec, southeast to NM. Native.

×*Elyhordeum macounii* comprises hybrids between *Elymus trachycaulus* and *Hordeum jubatum*. It backcrosses to *E. trachycaulus*. The backcrosses are usually identified as *E. trachycaulus* because the inflorescence axis may not disarticulate but has longer awns than *E. trachycaulus* ssp. *trachycaulus*. The long, narrow spikes of ×*Elyhordeum macounii* resemble those of ×*Elyleymus aristatus*, but the latter has only shorter awns.

× ***Elyhordeum stebbinsianum*** (Bowden) Bowden

Stebbins' wheatgrass

Elymus glaucus Buckley × *Hordeum brachyantherum* Nevski

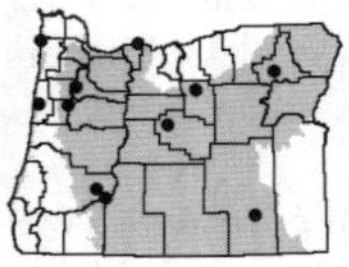

Culms 45–150 cm tall. **Leaves** blades (2)5–14 mm wide; lower surfaces smooth; upper surfaces minutely scabrous. **Inflorescences** spike-like, 6–16 cm, (1)3 spikelets per node; internodes 2.5–7 mm. **Spikelets** 8.5–11 mm; lateral spikelets sessile or on pedicels to 3 mm. **Glumes** 5.5–7(12) mm long without awns, 13–16 mm long with awns, 0.4–1 mm wide, linear to very narrowly lanceolate, 1(5)-veined, awns 4–10 mm. **Lemmas** 8–12 mm, glabrous to scabrous; lemma awns 5–27 mm. **Anthers** about 1.5 mm, indehiscent.

Disturbed moist to mesic grasslands. 0–1900 m. BR, BW, Casc, ECas, Est, Lava, WV. CA, WA. Native.

× *Elyhordeum stebbinsianum* is a hybrid between *Elymus glaucus* and *Hordeum brachyantherum*. It is completely sterile. Its thick spikes with spreading awns resemble those of *E. canadensis*.

× ELYLEYMUS Hybrid wheatgrass

Plants perennial. **Inflorescences** spikes or occasionally spike-like racemes, erect, 1–3 spikelets/node, pedicels absent or to 3 mm; disarticulation in the inflorescence axis and beneath the florets, tardy. **Spikelets** with 2–8 florets. **Glumes** linear to lanceolate, often awned. **Lemmas** 6–25 mm, glabrous or hairy, usually awned. **Anthers** indehiscent. **Caryopses** not developed.

North America. 11 taxa; 1 taxon in Oregon.

× *Elyleymus* taxa are hybrids between *Elymus* and *Leymus* species. They appear to be entirely sterile. These taxa are highly variable and may differ from the descriptions.

In the following description, measurements of spikelets, glumes, and lemmas exclude the awns.

× ***Elyleymus aristatus*** (Merrill) Barkworth & D.R. Dewey

Hybrid wheatgrass

× *Elysitanion aristatum* (Merr.) Bowden, *Elymus aristatus* Merr., *Elymus glaucus* Buckley var. *aristatus* (Merr.) Hitchc.

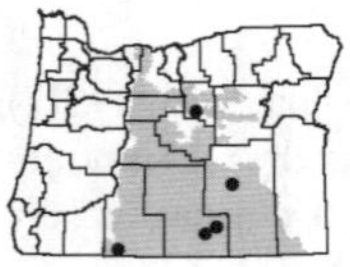

Plants 60–130 cm tall; cespitose or short-rhizomatous. **Leaves** evenly distributed; sheaths smooth, glabrous; auricles 0–0.5 mm; ligules 1–2.6 mm, rounded; blades 5–8 mm wide, smooth or scabrous; upper surface sometimes sparsely hairy. **Inflorescences** (4)6–15 cm × 3–5 mm excluding awns, 1–2(3) spikelets per node; internodes 2.5–4.5(7) mm, concealed by the overlapping spikelets. **Spikelets** 10–20 mm, sessile or nearly so, 1–3(4) florets. **Glumes** 5–7.5 mm long without awns, 6–11(15) mm long with awns, 0.3–1.1 mm wide, linear to narrowly lanceolate, 1(3)-veined, scabrous. **Lemmas** (5.6)7.5–12 mm, glabrous, smooth or scabrous, midveins prominent and scabrous distally; lemma awns 2–5 mm. **Anthers** (1.5)2.2–2.4 mm.

Grasslands and shrub steppe, often in disturbed areas. 900–1400 m. BR, ECas, Lava. CA, ID, NV, WA; CO, WY. Native.

× *Elyleymus aristatus* is the hybrid of *Elymus elymoides* with *Leymus triticoides* or *L. cinereus*. It is a relatively short-awned hybrid with dense spikes, similar to those of × *Elyhordeum macounii* except the latter has longer awns. The development of rhizomes, thickness of the spike, and development of pedicels may depend on which *Leymus* species is a parent.

ELYMUS Wildrye

Plants perennial; cespitose, rhizomatous, or stoloniferous. **Leaves** usually evenly distributed but sometimes mainly basal; sheaths open for most of their length; auricles often present; ligules membranous, often ciliolate; blades usually flat but sometimes folded or involute. **Inflorescences** terminal spikes with 1–3(5) spikelets per node; internodes 1.5–26 mm. **Spikelets** usually appressed to ascending but sometimes strongly divergent, 1–11 florets, the distal florets often reduced, the lowest florets usually bisexual but in some species sterile and glume-like; disarticulation usually above the glumes and beneath the florets but sometimes also below the glumes or in the inflorescence axis. **Glumes** usually 2 and subequal, usually linear-lanceolate to linear but sometimes bristle-like or oblanceolate to obovate, (0)1–7-veined, sometimes keeled over 1 vein, sometimes awned. **Lemmas** linear-lanceolate, faintly 5(7)-veined; tips acute or sometimes toothed; teeth 0.2 mm or extended as bristles to 10 mm, often awned; lemma awns terminal or from the sinus between the teeth, straight or curving and divergent, not bent. **Paleas** keels scabrous or ciliate, at least in part. **Anthers** 3. **Caryopses** with densely hairy tips.

Northern temperate regions. Approximately 150 species; 12 species treated in *Flora*.

Some species formerly treated in *Elymus* have been moved to *Leymus, Pascopyrum, Pseudoroegneria*, and *Thinopyrum*. Species now treated as members of *Elymus* include all the species formerly treated in *Sitanion* and some of those from *Agropyron*. See discussion under *Agropyron*. North Asian *Elymus ciliaris* and *E. tsukushiensis* were collected in Portland 1899–1903 but have not persisted. Both have one spikelet per node, lemmas with awns curving outwards at maturity and widest leaves to 10 mm wide when flat. Reports of *E. caninus* from Oregon are based on misidentifications of these specimens. *Elymus caninus* can be distinguished from all our other *Elymus* species by the hairy inner surface of its glumes.

When keying *Elymus*, measurements of spikes include the awns. Measurements of spikelets, glumes, and lemmas do not include the awns (unless specified). Rachis internodes are measured in the middle of the spike. Glume widths of lanceolate to linear glumes are measured at the widest point. Glume widths of linear to bristle-like glumes are measured about 5 mm above the base of the glumes.

Key to *Elymus* species:

1. Spikelets 1 at all or most nodes.
   2. Inflorescence disarticulating in the rachis when mature; anthers 1.5–2.6 mm . . . . **E. ×saundersii**
   2' Inflorescence not disarticulating in the rachis; anthers 2–6 mm.
      3. Lemmas with awns curving outwards at maturity.
         4. Glume veins 1–3, glumes narrowly lanceolate . . . . ***E. wawawaiensis***
         4' Glume veins (3)4–5(7), glumes lanceolate to oblanceolate . . . . ***Pseudoroegneria spicata*** (in part)
      3' Lemmas awnless or with straight awns.
         5. Leaf blades 6–10 mm wide, usually flat; glume midveins keeled and scabrous near the tip, relatively flat and smooth toward the base; anthers 4–7 mm . . . . ***E. repens*** (in part)
         5' Leaf blades 1–6(8) mm wide, involute or flat; glume midveins flat or keeled their entire length; anthers 0.7–6 mm.
            6. Plants strongly rhizomatous; lemmas glabrous to densely long-hairy . . . . ***E. lanceolatus***
            6' Plants cespitose, occasionally loosely so; lemmas glabrous to sparsely hairy.
               7. Anthers 3.5–8 mm; spikelets mostly 1–1.25 times as long as internodes . . . . ***Pseudoroegneria spicata*** (in part)
               7' Anthers 0.5–2.5 mm; spikelets mostly 2–3 times as long as internodes.
                  8. Glumes 3(5)-veined; hyaline margins of glumes unequal, usually widest in the distal third, the wider margins 0.3–1 mm wide . . . . ***E. violaceus***
                  8' Glumes 3–7-veined, hyaline margins of glumes equal, widest at or slightly below mid-length, 0.1–0.5 mm wide . . . . ***E. trachycaulus*** ssp. ***trachycaulus***

1' Spikelets 2–3(5) at all or most nodes.
   9. Both glumes awn-like or awl-like, 10–135 mm including awns, sometimes split lengthwise, flexuous to outcurving from near the base; inflorescence axis disarticulating at maturity.
      10. Glume awns split into 3–9 divisions; lemma awns about 0.2 mm wide at the base; inflorescence axis internodes 3–5.5 mm . . . . ***E. multisetus***
      10' Glume awns entire or split into 2(3) divisions; lemma awns about 0.4 mm wide at the base; inflorescence axis internodes 3–10(15) mm . . . . ***E. elymoides***
   9' One or both glumes not awn-like or awl-like, 0–43 mm including awns, entire, straight, or outcurving from well above the base; inflorescence axis not disarticulating at maturity.
      11. Inflorescence nodding.
         12. Distal lemma margins with hairs longer than those elsewhere on the lemma; glume awns 1–10 mm . . ***E. hirsutus***
         12' Lemma margins glabrous or with uniform hairs; glume awns (5)10–25(27) mm . . ***E. canadensis*** var. ***canadensis***
      11' Inflorescence stiffly erect.
         13. Inflorescence axis disarticulating tardily . . . . **E. ×hansenii**
         13' Inflorescence axis not disarticulating.
            14. Plants cespitose to more or less rhizomatous; glumes not keeled; lemma awns (0)1–30(35) mm; anthers 1.5–3.5 mm . . . . ***E. glaucus***
            14' Plants strongly rhizomatous; glumes keeled distally; lemma awns 0.2–4(10) mm; anthers 4–7 mm . . . . ***E. repens*** (in part)

Elymus canadensis Linnaeus var. ***canadensis***

Canada wildrye

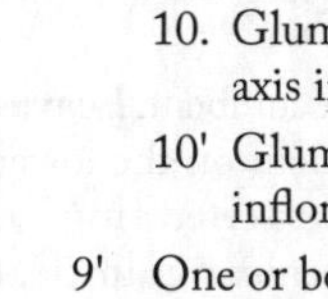

Plants (40)60–150(180) cm tall; cespitose, rarely with rhizomes to 4 cm long and 1–2 mm thick, often glaucous. **Culms** erect or decumbent. **Leaves** sheaths smooth or minutely scabrous, glabrous or hairy, often reddish brown; blades (3)4–15(20) mm wide; upper surfaces usually smooth or minutely scabrous and glabrous, rarely with sparse long hairs. **Inflorescences** (6)10–25(30) cm, usually nodding to almost pendent, often strongly glaucous, (1)3(5) spikelets per node; internodes 4–7 mm; disarticulation usually above the glumes and below each floret or rarely below the glumes. **Spikelets** 12–20 mm, more or less divergent, (2)3–5(7) florets. **Glumes** 11–40 mm including the awns, the basal 0–1 mm subterete and slightly hardened, glume bodies 6–13 × 0.5–1.6 mm, linear-lanceolate to almost awn-like, entire, widening or parallel-sided above the base, 3–5-veined, glabrous to scabrous-ciliate, rarely hairy on the veins; margins firm, not hyaline or scarious; glume awns (5)10–25(27) mm, straight to outcurved. **Lemmas** 8–15 mm; margins glabrous or with uniform hairs; lemma awns 15–40 mm, moderately to strongly outcurved. **Anthers** 2–3.5 mm.

Prairies, stream banks, ditches, roadsides. 100–1400 m. BW, Owy. ID, NV, WA; throughout most of North America. Native.

Elymus canadensis is a broad-leaved wildrye with a thick, nodding spike and conspicuous awns. It is scarce in Oregon and is found mainly in the Snake River drainage.

Elymus elymoides (Rafinesque) Swezey
Squirreltail

Plants perennial, 8–65(77) cm tall; cespitose. **Culms** erect. **Leaves** sheaths glabrous, scabrous, puberulent, or densely white-hairy; blades (1)2–4(6) mm wide, outer surfaces glabrous to puberulent, inner surfaces scabrous, short-hairy, hirsute or white-hairy. **Inflorescences** 3–20 cm, erect to slightly nodding, 2–3 spikelets per node or rarely 1 at some nodes; internodes 3–10(15) mm; disarticulation initially in the inflorescence axis, later beneath each floret. **Spikelets** 10–20 mm, divergent, sometimes glaucous, at least 1 spikelet at a node with 2–4(5) florets, 1–4(5) florets fertile, sometimes lowest floret sterile and glume-like or all florets sterile in the lateral spikelets. **Glumes** 20–135 mm including the awns, the bases hard and glabrous, the glume bodies 5–10 × 1–3 mm, linear to awn-like, 1–3-veined; margins firm; glume awns 15–125 mm, scabrous, sometimes split into 2(3) unequal divisions, flexuous to outcurving from near the base at maturity. **Lemmas** fertile lemmas 6–12 mm, glabrous, scabrous, or appressed-pubescent, 2 lateral veins extending into bristles to 10 mm; lemma awns 15–120 × 0.4 mm at the base, often reddish or purplish, scabrous, flexuous to curved near the base. **Anthers** 0.9–2.2 mm. **2*n***=28.

4 subspecies; 3 subspecies treated in *Flora*.

Similar *E. multisetus* has glumes divided well above the base into three to nine unequal segments. Some subspecies of *Elymus elymoides* have sterile florets that look like extra glumes. If a spikelet seems to have three glumes, one is the reduced lemma of a sterile floret. Three subspecies of *E. elymoides* occur in Oregon. *Elymus elymoides* ssp. *hordeoides* seems to have a coherent range and ecology, but the other two subspecies are widespread with overlapping distribution and habitats. The small anthers of *E. elymoides* suggest that self-pollination may be common in the species, creating complicated patterns of variation. *Elymus elymoides* forms hybrids with *Leymus triticoides* or *L. cinereus* called ×*Elyleymus aristatus*. The hybrid is sterile and has tardily dehiscent inflorescence axes, (2)3 spikelets per node, subulate glumes with relatively short awns and often short rhizomes if *L. triticoides* is one of the parents.

Key to *Elymus elymoides* subspecies:

1. Inflorescence with 3 spikelets per node, the central spikelet usually with 2 fertile florets, the florets of the lateral spikelets rudimentary to awn-like; lemma awns 15–30 mm . . . . ***E. e.*** ssp. ***hordeoides***

1' Inflorescence usually with 2 spikelets per node, each spikelet usually with (1)2–4(5) fertile florets; lemma awns 25–120 mm.

  2. One or more of the spikelets at most nodes appearing to have 3 glumes, 1 of these actually the lemma of a sterile lowermost floret; paleas usually with the veins extended as bristles; glume awns entire or split . . . . ***E. e.*** ssp. ***elymoides***

  2' No spikelets appearing to have 3 glumes, the lowermost floret of each spikelet well developed; paleas rarely with veins extended as bristles; glume awns entire . . . . ***E. e.*** ssp. ***brevifolius***

Elymus elymoides (Rafinesque) Swezey ssp. *brevifolius* (J.G. Smith) Barkworth
Longleaf squirreltail

Sitanion hystrix (Nutt.) J.G. Sm. var. *brevifolium* (J.G. Sm.) C.L. Hitchc.

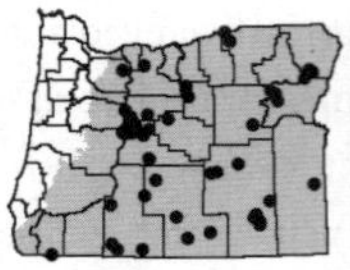

Culms 25–65(77) cm, erect. **Leaves** blades usually puberulent on the lower surface but sometimes glabrous. **Spikes** 7–20 cm, usually exserted from uppermost culm leaf sheath, usually with 2 spikelets per node. **Spikelets** with (1)2–4(5) florets; lowermost floret functional. **Glumes** awns 50–125 mm, entire. **Lemmas** awns 50–120 mm. **Paleas** rarely with the veins extended as bristles.

Dry, rocky soil, shrub steppe, grassland, and open forest, low elevation to alpine habitats. 50–2900 m. BR, BW, Casc, Col, ECas, Lava, Owy, Sisk. CA, ID, NV; east to KS, south to Mexico. Native.

In this subspecies, the lowermost floret in each spikelet is usually functional, so there are few or no reduced, awn-like lemmas.

Elymus elymoides (Rafinesque) Swezey ssp. *elymoides* (p. 401)
Common squirreltail

Sitanion hystrix (Nutt.) J.G. Sm. var. *hystrix*; *Sitanion hystrix* (Nutt.) J.G. Sm. var. *californicum* (J.G. Sm.) F.D. Wilson; *Elymus elymoides* (Raf.) Swezey ssp. *californicus* (J.G. Sm.) Barkworth

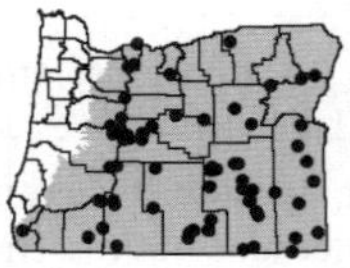

Culms 8–45 cm, erect to decumbent. **Leaves** blades glabrous or puberulent on the lower surface. **Spikes** 3–15 cm, exserted from or partly included in uppermost culm leaf sheath, usually with 2 spikelets per node. **Spikelets** with (1)2–3(4) fertile florets, lowest 1–2 florets sterile and glume-like. **Glumes** awns 15–85 mm, entire or split into 2 or 3 unequal divisions. **Lemmas** awns 25–75 mm, usually exceeded by those of the glumes. **Paleas** often with the veins extended as 1–2 mm bristles.

Dry, rocky soil, shrub steppe, grassland, and open forest, to alpine, often in disturbed sites. 100–2800 m. BR, BW, Casc, Col, ECas, Lava, Owy, Sisk. CA, ID, NV, WA; north to British Columbia, east to Saskatchewan, ND, and OK, south to Mexico. Native.

Elymus elymoides ssp. *elymoides* is characterized by the presence of awn-like reduced lemmas in one or more of the florets at each node, so that some of the spikelets appear to have three glumes.

Elymus elymoides (Rafinesque) Swezey ssp. *hordeoides* (Suksdorf) Barkworth (p. 401)
Barley squirreltail

Sitanion hystrix (Nutt.) J.G. Sm. var. *hordeoides* (Suksd.) C.L. Hitchc.

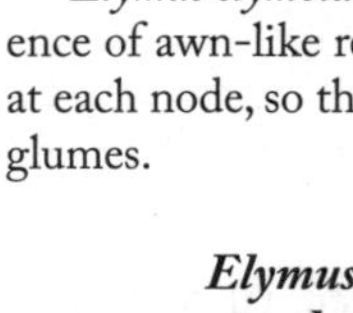

Culms 10–20 cm, erect. **Leaves** blades glabrous or puberulent on the lower surface. **Spikes** 3–6 cm, exserted or partly included, 3 spikelets per node. **Spikelets** in the central position usually with 2 fertile florets, the lateral spikelets usually with rudimentary to awn-like florets.

Glumes awns 15–50 mm, usually entire. **Lemmas** awns of the fertile florets 15–30 mm. **Paleas** with or without distinct bristles.

Dry, rocky areas, shallow soils. 400–1800 m. BW, Col, Lava. CA, ID, NV, WA. Native.

Plants of this subspecies are short and have three spikelets per node. They can be recognized by the six or more awn-like glumes and awn-like sterile lemmas at each node. The deeply divided glumes of *E. multisetus* can lead to confusion with *E. e.* ssp. *hordeoides*, but *E. multisetus* glumes are not divided all the way to the base. Dwarf alpine forms of other *E. elymoides* subspecies are sometimes misidentified as *E. e.* ssp. *hordeoides* because of their short stature. However, *E. e.* ssp. *hordeoides* grows at lower elevations, and alpine plants are usually *E. e.* ssp. *brevifolius*.

Elymus glaucus Buckley
Blue wildrye

Plants perennial, 30–140(200) cm tall; cespitose, sometimes somewhat rhizomatous, often glaucous. **Culms** erect or slightly decumbent. **Leaves** sheaths glabrous, scabrous, or short- to long-hairy, those of the lower leaves often purplish; blades 2–13(17) mm wide, usually lax and flat or sometimes involute; upper surfaces glabrous, scabrous or with appressed hairs on the veins, sometimes with longer often spreading hairs. **Inflorescences** 5–21 cm, usually erect, 2 spikelets per node or rarely 1 or 3; internodes 4–8(12) mm; disarticulation above the glumes and below each floret or rarely also below the glumes. **Spikelets** 8–25 mm, appressed to slightly divergent, (1)2–4(6) florets. **Glumes** 75% as long as or equaling the adjacent lemmas, bases often overlapping, usually flat and thin with evident veins, glume bodies (5)9–14(19) × 0.5–2 mm, linear-lanceolate, entire, widening above the base, (1)3–5(7)-veined, 2–3 veins extending to the tips, glabrous; veins smooth or evenly scabrous; margins 0.1–0.2 mm wide, whitish hyaline, tapering toward the tips; glume awns 0–5(9) mm, straight. **Lemmas** (8)9–14(16) mm, glabrous, scabrous, or short-hirsute; lemma awns, if present, (0)1–30(35) mm, usually straight to flexuous, sometimes slightly curving. **Anthers** 1.5–3.5 mm. **2*n*=28**.

3 subspecies; 2 subspecies treated in *Flora*.

Rare *E. glaucus* individuals with a single spikelet at each node can be confused with *E. trachycaulus*, which usually has shorter anthers, less tapered glumes that often lack awns, and stiffer, narrower leaves that are more basally concentrated. *Elymus glaucus* hybridizes with *E. elymoides*, *E. multisetus*, and *E. trachycaulus*, as well as species of *Leymus* and *Hordeum*. The intrageneric hybrids are often fertile.

Key to *Elymus glaucus* subspecies:

1. Lemma awns (5)10–25(35) mm; glume awns 0.5–5(9) mm; widely distributed . . . ***E. g.* ssp. *glaucus***

1' Lemma awns (0)1–5(7) mm; glume awns 0–2 mm; coastal . . . . ***E. g.* ssp. *virescens***

Elymus glaucus Buckley ssp. *glaucus* (p. 401)
Blue wildrye

Elymus glaucus Buckley ssp. *jepsonii* (Burtt Davy) Gould, *Elymus glaucus* Buckley var. *breviaristatus* Burtt Davy, *Elymus glaucus* Buckley var. *glaucus*, *Elymus glaucus* Buckley var. *jepsonii* Burtt Davy

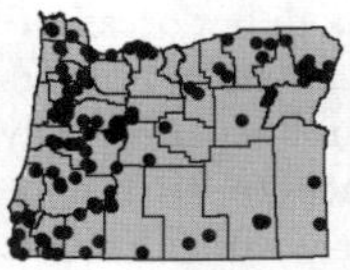

Leaves sheaths glabrous, scabrous or pubescent; blades usually 4–17 mm wide; upper surfaces glabrous or with appressed hairs or occasionally with spreading hairs, the hairs of fairly uniform length. **Glumes** awns 0.5–5(9) mm. **Lemmas** awns (5)10–25(35) mm.

Meadows and open woods. 0–2900 m. Throughout Oregon. CA, ID, NV, WA; north to AK, east to NY, southeast to TX, south to Mexico. Native.

This is the common subspecies of *E. glaucus* found throughout the range of the species. Its lemmas and glumes are nearly always awned.

Elymus glaucus Buckley ssp. *virescens* (Piper) Gould (p. 401)
Coastal blue wildrye

Elymus virescens Piper

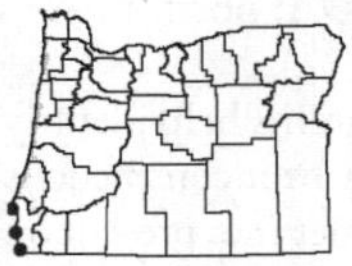

Leaves sheaths glabrous or scabrous; blades 2–10 mm wide, smooth or minutely scabrous to pubescent. **Glumes** awns 0–2 mm. **Lemmas** awns (0)1–5(7) mm.

Coastal cliffs, thickets, and forest edges. 50–1400 m. Est. CA, WA; north to AK. Native.

Elymus glaucus ssp. *virescens* is a coastal, awnless or short-awned form of the common *E. glaucus*. It grows inland as far as Josephine County in southwestern Oregon. Occasional awnless individuals of *E. glaucus* growing further inland are abnormal *E. g.* ssp. *glaucus*, hybrids involving that species, or misidentified *E. trachycaulus*. *Elymus trachycaulus* has a single spikelet at all or most nodes.

Elymus × *hansenii* Scribner
Hansen's wheatgrass

Sitanion anomalum J.G. Sm., *Sitanion hansenii* (Scribn.) J.G. Sm.

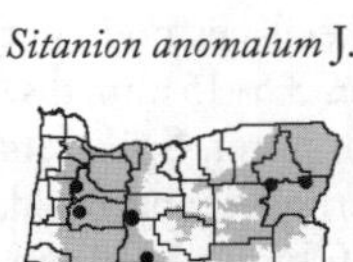

Plants perennial, 60–130 cm; cespitose, not rhizomatous. **Culms** erect to geniculate. **Leaves** sheaths smooth; blades 10–30 cm × (2)5–8 mm, flat or the margins involute, smooth or sometimes sparsely hairy. **Inflorescences** 5–20 cm × 0.5 mm without awns, 30–70 mm wide with awns, erect, with 2(3) spikelets per node; internodes to about 10 mm; disarticulation in the rachises, tardy. **Spikelets** 10–15 mm, (1)3–5 florets. **Glumes** 6.5–12 × 0.5–1 mm, narrowly lanceolate, 3(5)-veined; glume awns (0)25–35+ mm. **Lemmas** 8.5–14 mm, glabrous, scabrous or short-hairy; lemma awns 10–50 mm, outcurving. **Anthers** 1.1–2.6 mm.

Sagebrush steppe, open forest. 1000–2500 m. BW, Casc, ECas, Sisk, WV. CA, ID, NV, WA; east to CO. Native.

Elymus × *hansenii* is a hybrid of *E. glaucus* and *E. elymoides* or *E. multisetus*. Its inflorescence axis disarticulates, and it has two spikelets per node. This hybrid may be more variable than the description suggests.

Elymus hirsutus J. Presl

Northwestern wildrye

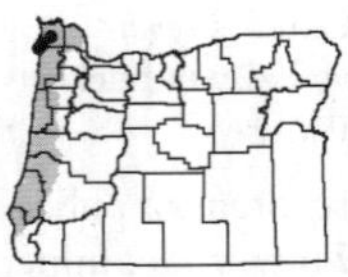

Plants perennial, 40–140 cm tall; cespitose, sometimes with short rhizomes. **Culms** usually somewhat decumbent. **Leaves** sheaths glabrous and smooth but occasionally minutely scabrous or hairy; blades 4–12 mm wide; upper surfaces pilose or villous but occasionally puberulent or minutely scabrous. **Inflorescences** 6–20 cm, nodding to pendent, 2 spikelets per node or rarely with 3 at some nodes; internodes 3–10(12) mm; disarticulation above the glumes and beneath each floret. **Spikelets** 12–20 mm, appressed to divergent, 2–4(7) florets. **Glumes** (4.5)7–10(11) × 0.7–1.5 mm, bases flat, occasionally hard for 0.5 mm; veins usually evident, linear-lanceolate, entire, widening or parallel-sided above the base, 3–5-veined, usually scabridulous to scabrous; veins occasionally hirsute beyond mid-length; margins hyaline or scarious; glume awns 1–10 mm, straight. **Lemmas** 7–14 mm, smooth or minutely scabrous; lateral veins hairy; margins hairy beyond mid-length; marginal hairs 0.5–1 mm and longer than those elsewhere on the lemma; lemma awns (2)8–30 mm, flexuous to moderately outcurving. **Anthers** 2–3.5 mm.

Forest edges, thickets and grasslands near the coast and in coastal mountains. 0–1000 m. CR, Est. WA; north to AK. Native.

The nodding spikes and ciliate lemma margins help identify *Elymus hirsutus*. The lemma margin hairs may be inconspicuous and should not be confused with the hairs sometimes present on the palea keels of *Elymus* species. *Elymus hirsutus* occasionally hybridizes with *Elymus glaucus*, *Hordeum brachyantherum*, and *Leymus mollis*.

Elymus lanceolatus (Scribner & J.G. Smith) Gould

Thickspike wheatgrass

Plants perennial, 22–130 cm, strongly rhizomatous; shoots often arising in clusters from the long slender rhizomes; foliage sometimes glaucous. **Culms** erect. **Leaves** sheaths glabrous or pubescent; blades 1.5–6 mm wide, outer surfaces usually glabrous, inner surfaces with appressed hairs. **Inflorescences** 3.5–26 cm, erect to slightly nodding, 1 spikelet per node but occasionally with 2 at a few nodes; internodes 3.5–15 mm; disarticulation above the glumes, beneath each floret. **Spikelets** 8–31 mm, 1.5–3 times longer than the internodes, appressed, 3–11 florets, glabrous or hairy, hairs to 1 mm. **Glumes** 5–14 mm, 50–75% the length of the adjacent lemmas, 0.7–1.3 mm wide, lanceolate, glabrous or hairy, smooth or scabrous, 3–5-veined, flat or weakly keeled; keels straight; margins narrow, tapering from the base or from beyond mid-length; tips acute to acuminate, sometimes mucronate or short-awned. **Lemmas** 7–12 mm, glabrous to densely long-hairy, all hairs similar, sometimes scabrous; tips acute to awned; lemma awns, if present, to 2 mm, straight. **Anthers** 2.5–6 mm.

3 subspecies.

Elymus lanceolatus is a strongly rhizomatous wheatgrass that lacks or nearly lacks awns. It is most similar to *Pascopyrum smithii*, which has more basally concentrated leaves and glumes that taper from near the base and have curved midveins. It is also similar to *E. trachycaulus*, which is cespitose, has consistently glabrous lemmas, and has glumes 75–100% as long as the lemmas.

Key to *Elymus lanceolatus* subspecies:

1. Lemmas densely hairy, hairs flexible, some hairs 1 mm or longer. . . . . ***E. l.*** ssp. ***psammophilus***

1' Lemmas glabrous or with stiff hairs shorter than 1 mm.

  2. Lemmas with hairs, not scabrous. . ***E. l.*** ssp. ***lanceolatus***

  2' Lemmas smooth, sometimes scabrous distally but mostly glabrous, sometimes the lemma margins hairy proximally. . . . ***E. l.*** ssp. ***riparius***

Elymus lanceolatus (Scribner & J.G. Smith) Gould ssp. *lanceolatus*

Thickspike wheatgrass

Agropyron dasystachyum (Hook.) Scribn. var. *dasystachyum*

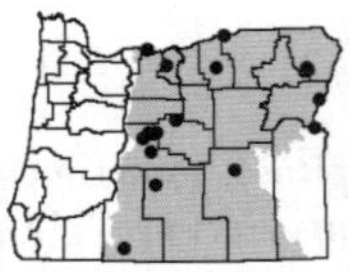

Culms 60–130 cm. **Spikes** 10–22 cm; internodes 7–15 mm, smooth, scabrous, or hairy distally. **Spikelets** 10–28 mm. **Lemmas** not scabrous, moderately hairy, hairs stiff, hairs shorter than 1 mm. **2*n*=28**.

Grasslands, sagebrush steppe, juniper woodlands, and dry forest openings on clay, sand, loam, and rocky soils. 50–1700 m. BR, BW, Col, ECas, Lava. CA, ID, NV, WA; north to AK, east to Ontario, southeast to NM. Native.

Elymus lanceolatus ssp. *lanceolatus* is similar to *Pascopyrum smithii*, which differs in having basally concentrated leaves, glumes that taper from mid-length or below, and glume midveins that "lean" to the side near the glume tip.

Elymus lanceolatus (Scribner & J.G. Smith) Gould ssp. *psammophilus* (J.M. Gillett & H. Senn) Á. Löve

Sand-dune wheatgrass

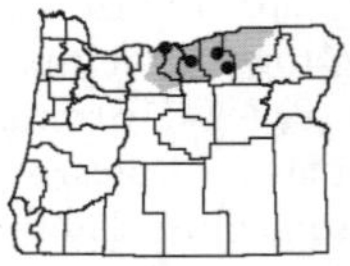

Culms 20–95 cm. **Spikes** 4–26 cm; internodes 3.5–13 mm, hairy at least distally. **Spikelets** 9–31 mm. **Lemmas** densely hairy, hairs flexible, usually many longer than 1 mm. **Paleas** hairy between the keels; keels hairy proximally.

Sandy soils in inland dunes and bunchgrass prairie. 0–1000 m. Col. WA; north to Yukon, northeast to Ontario. Native.

This rhizomatous wheatgrass of sandy soils is recognized by its densely hairy lemmas.

Elymus lanceolatus (Scribner & J.G. Smith) Gould ssp. *riparius* (Scribner & J.G. Smith) Barkworth

Streambank wheatgrass

Agropyron dasystachyum (Hook.) Scribn. var. *riparium* (Scribn. & J.G. Sm.) Bowden, *Agropyron riparium* Scribn. & J.G. Sm.

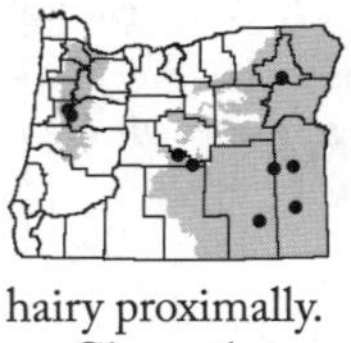

Culms usually 22–60 cm. **Spikes** 6–10 cm; internodes 3.5–10 mm, glabrous or sometimes scabrous. **Spikelets** 10–17 mm. **Lemmas** smooth or sometimes scabrous distally, mostly glabrous but margins sometimes hairy proximally.

Clay soils in mesic to moist habitats, alkaline meadows, riparian terraces, sagebrush steppe. 50–1800 m. BR, BW, Owy, WV. CA, ID, NV, WA; Widespread in western North America. Native.

Elymus lanceolatus ssp. *riparius* is a strongly rhizomatous wheatgrass that grows in moister habitats than the other *E. lanceolatus* subspecies. Other subspecies have more or less hairy lemmas.

Elymus multisetus (J.G. Smith) Burtt Davy

Big squirreltail

Sitanion jubatum J.G. Sm.

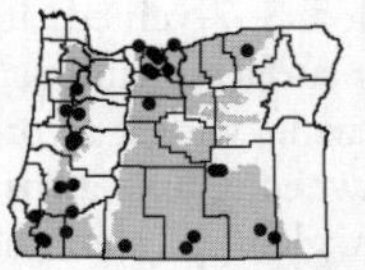

Plants perennial, 15–65 cm tall; cespitose. **Culms** erect to ascending. **Leaves** sheaths glabrous or white-villous; blades 1.5–4 mm wide; upper surfaces scabrous or hairy. **Inflorescences** 5–20 cm, erect, sometimes partially enclosed in the sheath of the uppermost leaf, 2 spikelets per node or rarely 3–4 at some nodes; internodes 3–5(8) mm; disarticulation initially at the rachis nodes, later beneath each floret. **Spikelets** 10–15 mm, divergent, 2–4 florets, lowest florets sterile and glume-like in 1 or both spikelets at each node. **Glumes** (10)30–100 mm including the awns, the bases hard and glabrous, glume bodies about (2)5–10 × 1–2 mm, setaceous, 2–3-veined; margins firm; glume awns (8)25–90 mm, each split above the base into 3–9 unequal divisions, scabrous, flexuous to outcurving from near the bases at maturity. **Lemmas** fertile lemmas 8–10 mm, smooth or scabrous near the tips, 2 lateral veins extending into bristles to 10 mm; lemma awns (10)20–110 × 0.2 mm at the base, divergent to arching. **Anthers** 1–2 mm. **2*n***=28.

Dry, often rocky grasslands and savannas. 50–2000 m. BR, Col, ECas, Lava, Sisk, WV. CA, ID, NV, WA; southeast to CO, south to Mexico. Native.

Elymus multisetus has inflorescences that fall apart at maturity and glumes that are divided into three to nine widely spreading awns. Very similar *E. elymoides* has glumes that are entire or unevenly split into two to three parts. Determining how much the glumes are split is complicated by the presence of glume-like sterile lemmas in both *E. multisetus* and some *E. elymoides* subspecies. *Elymus multisetus* glumes are divided above the base. If the glumes appear to be divided to the base, the plant is more likely to be *E. elymoides*.

Elymus repens (Linnaeus) Gould

Quackgrass

Agropyron repens (L.) P. Beauv., *Elytrigia repens* (L.) Desv. ex B.D. Jacks.

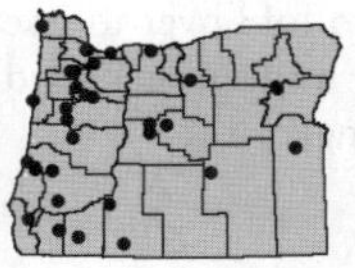

Plants perennial, 50–100 cm tall, strongly rhizomatous, sometimes glaucous. **Culms** erect to geniculate. **Leaves** sheaths with spreading hairs or glabrous proximally; blades 6–10 mm wide, usually flat; lower surfaces glabrous or sparsely pilose; upper surfaces usually with sparse spreading hairs over the veins but sometimes glabrous. **Inflorescences** 5–15 cm, 1 spikelet per node but occasionally with 2 at a few nodes; internodes 4–6(9.5) mm; disarticulation above the glumes, beneath each floret. **Spikelets** 10–27 mm, appressed to ascending, 4–7 florets. **Glumes** oblong, glabrous,

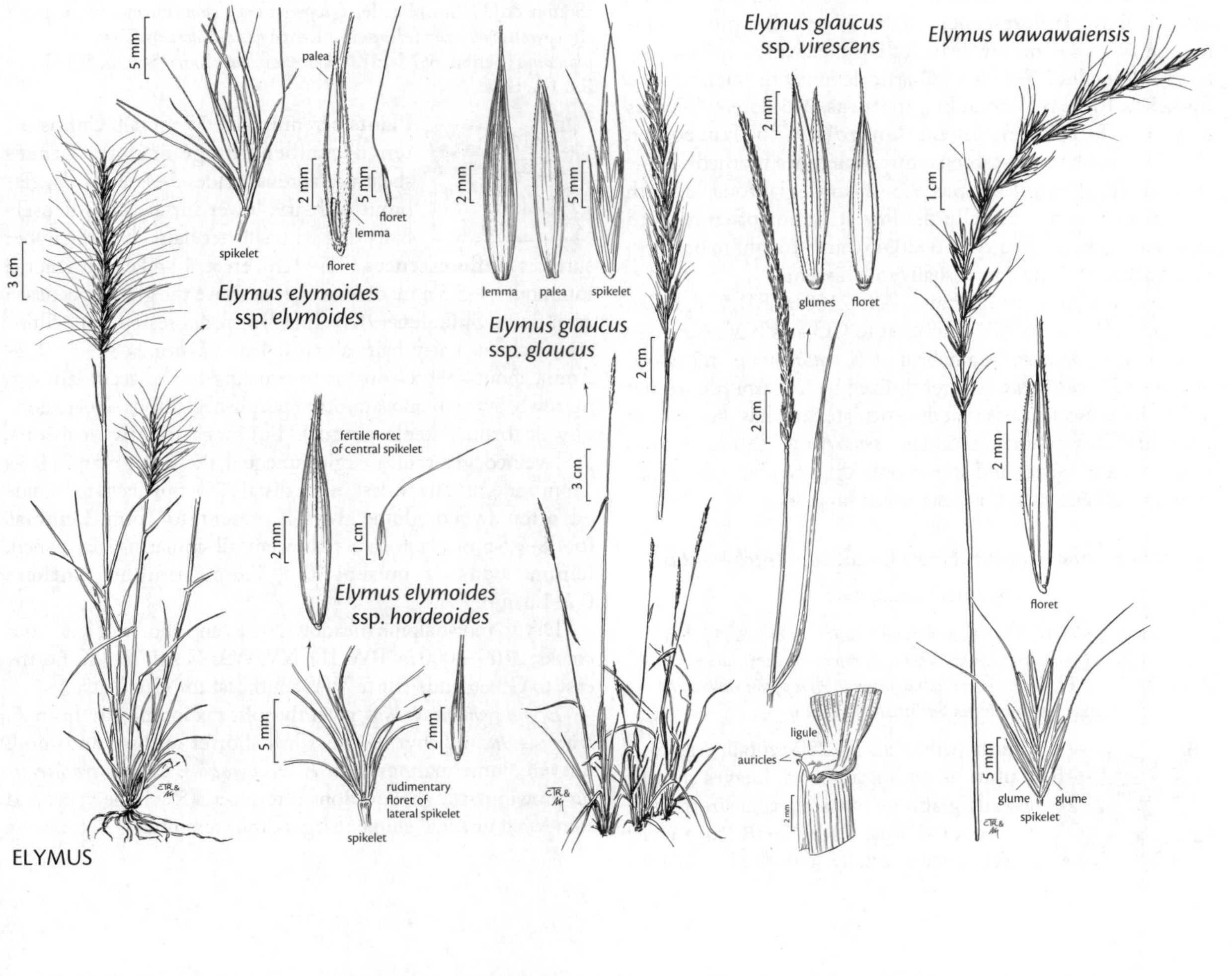

ELYMUS

the midveins scabrous and keeled near the tip but smooth and relatively flat near the base, hyaline margins present in the distal half; tips acute to acuminate, awnless or awned, to 3 mm; lower glumes 8.8–11.4 mm, 3–6-veined; upper glumes 7–12 mm, 5–7-veined. **Lemmas** 8–12 mm, glabrous, mostly smooth but sometimes minutely scabrous distally, awnless or awned; lemma awns 0–4(10) mm, straight. **Anthers** 4–7 mm. **2*n***=22, 42.

Disturbed areas, roadsides, ditches, streamsides. 0–2400 m. Throughout Oregon. CA, ID, NV, WA; throughout most of North America; cool temperate regions worldwide. Exotic.

Elymus repens is a strongly rhizomatous, nearly awnless wheatgrass most easily identified by its broad, flat leaves. Occasional individuals of *E. lanceolatus* and *E. trachycaulus* have leaf widths overlapping those of *E. repens,* but (ideally) an examination of the whole stand should reveal the difference. In addition, *E. trachycaulus* is cespitose. The midvein of the glume in *E. repens* is scabrous and keeled near the tip but smooth and not truly keeled near the base. In dried specimens, the keel may be observed as a tendency of the midvein to fold over to one side. In our similar native wheatgrasses, the midvein is keeled or not and scabrous or not throughout its length.

Elymus × *saundersii* Vasey

Saunder's wildrye

× *Agrositanion saundersii* (Vasey) Bowden

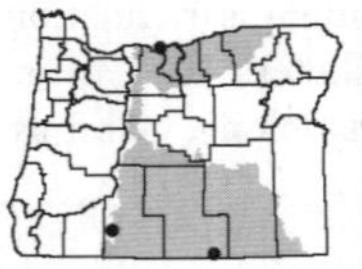

Plants perennial, 50–80 cm tall; cespitose, not rhizomatous. **Culms** erect, glabrous or pilose. **Leaves** somewhat basally concentrated; sheaths retrorsely hairy; blades 10–15 cm × 2.5–5 mm, flat, scabrous below, smooth above. **Inflorescences** (4)10–25 cm × 2–3 mm excluding the awns, 3–7 mm wide including the awns, 1 spikelet per node; internodes 2.5–11 mm; disarticulation at the rachis nodes. **Spikelets** 10–25 mm excluding the awns, 2–6 florets. **Glumes** 6–8 × 0.8–1.2 mm, linear-lanceolate to lanceolate, (1)3(5)-veined; veins scabrous; tips sometimes toothed; glume awns 0–5(13) mm. **Lemmas** 9.5–13 mm, glabrous, smooth proximally, scabrous distally, the lowest lemma often reduced and glume-like; lemma awns 6–20(45) mm, straight to outcurving. **Anthers** 1.5–2.6 mm, usually indehiscent.

Open, dry to mesic habitats. 1700–2100 m. BR, Col, ECas. CA, ID, NV; east to WY, southeast to CO and NM. Native.

Elymus × *saundersii* is a hybrid of *E. trachycaulus* and *E. elymoides* or *E. multisetus*. It is recognized by its cespitose habitat and an inflorescence axis that disarticulates and has one spikelet per node. This hybrid resembles *Pseudelymus saxicola* but has awns that are shorter and spread only slightly. This hybrid may be more variable than the description suggests.

Elymus trachycaulus (Link) Gould ssp. *trachycaulus*

Slender wheatgrass

Agropyron caninum (L.) P. Beauv. var. *andinum* (Scribn. & J.G. Sm.) C.L. Hitchc., *Agropyron tenerum* Vasey, *Agropyron trachycaulum* (Link) Malte ex H.F. Lewis var. *trachycaulum*, *Agropyron violaceum* (Hornem.) Lange var. *andinum* Scribn. & J.G. Sm.

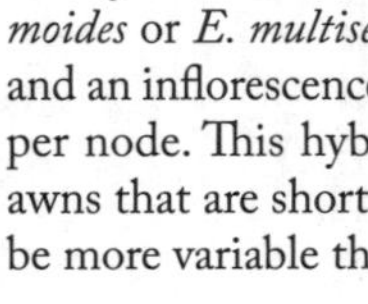

Plants perennial, 30–150 cm tall; cespitose. **Culms** ascending to erect. **Leaves** sheaths usually glabrous but sometimes retrorsely hirsute or villous; blades 2–5(8) mm wide; lower surfaces usually smooth and glabrous but sometimes hairy; upper surfaces usually glabrous but sometimes conspicuously hairy. **Inflorescences** (4)8–30 cm, erect, 1 spikelet at all or most nodes; internodes 8–15 mm; disarticulation above the glumes and beneath each floret. **Spikelets** 9–17(20) mm, usually at least twice as long as the internodes, 3–6 mm wide, appressed, 3–9 florets, their bases visible; rachillas glabrous or hairy, hairs to 0.3 mm. **Glumes** 5–17 mm, from 75% as long or longer than the adjacent lemmas, 1.8–3.2 mm wide, lanceolate to narrowly ovate, green but purple at higher latitudes and elevations, flat or asymmetrically keeled, 3–7-veined, at least 1 vein scabrous to near the base, inner surfaces glabrous; margins hyaline or scarious, 0.2–0.5 mm wide, widest at or slightly beyond mid-length; tips acute to awned; glume awns less than 2 mm, straight. **Lemmas** 6–13 mm, glabrous, smooth proximally but often minutely scabrous distally over the veins; tips acute, awnless or awned; lemma awns, if present, to 5 mm, straight. **Anthers** 1–3 mm. **2*n***=28.

Prairies, savannas, sagebrush steppe, and open forest. 50–2800 m. All ecoregions except Est. CA, ID, NV, WA; throughout North America; Eurasia. Native.

Herbarium specimens of *E. trachycaulus* can be confused with *E. glaucus* individuals that have a single spike at all the nodes, but *E. trachycaulus* usually has shorter anthers and less acuminate glumes. In the field, *E. trachycaulus* may be recognized by its stiffer leaves. *Elymus trachycaulus* growing at high elevations can resemble *E. violaceus* which has shorter anthers.

Elymus violaceus (Hornemann) Feilberg

High wheatgrass

Agropyron trachycaulum (Link) Malte ex H.F. Lewis var. *latiglume* (Scribn. & J.G. Sm.) Beetle, *Agropyron violaceum* (Hornem.) Lange, *Agropyron violaceum* (Hornem.) Lange var. *violaceum*, *Elymus alaskanus* (Scribn. & Merr.) Á. Löve ssp. *latiglumis* (Scribn. & J.G. Sm.) Á. Löve

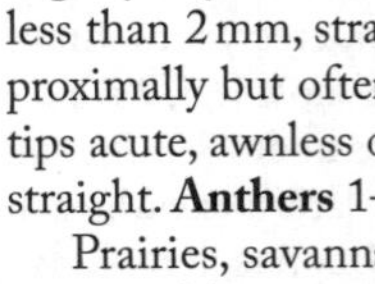

Plants perennial, 18–75 cm tall. **Culms** often decumbent or geniculate. **Leaves** sheaths glabrous; blades 3–4 mm wide, glabrous or hairy; lower surfaces less densely hairy and with shorter hairs than the upper surfaces. **Inflorescences** (3)5–9 cm, erect, 1 spikelet per node; internodes 4–5.5 mm; disarticulation above the glumes, beneath each floret. **Spikelets** (9)10–15(17) mm, appressed, (3)4–5 florets; rachillas hairy, hairs about 0.4 mm. **Glumes** 8–12 × 1.2–2 mm, about 75% as long as to equaling the adjacent lemmas, narrowly ovate to obovate, often purplish, glabrous or occasionally scabrous; keels smooth but occasionally scabrous, 3(5)-veined, glabrous; margins unequal, the wider margin 0.3–1 mm wide, usually widest in the distal 33%; tips acute to rounded, often awned; glume awns, if present, to 2 mm. **Lemmas** (6.4)8–9.5 mm, glabrous to pubescent, all similar, usually awned; lemma awns, if present, 0.5–3 mm, straight. **Anthers** 0.7–1.3 mm.

Dry to moist alpine meadows, talus, and shores of lakes and creeks. 2100–3000 m. BW. ID, NV, WA; North to AK, northeast to Greenland, east to WY, southeast to NM. Native.

Elymus violaceus was recently split taxonomically from *E. trachycaulus*. *Elymus violaceus* has shorter anthers and more uneven glume margins than *E. trachycaulus*. *Elymus trachycaulus* growing at high elevations often has 3(5) glume veins and somewhat unequal glume margins, thus resembling *E. violaceus*.

Elymus wawawaiensis (p. 401)
J.R. Carlson & Barkworth

Snake River wheatgrass

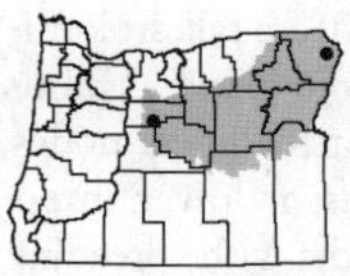

Plants perennial, (15)50–130 cm tall; cespitose, sometimes weakly rhizomatous. **Culms** erect. **Leaves** basal sheaths becoming glabrous; blades to 28 cm × 1.7–5 mm, inner surfaces densely pubescent or rarely sparsely pubescent. **Inflorescences** spikes, 5–20 cm, erect to slightly nodding, 1 spikelet per node; internodes 5–12 mm; disarticulation above the glumes, beneath each floret. **Spikelets** 10–22 mm long, about twice as long as the internodes, 2–8.5 mm wide, appressed, 4–10 florets. **Glumes** 4–10 × 0.5–1.3 mm, narrowly lanceolate, widest at or below mid-length, glabrous, often glaucous, 1–3-veined, flat or weakly keeled; margins 0.1–0.2 mm wide, widest near mid-length; tips usually acuminate, awned or awnless; glume awns 0–6 mm. **Lemmas** 6–12 mm, smooth or slightly scabrous; margins often sparsely pubescent near the base; longest lemma awns 9–28 mm, strongly divergent. **Anthers** 3.5–6 mm.

Dry, rocky canyon slopes, disturbed sagebrush steppe, roadsides. 800–1500 m. BW, Lava. ID, WA; MT, WY. Native.

Elymus wawawaiensis resembles a vigorous plant of *Pseudoroegneria spicata* with somewhat more overlapping spikelets. *Pseudoroegneria* differs in having wider glumes with (3)4–5(7) veins. *Elymus wawawaiensis* is endemic to the Salmon, Snake, and Yakima river drainages in Oregon, Idaho, and Washington. The "Secar" cultivar of *E. wawawaiensis* was developed for forage and erosion control and is planted widely in eastern Oregon and elsewhere in the arid West.

ERAGROSTIS Lovegrass

Plants annual or perennial; cespitose, stoloniferous or rhizomatous; sometimes with saucer-like glands on culms, leaves, inflorescence branches, lemmas and/or paleas. **Culms** sometimes rooting at the lower nodes, sometimes branched above the base. **Leaves** sheaths open with tufts of hairs at the tops, hairs 0.3–8 mm; ligules usually membranous and ciliate, the cilia sometimes longer than the membranous base, occasionally consisting of hairs only or of non-ciliate membranes; blades flat, folded, or involute. **Inflorescences** terminal and sometimes axillary open to contracted panicles. **Spikelets** laterally compressed, (1)2–60 florets; cleistogamous spikelets occasionally present on the axillary or terminal panicles; disarticulation below the fertile florets, the glumes sometimes deciduous, the paleas sometimes persistent, acropetal or irregular. **Glumes** 2, usually shorter than the adjacent lemmas, 1(3)-veined; tips obtuse to acute, awnless. **Calluses** glabrous or sparsely pubescent. **Lemmas** glabrous, (1)3(5)-veined, usually keeled; tips obtuse to acute, awnless or mucronate. **Paleas** 2-keeled, the keels usually ciliate. **Anthers** 2–3. **Caryopses** brown, often translucent.

Temperate, subtropical, and tropical regions worldwide. Approximately 350 species; 8 species treated in *Flora*.

Several of Oregon's *Eragrostis* species have glands on the culms, leaves, inflorescences branches, or spikelets. The glands may be discolored patches, or they may be "saucer-like" with raised edges. The glands give some species characteristic odors.

Key to *Eragrostis* species:

1. Perennial; anthers 0.6–1.2 mm . . . . ***E. curvula***
1' Annual; anthers usually 0.2–0.4 mm.
  2. Stems widely creeping, stoloniferous, forming mats . . . . ***E. hypnoides***
  2' Stems erect to decumbent, not creeping, not forming mats.
    3. Saucer-like glands present on the panicle branches, leaf blades, culms and/or on the keels of the glumes and lemmas.
      4. Leaf blade margin, glume keel, and lemmas without glands, other plant parts glandular.
        5. Inflorescence branches ascending to appressed; lower glumes (0.7)0.9–1.4 mm; upper glumes 1.2–1.8 mm . . . . ***E. lutescens***
        5' Inflorescence branches spreading; lower glumes 0.3–0.6(0.8) mm; upper glumes 0.7–1.2(1.4) mm . . . . ***E. pilosa*** var. ***pilosa*** (in part)
      4' Leaf blade margin with conspicuous glands; glume keel, lemmas, and other parts usually glandular.
        6. Spikelets 2–4 mm wide . . . . ***E. cilianensis***
        6' Spikelets usually less than 2 mm wide . . . . ***E. minor***
    3' Saucer-like glands lacking.
      7. Paleas deciduous; shorter glume 0.3–0.6(0.8) mm and less than 50% as long as lowest floret; lowest inflorescence branches usually whorled . . . . ***E. pilosa*** var. ***pilosa*** (in part)
      7' Paleas persisting after the lemma falls; shorter glume 0.5–1.7 mm and at least 50% as long as lowest floret; lowest inflorescence branches usually solitary or paired.
        8. Caryopses with a shallow groove on the side opposite the embryo . . . . ***E. mexicana*** ssp. ***virescens***
        8' Caryopses not grooved. . . . . ***E. pectinacea*** var. ***pectinacea***

Eragrostis cilianensis (Allioni) Vignolo ex Janchen (p. 405)

Stinkgrass

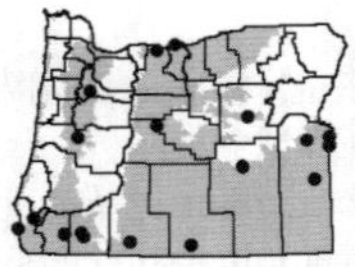

Plants annual, 15–45(65) cm tall, tufted. **Culms** erect or decumbent, sometimes with saucer-like glands below the nodes. **Leaves** sheaths glabrous or occasionally glandular, tops with hairs to 5 mm; blades (1)5–20 cm × (1)3–5(10) mm; lower surfaces glabrous or sometimes glandular, margins with conspicuous glands. **Inflorescences** (3)5–16(20) × 2–8.5 cm, oblong to ovate, condensed to open; primary branches 0.4–5 cm, appressed or diverging 20–80° from the inflorescence axis; pedicels 0.2–3 mm, stout, straight, stiff, usually divergent, occasionally appressed; disarticulation below the florets, each floret falling as a unit, rachillas persistent. **Spikelets** 6–20 × 2–4 mm, gray or greenish, 10–40 florets. **Glumes** broadly ovate to lanceolate, membranous, usually glandular; lower glumes 1.2–2 mm, usually 1-veined; upper glumes 1.2–2.6 mm, usually 3-veined. **Lemmas** 2–2.8 mm, broadly ovate, membranous, keels with 1–3 saucer-like glands, obtuse to acute. **Paleas** 1.2–2.1 mm, hyaline, keels scabrous or sometimes also ciliate. **Anthers** 3, 0.2–0.5 mm, yellow. **Caryopses** 0.5–0.7 mm, globose to broadly ellipsoid, not grooved. **2*n***=20.

Disturbed sites, roadsides. 0–1700 m. BR, Col, ECas, Est, Lava, Owy, Sisk, WV. CA, ID, NV, WA; throughout most of North America; Europe. Exotic.

Eragrostis cilianensis has large, dense inflorescences that seem too large for the small plants. As the common name suggests, it has an unpleasant odor. It also differs from the other weedy annual *Eragrostis* in having globose caryopses, yellow anthers, often ciliate paleas and wider spikelets.

Eragrostis curvula (Schrader) Nees

Weeping lovegrass

Eragrostis curvula (Schrad.) Nees var. *curvula*

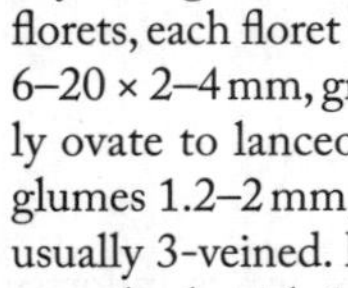

Plants perennial, (45)60–150 cm tall, cespitose, without glands. **Culms** erect, glabrous. **Leaves** mostly basal; sheaths with scattered hairs to 9 mm; blades 12–50(65) cm × 1–3 mm. **Inflorescences** 16–35(40) × (4)8–24 cm, ovate to oblong, open; primary branches 3–14 cm, diverging 10–80° from the inflorescence axis; pedicels 0.5–5 mm, appressed, flexible; disarticulation irregular or acropetal, basal rachilla internodes persistent. **Spikelets** 4–8.2(10) × 1.2–2 mm, gray to yellowish, with 3–10 florets. **Glumes** lanceolate, hyaline; lower glumes 1.2–2.6 mm; upper glumes 2–3 mm. **Lemmas** 1.8–3 mm, ovate, membranous, lateral veins conspicuous; tips acute. **Paleas** 1.8–3 mm, hyaline to membranous, obtuse. **Anthers** 3, 0.6–1.2 mm, reddish brown. **Caryopses** 1–1.7 mm, ellipsoid to obovoid, dorsally compressed, with or without a shallow, broad groove. **2*n***=40, 60, 80.

Roadsides, weedy meadows, disturbed areas. 0–900 m. BW, Casc, Col, ECas, Sisk, WV. CA, WA; east across southern US to VA and FL, south to Mexico; southern Africa. Exotic.

Eragrostis curvula is a densely cespitose plant that produces a nodding panicle in late summer. It slightly resembles a fine-leaved fescue.

Eragrostis hypnoides (Lamarck) Britton, Sterns & Poggenburg

Teal lovegrass

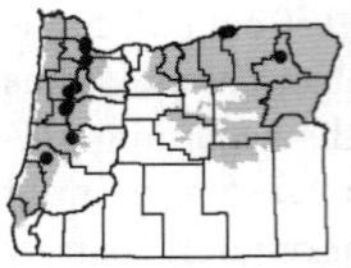

Plants annual, (2)5–12(20) cm tall, stoloniferous, mat-forming, without glands. **Culms** decumbent and rooting at the lower nodes, often branched, glabrous or lower internodes hairy. **Leaves** sheaths with spreading hairs on the margins, collars, and tips; hairs 0.1–0.6 mm; blades 0.5–2.5 cm × 1–2 mm. **Inflorescences** terminal and axillary, 1–3.5 × 0.7–2.5 cm, ovate, open to congested; primary branches 0.1–0.5 cm, appressed to strongly divergent, glabrous; pedicels 0.2–1 mm, ciliate; disarticulation acropetal. **Spikelets** 4–13 × 1–1.5 mm, loosely overlapping, greenish yellow to purplish, 12–35 florets. **Glumes** linear-lanceolate to lanceolate, hyaline; lower glumes 0.4–0.7 mm; upper glumes 0.8–1.2 mm. **Lemmas** 1.4–2 mm, ovate, strongly 3-veined, veins greenish; tips acuminate. **Paleas** 0.7–1.2 mm, hyaline, persistent, keels minutely scabrous, acute to obtuse. **Anthers** 2, 0.2–0.3 mm, brownish. **Caryopses** 0.3–0.5 mm, ellipsoid. **2*n***=20.

Muddy or sandy margins of rivers, lakes and ponds. 0–1400 m. BW, Col, CR, WV. CA, ID, NV, WA; throughout most of US, north to southeastern Canada, south to South America. Native.

Eragrostis hypnoides is a small, stoloniferous, late-season grass that forms mats on receding shorelines of streams and ponds.

Eragrostis lutescens Scribner

Sixweeks lovegrass

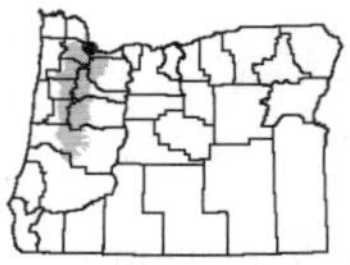

Plants annual, (2)6–25 cm tall, tufted. **Culms** glabrous with elliptical, yellowish, glandular pits below the nodes. **Leaves** sheaths with elliptical glandular pits and with sparse hairs to 2 mm; blades 2–12 cm × 1–3 mm, bases with glandular pits. **Inflorescences** 4–10(15) × 0.5–2 cm, narrowly elliptic, contracted, dense, glandular pits on inflorescence axis and branches; primary branches 1 per node, appressed or occasionally diverging to 30° from the inflorescence axis; pedicels 1.4–10 mm, appressed or divergent; disarticulation acropetal. **Spikelets** 3.6–7.5 × 1.2–2 mm, light yellowish or occasionally reddish purple mottled, 6–11(14) florets. **Glumes** ovate to lanceolate, hyaline; lower glumes (0.7)0.9–1.4 mm; upper glumes 1.2–1.8 mm. **Lemmas** 1.5–2.2 mm, ovate, hyaline, straw-colored, veins greenish and conspicuous; tips acute. **Paleas** 1.2–2 mm, hyaline, persistent, keels minutely scabrous. **Anthers** 3, 0.2–0.3 mm, purplish. **Caryopses** 0.5–0.8 mm.

Stream banks, lake shores, alkaline flats, and disturbed areas. 0–800 m. WV. CA, NV, WA; southeast to CO and NM. Native.

Eragrostis mexicana (Hornemann) Link ssp. ***virescens*** (J. Presl) S.D. Koch & Sánchez Vega

Mexican lovegrass

Eragrostis orcuttiana Vasey

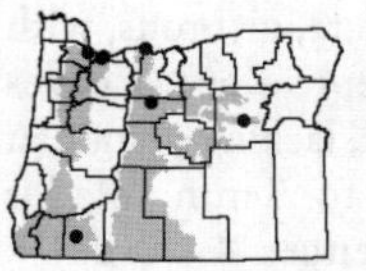

Plants annual, 10–130 cm tall, cespitose. **Culms** erect or sometimes geniculate, glabrous, without glands. **Leaves** sheaths lacking glands, with spreading hairs near the tips and on the collars, hairs to 4 mm, papillose-based; blades 5–25 cm × 2–7(9) mm. **Inflorescences** (5)10–40 × (2)4–18 cm, ovate; primary branches 3–12(15) cm, solitary or paired, appressed or diverging to 80° from the inflorescence axes; secondary branches somewhat appressed; pedicels 1–6(7) mm, almost appressed to narrowly divergent, stiff; disarticulation acropetal. **Spikelets** (4)5–10(11) × 0.7–1.5 mm, gray-green to purplish, 5–11(15) florets. **Glumes** 0.7–2.3 mm, ovate to lanceolate, membranous; the lowest glume at least 50% as long as the lowest floret. **Lemmas** 1.2–2.4 mm, ovate, membranous, glabrous or with a few hairs, gray-green; tips acute. **Paleas** 1–2.2 mm, hyaline, keels strongly scabrous; persisting after the lemma falls. **Anthers** 3, 0.2–0.5 mm, purplish. **Caryopses** 0.5–1 mm, ovoid to rectangular-prismatic, laterally compressed, shallowly to deeply grooved on the side opposite the embryo.

Disturbed, open areas including roadsides. 0–1300 m. ECas, Lava, Sisk, WV. CA, NV, WA; western US, scattered in eastern US; South America. Exotic.

Eragrostis mexicana ssp. *virescens* is a late-season, weedy grass with many gray florets in each spikelet. The more common *E. pectinacea* var. *pectinacea* is very similar but does not have grooved seeds.

Eragrostis minor Host

Little lovegrass

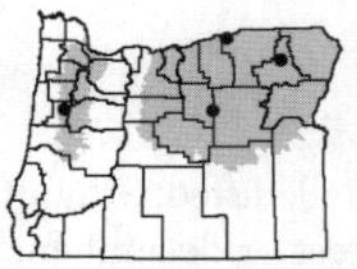

Plants annual, 10–45 cm tall, tufted. **Culms** erect to decumbent, sometimes with a ring of glandular tissue below the nodes. **Leaves** sheaths sometimes glandular on the midveins, tops with hairs to 4 mm; blades 1.5–10 cm × 1–3(4) mm, margins sometimes with saucer-like glands. **Inflorescences** 4–20 × 2.2–8(10) cm, ovate, open to contracted, rachises sometimes with glandular spots or pits below the nodes or rarely with a glandular ring, glands usually dull, greenish gray to straw-colored; primary branches 0.5–6 cm, diverging 20–100° from the inflorescence axis; pedicels 1–4 mm, stiff, straight, divergent, usually with a distal ring of saucer-like glands; disarticulation acropetal, paleas persistent. **Spikelets** 4–7(11) × 1.1–2.2 mm, mostly reddish purple to greenish to occasionally grayish, 7–12(20) florets. **Glumes** broadly ovate, membranous; lower glumes 0.9–1.4 mm; upper glumes 1.2–1.6 mm. **Lemmas** 1.4–1.8 mm, broadly ovate, membranous, keels occasionally with 1–2 saucer-like glands; tips acute to obtuse. **Paleas** 1.3–1.7 mm, hyaline, keels smooth or minutely scabrous. **Anthers** 2, 0.2–0.3 mm, reddish brown. **Caryopses** 0.4–0.7 mm, ellipsoid, not grooved. **2*n*=40.**

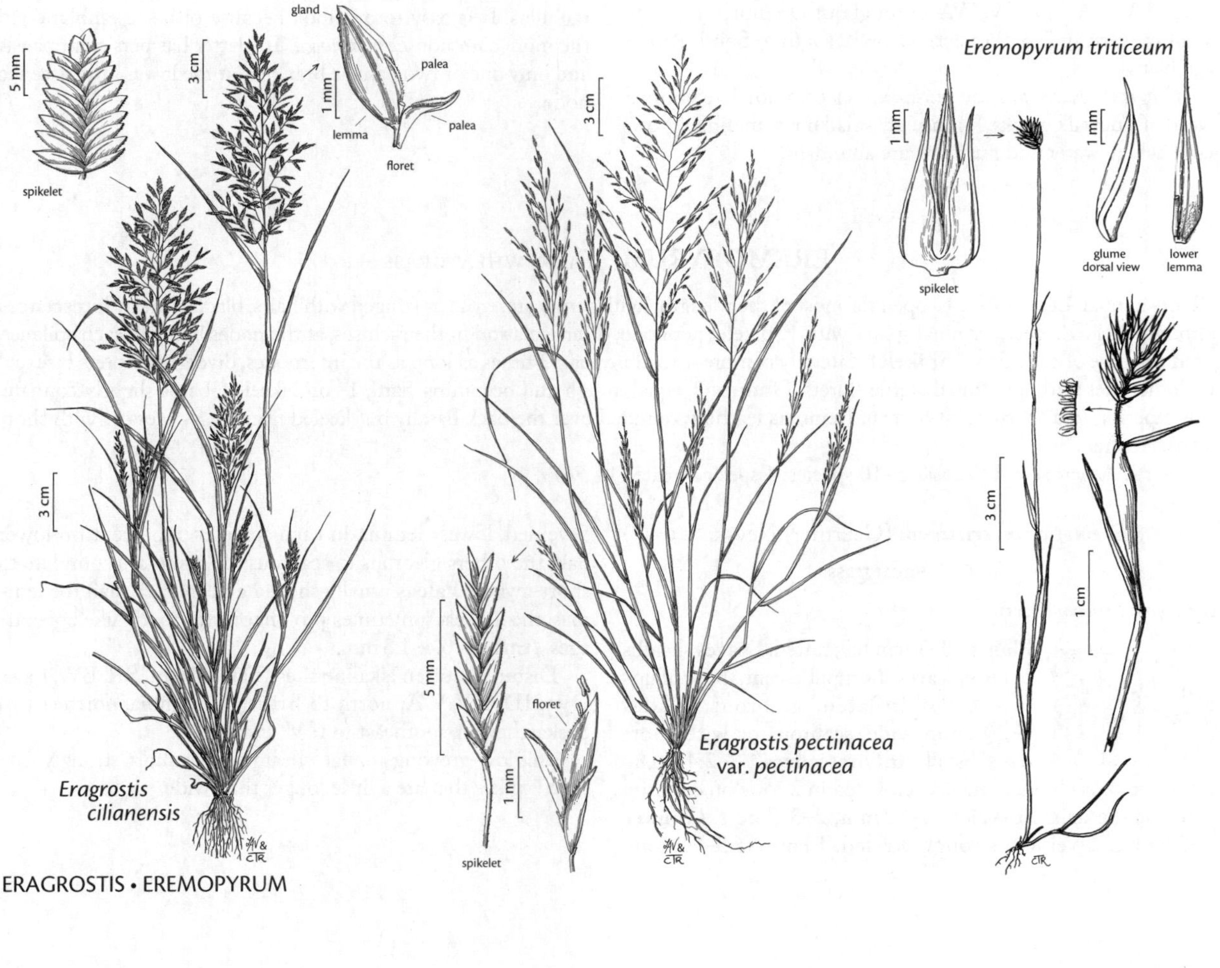

Disturbed places, pavement cracks. 50–900 m. BW, Col, Lava, WV. CA, ID, NV, WA; throughout most of Canada and US; Europe. Exotic.

Eragrostis minor closely resembles *E. pectinacea*, which is glandless. Similar *Eragrostis cilianensis* has denser inflorescences and wider spikelets.

Eragrostis pectinacea (Michaux) Nees var. **_pectinacea_** (p. 405)

Tufted lovegrass

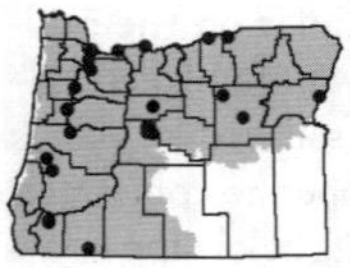

Plants annual, 10–80 cm tall, tufted, without glandular pits. **Culms** erect or decumbent, glabrous. **Leaves** sheaths with hairs to 4 mm at the tops; blades 2–20 cm × 1–4.5 mm. **Inflorescences** 5–25 × 3–12(15) cm, ovoid to pyramidal, usually open; primary branches 0.6–8.5 cm, appressed or diverging to 80° from the inflorescence axis, solitary or paired at the lowest 2 nodes; pedicels 1–7 mm, flexible, appressed or diverging up to 20° from the branch axes; disarticulation acropetal. **Spikelets** 3.5–11 × 1.2–2.5 mm, gray, yellowish brown or dark reddish purple, 6–22 florets. **Glumes** subulate to ovate-lanceolate, hyaline; lower glumes 0.5–1.5 mm, at least 50% as long as the adjacent lemmas; upper glumes 1–1.7 mm. **Lemmas** 1–2.2 mm, ovate-lanceolate, hyaline to membranous, grayish green proximally, reddish purple distally, acute. **Paleas** 1–2 mm, hyaline to membranous, persistent, keels minutely scabrous. **Anthers** 3, 0.2–0.4 mm, purplish. **Caryopses** 0.5–1.1 mm, pear-shaped, not grooved, slightly laterally compressed.

Disturbed sites. 0–1100 m. BW, Casc, Col, CR, ECas, Lava, Sisk, WV. CA, ID, NV, WA; throughout US, north just into southwestern and southeastern Canada, south to South America. Native.

Eragrostis pectinacea var. *pectinacea* is a common late summer weed of sidewalk cracks. It is usually small but sometimes grows large where water and nutrients are abundant.

Eragrostis pilosa (Linnaeus) P. Beauvois var. **_pilosa_**

India lovegrass

Eragrostis multicaulis Steud.

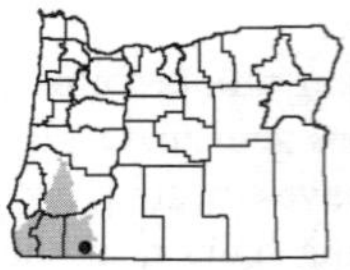

Plants annual, 8–45(70) cm tall, tufted. **Culms** erect or geniculate, glabrous, with few or no glandular depressions. **Leaves** sheaths mostly glabrous, lacking glandular pits, tips with hairs to 3 mm; blades 2–15(20) cm × 1–2.5(4) mm. **Inflorescences** 4–20(28) × 2–15(18) cm, ellipsoid to ovoid, diffuse; primary branches 1–10 cm, diverging 10–80(110)° from the inflorescence axis, capillary, whorled branches on the lowest 2 nodes, rarely glandular; pedicels 1–10 mm, flexible, appressed or divergent; disarticulation acropetal, rachillas persisting longer than the paleas. **Spikelets** (2)3.5–6(10) × 0.6–1.3 mm, gray, (3)5–17 florets. **Glumes** narrowly ovate to lanceolate, hyaline; lower glumes 0.3–0.6(0.8) mm, less than 50% as long as lowest floret; upper glumes 0.7–1.2(1.4) mm. **Lemmas** 1.2–1.8 mm, ovate-lanceolate, membranous to hyaline, grayish green proximally, reddish purple distally, lateral veins inconspicuous; acute. **Paleas** 1–1.6 mm, membranous to hyaline, tardily deciduous, keels scabrous. **Anthers** 3, 0.2–0.3 mm, purplish. **Caryopses** 0.5–0.9 mm, obovoid to prism-shaped, ventral surfaces flat. $2n$=40.

Disturbed sites, roadsides, lake shores. 0–300 m. Sisk. CA, ID, NV, WA; throughout US, north just into southwestern and southeastern Canada, south to Mexico; South America; worldwide. Exotic.

Eragrostis pilosa var. *pilosa* is recognized by its whorled lower panicle branches and its deciduous paleas that leave bare, zigzag rachillas. It is easy to overlook because of its resemblance to the more common *E. pectinacea*. The latter has persistent paleas and only one or two panicle branches at the lower inflorescence node.

EREMOPYRUM False wheatgrass

Plants annual. **Leaves** sheaths open for most of their length; ligules truncate, erose or fringed with hairs; blades flat. **Inflorescences** strongly 2-sided, erect, terminal spikes with 1 spikelet per node; disarticulation in the rachises, at the nodes beneath each spikelet, or at the base of each floret. **Spikelets** laterally compressed, more than 3 times as long as the internodes, divergent, florets bisexual or sometimes with additional sterile florets. **Glumes** 2, equal, tough and becoming hard, 1- or 2-keeled, bases slightly connate, tips tapering to a sharp point or awn. **Lemmas** leathery, rounded over the back basally but keeled near tip, awnless or with short awns. **Anthers** 3.

North America and Eurasia. 5–10 species; 1 species treated in *Flora*.

Eremopyrum triticeum (Gaertner) Nevski (p. 405)

Annual wheatgrass

Agropyron triticeum Gaertn.

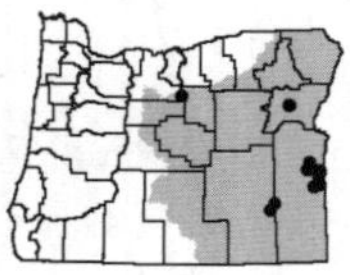

Plants 10–30 cm tall, tufted. **Leaves** sheaths of lower leaves short pubescent, those of upper leaves inflated, glabrous; blades 1–3(6) mm wide, scabrous or with short hairs distally. **Inflorescences** 1.3–2.4 × 0.8–2 cm, ovoid or oblong, spikelets crowded in 2 rows on opposite sides of the axis. **Spikelets** 6–12 mm, 2–3 florets. **Glumes** 4–7.5 mm, 1-veined, strongly keeled. **Lemmas** 5–7.5 mm, 5-veined, lowest lemma in each spikelet pubescent on lower half, the others glabrous to scabrous; tip acute or acuminate to short-awned. **Paleas** usually shorter and thinner than the lemmas, the 2 keels sometimes prolonged into tooth-like appendages. **Anthers** 0.4–1.3 mm.

Disturbed, often alkaline sites. 300–1800 m. BR, BW, Lava, Owy. ID, NV, WA; north to British Columbia, northeast to Saskatchewan, southeast to NM. Exotic.

This low-growing annual wheatgrass has short, strongly flattened spikes that are a little longer than wide.

FESTUCA Fescue

Plants perennial, densely to loosely cespitose, with or without rhizomes. **Culms** glabrous or sometimes scabrous or pubescent below the inflorescence; basal branching intra- or extravaginal. **Leaves** sheaths open or closed; auricles distinct upright flaps or absent; ligules membranous, usually ciliate; blades flat, folded, conduplicate or involute, glabrous, scabrous or pubescent or glaucous. **Inflorescences** open or contracted panicles, occasionally reduced to racemes; branches erect early in development, spreading to reflexed at flowering time and often returning to a nearly upright position after flowering. **Spikelets** laterally compressed, (1)2–10 florets, most bisexual but upper spikelets reduced or aborted; disarticulation above the glumes, beneath the florets. **Glumes** 2, usually exceeded by the florets, ovate to lanceolate, acute to acuminate; lower glumes 1(3)-veined; upper glumes 3(5)-veined. **Calluses** usually wider than long, glabrous, smooth, but sometimes elongated and/or pubescent. **Lemmas** membranous to leathery, generally keeled near the tip, 5(7)-veined; tips acute to attenuate, often minutely 2-toothed, usually awned; lemma awns, if present, terminal or subterminal, straight. **Anthers** 3. **Caryopses** glabrous or sometimes minutely hairy at the tip.

Temperate, arctic, and alpine regions worldwide. Approximately 450 species; 15 species treated in *Flora*.

Taxonomy of some taxa in this genus is controversial because of the diversity of species that look very similar but differ consistently in cryptic traits like ovary pubescence or leaf anatomy, or in habitat. The most severe problems involve the fine-leaved fescues, which have narrow, folded, mostly basal leaves. Some species formerly considered members of the genus *Festuca* have been placed in other genera. Tall and meadow fescues (formerly *F. arundinacea* and *F. pratensis*) are more closely related to *Lolium* than to *Festuca* and have been moved to their own genus, *Schedonorus*. The peculiar spike fescue (formerly *F. kingii*) is currently placed in the genus *Leucopoa*. Annual fescues with long, gradually tapered awns are now treated as *Vulpia*, but they may be returned to *Festuca* at some point. *Vulpia* species occasionally hybridize with fine-leaved fescues, mainly *F. rubra*.

When keying *Festuca*, lemma length is measured on the lowest floret of the spikelet, at or after anthesis. Lemma and spikelet lengths exclude awns. Assess leaf sheath closure on the second-from-the-top leaf of a vegetative shoot, not on culm leaves. In some fescues, such as *F. rubra*, the pale leaf sheaths soon dry out and die between the veins, leaving shredding, brown, somewhat shiny sheaths with contrasting white veins. In others, such as *F. idahoensis*, the pale leaf sheaths remain pale and intact, sometimes crumbling or cracking a bit, only belatedly dying and shredding. In circumstances in which their old sheaths remain in place, they might be confused with the first group except that the dead, brown leaves are obviously old, usually attached to dead or dying blades.

Leaf sclerenchyma under the epidermis on the outer/lower side is usually in one of four patterns: (a) three fibers, one at each margin and one at the midrib; (b) fibers near the margins, midrib, and some or all of the lateral veins; (c) sclerenchyma continuous, with fibers broad and some or all of them confluent into a broad band; or (d) sclerenchyma extending through the leaf from lower to upper surface. Leaf cross sections are cut from the middle third of fresh or dried leaves from sterile shoots. The leaf is broken near the middle and placed on a microscope slide, its broken end in a drop of water. Thin sections are sliced using a razor blade. Unstained sections are observed with a microscope. With experience, some traits can be observed in the field. The leaf is cut with scissors and observed with a 16× hand lens. Leaf cross-sectional shape is always visible, and the white sclerenchyma tissue may be visible.

Key to *Festuca* species:

1. Widest leaves more than 2 mm wide, flat or loosely rolled.
    2. Auricles present and represented by a laterally expanded, undulating flange or claw-like; lemma awns less than 4 mm, much shorter than the lemma bodies . . . . **SCHEDONORUS**
    2' Auricles absent or represented by inconspicuous swellings or small upright flaps; lemma awns absent to longer than the lemma bodies.
        3. Plants rhizomatous.
            4. Glumes not translucent, sometimes with narrow hyaline margins; sheaths of basal leaves brown with contrasting white veins, shredding lengthwise; leaf sheath margins fused; plants with bisexual florets . . . . ***F. rubra*** (in part)
            4' Glumes translucent; sheaths of basal leaves not shredding, pale with veins similar in color; leaf sheath margins overlapping; plants dioecious. . . . . **LEUCOPOA** (in part)
        3' Plants loosely to densely cespitose or growth habit various.
            5. Leaves mostly basal.
                6. Collars usually hairy at least on the edges; inflorescences 10–30 cm with branches 5–12 cm, spreading to drooping; apex of ovary and seed densely hairy. . . . . ***F. californica*** var. ***californica*** (in part)
                6' Collar glabrous; inflorescences (8)10–20(25) cm with branches 3–7 cm, appressed to reflexed; apex of ovary and seed glabrous. . . . . ***F. roemeri*** (in part)
            5' Leaves mostly cauline.
                7. Plants dioecious; inflorescence a spike-like condensed panicle . . . . **LEUCOPOA** (in part)
                7' Plants bisexual; inflorescence more or less open.
                    8. Callus longer than wide, approximately 0.4–0.6 mm; upper surface of leaves densely short-hairy; collar hairy. . . . . ***F. subuliflora***
                    8' Callus wider than long, approximately 0.1–0.3 mm; upper surface of leaves glabrous to scabrous or pubescent but not densely short-hairy; collar glabrous unless sheath is hairy.

9. Lemma veins 0-3; lemma awn often bent; leaves typically glabrous (or nearly so) on upper surface . . . . *F. subulata*

9' Lemma veins (3)5 (sometimes obscure); lemma awn straight; leaves scabrous or pubescent on upper surface. . . . . *F. elmeri*

1' Leaves less than 2 mm wide, folded or conduplicate.

10. Collars of lower leaves pubescent, rarely glabrous; panicles open, 10–30 cm long, branches lax, sometimes reflexed; plants 60–150 cm tall. . . . . ***F. californica*** var. ***californica*** (in part)

10' Collars of lower leaves glabrous; panicles open to contracted, 1–25(30) cm long, branches erect to reflexed, not lax; plants (5)8–120(140) cm tall.

11. Lower leaves with blades greatly reduced, sometimes to small points; apex of ovary and seed densely hairy. . ***F. viridula***

11' Lower leaves with blades normally developed; apex of ovary and seed glabrous or hairy.

12. Lemmas awnless or with a small awn less than 10% the length of the lemma.

13. Lemma 2.3–4(4.4) mm; plant epidermis smooth . . . . ***F. filiformis***

13' Lemma 4–11 mm; plant epidermis smooth to papillose.

14. Plants very densely cespitose; leaf sheaths usually pale, persistent; leaf surface papillose . . . ***F. campestris***

14' Plants cespitose to rhizomatous; leaf sheaths usually dark brown, translucent, shredding lengthwise, contrasting with white veins; leaf surface smooth or sometimes scabrous . . ***F. rubra*** (in part)

12' Lemmas awned, awns more than 10% the length of the lemma.

15. Plants rhizomatous or stoloniferous; vegetative shoots in part extravaginal . . . . ***F. rubra*** (in part)

15' Plants cespitose, lacking rhizomes; vegetative shoots usually intravaginal.

16. Anthers 0.4–1.8 mm; plants to 60 cm tall, often dwarfed; lemma bodies 2–5(5.6) mm, lemma awns 50% as long as the lemma bodies or less.

17. Anthers 0.4–0.8(0.9) mm when dry, 0.9–1.2 mm when fresh or rehydrated, usually retained long past anthesis; lemmas 2.5–5.5 mm . . . . ***F. brachyphylla***

17' Anthers 0.9–1.2 mm when dry, 1.2–1.8 mm when fresh or rehydrated, retained or lost after anthesis; lemmas 3.6–5.6 mm . . . . ***F. saximontana*** ssp. ***purpusiana***

16' Anthers 2 mm long or longer; plants to 100 cm tall, rarely dwarfed; lemma bodies (3)4–8 mm; lemma awns from less than 50% as long as to longer than the lemma body.

18. Some lemma awns longer than lemma bodies; apex of ovary and caryopses densely pubescent . . . . ***F. occidentalis***

18' Lemma awns approximately equal to or shorter than lemma bodies; apex of ovary and caryopses glabrous.

19. Mature lemmas more than 5.8 mm.

20. Foliage smooth; sheaths of basal leaves closed about 75% of their length, brown with contrasting whitish veins, splitting or shredding; leaf sclerenchyma bundles less than 2 times as broad as thick (except sometimes on serpentine substrates). . . . . ***F. rubra*** (in part)

20' Foliage rough, scabrous; sheaths of basal leaves open, usually pale (white, pink, light brown), usually lacking contrasting veins, usually crumbling or somewhat splitting but not shredding; leaf sclerenchyma bundles more than 2 times as broad as thick.

21. Leaves very narrow, readily spinning between the fingers . . . . ***F. idahoensis***

21' Leaves wider, not spinning between the fingers or spinning with angles that can be felt. . . . . ***F. roemeri*** (in part)

19' Mature lemmas less than 5.8 mm, usually less than 5.5 mm.

22. Foliage smooth; leaf sheath margins on vegetative shoots fused approximately 75% of their length; leaf sheath brown, usually shredding between white veins and usually contrasting in color with the pale veins exposed by the shredding sheaths; leaf sclerenchyma bundles less than 2 times as broad as thick . . . ***F. rubra*** (in part)

22' Foliage rough or smooth; leaf sheath margins overlapping, not fused; leaf sheaths whitish to light brown, sometimes pink or purplish, generally not shredding; leaf sclerenchyma bundles more than 2 times as broad as thick.

23. Hairs on inner leaf surface approximately as long as the leaf is thick; leaves often pubescent on outer surface; leaf sclerenchyma interrupted (forming discrete bundles) . . . . ***F. valesiaca***

23' Hairs on inner leaf surface much shorter than the leaf is thick; leaves glabrous or tomentose on outer surface; leaf sclerenchyma continuous or interrupted . . . . ***F. trachyphylla***

Festuca brachyphylla Schultes & Schultes f.

Alpine fescue

Festuca brachyphylla Schult. & Schult. f. ssp. *breviculmis* Fred., *Festuca brachyphylla* Schult. & Schult. f. ssp. *coloradensis* Fred., *Festuca ovina* L. var. *brachyphylla* (Schult. & Schult. f.) Piper ex Hitchc., *Festuca ovina* L. var. *brevifolia* (R. Br.) S. Watson

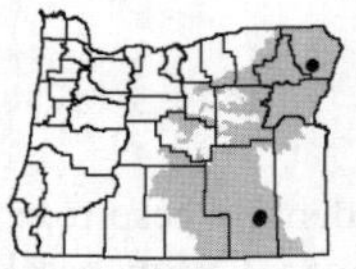

Plants 2.5–12 cm tall; densely cespitose, yellowish green or glaucous. **Culms** basal branching intravaginal. **Leaves** sheaths closed more than half their length, glabrous, not conspicuously splitting between the veins; collars glabrous; ligules 0.1–0.5 mm; blades 1.7–6 cm × 0.3–0.5(1) mm, conduplicate, glabrous or with minute hairs on inner surface; flag leaves 0.8–2.5 cm. **Leaf anatomy** cross section (0.25)0.3–0.5(0.8) mm by 0.35–0.95 mm, more or less round; veins 3–5; ribs 1(5); dorsal sclerenchyma bands less than 2 times as wide as thick (except sometimes at the midrib). **Inflorescences** (0.8)1.5–2.5 cm; branches 0.4–0.8 cm, appressed after anthesis. **Spikelets** 3.5–5.4 mm, 2–4 florets. **Glumes** glabrous or minutely scabrous at the tips; lower glumes 1.8–3 mm, 1-veined; upper glumes 2.3–3.5(4.3) mm, 3-veined. **Lemmas** 2.5–5.5 mm, 0(1)-veined, minutely scabrous at the tip; lemma awns 0.7–2 mm. **Anthers** 0.4–0.8(0.9) mm when dry, 0.9–1.2 mm when fresh or rehydrated, usually retained long past anthesis. **Ovaries** apex glabrous. **Caryopses** 1.8–2.5 mm, glabrous. ***2n***=28.

Alpine snowmelt basins. 1800–3000 m. BR, BW. CA, ID, NV, WA; north to AK, northeast to Greenland, east to MN, southeast to CO and AZ; circumpolar. Native.

This tiny alpine fescue is most likely to be confused with *F. saximontana*, which has longer anthers and leaf sclerenchyma bands that are more than twice as broad as thick. Oregon plants are morphologically intermediate between the named subspecies, fitting none well.

Festuca californica Vasey var. ***californica*** (p. 410)

California fescue

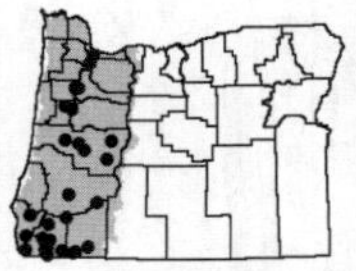

Plants 60–150 cm tall; densely cespitose. **Culms** with nodes exposed. **Leaves** mainly basal; sheaths usually papillose, sometimes hairy; collars with long hairs or puberulent, rarely glabrous; ligules 0.8–1.5 mm; blades 1.8–3.5 mm wide; folded or occasionally flat, somewhat stiff; upper surface with short hairs; lower surface glabrous; flag leaves 5–32 cm. **Inflorescences** 10–30 cm, open, lowest node with 1–2 spreading to drooping branches 5–12 cm. **Spikelets** 13–18 × (2.5)4.5–8 mm, 4–6 florets. **Glumes** glabrous or slightly scabrous on the keel; lower glumes 4.7–6.7 mm, 1-veined; upper glumes (5)6–8 mm, 3-veined. **Lemmas** 7.5–10.6 mm, 0(1)-veined, hairy or scabrous throughout; lemma awns 1.5–2.2 mm, much shorter than the lemma bodies. **Anthers** 4.5–5 mm. **Ovaries** apex densely hairy. **Caryopses** 4.5–5.5 mm, hairy at the tip. ***2n***=56.

Dry grasslands, grassy balds, savannas, open woodlands. 50–2000 m. Casc, CR, Sisk, WV. CA. Native.

When present, the hairy collars of *Festuca californica* are unique among Oregon fescues. This species is sometimes confused with *F. campestris,* a close relative with densely papillose leaves and sheaths that grows only east of the Cascades. Similar *Schedonorus arundinaceus* generally has tighter panicles and flatter, less papillose leaves, as well as flaring, ciliate auricles. *Bromus carinatus* and *B. sitchensis* are superficially similar in appearance and may grow in the same habitats but have longer lemma awns and hairier, closed leaf sheaths.

Festuca campestris Rydberg

Mountain rough fescue

Festuca altaica Trin., *Festuca scabrella* Torr. ex Hook.

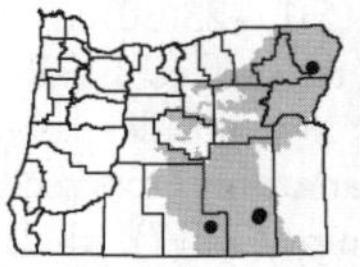

Plants (20)40–90 cm tall; densely cespitose. **Culms** with nodes hidden. **Leaves** mainly basal; sheaths pale, glabrous, papillose, conspicuous at the base of the plant and persisting for more than 1 year, entire, forming a dense tuft of split tubes at the base of the plant; collars glabrous; ligules 0.1–0.5 mm; blades 0.5–1.1 mm wide when tightly folded, 1.6–2.5(3) mm wide when flat or loosely rolled; somewhat stiff, outer surface papillose, inner surface with long hairs; flag leaves 2.5–7.5 cm. **Inflorescences** 5–18 cm, lowest node with 1–3 spreading branches 2.5–7(13) cm. **Spikelets** 8–12(16) × 2.5–7 mm, with (3)4–5(7) florets. **Glumes** glabrous or scabrous at the tips; lower glumes 4.5–7.5(8.5) mm, 1(3)-veined; upper glumes 5.3–8.2(9) mm, (1)3-veined. **Lemmas** (6.2)7–8.5(10) mm, (0)5-veined, slightly scabrous throughout, rarely glabrous; lemma awns 0.5–1.5 mm. **Anthers** (3.3)4.5–6 mm. **Ovaries** apex hairy. **Caryopses** 3.5–4.5 mm, hairy at the tip.

Dry grassland, sage-steppe. 600–3000 m. BR, BW. ID, WA; north to British Columbia, northeast to Saskatchewan and MT. Native.

The leaves of *Festuca campestris* are stiffer and more papillose than those of similar *F. idahoensis* or *Achnatherum* species. *Festuca campestris* is part of what was once called *F. scabrella,* which has been divided into three taxa: *F. altaica* (central Asia and North America south to British Columbia and Alberta), *F. hallii* (northern Great Plains), and *F. campestris.*

Festuca elmeri Scribner & Merrill

Elmer's fescue

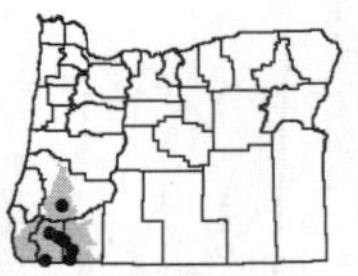

Plants 40–135 cm tall; loosely cespitose, green. **Culms** basal branching extravaginal. **Leaves** mainly cauline; sheaths open, glabrous, splitting between the veins toward the end of the season; collars glabrous; blades 1.8–6(7) mm wide, flat, lax; upper surface scabrous or pubescent; lower surface glabrous. **Inflorescences** 8–25 cm; branches more or less spreading. **Spikelets** 7–10.5 × (1)1.5–3 mm, 2–6 florets. **Glumes** glabrous or nearly so; lower glumes 1.7–4.2 mm, 1-veined; upper glumes 2.3–5.8 mm, 3-veined. **Calluses** 0.1–0.2(0.3) mm, wider than long, glabrous. **Lemmas** 4.8–7 mm, with (3)5 veins, scabrous or short-hairy throughout; lemma awns (1)2–5(8) mm, generally straight. **Anthers** 2–4 mm. **Ovaries** apex hairy. **Caryopses** 2.5–3.5 mm, hairy at the tip.

Dry open forest. 300–1400 m. Sisk. CA. Native.

Similar *Festuca subulata* and *F. subuliflora* have longer awns, obscure lemma veins, and usually smooth, glabrous lemmas. *Festuca subuliflora* has elongated, hairy calluses. Rare *F. subulata* plants have scabrous to hairy lemmas, leading to confusion with *F. elmeri.*

Festuca filiformis Pourret

Hair fescue, fine-leaved sheep fescue

Festuca capillata Lam., *Festuca ovina* L. var. *capillata* (Lam.) Alef.

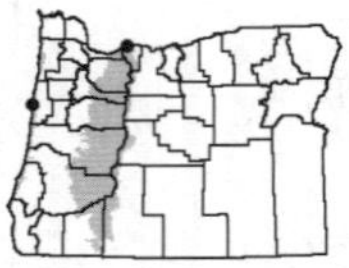

Plants 18–40(55) cm tall; densely cespitose, green to glaucous. **Culms** basal branching intravaginal. **Leaves** mainly basal, sheaths open, glabrous or becoming glabrous, not conspicuously splitting between the veins; collars glabrous; ligules 0.15–0.3 mm; blades (5)11–23(30) cm × 0.2–0.4(0.6) mm, conduplicate, very fine, outer surface usually glabrous, inner surface with very short hairs; lower blades normally developed; flag leaves 0.5–5 mm. **Leaf anatomy** cross sections 0.2–0.4(0.6) × 0.3–0.6 mm, elliptic; veins (3)5(7), rib 1; dorsal sclerenchyma in broad bands more than twice as wide as thick. **Inflorescences** 1–4(8) cm; branches 0.5–1.5(2) cm, appressed after anthesis. **Spikelets** 3–6(6.5) mm, 2–6 florets. **Glumes** glabrous or scabrous; lower glumes 1–2.5 mm, 1-veined; upper glumes 1.7–2.3(3.9) mm, 3-veined. **Lemmas** 2.3–4(4.4) mm, 0(1)-veined, scabrous on upper part only, awnless or short-awned; lemma awns 0–0.6 mm, less than 10% as long as the lemma body. **Anthers** 1.5–2.2 mm. **Ovaries** apex glabrous. **Caryopses** 1.5–2 mm, glabrous.

Lawns, roadsides, beaches, somewhat disturbed areas. 0–300 m. Casc, Est. WA; north to British Columbia; southeastern Canada and northeastern US; Europe. Exotic.

Festuca filiformis is identified by its small, virtually awnless lemmas and its very narrow leaves that have a thin, continuous to somewhat broken layer of sclerenchyma under the outer epidermis. This species rarely flowers and therefore is rarely collected, but it is planted as a lawn grass and can often be found in disturbed mesic sites.

Festuca idahoensis Elmer (p. 410)

Idaho fescue

Festuca idahoensis Elmer var. *idahoensis*

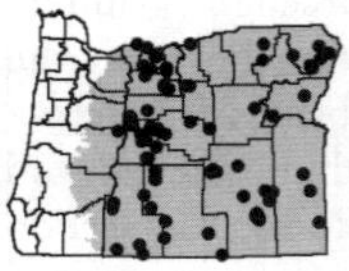

Plants 30–100 cm tall; densely cespitose, glaucous or less often green. **Culms** basal branching intravaginal. **Leaves** mostly basal; sheaths open, glabrous to hairy, usually pale (white, pink, light brown) and lacking contrasting veins, crumbling or somewhat splitting between the veins but not shredding; blades readily spinning between the fingers; collars glabrous; ligules 0.3–0.6 mm; blades (5)8–25(35) cm × 0.3–1 mm as folded, conduplicate; usually glaucous, sometimes green, outer surface scabrous, sometimes with long hairs, inner surface with hairs about as long as the leaf is thick; flag leaves (1.5)3–7(9.5) cm, usually conduplicate, sometimes loosely rolled. **Leaf anatomy** cross sections 0.35–0.6 × 0.55–0.7 mm, more or less round or hexagonal; veins (4)5(6); ribs (1)3(5); dorsal sclerenchyma in broad bands more than twice as wide as thick, occasionally continuous. **Inflorescences** (5)7–12(20) cm, open at anthesis, lowest node with 1–2 branches;

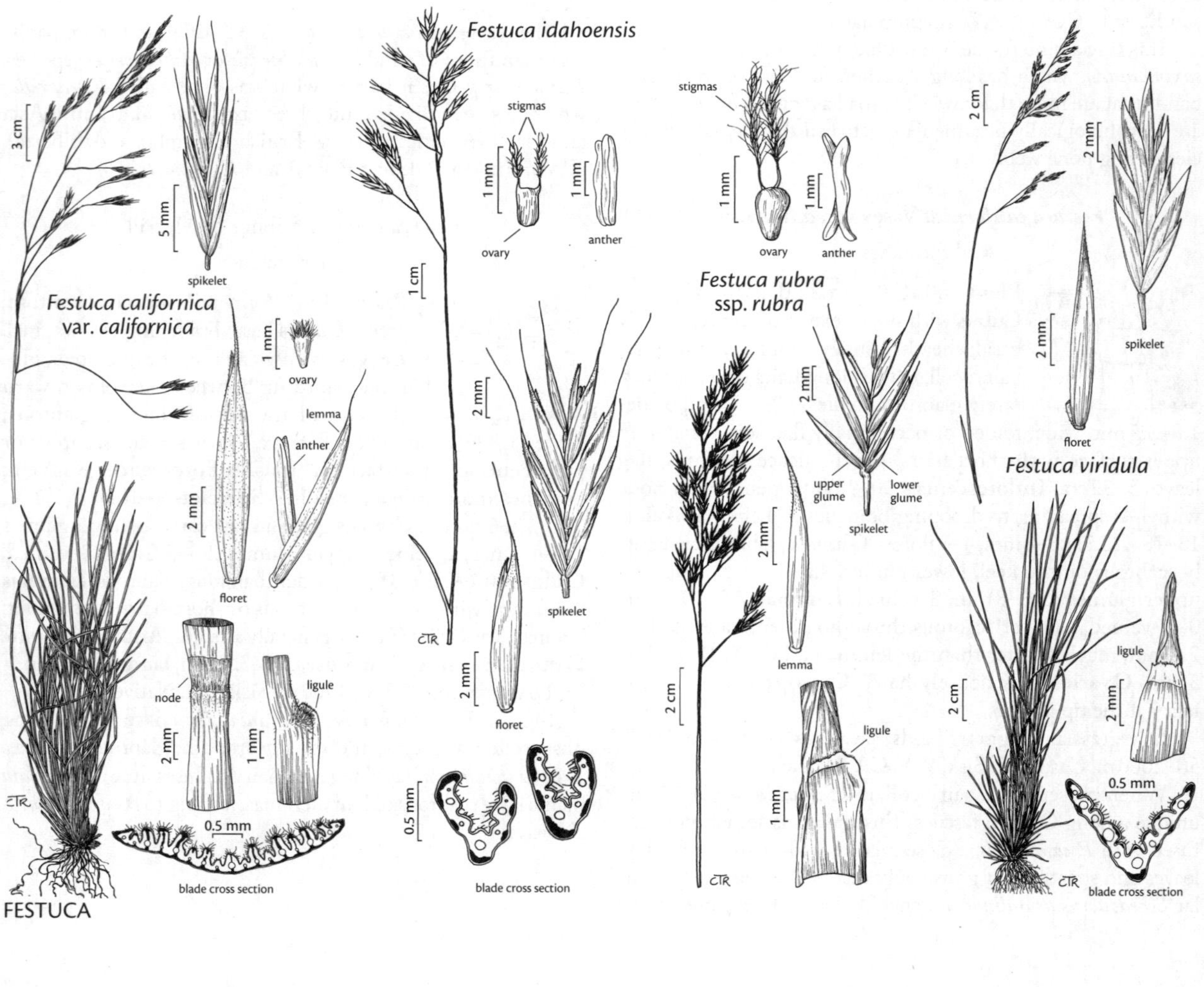

FESTUCA

branches (5.8)7.5–17(19) mm, ascending to reflexed at anthesis, appressed after anthesis or the lower branches sometimes remaining spreading to reflexed. **Spikelets** (5.8)7.5–17(19) mm, with (2)3–7(9) florets. **Glumes** glabrous or hairy at the tips only; lower glumes (2.5)3–5(6) mm, 1–3-veined; upper glumes (3)4–8 mm, 3–4-veined. **Lemma** (5.8)6–8(10) mm, 0(1)-veined, hairs on the distal part only; lemma awns (1)2–6(7) mm, shorter than to occasionally equaling the lemma body. **Anthers** (2.5)3.2–4(4.5) mm. **Ovaries** apex glabrous. **Caryopses** 4–4.5 mm, glabrous. ***2n***=28.

Prairie, grassland, shrub-steppe, open pine forest, often in partial shade or on north-facing slopes; also on south-facing slopes in subalpine and alpine zones. 100–3000 m. BR, BW, Casc, Col, ECas, Lava, Owy. CA, ID, NV, WA; north to British Columbia, northeast to Saskatchewan, SD, and NM. Native.

Festuca idahoensis is the most common fine-leaved fescue east of the Cascades. However, very similar fescues have been introduced. *Festuca trachyphylla* and *F. valesiaca,* planted on roadsides and used to renovate dry pastures, have shorter lemmas and denser inflorescences. Robust individuals of *F. occidentalis* can be similar to *F. idahoensis*, but in *F. occidentalis* many of the lemma awns are longer than the lemmas. In alpine sites, *F. idahoensis* can be confused with *F. brachyphylla* and *F. saximontana*, which have shorter lemmas. Plants west of the crest of the Cascades that were once included in *F. idahoensis* are now recognized as a distinct species, *F. roemeri,* which has broader leaves than *F. idahoensis.*

Festuca occidentalis Hooker
Western fescue

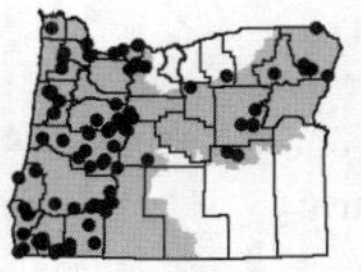

Plant (25)45–110 cm tall; densely cespitose, green. **Culms** basal branching intravaginal. **Leaves** mainly basal; sheaths open, with overlapping margins, not conspicuously splitting between the veins during the first year; collars glabrous; ligules 0.1–0.4(0.5) mm; blades (5)10–20(40) cm × 0.3–0.75 mm, conduplicate, more or less lax, green, inner surface with hairs, outer surface glabrous, rarely minutely scabrous, flag leaves 3–12 cm. **Leaf anatomy** cross sections 0.25–0.6 × 0.35–0.75 mm, diamond-shaped or triangular; veins (3)5–7, ribs 1(5); dorsal sclerenchyma bands less than twice as wide as thick. **Inflorescences** 5–20 cm; branches spreading, the lowest usually reflexed at anthesis and remaining reflexed thereafter. **Spikelets** 6–12 mm, with 2–5(7) florets. **Glumes** hairy at tips only; lower glumes 2–5 mm, 1-veined; upper glumes 3–6 mm, (1)3-veined. **Lemmas** 4–6.5(8) mm, 0(1)-veined, hairs near the tip only (rarely hairy throughout); lemma awns 3–12 mm, some longer than the lemma body. **Ovaries** apex densely hairy. **Caryopses** 3–4.5 mm, densely hairy at the tip.

Mesic to dry forest, in shade or partial shade. 0–2400 m. BW, Casc, CR, ECas, Est, Lava, Sisk, WV. CA, ID, WA; western US and Canada, Great Lakes region. Native.

Festuca occidentalis is a fine-leaved, forest-dwelling bunchgrass easily identified by the lemma awns that are longer than the lemma bodies. The only other fine-leaved fescue with densely hairy tips of the ovary and seed is subalpine, virtually awnless *F. viridula*. Similar *Festuca idahoensis* and *F. roemeri* have leaves that are oval, round, or hexagonal in cross section, with sclerenchyma bands more than twice as wide as thick. *Festuca rubra* has fused sheath margins and awns that are shorter than the lemmas.

Festuca roemeri (Pavlick) E.B. Alexeev
Roemer's fescue

Plants 35–100 cm tall; densely cespitose, glaucous to green. **Culms** basal branching intravaginal. **Leaves** mainly basal; sheaths open, glabrous but sometimes hairy, not conspicuously splitting between the veins, usually pale, becoming gray-brown with age; collars glabrous; ligules 0.1–0.4 mm; blades 15–35 cm × 0.5–1.2(2.5) mm, conduplicate or occasionally flat, outer surface usually glabrous, sometimes hairy, inner surface glabrous, scabrous, hairy; flag leaves 7–12(18) cm. **Leaf anatomy** cross sections usually 0.4–0.8 × 0.75–1.2 mm, elliptic or L-shaped; veins (5)7(10); ribs 5(9); dorsal sclerenchyma bands more than 2 times wider than thick, sometimes restricted to the midrib and margins only. **Inflorescences** (8)10–20(25) cm; branches 3–7 cm, ascending to reflexed at anthesis, appressed after anthesis or sometimes the lowest spreading to reflexed. **Spikelets** 9.5–12(13.5) mm, 3–6 florets. **Glumes** glabrous or scabrous to short-hairy at the tip only; lower glumes 2–5 mm, 1-veined; upper glumes 4–5.5(6.2) mm, 3-veined. **Lemmas** (5.8)6.5–8.2 mm, 0(5)-veined, with hairs on the distal part only; lemma awns 2–4(5) mm, shorter than or as long as the lemma body. **Anthers** 3–5 mm. **Ovaries** apex glabrous. **Caryopses** glabrous. ***2n***=28.

2 varieties.

Festuca roemeri is most often confused with the bunchgrass forms of *F. rubra*. *Festuca rubra* leaves are smooth, the sheath margins are fused, and bunchgrass forms of *F. rubra* usually have shorter lemmas. In cross section, leaves of sterile shoots of *F. rubra* are diamond shaped to triangular, angled next to the largest lateral veins. The leaf sclerenchyma bands are less than twice as wide as thick. *Festuca roemeri* is closely related to *F. idahoensis,* and intermediate plants can be found where the ranges of the two species come into contact at the east end of the Columbia Gorge, in the High Cascades, and in Klamath County. *Festuca idahoensis* has round to hexagonal leaves that spin readily between one's thumb and forefinger. *Festuca roemeri* leaves usually won't roll, though leaves from very dry serpentine sites will roll but with the angles easily felt. *Festuca trachyphylla* and *F. valesiaca* have smaller lemmas and denser inflorescences. Vegetative *F. valesiaca* cannot be distinguished from *F. roemeri.*

Key to *Festuca roemeri* varieties:

1. Hairs on inner surface of leaf few, short (approximately twice as long as the adjacent epidermal cells) . . . . ***F. r.*** var. ***roemeri***

1' Hairs on inner surface of leaf many, longer (much longer than the adjacent epidermal cells) and approximately as long as the dried leaf is thick . . . . . . ***F. r.*** var. ***klamathensis***

Festuca roemeri (Pavlick) E.B. Alexeev
var. **_klamathensis_** B.L. Wilson
Inland Roemer's fescue

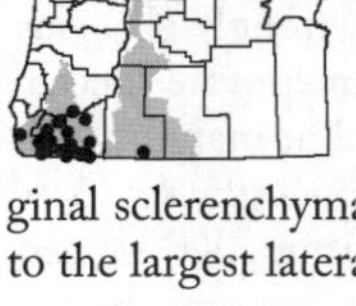

Leaves blades usually glaucous, inner surface pubescent with hairs approximately as long as the dried blades are thick. **Leaf anatomy** sclerenchyma usually present by lateral veins as well as at midrib and margins; the marginal sclerenchyma band sometimes fused with the band next to the largest lateral veins.

Grasslands, prairies, oak and pine savannas, serpentine and non-serpentine substrates. 100–2300 m. ECas, Sisk. CA. Native.

Festuca rubra growing on serpentine may have superficially similar leaf anatomy because the marginal sclerenchyma band and the one next to the largest lateral vein are so large that they grow into one another, creating a single broad band, but these plants can be identified as *F. rubra* by the fused leaf sheath margins on the vegetative shoots and their usually rhizomatous growth form. In eastern Jackson and western Klamath counties, plants intermediate between *F. roemeri* var. *klamathensis* and *F. idahoensis* can be found.

Festuca roemeri (Pavlick) E.B. Alexeev var. ***roemeri***

Roemer's fescue

Festuca idahoensis Elmer var. *oregona* (Hack. ex Beal) C.L. Hitchc., *Festuca idahoensis* Elmer var. *roemeri* Pavlick

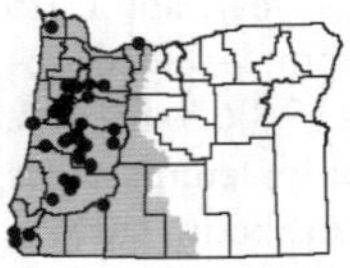

Leaves blades green or glaucous, inner surface sparsely scabrous, with few short hairs approximately twice as long as adjacent epidermal cells. **Leaf anatomy** sclerenchyma often limited to midrib and margins, sometimes with additional bands.

Grasslands. 50–1800 m. Casc, CR, ECas, Est, Sisk, WV. CA, WA; north to British Columbia. Native.

This taxon was once dominant in grasslands of northwestern Oregon but now survives in widely scattered locations. Though generally seen on non-serpentine substrates, it is capable of growing on serpentine.

Festuca rubra Linnaeus

Red fescue

Plants (8)10–120(130) cm tall, strongly rhizomatous to densely cespitose; shoots occasionally solitary, usually green, occasionally glaucous; basal branching intra- or extravaginal. **Leaves** mainly basal; sheaths closed for about 75% of their length; margins fused, readily splitting with age, with retrorse hairs but sometimes antrorsely hairy or glabrous, shredding lengthwise and becoming brown with contrasting whitish veins or remaining intact and gray-brown; ligules 0.1–0.5 mm; blades 4–40+ cm, generally 0.3–2.5 mm in diameter and conduplicate or less often 1.5–7 mm wide and flat; generally green, sometimes glaucous; outer/lower surface usually smooth and glabrous but sometimes scabrous; inner/upper surface scabrous or pubescent; flag leaves 3–15 cm; conduplicate leaves usually diamond-shaped or triangular in cross section; veins (5)7–9(13), ribs (3)5–7(9), usually conspicuous. **Leaf anatomy** dorsal sclerenchyma in 5–9(13) bands less than 2 times as wide as thick (on plants from serpentine soils sometimes with the marginal and adjacent bands confluent, creating a single large band that is more than 2 times as wide as thick); ventral sclerenchyma sometimes present in the tips of the ribs. **Inflorescences** (2)3.5–25(30) cm, usually open or loosely contracted, sometimes dense; lowest node with 1–3 branches; branches appressed after anthesis or sometimes the lowest spreading. **Spikelets** (6)7–17 mm, 3–10 florets, often green to reddish or brown, some forms strongly glaucous, marked with blue, turquoise or lavender. **Glumes** ovate-lanceolate to lanceolate, sometimes with narrow hyaline margins; lower glumes (1.5)3.5–5(7) mm, 1(3)-veined; upper glumes (3)4–6.5(8.5) mm, (1)3-veined. **Lemmas** 4–8(9.5) mm, 0(1)-veined, glabrous and smooth but sometimes scabrous toward the tips or densely pubescent throughout, attenuate or acuminate in side view, usually awned; lemma awns (0.1)0.4–4.5 mm, shorter than the lemma body. **Anthers** 1.8–4.5 mm. **Ovaries** apex glabrous or rarely with 1–4 hairs. **Caryopses** glabrous. **2*n*=**14, 28, 42, 56, 70, 128.

Approximately 10 subspecies; 6 subspecies treated in *Flora*.

Green, rhizomatous forms of *F. rubra* are easily identified, but cespitose plants are often confused with *F. roemeri*; see discussion under that species. Near timberline *F. rubra* can be confused with *F. viridula*, a cespitose species that has the lower leaf blades reduced to small points. *Festuca rubra* includes a diversity of intergrading taxa that are not easily delimited. The many diverse forms of Pacific Northwest *F. rubra* have been described as subspecies, but additional work is needed to clarify *F. rubra* taxonomy.

Key to *Festuca rubra* subspecies:

1. All shoots intravaginal; plant strictly cespitose.
    2. Anthers 1.8–2.2(3+) mm; lemma awns 1–3.3 mm; plants usually green . . ***F. r.*** ssp. ***commutata***
    2' Anthers 2.3–3.5 mm; lemma awns 0.1–3 mm; plants usually strongly glaucous . . . . ***F. r.*** ssp. ***juncea*** (in part)

1' Some shoots extravaginal; plants generally rhizomatous, the shoots arising in clumps from the buried rhizomes but sometimes cespitose in harsh environments.
    3. Leaf blades of vegetative shoots usually flat or loosely conduplicate, 1–3(7) mm wide; anthers 3.5–4.5 mm; plants strongly rhizomatous; sclerenchyma bands nearly always present on the ribs on the inner surface of the leaves . . . . . ***F. r.*** ssp. ***fallax***
    3' Leaf blades of vegetative shoots nearly always conduplicate (sometimes flat, especially in irrigated settings), 0.5–1.5(2) mm wide; anthers 1.8–3.5(4.5) mm; plants strongly or weakly rhizomatous; sclerenchyma usually absent on inner surface of the leaves.
        4. Plants growing east of the Cascades . . . . ***F. r.*** ssp. ***rubra*** (in part)
        4' Plants growing west of the Cascades.
            5. Inflorescences strongly glaucous, sometimes nearly white or marked with turquoise, green or purple . . . . . ***F. r.*** ssp. ***juncea*** (in part)
            5' Inflorescences not glaucous or whitish, often marked with purple but not turquoise.
                6. Plants usually of disturbed habitats; inflorescences well exserted from the uppermost leaf sheath; leaf sheath tightly enclosing the culm . . . . ***F. r.*** ssp. ***rubra*** (in part)

6' Plants of more natural habitats; inflorescences sometimes included in the uppermost leaf sheaths; leaf sheaths often loosely enclosing the culm.
   7. Lemmas 4.5–6 mm, acuminate in side view . . . . *F. r.* ssp. ***mediana***
   7' Lemmas 5.8–6.6(7) mm, attenuate in side view. . . . . *F. r.* ssp. ***secunda***

Festuca rubra Linnaeus ssp. ***commutata*** Gaudin

Chewing's fescue

Festuca rubra L. var. *commutata* Gaudin

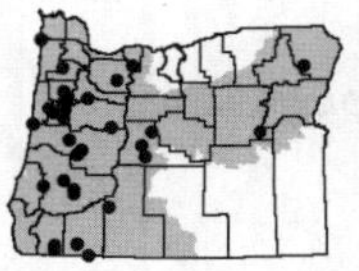

Plants densely cespitose, green or rarely glaucous. **Culms** basal branching intravaginal. **Leaves** sheaths not or only slowly becoming brown and shredding between the whitish veins; blades 0.3–0.7(1) mm wide, conduplicate, generally green. **Leaf anatomy** leaves usually diamond-shaped or triangular in cross section; veins usually 7, ribs (3)5; dorsal sclerenchyma generally in 5–7 bands less than 2 times as wide as thick; sclerenchyma girders absent. **Inflorescences** 4–13(30) cm; branches usually appressed after anthesis, lowest branches sometimes strongly spreading. **Spikelets** 7–11 mm, 3–9 florets, green to red or purple. **Glumes** ovate-lanceolate, acute; lower glumes 2.5–4 mm; upper glumes 3.2–5 mm. **Lemmas** 4.5–5.8(6) mm, green or reddish violet distally, glabrous; lemma awns 1–3.3 mm, shorter than the lemmas. **Anthers** 1.8–2.2(3+) mm.

Disturbed meadows, prairies, roadsides. 0–1900 m. BW, Casc, CR, ECas, Est, Lava, Sisk, WV. CA, ID, NV, WA; throughout most of Canada and US; temperate regions worldwide. Exotic.

Festuca rubra ssp. *commutata* is a common introduced bunchgrass. The leaf sheaths of *F. r.* ssp. *commutata* are not as obviously brown and shredding as are those of other *F. rubra* subtaxa. This subspecies can be confused with rhizomatous *F. rubra* subspecies that may sometimes have a clumped growth form and it is frequently confused with *F. roemeri*, which usually has longer lemmas, harsher, scabrous foliage and different leaf anatomy.

Festuca rubra Linnaeus ssp. ***fallax*** (Thuillier) Nyman

Flat-leaf red fescue

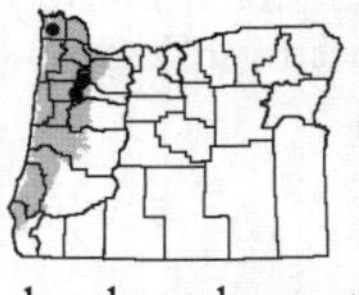

Plants 50–90(130) cm tall, strongly rhizomatous, sometimes also stoloniferous, sometimes seeming cespitose because shoots usually arise in dense clusters. **Culms** basal branching intra- or extravaginal. **Leaves** sheaths pubescent, shredding into fibers; blades usually flat, 1–3(7) mm wide, sometimes conduplicate and 0.8–2.5 mm in diameter, green. **Leaf anatomy** sclerenchyma near outer surface in 5–9 more or less broad strands, sclerenchyma generally present on the ribs on the inner surface. **Inflorescences** 9–15 cm, open. **Spikelets** 8–14(17) mm, (4)6–10 florets. **Glumes** ovate-lanceolate to lanceolate, acute to acuminate; lower glumes 3.5–4.5(7) mm; upper glumes 5–6.5(8.5) mm. **Lemmas** 5–7.5(9.5) mm, green or glaucous, often red-violet along the margins, glabrous; lemma awns 0.8–3.2 mm. **Anthers** 3.5–4.5 mm.

Disturbed moist to mesic habitats, sometimes a weed in grass seed fields. 50–700 m. CR, WV. CA, WA, north to British Columbia, northeast to Quebec; worldwide. Exotic.

This flat-leaved *F. rubra* is a weed of disturbed, often moist sites, including cultivated fields in which other types of fescue are being grown.

Festuca rubra Linnaeus ssp. ***juncea*** (Hackel) K. Richt

Coastal red fescue

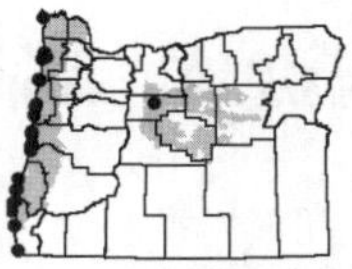

Plants 8–150 cm tall, usually strongly glaucous or occasionally green, often both in the same population; rhizomatous; shoots arising in clusters from rhizomes that may be deeply buried. **Culms** basal branching intra- and extravaginal. **Leaves** sheaths brown and shredding between the whitish veins; cauline sheaths often loose; blades 0.1–1.5 mm in diameter. **Leaf anatomy** sclerenchyma thick, the bands less than 2 times as wide as thick and on serpentine substrates sometimes large and confluent, thus producing some bands more than 2 times as wide as thick. **Inflorescences** 4.5–11 × (0.8)1–2 cm, unusually dense on coastal sands but not on inland sites, with blue and purple, sometimes with turquoise. **Glumes** glaucous but sometimes with a whitish coating; lower glumes 2.2–4.5 mm; upper glumes 3.2–6.5 mm; margins and tips scabrous; tips acute or short-pointed. **Lemmas** 4.4–6(6.5) mm, smooth, glabrous, often acute to short-pointed, sometimes awned; lemma awns 0.1–3 mm. **Anthers** 2.2–3.5 mm.

Shifting coastal sand dunes, coastal meadows and headlands and inland streamsides and roadsides on serpentine substrates. 0–1700 m. CR, Est, Lava. CA, WA; north to British Columbia, Atlantic Coast of eastern North America; Europe. Native.

Festuca rubra ssp. *juncea* is usually strongly glaucous. Coastal plants are short and remain short even when transplanted to more mesic habitats. They tend to have very dense inflorescences and very obscurely veined, glaucous lemmas. Inland plants are taller and have proportionately narrower inflorescences. Three somewhat different forms are provisionally included in this subspecies. The first is *Festuca ammobia*, also called *F. rubra* ssp. *arenicola*, which grows on shifting sands. The second is a small plant of coastal headlands, considered the same as coastal *F. r.* ssp. *pruinosa* of Europe. The third is a tall, strongly rhizomatous plant of rocky streamsides on serpentine substrates; it may correspond to *F. r.* ssp. *juncea*. Botanists in Great Britain treat *F. r.* ssp. *pruinosa* as a synonym of inland *F. r.* ssp. *juncea*.

Festuca rubra Linnaeus ssp. ***mediana*** (Pavlick) Pavlick

Median red fescue

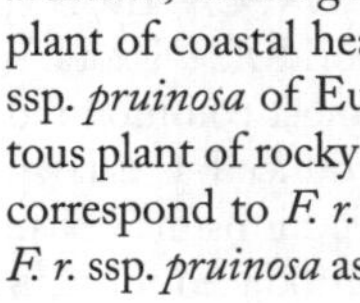

Plants 20–30(60) cm tall; rhizomatous; shoots arising in dense clusters. **Culms** basal branching extravaginal. **Leaves** sheaths glabrous or hairy but sometimes glaucous, shredding into fibers; cauline sheaths often loosely enclosing the culms; blades 0.7–1.5 mm in diameter, conduplicate. **Leaf anatomy** sclerenchyma in narrow strands; sometimes present on ribs on inner surface. **Inflorescences** generally congested, usually secund, often partially included in the uppermost sheaths at maturity; branches scabrous. **Spikelets** to 12 mm, green to violet, sometimes glaucous, 4–8 florets; lower glumes 2.5–3.2 mm; upper glumes 3.5–4.6(5) mm. **Lemmas** 4.5–6 mm, acuminate in side view; lemma awns 0.5–2 mm. **Anthers** 2.5–3 mm.

Sand dunes and beaches, salt marshes, meadows and prairies. 0–100 m. Est. CA, WA; north to British Columbia. Native?

Festuca rubra ssp. *mediana* is the name published for presumed native coastal plants from the Pacific Northwest with relatively small floral parts and acuminate lemmas. Viewed from the side, the lemma bodies taper with somewhat concave margins. Compare with *F. rubra* ssp. *secunda*. The validity of the taxon and its origin are both uncertain, and more research is needed.

Festuca rubra Linnaeus ssp. ***rubra*** (p. 410)

Red fescue

Festuca rubra L. var. *rubra*

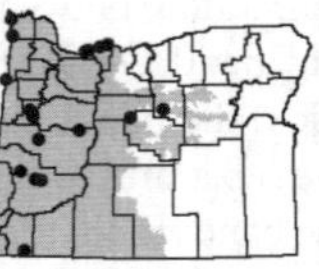

Plants 15–90(120) cm tall, more or less rhizomatous, green, yellow-green or glaucous; shoots extravaginal. **Culms** basal branching intra- and extravaginal. **Leaves** mainly basal; sheaths usually pubescent but sometimes glabrous, tightly enclosing the culm, the hairs retrorse, tending to turn brown early, usually conspicuously splitting between the veins producing a contrast of brown sheaths with whitish veins; blades 1–4 mm wide; lax, leaf blades of vegetative shoots almost always conduplicate unless under irrigation, green or glaucous. **Leaf anatomy** cross sections 0.4–1.1 × 0.65–1.25 mm, diamond-shaped; veins 5–7; ribs 5; dorsal sclerenchyma bands less than 2 times as broad as thick; sclerenchyma bands only occasionally present at the tops of the ribs on the inner surface. **Inflorescences** (2)5–14(20) × 1–8 cm, well exserted from uppermost leaf sheath; branches appressed after anthesis or sometimes the lower branches spreading. **Spikelets** (6)9–12(13) × 3–4.5 mm, (2)4–6(10) florets. **Glumes** glabrous or hairy at tip only or throughout; lower glumes (2)2.5–3.5(4.5) mm, 1(3)-veined; upper glumes 3.5–5.5(6) mm, 3-veined. **Lemmas** (4)5–7(8) mm, faintly 0–3(5)-veined, glabrous or pubescent at tip only or throughout; tip entire, awned; lemma awns (0.3)1–3.3(5) mm. **Anthers** (2.2)3–4(4.5) mm.

Disturbed meadows, swales, roadsides, lawns, pastures. 0–2000 m. Casc, CR, ECas, Est, Lava, Sisk, WV. CA, ID, NV, WA; throughout most of North America; worldwide. Exotic.

Festuca rubra ssp. *rubra* is probably the most common red fescue in Oregon. It has been widely introduced as a lawn grass, soil stabilizer, and pasture grass. Native forms may exist, but distinguishing them from the diverse introduced forms is impossible.

Festuca rubra Linnaeus ssp. ***secunda*** Pavlick

Secund red fescue

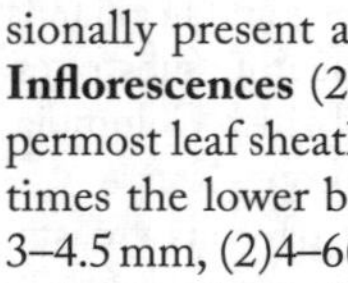

Plants (20)30–70(80) cm tall; rhizomatous, often with shoots arising in loose clusters. **Culms** basal branching extravaginal. **Leaves** sheaths glabrous or hairy or sometimes glaucous, shredding into fibers; cauline sheaths often loose; blades of vegetative shoots 0.7–1.5 mm in diameter, conduplicate, sometimes glaucous, outer surfaces more or less evenly scabrous, inner surfaces pilose on the ribs; cauline blades to 2.5 mm wide, sometimes flat. **Leaf anatomy** sclerenchyma bands less than 2 times as wide as thick; sclerenchyma rarely present on ribs on leaf inner surface. **Inflorescences** 7.5–12 cm, generally open, usually nodding, often partially included in the uppermost sheaths at maturity. **Spikelets** 9.5–13 mm, green to violet, sometimes glaucous, 4–7 florets. **Lower glumes** 3.1–4.5 mm; upper glumes (4.5)5–5.6 mm. **Lemmas** 5.8–6.6(7) mm, attenuate in side view, green but sometimes with violet margins, glabrous or uniformly hairy; margins and tips scabrous; lemma awns 1–5 mm. **Anthers** 2.8–3.9 mm.

Coastal salt marshes. 0–100 m. Est. WA; north to AK. Native?

This name was published for presumed native coastal plants with relatively long floral parts and attenuate lemmas. Viewed from the side; the margins of the lemmas taper straight from the widest part toward the awn. Hairy lemmas occur more often in *F. rubra* ssp. *secunda* than in our other *F. rubra* subtaxa. See *F. rubra* ssp. *mediana*. The validity of the taxon and its origin are both uncertain. More research is needed.

Festuca saximontana Rydberg ssp. ***purpusiana*** (St.-Yves) Tzvelev

Rocky Mountain fescue

Festuca saximontana Rydb. var. *purpusiana* (St.-Yves) Fred. & Pavlick

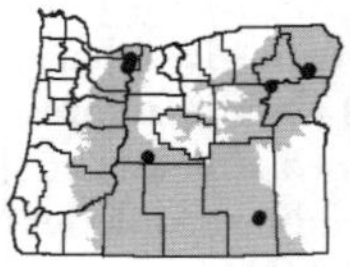

Plants 5–25 cm tall; densely cespitose, usually glaucous. **Culms** basal branching intravaginal. **Leaves** mainly basal; sheaths usually open half or more of their length, not splitting between the veins; blades 2–6.5 cm × 0.3–0.9 mm, conduplicate, inner surface hairy, outer surface glabrous; flag leaves 1–2.5 cm. **Leaf anatomy** cross sections 0.3–0.5 × 0.55–0.8 mm, generally round; veins 5; ribs 1 (often with an additional 2–4 poorly defined ribs), dorsal sclerenchyma in bands more than 2 times as wide as thick or continuous. **Inflorescences** 1–5 cm; branches 0.5–1.2 cm, appressed after anthesis. **Spikelets** (4.8)5.5–6.5(7.6) mm, 2–5 florets. **Glumes** glabrous or sometimes scabrous distally; lower glume 2–3.4 mm, 1-veined; upper glume 3–4.8 mm, 3-veined. **Lemmas** 3.6–5.6 mm, 0(1)-veined, glabrous to distally scabrous; lemma awns 0.8–1.8 mm. **Anthers** 0.9–1.2 mm when dry, 1.2–1.8 mm when fresh or rehydrated, retained or lost after anthesis. **Ovaries** apex glabrous. **Caryopses** 2–2.5 mm, glabrous.

Alpine. 1100–3000 m. BR, BW, Casc, ECas. CA, ID, NV, WA; western Canada and US; eastern Russia. Native.

Festuca saximontana is an alpine species that is most likely to be confused with *F. idahoensis*, which differs in having longer awns and anthers. *Festuca brachyphylla* is found in the similar alpine habitats but has shorter anthers and proportionately narrower leaf sclerenchyma bands. Its leaf sheaths are less conspicuous and persistent.

Festuca subulata Trinius

Bearded fescue

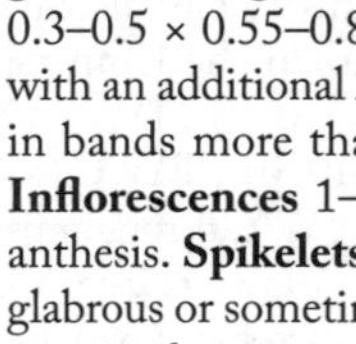

Plants 35–120 cm tall; loosely cespitose, green. **Culms** branching extravaginal. **Leaves** mostly cauline; sheaths glabrous to minutely scabrous, splitting between the veins; collars glabrous; ligules 0.2–0.7(1) mm, ciliate, the cilia much shorter than the membranous part; blades 3–8(10) mm wide, flat, lax; upper surface glabrous to sparsely hairy; lower surface glabrous or soft hairy. **Inflorescences** (10)15–30(40) cm, lowest node with (1)2(3) branches; branches 6–10(15) cm, spreading or drooping. **Spikelets** 6–11(12) × 1.5–3.5 mm, 3–4(5) florets. **Glumes** glabrous or with short hairs at the apex only; lower glumes (1.8)2–3.2(4) mm, 1-veined; upper glumes (2)3.5–5.5(6) mm, 1–3-veined. **Calluses** 0.1–0.2 mm, wider than long, glabrous. **Lemmas** (5)6–7.5(8.8) mm, 0–3-veined, glabrous or sometimes sparsely to densely scabrous or short-hairy; awns (2.5)5–10(17) mm, sometimes curving, longest awns longer than the lemma body. **Ovaries** apex hairy. **Caryopses** 4–5.5 mm; tips hairy. **2*n***=14, 28.

Moist forest, rarely in meadows. 0–2000 m. BW, Casc, CR, ECas, Est, Sisk, WV. CA, ID, NV, WA; north to British Columbia, east to WY. Native.

Festuca subulata is a common, leafy-stemmed grass of forests, with a nodding inflorescence. Similar *Melica* species and *Bromus vulgaris* have closed leaf sheaths. *Trisetum canescens* and *T. cernuum* have lemma awns arising from the back. *Festuca subuliflora* is similar but has long calluses, long hairs on the collar and velvety pubescence on the upper surface of the leaves. *Festuca elmeri* has consistently scabrous to short-hairy lemmas with distinctly visible veins.

Festuca subuliflora Scribner

Crinkle-awn fescue

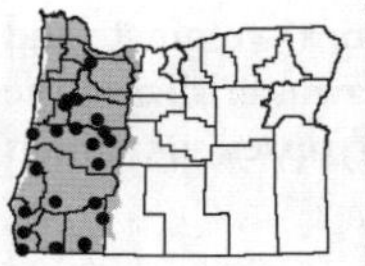

Plants (40)60–100(125) cm tall; loosely cespitose, green. **Culms** glabrous, 2–4 exposed nodes; basal branching extravaginal. **Leaves** mainly cauline; sheaths glabrous to abundantly hairy, splitting between the veins toward the end of the growing season; collars with sparse long hairs or rarely glabrous; ligules 0.1–0.5 mm, ciliate, hairs equaling or longer than the membranous portion; blades 2–6(10) mm wide, flat, lax; upper surface with dense, soft, velvety hairs; lower surface usually glabrous. **Inflorescences** (7)10–20 cm, lowest node 1(2)-branched; branches 3–16 cm, spreading to drooping. **Spikelets** 8–12.5 × 1–5 mm, (2)3–4(5) florets. **Glumes** glabrous or scabrous on the midvein and minutely scabrous toward the tip; lower glumes 2–4 mm, 1-veined; upper glumes 3.5–5(6) mm, 1–3-veined. **Calluses** (0.3)0.4–0.6 mm, longer than wide, hairy. **Lemmas** 6–9 mm, (3)5-veined, scabrous over the entire surface or more so on the veins; lemma awns 10–15 mm, crinkled, curved, longest awns much longer than the lemma bodies. **Anthers** 2.5–4 mm. **Ovaries** apex hairy. **Caryopses** 4–5 mm; tips hairy. **2*n***=28.

Mesic to dry forests. 0–1500 m. Casc, CR, Sisk, WV. CA, WA; north to British Columbia. Native.

Festuca subuliflora is a leafy-stemmed forest grass with a nodding inflorescence, similar to *F. subulata*, which tends to grow in moister microsites. The unusual elongated calluses of *F. subuliflora* probably aid in dispersal. See *F. subulata* for a discussion of similar species.

Festuca trachyphylla (Hackel) Krajina

Hard fescue, sheep fescue

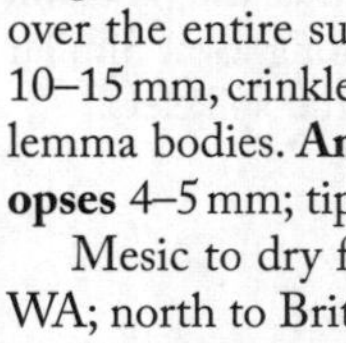

Plants (20)30–50(70) cm tall; densely cespitose, green or glaucous; basal branching intravaginal. **Leaves** mainly basal; sheaths open, somewhat flattened, not conspicuously splitting between the veins; blades 8–30 cm × 0.3–1.2 mm, conduplicate; inner surface with minute hairs that are much shorter than the leaf is thick, outer surface glabrous or tomentose. **Leaf anatomy** cross sections 0.4–0.6 × 0.7–1.2 mm, elliptic or ovate; veins (5)7+; ribs 1–5; dorsal sclerenchyma bands more than 2 times as wide as thick, interrupted or continuous. **Inflorescences** 3–10(30) cm; branches 1.2–3.6 cm, appressed after anthesis. **Spikelets** 5.5–9(10) mm, 4–7(8) florets. **Glumes** scabrous at the tip and often along the midrib; lower glumes 2–4 mm, 1–3-veined; upper glumes 3–5.5 mm, 3-veined. **Lemmas** (3.8)4–5(5.5) mm, 0(1)-veined, glabrous, scabrous, or with long hairs on margins near tip; lemma awns 0.5–2.5 mm, usually less than half as long as the lemma body. **Anthers** (1.6)2–3(3.4) mm. **Ovaries** apex glabrous. **Caryopses** 2.5–3.5 mm, glabrous. **2*n***=28, 42.

Roadsides, disturbed areas, ski slopes, persisting from cultivation or where planted for erosion control, occasionally naturalizing. 0–2800 m. BR, BW, Casc, Lava, Sisk, WV. CA, WA; scattered throughout North America; worldwide. Exotic.

Festuca trachyphylla is a fine-leaved bunchgrass with small lemmas and dense inflorescences. *Festuca idahoensis* and *F. roemeri* have longer lemmas and more open inflorescences, at least at anthesis. *Festuca rubra* ssp. *commutata* has fused leaf sheath margins and leaves that are diamond-shaped in cross section with sclerenchyma bundles less than twice as broad as thick. *Festuca valesiaca* is very similar but has tighter, less conspicuous leaf sheaths; *F. trachyphylla* and *F. valesiaca* are best distinguished by leaf anatomy. The glaucous forms of *Festuca trachyphylla* are often called sheep fescues and labeled *F. ovina*, but the name *F. ovina* is best limited to a species of northern Europe and few cultivars are derived from it. The hard and sheep fescues are sometimes considered part of *F. longifolia,* but current usage limits that name to a British species, which is rarely cultivated.

Festuca valesiaca Schleicher ex Gaudin

Valais fescue, Covar sheep fescue

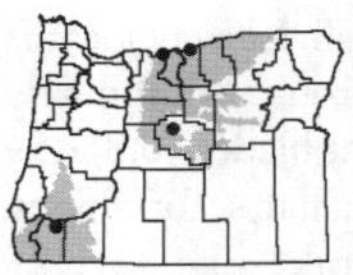

Plants 20–50(75) cm tall; densely cespitose, green or glaucous. **Culms** basal branching intravaginal. **Leaves** mainly basal; sheaths open, glabrous, not conspicuously splitting between the veins; blades (4)10–30 cm × 0.3–0.9 mm, conduplicate, flag leaves 1–8 cm; hairs on inner leaf surface approximately as long as the leaf is thick; leaves often pubescent on outer surface. **Leaf anatomy** cross sections 0.2–0.5 × 0.4–0.9 mm, elliptic; veins 5–7; ribs 1–3; dorsal sclerenchyma bands more than 2 times as wide as thick, 3 at midrib and margins and sometimes with smaller bands between these. **Inflorescences** (2)5–9 cm; branches 0.5–1.5 cm, appressed after anthesis. **Spikelets** (4.5)5–7.5(8.5) mm, 2–5 florets. **Glumes** sparsely scabrous at the tip, acute to mucronate; lower glumes 1.9–2.6(3) mm, 1-veined; upper glumes (2.2)2.5–4(4.3) mm, 3-veined. **Lemmas** (2.6)4–5(5.2) mm, 0(1)-veined, glabrous throughout or scabrous or ciliate near the tip; lemma awns (0.5)1.5–2.2 mm, usually less than half as long as the lemma body. **Anthers** 2.1–2.6 mm. **Ovaries** apex glabrous. **Caryopses** 2.8–3.2 mm, glabrous.

Dry roadsides, dry pastures and shrub-steppe. 50–2000 m. Col, Lava, Sisk. WA; southeast to AZ; Eurasia. Exotic.

Similar *Festuca idahoensis* and *F. roemeri* have longer lemmas and more open inflorescences, at least at anthesis. *Festuca rubra* ssp. *commutata* has fused leaf sheath margins and leaves diamond-shaped in cross section with sclerenchyma bundles less than two times as broad as thick. *Festuca trachyphylla* is very similar but has looser, more conspicuous sheaths and very short hairs on the inner leaf surface.

Festuca viridula Vasey (p. 410)

Green fescue

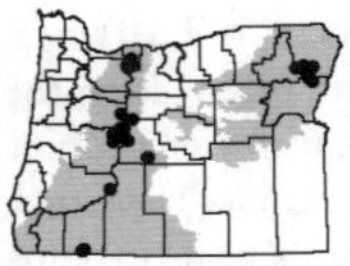

Plants (35)50–100 cm tall; densely cespitose, green. **Culms** glabrous, 1–3 exposed nodes; branching intravaginal. **Leaves** mainly basal; sheaths open, glabrous, sometimes turning brown with contrasting white veins, usually not conspicuously splitting between the veins; blades mostly 10–25 cm but the lower few blades greatly reduced, mostly 1–4 cm × 0.8–2(2.5) mm, conduplicate, flat, or loosely rolled, green; inner surface with very short hairs, outer surface glabrous, flag leaves 4–8(12) cm; collars glabrous. **Leaf anatomy** cross sections 0.4–0.9 × 0.6–1.3 mm, elliptic; veins 7–12, ribs 5–10, some of the dorsal sclerenchyma bands more than 2 times as wide as thick; ventral sclerenchyma bands often present on the tops of the ribs. **Inflorescences** 4–12(15) cm; branches 1.5–3 cm, appressed after anthesis. **Spikelets** 9–12 mm, (2)3–6(7)florets. **Glumes** scabrous at the tips; lower glumes (2.4)3.5–5 mm, 1-veined; upper glumes 4.5–6(7.5) mm, 3-veined. **Lemmas** (5.5)6.5–8 mm, 0(1)-veined, glabrous or minutely scabrous on the tips and/or along the margins, awnless or awned; lemma awns 0–0.5(1.4) mm. **Anthers** (2)3–4(5) mm. **Ovaries** apex densely hairy. **Caryopses** 3.5–5 mm; tips densely hairy. **2*n*=**28.

Montane to alpine, dry grasslands, meadows and open forest. 1000–2700 m. BW, Casc, ECas, Sisk. CA, ID, NV, WA; north to British Columbia, northeast to MT. Native.

Festuca viridula is relatively easy to identify because few similar species are found in its habitat. It might be confused with *F. idahoensis*, which has finer leaves, or *F. rubra*, which is usually found in moister habitats, but those species have normally developed blades on their lower leaf sheaths. *Festuca viridula* ovaries and seeds are densely hairy at the tip. The only other fine-leaved fescue sharing that trait is *F. occidentalis*, a delicate forest species with awns longer than the lemma bodies.

GASTRIDIUM Nit grass

Plants annual. **Leaves** sheaths open; auricles vestigial or absent; ligules membranous; blades flat. **Inflorescences** spike-like panicles with more than 1 spikelet at node; branches appressed to ascending. **Spikelets** laterally compressed with 1 bisexual floret; disarticulation above the glumes. **Glumes** 2, unequal, much longer than the floret, 1-veined, swollen at the base; tips acuminate and awnless. **Calluses** short, blunt, glabrous. **Lemmas** 5-veined, the lateral veins faint, awned or awnless; lemma awns arising in the upper 33% of the lemma, sometimes subterminal, shorter or longer than the lemmas, bent. **Anthers** 3. **Caryopses** sometimes adhering to the lemma or palea.

Cosmopolitan. 2 species; 1 species treated in *Flora*.

Gastridium phleoides (Nees & Meyen) C.E. Hubbard (p. 418)

Nit grass

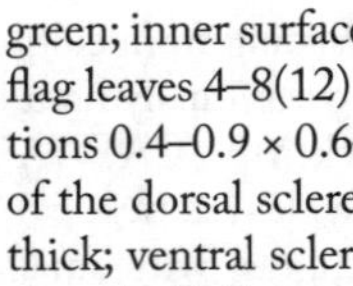

Plants annual, 7–70 cm tall. **Leaves** sheaths smooth or minutely scabrous and minutely papillose; ligules 1–7 mm; blades (0.8)1.5–20 cm × (0.4)1.5–4(6) mm, flat, scabridulous. **Inflorescences** dense, spike-like, (1.1)2–16.5 × 0.15–3.7 cm. **Spikelets** 3–7 mm. **Glumes** much longer than the lemma, strongly keeled or winged, 1-veined, scabrous, swollen near the base to accommodate the lemma, gradually narrowed into an awn-like point; lower glumes 3–7 mm; upper glumes 2.7–5.5 mm. **Lemmas** 1–1.5 mm, 5-veined, with short appressed hairs; tips rounded or truncate, awnless or awned; lemma awns to 6 mm, arising above mid-length to just below apex, twisted. **Anthers** 0.5–0.9 mm. **Caryopses** 0.8–1 × 0.4–0.5 mm. **2*n*=**14.

Dry or seasonally dry disturbed areas including roadsides. 0–1100 m. Casc, CR, Est, Sisk, WV. CA; AZ, MA, SC, TX, South America; Africa, southwestern Asia, Australia. Exotic.

Gastridium phleoides resembles an annual *Agrostis* or *Apera*. The glumes have a distinctive swollen pouch at the base and enclose a very small lemma usually with a long awn. Similar *Apera interrupta* has a looser panicle and shorter spikelets.

GLYCERIA Mannagrass

Plants usually perennial; rhizomatous. **Culms** rooting at the lower nodes. **Leaves** cauline; sheaths closed for at least 75% of their length; auricles sometimes present; ligules scarious, glabrous, erose to lacerate, those of lower leaves shorter but otherwise similar to those of upper leaves; blades flat or folded. **Inflorescences** panicles or racemes; branches appressed to widely spreading or reflexed. **Spikelets** cylindrical and terete or oval and laterally compressed, 2–16 florets, terminal floret in each spikelet reduced and sterile; disarticulation above the glumes and beneath the florets. **Glumes** 2, much smaller than to equaling the adjacent lemmas, 1-veined, obtuse or acute, often erose. **Calluses** glabrous. **Lemmas** membranous to thinly leathery, rounded over the back, glabrous or with hairs to 0.1 mm, 5–11-veined; veins often conspicuous and ridged, not converging distally; margins hyaline and transparent at and near the tips, sometimes with a purplish band below the hyaline portion; tips acute to rounded or truncate, entire or erose, awnless. **Anthers** (1)2–3.

Circumboreal. Approximately 35 species; 9 species treated in *Flora*.

Glyceria can be confused with *Torreyochloa pallida* and *Puccinellia* species, which have open leaf sheaths. The native *Glyceria* species with long, terete spikelets (section *Glyceria*) are distinctly different from each other, but introduced *G. fluitans* hybridizes with *G. leptostachya* to produce *G.* × *occidentalis*. Due to a series of hybrids and backcrosses, the morphology of the parental species can be difficult to discern. Some individuals are difficult to distinguish from *G. borealis*.

When keying within this genus, it is helpful to measure the culm thickness near mid-length of the basal internode; do not include the leaf sheath. Ligule measurements include both basal and upper leaves unless noted otherwise. Lemma lengths are measured on the lowest lemmas of most spikelets in a panicle. Additionally, the number of spikelets on a branch is counted on the longest primary branches and includes all the spikelets on the secondary (and higher order) branches of that primary branch.

Key to *Glyceria* species:

1. Spikelets short and rounded to elliptic, laterally compressed, length 1–4 times width, palea keels not winged.
   2. Veins of 1 or both glumes extending to the tips; lemma tips nearly flat; anthers 3 . . . . ***G. grandis*** var. ***grandis***
   2' Veins of glumes terminating below the tips; lemma tips prow-tipped; anthers 2.
      3. Spikelets (2.5)3–5 mm wide; lemma veins evident but not raised distally; palea lengths 1.5–1.8 times width . . . . ***G. canadensis*** var. ***canadensis***
      3' Spikelets 1.2–2.9 mm wide; lemma veins distinctly raised throughout; palea lengths 1.5–3.5 times width.
         4. Leaf blades 2–6 mm wide; anthers 0.2–0.6 mm; culms 1.5–3.5 mm thick . . . . ***G. striata***
         4' Leaf blades 6–15 mm wide; anthers 0.5–0.8 mm; culms 2.5–8 mm thick. . . . . ***G. elata***

1' Spikelets long and narrow, length more than 5 times width, cylindrical and round in cross section, except at anthesis when slightly laterally compressed; palea keels usually winged toward the tip.
   5. Lemma tips with 1 strongly developed lobe on 1 or both sides, entire to crenulated between the lobes; leaf blades 3–12 cm; primary panicle branches 1.5–9.5 cm . . . . ***G. declinata***
   5' Lemma tips entire or crenulated but not with 1 strongly developed lobe on 1 or both sides; leaf blades 5–30 cm; primary panicle branches 3–18 cm.
      6. Lemmas 5–8 mm.
         7. Anthers 0.6–1.6 mm; lemma tips usually slightly lobed or irregularly crenate . . . . ***G. ×occidentalis*** (in part)
         7' Anthers 1.5–3 mm; lemma tips usually entire . . . . ***G. fluitans***
      6' Lemmas 2.5–5.9 mm.
         8. Lemmas smooth or scabridulous only on the veins at 20× magnification . . . . ***G. borealis***
         8' Lemmas scabridulous both on and between the veins at 20× magnification
            9. Lemmas usually 4.5–5.9 mm, tips acute; anthers 2 . . . . ***G. ×occidentalis*** (in part)
            9' Lemma 2.5–4.5 mm, tips truncate to obtuse; anthers 3. . . . . ***G. leptostachya***

Glyceria borealis (Nash) Batchelder (p. 418)

Boreal mannagrass

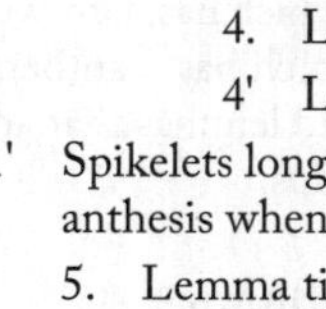

Plants perennial, 60–100 cm tall; rhizomatous. **Culms** 1.5–5 mm thick, often decumbent and rooting at the lower nodes. **Leaves** sheaths glabrous, keeled; blades 9–25 cm × 2–7 mm wide. **Inflorescences** 18–40(50) × 0.5–2(5) cm, arching, usually narrow, open at anthesis, bases often enclosed in the upper sheaths at maturity; branches 5–10(15) cm, usually 1–3(5) per node, appressed to strongly ascending, occasionally spreading, longer branches with 3–6 spikelets. **Spikelets** 2–22 mm, cylindrical and round in cross section. **Glumes** lower glumes 1.2–2.2 mm; upper glumes 2–3.8 mm, rounded. **Lemmas** 2.7–5.4 mm; veins raised, smooth or minutely scabrous, the regions between the veins smooth, rarely minutely scabrous, midvein terminating about (0.1)0.2 mm short of the apical margin; tips acute, sometimes obtuse, entire or nearly so. **Paleas** keels narrowly winged. **Anthers** 3, 0.4–1.5 mm. **Caryopses** 1.2–2 mm. $2n$=20.

Emergent in ponds, streams, and sloughs. 0–2600 m. BR, BW, Casc, CR, ECas, Est, Lava, Owy, WV. CA, ID, NV, WA; throughout Canada, western and northern US. Native.

Glyceria borealis differs from other species with long spikelets in having smooth (occasionally minutely scabrous) lemmas.

Glyceria canadensis (Michaux) Trinius var. ***canadensis***

Canadian mannagrass

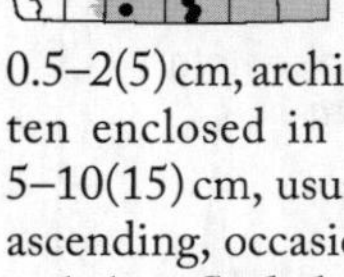

Plants perennial, 60–150 cm tall; rhizomatous. **Culms** 2.5–5 mm thick, erect or the bases decumbent. **Leaves** sheaths retrorsely scabrous to minutely scabrous; blades 3–8 mm wide. **Inflorescences** 10–30 × 10–20 cm, pyramidal, open, nodding; branches 7–20 cm, lax, divergent, often drooping, with 15–60+ spikelets. **Spikelets** 5–8 × (2.5)3–5 mm, laterally flattened, oval in side view, 4–10 florets. **Glumes** narrowing from mid-length or above to the broadly (more than 45°) acute or rounded tips, 1-veined; veins terminating below the tips; lower glumes 1.6–2.4 mm, ovate to rectangular; upper glumes 1.8–2.5 mm, lanceolate, acute. **Lemmas** 2.4–4 mm, ovate when viewed from the back, 5–7-veined; veins evident but not raised distally, smooth; tips acute, prow-shaped. **Paleas** length 1.5–1.8 times the width; keels not winged. **Anthers** 2, 0.4–0.5 mm, dehiscent at maturity.

Emergent in bogs, wet ditches, lake shores, cranberry farms. 0–50 m. Est. WA; north to British Columbia; midwestern and eastern Canada and US. Exotic.

Glyceria canadensis is characterized by large, more or less round spikelets. Its lemma veins are raised in the lower part of the lemma but merge into the lemma surface distally. *Glyceria striata* and *G. elata* have smaller spikelets and lemmas, and the lemmas have strong, raised veins nearly to the tip. *Glyceria canadensis* forms sterile hybrids with *G. striata*.

Glyceria declinata Brébisson

Low glyceria

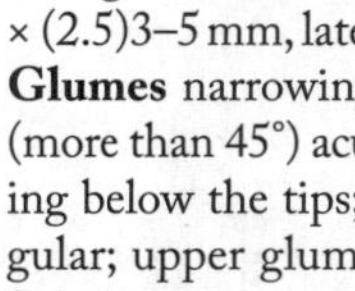

Plants usually perennial, (10)20–92 cm tall, sprawling. **Culms** 1.5–2.5 mm thick, ascending to erect from a decumbent, branching base. **Leaves** sheaths glabrous, keeled; blades 3-12 cm × 4–8 mm. **Inflorescences** 6–30 × 1–2.5 cm, narrow; primary branches 1.5–9.5 cm, ascending, 1–5 spikelets. **Spikelets** 11–24 × 1.3–3 mm, cylindrical and terete, except slightly laterally compressed at anthesis, 8–15 florets. **Glumes** lower glumes 1.4–3.5 mm; upper glumes 2.5–4.9 mm. **Lemmas** 3.4–6 mm, 7-veined; veins and the regions between them minutely scabrous, midveins extending to within 0.1 mm of the apical margins; tips acute with a well-developed lobe on one or both sides opposite the lateral veins, entire to crenulate between the lateral lobes. **Paleas** keels winged. **Anthers** 0.5–1.4 mm, usually purple. **Caryopses** 1.8–2.5 mm. **2*n***=20.

Marshes, wet ditches, disturbed wet meadows, and vernal pools. 50–300 m. Casc, Sisk, WV. CA, NV; widely scattered in North America; Europe. Exotic.

Glyceria declinata is most similar to *G.* × *occidentalis*, which has crenate but not truly lobed lemma tip margins. Some intermediate plants can only be identified by chromosome number; *Glyceria declinata* is diploid and *G.* × *occidentalis* is tetraploid.

Glyceria elata (Nash) M.E. Jones
Tall mannagrass

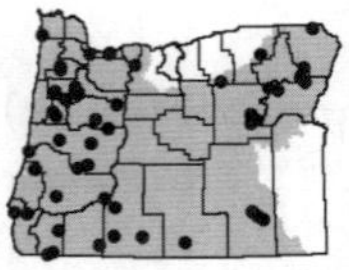

Plants perennial, 75–150 cm tall; rhizomatous. **Culms** 2.5–8 mm thick, spongy, decumbent. **Leaves** sheaths minutely scabrous or minutely pubescent, only weakly or not keeled; blades 6–12(15) mm wide. **Inflorescences** 15–30 × 12–30 cm, pyramidal, open; branches 12–17 cm, divergent to drooping, lax, with 30–50+ spikelets. **Spikelets** 3–6 × 1.5–2.8 mm, laterally compressed, oval in side view, 3–4(6) florets. **Glumes** 1–1.5(2) times longer than wide, narrowing beyond mid-length; veins terminating below the apical margins; tips obtuse to rounded; lower glumes 0.7–1.5 mm; upper glumes 1–1.5 mm. **Lemmas** 1.7–2.2 mm, oval when viewed from the back, 5–7-veined; veins raised throughout, minutely scabrous, green with purple tips, or sometimes entirely purple; tips rounded, prow-shaped. **Paleas** length 2.4–3 times the width; keels not winged. **Anthers** 2, 0.5–0.8 mm. **2*n***=20.

Ponds, stream banks, wet meadows, and moist woods. 0–2500 m. BR, BW, Casc, CR, ECas, Est, Lava, Sisk, WV. CA, ID, NV, WA; north to British Columbia, northeast to Alberta;NM. Native.

Glyceria elata is very similar to *G. striata*, which has narrower leaves. It is also confused with *G. grandis*, which has 3 anthers, glume veins that extend to the glume tips and lemmas that are usually entirely purple and have almost flat tips.

Glyceria fluitans (Linnaeus) R. Brown
Water mannagrass

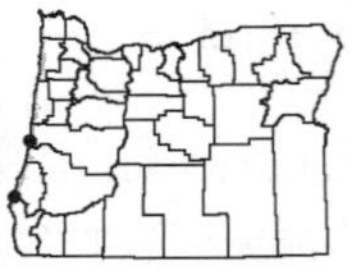

Plants perennial, 20–150 cm tall; rhizomatous. **Culms** erect or spreading, sometimes decumbent, distal portion sometimes floating in shallow water. **Leaves** sheaths glabrous, keeled; blades 5–25 cm × 3–10 mm wide. **Inflorescences** 10–50 × 2–3 cm; branches 3–5 cm, paired or solitary, usually appressed to ascending, divergent at anthesis, 1–4 spikelets. **Spikelets** (15)18–39 × 1.7–3.3 mm, cylindrical and terete, slightly laterally flattened at anthesis, rectangular in side view, 8–16 florets. **Glumes** lower glumes 1.3–3.9 mm; upper glumes 2.7–5 mm. **Lemmas** 5.2–8 mm, slightly scabrous over and between the veins, sometimes almost smooth, midveins extending to within 0.1 mm of the apical margins; tips

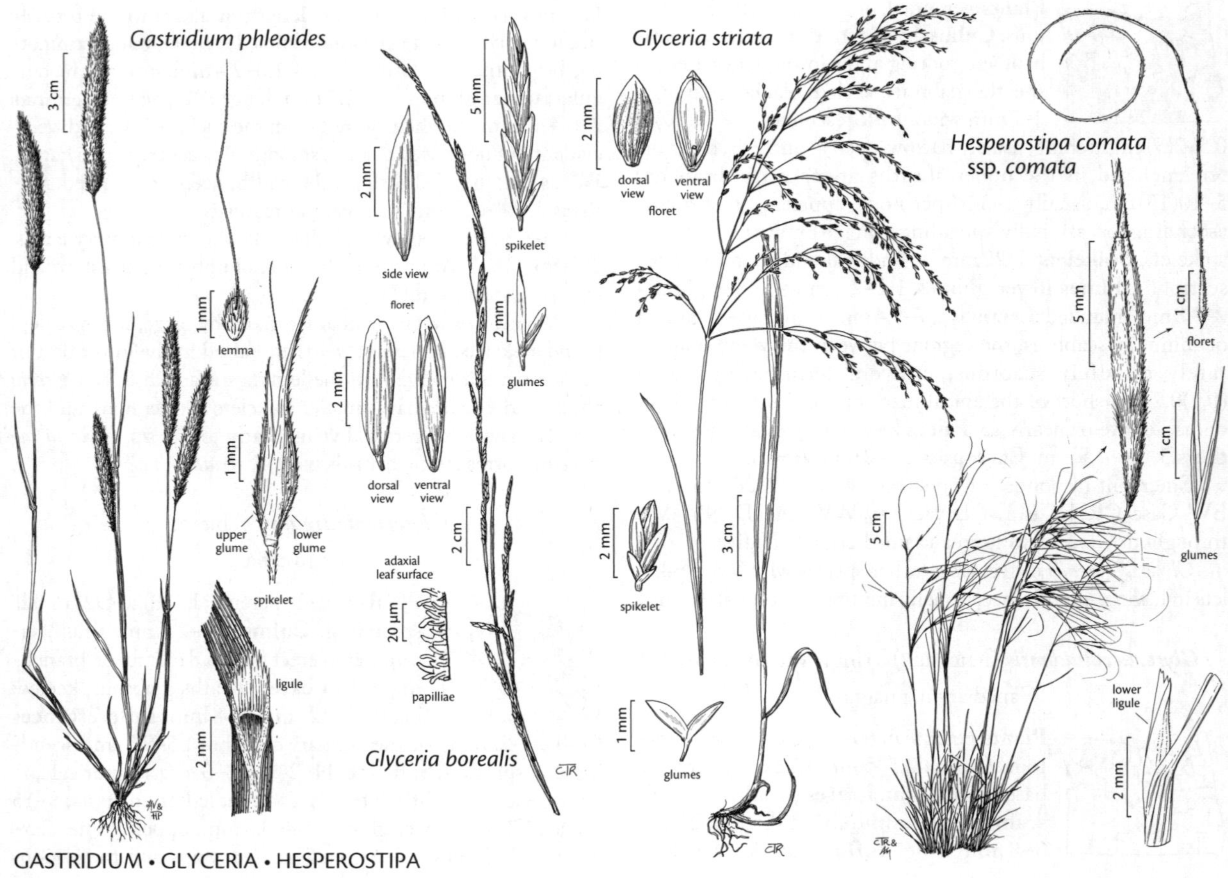

GASTRIDIUM • GLYCERIA • HESPEROSTIPA

acute, usually entire. **Paleas** keels winged. **Anthers** 1.5–3 mm, usually purple. **Caryopses** 2.6–3 mm. **2*n*=40.**

Emergent in freshwater wetlands near the coast. 0–1800 m. Est. CA, WA; SD, north to British Columbia, eastern North America from Newfoundland to PA; Eurasia. Exotic.

Glyceria fluitans is an introduced, emergent grass, with long, narrow spikelets, recognized by its very long lemmas. It is most often confused with its hybrid descendent, *G.* × *occidentalis,* which may have lemmas up to 5.9 mm but differs in having shorter anthers.

Glyceria grandis S. Watson var. *grandis*

American mannagrass

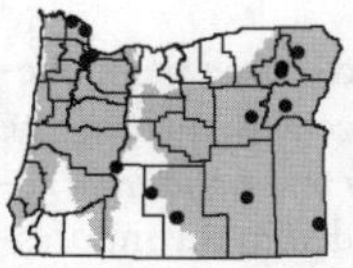

Plants perennial, 50–150(200) cm tall; rhizomatous. **Culms** 8–12 mm thick, erect or decumbent. **Leaves** sheaths smooth or minutely scabrous, keeled; blades 4.5–15 mm wide. **Inflorescences** 16–42 × 12–20 cm, open; branches (7)10–18 cm, lax, widely divergent to drooping, with 35–80+ spikelets. **Spikelets** 3.2–6.4 × 2–3 mm, somewhat laterally flattened, oval to elliptic in side view, with 4–8 florets. **Glumes** mostly hyaline, usually the midvein of 1 or both glumes extending to the tips; tips acute; lower glumes 1–2.3 mm; upper glumes 1.5–2.7 mm. **Lemmas** 1.8–3 mm, prominently (5)7-veined; veins often minutely scabrous, regions between the veins smooth; tips rounded to truncate, sometimes erose, almost flat at maturity. **Paleas** length more than 3 times the width; keels not winged. **Anthers** 3, 0.5–1.2 mm. **2*n*=20.**

Emergent in ponds, stream banks, ditches. 0–1700 m. BR, BW, Casc, CR, Lava, Owy, WV. CA, ID, NV, WA; most of Canada and US. Native.

Glyceria grandis is superficially similar to *G. elata* but has hyaline glumes with midveins extending to the tips and nearly flat lemma tips. *Glyceria grandis* usually has purple lemmas; those of *G. elata* are often green with purple bands near the tip. *Glyceria grandis* is sometimes confused with *Torreyochloa pallida*, which has open sheaths and 3-veined upper glumes.

Glyceria leptostachya Buckley

Narrow mannagrass

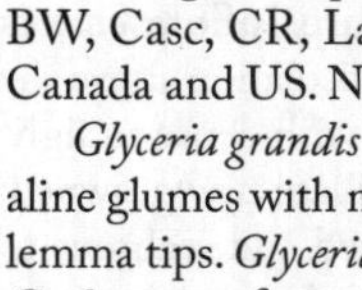

Plants perennial, 50–100(150) cm tall; rhizomatous. **Culms** 3–8 mm thick, spongy, erect to decumbent. **Leaves** sheaths finely, minutely scabrous, not or weakly keeled; blades 12–30 cm × 3.5–11 mm wide. **Inflorescences** 20–40 × 2.5–8 cm; branches 4.2–14.7 cm, appressed to ascending, with 3–8(10) spikelets. **Spikelets** 9–20 × 0.4–3 mm, cylindrical and terete except at anthesis when slightly laterally flattened, 6–15 florets. **Glumes** lower glumes 0.5–2.1 mm; upper glumes 1.4–3.4 mm. **Lemmas** 2.5–4.5 mm, somewhat indented below the apical margins at maturity; veins raised, scabrous to scabridulous over and between the veins, midveins extending to within 0.1 mm of the apical margins; tips truncate to obtuse, minutely crenate. **Paleas** keels winged. **Anthers** 3, 0.3–0.9 mm. **Caryopses** 1.8–2.4 mm. **2*n*=40.**

Emergent in swamps, lake shores and stream banks. 0–1500 m. Casc, CR, ECas, Est, Owy, Sisk, WV. CA, WA; north to AK. Native.

Glyceria leptostachya has long, narrow spikelets and scabrous lemmas. It is most often confused with its hybrid descendent, *G.* × *occidentalis,* which usually has longer lemmas and glumes and more acute lemma tips.

Glyceria × *occidentalis* (Piper) J.C. Nelson

Western mannagrass

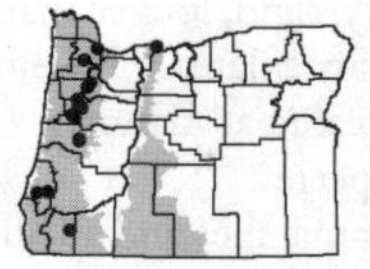

Plants perennial, 60–160 cm tall; rhizomatous. **Culms** 2.5–5 mm thick, erect or decumbent. **Leaves** sheaths smooth to minutely scabrous, keeled, sometimes weakly so; blades 20–30 cm × (2.5)4–12 mm wide. **Inflorescences** 20–50 × 2–15 cm, usually narrow, open at anthesis; primary branches 4.5–18 cm, somewhat lax, usually ascending but strongly divergent at anthesis, 2–8 spikelets. **Spikelets** 13–23 × 1.5–3.5 mm, terete except at anthesis when slightly laterally flattened, 6–13 florets. **Glumes** lower glumes 1.1–2.8 mm; upper glumes 2.9–3.7 mm. **Lemmas** 4.5–5.9 mm, minutely scabrous, midveins extending to within 0.1 mm of the apical margins; tips acute, slightly lobed or irregularly crenate, occasionally entire. **Paleas** keels winged. **Anthers** 2, 0.6–1.6 mm. **Caryopses** 2.1–2.7 mm. **2*n*=40.**

Emergent in ponds, lake shores, stream banks and marshy areas. 0–400 m. CR, ECas, Sisk, WV. CA, ID, WA. Native and exotic.

Glyceria × *occidentalis* was long thought to be an uncommon native species, but it was recently recognized as a series of hybrids and backcrosses between native *G. leptostachya* and introduced *G. fluitans.*

Glyceria striata (Lamarck) Hitchcock (p. 418)

Fowl mannagrass

Glyceria nervata (Willd.) Trin., *Glyceria striata* (Lam.) Hitchc. var. *stricta* (Scribn.) Fernald

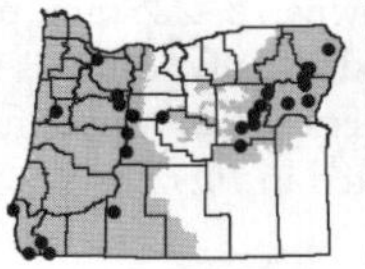

Plants perennial, 20–80(100) cm tall; rhizomatous. **Culms** 1.5–3.5 mm thick, not or only slightly spongy. **Leaves** sheaths smooth to minutely scabrous, keeled, sometimes only weakly so; blades 2–6 mm wide. **Inflorescences** 6–25 × 2.5–21 cm, pyramidal, open, nodding; branches 5–13 cm, straight to lax, sometimes ascending, 15–50 spikelets; lower branches usually strongly divergent to drooping at maturity. **Spikelets** 1.8–4 × 1.2–2.9 mm, laterally flattened, oval in side view, with 3–7 florets. **Glumes** ovate, 1–1.5 times longer than wide, narrowing from mid-length or above; veins terminating below the apical margins; tips often splitting with age; lower glumes 0.5–1.2 mm, rounded to obtuse; upper glumes 0.6–1.2 mm, acute or rounded. **Lemmas** 1.2–2 mm, ovate when viewed from the back; veins raised, minutely scabrous over and between the veins; tips acute, prow-shaped. **Paleas** lengths 1.5–3 times widths; keels not winged. **Anthers** 2, 0.2–0.6 mm, purple or yellow. **2*n*=40.**

Emergent in ponds, bogs, lakesides, stream banks, wet meadows, wet forests. 50–2500 m. BW, Casc, CR, ECas, Est, Sisk, WV. CA, ID, NV, WA; throughout North America. Native.

Glyceria striata is a common wetland grass with an open panicle. The strongly veined lemmas readily fall from the panicle branches. *Glyceria elata* is very similar but has wider leaves, thicker culms, and slightly larger spikelet parts. The two species have sometimes been considered to be conspecific, but there is molecular evidence that suggests they do not hybridize.

HAINARDIA Thintail

Plants annual. **Culms** branched above the base; internodes solid. **Leaves** sheaths open, the uppermost expanded and partially enclosing the inflorescence; ligules membranous; blades flat or convolute. **Inflorescences** cylindrical spikes with spikelets solitary and embedded in the rachis; the upper glumes exposed. **Spikelets** oriented radially to the inflorescence axis, dorsiventrally flattened, 1–2 florets, the second floret (if present) reduced and sterile; disarticulation at the nodes of the inflorescence axis. **Glumes** 1(2 on the terminal spikelet); upper glumes leathery, rigid, longer than the lemmas, 3–7(9)-veined, acute, awnless but sometimes with a short point. **Lower lemmas** membranous, lanceolate, 3(5)-veined, awnless. **Anthers** 1–3. **Caryopses** oblong, somewhat dorsally compressed, with an apical appendage, concealed at maturity.

North America, Africa, and Eurasia. 1 species.

Hainardia is a monospecific genus of salt marshes and alkaline soils. Its spikelets are embedded in the axis of the inflorescence.

Hainardia cylindrica (Willdenow) Greuter

Thintail

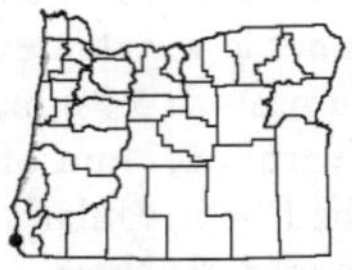

Plants 5–40(45) cm tall. **Culms** glabrous. **Leaves** sheaths 1–6 cm, glabrous, the uppermost enclosing the bottom of the inflorescence; ligules 0.2–1 mm, denticulate; blades to 7 cm × 1.5–2.5 mm; upper surfaces scabrous. **Inflorescences** spikes, 8–25 mm, rigid, straight or somewhat curved; spikelets embedded in the axis. **Spikelets** 5–8 mm, slightly longer than the adjacent internodes. **Glumes** ovate-lanceolate; upper glumes glabrous, midveins prominent; tips acute, sometimes mucronate. **Lower lemmas** 4–6 mm, awnless. **Anthers** 2–3.5 mm. ***2n***=14, 26, 52.

Disturbed coastal areas, roadsides, pond margins, vernal pools. 0–50 m. Est. CA; south to Mexico, TX, LA, SC; Europe. Exotic.

Hainardia cylindrica has one stiff, pointed glume per spikelet. The glumes diverge from the inflorescence axis, making the inflorescence look barbed. *Parapholis incurva* and *Scribneria bolanderi* are similar in having spikelets embedded in the inflorescence axis, but in both species two glumes are present on each spikelet. In addition, *S. bolanderi* is a more delicate plant with spikes only 1–2.5 mm wide. *Lolium rigidum* can have spikelets sunken in the inflorescence axis, but it has two or more florets per spikelet.

HESPEROSTIPA Needle-and-thread

Plants perennial; cespitose. **Culms** erect. **Leaves** not overwintering, not basally concentrated; sheaths smooth; ligules membranous, often ciliate; blades usually tightly involute; ventral surfaces conspicuously ridged; tips narrowly acute but not sharp. **Inflorescences** panicles. **Spikelets** with 1 narrowly cylindrical floret, rachilla not prolonged beyond the base of the floret; disarticulation above the glumes and beneath the floret. **Glumes** tapering from near the base to a hair-like tip. **Calluses** sharp, 2–6 mm, with dense straight hairs distally. **Lemmas** hard; margins slightly overlapping at maturity, the upper portion fused into a papillose, ciliate crown, awned; the lemma-awn junction distinct; lemma awns 75–225 mm, persistent, bent twice, the lower segments twisted and scabrous to long-hairy, the terminal segment not twisted, usually scabrous to long-hairy. **Paleas** equal to the lemmas, flat, pubescent, leathery, 2-veined; tips hard and prow-shaped. **Anthers** 3. **Caryopses** fusiform.

North America. 5 species; 1 species treated in *Flora.*

Hesperostipa comata (Trinius & Ruprecht) (p. 418) Barkworth ssp. *comata*

Needle-and-thread

Stipa comata Trin. & Rupr. var. *comata, Stipa comata* Trin. & Rupr. var. *intosa* Piper

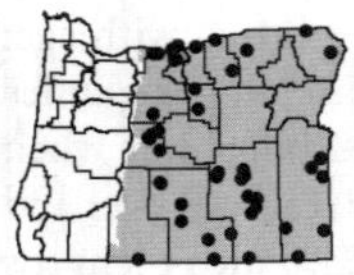

Plants 12–110 cm tall. **Leaves** mainly basal; sheaths glabrous; ligules 1–6.5 mm, those of upper leaves to 7 mm, scarious, acute; blades 0.5–4 mm wide, usually involute, inner surfaces conspicuously ridged. **Inflorescences** panicles 10–32 × 1–10 cm, contracted, often partially enclosed in the uppermost sheath at maturity; branches ascending. **Glumes** subequal to unequal, longer than the adjacent lemma, 3–5-veined, glabrous, entire; tips sometimes awned; lower glumes 18–35 mm; upper glumes 15–34 mm. **Florets** 7–13 mm long, 0.9–1.1 mm thick. **Calluses** 2–4 mm, with long, dense, silky hairs, sharp. **Lemmas** 5-veined, hard, evenly pubescent with white hairs about 1 mm, sometimes glabrous immediately above the callus; awned, lemma awns 75–225 mm, wavy or curled. **Paleas** about as long as the lemma, leathery, pubescent. **Anthers** 3 mm, with a tuft of hairs near the tip. ***2n***=38, 44, 46.

Grasslands, shrub steppe, and open forests, often in disturbed microsites. 50–1700 m. BR, BW, Col, ECas, Lava, Owy. CA, ID, NV, WA; north to AK, east to NY, southeast to TX. Native.

Hesperostipa comata is striking for its unusually long awns. It is palatable early in the season, but livestock avoid it once the sharp florets and long awns develop.

HIEROCHLOË Sweetgrass

Plants perennial; loosely cespitose. **Leaves** sheaths open; auricles absent; ligules membranous, sometimes shortly ciliate; blades flat or rolled. **Inflorescences** more or less open panicles. **Spikelets** laterally compressed, with 3 florets, the 2 lowest florets staminate, about as long as the distal bisexual floret; disarticulation above the glumes, the florets falling together. **Glumes** 2, unequal or subequal, equaling or exceeding the florets, lanceolate to ovate, keeled. **Calluses** blunt, glabrous or pubescent. **Lemmas** lower lemmas strongly compressed, 3-veined with appressed brown hairs; tips entire or bilobed, awnless or awned from between the teeth; upper lemmas somewhat hard, shiny, inconspicuously 3–7-veined, awnless. **Anthers** 2 or 3. **Caryopses** tightly enclosed by the florets.

Temperate regions of North America and Eurasia. Approximately 30 species; 2 species treated in *Flora*.

The foliage of *Hierochloë* produces a sweet smell when burned or crushed due to the presence of coumarins. *Hierochloë* is closely related to *Anthoxanthum* and sometimes is included in that genus. The traditional treatment is followed here because it can be argued that species that appear intermediate between the two genera are so only superficially and because the native *Hierochloë* are easily distinguished from introduced *Anthoxanthum* species.

Key to *Hierochloë* species:

1. Uppermost culm leaves 0.5–3 cm; leaf blades 2.5–5.5 mm wide; glumes longer than the florets; lemmas acute, entire; fertile lemmas 2.9–3.5 mm . . . . ***H. odorata***

1' Uppermost culm leaves (1.5)3–13.5 cm; leaf blades (3)5–15 mm wide; glumes as long as or slightly shorter than the florets; lemmas rounded and shallowly bilobed; fertile lemmas 3.5–4.5 mm . . . . ***H. occidentalis***

Hierochloë occidentalis Buckley

California sweetgrass

Anthoxanthum occidentale (Buckley) Veldkamp, *Hierochloë macrophylla* Thurb.

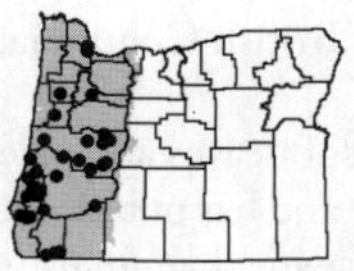

Plants perennial, (40)60–90 cm tall; loosely cespitose, rhizomes 2–4 mm thick. **Leaves** blades 20–40 cm × (3)5–15 mm, narrowed at the base, flat, glaucous on the ventral side and shiny dark green on the dorsal side, twisting at the base so that the dark green side is up, glabrous; flag leaves (1.5)3.2–11.5(13.5) cm. **Inflorescences** 8–13 cm, open. **Spikelets** 4.5–6 mm. **Glumes** 3.5–4.5 mm, ovate, equaling or slightly shorter than florets; glabrous, green at the back and with very wide hyaline margins, awnless. **Lemmas** lower lemmas staminate, 4–5 × 0.7–1 mm; usually mostly glabrous on the body, sometimes with a few scattered hairs, minutely scabrous near the tips; margins usually with longer hairs; tips rounded or bilobed, awnless or awned; awns of lower lemmas to 1 mm; upper lemmas bisexual, 3.5–4.5 mm (where fertile), shiny, brown; margins pilose especially toward the tips, acute. **Anthers** 2–3.5 mm. **2*n***=42.

Moist to fairly dry coniferous, deciduous, or mixed forest. 0–1400 m. Casc, CR, Est, Sisk, WV. CA, WA. Native.

Hierochloë occidentalis is recognized by its open panicles, rounded spikelets, and broad, pleasant-smelling leaves that twist at the base so that the dark green dorsal surfaces are on the upper side. Similar *H. odorata* has much reduced flag leaves and denser panicles with shorter spikelets.

Hierochloë odorata (Linnaeus) P. Beauvois (p. 422)

Vanilla grass, hairy sweetgrass

Anthoxanthum hirtum (Schrank) Y. Schouten & Veldkamp

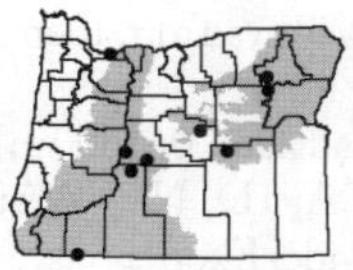

Plants perennial, 40–85(110) cm tall; loosely cespitose, rhizomes 0.7–2 mm thick. **Leaves** blades to 25 cm on sterile shoots, much reduced on culms, 2.5–5.5 mm wide, narrowed at the base, dorsal surfaces glabrous and shiny, ventral surfaces hairy; flag leaves 1–3(6) cm × 3–4.5 mm. **Inflorescences** (5)7.5–15 cm, open. **Spikelets** 4–6.3 mm. **Glumes** 4–6.3 mm, ovate, exceeding the florets, glabrous, hyaline and more or less transparent, white to light brown, tinged with green or purple on the back near the base. **Lemmas** lower lemmas staminate, 3–5 × 1.1–1.5 mm, length/width ratio usually less than 4, more or less hairy all over, with hairs 0.5–1 mm on margins, to 0.5 mm toward the tips; tips acute to awned; awns of lower lemmas 0.1–1 mm; upper lemmas bisexual, 2.9–3.5 mm, shiny, hard, hairy distally, the hairs 0.5–1 mm and with bases strongly divergent from the lemma surface, acute. **Anthers** 1.6–2.1 mm in lower, staminate florets, 1.2–1.3 mm in upper, bisexual florets.

Wet meadows and freshwater marshes. 600–2200 m. BW, Casc, ECas, Sisk. CA, ID, NV, WA; north to AK, east to Greenland; northeastern US. Native.

Hierochloë odorata is recognized by its more or less round spikelets with almost transparent glumes surrounding hairy brown lemmas. Its culm leaves are disproportionately small. It is very similar to *H. occidentalis,* which has broader leaves, longer flag leaves, and more open panicles.

HOLCUS Velvetgrass

Plants usually perennial; cespitose or rhizomatous. **Culms** glabrous or pubescent, with nodes glabrous or retrorsely pubescent. **Leaves** sheaths open; ligules membranous, glabrous or puberulent; blades flat, pubescent. **Inflorescences** terminal panicles. **Spikelets** laterally compressed, with 2(3) florets; lower florets bisexual; upper florets staminate or sterile, rachillas curved below the lowest florets, sometimes prolonged beyond the base of the distal florets; disarticulation below the glumes. **Glumes** equaling or exceeding the florets, strongly keeled, unawned; lower glume 1-veined; upper glume 3-veined. **Calluses** glabrous or pubescent. **Lemmas** firm, shiny, glabrous or pubescent, obscurely 3–5-veined, often bidentate; lower lemmas unawned; upper lemmas awned; lemma awns arising from below the lemma tip, hooked or bent at maturity. **Anthers** 3.

Cosmopolitan. 8 species; 2 species treated in *Flora*.

Key to *Holcus* species:

1. Plants cespitose; culms usually pubescent on the internodes as well as the nodes; lemma awns less than 2 mm, hooked at maturity, included or exserted sideways from the glumes . . . . ***H. lanatus***

1' Plants rhizomatous; culms glabrous (or nearly so) except at the nodes; lemma awns more than 2.5 mm, straight to curved at maturity, exserted more than 1 mm beyond the glumes. . . . . ***H. mollis***

Holcus lanatus Linnaeus (p. 422)

Velvetgrass

Plants perennial, 20–100 cm; cespitose. **Culms** usually erect; lower internodes densely soft hairy with hairs to 1 mm; upper internodes often glabrous; nodes pubescent. **Leaves** sheaths densely soft hairy; blades (3)5–10 mm wide, flat, densely soft pubescent. **Inflorescences** 3–15(20) × 1–8 cm; branches and pedicels soft hairy. **Spikelets** 3–6 mm, 2 florets. **Glumes** 4–5 mm, whitish green; keels ciliate; veins ciliate and often purple; areas between the veins usually scabrous, puberulent or villous; lower glumes lanceolate, acute; upper glumes ovate, obtuse and often with an awn to 1.5 mm. **Lemmas** 1.7–2.5 mm, glabrous; tips of lower lemmas acute and awnless, those of upper lemmas bifid and awned; lemma awns on upper florets forming a hook when mature (straight when young), hidden by the glumes or exserted sideways from them, less than 2 mm. **Anthers** (1.2)2–2.5 mm. ***2n***=14.

Moist soil, disturbed sites, grasslands, roadsides. 0–1600 m. Casc, CR, ECas, Est, Lava, Sisk, WV. CA, ID, NV, WA; throughout most of North America; worldwide. Exotic.

Holcus lanatus can be recognized by feel because its foliage is densely covered with soft, velvety hairs. Its pale inflorescences change in color and shape seasonally as the branches spread during anthesis and then clump back together. Somewhat similar *H. mollis* is rhizomatous, less densely hairy and has more or less glabrous lower internodes contrasting with its hairy nodes.

Holcus mollis Linnaeus

Creeping velvetgrass

Plants perennial, 20–100(150) cm; rhizomatous. **Culms** usually decumbent; internodes glabrous or sparsely pubescent; nodes pubescent. **Leaves** sheaths glabrous or pubescent; blades 3–10 mm wide, flat, pubescent. **Inflorescences** 4–20(22) × up to 3 cm; branches puberulent or ciliate, pedicels with spreading hairs. **Spikelets** 4–6(7) mm, 2 florets. **Glumes** 5–6 mm, whitish green when young, straw-colored with age; veins ciliate and often purple, area between the veins scabrous; ovate. **Lemmas** 2–2.5 mm, glabrous; tips of lower lemmas acute, awnless, those of upper lemmas bifid and awned; lemma awns on upper florets straight to bent, not hook-shaped, more than 2.5 mm, exserted more than 1 mm from the glumes. **Anthers** about 2 mm. ***2n***=28, 35, 42, 49.

Disturbed sites, roadsides, forest edge, prairie, oak savanna. 0–1500 m. CR, Est, WV. CA, WA; north to British Columbia; Europe. Exotic.

Holcus mollis is a rhizomatous grass with soft leaves and pale inflorescences. Similar *H. lanatus* is cespitose and has pubescent lower internodes and strongly hook-shaped awns that are usually hidden by the glumes.

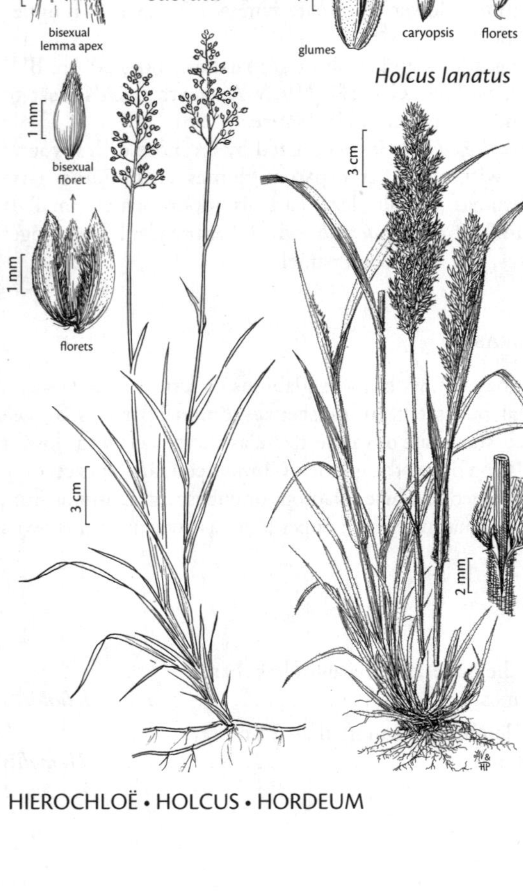

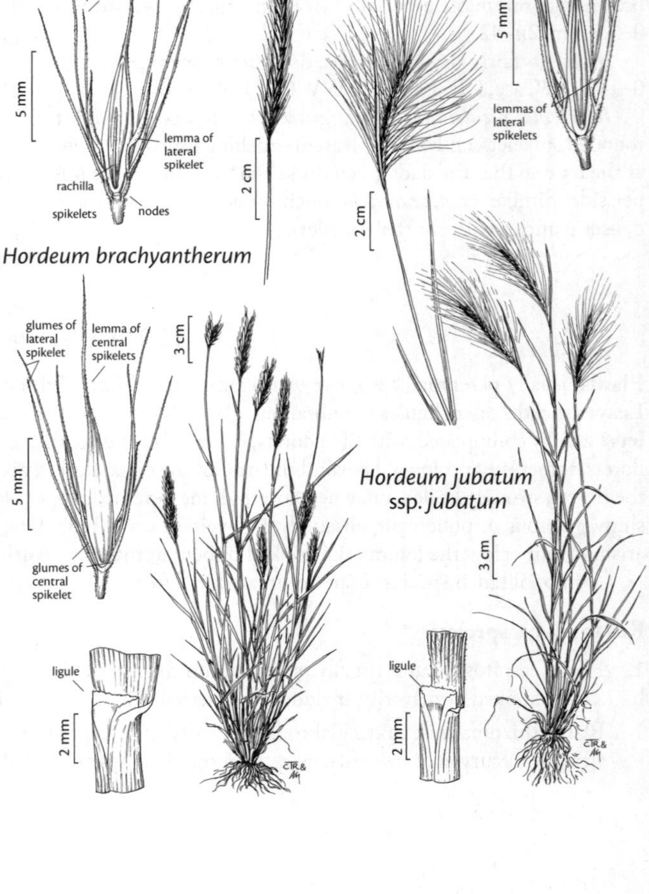

HORDEUM Barley

Plants annual or perennial; usually cespitose, sometimes rhizomatous. **Leaves** sheaths open; auricles present or absent; ligules hyaline, truncate; blades flat to involute, more or less pubescent on both sides. **Inflorescences** spikes or spike-like racemes with 3 spikelets at each node, the central spikelet usually sessile, sometimes on a pedicel to 2 mm; lateral spikelets usually on pedicels, the pedicels straight or curved; all spikelets often sessile in cultivated taxa; disarticulation in the inflorescence axis, spikelets falling in 3's, cultivated forms usually not disarticulating. **Spikelets** with 1 floret; glumes awn-like, usually longer than the floret; lateral spikelets usually sterile or staminate, often bisexual in cultivated forms, the lemmas awned or awnless; central spikelets bisexual, with sessile florets, rachillas prolonged beyond the floret; lemmas ovate, 5-veined, usually awned. **Anthers** 3. **Caryopses** usually tightly enclosed in the lemmas and paleas at maturity.

Temperate and subtropical regions. Approximately 32 species; 6 species treated in *Flora*.

As an aid to keying within this genus, break the spike to isolate a node and reveal the set of three spikelets. Spike and lemma lengths exclude awns unless noted otherwise; glume lengths include awns. Disarticulation is in the rachis unless noted otherwise.

Key to *Hordeum* species:

1. Auricles well developed, at least some more than 1 mm.
    2. Leaf blades 5–15 mm wide; lemma awns 30–180 mm, or if awnless then prominently 3-lobed near the tip; ligules 0.5–1 mm; lemmas more than 2 mm wide . . . . *H. vulgare* ssp. *vulgare*
    2' Leaf blades mostly less than 5(7) mm wide; lemma awns equal to or less than 5 cm; lemmas never 3-lobed near the tip; ligules 1–4 mm; lemmas of the central spikelets equal to or less than 2 mm wide . . . . *H. murinum*

1' Auricles lacking or less than 1 mm.
    3. Inflorescences arching to horizontal at anthesis. . . . . *H. jubatum* ssp. *jubatum*
    3' Inflorescences more or less erect.
        4. Plants perennial, (20)40–80(100) cm tall; leaf blades 2–6(9) mm wide . . . . *H. brachyantherum*
        4' Plants annual, usually less than 30 cm tall; leaf blades mostly 2–4 mm wide.
            5. Lemmas of the lateral spikelets with awns 3–8 mm long . . . . *H. marinum* ssp. *gussoneanum*
            5' Lemmas of the lateral spikelets awnless or with awns to 3 mm long . . . . *H. depressum*

Hordeum brachyantherum Nevski (p. 422)

Meadow barley

Hordeum brachyantherum Nevski ssp. *brachyantherum*, *Hordeum brachyantherum* Nevski ssp. *californicum* (Covas & Stebbins) Bothmer, N. Jacobsen & Seberg, *Hordeum nodosum* L.

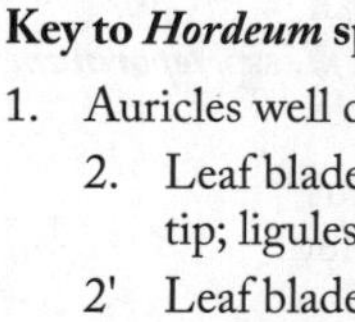

Plants perennial, (20)40–80(100) cm, loosely to densely cespitose. **Leaves** sheaths glabrous or densely pubescent; auricles absent; blades to 19 cm × 2–6(9) mm. **Inflorescences** erect spikes, 3–8.5 × 0.7–1.5 cm. **Central spikelets** glumes 9–19 × 0.2 mm, usually awn-like throughout, rarely flattened near the base; lemmas 5.5–10 mm, glabrous, rarely pubescent; lemma awns 3.5–14 mm; anthers 0.8–4 mm. **Lateral spikelets** staminate; glumes 7–19 mm, awn-like; lower glumes sometimes flattened near the base; lemmas rudimentary to well developed; lemma awns to 7.5 mm. **Glumes** ascending to slightly divergent at maturity. **Anthers** 0.8–4 mm. **2*n***=14, 28, 42.

Marshes, moist meadows, wet prairies, shores. 0–2400 m. Throughout Oregon. CA, ID, NV, WA; throughout Canada and US except midwest and southeast; eastern Russia. Native.

Hordeum brachyantherum is a sod-forming perennial with neat, narrow spikes. Similar *H. pusillum* is annual, and its glumes are somewhat flattened. It has been reported in Oregon, but its presence has not been confirmed. *Hordeum brachyantherum* has been divided into two subspecies that we do not recognize, the diploid *H. b.* ssp. *californicum* and the polyploid *H. b.* ssp. *brachyantherum*. The two overlap morphologically and cannot be identified in the absence of a chromosome count.

Hordeum depressum (Scribner & J.G. Smith) Rydberg

Low barley

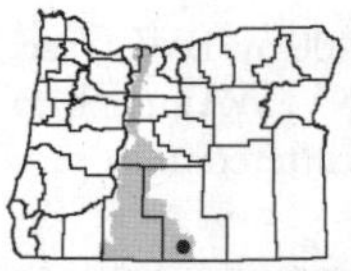

Plants annual, 10–30(55) cm tall; cespitose. **Leaves** sheaths pubescent; auricles absent; ligules 0.3–0.8 mm; blades to 7.5(13.5) cm long and to 4.5 mm wide. **Inflorescences** 2.2–7 × 0.4–0.8 cm. **Central spikelets** with glumes 5.5–20.5 mm, to 0.5 mm wide, awn-like throughout, or slightly flattened at base; lemmas 5–9 mm, glabrous, awned; lemma awns 3–12 mm, straight to ascending; anthers 0.5–1.5 mm. **Lateral spikelets** staminate or sterile, occasionally bisexual, with glumes 5–20 mm; upper glumes awn-like throughout; lower glumes slightly flattened near base; lemmas 1.8–8.5 mm; lemma awns 0–3 mm. **Glumes** straight, ascending. **2*n***=28.

Seasonally moist, often alkaline sites, including vernal pools and roadsides. 50–500 m. ECas. CA, NV, ID, WA. Native.

Hordeum jubatum Linnaeus ssp. *jubatum* (p. 422)

Squirreltail barley

Hordeum jubatum L. var. *jubatum*

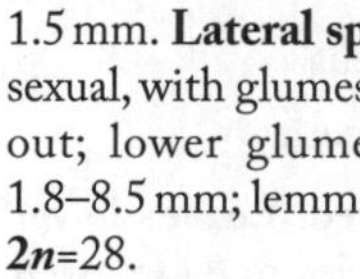

Plants perennial, sometimes blooming the first year and appearing annual, 20–80 cm tall; cespitose. **Leaves** sheaths glabrous or pubescent; auricles absent or rarely to 0.5 mm; ligules to 0.8 mm; blades to 15 cm long and to 5 mm wide. **Inflorescences** 3–15 × 0.7–1.2 cm, usually nodding at and after anthesis. **Central spikelets** with glumes 35–85 mm, awn-like throughout, strongly spreading at maturity; lemmas 4–8.5 mm, glabrous, awned; lemma awns 35–90 mm, straight to ascending; anthers 0.6–1.2 mm. **Lateral spikelets** staminate or sterile, with glumes 17–83 mm, awn-like; lemmas 4–6.5 mm; lemma awns 2–15 mm, divergent. **Glumes** bent, divergent to strongly divergent at maturity. **Anthers** 1–1.5 mm. **2*n***=28.

Seasonally moist disturbed sites, including roadsides, seasonal lakes, pastures, meadows, and coastal salt marshes. 0–2300 m. BR, BW, Col, ECas, Est, Lava, Owy, WV. CA, ID, NV, WA; throughout North America; South America; central and eastern Asia, Europe. Native.

Mature *H. jubatum* may be confused with *Elymus elymoides* and *E. multisetus,* which are usually hairier and have sessile lateral spikelets in erect spikes.

Hordeum marinum Hudson ssp. *gussoneanum* (Parlatore) Thellung

Mediterranean barley, geniculate barley

Hordeum geniculatum All., *Hordeum gussoneanum* Parl., *Hordeum hystrix* Roth

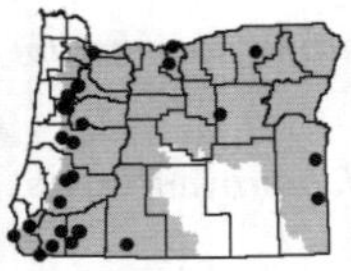

Plants annual, 10–30(50) cm tall, tufted. **Leaves** sheaths somewhat pubescent; auricles usually absent or to 0.3 mm; ligules 0.2–0.5 mm; blades to 8 cm × 1.5–4 mm. **Inflorescences** 1.5–7 × 0.5–1(2) cm. **Central spikelets** with glumes 14–26 × 0.2–1.1 mm at the base, setaceous distally; lemmas 5–8 mm, smooth to scabrous, sometimes pubescent; lemma awns 6–18 mm; anthers 0.8–1.3 mm. **Lateral spikelets** sterile; lower glumes setaceous or occasionally slightly widened, not winged; upper glumes setaceous; lemmas 4–6 mm; lemma awns 3–8 mm. **Glumes** straight, ascending to slightly divergent at maturity, not ciliate. **2*n***=28.

Seasonally moist, disturbed areas. 50–1400 m. BW, Casc, Col, ECas, Lava, Owy, Sisk, WV. CA, ID, NV, WA; north to British Columbia, northeast to MT; widely scattered in eastern US; worldwide. Exotic.

Hordeum marinum ssp. *gussoneanum* is a common, weedy, annual barley usually without or with very short auricles. Similar *H. murinum* grows in similar habitats but has longer auricles.

Hordeum murinum Linnaeus

Mouse barley, hare barley, wall barley

Plants annual, to 110 cm tall, loosely tufted. **Leaves** lower sheaths glabrous or somewhat pilose; auricles to 8 mm, well developed; ligules 1–4 mm; blades to 28 cm × up to 5(7) mm. **Inflorescences** 3–8 × 0.7–1.6 cm. **Central spikelets** sessile, florets sessile or pedicellate, pedicels, if present, to 2 mm; glumes 11–25 × 0.8–1.8 mm; lemmas 8–14 mm long and up to 2 mm wide, more or less smooth; lemma awns 20–40 mm; anthers 0.2–3.2 mm, gray to yellow, sometimes with purple spots. **Lateral spikelets** staminate, florets sessile; glumes similar to those of central spikelet; lemmas 8–15 mm; lemma awns 20–50 mm. **Glumes** erect to slightly spreading, flattened; margins ciliate. **Anthers** 2.5–6.5 mm.

3 subspecies.

Hordeum murinum is readily distinguished from the other annual barleys by the well-developed auricles on its leaves.

Key to *Hordeum murinum* subspecies:

1. Floret of central spikelet sessile or subsessile; palea of lateral floret more or less glabrous . . . . ***H. m.*** ssp. ***murinum***

1' Floret of central spikelet pedicellate; palea of lateral floret pubescent.

 2. Internodes of the inflorescence axis rarely as much as 2 mm long; leaf sheath glabrous; central floret slightly shorter than lateral florets. . . . . ***H. m.*** ssp. ***glaucum***

 2' Internodes of inflorescence axis 2 mm long or greater; leaf sheath glabrous or hairy; central floret much shorter than lateral florets. . . . . ***H. m.*** ssp. ***leporinum***

Hordeum murinum Linnaeus ssp. *glaucum* (Steudel) Tzvelev

Smooth barley

Hordeum glaucum Steud.

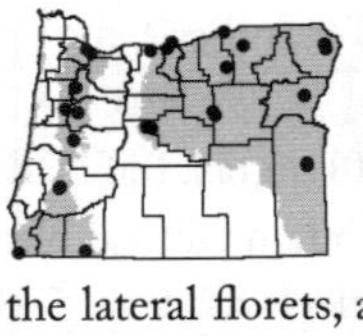

Plants summer annuals, 15–40 cm tall. **Leaves** sheaths glabrous. **Inflorescences** axis internodes up to 2mm. **Central spikelets** florets pedicellate, slightly shorter than lateral floret; lemmas subequal to those of the lateral florets, awns as long as or longer than those of the lateral florets; anthers 0.2–0.6 mm, much shorter than those of the lateral florets, more or less covered with purple spots. **Lateral spikelets** paleas more or less densely pilose, especially on the lower 50%. **Anthers** 1.2–1.8 mm. **2*n***=14.

Disturbed sites and fields, often in arid areas. 0–1600 m. BW, Col, Est, Lava, Owy, Sisk, WV. CA, ID, NV, WA; western Canada and US; Mediterranean. Exotic.

Hordeum murinum Linnaeus ssp. *leporinum* (Link) Arcangeli

Hare barley

Hordeum leporinum Link

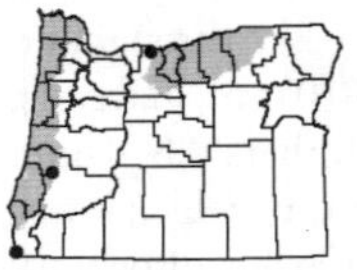

Plants winter annuals, 30–110 cm tall. **Leaves** sheaths glabrous to pilose. **Inflorescences** internodes of the axis at least 2(3) mm. **Central spikelets** florets pedicellate, shorter than those of the lateral florets; anthers 0.9–3 mm. **Lateral spikelets** paleas scabrous on the lower 50%. **Anthers** 1.2–3.2 mm. **2*n***=28, 42.

Disturbed sites, often in arid areas. 0–1400 m. Col, CR, Est. CA, ID, NV, WA; north to British Columbia; western and eastern US; Mediterranean. Exotic.

Hordeum murinum Linnaeus ssp. *murinum*

Wall barley, way barley

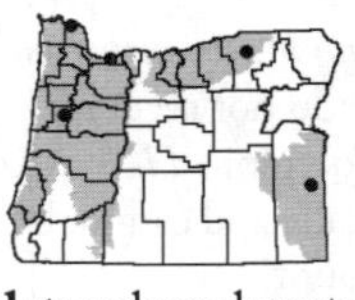

Plants winter annuals, 30–60 cm tall. **Central spikelets** florets sessile to subsessile; lemmas subequal to those of the lateral florets, awns longer than those of the lateral florets; anthers 0.8–1.4 mm. **Lateral spikelets** paleas almost glabrous. **Anthers** 0.8–1.4 mm. **2*n***=28.

Disturbed, often moist sites. 0–1200 m. Casc, Col, CR, Owy, WV. CA, WA; north to British Columbia, southeast to AZ, eastern US from ME to VA; Europe. Exotic.

Hordeum vulgare **Linnaeus ssp.** ***vulgare***
Barley

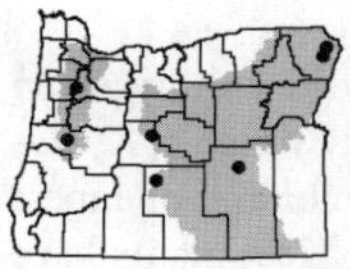

Plants summer or winter annuals, to 100(150) cm tall, tufted. **Leaves** lower sheaths pilose; upper sheaths glabrous; auricles to 6 mm, well developed; ligules 0.5–1 mm; blades to 30 cm × 5–15 mm. **Inflorescences** 5–10 × 0.8–2 cm, often nodding, usually not disarticulating at maturity. **Central spikelets** sessile, glumes straight to curved, 10–30 mm, pubescent, flattened near the base, awned; lemmas 6–12 × 3+ mm, glabrous, sometimes scabrous, awned or if awnless then prominently 3-lobed; lemma awns, 30–180 mm, usually scabrous; anthers 6–10 mm. **Lateral spikelets** sterile or one or both fertile, if sterile usually raised on stalks to 3 mm, if fertile usually sessile; glumes similar to those of central spikelets; lemmas usually 6–15 mm, obtuse to acute and awnless if sterile, awned if fertile; lemma awns 0–15 mm. **2*n***=14, 28.

Cultivated fields, disturbed areas, roadsides, and railroads. 100–1700 m. BR, BW, Lava, WV. CA, ID, NV, WA; throughout North America; worldwide. Exotic.

Hordeum vulgare ssp. *vulgare* is a cultivated crop that occasionally grows on roadsides from seed lost from grain trucks. Archaeological records suggest that *H. vulgare* was one of the earliest domesticated crops.

KOELERIA Junegrass

Plants perennial; usually cespitose. **Leaves** sheaths open; ligules membranous; blades flat to involute, pubescent or glabrous. **Inflorescences** usually dense, spike-like terminal panicles; main axis and branches softly pubescent. **Spikelets** laterally compressed, 2–4 florets, rachillas usually prolonged beyond the base of the distal floret or with a vestigial floret; disarticulation above the glumes, beneath the florets. **Glumes** subequal to or slightly exceeding the lemmas, awnless; lower glume 1-veined and somewhat narrower and shorter than the upper glume; upper glume faintly 3(5)-veined. **Calluses** glabrous or pubescent. **Lemmas** 5-veined; margins shiny, acute, sometimes awned. **Anthers** 3.

Temperate regions worldwide and on mountains in the tropics. 35 species; 1 species treated in *Flora.*

Trisetum wolfii individuals with narrow inflorescences resemble *Koeleria* but have hairy rachillas prolonged beyond the upper floret.

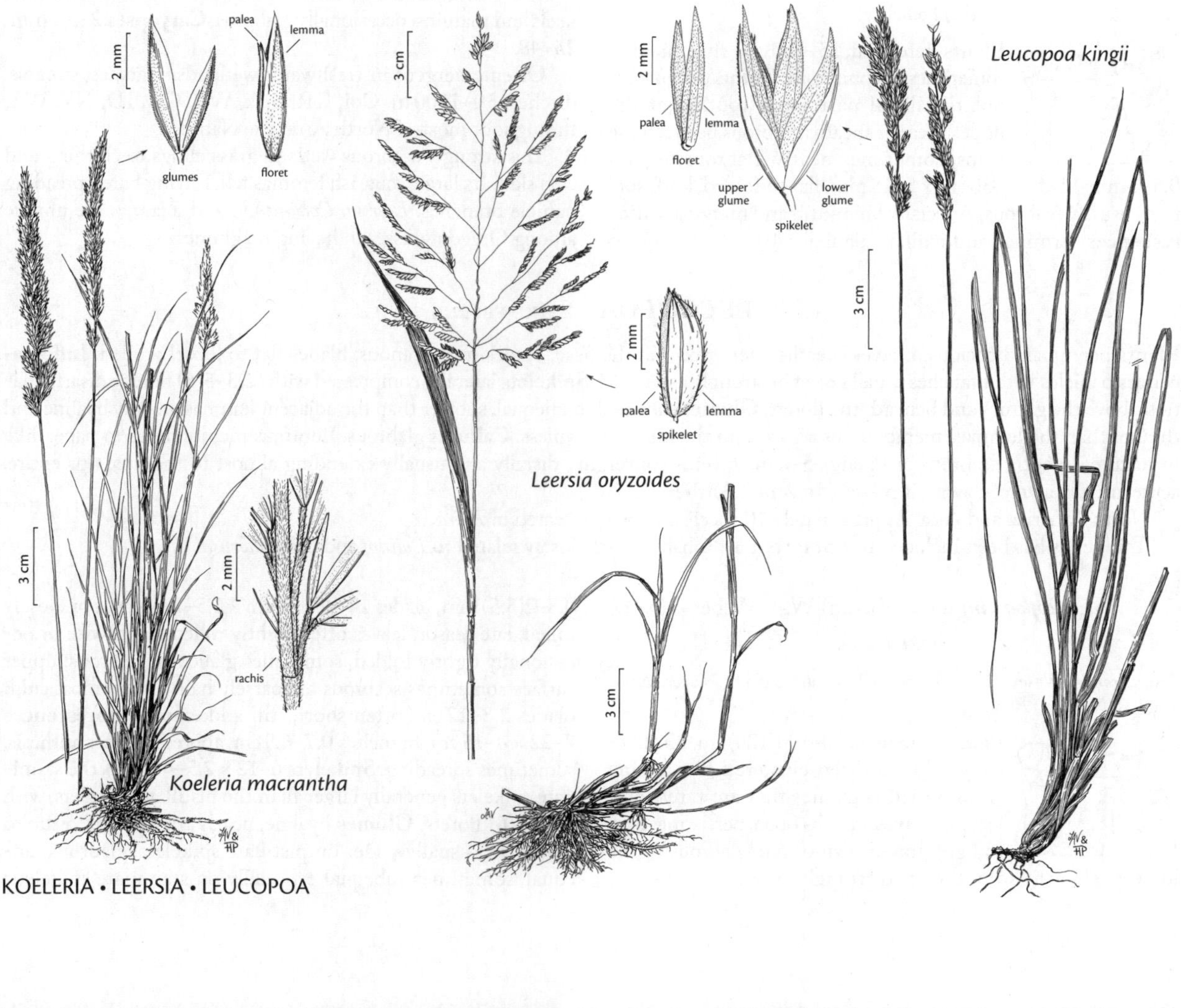

KOELERIA • LEERSIA • LEUCOPOA

Koeleria macrantha (Ledebour) Schultes (p. 425)

Junegrass

Koeleria cristata (L.) Pers., *Koeleria cristata* (L.) Pers. var. *cristata*, *Koeleria cristata* (L.) Pers. var. *longifolia* Vasey ex Burtt Davy, *Koeleria nitida* Nutt.

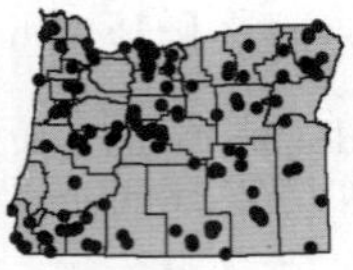

Plants perennial, (10)20–85(130) cm tall, densely to loosely cespitose. **Culms** glabrous but pubescent below the panicles and near the nodes. **Leaves** basal to distinctly cauline; sheaths pubescent or glabrous; ligules 0.5–2 mm; blades 2–20 cm × 0.5–3(4.5) mm, flat to involute, minutely scabrous to densely pubescent or occasionally glabrous. **Inflorescences** condensed panicles 4–27 × 0.5–2 cm, dense except often interrupted near the base, axis and branches densely soft-pubescent, sometimes glabrous or puberulent. **Spikelets** 2.5–6.5 mm, 2(3) florets. **Glumes** 2.5–5 mm, ovate, 1-veined; the upper glume wider and of a different shape than the lower, glabrous to scabrous, acute. **Lemmas** 2.5–6.5 mm, 5-veined, glabrous, sometimes scabrous, acuminate, usually awnless; lemma awns when present 1 mm. **Anthers** 1–3 mm. **2*n***=14.

Grasslands, meadows, upland prairies, forest openings, savannas. 0–2900 m. Throughout Oregon. CA, ID, NV, WA; throughout North America except Atlantic coast; Eurasia. Native.

Koeleria macrantha is a cespitose grass with a shiny, often silvery, condensed panicle. Because it is highly variable, *K. macrantha* is difficult to identify. The panicle's color and openness vary from plant to plant and with maturity. Most plants have broad leaves more or less evenly distributed along the culm, but those from dry, alpine sites produce narrow, involute, basally concentrated leaves like fine-leaved fescues. Plants from windy coastal bluffs may be dwarfed, with the panicle half as long as the entire plant.

LEERSIA Cutgrass

Plants usually perennial; rhizomatous or cespitose. **Culms** erect or decumbent, often rooting at the nodes, branched or unbranched. **Leaves** sheaths open; ligules membranous; blades flat or folded, sometimes involute when dry. **Inflorescences** terminal panicles, usually exserted, sometimes axillary panicles also present. **Spikelets** 1 floret, laterally compressed; disarticulation beneath the florets. **Glumes** absent. **Calluses** glabrous. **Lemmas** 5-veined, usually awnless. **Anthers** 1, 2, 3, or 6. **Caryopses** laterally compressed.

Tropical and warm-temperate regions of the world. 17 species; 1 species treated in *Flora*.

Leersia oryzoides (Linnaeus) Swartz (p. 425)

Rice cutgrass

Homalocenchrus oryzoides (L.) Pollich

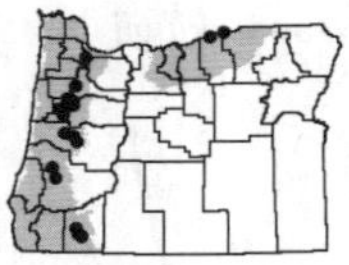

Plants perennial, 35–150 cm; rhizomatous, sometimes stoloniferous. **Culms** decumbent but the distal part erect, rooting at the nodes. **Leaves** sheaths scabrous between the veins, sometimes nearly glabrous; ligules 0.5–1 mm; blades 7–30 cm × 5–15 mm, flat to folded, both surfaces usually scabrous, especially on midrib and margins. **Inflorescences** terminal and axillary panicles 10–30 cm × up to 15 mm wide, often partially enclosed in uppermost leaf sheath; distal 67% of branches spikelet-bearing. **Spikelets** elliptic, 4–6.5 × 1.3–1.7 mm. **Lemmas** 4–6.5 mm, 5-veined, hairy on the keels and margins, occasionally glabrous. **Caryopses** 2–3.5 mm. **2*n***=48.

Often emergent in freshwater wetlands, marshes, streams, ditches. 50–1100 m. Col, CR, Sisk, WV. CA, ID, NV, WA; throughout most of North America. Native.

This strongly scabrous wetland grass clings to clothing and cuts skin. Its large, squarish lemmas fall, leaving bare, spreading panicle branches. *Leersia, Coleanthus,* and *Zizania* are unique among Oregon grasses in having no glumes.

LEUCOPOA Spike fescue

Plants perennial, dioecious. **Leaves** sheaths open except at the base; ligules membranous; blades flat to loosely rolled. **Inflorescences** panicles with branches usually erect or strongly ascending. **Spikelets** laterally compressed with (2)3–5(6) florets; disarticulation above the glumes and beneath the florets. **Glumes** subequal to unequal, shorter than the adjacent lemmas, mostly hyaline and thinner than the lemmas, membranous adjacent to the midvein, awnless. **Calluses** glabrous. **Lemmas** membranous to paper-like in texture, smooth, scabrous or hirsute, 5-veined; veins converging distally and usually extending almost to the tips; tips entire, acute, usually awnless; awns, if present, to 2 mm. **Anthers** 3.

North America and Asia. Approximately 10 species; 1 species treated in *Flora*.

This genus has been included in *Festuca* but is probably more closely related to *Lolium* and *Schedonorus*.

Leucopoa kingii (S. Watson) W.A. Weber (p. 425)

Spike fescue

Festuca confinis Vasey, *Festuca kingii* (S. Watson) Cassidy, *Hesperochloa kingii* (S. Watson) Rydb.

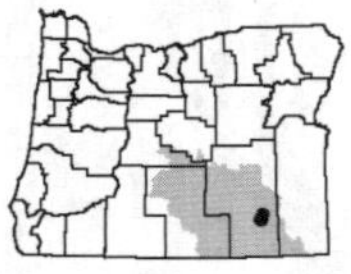

Plants perennial, 30–100(120) cm tall, dioecious; shoots often clustered but arising from buried rhizomes that are rarely collected. **Leaves** sheaths open, persisting, dark to light gray-brown or tan, glabrous or the lowest sheath sometimes retrorsely scabrous; ligules 0.8–2(3.5) mm; blades 14–40(50) cm × 1.5–7 mm, flat or loosely rolled, late season leaves often tightly rolled lengthwise or occasionally tightly folded, sometimes glaucous, glabrous; upper surface sometimes scabrous or sparsely hairy; uppermost culm blades 2.3–12 cm, often short, stiff and erect. **Inflorescences** 7–22 × 4–15 cm; branches 0.7–6.5 cm, appressed after anthesis, sometimes spreading. **Spikelets** 6–12 × 2.5–4.5 mm, the staminate spikelets generally larger than the pistillate spikelets, with (2)3–4(6) florets. **Glumes** hyaline, nearly transparent, ovate to lanceolate, usually wider on pistillate spikelets, glabrous, unequal, sometimes subequal especially on staminate spikelets;

lower glume 3–5.5(6.5) mm, 1-veined; upper glume 4–6.5(7.5) mm, 1–3-veined. **Lemmas** 4.5–8(10) mm, 1–5-veined; veins often faint, more or less glabrous or scabrous to short-hairy throughout or only on the midvein; tip entire, acute, awnless or mucronate. **Anthers** of staminate florets (2.5)3.5–5(6) mm; vestigial in pistillate florets. **Caryopses** 3.5–5 × 1.3–1.7 mm, with hairy tips. **2*n***=56.

Grasslands, dry sagebrush plains to subalpine meadows. 2400–2900 m. BR. CA, ID, NV; northeast to MT, east to SD, southeast to KS. Native.

In Oregon, *Leucopoa kingii* is known only from Steens Mountain in Harney County, although there are unconfirmed reports from Hart Mountain and the Wallowas.

LEYMUS Wildrye

Plants perennial; rhizomatous, sometimes cespitose. **Culms** erect. **Leaves** sheaths open; auricles usually present; ligules membranous; blades often stiff; the upper surfaces usually with closely spaced, prominent veins, sometimes with unequal, widely spaced veins. **Inflorescences** 2-sided spikes, racemes, or condensed panicles with 1–8(35) spikelets associated with each node. **Spikelets** 0.5–3.75 times the length of the inflorescence internodes, usually sessile, sometimes on pedicels up to 2(15) mm, appressed or ascending, 2–12 florets, the terminal floret usually reduced; disarticulation above the glumes, beneath the florets. **Glumes** usually 2, subequal but occasionally one or both reduced, lanceolate and narrowing in the distal quarter, or lanceolate to subulate and tapering from below mid-length, glabrous or hairy, 0–3(7)-veined; veins apparent at least at mid-length; keels, if present, straight or almost so; tips acute, acuminate, or tapering to an awn-like tip; awns, if present, to 4 mm. **Lemmas** inconspicuously 5–7-veined, rounded over the back at least toward the base, sometimes keeled distally; tips acute, awned or not; lemma awns, if present, to 7 mm, straight. **Anthers** 3. **Caryopses** with hairy tips.

Temperate regions of the northern hemisphere. Approximately 50 species; 6 species treated in *Flora*.

Leymus species are strictly outcrossing polyploids. In most *Leymus*, the spikelets are on pedicels and therefore the inflorescence is technically a raceme, though it is nearly always called a spike for consistency with other genera of wheatgrasses. In *L. condensatus*, the inflorescence is actually a condensed panicle, though still called a spike. *Leymus racemosus*, mammoth wildrye, has been reported from Oregon, but no specimens have been seen. It is a strongly rhizomatous plant with unusually thick culms, 8–12 mm thick at the lower nodes. Its leaves are 8–20 mm wide. It has been introduced to dry, sandy soils in southern ID.

When keying *Leymus*, culm thickness is measured on the lower internodes and does not include the surrounding leaf sheath. Descriptions of the inflorescence nodes and internodes apply to the middle of the spike.

Key to *Leymus* species:

1. Glumes flat or rounded on the back, tapering from mid-length or above, flexible . . . . . *L. mollis* ssp. *mollis*

1' Glumes keeled, at least distally, tapering from below mid-length, stiff.
- 2. Plants cespitose; culms (70)100–350 cm tall.
  - 3. Plants green, not glaucous; blades 3–12 mm wide; inflorescences unbranched . . . . . *L. cinereus*
  - 3' Plants often glaucous; blades 10–28 mm wide; inflorescences often branched, especially at the lower nodes (bend the inflorescence to find the branches) . . . . . *L. condensatus*
- 2' Plants rhizomatous; culms solitary or in small groups, 40–125 cm tall.
  - 4. Glumes glabrous or scabrous distally . . . . . *L. triticoides*
  - 4' Glumes hairy at least distally.
    - 5. Lemmas densely hairy throughout; anthers dehiscent; glumes with 0-1(3) veins . . . . . *L. flavescens*
    - 5' Lemmas glabrous toward the base; anthers usually indehiscent; glumes with 3(5) veins . . . . *L.* × *vancouverensis*

Leymus cinereus (Scribner & Merrill) Á. Löve
Great Basin wildrye

Elymus cinereus Scribn. & Merr., *Elymus cinereus* Scribn. & Merr. var. *cinereus*, *Elymus cinereus* Scribn. & Merr. var. *pubens* (Piper) C.L. Hitchc.

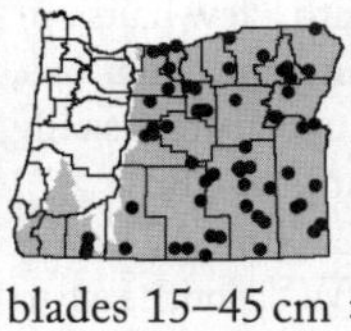

Plants 70–270 cm tall, usually green, not glaucous; strongly cespitose, rhizomes, if present, small and inconspicuous. **Culms** 2–5 mm thick. **Leaves** sheaths glabrous or hairy; auricles to 1.5 mm; ligules 1.5–8 mm; blades 15–45 cm × 3–12 mm, strongly involute to flat; lower surfaces glabrous; upper surfaces scabrous, 11–25-veined. **Inflorescences** true (unbranched) spikes 10–29 cm × 8–17 mm, 2–7 spikelets per node; internodes 4–9 mm. **Spikelets** 9–25 mm. **Glumes** 8–18 × 0.5–2.5 mm, stiff, keeled, the central portion thicker than the margins, tapering from below mid-length, smooth or scabrous, 0–1(3)-veined; veins inconspicuous at mid-length. **Lemmas** 6.5–12 mm, glabrous or hairy, hairs 0.1–0.3 mm; tips acute or awned; lemma awns to 3 mm. **Anthers** 4–7 mm. **2*n***=28, 56.

Streamside terraces, toe slopes, and roadsides in sagebrush steppe and juniper woodland. 0–2300 m. BR, BW, Col, ECas, Lava, Owy, Sisk. CA, ID, NV, WA; north to British Columbia, northeast to Manitoba, southeast to NM. Native.

Leymus condensatus, sparingly introduced, is often glaucous, has broader leaves, and usually has inconspicuously branched inflorescences. Rhizomes observed in some *L. cinereus* individuals may result from introgression following hybridization with *L. triticoides*. Hybrids with *Elymus* are called × *Elyleymus aristatus*. This hybrid is sterile and has tardily dehiscent inflorescence axes, (2)3 spikelets per node, and subulate glumes with relatively short awns.

Leymus condensatus (J. Presl) Á. Löve
Giant wildrye

Elymus condensatus J. Presl var. *condensatus*

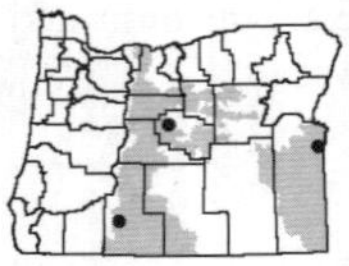

Plants 115–350 cm tall; cespitose or short-rhizomatous, often glaucous. **Culms** 6–10 mm thick. **Leaves** with ligules 0.7–6 mm on basal leaves, 4–7.5 mm on uppermost culm leaves; blades 10–28 mm wide, often glaucous; lower surfaces glabrous; upper surfaces minutely scabrous. **Inflorescences** dense, spike-like panicles, 17–44 × 2–6 cm, usually branched at the lower nodes; internodes 3.5–10 mm, glabrous. **Spikelets** 9–25 mm, sessile or stalked, the stalks, if present, 0.8–2 mm. **Glumes** narrowly lanceolate, tapering below mid-length, 6–16 × 0.5–2.5 mm, stiff, keeled, smooth near the base, scabrous near the tip, 0–1(3)-veined; tips tapering gradually into an awn approximately as long as the glume body. **Lemmas** 7–14 mm, usually glabrous; tips acute to awned; lemma awns 0–4 mm. **Anthers** 3.5–7 mm. ***2n***=28, 56.

Disturbed sagebrush steppe. 700–1600 m. ECas, Lava, Owy. CA. Exotic.

Leymus condensatus is a big, broad-leaved, often glaucous bunchgrass of dry areas. Its inflorescences look like dense spikes, but bending them may reveal the branches, especially near the base. An individual plant may have both clearly branched and apparently unbranched inflorescences. *Leymus cinereus* has narrower leaves and unbranched spikes.

Leymus flavescens (Scribner & J.G. Smith) Pilger
Yellow wildrye

Elymus arenicola Scribn. & J.G. Sm., *Elymus flavescens* Scribn. & J.G. Sm.

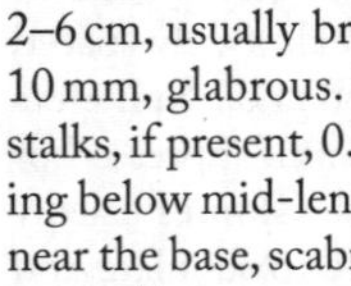

Plants 40–120 cm tall, strongly rhizomatous, rarely cespitose. **Culms** 2–4 mm thick. **Leaves** sheaths glabrous; auricles absent, sometimes with a few hairs in the auricle position; ligules 0.3–1.5 mm; blades 3–4 mm wide, usually involute, inner surfaces scabrous, sometimes with scattered hairs to 1 mm, approximately 15-veined. **Inflorescences** spikes or dense, spike-like racemes, 10–20 × 1.2–20 cm, with 2 spikelets per node; internodes 7–10 mm, densely hairy. **Spikelets** 13.5–25 mm, paired, one of each pair sessile and the other pedicellate, pedicels up to 15 cm. **Glumes** 8.5–16 × 0.5–2.5 mm, stiff, keeled distally, the central portion thicker than the margins, tapering from below mid-length, hairy, 0–1(3)-veined; veins inconspicuous at mid-length; lower glumes 8.5–13.5 mm; upper glumes 10–16 mm. **Lemmas** 10.5–15 mm, densely hairy, hairs 2–3 mm; tips awnless or awned; lemma awns to 2 mm. **Anthers** 4.5–7 mm, dehiscent.

Unstable sandy soils, dunes. 50–700 m. Col, ECas, Est, Owy. ID, WA. Native.

Hybrids between *L. flavescens* and *L. triticoides* seem to be nearly as common as true *L. flavescens*. They have sparsely hairy lemmas and grow within the range of *L. flavescens*.

Leymus mollis (Trinius) Pilger ssp. ***mollis***
American dunegrass

Elymus mollis Trin., *Elymus mollis* Trin. ssp. *mollis*

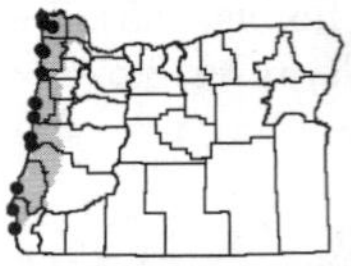

Plants (50)70–170 cm tall, often glaucous, strongly rhizomatous with rhizomes 4–6 mm thick. **Culms** 3–6 mm thick, usually densely pubescent below the spikes for 10–40 mm. **Ligules** 0.2–2.5 mm; basal blades 22–94 cm × 5–15 mm; upper surfaces minutely to definitely scabrous, 20–40-veined. **Inflorescences** spikes, 12–34 × 1–2 cm, usually with 2 spikelets per node; internodes 4.5–9.5 mm, with soft hairs 0.1–0.7 mm. **Spikelets** 15–34 mm. **Glumes** lanceolate, tapering from mid-length or above, 13–34 × 3–4 mm, often longer than the lemmas, flat or rounded on the back, flexible, pilose, sometimes almost glabrous, the central portion scarcely thicker than the margins, 3(5)-veined at mid-length; tips acute. **Lemmas** 11–20 mm, densely hairy, hairs 0.5–1 mm, soft; tips acute, awnless. **Anthers** 4–9 mm. ***2n***=28.

Coastal sands. 0–50 m. CR, Est. CA, WA; north to AK, New Brunswick to Greenland, south into northeastern US; eastern Asia, Iceland. Native.

Before European settlement, *Leymus mollis* was one of the dominant native plants on the small dunes that typically formed inland of sandy beaches. The introduction of *Ammophila arenaria* for dune stabilization severely modified coastal sandy habitats and displaced most *Leymus mollis* populations.

Leymus triticoides (Buckley) Pilger (p. 430)
Beardless wildrye

Elymus triticoides Buckley var. *pubescens* Hitchc., *Elymus triticoides* Buckley var. *triticoides*

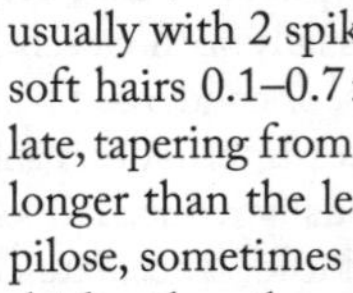

Plants 45–125 cm tall, strongly rhizomatous. **Culms** 1.8–3 mm thick. **Leaves** often basally concentrated; sheaths glabrous or hairy, hairs 0.5–1 mm; auricles to 1 mm; ligules 0.2–1.3 mm, truncate, erose; blades 10–35 cm × 3.5–10 mm, flat to involute, usually stiffly ascending; upper surfaces scabrous or sparsely hairy, hairs to 0.8 mm, more abundant toward the base; veins 11–27. **Inflorescences** spikes, 5–20 × 0.5–1.5 cm, 2 spikelets at each node near the middle of the spike, sometimes 1 or 3 at other nodes; internodes 5–11.5 mm, smooth and glabrous, sometimes strigillose distally, edges ciliate, cilia to 0.4 mm. **Spikelets** 10–22 mm, sessile, sometimes with pedicels, the pedicels to 1.5(5) mm. **Glumes** awl-like but sometimes flattened, 5–16 × 0.5–1.2 mm, bases not overlapping, glabrous and smooth proximally, scabrous distally, tapering from below mid-length to the subulate tips, stiff, keeled, the central portion thicker than the margins, 1(3)-veined; veins inconspicuous at mid-length. **Calluses** glabrous, occasionally with a few hairs approximately 0.1 mm. **Lemmas** 5–12 mm, shiny and glabrous, occasionally sparsely hairy, hairs to 0.3 mm; tips acute, usually awned; lemma awns to 3 mm. **Anthers** 3–6 mm. ***2n***=28.

Dry to moist, often alkaline meadows. 0–2200 m. All ecoregions except Est. CA, ID, NV, WA; east to WY, southeast to NM. Native.

The "Shoshone" cultivar distributed as *L. triticoides* is actually *L. multicaulis*. *Leymus triticoides* hybridizes with other members of the Triticeae tribe, including *Elymus elymoides*, *E. lanceolatus*, *Hordeum jubatum*, *Leymus cinereus*, *L. condensatus*, *L. flavescens*, and *L. mollis*.

Leymus* × *vancouverensis (Vasey) Pilger

Vancouver wildrye

Elymus vancouverensis Vasey

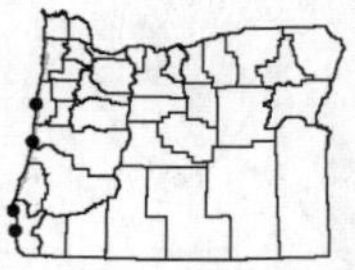

Plants 60–122 cm tall, green or slightly glaucous; rhizomatous. **Culms** 2–3.5 mm thick. **Leaves** auricles to 1 mm; ligules 0.4–1.2 mm; blades to 9 mm wide. **Inflorescences** spikes, 7–32 × 0.7–1.1 cm, sometimes glaucous, often purplish or green with traces of purple, with 1–2 spikelets per node; internodes 8–12 mm; surfaces hairy distally, hairs 0.1–0.3 mm; edges ciliate, hairs to 1 mm, coarser than the surface hairs. **Spikelets** 15–20 mm. **Glumes** lanceolate, 9–28 × 1.5–4 mm, glabrous towards the base, pubescent toward the tip, midveins glabrous or with hairs to approximately 1.3 mm, stiff, keeled or flat near the base, keeled toward the tip, the central portion thicker than the margins, 3(5)-veined at mid-length, tapering from below mid-length to an awn; awns to 4 mm. **Lemmas** 8–10 mm, usually glabrous toward the base; margins and tips sometimes pubescent; tips tapering to an awn; lemma awns to 4 mm. **Anthers** 3.3–7.3 mm, usually indehiscent. **2*n***=28, 42.

Coastal stream banks and beaches. 0–50 m. Est. CA, WA; north to British Columbia. Native.

Leymus × *vancouverensis* is a hybrid of *L. mollis* and *L. triticoides* that occurs along the coast. Although it seems to be sterile, it persists in areas where *L. triticoides* is now absent.

LOLIUM Ryegrass

Plants annual or perennial; cespitose to short-rhizomatous. **Culms** erect to decumbent. **Leaves** sheaths open; auricles often present, thin, clasping; ligules membranous; blades flat. **Inflorescences** 2-sided spikes; spikelets oriented edge-wise to the inflorescence axis and sometimes sunken into the inflorescence axis. **Spikelets** laterally flattened, 2–22 florets, distal florets reduced; disarticulation above the glumes and beneath the florets. **Glumes** 1(2 in the terminal spikelet), 3–9-veined, awnless. **Calluses** short, blunt, glabrous. **Lemmas** lanceolate, ovate, or oblong, rounded over the back, 3–7-veined; tips sometimes hyaline, awned or awnless; lemma awns subterminal, more or less straight. **Anthers** 3. **Caryopses** dorsally compressed, oblong, broadly elliptic or ovate.

North America, Eurasia, and North Africa. Approximately 5 species; 3 species treated in *Flora.*

Except for its inflorescence structure, *Lolium* is very similar to *Schedonorus*. It would be logical to merge the two genera. Hybrids between *Lolium* and *Schedonorus* are recognized by spikelets edge-wise to the sparsely branched inflorescence axis. Both F1 hybrids and various backcrosses can sometimes be found.

Boundaries between the *Lolium* species are blurry. Hybrids can occur between all pairs of species, though some hybrids are more common than others. *Lolium rigidum* was collected in Portland in 1916. It is an annual with glumes about as long as the spikelets and caryopses more than 3 times as long as wide. The spikelets of *L. rigidum* are often sunken into the inflorescence axis.

Key to *Lolium* species:

1. Plants annual and longest glumes approximately 1.5 times as long as the lemmas .... ***L. temulentum*** ssp. ***temulentum***

1' Plants annual or perennial, if annual then with glume not longer than the mature spikelet.

  2. Plants long-lived perennials, with vegetative shoots at flowering and fruiting time; florets 2–10 per spikelet; leaves folded in bud; lemmas awnless or with awns to approximately 8 mm .... ***L. perenne***

  2' Plants annuals or short-lived perennials, often lacking vegetative shoots at flowering and fruiting; florets 10–22 per spikelet; leaves rolled in bud; lemmas with awns to 15 mm or rarely awnless .... ***L. multiflorum***

Lolium multiflorum Lamarck

Annual ryegrass

Festuca perennis (L.) Columbus & J.P. Sm.

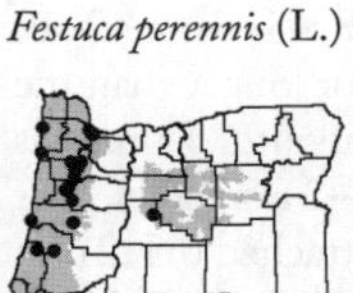

Plants annual or short-lived perennials, to 150 cm tall; cespitose; often lacking vegetative shoots at flowering and fruiting. **Leaves** sheaths open; blades usually 10–30 cm × (2)3–8(13) mm, rolled in the bud. **Inflorescences** spikes, 15–45 × 0.8–2 cm, 5–38 spikelets, main axis 0.8–2 mm thick at nodes, not flexuous. **Spikelets** 8–31 × 2–10 mm, with (10)11–22 florets. **Glumes** 5–18 mm, usually 25–50% as long as the spikelet, membranous to somewhat hard. **Lemmas** 4–8.2 × 1–2 mm, usually awned; lemma awns to 15 mm, attached 0.2–0.7 mm below the lemma tip. **Anthers** 2.5–4(5) mm. **Caryopses** 2.5–4 × 0.7–1.5 mm, 3 or more times longer than wide. **2*n***=14.

Disturbed sunny areas, prairies, meadows, savannas, roadsides. 0–1400 m. CR, Est, Lava, Sisk, WV. CA, ID, NV, WA; throughout North America; temperate regions worldwide. Exotic.

Lolium multiflorum is more common west of the Cascades. It invades disturbed upland prairies and savannas, including those on mild serpentine substrates. See also discussion under *L. perenne*, with which this species hybridizes.

Lolium perenne Linnaeus (p. 430)

Perennial ryegrass

Lolium perenne L. var. *cristatum* (Pers.) Döll, *Festuca perennis* (L.) Columbus & J.P. Sm.

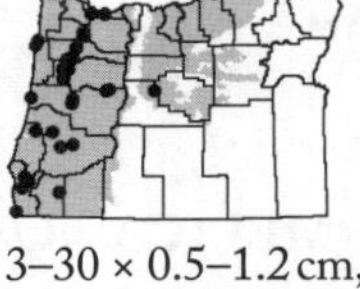

Plants long-lived perennials, to 100 cm tall; cespitose; with vegetative shoots at flowering and fruiting time. **Leaves** sheaths open; blades usually 10–30 cm × (1)2–4(6) mm, folded in the bud. **Inflorescences** spikes, 3–30 × 0.5–1.2 cm, 5–37 spikelets, main axis 0.5–2.5 mm thick at the nodes, often flexuous. **Spikelets** 5–22 × 1–7 mm, (2)5–9(10) florets. **Glumes** 3.5–15 mm, 50–75% as long as the spikelet, rarely slightly exceeding it, membranous to somewhat hard. **Lemmas** 3.5–9 × 0.8–2 mm, awnless or awned; lemma awns

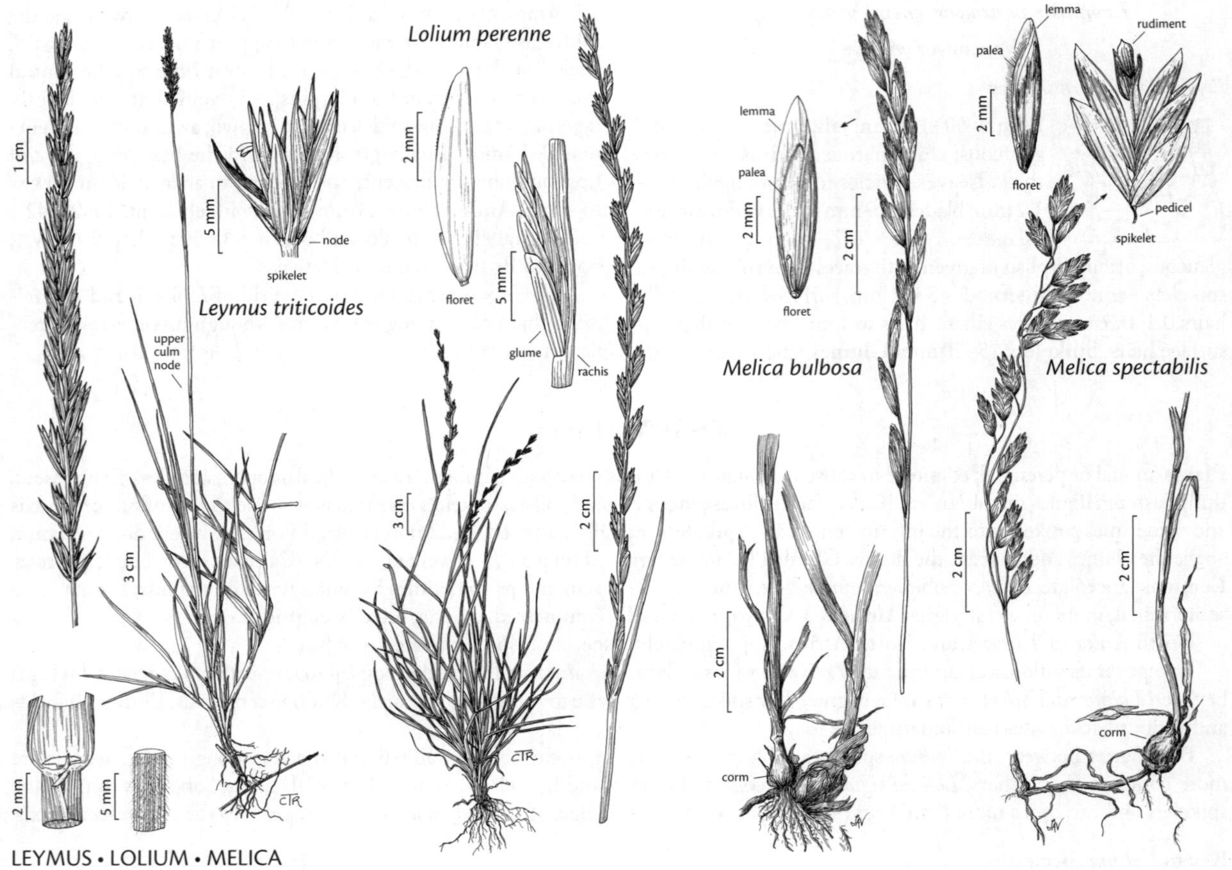

LEYMUS • LOLIUM • MELICA

0–8 mm, attached 0.2–0.7 mm below the lemma tip. **Anthers** 2–4.2 mm. **Caryopses** 3–5.5 × 0.7–1.5 mm, 3 or more times longer than wide. **2*n***=14.

Disturbed open areas, meadows, roadsides. 0–1300 m. Casc, Col, CR, Est, Lava, Sisk, WV. CA, ID, NV, WA; throughout North America; temperate regions worldwide. Exotic.

Typical *L. perenne* is perennial with remnants of last year's leaves at the base, few florets, and awnless or short-awned lemmas. Typical *L. multiflorum* is annual with many florets per spikelet and awned lemmas. The two species hybridize and intergrade. Plants that combine traits of both species can be called *L.* × *hybridum*. Alternatively, the closely related *L. perenne* and *L. multiflorum* might be treated as subspecies within *L. perenne.*

Lolium temulentum Linnaeus ssp. *temulentum*

Darnel

Festuca temulenta (L.) Columbus & J.P. Sm.

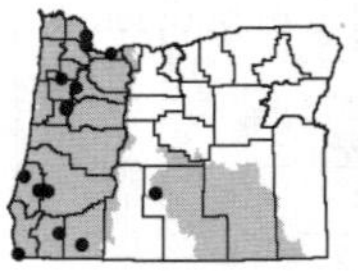

Plants annual to 120 cm tall. **Leaves** sheaths open; blades to 27 cm × (1.5)3–10(12) mm. **Inflorescences** spikes, 5–40 × 0.8–2.2 cm, 5–26 spikelets, main axis 0.5–3.5 mm thick at the nodes; spikelets not sunken in the axis and not hidden by the glume. **Spikelets** 5–28 × 1–8 mm, 2–10 florets. **Glumes** (5.5)7–28 mm, 75% as long or longer than the florets, somewhat hard; the longest glumes about 1.5 times as long as the lemmas. **Lemmas** 4.5–8.5 × 1.5–3 mm, awned or awnless; lemma awns, if present, to 23 mm, attached 0.5–2 mm below the tips. **Anthers** 1.5–4 mm. **Caryopses** 3.8–7 × 1–3 mm, 2–3 times longer than wide. **2*n***=14.

Disturbed areas. 0–1400 m. BR, Casc, CR, Est, Sisk, WV. CA, ID, WA; throughout most of North America; temperate regions worldwide. Exotic.

MELICA Oniongrass, melic

Plants cespitose or rhizomatous. **Culms** bases often thickened, forming bulb-like corms. **Leaves** sheaths closed or splitting toward the top or remaining closed unless mechanically torn, auricles absent; ligules membranous, sometimes with a sparse fringe of hairs, sometimes closed in front; blades flat to involute. **Inflorescences** open to contracted panicles, occasionally racemes. **Spikelets** 1–7 bisexual florets proximally and 1–4 sterile florets near the tip forming a cluster (rudiment), florets of the rudiment similar to but smaller than the bisexual florets or reduced, truncate, club-shaped, or more or less round; disarticulation above or below the glumes. **Glumes** shorter than the lemmas, hyaline or scarious in texture throughout or the outer 33% and margins usually marked with purple; lower glumes 1–9-veined; upper glumes with 3–11 veins. **Lemmas** with 7–11 veins that do not converge toward tip, hyaline in the distal 33–50%, awned or not; lemma awns, if present, straight. **Paleas** subequal to lemmas or little more than 50% of their length. **Lodicules** short, truncate. **Anthers** (2)3. **Ovaries** glabrous. **Caryopses** rounded-cylindrical, dark brown, smooth, shiny.

Temperate regions worldwide. Approximately 80 species; 9 species treated in *Flora*.

In species with corm-like bases, the bases may be obviously thick, rounded, and bulb-like, or they may be somewhat broad but gradually tapered. Many species have upper, sterile lemmas that are reduced to truncate or club-like clusters called rudiments. Specimens that lack these two unusual *Melica* traits can be identified to genus based on the closed leaf sheaths, the glumes that are hyaline at least distally, and the relatively large number of veins on the glumes and lemmas. Specimens should be collected with the stem bases and roots, structures essential to confident identification. *Melica californica* has been reported repeatedly from southwestern Oregon, generally based on specimens of *M. bulbosa* collected without their bulb-like culm bases. *Melica californica* has a pale, narrow inflorescence. Its rudiments are 1.4–3 mm and club-like.

When keying within this genus, glume width is the distance from one margin to the other. Lemma lengths are measured on the fertile (lower) lemmas.

Rudimentary scales at the base of the ovary are called *lodicules*, which are sometimes used in describing *Melica*.

Key to *Melica* species:

1. Lemmas awned; culms lacking thickened, bulb-like bases.
  2. Lemma awns 0–3(4) mm; lemmas usually strongly long-ciliate-pilose at base . . . . ***M. harfordii*** (in part)
  2' Lemma awns (3)5–12 mm; lemmas not long-ciliate-pilose at base.
    3. Leaf blades 2–6 mm wide; lower panicle branches usually paired . . . . ***M. aristata***
    3' Leaf blades 5–13 mm wide; panicle branches usually solitary . . . . ***M. smithii***

1' Lemmas awnless; culms bulbous-based or not.
  4. Panicle branches spreading or pedicels sharply bent so that the spikelets point sideways or down.
    5. Pedicels sharply bent so that spikelets point sideways or down; inflorescence a narrow raceme-like panicle; glumes subequal to equal the spikelet length; disarticulation below glumes . . . . ***M. stricta*** var. ***stricta***
    5' Pedicels erect to flexuous, not sharply bent; inflorescence clearly a panicle; glumes less than 80% of spikelet length; disarticulation above glumes.
      6. Inflorescence branches appressed to ascending; rachilla internodes swollen when fresh, longitudinally wrinkled and dull orange-brown when dry . . . . ***M. fugax*** (in part)
      6' Inflorescence branches flexuous or drooping; rachilla internodes not swollen when fresh, not wrinkled or orange-brown when dry . . . . ***M. geyeri*** var. ***geyeri***
  4' Panicle branches appressed and pedicels erect so that the entire inflorescence is narrow.
    7. Lemmas acuminate, 9–13 mm, usually with sparse long hairs at least on veins near the base; culms usually with bulb-like bases . . . . ***M. subulata***
    7' Lemmas obtuse, 4–16 mm, glabrous, scabrous, or hairy; culms with or without bulb-like bases.
      8. Rachilla internodes obvious, swollen when fresh, longitudinally wrinkled and dull orange-brown when dry . . . . ***M. fugax*** (in part)
      8' Rachilla internodes not swollen when fresh, not wrinkled and orange-brown when dry.
        9. Lemmas hairy; culms without bulb-like bases . . . . ***M. harfordii*** (in part)
        9' Lemmas glabrous or scabrous; culms usually with bulb-like bases.
          10. Ligules 0.1–3 mm; glumes usually less than 50% as long as the spikelets; bulb-like bases connected to the rhizome by a root-like structure 10–30 mm long . . . . ***M. spectabilis***
          10' Ligules 2–6 mm; glumes (50)67% to as long as the spikelets; bulb-like bases sessile on the rhizome . . . . ***M. bulbosa***

Melica aristata Thurber ex Bolander

Awned melic

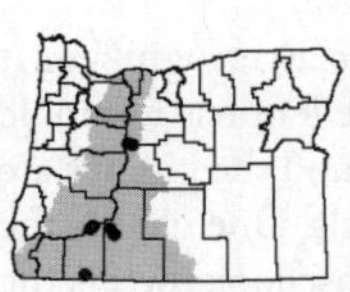

Plants 40–120 cm tall; cespitose. **Culms** bases not bulb-like. **Leaves** sheaths glabrous or scabrous, sometimes sparsely pilose, closed; blades 2–6 mm wide, flat, often sparsely pilose on both surfaces. **Ligules** 3–5 mm, often closed in front. **Inflorescences** 10–26 × 0.5–1.5 cm with branches appressed or strongly ascending; lower branches usually paired; pedicels not sharply bent. **Spikelets** 11–21 mm, (2)3–5 bisexual florets and 1–4 sterile florets; rachilla internodes 3.4–3.8 mm; disarticulation above glumes. **Glumes** lower glumes 8.5–11 × 1.5–2.5 mm, with 3–5 veins; upper glumes 9–12 × 2–3 mm, with 5–7 veins. **Lemmas** of fertile florets 8–13 mm, glabrous or with hairs on the marginal veins, with 5–7 veins, bifid to emarginate, awned; lemma awns (3)5–12 mm, arising from sinus of the bifid apex. **Anthers** 2, 2–3 mm. **2*n***=18.

Dry open forests, clearcuts, sunny slopes, rock outcrops, ridges, sometimes more moist or shady sites. 1000–2000 m. Casc, ECas, Sisk. CA, WA; KY. Native.

The 5–12 mm awns distinguish this species from all other *Melica* species except *M. smithii*, which has wider leaves.

Melica bulbosa Geyer ex Porter & J.M. Coulter (p. 430)

Oniongrass

Melica bella Piper ssp. *intonsa* Piper, *Melica bulbosa* Geyer ex Porter & J.M. Coult. var. *bulbosa*

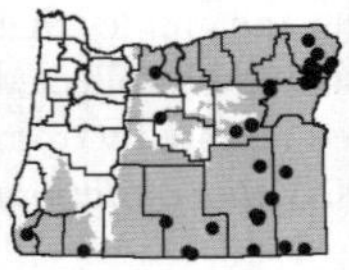

Plants 29–100 cm tall; loosely cespitose, rhizomatous. **Culms** bases modified as corms almost sessile on rhizome. **Leaves** sheaths usually minutely scabrous, sometimes sparsely pilose; blades 1.5–5 mm wide, flat to somewhat involute, minutely scabrous to crisp-pubescent; uppermost culm leaf not exceeding the inflorescence. **Ligules** glabrous, (2)3–4(6) mm, usually open in front. **Inflorescences** 7–30 × 0.7–6 cm with branches appressed; pedicels erect. **Spikelets** (6)9–18(24) mm, 4–7 bisexual florets and 1–4 sterile florets; rachilla internodes 1–2 mm; disarticulation above glumes. **Glumes** (50)67% to as long as the spikelets; lower glumes (4)5.5–10.5 × 2–3 mm, with 3–5 veins; upper glumes (5)6–11(14) × 2.3–3.5 mm, with 5–7 veins. **Lemmas** of fertile florets (6)7–10(12) mm, glabrous, with 7–11 veins, emarginate to acute, awnless. **Anthers** 3, 1.5–4 mm. **2*n*=18.**

Dry, rocky slopes, usually grasslands or sagebrush steppe, sometimes on more mesic sites with aspen. 200–2600 m. BR, BW, Col, ECas, Owy, Sisk. CA, ID, NV, WA; north to British Columbia, northeast to Saskatchewan, southeast to CO. Native.

Sessile, onion-like bulbs and more or less acute lemmas typify *M. bulbosa. Melica spectabilis* is similar but has bulbs connected to the rhizome by a root-like structure. *Melica subulata* has longer, acuminate lemmas.

Melica fugax Bolander

Little melic

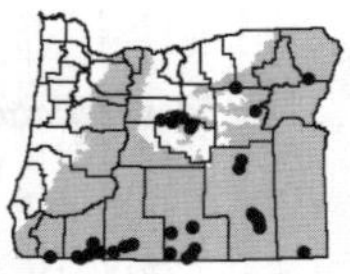

Plants 10–60 cm tall; cespitose. **Culms** bases modified as bulb-like corms. **Leaves** sheaths minutely glabrous to scabrous; blades 1.2–5 mm wide, flat, glabrous or sometimes pilose on both surfaces; flag leaf not exceeding the inflorescence. **Ligules** glabrous, 1–3 mm, usually open in front. **Inflorescences** clearly paniculate, 4.5–18 × 5–7 cm with branches appressed to ascending; pedicels straight. **Spikelets** (4)8–12(17) mm, 2–5 bisexual florets and 1–4 sterile florets; rachilla internodes 2.1–2.3 mm, swollen when fresh, longitudinally wrinkled and dull orange-brown when dry; disarticulation above glumes. **Glumes** lower glumes 2.5–5(6) × 1.5–2.5 mm, with 1–3 veins; upper glumes (3.5)4.5–7 × 2.5–3.5 mm, with 5 veins. **Lemmas** of fertile florets 4–7 mm, glabrous or scabrous, with 4–11 veins, rounded to acute, awnless. **Anthers** 3, 1–2 mm. **2*n*=18.**

Dry sunny slopes and savannas, often with *Pinus ponderosa*, sometimes with sagebrush, or on serpentine. 600–2800 m. BR, BW, Casc, ECas, Owy, Sisk. CA, ID, NV, WA; north to British Columbia. Native.

The swollen rachilla internodes of *M. fugax* are inconspicuous but utterly distinctive once observed. *Melica bulbosa* and *M. spectabilis* are also bulbous-based but do not have the swollen rachilla nodes. The *M. fugax* panicle matures with the panicle branches spread out stiffly perpendicular to the inflorescence, unlike other Oregon *Melica* species.

Melica geyeri Munro var. *geyeri*

Geyer's oniongrass

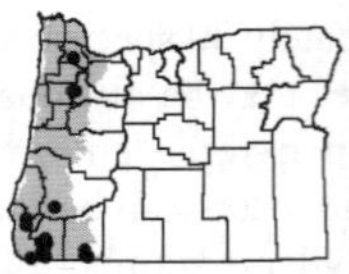

Plants 65–200 cm tall; cespitose, rhizomatous. **Culms** bases modified as bulb-like corms sessile on the rhizome. **Leaves** sheaths scabrous or minutely so, sometimes sparsely pilose, particularly at the throat and collar; blades 2–8 mm wide, flat to involute; lower surfaces minutely scabrous; upper surfaces hairy. **Ligules** glabrous, 2–3.5 mm, mechanically split in front. **Inflorescences** 10–30 × 0.5–15 cm with branches divergent to reflexed, flexuous; pedicels straight. **Spikelets** (8)12–24 mm, 4–7 bisexual florets and 1–4 sterile florets per spikelet; rachilla internodes 2–3 mm; disarticulation above glumes. **Glumes** usually purple; lower glumes 3.5–7 × 1.5–2 mm, with 5–9 veins; upper glumes 6–12 × 2–2.5 mm, with 5–11 veins. **Lemmas** of fertile florets 7.5–12.5 mm, glabrous or scabrous, 7 veins, acute to more or less obtuse, awnless. **Anthers** 3, 2.5–4 mm. **2*n*=18.**

Dry forests, common on serpentine. 50–1300 m. CR, Sisk, WV. CA. Native.

Melica geyeri is identified by its spreading panicle branches and relatively long spikelets with contrasting dark purple glumes and paler lemmas. Rare individuals with appressed panicle branches can be confused with the other bulb-bearing *Melica* species, though the panicle branches of *M. geyeri* tend to be flexuous even if erect.

Melica harfordii Bolander

Harford melic

Melica harfordii Bol. var. *minor* Vasey

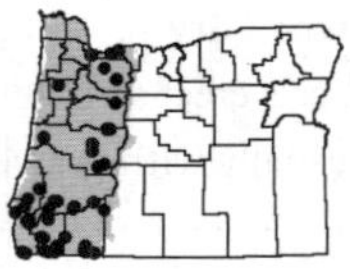

Plants 35–120 cm tall; cespitose. **Culms** bases not bulb-like. **Leaves** sheaths glabrous or pilose, often most pilose at the throat and collar; blades 1.5–4.5 mm wide, flat to involute; lower surfaces smooth; upper surfaces minutely scabrous, glabrous or short-hairy. **Ligules** glabrous or with a fringe of hairs, hairs sometimes as long as the membranous portion, 1–3 mm, usually closed and projecting upward in front in a broad tooth. **Inflorescences** 6–25 × 0.8–2 cm with branches appressed; pedicels straight. **Spikelets** 7–20 mm, 2–6 bisexual florets and 1–4 sterile florets per spikelet; rachilla internodes 2–2.4 mm; disarticulation above glumes. **Glumes** lower glumes 4–10 × 1.5–2.5 mm, with 3–5 veins; upper glumes 5–11.5 × 1.8–2.5 mm, 5–7 veins. **Lemmas** of fertile florets 6–16 mm, hairy, long-ciliate-pilose at the base, hairs to 0.75 mm on the back, 0.7–1.3 mm on the margins, 9–11 veins, obtuse or rounded, mucronate or more often awned; lemma awns 0–3(4) mm, arising from the apex. **Anthers** 3, 2.2–4 mm. **2*n*=18.**

Sunny, rocky openings on slopes; dry open forests. 50–2300 m. Casc, CR, Sisk, WV. CA, WA. Native.

The challenge in identifying *M. harfordii* is recognizing it as a *Melica*. It lacks bulb-like bases, its sterile lemmas resemble the fertile lemmas (but are slightly smaller), and the glumes are green and normal in texture in the lower half. One clue is the purple coloration on the green lemmas. A close look reveals the 9–11 lemma veins and the closed leaf sheath.

Melica smithii (Porter ex A. Gray) Vasey

Smith's melic

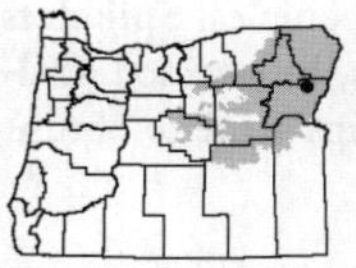

Plants 60–160 cm tall; loosely cespitose. **Culms** bases thickened basally, rarely forming bulb-like corms. **Leaves** sheaths glabrous, sometimes pilose or retrorsely scabrous particularly at the throat, closed in front unless mechanically torn; blades 5–12 mm wide, flat, both surfaces minutely scabrous, sometimes the upper surface hairy. **Ligules** usually glabrous except at the base, 3–9 mm, closed in front unless mechanically split. **Inflorescences** 12–30 × 1.2–21 cm; branches solitary, spreading to reflexed; pedicels straight. **Spikelets** 12–18 mm, 3–5 bisexual florets and 1–4 sterile florets per spikelet; rachilla internodes 2.5–3 mm; disarticulation above glumes. **Glumes** lower glumes 4.5–7 × 1–1.5 mm, 1–3 veins; upper glumes 6.5–9 × 1–1.5 mm, 3–5 veins. **Lemmas** of fertile florets 9.5–12 mm, glabrous or scabrous, 7-veined, acuminate, bifid to emarginate, awned; lemma awns 3–10(12) mm, arising from the apex or from the sinus of a bifid apex. **Anthers** 3, 1.3–2.5 mm.

Cool, moist forests in mountains. 700–2400 m. BW. ID, WA; north to British Columbia, northeast to Alberta, east to WY; Great Lakes region and northeastern Canada. Native.

The wide, flat leaves make *M. smithii* unique among Oregon *Melica*. In addition, it has long lemma awns, a trait shared only with *M. aristata*. It is most often confused with *Bromus vulgaris*, which differs in having hairy sheaths and blades, acute lodicules, and densely hairy tips on the ovaries and caryopses.

Melica spectabilis Scribner (p. 430)

Purple oniongrass

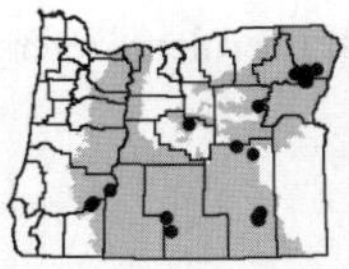

Plants 45–100 cm tall; loosely cespitose, rhizomatous. **Culms** bases modified as bulb-like corms attached to the rhizomes by a root-like, 10–30 mm structure; sheaths glabrous, often pilose at the throat and collar. **Leaves** sheaths open above; blades 2–5 mm wide, flat to slightly involute; lower surfaces minutely scabrous over the veins; upper surfaces usually glabrous. **Ligules** 0.1–3 mm, glabrous, mostly 1–3 mm, open in front. **Inflorescences** 5–26 × 0.8–2.5 cm; branches usually appressed, sometimes divergent and flexuous; pedicels straight. **Spikelets** (7)8.5–12.5(19) mm, 3–7 bisexual florets and 1–4 sterile florets per spikelet; rachilla internodes 1–2 mm; disarticulation above glumes. **Glumes** usually less than 50% as long as the spikelets; lower glumes 3.5–6.4 × 1.5–3 mm, 1–3 veins; upper glumes 5–7 × 2.3–3.5 mm, 5–7 veins. **Lemmas** of fertile florets 6–9 mm, glabrous to scabrous, 5–11 veins, rounded to acute, awnless. **Anthers** 3, 1.5–3 mm. ***2n***=18.

Moist montane meadows and open forest. 200–2800 m. BR, BW, Casc, ECas. CA, ID, NV, WA; north to British Columbia, northeast to Alberta, southeast to CO. Native.

The bulb-like culm bases, connected to the rhizome by a little root-like structure, identify *M. spectabilis*. Otherwise, this species is easy to confuse with *M. bulbosa*.

Melica stricta Bolander var. ***stricta***

Rock melic

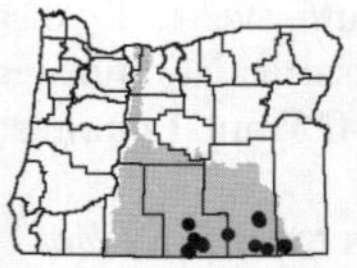

Plants 9–85 cm tall; densely cespitose. **Culms** bases not bulb-like. **Leaves** sheaths mostly open above, minutely scabrous, rarely densely hairy with spreading hairs; blades 1.5–5 mm wide, flat to involute; lower surfaces glabrous, minutely scabrous, rarely densely hairy with spreading hairs; upper surfaces strigose, glabrous or minutely scabrous. **Ligules** glabrous, 2.5–3.5(4.3) mm, closed or open in front. **Inflorescences** narrow raceme-like panicles, 3–30 × 0.8–3.7 cm, sometimes nearly racemose, with branches appressed; pedicels sharply bent below the spikelets which hang downward. **Spikelets** (6)10–20(23) mm, 1–7 bisexual florets and 1–4 sterile florets per spikelet; rachilla internodes 1.8–2.1 mm; disarticulation below glumes. **Glumes** subequal to as long as the spikelet; lower glumes (6)9.5–16 × 3.5–5 mm, 4–7 veins; upper glumes (6)9.5–16(18) × 3–5 mm, 5–9 veins. **Lemmas** of fertile florets 6–16 mm, glabrous or minutely scabrous, 5–9 veins, acute, awnless. **Anthers** 3, 1–2 mm. ***2n***=18.

Dry, open slopes in mountains. 1400–2600 m. BR, ECas. CA, ID; southeast to UT. Native.

This *Melica* has distinctive inflorescences with the spikelets spreading sideways or hanging down on the strongly curved pedicels. The inflorescences often look like racemes, although closer examination reveals that the lower nodes have delicate appressed branches.

Melica subulata (Grisebach) Scribner

Alaska oniongrass, tapered oniongrass

Melica subulata (Griseb.) Scribn. var. *pammelii* (Scribn.) C.L. Hitchc., *Melica subulata* (Griseb.) Scribn. var. *subulata*

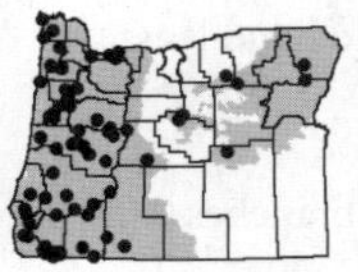

Plants 55–125 cm tall; cespitose, rhizomatous. **Culms** bases modified as bulb-like corms, sessile on rhizomes. **Leaves** sheaths minutely scabrous, sometimes glabrous or pilose; mostly open above; blades 2–10 mm wide, flat, occasionally somewhat involute; lower surfaces smooth or minutely scabrous; upper surfaces minutely scabrous, glabrous or hairy. **Ligules** glabrous, mostly 1–5 mm, often closed in front. **Inflorescences** 8–25 × 0.5–10.4 cm with branches usually appressed to ascending, occasionally divergent; pedicels erect. **Spikelets** (10)12–21(28) mm, 2–5 bisexual florets and 1–4 sterile florets per spikelet; rachilla internodes 1.8–2 mm; disarticulation above glumes. **Glumes** usually purple; lower glumes 4–6(8) × 1.3–2.2 mm, 1–3 veins; upper glumes 5.5–8(11.5) × 2–3 mm, 3–5 veins. **Lemmas** of fertile florets 5.5–18 mm, usually with appressed hairs over the veins, hairs longest towards the base, with 7–9 veins, strongly tapering and acuminate, awnless. **Anthers** 3, 1.5–2.5 mm. ***2n***=18.

Common in mesic forest. 0–2200 m. BW, Casc, CR, ECas, Est, Sisk, WV. CA, ID, WA; north to AK, east to SD. Native.

The bulbous culm bases and acuminate lemmas identify this common forest oniongrass.

MOLINIA Moorgrass

Plants perennial; cespitose. **Culms** often disarticulating at the first node; basal internodes persistent, often swollen. **Leaves** mostly basal; ligules of hairs; blades flat or convolute, eventually disarticulating from the sheaths. **Inflorescences** panicles. **Spikelets** laterally compressed, with (1)2–5 florets; disarticulation beneath the florets. **Glumes** shorter than the florets, 1–3-veined. **Calluses** blunt, glabrous or with sparse hairs to 0.5 mm. **Lemmas** glabrous, 3(5)-veined, rounded over the back; tips acute to obtuse, awnless. **Anthers** 3.

Temperate regions. 2–5 species; 1 species treated in *Flora.*

Molinia caerulea (Linnaeus) Moench

Purple moorgrass

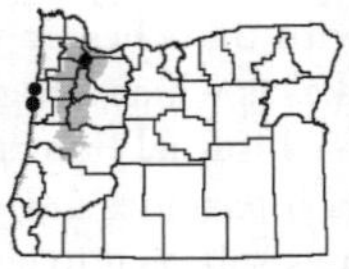

Plants 15–150(250) cm tall; cespitose. **Culms** erect, rigid; basal internodes 2–6 cm, usually swollen and somewhat club-shaped. **Leaves** sheaths smooth, glabrous except for sparse marginal hairs; collars often marked by a narrow crease; ligules a fringe of hairs, obvious; blades 2–5(10) cm wide, flat or convolute. **Inflorescences** panicles, 5–40 cm × 0.4–2 mm, interrupted. **Spikelets** 4–9 mm with (1)2(5) florets, usually purple, ripening brown, lowest floret more or less sessile, second raised on a conspicuous rachilla internode; 3rd to 5th florets, if present, much reduced. **Glumes** 1–3-veined; lower glumes 1.5–2.5 mm; upper glumes 2–3 mm. **Lemmas** 2.5–4.5 mm, faintly 3–5-veined, smooth, shiny and leathery; tips acute to obtuse, awnless. **Paleas** expanded and more strongly bowed than the lemmas. **Anthers** 1.5–3 mm.

Marshy flats in stabilized coastal dunes, other wetlands. 0–100 m. Est, WV. southeastern Canada, northeastern US; Eurasia. Exotic.

This coastal grass has narrow, interrupted panicles. The purple to brown spikelets contain a sessile lower floret and a second floret raised on a stalk. The paleas are expanded and more strongly bowed than the lemmas.

MUHLENBERGIA Muhly

Plants 2–100(300) cm tall, annual or perennial; rhizomatous; cespitose or mat-forming. **Culms** erect to decumbent. **Leaves** sheaths open; ligules membranous or hyaline, sometimes with minute hairs on the margins; blades flat, folded or involute. **Inflorescences** terminal, sometimes axillary, open to contracted panicles. **Spikelets** 0.8–4 mm, flattened, 1(3) florets; disarticulation above the glumes, occasionally below the pedicels. **Glumes** 2, (0)1(3)-veined; tips truncate to acuminate, sometimes mucronate or awned; lower glumes sometimes rudimentary or absent, occasionally bifid; upper glumes shorter or longer than the florets. **Calluses** short, glabrous or pubescent. **Lemmas** 3-veined; tips awnless to awned; lemma awns if present, straight or wavy to curled, terminal or arising between 2 minute teeth. **Anthers** (1)3. **Caryopses** elongate, fusiform to elliptic, slightly dorsally compressed.

Tropical and temperate regions. Approximately 155 species; 7 species treated in *Flora.*

In the following key, "nodulose" refers to pale, blunt bumps that are thickenings of the fibers in the culm. "Puberulent" refers to hairs so small they can only be seen with a hand lens.

Key to *Muhlenbergia* species:

1. Panicles open, 1.5–20 cm wide; panicle branches 0.05–0.1 mm thick, diverging 10–100° from the inflorescence axes at maturity.
   2. Glumes minutely hairy at least on the margins or toward the tips; lemmas 0.8–1.5 mm . . . . ***M. minutissima***
   2' Glumes glabrous or minutely scabrous on the veins; lemmas 1.2–2.1 mm.
      3. Plants rhizomatous; habitat moist, alkaline meadows east of the Cascades . . . . ***M. asperifolia***
      3' Plants cespitose; habitat coastal cranberry bogs . . . . ***M. uniflora***

1' Panicles contracted, 0.1–3 cm wide; panicle branches more than 0.1 mm thick, appressed or diverging up to 30(40)° from the inflorescence axes at maturity.
   4. Leaf blades 0.2–2 mm wide, flat, involute, or folded at maturity.
      5. Culms usually nodulose throughout; glume tips acute, sometimes mucronate . . . . ***M. richardsonis*** (in part)
      5' Culms not nodulose; glume tips rounded to subacute. . . . . ***M. filiformis***
   4' Leaf blades 2–15 mm wide, flat at maturity.
      6. Glumes approximately 50% as long as the lemmas; lemmas glabrous; culms nodulose . . . . ***M. richardsonis*** (in part)
      6' Glumes approximately as long as or longer than the lemmas; lemmas with hairs, these sometimes restricted to the callus; culms not nodulose.
         7. Lemma bases with hairs approximately as long as the florets, usually 2–3.5 mm. . . . . ***M. andina***
         7' Lemma bases with hairs shorter than the florets, usually shorter than 1.5 mm . . . . ***M. mexicana***

Muhlenbergia andina (Nuttall) Hitchcock
Foxtail muhly

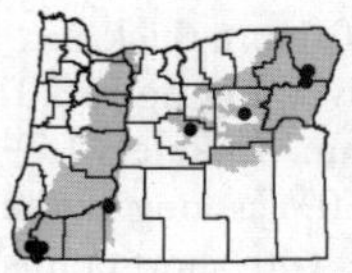

Plants perennial, 25–85 cm tall; rhizomatous. **Culms** 0.9–1.7 mm thick, not nodulose; internodes glabrous for most of their length, scabrous to short-hairy below the nodes. **Leaves** blades 4–16 cm × 2–4(5) mm, flat at maturity; lower surface scabrous; upper surface pubescent. **Inflorescences** 2–15 × 0.5–2.8 cm, contracted, dense; primary branches 0.5–5 cm × over 0.1 mm, appressed or strongly ascending, diverging up to 30(40)° at maturity; pedicels 0.5–1.5 mm, appressed, strigose. **Spikelets** 2–4 mm. **Glumes** equal to subequal, 2–4 mm, subequal to or longer than the florets and lemmas, 1-veined; veins minutely scabrous; tips acuminate to awn-tipped. **Lemmas** 2–3.5 mm, lanceolate, grayish green, hairy on the calluses and lemma bases, hairs 2–3.5 mm and 75% as long or longer than the lemmas; tips acuminate; lemma awns 1–10 mm. **Anthers** 0.4–1.5 mm, yellow. **2*n***=20.

Stream banks, marshes, lake margins, moist meadows. 100–2000 m. BW, Casc, Sisk. CA, ID, NV; north to British Columbia, northeast to Manitoba, southeast to TX. Native.

Muhlenbergia andina is an uncommon plant of moist, montane habitats, with narrow inflorescences and noticeably hairy, awned lemmas. The common *M. mexicana* is similar but has shorter lemma hairs.

Muhlenbergia asperifolia (Nees & Meyen ex Trinius) Parodi
Alkali muhly, scratchgrass

Sporobolus asperifolius Nees & Meyen ex Trin.

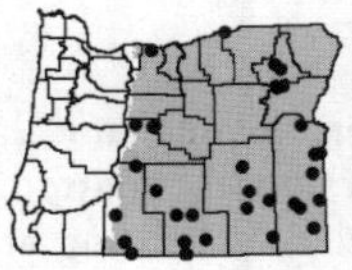

Plants perennial, 10–60(100) cm tall; rhizomatous, occasionally stoloniferous. **Culms** with bases somewhat compressed-keeled; internodes glabrous, shiny below the nodes. **Leaves** blades 2–7(11) cm × 1–2.8(4) mm, flat, occasionally conduplicate; lower surface smooth or minutely scabrous; upper surface minutely scabrous; margins and midveins not conspicuously thickened, greenish; tips acute. **Inflorescences** 6–21 × 4–16 cm, broadly ovoid, open; branches whitish; primary branches 3–12 cm × 0.05–0.1 mm, capillary; lower branches spreading, not appearing clustered; pedicels 3–14 mm, longer than the spikelets. **Spikelets** 1.2–2.1 mm. **Glumes** equal, 0.6–1.7 mm, 1-veined, purplish, minutely scabrous particularly on the veins; tips acute. **Lemmas** 1.2–2.1 mm, lanceolate to oblong-elliptic, somewhat gray, glabrous, smooth, occasionally minutely scabrous near the tips; tips acute, awnless or mucronate, the mucros to 0.3 mm. **Anthers** 1–1.3 mm, greenish yellow to purplish at maturity. **2*n***=20, 22, 28.

Moist, alkaline meadows. 100–1500 m. BR, BW, Col, ECas, Lava, Owy. CA, ID, NV, WA; north to British Columbia, northeast to Ontario, southeast to TX, south to Mexico. Native.

Muhlenbergia filiformis (p. 437)
(Thurber ex S. Watson) Rydberg
Pull-up muhly

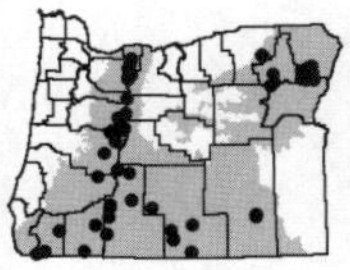

Plants annual but sometimes appearing perennial, (3)5–20(35) cm tall. **Culms** often rooting at the lower nodes; not nodulose; internodes glabrous. **Leaves** blades 1–4(6) cm × 0.6–1.6 mm, flat or involute; lower/outer surface smooth or minutely scabrous; upper/inner surface scabrous or pubescent. **Inflorescences** spike-like, 1.6–6 × 0.2–0.5 cm, interrupted near the base, long-exserted from the upper leaf sheath; primary branches 0.9–1.2 cm × over 0.1 mm, appressed at maturity or diverging 30(40)°; pedicels 1–3 mm, scabrous. **Spikelets** 1.5–3.2 mm. **Glumes** greenish gray, glabrous, 1-veined, rounded to subacute; lower glumes 0.6–1.4 mm; upper glumes 0.7–1.7 mm; tips rounded to subacute. **Lemmas** (1.5)1.8–2.5(3.2) mm, lanceolate, dark greenish, appressed-pubescent on the margins and midveins, hairs shorter than 0.3 mm; tips minutely scabrous, acute to acuminate, awnless, sometimes with small points at the tip. **Anthers** 0.5–1.2 mm, purplish. **2*n***=18.

Moist meadows, lake shores, stream banks. 100–2800 m. BR, BW, Casc, ECas, Sisk. CA, ID, NV, WA; north to British Columbia, east to NE, south to Mexico. Native.

Muhlenbergia filiformis is a tiny, delicate annual usually about 5 cm tall, with a narrow, purplish inflorescence. It may form dense, turf-like stands. It often grows with similar *M. richardsonis,* a more robust, more or less rhizomatous perennial with nodulose stems and glabrous lemmas. This species gets its common name of "pull-up muhly" because cattle pull up clumps of the short plants but cannot manipulate them to eat them and thus drop the tufts on the meadows.

Muhlenbergia mexicana (Linnaeus) Trinius
Wirestem muhly

Plants perennial, 30–90 cm tall; rhizomatous. **Culms** 0.5–2 mm thick, much branched above the base, not nodulose; internodes dull, puberulent or glabrous for most of their length, sometimes strigose immediately below the nodes. **Leaves** blades 2–20 cm × 2–6 mm, flat, scabrous or smooth, those of the secondary branches similar in length and width to those of the main branches. **Inflorescences** terminal and axillary, 2–21 × 0.3–3 cm, dense; appressed or diverging up to 30° from the inflorescence axis; primary branches 0.3–5.5 cm × over 0.1 mm. **Spikelets** 1.5–3.8 mm, often purple-tinged. **Glumes** subequal, 1.5–3.7 mm, approximately as long as or slightly shorter than the lemmas, 1-veined, tapering from the bases to the acuminate tips; awnless or awned, awns, if present, to 2 mm. **Lemmas** 1.5–3.8 mm, lanceolate, pubescent on calluses and lower portion of the midveins and margins, hairs shorter than the florets, less than 1.5 mm long; tips minutely scabrous, acuminate, awnless or awned; lemma awns 0–10 mm. **Anthers** 0.3–0.5 mm, yellowish to purplish. **2*n***=40.

Throughout North America. 2 varieties.

Muhlenbergia mexicana flowers relatively late in summer after river levels drop. It is seldom collected. The similar *M. andina* has lemma hairs about as long as the lemma body.

Key to *Muhlenbergia mexicana* varieties:

1. Lemmas awned, awns 3–10 mm . . . . . ***M. m.*** var. ***filiformis***
1' Lemmas awnless or with awns less than 3 mm . . . . ***M. m.*** var. ***mexicana***

Muhlenbergia mexicana (Linnaeus) Trinius var. ***filiformis*** (Torrey) Scribner

Muhlenbergia foliosa (Roem. & Schult.) Trin. ssp. *ambigua* (Torr.) Scribn.

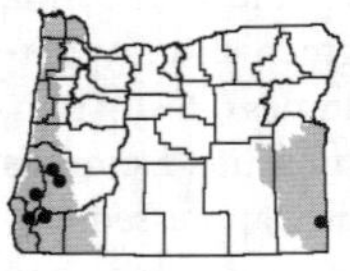

Lemmas awned; lemma awns 3–10 mm.

Rocky stream banks below the high water line. 100–1200 m. CR, Owy, Sisk. Native.

Muhlenbergia mexicana (Linnaeus) Trinius var. ***mexicana***

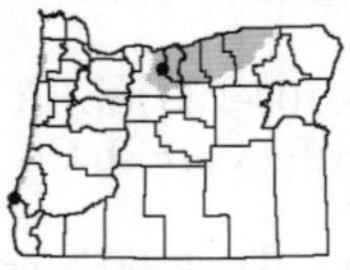

Lemmas awnless or nearly so; lemma awns (if present) less than 3 mm.

Cranberry bogs, stream banks, moist meadows. 300–700 m. Col, Est. CA, ID, NV, WA; throughout most of North America. Exotic.

Muhlenbergia mexicana var. *mexicana* has been collected in the Columbia Gorge and in a cranberry bog in Coos County. The Coos County specimen may have been introduced with cranberry stock from Wisconsin.

Muhlenbergia minutissima (Steudel) Swallen

Annual muhly

Sporobolus microspermus (Lag.) Hitchc.

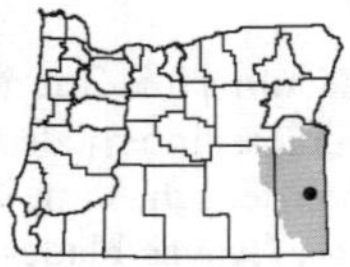

Plants annual, 5–40 cm tall, tufted. **Culms** slender; internodes mostly glabrous, minutely scabrous or smooth; minutely scabrous or with tiny, appressed hairs below the nodes. **Leaves** blades 0.5–4(10) cm × 0.8–2 mm, flat or involute; lower surface scabrous; upper surface shortly pubescent. **Inflorescences** 5–16(21) × 1.5–6.5 cm, open; primary branches 8–42 × 0.05–0.1 mm, often capillary, diverging 25–80° from the inflorescence axes; pedicels 2–7 mm, straight or curved but rarely curved as much as 90°. **Spikelets** 0.8–1.5 mm. **Glumes** with sparse, tiny, appressed hairs at least near the tips, 1-veined; lower glumes 0.5–0.8 mm, obtuse to acute; upper glumes 0.6–0.9 mm, broader than the lower glumes, obtuse. **Lemmas** 0.8–1.5 mm, lanceolate, brownish to purplish, glabrous or the midveins and margins appressed-pubescent; tips obtuse to subacute, awnless. **Anthers** 0.2–0.7 mm, purplish. **2*n***=60, 80.

Shores of lakes, ponds, reservoirs and streams, in sandy or gravelly soil. 1100–1500 m. Owy. CA, ID, NV, WA; northeast to MT, east to SD, southeast to TX, south to Mexico. Native.

This rare, tufted annual has a diffuse but narrow panicle. It is known from one Oregon location near Jordan Craters in Malheur County. Similar *M. asperifolia* is a rhizomatous perennial with a more open panicle.

Muhlenbergia richardsonis (Trinius) Rydberg (p. 437)

Mat muhly

Muhlenbergia squarrosa Rydb.

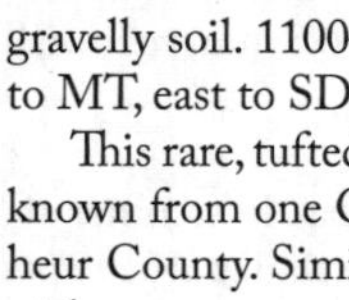

Plants perennial, 5–30 cm tall; rhizomatous. **Culms** 0.4–1 mm thick; internodes usually nodulose for most of their length, puberulent or nodulose below the nodes. **Leaves** blades 0.4–6.5 cm × 0.5–4.2 mm, flat or involute, straight or curving; lower/outer surface glabrous; upper/inner surface minutely hairy. **Inflorescences** spike-like, 1–1.5 × 0.1–1.7 cm, inflorescence axes usually concealed by the branches; primary branches 0.4–5 cm × over 0.1 mm thick, usually closely appressed at maturity, rarely diverging up to 20° from the inflorescence axes; pedicels 0.2–2 mm. **Spikelets** 1.7–3.1 mm. **Glumes** subequal, 0.6–2 mm, 33–50% as long as the lemmas, green, 1(2)-veined, acute, sometimes with short points less than 0.2 mm; tips acute, sometimes mucronate. **Lemmas** 1.7–3.1 mm, lanceolate, dark greenish, gray or mottled, glabrous; tips minutely scabrous, acute to acuminate, sometimes mucronate, mucros to 0.5 mm. **Anthers** 0.9–1.6 mm, yellow to purplish. **2*n***=40.

Seasonally moist alkaline meadows, river banks. 50–2700 m. BR, BW, Casc, Col, CR, ECas, Lava, Owy, WV. CA, ID, NV, WA; north to Yukon, east to Nova Scotia, south to northern Mexico. Native.

Muhlenbergia richardsonis is a small rhizomatous perennial with narrow purplish inflorescences much like those of *M. filiformis*. The nodulose culms are unique among Oregon grasses when present but can be hard to see. Young *M. richardsonis* culms are green with slightly thickened, straw-colored ribs; older culms are straw-colored throughout. The nodules are similar in thickness and color to the ribs but extend across the space between the ribs. Occasional *M. richardsonis* that lack the nodules can be difficult to distinguish from *M. filiformis*, an annual that has culms glabrous even below the nodes and glumes erose and rounded to subacute. *Muhlenbergia richardsonis* populations increase under heavy grazing.

Muhlenbergia uniflora (Muhlenberg) Fernald

Bog muhly

Plants perennial but often seeming annual; 5–45 cm tall, tufted or loosely matted. **Culms** compressed-keeled, branching at lower nodes; internodes mostly glabrous or sometimes minutely puberulent below the nodes. **Leaves** blades 1–15 cm × 0.8–2 mm, flat to conduplicate; lower surface smooth or minutely scabrous; upper surface minutely hairy. **Inflorescences** 2–20 × (0.2)2.5–6 cm, diffuse; primary branches approximately 0.1 mm thick, ascending, diverging 10–60° from the inflorescence axes, naked basally; pedicels 0.2–7 mm, glabrous. **Spikelets** 1.3–2.1 mm, dark purplish to gray. **Glumes** equal, 0.4–1.3 mm, glabrous on the veins, 1-veined; tips scabrous, acute to obtuse, sometimes erose, awnless. **Lemmas** 1.2–2 mm, oblong-elliptic, dark purplish to gray, glabrous, faintly 3-veined; tips acute to obtuse, awnless. **Anthers** 0.6–0.9 mm, purple.

Bogs and lakeshores. 0–100 m. Est. British Columbia, northeastern North America. Exotic.

Muhlenbergia uniflora is a diffuse-panicled bog grass that resembles *M. minutissima*, which has smaller spikelets with minutely hairy glumes and lives in alkaline habitats. *Muhlenbergia uniflora* was first collected in a Curry County cranberry bog in 1997 and was likely introduced with planting stock from eastern North America.

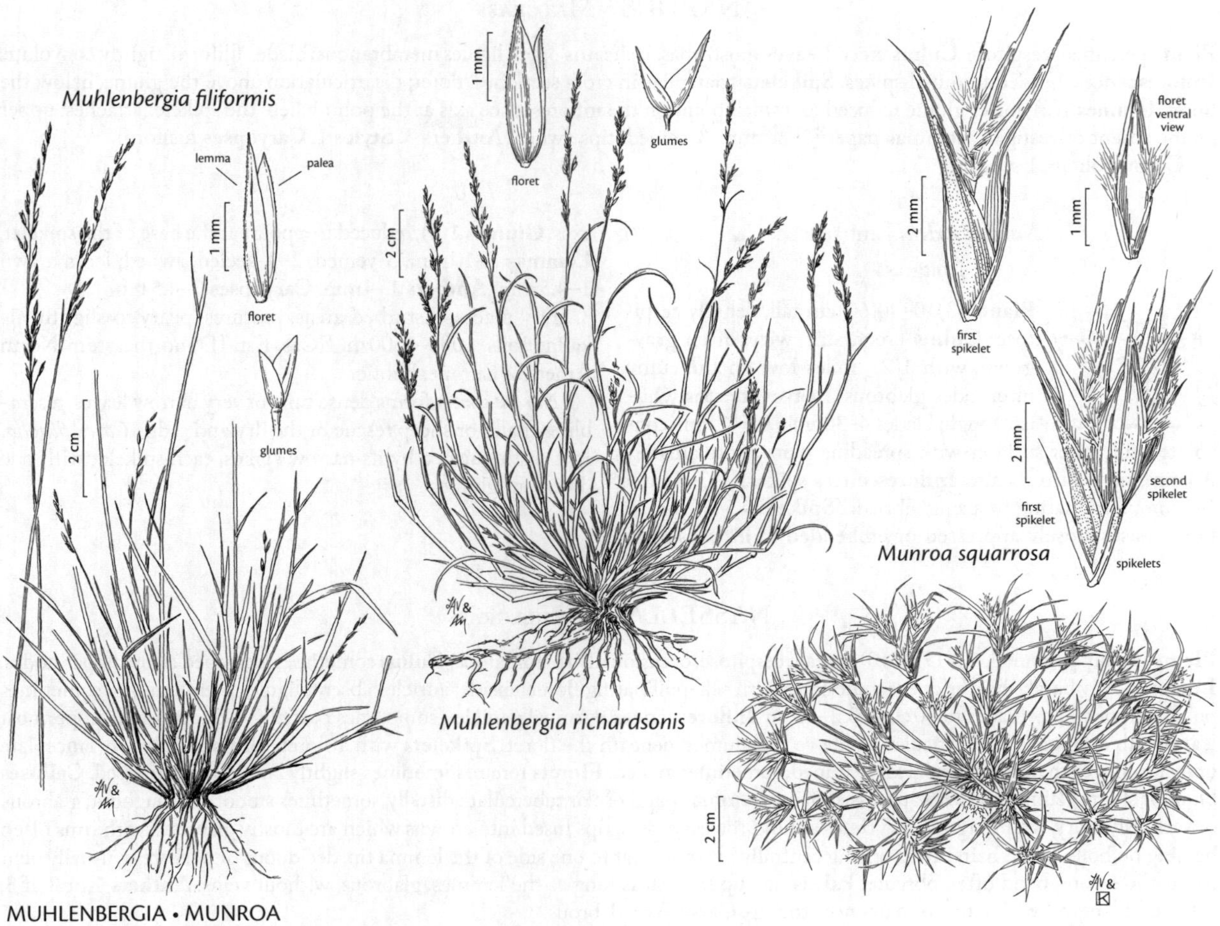

MUHLENBERGIA • MUNROA

MUNROA False buffalograss

Plants annual; stoloniferous, mat-forming. **Stolons** terminating in fascicles of leaves from which new culms arise. **Leaves** mostly basal, sometimes with a purple tint; sheaths open; ligules of hairs; blades usually involute with thick white margins and sharply pointed tips. **Inflorescences** terminal head-like panicles with spike-like branches, partly hidden in the tuft of leaves. **Spikelets** laterally compressed, 2–10 florets, lower florets bisexual or pistillate and the terminal florets sterile; disarticulation above the glumes or beneath the leaves that subtend the panicles. **Glumes** (1)2, shorter than the spikelets, keeled, 1-veined, awnless; lower glumes present on all spikelets; upper glumes reduced or absent on the terminal spikelet. **Lemmas** with a tuft of hairs on the margins at mid-length, membranous or leathery, 3-veined; tips emarginate or 2-lobed, sometimes awned. **Anthers** 2 or 3. **Caryopses** dorsally compressed.

North America and South America. 5 species; 1 species treated in *Flora*.

Munroa squarrosa (Nuttall) Torrey (p. 437)

False buffalograss

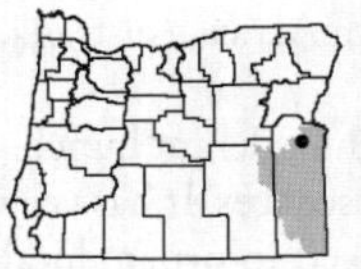

Plants 3–15(30) cm tall, annual; stoloniferous. **Culms** decumbent, much-branched, scabrous to minutely hairy. **Leaves** arising in bundles with sharp tips; sheaths with a tuft of hairs at the throat; ligule of hairs 0.5(1) mm; blades 1–5 cm × 1–2.5 mm, involute, sometimes flat or folded, glabrous, with a conspicuous midrib and white margins; uppermost blades exceeding the inflorescence. **Inflorescences** head-like, 5–13 × 0.4–1.3 cm, partially hidden in leaf sheaths. **Spikelets** 6–8(10) mm, 3–5 florets. **Glumes** of first 1–2 spikelets subequal, 2.5–4.2 mm, narrow, 1-veined, acute; on the upper spikelets the lower glume reduced or absent. **Lemmas** with a tuft of conspicuous hairs at mid-length, 5–8 mm, those of lower spikelets minutely scabrous, lanceolate; lateral veins excurrent; lemma awns 0.5–2 mm, stout, scabrous. **Anthers** 1–1.5 mm.

Dry, open areas, usually in sandy soil or disturbed sites, often around ant mounds. 700–1100 m. Owy. CA, NV; northeast to Alberta and Manitoba, east to NE, southeast to TX and northern Mexico. Native.

Munroa squarrosa forms mounds of curved, sharp-pointed leaves partially hiding the dense inflorescences. It strongly resembles *Crypsis vaginiflora*, a species not yet reported from Oregon (but present in surrounding states), which has less sharp leaf tips and shorter spikelets with only one floret each.

NARDUS Matgrass

Plants perennial, cespitose. **Culms** erect. **Leaves** mostly basal; sheaths open; ligules membranous; blades filiform, tightly convolute. **Inflorescences** 1-sided terminal spikes. **Spikelets** triangular in cross section, 1 floret; disarticulation above the glume, below the floret. **Glumes** 1(2); lower glume reduced to a small point on the inflorescence axis at the point where the spikelet attaches; upper glume absent or vestigial. **Lemmas** papery in texture, 3-veined; tips awned. **Anthers** 3. **Styles** 1. **Caryopses** fusiform.

Cosmopolitan. 1 species.

Nardus stricta Linnaeus (p. 439)

Matgrass

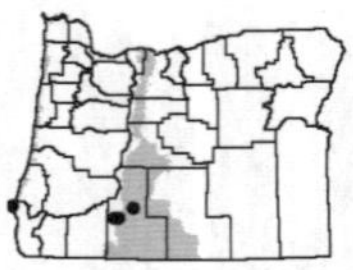

Plants (3)10–40(60) cm tall, densely cespitose. **Culms** erect, stiff, wiry, often gray-green, with 1(2) nodes low on the culm; internodes glabrous. **Leaves** sheaths whitish, tough; blades 4–30 cm × 0.5–1 mm, convolute, stiff, outer surfaces with spreading hairs approximately 0.3 mm; tips sharply acute. **Inflorescences** spikes, (1)3–8 cm × 1–3 mm, terminating in a single bristle. **Spikelets** 5–10 mm, 1 floret, bases closely appressed or embedded in inflorescence axes. **Glumes** 1(2), reduced to a point at the base of the spikelet. **Lemmas** 5–10 mm, 3-veined, 2–3-keeled, awned; lemma awn 1–4.5 mm. **Anthers** 1–4 mm. **Caryopses** 3–4.5 mm.

Dry places, disturbed areas, pastures, peaty to slightly alkaline soils. 1000–1500 m. ECas, Est. ID; northeastern North America; Europe. Exotic.

Nardus stricta forms dense tufts of very narrow leaves, resembling Idaho or sheep fescue or the dryland sedge *Carex filifolia.* It is distinguished by its narrow spikes, each spikelet with one divergent glume.

NASSELLA Needlegrass

Plants usually perennial, 10–175(210) cm tall, cespitose, occasionally rhizomatous. **Culms** sometimes branched at the upper nodes. **Leaves** mostly basal; sheaths open, sometimes with self-pollinating florets inside; auricles absent; ligules membranous, sometimes pubescent or ciliate; blades to 60 cm × 0.2–8 mm. **Inflorescences** terminal panicles, sometimes partially included in the uppermost leaf sheath at maturity; disarticulation above the glumes, beneath the floret. **Spikelets** with 1 floret. **Glumes** narrowly lanceolate or ovate, longer than the floret, (1)3–5-veined, sometimes awned. **Florets** terete, sometimes slightly laterally compressed. **Calluses** blunt or sharp, glabrous or with appressed hairs. **Lemmas** papillose or tuberculate distally, sometimes smooth throughout, glabrous or variously hairy, wrapping 1.2–1.5 times around the caryopses, tips fused into crowns which are mostly glabrous, with rims often bearing bulbous-based hairs; awns arising centrally or somewhat to one side of the lemma tip, deciduous or persistent, usually bent twice, the second bend often obscure. **Paleas** flat, up to 50% as long as the lemmas, glabrous, without veins. **Anthers** 1 or 3, if 3, often of 2 lengths, with a tuft of hairs near the tip. **Caryopses** glabrous.

Mainly warm and tropical areas. Approximately 116 species; 1 species treated in *Flora.*

The genus *Nassella* includes many species that were formerly included in *Stipa*. It is characterized by strongly overlapping lemma margins and reduced, veinless paleas. The lemma tips are fused into a crown, formed by a membrane surrounding the base of the awn. The crown is often edged with hairs. Many *Nassella* species form both cross-pollinating and self-pollinating florets in the terminal panicle. The self-pollinating florets have 1–3 small anthers; the cross-pollinating florets have three longer anthers. Some species have self-pollinating inflorescences hidden in their basal leaf sheaths. These hidden inflorescences lack glumes and usually lack awns.

Nassella tenuissima (Trinius) Barkworth

Mexican feathergrass, Mexican needlegrass

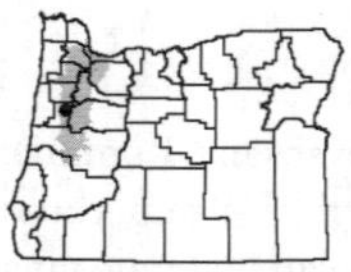

Plants perennial, 25–100 cm tall, densely cespitose. **Culms** 0.4–0.7(1.1) mm thick; internodes mostly glabrous, pubescent just below the lower nodes; nodes glabrous. **Leaves** sheaths glabrous, sometimes minutely scabrous; ligules 1–5 mm, acute, glabrous; blades 7–60 cm × 0.2–1.5 mm, usually convolute, stiff, glabrous, minutely scabrous. **Inflorescences** loosely contracted, 8–50 cm, often partly enclosed by the uppermost leaf sheath at maturity; branches ascending, 2–8 cm; pedicels 1–11 mm. **Glumes** subequal, narrowly lanceolate, 5–13 × 0.5–1.2 mm, glabrous, awned. **Florets** somewhat laterally compressed. **Calluses** blunt, 0.2–0.5 mm, with appressed hairs to approximately 25–33% as long as the lemmas. **Lemmas** (1.5)2.5–3 × approximately 0.5 mm, widest at about mid-length, covered with minute bumps, midveins pubescent on the lower 75%; crowns 0.1–0.2 × 0.2–0.25 mm, rims with hairs less than 0.5 mm; awns arising near the center of the lemma tip, more or less wavy, obscurely bent twice, 45–100 mm. **Anthers** 3, 1.2–1.5 mm.

Disturbed urban areas, pine and oak savannas and dry grasslands. 0–100 m. WV. NM, TX, Mexico; South Africa, Australia, New Zealand. Exotic.

Nassella tenuissima is an attractive, drought-tolerant bunchgrass with narrow leaves and feathery inflorescences. It has very long, wavy lemma awns. It is frequently used in ornamental plantings and occasionally escapes into disturbed habitats.

PANICUM Panic grass, witchgrass

Plants annual or perennial, cespitose. **Leaves** cauline and basal, not forming an overwintering basal rosette; ligules membranous with cilia; blades flat or folded, glabrous to spreading pubescent. **Inflorescences** terminal and often axillary panicles; disarticulation below the glumes, sometimes at the base of the upper florets. **Spikelets** dorsiventrally compressed, sometimes subterete or laterally compressed, 1–8 mm, awnless. **Glumes** unequal, ours glabrous to scabrous; lower glumes absent to almost as long as the spikelets, 1–9-veined; upper glumes approximately as long as the spikelets to much longer, 3–13(15)-veined, membranous. **Lower florets** sterile or staminate, the lower lemma similar to the upper glume. **Upper florets** bisexual, the upper lemma hard and leathery, usually shiny, glabrous, usually smooth, margins involute and usually clasping the palea. **Upper paleas** striate, rarely transversely rugose. **Anthers** 3. **Caryopses** smooth.

Tropical and warm-temperate regions. Approximately 500 species; 4 species treated in *Flora*.

Panicum species are easily confused with *Dichanthelium*, hence the inclusion of the latter genus in the key below. *Panicum* species are warm-season grasses that use C4 photosynthesis, while members of the genus *Dichanthelium*, formerly placed in *Panicum*, are cool-season grasses that use C3 photosynthesis.

Key to *Panicum* species:

1. Pedicels less than 2 mm; panicle branches 1-sided . . . . ***P. rigidulum*** ssp. ***rigidulum***

1' At least some pedicels 2–20 mm; panicle branches not 1-sided.

   2. Plants perennial; leaves of 2 forms, the basal ones smaller and usually wider than the cauline; rosette well developed, usually persistent . . . . **DICHANTHELIUM**

   2' Plants annual or short-lived perennial; leaves not forming a basal rosette.

      3. Spikelets 4–6 mm long . . . . ***P. miliaceum***

      3' Spikelets 2–3.5 mm long.

         4. Foliage with long, straight, spreading, papillose-based hairs; stem nodes short-hairy; ultimate inflorescence branches widely spreading . . . . ***P. capillare*** ssp. ***capillare***

         4' Foliage glabrous or nearly so, hairs, if present, not papillose-based; stem nodes glabrous; ultimate inflorescence branches more or less appressed . . . . ***P. dichotomiflorum*** ssp. ***dichotomiflorum***

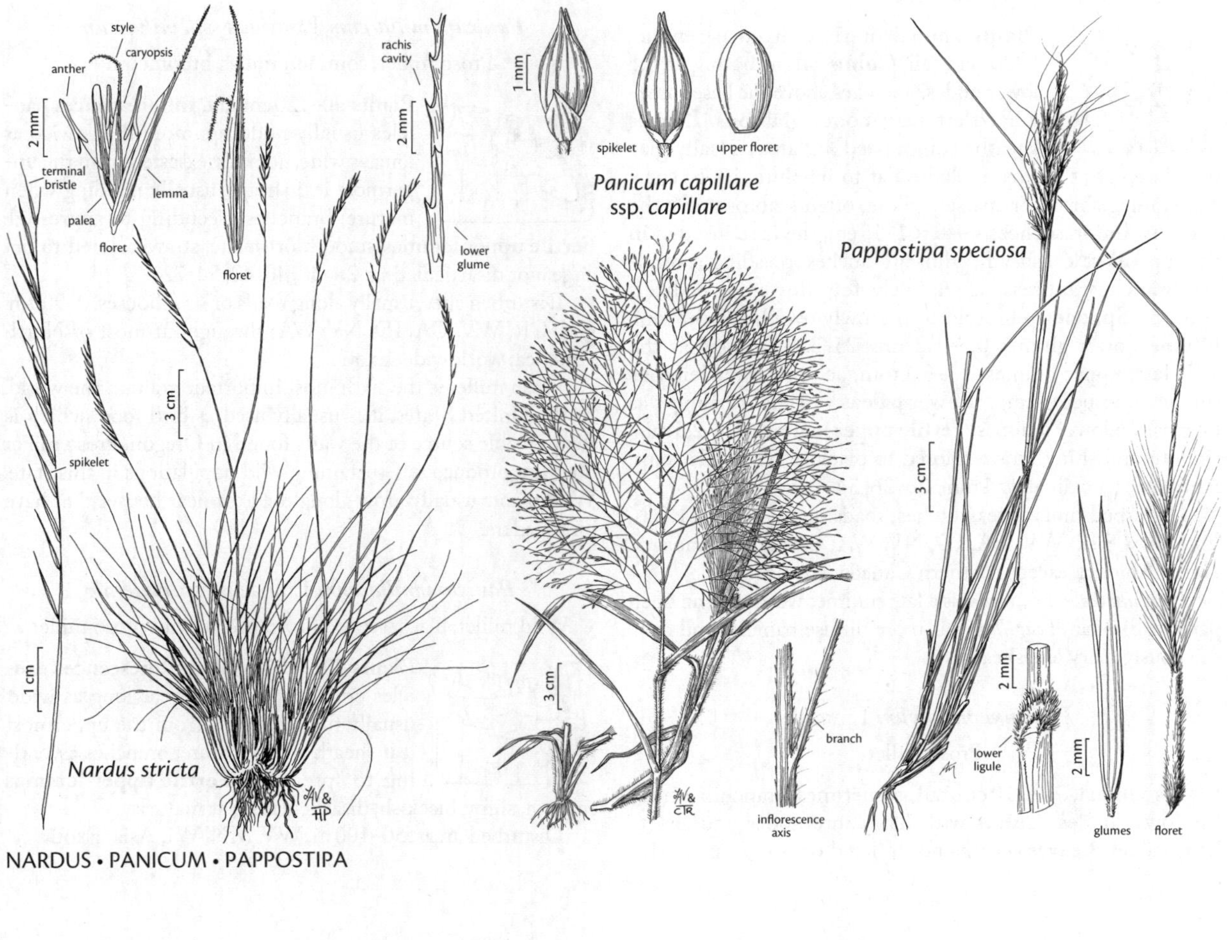

NARDUS • PANICUM • PAPPOSTIPA

Panicum capillare Linnaeus ssp. *capillare* (p. 439)

Witchgrass

Panicum barbipulvinatum Nash, *Panicum capillare* L. var. *capillare*

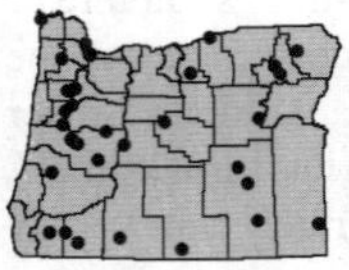

Plants annual, 15–130 cm tall. **Culms** often branching above the base; nodes hairy. **Leaves** sheaths round in cross section, with dense, spreading, papillose-based hairs; blades flat or folded, 5–40 cm × 3–18 mm, hairy. **Inflorescences** diffuse, 13–50 cm, usually more than half as long as the plants, 7–24 cm wide, the base included in the uppermost leaf sheath or exserted, disarticulating at the base at maturity and becoming a tumbleweed; branches spreading; pedicels 0.5–2.8 mm. **Spikelets** elliptical to lanceoloid, 1.9–4 mm. **Glumes** lower glumes 33–50% as long as the spikelets, 1–3-veined; upper glume 1.8–3.1 mm, with 7–9 veins, acuminate. **Lower lemmas** approximately as long as the upper glume, longer than the fertile lemma, acuminate. **Lower paleas** absent. **Fertile upper lemmas** 1.5–1.8 mm, shorter than the spikelet, straw-colored or tan, sometimes blackish, shiny, smooth, hard. **Anthers** 0.5–1 mm. ***2n***=18.

Dry to seasonally wet, often disturbed sites. 0–1600 m. Throughout Oregon. CA, ID, NV, WA; throughout most of North America; South America Eurasia. Native.

This common weedy grass blooms in late summer and fall in disturbed sites. It is identified by its widely spreading panicle branches and spreading-hairy leaf sheaths.

Panicum dichotomiflorum Michaux ssp. *dichotomiflorum*

Fall panicum

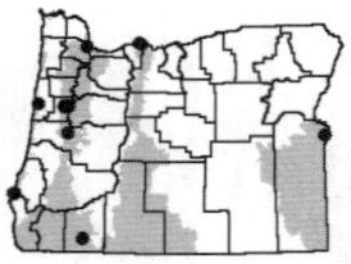

Plants annual or short-lived perennial, 5–200 cm tall. **Culms** often rooting at the lower nodes, branches above the base somewhat flattened; nodes glabrous. **Leaves** sheaths compressed, inflated, usually glabrous except at the base; blades flat to involute, 10–65 cm × 3–25 mm, glabrous or sparsely pilose, often scabrous near the margins. **Inflorescences** 4–40 × 1–15 cm, the base inserted in the uppermost leaf sheath, primary branches spreading, secondary branches appressed, relatively few-flowered; pedicels 1–6 mm. **Spikelets** ellipsoid to narrowly ovoid, 2.3–3.8 mm. **Glumes** lower glumes 0.6–1.2 mm, 25–33% as long as the spikelets; upper glumes 2.5–3.1 mm, acute. **Lower lemmas** similar to upper glumes. **Lower paleas** vestigial to almost as long as the lower lemmas. **Fertile upper lemmas** 1.4–2.5 mm, smooth and shiny, straw-colored to blackish with pale veins. **Anthers** approximately 1 mm. ***2n***=36, 54.

Disturbed moist sites, ditches, roadsides. 0–800 m. ECas, Est, Owy, Sisk, WV. CA, ID, NV, WA; throughout most of North America except northern Canada. Exotic.

Panicum dichotomiflorum is a late summer weed with an open panicle. Similar *P. capillare* has more diffuse panicles and conspicuously hairy leaf sheaths.

Panicum miliaceum Linnaeus

Proso millet

Plants annual, 20–210 cm tall, sometimes branching from the lower nodes. **Culms** with hairy internodes and short-hairy nodes. **Leaves** sheaths not inflated or compressed, with spreading papillose-based hairs; blades 15–40 cm × 7–25 mm. **Inflorescences** dense, usually nodding, 6–20 × 4–11 cm, included in or shortly exserted from the uppermost sheaths at maturity; branches appressed to spreading, spikelets confined to the distal portions; pedicels 1–9 mm. **Spikelets** ovoid, 4–6 mm, usually glabrous. **Glumes** lower glumes 2.8–3.6 mm, 50–75% as long as the spikelets, 5–7-veined; upper glumes 4–5.1 mm, slightly exceeding the upper florets, 11–13(15)-veined. **Lower lemmas** 4–4.8 mm, 9–13-veined, similar to the upper glumes. **Lower paleas** 50% or less the length of the upper florets. **Fertile upper lemmas** 3–3.8 × 2–2.5 mm, leathery or hard, smooth or striate, more or less shiny. **Anthers** 1–2 mm.

2 subspecies.

Panicum miliaceum has larger spikelets than Oregon's other *Panicum* species. It has been cultivated as a grain crop for thousands of years. In North America, it is mainly grown for birdseed.

Key to *Panicum miliaceum* subspecies:

1. Mature upper florets blackish, disarticulating at maturity; culms 70–210 cm tall; panicles erect, more or less exserted at maturity; panicle branches ascending to spreading . . . . ***P. m.*** ssp. ***ruderale***

1' Mature upper florets straw-colored to orange, not disarticulating; culms 20–120 cm tall; panicles usually nodding, not fully exserted; panicle branches ascending to appressed . . . . ***P. m.*** ssp. ***miliaceum***

Panicum miliaceum Linnaeus ssp. *miliaceum*

Proso millet, common millet, broomcorn

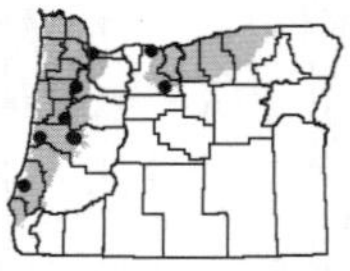

Plants 20–120 cm tall. **Inflorescences** panicles usually nodding, more than twice as long as wide, not fully exserted from the uppermost leaf sheath, usually nodding when mature; branches ascending to appressed. **Fertile upper lemmas** smooth or striate, straw-colored to orange, not disarticulating. ***2n***=36, 40, 42, 54, 72.

Disturbed sites, usually along roads or near houses. 0–900 m. Col, CR, WV. CA, ID, NV, WA; throughout most of North America; worldwide. Exotic.

Proso millet is the sixth most important grain in the world. In the United States, it is usually used as bird food, which is the probable source of the waifs found in Oregon. Proso millet does not produce self-sustaining wild populations in this state. Wild plants usually grow along roads or near houses that have bird feeders.

Panicum miliaceum Linnaeus ssp. *ruderale*

Wild millet, black-seeded proso millet, wild proso millet

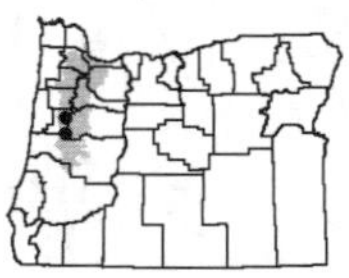

Plants 70–210 cm tall. **Inflorescences** panicles erect, approximately as long as wide, usually fully exserted from the uppermost leaf sheath when mature; branches ascending to spreading. **Fertile upper lemmas** smooth, shiny, blackish, disarticulating at maturity.

Disturbed areas. 50–100 m. WV. MN, WI; Asia. Exotic.

Panicum miliaceum ssp. *ruderale* is similar to cultivated *P. m.* ssp. *miliaceum,* but the panicle is more open, and the mature fertile florets are shiny dark brown to black and readily fall from the inflorescence.

Panicum rigidulum Bosc ex Nees ssp. ***rigidulum***

Redtop panic grass, redtop panicum

Coleataenia rigidula (Bosc ex Nees) LeBlond ssp. *rigidula*

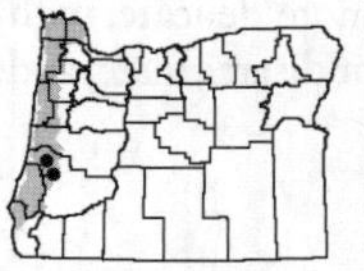

Plants perennial, 40–150 cm tall, cespitose. **Culms** stout, compressed; nodes glabrous. **Leaves** sheaths strongly compressed or keeled, glabrous or sparsely hairy near the top; blades flat or folded, 8–50 cm × 5–12 mm, glabrous or minutely scabrous. **Inflorescences** terminal and axillary, 9–40 × 3–30 cm, 33–75% as wide as long, ultimate branches 1-sided and more or less appressed; pedicels 0.5–1.5 mm. **Spikelets** lanceolate, 1.6–2.5 mm, often purple. **Glumes** lower glumes 67–75% as long as the spikelets, 3-veined; upper glumes 1.6–2.5 mm, rounded to attenuate. **Lower lemmas** like upper glumes. **Lower paleas** up to 67% as long as the lower lemmas. **Fertile upper lemmas** 1.4–2 × 0.6–0.8 mm, shiny, thick, hard, with a tuft of minute hairs at the tip. **Anthers** 0.5–0.75 mm. ***2n***=18.

Disturbed areas, cultivated fields, along rivers in sandy soil or gravel bars. 50–100 m. CR. CA; Great Lakes region to ME, south to Central America. Exotic.

Among our *Panicum* species, *P. rigidulum* is unique because of its one-sided panicle branches, perennial habit, and lanceolate spikelets.

PAPPOSTIPA South American needlegrass

Plants perennial, cespitose. **Leaves** sheaths open; ligules membranous, 0.5–3 mm, glabrous or hairy; blades usually convolute, 0.5–2.5 mm wide, tips narrowly pointed. **Inflorescences** panicles, often partly included in the sheath of the uppermost leaf; disarticulation above the glumes. **Spikelets** 5.5–24 mm, 1 floret. **Glumes** longer than the floret, hyaline, 1–3(5)-veined. **Calluses** acute, 1–4 mm, with antrorsely appressed hairs distally, hairs white. **Lemmas** fusiform, thickly membranous, evenly hairy or glabrous on the distal part of the back; tips not fused into a crown, lobes to 1 mm, awned; lemma awns bent once, 20–70(200) mm, persistent; lower segment 15–33 mm, densely hairy, hairs 4–9 mm; terminal segment 12–45(170) mm. **Paleas** approximately as long as the lemmas, sometimes half as long. **Anthers** 3. **Caryopses** fusiform.

North America and South America. Approximately 14 species; 1 species treated in *Flora.*

Pappostipa is a genus of mostly South American needlegrasses that has been previously included in the genera *Achnatherum, Jarava,* and *Stipa.* The taxonomy of this group is not settled, and the generic name for this taxon may change again in the future.

Pappostipa speciosa (Trinius & Ruprecht) Romaschenko, P.M. Peterson & Soreng (p. 439)

Desert needlegrass

Achnatherum speciosum (Trin. & Rupr.) Barkworth, *Jarava speciosa* (Trin. & Rupr.) Peñail., *Stipa speciosa* Trin. & Rupr.

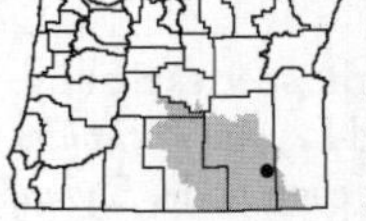

Plants perennial, 30–60 cm tall, cespitose. **Culms** orange-brown at the base. **Leaves** sheaths mostly glabrous, throats densely hairy, basal sheaths reddish brown, becoming flat with age; lower ligules 0.3–1 mm, densely hairy, hairs 0.2–1 mm; upper ligules to 2.5 mm, hyaline to scarious, glabrous or hairy; blades 10–30 cm × 0.5–2 mm when flat but usually rolled or convolute and to 1 mm in diameter; outer surfaces glabrous; inner surfaces with spreading hairs. **Inflorescences** congested, 10–15 cm; branches ascending. **Spikelets** terete or slightly laterally compressed, 16–24 mm. **Glumes** tapering from below mid length, glabrous; tip narrowly acute, awnless; lower glumes 16–24 mm, 1-veined; upper glumes 13–19 mm, 3–5-veined. **Florets** (6)8–10 mm, 0.6–0.8 mm thick. **Calluses** sharp, 0.8–1.6(3) mm, hairy. **Lemmas** densely and evenly hairy, hairs approximately 0.5 mm; lemma awns bent once, 35–45(80) mm, first segment with spreading hairs 3–8 mm, terminal segment glabrous, smooth. **Paleas** 3.2–5.1 mm, approximately 50% as long as the lemma. **Anthers** 1–3 mm long, a tuft of hairs at the tip. ***2n***=66, 68, ±74.

Rocky slopes, in well-drained soils, sagebrush steppe. 1300–1500 m. BR. CA, NV; east to CO, south to Mexico; southern South America. Native.

Pappostipa speciosa is a needlegrass with reddish brown leaf sheaths, densely hairy ligules on the lower leaves, and long hairs on the lower awn segment. The long-awned species of *Achnatherum* have the awn bent twice and lack densely hairy ligules.

PARAPHOLIS Sicklegrass

Plants annual. **Culms** erect to prostrate, sometimes branched above the base. **Leaves** sheaths not keeled; ligules membranous; blades flat to convolute, 0.3–3.5 mm wide. **Inflorescences** terminal and axillary spikes, the spikelets closely appressed or embedded in the inflorescence axis; disarticulation in the rachis below each spikelet. **Spikelets** cylindrical, sessile, with 1 floret. **Glumes** 2, subequal, side by side, curved and asymmetrical, usually longer than the floret and covering the rachis cavity, leathery, stiff, veins conspicuous, margins translucent, tip awnless. **Lemmas** rounded on the back, membranous, translucent, 1–3-veined, awnless. **Anthers** 3. **Caryopses** narrowly ovoid to ellipsoid.

Coasts and salt marshes worldwide. 6 species; 1 species treated in *Flora.*

Parapholis incurva (Linnaeus) C.E. Hubbard
Curved sicklegrass

Pholiurus incurvus (L.) Schinz & Thell.

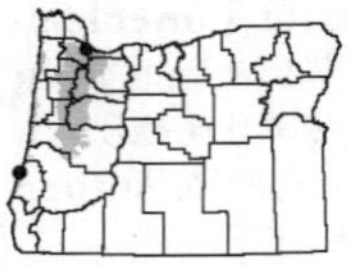

Plants annual, 2–35 cm tall, tufted. **Leaves** upper sheaths strongly inflated, enclosing the lower spikelets; ligules to 1.5 mm; blades 1–3 mm wide, outer surface glabrous, inner surface scabrous. **Inflorescences** spikes, usually curved, 1–15 cm × 1–2 mm, spikelets usually embedded in the inflorescence axis. **Spikelets** 4.5–7.5 mm. **Glumes** 6–8 mm, (3)4–7-veined, longer than and hiding the floret. **Lemmas** 3.5–5.5 mm, 3-veined; tips rounded. **Anthers** 0.5–1.3 mm. ***2n***=32, 36, 38, 42.

Coastal salt marshes, disturbed areas. 0–50 m. Est, WV. CA; south to Mexico; British Columbia, Atlantic coast of North America. Exotic.

Parapholis incurva has spikelets sunken into the inflorescence axis, guarded by a pair of glumes. *Hainardia cylindrica* and certain *Lolium rigidum* individuals have cylindrical inflorescences with sunken spikelets, but they have only one glume per spikelet. *Scribneria bolanderi* is similar but more delicate, with awned lemmas, inflorescence axes that do not disarticulate, and it does not grow in coastal environments.

PASCOPYRUM Wheatgrass

Plants perennial, rhizomatous. **Culms** 20–100 cm. **Leaves** mostly basal; sheaths glabrous, rarely hairy; auricles present; ligules membranous. **Inflorescences** 2-sided spikes usually with 1 spikelet per node, spikelets at lower 4–6 nodes usually sterile, lowest internode to 26 mm, to 2 times as long as the middle internodes; disarticulation above the glumes, beneath the florets. **Spikelets** straight, 1–3 times the length of the internodes, 2–12 florets. **Glumes** usually narrowly lanceolate, tapering from mid-length or below, slightly curving to the side distally, not keeled, 50–67% the length of the spikelets, stiff, 3–5-veined basally, 1-veined distally, tips acuminate. **Lemmas** lanceolate, rounded on the back, acute to awned; lemma awns straight, 0–5 mm, arising from the tip. **Anthers** 3. **Caryopses** 4–5 mm, falling with the lemmas.

North America. 1 species.

Pascopyrum is a monotypic genus that for many years was included in *Agropyron*. It is octoploid and originated from hybridization of *Elymus* and *Leymus* species.

Pascopyrum smithii (Rydberg) Barkworth & D.R. Dewey (p. 443)
Western wheatgrass

Agropyron elmeri Scribn., *Agropyron smithii* Rydb., *Agropyron smithii* Rydb. var. *molle* (Scribn. & J.G. Sm.) M.E. Jones, *Agropyron smithii* Rydb. var. *smithii*, *Elymus smithii* (Rydb.) Gould

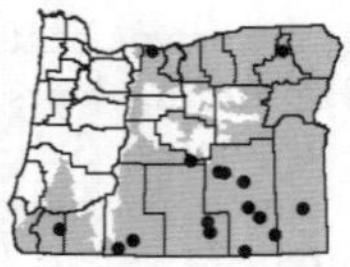

Plants 20–100 cm tall, rhizomatous. **Leaves** sheaths glabrous or sometimes with spreading hairs; auricles 0.2–1 mm, often purple; ligules approximately 0.1 mm; blades flat to involute. 2–26 cm × 1–4.5 mm; lower surfaces glabrous; upper surfaces with prominent veins, scabrous or sparsely hairy. **Inflorescences** spikes, 5–17 cm × 5–12 mm; spikelets 1(2) per node, spikelets at lower 4–6 nodes often sterile; middle internodes 4.5–11 mm, lowest internodes to 26 mm. **Spikelets** 12–26(30) mm, 1–3 times the length of the internodes. **Glumes** 5–15 mm, smooth or scabrous along the keel, midveins tending to curve to one side, the upper glumes usually longer than the lower glumes; lower glumes 0.15–0.8 mm wide at 75% length. **Lemmas** 6–14 mm, 5-veined, leathery, glabrous to moderately pubescent, tips acute, awns 0–5 mm. **Anthers** 2.5–6 mm. ***2n***=56.

Mesic to moist meadows in sagebrush steppe, in clay and sandy, often alkaline soils. 100–1600 m. BR, BW, Col, ECas, Owy, Sisk. CA, ID, NV, WA; north to British Columbia, east to NH, south to Mexico. Native.

Pascopyrum smithii can be confused with the species that hybridized to produce it, *Elymus lanceolatus* and *Leymus triticoides*. *Leymus triticoides* usually has two spikelets at each node. *Elymus lanceolatus* has leaves more evenly distributed along the culm.

PASPALUM Paspalum

Plants annual or perennial, up to 4 m tall, cespitose or rhizomatous. **Culms** solid or hollow, erect, spreading or prostrate. **Leaves** sheaths open; auricles sometimes present; ligules membranous; blades flat or folded. **Inflorescences** terminal or axillary, panicles with 1 to many spike-like branches; branches spreading or erect, flattened, usually winged, usually terminating in a spikelet; disarticulation below the glumes. **Spikelets** dorsiventrally flattened, plano-convex, subsessile to short-stalked, solitary or paired, in 2 rows along one side of the branches, with 2 florets, the lower sterile, the upper bisexual. **Glumes** lower glumes absent, or present on only some spikelets in each branch, veinless or 1-veined, awnless; upper glumes membranous with rounded tips, awnless. **Lemmas** lower lemmas similar to upper glumes; upper lemmas convex, involute, clasping the paleas, straw-colored to dark brown, hard, smooth to slightly rugose. **Caryopses** orbicular to elliptical, plano-convex or flattened, white, yellow, or brown.

Tropical and warm temperate regions worldwide. 300–400 species; 2 species treated in *Flora*.

Key to *Paspalum* species:

1. Inflorescences with (2)4–7 branches; primary branches 4–13 cm; spikelets paired; upper glumes pilose on margins, sometimes on lateral veins, glabrous over the back. . . . . *P. dilatatum*

1' Inflorescences with 2(3) branches; primary branches 1.5–5.5 cm; spikelets usually solitary; upper glumes glabrous on margins, sparsely pubescent over the back, the hairs white, inconspicuous . . . . *P. distichum*

Paspalum dilatatum Poiret (p. 443)

Dallisgrass

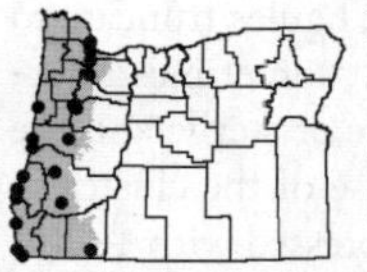

Plants perennial, 50–175 cm tall, loosely cespitose or short rhizomatous. **Culms** bent at the base or decumbent. **Leaves** sheaths glabrous or pubescent; blades flat, to 35 cm × 2–16.5 mm, mostly glabrous, the upper surfaces with a few long hairs near base. **Inflorescences** 1.5–12 cm, branches 2–7, 4–13 cm, spikelets in rows on one side of rachis. **Spikelets** dorsiventrally compressed, paired, 2.3–4 × 1.7–2.5 mm, 2 florets. **Glumes** 1; lower glume absent; upper glumes 2.3–4 mm, 5–7-veined, with long hairs on the margins. **Lemmas** lower lemmas like upper glumes; upper lemmas 2–3 mm, straw-colored, hard, tips acuminate, awnless, glabrous. **Anthers** 1–1.2 mm. **Caryopses** 2–2.3 mm, white to brown. **2*n***=20, 40, 50.

Ditches, seepy road cuts, disturbed moist sites. 0–700 m. CR, Est, Sisk, WV. CA, NV; south to Mexico; tropical and warm-temperate areas worldwide. Exotic.

This is a tall grass with spike-like, one-sided branches scattered well down the culm and paired oval spikelets. It stays green after most grasses in its habitats have turned brown. *Digitaria* species are superficially similar, but the inflorescence branches arise close together.

Paspalum distichum Linnaeus

Knotgrass

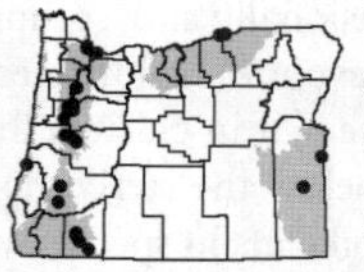

Plants perennial, 5–65 cm tall, strongly rhizomatous. **Culms** erect or ascending; late-season shoots bulbous at the base. **Leaves** sheaths mostly glabrous, sparsely long-pubescent distally, those growing under water densely hairy; blades flat or folded, to 14 cm × 1.8–11.5 mm, glabrous or sparsely long-pubescent. **Inflorescences** V-shaped panicles, 1.4–7 cm; branches usually 2(3), 1.5–5.5 cm, spikelets in rows on one side of branch. **Spikelets** solitary or rarely paired, 2.4–3.2 × 1.1–1.6 mm, 2 florets. **Glumes** 1 or 2; lower glumes absent or triangular and to 1 mm; upper glumes 2.4–3.2 mm, 3-veined, sparsely short-pubescent on the back. **Lemmas** lower lemmas as long as upper glumes but glabrous; upper lemmas 2.5–3 mm, obscurely 3-veined, hard, straw-colored, shiny, glabrous; tips acute, awnless. **Anthers** 1–1.5 mm. **Caryopses** 1.9–2.1 mm, yellow. **2*n***=20, 30, 40, 48, 60.

Wet ditches, marshes, river banks. 0–1200 m. Col, Est, Owy, Sisk, WV. CA, ID, NV, WA; northwestern and southern US; tropical and warm-temperate areas worldwide. Native.

Paspalum distichum is a strongly rhizomatous grass usually with two short panicle branches. *Cynodon dactylon* also has spike-like inflorescence branches and is equally rhizomatous, but it usually has more inflorescence branches and laterally flattened spikelets with two glumes. *Digitaria* species are loosely cespitose and often have smaller spikelets.

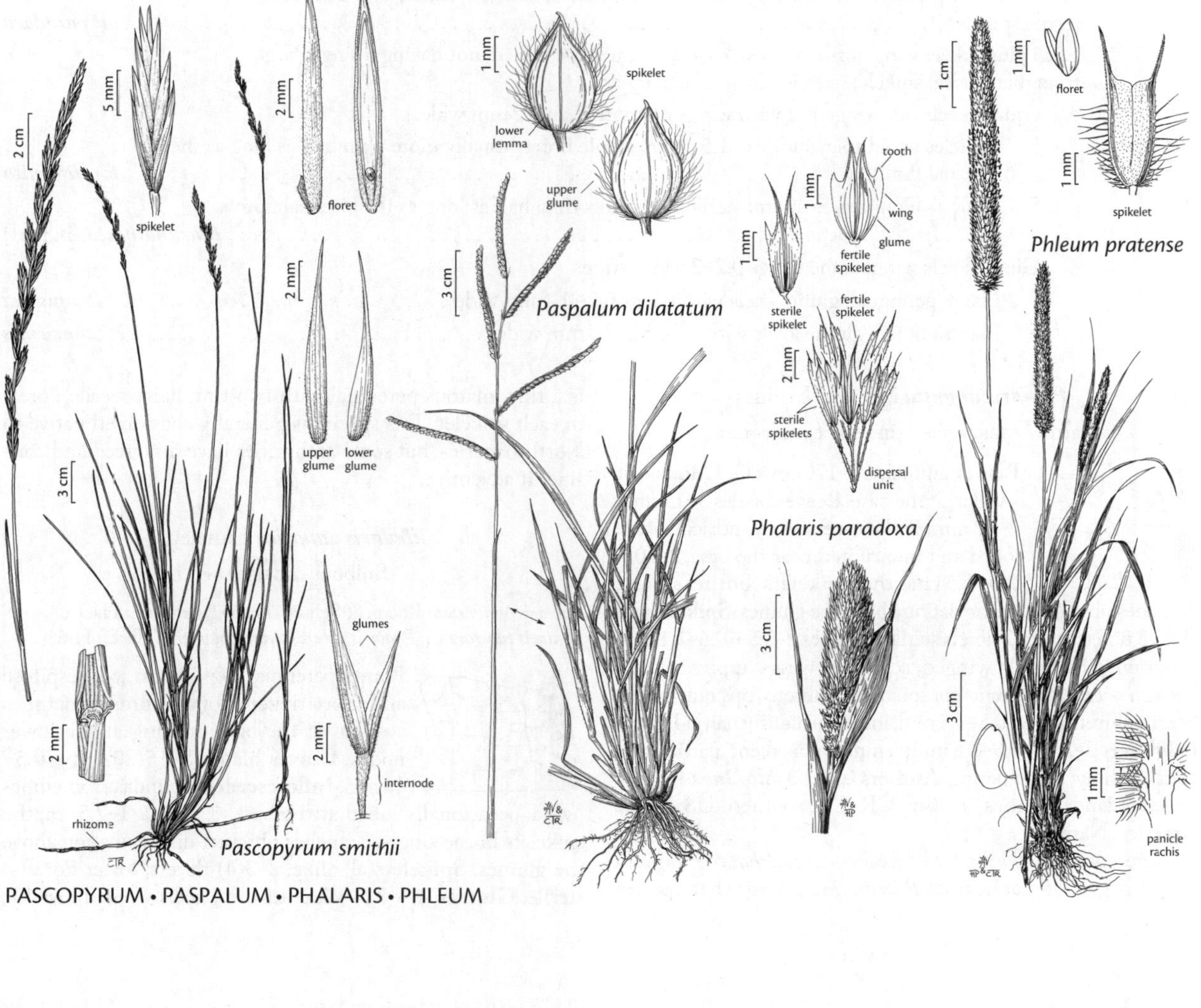

PASCOPYRUM • PASPALUM • PHALARIS • PHLEUM

PHALARIS Canarygrass

Plants annual or perennial, 4–230 cm tall, cespitose or rhizomatous. **Culms** sometimes swollen at the base. **Leaves** more or less evenly distributed; sheaths open except occasionally at base, uppermost sheaths sometimes somewhat inflated; ligules truncate to acuminate, hyaline; blades flat, sometimes revolute. **Inflorescences** dense terminal panicles, ovoid to cylindric, sometimes interrupted, the spikelets borne singly or in clusters, if in clusters then the spikelets of 2 forms, the lower staminate or sterile and the upper bisexual or pistillate; disarticulation below the sterile florets in species with solitary spikelets; at the base of the cluster or sometimes below the bisexual or pistillate spikelets in species with clustered spikelets. **Spikelets** laterally compressed with 1–3(4) florets, the terminal floret sexual, lower florets, if present, sterile. **Glumes** subequal, longer than the florets, 1–5-veined, keeled, the keels often winged. **Florets** sterile florets reduced to tiny knobs, linear, bristle-like structures or lanceolate lemmas to 75% as long as the bisexual florets; terminal florets bisexual or in spikelet clusters sometimes staminate or sterile). **Lemmas** of terminal florets leathery to hard, acute to acuminate or beaked, awnless, shiny, glabrous or hairy, inconspicuously 5-veined. **Paleas** similar in texture to the lemmas, enclosed by the lemmas at maturity, 1-veined. **Anthers** 3.

Temperate regions. 22 species; 6 species treated in *Flora*.

The sterile florets in the spikelet of many *Phalaris* species are often misinterpreted as tufts of hairs, but they are actually small, linear, hairy, vestigial lemmas. Two annual *Phalaris* were introduced to western Oregon but have not been found in more than 60 years. *Phalaris caroliniana* has glume wing margins entire and 0.8–1.5 mm wide, two sterile florets 1.5–2.5 mm, and a somewhat beaked fertile floret. *Phalaris minor* has glume wing margins often irregularly toothed or crenate and 0.3–0.5 mm wide and one sterile floret 0.7–1.8 mm.

Key to *Phalaris* species:

1. Inflorescences branched or at least lobed at flowering time; glume keels not winged or with a wing up to 0.2 mm wide . . . . *P. arundinacea* (in part)

1' Inflorescences smooth cylinders or ovals with obscure branches, or somewhat lobed near the base; glumes not to broadly winged (if inflorescence basally lobed, then glume keels winged).

  2. Inflorescences approximately 10 times as long as wide; glume keel wings approximately 0.4 mm wide . . . . ***P. angusta***

  2' Inflorescences rarely more than 6 times as long as wide; glume keel wings 0–0.6 mm wide.

    3. Each glume keel wing with 1 large, triangular tooth; spikelets in clusters and of 2 obviously different kinds, each cluster with 4–7 narrow staminate or sterile spikelets and 1 broader fertile spikelet . . . . ***P. paradoxa***

    3' Each glume keel wing entire to more or less irregularly toothed, not having a single large triangular tooth; spikelets not in clusters, all similar.

      4. Glume keels not winged or with wings no more than 0.2 mm wide.

        5. Panicles ovoid to cylindrical, 1.5–6 cm; sterile florets usually more than half as long as the bisexual florets . . . . ***P. californica***

        5' Panicles elongate, 5–40 cm; sterile florets less than half as long as the bisexual florets . . . . ***P. arundinacea*** (in part)

      4' Glume keels winged, the wings 0.2–2.5 mm wide.

        6. Plants perennial; glume keels with wings 0.2–0.4 mm wide . . . . ***P. aquatica***

        6' Plants annual; glume keels with wings 2–2.5 mm wide . . . . ***P. canariensis***

Phalaris angusta Nees ex Trinius

Narrow canarygrass, timothy canarygrass

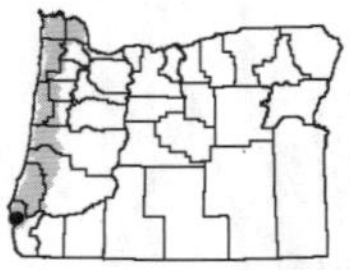

Plants annual, 10–170 cm tall. **Culms** not swollen at the base. **Leaves** blades 3–15 cm × 2–12 mm. **Inflorescences** cylindrical, often lobed and interrupted near the base, 2–20 × 0.6–1.5 cm, the spikelets borne singly; branches obscure; disarticulation above the glumes. **Spikelets** all alike, 3 florets; lower 2 florets sterile. **Glumes** 2–5.5 × 0.6–1.1 mm, often purplish; keels winged, scabrous, wings approximately 0.4 mm wide; lateral veins conspicuous, scabrous, tips mucronate. **Sterile florets** linear, 0.5–1.5 mm, inconspicuously hairy. **Terminal florets** 2–3.8 × 0.9–1.5 mm; lemmas pubescent, particularly distally, shiny; tips tapering. **Anthers** 0.5–1.3 mm. **2*n***=14.

Grasslands, prairies. 0–50 m. CR. CA; southern US; South America. Native?

The somewhat lobed inflorescence of *Phalaris angusta* might suggest an abnormal *P. arundinacea*, but that species is a rhizomatous perennial with two tiny, hairy sterile florets in each spikelet. This species is generally considered native to North America, but some authorities have disagreed and consider it adventive.

Phalaris aquatica Linnaeus

Bulbous canarygrass

Phalaris commutata Roem. & Schult., *Phalaris stenoptera* Hack., *Phalaris tuberosa* L., *Phalaris tuberosa* var. *stenoptera* (Hack.) Hitchc.

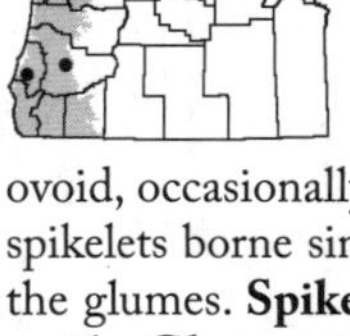

Plants perennial, 60–200 cm tall, cespitose and short-rhizomatous. **Culms** sometimes swollen at the base, rooting at the lower nodes. **Leaves** blades 5–15(20) cm × 0.5–10 mm. **Inflorescences** cylindric, sometimes ovoid, occasionally lobed at the base, 1.5–15 × 1–2.5 cm, the spikelets borne singly; branches obscure; disarticulation above the glumes. **Spikelets** all alike, 2–3(4) florets; lower floret(s) sterile. **Glumes** 4.4–7.5 × 1.2–1.5 mm; keels winged distally;

wings 0.2–0.4 mm wide, usually entire; lateral veins conspicuous, not scabrous; tips acute. **Sterile florets** upper sterile floret 1–3 mm and pubescent, nearly as long as the fertile floret; lower sterile floret, if present, to 0.7 mm. **Terminal florets** 3.1–4.6 × 1.2–1.5 mm; lemmas straw-colored, pubescent, acute. **Anthers** 3–3.6 mm. **2*n*=28.**

Mesic to seasonally wet areas, disturbed areas, roadsides, ditch banks, prairies, often in clay soils. 0–400 m. CR, Sisk, WV. CA; MT; nearly worldwide. Exotic.

Phalaris aquatica is a tall grass with a dense, oval inflorescence. It is the only Oregon *Phalaris* that does not consistently have two sterile florets in each spikelet. It often grows near *P. arundinacea*, which is strongly rhizomatous, has a lobed inflorescence with longer, more spreading branches, and usually grows in wetter microsites. *Phalaris aquatica* hybridizes with both *P. minor* and *P. arundinacea*. These hybrids have been cultivated for forage.

Phalaris arundinacea Linnaeus

Reed canarygrass

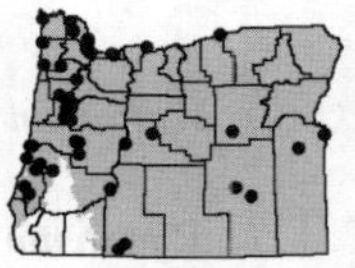

Plants perennial, 40–245 cm tall, rhizomatous. **Culms** not swollen at the base. **Leaves** blades 10–30 cm × 5–20 mm. **Inflorescences** elongate, branched and lobed in shape when mature at least near the base, 5–40 × 1–4 cm, the spikelets borne singly; branches appressed but spreading during anthesis, to 5(8) cm; disarticulation above the glumes. **Spikelets** all alike, 3 florets; lower 2 florets sterile. **Glumes** 4–8.1 × 0.8–1 mm; keels smoothly curved, usually scabrous, not winged or only narrowly winged distally; wings to 0.2 mm wide; lateral veins inconspicuous; tips acute. **Sterile florets** 1.5–2 mm, pubescent. **Terminal florets** 2.5–4.2 mm; lemmas dull yellow when immature, shiny gray-brown to brown at maturity, becoming glabrous on the lower portion, pubescent in the upper portion and on the margins; tips acute. **Anthers** 2.5–3 mm. **2*n*=28.**

Disturbed wet areas, marshes, shores, wet prairies, meadows, clearcuts. 0–1800 m. All ecoregions except Sisk. CA, ID, NV, WA; throughout North America; circumboreal. Native and exotic.

Phalaris arundinacea is an invasive grass that forms dense monocultures, excluding many native species. Exotic populations are present in Oregon, and native populations may exist as well. Eurasian forms have been planted extensively for forage and erosion control. It seems likely that the most aggressive populations are introduced, though it is difficult to be sure in the absence of morphological traits distinguishing the two. A usually sterile, variegated form with white-striped leaves has been called *P. arundinacea* var. *picta*. It is grown ornamentally and occasionally escapes near gardens or where yard waste is dumped.

Phalaris californica Hooker & Arnott

California canarygrass

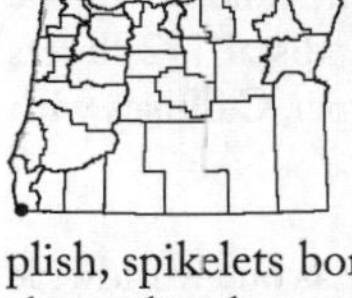

Plants perennial, 60–160 cm tall, rhizomatous. **Culms** swollen at the base. **Leaves** blades 5–35(40) cm × 3–12(18) mm. **Inflorescences** ovoid to cylindrical, often truncate at the base, 1.5–6 × 1–3 cm, often purplish, spikelets borne singly; branches obscure; disarticulation above the glumes. **Spikelets** all alike, 3 florets; lower 2 florets sterile. **Glumes** 5–8 × 0.9–1.6 mm; keels not winged or only narrowly winged distally, scabrous; lateral veins conspicuous, smooth; tips acute to acuminate. **Sterile florets** 1.8–3.5 mm, usually more than half as long as the bisexual florets, densely pubescent. **Terminal florets** 3.5–5 mm; lemmas straw-colored, becoming darker at maturity, sparsely pubescent, shiny; tips acute to weakly acuminate. **Anthers** 3–3.5 mm. **2*n*=28.**

Coastal ravines and moist open areas. 0–50 m. Est. CA. Native.

Phalaris californica has a dense, unlobed, ovoid to cylindrical inflorescence and narrow, wingless glumes. The sterile florets are relatively large, usually more than half as long as the terminal floret.

Phalaris canariensis Linnaeus

Annual canarygrass

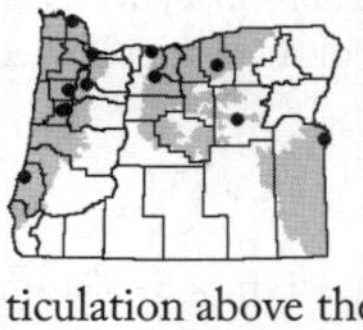

Plants annual, 30–100 cm tall. **Leaves** blades 3–25 cm × 2–10 mm. **Inflorescences** ovoid to oblong-ovoid, not lobed, truncate at the base, 1.5–5 × 1.5–2 cm, spikelets borne singly; branches not evident; disarticulation above the glumes. **Spikelets** all alike, 3 florets; lower 2 florets sterile. **Glumes** 7–10 × 2–2.5 mm, smooth, glabrous, sometimes sparsely pilose between the veins; keels winged, wings to 0.6 mm wide, widening toward the tip; lateral veins inconspicuous, smooth; tips rounded, short-pointed. **Sterile florets** lanceolate, 2–4.5 mm, 33% or more the length of the bisexual floret, acute, sparsely pubescent. **Terminal florets** 4.5–6.8 mm; lemmas ovate, straw-colored to gray-brown, densely pubescent, shiny. **Anthers** 2–4 mm. **2*n*=12.**

Disturbed areas, often around bird feeders. 0–900 m. Col, CR, Est, Lava, Owy, WV. CA, ID, WA; north to Yukon, scattered east throughout the US and southern Canada; nearly worldwide. Exotic.

Phalaris canariensis has larger spikelets than Oregon's other *Phalaris* species, with distinctive glumes so widely winged that they are almost semicircular near the tips.

Phalaris paradoxa Linnaeus (p. 443)

Hooded canarygrass

Phalaris paradoxa L. var. *praemorsa* (Lamb.) Coss. & Durieu

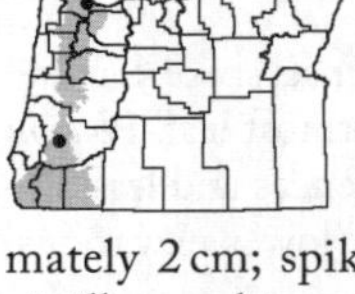

Plants annual, 20–100 cm tall. **Culms** not swollen at the base. **Leaves** blades 5–10(15) cm × 2–5 mm. **Inflorescences** obovoid to clavate, tapering at the base, rounded to truncate at the top, 3–9 × approximately 2 cm; spikelets in clusters consisting of 1 terminal pistillate or bisexual spikelet surrounded by 5–6 staminate (or rarely sterile) spikelets; branches obscure, pedicels with stiff spreading hairs; disarticulation beneath the spikelet cluster. **Spikelets** of 2 types, some staminate or sterile, others bisexual or pistillate; florets 3; lower 2 florets sterile. **Glumes** of staminate or sterile spikelets usually narrowly winged and to 9 mm, widest near the tip, those of the spikelets at the base of the panicle reduced to knobs of tissue terminating the pedicels; glumes of pistillate or bisexual spikelets 5–8 × approximately 1 mm; keels winged, wings 0.2–0.4 mm wide and forming a single prominent tooth; tips unwinged, acuminate to awned, awns approximately 0.5 mm. **Sterile florets** knob-like projections, 0.2–0.4 mm on the callus of the fertile floret, often with 1–2 hairs. **Terminal florets** of all spikelets 2.5–3.5 mm; lemmas hard, shiny, glabrous or with a few short hairs near the tip. **Anthers** 1.5–2.5 mm. **2*n***=14.

Disturbed areas. 0–200 m. Sisk, WV. CA, ID, WA; sparsely scattered in North America; Mediterranean and worldwide around harbors and ballast dumps. Exotic.

Phalaris paradoxa has a strong superficial resemblance to *P. aquatica,* but the spikelets are borne in clusters and each glume has a single large tooth. The plants superficially resemble *Phleum* or *Alopecurus.*

PHLEUM Timothy

Plants annual or perennial, cespitose, sometimes rhizomatous or stoloniferous. **Leaves** sheaths open; ligules membranous; blades usually flat. **Inflorescences** dense, cylindrical, spike-like panicles with more than 1 spikelet associated with each node; disarticulation above the glumes or later in the season beneath the glumes. **Spikelets** strongly laterally compressed, 1 floret. **Glumes** 2, equal, longer than the floret, stiff, strongly keeled, keels ciliate, 3-veined, tips truncate to tapered, often with stiff, awn-like tips. **Calluses** blunt, glabrous. **Lemmas** white, often translucent, 5–7-veined; tips acute, entire, awnless or with a weak awn arising below the apex. **Paleas** subequal to the lemmas, 2-veined. **Anthers** 3. **Caryopses** elongate-ovoid.

Cosmopolitan. Approximately 15 species; 2 species treated in *Flora.*

Phleum species are often confused with *Alopecurus,* which has softer inflorescences and unawned glumes. *Phleum arenarium,* an annual with glumes tapering to minute points, was collected on ballast in Portland 1907–1910. It is apparently not persisting in Oregon.

Key to *Phleum* species:

1. Culm bases not bulbous; inflorescences 1–6 cm × 5–12 mm; glume awns usually more than 2 mm; upper leaf sheaths more or less inflated . . . . ***P. alpinum***

1' Culm bases bulbous; inflorescences (3)5–10(16) cm × 5–7.5(10) mm; glume awns less than 2 mm; upper leaf sheaths not inflated . . . . ***P. pratense***

Phleum alpinum Linnaeus

Alpine timothy

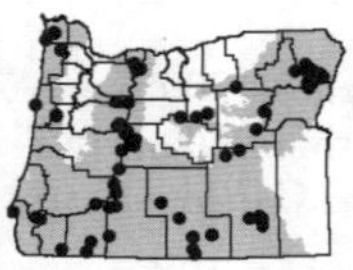

Plants perennial, 15–50 cm tall, cespitose or short-rhizomatous. **Culms** with lower internodes not enlarged or bulbous. **Leaves** sheaths of uppermost culm leaves inflated; blades flat, to 17 cm × 4–7 mm. **Inflorescences** ovoid, lanceolate, or oblong, 1.5–3 times as long as wide, 1–6 cm × 5–12 mm. **Glumes** 2.5–4.5 mm, 3-veined, sides scabrous, keels with spreading hairs, tips awned, awns 1.5–2.5(3.2) mm. **Lemmas** 1.7–2.5 mm, 5–7-veined, mostly glabrous, keels with spreading hairs; tips acute, awnless. **Anthers** 1–2 mm. **2*n***=14, 28.

Moist mountain meadows, streamsides, coastal headlands. 50–2800 m. BR, BW, Casc, CR, ECas, Est, Sisk. CA, ID, NV, WA; western and northern North America; South America, circumboreal. Native.

Phleum alpinum is a small grass with a dense, spike-like inflorescence and inflated sheaths on the uppermost leaf. *Phleum pratense* is a taller grass with longer inflorescences and less distinctly inflated sheaths. It usually grows at lower elevations, though there is an overlap in range.

Phleum pratense Linnaeus (p. 443)

Timothy

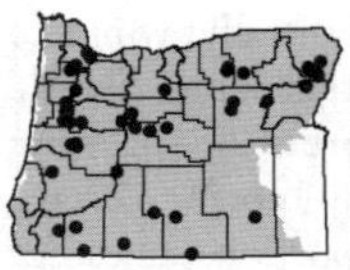

Plants perennial, (20)50–150 cm tall, loosely to densely cespitose. **Culms** with lower internodes often enlarged or bulbous. **Leaves** sheaths of uppermost culm leaves not inflated; blades flat, to 45 cm × 4–8(10) mm, glabrous. **Inflorescences** cylindric, 5–20 times as long as wide, (3)5–10(16) cm × 5–7.5(10) mm. **Glumes** 3–4 mm, 3-veined, sides usually puberulent, keels with spreading hairs; tips awned, awns 1–2 mm. **Lemmas** 1.2–2 mm with 5–7 veins, membranous, usually puberulent; tips acute, awnless. **Anthers** 1.6–2.3 mm. **2*n***=14, 28, 42.

Old fields, pastures, rangelands, and disturbed sites. 0–2300 m. BR, BW, Casc, Col, CR, ECas, Lava, Sisk, WV. CA, ID, NV, WA; throughout North America; temperate regions worldwide. Exotic.

Phleum pratense is a common grass with a dense, spike-like inflorescence. It is similar to *Alopecurus pratensis,* a plant of generally wetter habitats, which has a softer inflorescence with tapered, awnless glumes. In *Alopecurus,* the spikelets disarticulate from the inflorescence axis when mature; those of *P. pratense* are persistent on the plant well into the fall.

PHRAGMITES Reed

Plants perennial, 100–400 cm tall, rhizomatous or stoloniferous, often forming dense stands. **Leaves** sheaths open; ligules membranous, ciliate; blades flat or folded. **Inflorescences** terminal, plumose panicles; disarticulation above the glumes and below the florets. **Spikelets** weakly laterally compressed, 2–8 florets, lower 1–2 florets staminate, middle florets bisexual, distal 1–2 florets rudimentary. **Glumes** shorter than the florets, 1–3-veined, glabrous, the lower glumes much shorter than the upper. **Calluses** with hairs 6–12 mm. **Lemmas** (1)3–5(7)-veined, glabrous, awnless. **Anthers** 1–3. **Caryopses** rarely maturing.

Cosmopolitan. 1 species.

The genus *Phragmites* is subtly variable, and some botanists divide it into three or four species. Here we treat it as one worldwide species with two subspecies in Oregon, one native and one introduced.

Phragmites australis (Cavanilles) Trinius ex Steudel
Common reed

Plants perennial, 100–400 cm tall, forming dense stands from rhizomes. **Culms** erect or ascending, 0.5–1.5 cm thick. **Leaves** blades flat to folded, 15–40 cm × 20–40 mm, glabrous. **Inflorescences** panicles plumose, ovoid or lanceoloid, 15–35 cm × 8–20 mm. **Spikelets** 8–16 mm, with 2–10 florets; rachilla internodes with long, silky hairs. **Glumes** lower glumes 3–7 mm, upper glumes (4)5–10 mm. **Lemmas** linear, 8–15 mm, margins inrolled; tips long-acuminate, awnless. **Anthers** 1.5–2 mm. **Caryopses** 2–3 mm.
3 subspecies; 2 subspecies in Oregon.

Key to *Phragmites australis* subspecies:

1. Lower culm internodes reddish purple (remove leaf sheath); lower glumes 3–7 mm; membranous portion of ligules of middle leaves (0.2)0.4–0.9 mm. . . . ***P. a.*** ssp. ***americanus***

1' Lower culm internodes yellowish or yellowish brown; lower glumes 2.5–4.2(4.8) mm; membranous portion of ligules of middle leaves 0.1–0.4 mm . . . . ***P. a.*** ssp. ***australis***

Phragmites australis (Cavanilles) Trinius ex Steudel ssp. ***americanus*** Saltonstall, P.M. Peterson & Soreng (p. 448)
American common reed

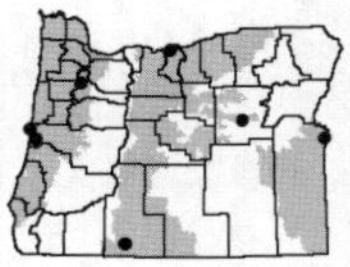

Plants rarely forming monocultures. **Culms** exposed, smooth, shiny; lower internodes reddish purple to chestnut brown. **Leaves** sheaths deciduous with age exposing the culms; ligules on middle leaves with membranous portion (0.2)0.4–0.9 mm; blades usually yellow-green. **Glumes** lower glumes 3–7 mm; upper glumes 5.5–11 mm. **Lemmas** 8–13.5 mm.

600–1400 m. Col, CR, ECas, Est, Lava, Owy, WV. CA, ID, NV, WA; Scattered throughout North America. Native.

This native appears to be the common subspecies of inland marshes, including the extensive wetlands of Klamath County; there are a few coastal records as well. It has been proposed that this subspecies is reproductively isolated from the introduced *P. a.* ssp. *australis* and should be recognized as a species.

Phragmites australis (Cavanilles) Trinius ex Steudel ssp. ***australis*** (p. 448)
Old World common reed

Phragmites communis (L.) Trin., *Phragmites phragmites* (L.) H. Karst.

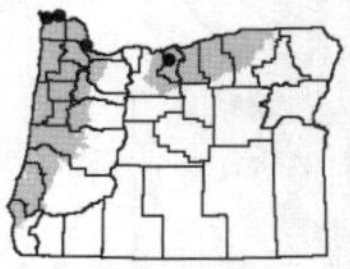

Plants often forming monocultures. **Culms** mostly hidden, ridged, dull; lower internodes yellowish, tan, or yellowish brown. **Leaves** sheaths persistent, not exposing the culm; ligules on middle leaves with membranous portion 0.1–0.4 mm; blades usually dark green or glaucous, sometimes yellow-green near the coast. **Glumes** lower glumes 2.5–4.2(4.8) mm; upper glumes 4.5–7.5 mm. **Lemmas** 7.5–12 mm.

0–1400 m. Col, CR, Est, WV. CA, ID, NV, WA; north to British Columbia; throughout southern Canada and US; Africa, Australia, Eurasia. Exotic.

PIPTATHEROPSIS Ricegrass

Plants perennial, cespitose to rhizomatous. **Leaves** usually mostly basal; sheaths open; ligules membranous to hyaline; blades flat, involute, or folded, tips acute to acuminate but not sharp. **Inflorescences** terminal panicles; disarticulation above the glumes, below the floret. **Spikelets** with 1 floret. **Glumes** shorter than to exceeding the floret, subequal or the lower glume longer, membranous, 1–9-veined, tips obtuse to acute or acuminate. **Calluses** blunt, 0.1–0.6 mm, glabrous or hairy. **Lemmas** dorsally compressed or sometimes terete, smooth, stiffly membranous or leathery, 3–7-veined, margins separated, tips lobed or not, glabrous or with short hairs that do not spread; lemma awns nearly straight to bent once or twice, 1–18(20) mm, deciduous or persistent. **Anthers** 3. **Caryopses** ovoid to obovoid, glabrous.

North America and Eurasia. Approximately 30 species; 1 species treated in *Flora*.

Piptatheropsis exigua (Thurber) Romaschenko, P.M. Peterson & Soreng (p. 448)
Little ricegrass

Piptatherum exiguum (Thurb.) Dorn, *Oryzopsis exigua* Thurb. *Stipa exigua* (Thurb.) Columbus & J.P. Smith

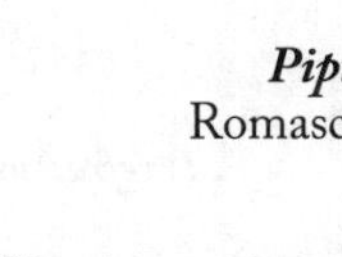

Plants 12–40 cm tall, densely cespitose. **Leaves** sheaths mostly smooth, sometimes minutely scabrous distally; ligules acute, 1.2–3.5 mm; blades usually folded and 0.4–0.8 mm wide, 30–90 cm × 0.6–1.4 mm when flat, both surfaces scabrous. **Inflorescences** panicles 3.5–9 × 1–5 cm; branches straight, appressed, lower branches 1–2 cm. **Spikelets** terete, 3.5–6 mm, 1 floret. **Glumes** 3.5–6 × 1.8–2.2 mm, with 3–5(7) veins; tips acute. **Florets** terete, 3–6 mm. **Calluses** blunt, 0.2–0.5 mm, hairy. **Lemmas** tan or gray-brown at maturity, leathery or stiffly membranous, shiny, hairy especially in the upper half, margins not overlapping at maturity, 3-veined, tips with 2 thick lobes 0.3–0.4 mm, awned; lemma awns twisted, strongly bent once, 3.9–7 mm, persistent, arising from between the lobes. **Paleas** similar to lemmas in length, texture and hairiness. **Anthers** 1.5–3 mm. **Caryopses** approximately 2.5 mm long, 1 mm thick.

Rocky slopes and outcrops in upper montane regions. 1200–2800 m. BW. CA, ID, NV, WA; north to British Columbia, northeast to Alberta. Native.

This densely tufted grass has narrow inflorescences, short, spindle-shaped lemmas, and persistent short awns bent almost 90° when mature.

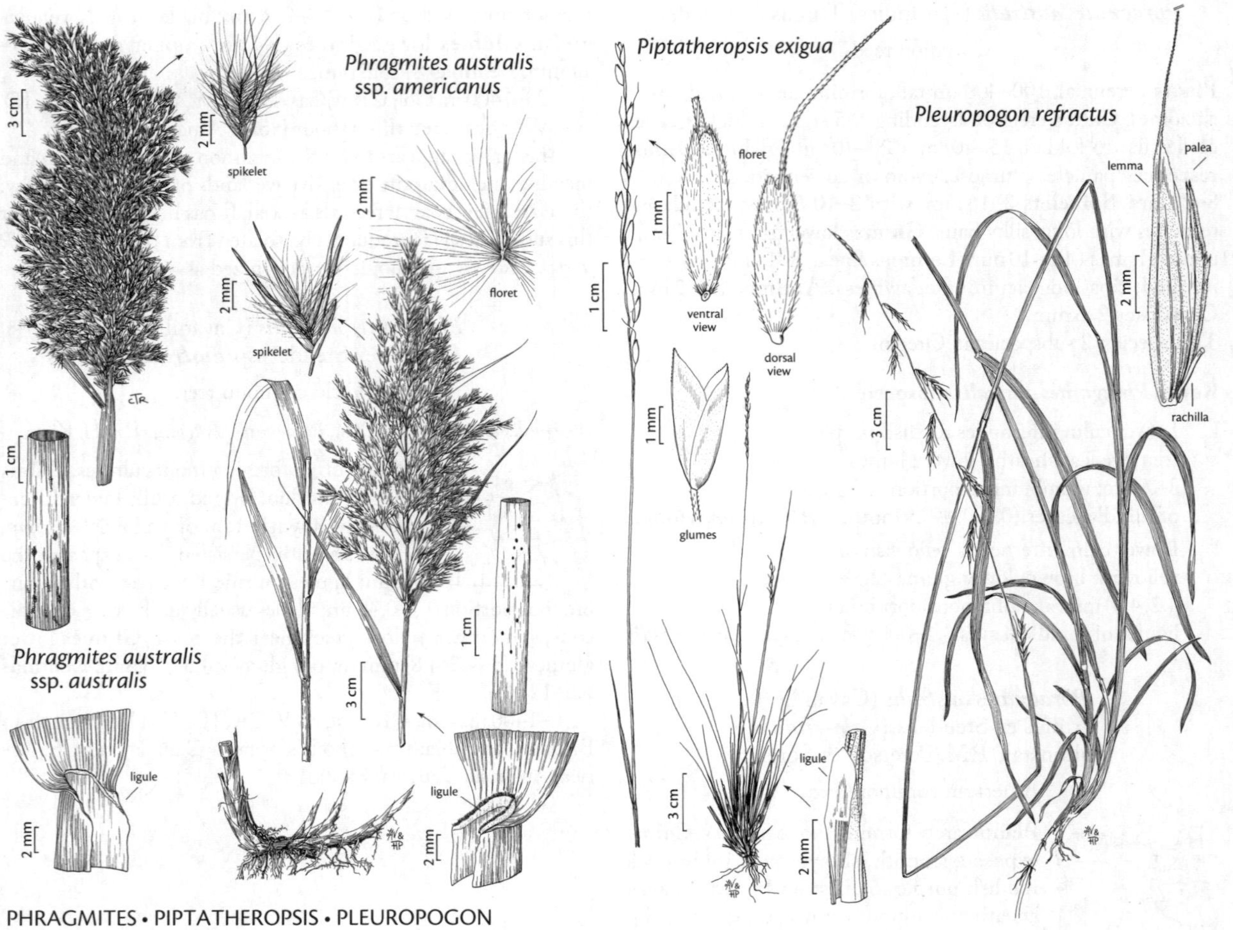

PHRAGMITES • PIPTATHEROPSIS • PLEUROPOGON

PLEUROPOGON Semaphoregrass

Plants perennial or annual, cespitose or rhizomatous. **Leaves** cauline; sheaths closed; ligules membranous; blades flat to folded, tips mucronate to awned. **Inflorescences** terminal racemes; disarticulation above the glumes and beneath the florets. **Spikelets** laterally compressed, with 5–20(30) florets. **Glumes** unequal to subequal, shorter than the adjacent lemmas, lower glumes 1-veined, upper glumes 1–3-veined. **Calluses** rounded, glabrous. **Lemmas** 7(9)-veined, veins parallel, margins and tips scarious, tips entire or emarginate, sometimes awned; lemma awns, if present, straight. **Paleas** with 2 keels, the keels winged and with 1–2 scabrous awns or smooth, flat, triangular appendages. **Anthers** 3.

Circumboreal. 5 species; 2 species treated in *Flora*.

Semaphore grasses are recognized by their long spikelets attached directly to the inflorescence axis. The palea keels are winged and have awns or appendages. The leaf tips may have small points or even small awns.

Key to *Pleuropogon* species:

1. Mature spikelets dangling; lemmas 8–10 mm; lower glume 3–6 mm; upper glume 4–7(8.3) mm; palea keels awnless . . . . ***P. refractus***

1' Mature spikelets ascending to erect; lemmas 4.5–7.5 mm; lower glume 2–3 mm; upper glume 2.5–4.5 mm; palea keels with 2 awns 3–9 mm . . . . ***P. oregonus***

Pleuropogon oregonus Chase

Oregon semaphoregrass

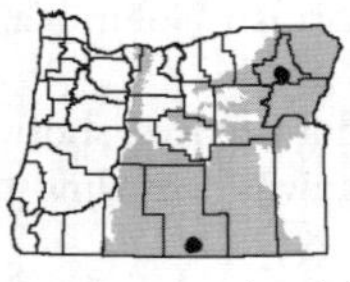

Plants perennial, 40–95 cm tall, rhizomatous. **Leaves** sheaths glabrous or minutely scabrous; blades flat, 5–17 cm × 4–9 mm, glabrous or scabrous over the veins; tips with short points or with awns to 4.5 mm; uppermost culm blades 1.5–7 cm. **Inflorescences** racemes 13–20 cm × 10–33 mm, 6–7 spikelets that are ascending to erect at maturity. **Spikelets** 20–40(50) mm, 7–14 florets, ascending, lower florets bisexual, upper florets pistillate, terminal floret usually sterile. **Glumes** lanceolate to ovate, acute, erose; lower glumes 2–3 mm; upper glumes 2.5–4.5 mm. **Lemmas** 4.5–7 mm, 7-veined; tips truncate; awns 5–12 mm. **Paleas** keels each with an awn 3–9 mm, arising 33–50% above the base. **Anthers** approximately 4 mm. **Caryopses** 2.5–3 mm.

Wet meadows, stream banks. 900–1700 m. BR, BW, ECas. Native.

From the long spikelets to awned palea keels and often awned leaf tips, *Pleuropogon oregonus* is an odd grass. The only similar Oregon species is *P. refractus,* which has dangling spikelets, paleas with triangular appendages, and reduced flag leaves. *Pleuropogon oregonus* is endemic to Oregon and is known from only two populations.

Pleuropogon refractus (A. Gray) Benth. ex Vasey (p. 448)

Nodding semaphoregrass

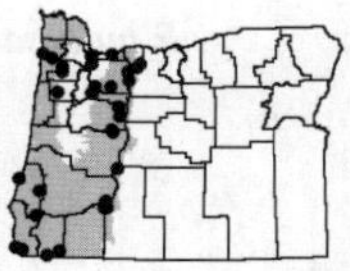

Plants perennial, (85)100–150 cm tall, rhizomatous. **Leaves** sheaths glabrous or pubescent; blades flat, 10–40 cm × 5–14 mm, glabrous; tips acute or acuminate and with a short point, rarely awned to 0.3 mm; uppermost culm blades often absent or reduced, 0–0.6(1) cm and shorter than the associated ligule, sometimes normally developed and up to 7(17) cm. **Inflorescences** racemes (10)20–35 cm × 12–33 mm, 6–14 spikelets reflexed and dangling at maturity. **Spikelets** (20)25–28 mm, 7–14 florets, usually all but the terminal floret bisexual. **Glumes** linear to ovate, often scabrous, apexes obtuse, sometimes notched; lower glumes 3–6 mm; upper glumes 4–7(8.3) mm. **Lemmas** 8–10 mm, 7(9)-veined; tips truncate, sometimes toothed; awns 2–15(21) mm. **Palea** keels with a triangular appendage 0.2–1 mm. **Anthers** 3.5–4 mm. **Caryopses** 3.5–4.5 mm. **2*n***=32, 36.

Wet meadows, streamsides, moist shaded woods. 50–1700 m. Casc, CR, Est, Sisk. CA, WA; north to British Columbia. Native.

Pleuropogon refractus is an uncommon wetland grass with long, distant spikelets that hang down at an angle. The palea keels have triangular appendages, and the uppermost leaf blades are so reduced that they may be shorter than the ligule.

POA Bluegrass

Robert J. Soreng, Barbara L. Wilson, Richard Brainerd, and Nick Otting

Plants annual or perennial; cespitose or culms solitary; with or without rhizomes or stolons; usually with bisexual florets but sometimes monoecious or dioecious; basal branching intra- or extravaginal. **Culms** usually unbranched above the base. **Leaves** sheaths mostly closed to mostly open; ligules membranous, truncate to acuminate; blades flat, folded, or involute, the upper side with 2 longitudinal grooves; tips often prow-shaped. **Inflorescences** panicles, occasionally reduced and raceme-like; disarticulation above the glumes and beneath the florets. **Spikelets** usually laterally compressed, occasionally nearly round in cross section, usually lanceolate, sometimes ovate, 2–17 mm long, with (1)2–6(13) florets, occasionally forming bulblets. **Glumes** 2, usually shorter than the lowest lemma of the spikelets, usually keeled, 1–3(5)-veined, awnless. **Calluses** blunt, glabrous or with a tuft of cobwebby hairs, or with a crown of straight to slightly sinuous hairs. **Lemmas** usually keeled, sometimes rounded over the back, 5(7–11)-veined; margins scarious-hyaline distally; tips truncate or obtuse to acuminate, awnless. **Functional anthers** (1–2)3. **Ovaries** glabrous. **Caryopses** ellipsoidal, 1–4 mm, often shallowly grooved on the ventral side.

Cosmopolitan. Approximately 500 species; 35 species treated in *Flora.*

Poa is the most diverse grass genus in Oregon. Four species have two or more Oregon subspecies, bringing the total to 42 taxa. Bluegrasses are community dominants or important components of dry to moist grasslands, forests, and alpine zones. They are planted for forage as well as for soil stabilization and lawns. *Poa annua, P. bulbosa,* and some other *Poa* species may be economically important weeds in crops. *Poa pratensis, P. compressa,* and *P. bulbosa* are significant invaders of native ecosystems.

The genus *Poa* exhibits great diversity in breeding systems, though each *Poa* species is usually consistent in its method of reproduction. Plants may be synoecious (with bisexual florets), gynomonoecious (with both bisexual and pistillate florets on the same plant), sequentially gynomonoecious (with a higher proportion of pistillate florets as the season progresses), gynodioecious (some plants with only pistillate florets and others with bisexual florets), or dioecious (with separate male and female plants). Self-fertilization is common. Some plants set seed apomictically (asexually). One introduced Oregon species produces bulblets asexually in the inflorescence. Many species reproduce vegetatively through spread of rhizomes or stolons. The prevalence of asexual reproduction and selfing in combination with at least occasional outcrossing produces a complex pattern of variation in *Poa* that can lead to disagreement about species limits. This *Poa* treatment follows that of Soreng (2007). The most widespread and abundant bluegrasses in Oregon are introduced *P. annua, P. bulbosa,* and *P. pratensis* and native *P. secunda.* Other relatively common species include introduced *Poa compressa, P. palustris,* and *P. trivialis* and native *P. cusickii* and *P. wheeleri.* All other species are uncommon or rare in Oregon.

The following identification hints may be helpful when keying within this genus:

Leaf sheath closure is measured on the uppermost leaf of the culm (flag leaf). Ligule length is measured on the upper one to two culm leaves unless otherwise specified. Floret pubescence is evaluated on the lower florets within the spikelet and is checked on several spikelets. If hairs are found in a particular position on any florets, hairs should be considered present at that position for keying purposes, even if other florets lack such hairs. The length of callus hairs is measured when they are stretched out. Anther measurements are based on functional anthers, which produce pollen, are plump before anthesis, and open at and after anthesis.

In general, inflorescence parts of similar *Poa* species overlap in length but have distinctive surface features, necessitating precise terms for hairs. (1) Scabrous: rough with tiny protrusions scarcely longer than wide. (2) Smooth: not scabrous. (3) Glabrous: not hairy (and thus, smooth and glabrous are not synonyms). (4) Hispidulous: rough with very short, firm, stiff hairs. (5) Muriculate: rough with tiny, almost undetectable points.

Terms related to hair length are defined as follows, but these lengths are only approximations. (1) Puberulent: with hairs up to about 0.15 mm long (giving the surface a sparkly appearance at 10× magnification although distinct hairs may be difficult to see.) (2) Short-villous: with hairs up to about 0.3 mm long. (3) Long-villous: with hairs about 0.3–0.4+ mm long.

Callus hairs in *Poa* can follow one of these patterns. (1) Glabrous: no hairs. (2) Cobwebby hairs: long, flexible, woolly, crinkled, or pleated hairs that may be produced in a tuft below the lemma keel (and, in some species, below the marginal veins as well) or diffusely, meaning they are distributed around the lemma base, though somewhat concentrated and longer toward the base of the keel. (3) Crown of hairs: straight to slightly sinuous hairs distributed more or less evenly around the lemma base.

Three patterns of basal branching occur in *Poa*. (1) Extravaginal: new shoots grow sideways from the mother shoot, breaking through sheaths, and the very first leaves consist only of bladeless sheaths. (2) Intravaginal: new shoots grow up within the sheath, diverging only gradually from the mother shoot, and the very first leaves have well-developed blades. (3) Pseudointravaginal: branching is intravaginal, and the very first leaves have rudimentary blades.

Key to *Poa* species:

1. Florets usually producing bulblets rather than seeds; vegetative shoots with bulbous bases. . . . . ***P. bulbosa***
1' Florets producing seeds rather than bulblets; shoots lacking bulbous bases.
 2. Plants annual (rarely biennial).
  3. Calluses glabrous.
   4. Anthers 0.6–1.1 mm; lemmas 2.5–4 mm; lower inflorescence branches usually spreading to reflexed with spikelets relatively uncrowded. . . . . ***P. annua***
   4' Anthers 0.1–0.5(0.6) mm; lemmas 2–2.5 mm; lower inflorescence branches usually ascending with spikelets crowded . . . . ***P. infirma***
  3' Calluses with a tuft of cobwebby hairs on dorsal side (sometimes sparse).
   5. Lemmas without hairs; rachilla internodes usually 1–1.2 mm long or longer . . . . ***P. bolanderi***
   5' Lemmas puberulent; rachilla internodes about 1 mm long, some of them shorter. . . . . ***P. howellii***
 2' Plants perennial.
  6. Lemmas (6)7.5–11 mm long and plants strongly rhizomatous. . . . .***P. macrantha***
  6' Lemmas 2–8 mm long, if more than 7 mm long, then plants cespitose.
   7. Culms and nodes distinctly compressed, wiry, several nodes exposed; plants strongly rhizomatous . . ***P. compressa***
   7' Culms and nodes terete, or indistinctly compressed, wiry or not, zero to several nodes exposed; plants rhizomatous or not.
    8. Lemmas rounded on back at least in the lower half; spikelets subterete; calluses never with cobwebby hairs; sheaths of upper culm leaves closed less than 25% of their length.
     9. Plants rhizomatous . . . .***P. × limosa***
     9' Plants cespitose (sometimes becoming somewhat rhizomatous on unstable slopes).
      10. Panicles more or less dense, branches little spreading; lemmas glabrous or scabrous but not hairy . . . . ***P. secunda*** (in part)
      10' Panicles open at flowering time, the primary and secondary branches spreading; lemmas hairy or long-scabrous on keels, marginal veins and/or base of lemma.
       11. Plants densely cespitose; ligules of culm leaves 0.5–6(10) mm . . . . ***P. secunda*** (in part)
       11' Plants loosely cespitose; ligules 0.3–1 mm. . . . . ***P. × multnomae***
    8' Lemmas keeled on back; spikelets laterally compressed; calluses with or without cobwebby hairs; sheaths of upper culm leaves closed more or less than 25% of their length.
     12. Spikelets broadly ovoid, compact, with rachilla internodes not exceeding 1 mm; lemmas with prominent silky hairs on the keel and marginal veins and sparse silky hairs between the veins; callus glabrous; plants cespitose . . . . ***P. alpina* ssp. *alpina***
     12' Plants without the above combination of characters; spikelets ovoid to lanceoloid, with rachilla internodes usually exceeding 1 mm between at least the upper florets; lemmas variously glabrous or pubescent; callus glabrous or variously hairy; plants cespitose or rhizomatous.
      13. Sheaths of upper culm leaves closed less than 25% of their length; spikelets all perfect flowered (except dioecious *P. pringlei*); plants without rhizomes, cespitose or more or less stoloniferous. . . . . **Subkey 1**
      13' Sheaths of upper culm leaves closed more than 25% of their length; spikelets perfect-flowered or all pistillate; plants rhizomatous or cespitose . . . . **Subkey 2**

Subkey 1: Perennials with sheaths of culm leaves mostly open; plants cespitose or sometimes rooting from lower nodes.

1. Plants with half or more of the shoots vegetative at flowering time; calluses usually with a crown of hairs 0.2–2 mm long on at least some florets, lacking cobwebby hairs, sometimes glabrous.
 2. Inflorescences very densely ovoid with almost smooth outlines; longest branches 0.5–1.5 cm long .... ***P. unilateralis*** ssp. ***unilateralis***
 2' Inflorescences contracted to open, elongate, with somewhat irregular outlines; longest branches (0.5)1–15 cm long.
  3. Glumes broadly hyaline except for the veins (pull glume away from lemma to check), shiny; plants dioecious .... ***P. pringlei***
  3' Glumes hyaline only on the margins, shiny or dull; plants bisexual.
   4. Panicles open at flowering time, the primary and secondary branches spreading.
    5. Upper glumes (3.7)4.1–6.5 mm; lemmas 4–6 mm and hairy on keels and often the intermediate veins but not between them; ligules 2–6 mm; widespread .... ***P. stenantha***
    5' Upper glumes 2.5–3.8 mm; lemmas 3 (3.5)–5 mm and long-scabrous or very short hairy on keels and marginal veins; ligules 0.3–1(1.5) mm; Columbia Gorge .... ***P. × multnomae***
   4' Panicles more or less dense, branches little spreading.
    6. Calluses and/or lemmas scabrous or with hairs, at least on nerves.
    6' Calluses glabrous; lemmas usually glabrous.
     7. Calluses with a crown of hairs 0.1–0.2 mm long; lemmas short-hairy on keels and marginal veins for the proximal 33–50% of their length, intercostal area glabrous or with very short hairs near the base .... ***P. unilateralis*** ssp. ***pachypholis***
      8. Lemmas 2.5–3 mm long and usually glabrous; keels and marginal veins rarely sparsely puberulent near the base, usually glabrous; glumes often equaling or exceeding upper florets; anthers 0.2–0.8 mm .... ***P. lettermanii***
      8' Lemmas 3–5.8 mm long, if less than 3 mm long lemmas short- to long-villous on the keels and marginal veins; glumes rarely equaling the upper florets; anthers (0.6)0.8–1.2(1.7) mm.
     7' Calluses glabrous or with a crown of hairs 0.1–0.5 mm long; lemmas glabrous or the keels and marginal veins scabrous to short-hairy, intercostal regions similar to veins. .... ***P. secunda***
      9. Lemmas 4.1–5.8 mm, glabrous to puberulent, narrowly lanceolate .... ***P. suksdorfii***
      9' Lemmas 3–4.6 mm, short- to long-villous on keels and marginal veins, lanceolate. .... ***P. laxa***

1' Plants with nearly all the shoots of the current season fertile at flowering time (late-season plants may have a post-flowering flush of new shoots for the next year); calluses with cobwebby hairs which may be sparse (glabrous in alpine *P. glauca* ssp. *rupicola*).
 10. Upper 67% of culm lacking nodes; lemmas fairly evenly but sometimes sparsely pubescent across the lower back between the keel and lateral veins; calluses glabrous .... ***P. glauca*** ssp. ***rupicola***
 10' Upper 67% of culm with nodes; lemmas pubescent only on keel and marginal veins; calluses with cobwebby hairs, these well developed or minute.
  11. Ligules 0.2–0.8(1) mm, truncate; cobwebby hairs of callus to 33% as long as lemmas .... ***P. nemoralis***
  11' Ligules 1–6 mm, truncate to acute; cobwebby hairs of callus (50)67+% as long as lemma .... ***P. palustris***

Subkey 2: Perennials with sheaths of culm leaves closed more than 25% of their length; plants rhizomatous or cespitose.

1. Calluses with well-developed cobwebby hairs, sometimes sparse, extending 2+ mm when stretched.
 2. Habitat forests west of Cascades, below 1300 m elevation; inflorescences (6)12–30 cm long, sparsely flowered, nodding or with drooping branches; leaf sheaths closed 25% to most of their length; plants loosely cespitose to rhizomatous; substrate not strongly ultramafic.
  3. Lemmas glabrous; panicle branches strongly ascending .... ***P. marcida***
  3' Lemmas hairy on keels and major veins; panicle branches ascending to broadly spreading. .... ***P. laxiflora***
 2' Differing from the above in some way.
  4. Lemmas 4–5.6(7) mm; substrate strongly ultramafic.
   5. Inflorescences loosely contracted; lemmas smooth to scabrous; leaf blades firm, involute; substrate serpentine .... ***P. piperi***
   5' Inflorescences open; lemmas with silky hairs on keel and marginal veins; leaf blades softer, mostly flat; substrate rocky gabbro or peridotite. .... ***P. rhizomata***
  4' Lemmas 2–4.3(6) mm; substrate not strongly ultramafic.
   6. Sheaths densely scabrous; habitat low and moderate elevations. .... ***P. trivialis***

6' Sheaths smooth or nearly so; habitat low to high elevations.
7. Habitat coastal sands.
8. Lemmas glabrous or sparsely hairy; cobwebby hairs originating from diffuse locations on the calluses; plants gynodioecious . . . . ***P. confinis*** (in part)
8' Lemmas with prominent silky hairs on keel and marginal veins; cobwebby hairs originating from discrete tufts on callus backs and sometimes bases of marginal veins; spikelets with bisexual flowers. . . . . ***P. pratensis*** (in part)
7' Habitat various, not coastal sands.
9. Plants rhizomatous; inflorescences relatively dense, branches generally not reflexed; anthers 1.2–2 mm . . . . ***P. pratensis*** (in part)
9' Plants loosely to densely cespitose, rarely with short, slender rhizomes; branches widely spreading to reflexed at anthesis; anthers 0.2–1.1 mm.
10. Habitat mesic to wet; lemma keels with sparse hairs on the basal 50%; anthers 0.2–0.6(1.1) mm long; panicle branches usually moderately scabrous . . . . ***P. leptocoma***
10' Habitat mesic to dry; lemma keels with dense hairs on the basal 75%; anthers 0.6–1 mm long; panicle branches smooth . . . . ***P. reflexa***

1' Calluses glabrous or infrequently with a tuft of short cobwebby hairs less than 2 mm long when stretched.
11. Upper culm leaf blades absent or vestigial, usually much less than 1 cm long, of the same texture as the relatively firm, involute vegetative shoot blades; lemmas evidently silky-hairy on the keel and marginal veins for about half the length; ligules (1.5)1.8–18 mm. . . . . ***P. fendleriana*** ssp. ***longiligula***
11' Upper culm leaf blades well developed usually exceeding 1 cm long, or if shorter then soft and withering, vegetative shoot blades flat or folded and involute on margins; lemmas usually sparsely pubescent on the keel and marginal veins for less than half the length, sometimes glabrous; ligules 0.2–6 mm long.
12. Plants rhizomatous, producing turf, isolated tufts connected by rhizomes, or solitary shoots.
13. Habitat strictly coastal; leaf blades 0.5–1(1.5) mm wide, involute; lemmas 2.5–4(4.5) mm . . . ***P. confinis*** (in part)
13' Habitat inland, usually montane; leaf blades 2–5 mm wide, sometimes flat; lemmas 2.5–8 mm.
14. Longer glumes 3.5–7.5 mm long; plants entirely pistillate; glumes shiny, broadly hyaline except near the veins (pull glumes away from lemmas to check). . . . . ***P. pringlei***
14' Longer glumes 2.5–3.5 mm long; plants entirely pistillate or bisexual or staminate; glumes not so shiny, hyaline only on the margins, which may be broad.
15. Leaf sheaths, including collars, glabrous and not scabrous, even on the lower leaves (check collars) . . . . ***P. cusickii*** (in part)
15' Leaf sheaths hairy or retrorsely scabrous on or near the collar (at least on the lower leaves).
16. Panicles dense; anthers lacking or aborted; plants densely cespitose
17. Lemmas 5–7 mm long; habitat montane to alpine . . . . ***P. chambersii***
17' Lemmas 2.5–4.5 mm long; habitat wet to moist alkaline meadows, usually in sagebrush steppe . . . . ***P. × limosa***
16' Panicles loosely contracted; anthers usually well developed on at least some plants in the population, occasionally aborted; plants distinctly rhizomatous.
18. Collars hairy, at least near the front margins, collar margin hairs distinctly longer than those elsewhere on the sheaths; plants never all pistillate . . . . ***P. nervosa***
18' Collars glabrous or hairy, collar margin hairs, if present, equaling or shorter than those elsewhere on the sheaths; plants all pistillate (rarely some individuals bisexual). . . . . ***P. wheeleri***
12' Plants without rhizomes, forming dense tufts.
19. Plants in or east of the Cascades or in the Columbia Gorge.
20. Lemmas slightly to densely scabrous to hairy between the veins; panicles open; ligules 0.3–1(2.8) mm . . . . ***P. × multnomae***
20' Lemmas glabrous between the veins, or if hairy or scabrous then panicles dense; panicles open or dense, if open, then ligules (1)2–4 mm.
21. Anthers (0.6)0.8–1.1 mm and well developed; lemmas glabrous between the veins; habitat alpine. . . . . ***P. laxa***
21' Anthers 1.3–3.5 mm and well developed or 0.1–0.2 mm and aborted (rarely longer and aborted); lemmas glabrous, scabrous, or hairy between the veins; habitat various, from moderate elevations to alpine.
22. Panicles open, with (1)6–17(22) spikelets; ligules of both vegetative shoots and cauline leaves hyaline, smooth, (1)2–4 mm. . . . . ***P. leibergii***

22' Panicles dense, with 10–100 spikelets; ligules usually shorter, dorsally somewhat scabrous at least on the lower leaves (except in *P. c.* ssp. *cusickii*), those of vegetative shoots 0.2–0.5(2.5) mm long, those of cauline leaves 1–3(6) mm long . . . . **P. cusickii** (in part)

19' Plants of southwestern Oregon.

23. Plants staminate, functional anthers present . . . . **P. pringlei** (in part)

23' Plants pistillate, functional anthers lacking or presence of functional anthers unknown.

24. Glumes shiny; all lemmas glabrous; calluses glabrous; sheaths of upper culm leaves closed to 33% of their length; panicles with longest branches 0.5–1.5(2) cm long with 1–3 spikelets; plants of rocky subalpine slopes . . . . **P. pringlei** (in part)

24' Glumes not so shiny; lemmas of the lowest florets in each spikelet usually sparsely pubescent at least on the lower marginal veins, upper lemmas usually glabrous; calluses sometimes with sparse, short hairs; sheaths of upper culm leaves closed 50–75% of their length; panicles with longest branches 1–3(4) cm long, with 1–8 spikelets; plants of riparian meadows to alpine slopes . . . . ***P. cusickii*** ssp. ***purpurascens***

Poa alpina Linnaeus ssp. ***alpina*** (p. 454)

Alpine bluegrass

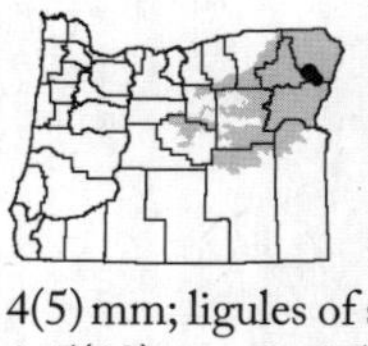

Plants perennial, 10–40 cm tall, not glaucous; densely cespitose. **Basal branching** strictly intravaginal. **Culms** nodes terete. **Leaves** sheaths closed to 33% of their length; ligules of upper cauline leaves to 4(5) mm; ligules of sterile shoots 1–2(3) mm; cauline blades flat, 1–5(12) cm × 2–4.5 mm, smooth; upper blades much reduced in length. **Inflorescences** erect, ovoid to pyramidal, open or loosely contracted at maturity, 2–6(8) cm; branches 1–2 per node and ascending to spreading, straight, 1–3(4) cm long. **Spikelets** broadly ovoid, laterally compressed, 1.5–2.5 times as long as wide, 3.9–6.2 mm; florets 3–7; rachilla internodes 0.5–0.8 mm. **Glumes** broadly lanceolate to narrowly ovate; keels sparsely scabrous; lower glumes (2.3)2.7–3.2 mm, 3-veined; upper glumes (2.6)2.8–3.6 mm, shorter than or subequal to the lowest lemma. **Calluses** glabrous. **Lemmas** broadly lanceolate, 3–5 mm long, distinctly keeled; keels and marginal veins short- to long-villous, area between veins sparsely to moderately short-villous; tips acute. **Anthers** 1.3–2.3 mm.

Subalpine and alpine habitats, often in more or less disturbed areas, usually on limestone. 1500–3000 m. BW. ID, NV, WA; north to AK, east to Newfoundland, southeast to CO and NM; circumboreal. Native.

Poa alpina is a densely cespitose alpine grass. It has relatively wide leaves and compact, ovoid spikelets with broad, ovate glumes. *Poa glauca* and *P. laxa* are similar but have less broadly ovoid spikelets, mostly extravaginal basal branching, and narrower leaves. *Poa laxa* has shorter anthers, and *P. glauca* has longer rachilla internodes. All three species occur in the Wallowa Mountains.

Poa annua Linnaeus (p. 454)

Annual bluegrass

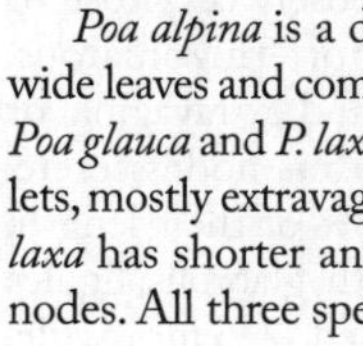

Plants annual, rarely biennial, 2–20(45) cm tall; cespitose. **Basal branching** intravaginal. **Culms** nodes terete or weakly compressed. **Leaves** sheaths closed about 33% of their length; ligules 0.5–3(5) mm; blades flat or weakly folded, 1–10 cm × 1–3(6) mm, smooth. **Inflorescences** 1–7(10) cm; branches ascending to spreading or reflexed. **Spikelets** 3–5 mm long, florets 2–6; rachilla internodes smooth, glabrous, more or less concealed, occasionally exposed; distal rachilla internode less than 50(75)% of the length of distal lemma. **Glumes** lower glumes smooth, 1-veined. **Calluses** glabrous. **Lemmas** lanceolate, 2.5–4 mm, keeled, smooth; keels, marginal veins, and usually intermediate veins crisply puberulent to long-villous, rarely glabrous, area between veins glabrous; tips obtuse to acute. **Anthers** 0.6–1.1 mm, oblong. **2*n***=28.

Disturbed areas, parking lots, lawns, shores. 0–1700 m. BR, BW, Casc, Col, CR, Est, Lava, Sisk, WV. CA, ID, NV, WA; throughout southern Canada and US; worldwide. Exotic.

Poa annua is a small but abundant annual with spreading inflorescence branches and oblong anthers. *Poa infirma* is very similar but has ascending inflorescence branches and shorter, nearly spherical anthers. *Poa supina*, supine bluegrass, has inflorescences similar to *P. annua*, but it is a stoloniferous perennial with longer anthers. It is planted on shady golf courses and lawns where it forms dense, spongy turf. It has not been found in the wild in Oregon but is expected to escape here.

Poa bolanderi Vasey

Bolander's bluegrass

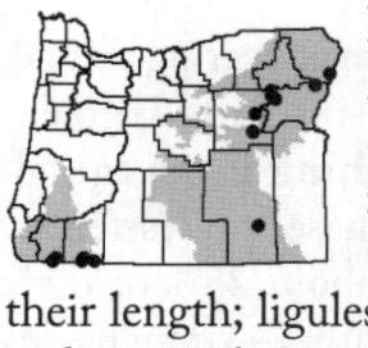

Plants annual, rarely short-lived perennial, 20–60(70) cm tall, often glaucous; densely cespitose, sterile shoots few. **Basal branching** intra- and extravaginal. **Culms** nodes terete. **Leaves** sheaths closed 50–75% of their length; ligules 2.5–7 mm; blades flat (folded), 1.5–5 mm wide, smooth or sparsely scabrous; cauline blades 3–15 cm; uppermost blades 1–4 cm. **Inflorescences** erect or occasionally slightly nodding, usually eventually open, sometimes interrupted, sparse, (5)10–15(25) cm, 25–50% of the plant height; branches initially erect, usually some eventually spreading or reflexed. **Spikelets** (3)4–7 mm, florets 2–3(4); rachilla internodes mostly exceeding 1 mm, 2nd internodes exceeding (1)1.2 mm, smooth or sparsely scabrous, glabrous. **Glumes** unequal, distinctly shorter than the adjacent lemmas; keels smooth or sparsely scabrous; lower glumes 1–3-veined, 50–67% as long as the lowest lemma. **Calluses** with sparse cobwebby hairs on at least some florets. **Lemmas** lanceolate to narrowly lanceolate, 2.5–4 mm long, smooth or scabrous throughout, glabrous; tips narrowly acute, usually purple near the tip. **Anthers** 0.5–1(1.8) mm. **2*n***=28.

Openings in montane conifer forests. 1500–2700 m. BR, BW, Sisk. CA, ID, NV, WA; east to UT. Native.

Poa bolanderi is a tall, often glaucous annual with long, smooth lemmas. Similar *Poa howellii* has hairy lemmas and usually grows at lower elevations.

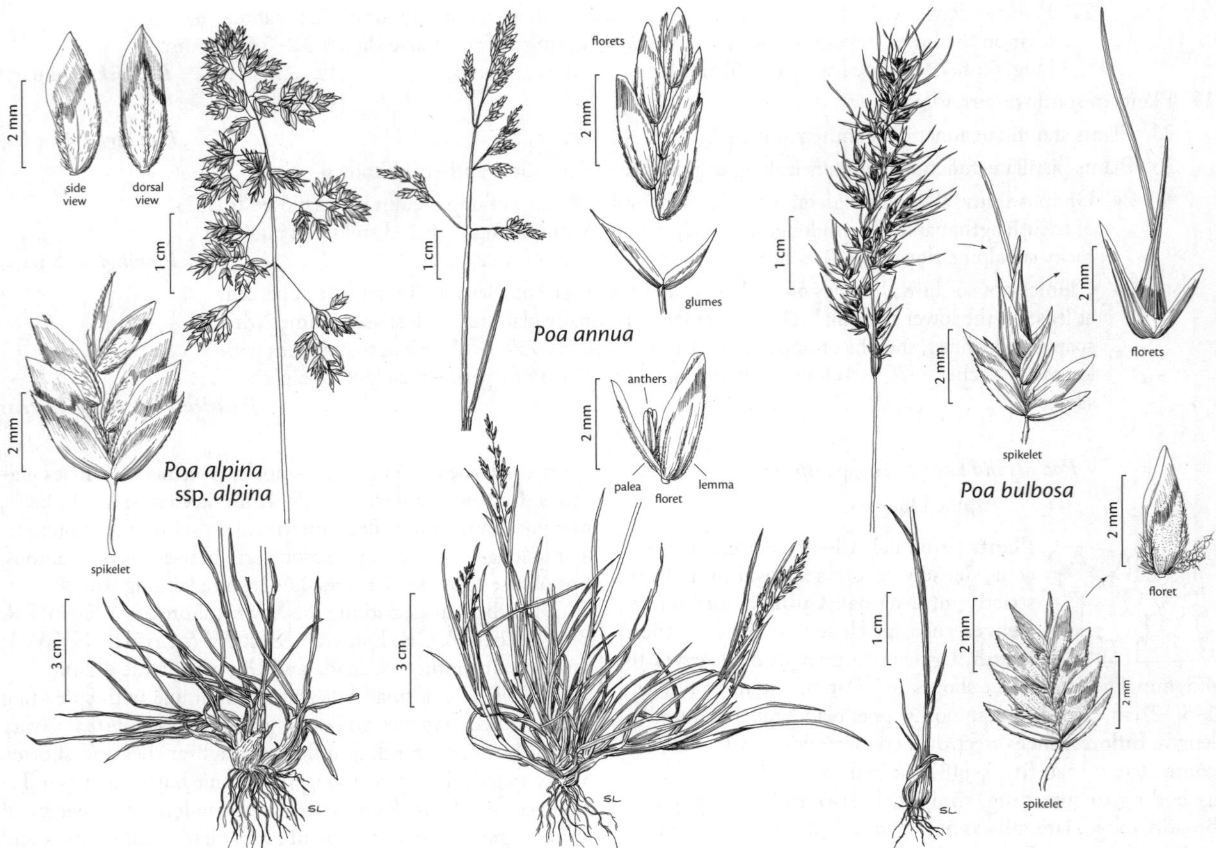

POA

Poa bulbosa Linnaeus (p. 454)

Bulbous bluegrass

Poa bulbosa L. ssp. *vivipara* (Koeler) Arcang.

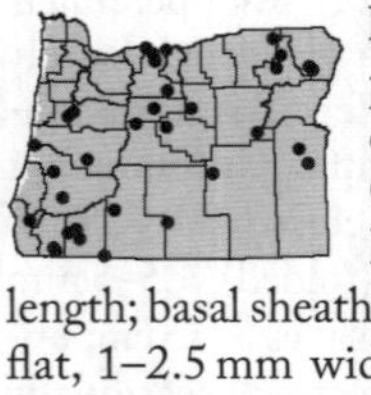

Plants perennial but often appearing annual, producing bulblets, 15–60 cm tall, green; cespitose. **Basal branching** intravaginal. **Culms** bulbous at the base; nodes terete. **Leaves** sheaths closed about 25% of their length; basal sheaths swollen at the base; ligules 1–3 mm; blades flat, 1–2.5 mm wide, soon withering. **Inflorescences** ovoid, 3–12 cm; branches 2–5 per node, ascending to spreading. **Spikelets** 3–5 mm, florets 3–7, commonly all or all but the lowest producing bulblets, occasionally all producing seeds. **Glumes** keels scabrous; lower glumes 3-veined; upper glumes shorter than or subequal to lowest lemma. **Calluses** with cobwebby hairs or glabrous. **Lemmas** lanceolate, 3–4 mm, distinctly keeled, glabrous throughout, or keels and marginal veins short- to long-villous, area between veins glabrous or softly puberulent; tips acute. **Anthers** developed and 1.4–2 mm, or aborted late in development, or not developed. ***2n***=21–42.

Sunny to partly shady disturbed grasslands, shrub steppe, roadsides. 0–2300 m. All ecoregions except Est. CA, ID, NV, WA; throughout US; Mexico; temperate regions worldwide. Exotic.

Poa bulbosa is usually recognized by the bulblets produced in the inflorescence. Occasional non-bulblet-producing plants can be recognized by their bulb-like shoot bases and the presence of bulb-bearing individuals nearby. Some botanists recognize two subspecies, *P. b.* ssp. *bulbosa* producing only normal caryopses and *P. b.* ssp. *vivipara*, producing bulblets. In Oregon, populations either are entirely viviparous or include both viviparous and seed-producing plants.

Poa chambersii Soreng

Chambers' bluegrass

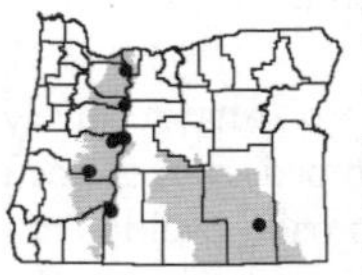

Plants perennial, dioecious or gynodioecious, 10–50 cm long; loosely cespitose or with solitary shoots, short-rhizomatous. **Basal branching** intra- and extravaginal or mostly intravaginal. **Culms** nodes terete, 0–1 exserted. **Leaves** sheaths closed 33–88% of their length, bases of basal sheaths glabrous; collars smooth, glabrous; ligules 0.5–2(2.5) mm; cauline blades flat or folded, 2–5 mm wide, smooth or upper surface sparsely scabrous, mainly over the veins, blades gradually reduced up the culm; uppermost blade 0.7–6 cm. **Inflorescences** erect, lanceoloid to ovoid, tightly to loosely contracted, 2–9 cm; spikelets 15–30, proximal internodes less than 2 cm; branches erect to ascending to slightly spreading, 1–2 per node, 0.9–3.2 cm, with 1–4 spikelets. **Spikelets** to 3 times as long as wide, 6–12 mm, florets 2–7; rachilla internodes 0.8–1.5 mm long. **Glumes** 60–80% as long as adjacent lemmas; lower glumes 3-veined. **Calluses** of at least some lower florets with sparse cobwebby hairs, rarely all florets glabrous, hairs 1–2 mm. **Lemmas** lanceolate, 5–7 mm, distinctly keeled, smooth or sparsely finely scabrous, glabrous throughout or keels and marginal veins sparsely soft-puberulent over the proximal 20%, area between the veins glabrous, 5–7-veined; tips acute. **Anthers** vestigial and 0.1–0.2 mm, aborted late in development, or 1.8–3.7 mm.

Openings in upland forest and snowmelt pockets in alpine areas. 1700–2800 m. BR, Casc. Native.

Although some plants may be pistillate, populations always include individuals with functional anthers. *Poa chambersii* is similar to *P. wheeleri* and *P. cusickii* ssp. *purpurescens*, but the latter have denser inflorescences and only pistillate spikelets.

Poa chambersii populations in the Cascades are dioecious; those on Steens Mountain are gynodioecious and will be described as a new species in the future.

Poa compressa Linnaeus

Canada bluegrass

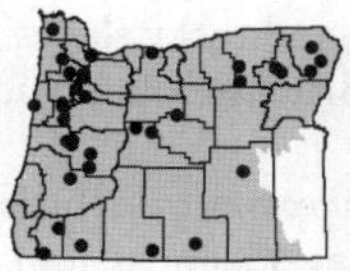

Plants perennial, 15–60 cm tall; loosely cespitose or with solitary shoots, extensively rhizomatous. **Basal branching** intra- and extravaginal or mostly extravaginal. **Culms** wiry, strongly compressed; nodes strongly compressed, some proximal nodes usually exserted. **Leaves** sheaths closed to 20% of their length; ligules 1–3 mm; blades flat, 1.5–4 mm wide; cauline blades subequal. **Inflorescences** erect, lanceoloid to ovoid, often interrupted, sparse to congested, 2–10 cm; spikelets 15–80; branches erect to ascending, or infrequently spreading, 0.5–3 cm, 1–3 per node, with 1–15 spikelets. **Spikelets** (2.3)3.5–7 mm, florets (2)3–7(8). **Glumes** lower glumes 3-veined. **Calluses** with cobwebby hairs, less often glabrous. **Lemmas** lanceolate, 2.3–3.5 mm, distinctly keeled; keels and marginal veins short-villous, area between veins glabrous; tips acute. **Anthers** 1.3–1.8 mm. $2n$=42.

Roadsides, moist or mesic meadows, disturbed areas, pavement cracks. 0–2000 m. All ecoregions except Owy. CA, ID, NV, WA; throughout North America; nearly worldwide. Exotic.

Poa compressa is characterized by its compressed culms and nodes, rhizomatous habit, lower culm nodes exserted from the leaf sheaths, and proportionately small inflorescences with scabrous panicle branches.

Poa confinis Vasey

Coastline bluegrass

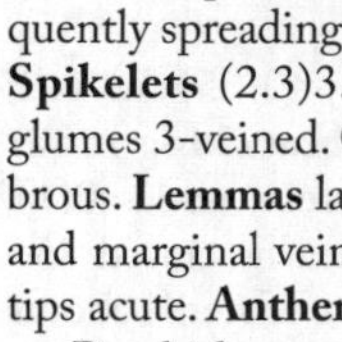

Plants perennial, gynodioecious, 7–30(35) cm tall, densely to loosely tufted; rhizomatous and stoloniferous, rhizomes and stolons to 100 cm long. **Basal branching** mainly intravaginal, some extravaginal. **Culms** slender; nodes terete, 0–1 exserted. **Leaves** sheaths closed 33–50% of their length, smooth, bases of basal sheaths glabrous; collars smooth, glabrous; ligules 0.5–1.5(2.2) mm; blades of tillers with upper surface moderately to densely scabrous or very short-hairy on and between the veins; cauline blades involute, mostly filiform, 0.5–1(1.5) mm wide; lower surfaces smooth; upper surfaces sparsely scabrous on and between the veins; uppermost blade (0.5)1–5 cm long. **Inflorescences** erect, ovoid, moderately to loosely congested, 1–5(7) cm; spikelets fewer than 50; branches erect to ascending, slightly lax, 0.5–3 cm, 1–2 per node. **Spikelets** to 3 times as long as wide, compact, 3–6(8) mm, florets 2–5; rachilla internodes glabrous or sparsely softly puberulent. **Glumes** slightly unequal; keels smooth or scabrous; lower glumes 2–4 mm long, 1–3-veined, about 67% as long as the lowest lemma; upper glumes 2.9–5 mm long. **Calluses** with sparse, diffuse, cobwebby hairs 1–2 mm, infrequently glabrous. **Lemmas** lanceolate, 2.5–4(4.5) mm, distinctly keeled, moderately to densely finely scabrous, glabrous throughout or keels and sometimes marginal veins proximally sparsely softly puberulent; margins narrowly scarious; tips acute. **Anthers** vestigial and 0.1–0.2 mm or functional and 1.5–2 mm.

Coastal sand dunes and stabilized sandy soil, occasionally edges of coastal forests. 0–50 m. Est. CA, WA; north to British Columbia. Native. $2n$=42.

Poa confinis is a rhizomatous grass that often puts up shoots in clusters. It grows in coastal sandy habitats. *Poa macrantha* grows in the same habitat but has much larger lemmas and inflorescences. *Poa pratensis* has long hairs on the lemma keels and marginal veins and cobwebby callus hairs that arise as a tuft on the dorsal surface, sometimes also with two additional tufts by the marginal veins.

Poa cusickii Vasey

Cusick's bluegrass

Plants perennial, gynodioecious, or all pistillate, 10–60(70) cm long, usually densely cespitose with short rhizomes. **Basal branching** intravaginal or intra- and extravaginal. **Culms** 0.5–1.8 mm thick; nodes terete, 0–2 exserted. **Leaves** sheaths closed 25–75% of their length, bases of basal sheaths glabrous; collars smooth or scabrous, glabrous; ligules of cauline leaves 1–3(6) mm; tips truncate to acute; ligules of sterile shoots 0.2–0.5(2.5) mm; tips usually truncate; blades involute to flat, 0.2–3.5 mm wide, all about equal or the upper ones slightly reduced in length; upper surface usually densely scabrous or hispidulous to softly puberulent, infrequently nearly smooth and glabrous; lower surface smooth or scabrous; uppermost blade 0.5–5(6) cm. **Inflorescences** usually erect, tightly or loosely contracted, narrowly lanceoloid to ovoid, 2–10(12) cm; spikelets 10–100; branches erect or steeply ascending, 0.5–4(5) cm long, 1–3(5) per node, with 1–15 spikelets. **Spikelets** broadly lanceolate to narrowly ovate, to 3 times as long as wide, (3)4–10 mm, florets 2–6; rachilla internodes 0.5–1.2 mm long. **Glumes** lanceolate, lower glumes 3-veined, distinctly shorter than the lowest lemma. **Calluses** glabrous or with short, sparse, sometimes diffuse cobwebby hairs less than 25% of the lemma length. **Lemmas** lanceolate to broadly lanceolate, (3)4–7 mm, distinctly keeled, membranous, smooth or sparsely to densely scabrous, glabrous throughout, or keels and/or marginal veins softly puberulent near the base; tips acute. **Anthers** vestigial and 0.1–0.2 mm, aborted late in development, or well developed and 2–3.5 mm.

4 subspecies.

Poa cusickii is a common upland bluegrass with a dense inflorescence. It is usually densely cespitose, though some plants may have short rhizomes. Similar *P. fendleriana* has reduced uppermost flag blades, and lemmas with silky hairs on the keels and marginal veins. *Poa leibergii* has narrower leaves and usually longer ligules. *Poa wheeleri* can be misidentified as *P. cusickii* if the specimen is collected without its distinctive scabrous leaf sheaths. *Poa pringlei*, restricted to southwestern Oregon near the California border, has longer ligules, sheaths closed to about a third their length, and glabrous (to scabrous) lemmas. It is dioecious, whereas *P. cusickii* plants are bisexual or entirely pistillate.

Key to *Poa cusickii* subspecies:

1. Cauline blades usually less than 1.5 mm wide; panicle branches moderately to densely scabrous; all vegetative shoots intravaginal; lemmas totally glabrous; calluses glabrous; functional anthers present or absent.
    2. Panicle branches usually 1.7–4(5) cm long, stout to slender, with 2–15 spikelets; functional anthers sometimes present . . . . *P. c.* ssp. ***cusickii***
    2' Panicle branches 0.5–1.7 cm long, stout, with 2–5 spikelets; functional anthers absent . . . . *P. c.* ssp. ***pallida***

1' Some cauline blades 1.5–3 mm wide; panicle branches smooth to moderately scabrous; some vegetative shoots extravaginal; lemmas glabrous or sparsely pubescent on the keel and marginal veins; calluses sometimes with a few short hairs; functional anthers absent.
    3. Lemmas and calluses glabrous; panicles erect, contracted; panicle branches smooth or very sparsely scabrous . . . . *P. c.* ssp. ***epilis***
    3' Lemmas of lower florets in each spikelet usually sparsely pubescent at least on the lower marginal veins, lemmas of upper florets glabrous; calluses sometimes with sparse, short hairs; panicles somewhat lax and loosely contracted; panicle branches usually sparsely scabrous . . . . . .*P. c.* ssp. ***purpurascens***

Poa cusickii Vasey ssp. ***cusickii*** (p. 459)

Cusick's bluegrass

Poa cusickii Vasey var. *cusickii*

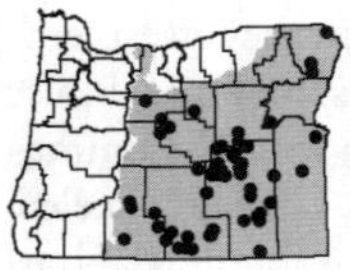

Plants gynodioecious and dioecious, 10–60(70) cm; densely cespitose. **Basal branching** intravaginal. **Culms** mostly erect; nodes 0–1, well exserted. **Leaves** sheaths closed 25–67% of their length; blades of sterile shoots flat, folded, or involute, 0.5–1 mm wide; cauline blades less than 1.5 mm wide; uppermost blades (0.5)1.5–5 cm long, to 33% as long as their sheaths. **Inflorescences** contracted, usually 5–10(12) cm; spikelets 20–100; branches 1.7–4(5) cm, moderately to densely scabrous, with 2–15 spikelets. **Spikelets** 4–10 mm. **Calluses** glabrous. **Lemmas** 4–7 mm, glabrous. **Anthers** vestigial and 0.1–0.2 mm, or functional and 2–3.5 mm. **2*n***=28.

Riparian meadows and swales in upper sagebrush zone to alpine slopes and ridges. 300–3000 m. BR, BW, ECas, Lava, Owy. CA, ID, NV, WA. Native.

This is the most widespread *Poa cusickii* subspecies, occurring from low elevations to the alpine zone. It is the only subspecies that sometimes produces functional anthers. It can be confused with *Poa leibergii,* which has longer ligules on the vegetative shoots and more spikelets in the inflorescence.

Poa cusickii Vasey ssp. ***epilis*** (Scribner) W.A. Weber

Skyline bluegrass

Poa cusickii Vasey var. *epilis* (Scribn.) C.L. Hitchc., *Poa epilis* Scribn.

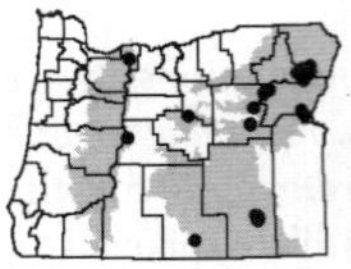

Plants all pistillate, 20–45 cm tall; densely cespitose. **Basal branching** intra- and extravaginal. **Culms** mostly erect; nodes 1–2, well exserted. **Leaves** sheaths closed 33–75% of their length; blades of sterile shoots 0.7–1 mm wide; blades of cauline leaves flat or folded, more than 1.5 mm wide; uppermost blades 1.5–5 cm long, about 40% as long as their sheaths. **Inflorescences** usually contracted, usually 2–7 cm; spikelets 20–70; branches 1–3 cm, smooth to sparsely scabrous, with 1–8 spikelets. **Spikelets** (3)4–8 mm. **Calluses** glabrous. **Lemmas** 3–6 mm, glabrous or scabrous. **Anthers** usually aborted late in development. **2*n***=56.

Subalpine and alpine meadows, thickets, and rocky slopes. 1000–3000 m. BR, BW, Casc. North to British Columbia, southeast to CO. Native.

Poa cusickii ssp. *epilis* is similar to *P. c.* ssp. *purpurascens*, which has some hairs on the lemmas and calluses, longer lemmas, fewer and longer spikelets, and both intra- and extravaginal branching. Both are coarser plants than *P. c.* sspp. *cusickii* and *pallida.*

Poa cusickii Vasey ssp. ***pallida*** Soreng

Pale Cusick's bluegrass

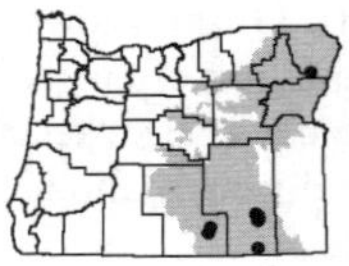

Plants all pistillate, 10–40(55) cm tall; densely cespitose. **Basal branching** intravaginal. **Culms** mostly erect; nodes 0(1), little exserted. **Leaves** sheaths closed 25–67% of their length; blades of sterile shoots 0.5–1 mm wide; cauline blades flat, folded, or involute, usually less than 1.5 mm wide; uppermost blades 0.5–2(3) cm long, 10–30% as long as their sheaths. **Inflorescences** contracted, usually 2–6 cm; spikelets 10–40; branches 0.5–1.7 cm long, moderately to densely scabrous, with 2–5 spikelets. **Spikelets** 4–10 mm. **Calluses** glabrous. **Lemmas** 4–7 mm long, glabrous. **Anthers** vestigial and 0.1–0.2 mm, or 2–3.5 mm. **2*n***=56, 59.

Alpine grasslands and ridges. 1900–3000 m. BR, BW. CA, ID, NV, WA; north to Yukon, southeast to CO. Native.

Poa cusickii ssp. *pallida* is a fine-leaved bunchgrass with a narrow, dense panicle and scabrous branches. In Oregon, its range is limited to alpine habitats in the southeastern part of the state. *Poa cusickii* ssp. *cusickii* has longer panicle branches. Hybridization occurs with *P. fendleriana*, forming *P.* × *nematophylla,* which may have hairy lemmas.

Poa cusickii Vasey ssp. ***purpurascens*** (Vasey) Soreng

Purple Cusick's bluegrass

Poa cusickii Vasey var. *purpurascens* (Vasey) C.L. Hitchc.

Plants all pistillate, 25–50 cm tall, densely to moderately cespitose. **Basal branching** mostly extravaginal. **Culms** decumbent; nodes 1–2, well exserted. **Leaves** sheaths closed 50–75% of their length; blades of sterile shoots mostly 1–2 mm wide; blades of culm leaves flat or folded, more than 1.5 mm wide; uppermost blades 3–6 cm long, 20–60% as long as their sheaths. **Inflorescences** slightly lax, ovate, loosely contracted, usually 4–7 cm; spikelets 13–50; branches 1–3(4) cm long, smooth to moderately scabrous, with 1–8 spikelets. **Spikelets** 7–10 mm. **Calluses** of lower lemmas in each spikelet usually with sparse, short cobwebby hairs, hairs less than 25% of the lemma in length, sometimes glabrous, calluses of upper lemmas glabrous. **Lemmas** 4–7 mm; keels and marginal veins of some lower lemmas in each spikelet usually sparsely softly puberulent near the base, sometimes glabrous; upper lemmas glabrous. **Anthers** usually aborted late in development. **2*n***=28.

Subalpine to alpine scree, meadows, and thickets. 1300–2800 m. Casc, Sisk. CA, WA; north to British Columbia. Native.

Poa cusickii ssp. *purpurascens* grows at high elevations in the Cascades and Siskiyou Mountains. The proximal lemmas of *P. c.* ssp. *purpurascens* usually have short cobwebby hairs, and the distal lemmas have glabrous calluses. This can lead to confusion with *P. rhizomata*, which has well-developed cobwebby hairs and open panicles. The other subspecies of *P. cusickii* have glabrous lemmas.

Poa fendleriana (Steudel) Vasey ssp. ***longiligula*** (Scribner & T.A. Williams) Soreng

Longtongue muttongrass

Poa longiligula Scribn. & T.A. Williams

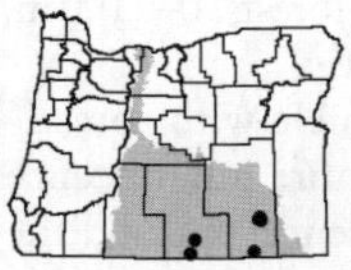

Plants all pistillate, rarely dioecious, perennial, 15–70 cm tall, densely to loosely cespitose, often weakly rhizomatous. **Basal branching** mainly intravaginal, some extravaginal. **Culms** sometimes stout; nodes terete, 0–1 exserted. **Leaves** sheaths closed about 33% of their length, bases of basal sheaths glabrous; collars smooth or scabrous near the throat; ligules of middle cauline leaves (1.5)1.8–18 mm long; blades of tillers usually moderately to densely scabrous or very short-hairy on and between the veins, infrequently nearly smooth and glabrous; cauline blades usually involute, (0.5)1–3(4) mm wide; lower surfaces smooth or infrequently scabrous; upper surfaces usually scabrous or softly puberulent; cauline blades greatly reduced in length higher on the culm; uppermost blades absent or reduced, 0–1 cm, to about 10(20)% as long as their sheaths. **Inflorescences** erect, contracted, narrowly lanceoloid to ovoid, 2–12(30) cm, frequently with more than 100 spikelets; branches erect, 1–8 cm, 1–2 per node, with 3–15(25) spikelets. **Spikelets** broadly lanceolate to ovate, to 3 times long as wide, (3)4–8(12) mm; florets 2–7(13); rachilla internodes usually 0.8–1.3 mm long. **Glumes** lanceolate; lower glumes 1–3-veined, distinctly shorter than the lowest lemma. **Calluses** glabrous. **Lemmas** lanceolate, 3–6 mm, distinctly keeled, long-villous on keels and marginal veins, area between veins glabrous or infrequently softly puberulent; tips acute. **Anthers** vestigial and 0.1–0.2 mm, or 2–3 mm. **2*n***=56.

Rocky slopes in sagebrush steppe, subalpine to alpine grasslands. 700–2600 m. BR, ECas. CA, ID, NV, WA; north to British Columbia, southeast to CO, south to Mexico. Native.

Poa fendleriana ssp. *longiligula* is a cespitose grass of dry grasslands and shrub steppe, with a narrow, dense inflorescence, long, acute lemmas with hairy keels and marginal veins, and long, acute ligules. The blade of the uppermost culm leaf is reduced to an inconspicuous point or is occasionally up to 1 cm long.

Poa glauca Vahl ssp. ***rupicola*** (Nash) W.A. Weber

Timberline bluegrass

Poa rupicola Nash

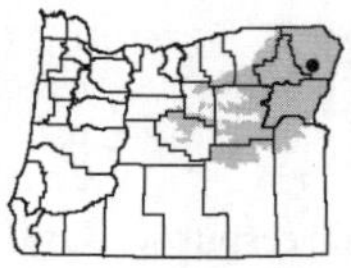

Plants perennial, 5–15 cm tall, usually glaucous; densely cespitose; all or most shoots flowering at anthesis, vegetative shoots appearing late in the growing season. **Basal branching** mostly extravaginal. **Culms** wiry; nodes terete or slightly compressed, the uppermost usually in proximal to 33% of culm. **Leaves** sheaths closed to 20% of their length, bases of basal sheaths glabrous or with sparse minute hairs; ligules 1–4(5) mm; blades flat or folded, 0.8–2.5 mm wide, appressed or abruptly ascending to spreading, straight. **Inflorescences** narrow, generally 20(33)% as broad as long at maturity, rarely raceme-like, erect, usually narrowly lanceoloid to ovoid, contracted or somewhat open, sparse, 1–5 cm; branches erect, ascending, or weakly spreading, short, 2–3(5) per node, glaucous. **Spikelets** 3–7(9) mm, 2–3 times long as wide, usually glaucous, florets 2–5; rachilla internodes up to 1.2 mm long, smooth and glabrous to densely short-hairy. **Glumes** subequal, narrowly to broadly lanceolate; keels smooth or sparsely scabrous; tips acute; lower glumes 3-veined; upper glumes 2–3.8(5.2) mm long, distinctly shorter than to subequal to lowest lemmas. **Calluses** glabrous. **Lemmas** lanceolate to broadly lanceolate, 2.5–4 mm, distinctly keeled; keels and marginal veins short-villous; intermediate veins usually sparsely softly puberulent to short-villous; areas between veins at least sparsely crisply puberulent to softly puberulent; tips obtuse or acute, usually bronze-colored in part. **Anthers** (1)1.2–2.5 mm. **2*n***=42–±100.

Dry, often disturbed, alpine slopes and ridges. 2000–2800 m. BW. CA, ID, NV, WA; western North America. Native.

Poa glauca is a small, glaucous, cespitose grass of dry alpine habitats. Very similar *P. laxa* grows in moister microsites and has lemmas that are glabrous between the veins and sparsely scabrous over the keels and smaller anthers. *Poa nemoralis* is taller and has shorter ligules. *Poa interior,* not confirmed in Oregon, usually has cobwebby callus hairs.

Poa howellii Vasey & Scribner

Howell's bluegrass

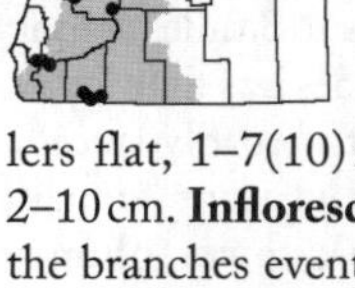

Plants annual, rarely perennial, (10)25–80(120) cm tall; densely cespitose. **Basal branching** intravaginal. **Culm** nodes terete. **Leaves** sheaths closed 50% or more of their length; ligules 1.5–5(10) mm; blades of tillers flat, 1–7(10) mm wide, finely scabrous; cauline blades 2–10 cm. **Inflorescences** erect, eventually open, 10–25(30) cm, the branches eventually spreading or reflexed. **Spikelets** (2)4–6 mm, florets 2–5; rachilla internodes about 1 mm, smooth, softly puberulent or occasionally glabrous. **Glumes** slightly unequal; keels and sometimes lateral veins scabrous; lower glumes 1–3-veined. **Calluses** with sparse cobwebby hairs on at least some florets. **Lemmas** lanceolate to narrowly lanceolate, 2.5–3.5 mm, distinctly keeled, evenly crisply puberulent near the base, finely scabrous distally; margins narrowly hyaline; tips narrowly acute, rarely purple. **Anthers** 0.2–1 mm.

Open, often moist woods and bottomlands, often with moss or boulders. 0–1100 m. Casc, ECas, Sisk, WV. CA, WA; north to British Columbia. Native.

Poa howellii is an erect, green, annual bluegrass with a long inflorescence. Very similar *P. bolanderi* has glabrous to scabrous lemmas and longer rachilla internodes and usually grows at higher elevations.

Poa infirma Kunth

Weak bluegrass

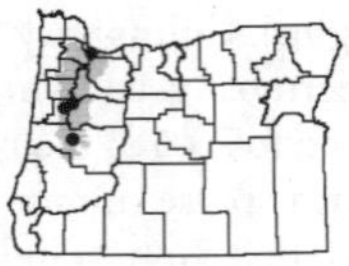

Plants annual, 2–15 cm tall; cespitose. **Basal branching** intravaginal. **Culms** nodes terete. **Leaves** sheaths closed about 33% of their length; ligules 0.5–3 mm; blades flat, 1–3(4) mm wide, smooth. **Inflorescences** 1–6 cm; branches ascending; spikelets crowded. **Spikelets** 3–5 mm, florets 2–6; rachilla internodes smooth, glabrous, usually exposed in side view, distal internode 50–75% the length of the distal lemma. **Glumes** smooth; lower glumes 1-veined. **Calluses** glabrous. **Lemmas** lanceolate, 2–2.5 mm, distinctly keeled, smooth throughout; keels, marginal and intermediate veins crisply puberulent to long-villous, area between veins glabrous; intermediate veins prominent; tips obtuse to acute. **Anthers** 0.1–0.5(0.6) mm; spherical to short-elliptical. ***2n***=14.

Disturbed areas, gravel parking areas, railroads. 50–1100 m. WV. CA; British Columbia, SC; South America; Europe. Exotic.

Poa infirma is a small annual bluegrass very similar to *P. annua* but with ascending inflorescence branches and smaller, nearly spherical anthers. It has been collected rarely in Oregon.

Poa laxa Haenke

Lax bluegrass

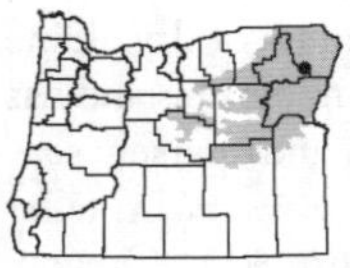

Plants perennial, 8–35 cm tall, not or only slightly glaucous; densely cespitose. **Basal branching** mixed, mainly extravaginal, or mainly pseudointravaginal, sometimes also intravaginal. **Culms** slender, 0.5–0.9 mm thick; nodes terete. **Leaves** sheaths closed to 33% of their length, bases of basal sheaths glabrous; collars smooth or scabrous, glabrous; ligules 2–4 mm; blades of tillers similar to cauline blades; cauline blades flat, 1–2(3) mm wide, thin, soft, smooth, blades not much smaller higher on the culm. **Inflorescences** usually loosely contracted, sparse, or rarely contracted and dense, 2–8 cm; branches ascending, or weakly spreading, infrequently erect, fairly straight or flexuous, slender, 1–3(4) cm, 1–3(5) per node, with 1–8 spikelets. **Spikelets** to 3 times as long as wide, 4–6 mm, florets 2–5; rachilla internodes less than 1 mm long, smooth, glabrous. **Glumes** lanceolate to broadly lanceolate, nearly equaling or slightly longer than adjacent lemmas, thin; keels smooth or sparsely scabrous; lower glumes 1–3-veined; upper glumes shorter than or subequal to lowest lemma. **Calluses** with cobwebby hairs usually less than 25% as long as the lemma, sparse, or glabrous. **Lemmas** lanceolate to broadly lanceolate, 3–4.6 mm, distinctly keeled, thin; keels and marginal veins short- to long-villous; intermediate veins glabrous to sparsely softly puberulent, area between veins glabrous; tips acute. **Anthers** (0.6)0.8–1.1(1.3) mm.

Mesic alpine habitats. 2900–3100 m. BW. Widely scattered in North America; Europe. Native.

Poa laxa is a tiny, densely tufted, alpine bluegrass. Similar *P. glauca* has lemmas that are hairy between the veins, villous lemma keels, and larger anthers. North American *P. laxa* are divided into two subspecies, *P. l.* ssp. *banffiana* and *P. l.* ssp. *fernaldiana*. Oregon plants do not fit into either one easily.

Poa laxiflora Buckley (p. 459)

Looseflower bluegrass

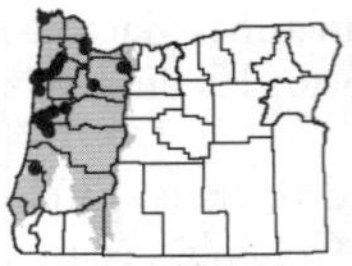

Plants perennial, 50–120 cm; loosely cespitose or with solitary shoots, long-rhizomatous. **Basal branching** extravaginal. **Culms** retrorsely scabrous. **Leaves** sheaths closed 25–50% of their length, retrorsely scabrous; ligules 2–3.5 mm; blades flat, 3–8 mm wide. **Inflorescences** open, sparse, 14–30 cm; branches widely spreading, fairly straight, (5.5)8–12(15) cm, 1–3(4) per node, antrorsely scabrous, with 3–13 spikelets. **Spikelets** 4–8 mm, florets 2–4; rachilla internodes about 1 mm long. **Glumes** keels scabrous; lower glumes lanceolate, 3-veined but margins tending to roll up, often concealing the marginal veins and appearing 1-veined; upper glumes shorter than or subequal to the lowest lemma. **Calluses** with cobwebby hairs. **Lemmas** lanceolate, 3.2–6 mm, distinctly keeled, smooth or sparsely finely scabrous; keels long-villous to 67–75% of their length; marginal vein hairs sparse; intermediate veins usually glabrous, rarely sparsely puberulent, glabrous between the veins; tips acute. **Anthers** 0.5–1.1 mm.

Moist, shady, riparian terraces in conifer forests. 0–1100 m. Casc, CR, Est, WV. WA; north to AK. Native.

Poa laxiflora has an extremely open panicle with spreading, nodding branches. In *P. marcida,* the entire inflorescence nods, but the branches parallel the inflorescence axis, and the lemmas taper more gradually. *Poa nervosa* has shorter panicles, usually glabrous calluses, and longer anthers. *Poa leptocoma*, usually found at higher, more eastern sites, has one-veined lower glumes and a more erect inflorescence.

Poa leibergii Scribner (p. 459)

Leiberg's bluegrass

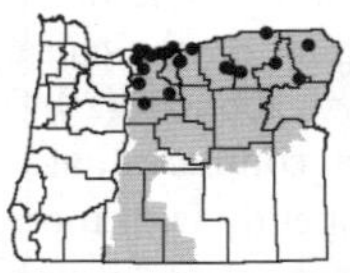

Plants perennial, gynodioecious, 5–35 cm tall; densely cespitose. **Basal branching** intravaginal. **Culms** 0.5–0.7 mm thick; nodes 0–1 exserted. **Leaves** sheaths closed 40–60% of their length, bases of basal sheaths glabrous; collars smooth, glabrous; ligules (1)2–4 mm long, hyaline, smooth; sterile shoot ligules like those of the culm leaves; blades of tillers with lower surfaces smooth or sparsely scabrous; cauline blades flat, folded or involute, 0.5–1(1.5) mm wide, usually soon withering, smooth or sparsely scabrous. **Inflorescences** erect to lax, lanceoloid to ovoid or pyramidal, contracted to open, 1–5(8) cm; spikelets (1)6–17(22); branches erect to spreading, slender, 1–4 cm, 1–2 per node, with 1–2(3) spikelets. **Spikelets** broadly lanceolate to broadly ovate, to 3 times as long as wide, 4–8 mm; florets 2–3; rachilla internodes glabrous. **Glumes** thin, somewhat shiny; lower glumes 3-veined, distinctly shorter than the lowest lemma. **Calluses** glabrous. **Lemmas** lanceolate, 3.5–7 mm, distinctly keeled, thinly membranous, smooth or scabrous, glabrous; tips acute to truncate and erose. **Anthers** vestigial and 0.1–0.2 mm, or 1.3–3 mm.

Wet rock ledges, vernal pools, edges of wet meadows, in sagebrush steppe to alpine habitats. 50–2900 m. BW, Col, ECas, Lava. ID, NV, WA. Native.

Poa leibergii is a cespitose upland species, lacking cobwebby callus hairs. It is most likely to be confused with *P. cusickii* ssp. *cusickii,* which has shorter ligules on the leaves of vegetative shoots and more condensed panicles with more spikelets.

Poa leptocoma Trinius

Bog bluegrass

Poa leptocoma Trin. ssp. *leptocoma*, *Poa leptocoma* Trin. var. *leptocoma*

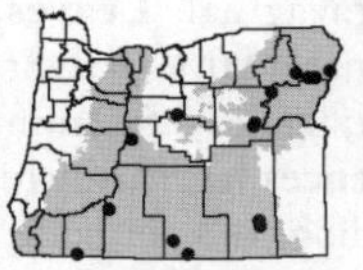

Plants perennial, 15–100 cm, often partially purplish, more or less cespitose. **Basal branching** mostly extravaginal. **Leaves** sheaths closed 25–67% of their length, smooth or sparsely scabrous; ligules 1.5–4(6) mm; blades flat, 1–4 mm wide. **Inflorescences** lax, open, sparse, 5–15 cm; branches spreading to reflexed, capillary, (2)3–8 cm, 1–3(5) per node, usually moderately scabrous, with (3)4–15 spikelets. **Spikelets** lanceolate or narrowly lanceolate, 4–8 mm, green, or partly purple to dark purple, florets 2–5. **Glumes** tapered from the base or lanceolate, thin; keels usually scabrous; lower glumes 1-veined; upper glumes distinctly shorter than to nearly equaling lowest lemma. **Calluses** with sparse cobwebby hairs. **Lemmas** lanceolate, 3–4 mm, often partly purple, distinctly keeled, thin, smooth, or sparsely scabrous near the tip; keels softly puberulent to long-villous on 25–67% the length; marginal veins softly puberulent to long-villous, hairs frequently sparse; lateral veins and area between veins glabrous; margins infolded; tips sharply acute to acuminate, usually bronze-colored. **Anthers** 0.2–1.1 mm. ***2n***=42.

Montane to alpine streamsides, lakeshores, and wet meadows. 1100–2800 m. BR, BW, Casc, ECas, Sisk. CA, ID, NV, WA; western Canada and US; Asia. Native.

Poa leptocoma is a loosely cespitose grass with short anthers and somewhat open panicles. It is most similar to *P. reflexa*, a plant of drier habitats that has less scabrous panicle branches, longer anthers, and sparsely hairy lateral lemma veins, at least on one side of the lemma.

Poa lettermanii Vasey

Letterman's bluegrass

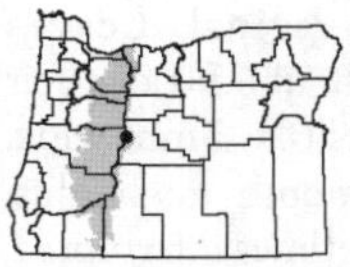

Plants perennial, 1–12 cm tall; densely cespitose. **Basal branching** intra- and extravaginal or mainly intravaginal. **Leaves** sheaths closed to 25% of their length; ligules 1–3 mm; blades flat or folded, or slightly inrolled, 0.5–2 mm wide. **Inflorescences** erect, contracted, 1–3 cm long, usually exserted from the sheaths; branches erect to steeply ascending, slender, to 1.5 cm. **Spikelets** 3–4 mm, green or purple, florets 2–3; rachilla internodes less than 1 mm long, smooth. **Glumes** lanceolate to broadly lanceolate, 2.4–3.6(4) mm, usually equaling or exceeding the lowest lemma, frequently exceeding the upper florets; lower glumes 3-veined. **Calluses** glabrous. **Lemmas** lanceolate, 2.5–3 mm long, distinctly keeled, thin, glabrous; keels and marginal veins rarely sparsely puberulent; tips acute. **Anthers** 0.2–0.8 mm. ***2n***=14.

Rocky alpine ridges and ledges. 2900–3100 m. Casc. CA, ID, NV, WA; north to British Columbia, east to CO. Native.

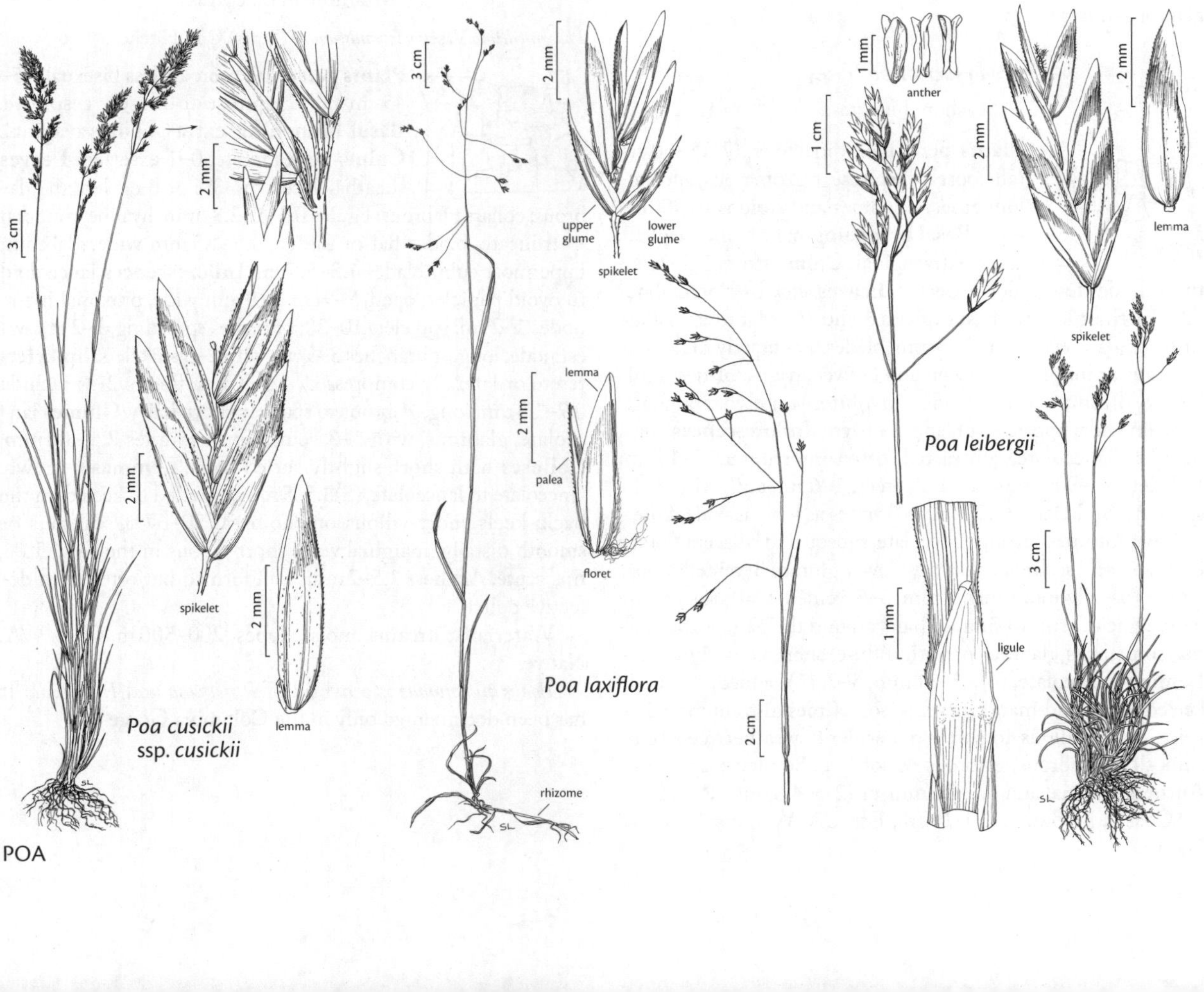

POA

This is a small, cespitose, alpine bluegrass, most similar to *P. suksdorfii*. Both have subequal glumes that tend to be longer than the lowest lemma, but *P. suksdorfii* has larger spikelets with longer glumes. In Oregon, *P. lettermanii* is known only from high elevation on South Sister in the central Cascades.

Poa* × *limosa Scribner

Hybrid bluegrass

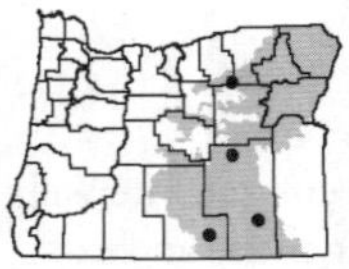

Plants perennial, sterile, pistillate, or bisexual, 20–80 cm tall; rhizomatous to loosely cespitose. **Basal branching** intra- and extravaginal or partly extravaginal. **Leaves** sheaths closed usually about 15% of their length; ligules 1–4 mm long; blades of tillers 0.5–2 mm wide; cauline blades flat, folded, 0.5–5 mm wide, smooth or scabrous. **Inflorescences** erect, usually contracted, sometimes interrupted, 5–15 cm; branches erect, less than 4 cm. **Spikelets** somewhat laterally compressed, 4–7 mm long; florets 2–5; rachilla internodes smooth. **Glumes** 2.5–3.5 mm long; lower glumes 3-veined. **Calluses** glabrous or with cobwebby hairs to 25% as long as the lemma. **Lemmas** narrowly lanceolate, 2.5–4.5 mm, distinctly to weakly keeled, glabrous throughout or keels and marginal veins sparsely long-villous; tips acute. **Anthers** aborted late in development or 1.3–2.2 mm.

Wet alkaline meadows in sagebrush steppe. 1300–2200 m. BR, BW. CA, NV; British Columbia, Alberta, Saskatchewan. Native and exotic.

Poa × *limosa* is a hybrid of *P. pratensis* and *P. secunda* ssp. *juncifolia*. It is more or less rhizomatous like *P. pratensis*, but the spikelets are more nearly terete, and the lemma veins are glabrous or less hairy.

Poa macrantha Vasey

Seashore bluegrass

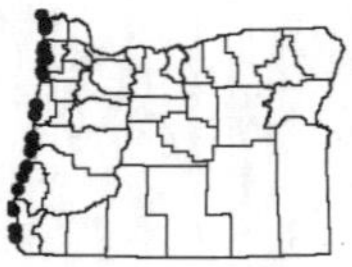

Plants perennial, dioecious, (7)15–60 cm tall; loosely cespitose, rhizomatous, and stoloniferous, rhizomes and stolons to 400 cm long. **Basal branching** mainly intravaginal, some extravaginal. **Culms** stout, 1.5–2 mm thick; nodes terete, 0(1) exserted. **Leaves** sheaths closed about 50% of their length, bases of basal sheaths glabrous; collars smooth, glabrous; ligules 1–5 mm; blades moderately to densely scabrous or minutely hairy on and between veins; blades of tillers to 30 cm; cauline blades involute, somewhat arcuate, 2–4 mm wide; uppermost blades 1–10 cm. **Inflorescences** erect, ovoid to lanceolate, contracted, often interrupted, 3–15 cm; spikelets 15–80; branches 1-2, erect, 1–6 cm, stiff, with 3–17 spikelets. **Spikelets** 9–17 mm, to 3 times as long as wide, florets 3–6(10). **Glumes** broadly lanceolate, subequal to adjacent florets; keels sparsely scabrous near tips; lower glumes 3-veined; upper glumes usually more than 7 mm, 3–5-veined. **Calluses** usually with a line of crisp to sinuous hairs around the base of the lemma, sometimes glabrous or with diffuse, short, cobwebby hairs. **Lemmas** lanceolate, (6)7.5–11 mm, 5–7(11)-veined, distinctly keeled; keels and marginal veins, sometimes also intermediate veins, short-villous to softly puberulent; area between veins smooth to scabrous, glabrous or softly puberulent; tips acute. **Anthers** vestigial and 0.1–0.2 mm, or (2)3–4(5) mm. **2*n*=28**.

Coastal sand dunes. 0–100 m. Est. CA, WA; north to AK. Native.

Poa macrantha is a strongly rhizomatous, dioecious plant growing in coastal sand dunes. It has broad, dense inflorescences and unusually long lemmas. No other Oregon species is similar.

Poa marcida Hitchcock

Withered bluegrass

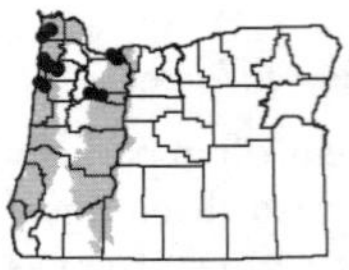

Plants perennial, dioecious, 20–80 cm tall, loosely to densely cespitose. **Basal branching** mainly pseudointravaginal. **Leaves** sheaths closed approximatly 90% of their length, to near the collar; ligules 0.5–2 mm long; blades flat, 1.5–5 mm wide. **Inflorescences** lax, narrowly lanceolate, 6–22 cm, sparse; branches ascending, 1–3 per node. **Spikelets** 3.5–7 mm, florets (1)2(4); rachilla internodes about 1 mm long. **Glumes** keels scabrous; lower glumes 1-veined; upper glumes shorter than or subequal to lowest lemma. **Calluses** with sparse cobwebby hairs. **Lemmas** narrowly lanceolate, 3.2–5 mm long, distinctly keeled, glabrous, smooth; tips acuminate. **Anthers** 0.5–1.2 mm.

Moist, late successional conifer forests. 0–1100 m. Casc, CR, Est. WA; north to British Columbia. Native.

Poa marcida has narrow, nodding, few-branched inflorescences with branches that parallel the inflorescence axis. The lemmas are long and taper more gradually than other *Poa* species. *Poa laxiflora* also has nodding inflorescences, but its branches are strongly spreading, and its lemma tips are more rounded.

Poa* × *multnomae Piper

Multnomah bluegrass

Poa gracillima Vasey var. *multnomae* (Piper) C.L. Hitchc.

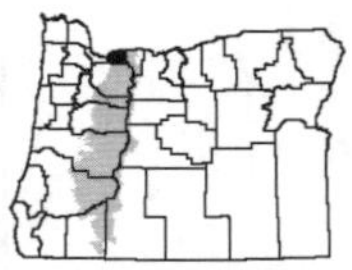

Plants perennial, more or less bisexual, 15–45 cm tall; cespitose to loosely cespitose. **Basal branching** extra- or intravaginal. **Culms** nodes terete, 0–1 exserted. **Leaves** sheaths closed to 33% of their length, glabrous; collars glabrous; ligules 0.3–1 (2.8) mm, hyaline, rounded to truncate; blades flat or folded, 0.5–1.5 mm wide, glabrous; uppermost culm blades 1.3–5.5 cm. **Inflorescences** lanceoloid to ovoid panicles, open, 5–9 cm, 3–9 mm wide, proximal internodes 1–2 cm; spikelets 10–30; branches spreading, 1–2 at lowest node; longest branches 3–5 cm, with 2–5 spikelets. **Spikelets** terete or laterally compressed, 4.5–8 mm, florets 2–4; rachilla 0.7–1.5 mm long, glabrous to sparsely short-hairy. **Glumes** lanceolate, glabrous, with 1–3 veins; upper glumes 2.5–3.8 mm. **Calluses** with short, slightly curled hairs. **Lemmas** narrowly lanceolate to lanceolate, (3)3.5–5 mm, rounded to keeled on the back; keels short-villous on the basal 50–67%, scabrous or smooth distally; marginal veins short-villous in the basal 50%; tips acute. **Anthers** 1.5–2 mm, well formed but often with defective pollen.

Waterfalls, streams, moist slopes. 200–500 m. Casc. WA. Native.

Poa × *multnomae* is a hybrid of *P. nervosa* and *P. secunda*. It has been documented only in the Columbia Gorge.

Poa nemoralis Linnaeus

Woodland bluegrass, wood bluegrass

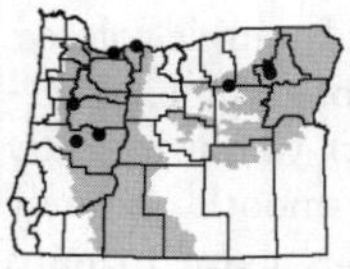

Plants perennial, 30–80 cm long; densely cespitose; shoots all or most flowering at anthesis, vegetative shoots appearing late in the growing season. **Basal branching** intra- or extravaginal or mostly extravaginal. **Culms** nodes slightly compressed, 2–5 exserted; the uppermost node in upper 25–50% of culm. **Leaves** sheaths closed to 20% of their length, bases of basal sheaths glabrous; ligules 0.2–0.8(1) mm long, truncate; blades mostly flat, 0.8–3 mm wide. **Inflorescences** narrowly lanceoloid to ovoid, generally 25–40% as broad as long at maturity, erect, or nodding in shade forms, sparsely to moderately congested, 7–16(20) cm long; branches ascending to widely spreading, 2–5 per node, fairly straight. **Spikelets** narrowly lanceolate to lanceolate, to 2.5–3.5 times as long as wide, 3–8 mm, usually not glaucous, florets (1)2–5; rachilla internodes mostly less than 1 mm. **Glumes** tapering from the base or narrowly lanceolate; keels smooth or sparsely scabrous; lower glumes 3-veined, 6.4–11 times as long as wide; upper glumes shorter than or subequal to lowest lemma. **Calluses** with sparse, often short, cobwebby hairs. **Lemmas** narrowly lanceolate to lanceolate, 2.4–4 mm, lower lemmas less than 20% as wide as long, distinctly keeled; keels and marginal veins short-villous; intermediate veins glabrous, area between veins smooth or muriculate, glabrous; tips acute, usually bronze-colored in part. **Anthers** 0.8–1.9 mm. ***2n***=28, 42.

Moist, low elevation forests, streamsides, meadows. 50–2000 m. BW, Casc, ECas, WV. CA, WA; throughout most of North America; Eurasia. Exotic.

Poa nemoralis is a cespitose forest grass with short ligules and mostly open leaf sheaths. Similar *P. interior* has been reported from Oregon, but its presence here has not been confirmed. It has longer ligules and wider glumes and lemmas. *Poa palustris* has a more spreading growth form, culms branched above the base, well-developed callus hairs, narrow hyaline lemma margins, and incurving lemma keels.

Poa nervosa (Hooker) Vasey

Veiny bluegrass

Poa nervosa (Hook.) Vasey var. *nervosa*, *Poa vaseyana* Scribn. ex Beal

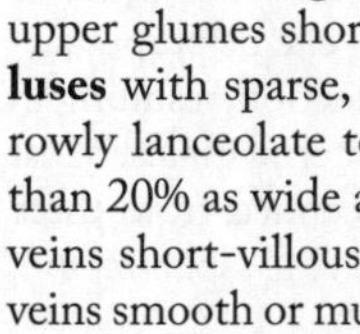

Plants perennial, sequentially gynomonoecious, 20–65 cm tall; loosely cespitose or with solitary shoots, short-rhizomatous. **Basal branching** mainly extravaginal. **Culms** nodes terete, 1–2 exserted. **Leaves** sheaths closed over 67% their length, bases of basal sheaths glabrous; collars of proximal leaves usually pubescent on and near the margins, collar margin hairs longer than those of the sheaths; ligules 0.5–1.5 mm, ligules of lower culm and sterile shoot leaves to 1 mm; blades of tillers similar to cauline blades or sometimes longer; cauline blades usually flat, 2–4.5 mm wide, smooth or upper surface sparsely scabrous mainly over the veins, blades gradually reduced in length higher on the culm or middle blades longest; uppermost blades mostly 3–8 cm long. **Inflorescences** erect or lax, ovoid to pyramidal, open or loosely contracted, sparse, 8–15 cm; spikelets 25–80; branches ascending to spreading, lax, 2.5–8 cm long, 3–5 per node, with 2–8 spikelets. **Spikelets** 4–7 mm, to 3.5 times as long as wide, florets 3–8; rachilla internodes smooth or scabrous, glabrous or sparsely hispidulous. **Glumes** lanceolate, over 67% as long as adjacent lemmas; lower glumes 1–3(5)-veined. **Calluses** glabrous or rarely with minute cobwebby hairs. **Lemmas** lanceolate, 3–4.5 mm, distinctly keeled; keels and marginal veins glabrous or infrequently sparsely short-hairy, area between veins glabrous or hispidulous, smooth or finely scabrous; tips acute. **Anthers** vestigial and 0.1–0.2 mm or 2.5–4 mm.

Moist conifer or mixed conifer-deciduous forests, waterfalls, seeps. 50–300 m. Casc, WV. WA. Native.

Poa wheeleri, once considered part of *P. nervosa*, has sheaths of lower leaves that are densely scabrous at least in the upper half. Reports of *P. nervosa* east of the Cascades date from the time before *P. wheeleri* was recognized as a separate species. *Poa nervosa* specimens with very inconspicuous collar hairs might be misidentified as *P. chambersii*, which has a denser inflorescence. *Poa laxiflora* has a longer panicle with longer, more nodding branches. *Poa* × *multnomae*, a hybrid of *P. nervosa* × *P. secunda*, occurs in the Columbia Gorge. It is cespitose and lacks leaf sheath hairs.

Poa palustris Linnaeus

Fowl bluegrass

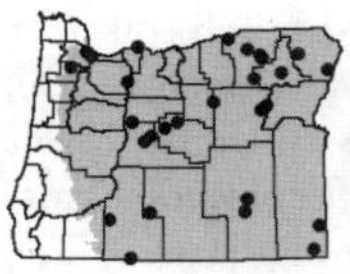

Plants perennial, 25–120 cm tall, loosely to densely cespitose and often stoloniferous; shoots all or most flowering at anthesis, vegetative shoots appearing late in the growing season. **Basal branching** extra- and intravaginal. **Culms** nodes terete to slightly compressed, proximal nodes often slightly swollen; the uppermost node usually at or above the middle of the culm. **Leaves** sheaths closed to 20% of their length, bases of basal sheaths glabrous; ligules (1)1.5–6 mm, truncate to acute; blades flat, 1.5–8 mm wide, usually several on the culm, steeply ascending or spreading to 80°; upper blades often lax. **Inflorescences** lax, eventually open, sparsely to moderately congested, (9)13–30(41) cm, generally 33–50% as broad as long at maturity; spikelets 25–100+; branches initially erect, eventually widely spreading to slightly reflexed, fairly straight, slender, 4–15 cm, 33–50% of the panicle length, 2–9 per node. **Spikelets** narrowly to broadly lanceolate, 3–3.5 times as long as wide, 3–5 mm, florets (1)2–5; rachilla internodes mostly less than 1 mm. **Glumes** tapering from the base or lanceolate, distinctly keeled; keels smooth or sparsely scabrous; lower glumes 3-veined, long tapered to a slender point, 6.4–10 times as long as wide. **Calluses** with moderately dense cobwebby hairs that are at least (50)67% as long as the lemma. **Lemmas** narrowly lanceolate to lanceolate, 2–3 mm, distinctly keeled; keels straight or gradually arched, usually abruptly inwardly arched at the junction of the scarious tips; keels and marginal veins short-villous, area between veins muriculate, glabrous; intermediate veins obscure; margins distinctly inrolled; tips obtuse or acute, usually bronze-colored in part, frequently incurved and blunt with a short hyaline margin. **Anthers** 1–1.8 mm. ***2n***=28.

Moist meadows, shorelines, ditches, forests, disturbed areas. 0–2600 m. BR, BW, Casc, Col, ECas, Lava, Owy, WV. CA, ID, NV, WA; throughout most of Canada and US; circumboreal. Exotic?

Poa palustris has an open inflorescence, small lemmas and an open growth form produced by culms that often branch above the base. *Poa nemoralis* differs in having short, truncate ligules. *Poa palustris* from drier sites can be similar to *P. interior*, which has more lax leaf blades, calluses with sparse, short hairs, wider hyaline lemma margins and straight to slightly curved lemma keels. *Poa interior* has not been documented in Oregon. Both native and introduced populations of *Poa palustris* exist in North America, and Oregon plants have generally be considered introduced. However, some populations may be native.

Poa piperi Hitchcock

Piper's bluegrass

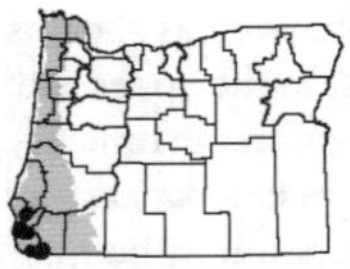

Plants perennial, dioecious, 20–55 cm; loosely cespitose, rhizomatous. **Basal branching** extra- and intravaginal. **Culms** nodes terete, 0–1 exserted. **Leaves** sheaths closed 33–50% of their length, bases of basal sheaths glabrous; collars of at least some leaves usually sparsely short-hairy; ligules 1–2 mm; blades of tillers to 40 cm long; upper surface moderately to densely scabrous or minutely hairy on and between the veins; cauline blades involute, 1–3 mm wide; lower surfaces smooth; upper surfaces sometimes sparsely hairy; blades much reduced higher on the culm; uppermost blades 1–4.5 cm. **Inflorescences** erect to nodding, lanceoloid to ovoid, loosely contracted, sparse, 4–8 cm; spikelets 18–60; branches ascending, lax, 3–8 cm long, 1–2 per node, with 3–8 spikelets. **Spikelets** 6–9(11) mm, to 3 times as long as wide, florets 2–5(7); rachilla internodes 1–2 mm long. **Glumes** subequal; lower glumes 3-veined. **Calluses** with diffuse, soft cobwebby hairs about 50% as long as the lemma. **Lemmas** lanceolate, 4–6(7) mm, distinctly keeled, glabrous throughout, smooth or sparsely to moderately finely scabrous; keels scabrous; tips acute. **Anthers** vestigial and 0.1–0.2 mm, or 2–3 mm. **2*n***=28.

Forest openings on serpentine soils. 300–1600 m. CR, Sisk. CA. Native.

Poa piperi is a dioecious, usually glaucous grass with nearly transparent glumes and cobwebby callus hairs.

Poa pratensis Linnaeus

Kentucky bluegrass

Plants perennial, 5–70(100) cm tall, densely to loosely cespitose and extensively rhizomatous. **Basal branching** mainly extravaginal or equally extra- and intravaginal. **Culms** nodes terete or slightly compressed, 1–2(3) nodes exposed. **Leaves** sheaths closed 25–50% of their length, bases of basal sheaths glabrous; collars smooth, glabrous; ligules truncate to rounded, 0.9–2(3.1) mm; blades of extravaginal shoots like cauline blades, those of intravaginal shoots, when present, sometimes distinctly narrower, flat to involute, 0.4–1 mm wide; cauline blades flat or folded, to involute on the margins, 0.4–4.5 mm wide; lower surfaces smooth, glabrous; upper surfaces smooth or sparsely scabrous, frequently with sparse, slender hairs 0.2–0.8 mm long; uppermost blades 1.5–10 cm. **Inflorescences** narrowly ovoid to narrowly or broadly pyramidal, loosely contacted to open, sparsely to moderately congested, 2–15(20) cm; spikelets (25)30–100+; branches spreading, (1)2–9 cm long, (1)2–7(9) per node; spikelets 4–30(50), usually fairly crowded in distal half. **Spikelets** 3.5–6(7) mm, to 3.5 times as long as wide, florets 2–5; rachilla internodes mostly less than 1 mm, smooth, glabrous. **Glumes** narrowly lanceolate to lanceolate (broadly lanceolate), unequal to subequal, usually distinctly shorter than the adjacent lemmas; keels sparsely to densely scabrous (smooth); lower glumes usually lanceolate and 3-veined, occasionally sickle-shaped and 1-veined, 1.5–4(4.5) mm; upper glumes 2–4.5(5) mm. **Calluses** with cobwebby hairs arising as a dorsal tuft only, or with secondary tufts under each marginal vein, hairs from half as long as to longer than the lemma. **Lemmas** lanceolate, 2–4.3(6) mm, green or strongly purple colored, distinctly keeled; keels and marginal veins long-villous; intermediate veins glabrous or infrequently short-villous to softly puberulent, area between veins glabrous, smooth or finely muriculate below, smooth or sparsely scabrous above; margins narrowly to broadly hyaline; tips acute. **Anthers** 1.2–2 mm, occasionally aborted late in development.

6 subspecies; 5 subspecies treated in *Flora*.

Poa pratensis is a common grass with slightly nodding inflorescences and copious cobwebby callus hairs. *Poa rhizomata*, of rocky ultramafic slopes in southwestern Oregon, is similar but has sparse inflorescences, acute ligules, and unisexual inflorescences. *Poa confinis* of coastal sands has glabrous (to sparsely hairy) lemmas and diffuse, not tufted callus hairs. *Poa wheeleri*, a common eastern Oregon species, lacks the lemma hairs and has denser inflorescences and more strongly scabrous leaf sheaths.

Poa pratensis is common, polyploid, and highly variable. It hybridizes with related taxa including *P. alpina*, *P. secunda*, and *P. wheeleri*. The resulting variation in *P. pratensis* can be perpetuated through both sexual and asexual seed production, resulting in many recognizable forms that are linked by intermediates. Most or all *Poa pratensis* found in Oregon are descended from three introduced subspecies (*P. p.* ssp. *angustifolia*, *P. p.* ssp. *pratensis*, and *P. p.* ssp. *irrigata*).

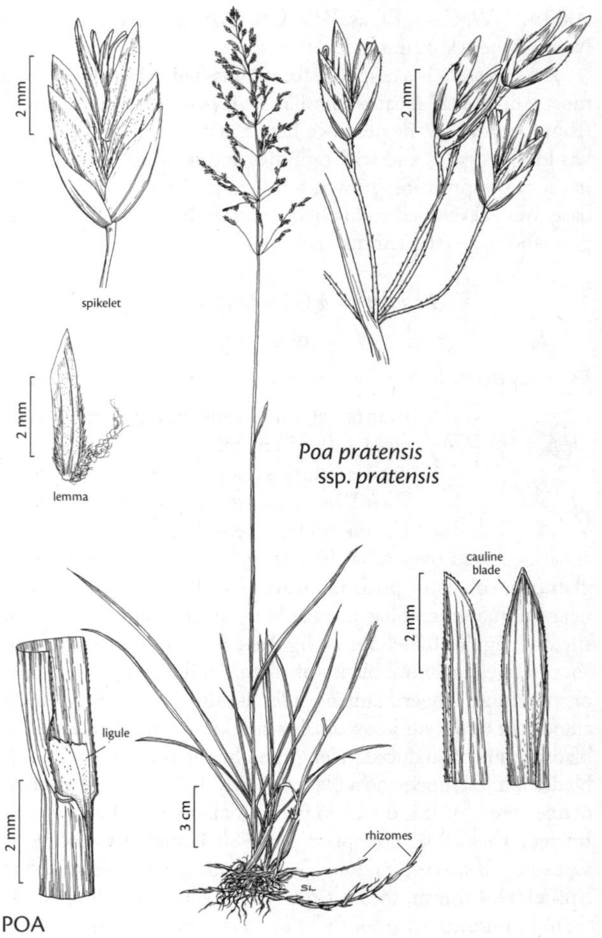

POA

Key to *Poa pratensis* subspecies:

1. Panicle branches smooth or almost smooth.
   2. Basal branching primarily extravaginal; blades flat or folded, soft, adaxial surfaces usually glabrous, sometimes sparsely hairy; plants of alpine and tundra regions. . . . . *P. p.* ssp. ***alpigena***
   2' Basal branching both intra- and extravaginal; blades folded or involute, somewhat firm, adaxial surfaces often sparsely hairy; plants widely distributed, but not in alpine or tundra regions . . . . *P. p.* ssp. ***agassizensis***

1' Panicles branches more or less scabrous.
   3. Intravaginal innovation shoots present, intra- and extravaginal blades alike, 0.4–1 mm wide, folded to involute, somewhat firm, adaxial surfaces often sparsely and softly hairy. . . . . . *P. p.* ssp. ***angustifolia***
   3' Intravaginal innovation shoots present or absent, if present, then differentiated or alike, at least some with blades 1.5–4.5 mm wide, flat or folded, adaxial surfaces rarely hairy.
      4. Culms 8–30(50) cm tall, often somewhat glaucous; blades flat; intravaginal shoots absent or present and with blades similar to those of the extravaginal shoots; panicles with 4-8 spikelets per branch and 1–2(5) branches per node. . . . . *P. p.* ssp. ***irrigata***
      4' Culms 8–100 cm tall, not glaucous; blades flat or folded; intravaginal shoots present, with blades similar to those of the extravaginal shoots or distinctly narrower; panicles with several to many spikelets per branch and 3–5(7) branches per node . . . . . *P. p.* ssp. ***pratensis***

Poa pratensis Linnaeus ssp. ***agassizensis*** (B. Boivin & D. Löve) Roy L. Taylor & MacBryde

Agassiz's Kentucky bluegrass

Poa agassizensis B. Boivin & D. Löve

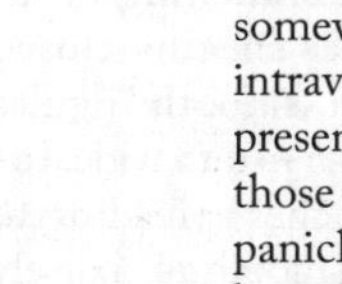

Plants moderately to densely tufted; shoots clustered. **Basal branching** intra- and extravaginal. **Culms** 20–40(50) cm. **Leaves** blades of tillers all involute or folded, usually shorter than 15 cm, 0.8–2 mm wide, adaxial surfaces sparsely pubescent. **Inflorescences** ovoid, narrowly pyramidal or loosely contracted, 4–8 cm, with 2–5 branches per node; branches ascending, smooth or sparsely to moderately densely scabrous, with several spikelets per branch. **Spikelets** lanceolate. **Glumes** with keels strongly compressed, sparsely to moderately scabrous; upper glumes shorter than to nearly equaling the lowest lemmas. **Lemmas** 2–4 mm, finely muriculate; lateral veins glabrous.

Prairies and mountain grasslands. 1800–2100 m. Owy. CA, ID, NV, WA; western and central North America; cool temperate regions worldwide. Native?

Poa pratensis ssp. *agassizensis* grows throughout the drier, cool-temperate range of the species in North America. It may consist of ecotypes derived from cultivated material, or be a native form that has adapted to xeric conditions. The least distinctive of the subspecies treated here, it closely approaches ssp. *angustifolia* in having involute leaves and small spikelets, but has shorter and broader leaves and more condensed panicles.

Poa pratensis Linnaeus ssp. ***alpigena*** (Lindman) Hiitonen

Alpine Kentucky bluegrass

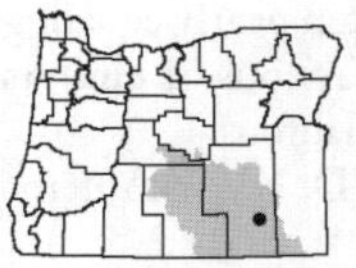

Plants strongly anthocyanic; moderately to loosely tufted; shoots usually solitary. **Basal branching** mainly extravaginal. **Culms** 15–70 cm. **Leaves** blades of tillers flat or folded, shorter than 15 cm, 1–3.6 mm wide, soft, adaxial surfaces usually glabrous, sometimes sparsely pubescent; cauline blades flat or folded. **Inflorescences** narrowly pyramidal or contracted, expanding well after emergence from the sheath, 3–13(20) cm, with (1)2–5(7) branches per node; branches steeply ascending to eventually spreading or somewhat reflexed, 1–6 cm, smooth or sparsely scabrous, with 5–15 spikelets. **Spikelets** narrowly lanceolate, 3.3–5.5 mm. **Glumes** with keels distinct, smooth or sparsely scabrous near the tips; upper glumes nearly equaling the lowest lemmas. **Lemmas** 2.5–3.5 mm, smooth or finely muriculate; lateral veins frequently short-villous to softly puberulent.

Mesic to moist arctic and alpine meadows. 2000–2900 m. BR. ID, WA; north to Alaska, throughout Canada; rare in Rocky Mountains and New England; southern South America; circumboreal. Native.

In Oregon, this is a high alpine taxon, with small spikelets, and is known from cirques on Steens Mountain.

Poa pratensis Linnaeus ssp. ***angustifolia*** (Linnaeus) Lejeune

Thin-leaved Kentucky bluegrass

Plants moderately to densely tufted. **Basal branching** intra- and extravaginal, intravaginal shoots clustered. **Culms** 25–80 cm. **Leaves** blades of tillers all involute, 10–45 cm × 0.4–1 mm, sometimes narrower than the cauline blades, adaxial surfaces sparsely pubescent; cauline blades involute or folded, somewhat firm, adaxial surfaces sparsely pubescent. **Inflorescences** narrowly pyramidal or loosely contracted, 8–18 cm; branches ascending to spreading, smooth or sparsely to densely scabrous, with several to many spikelets per branch. **Spikelets** narrowly lanceolate. **Glumes** with keels strongly compressed, sparsely to moderately scabrous; upper glumes shorter than to nearly equaling the lowest lemmas. **Lemmas** 2.5–3.5 mm, finely muriculate; lateral veins glabrous.

Mesic grasslands. CA, ID, NV, WA; Eurasia. Exotic.

Poa pratensis ssp. *angustifolia* is a western Eurasian taxon characterized by the predominance of fascicles of elongate, narrow, involute blades on the intravaginal vegetative shoots and slender panicles with small spikelets.

Poa pratensis Linnaeus ssp. ***irrigata*** (Lindman) H. Lindberg

Dryland Kentucky bluegrass

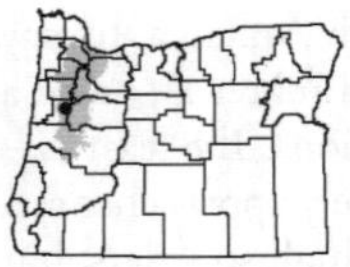

Plants moderately densely to loosely tufted, sometimes forming turf. **Basal branching** mainly extravaginal. **Culms** solitary, 8–30(50) cm. **Leaves** blades of tillers usually shorter than 15 cm, 2–4.5 mm wide, adaxial surfaces usually glabrous; cauline blades flat. **Panicles** broadly pyramidal, open, 2–10 cm, with 1–2(5) branches per node; branches widely spreading, 1.5–6 cm, smooth or sparsely to moderately scabrous, with 4–8 spikelets. **Spikelets** lanceolate to broadly lanceolate. **Glumes** with keels distinct, sparsely to moderately scabrous; lower and upper glumes often nearly equaling the lowest lemmas, often noticeably glaucous. **Lemmas** 3–5(6) mm, finely muriculate; lateral veins glabrous.

Mesic grasslands. 50–200 m. WV. CA, ID, NV, WA; temperate regions worldwide. Exotic.

Poa pratensis ssp. *irrigata* includes subarctic to boreal, coastal, and lowland plants that differ from those of ssp. *pratensis* in their primarily extravaginal branching, shorter stature, wide, flat blades, short, stout panicles with fewer branches, larger spikelets, and relatively longer glumes that are often pruinose.

Poa pratensis Linnaeus ssp. ***pratensis*** (p. 462)

Typical Kentucky bluegrass

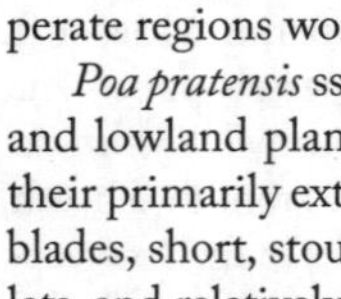

Plants densely to loosely tufted, often forming turf, culms clustered. **Basal branching** intra- and extravaginal. **Culms** 8–100 cm. **Leaves** blades of tillers all flat or some involute, 10–45 cm × 0.4–4 mm, some distinctly narrower than the cauline blades, usually soft, adaxial surfaces sparsely pubescent; cauline blades flat or folded. **Panicles** broadly pyramidal, open or somewhat contracted, 5–18 cm, with 3–5(7) branches per node; branches spreading to somewhat reflexed, smooth or sparsely to fairly densely scabrous, with several to many spikelets per branch. **Spikelets** lanceolate to broadly lanceolate. **Glumes** with keels strongly compressed, sparsely to moderately scabrous; upper glumes shorter than or nearly equaling the lowest lemmas. **Lemmas** 2.8–4.3 mm, finely muriculate; lateral veins glabrous. ***2n***=21–144.

Mesic grasslands; 0–2200 m. Casc, ECas, Est, Lava, Owy, WV. CA, ID, NV, WA; temperate and subarctic areas worldwide. Exotic.

Poa pratensis ssp. *pratensis* grows throughout most of the range of the species. It usually has a few narrow, flat or involute, intravaginal shoot leaves, in addition to some broader, extravaginal shoot leaves and is intermediate between sspp. *angustifolia* and *irrigata*.

Poa pringlei Scribner

Pringle's bluegrass

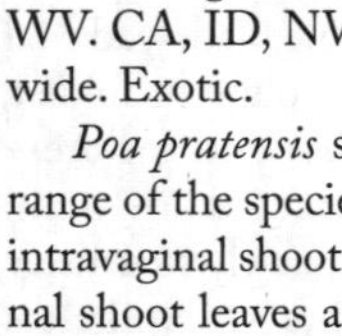

Plants perennial, dioecious, 5–35 cm tall; densely cespitose. **Basal branching** intravaginal. **Culms** 0.5–0.9 mm thick; nodes 0(1) exserted. **Leaves** sheaths terete, closed to 33% of their length, bases of basal sheaths glabrous; collar margins smooth or scabrous to hispidulous; ligules (0.5)1–6 mm long, translucent; tips truncate to acute, ligules of sterile shoots 1–2.5 mm long; blades involute, frequently somewhat arcuate, 1.5–3 mm wide; lower surfaces smooth; upper surfaces densely scabrous or hispidulous; cauline blades only slightly reduced higher on the culm. **Inflorescences** erect, narrowly lanceoloid to ovoid, moderately congested, 1–6 cm; spikelets 6–20(25); branches erect, 0.5–1.5(2) cm, 1–2 per node, with 1–3 spikelets. **Spikelets** broadly lanceolate, to 3.5 times as long as wide, 6–8(12) mm, shiny, florets 2–5; rachilla internodes smooth. **Glumes** lanceolate to broadly lanceolate, 3.5–7.5 mm, those of male plants shorter and with rounder tips than those of female plants, thin, shiny, almost transparent; keels smooth or sparsely scabrous; lower glumes slightly shorter than adjacent lemma, 3-veined. **Calluses** glabrous. **Lemmas** lanceolate, 5–8 mm, thinly membranous, distinctly keeled, smooth or sparsely finely scabrous, glabrous; tips acute. **Anthers** vestigial and 0.1–0.2 mm on female plants, 2–4 mm on male plants.

Rocky, subalpine and alpine slopes. 1200–2300 m. Sisk. CA. Native.

Poa pringlei is a dioecious grass with relatively long, nearly transparent glumes. It is cespitose, but the culms spread horizontally at the base, making it appear rhizomatous.

Poa reflexa Vasey & Scribner

Nodding bluegrass

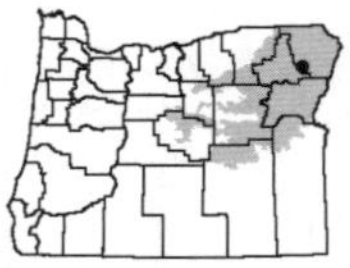

Plants perennial, short-lived, 10–60 cm tall; densely cespitose. **Basal branching** intra- and extravaginal. **Leaves** sheaths closed 33–50% of their length, smooth; ligules 1.5–3.5 mm; blades flat, 1.5–4 mm wide. **Inflorescences** nodding, open, 4–15 cm; branches spreading to reflexed, flexuous, (2)3–7 cm, 1–2 per node, smooth to sparsely scabrous, with (3)6–18 spikelets. **Spikelets** lanceolate to broadly lanceolate, 4–6 mm, usually partly to wholly purplish, florets 3–5. **Glumes** narrowly to broadly lanceolate; keels smooth or nearly so; lower glumes 1-veined; upper glumes shorter than or subequal to lowest lemma. **Calluses** with cobwebby hairs. **Lemmas** lanceolate, 2–3.5 mm, partly to mostly purple colored, distinctly keeled; keels short- to long-villous for over 67% of their length; marginal veins short- to long-villous; lateral veins usually sparsely softly puberulent at least on one side, area between veins smooth; tips acute, slightly bronze colored or not. **Anthers** 0.6–1 mm.

Subalpine and alpine meadows, streamsides, talus slopes. 1000–2600 m. BW. ID, NV; northeast to MT, southeast to NM. Native.

Poa reflexa is a rare subalpine bluegrass similar to *P. leptocoma*. *Poa leptocoma* grows in wetter habitats and has glabrous lateral lemma veins and scabrous panicle branches.

Poa rhizomata Hitchcock

Rhizome bluegrass

Plants perennial, subdioecious (many plants either staminate or pistillate, but others with bisexual florets), 20–65 cm long; loosely cespitose or with solitary shoots, short-rhizomatous. **Basal branching** intra- and extravaginal or mainly extravaginal. **Culm** nodes terete, 1–2 exserted. **Leaves** sheaths closed 50–67% of their length, bases of basal sheaths glabrous; collars smooth, glabrous; ligules of cauline leaves acute to acuminate, 2–8 mm long; ligules of sterile shoots 2–5 mm long; blades of tillers to 20 cm long, otherwise similar to cauline blades; cauline blades usually flat or V-shaped, occasionally folded, 1–3.5 mm wide, smooth or sparsely scabrous mainly over the veins and margins, distinctly keeled, blades gradually reduced higher on the culm; uppermost blades (1.4)3–6(8) cm long. **Inflorescences** nodding at the tips, ovoid, sparse, (2)4–10 cm; spikelets 20–50; branches ascending to spreading, lax, 1.5–4.5 cm long, 1–2(4) per node, with 2–7 spikelets. **Spikelets** (4)6–9(12) mm, to 3.5 times as long as wide, florets 3–8; rachilla internodes smooth or sparsely scabrous, glabrous or infrequently sparsely softly puberulent. **Glumes** narrowly lanceolate to lanceolate, over 50% as long as adjacent lemmas; keels scabrous; lower glumes 1–3(5)-veined. **Calluses** with cobwebby hairs more than half as long as the lemma. **Lemmas** lanceolate, 4–6.5 mm, distinctly keeled; keels and marginal veins sparsely short- to long-villous, area between veins sparsely scabrous, glabrous, 5–7-veined; tips acute. **Anthers** vestigial and 0.1–0.2 mm, or 2.5–4 mm. ***2n***=28.

Rocky gabbro or peridotite soils in montane, mixed conifer forests. 500–2200 m. Sisk. CA. Native.

Poa rhizomata resembles *P. pratensis* with sparse inflorescences. *Poa pratensis* has truncate to rounded ligules and bisexual, usually smaller florets. *Poa chambersii* is somewhat similar but has more closed sheaths, shorter ligules, glabrous or less hairy lemmas, and calluses glabrous or with sparser, shorter cobwebby hairs.

Poa secunda J. Presl

Secund bluegrass

Plants perennial, (10)15–120 cm tall; densely cespitose; basal leaf tufts (excluding culms) 2–20+ cm tall, glaucous or not. **Basal branching** intra- and extravaginal. **Culms** slender to stout; nodes terete, 0–2 exserted. **Leaves** sheaths closed to 25% of their length, bases of basal sheaths glabrous; collars smooth or scabrous, glabrous; ligules 0.5–6(10) mm; blades flat, folded, or involute, 0.4–3(5) mm wide, green to strongly glaucous, thin, soft and soon withering, or thick, firm and persisting, smooth, or scabrous mainly over the veins, blades gradually reduced distally or the middle blades longest; uppermost blades 0.8–10(17) cm. **Inflorescences** erect or somewhat lax, narrowly lanceoloid to ovoid, usually contracted, more or less open in flower, infrequently remaining open at maturity, usually moderately congested, 2–25(30) cm; spikelets 10–100+; branches usually erect or ascending, infrequently spreading at maturity, (0.5)1–8(10) cm, 1–3 per node, with (1)2–20(60+) spikelets in distal 50–67%. **Spikelets** usually narrowly lanceolate, more or less terete to slightly laterally compressed, (3.8)4–5 times as long as wide, (4)5–10 mm, florets (2)3–5(10); rachilla internodes usually 1–2 mm long, smooth to scabrous. **Glumes** broadly lanceolate; keels indistinct; lower glumes 3-veined. **Calluses** glabrous, or with a line of crisp or slightly sinuous hairs 0.1–0.5(2) mm long around the base. **Lemmas** lanceolate to narrowly lanceolate or slightly oblanceolate, 3.5–6 mm, usually weakly keeled, glabrous throughout or keels and marginal veins softly puberulent to short-villous, area between veins smooth or scabrous, glabrous, short-villous, crisply puberulent, or loosely softly puberulent over the basal 67%, hairs usually 0.1–0.5 mm long; margins strongly inrolled below, broadly scarious above, glabrous; tips obtuse to broadly acute, blunt or pointed. **Anthers** 1.5–3 mm.

2 subspecies.

Poa secunda is a common, highly variable, densely cespitose grass mainly of uplands. The spikelets are terete to slightly compressed, and the lemmas are more or less rounded over the back and usually hairy to scabrous throughout. *Poa stenantha* can be very similar, but its lemmas are a bit more strongly keeled and have hairs only on the veins. Callus hairs, if present, are 0.2–2 mm long. Hybrids between *P. secunda* and *P. pratensis* (*P.* × *limosa*) occur in moist, alkaline meadows in southeastern Oregon. They resemble *P. secunda* but are more or less rhizomatous and may have more strongly keeled lemmas. Hybrids with *P. nervosa,* called *P.* × *multnomae,* occur in the Columbia Gorge. They have more open panicles and generally grow in moister habitats than *P. secunda*. *Puccinellia lemmonii* and *Poa secunda* are superficially similar, but *Puccinellia lemmonii* has lower glumes that are less than 50% as long as the lowest lemma, and it grows in more strongly alkaline meadows and flats that are generally wet in spring. *Poa secunda* is a highly variable taxon. Its combination of asexual and sexual seed production produces multiple widespread, morphologically consistent forms that are sympatric. Many of these variants have been treated as species. However, intermediates can be found between all of them. Here we gather these forms into two categories treated as subspecies.

Key to *Poa secunda* subspecies:

1. Lemmas glabrous, sometimes scabrous, rarely sparsely short-hairy; ligules of vegetative shoots and often culms less than 2 mm long, more or less truncate, scabrous; leaf blades more or less firm, retaining shape; inflorescence branches appressed . . . . *P. s.* ssp. ***juncifolia***

1' Lemmas more or less evenly short-hairy on keel and sides across base, sometimes sparsely so; ligules of vegetative shoots and culms usually more than 2 mm long, acute or acuminate, smooth or sparsely scabrous; leaf blades usually more or less soft, soon withering; inflorescence branches appressed or spreading (if spreading and with short truncate ligules, see *P.* × *multnomae*) . . . . *P. s.* ssp. ***secunda***

Poa secunda J. Presl ssp. *juncifolia* (Scribner) Soreng

Alkali bluegrass, big bluegrass, Nevada bluegrass

Poa ampla Merr., *Poa brachyglossa* Piper, *Poa juncifolia* Scribn., *Poa nevadensis* Vasey ex Scribn.

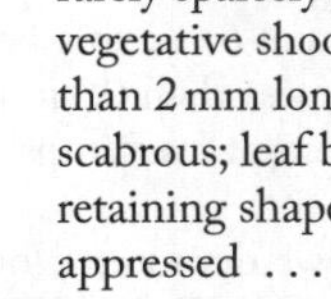

Plants 30–120 cm tall, usually medium-sized to robust, or rarely tiny. **Basal branching** mainly extravaginal. **Ligules** of sterile vegetative shoots mostly 0.5–2 mm long, scabrous; tips truncate to obtuse; blades 1–3(5) mm wide, slightly lax to firm, tending to hold their form and persist. **Inflorescences** narrowly lanceoloid, contracted at maturity, (4)10–25(30) cm; branches erect, scabrous. **Spikelets** (4)7–10 mm. **Calluses** glabrous. **Lemmas** more or less scabrous, usually glabrous throughout, or less often keel and marginal veins crisply puberulent over the basal 25%, hairs usually less than 0.2 mm. **Paleas** glabrous. ***2n***=±63.

Montane and wet meadows, riparian terraces. 0–2900 m. BR, BW, Casc, Col, ECas, Lava, Owy, Sisk, WV. CA, ID, NV, WA; north to AK, east to ND; South America. Native.

Usually, *Poa secunda* ssp. *juncifolia* is more robust than *P. s.* ssp. *secunda* and grows in moister and sometimes more saline habitats. It has short ligules on vegetative shoot leaves, nonwithering leaves and glabrous lemmas.

Poa secunda J. Presl ssp. *secunda* (p. 467)

Canby's bluegrass, Pacific bluegrass, pine bluegrass, Sandberg's bluegrass

Poa gracillima Vasey var. *gracillima*, *Poa incurva* Scribn. & T.A. Williams, *Poa sandbergii* Vasey, *Poa scabrella* (Thurb.) Benth. ex Vasey

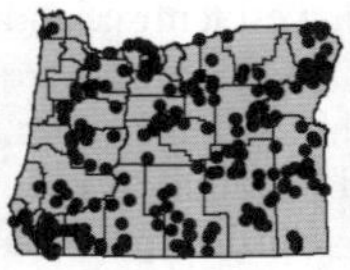

Plants (10)15–100 cm tall, tiny to medium-sized, rarely robust. **Basal branching** mixed intra- and extravaginal or mainly intravaginal. **Ligules** of vegetative shoots mostly 2–6 mm long, smooth or scabrous; tips obtuse to acuminate; blades 0.4–3 mm wide, usually thin, lax and soon withering, to less often moderately thick, moderately firm and somewhat persistent. **Inflorescences** usually narrowly lanceoloid to ovoid, contracted at maturity or occasionally pyramidal, remaining open at maturity and sparse, 2–15(20) cm; branches erect or ascending, less often widely spreading at maturity. **Spikelets** (4)5–8 mm. **Calluses** glabrous or pubescent. **Lemmas** with keels and marginal veins long-villous to crisply puberulent or softly puberulent over proximal 67%, area between veins usually with some hairs usually less than 0.5 mm. **Paleas** often short-villous to softly puberulent at least in part. $2n$=±84.

Sagebrush steppe, upland prairies, meadows, dry forests, low alpine slopes. 0–3000 m. Throughout Oregon. CA, ID, NV, WA; western and northern North America; South America. Native.

Poa secunda ssp. *secunda* is usually less robust than *P. s.* ssp. *juncifolia*, with leaves that may wither early in the growing season, longer ligules, and puberulent lemmas. It usually grows in drier, more upland sites than *P. s.* ssp. *juncifolia*.

Poa stenantha Trinius var. *stenantha*

Northern bluegrass

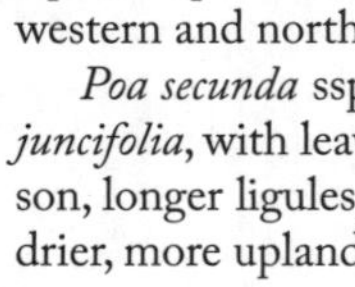

Plants perennial, 20–60(100) cm tall; densely to loosely cespitose. **Basal branching** mostly extravaginal, some intravaginal. **Culms** nodes 1–2 exserted. **Leaves** sheaths closed to 25% of their length, bases of basal sheaths glabrous; ligules 2–6 mm; blades flat or folded, 1.5–4(5) mm wide, smooth or sparsely scabrous, blades not much reduced higher on the culm. **Inflorescences** lax, loosely contracted to open, 5–18(25) cm; spikelets 20–65; branches ascending to spreading, 3–15 cm long, usually 2(7) per node, with 3–10(15) spikelets in the distal half. **Spikelets** lanceolate to narrowly ovate, 3–3.6 times long as wide, 6–10 mm, often slightly glaucous, florets 3–4(7); rachilla internodes 1.2–2 mm long, smooth or sparsely muriculate. **Glumes** subequal, lanceolate to broadly lanceolate, dull, frequently glaucous; tips obtuse to acute; lower glumes 3-veined; upper glumes (3.7)4.1–6.5 mm long. **Calluses** usually with a crown of hairs 0.2–2 mm long, sometimes glabrous. **Lemmas** lanceolate, distinctly compressed and keeled, 4–6 mm; keels short- to long-villous to 75% of their length; marginal and sometimes intermediate veins short- to long-villous; areas between veins glabrous or sparsely puberulent or hispidulous only near the base with hairs distinctly shorter than those of the keel and veins; margins weakly inrolled, broadly scarious, glabrous; tips acute. **Anthers** 1.2–2 mm.

Montane forests, cliffs, coastal meadows. 100–3000 m. BW, Casc, CR, Sisk, WV. ID, WA; north to AK, southeast to CO; South America. Native.

Poa stenantha usually has an open panicle like *P. pratensis* but lacks long cobwebby callus hairs and is usually more cespitose. *Poa secunda* has lemmas that are more rounded over the back and scabrous or hairy between the lemma veins.

Poa suksdorfii (Beal) Vasey ex Piper

Western bluegrass

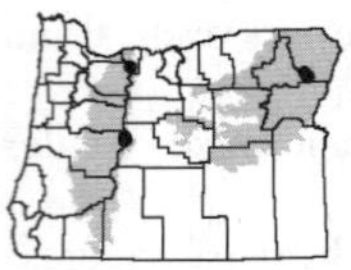

Plants perennial, 7–25 cm tall; densely cespitose. **Basal branching** intravaginal and extravaginal or mainly intravaginal. **Leaves** sheaths closed up to 33% of their length; ligules of cauline leaves 1–3 mm, ligules of uppermost tiller leaves 0.5–2.5 mm long; blades of tillers with upper surface scabrous, minutely hairy, or puberulent on and between the veins; cauline blades folded to involute; uppermost blades 1–2 mm wide. **Inflorescences** erect, narrowly lanceoloid, contracted, moderately congested, 3–6 cm; branches erect, slender, to 1.5 cm, 1–2 per node. **Spikelets** 4.2–7 mm, often purple, florets 2–4; rachilla internodes 1–1.5 mm long, smooth, sometimes sparsely hispidulous. **Glumes** lanceolate; (3.3)4.3–5.7 mm; lower glumes shorter than to equal to the lowest lemma, 3-veined; upper glumes frequently longer than the lowest lemma, 3–5-veined, not exceeding the upper florets. **Calluses** glabrous. **Lemmas** narrowly lanceolate, 4.1–5.8 mm, more or less keeled, thin, glabrous or sometimes puberulent; tips acute. **Anthers** usually 0.8–1.2(1.7) mm, occasionally aborted late in development.

Dry, rocky, alpine slopes. 1800–3000 m. BW, Casc. WA; north to British Columbia. Native.

Poa suksdorfii is a small, cespitose, alpine bluegrass with the upper glume longer than the lowest lemma. It is coarser than similar *P. lettermanii*, which has shorter glumes. *Poa glauca* and *P. laxa* have shorter glumes that are shorter than the lowest lemma.

Poa trivialis Linnaeus

Rough bluegrass

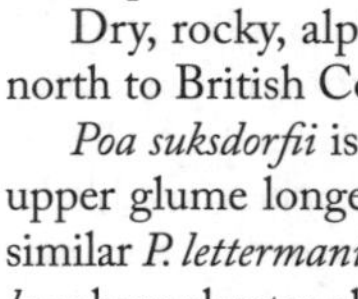

Plants perennial, short-lived, 25–120 cm long, loosely to densely cespitose, usually weakly stoloniferous. **Basal branching** intravaginal. **Culms** sometimes trailing and rooting at the nodes; nodes terete or slightly compressed, (0)1–3 exserted. **Leaves** sheaths closed about 25–50% of their length, usually densely scabrous, bases of basal sheaths glabrous; collars smooth or scabrous, glabrous; ligules 3–10 mm; blades flat, 1–5 mm wide, scabrous. **Inflorescences** erect or lax, pyramidal, open, 8–25 cm; spikelets 35–100+; branches ascending to spreading, flexuous to fairly straight, 2–8(10) cm, 3–7 per node, with 5–35 spikelets in distal 50–75%. **Spikelets** to 3 times as long as wide, 2.5–4(5) mm, florets 2–4; rachilla internodes smooth or muriculate. **Glumes** keels scabrous; lower glumes subulate to narrowly lanceolate, usually arched to crescent-shaped, 1-veined, distinctly shorter than lowest lemma. **Calluses** with cobwebby hairs more than 67% as long as the lemma. **Lemmas** lanceolate, distinctly keeled, 2.3–3.5 mm; keels with usually sparse, inconspicuous, soft, short hairs in the proximal 67%; marginal veins glabrous or infrequently with short hairs on the proximal 25%, area between the veins smooth, glabrous; upper lemmas sometimes glabrous throughout; margins glabrous; tips acute. **Anthers** 1.3–2 mm. **2*n***=14.

Disturbed areas, riparian areas, moist deciduous forest, pastures, lawns, roadsides. 0–1700 m. BR, BW, Casc, Col, CR, Est, Sisk, WV. CA, ID, NV, WA; throughout most of North America; Europe. Exotic.

Poa trivialis is identified by rough, scabrous foliage and 1-veined, usually sickle-shaped lower glumes. *Poa pratensis* is more strongly rhizomatous and has smooth foliage, and has lower glumes that are usually 3-veined and lanceolate.

Poa unilateralis Scribner ex Vasey
San Francisco bluegrass

Plants perennial, 5–40 cm long, frequently glaucous; densely cespitose. **Basal branching** intra- and extravaginal or mainly intravaginal. **Leaves** sheaths closed 10(20)% of their length, bases of basal sheaths glabrous; ligules (0.8)2–6 mm long; blades of tillers usually 1–1.5 mm wide, thin, soon withering, distinctly narrower than those of the culm, infrequently broad and flat, or involute; cauline blades flat or folded, 2–5 mm wide, soft, thin and soon withering or moderately thick and somewhat fleshy and retaining their form, smooth, blades gradually reduced in length higher on the culm. **Inflorescences** erect, nearly cylindrical, contracted, 3–7 cm; spikelets (20)30–80(120); branches 0.5–1.5(4.5) cm long, 3–7 per node, densely papillose and sparsely scabrous. **Spikelets** lanceolate to narrowly ovate, to 3.5 times as long as wide, 4.5–7 mm, florets 3–5. **Glumes** lanceolate, slightly unequal; keels smooth and papillate or scabrous; lower glumes 3-veined. **Calluses** glabrous or with a crown of tiny crisp hairs around the base, hairs 0.1–0.2 mm long. **Lemmas** lanceolate, distinctly keeled, 3–4.5 mm, glabrous, or keels and marginal veins short-villous on proximal half, sparsely crisply puberulent near the base; margins glabrous; tips acute. **Anthers** 1.5–3 mm.

2 subspecies.

Poa unilateralis is a cespitose grass with dense inflorescences that grows on coastal headlands. Its leaf sheaths are open most of their length, and its calluses are glabrous or nearly so.

Key to *Poa unilateralis* subspecies:

1. Inflorescences contracted to open, elongate, with somewhat irregular outlines; longest branches (1.2)2.5–4 cm long; northern coast of Oregon .... ***P. u.*** ssp. ***pachypholis***

1' Inflorescences very densely ovoid, with almost smooth outlines; longest branches 0.5–1.5 cm long; southern coast of Oregon .... ***P. u.*** ssp. ***unilateralis***

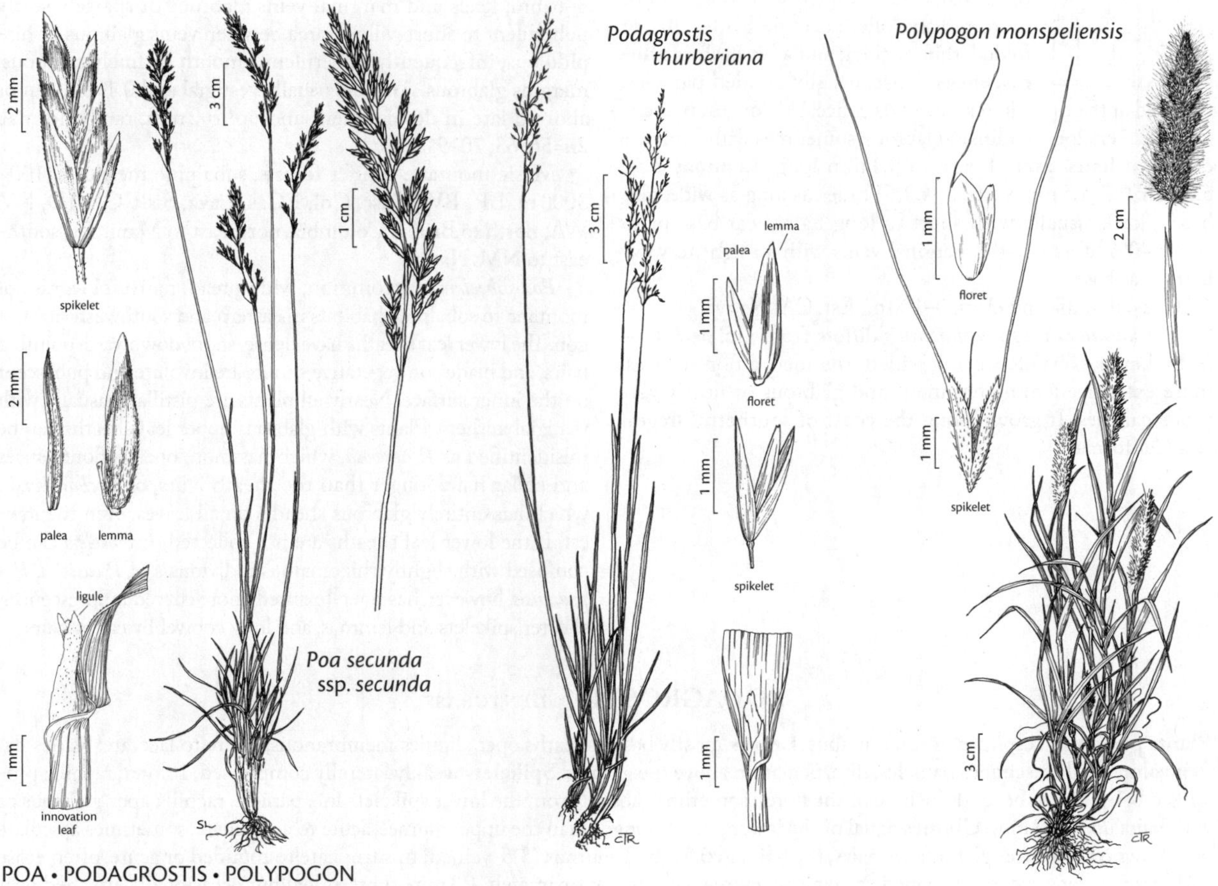

POA • PODAGROSTIS • POLYPOGON

Poa unilateralis Scribner ex Vasey ssp. *pachypholis* (Piper) D.D. Keck ex Soreng

Seacliff bluegrass

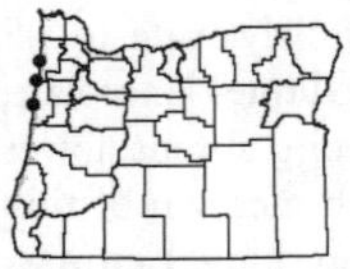

Leaves blades of sterile shoots similar to those of the culm, not fleshy. **Inflorescences** contracted panicles, with somewhat irregular outlines, tapering to the somewhat pointed tip; inflorescence axis somewhat hidden but easily exposed by moving the spikelets; longest branches (1.2)2.5–4 cm long. **Calluses** with a short crown of hairs around base, hairs 0.1–0.2 mm long, occasionally with a few longer straight or curly hairs to 0.5 mm long, rarely glabrous. **Lemmas** 3.5–6.2 times as long as wide, 2.8–5 mm long, 0.7–1.1 mm wide; glabrous or with sparse hairs between the veins near the base, rarely very short-hairy throughout; keels and marginal veins short-hairy for at least 50% of their length; tips acute to truncate.

Coastal bluffs and cliffs. 50–200 m. Est. WA. Native.

Poa unilateralis ssp. *pachypholis* is distinguished from ssp. *unilateralis* by having less dense, more irregularly shaped inflorescences with more tapering tips, less extensive hairiness on the lemmas and calluses with short hairs. It occurs along the coast from northern Oregon to southern Washington.

Poa unilateralis Scribner ex Vasey ssp. *unilateralis*

San Francisco bluegrass

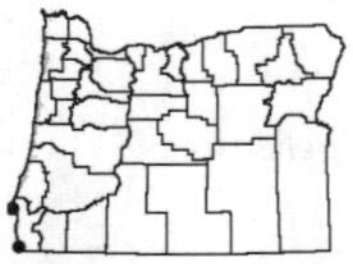

Leaves blades of sterile shoot blades sometimes narrower than those of the culm or more tightly rolled; cauline blades, flat or folded, thin or thick and a bit fleshy. **Inflorescences** dense, usually 1-sided panicles, rounded at the tip; inflorescence axis concealed; longest branches 0.5–1.5 cm long. **Calluses** glabrous, sometimes with a minute crown of hairs around base to 0.1 mm long. **Lemmas** 3.5–5.5 mm, 0.7–1.5 mm wide, 3–4.2(5) times as long as wide, glabrous; keels usually with short to long hairs near base or to about 40% of its length; marginal veins with straight to wavy hairs near base.

Coastal bluffs and cliffs. 0–100 m. Est. CA. Native.

Poa unilateralis ssp. *unilateralis* differs from ssp. *pachypholis* by having very dense, one-sided, rounded inflorescences, more extensive lemma hairiness, and glabrous or nearly glabrous calluses. It grows along the coast of southern Oregon and California.

Poa wheeleri Vasey

Wheeler's bluegrass

Poa curta Rydb., *Poa nervosa* (Hook.) Vasey var. *wheeleri* (Vasey) C.L. Hitchc.

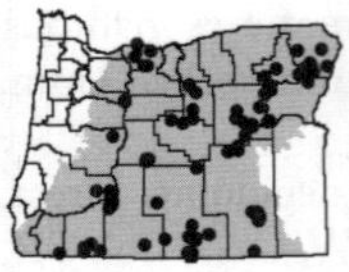

Plants perennial, pistillate, 35–80 cm tall, densely to loosely tufted, or with solitary shoots, short-rhizomatous. **Basal branching** mainly extravaginal. **Culms** nodes terete, 1–2 exserted. **Leaves** sheaths closed 33–75% of their length, bases of basal sheaths glabrous, at least some basal sheaths with the distal 25% retrorsely hairy; collars of proximal leaves glabrous or with hairs of the same length as those of sheaths; ligules 0.5–2 mm, ligules of sterile shoots of lower culm and sterile shoot leaves to 1.5 mm; blades of tillers folded or involute, infrequently flat, usually with upper surface densely scabrous to minutely hairy on and between the veins; cauline blades flat or folded, 2–3.5 mm wide, smooth or sparsely scabrous, glabrous or hispidulous, blades gradually reduced distally or middle blades longest; uppermost 1–10 cm long. **Inflorescences** erect or nodding at the tips, ovoid to pyramidal, loosely contracted to open, 5–12(18) cm, with 20–70 spikelets, with 2–5 branches per node; branches ascending to spreading or reflexed, lax, (1)1.7–6.5 cm long, with 2–8(12) spikelets. **Spikelets** to 3.5 times as long as wide, 5.5–10 mm, with 2–7 florets; rachilla internodes smooth or scabrous, glabrous or sparsely to densely hispidulous. **Glumes** lanceolate; lower glumes 1–3-veined, less than half as long as adjacent lemmas. **Calluses** glabrous. **Lemmas** lanceolate, distinctly keeled, 3–6 mm; keels and marginal veins glabrous or sparsely softly puberulent to short-villous, area between veins glabrous or hispidulous, infrequently puberulent, smooth or finely scabrous; margins glabrous. **Anthers** usually vestigial and 0.1–0.2 mm or aborted late in development and up to 2 mm, rarely normal. $2n$=56, 63, 70–91.

Mesic montane conifer forests, subalpine meadows. 100–3000 m. BR, BW, Casc, Col, ECas, Lava, Sisk. CA, ID, NV, WA; north to British Columbia, northeast to Manitoba, southeast to NM. Native.

Poa wheeleri is a common, widespread, native bluegrass of montane to subalpine habitats in eastern and southwestern Oregon. The lower leaf sheaths have dense, short, downward-pointing hairs, and blades on vegetative shoots are involute and pubescent on the inner surface. Nearly all plants are pistillate, usually with vestigial anthers. Plants with glabrous upper leaf sheaths can be misidentified as *P. nervosa,* which has more open inflorescences and collar hairs longer than the sheath hairs, or *P. chambersii,* which has entirely glabrous sheaths on all leaves, even the lowest. If the lower leaf sheaths are not collected, *P. wheeleri* can be confused with slightly rhizomatous individuals of *P. cusickii*. *Poa pratensis*, however, has finer-textured, more curved inflorescences, shorter spikelets and lemmas, and long cobwebby callus hairs.

PODAGROSTIS Bentgrass

Plants perennial; cespitose or rhizomatous. **Leaves** mostly basal; sheaths open; ligules membranous, entire to lacerate; blades flat to involute. **Inflorescences** panicles; disarticulation above the glumes. **Spikelets** weakly laterally compressed, 1 floret; rachilla prolonged 0.1–1.9 mm beyond the base of the floret, sometimes absent from the lower spikelets in a panicle, rachilla apex glabrous or with hairs up to 0.3 mm. **Glumes** equal or the lower glumes longer than the upper glumes, acute to acuminate, sometimes apiculate but unawned. **Calluses** glabrous or hairy, the hairs to 0.5 mm. **Lemmas** (3)5-veined; tips truncate to rounded or acute, often erose with lemma veins excurrent, awned or awnless; lemma awns to approximately 1.3 mm, subterminal but occasionally attached near mid-length. **Paleas** more than half as long as the lemmas, 2-veined, thinner than the lemmas. **Anthers** 3.

North America and South America. Approximately 6 species; 2 species treated in *Flora*.

The *Podagrostis* species have previously been treated within *Agrostis*, but recent work indicates they are not closely related to that genus. In *Podagrostis*, the rachilla is usually prolonged beyond base of the floret; *Agrostis* does not have a prolonged rachilla. In addition, *Podagrostis* has opaque lemmas with relatively conspicuous veins and well-developed paleas.

Key to *Podagrostis* species:

1. Plants often less than 15 cm tall; leaves 0.3–0.8(1.5) mm wide; paleas 67–75% as long as the lemmas; rachillas prolonged 0–0.1 mm beyond the floret base . . . . ***P. humilis***

1' Plants usually more than 15 cm tall; leaves 1–3 mm wide; paleas almost as long as the lemmas; rachillas prolonged 0.1–0.5 mm beyond the floret base . . . . ***P. thurberiana***

Podagrostis humilis (Vasey) Björkman

Alpine bentgrass

Agrostis humilis Vasey

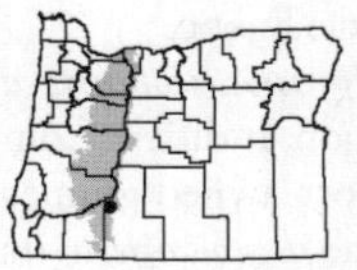

Plants 3–15(26) cm tall; cespitose, sometimes rhizomatous. **Leaves** blades flat or folded, 1.5–4(5) cm × 0.3–0.8(1.5) mm; flag leaf blade usually below mid-culm, sometimes near mid-culm. **Inflorescences** linear to narrowly oblong, 1.5–6 × 0.2–0.7(1.5) cm, sometimes wider at anthesis; branches appressed to erect, ascending at anthesis. **Spikelets** usually purple; rachilla prolongations 0–0.1 mm. **Glumes** ovate to elliptic, 1.6–2 mm, usually exceeding the lemma by 0.2–0.4 mm. **Calluses** glabrous or sparsely hairy, hairs to 0.5 mm. **Lemmas** 1.5–2 mm, usually purplish, opaque, smooth, 5-veined; tips usually awnless; awns, if present, to approximately 1.3 mm. **Paleas** 67–75% as long as the lemma.

Wet alpine and subalpine meadows. 1100–2700 m. Casc. CA, ID, NV, WA; north to AK, northeast to Alberta and MT, southeast to CO. Native.

Podagrostis humilis is most similar to *Agrostis variabilis*, which has paleas usually less than a third as long as the lemmas.

Podagrostis thurberiana (Hitchcock) Hultén (p. 467)

Thurber's bentgrass

Agrostis thurberiana Hitchc.

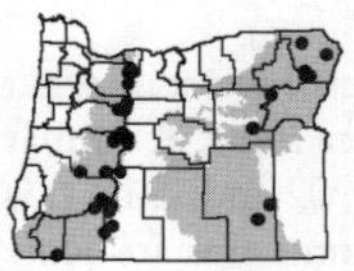

Plants 10–40(50) cm tall; cespitose, sometimes rhizomatous. **Leaves** blades flat or sometimes folded, 3–10(15) cm × 1–3 mm; flag leaf blade usually near mid-culm. **Inflorescences** linear to narrowly oblong, (2.5)5–12 × 0.5–0.7(2.5) cm; branches erect to ascending. **Spikelets** purple or sometimes greenish; rachilla prolongations 0.1–0.5 mm, glabrous or with a tuft of short hairs at the apex. **Glumes** elliptic, 2–2.5 mm, as long as the lemmas or up to 0.1 mm longer. **Calluses** glabrous. **Lemmas** 2–2.5 mm, greenish or partly purplish, smooth, opaque, 5-veined; tips usually awnless; awns, if present, to approximately 1.3 mm. **Paleas** approximately as long as the lemma.

Montane to alpine wet meadows. 900–2600 m. BR, BW, Casc, Sisk. CA, ID, NV, WA; north to AK, southeast to CO. Native.

Podagrostis thurberiana looks like *Agrostis*, but its leaves are a bit wide, its lemmas are more opaque and conspicuously veined and its rachilla is prolonged beyond the floret. Occasional depauperate *P. thurberiana* plants might look somewhat like *P. humilis*, but they can always be distinguished by the rachilla prolongation, which is nearly absent in *P. humilis*.

POLYPOGON Rabbitsfoot grass, beardgrass

Plants annual or perennial, 4–120 cm tall, not rhizomatous. **Culms** rooting at the lower nodes, branched near the base. **Leaves** sheaths open; ligules membranous or hyaline, erose, ciliate; blades flat to convolute. **Inflorescences** dense panicles, continuous or interrupted below; disarticulation at the base of the stipe. **Spikelets** borne on a stipe connected to a pedicel or directly to the panicle branch; laterally compressed, 1 floret; rachilla not prolonged beyond the floret. **Glumes** lanceolate, 2, longer than the floret, bases not fused; tips entire to emarginate or bilobed, usually awned; glume awns arising from between the lobes, sometimes at the tip, curved, glabrous. **Lemmas** 1–3(5)-veined, often awned; lemma awns terminal, subterminal, or arising from just above mid-length. **Paleas** 33% to as long as the lemmas. **Anthers** 3. **Caryopses** slightly flattened, broadly ellipsoid to oblong-ellipsoid.

Tropical and warm-temperate regions worldwide. Approximately 18 species; 4 species treated in *Flora*.

Polypogon species resemble *Agrostis* but have strongly scabrous, usually awned glumes that are supported on a stipe. *Polypogon* occasionally hybridizes with *Agrostis*. The hybrids look like *Agrostis*, but the spikelets disarticulate below the glumes, often at the base of a stipe. *Polypogon fugax* was collected in Portland in the early 1900s but apparently has not persisted. It is an annual with glume awns similar to *P. interruptus*, but its glumes have lobes 0.1–0.2 mm.

Key to *Polypogon* species:

1. Glumes with awns 4–12 mm.
   2. Glume lobes 0.1–0.2 mm long, to 10% as long as the glume bodies . . . . ***P. monspeliensis***
   2' Glume lobes 0.3–1.2 mm long, more than 17% as long as the glume bodies. . . . . ***P. maritimus***

1' Glumes unawned or with awns to 3.2 mm.
   3. Glumes unawned. . . . . ***P. viridis***
   3' Glumes awned, the awns 1.5–3.2 mm . . . . ***P. interruptus***

Polypogon interruptus Kunth

Ditch beardgrass

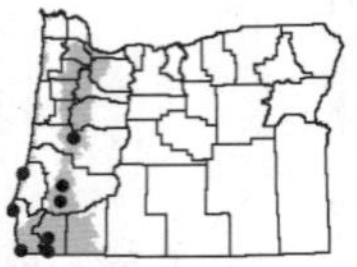

Plants perennial, often flowering the first year, 20–80(90) cm tall. **Leaves** sheaths smooth; ligules 2–6 mm; blades 5–9 cm × 3–6 mm. **Inflorescences** usually interrupted or lobed, 3–15 × 0.5–3 cm; pedicels not developed; stipes 0.2–0.7 mm. **Glumes** 2–3 mm, scabrous, larger prickles extending up the keel beyond mid-length; glumes not tapering to the tips; tips acute to truncate, unlobed or the lobes to 0.1 mm, awned; glume awns 1.5–3.2 mm. **Lemmas** 0.8–1.5 mm, glabrous, smooth; tips obtuse, not emarginate, awned; lemma awns 1–3.2 mm. **Anthers** 0.5–0.7 mm. **2*n*=42.**

Wet, often disturbed sites. 0–500 m. Est, Sisk, WV. CA, ID, NV, WA; east to NE, southeast to LA, south to South America; NY, WI. Native.

Polypogon interruptus has a lobed inflorescence somewhat like a sparse *P. monspeliensis* but with shorter awns. A similar grass, *P. fugax*, was collected in Portland in 1902 and 1916. It is an annual with leaves 2–11 mm wide. Its glume tips are acute to rounded and have lobes 0.1–0.2 mm. Its lemma awns are up to 2 mm.

Polypogon maritimus Willdenow

Mediterranean beardgrass

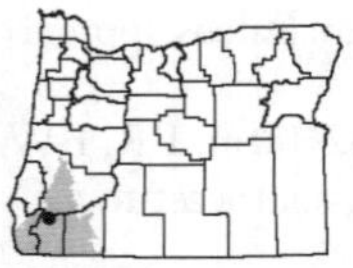

Plants annual, (5)20–40 cm tall. **Leaves** sheaths glabrous; the uppermost sheaths sometimes inflated; ligules to 7 mm; blades (1)3–9(14) cm × 0.5–5 mm. **Inflorescences** narrowly ellipsoid, sometimes lobed, (1)2–8(15) cm; pedicels to 0.5 mm; stipes 0.1–1.2 mm. **Glumes** 1.8–8.2 mm, ciliate, with large hairs or protrusions in the lower half; tips lobed, lobes 0.3–1.2 mm, about 17% as long as the glume body, awned; glume awns 4–10 mm, yellowish. **Lemmas** 0.5–1.5 mm, glabrous, awned or awnless; lemma awns 0–1 mm. **Anthers** 0.4–0.5 mm. **2*n*=14, 42.**

Disturbed wet sites, streamsides. 100–300 m. Sisk. CA, NV; FL, GA, SC; Eurasia. Exotic.

This rare or overlooked introduction has lobed, ciliate glumes that look almost spiny near the base.

Polypogon monspeliensis (Linnaeus) Desfontaines (p. 467)

Rabbitsfoot grass, annual beardgrass

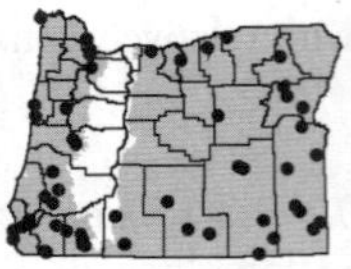

Plants annual, 5–65 cm tall. **Leaves** sheaths glabrous; the uppermost sheaths sometimes inflated; ligules 2.5–16 mm; blades 1–20 cm × 1–7 mm. **Inflorescences** narrowly ellipsoid, sometimes lobed, 1–17 cm; pedicels absent or to 0.2 mm; stipes 0.1–0.2 mm. **Glumes** 1–2.7 mm, ciliate, minutely hairy throughout, largest hairs restricted to the lower half; tips rounded, lobed, lobes 0.1–0.2 mm, 10% or less as long as the glume body, awned; glume awns 4–10 mm, yellowish. **Lemmas** 0.5–1.5 mm, glabrous, awned; lemma awns 0.5–1(4.5) mm. **Anthers** 0.2–1 mm. **2*n*=28, 35.**

Disturbed wet sites, ditches, roadsides, often on alkaline soils. 0–1700 m. All ecoregions except Casc. CA, ID, NV, WA; throughout most of North America; worldwide. Exotic.

A hybrid of *P. monspeliensis* and *Agrostis stolonifera* (×*Agropogon lutosus*) has been found in Oregon. It differs from *P. monspeliensis* in having more tapered, short-awned glumes and lemmas with subterminal awns. Unlike *A. stolonifera*, its spikelets are on short stipes.

Polypogon viridis (Gouan) Breistroffer

Water beardgrass

Agrostis semiverticillata (Forssk.) C. Chr., *Agrostis verticillata* Vill.

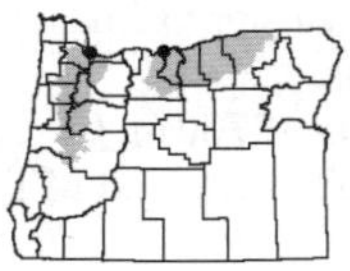

Plants perennial, often flowering the first year, 10–90 cm tall. **Leaves** sheaths glabrous, smooth; ligules to 5 mm; blades 2–13 cm × 1–6 mm. **Inflorescences** ovoid-oblong to pyramidal, interrupted, 2–10 cm; pedicels not developed; stipes 0.1–0.6 mm. **Glumes** 1.5–2 mm, scabrous on the back and keel; tips obtuse to truncate, awnless. **Lemmas** approximately 1 mm; tips erose, awnless. **Anthers** 0.3–0.5 mm. **2*n*=28.**

Mesic stream banks, rivers, ditches. 0–100 m. Col, WV. CA, NV, WA; north to British Columbia, southeast to TX; Eurasia. Exotic.

This small grass resembles an *Agrostis* with unusually scabrous glumes, but its spikelets disarticulate below the glumes. Unlike other *Polypogon* species, it lacks glume awns. It has been collected from two sites along the Columbia River.

×PSEUDELYMUS Hybrid wheatgrass

Plants perennial; cespitose or short-rhizomatous. **Culms** glabrous. **Leaves** not concentrated at the base, auricles present; ligules truncate. **Inflorescences** 2-sided spikes with 1(2) spikelet per node; disarticulation below the florets and usually in the inflorescence axis. **Spikelets** with 3–5 florets. **Glumes** linear-lanceolate to lanceolate. **Lemmas** awned or awnless. **Anthers** indehiscent.

Western North America. 1 species.

×***Pseudelymus saxicola*** (Scribner & J.G. Smith) Barkworth & D.R. Dewey

Hybrid wheatgrass

×*Agrositanion saxicola* (Scribn. & J.G. Sm.) Bowden, *Agropyron saxicola* (Scribn. & J.G. Sm.) Piper

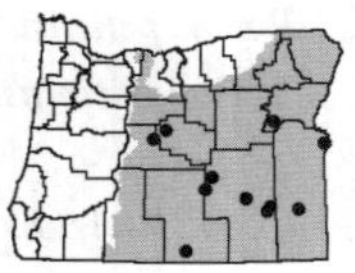

Plants 50–80 cm tall; cespitose. **Culms** 0.8–1 mm thick. **Leaves** sheaths glabrous to puberulent; ligules truncate, 0.2–0.4 mm; blades 1.5–2 mm wide. **Inflorescences** 7–14 cm, with 1(2) spikelet per node; internodes 7–9 mm. **Glumes** 6–8 × 0.7–1 mm, (1)3–4-veined; awns 7–15 mm, spreading. **Lemmas** 10–11 mm; lemma awns (4)8–26+ mm, straight to spreading. **Anthers** approximately 2.5 mm. **2*n*=14, 28.**

Dry grassland, shrub steppe. 600–2500 m. BR, BW, ECas, Lava, Owy. CA, WA; east to MT, KS, and NM. Native.

×*Pseudelymus saxicola* is a hybrid of *Pseudoroegneria spicata* and *Elymus elymoides*. It has narrower leaves, longer internodes, and less overlapping spikelets than most wheatgrass hybrids. It is most likely to be confused with *Elymus* ×*saundersii* but has longer glumes and lemma awns.

PSEUDOROEGNERIA Bluebunch wheatgrass

Plants perennial; cespitose, sometimes rhizomatous. **Leaves** basal and cauline; sheaths open; auricles well developed; ligules membranous; blades flat to loosely involute. **Inflorescences** terminal spikes with 1 spikelet per node; internodes (7)10–20(28) mm at mid-length; lower internodes often longer than those at mid-length; disarticulation above the glumes and beneath the florets. **Spikelets** (8)12–25 mm, 1.1–1.5(2) times as long as the internodes, usually appressed, 4–9 florets. **Glumes** lanceolate to oblanceolate, narrowing beyond mid-length, from shorter than to slightly longer than the lowest lemma; veins prominent, (3)4–5(7)-veined, acute to obtuse, occasionally truncate. **Lemmas** inconspicuously 5-veined, awned or awnless; lemma awns, if present, straight or strongly bent and divergent. **Paleas** subequal to lemmas. **Anthers** 3.

North America and Eurasia. Approximately 15–20 species; 1 species treated in *Flora.*

Pseudoroegneria spicata has previously been treated in the genera *Agropyron, Elymus, Elytrigia,* and *Roegneria.* It may be conspecific with up to three named East Asian species of *Pseudoroegneria.* Bluebunch wheatgrass exhibits variation in awn presence/absence and pubescent vs. glabrous forms.

Pseudoroegneria spicata (Pursh) Á. Löve (p. 473)
Bluebunch wheatgrass

Agropyron inerme (Scribn. & J.G. Sm.) Rydb., *Agropyron spicatum* (Pursh) Scribn. & J.G. Sm., *Agropyron spicatum* (Pursh) Scribn. & J.G. Sm. var. *pubescens* Elmer, *Agropyron spicatum* (Pursh) Scribn. & J.G. Sm. var. *spicatum, Pseudoroegneria spicata* (Pursh) Á. Löve ssp. *spicata*

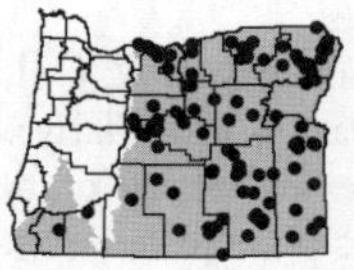

Plants 30–100 cm tall; cespitose, sometimes rhizomatous. **Culms** erect or ascending, nodes glabrous. **Leaves** sheaths glabrous; auricles well developed; ligules membranous; blades flat to loosely involute, 2–6 mm wide; lower surfaces smooth, glabrous; upper surfaces scabrous or long hairy. **Inflorescences** spikes 8–15 × 0.3–0.8(1) cm. **Spikelets** 8–22(25) mm, 4–9 florets, 1.1–1.5(2) times the length of the internodes. **Glumes** unequal, 6–13 mm, 4–7-veined, glabrous, sometimes scabrous over the veins. **Lemmas** 9–14 mm, 5-veined, glabrous to scaberulous or very shortly puberulent; tips acuminate, awned or not; lemma awns 0–25 mm, terminal, straight to strongly divergent. **Anthers** 4–8 mm. **2*n***=14, 28.

Medium-textured soils in arid and semi-arid steppe, shrub-steppe, and open woodland communities. 50–2900 m. BR, BW, Col, ECas, Lava, Owy, Sisk. CA, ID, NV, WA; north to AK, east to NE, south to northern Mexico. Native.

Elymus wawawaiensis is very similar but is usually more vigorous and has narrowly lanceolate glumes with 1–3 veins, shorter internodes, and more overlapping spikelets. Some *Pseudoroegneria spicata* are awnless due to a single, genetically dominant gene. Awnless plants can be common in some parts of eastern Oregon, and mixed populations exist. In moist habitats, *P. spicata* is sometimes rhizomatous but rarely flowers. Usually the leaves are scabrous, but some northeastern Oregon plants have hairy leaves. *Pseudoroegneria spicata* hybridizes with several species of *Elymus,* including *E. canadensis, E. elymoides, E. lanceolatus,* and *E. multisetus.* Hybrids are mostly sterile, but a few seeds may be produced, allowing introgression to occur between the parental populations. Hybrids with *E. elymoides,* called *Pseudelymus* × *saxicola,* are encountered fairly often. They resemble *Pseudoroegneria spicata* but disarticulate in the inflorescence axis and are usually sterile.

PUCCINELLIA Alkali grass

Plants perennial; cespitose or sometimes more or less stoloniferous, sometimes mat-forming. **Culms** erect or decumbent. **Leaves** sheaths open to the base or nearly so; ligules membranous, entire or erose; blades flat, folded or involute. **Inflorescences** panicles; branches spreading to appressed; disarticulation above the glumes, beneath the florets. **Spikelets** more or less terete or somewhat laterally flattened. **Glumes** 2, usually unequal, usually much shorter than the lowest lemma, rounded or weakly keeled on the back, awnless. **Calluses** blunt, glabrous or pubescent. **Lemmas** membranous, glabrous or more or less pubescent near the base, mainly on the veins, rounded over the back, (3)5(7)-veined; veins prominent or faint, more or less parallel, not converging at the tip; tips acute to truncate, occasionally acuminate, awnless. **Paleas** 2-veined; veins terminating at or beyond the midpoint. **Anthers** 3. **Caryopses** oblong, terete to dorsally flattened, falling free or with the palea attached.

Temperate to arctic regions of the Northern Hemisphere. Approximately 120 species; 5 species treated in *Flora.*

Most *Puccinellia* species live in saline or alkaline environments. Self-pollination, hybridization, and polyploidy have combined to create a confusing pattern of variation. The parallel lemma veins can cause confusion with *Poa.* Most *Poa* species have more strongly laterally flattened spikelets, with lower glumes that are usually more than half as long as the lower lemma. Lower glumes of *Puccinellia* are often a third or less as long as the lower lemma.

The angle of the panicle branches (erect, descending, etc.) should be judged when the seeds are mature. Lemma lengths should be measured on the lowest lemma in the spikelet. Tumid cells are swollen, so the pedicel surface looks somewhat beaded. In pedicels with tumid cells, a small part of the pedicel just below the glumes lacks scabers and is somewhat thickened; very short pedicels may be entirely free of scabers and have all their surface cells tumid. The absence, presence, and density of scabrules on the lemma margins are important features for identification. In this genus, scabrules are short pointed hairs or very fine teeth, seen with 20× or stronger magnification. In some species, e.g., *P. nutkaensis* and *P. nuttalliana,* the scabrules are very dense, with little or no space between them. In others, such as *P. pumila,* the lemma margin is mostly smooth or may have a few scattered scabrules, widely separated or a few together. The lemma margins may be rolled under, hiding the scabrules, but examining several lemmas under the microscope will reveal some partly exposed lemma margins on which this trait can be assessed.

Key to *Puccinellia* species:

1. Plants coastal.
    2. Plants shorter (to 40 cm); pedicels smooth, not scabrous; lemma margins lacking scabrules or with a few scabrules at and near the tips. . . . . *P. pumila*
    2' Plants usually taller (10–100 cm); pedicels scabrous; lemma margins with dense scabrules at and near the tips.
        3. Pedicel epidermal cells often tumid; lower branches of the panicles usually erect to ascending, occasionally spreading to descending; lemmas (2.2)2.8–4.5(5) mm; lower glumes 1–1.6 mm . . . . *P. nutkaensis*
        3' Pedicel epidermal cells not tumid; lower branches of the panicles erect to descending; lemmas (2)2.2–3(3.5) mm; lower glumes 0.5–1.5 mm . . . . *P. nuttalliana* (in part)

1' Plants east of the Cascade Range.
    4. Lemmas 1.5–2.2 mm, tips widely obtuse to truncate; anthers 0.4–0.8 mm; lower panicle branches horizontal to descending . . . . *P. distans*
    4' Lemmas 2–5 mm, tips usually acute to obtuse, occasionally acuminate or rounded; anthers 0.5–2.2 mm; lower panicle branches erect to descending.
        5. Leaves involute, very narrow, 1.2–1.9 mm wide when flattened; lemma tips acute; lemma apical margins smooth to scabrous . . . . *P. lemmonii*
        5' Leaves involute or flat, usually wider, 1–4 mm wide when flat; lemma tips usually obtuse, sometimes acute; lemma apical margins densely scabrous . . . . *P. nuttalliana* (in part)

Puccinellia distans (Jacquin) Parlatore

European alkali grass

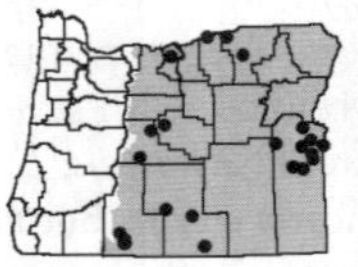

Plants perennial, 5–60 cm tall; cespitose. **Leaves** blades flat to involute, 1–7 mm wide. **Inflorescences** panicles open at maturity, 2.5–20 cm; lower branches horizontal to descending; spikelets usually confined to distal 67% of the branches. **Spikelets** 2.5–7 mm, 2–7 florets. **Glumes** rounded over the back; veins obscure; tips acute to truncate; lower glumes 0.4–1.3 mm; upper glumes 0.9–1.8 mm. **Lemmas** 1.5–2(2.2) mm, 5-veined; veins obscure, not extending to the margins; margins near the tip hyaline and often yellowish, uniformly and densely scabrous; tips widely obtuse to truncate, entire, awnless. **Anthers** 0.4–0.8 mm. **2*n***=14, 28, 42.

Disturbed alkaline sites, lake margins, stream banks, alkaline meadows, roadsides. 100–1400 m. BR, BW, Col, ECas, Lava, Owy. CA, ID, NV, WA; throughout northern North America; Eurasia. Exotic.

Puccinellia distans has the shortest lemmas of any Oregon *Puccinellia*. Its lower inflorescence branches angle downward. It has not been found at the coast but should be looked for there.

Puccinellia lemmonii (Vasey) Scribner

Lemmon's alkali grass

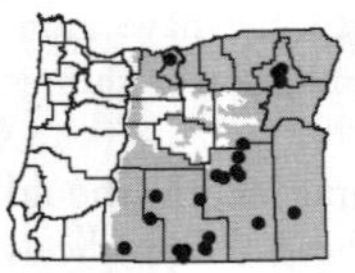

Plants perennial, 5–40 cm tall; cespitose. **Leaves** blades involute, 1.2–1.9 mm wide when flat. **Inflorescences** panicles compact to open at maturity, 2–18 cm, lower branches ascending to descending, usually spikelet-bearing to the base; pedicels scabrous, lacking tumid epidermal cells. **Spikelets** 3.5–8 mm, 2–6 florets. **Glumes** rounded over the back; veins obscure; tips acute to obtuse; lower glumes 0.7–1.5 mm; upper glumes 1.4–3 mm. **Lemmas** 2.4–4 mm, sometimes weakly keeled, 5-veined; veins obscure, midveins often slightly scabrous and prominent in the distal half, often extending to the margins near the tip; lateral veins not extending to the margins, apical margins; smooth to scabrous, entire; tips acute, entire. **Anthers** 1–2 mm. **2*n***=14.

Alkaline flats and meadows. 800–1700 m. BR, BW, Col, ECas, Owy. CA, ID, NV; northeast to MT, east to WY. Native.

This grass is unusual among *Puccinellia* species because of its dense basal tuft of very narrow leaves and its pointed lemmas. *Puccinellia nuttalliana* may have leaves concentrated at the base but usually has leaves scattered up the culm as well. Additionally, its lemmas are usually obtuse. Fine-leaved *Poa secunda* ssp. *secunda* has a superficially similar appearance but has relatively longer glumes and tends to live in drier, less strongly alkaline habitats.

Puccinellia nutkaensis (J. Presl) Fernald & Weath.

Alaska alkali grass

Puccinellia grandis Swallen, *Puccinellia lucida* Fernald & Weath.

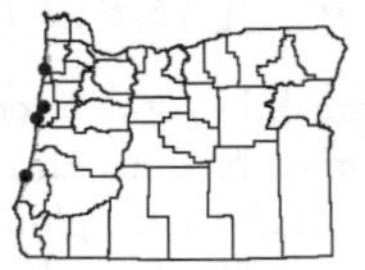

Plants perennial, 10–90 cm tall; cespitose, sometimes rooting at the nodes of buried stems and appearing rhizomatous or stoloniferous. **Leaves** blades flat to involute, 1.5–6 mm wide when flat. **Inflorescences** panicles compact to open at maturity, 5–30 cm, lower branches erect to ascending, occasionally spreading to descending; spikelet-bearing from near the base, sometimes the spikelets confined to the distal half; pedicels sparsely to densely scabrous, often subtended by a small bract; epidermal cells often tumid. **Spikelets** 3.5–12 mm, 3–7 florets. **Glumes** rounded over the back; veins obscure; tips acute to truncate; lower glumes 1–1.6 mm; upper glumes 2–3 mm. **Lemmas** (2.2)2.8–4(5) mm, 5-veined; veins obscure, not extending to the margins, midveins smooth distally; margins near the tips uniformly and densely scabrous; tips usually acute to obtuse, sometimes acuminate, entire. **Anthers** 0.5–1.4 mm. **2*n***=42, 56.

Coastal salt marshes. 0–50 m. Est. CA, WA; north to AK, most of Canada east to Greenland. Native.

Similar *Puccinellia pumila* lacks the dense scabrules along the lemma margins near the tip. Larger *P. nutkaensis* plants, with more open panicles, can be confused with *P. nuttalliana*, which is rare at the coast. *Puccinellia nuttalliana* lacks tumid cells on the pedicels, tends to be more strongly scabrous on inflorescence branches and pedicels, with the pedicels scabrous very nearly to the top, and has lemma margins that are entire except for the dense scabrules.

Puccinellia nuttalliana (Schultes) Hitchcock (p. 473)

Nuttall's alkali grass

Puccinellia airoides S. Watson & J.M. Coult., *Puccinellia cusickii* Weath.

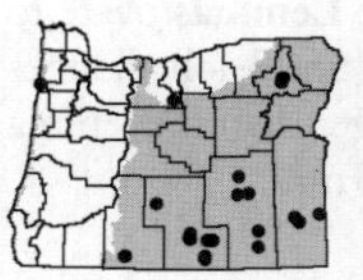

Plants perennial, 10–100 cm tall; cespitose. **Leaves** blades flat to involute, 1–4 mm wide. **Inflorescences** panicles compact to open at maturity, 5–30 cm, usually distinctly exserted from the sheaths, lower branches erect to diverging or occasionally descending, spikelet-bearing from the base or in the distal 67%; pedicels scabrous, lacking tumid epidermal cells. **Spikelets** 3.5–9 mm, 2–7 florets. **Glumes** rounded over the back; veins obscure; tips acute to obtuse; lower glumes 0.5–1.5 mm, usually less than half as long as the adjacent lemmas; upper glumes 1–2.8 mm. **Lemmas** (2)2.2–3(3.5) mm, 5-veined; veins obscure, not extending to the margins, smooth distally; lateral margins inrolled or not; margins near the tips uniformly and densely scabrous; tips acute to obtuse, entire. **Anthers** 0.6–2 mm. **2*n***=28, 42, 56.

Alkaline and saline areas. 800–1400 m. BR, BW, ECas, Est, Lava, Owy. CA, ID, NV, WA; western and northern North America. Native.

Similar *Puccinellia lemmonii* has a dense tuft of very narrow basal leaves and acute lemmas. See also discussion for *P. nutkaensis*.

Puccinellia pumila (Vasey) Hitchcock

Smooth alkali grass

Puccinellia paupercula (Holm) Fernald & Weath. var. *alaskana* (Scribn. & Merr.) Fernald & Weath.

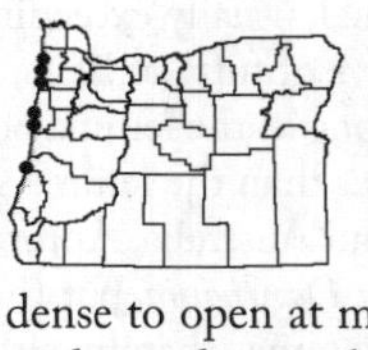

Plants perennial, 8–40 cm tall; cespitose or sometimes rooting at the nodes of buried stems and appearing rhizomatous or stoloniferous. **Leaves** blades flat to involute, 1–3 mm wide. **Inflorescences** panicles, dense to open at maturity, 3–20 cm, lower branches ascending to descending, spikelets borne from near the base or confined to the distal 67%; pedicels smooth or with a few scattered scabers, often with tumid epidermal cells distally. **Spikelets** 4–9 mm, 3–7 florets. **Glumes** rounded over the back; veins obscure to distinct; tips acute to obtuse; lower glumes 1.4–2(4) mm; upper glumes 2–3(9) mm. **Lemmas** 2.5–3.5(4.6) mm, 5-veined; veins obscure, not extending to the margins; margins near the tip usually smooth, occasionally with a few scattered scabrules, entire; tips acute to obtuse, entire. **Anthers** 0.7–1.2 mm.

Coastal salt marshes. 0–50 m. Est. Native. CA, WA; northwestern and northeastern North America. Native.

Puccinellia pumila is a short grass of coastal salt marshes, with nearly smooth inflorescence branches and pedicels. Its lemma margins lack the pointed hairs and scabrules that characterize *P. nutkaensis*.

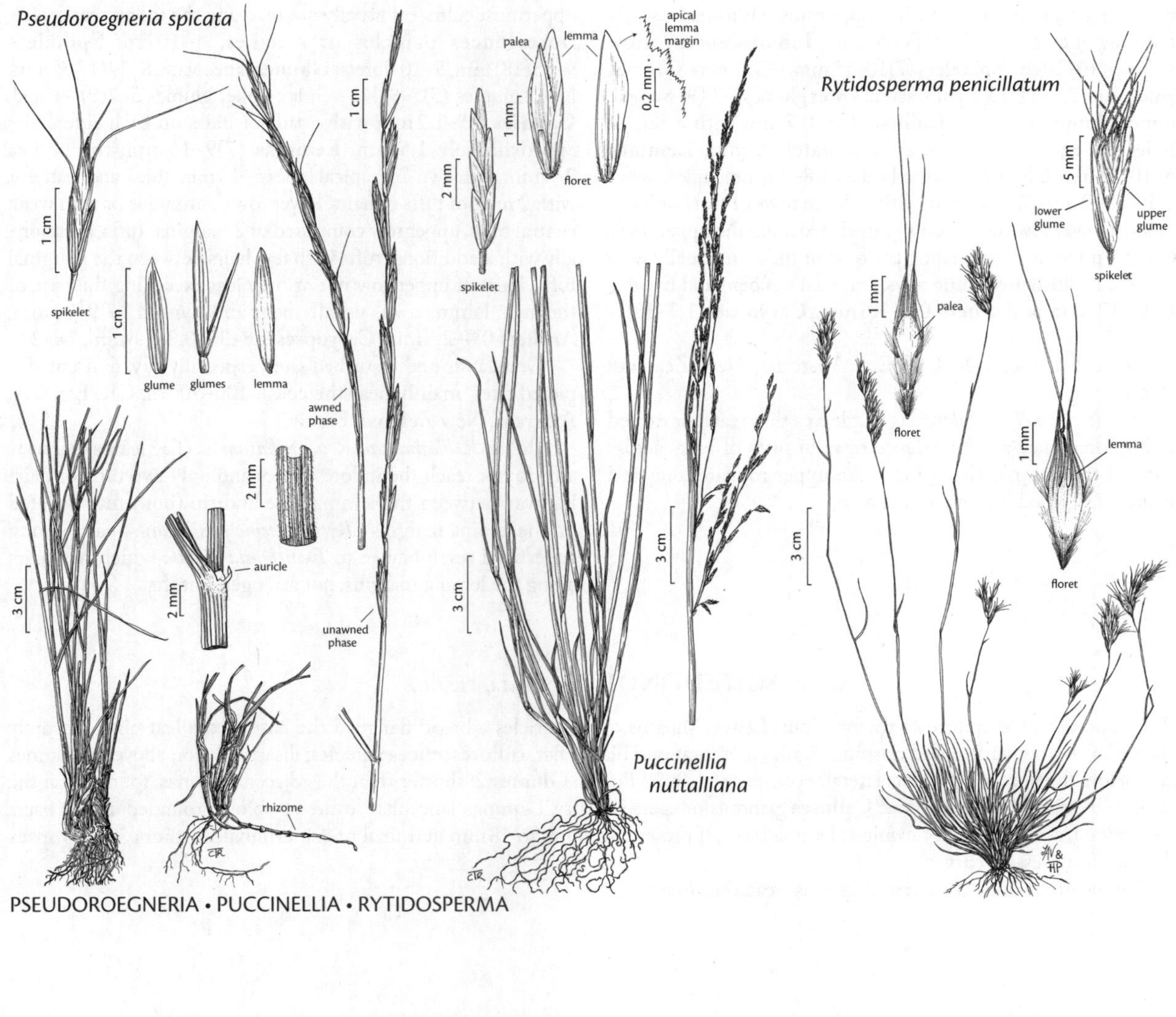

PSEUDOROEGNERIA • PUCCINELLIA • RYTIDOSPERMA

RYTIDOSPERMA Wallaby grass

Plants perennial; cespitose or somewhat spreading on short rhizomes. **Leaves** sheaths open, glabrous or hairy, often with a tuft of spreading hairs at the top; ligules a fringe of hairs; blades flat or involute; basal leaves much shorter than the culms. **Inflorescences** panicles or racemes; disarticulation above the glumes and below the florets. **Spikelets** terete, 3–10 florets, the florets bisexual, the terminal floret reduced. **Glumes** subequal, usually exceeding the spikelets (excluding awns), 5–11-veined. **Lemmas** ovate to lanceolate, 5–9-veined with 2 transverse rows of tufts of hairs, the rows sometimes reduced to marginal tufts; tips 2-lobed, lobes drawn out into awn-like bristles at least as long as the lemma body; awns present, originating between and longer than the lemma lobes, twisted and usually bent. **Paleas** longer than the lemma. **Anthers** 3. **Caryopses** obovate to elliptic, 1.2–3 mm.

North America, South America, Asia, and Australia. Approximately 45 species; 2 species treated in *Flora.*

Rytidosperma has often been included in *Danthonia*, but that genus is now restricted to Northern Hemisphere plants having hair tufts only at the base of the callus and lemmas that are either glabrous or with evenly distributed hairs.

Key to *Rytidosperma* species:

1. Uppermost culm leaf usually reaching or exceeding the inflorescence; upper row of lemma hairs in a more or less continuous row of tufts, the hairs much exceeding the base of the awn. . . . . ***R. biannulare***

1' Uppermost culm leaf usually not reaching the inflorescence; upper row of lemma hairs in isolated tufts or only at the margins, the hairs not or only just exceeding the base of the awn . . . . ***R. penicillatum***

Rytidosperma biannulare (Zotov) Connor & Edgar
Tasmanian wallaby grass

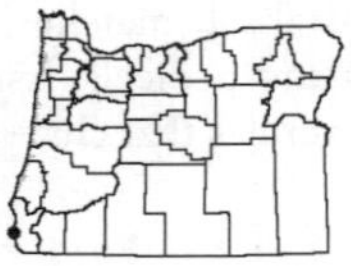

Plants 30–85 cm tall; cespitose. **Leaves** sheaths glabrous or hairy, the top with tufts of hair to 5 mm, these sometimes extending across the collar; blades usually involute, 30–40 cm × 5 mm; margins and sometimes upper surfaces scabrous; young blades sparsely spreading-hairy, becoming glabrous at maturity; uppermost culm leaf usually reaching or exceeding the inflorescence. **Inflorescences** narrow panicles 10–20 cm. **Spikelets** (7)10–15 mm, 6–7 florets. **Glumes** lanceolate, 7.5–11(13.2) mm, acute; lower glumes 5–7(9)-veined; upper glumes 5-veined. **Calluses** 0.5–0.7 mm with a tuft of hairs on each side, hairs approximately 1 mm. **Lemmas** 6–10(12.5) × 0.5–0.7 mm, the bodies 1.8–2.8 mm, apical lobes 3.5–5(8.5) mm, 7–9-veined, with 2 dense rows of tufts of hairs, hairs of lower rows not or only scarcely reaching the upper rows, hairs of upper rows exceeding the base of the awn, usually with scattered hairs between the rows; lemma awns bent and twisted, 6–10(12.5) mm. **Anthers** 0.8–1.6 mm. **Caryopses** 1.2–1.9 × 0.6–0.8 mm.

Disturbed areas. 0–50 m. Est. Australia, New Zealand. Exotic.

Rytidosperma biannulare has flag leaves that reach or exceed the inflorescence and two dense rows of tufts of hairs across back of the lemma. The hairs of the upper row are long and white and exceed the base of the awn.

Rytidosperma penicillatum (p. 473)
(Labillardière) Connor & Edgar
Hairy wallaby grass, hairy oatgrass, poverty grass

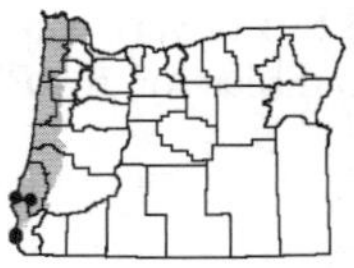

Plants 30–90 cm tall; cespitose to spreading on short rhizomes. **Leaves** sheaths hairy or glabrous, the top with tufts of hairs 1–3.5 mm; blades rolled, folded or flat, to 30 cm × up to 5 mm, glabrous or pubescent; uppermost culm leaf usually not reaching the inflorescence. **Inflorescences** panicles or racemes, 4–10 cm. **Spikelets** 9–15(18) mm, 5–10 florets. **Glumes** lanceolate, 8–14(17.5) mm; lower glumes (5)7–9(11)-veined; upper glumes 5–7(9)-veined. **Calluses** 0.5–1.2 mm with a tuft of hairs on each side, hairs approximately 1.5 mm. **Lemmas** (7)9–16 mm, the bodies 2–4 mm with awn-like apical lobes 5–13 mm, thick and leathery, with 2 rows of tufts of hairs, lower row continuous or with weak central tufts, upper row composed of 2 marginal tufts, occasionally with 2 additional tufts with few hairs between the marginal tufts, hairs of upper row not or only just exceeding the base of the awn; lemma awns usually bent and twisted, (7)9–16 mm. **Anthers** 0.4–2.5 mm. **Caryopses** 1.8–3 × 0.8–1.6 mm. **2*n*=24.**

Grasslands and disturbed sites, especially dry, nutrient-depleted sites, mainly near the coast. 100–500 m. CR, Est. CA; Australia, New Zealand. Exotic.

Unlike *R. biannulare, R. penicillatum* has flag leaves that usually do not reach the inflorescences and only two tufts of hairs halfway between the lemma base and tip (both tufts located on the lemma margins). *Rytidosperma penicillatum* has a strong superficial resemblance to *Danthonia spicata* which has hairs along the lemma margins, not arranged in tufts.

SCHEDONORUS Tall fescue

Plants perennial; cespitose or rhizomatous. **Leaves** sheaths open; auricles a broad flange at the base of the leaf blade or rarely claw-like, narrow, curved and clasping; ligules membranous; blades flat. **Inflorescences** panicles; disarticulation above the glumes and below the florets. **Spikelets** laterally compressed, 2–22 florets. **Glumes** 2, shorter than the adjacent lemmas, rounded on the back, 3–9-veined, acute, awnless. **Calluses** glabrous or sparsely hairy. **Lemmas** lanceolate, ovate or oblong, rounded on the back, 3–7-veined, acute, awned or awnless; lemma awns, if present, straight, to 18 mm, terminal or subterminal. **Anthers** 3. **Caryopses** dorsally compressed, curved.

Temperate regions. 3 species; 2 species treated in *Flora.*

The species of *Schedonorus* have long been placed in *Festuca* but are clearly more closely related to *Lolium.* Hybrids between *Schedonorus* and *Lolium,* × *Schedololium,* are occasionally found in Oregon. The hybrids usually have a branched inflorescence with many of the spikelets oriented edgewise to the rachis and having only one glume.

Key to *Schedonorus* species:

1. Auricles and flared, rounded area at base of blade with ciliate margins (sometimes just a single hair can be found; hairs wear off old leaves); lemma awns usually 0–1.5(4) mm; leaves 4–12 mm wide . . . . ***S. arundinaceus***

1' Auricles and flared, rounded area at base of blade lacking cilia; lemma awns 0–0.5 mm; leaves 3–5(7) mm wide . . . . ***S. pratensis***

Schedonorus arundinaceus (Schreber) Dumortier (p. 477)

Tall fescue

Festuca arundinacea Schreb., *Festuca elatior* L. var. *arundinacea* (Schreb.) Weinm.

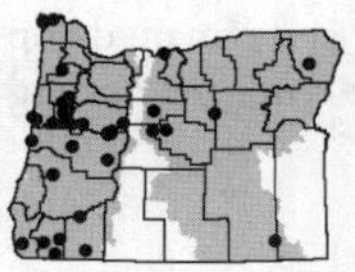

Plants 50–210 cm tall; usually cespitose or rarely rhizomatous. **Auricles** with hairs on the margins or sometimes with just 1 hair, hairs wearing off old leaves; ligules 1(2) mm; blades flat or convolute on young shoots, 4–12 mm wide. **Inflorescences** 10–35 × 8–25 cm. **Spikelets** 8–15.5 × 2–3.5 mm, 3–6(9) florets. **Glumes** lower glumes 3–5 mm; upper glumes 4.5–7(9) mm. **Lemmas** (4)5–9(11.5) mm, usually scabrous or minutely hairy at least distally or rarely smooth, awnless or awned; lemma awns 0–(1.5)4 mm, arising from the tip or to 0.4 mm below the tip. **Anthers** 2.5–4 mm. **Caryopses** 2–4 × 0.9–1.6 mm. **2*n***=28, 42, 56, 63, 70.

Grasslands, meadows, pastures, roadsides, lawns. 0–2000 m. BR, BW, Casc, Col, CR, Est, Lava, Sisk, WV. CA, ID, NV, WA; throughout most of North America; temperate regions worldwide. Exotic.

Schedonorus arundinaceus is a coarse, usually cespitose grass with a nodding inflorescence and short awns. Its flange-like auricles are ciliate on the margins. The cilia can wear off older leaves, so finding even one hair defines the plant as tall fescue. *Schedonorus pratensis* is superficially similar but rhizomatous, and lacks cilia on the auricles.

Schedonorus pratensis (Hudson) P. Beauvois

Meadow fescue

Festuca elatior L. var. *elatior*, *Festuca pratense* Huds., *Lolium pratense* (Huds.) Darbysh.

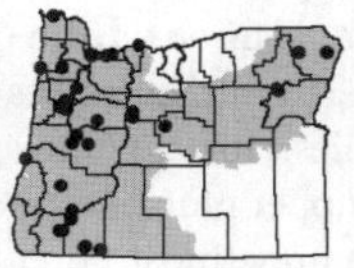

Plants to 130 cm; rhizomatous. **Auricles** lacking hairs; blades flat, folded or convolute on young shoots, 2–7 mm wide. **Inflorescences** (6)10–25 × 9–15 cm. **Spikelets** (8.5)12–15.5(17) × 2–5 mm, (2)4–10(12) florets. **Glumes** lower glumes 2–4.5 mm; upper glumes 3–5 mm. **Lemmas** 5–8 mm, smooth or sometimes slightly scabrous distally; tips awnless but sometimes with short points to 0.5 mm. **Anthers** 1.5–4.6 mm. **Caryopses** 3–4 × 1–1.5 mm.

Moist meadows, pastures, ditches. 100–1800 m. BW, Casc, CR, ECas, Est, Lava, Sisk, WV. CA, ID, NV, WA; throughout most of North America; temperate regions worldwide. Exotic.

Schedonorus pratensis closely resembles *S. arundinaceus* but lacks cilia on the auricles, lacks lemma awns, and is rhizomatous. In the past, *S. pratensis* was widely planted in Oregon as a forage grass, but it is no longer used for that purpose.

SCLEROCHLOA Hardgrass

Plants annual. **Culms** short. **Leaves** sheaths open or closed; ligules membranous; blades flat or folded, exceeding the inflorescences. **Inflorescences** terminal, 1-sided racemes or occasionally panicles; disarticulation tending to be late, variable in location. **Spikelets** laterally compressed. **Glumes** unequal, shorter than the lowest lemmas, with wide hyaline margins; tips obtuse to emarginate, awnless. **Calluses** blunt, glabrous. **Lemmas** hard at maturity. **Anthers** 3. **Caryopses** with a beak formed by the persistent style base.

Temperate regions. 2 species; 1 species treated in *Flora.*

Sclerochloa dura (Linnaeus) P. Beauvois (p. 477)

Fairgrounds grass, hardgrass

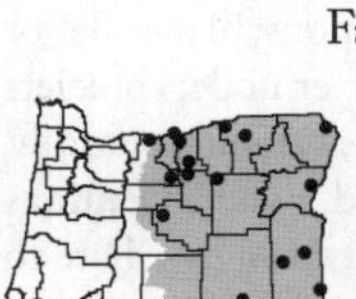

Plants annual, 2–15(30) cm tall; cespitose, becoming straw-colored with age. **Culms** spreading or prostrate, compressed. **Leaves** strongly overlapping; sheaths open or closed up to half their length, keeled, glabrous; ligules acute, (0.3)0.7–2(3.5) mm; blades flat to folded, 0.5–5(7) cm × 1–4 mm, glabrous; tips prow-shaped. **Inflorescences** congested racemes or panicles, 1–4(5) cm × 0.4–2 mm, often partly enclosed in the uppermost leaf sheath. **Spikelets** laterally compressed, (3.4)5–12 mm with (2)3–4(7) florets. **Glumes** lower glumes 1.4–3(3.7) mm, (1)3–5-veined; upper glumes 2.6–5.4(6.2) mm, (3)5–7-veined. **Lemmas** (3.4)4.5–6(7) mm (5)7–9 veins, glabrous but midveins minutely scabrous; tips rounded, awnless. **Anthers** 0.8–1.5 mm. **2*n***=14.

Disturbed, trampled areas with compacted soils such as parking lots, fairgrounds, campsites, roadsides. 50–1400 m. BR, BW, Col, Lava, Owy. CA, ID, WA; scattered throughout North America; temperate regions worldwide. Exotic.

Sclerochloa dura is a short, tough grass of parking lots. It is easily confused with *Poa annua* which has more openly branched inflorescences, hairs on the lemma veins, and obtuse to acute lemma tips.

SCOLOCHLOA Rivergrass

Plants perennial, strongly rhizomatous. **Culms** erect or ascending, sometimes rooting at the lower nodes; nodes glabrous. **Leaves** flat or slightly involute; sheaths open; auricles lacking; ligules membranous, truncate to rounded; blades rolled in bud. **Inflorescences** terminal, open panicles; disarticulation above the glumes, beneath the florets. **Spikelets** laterally compressed, 3–4 florets. **Glumes** unequal, acute to acuminate, awnless. **Lemmas** 3–9-veined; veins tending to protrude from the tips; tips erose or slightly 3-lobed, awnless. **Anthers** 3. **Caryopses** dorsiventrally compressed.

Temperate regions of North America and Eurasia. 1 species.

Scolochloa festucacea (Willdenow) Link (p. 477)

Common rivergrass

Fluminea festucacea (Willd.) Hitchc.

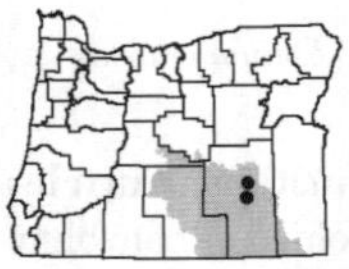

Plants 70–200 cm tall; strongly rhizomatous, rhizomes with thick corky epidermis. **Leaves** sheaths inflated, glabrous; blades flat, 20–45 cm × 4–12 mm, glabrous. **Inflorescences** 15–30 × 10–22 cm. **Spikelets** 7–11 mm, 3(4) florets. **Glumes** glabrous, awnless; lower glumes 5–8 mm, exceeded by the uppermost floret; upper glumes 6.5–10 mm, 3–7 veins, equaling or exceeding the uppermost floret. **Lemmas** 4–9 mm, 3–9 veins, papery in texture, glabrous; tips indistinctly 3-lobed or toothed; awnless. **Anthers** 3, 2–4 mm.

Ponds, marshes, seasonally flooded areas, shores of freshwater to moderately saline lakes and streams. 1100–1300 m. BR. ID; north to from British Columbia to AK, northeast to Manitoba, east to IA and KS; Eurasia. Native.

Scolochloa festucacea is a rhizomatous grass with an open panicle, often forming dense stands. In Oregon, it grows in moderately alkaline wetlands that are wet in spring and remain damp through the summer. It provides important cover for nesting waterfowl and shorebirds.

SCRIBNERIA Scribner's grass

Plants annual, tufted. **Culms** often branched at the lower nodes; nodes purple. **Leaves** sheaths open; ligules membranous; blades involute, very narrow. **Inflorescences** spikes with 1 spikelet per node or rarely 2–4 spikelets per node; lower spikelets sessile; upper spikelets with pedicels less than 3 mm; disarticulation above the glumes, beneath the florets. **Spikelets** tangential to and partially embedded in the inflorescence axis, laterally compressed, 1 floret. **Glumes** 2, longer than the floret, tinged with red or purplish, leathery, 2-keeled, awnless; lower glumes longer and narrower than the upper glumes, 2–3-veined; upper glumes 3–4-veined. **Lemmas** inconspicuously 5-veined, bifid, awned; lemma awns arising from the sinus between the short teeth. **Paleas** tightly clasped by the lemmas. **Anthers** 1. **Caryopses** fusiform.

Pacific states of North America. 1 species.

Scribneria bolanderi (Thurber) Hackel (p. 477)

Scribner's grass

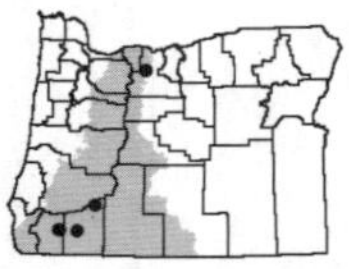

Plants 3–15(45) cm tall. **Culms** ascending to erect. **Leaves** blades 1–3 cm × 0.8–1.6 mm, outer surfaces scabrous on the midveins. **Inflorescences** (2)4–12 cm × 1–3 mm. **Spikelets** slightly longer than the adjacent internodes. **Glumes** (3)4–7 mm. **Lemmas** hidden by the glumes; lemma awns 2–4 mm. ***2n***=26.

Vernal pools, seasonally wet spots with compacted soils such as old road beds, areas with shallow, rocky soils. 400–1100 m. Casc, ECas, Sisk. CA, WA; south to Mexico. Native.

Scribneria bolanderi is a rare, inconspicuous grass of shallow, seasonal wetlands. The spikelets are sunken into the inflorescence axis and oriented so that both glumes are readily visible. Similar *Hainardia cylindrica* has spikelets oriented edgewise to the inflorescence axis, and each spikelet has only one glume. Also similar *Parapholis incurva* is a coarser plant that lacks lemma awns. Its inflorescence axis breaks apart as the seeds fall; that of *S. bolanderi* remains intact. Immature, depauperate *Lolium* species may resemble *Scribneria* but have only one glume per spikelet.

SECALE Rye

Plants annual, biennial, or short-lived perennials. **Culms** hollow. **Leaves** cauline; sheaths open; ligules membranous; blades flat or involute. **Inflorescences** terminal, laterally compressed, 2-sided spikes; middle internodes 2–4 mm, 1 spikelet per node; spikelets strongly ascending; disarticulation in the rachises below the spikelets, or in cultivated forms not disarticulating or doing so tardily. **Spikelets** sessile, 2 florets. **Glumes** linear to subulate, shorter than the adjacent lemmas, scabrous, 1-veined, keeled, awned. **Lemmas** strongly compressed, flattened over the back; midveins smooth to scabrous; marginal veins strongly keeled and strongly ciliate to scabrous; tips tapering to an awn; lemma awns 2–50 mm, scabrous. **Anthers** 3.

Temperate regions. 3 species; 1 species treated in *Flora*.

Secale strictum has been occasionally collected in Oregon but does not appear to persist. It is a perennial with inflorescence axes that readily disarticulate and lemmas that are 8–16 mm.

Secale cereale Linnaeus (p. 479)

Rye

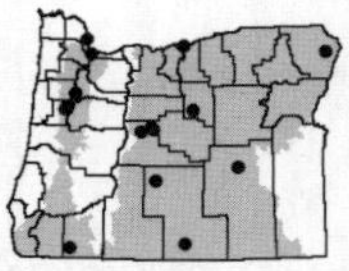

Plants annual or biennial, (35)50–120(300) cm tall. **Leaves** blades (3)4–12 mm wide, glabrous. **Inflorescences** spikes usually nodding at maturity, (2)4.5–12(19) × approximately 1 cm; spikelets in 2 rows on opposite sides of the axis; disarticulation tardy or not occurring. **Spikelets** 10–18 mm. **Glumes** 8–20 mm, awns 1–3 mm. **Lemmas** 14–18 mm, 5-veined, tough; marginal veins keeled and ciliate with conspicuous teeth or broad hairs; lemma awns straight, 7–50 mm. **Anthers** approximately 7 mm. **2*n***=14.

Disturbed grasslands, roadsides, cultivated fields. 50–1600 m. BR, BW, Col, ECas, Lava, Sisk, WV. CA, ID, NV, WA; throughout North America; temperate regions worldwide. Exotic.

Secale cereale is a large, pale green grass with a usually nodding spike and coarse hairs on the keeled lemma margins. Closely related *Triticum aestivum* does not have hairs on the lemma margins.

SETARIA Foxtail

Plants annual or perennial; cespitose, rarely rhizomatous. **Culms** erect or decumbent, 10–600 cm tall. **Leaves** sheaths open; ligules membranous and ciliate or comprised of hairs; blades flat or folded. **Inflorescences** terminal panicles, usually dense and spike-like; disarticulation usually below the glumes spikelets falling intact but leaving the bristles behind (disarticulation between the lower and upper florets in cultivated *S. italica*). **Spikelets** dorsiventrally flattened or round in cross section, ellipsoid, rarely globose, subsessile or on short pedicels, in clusters or single on short branches, some or all spikelets subtended by 1–several bristles. **Glumes** lower glumes less than half as long as the spikelets, 1–7-veined; upper glumes half to nearly equal the upper lemmas in length, 3–9-veined. **Florets** lower florets staminate or sterile; lower lemmas equaling or rarely exceeding the upper lemmas, rarely absent, 5–7-veined; upper florets bisexual; upper lemmas and paleas hard and transversely rugose or rarely smooth. **Anthers** 3. **Caryopses** small, ellipsoid to subglobose, dorsiventrally compressed.

Tropical and warm-temperate regions worldwide. Approximately 140 species; 3 species treated in *Flora*.

Most *Setaria* species are weeds of disturbed areas, including crop fields and are of major economic importance in some areas. One species, *S. italica,* foxtail millet, is cultivated for grain. It is similar to *S. viridis* but its inflorescences are usually wider and its spikelets are about 3 mm. The common name “foxtail” is shared with the genus *Alopecurus*, which has awns in the inflorescence but no bristles.

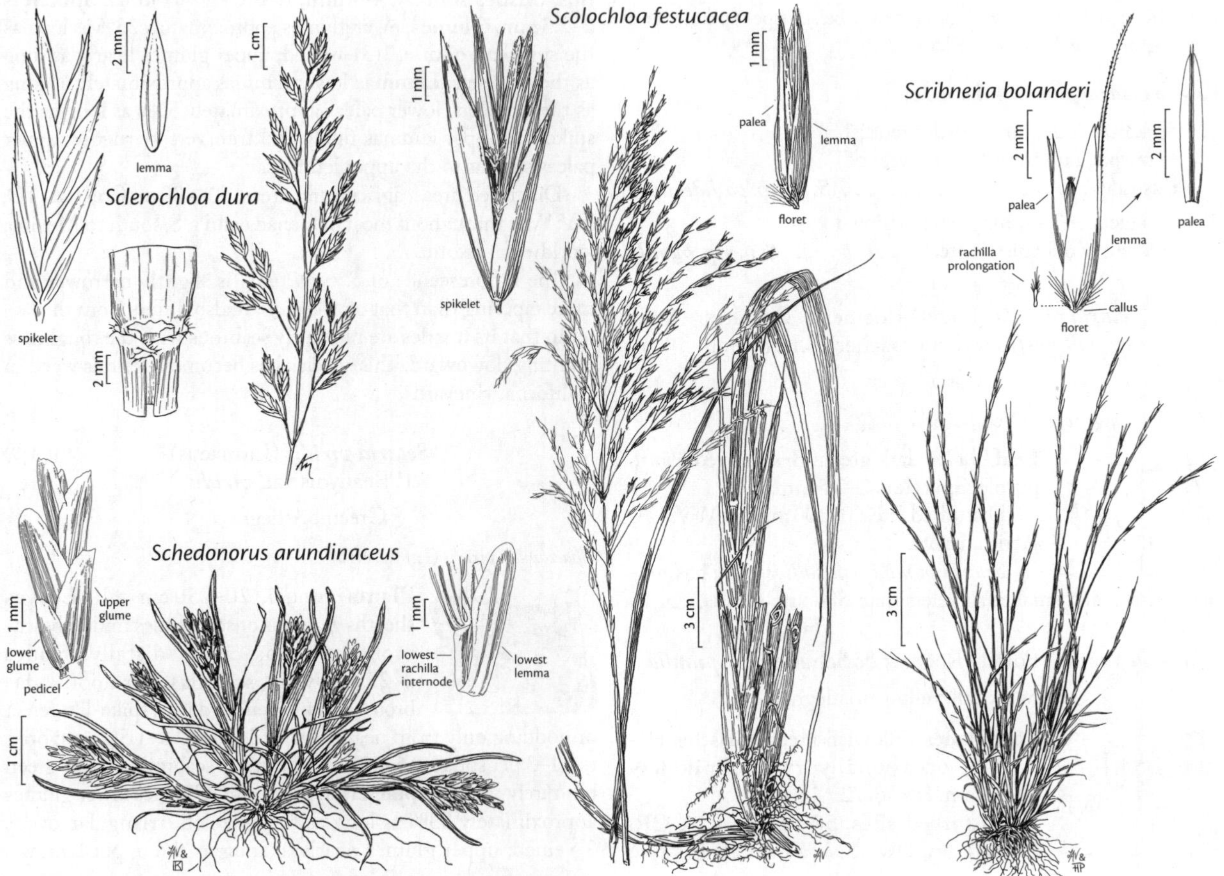

SCHEDONORUS • SCLEROCHLOA • SCOLOCHLOA • SCRIBNERIA

Setaria parviflora was collected in Portland in 1916 but did not persist. It is native to the southern half of the United States, Central America and the West Indies. It is a rhizomatous perennial with scabrous leaves that lack hairs and upper glumes half to two-thirds as long as the spikelets. *Setaria viridis* var. *major* was collected in Portland in 1920, but apparently it has not persisted. It is 100–250 cm tall with leaf blades 10–25 mm wide and panicles 10–20 cm.

Key to *Setaria* species:

1. Inflorescence bristles retrorsely scabrous (the points directed toward the base of the bristle); inflorescences tending to stick to clothing. . . . . ***S. verticillata***

1' Inflorescence bristles antrorsely scabrous (the points directed away from the base of the bristle); inflorescences not sticking to clothing

 2. Upper glume approximately as long as upper lemma; spikelet or cluster of spikelets subtended by 1–3 bristles. . . . . ***S. viridis*** var. ***viridis***

 2' Upper glume 50–67% as long as upper lemma; spikelet or cluster of spikelets subtended by 4–12 bristles. . . . . ***S. pumila***

Setaria pumila (Poiret) Roemer & Schultes
Yellow foxtail

Plants annual, 30–130 cm tall. **Leaves** sheaths glabrous; blades 4–10 mm wide; basal portion of upper surfaces with papillose-based hairs. **Inflorescences** dense, spike-like, 1–7(15) × approximately 1 cm, uniform in width throughout; bristles 4–12 per spikelet, 3–8 mm, antrorsely scabrous. **Spikelets** 2–3.4 mm. **Glumes** lower glumes approximately 33% as long as the spikelets, 3-veined, acute; upper glumes approximately 50% as long as the spikelets and 50–67% as long as the upper lemmas. **Lemmas** lower lemmas as long as the upper lemmas and lower paleas; upper lemmas exposed, strongly transversely rugose.

2 subspecies.

Setaria pumila inflorescences are usually smaller and paler than those of *S. viridis,* which has fewer bristles per spikelet and upper glumes as long as the lemmas.

Key to *Setaria pumila* subspecies:

1. Spikelets 2–2.5 mm; bristles reddish to purple; rare introduction west of Cascades . . . . ***S. p.*** ssp. ***pallide-fusca***

1' Spikelets 3–3.4 mm; bristles yellow to reddish; widely distributed . . . . ***S. p.*** ssp. ***pumila***

Setaria pumila (Poiret) Roemer & Schultes ssp. ***pallide-fusca*** (Schumacher) B.K. Simon
Cattail grass

Chaetochloa nigrirostris (Nees) Skeels

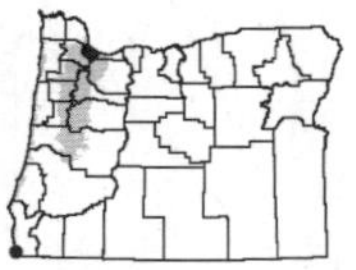

Leaf blades dark green. **Bristles** reddish to purple. **Spikelets** 2–2.5 mm.

Disturbed sites. 0–100 m. Est, WV. LA; Africa. Exotic.

Setaria pumila ssp. *pallide-fusca* has reddish bristles and smaller spikelets than *S. p.* ssp. *pumila.*

Setaria pumila (Poiret) Roemer & Schultes ssp. ***pumila***
Yellow foxtail, yellow bristlegrass

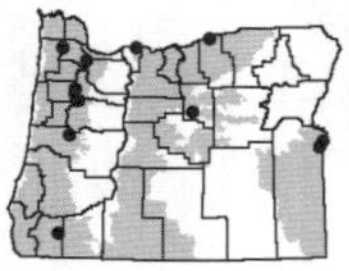

Leaf blades yellowish green. **Bristles** yellowish, occasionally rusty. **Spikelets** 3–3.4 mm. **2*n***=36, 72.

Disturbed sites. 50–700 m. Col, CR, ECas, Lava, Owy, Sisk, WV. CA, ID, NV, WA; throughout most of North America; worldwide. Exotic.

Setaria pumila ssp. *pumila* is widespread but uncommon in Oregon.

Setaria verticillata (Linnaeus) P. Beauvois
Hooked bristlegrass

Chaetochloa verticillata (L.) Scribn., *Setaria verticillata* (L.) P. Beauv. var. *verticillata*

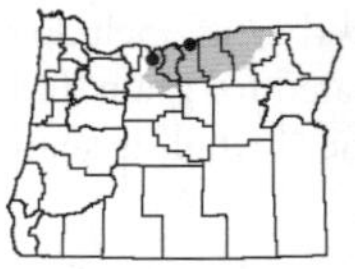

Plants annual, 30–100 cm tall. **Leaves** sheaths glabrous; margins ciliate distally; blades 5–15 mm wide, glabrous to scabrous. **Inflorescences** dense, spike-like, 5–15 cm, approximately 0.8 cm wide, tapering to the tips; bristles solitary, 4–7 mm, retrorsely scabrous. **Spikelets** 2–2.3 mm. **Glumes** lower glumes approximately 33% as long as the spikelets, obtuse, 1(3)-veined; upper glumes nearly as long as the spikelets. **Lemmas** lower lemmas approximately as long as the spikelets; lower paleas approximately 50% as long as the spikelets; upper lemmas finely and transversely rugose; upper paleas similar to the upper lemmas.

Disturbed areas, agricultural areas. 50–400 m. Col. CA, ID, NV, WA; throughout most of Canada and US, south to Mexico; worldwide. Exotic.

The inflorescence of *S. verticillata* is slightly narrower and more tapering than that of *S. viridis*. It also differs from the latter in that its bristles are retrorsely scabrous with the roughness pointing downward. This species has become a serious weed in California vineyards.

Setaria viridis (Linnaeus) P. Beauvois var. ***viridis***
(p. 479)

Green bristlegrass

Chaetochloa viridis (L.) Scribn.

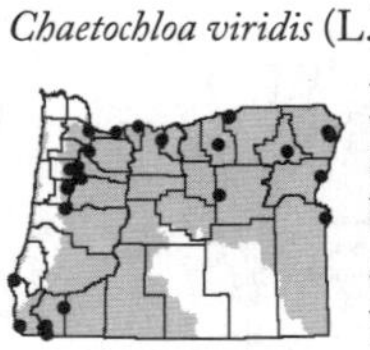

Plants annual, 20–250 cm tall. **Leaves** sheaths glabrous or sometimes minutely scabrous; margins ciliate distally; blades 4–25 mm wide, scabrous or smooth, glabrous. **Inflorescences** dense, spike-like, erect or nodding only from near the tips, 3–20 × 0.6–1.5(2) cm; bristles 1–3 per spikelet, 5–10 mm, antrorsely scabrous, usually green but rarely purple. **Spikelets** 1.8–2.2 mm. **Glumes** lower glumes approximately 33% as long as the spikelets, triangular-ovate, 3-veined; upper glumes nearly as long as the upper lemmas.

Lemmas lower lemmas slightly longer than the upper lemmas; lower paleas approximately 33% as long as the lower lemmas; upper lemmas pale green, very finely and transversely rugose, 5–6-veined; upper paleas similar to the upper lemmas. **2*n***=18.

Disturbed sites, cultivated fields. 0–900 m. BW, Casc, Col, ECas, Est, Lava, Owy, Sisk, WV. CA, ID, NV, WA; throughout most of Canada and US, south to Mexico; warm-temperate regions worldwide. Exotic.

Setaria viridis has an erect, bristly, green, spike-like panicle. *Setaria pumila* has smaller, paler panicles with more bristles under each spikelet, and the upper glumes are only about half as long as the upper lemmas. *Setaria verticillata* has narrower, slightly tapering inflorescences, and the bristles are retrorsely scabrous.

SORGHUM Sorghum

Plants annual or perennial, 0.5–5 m tall. **Culms** with solid internodes. **Leaves** sheaths open; ligules membranous, ciliate, or comprised of hairs; blades usually flat. **Inflorescences** panicles, primary branches whorled; spikelets in clusters with 1 sessile spikelet and 1–2 stalked spikelets; disarticulation in the panicle branches, sometimes below the pedicels, some cultivated forms not disarticulating. **Spikelets** sessile spikelets fertile, dorsiventrally compressed, 2 florets, the lower 1 sterile and reduced to a hyaline lemma; stalked spikelets staminate or sterile, about as long as the sessile spikelets or somewhat shorter. **Glumes** 2, tough, concealing the florets; lower glumes dorsally compressed, rounded basally, 2-keeled or winged distally, 5–15-veined, usually awnless; upper glumes 2-keeled, sometimes awned. **Lemmas** hyaline, sometimes awned.

Warmer regions worldwide. Approximately 25 species; 2 species treated in *Flora*.

Sorghum species are cultivated for forage, grain, and sugars. Although the two species of *Sorghum* present in Oregon are distinct, they are part of a confusingly polymorphic complex of hybridizing species.

Key to *Sorghum* species:

1. Annual or short-lived perennial, not rhizomatous; spikelets not or tardily disarticulating at maturity .... ***S. bicolor***

1' Perennial from deep rhizomes; spikelets disarticulating at maturity .... ***S. halepense***

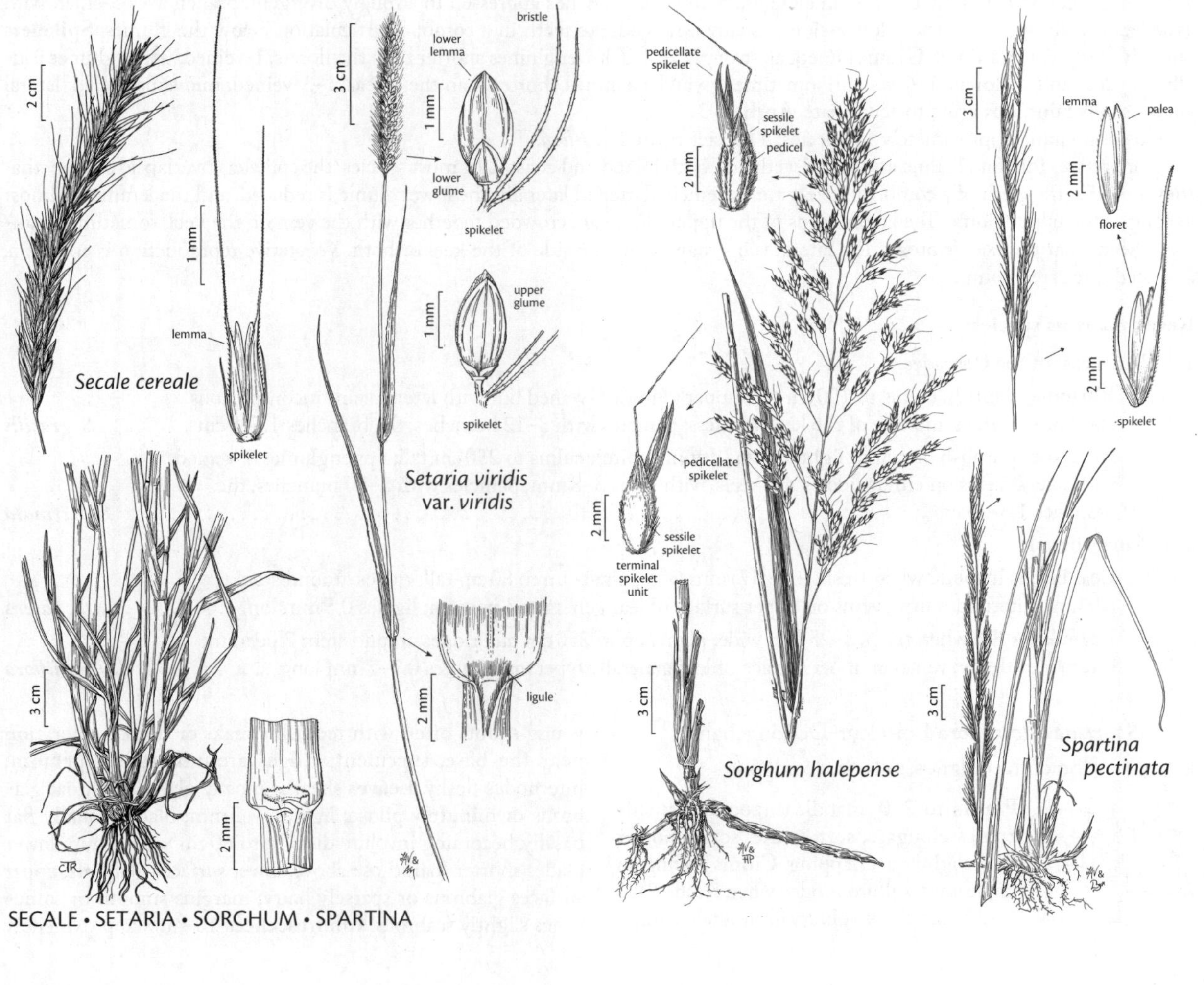

SECALE • SETARIA • SORGHUM • SPARTINA

Sorghum bicolor (Linnaeus) Moench

Sorghum, milo, shattercane

Sorghum bicolor (L.) Moench ssp. *bicolor*

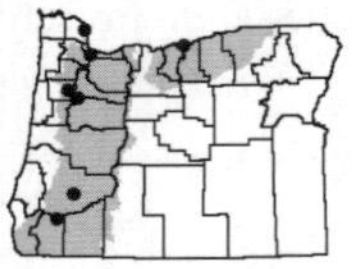

Plants annual or sometimes short-lived perennials, 50–500 cm tall, culms solitary or clustered, not rhizomatous. **Culms** 1–5 cm thick. **Leaves** blades flat, 5–100 mm wide, glabrous or slightly scabrous. **Inflorescences** open or dense, 5–60 × 3–30 cm, brown; disarticulation tardy or not occurring. **Sessile spikelets** 3–9 mm, bisexual, dorsiventrally compressed. **Pedicellate spikelets** 3–6 mm, staminate or sterile. **Glumes** hard or leathery to membranous, glabrous to densely hairy; keels winged. **Upper lemmas** awnless or awned; lemma awns, if present, 5–30 mm, bent and twisted. **Anthers** 2–2.8 mm. **Caryopses** exposed at maturity or not. ***2n***=20, 40.

Disturbed areas, cultivated fields. 50–600 m. Casc, Col, Sisk, WV. CA, ID, NV, WA; eastern Canada, throughout US; tropical and temperate regions worldwide. Exotic.

Sterile *Sorghum bicolor* plants may resemble corn.

Sorghum halepense (Linnaeus) Persoon (p. 479)

Johnson grass

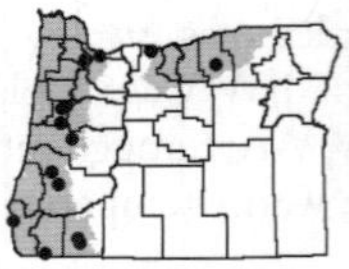

Plants perennial, 50–200 cm tall; rhizomatous. **Culms** 0.4–2 cm thick. **Leaves** blades flat, 8–40 mm wide, glabrous to slightly scabrous. **Inflorescences** open, 10–50 × 5–25 cm, dark reddish brown at maturity; disarticulation below the sessile spikelets, or below the pedicellate spikelets. **Sessile spikelets** 3.8–6.5 × 1.5–2.3 mm, bisexual. **Pedicellate spikelets** 3.6–5.6 mm, staminate. **Glumes** hard, shiny, appressed pubescent. **Upper lemmas** awnless or awned; lemma awns, if present, bent, twisted, to 13 mm. **Anthers** 1.9–2.7 mm. **Caryopses** not exposed at maturity. ***2n***=20, 40.

Disturbed, often moist, grassy areas, roadsides. 50–600 m. Col, CR, Est, Sisk, WV. CA, ID, NV, WA; south to Mexico; tropical and temperate regions worldwide. Exotic.

Sorghum halepense is a weedy, perennial grass that spreads from deeply buried rhizomes. At maturity, it can be recognized by the large, reddish brown inflorescences. *Sorghum bicolor* is usually annual, more robust, with a denser inflorescence.

SPARTINA Cordgrass

Plants perennial, 10–350 cm tall; cespitose or rhizomatous. **Leaves** sheaths open, smooth; ligules appearing to consist of hairs only (actually consisting of a short, inconspicuous membrane and longer hairs); blades flat or involute. **Inflorescences** panicles with 1–75 spike-like branches attached to an elongate main axis; branches appressed to strongly divergent, branch axes 3-sided with spikelets usually sessile on the 2 lower sides, usually arranged like teeth in a comb; disarticulation below the glumes. **Spikelets** laterally compressed, 1 floret. **Glumes** unequal, strongly keeled; lower glumes shorter than the florets, 1-veined; upper glumes usually longer than the florets, 1–6-veined, sometimes awned. **Lemmas** shorter than the paleas, 1–3-veined; midveins keeled; lateral veins usually faint; tips acute to acuminate. **Anthers** 3.

Cosmopolitan. Approximately 15 species; 4 species treated in *Flora*.

Cordgrasses inhabit alkaline and saline wetlands, both inland and coastal. In most species, the spikelets overlap so densely that they look like the teeth of a comb. Spikelets are extremely flattened laterally, the lower glume is reduced, and the lemma is almost as long as the upper glume. The lateral veins of the upper glume are crowded together with the vein on the keel, sometimes making it seem that the keel is broad. The lateral veins may be on one side of the keel or both. Vegetative reproduction is common, and seeds are rarely found.

Key to *Spartina* species:

1. Plants east of the Cascades.
    2. Rhizomes whitish; culms to 100 cm tall; upper glumes 3-veined but with lateral veins inconspicuous and both on the same side of the keel, awnless; panicles with 3–12 branches, the branches 1.5–8 cm . . . . ***S. gracilis***
    2' Rhizomes purplish brown or light brown (drying white); culms to 250 cm tall; upper glumes 3-veined, with lateral veins on either side of the keels, with awns 3–8 mm; panicles with 5–50 branches, the branches 1.5–15 cm . . . . ***S. pectinata***

1' Plants coastal.
    3. Leaf blades involute when fresh, 0.5–4(7) mm wide; plants up to 80 cm tall; ridges around stem 6 per mm; internodes firm; veins on upper surface of leaf generally 3 per mm; ligules 0.5 mm long. . . . . ***S. patens***
    3' Leaf blades flat when fresh, 3–25 mm wide; plants up to 250 cm tall; ridges around stem 2 per mm; internodes fleshy; veins on upper surface of leaf generally 6 per mm; ligules 0.7–2 mm long . . . . ***S. alterniflora***

Spartina alterniflora Loiseleur-Deslongchamps

Smooth cordgrass, saltwater cordgrass

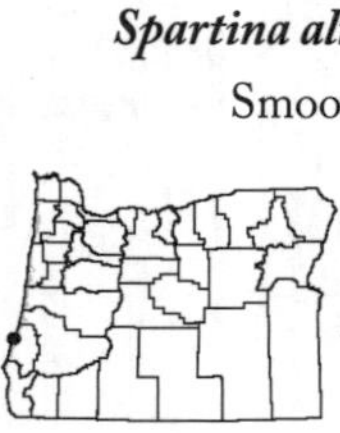

Plants to 250 cm tall; rhizomatous; rhizomes elongate, soft, white, scales not or only slightly overlapping. **Culms** having an unpleasant, sulfurous odor when fresh, (3)5–15(20) mm thick, solitary or in small clumps, young shoots often with reddish streaks or diffuse coloration near the base, succulent; ridges around stem 2 per mm; internodes fleshy. **Leaves** sheaths mostly glabrous, throat glabrous or minutely pilose; ligules 1–2 mm; blades usually flat basally, becoming involute distally, to 60 cm × 3–25 mm; lower blades shorter than those above; lower surfaces glabrous; upper surfaces glabrous or sparsely hairy; margins smooth or sometimes slightly scabrous. **Inflorescences** 10–40 × approximately

2 cm, 3–25 branches, often partially enclosed in the uppermost sheath; branches loosely appressed, 5–15 cm, generally remote to moderately overlapping throughout the panicles, 20–30 spikelets. **Spikelets** 8–14 mm, usually divergent, equally overlapping on all the branches. **Glumes** acuminate to obtuse, straight, sides usually glabrous, sometimes pilose near the base or appressed-pubescent, occasionally with a minute point; lower glumes 4–10 mm, generally 50–67% as long as the upper glume; upper glumes 8–14 mm; keels glabrous to softly shaggy-hairy; lateral veins absent. **Lemmas** approximately 8–11 mm, glabrous or with sparse spreading hairs; tips usually acuminate. **Anthers** 3–6 mm. **2*n***=56, 70.

Intertidal mudflats. 0–50 m. Est. CA, WA; coasts of eastern North America; South America; Asia, Europe. Exotic.

Spartina alterniflora forms dense stands in estuaries, choking out native salt marsh species and converting mudflats into *Spartina*-dominated marshes. Infestations have occurred in Oregon near the mouth of the Siuslaw River and in Coos Bay. Both may have been eradicated by weed control agencies. Invasive *S. anglica,* known from the Seattle, Washington area, is similar but has firmer internodes, glumes with appressed hairs, and up to 12 panicle branches.

Spartina gracilis Trinius

Alkali cordgrass

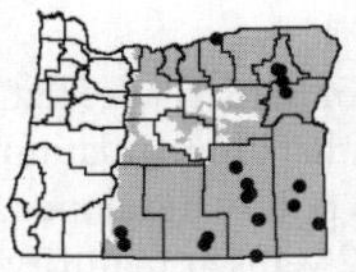

Plants 40–100 cm tall, strongly rhizomatous; rhizomes elongate, 1.5–5.5 mm thick, whitish, closely overlapping. **Culms** 2–3.5 mm thick, firm. **Leaves** sheaths glabrous but throats sometimes hairy; ligules 0.5–1 mm; blades flat at base, becoming involute, 6–30 cm × 2–4 mm where flat; lower surfaces glabrous; upper surfaces and margins scabrous. **Inflorescences** 8–25 cm, 3–12 branches; branches usually appressed but rarely spreading, 1.5–8 cm, alternate, differing only slightly in length and spacing within the panicle, 10–30 spikelets. **Spikelets** ovate to lanceolate, 6–11 mm. **Glumes** glabrous or with sparsely hairy margins, acute or short-pointed; lower glumes curved, 3–7 mm, glabrous or sparsely pubescent, 50–67% as long as the upper glumes; keels glabrous or hairy; upper glumes 6–10 mm, with short stiff hairs on the surface and veins; keels with spreading hairs 0.2–0.5 mm; lateral veins 2, inconspicuous, both on the same side of the keel. **Lemmas** about 6–7 mm, glabrous or sparsely hairy; keels hairy at least distally, hairs 0.3–1 mm; margins sparsely hairy; tips obtuse to rounded, sometimes obscurely lobed. **Anthers** 2.5–5 mm. **2*n***=40, 42.

Margins of alkaline lakes and streams. 100–1500 m. BR, BW, Col, ECas, Owy. CA, ID, NV, WA; north to Northwest Territory, east to KS, south to central Mexico. Native.

Spartina gracilis is the more delicate of our two native, inland cordgrasses. *Spartina pectinata* has thicker rhizomes that are darker when fresh, thicker culms, wider, flatter leaves and distinctly awned glumes.

Spartina patens (Aiton) Muhlenberg

Saltmeadow cordgrass

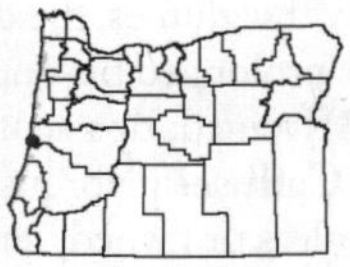

Plants 15–150 cm tall, strongly rhizomatous; rhizomes elongate, 1–6 mm thick, whitish, scales not overlapping. **Culms** 1–6 mm thick, usually solitary, hard, ridges around stem 6 per mm. **Leaves** sheaths glabrous or mostly glabrous, throats occasionally hairy; ligules 0.5–1 mm; blades involute except sometimes near base, 10–50 cm × (0.5)1–4(7) mm where flat; lower surfaces glabrous; upper surfaces scabrous; margins strongly scabrous. **Inflorescences** 3–15 cm, 2–15 branches; branches appressed to strongly divergent, 1–7 cm, alternate, differing only slightly in length and spacing within a panicle, 10–30 spikelets. **Spikelets** 7–12 mm. **Glumes** glabrous or sparsely hairy on the sides; keels scabrous or with stiff short hairs 1–2.5 mm; tips awnless; lower glumes linear, 3–8 mm, generally 67% as long as the upper glumes; upper glumes 7–12 mm; lateral veins 2, these on the same side of the keel, usually hairy. **Lemmas** approximately 7–8 mm, glabrous or sparsely hairy; keels hispid distally; tips obtuse, rounded or obscurely lobed. **Anthers** 3–5 mm. **2*n***=28.

Upper parts of coastal salt marsh. 0–50 m. Est. CA, WA; coasts of eastern North America, Central America, Caribbean; South America. Exotic.

Like *S. alterniflora, S. patens* forms dense stands that exclude native plants. In eastern North America, it is expanding its range along roads that are salted in winter. *Spartina densiflora,* introduced to coastal California and Washington, is similar but cespitose and has about two ridges per millimeter on its culms.

Spartina pectinata Link (p. 479)

Prairie cordgrass, freshwater cordgrass

Spartina michauxiana Hitchc.

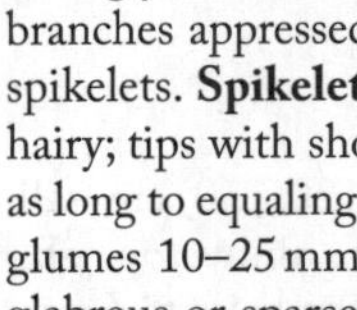

Plants to 250 cm tall, strongly rhizomatous; rhizomes elongate, (2)3–8 mm thick, purplish brown or light brown, drying white, scales closely overlapping. **Culms** 2.5–11 mm thick, solitary or in small clumps, hard. **Leaves** sheaths mostly glabrous but throats often hairy; ligules 1–3 mm; blades flat when fresh, becoming involute when dry, 20–96 cm × 5–15 mm, both surfaces glabrous; margins strongly scabrous. **Inflorescences** 10–50 cm, 5–50 branches; branches appressed to somewhat spreading, 1.5–10 cm, 10–80 spikelets. **Spikelets** 10–25 mm. **Glumes** glabrous to sparsely hairy; tips with short awns; lower glumes 5–10 mm, from 75% as long to equaling the adjacent lemmas; keels with hairs; upper glumes 10–25 mm (including the awn), exceeding the florets, glabrous or sparsely hairy; keels scabrous to hairy with hairs approximately 0.3 mm; lateral veins usually glabrous or rarely with hairs on either side of and close to the keels; glume awns 3–8 mm. **Lemmas** approximately 10 mm, glabrous; keels with short hairs distally; tips bilobed, lobes 0.2–0.9 mm. **Anthers** 4–6 mm.

Wet to seasonally dry marshes, ponds, ditches, and roadsides. 50–1000 m. BR, BW, Col, Lava. ID, WA; east to Labrador, southeast to NM. Native.

Spartina pectinata is the more robust of our inland cordgrasses. It is Oregon's only *Spartina* with distinctly awned glumes.

SPHENOPHOLIS Wedgescale

Plants usually perennial, (5)20–130 cm tall; usually cespitose. **Leaves** sheaths open; ligules membranous; blades flat or involute. **Inflorescences** panicles; disarticulation below the glumes, the distal floret occasionally disarticulating before the others. **Spikelets** laterally compressed, 2–3 florets; the rachillas prolonged beyond the base of the distal floret. **Glumes** approximately as long as the lowest floret; lower glumes narrower and 1(3)-veined; tips acute; upper glumes elliptical to oblanceolate or obovate, 3(5)-veined; tips rounded, truncate, acute or acuminate. **Calluses** glabrous. **Lemmas** not hard, rounded on lower back, usually keeled near the tips, 3(5)-veined but veins very faint; awnless or awned; lemma awns terminal or arising just below the tips, straight or bent. **Anthers** 3. **Caryopses** linear-ellipsoid.

North America and the Caribbean. Approximately 6 species; 1 species treated in *Flora*.

Sphenopholis obtusata (Michaux) Scribner (p. 483)

Prairie wedgescale

Sphenopholis obtusata (Michx.) Scribn. var. *lobata* (Trin.) Scribn. ex B.L. Rob., *Sphenopholis obtusata* (Michx.) Scribn. var. *obtusata*

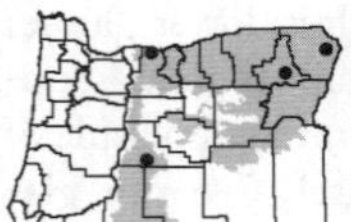

Plants perennial, sometimes flowering the first year, (9)20–130 cm tall; cespitose. **Leaves** sheaths glabrous or hairy; ligules 1–2.5 mm; blades usually flat or rarely slightly involute, 5–14 cm × (1)2–8 mm, scabrous or pubescent. **Inflorescences** spike-like, dense, (2)5–15(25) cm × 0.5–2 mm, pale, shiny. **Spikelets** 2.2–3.6 mm. **Glumes** awnless, scabrous; lower glumes less than 33% as wide as the upper glumes; upper glumes 1.5–2.5 mm, 2–3 times as long as wide, expanded beyond mid-length, 3(5)-veined; tips rounded to truncate. **Lemmas** 2.4–2.7 mm, 3-veined; tips acute, awnless. **Anthers** 0.2–1 mm. **2*n***=14.

Moist grasslands, pond shores, marshes, dunes and disturbed sites. 100–2100 m. BW, Col, ECas. CA, ID, NV, WA; throughout most of North America. Native.

The narrow, spike-like, shining panicles of *Sphenopholis obtusata* resemble those of *Koeleria macrantha* and, like that species, it has one glume distinctly wider than the other. However, *K. macrantha* can grow in drier habitats, its glumes are acute, and its spikelets disarticulate above the glumes.

SPOROBOLUS Dropseed

Plants annual or perennial, usually cespitose. **Leaves** sheaths open, usually with a tuft of hairs at the tips; ligules of hairs; blades flat, folded, involute, occasionally terete. **Inflorescences** terminal panicles, sometimes hidden in the upper leaf sheaths; disarticulation above the glumes. **Spikelets** rounded to laterally compressed with 1(3) floret. **Glumes** 0–1-veined. **Calluses** poorly developed, glabrous **Lemmas** membranous or papery in texture, 1(3)-veined, awnless. **Paleas** often splitting between the 2 veins at maturity. **Anthers** (2)3. **Fruits** achenes, often translucent, the outer layer free from the seed, in most species becoming mucilaginous when wet and sometimes forcibly ejecting the seed when it dries.

Tropical, subtropical, and warm-temperate regions. Approximately 160 species. 2 species treated in *Flora*.

Sporobolus indicus was collected in Portland in 1900 but is not persisting. It has ascending upper culm leaf blades and very narrow inflorescences.

Key to *Sporobolus* species:

1. Leaf sheath tops glabrous or with a few scattered hairs; upper culm leaf blades ascending .... ***S. airoides***

1' Leaf sheath tops with a conspicuous tuft of white hairs; upper culm leaf blades nearly perpendicular to the culms .... ***S. cryptandrus***

Sporobolus airoides (Torrey) Torrey

Alkali sacaton

Sporobolus airoides (Torr.) Torr. var. *airoides*

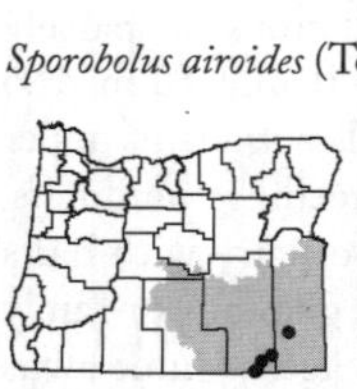

Plants perennial, 35–120(150) cm tall; cespitose. **Leaves** sheaths shiny, tops glabrous or sparsely hairy, the hairs to 6 mm; ligules of hairs 0.1–0.3 mm; blades (1)2–5(6) mm wide; upper culm leaf blades ascending. **Inflorescences** diffuse, (10)15–45 × 15–25 cm, bases often included in uppermost sheath; primary and secondary branches spreading, brownish. **Spikelets** 1.3–2.8 mm. **Glumes** lanceolate to ovate; lower glumes 1.5–1.8 mm; upper glumes 1.1–2.8 mm. **Lemmas** 1.2–2.5 mm, acute. **Anthers** 1.1–1.8 mm. **Achenes** 1–1.4 mm, reddish brown. **2*n***=80, 90, 108, 126.

Dry, alkaline, sandy soils. 400–1300 m. BR, Owy. CA, ID, NV, WA, north to British Columbia, east to MO, south to Mexico. Native.

Sporobolus airoides is a cespitose perennial of sandy, alkaline soils. It has a large, open panicle. The common name alkali sacaton is sometimes applied to *Muhlenbergia asperifolia*, a common rhizomatous perennial of alkaline, often fine-textured soils, which has a smaller, whitish inflorescence. *Panicum capillare* and *P. dichotomiflorum* have diffuse panicles like *S. airoides* and can grow in fairly alkaline habitats. They are both annuals, and their spikelets are dorsiventrally compressed.

Sporobolus cryptandrus (Torrey) A. Gray (p. 483)

Sand dropseed

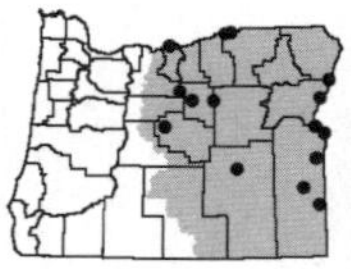

Plants perennial, 30–100(120) cm tall; cespitose. **Leaves** sheaths glabrous or minutely scabrous; margins sometimes ciliate distally, tops with conspicuous tufts of hairs, the hairs to 4 mm; ligules of hairs 0.5–1 mm; blades 2–6 mm wide; upper culm leaf blades nearly perpendicular to the culms. **Inflorescences** contracted and spike-like early in development, narrowly pyramidal later, 15–40 cm × 2–12(14) mm; primary branches appressed, spreading or reflexed; secondary branches appressed. **Spikelets** 1.5–2.7 mm. **Glumes** linear-lanceolate to ovate; lower glumes 0.6–1.1 mm; upper glumes 1.5–2.7 mm. **Lemmas** 1.4–2.7 mm, acute. **Anthers** 0.5–1 mm. **Achenes** 0.7–1.1 mm, light brown to reddish orange. **2*n***=36, 38, 72.

Sandy soils and washes, rocky slopes and calcareous ridges, roadsides in sagebrush steppe, dry grasslands, and open forests. 50–1300 m. BR, BW, Col, Lava, Owy. Throughout most of North America. Native.

Sporobolus cryptandrus is a common but inconspicuous perennial that blooms later in summer than most of our other grasses. It is recognized by the conspicuous tuft of hairs at the top of the leaf sheath, just below the blade. Its upper culm leaf blades jut out from the culm at a nearly perpendicular angle.

TAENIATHERUM Medusahead

Plants annual. **Culms** erect. **Leaves** sheaths open; auricles usually present; ligules truncate, membranous; blades flat to involute. **Inflorescences** terminal spikes with 2(4) spikelets per node; internodes 0.5–3.5 mm; disarticulation above the glumes, between the florets. **Spikelets** with 2(3) florets, the lowest floret bisexual, the other(s) highly reduced, sterile. **Glumes** equal, awn-like, erect to spreading or reflexed, bases connate. **Lemmas** 3-veined, awned; lemma awns divergent, becoming bent. **Anthers** 3. **Caryopses** narrowly elliptic; tips pubescent.

North America, Australia, and Eurasia. 1 species.

Taeniatherum caput-medusae (Linnaeus) Nevski (p. 483)

Medusahead

Elymus caput-medusae L.

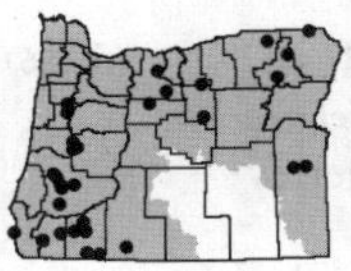

Plants annual, (5)10–55(70) cm tall, tufted. **Culms** erect to decumbent. **Leaves** sheaths glabrous; blades flat to involute, 0.5–2.5 mm wide, glabrous or scabrous. **Inflorescences** terminal spikes 1.2–6 cm; spikelets 2(4) per node. **Spikelets** 6–45 mm, with 2(3) florets. **Glumes** awn-like, erect to reflexed, (5)7–28 mm, 1-veined, scabrous with a few scattered hairs. **Lemmas** 5.5–8 mm, 5-veined, glabrous or scabrous; tips awned; lemma awns divergent, straight when young but often becoming twisted, (20)30–110 mm, arising from the lemma apex, the longest longer than the inflorescence axis. **Anthers** 0.8–1 mm. **Caryopses** 4–5.2 mm. **2*n***=14.

Disturbed areas, fields, dry roadsides, degraded sage steppe. 0–1900 m. All ecoregions except BR. CA, ID, NV, WA; north to British Columbia, southeast to UT; Australia, Eurasia. Exotic.

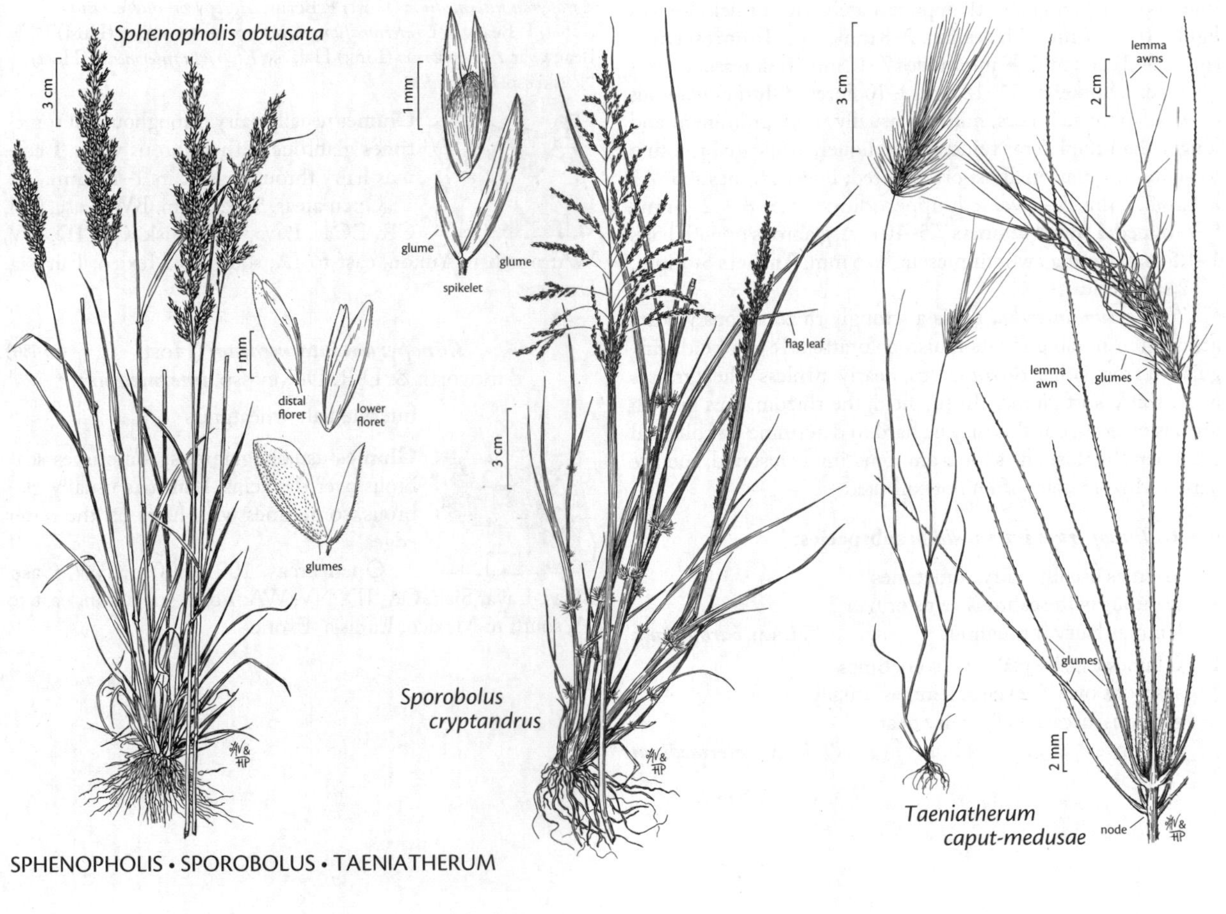

SPHENOPHOLIS • SPOROBOLUS • TAENIATHERUM

Taeniatherum caput-medusae has long, twisted awns that are longer than the inflorescence axis. It may be confused with *Elymus elymoides*, a long-awned, perennial bunchgrass with an inflorescence axis that disarticulates when mature. This grass is an aggressive invader of grasslands and shrub steppe in the arid west. It forms a dense thatch that is slow to decompose and forms monocultures that exclude both native and introduced plants, including other invasive annual grasses such as *Bromus tectorum*. Its forage value is limited due to its low stature and its long, irritating awns.

THINOPYRUM Wheatgrass

Plants perennial; cespitose or rhizomatous. **Leaves** sheaths open, glabrous or ciliate; auricles present or absent; ligules membranous; blades convolute or flat. **Inflorescences** terminal 2-sided spikes with 1 spikelet at all or most nodes; internodes 5–30 mm; disarticulation tardy, usually beneath the florets, sometimes in the rachis. **Spikelets** appressed to ascending, often diamond-shaped in outline and arching outwards at maturity, 1–3 times as long as the middle internodes. **Glumes** rectangular to lanceolate, narrowing beyond mid-length, stiff, hard to leathery, keeled or rounded at the base, usually more strongly keeled distally, 4–9-veined; tips truncate to acute, lacking lateral teeth, sometimes short-pointed. **Lemmas** 5-veined, with leathery tips, truncate, obtuse or acute, sometimes short-pointed or awned; lemma awns, if present, up to 5 mm. **Anthers** 3.

Temperate regions. Approximately 10 species; 2 species treated in *Flora*.

Thinopyrum pycnanthum was collected in Portland in 1915–1919 but apparently has not persisted.

Within this genus, measurements and descriptions of the inflorescence internodes refer to those in the middle of the spike. Culm width measurements include the surrounding leaf sheath.

Key to *Thinopyrum* species:

1. Plants strongly rhizomatous; glume tips usually acute to obtuse with veins unequal in length . . . . *T. intermedium*

1' Plants lacking rhizomes; glume tips truncate with all veins approximately equal in length . . . . *T. ponticum*

Thinopyrum intermedium (Host) Barkworth & D.R. Dewey

Intermediate wheatgrass

Plants perennial, 50–115 cm tall; rhizomatous, glaucous. **Culms** glabrous or hairy or occasionally hairy only at the nodes, lowest internode including sheath approximately 3 mm thick. **Leaves** ligules 0.1–0.8 mm; blades flat, 2–8 mm wide. **Inflorescences** spikes 8–21 cm; middle internodes 7–12 mm; inflorescence axes pubescent. **Spikelets** 11–18 mm, 3–10 florets. **Glumes** glabrous or with appressed hairs, midvein usually more prominent and longer than the lateral veins; tips obliquely truncate or obtuse to acute or occasionally short-pointed; lower glumes 4.5–7.5 × 1.5–2.5 mm, 5–6-veined; upper glumes 5.5–8 × 2–3 mm, 5–7-veined. **Lowest lemmas** 7.5–10 mm, glabrous or with hairs 1–1.5 mm; lemma awns, if present, to 5 mm. **Anthers** 5–7 mm.

2 subspecies.

Thinopyrum intermedium is a strongly rhizomatous wheatgrass often notable for its bluish coloration. It can be distinguished from other rhizomatous, nearly awnless wheatgrasses by its thick, stiff glumes. In the field, the rhizomatous growth form is obvious, but this may be hard to determine on collected specimens because the shoots are sometimes clustered, and the narrow rhizomes are often not collected.

Key to *Thinopyrum intermedium* subspecies:

1. Glumes usually hairy, sometimes glabrous with scabrous veins or hairy; lemmas hairy throughout. . . . . *T. i.* ssp. ***barbulatum***

1' Glumes usually glabrous, sometimes scabrous over the veins; lemmas usually glabrous, occasionally hairy near margins . . . . *T. i.* ssp. ***intermedium***

Thinopyrum intermedium (Host) Barkworth & D.R. Dewey ssp. *barbulatum* (Schur) Barkworth & D.R. Dewey

(p. 486)

Pubescent wheatgrass

Agropyron glaucum (Desf. ex DC. & Lam.) Roem. & Schult., *Agropyron intermedium* (Host) P. Beauv., *Agropyron intermedium* (Host) P. Beauv. var. *intermedium*, *Agropyron intermedium* (Host) P. Beauv. var. *trichophorum* (Link) Halácsy, *Elytrigia intermedia* (Host) Nevski ssp. *intermedia*

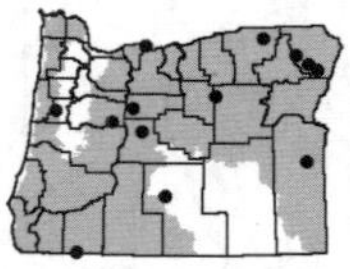

Glumes usually hairy throughout but sometimes glabrous with scabrous veins. **Lemmas** hairy throughout, hairs 1–1.5 mm.

Open areas. 50–2200 m. BW, Casc, Col, CR, ECas, Lava, Owy, Sisk. CA, ID, NV, WA; north to Yukon, east to IA, south to Mexico; Eurasia. Exotic.

Thinopyrum intermedium (Host) Barkworth & D.R. Dewey ssp. *intermedium*

(p. 486)

Intermediate wheatgrass

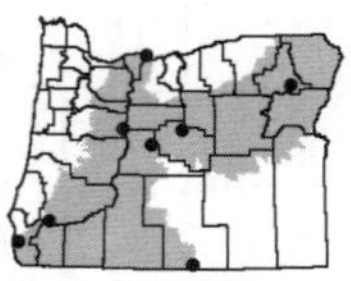

Glumes usually glabrous, sometimes scabrous over the veins. **Lemmas** usually glabrous, sometimes with hairs on the outer edges.

Open areas. 100–1800 m. BW, Casc, ECas, Lava, Sisk. CA, ID, NV, WA; north to Yukon, east to IA, south to Mexico; Eurasia. Exotic.

Thinopyrum ponticum Barkworth & D.R. Dewey

Tall wheatgrass

Agropyron elongatum (Host) P. Beauv., *Elytrigia elongata* (Host) Nevski, *Elytrigia pontica* (Podp.) Holub ssp. *pontica*

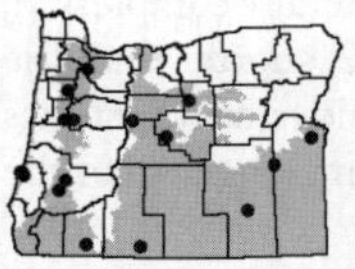

Plants perennial, 50–200 cm tall; cespitose, green. **Culms** glabrous; lower internode including sheath approximately 3.5 mm thick. **Ligules** 0.3–1.5 mm; blades 2–6.5 mm wide. **Inflorescences** spikes 10–42 cm; lower internodes 16–42 mm; middle internodes 9–19 mm; inflorescence axes glabrous. **Spikelets** often diamond-shaped and arched away from the inflorescence axis, 13–30 mm, 6–12 florets. **Glumes** glabrous, midveins approximately equal in length and prominence to the lateral veins; tips truncate; lower glumes 6.5–10 mm; upper glumes 7–10 mm. **Lemmas** 9–12 mm, glabrous, awnless. **Anthers** 4–6 mm.

Roadsides, grasslands, often in alkaline areas. 0–1400 m. BR, ECas, Est, Lava, Owy, Sisk, WV. CA, ID, NV, WA; western North America, scattered in eastern North America; Eurasia. Exotic.

Thinopyrum ponticum is a very large, cespitose grass with spikes that have a single spikelet at each node. *Leymus cinereus* and its relatives have two to eight spikelets per node and more or less acute glumes. *Thinopyrum intermedium* is strongly rhizomatous, has narrower culms, and has the glume midvein slightly longer than the other veins.

TORREYOCHLOA False mannagrass

Plants perennial; rhizomatous. **Culms** sometimes rooting at the lower nodes. **Leaves** sheaths open; ligules membranous; blades flat. **Inflorescences** panicles with scabrous branches; disarticulation above the glumes and beneath the florets. **Spikelets** laterally compressed to terete, 2–8 florets. **Glumes** shorter than the lowest lemmas, rounded across the back or weakly keeled, awnless; lower glumes 1(3)-veined; upper glumes (1)3(5)-veined. **Calluses** blunt, glabrous. **Lemmas** rounded to weakly keeled, prominently (5)7–9-veined; veins generally parallel; tips minutely scabrous or entire to serrate-erose, awnless. **Anthers** usually 3. **Caryopses** oblong, flattened dorsally, falling free.

North America and northeastern Asia. 4 species; 2 species treated in *Flora*.

Torreyochloa is similar in appearance but not closely related to *Glyceria* and *Puccinellia*. It was long placed in *Glyceria* although that genus has closed leaf sheaths. *Puccinellia* differs from *Torreyochloa* in having fewer, fainter lemma veins. *Puccinellia* grows in saline to alkaline habitats, while *Torreyochloa* grows only in non-saline, non-alkaline locations.

Key to *Torreyochloa* species:

1. Inflorescences linear to narrowly elliptic, 5.5–19 times as long as wide, less than 1 cm wide; panicle branches appressed (or rarely slightly spreading); leaf blades 3.4–7.2 mm wide . . . . ***T. erecta***

1' Inflorescences narrowly to widely conic, ovoid, or obovoid, 1.2–7.5 times as long as wide, 1–12 cm wide; panicle branches spreading; leaf blades 3.6–18 mm wide . . . . ***T. pallida*** var. ***pauciflora***

Torreyochloa erecta (Hitchcock) G.L. Church

Spiked false mannagrass, erect mannagrass

Glyceria erecta Hitchc.

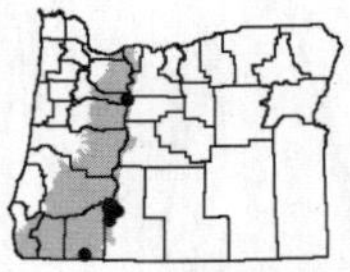

Plants perennial, 20–62 cm tall. **Culms** 1–1.4 mm thick. **Leaves** ligules of larger cauline leaves truncate or rounded to acute, 2.6–6.5 mm; blades of widest cauline leaves 3.4–6(7.2) mm wide. **Inflorescences** linear to narrowly elliptic, 5.5–11 × 0.3–1 cm, (5.5)6–19 times as long as wide; branches appressed or rarely slightly spreading. **Spikelets** 4.2–6.4 mm, 4–6 florets. **Glumes** lower glumes 0.8–1.6 mm; upper glumes 1–2.1 mm. **Lemmas** 2.3–3.1 mm, obtuse to acute, awnless. **Anthers** 0.6–0.8 mm. **2*n***=14.

Subalpine and alpine lakes and streams. 1500–2400 m. Casc, Sisk. CA, ID, NV. Native.

Torreyochloa pallida (Torrey) G.L. Church (p. 486)
var. ***pauciflora*** (J. Presl) J.I. Davis

Weak mannagrass, pale false mannagrass

Glyceria otisii Hitchc., *Puccinellia pauciflora* (J. Presl) Munz, *Puccinellia pauciflora* (J. Presl) Munz var. *holmii* (Beal) C.L. Hitchc., *Puccinellia pauciflora* (J. Presl) Munz var. *microtheca* (Buckley) C.L. Hitchc., *Puccinellia pauciflora* (J. Presl) Munz var. *pauciflora*

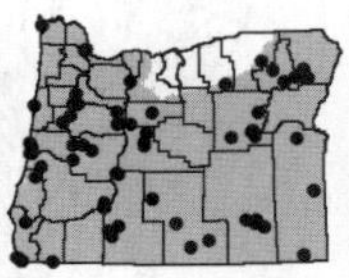

Plants perennial, 20–145 cm tall. **Culms** 1.3–4.8 mm thick near the base. **Leaves** ligules of larger cauline leaves truncate or acute to attenuate, 3–9 mm; blades of widest cauline leaves 3.6–18 mm wide. **Inflorescences** narrowly to widely conic, ovoid, or obovoid, (3)5–25 × (1)2–14 cm, 1.2–5.8(7.5) times as long as wide; branches spreading. **Spikelets** 3.6–6.9 mm, with (3)4–8 florets. **Glumes** lower glumes 0.7–2.1 mm; upper glumes 0.9–1.8 mm. **Lemmas** 2.2–3.3 mm, truncate to acute, awnless. **Anthers** of lowest floret 0.5–0.7 mm. **2*n***=14.

Wetlands, including swamps, marshes, bogs, and streams. 0–2500 m. All ecoregions except Col. CA, ID, NV, WA; north to AK, east to SD. Native.

Torreyochloa pallida grows emergent in water and looks like an unusually wide-leaved *Glyceria*. However, *Glyceria* species have closed leaf sheaths.

TRIPLASIS Sandgrass

Plants perennial or annual; cespitose or occasionally rhizomatous. **Culms** ascending to erect. **Leaves** sheaths open; ligules of hairs or membranous and ciliate; blades flat or involute, 1–5 mm wide. **Inflorescences** both terminal and axillary panicles, the terminal panicles exserted from or partially included in the uppermost leaf sheath, the axillary usually self-pollinating and hidden in the upper sheaths; disarticulation above the glumes and beneath the florets and later at the culm nodes. **Spikelets** laterally compressed, 2–5 florets; lower florets fertile; upper florets sterile; rachilla prolonged beyond the terminal floret. **Glumes** 2, shorter than the first lemma, 1-veined, keeled. **Calluses** with a tuft of soft hairs. **Lemmas** 3-veined; veins villous; tips bilobed, midveins sometimes extending into an awn; lemma awns 0–11 mm. **Paleas** bowed out, hairy on the keels, the distal hairs 0.5–2 mm, longer than the proximal ones. **Anthers** 3. **Caryopses** dorsiventrally compressed.

North America. 2 species; 1 species treated in *Flora*.

Triplasis purpurea (Walter) Chapman (p. 486)

Purple sandgrass

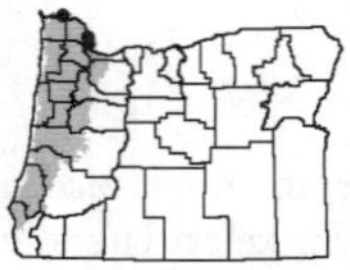

Plants annual or occasionally perennial, 14–100 cm tall; cespitose or occasionally rhizomatous. **Culms** nodes with hairs appressed and very inconspicuous to spreading. **Leaves** sheaths inflated, with hairs at the throat; ligules of hairs to 1 mm; blades flat to involute, 1–5 mm wide, scabrous with a few scattered, long, spreading, papillose-based hairs. **Inflorescences** 3–7 cm × 1–6 mm, partially hidden in leaf sheaths; disarticulation in the culm and between florets, above the glumes. **Spikelets** 6.5–9 mm, 3–4 florets. **Glumes** approximately 2 mm, glabrous or scabrous; tips erose. **Calluses** hairy. **Lemmas** 3–4 mm, 3-veined; veins with dense, white hairs; tips truncate with 2 short, rounded lobes; lemma awns straight, to 2 mm. **Paleas** about 2.5 mm; keels densely hairy. **Anthers** 2 mm, reddish purple.

Sandy soils, river shores. 0–50 m. CR, WV. central and eastern US, Ontario. Exotic.

This odd grass has tufts of palea hairs easily visible in the spikelets. Most of its inflorescences are hidden in the leaf sheaths although the terminal one is exposed. Late in the season, the grass disarticulates at the culm nodes. *Sporobolus cryptandrus* is similar in having partly hidden inflorescences, but it has dense tufts of hairs at the top of the leaf sheath and has glabrous lemmas and paleas. *Triplasis purpurea* grows along the lower Columbia River, where it likely was introduced from ship ballast.

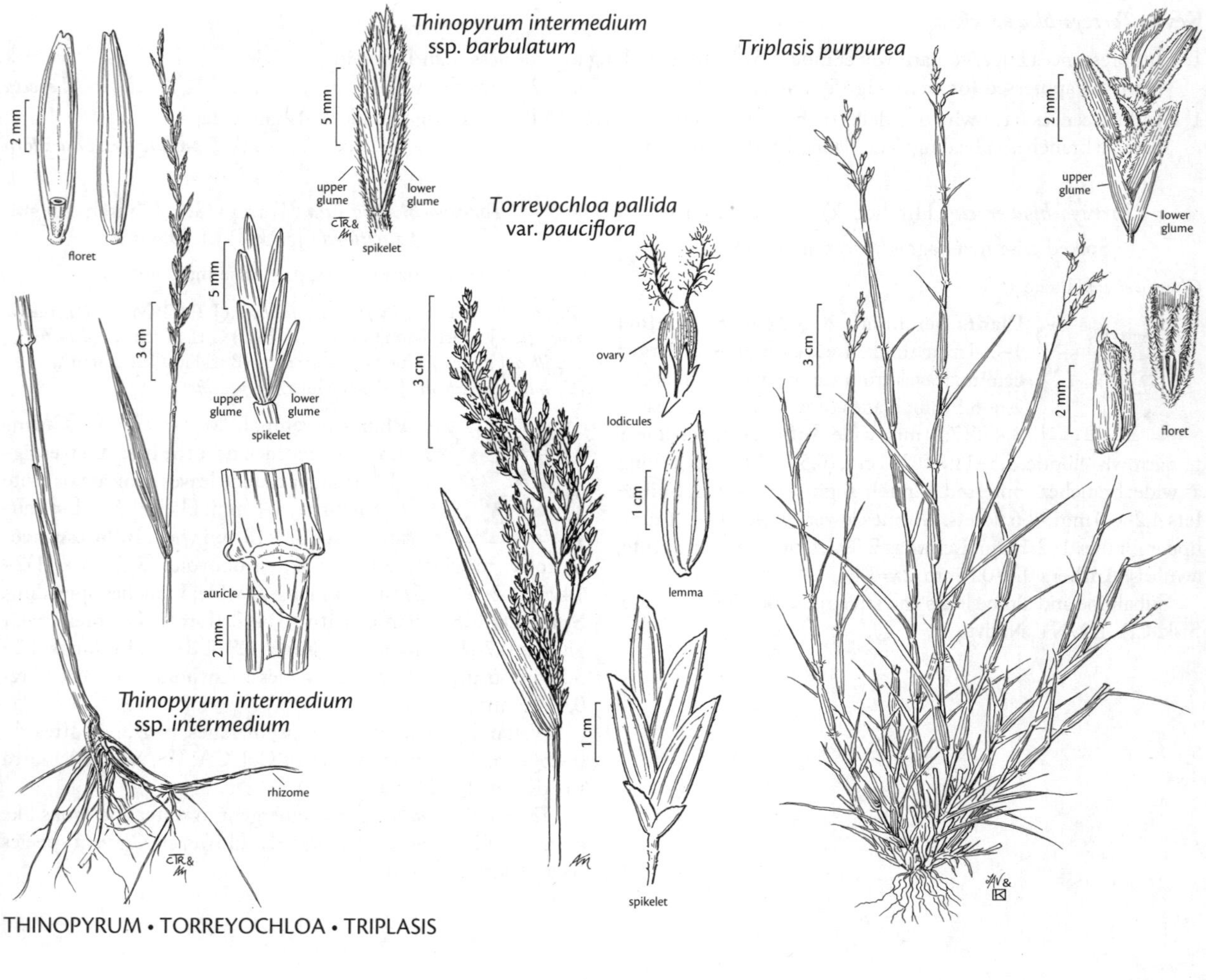

THINOPYRUM • TORREYOCHLOA • TRIPLASIS

TRISETUM Trisetum, false oat, oatgrass

Plants annual or perennial; cespitose or rhizomatous. **Leaves** sheaths open or with margins fused only at the base; ligules membranous, erose to lacerate or sometimes minutely ciliate; blades flat. **Inflorescences** terminal, open to condensed panicles but sometimes spike-like, green, tan, often purple-tinged, shiny; disarticulation above the glumes and beneath the florets, later below the glumes. **Spikelets** on pedicels or rarely sessile, laterally compressed, 2–5 florets, reduced florets sometimes present at the tip. **Glumes** keels scabrous; tips usually acute, awnless but often with an apiculus. **Calluses** hairy. **Lemmas** 3–7-veined; margins hyaline; tips usually bifid but sometimes entire, awnless or awned; lemma awns, if present, arising from above the middle. **Paleas** 2-veined; the veins sometimes excurrent as bristle-like tips. **Anthers** 3. **Caryopses** elongate-fusiform, compressed, brown; endosperm milky.

Temperate, subalpine, and arctic regions worldwide and high mountains in the tropics. Approximately 75 species; 4 species treated in *Flora*.

The wide hyaline lemma margins give *Trisetum* inflorescences a silvery or shining-tan glint.

Key to *Trisetum* species:

1. Lemmas awnless or with awns to 2 mm . . . . *T. wolfii*

1' Lemmas with awns 4–14 mm.

2. Inflorescences usually dense, spike-like; leaves usually in a basal tuft; glumes subequal, both glumes lanceolate; upper glumes less than 2 times as wide as the lower glumes . . . . *T. spicatum*

2' Inflorescences usually panicle-like, dense to open; leaves basal and cauline; glumes unequal (rarely subequal); upper glumes broadly lanceolate to ovate or obovate, 2 or more times as wide as the lower glumes.

3. Most panicle branches, except occasionally the lowermost, spikelet-bearing nearly to the base; panicles erect or nodding at the tips; branches ascending to somewhat divergent; lower glumes 2.8–5 mm; ligules (1.5)3–6 mm . . . . *T. canescens*

3' Most panicle branches, except occasionally the uppermost, spikelet-bearing only towards the tips; panicles nodding; branches of at least the lowest 1–3 whorls spreading or drooping; lower glumes 0.75–2.5(3.5) mm; ligules 1.5–3(3.4) mm . . . . *T. cernuum*

Trisetum canescens Buckley (p. 488)

Tall trisetum

Trisetum cernuum Trin. ssp. *canescens* (Buckley) Calder & R.L. Taylor, *Trisetum cernuum* Trin. var. *canescens* (Buckley) Beal

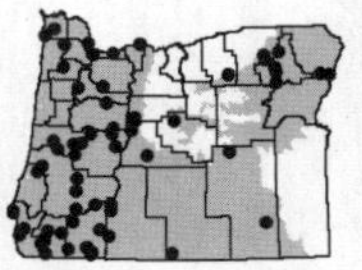

Plants perennial, 40–120 cm tall; cespitose. **Leaves** basal and cauline; ligules (1.5)3–6 mm; blades (3)7–10 mm wide; margins and surfaces sometimes with scattered hairs 1–3 mm. **Inflorescences** erect or nodding only near the tips, 10–25 × (0.75)1–3(4) cm, green or tan, occasionally purple-tinged; branches ascending to somewhat divergent, 1–5.5 cm, most spikelet-bearing nearly to the base, sometimes the lowermost branches naked below. **Spikelets** 7–9 mm, 2–4 florets. **Glumes** lower glumes narrow, lanceolate to subulate, acute or long-tapered, 2.8–5 mm; upper glumes broadly lanceolate to obovate, tapering to the tips, acute, (3.4)5–7(9) mm, usually shorter than the lowest florets, at least twice as wide as the lower glumes. **Lemmas** 5–7 mm, glabrous; tips with 2 awn-like teeth to 2.5(3.2) mm, awned; lemma awns bent, 7–14 mm, arising mostly in the upper 33% of the lemmas. **Caryopses** glabrous or finely hairy distally. ***2n***=28, 42.

Dry to moist forests and stream banks. 0–2500 m. BR, BW, Casc, CR, ECas, Est, Sisk, WV. CA, ID, NV, WA; north to British Columbia, southeast to CO. Native.

Trisetum canescens is most similar to *T. cernuum*, which has a nodding inflorescence, longer panicle branches that lack spikelets in the lower half, consistently short ligules, and proportionately smaller lower glumes. *Trisetum cernuum* grows in slightly moister habitats. The two species are most easily distinguished in the field where the panicle characteristics can be easily assessed.

Trisetum cernuum Trinius

Nodding trisetum, nodding oatgrass

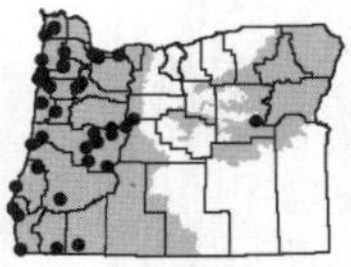

Plants perennial, (30)50–110 cm tall; cespitose. **Leaves** basal and cauline; ligules 1.5–3(3.4) mm; blades (3)7–12 mm wide, often minutely scabrous. **Inflorescences** open, nodding throughout, 10–30 × (1)2–9 cm, green or tan, occasionally purple-tinged; branches filiform, flexuous, at least the lowest 1–3 whorls spreading or drooping, 2–12+ cm, most spikelet-bearing in the outer 50–67%, the uppermost sometimes spikelet-bearing to the base. **Spikelets** 6–12 mm, 3–5 florets; upper 1–2 florets usually reduced. **Glumes** lower glumes subulate, 0.75–2.5(3.5) mm; upper glumes ovate or obovate, 3.5–5 mm, usually shorter than the lowest florets, 2–3 times as wide as the lower glumes, rounded to the acuminate tips. **Lemmas** broadly lanceolate, 5–6 mm, glabrous; tips with 2 teeth to 1.3 mm, awned; lemma awns arcuate to flexuous, (7)9–14 mm, arising from above mid-length to just below the teeth, conspicuous. **Anthers** about 1 mm. **Caryopses** densely to sparsely pubescent. ***2n***=42.

Moist forests and streamsides. 0–1600 m. BW, Casc, CR, ECas, Est, Sisk, WV. CA, ID, NV, WA; north to AK, northeast to MT. Native.

Trisetum cernuum is a flat-leaved grass with a nodding inflorescence. *Trisetum canescens* is very similar but has an erect inflorescence with shorter branches that are usually spikelet-bearing to the base and it usually grows in slightly drier habitats. It has longer ligules, on at least some leaves, and proportionately larger lower glumes. Occasional plants of *T. cernuum* have lower glumes more than 3 mm but otherwise have the traits of the species. *Trisetum canescens* has been treated as a subspecies of *T. cernuum*.

Trisetum spicatum (Linnaeus) K. Richter

Spike trisetum, spike false oat, narrow false oat

Trisetum sesquiflorum Trin., *Trisetum spicatum* (L.) K. Richt. var. *congdoni* (Scribn. & Merr.) Hitchc., *Trisetum spicatum* (L.) K. Richt. var. *spicatum*

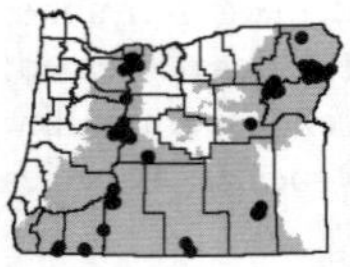

Plants perennial, 10–120 cm tall; cespitose. **Leaves** mostly basal or evenly distributed; ligules 0.5–4 mm; blades 1–5 mm wide. **Inflorescences** dense, spike-like, often interrupted basally, (5)20–30(50) × (0.5)1–2.5(5) cm, green to purplish or tawny, usually shiny and silvery; branches less than 2(3.5) mm, spikelet-bearing to the base. **Spikelets** 5–7.5 mm, 2(3) florets. **Glumes** lanceolate, smooth or sometimes sparsely scabrous or pilose; margins widely scarious; tips acute to acuminate or apiculate; lower glumes 3–4(5.5) mm; upper glumes 4–7 mm, as long as or longer than the lowest florets, less than twice as wide as the lower glumes. **Lemmas** narrowly to broadly lanceolate, 3–6(7) mm, glabrous, minutely scabrous, or pilose; tips bifid; teeth usually less than 1 mm; awned; awns bent and basally twisted, 3–8 mm, arising in the upper 33% of the lemmas. **Anthers** 0.7–1.4 mm. **Caryopses** glabrous. **2*n***=14, 28, 42.

Moist montane meadows, forests, and rocky slopes. 900–3100 m. BR, BW, Casc, ECas, Sisk. CA, ID, NV, WA; throughout North America except midwestern and southern US; South America; Eurasia. Native.

Trisetum spicatum is recognized by its dense, shiny, tan inflorescence with divergent awns. The shiny inflorescence of *Koeleria macrantha* is narrower and lacks awns.

Trisetum wolfii Vasey

Wolf's trisetum, beardless trisetum, Wolf's false oat

Graphephorum wolfii (Vasey) ex J.M. Coult., *Trisetum muticum* Scribn.

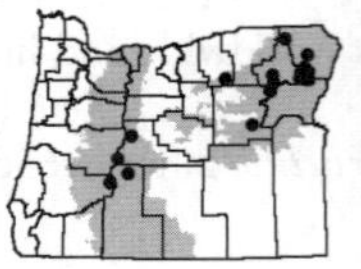

Plants perennial, 20–80(100) cm tall, short-rhizomatous. **Culms** glabrous or retrorsely pubescent below the nodes. **Leaves** usually concentrated on the lower 33% of the culms; ligules (1.2)2.5–4(5) mm; blades 2–7 mm wide, smooth, scabrous or sparsely pilose; tips sometimes prow-like. **Inflorescences** dense, spike-like, stiffly erect, (10)20–40(50) × usually 1–1.5 cm, green, tan or purple-tinged; branches appressed-ascending, spikelet-bearing to the base. **Spikelets** 4–7(8) mm, 2(3) florets. **Glumes** usually exceeding the lowest florets; lower glumes 4–7 mm; upper glumes 4–6.5 mm, slightly wider than the lower glumes. **Lemmas** lanceolate, 4–6.5 mm, firmer than the glumes, minutely scabrous-puberulent; tips obscurely bifid, awnless or awned; lemma awns, if present, to 2 mm, arising just below and rarely exceeding the lemma tips. **Anthers** (0.6)1(1.5) mm. **Caryopses** pubescent. **2*n***=14.

Moist montane meadows and riparian areas. 800–2700 m. BW, Casc, ECas. CA, ID, NV, WA; north to British Columbia, northeast to Saskatchewan, southeast to NM. Native.

Trisetum wolfii is a fairly nondescript grass with a dense, narrow, shiny panicle.

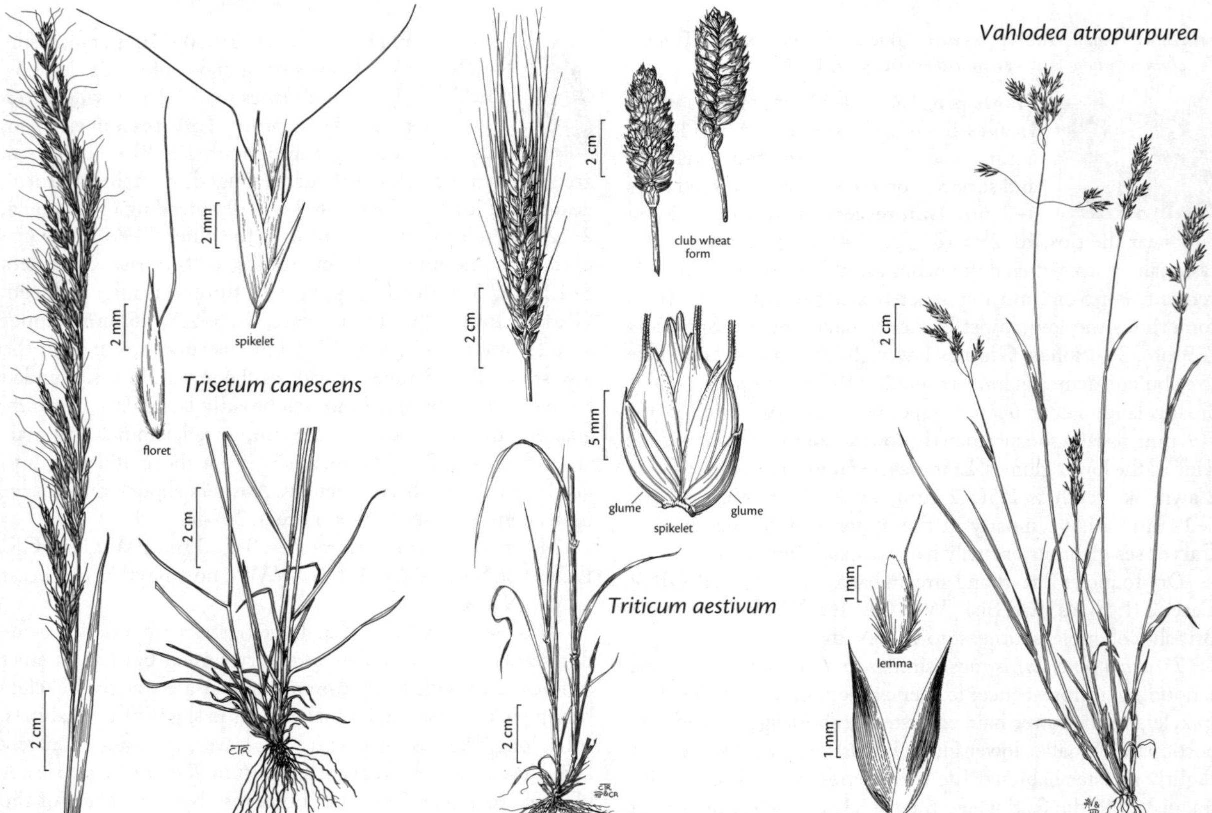

TRISETUM • TRITICUM • VAHLODEA

TRITICUM Wheat

Plants annual. **Culms** solitary or branched at the base. **Leaves** sheaths open; auricles present, often deciduous at maturity; ligules truncate, membranous; blades flat. **Inflorescences** terminal, 2-sided spikes with 1 spikelet per node; internodes (0.5)1.4–8 mm; disarticulation in the rachis; spikelets usually falling with the internode below or sometimes falling with the internode above, cultivated taxa usually not disarticulating. **Spikelets** usually 1–3 times the length of the internodes, appressed to ascending, 2–9 florets, the distal florets usually sterile. **Glumes** subequal, ovate, rectangular, or lanceolate, thick, usually stiff, keeled distally, the keels often winged and ending in a tooth or awn, some taxa with a second keel or prominent lateral vein. **Lemmas** keeled, papery or leathery in texture, the 2 lowest lemmas usually awned, distal lemmas awned or awnless; lemma awns 3–23 mm on lower lemmas, to 20 mm on upper lemmas, arising from the apex. **Anthers** 3. **Caryopses** tightly or loosely enclosed by the glumes and lemmas.

Cosmopolitan. Approximately 25 species; 1 species treated in *Flora*.

Triticum aestivum Linnaeus (p. 488)
Wheat

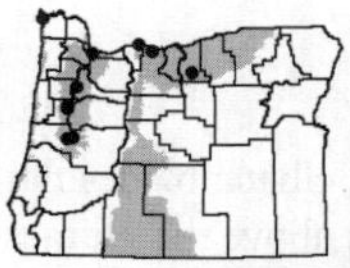

Plants annual, 14–150 cm tall; cespitose. **Leaves** sheaths glabrous or hairy; blades 6–15(20) mm wide, glabrous or pubescent. **Inflorescences** spikes (3.5)6–18 × 0.8–2.1 cm; disarticulation tardy or not disarticulating. **Spikelets** 10–15 mm, 2 florets. **Glumes** subequal, shorter than adjacent lemmas, hairy, short-awned; upper glumes 6–12 mm, 4–7-veins. **Lemmas** 10–15 mm, 0–several veins, mostly glabrous, scabrous, hairs near apex; tips toothed or awned; lemma awns, if present, straight or curved, to 8(12) mm, arising from the apex. **Anthers** 2.5 mm. **2*n***=42.

Disturbed sites, cultivated fields, roadsides. 0–1700 m. Col, ECas, Est, WV. CA, ID, NV, WA; throughout North America; worldwide. Exotic.

Wheat is the most widely grown grain crop. It has a thick spike that may be erect or nodding. Awnless forms, called club wheats, commonly grown in the Pacific Northwest, appear startlingly different from awned forms.

VAHLODEA Mountain hairgrass

Plants perennial; loosely cespitose. **Leaves** sheaths open; ligules membranous; blades flat. **Inflorescences** terminal panicles; branches very thin, flexuous; spikelets tending to be concentrated in outer parts of branches; disarticulation above glumes, beneath florets. **Spikelets** laterally compressed, usually 2 florets or sometimes with additional distal vestigial florets; rachillas prolonged beyond the distal florets for 0.5 mm or less, usually glabrous. **Glumes** as long as or longer than the florets; lower glumes 1(3)-veined; upper glumes 3-veined. **Calluses** obtuse, hairs about half as long as the lemmas. **Lemmas** ovate, 5(7) faint veins, awned; lemma awns twisted and bent, attached near the middle of the lemmas, visible between the glumes. **Anthers** 3. **Caryopses** ellipsoid, irregularly triangular or ovate in cross section, deeply grooved.

Circumboreal and southern South America. 1 species.

Vahlodea was formerly placed in *Deschampsia* but was removed based on its loosely cespitose habit, mostly cauline leaves, short, usually hairless rachilla prolongation, and caryopses traits.

Vahlodea atropurpurea (Wahlenb.) Fries ex Hartman (p. 488)
Mountain hairgrass

Aira atropurpurea Wahlenb., *Deschampsia atropurpurea* (Wahlenb.) Scheele, *Deschampsia atropurpurea* (Wahlenb.) Scheele var. *latifolia* (Hook.) Scribn. ex Macoun

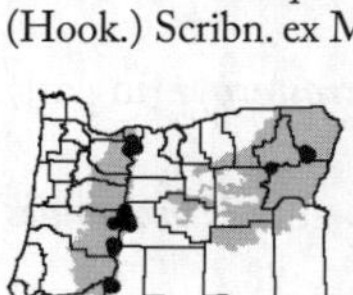

Plants 15–80 cm tall; cespitose. **Lower** blades to 30 cm × 1–8.5 mm; uppermost culm blades 1–5 mm wide, glabrous. **Inflorescences** 3–20 × 0.8–4 cm. **Spikelets** 4–7 mm. **Glumes** strongly keeled, awnless; lower glumes 4–5(6.5) mm; upper glumes 4–5.5(7) mm, 3-veined. **Lemmas** 1.8–3 mm; tips scabrous, ciliate; lemma awns bent, twisted, 2–4 mm, arising from mid-length. **Anthers** 0.5–1.2 mm. **2*n***=14.

Montane, moist to wet, open woods, streamsides, snowbeds, meadows. 1100–2800 m. BW, Casc. CA, ID, WA; northern and western North America; southern South America, circumboreal; circumboreal, southern South America. Native.

Vahlodea atropurpurea resembles *Deschampsia cespitosa*, but *V. atropurpurea* leaves are flat, relatively broad, and not concentrated at the base of the plant.

VENTENATA Ventenata

Plants annual, tufted. **Culms** puberulent below the nodes, nodes glabrous. **Leaves** sheaths open; ligules hyaline; blades flat, becoming involute with age. **Inflorescences** panicles, spikelets borne on pedicels near the ends of the branches; disarticulation above glumes and between the florets. **Spikelets** laterally compressed, 2–10 florets. **Glumes** unequal, lanceolate, acuminate; lower glumes 3–7-veined, upper glumes 3–9-veined. **Calluses** of lower florets shorter than those of the upper florets, sparsely hairy; calluses of the upper florets with a dense tuft of white hairs. **Lemmas** lanceolate, chartaceous, 5-veined, tips entire or bifid, awned or awnless; awn of lowest lemma within a spikelet straight, terminal, sometimes absent; awns of distal lemmas bent, dorsal, always present. **Anthers** 3. **Caryopses** shorter than lemmas, concealed at maturity, glabrous.

North America, North Africa, western Asia, and Europe. 5 species; 1 species treated in *Flora*.

Ventenata dubia (Leers) Cosson (p. 492)

Ventenata, North Africa Grass

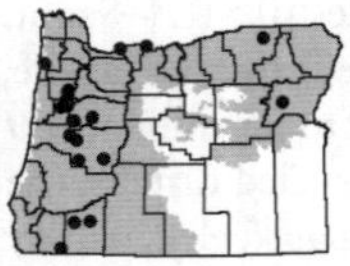

Plants 15–75 cm tall. **Leaves** sheaths glabrous or sparsely pubescent; ligules obtuse, usually lacerate, 1–8 mm; blades flat initially, involute with age, 2–7(12) cm × 0.8–2.5 mm, lower surfaces glabrous, upper surfaces scabrous. **Inflorescences** diffuse, (7)15–20 × 2–12 cm; branches 2–5 at lower nodes. **Spikelets** 9–15 mm, 2–3 florets, lowest floret staminate, upper floret(s) bisexual. **Glumes** 5–9-veined, lower glumes 4.5–6 mm; upper glumes 6–8 mm. **Lemmas** 5–7.5 mm, 5-veined, awned, tips of lower lemmas entire, those of upper lemmas bifid; lemma awns of lowest lemma in each spikelet straight, to 4 mm, arising from the near the tip; upper lemma awns bent and twisted, 10–16 mm, arising from the middle of the back. **Anthers** 1–2 mm. **Caryopses** about 3 mm.

Disturbed sites, roadsides, pastures, grasslands. 50–1900 m. BW, Casc, Col, CR, ECas, Sisk, WV. CA, ID, WA; scattered in northwestern and northeastern North America; North Africa, Europe. Exotic.

Ventenata dubia resembles the delicate, annual grass *Deschampsia danthonioides*, which has 1–3-veined glumes. Long, bent awns arise from near the base of the *D. danthonioides* lemma backs. The first North American specimen of *Ventenata dubia* was collected in Washington in 1952. Now it is widespread and abundant in northwestern North America.

VULPIA Annual fescue

Plants usually annual, rarely perennial. **Leaves** sheaths open; ligules membranous, usually shorter than 1 mm, ciliate; blades flat or rolled, usually glabrous. **Inflorescences** panicles or racemes, branches ascending to spreading; disarticulation above the glumes and beneath the florets, occasionally below the pedicels. **Spikelets** laterally compressed, with 1–11(17) florets, the distal florets reduced. **Glumes** 2, shorter than the adjacent florets; tips acute to acuminate, awnless or awned; lower glume usually much shorter than the upper glume, 1-veined; upper glume 3-veined. **Calluses** blunt, glabrous. **Lemmas** lanceolate, 3–5-veined; tips entire, acute to acuminate, usually awned; lemma awns terminal, straight. **Anthers** 1(3). **Caryopses** elongate, dorsally compressed, curved in cross section.

Cosmopolitan. Approximately 30 species; 4 species treated in *Flora*.

Vulpia species are small annuals of disturbed sites. Sterile hybrids are formed with some species of *Festuca* and *Lolium*. *Vulpia* has been included in *Festuca* in the past and currently is in some references. The species in this genus are usually self-fertilizing. Outcrossing individuals (which have 3 anthers), occur occasionally in all our species, usually when environmental conditions are especially favorable.

Culm width measurements include the surrounding leaf sheath. Spikelet and lemma measurements exclude the awns. Relative lengths of upper and lower glumes should not be assessed on the terminal spikelet of a branch.

Key to *Vulpia* species:

1. Inflorescence branches and/or spikelets spreading to reflexed . . . . ***V. microstachys*** (in part)

1' Inflorescence branches and spikelets ascending to erect.

  2. Lower glumes less than 50% as long as the upper glumes, 0.5–2 mm, sometimes seemingly absent. . . . . ***V. myuros***

  2' Lower glumes 40% or more the length of the upper glumes, usually more than 3 mm.

    3. Florets usually (4)5–11(17) per spikelet; awns usually less than 4(9) mm, shorter than or equaling lemmas; caryopses 1.7–3.7 mm . . . . ***V. octoflora***

    3' Florets usually 1–5(8) per spikelet; awns usually more than 4 mm, equaling or longer than the lemmas; caryopses 3.5–6.5 mm.

      4. Paleas usually slightly longer than the lemmas; pulvini present at base of panicle branches and often the spikelets; plants immature (mature individuals will key to lead 1). . . . . . . ***V. microstachys*** (in part)

      4' Paleas shorter than or as long as the lemmas; pulvini lacking in the inflorescence; plants immature or mature . . . . ***V. bromoides***

Vulpia bromoides (Linnaeus) Gray

Brome fescue, rattail fescue

Festuca bromoides L., *Festuca dertonensis* (All.) Asch. & Graebn.

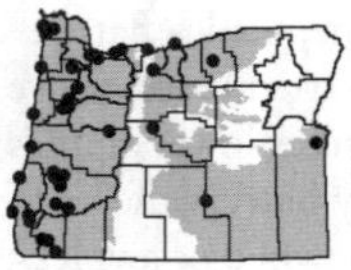

Plants annual, 5–75 cm tall. **Leaves** blades usually 2–10 cm × 0.5–2.5 mm, glabrous or puberulent. **Inflorescences** panicles 1.5–15 × 0.5–3 cm; conspicuously exserted from the uppermost leaf sheath; branches usually appressed to erect at maturity, without pulvini in the axils of the branches. **Spikelets** 5–10 mm, 4–8 florets, not closely overlapping. **Glumes** lower glumes 3.5–5 mm, 50–80% as long as the upper glumes; upper glumes 4.5–9.5 mm, midveins scabrous distally. **Lemmas** 4–8 mm, 5-veined, scabrous distally, awned; lemma awns of lowermost lemma in each spikelet 2–13 mm. **Paleas** equal to or shorter than the lemmas. **Anthers** 0.4–0.6(1.5) mm. **Caryopses** 3.4–5 mm. **2*n***=14.

Disturbed areas including cultivated fescue and ryegrass fields, roadsides. 0–1700 m. BR, Casc, Col, CR, Est, Lava, Owy, Sisk, WV. CA, ID, NV, WA; scattered throughout North America; South America; Europe. Exotic.

Vulpia bromoides closely resembles *V. myuros*, but, in the latter species, the lower glumes are often tiny and always half or less as long as the upper glumes, and the inflorescences are looser. Immature *Vulpia microstachys* can be confused with *V. bromoides*.

Vulpia microstachys (Nuttall) Munro

Small fescue, desert fescue

Festuca arida Elmer, *Festuca confusa* Piper, *Festuca eastwoodae* Piper, *Festuca grayii* (Abrams) Piper, *Festuca microstachys* Nutt., *Festuca pacifica* Piper, *Festuca reflexa* Buckley, *Vulpia microstachys* (Nutt.) Munro ex Benth. var. *ciliata* Beal, *Vulpia microstachys* (Nutt.) Munro ex Benth. var. *confusa* (Piper) Lonard & Gould, *Vulpia microstachys* (Nutt.) Munro ex Benth. var. *microstachys*, *Vulpia microstachys* (Nutt.) Munro ex Benth. var. *pauciflora* (Scribn. ex Beal) Lonard & Gould, *Vulpia pacifica* (Piper) Rydb.

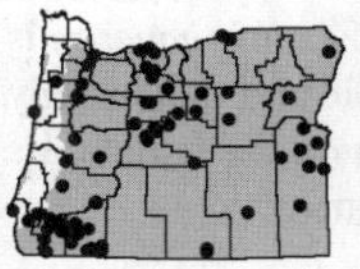

Plants annual, 15–75 cm tall. **Leaves** blades usually less than 10 cm × 0.5–1 mm, glabrous or pubescent. **Inflorescences** open panicles or sometimes spike-like racemes 2–24 × 0.8–8 cm; branches appressed to erect when immature, spreading to reflexed at maturity, pulvini in the axils of the branches. **Spikelets** 4–10 mm, 1–6 florets. **Glumes** unequal to subequal, smooth, scabrous, or pubescent; lower glumes 1.7–6.5 mm, 50–97% as long as the upper glumes; upper glumes 3.5–7.5 mm. **Lemmas** 3.5–9.5 mm, 5-veined, smooth, scabrous or evenly pubescent; lemma awns of lowest lemma in each spikelet (3)6–20 mm. **Paleas** usually slightly longer than the lemmas. **Anthers** 0.7–3 mm. **Caryopses** 3.5–6.5 mm. **2*n*=42.**

Grasslands, scablands, sagebrush steppe and disturbed places. 0–1800 m. All ecoregions except CR. CA, ID, NV, WA; north to British Columbia, northeast to MT, southeast to NM, south to Mexico. Native.

When mature, the inflorescence branches of *Vulpia microstachys* are usually spread 90° or more, and often the spikelets spread similarly from the branches. It is unique among our *Vulpia* species in having pulvini (swellings that can expand to move the branches) at the base of inflorescence branches and spikelets. Immature plants are easily confused with *V. bromoides*, which has consistently glabrous (or very short puberulent) leaf sheaths and consistently unequal glumes; many immature plants cannot be identified to species. Numerous intraspecific taxa have been described, but their usefulness is questionable.

Vulpia myuros (Linnaeus) C.C. Gmelin

Rattail fescue, rat-tail six-weeks grass

Festuca megalura Nutt., *Festuca myuros* L., *Vulpia myuros* (L.) C.C. Gmel. var. *hirsuta* (Hack.) Asch. & Graebn., *Vulpia myuros* (L.) C.C. Gmel. var. *myuros*

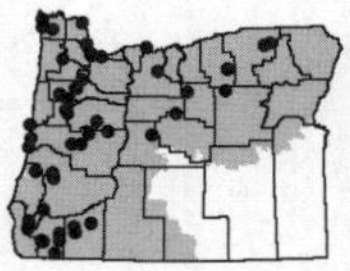

Plants annual, 10–75(90) cm tall. **Leaves** blades 2.4–10.5(17) cm × 0.4–3 mm, usually glabrous. **Inflorescences** dense panicles or spike-like racemes 3–25 × 0.5–1.5(2) cm, often enclosed in the uppermost leaf sheath at maturity; pulvini absent; branches usually appressed to ascending. **Spikelets** 5–12 mm, 3–7 florets. **Glumes** glabrous; lower glumes 0.5–2 mm, 20–50% as long as the upper glumes, (except occasionally on the uppermost spikelet of a branch); upper glumes 2.5–5.5 mm. **Lemmas** 4.5–7 mm, 5-veined, usually scabrous distally, glabrous but the margins sometimes strongly ciliate; lemma awns 5–15(22) mm. **Anthers** 0.5–1(2) mm. **Caryopses** fusiform, 3–5 mm, glabrous. **2*n*=42.**

Disturbed sites with well-drained soil. 0–1700 m. BW, Casc, Col, CR, ECas, Est, Lava, Sisk, WV. CA, ID, NV, WA; scattered throughout North America; North Africa, Europe. Exotic.

With its narrow panicles, *V. myuros* closely resembles *V. bromoides*, but the latter species has lower glumes at least half as long as the upper glumes and consistently dense inflorescences with individual spikelets that are hard to distinguish. Some *V. myuros* plants have dense inflorescences like *V. bromoides*, but others have looser inflorescences with the individual spikelets easily visible. Plants with strongly ciliate lemmas have been called *V. megalura*, but there seems to be no other trait distinguishing them from typical *V. myuros*.

Vulpia octoflora (Walter) Rydberg (p. 492)

Sixweeks fescue, six-weeks grass

Festuca octoflora Walter var. *hirtella* (Piper) Hitchc., *Festuca octoflora* Walter var. *octoflora*, *Vulpia octoflora* (Walter) Rydb. var. *hirtella* (Piper) Henrard, *Vulpia octoflora* (Walter) Rydb. var. *octoflora*

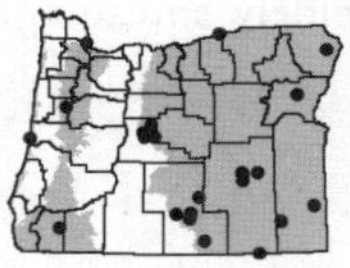

Plants annual, 4–60 cm tall. **Leaves** blades to 10 cm × 0.5–1 mm wide, glabrous or pubescent. **Inflorescences** panicles 1–7 × 0.5–1.5 cm; branches appressed to spreading; pulvini absent. **Spikelets** (4)5–11(13) mm, (4)5–11(17) florets, closely overlapping. **Glumes** lower glumes 1.7–4.5 mm, 50–67% as long as the upper glumes; upper glumes 2.5–7.2 mm. **Lemmas** 2.7–6.5 mm, 5-veined, smooth, scabrous or pubescent; tips entire, awned; lemma awns of lowermost lemma in each spikelet 0.2–9 mm. **Anthers** 0.3–1.5 mm. **Caryopses** 1.7–3.7 mm. **2*n*=14.**

Found on sandy soil, disturbed sites in grasslands, shrub steppe, open forests, roadsides, and ditches. 0–1500 m. BR, BW, Col, Est, Lava, Owy, Sisk, WV. CA, ID, NV, WA; throughout most of North America; South America; Asia, Europe. Native.

Vulpia octoflora has more florets per spikelet and usually has shorter awns than the other *Vulpia* species. Its branches never spread widely as do those of *V. microstachys*. Dwarfed individuals of both *V. octoflora* and *V. microstachys* may occur together in shrub steppe and can be difficult to distinguish until mature. Both our native *Vulpia* species have been displaced by introduced *Bromus tectorum*, with *V. octoflora* suffering the greatest decline. Three varieties have been described based on spikelet and lemma length and pubescence. These characters, however, vary independently and do not seem useful taxonomically.

ZIZANIA Wild rice

Plants annual or perennial. **Culms** to 5 m tall, erect and emergent. **Leaves** sheaths open; ligules membranous or scarious; blades flat. **Inflorescences** panicles, lower branches staminate, upper branches pistillate, occasionally the middle branches with staminate and pistillate spikelets intermingled; disarticulation beneath the spikelets. **Spikelets** unisexual, terete, staminate spikelets pendent. **Glumes** absent. **Calluses** inconspicuous. **Lemmas** 5-veined, those of staminate spikelets membranous, awnless, those of pistillate spikelets papery or leathery, margins clasping the palea margins, tips acute to acuminate or sometimes awned; lemma awns terminal, slender. **Paleas** 3-veined. **Anthers** 6. **Caryopses** cylindrical.

North America and Asia. 4 species; 1 species treated in *Flora*.

Two other *Zizania* species may be introduced to Oregon in the future. *Zizania latifolia* is a rhizomatous perennial from Asia that is sold horticulturally. It is invasive in Hawaii and host to a fungus that could potentially devastate North American *Zizania* stands. The middle branches of its inflorescences have both pistillate and staminate spikelets mixed together. *Zizania aquatica* is an annual that is sometimes planted for wildlife food. It is very similar to *Z. palustris*, but its lemmas are flexible and dull or only slightly shiny and bear short, scattered hairs that are not more densely arranged near the lemma tip than lower on the lemma. Its aborted pistillate spikelets are 0.4–1 mm wide. Its pistillate inflorescence branches are usually spreading at maturity.

Zizania palustris Linnaeus var. ***palustris*** (p. 492)
Wild rice

Zizania aquatica L. ssp. *angustifolia* (Hitchc.) Tzvelev, *Zizania aquatica* L. var. *angustifolia* Hitchc.

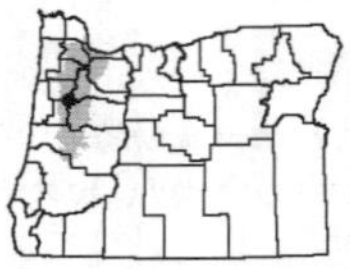

Plants annual, to 3 m tall, usually emergent. **Ligules** 3–16 mm. **Leaves** blades 20–60 cm × 3–21(40+) mm, glabrous. **Inflorescences** 24–60 × 1–20(40) cm, staminate branches ascending to divergent, pistillate branches mostly appressed or sometimes a few widely spreading. **Spikelets** staminate spikelets lanceolate, 6–17 mm, acuminate or awned, lemma awns to 2 mm; pistillate spikelets 8–33 × 1–2.6 mm, leathery or hard, shiny, glabrous or with lines of short hairs, tips usually hairy and abruptly narrowed, lemma awns to 10 cm; aborted lemmas 0.6–2.6 mm wide. **Caryopses** 6–30 × 0.6–2 mm.

Lakes, marshes, streams. 50–100 m. WV. ID, WA; throughout mid-latitude North America. Exotic.

Zizania palustris is a large aquatic grass harvested for human food and is sometimes planted to provide cover and food for wildlife.

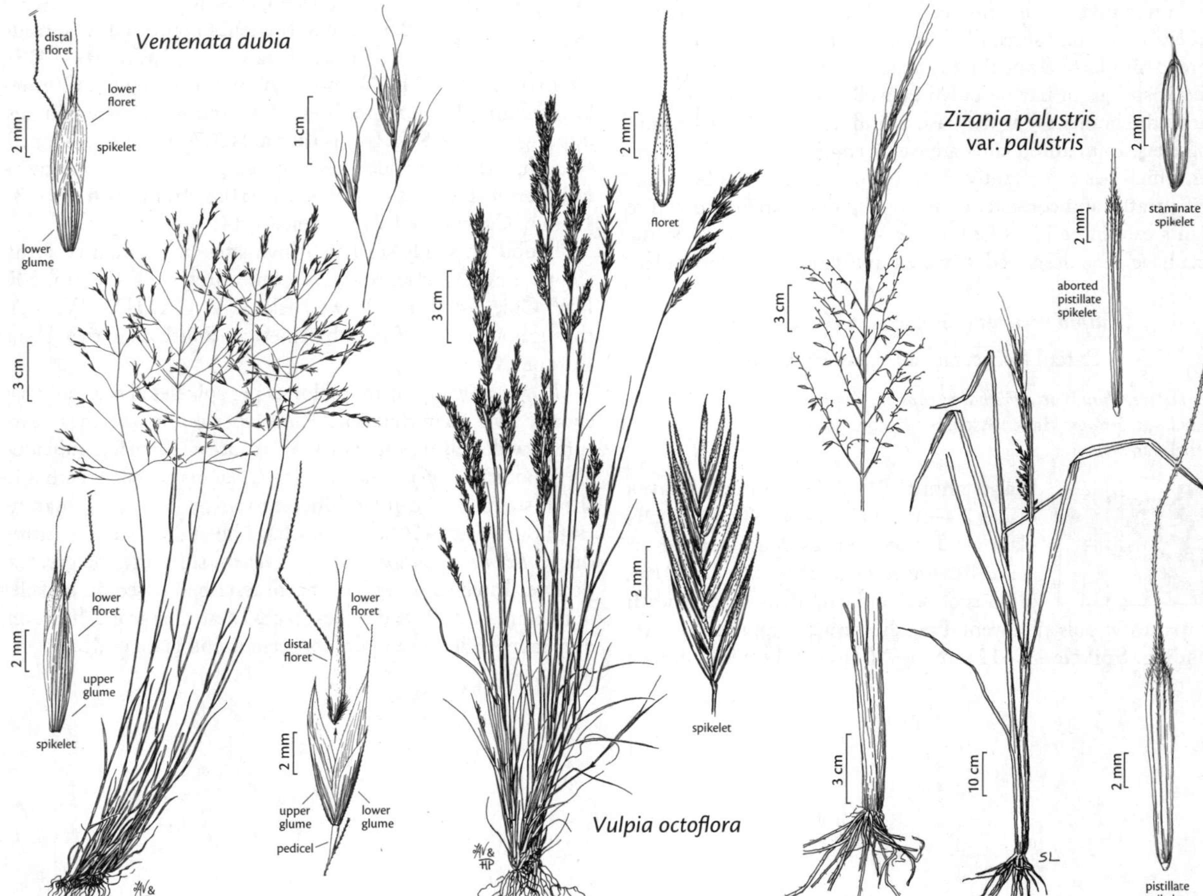

VENTENATA • VULPIA • ZIZANIA

PONTEDERIACEAE Pickerel-weed family

Tamra Dickson Prior

Herbs aquatic, rooting in mud or occasionally free floating. **Stems** vegetative and flowering; vegetative stems indeterminate, with many leaves, glabrous; flowering stems determinate, submersed or emersed, with a single leaf, spathe, and terminal inflorescence. **Leaves** petiolate and sessile, forming basal rosettes or alternate on elongated stems, stipules retained after withering; petiolate leaf blades cordate to ovate; sessile leaf blades linear to oblanceolate, base sheathing, margins entire. **Inflorescences** spicate or 1-flowered; spathes bract-like, folding or clasping. **Flowers** sessile, 3-merous; tepals connate, yellow, blue, mauve, or white, limb 6-lobed; stamens 3 or 6, adnate to tepals; pistils 3-locular, 1 or 3 developing to maturity. **Fruits** capsules. **Seeds** smooth or winged.

Cosmopolitan, concentrated in tropical and subtropical regions. 6 genera; 2 genera treated in *Flora*.

Eichhornia crassipes, a highly invasive introduced species, has been collected once, in Wasco County, in 1958. Since that time no other occurrences have been reported.

Key to genera of Pontederiaceae:

1. Inflorescences with fewer than 30 flowers; fruits with 10–200 seeds . . . . **HETERANTHERA**

1' Inflorescences generally with more than 50 flowers; fruits with a single seed . . . . **PONTEDERIA**

HETERANTHERA Mud plantain

Herbs perennial; rooting in mud. **Stems** vegetative submersed; flowering stems submersed and growing to water surface. **Leaves** floating or submersed; blades reniform, cordate, or oblong, apex membranous and acute to obtuse. **Inflorescences** elongating over 1–several days. **Flowers** open 1 day only; tepals connate half or more of their length, limb lobes obtuse to acuminate; stamens 3, filaments yellow or purple, anthers yellow or purple; style 3-lobed. **Fruits** elongate. **Seeds** 10–200, ovoid, testa with longitudinal wings.

Western Hemisphere, Africa. 12 species; 1 species treated in *Flora*.

Heteranthera dubia (Jacquin) MacMill (p. 496)

Water stargrass

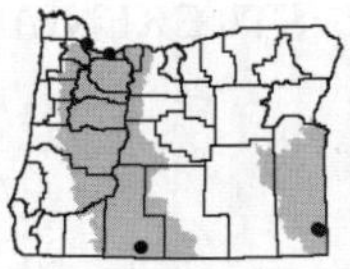

Herbs perennial; rooting in mud. **Leaves** petiolate leaves not produced; sessile leaves alternate or appearing opposite at nodes with inflorescences; blades linear, thin, 2–10 cm × 1–5 mm; stipules absent. **Inflorescences** 1-flowered; spathes 12–45 mm, glabrous; peduncle absent. **Flowers** opening within 2 hours after dawn, wilting by dusk; perianth salverform, yellow, tube 1.5–11 cm, limb lobes equal, linear, 4–11 mm, of uniform color; stamens equal, 2–10 mm, filaments centrally inflated, glabrous, anthers coiling by early afternoon after anthesis; style glabrous or scarcely pubescent. **Seeds** 7–30, 12–22-winged, 1–1.6 × 0.5–1.3 mm. $2n$=15.

Streams, rivers, ponds, lakes. Flowering Jul–Sep. 0–1400 m. Casc, ECas, Owy, WV. CA, ID, WA; north to British Columbia, east to the Atlantic Ocean, south to AZ and TX. Native.

The morphology of *H. dubia* varies in relation to its habitat. The leaves of plants found in ponds and lakes tend to be longer and narrower than those of plants found in streams and rivers. This species grows best in alkaline waters and can cause an increase in water pH when growing in high densities.

PONTEDERIA Pickerel-weed

Herbs, rooted in mud. **Stems** vegetative submersed and growing to surface, or rhizomatous; flowering stems submersed and growing to surface or constricted below first node and emersed. **Leaves** petiolate and sessile, cordate to reniform; apex obtuse to acuminate; petiolate leaves floating or submersed; sessile leaves forming basal rosettes, blades thin. **Inflorescences** spicate, elongating over several days; spathes folded, apex acute to acuminate; peduncle glandular-pubescent or pilose. **Flowers** open one day only; perianths funnel-shaped, purple to blue or white, limb lobes oblong to oblanceolate, glandular-pubescent or pilose, apex obtuse to acute; stamens 6, distal 3 shorter than proximal 3, filaments ovate to oblong, purple with glandular-pubescence, anthers yellow; styles 1 or sometimes more. **Fruits** ovoid. **Seeds** 1, ovoid, smooth.

Western Hemisphere. 6 species; 1 species treated in *Flora*.

Pontederia cordata Linnaeus (p. 496)

Pickerel-weed

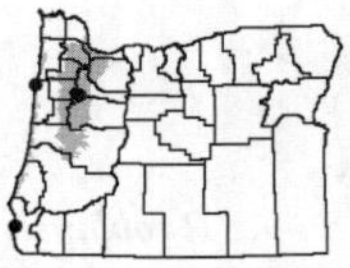

Herbs perennial. **Roots** rhizomatous. **Stems** contracted when vegetative, erect when flowering; flowering stems to 120 cm. **Leaves** petiolate leaves emersed, petioles constricted below blade, to 60 cm, stipules 7–29 cm, blades lanceolate to cordate, 6–22 × 0.7–12 cm; sessile leaves blades linear. **Inflorescences** spikes with up to several hundred flowers, 2–15 cm; spathes 5–17 cm. **Flowers** tubes 3–9 mm, limb lobes 5–8 mm, distal central lobe with 2-lobed yellow spot; proximal stamens 7–13 mm, distal stamens 1.5–7 mm; style 3-lobed. **Fruits** 4–6 × 2–3 mm, with dentate ridges.

Streams, rivers, ponds, lakes. Flowering Jun–Sep. 0–50 m. Est, WV. East from Ontario, TX, and KS to the Atlantic Ocean. Exotic.

This species is considered a noxious weed in areas where it has been introduced. Like *Heteranthera dubia*, in high quantities this plant can cause an increase in local water pH.

POTAMOGETONACEAE Pondweed family

Nick Otting, Richard Brainerd, and Barbara L. Wilson

Herbs perennial or rarely annual, rhizomatous or not; turions sometimes present. **Leaves** alternate or nearly opposite, submersed or both submersed and floating, petiolate or sessile; stipules well developed, sometimes connate and sheathing, sometimes fused to the leaf, if fused to the leaf then often ligulate (extending beyond the base of the blade as a ligule). **Inflorescences** terminal or axillary spikes, cylindric or capitate, rarely paniculate, pedunculate; bracts 0. **Flowers** bisexual; tepals 4 in 1 series; stamens (2)4, attached to the petals, in 1 series, anthers distinct, dehiscing longitudinally; pistils 1 or 4, sessile or rarely short-stipitate. **Fruits** drupaceous, 1-seeded.

Cosmopolitan in fresh water (rarely brackish) habitats. 3 genera; 2 genera treated in *Flora.*

Leaves of Potamogetonaceae species are stipular. The submersed leaves are attached to stem nodes or to the stipules. If the leaves are attached to the stipules, they appear to have a ligule above the point of attachment. Reproductive features are important in separating taxa. We strongly recommend that no one collect specimens of Potamogetonaceae that are lacking reproductive structures.

North American pondweeds have often been classified in a single genus, *Potamogeton* in the broad sense. However, the genus includes two lineages that we recognize at the generic level, *Potamogeton* in the strict sense and *Stuckenia.* In *Stuckenia,* leaf blades are attached to most of the stipule, diverging from it in the upper third, whereas in *Potamogeton,* blades are free from stipules or attached only in the lower fourth. Submersed leaves of *Stuckenia* are opaque, channeled, and turgid, whereas those of *Potamogeton* are translucent, flat, and without grooves or channels.

Key to Potamogetonaceae genera:

1. Some leaves floating . . . . **POTAMOGETON** (in part)

1' All leaves submersed.

  2. Submersed leaf blades not attached to stipules, attached at the stem nodes or petiolate, or attached to the stipule below its middle, the stipule not sheathing, or forming a short sheath around the stem below the point of attachment; leaf blades translucent, not channeled, flattened; peduncle stiff and usually holding inflorescence above water surface . . . . **POTAMOGETON** (in part)

  2' Submersed leaf blades attached to distal 33% of stipule, the stipule fused, forming a sheath around the stem below the point of attachment; leaf blades opaque, channeled, turgid; peduncle flexible, not holding inflorescence above water surface . . . . **STUCKENIA**

POTAMOGETON Pondweed

Herbs rhizomes present or absent; tubers absent; turions present or absent. **Stems** terete or compressed, nodes occasionally with small, round, raised oil glands. **Leaves** submersed or both submersed and floating, alternate to nearly opposite; submersed leaves petiolate or sessile, blade translucent, linear to broadly elliptic, channeled or not, flattened; margins entire or serrate, rarely wavy-margined; tip acute to subulate to obtuse; veins 1–35, stipules free from base of leaf blade, or adnate to base of leaf blade for less than 50% of length of stipule, or adnate to base of leaf blades for 67% or more length of stipule, if adnate, then extending past the adnation as free ligule; floating leaves petiolate, rarely nearly sessile, blade elliptic to ovate, leathery; base cuneate to rounded or cordate; margins entire; tip acute to obtuse; veins 1–51; stipules free from base of leaf blade. **Inflorescences** spikes, submersed or emersed, capitate or cylindric; peduncles flexible or stiff, sometimes projecting inflorescence above surface of water. **Flowers** pistils 1 or 4. **Fruits** dorsally rounded or keeled, flattened to turgid, beaked or not.

Approximately 100 species; 17 species treated in *Flora.*

Potamogeton species provide significant habitats and food for aquatic animals. They help stabilize the substrate.

Hybridization is common among members of the genus. Most are recognized as hybrids by their infertility and their intermediate morphology, usually between parental species which are found growing with them.

Many species have paired oil glands on the stem at the node of submersed leaves. These glands are especially common on species with sessile leaves. Low circular bumps ranging from green to golden to white, they are usually present at most nodes, but sometimes are lacking at most nodes. The glands are best observed with dried specimens, a good light source, and magnification of at least 15×.

Important features of the fruit include presence or absence of lateral and abaxial wings, ribs, ridges, or keels. Here, "ribbed" indicates a raised "vein" on a rounded surface; "ridged," a ridge with an obtuse angle; "keeled," a ridge with an acute angle; and "winged," a ridge that appears to have a wing distally. Fruit lengths do not include the beak length.

Key to *Potamogeton* species:

1. Submersed leaf blades attached to stipules.

  2. All leaves submersed, somewhat stiff, conspicuously 2-ranked, with 20–60 veins . . . . ***P. robbinsii***

  2' Floating leaves sometimes present; submersed leaves lax, not conspicuously 2-ranked, with 20 or fewer veins . . . . ***P. diversifolius***

1' Submersed leaf blades not attached to stipules, attached at stem nodes or petiolate.
 3. Submersed leaf blades elliptic to lanceolate to broadly linear, 3–58 mm wide.
  4. Leaves conspicuously serrate; stem flattened; fruit beaks 2–3 mm . . . . ***P. crispus***
  4' Leaves entire or microscopically serrate at tip; stem terete; fruit beaks 1 mm or less.
   5. Submersed leaves clasping stem; floating leaves absent.
    6. Leaves 5–20 cm long, tips hood-like, often splitting when pressed; stipules 2–10 cm; fruits 4–5.7 mm . . . . ***P. praelongus***
    6' Leaves 1.5–12 cm long, tips flat, not splitting when pressed; stipules less than 2 cm; fruits 2.2–4.2 mm. . . . . ***P. richardsonii***
   5' Submersed leaves petiolate or sessile, not clasping; floating leaves absent or present.
    7. Submersed leaf blades folded and strongly falcate, usually more than 20 mm wide, with 19 or more veins . . . . ***P. amplifolius***
    7' Submersed leaf blades usually not folded and falcate, often less than 20 mm wide, with 19 or fewer veins.
     8. Submersed leaves with petioles 5–130 mm.
      9. Submersed leaves (1.5)2–4(5) cm broad, tips often microscopically serrate, petioles absent, rarely as much as 2 cm long; petioles of floating leaves usually shorter than the blade; fruits grayish green to olive-green . . . . ***P. illinoensis*** (in part)
      9' Submersed leaves rarely over 2 cm broad, tips entire, petioles usually over 2 cm long; petioles of floating leaves usually exceeding the blade; fruits red to reddish brown. . . . . ***P. nodosus***
     8' Submersed leaves sessile.
      10. Fruits plump, concealing embryo shape, short-pedicellate; stipules blunt; submersed leaf blades 7–9-veined, entire . . . . ***P. alpinus***
      10' Fruits compressed, often revealing embryo shape when mature, sessile; stipules acute to obtuse; submersed leaf blades 3–19-veined, often minutely serrate near the tip.
       11. Fruits 1.9–2.3 × 1.8–2 mm; submersed leaf blades with 3–7(9) veins, 20–100 × (3)10–27 mm . . . . ***P. gramineus***
       11' Fruits 2.5–3.6 × 2.1–3 mm; submersed leaf blades with 7–19 veins, 60–200 × 15–50 mm . . . . ***P. illinoensis*** (in part)
 3' Submersed leaf blades ribbon-like or thread-like, 0.1–10 mm wide.
  12. Floating leaves present at least in some plants in the population.
   13. Floating leaf blades acute at base; submersed leaves ribbon-like, 3–10 mm wide, with lacunae forming a prominent band along the midrib . . . . ***P. epihydrus***
   13' Floating leaf blades rounded to cordate at base; submersed leaves subterete, 1–2 mm wide, without a prominent band of lacunae along midrib . . . . ***P. natans***
  12' Floating leaves absent.
   14. Nodal glands absent.
    15. Leaves 15–35-veined, more than 2 mm wide; stems conspicuously flattened; peduncles usually terminal . . . . ***P. zosteriformis***
    15' Leaves 3–5-veined, less than 2 mm wide; stems terete; peduncles axillary.
     16. Fruits with dorsal surface rounded; stipules 3.1–9.2 mm . . . . ***P. berchtoldii*** (in part)
     16' Fruits with dorsal keel; stipules 2–22 mm . . . . ***P. foliosus*** (in part)
   14' Nodal glands present, often obscure, low, greenish bumps at only a few nodes.
    17. Fruits with dorsal surface rounded; stipules 3.1–9.2 mm.
     18. Mature fruits obovoid, sides concave; beaks displaced ventrally, rarely at middle of fruit tip; inflorescences interrupted; stipules connate but often rupturing . . . . ***P. pusillus***
     18' Mature fruits widest at middle or ovoid, sides rounded; beaks located in middle of fruit tip; inflorescences continuous; stipule margins free . . . . ***P. berchtoldii*** (in part)
    17' Fruits with dorsal surface keeled; stipules 2–22 mm.
     19. Stipules strongly fibrous, the fibers free at stipule tips as prominent cilia 2–3 mm long; veins persistent as the rest of the stipule disintegrates; inflorescences interrupted; fruits pale green. . . . . ***P. fibrillosus***
     19' Stipules not strongly fibrous, often deciduous; veins not persisting; inflorescences usually not interrupted; fruits olive to greenish brown . . . . ***P. foliosus*** (in part)

Potamogeton alpinus Balbis (p. 496)

Northern pondweed, reddish pondweed

Potamogeton alpinus Balb. ssp. *tenuifolius* (Raf.) Hultén, *Potamogeton alpinus* Balb. var. *subellipticus* (Fernald) Ogden, *Potamogeton alpinus* Balb. var. *tenuifolius* (Raf.) Ogden

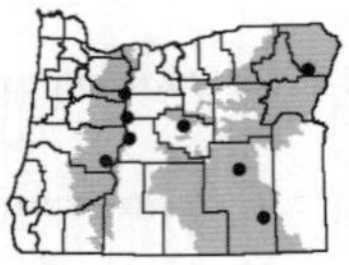

Stems terete, to 200 cm; nodal glands absent. **Leaves** submersed and floating or submersed only. **Submersed leaves** sessile, attached to stem nodes, not attached to stipules; stipules (12)15–25(40) mm, tip blunt; blades oblong-linear to linear-lanceolate, 45–180(250) × 5–20 mm; base rounded to cuneate; margins entire; tip obtuse or acute; lacunae in 0–6 rows on each side of midvein; veins 7–9. **Floating leaves** elliptic or oblanceolate to obovate or oblong-linear, 40–70(100) × 10–25(40) mm, reddish green; base gradually tapering to petiole; tip obtuse or acute; veins (7)9–13(15); petioles 0–12 mm. **Inflorescences** emersed; spikes cylindric, 10–35 mm; peduncles terminal or axillary, erect, 30–100(160) mm. **Fruits** pedicellate, obovoid, turgid, (2.5)3–3.5 × (1.7)2–2.4 mm, dorsally keeled, laterally keeled or not; beaks dorsally curved, 0.5–0.9 mm.

Ponds, lakes, and slow-moving streams. 1200–2200 m. BR, BW, Casc. CA, ID, NV, WA; north to British Columbia, east to Greenland; Eurasia. Native.

Potamogeton alpinus is often red when fresh. This species hybridizes with *P. nodosus, P. gramineus,* and *P. praelongus.*

Potamogeton amplifolius Tuckerman

Large-leaved pondweed, broad-leaved pondweed

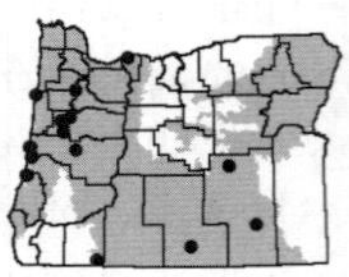

Stems terete, 6–110 cm; nodal glands absent. **Leaves** submersed and floating or rarely submersed only. **Submersed leaves** petiolate, attached to stem nodes, not attached to stipules; stipules 15–117 mm, tip acute; blades ovate to oblanceolate, distinctly falcate, 50–125 × 15–58 mm; base rounded to acute; margins entire and recurved, wavy; tip acute to round-apiculate; veins 19–49; petioles terete, 0.9–11.5 cm. **Floating leaves** lanceolate to round elliptic, 43–92 × 25–38 mm, light green; base rounded to cordate; tip acute to rounded; veins 27–49; petioles 23–226 mm. **Inflorescences** emersed; spikes cylindric, 34–65 mm; peduncles terminal or axillary, erect, cylindric, 45–223 mm. **Fruits** sessile, obovoid, turgid, 5–6.7 × 4.5–5.2 mm, reddish brown, dorsally keeled, laterally ridged; beaks erect, 0.5–0.8 mm.

Lakes, ponds, streams. 0–2100 m. BR, BW, Casc, CR, ECas, Est, WV. CA, ID, WA; north to British Columbia, east to Newfoundland. Native.

Potamogeton amplifolius has large submersed leaves that are folded and falcate and have numerous veins. This species hybridizes with *P. illinoensis.*

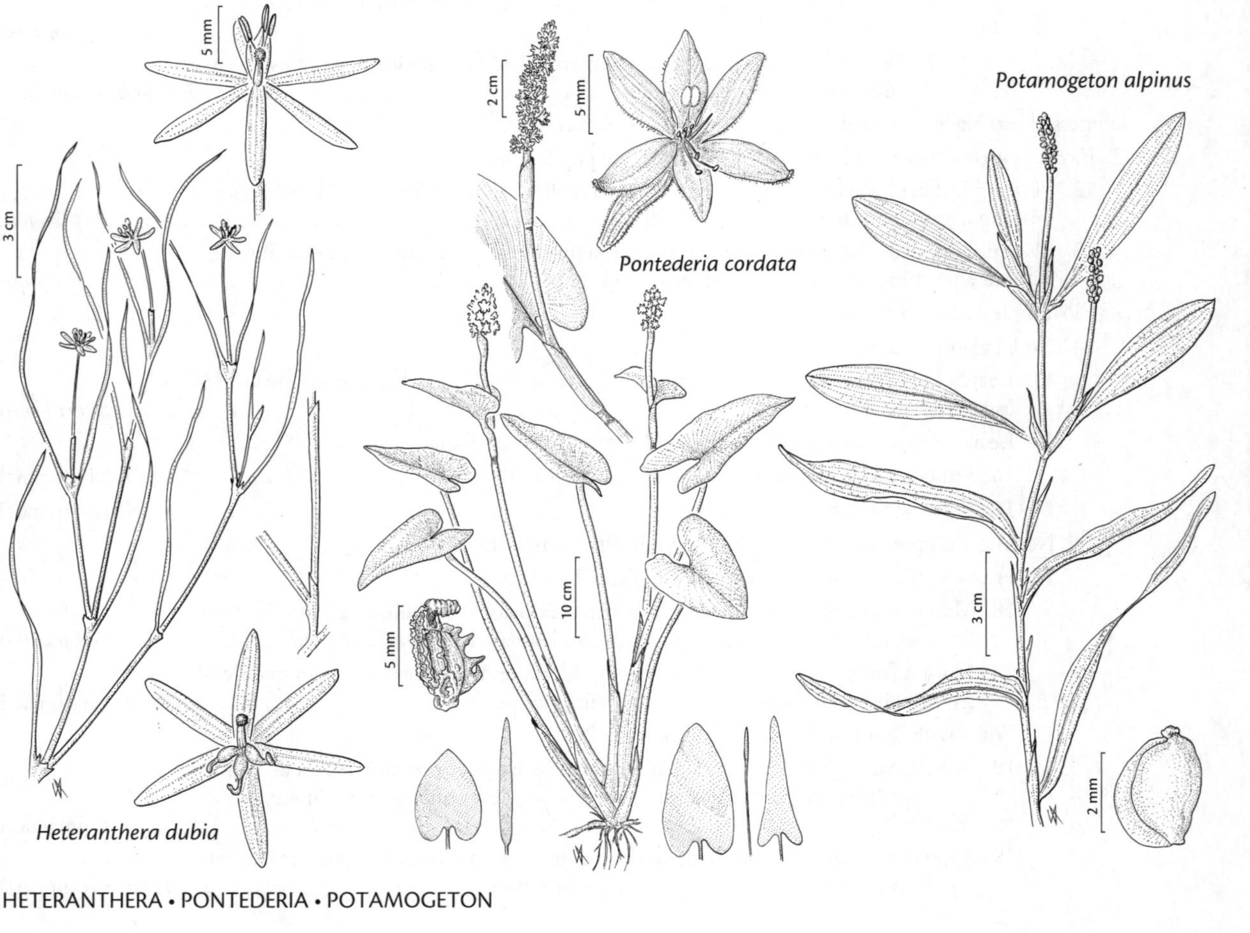

HETERANTHERA • PONTEDERIA • POTAMOGETON

Potamogeton berchtoldii Fieber

Berchtold's pondweed

Potamogeton pusillus L. ssp. *tenuissimus* (Mert. & Koch) R.R. Haynes & Hellq., *Potamogeton pusillus* L. var. *tenuissimus* Mert. & Koch

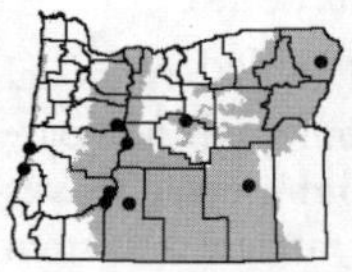

Stems terete to slightly compressed, 18–150 cm; nodal glands present on at least some nodes, to 0.5 mm diameter, green, gold, brown, or rarely white. **Leaves** submersed, attached to stem nodes, not attached to stipules; stipules convolute, 3.1–9.2 mm, rarely fibrous, tip acute to obtuse; blades linear, 9–54 × 0.2–2.5 mm; base slightly tapering; margins entire; tip subulate to obtuse; lacunae in 1–5 rows on each side of midrib; veins 1–3(5). **Inflorescences** submersed or emersed; spikes capitate to cylindric, continuous to interrupted, 1.5–10.1 mm; peduncles more than 3 per plant, axillary or terminal, cylindric to slightly clavate, erect, rarely recurved, 5–66 mm. **Fruits** sessile, ovoid, sides rounded, rarely concave, 1.5–2.2 × 1.2–1.6 mm, green to brown, not keeled; beaks erect, located at middle of fruit tip, 0.1–0.6 mm.

Shallow waters of lakes and streams. 0–1700 m. BR, BW, Casc, ECas, Est. CA, ID, NV, WA; north to Northwest Territories, east to Newfoundland, southeast to FL; Eurasia. Native.

This pondweed is sometimes treated as a subspecies of *P. pusillus* (*P. p.* ssp. *tenuissimus*). Immature plants usually cannot be distinguished from *P. foliosus* or *P. pusillus*.

Potamogeton crispus Linnaeus

Curled pondweed

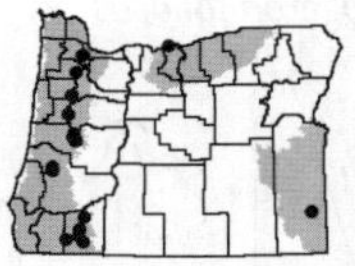

Stems flattened, to 100 cm; nodal glands absent. **Leaves** submersed, sessile, attached to stem nodes, not attached to stipules; stipules persistent to deliquescent, to 0.5 cm, tip obtuse; blades linear, 12–90 × 4–10 mm, base obtuse to rounded, not to slightly clasping; margins conspicuously serrate, wavy; tip round to acute; lacunae in 2–5 rows on each side of midrib; veins 3–5. **Inflorescences** emersed, terminal or rarely axillary; spikes cylindric, 10–15 mm; peduncles erect to ascending, 25–40 mm. **Fruits** sessile, obovoid, turgid to slightly concave, 6 × 2.5 mm, yellow to red or reddish brown, not dorsally or laterally keeled; beaks apically recurved, 2–3 mm.

Quiet waters, especially brackish, alkaline, or eutrophic waters of ponds, lakes, and streams. 0–1300 m. Col, CR, Owy, Sisk, WV. CA, NV, WA; north to British Columbia, east to CT, south to FL and Costa Rica, South America; Australia, Eurasia. Exotic.

Potamogeton cripsus is readily identified by its wavy-margined, serrate leaves. This species is native to Eurasia and has been introduced to quiet waters nearly worldwide. This species hybridizes with *P. praelongus*.

Potamogeton diversifolius Rafinesque (p. 498)

Water-thread pondweed, diverse-leaved pondweed

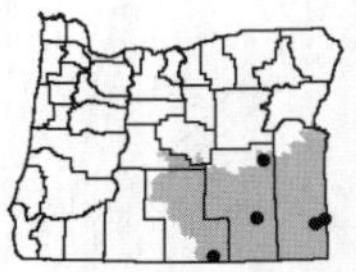

Stems compressed, 10–35 cm; nodal glands absent. **Leaves** both submersed and floating or submersed only. **Submersed leaves** sessile with blade attached to stipule near middle; stipules persistent to decomposing, ligulate, 15–23 mm, tip obtuse; blades linear, often falcate, 20–40 × 0.5–1(1.5) mm; base slightly tapering; margins entire; tip acute; lacunae in 1–2 rows on each side of midrib; vein 1. **Floating leaves** obovate to elliptic, (10)15–30 × 3–8.5 mm, light green, base acute, tip round to acute, veins 3–7; petioles 7–8 mm. **Inflorescences** submersed spikes capitate, 2–3 mm, emersed spikes cylindric, 5–9.7 mm; peduncles axillary, recurved, 3–5 mm, emersed peduncles axillary or terminal, erect to slightly recurved, 6–15 mm. **Fruits** sessile, orbicular, compressed, 1–1.5 × 0.9–2 mm, greenish brown, dorsally and laterally winged, lateral wings with sharp points; beaks erect, 0.1 mm.

Ponds, lakes, streams. 1100–1800 m. BR, Owy. CA, ID, NV; east to CT, Mexico. Native.

Potamogeton epihydrus Rafinesque (p. 498)

Ribbon-leaf pondweed, ribbon-leaved pondweed

Potamogeton epihydrus Raf. ssp. *nuttallii* (Cham. & Schltdl.) Calder & Roy L. Taylor, *Potamogeton epihydrus* Raf. var. *epihydrus*, *Potamogeton epihydrus* Raf. var. *nuttallii* (Cham. & Schltdl.) Fernald

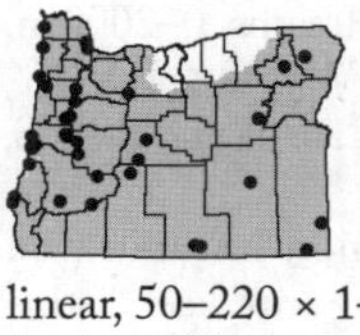

Stems flattened, 10–90 cm; nodal glands absent. **Leaves** submersed and floating or submersed only. **Submersed leaves** sessile, attached to stem nodes, not attached to stipules; stipules 10–30 mm, tip obtuse; blades linear, 50–220 × 1–10 mm; margins entire; tip blunt to acute; lacunae forming a prominent band along the midvein; veins 3–13. **Floating leaves** narrowly oblanceolate to elliptic, 20–80 × 4–20 mm, base acute, tip rounded or bluntly cuspidate; veins 11–41; petioles 20–125 mm. **Inflorescences** emersed; spikes cylindric, 8–40 mm; peduncles axillary, erect, 15–50(160) mm. **Fruits** sessile, round to obovoid, flattened, dorsally and laterally keeled, 2.5–4.5 × 2–3.6 mm, greenish brown; beaks erect, 0.5 mm.

Still or flowing waters of lakes, ponds and streams. 0–2000 m. All ecoregions except Col. CA, ID, WA; north to AK, east to Newfoundland, southeast to FL; Europe. Native.

Potamogeton epihydrus is distinguished by submersed linear leaves with a prominent band of lacunae along the midvein. This species hybridizes with *P. gramineus* and *P. nodosus*.

Potamogeton fibrillosus Fernald

Fibrous-stipuled pondweed

Potamogeton foliosus Raf. ssp. *fibrillosus* (Fernald) R.R. Haynes & Hellq., *Potamogeton foliosus* Raf. var. *fibrillosus* (Fernald) R.R. Haynes & Reveal

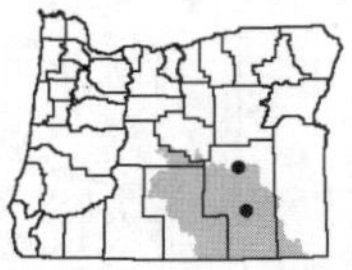

Stems slightly compressed, 4–75 cm; nodal glands common. **Leaves** submersed, sessile, attached to stem nodes, not attached to stipules; stipules 2–22 mm, brown (rarely white), fibrous, shredding at tip, veins persisting as fibers; blades linear, 13–82 × 0.3–2.3 mm; margins entire; tip acute to apiculate; lacunae rarely present; veins 1–3(5). **Inflorescences** emersed; spikes interrupted, 1.5–7 mm; peduncles axillary, recurved, 3–11(37) mm. **Fruits** sessile, round to obovoid, flattened, 1.4–2.7 × 1.1–2.2 mm, pale green, dorsally and laterally keeled; beaks erect, 0.2–0.4 mm.

Warm waters of shallow lakes, springs, streams and rivers. 1000–1400 m. BR. CA, ID, WA; east to WY. Native.

Potamogeton fibrillosus is an uncommon species of the western United States. It is sometimes included in *P. foliosus* as *P. f.* ssp. *fibrillosus*. It differs from *P. foliosus* by the stipular tissue between the veins decomposing, leaving only strands formed by the fibrous veins. In addition, nodal glands are common.

Potamogeton foliosus Rafinesque (p. 498)

Leafy pondweed

Potamogeton foliosus Raf. var. *foliosus*, *Potamogeton foliosus* Raf. var. *macellus* Fernald

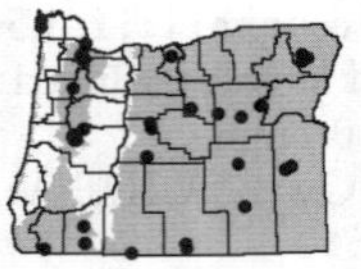

Stems flattened, 4–75 cm; nodal glands rarely present. **Leaves** submersed, sessile, attached to stem nodes, not attached to stipules; stipules 2–22 mm, greenish to brown, membranous to slightly fibrous, veins not persisting; blades linear, 13–82 × 0.3–2.3 mm; margins entire, tip acute to apiculate; lacunae rarely present; veins 1–3(5). **Inflorescences** emersed; spikes continuous (rarely interrupted), 1.5–7 mm; peduncles axillary, recurved, 3–11(37) mm. **Fruits** sessile, round to obovoid, flattened, 1.5–2.7 × 1.2–2.2 mm, olive-green to green-brown, dorsally and laterally keeled; beaks erect, 0.2–0.6 mm.

Ponds, lakes, and slow or fast-moving streams. 0–2000 m. BR, BW, Col, ECas, Est, Lava, Owy, Sisk, WV. CA, ID, NV, WA; north to AK, east throughout most of Canada and the US, south to Central America. Native.

Potamogeton foliosus is the most common linear-leaved pondweed in North America. The undulating fruit keel distinguishes this species and *P. fibrillosus* from our other linear-leaved pondweeds. Immature plants usually cannot be distinguished from *P. berchtoldii* or *P. pusillus*.

Potamogeton gramineus Linnaeus

Variable pondweed

Potamogeton gramineus L. var. *gramineus*, *Potamogeton gramineus* L. var. *maximus* Morong ex A. Benn., *Potamogeton gramineus* L. var. *myriophyllus* J.W. Robbins, *Potamogeton heterophyllus* Schreb.

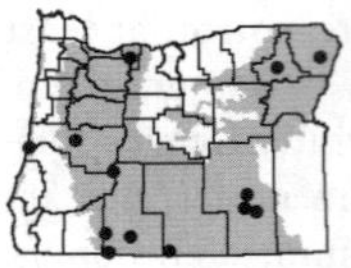

Stems terete to flattened, to 150 cm; nodal glands absent. **Leaves** submersed and floating or submersed only. **Submersed leaves** sessile or rarely petiolate, attached to stem nodes, not attached to stipules; stipules 10–16 mm, tip acute to obtuse; blades linear to narrowly lanceolate or oblanceolate, 31–91 × 3–27 mm; base attenuate, margins entire, rarely wavy; tip acuminate; lacunae in 1–2 rows on each side of midvein; veins 3–7(9); petioles to 30 mm. **Floating leaves** elliptic to ovate, 35–40 × 16–20 mm, yellow-green to dark green, base rounded, tip acuminate, veins 11–13; petioles 30–45 mm. **Inflorescences** emersed; spikes cylindric, 15–35 mm; peduncles axillary and terminal, erect to ascending, cylindric, 32–77 mm. **Fruits** sessile, ovoid, laterally compressed, 1.9–2.3 × 1.8–2 mm, greenish brown, dorsally and laterally keeled; beaks erect, 0.3–0.5 mm.

Ponds, lakes and streams. 50–2300 m. BR, BW, Casc, ECas, Est, WV. CA, ID, NV, WA; north to Northwest Territories, east to Newfoundland, southeast to AZ; Eurasia. Native.

This species hybridizes with *P. alpinus*, *P. illinoensis*, *P. natans*, *P. nodosus*, *P. praelongus*, and *P. richardsonii*.

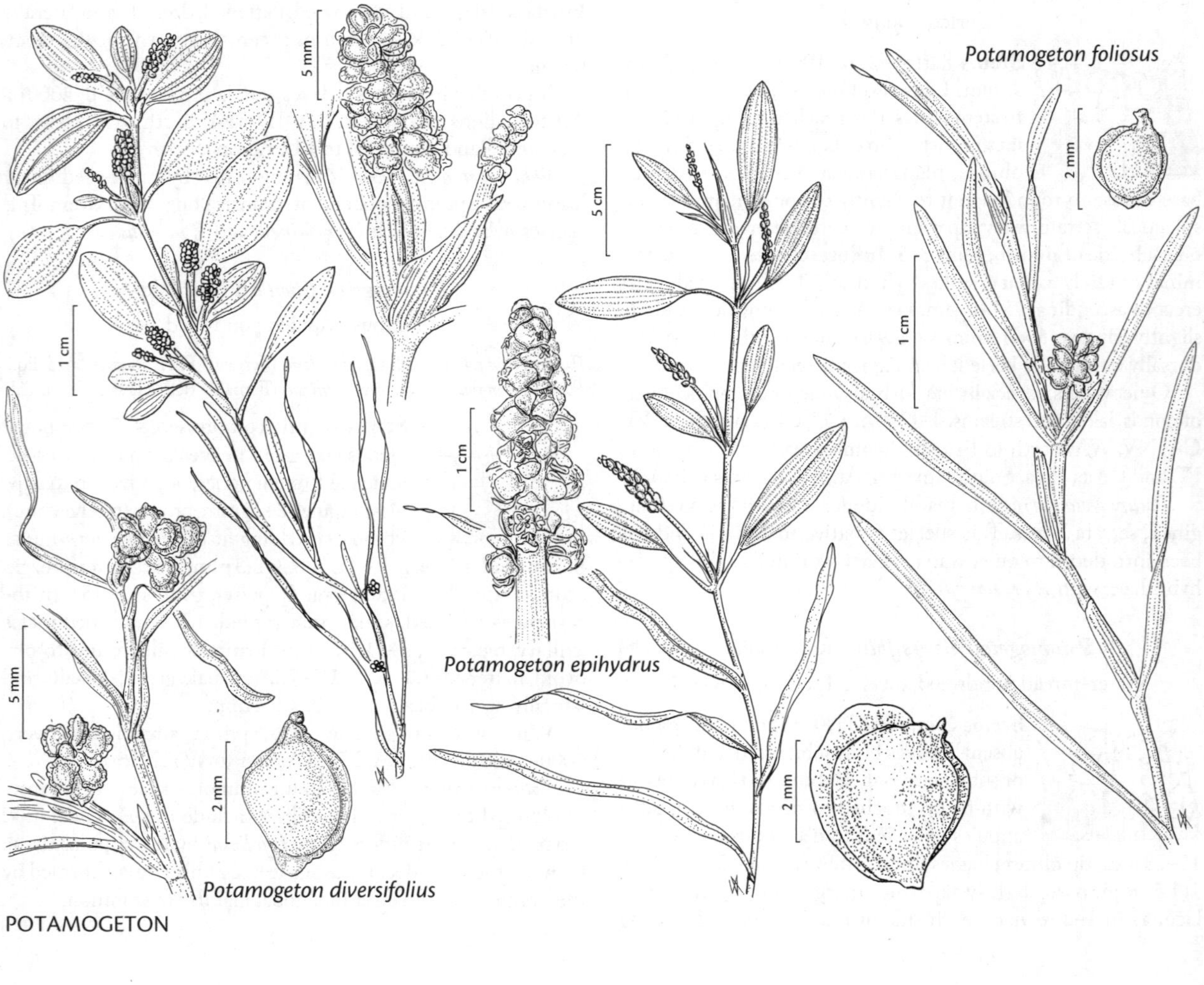

POTAMOGETON

Potamogeton illinoensis Morong

Illinois pondweed

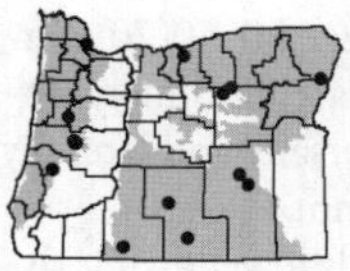

Stems terete, 28–120 cm; nodal glands absent. **Leaves** both submersed and floating or submersed only. **Submersed leaves** petiolate or sessile, attached to stem nodes, not attached to stipules; stipules 10–80 mm, tip acuminate; blades elliptic to lanceolate (linear), often falcate, 50–200 × 2–45 mm; base acute; margins entire, often wavy; tip acute-mucronate; lacunae in 2–5 rows on each side of midrib; veins 7–19; petioles 0–40 mm. **Floating leaves** elliptic to oblong-elliptic, 40–190 × 20–65 mm, light green, base cuneate, tip round-mucronate, veins 13–29; petioles 20–90 mm. **Inflorescences** emersed; spikes cylindric, 25–70 mm; peduncles terminal or axillary, erect to ascending, cylindric, 4–30 cm. **Fruits** sessile, obovoid to ovoid, laterally compressed, dorsally keeled, 2.5–3.6 × 2.1–3 mm, grayish green to olive green, laterally ridged, dorsal keel well developed; beaks erect to slightly recurved, 0.5–0.8 mm.

Streams, lakes, ponds, and sloughs, often in alkaline waters. 50–2300 m. BR, BW, Col, CR, ECas, WV. CA, WA; north to Northwest Territories, east to CT, south to South America. Native.

This species hybridizes with *P. amplifolius, P. gramineus,* and *P. nodosus.*

Potamogeton natans Linnaeus

Floating-leaf pondweed

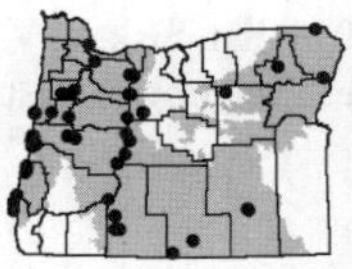

Stems terete, 30–90 cm; nodal glands absent. **Leaves** submersed and floating. **Submersed leaves** sessile, attached to stem nodes, not attached to stipules; stipules 45–100 mm, whitish, fibrous, tip obtuse; blades rigid, linear, subterete, 9–20 cm × 0.7–2.5 mm; base slightly tapering, margins entire; tip obtuse; veins 3–5, obscure. **Floating leaves** elliptic to ovate, 35–110 × 15–60 mm, base rounded to cordate, tip acute to rounded, veins 17–37; petioles 55–290 mm. **Inflorescences** emersed; spikes cylindric, 25–50 mm; peduncles terminal, erect to ascending, 45–95 mm. **Fruits** sessile, obovoid, turgid, 3.5–5 × 2–3 mm, green to greenish brown, not keeled; beaks erect to apically recurved, 0.4–0.8 mm.

Ponds, lakes, and slow-moving streams. 0–2000 m. BR, BW, Casc, CR, ECas, Est, WV. CA, ID, NV, WA; north to Northwest Territories, east to Newfoundland, southeast to NM; Eurasia. Native.

Potamogeton natans is a common pondweed of still or slow moving water. Its leathery floating leaves are often cordate at the base. Its petioles have reflexed tips with a light-colored band that contrasts with darker tissue below. Its submersed leaves are rigid and subterete. This species hybridizes with *P. nodosus.*

Potamogeton nodosus Poiret

Long-leaf pondweed, long-leaved pondweed, Loddon's pondweed

Potamogeton americanus Cham. & Schltdl.

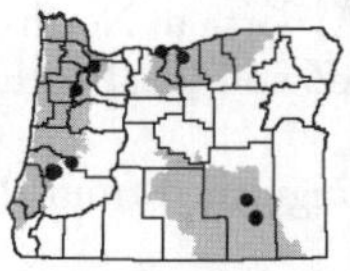

Stems terete, to 100 cm; nodal glands absent. **Leaves** submersed and floating, or submersed only. **Submersed leaves** attached at stem node, not attached to stipules; stipules 30–90 mm, tip round to acute; petioles 20–130 mm; blades linear-lanceolate to lance-elliptic, 90–200 × 10–35 mm; base acute; margins entire; tip acute; lacunae in 2–5 rows each side of midrib; veins 7–15. **Floating leaves** lenticular to elliptic, 30–110 × 15–45 mm, green, base cuneate to rounded, tip acute to rounded, veins 9–21; petioles 3.5–26 cm. **Inflorescences** emersed; spikes cylindric, 20–70 mm; peduncles terminal, erect to ascending, cylindric, 30–150 mm. **Fruits** sessile, obovoid, 2.7–4.3 × 2.5–3 mm, red to reddish brown, dorsally keeled, laterally ridged, lateral ridges with blunt to sharp tips; beaks erect.

Clear to turbid waters of lakes, streams, and sloughs. 0–2300 m. BR, Col, CR, WV. CA, ID, NV, WA; north to British Columbia, east to CT, south to Mexico, Central America, South America; Eurasia. Native.

This species hybridizes with *P. alpinus, P. epihydrus, P. gramineus, P. illinoensis, P. natans,* and *P. richardsonii.*

Potamogeton praelongus Wulfen

White-stemmed pondweed, white-stalked pondweed

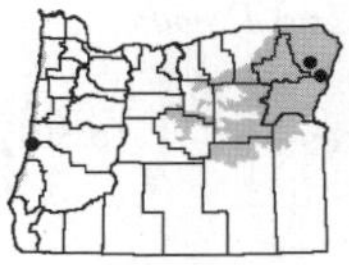

Stems terete, to 210 cm; nodal glands absent. **Leaves** submersed, sessile, attached at stem nodes, not attached to stipules; stipules 30–81 mm, fibrous, shredding at tip; blades linear-lanceolate, 50–200 × 10–30 mm; base clasping but without basal lobes, rarely wavy; tip obtuse; veins 11–33. **Inflorescences** emersed; spikes cylindric, 34–75 mm; peduncles terminal or axillary, erect to spreading, 10–30 cm. **Fruits** sessile, obovoid, turgid, 4–5.7 × 3.2–4 mm, greenish brown, dorsally keeled, occasionally laterally keeled; beaks erect, 0.6–1 mm.

Neutral to alkaline waters of lakes and streams. 0–1900 m. BW, Est. CA, ID; north to British Columbia, east to Newfoundland, south to Mexico; Eurasia. Native.

This species hybridizes with *P. alpinus, P. crispus* and *P. gramineus.*

Potamogeton pusillus Linnaeus

Small pondweed

Potamogeton pusillus L. var. *pusillus*

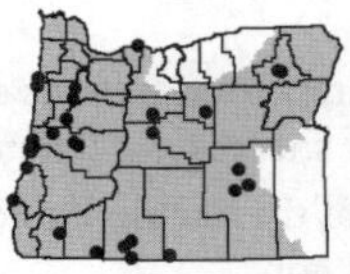

Stems terete to slightly compressed, 18–150 cm; nodal glands present on at least some nodes, to 0.5 mm diameter, green, gold, brown, or rarely white. **Leaves** submersed, attached to stem nodes, not attached to stipules; stipules connate, 3.1–9.2 mm, rarely fibrous, tip obtuse; blades linear, 14–65 × 0.5–1.9 mm; base slightly tapering; margins entire; tip acute or rarely apiculate, rarely with a bristle; lacunae in 0–2 rows on each side of midrib; veins 1–3. **Inflorescences** submersed or emersed; spikes cylindric, interrupted, 1.5–10.1 mm; peduncles 1–3 per plant, axillary or terminal, filiform to cylindric, erect, rarely recurved, 5–66 mm. **Fruits** sessile, obovoid, sides centrally concave, 1.5–2.2 × 1.2–1.6 mm, green to brown, not keeled; beaks erect, displaced ventrally or rarely at middle of fruit tip, 0.1–0.6 mm.

Streams, lakes, marshes. 0–1600 m. BR, BW, Casc, CR, ECas, Est, Lava, Sisk, WV. CA, ID, NV, WA; north to Northwest Territories, east to CT, southeast to AZ and FL, South America; Africa, Eurasia. Native.

Immature plants usually cannot be distinguished from *P. foliosus* or *P. berchtoldii*.

Potamogeton richardsonii (A. Bennett) Rydberg (p. 501)

Clasping-leaved pondweed, Richardson's pondweed

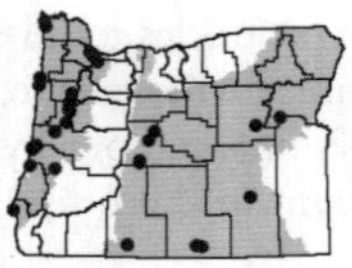

Stems terete, to 100 cm; nodal glands absent. **Leaves** submersed, sessile, attached at stem nodes, not attached to stipules; stipules 1.2–1.7 mm, fibrous, disintegrating to persistent fibers, even on proximal portion of stem, tip shredding, obtuse; blades ovate-lanceolate to narrowly lanceolate, 1–12 cm, 10–20 mm wide; base rounded, clasping; margins entire to wavy; tip acute to obtuse; lacunae absent; veins 3–35. **Inflorescences** emersed; spikes cylindric, 13–37 mm; peduncles terminal or axillary, erect to rarely recurved, 15–148 mm. **Fruits** sessile, obovoid, turgid to concave, 2.2–4.2 × 1.7–2.9 mm, greenish brown, rarely dorsally keeled; beaks erect, 0.4–0.7 mm.

Alkaline lakes, streams, and rivers. 0–2400 m. BR, BW, CR, ECas, Est, Lava, WV. CA, ID, NV, WA; north to Northwest Territories, east to CT. Native.

This species hybridizes with *P. gramineus* and *P. nodosus*.

Potamogeton robbinsii Oakes (p. 501)

Robbins' pondweed

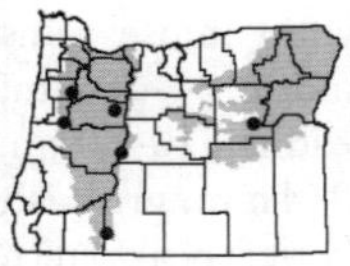

Stems terete, to 100 cm; nodal glands absent. **Leaves** submersed, conspicuously 2-ranked, sessile, more or less stiff, blades attached to lower 25% of petiole; stipules ligulate, 0.5–2 cm, fibrous, shredding at tip, tip obtuse; blades linear to lanceolate, 20–70(120) × 3–4(8) mm; base rounded, with basal lobes (auricles); margins minutely spinulose to serrulate; tip acute; lacunae absent; veins 20–60. **Inflorescences** often branched; peduncles axillary, erect, 30–50(70) mm; spikes moniliform, 7–20 mm. **Fruits** stipitate, obliquely obovoid, turgid, 3–4(5) × 2(3.3) mm, brown, dorsally and laterally keeled; beaks erect, recurved at tip, 0.7–0.9 mm.

Shallow to deep water of ponds, lakes and slow-flowing rivers. 50–2400 m. BW, Casc, WV. CA, ID, WA; north to AK, northeast to Nunavut, east to Prince Edward Island. Native.

Potamogeton robbinsii is our only species with branched inflorescences and auriculate leaf blades. It flowers only rarely.

Potamogeton zosteriformis Fernald

Flatstem pondweed, eel grass pondweed

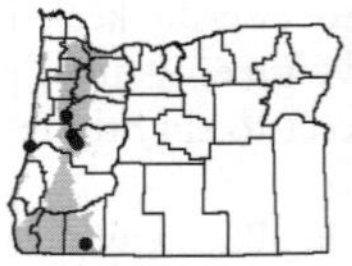

Stems conspicuously flattened, 60–120 cm; nodal glands absent (rarely present), 0.3 mm diameter, gold. **Leaves** submersed, sessile, rigid, with blades attached at stem nodes, not attached to stipules; stipules 15–35 mm, fibrous, shredding at tip, tip obtuse to acuminate; blades linear, 100–200 × 2–5 mm; base rounded; margins entire; tip blunt, acuminate, or bristle-tipped; lacunae absent (but prominent midvein may sometimes appear lacuna-like); veins 15–35. **Inflorescences** emersed; spikes cylindric, 15–30 mm; peduncles usually terminal, erect to ascending, occasionally recurved, 20–50 mm. **Fruits** quadrate-oblong or nearly orbicular, turgid, 4–5 × 3–3.5 mm, light green to olive-green, dorsally keeled, keel wing-like; beaks erect, 0.6–1 mm.

Lakes, ponds, and slow streams. 100–1100 m. Est, Sisk, WV. CA, ID, NV, WA; north to AK and Northwest Territories, east to Newfoundland. Native.

STUCKENIA Pondweed

Plants rhizomatous; tubers absent or present; turions absent. **Stems** more or less terete, nodes without oil glands. **Leaves** submersed, alternate, opaque, sessile, linear, channeled, turgid; base acute; margins entire; tip obtuse to acuminate; veins 1–5; blades attached to stipules for 67% or more the length of the stipules, thus diverging from the stipules in the upper 33% of stipule length; stipules extending past attachment as free ligules. **Inflorescences** spikes, capitate, cylindric, or verticillate (arranged in whorls), submersed; peduncles flexible, not projecting inflorescence above surface of water. **Flowers** pistils 4. **Fruits** dorsally rounded, beaked or not, turgid.

Cosmopolitan. Approximately 6 species. 3 species treated in *Flora*.

Key to *Stuckenia* species:

1. Leaf tip acute, apiculate, or cuspidate, rarely round; stems abundantly branched, especially on distal portions; fruit distinctly beaked; stipular sheaths not inflated.
   2. Leaves (0.4)1.5–3(8.5) mm wide; leaf tip rounded-apiculate (round); fruit beaks 0.2–0.3 mm; plants of alkaline habitats in SE Oregon . . . . ***S. striata***

   2' Leaves 0.2–1 mm wide; leaf tip acute to mucronate or rarely apiculate; fruit beaks 0.5–1.1 mm; widespread . . . . ***S. pectinata***

1' Leaf tip notched, obtuse or round, rarely apiculate on proximal part of stem; stems sparsely branched on distal portions; fruits without a beak; stipular sheath often inflated . . . . ***S. filiformis***

Stuckenia filiformis (Persoon) Börner
Threadleaf pondweed

Stems freely branched proximally, sparsely branched distally, subterete, (10)20–60(100) cm. **Leaves** stipules near base of stem often with sheaths inflated, 10–40(95) mm, midstem stipules tight to stem, more or less the same width as stem, ligule 2–20 mm; blade filiform or linear, 10–150 × 0.2–2(3.7) mm; margins entire; tip notched, blunt, or rarely apiculate; veins 1–3. **Inflorescences** peduncles terminal, erect, filiform to slender, 20–100(150) mm; spikes cylindric to verticillate, 5–55 mm. **Fruits** obovoid, turgid, 2–3 × 1.5–2.4 mm, tan to dark brown, without a beak.

Approximately 5 subspecies; 2 subspecies treated in *Flora*.

Key to *Stuckenia filiformis* subspecies:

1. Plants 20–100 cm; stipules on proximal portion of stem inflated, disintegrating with age; fruits often absent . . . . *S. f.* ssp. ***occidentalis***

1' Plants 10–30 cm; stipules on proximal portion of stem tightly clasping or slightly enlarged, persistent; fruits common . . . . *S. f.* ssp. ***alpina***

Stuckenia filiformis ssp. ***alpina*** (Blytt) R.R. Haynes, Les, & M. Král
Slenderleaf pondweed

Potamogeton filiformis Pers. var. *alpinus* (Blytt) Asch. & Graebn., *Potamogeton filiformis* Pers. var. *borealis* (Raf.) H. St. John, *Potamogeton filiformis Pers.* var. *macounii* (Morong) Morong

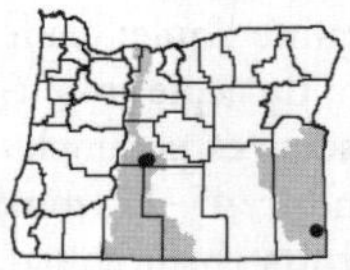

Stems 10–30 cm. **Leaves** stipules persistent, those near base of stem tightly clasping to slightly enlarged, to 1 mm wide.

Shallow to deep waters of ponds, lakes, and streams. 1200–2400 m. ECas, Owy. CA, ID, NV, WA; north to Northwest Territories, east to Newfoundland, southeast to NM; Asia. Native.

Stuckenia filiformis ssp. ***occidentalis*** (J.W. Robbins) R.R. Haynes, Les, & M. Král
Western pondweed

Potamogeton filiformis var. *occidentalis* (J.W. Robbins) Morong

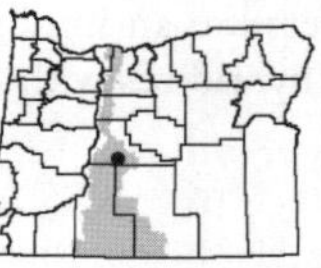

Stems 20–100 cm. **Leaves** stipules disintegrating with age, those on proximal portion of stem inflated; 0.2–5 mm wide.

Mainly in cold streams and rivers, occasionally in ponds and lakes. 900–2100 m. ECas. ID, WA; north to Northwest Territories, east to NH, southeast to NM. Native.

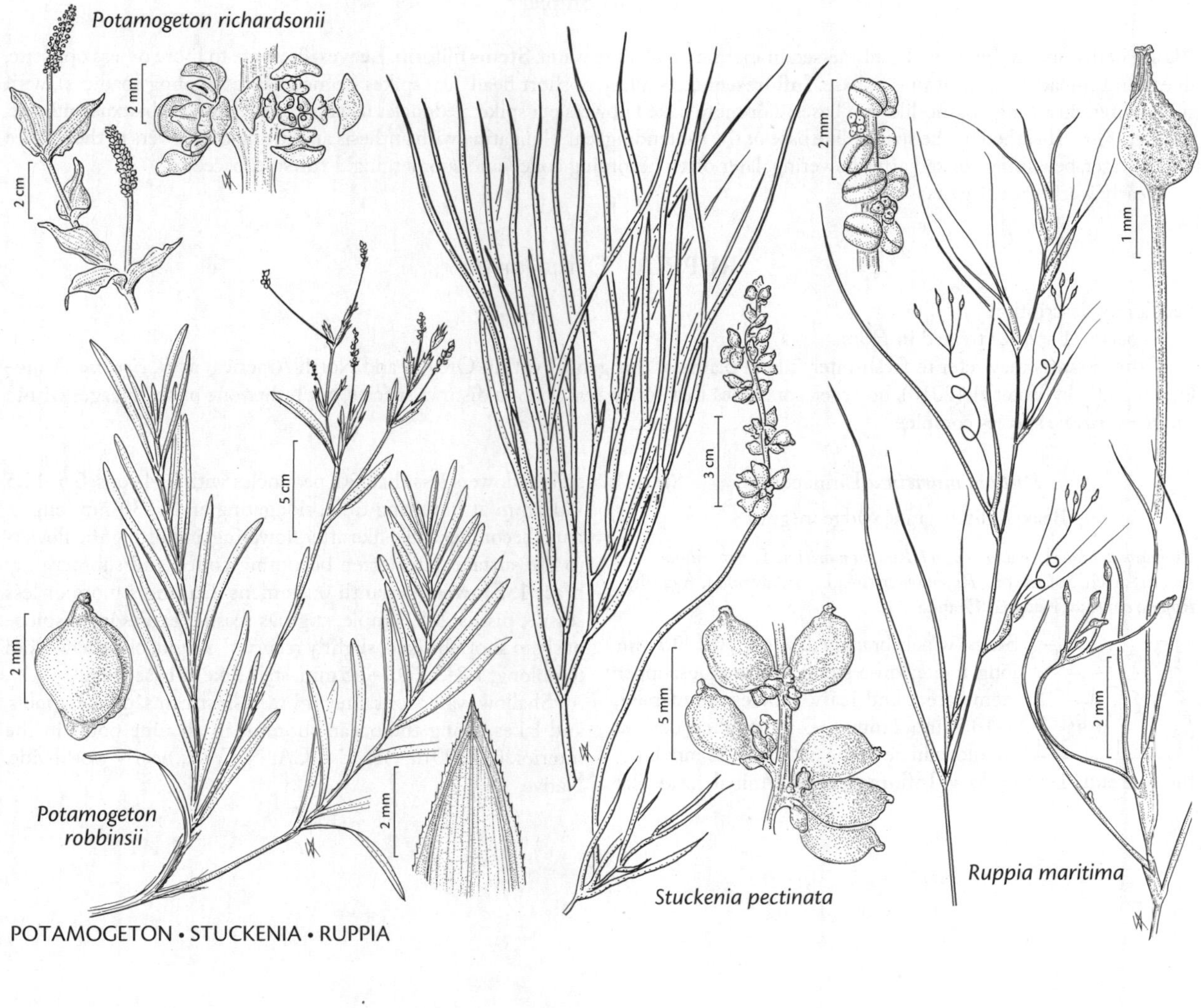

POTAMOGETON • STUCKENIA • RUPPIA

Stuckenia pectinata (Linnaeus) Börner (p. 501)

Sago pondweed, fennel-leaved pondweed

Potamogeton pectinatus L.

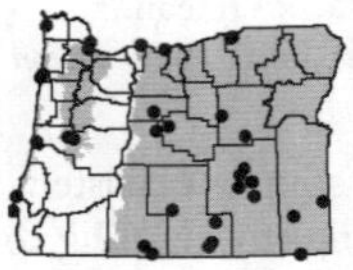

Stems abundantly branched, especially distally, terete to slightly compressed, to 75 cm. **Leaves** stipules with sheaths not inflated, 8–11 mm, with short hyaline ligule; blade linear, 56–92 × 0.2–1 mm; tip acute to acuminate, rarely apiculate; veins 1–3. **Inflorescences** peduncles terminal or axillary, erect to ascending, cylindric, 45–114 mm; spikes verticillate to cylindric, 14–22 mm. **Fruits** oblanceoloid, turgid, 3.8–4 × 2.5–3.1 mm, yellow-brown to brown; beaks erect, 0.5–1.1 mm.

Lakes, streams, and estuaries, often in brackish or alkaline waters. 0–1500 m. BR, BW, Col, ECas, Est, Lava, Owy, WV. CA, ID, NV, WA; north to Northwest Territories, east to Newfoundland, south to South America; Australia, Eurasia. Native.

When floating, the distinctive palm-like arrangement of leaves can be seen. *Stuckenia pectinata* tubers are an important food source for waterfowl. This species hybridizes with *S. filiformis*.

Stuckenia striata (Ruiz & Pavón) Holub

Striped pondweed, Nevada pondweed

Potamogeton striatus Ruiz & Pavón, *Potamogeton filiformis* Pers. var. *latifolius* (J.W. Robbins) Reveal, *Potamogeton latifolius* (J.W. Robbins) Morong

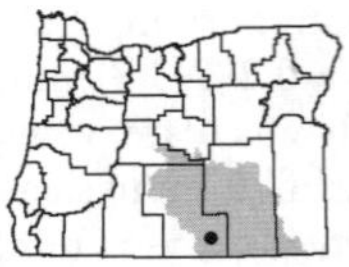

Stems abundantly branched distally, terete to 5-ridged, to 200 cm. **Leaves** stipules with sheaths not inflated, 12–34 mm, ligule 2–11 mm; blades linear, 50–210 × (0.4)1.5–3(8.5) mm; tip rounded-apiculate (round); veins 3–5. **Inflorescences** peduncles axillary, rarely terminal, erect to ascending, cylindric, 12–52 mm; spikes cylindric, rarely verticillate, 13–45 mm. **Fruits** obovoid to oblanceoloid, turgid, 3–3.9 × 2.8–3 mm, brown to reddish brown; beaks, erect or rarely recurved to tip, 0.2–0.3 mm.

Alkaline rivers, canals, and ponds. 1100–1400 m. BR. CA, ID, NV; east to CO, southeast to TX, south to Mexico, Central America, South America. Native.

RUPPIACEAE Ditch-grass family

Lareina Shipler

Plants herbs, annual (perennial), submersed in marine or alkaline water. **Stems** filiform. **Leaves** alternate to more or less opposite, linear and thread-like, sheathing at base. **Inflorescences** axillary in short head-like spikes (somewhat resembling spadices), with subtending, sheathing, spathe-like leaf bases, 2 or more naked flowers per spike; peduncles usually coiled. **Flowers** bisexual, minute, peltate, at first sheathed by the spathe-like base of the subtending leaf, elongating with anthesis and lifting the flowers to the surface of the water, becoming stalked after flowering, later often becoming coiled and submerging. **Fruits** drupaceous.

Nearly worldwide. 1 genus.

RUPPIA Ditch-grass

See family description.

4 species; 1 species treated in *Flora*.

Other sources may refer to freshwater, inland plants of this genus (within Oregon and North America) as *R. cirrhosa*. A molecular study by Ito et al. (2010), however, concludes that *R. cirrhosa* is not a distinct species, but rather one of six lineages within the *R. maritima* species complex.

Ruppia maritima Linnaeus (p. 501)

Beaked ditch-grass, widgeon-grass

Ruppia maritima L. var. *maritima*, *Ruppia maritima* L. var. *obliqua* (Schur) Asch. & Graebn., *Ruppia maritima* L. var. *rostrata* J. Agardh, *Ruppia cirrhosa* (Petagna) Grande

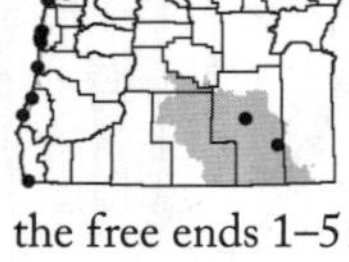

Stems widely branched, 50 cm × 0.1–0.7 mm long, terete; lower stems rhizomatous; upper stems erect and leafy. **Leaves** submersed, 3–10.5 cm × 1 mm, apex more or less obtuse; stipules adnate to the petiole, 5–15 mm long, the free ends 1–5 mm long. **Inflorescences** terminal, head-like spikes, flowers less than 20; peduncles with 0–4 coils, 0.5–16.5 × 0.5 mm at flowering time, elongating and 5–30 mm long in fruit, becoming stalk-like after flowering and lifting the flowers to the surface, later often becoming coiled and submerging; bract 1. **Flowers** perianth 0; stamens 2, anthers more or less sessile; pistils 2–8, simple, stigmas sessile. **Fruits** inconspicuously to short-beaked, slightly recurved, turgid, obliquely ovoid to oblong, 1.8–3 × 0.7–1.5 mm, stalk-like at base. **2*n***=20.

Shallow waters of saline or brackish streams, ditches, pools, and lakes, along the ocean shore or in alkaline pools in the interior. 0–1300 m. BR, Est. CA, NV, WA; nearly worldwide. Native.

SCHEUCHZERIACEAE Scheuchzeria family

Tamra Dickson Prior and Lareina Shipler

Herbs perennial, aquatic; either emergent, terrestrial, or on floating mats; rhizomatous. **Stems** erect, generally simple, slender; usually sheathed at base by old leaves. **Leaves** basal and cauline, alternate, simple, linear, nearly cylindrical, auriculate. **Inflorescences** terminal racemes; pedicels elongating after flower; bracts leaf-like. **Flowers** bisexual, radial; tepals 6, in 2 more or less dissimilar whorls; stamens 6, anthers united at base, turned outward, dehiscing longitudinally; ovary superior; pistils 3(6), distinct to slightly connate at base. **Fruits** follicles. **Seeds** 1–3.

Colder parts of the Northern Hemisphere. 1 genus.

SCHEUCHZERIA Rannoch-rush

Family comprised of a single genus, *Scheuchzeria*, and a single species *S. palustris*.

Scheuchzeria palustris Linnaeus (p. 503)

Rannoch-rush, podgrass, American scheuchzeria

Scheuchzeria palustris L. var. *americana* Fernald

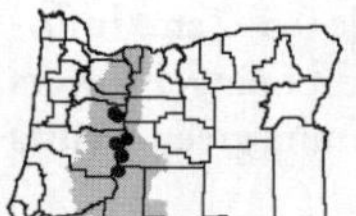

Plants glabrous. **Rhizomes** elongated and creeping. **Stems** 10–40 cm in a Z-like pattern. **Leaves** grass-like, 2–35 cm × 1–2 mm; cauline leaves reduced distally; leaf sheaths pubescent. **Inflorescences** 2–10-flowered (usually few), 2–10 cm; bracts sheathing; pedicels 5–25 mm. **Flowers** perianth segments 2–4 mm long, apex acute, greenish white to yellowish green; anthers and pistils longer than perianth segments. **Fruits** ovoid, 5–9 mm long, beak curved, light green to brown. **Seeds** narrowly ellipsoid, 4–5 mm, brown to black. ***2n***=22.

Bogs, marshes, and pond margins. Flowering Jul–Aug. 900–1700 m. Casc, ECas. CA, ID, WA; north to AK, east to Newfoundland and NY; NM; Eurasia. Native.

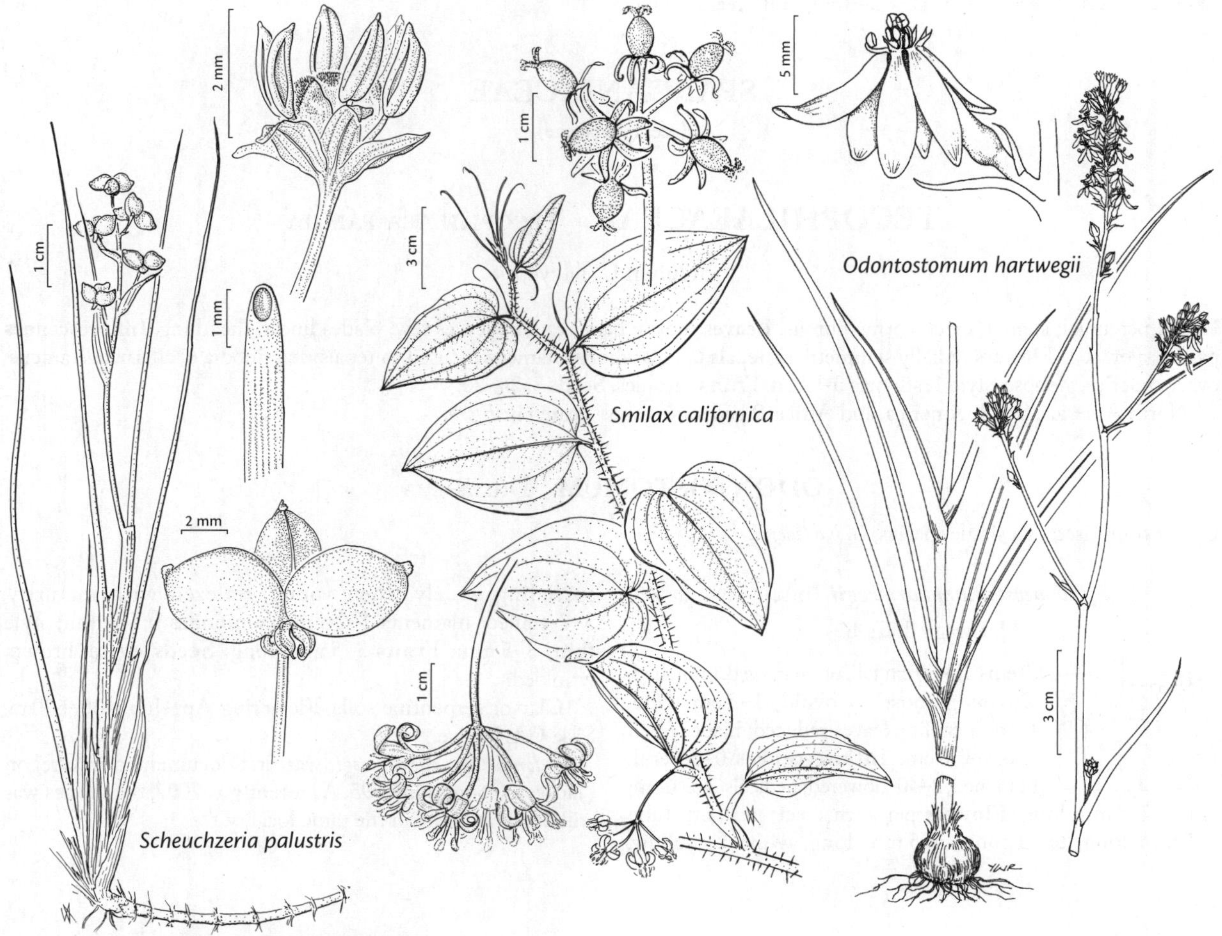

SCHEUCHZERIA • SMILAX • ODONTOSTOMUM

SMILACACEAE Catbrier family

Tamra Dickson Prior

Plants perennial; from rhizomes, stolons, or caudices; dioecious. **Stems** erect or climbing, usually prickly, scaly below leaves. **Leaves** simple, leathery, with 3 prominent veins, reticulate venation between prominent veins. **Inflorescences** umbels. **Flowers** unisexual, tepals 6, in 2 whorls, free. **Staminate flowers** stamens 2–3-whorled. **Pistillate flowers** bearing staminodia; ovary superior; pistils 3. **Fruits** berries. **Seeds** 1–3.

Cosmopolitan. 4–12 genera; 1 genus treated in *Flora.*

This family has been included in Liliaceae *sensu lato*. Its main morphological delineators are that its members are dioecious and have reticulate venation, both uncharacteristic of monocotyledons.

SMILAX Greenbrier

Plants woody. **Roots** filamentous. **Leaves** alternate, ovate to broadly ovate, apex acute, base cordate to subcordate. **Inflorescences** loose to dense; pedunculate. **Staminate flowers** stamens 6, free, 2–3-whorled. **Pistillate flowers** with 6 staminodes; style absent or 1, short; stigmas 3, recurved.

Cosmopolitan, mainly in the tropics and subtropics. Approximately 350 species; 1 species treated in *Flora.*

Smilax californica (A. de Candolle) A. Gray (p. 503)

California smilax, greenbrier

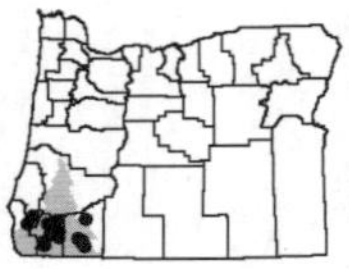

Vines with short rhizomatous caudices. **Stems** 2–11 m long, glabrous; no central pith; prickles 3–11 mm long, bristle-like, flexible, may be lacking distally. **Leaves** evergreen, 4–12 × 2–8 cm, dull green, drying grayish green, glabrous, margins entire; petiole 0.5–2 cm. **Inflorescences** 3–20-flowered; peduncles 2–5 cm, drooping. **Flowers** tepals linear, reflexed from middle, 3–6 × 1–2 mm, green. **Fruits** ovoid, 6–9 mm, black.

Stream banks, rivers, and springs in mixed evergreen and ponderosa pine forests. Flowering May–Jul. 50–800 m. Sisk. CA. Native.

SPARGANIACEAE

(see Typhaceae)

TECOPHILAEACEAE Tecophilaea family

Stephen C. Meyers

Herbs perennial; stem 1, erect; corms fibrous. **Leaves** mostly basal, cauline leaves few; blades linear, sheathing. **Inflorescences** panicles; bract 1. **Flowers** radially symmetrical; tepals 6, in 2 whorls; stamens 6, fused to tepals; staminodia 6, alternate stamens; ovary superior, globose; style 1, stigmas 3-lobed. **Fruits** capsules. **Seeds** 3. **2*n***=20.

North America, South America, and Africa. 9 genera; 1 genus treated in *Flora.*

ODONTOSTOMUM Doll's-lily

Genus comprised of a single species, *O. hartwegii.*

Odontostomum hartwegii Torrey (p. 503)

Hartweg's doll's-lily

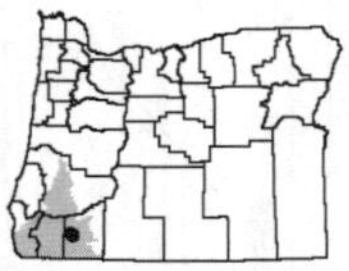

Stems 10–50 cm tall, often curved, branched. **Corms** globose to ovoid, 1–3 cm wide. **Leaves** cauline leaves reduced; blades 6–25 × 0.3–0.8 cm. **Inflorescences** 0–several, branched, 2–40-flowered; pedicels 2–20 mm; bracts 2–8 mm long. **Flowers** perianth erect; perianth tube 4–5 mm long; tepal lobes 4–5 mm long, white to yellow, spreading widely when young, reflexed with maturity, 3–7-veined; filaments 1–2 mm, staminodes 0.5–1 mm; style thin, 3–5 mm. **Fruits** 3–5 mm long. **Seeds** ovate, brown, wrinkled.

Clay or serpentine soils. Flowering Apr–Jun. 400–600 m. Sisk. CA. Native?

Odontostomum hartwegii was first documented in Oregon (Jackson County) in 1995. As recently as 2007, the species was still known to exist in the same locality.

TOFIELDIACEAE False-asphodel family

Stephen C. Meyers

Herbs perennial; stems erect, glabrous or glandular-hairy. **Leaves** basal, 2-ranked; blades linear. **Inflorescences** terminal racemes, usually head-like, sometimes spike-like; bracts present, subtending flower clusters, bracteoles present. **Flowers** tepals 6, in 2 whorls, distinct; stamens 6, filaments flattened, wider at base, anthers attached at base; ovary superior; styles 3. **Fruits** capsules, 3-beaked. **Seeds** appendaged.

Temperate regions worldwide. 3 genera; 1 genus treated in *Flora.*

TRIANTHA Tofieldia, false-asphodel

Plants 10–70 cm; stems glandular below inflorescence or glandular-pubescent, glands 0.5–6 times longer than wide. **Leaves** 1–3, 5–50 cm × 3–8 mm. **Inflorescences** 3–40-flowered, globose to cylindrical-ovate, 1–8 cm, glandular-pubescent; bracteoles 3, resembling a calyx. **Flowers** usually in clusters of 3; tepals 3–7 mm, white to yellowish, those of inner whorl longer; stamens 3–6 mm; styles 0.5–3 mm. **Fruits** 4–9 mm, bright red. **Seeds** approximately 1 mm, whitish to brown; seed coat present or not. **2*n***=30.

North America. 4 species; 2 species treated in *Flora.*

The morphological characters separating the two *Triantha* species present in western North America are minor, difficult to distinguish, and often overlapping. The morphological characters separating subspecific taxa are even more esoteric. More study (preferably molecular) is needed to determine if these taxa should remain distinct. Here, we conservatively include the taxa recent references list as present in our state.

Key to *Triantha* species:

1. Seed coats absent; seeds brown; inflorescences cylindrical-ovate . . . . *T. glutinosa*

1' Seed coats present; seeds whitish; inflorescences globose to cylindrical-ovate . . . . *T. occidentalis*

Triantha glutinosa (Michaux) Baker

Sticky tofieldia, sticky false-asphodel

Tofieldia glutinosa (Michx.) Pers.

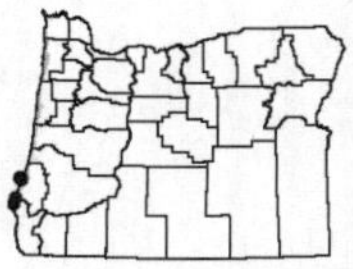

Inflorescences cylindrical-ovate, 3–40-flowered. **Flowers** tepals 3.5–5 mm, white to yellowish; stamens 3–6 mm; styles 0.5–1.5 mm. **Seeds** brown; seed coats absent.

Wet meadows, marshes. Flowering Jun–Aug. 0–50 m. Est. WA; northern North America. Exotic.

Where their ranges overlap, hybrids between *T. glutinosa* and *T. occidentalis* have been documented.

Triantha occidentalis (S. Watson) R.R. Gates

Western false-asphodel

Inflorescences globose to cylindrical-ovate, 3–30-flowered. **Flowers** tepals 3–7 mm, white to yellowish; stamens 3–4 mm; styles 0.5–3 mm. **Seeds** whitish; seed coats present.

Western North America. 3 subspecies; 2 subspecies treated in *Flora.*

Key to *Triantha occidentalis* subspecies:

1. Inflorescences generally cylindrical-ovoid; styles 0.5–1.5 mm. . . . . *T. o.* ssp. ***brevistyla***

1' Inflorescences generally globose; styles 1–3 mm . . . . *T. o.* ssp. ***occidentalis***

Triantha occidentalis (S. Watson) R.R. Gates ssp. *brevistyla* (C.L. Hitchcock) Packer

(p. 506)

Short-style tofieldia, western false-asphodel

Tofieldia glutinosa (Michx.) Pers. ssp. *brevistyla* C.L. Hitchc., *Tofieldia glutinosa* (Michx.) Pers. var. *brevistyla* (C.L. Hitchc.) C.L. Hitchc., *Tofieldia glutinosa* (Michx.) Pers. var. *intermedia* (Rydb.) B. Boivin

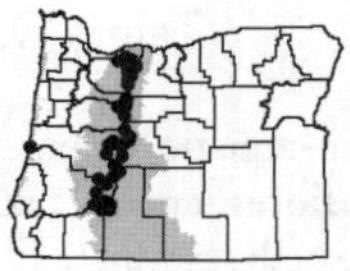

Inflorescences generally cylindrical-ovoid. **Flowers** styles 0.5–1.5 mm.

Wet meadows, marshes, bogs. Flowering Jun–Aug. 400–2200 m. Casc, ECas, Est. WA; north to AK. Native.

Triantha occidentalis (S. Watson) R.R. Gates ssp. *occidentalis*

Western false-asphodel

Tofieldia glutinosa (Michx.) Pers. ssp. *occidentalis* (S. Watson) C.L. Hitchc., *Tofieldia glutinosa* (Michx.) Pers. var. *occidentalis* (S. Watson) C.L. Hitchc., *Tofieldia occidentalis* S. Watson ssp. *occidentalis*

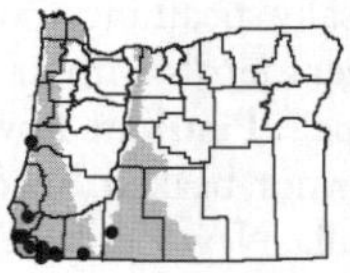

Inflorescences generally globose. **Flowers** styles 1–3 mm.

Wet meadows, marshes, bogs. Flowering Jul–Aug. 300–2200 m. CR, ECas, Sisk. CA. Native.

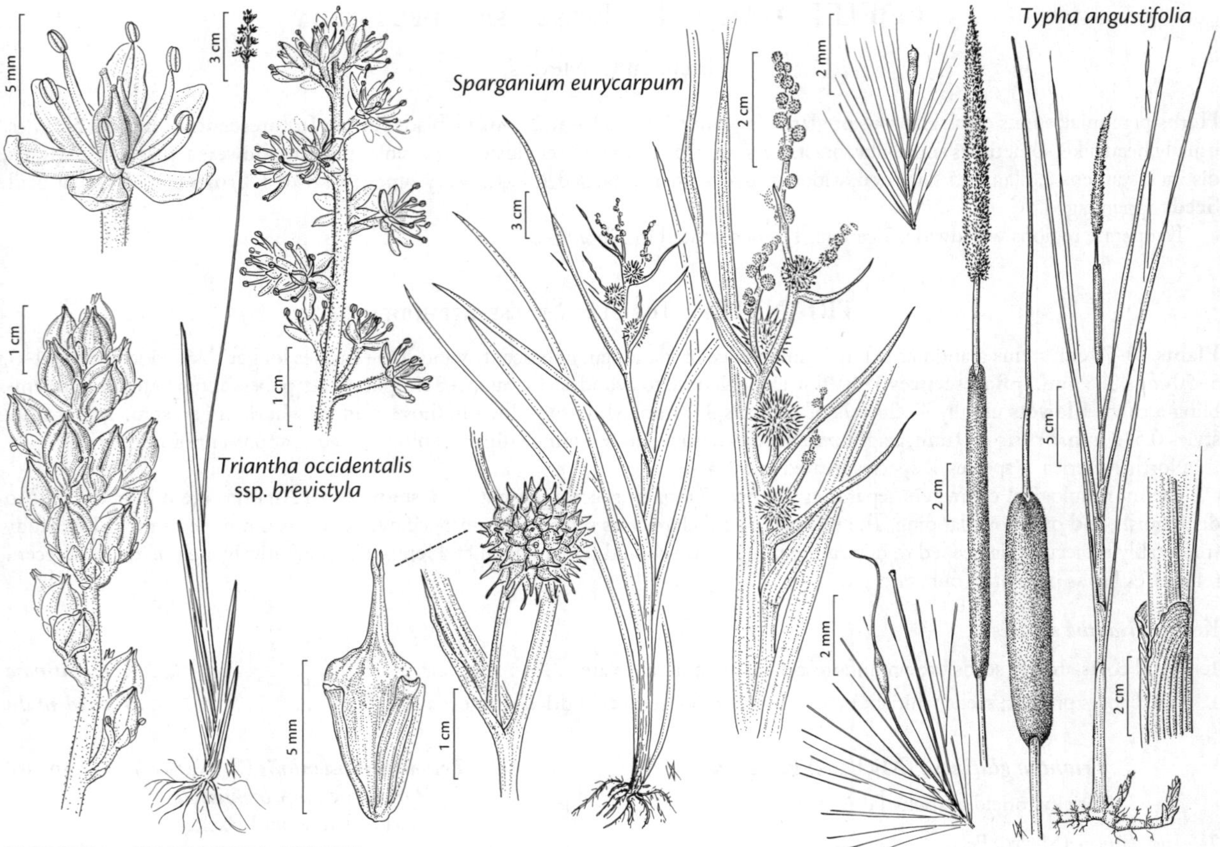

TRIANTHA • SPARGANIUM • TYPHA

TYPHACEAE Cattail family

Tamra Dickson Prior

Herbs perennial, monoecious, aquatic or semi-aquatic, emergent, floating or erect; stems round; rhizomatous. **Leaves** basal and cauline, 2-ranked, ascending; blades linear. **Inflorescences** 1, axillary or terminal, erect, spike-like and cylindrical or with multiple spherical heads; bracts present. **Flowers** unisexual; staminate flowers deciduous, stamens 2–several; ovary superior; pistil 1, style 1, stigmas 1–2(4). **Fruits** follicles or achene-like drupes.

Cosmopolitan. 2 genera.

1. Inflorescences with multiple globose heads . . . . **SPARGANIUM**
1' Inflorescences cylindrical, spike-like . . . . **TYPHA**

SPARGANIUM Bur-reed

Plants emergent, floating, or erect. **Leaves** basally sheathing, spongy; margins entire. **Inflorescences** 2–50+-flowered, heads axillary or terminal, usually unisexual, globose. **Flowers** tepals with or without dark spot near tip, persistent in fruit. **Staminate flowers** white; stamens 2–5(8), much longer than tepals. **Pistillate flowers** white or green; pistil longer than tepals; stigmas 1(2). **Fruits** achene-like drupes, narrowing at midpoint or not, beaked. **Seeds** 1(2), narrowly-ovoid, thin-coated.

Temperate Northern Hemisphere, Australia, New Zealand. Approximately 14 species; 4 species treated in *Flora*.

Key to *Sparganium* species:

1. Inflorescences branched; staminate heads 10–50+; stigmas usually 2. . . . . ***S. eurycarpum***
1' Inflorescences unbranched; staminate heads 10 or less; stigma 1.
    2. Staminate heads 1–2; fruit beaks 1.5 mm long or less . . . . ***S. natans***
    2' Staminate heads 3–7(10); fruit beaks 1.5 mm long or greater.
        3. Leaves not keeled, 2–5 mm wide; fruit beaks 2 mm long or less . . . . ***S. angustifolium***
        3' Leaves keeled near base, 4–10 mm wide; fruit beaks 2 mm long or greater. . . . . ***S. emersum***

Sparganium angustifolium Michaux
Floating bur-reed, narrow-leaf bur-reed

Sparganium angustifolium Michx. var. *multipedunculatum* (Morong) Brayshaw, *Sparganium emersum* Rehmann var. *multipedunculatum* (Morong) Reveal, *Sparganium simplex* Huds. var. *multipedunculatum* Morong

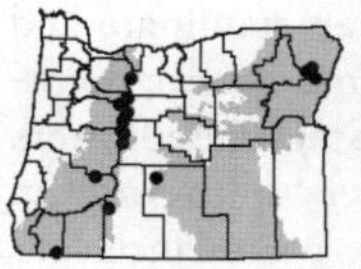

Plants terrestrial to emergent, to 2.5 m. **Leaves** 0.2–1 m × 2–5(8) mm, not keeled, uppermost leaf extending above inflorescence. **Inflorescences** emergent, unbranched; staminate heads 3–4(5); pistillate heads 1–4, 1–3 cm diameter in fruit. **Flowers** tepals lance-ovate, without spot near tip, tips toothed; stigma 1. **Fruits** red to brown, dull; body 3–4 × 1.5–2 mm, narrower at midpoint; beaks straight, 1.5–2 mm long. **2*n***=30.

Lakes, ponds, and streams, occasionally shorelines. Flowering Jul–Aug. 900–2800 m. BR, BW, Casc, Sisk. CA, ID, NV, WA; east to Newfoundland, southeast to NM; circumboreal. Native.

Sparganium angustifolium is known to hybridize with *S. emersum* in habitats where both species are present.

Sparganium emersum Rehmann
Simple-stem bur-reed

Sparganium emersum Rehmann ssp. *emersum*, *Sparganium emersum* Rehmann var. *emersum*, *Sparganium simplex* Huds.

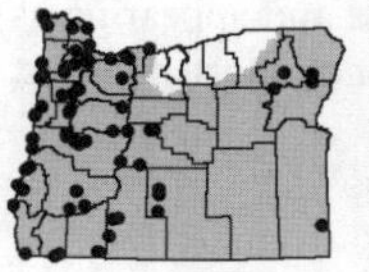

Plants terrestrial to emergent, to 2 m. **Leaves** 0.1–0.8 m × 4–10 mm, keeled near base, uppermost leaf extending slightly or not above inflorescence. **Inflorescences** erect, unbranched; staminate heads 3–7(10); pistillate heads 1–6, 1.5–3 cm diameter in fruit. **Flowers** tepals linear-lanceolate, without spot near tip, tips toothed; stigma 1. **Fruits** green to red-brown, shiny; body 3–4 × 1.5–2 mm, narrower at midpoint; beaks straight to curved, 2–4.5 mm long. **2*n***=30.

Fresh water habitats. Flowering Jun–Aug. 0–2300 m. All ecoregions except Col. CA, ID, NV, WA; throughout North America; Eurasia. Native.

This species can be easily confused with *S. angustifolium*. *Sparganium emersum*, however, usually has more staminate heads and fruits with longer beaks, and its uppermost leaf extends only slightly or not above the inflorescence. See also *S. angustifolium*.

Sparganium eurycarpum Englemann (p. 506)
Broad-fruit bur-reed

Sparganium californicum Morong, *Sparganium erectum* L. ssp. *stoloniferum* (Graebn.) C.S. Cook & M.S. Nicholls, *Sparganium eurycarpum* Engelm. ssp. *eurycarpum*, *Sparganium greenei* Morong

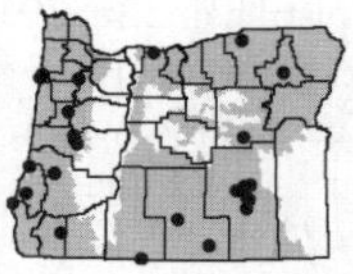

Plants emergent, to 2.5 m. **Leaves** 0.2–2.5 m × 5–20 mm, keeled near base. **Inflorescences** emergent, branched; staminate heads 10–50+; pistillate heads (1)2–8, 1.5–4 cm diameter in fruit. **Flowers** tepals linear, with spot near tip, tips usually entire; stigmas (1)2. **Fruits** tan, darkening with age, dull or shiny; body 5–10 × 3–7 mm, not narrower at midpoint; beaks straight, 2–4 mm long. **2*n***=30.

Ditches, marshes, brackish waters. Flowering Jun–Aug. 0–1700 m. BR, BW, Col, CR, ECas, Est, Sisk, WV. CA, ID, NV, WA; northeast to Northwest Territories, east to New Brunswick, southeast to NM; eastern Asia. Native.

Sparganium natans Linnaeus
Small bur-reed

Sparganium minimum (Hartm.) Fr.

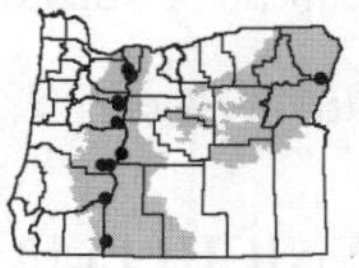

Plants aquatic, to 0.6 m. **Leaves** 0.1–0.6 m × 1–5 mm, not keeled. **Inflorescences** unbranched; staminate heads 1(2); pistillate heads 1–3, 0.5–1.5 cm diameter in fruit. **Flowers** tepals lance-ovate, without spot near tip, tips toothed; stigma 1. **Fruits** dark green to brown, shiny or dull; body 2–4 × 1–1.5 mm; beaks curved, 0.5–1.5 mm long. **2*n***=30.

Shallow water; acid to slightly basic waters of bays, pools, ditches, and peat bogs. Flowering Jul–Aug. 700–2000 m. BW, Casc, ECas. CA, ID, WA; northern North America; circumboreal. Native.

TYPHA Cattail

Plants aquatic to semi-aquatic, to 3 m tall; stems cylindrical, pithy; rhizomes to 50+ cm long, starchy. **Leaves** alternate, linear, spongy, sheathing. **Inflorescences** 1000+-flowered, spike-like, cylindrical, terminal, densely crowded, with numerous papery scales; staminate flowers on narrow stalk above pistillate flowers; bare stalk between staminate and pistillate flowers 0–10 cm. **Staminate flowers** stamens 2–5, anthers much longer than filaments, yellow. **Pistillate flowers** bractlets present or not. **Fruits** dry, thin-walled, nearly transparent, opening in water. **2*n***=30.

Approximately 15 species; 2 species treated in *Flora*.

Key to *Typha* species:

1. Bare stalk between staminate and pistillate flowers 1–10 cm; stems 2–3 mm wide near inflorescence; pistillate bractlets present . . . . *T. angustifolia*

1' Bare stalk between staminate and pistillate flowers absent; stems 3–7 mm wide near inflorescence; pistillate bractlets absent. . . . . *T. latifolia*

Typha angustifolia Linnaeus (p. 506)

Lesser cattail, narrow-leaf cattail

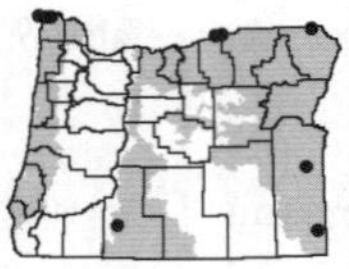

Stems 5–12 mm wide, 2–3 mm wide near inflorescence. **Leaves** 3–12 mm wide, deep green, distal blades usually exceeding inflorescence. **Inflorescences** bare stalk between staminate and pistillate flowers 1–10 cm approximately equal in length; staminate scales hair- to strap-like, straw-colored to brown. **Staminate flowers** 4–6 mm; anthers 1.5–2 mm. **Pistillate flowers** up to 2 mm in flower, up to 7 mm in fruit; stigmas linear, white, drying to brown, elongating with age; bractlets equal to or slightly shorter than pistil hairs. **2*n***=30.

Wet soil, quiet to slow-moving often brackish water. Flowering Jun–Sep. 0–1400 m. BW, Col, CR, ECas, Est, Owy. CA, ID, NV, WA; throughout North America; Eurasia. Exotic.

In California, hybrids between *T. angustifolia* and *T. latifolia* have been reported. No hybrids have yet been reported in Oregon.

Typha latifolia Linnaeus

Broad-leaf cattail, common cattail

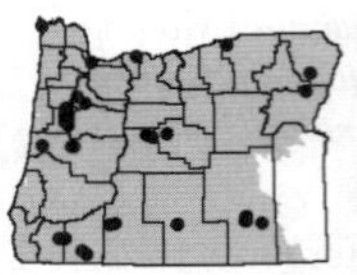

Stems 1–2 cm wide, 3–7 mm wide near inflorescence. **Leaves** 5–25(35) mm wide, grayish green, distal blades approximately equal in height to inflorescence. **Inflorescences** bare stalk between staminate and pistillate flowers absent (rarely to 4 cm in some clones); staminate scales hair-like, clear to tan-colored. **Staminate flowers** 5–12 mm; anthers 1–3 mm. **Pistillate flowers** up to 3 mm in flower, up to 15 mm in fruit; stigmas ovate to ovate-lanceolate, green, drying to reddish brown or black; bractlets absent. **2*n***=30.

Wet soil, quiet to slow-moving fresh or brackish water. Flowering Jun–Aug. 0–1700 m. All ecoregions except Owy. CA, ID, NV, WA; throughout North America; worldwide. Native.

See also *T. angustifolia.*

ZANNICHELLIACEAE Horned-pondweed family

Tamra Dickson Prior

Herbs monoecious, annual or perennial, aquatic, submersed, glabrous; rhizomes creeping. **Stems** thread-like, fragile. **Leaves** alternate, opposite and/or somewhat whorled, simple, sessile; stipules fused to blade or free and sheath-like; ligules not auriculate; blades linear. **Inflorescences** axillary cymes, sessile; bract 1, cup-shaped; staminate flowers below pistillate flowers. **Staminate flowers** perianth parts generally 0, if 3 then minute and scale-like; stamen generally 1 (but may be 2–3 united and appearing as 1), anther dehiscing longitudinally. **Ovaries** superior. **Pistillate flowers** perianth parts 0; pistils (1)4–5(9), distinct, short-stipitate, style 1. **Fruits** achene-like drupes. **Seeds** 1.

Nearly worldwide. 4 genera; 1 genus treated in *Flora.*

ZANNICHELLIA Horned-pondweed

Roots single or in pairs at nodes. **Stems** usually few-branched. **Leaves** entire, veins 1(3), almost terete. **Inflorescences** staminate flowers usually 1; pistillate flowers 2–4(6). **Fruits** compressed, abaxially curved, ridged and toothed, adaxially more or less straight, base stalk-like, tip beaked; margin convex and minutely dentate.

Cosmopolitan. Approximately 5 species; 1 species treated in *Flora.*

Zannichellia palustris Linnaeus (p. 509)

Horned-pondweed

Zannichellia palustris L. var. *stenophylla* Asch. & Graebn.

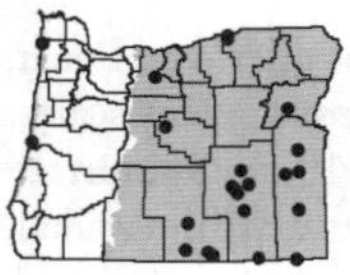

Stems 30–600 cm × 0.2–0.6 mm. **Leaves** 3–6 cm × 0.2–1 mm, tips acute. **Staminate flowers** filament 1.5–2 mm, connective tissue extending beyond anther sacs to form a blunt tip. **Pistillate flowers** style 0.4–0.7 mm. **Fruits** 1.5–3 × 0.5–1 mm; beak 1–1.5 mm. **2*n***=24, 36.

Brackish or fresh water streams, ponds, lakes, and estuaries. Flowering Mar–Nov. 0–1800 m. BR, BW, Col, ECas, Est, Lava, Owy. CA, ID, NV, WA; worldwide. Native.

Currently, only one species of *Zannichellia* is recognized within North America. Species delimitations within *Zannichellia* are in debate; more than one species may exist. Further study is needed.

ZOSTERACEAE Eel-grass family

Rory Nichols

Herbs perennial (or rarely annual), marine aquatic, intertidal and subtidal, grass-like. **Roots** rhizomatous. **Stems** leafy. **Leaves** alternate, 2-ranked or tufted, less than 2 m long; sheath open or closed; blades linear, ribbon-like, apex obtuse or notched; stipules absent. **Inflorescences** spikes terminal or axillary, enclosed by membranous spathe (leaf-like bract) in flower and in fruit or not; flowers many, in 2 rows on 1 side of axis, pedunculate. **Flowers** unisexual, staminate and pistillate on same plant or different plants; retinacula (subtending bracts) present or not; perianth absent; staminate flowers stamen 1, generally sessile anthers dehiscing longitudinally, pollen linear; pistillate flowers perianth absent; ovary 1, ovule 1; style 1, stigmas 2. **Fruits** achene-like. **Seeds** 1.

Widely distributed in temperate coastal oceans worldwide. 3 genera; 2 genera treated in *Flora.*

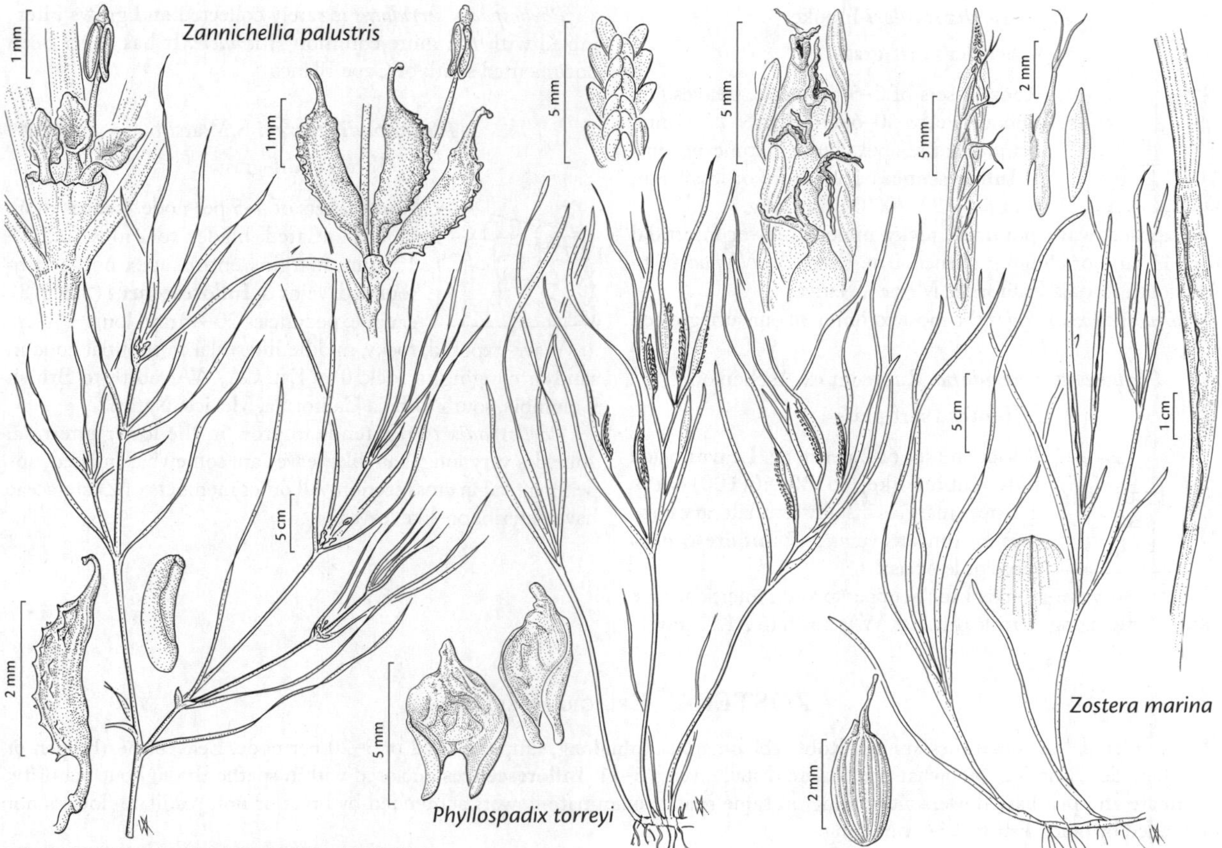

ZANNICHELLIA • PHYLLOSPADIX • ZOSTERA

Eel-grasses are found in marine environments such as tide pools and bays. Tangled clumps of eel-grass often wash up on Oregon's beaches due to strong waves uprooting plants. The eel-grasses are grass-like in appearance, growing in beds or intermixed with marine algae. Zosteraceae is most closely related to Potamogetonaceae and Zannichelliaceae. Eel-grasses are ecologically very important, providing food and/or shelter to numerous organisms, including fish, invertebrates, and waterfowl. Some species of algae grow epiphytically on eel-grasses. Generally, species of this family are under-collected, and their subtidal distribution is poorly documented. In Oregon, Zosteraceae is unique in that it is the only group of vascular plants that inhabit our marine waters, never occurring above high tide.

Key to Zosteraceae genera:

1. Plants dioecious; rhizomes short, thick; spadices enclosed in spathes at maturity; growing in exposed rocky intertidal areas . . . . **PHYLLOSPADIX**

1' Plants monoecious; rhizomes long, slender; spadices protrude from spathes when mature; growing in protected estuaries and bays . . . . **ZOSTERA**

PHYLLOSPADIX Surf-grass, surfgrass

Herbs perennial, tufted. **Roots** tuber-like, stout, thick, simple, in 2 sets of 3–5 per node or 1 set of 2 per node. **Leaves** prior years' decayed leaf fibers persistent; sheaths open; blade margins entire or serrulate; veins 3–7. **Inflorescences** spadices enclosed by spathes when young, projecting when mature, dioecious; flowers subtended by a bract; ovary tapering to a short style. **Fruits** 4–5 mm long.

Temperate coastal waters of the northern Pacific Ocean. 5 species; 3 species treated in *Flora*.

Key to *Phyllospadix* species:

1. Leaf blades more or less inflated (hair-like), width (0.5)1–1.5 mm . . . . ***P. torreyi***

1' Leaf blades not inflated (flat, ribbon-like), width 1.5–3 mm.

  2. Roots in 2 sets of 3–5 per node; leaf veins 3; margins entire . . . . ***P. scouleri***

  2' Roots 2 per node; leaf veins 5–7; margins serrulate distally (run fingers from apex toward base to feel teeth) . . . . ***P. serrulatus***

Phyllospadix scouleri Hooker

Scouler's surf-grass

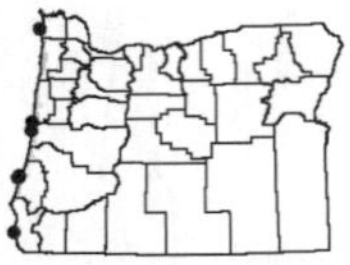

Roots 2 sets of 3–5 per node. **Leaves** flat, ribbon-like, to 30–60(100) cm × 1.5–3 mm, margins entire, apex obtuse to truncate, veins 3. **Inflorescences** 1(2), more or less basal; peduncles 20–70(100) mm long.

Exposed, wave-pounded, rocky, upper to lower intertidal zone in sand or clinging to rock. 0 m. Est. CA, WA; north to AK, south to Baja California, Mexico. Native.

Phyllospadix scouleri is the most common of our surf-grasses.

Phyllospadix serrulatus Ruprecht ex Ascherson

Toothed surf-grass

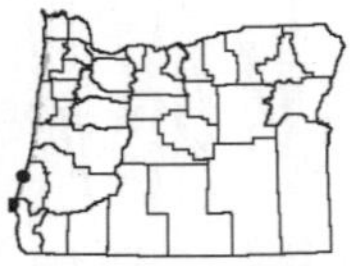

Roots one set of 2 per node. **Leaves** blade flat, ribbon-like, to 30–60(100) cm × 2–3 mm, margins distally serrulate, apex obtuse to truncate, veins 5–7. **Inflorescences** more or less basal.

Exposed, wave-pounded, rocky, upper to lower intertidal zone in sand or clinging to rock. 0 m. Est. WA; north to AK. Native.

Phyllospadix serrulatus is rarely collected and grows intermixed with the more common *P. scouleri*. It has never been documented south of Cape Blanco.

Phyllospadix torreyi S. Watson (p. 509)

Torrey's surf-grass

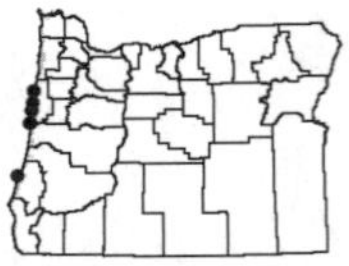

Roots 2 sets of 3–5 per node. **Leaves** more or less inflated, blades to 2 m × (0.5)1–1.5 mm, margins entire, apex notched to rounded, veins 3. **Inflorescences** (2)3–5(8), cauline; peduncles 20–40 mm long.

Wave-exposed, rocky, middle intertidal to subtidal zone in sand or clinging to rock. 0 m. Est. CA, WA; north to British Columbia, south to Baja California, Mexico. Native.

Phyllospadix torreyi tends to grow in the lower intertidal zone. Its very long, hair-like leaves are somewhat inflated, appearing oval in cross section. All other members of Zosteraceae have flat, ribbon-like leaves.

ZOSTERA Eel-grass, eelgrass

Herbs perennial or sometimes annual. **Roots** rhizomatous, thin, long, simple roots 2 or 5–20 per node. **Leaves** sheath open or closed; blades entire or somewhat denticulate distally; veins 3–11. **Inflorescences** enclosed within spathe throughout maturity; staminate and pistillate flowers alternating on same spadix; staminate flowers subtended by bract or not; pistillate flowers not subtended by bract. **Fruits** 2.5–5 mm long.

Temperate coastal waters worldwide. 12 species; 2 species treated in *Flora*.

Both species of *Zostera* reproduce sexually but also rely on spreading vegetatively by being uprooted by winter storms. In Oregon, *Z. marina* is native and *Z. japonica* is exotic. Their native ranges, however, overlap in the western Pacific.

Key to *Zostera* species:

1. Roots 2 per node; leaf sheaths open, with 2 membranous flaps; leaf blades to 30 cm × 1–1.5 mm, veins 3 . . . . ***Z. japonica***

1' Roots 5–20 per node; leaf sheaths closed, without membranous flaps; leaf blades to 150 cm × (2)3–11 mm, veins 5–11. . . . . ***Z. marina***

Zostera japonica Ascherson & Graebner

Japanese eel-grass

Zostera nana Roth

Herbs perennial or sometimes annual. **Roots** 2 per node. **Leaves** sheath open, with 2 membranous flaps; blades to 20(30) cm × 1–1.5 mm, apex obtuse or retuse, veins 3. **Inflorescences** 2–5; staminate flowers each subtended by bract. **Fruits** 2.5 mm long.

Protected waters in bays or estuaries on tidal flats in high to low intertidal zone in mud or sand. 0 m. Est. CA, WA; north to AK; Africa, eastern Asia, Europe. Exotic.

Zostera japonica was inadvertently introduced to the western coast of North America in the early 1900s, likely in a shipment of Pacific oysters (*Crassostrea gigas*) from Japan. It has spread very rapidly since and is likely present in all of Oregon's bays and estuaries. This species competes with *Z. marina* in the low intertidal zone, while colonizing the typically unvegetated tidal flats in the middle to high intertidal zone.

Zostera marina Linnaeus (p. 509)

Eel-grass, grass wrack

Zostera marina L. var. *stenophylla* Asch. & Graebn., *Zostera pacifica* S. Watson

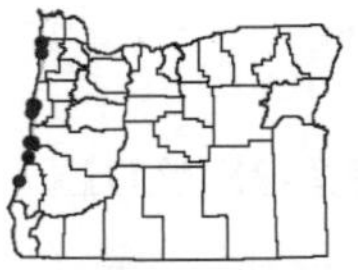

Herbs perennial. **Roots** 5–20 per node. **Leaves** sheath closed, 5–20 cm, membranous flaps absent; blades to 150 cm × (2)3–11 mm, apex obtuse, round-obtuse, or slightly mucronate; veins 5–11. **Inflorescences** 10 or more; staminate flower subtending bracts absent or rarely 1 subtending the most proximal flower. **Fruits** 2–5 mm long.

Protected waters in bays or estuaries in lower intertidal zone to subtidal in mud or sand. 0 m. Est. CA, WA; north to AK, south to Baja California, Mexico; Eurasia. Native.

In comparison to the dainty *Z. japonica*, *Z. marina* is a much larger, more robust plant with considerably wider leaves.

GLOSSARY

For terms that are illustrated, the page numbers of the illustration can be found at the end of the entry. The references *Plant identification terminology: An illustrated glossary* by Harris and Harris (2001) and *Merriam-Webster's collegiate dictionary* (2000) were consulted in the preparation of these definitions.

A

abaxial The side facing away from the axis or stem (compare to adaxial).

abort To fail to develop fully.

abrupt Terminating suddenly. (p. 515)

abscission The natural detachment of an organ or structure caused by the breakdown of cells at the point of attachment.

acaulescent Lacking a distinct stem.

achene Dry, indehiscent fruit containing one seed attached to the ovary wall at a single point. (p. 517)

achlorophyllous Lacking chlorophyll (not green).

acid Having a pH less than 7 (compare to alkaline).

acropetal Proceeding from the base upwards toward the apex.

acroscopic Facing the tip or apex (compare to basiscopic).

acuminate Gradually tapering to a narrow point with concave sides. (p. 515)

acute Tapering to a pointed apex with more or less straight sides. (p. 515)

adaxial The side facing the axis or stem (compare to abaxial).

adnate Fused to unlike parts (compare to connate).

adventitious Developing or arising in an unusual position, usually said of roots on a stem.

adventive Not native.

aerenchyma A tissue of thin-walled cells with many air spaces, especially common in aquatic plants.

aerial Occurring in the air, above ground or water.

aggregate To cluster densely.

alkaline Basic; having a pH greater than 7 (compare to acid).

allopolyploid Having more than two complete sets of chromosomes, from two or more different species.

allotetraploid Having four complete sets of chromosomes, from two or more different species.

alluvial Composed of unconsolidated sediment which has been eroded and deposited by water.

alpine Above timberline; see *Ecology* chapter.

alternate Borne singly at a node, as leaves on a stem; borne between other organs.

amphibious Living both in water and on land.

amplexicaul Clasping the stem. (p. 515)

anastomose To join together after branching.

androgynous With staminate flowers above the pistillate flowers in an inflorescence (compare to gynaecandrous).

annual A plant that germinates from seed, flowers, sets seed, and dies in one year or less (compare to biennial, perennial).

annular In the form of a ring.

annulus A row of thick-walled cells along one side of a fern sporangium that aids in spore dispersal.

anterior Towards the front; on the side away from the main axis.

anther The expanded, apical, pollen bearing portion of the stamen. (p. 517)

anthesis The flowering period, when a flower is fully expanded and functioning.

anthocyanin Water soluble plant pigment, ranging in color from blue to purple to red.

antrorse Directed forward or towards the apex (compare to retrorse).

apex (pl. **apices**) The tip; point farthest from the point of attachment. (pp. 513, 515)

aphyllopodic Having the lowest leaves reduced to scales.

apical At or towards the apex or tip (compare to basal).

apiculate Ending abruptly with a short, slender point. (p. 515)

apiculus A small, slender point.

apophysis (pl. **apophyses**) The part of a cone scale that is exposed when the cone is closed; a swelling or protuberance.

appendage An attached, supplementary or secondary part.

appressed Pressed flat against another organ.

aquatic Growing in water.

archegonium (pl. **archegonia**) The female reproductive structure that produces female gametes in moss and fern gametophytes.

arcuate Moderately curved like a bow.

areola (pl. **areolae**) An area defined on a surface, such as the angular space between leaf veins.

aril Fleshy, corky, or hard appendage arising at or near the point of seed attachment, sometimes covering the entire seed. (p. 521)

aristate Awn- or bristle-tipped.

articulate Jointed; separating freely along a well-defined line.

ascending Rising obliquely or curving upwards.

asexual Reproducing without fertilization.

asymmetric Irregular in shape, not divisible into identical mirror-image halves.

attenuate Gradually narrowed or tapered, more gradual than acuminate. (p. 515)

auricle Ear-shaped appendage or lobe. (p. 523)

auriculate With auricles.

awl-like Narrow throughout but with a broader base tapering to sharp tip. (p. 521)

awn A stiff, narrow, bristle-like appendage from the tip or dorsal surface. (p. 523)

axil The upper angle between a stem and a leaf or branch of the stem.

axile On or pertaining to the axis.

axillary Pertaining to the axil; in the axil.

axis (pl. **axes**) The main or central line or stem about which the lateral organs or parts are arranged.

B

backcross Cross of a hybrid with one of the parental species.

bald Lacking the usual covering; usually an area without trees.

ballast Heavy material placed low in a vessel such as a ship to improve stability.

band Stripe.

barbed Bearing sharp, rigid, reflexed points like barbs of a fishhook.

bark The outermost layers of a woody stem.

basal Related to or situated at the base (compare to apical).

basalt A dark gray to black, dense to fine-grained, igneous rock.

basic Alkaline; with a pH greater than 7 (compare to acid).

basifixed Attached at the base.

basin A depression; watershed.

basiscopic Facing toward the base (compare to acroscopic).

beak A prolonged, firm tip. (p. 517)

bearded Bearing one or more tufts of long hair.

berry A fleshy fruit with more than one seed, developing from a single pistil.

bicolored Of two distinct colors.

biconvex Convex on both sides.

bidentate Having two teeth.

biennial A plant which lives for more than one but less than two years after germination (compare to annual, perennial).

bifid Forked or divided for about half its length.

bilateral Arranged on opposite sides.

bilobed Divided into two lobes.

bipinnate Twice or doubly pinnate.

biradial Having both bilateral and radial symmetry.

bisexual A flower with both male and female reproductive organs.

blade The broad part of a leaf or petal. (pp. 513, 519, 523)

bloom A whitish or glaucous powdery coating of a surface; a flower; to produce flowers.

bog A wetland with spongy peat deposits, acidic water (pH less than 5.5), and low nutrient levels; see *Ecology* chapter.

boreal Of, relating to, or located in northern regions.

bottomland Low lying, alluvial land by a river or stream.

brackish Water that is saltier than fresh water, but not as salty as seawater.

bract A modified leaf or leaf-like structure at the base of a flower or inflorescence.

bracteate With bracts.

bracteole A small bract.

bractlet A small bract.

branchlet A small branch growing from a larger branch.

bristle A short, stiff, hair-like structure.

bud An undeveloped shoot or flower, often covered with protective scales.

bulb An underground bud with thickened fleshy scales, such as an onion.

bulbil A small bulb or bulb-like structure.

bulblet See bulbil.

bur A structure armed with hooks or barbs.

buttressed With props or supports.

C

caducous Falling off early compared to similar structures in other plants.

caespitose See **cespitose**.

calcareous Derived from limestone; containing lime.

callous Hardened or thickened.

callus A hard thickening or protuberance; in grasses, the thickened basal extension of the lemma. (p. 523)

calyx (pl. **calyces**) The outer whorl of the perianth; collective term for all of the sepals of a flower.

campanulate Bell-shaped.

capillary Slender and fine; hair-like.

capitate Head-like or in a rounded cluster. (p. 516)

capsular Pertaining to a capsule; capsule-like.

capsule A dry, dehiscent fruit composed of more than one locule.

carinate Longitudinally keeled.

carpel A simple pistil or one unit of a compound pistil.

cartilaginous Tough and firm yet flexible, like cartilage; without chlorophyll.

caruncle A protuberance at or surrounding the hilum (scar at point of attachment) of a seed.

caryopsis (pl. **caryopses**) A dry, one-seeded indehiscent fruit with the seed coat fused to the pericarp, as in the fruits of the grass family; a grain.

caudex (pl. **caudices**) Persistent, often woody base of an herbaceous perennial; compact stem or trunk of a fern.

caulescent With an obvious leafy stem above the ground.

cauline Of or relating to the stem.

cell A hollow cavity or compartment within a structure; the basic structural unit of which plant tissue is composed.

cellular Made of small cavities or compartments.

cespitose Growing in dense tufts.

chaffy Bearing thin, dry scales or bracts.

chamber A hollow space.

channel A deep groove.

chaparral A vegetation type consisting of dense thickets of evergreen shrubs.

chartaceous With a thin, papery texture, usually not green.

chasmogamous Flowers that open before fertilization and are usually cross-pollinated (compare to cleistogamous).

chlorophyll The green pigment in plants.

ciliate With a fringe of hairs along the margin.

ciliolate With a fringe of minute hairs along the margin.

cilium (pl. **cilia**) Small hair, usually along the margin.

circinate Coiled from the tip downward.

circumboreal Throughout the boreal (northern) regions of North America and Eurasia.

circumpolar Throughout the polar regions of North America and Eurasia.

cirque An amphitheater-shaped valley head.

clade A group composed of an ancestor taxon and all of its descendants (a monophyletic group).

clasp To surround, entirely or in part.

clavate Club-shaped, widening gradually towards the apex.

clavellate Of a diminutive club-shape.

claw The narrow base of some petals or sepals.

cleft Cut or split about halfway to the middle or base.

cleistogamous Flowers that self-fertilize without opening (compare to chasmogamous).

cleistogene A plant or plantlet that bears cleistogamous flowers.

cliff A steep, vertical or over-hanging face of rock or earth; see *Ecology* chapter.

climbing Growing more or less erect by leaning or twining on a support.

clone A set of genetically identical individuals that are produced by vegetative reproduction from a single parent.

coalesce Unite to form a single unit.

coat The outer covering of a structure, usually a seed.

cobbles Rocks that are between gravel and boulders in size, or about 6–25 cm in diameter.

collar The area on the outside of a grass leaf at the junction of the sheath and blade. (p. 523)

colony A group of plants, usually connected by underground organs.

column A structure formed by united filaments or by united style and filaments. (p. 517)

columnar Shaped like a column.

compound Composed of two or more like parts.

compressed Flattened.

concave Hollowed out or curved inward (compare to convex).

concolor Of uniform color.

conduplicate Folded lengthwise, with the upper surface within. (p. 513)

Leaves

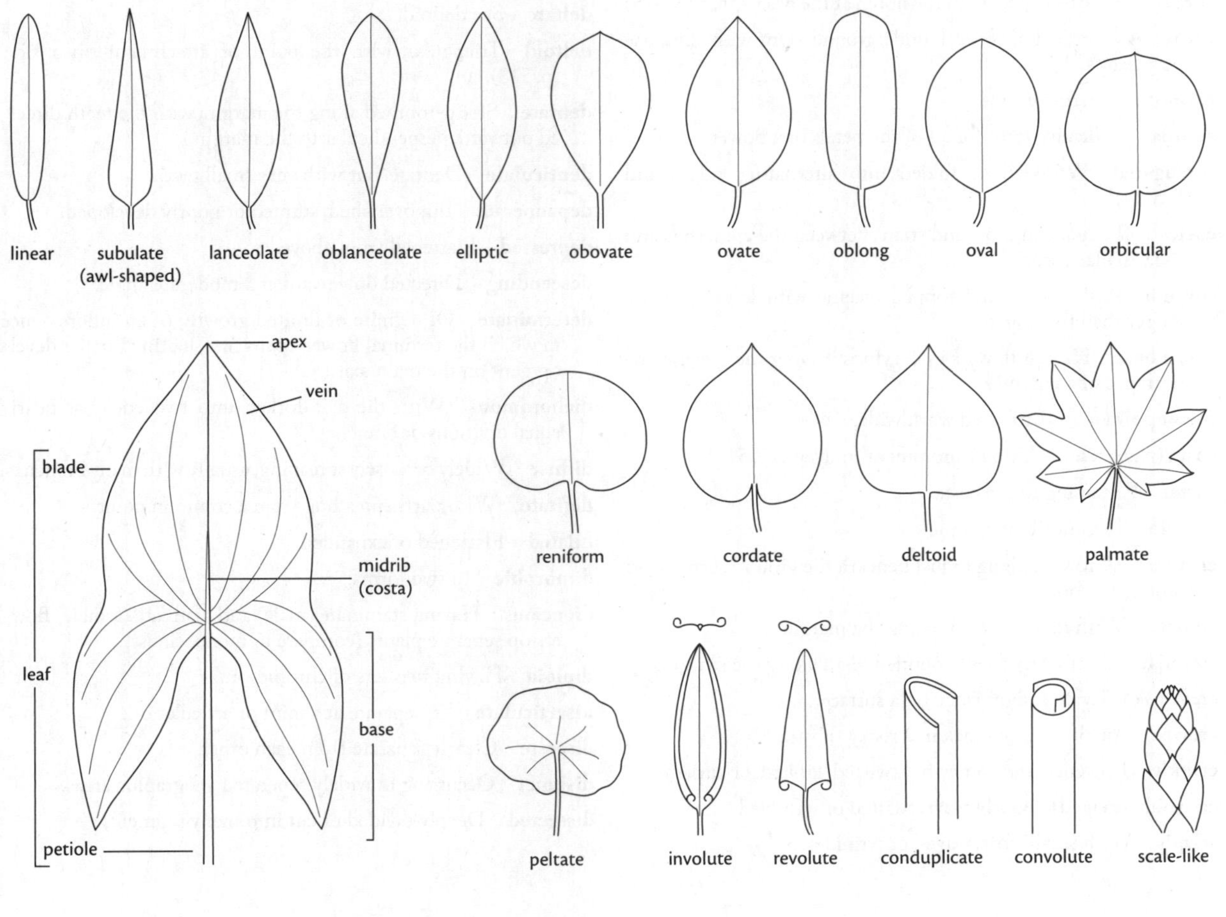

cone A dense series of spirally arranged woody bracts or sporophylls. (p. 521)

confluent Blending or coming together.

congested Densely crowded.

conic Cone-shaped, with the point of attachment at the broad end.

conifer A plant that bears cones.

connate Of like parts fused together (compare to adnate, distinct).

connective The part of the stamen that connects the two pollen sacs of the anther.

connivent Converging, but not fused or united, usually with bases separate and apices approaching each other.

conspecific Belonging to the same species.

constricted Narrowed.

contiguous Adjoining or touching but not united.

contorted Twisted or bent.

contracted Narrowed; shortened.

contractile Having the ability to contract.

contraction The process of making something or of becoming smaller or narrower.

convergent Approaching or growing towards one another.

convex Rounded and curved outward on the surface (compare to concave).

convolute Rolled together longitudinally. (p. 513)

copious Large in number or quantity; abundant.

cordate Heart-shaped, with the notch at the base. (pp. 513, 515)

corm A short, solid, vertical, underground stem with thin, papery leaves.

cormose Bearing corms.

corolla Collective term for all of the petals of a flower.

corrugated Wrinkled or folded into alternating ridges and furrows.

cortex The tissue in roots and stems between the epidermis and the vascular tissue.

corymb A flat- or round-topped raceme with lower pedicels longer than the upper.

corymbose Having flowers in corymbs; having the general appearance of a corymb.

cosmopolitan Distributed worldwide.

costa (pl. **costae**) A rib or prominent midvein. (p. 513)

costal Pertaining to the costa.

costule The midrib of a pinnule.

creeping Growing along or just beneath the ground surface and producing roots.

crenate With rounded teeth along the margin.

crenulate With very small, rounded teeth along the margin.

crest An elevated rib or ridge on a surface.

crevice An opening or crack in a rocky substrate.

crinkle Flattened and somewhat twisted, kinked, or curled.

crisp (alt. **crisped**) Curly, wavy, twisted or wrinkled.

crisply With small curls, waves, or wrinkles.

cristate Crested; with a tuft or crest at the tip.

crown The top part of a tree or shrub; the persistent base of an herbaceous perennial; a corona.

culm An above ground stem. (p. 523)

cultivar A form of plant originating under cultivation.

cuneate Wedge-shaped, narrowly triangular and attached at the narrow end; with the sides coming together at an acute angle at the base. (p. 515)

cusp A short, sharp, rigid point.

cuspidate Tipped with a short, sharp, abrupt point.

cylindric Cylinder-shaped; elongate and round in cross section.

cyme A branched, flat- or round-topped, determinate inflorescence in which the terminal (often central) flower matures first. (p. 516)

cymose With flowers arranged in a cyme.

D

deciduous Falling off; not persistent (compare to evergreen).

decumbent Lying on the ground but with the tip ascending.

decurrent Extending downward from the point of insertion. (p. 515)

deflation plain Part of a sand dune landscape immediately behind the foredune where the sand has been wind-scoured down to the summer water table level.

deflexed Bent abruptly downward.

dehisce (adj. **dehiscent**) To split open, generally at maturity.

deliquescent Having a tendency to melt or dissolve.

deltate See deltoid.

deltoid Triangular, with the point of attachment on a side. (p. 513)

dentate Sharp-toothed along the margin, with the teeth directed outward, perpendicular to the margin.

denticulate Dentate, but with very small teeth.

depauperate Impoverished, stunted or poorly developed.

depressed Flattened from above.

descending Directed downward at a moderate angle.

determinate Of definite or limited growth; of an inflorescence in which the terminal flower opens first, halting further development on the main axis.

dichotomous With the axis forked into two equal or nearly equal divisions or branches.

diffuse Widely or loosely spreading, usually with many branches.

digitate With parts spreading from a common point.

dilated Flattened or expanded.

dimorphic In two forms.

dioecious Having staminate (male) and pistillate (female) flowers on separate plants (compare to monoecious).

diploid Having two sets of chromosomes.

disarticulate To separate at a joint or articulation.

discrete Clearly separate from each other.

disjunct Occurring in widely separated geographic areas.

dissected Deeply divided or cut into many segments.

distal Toward the tip, or the end of the organ opposite the end of attachment (compare to proximal).

distended Stretched or swollen.

distichous Arranged in two vertical rows (ranks) on opposite sides of the axis.

distinct Separate, not attached to like parts (compare to connate).

diurnal Opening or occurring in the daytime.

divaricate Spreading at a very wide angle, extremely divergent.

divided Cut or lobed to the base or to the midrib.

dominant Of an organism or taxon that by its size, abundance, or coverage exerts considerable influence on a community's biotic and abiotic conditions.

dorsal Pertaining to the back or outward surface of an organ in relation to the axis, as in the lower surface of a leaf; abaxial (compare to ventral).

dorsiventral With an upper and lower surface; from top to bottom or back to front (not side to side).

drought A period of abnormally dry weather.

drupaceous Bearing drupes, resembling a drupe, or consisting of drupes.

drupe A fleshy indehiscent fruit with a stone or pit surrounding usually a single seed, as in a peach or cherry.

duff Organic matter in various stages of decomposition on the floor of a forest.

dune (adj. **dunal**) A mound or ridge of sand or other loose sediment, formed by wind.

E

echinate Bearing prickles or spines.

ecoregion A geographically defined area that has similar climate, soils and geology, and a distinctive composition of species.

ecotone A transition area between two biomes or plant communities.

elaiosome A lipid- and protein-rich structure on the seeds of some plants that is attractive to ants, which disperse the seeds.

ellipsoid Elliptic in long section or outline and circular in cross section.

elliptic Oval-shaped; of a plane with the shape of an ellipse, longer than wide and rounded at both ends, the widest part near the middle. (p. 513)

elongate Drawn out in length.

emarginate Notched at the apex. (p. 515)

embryo The new plant in a seed or archegonium.

emergent Emersed.

emersed Rising above the surface of the water.

endemic Confined to a given region.

endemism The state of being confined to a given region.

endosperm The nutritive tissue surrounding the embryo of a seed.

endosporic Of a plant gametophyte that grows within the confines of microspore and megaspore walls.

enfolding Wrapping up or enveloping.

entire With a margin that is not toothed, notched, or divided.

ephemeral Lasting a very short time.

epidermis Outermost layer of cells covering the plant, beneath the cuticle.

epipetalous Attached to the petals.

epiphyte (adj. **epiphytic**) A plant not connected to the ground, growing upon another plant but not deriving water or nutrients from it (compare to lithophyte).

episporic Pertaining to the epispore, the principal outer covering of a spore.

equilateral With sides of equal length.

equitant Conduplicate and overlapping in two ranks.

erect Upright.

erose With irregular teeth on the margin.

estuary A partially enclosed, brackish body of water forming a transition between the fresh water of a river or stream and the ocean; see *Ecology* chapter.

eutrophic Of a body of water rich in nutrients for plant growth.

evergreen Having leaves throughout the year (compare to deciduous).

excurrent Extending beyond what is typical, such as a vein that proceeds beyond the margin.

Apex and Base Shapes

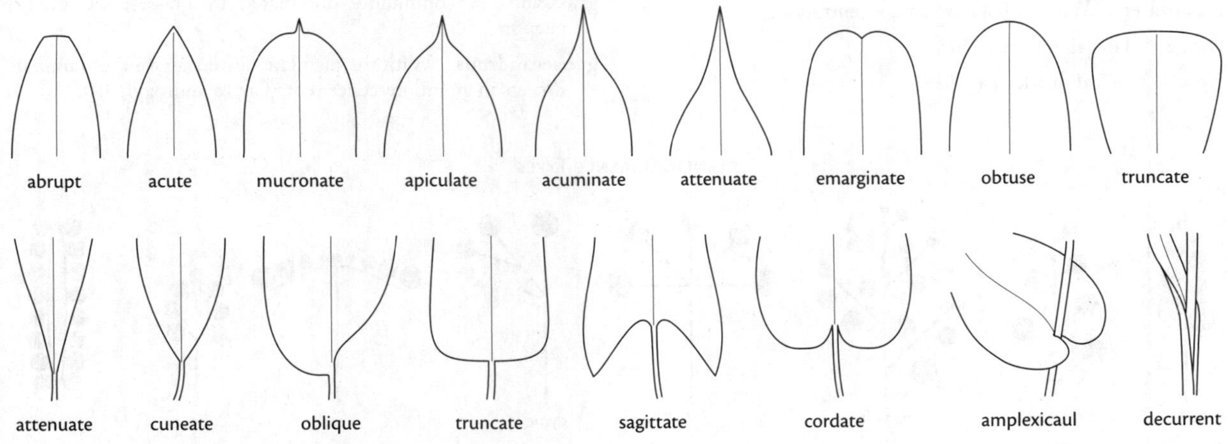

excurve To curve outward, away from the axis (compare to incurve).

exfoliate To peel off in scales or flakes.

exotic Not native; post-colonial arrivals from distant parts of North America or from other continents.

exserted Projecting out or beyond surrounding parts (compare to included).

extravaginal Outside an enclosing sheath; a type of basal branching in grasses in which new shoots grow laterally from the mother shoots, breaking through sheaths, and the very first leaves are bladeless sheaths.

F

falcate Strongly curved, sickle-like.

family A taxonomic grouping of closely related genera.

fascicle A close cluster or bundle. (p. 521)

fen A wetland fed by groundwater and dominated by mosses, herbs, and shrubs, with a pH generally higher than 5.5; see *Ecology* chapter.

fertile Producing seeds, spores, or pollen capable of germination; of a plant with reproductive organs.

fibrillose Bearing thread-like structures.

filament Any thread-like body, especially the stalk of stamens. (p. 517)

filiform Thread-like.

fimbriate Fringed, with slender processes along the margin.

fissured Cracked or fractured.

flaccid Limp, not rigid or turgid.

flag leaf The uppermost leaf on a grass stem.

flange A projecting rim or edge.

flexuous With curves, bends, or zigzags.

floodplain A low-lying area near a river or stream subject to flooding.

floret A small flower. (p. 523)

fluvial Relating to a stream or river.

follicle A dry, dehiscent fruit of one carpel opening along one side.

forb A non-grass-like herb.

foredune A dune ridge that runs parallel to the shore.

free Not fused or united.

freshwater Water with a low salt concentration.

fringe Hairs along a margin.

frond The leaf of a fern. (p. 519)

fruit A mature ovary along with any other structures that are attached and ripen with it as a unit.

funnelform Funnel-shaped.

furrow A groove or depression in a surface.

fused Joined and growing together.

fusiform Spindle-shaped, widest in the middle and tapering at both ends.

G

gamete A male or female sexual cell that combines with another at fertilization to form a zygote; in pteridophytes, gametes are produced on the prothallus (gametophyte) by antheridia (male) and archegonia (female).

gametophyte The sexual or haploid stage in the life cycle of land plants and algae producing the ova and sperm; in pteridophytes it is a separate, inconspicuous, and short-lived generation.

gemma (pl. **gemmae**) A vegetative bud used for propagation and sometimes dispersal.

gemmiferous Producing gemmae.

geniculate With a knee-like bend.

genus (pl. **genera**) A taxonomic grouping of closely related species.

gibbous Somewhat swollen on one side.

glabrate Nearly glabrous.

glabrescent Becoming glabrous; almost glabrous.

glabrous Naked, lacking hairs or scales.

glacier A persistent body of ice on land that is constantly in motion.

gland A structure that secretes oily or sticky substances.

glandular Bearing glands.

glaucous With a distinct bluish, waxy bloom or powder on the surface.

globose Almost spherical in shape.

globular Globe-shaped.

glomerule A head or dense cluster of flowers.

glume One of the paired bracts at the base of a grass spikelet. (p. 523)

graduated Arranged in a series.

granular With a sand-like texture.

grassland A community dominated by grasses; see *Ecology* chapter.

gynaecandrous With the pistillate flowers above the staminate flowers in an inflorescence (compare to androgynous).

Inflorescence Types

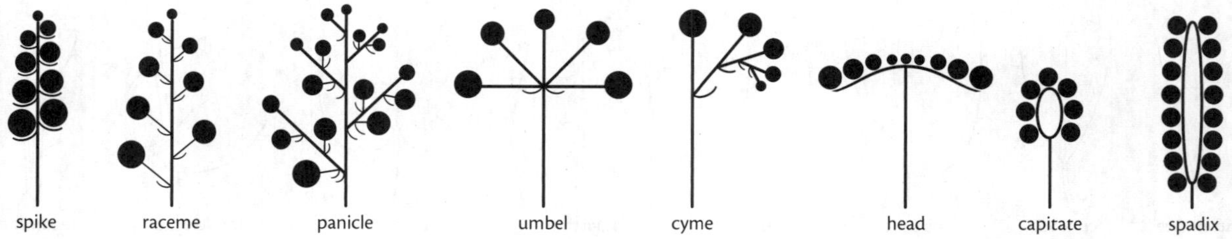

gynodioecious Having pistillate flowers on one plant and bisexual flowers on another plant.

gynomonoecious Having both pistillate and bisexual flowers on the same plant.

H

habit The general appearance or characteristic mode of growth of a plant.

habitat The kind of place in which a plant grows.

hair A fine epidermal outgrowth, usually elongate.

hastate Arrowhead-shaped with the basal lobes pointed, narrow, and spreading at a wide angle.

head An inflorescence of densely clustered, sessile, or nearly sessile flowers. (p. 516)

heath A shrubland characterized by open, low-growing woody vegetation.

hemispheric In the shape of half of a sphere.

herb A vascular plant which is not woody.

heterosporous Bearing spores of distinctly different types.

heterotrophic Dependent on external sources for organic nutrients.

hexagonal With six sides.

hirsute With distinct hairs, often rather stiff or bristly.

hispid With stiff, spreading hairs or bristles.

hispidulous With very small, stiff, spreading hairs or bristles.

homogeneous Of similar characteristics or morphology; similar throughout.

hood A hollow arched covering; the arched upper petals or sepals in some species.

hummock A hill or mound.

humus The dark-colored, organic material in soil derived from the decomposition of biomass.

hyaline Thin, membranous, and translucent or transparent.

hybrid The progeny from a cross between different taxa (genera, species, or subspecific taxa).

hydathode A water-secreting gland, usually on the margin of a leaf at the end of a vein.

hypogynous Having floral parts (stamens, petals, or sepals) borne on the receptacle beneath the pistil and free of the ovary.

I

igneous Of rock from lava or magma.

imbricate With overlapping edges.

immersed Embedded in another organ; under water.

impressed Sunk below the surface.

incised Cut sharply and deeply.

incision A sharp, deep cut.

included Enclosed, not protruding (compare to exserted).

incomplete Lacking an expected part.

incurve To curve inward (compare to excurve).

indehiscent Not opening at maturity along definite lines or by pores.

indeterminate An inflorescence in which the lower flowers bloom first, permitting indefinite growth of the main axis.

indumentum The covering of a plant surface, especially hairs and scales.

indusium (pl. **indusia**) A thin, epidermal outgrowth of a fern leaf covering a sorus. (p. 519)

inferior Situated below another organ or part; an inferior ovary has other floral parts attached above its point of attachment (compare to superior).

inflated Swollen, expanded, or distended.

inflorescence The flowering part of the plant. (pp. 516, 523)

inframarginal Inside of the margin.

infraspecific Of taxonomic ranks below species, such as subspecies and variety.

innovation A basal offset from the base of a stem.

inrolled Rolled inwards; involute or revolute.

intercostal Between the ribs or nerves.

intergrade To merge into one another gradually.

internode Portion of stem between two nodes.

interrupted Discontinuous.

intertidal Of the area of seashore that is covered with water at high tide and exposed at low tide.

intramarginal Within the margins.

intravaginal Branching in which the young shoot grows up inside the leaf sheath, emerging at the sheath mouth.

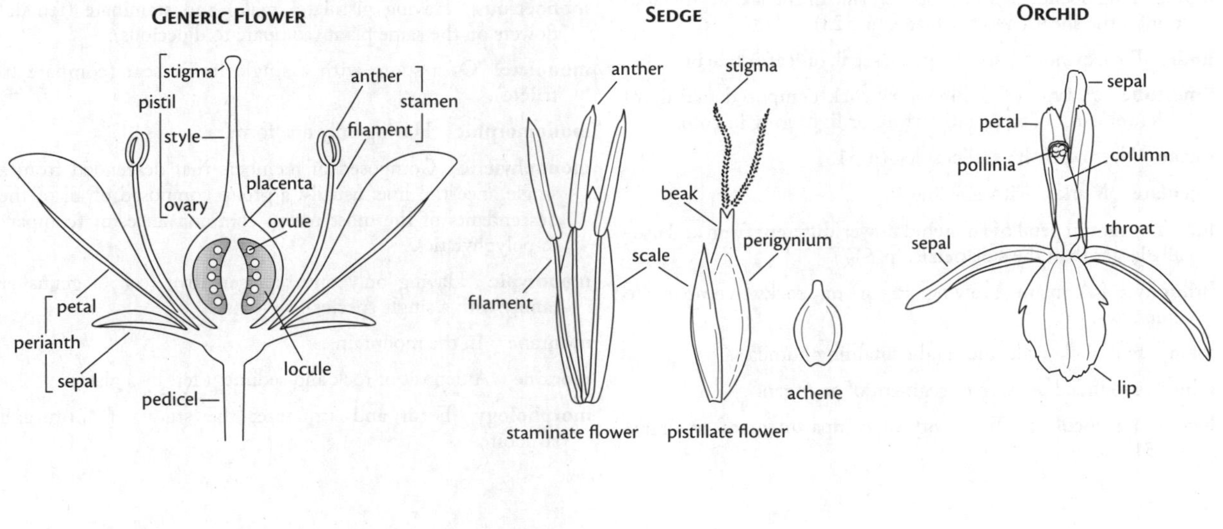

introgression The modification of species by the repeated backcrossing of hybrids with the parent species.

involucrate With an involucre.

involucre One or more whorls of bracts subtending an inflorescence.

involute Rolled inwards or to the adaxial (upper) side (compare to revolute). (p. 513)

irregular Not similar in size or arrangement.

isodiametric Nearly spherical.

J

joint A node; the section of the stem from which a leaf or branch arises.

jointed Having one or more nodes or points of articulation.

K

keel A ridge, usually on the back, like the keel of a boat.

L

lacerate With an irregular, ragged margin.

laciniate Irregularly divided into very narrow, pointed segments.

lacuna (pl. **lacunae**) A gap or space within a tissue.

lamina Expanded portion of an organ such as a leaf or petal; a thin, flat part.

lanceolate Lance-shaped, several times longer than wide, widest below the middle and tapering toward the ends, applied to flat structures. (p. 513)

lanceoloid Lance-shaped; lanceolate in long section or outline and circular in cross section, thickest below the middle and tapering toward the ends, applied to three-dimensional structures.

lateral To the side.

lax Loosely arranged; open and spreading.

leaf An expanded, usually photosynthetic organ of a plant. (pp. 513, 521)

leaflet A segment of a compound leaf.

lemma The lower of the two bracts that subtend a grass floret. (p. 523)

lenticular Shaped like a biconvex lens.

ligulate With ligules.

ligule An appendage on the adaxial side of the leaf at the junction of the sheath and the blade. (p. 523)

limb The expanded part of a petal, sepal, or leaf; a tree branch.

limestone A type of sedimentary rock composed mainly of calcium carbonate, usually white or light gray in color.

linear Narrow, with parallel sides. (p. 513)

lineolate Marked with fine lines.

lip The central petal of an orchid flower, different from and usually larger than the side petals. (p. 517)

lithophyte A plant growing in or on rocks (compare to epiphyte).

loam Soil of clay, silt, and sand containing humus.

lobe A rounded division or segment of an organ.

locule (adj. **locular**) The cavity or compartment of an organ. (p. 517)

loculicidal Dehiscing through the locules, usually near the center of the locule on the ovary wall (compare to septicidal).

lodicule One of the (usually paired) small scales at the base of the ovary in grasses.

lowland Low elevation areas; not montane.

lunate Crescent-shaped.

M

mammillate Nipple-like.

margin The edge.

marsh A wetland that is dominated by herbaceous vegetation and flooded most or all of the year; see *Ecology* chapter (compare to swamp).

meadow A vegetation type dominated by grasses and forbs; see *Ecology* chapter.

medial Of the middle.

megasporangium (pl. **megasporangia**) The structure producing female spores.

megaspore A female spore that gives rise to a female gametophyte.

megasporocarp A walled body enclosing one or more megasporangia.

membrane A thin, flexible, often translucent piece of tissue.

mentum A chin-like or spur-shaped projection that is formed by the sepals and base of the column in some orchids.

meristem Actively dividing tissue at the growing tips of roots and shoots.

mesic Moist (compare to xeric).

microsporagium (pl. **microsporangia**) A structure that produces microspores.

microspore A male spore that grows into a male gametophyte.

microsporocarp A walled body enclosing one or more microsporangia.

midrib The central vein of a leaf or other organ. (p. 513)

midvein The main or central vein of a leaf.

moniliform Cylindrical and constricted at regular intervals, giving a beaded necklace-like appearance.

monochromatic Of one color.

monocotyledon (alt. **monocot**) A flowering plant with a single cotyledon (seed leaf) in the embryo.

monoecious Having pistillate (male) and staminate (female) flowers on the same plant (compare to dioecious).

monolete Of a spore with a single linear scar (compare to trilete).

monomorphic Having only one form.

monophyletic Composed of members that descended from a single ancestral line; usually, a group composed of all of the descendants of the most recent common ancestor (compare to polyphyletic).

monotypic Having only one representative (e.g., a genus or family with a single species).

montane In the mountains.

moraine A deposit of rock and sediment left by a glacier.

morphology Form and structure; the study of form and structure.

motile Capable of self-propelled movement.

mucilaginous Slimy.

mucro A short, sharp tip.

mucronate With a mucro. (p. 515)

mucronulate Tipped with a small, sharp point.

muriculate With fine, small, sharp projections.

mycoheterotroph A plant that gets food from a fungus rather than photosynthesis, either in whole or in part.

mycorrhizal Pertaining to the symbiotic relationship between a fungus and the root system of a vascular plant.

mycotrophic Having a symbiotic or parasitic relationship with a fungus, obtaining some or all of its water or food from fungi.

N

naked Lacking appendages.

nectary Gland that secretes nectar, a sweet fluid, commonly found in insect-pollinated flowers.

needle-like Linear and usually stiff. (p. 521)

nerve A strand of vascular and supporting tissue.

nocturnal Opening or occurring at night.

node A point on the stem where leaves or branches are inserted.

nodulose With minute knobs or nodules.

nut Dry fruit with a hard exterior that typically contains a single seed.

nutlet A small, dry fruit.

O

obconic Cone-shaped, with the narrow end at the point of attachment.

obcordate Heart-shaped, with the pointed end at the point of attachment.

oblanceolate Inversely lanceolate, with attachment at narrow end. (p. 513)

oblanceoloid Oblanceolate in outline or long section and circular in cross section.

oblique With unequal sides; slanting. (p. 515)

oblong 2 to 4 times longer than broad with the sides mostly parallel. (p. 513)

obovate Of a flat structure that is egg-shaped in outline and attached at the narrower end. (p. 513)

obovoid Egg-shaped, with the point of attachment at the narrower end.

obpyriform Pear-shaped, with the narrower end at the point of attachment.

obtuse Blunt or rounded, with the sides coming together at an angle wider than 90°. (p. 515)

opposite Having two leaves attached at each node, directly across from each other; borne on the same radius as another part (compare to alternate).

orbicular Rounded or circular in outline. (p. 513)

orbiculate Orbicular.

outcrop A rock formation that is visible on the surface; see *Ecology* chapter.

oval Broadly elliptic, narrowing somewhat from the middle to rounded ends. (p. 513)

ovary The basal portion of the pistil that develops into fruit after fertilization and contains the seeds. (p. 517)

ovate Of a flat structure with an egg-shaped outline, attached at the broad end. (p. 513)

ovoid Egg-shaped, with the point of attachment at the broad end.

ovule An immature seed. (p. 517)

P

palea The uppermost bract subtending a grass floret. (p. 523)

palmate Radiating from a common point, like the fingers of a hand. (p. 513)

panicle A branched inflorescence maturing from the bottom upwards. (p. 516)

papilla (pl. **papillae**) A small, rounded, superficial projection.

papillate With papillae.

parietal Pertaining to the wall of an organ.

Fern

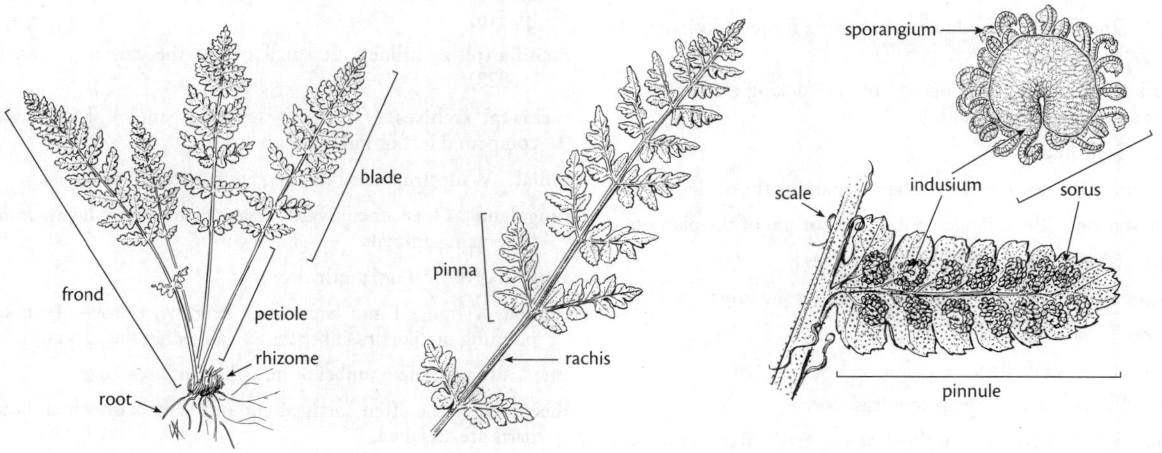

peat Soft, soil-like, acidic material made of partially decomposed plant matter.

pedicel The stalk of an individual flower in an inflorescence. (p. 517)

pedicellate With a pedicel.

peduncle Stalk of a solitary flower or an inflorescence.

pedunculate Having a peduncle.

peltate Shield-shaped; of leaves, with the stalk attached on the lower surface, not at the margin. (p. 513)

pendent Drooping or hanging down.

pendulous See pendent.

pentagonal With five sides.

perennial With a life cycle lasting two or more years (compare to annual, biennial).

perfect With both male and female parts in the same flower.

perianth The sepals and petals of a flower, collectively. (p. 517)

pericarp The wall of a fruit.

peridotite Coarse-grained igneous rock.

perigynium (pl. **perigynia**) A specialized bract that encloses the pistillate flower of *Carex*. (p. 517)

petal An individual segment of a corolla, often the showy part of a flower. (p. 517)

petaloid Petal-like.

petiolate Having a petiole.

petiole Stalk at base of leaf blade that attaches to the stem. (pp. 513, 519)

petiolule The stalk of a leaflet.

phyllopodic With the lowest leaves well developed.

phylogeny The development or evolutionary history of a group of organisms and the relationships among its members.

pilose With long, soft, straight, distinct hairs.

pinna (pl. **pinnae**) The leaflet of a pinnate leaf, or the primary division of a bi- or tripinnate leaf. (p. 519)

pinnate Resembling a feather, as in a compound leaf with leaflets arranged on opposite sides.

pinnatifid Pinnately lobed half the distance, or more, to the midrib (but not reaching the midrib).

pinnule The second or third order divisions of a bi- or tripinnate leaf. (p. 519)

pistil The female reproductive organ of a flower, consisting of a stigma, style, and ovary. (p. 517)

pistillate Having pistils or a pistil and lacking stamens (compare to staminate). (p. 517)

pith Cork-like tissue.

placenta The portion of the ovary containing the ovules. (p. 517)

placentation The arrangement of the portion of the placenta.

plains An area of flat land with few trees.

plano-convex Flat on one side, convex on the other.

plicate Folded into pleats.

plumose Feather-like.

pod A dry fruit that opens to release seeds.

pollen The microspores or the dust-like fertilizing cells of gymnosperms and angiosperms.

pollinate To transfer pollen from anther to stigma.

pollinium (pl. **pollinia**) A waxy cluster of pollen grains transported as a unit during pollination. (p. 517)

polygonal With three or more sides, but typically used when there are five or more sides.

polymorphic Existing in many forms.

polyphyletic Composed of members that descended independently from two or more ancestral lines (compare to monophyletic).

polyploid With more than two sets of the basic chromosome number.

prairie A large open area of grassland; see *Ecology* chapter.

prickle A sharp outgrowth of epidermis or bark. (p. 521)

primary The first division of a leaf or inflorescence.

process A projecting appendage.

prostrate Lying flat along the ground.

protuberance A rounded swelling or bump on the surface.

proximal Toward the base or attached end of an organ (compare to distal).

pruinose With a waxy, powdery coating.

pseudolateral Terminal, but appearing lateral.

pteridophyte A fern or fern ally; vascular plants reproducing by means of spores.

puberulent Minutely pubescent.

pubescence Hairiness; with short, soft hairs.

pubescent Covered with hairs, especially short, soft hairs.

pulvinus (pl. **pulvini**) A swelling or enlargement at the base of the petiole or inflorescence branch.

pumice Volcanic rock with a rough surface that is full of cavities.

pustulose With small blisters.

pyramidal Pyramid-shaped.

Q

quadrangular Four-angled.

quadrate Square or rectangular.

R

raceme An inflorescence with pedicellate flowers on an elongate, unbranched axis maturing from the bottom upwards. (p. 516)

rachilla (pl. **rachillae**) A small rachis; the axis of a spikelet. (p. 523)

rachis (pl. **rachises**) The axis (excluding petiole) of a pinnately compound leaf or inflorescence. (p. 519)

radial With structures radiating from a central point.

rangeland Open, uncultivated areas capable of producing forage for grazing animals.

rank A (vertical or lengthwise) row.

ranked Arranged into (vertical or lengthwise) rows; 1-ranked: pointing up like brush bristles; 2-ranked: in flat sprays.

ray A branch of an umbel or head-like inflorescence.

receptacle The often enlarged end of the stalk on which flower parts are attached.

recurve To curve backward (compare to excurve, incurve).

reflex To bend backward.

regular Uniform or symmetrical in shape and structure.

reniform Kidney-shaped. (p. 513)

resin A thick, sticky liquid that hardens into a solid.

reticulate In the form of a network; net-veined.

reticulum A network formed by repeated branching and anastamosing, often pertaining to veins in leaves.

retrorse Directed downwards or backwards, away from the apex (compare to antrorse).

retuse A rounded apex with a small notch.

revolute With margins rolled downwards or to the abaxial side (compare involute). (p. 513)

rhizomatous With rhizomes.

rhizome A modified underground stem, usually growing horizontally. (pp. 519, 523)

rhombic Diamond-shaped.

riparian Relating to areas adjacent to rivers and streams; see *Ecology* chapter.

root The typically underground extension of the plant axis that tends to grow downwards and is without leaves or nodes. (p. 519)

rootstock A rhizome; a swollen, very short, erect, underground stem.

rosette A group of organs radiating from the center, usually leaves at ground level.

rotate Flat and circular or disc-shaped.

rudimentary Not fully developed.

rugose Wrinkled.

rugulate See rugulose.

rugulose Slightly or finely wrinkled.

S

sac A pouch-like structure.

saccate Sac-shaped; with a sac.

sagittate Arrowhead-shaped. (p. 515)

saline Salty.

salverform With a slender tube and an abruptly spreading, flattened limb.

savanna (alt. **savannah**) A grassland region with scattered trees.

scaber A rough, minute, hard projection.

scaberulous Slightly rough to the touch due to minute, hard projections or to the presence of short, stiff hairs; diminutive of scabrous.

scabland Barren, rocky plains with little or no soil.

scabrid Roughened.

scabridulous Minutely roughened.

scabrous Rough to the touch due to minute, hard projections or to the presence of short, stiff hairs.

scabrule Small scaber.

scale A small, often leaf-like organ, often dry and membranous. (pp. 519, 521)

scape A leafless flowering stalk rising from ground level.

scapose With flowers on a scape; scape-like.

scar A mark left by the separation of two structures, as a leaf scar on a stem or on a seed after detachment from the placenta. (p. 521)

scarious Thin, dry, and translucent.

schizocarp A dry, indehiscent fruit that separates into one-seeded pieces at maturity.

sclerenchyma Supporting or protective tissue composed of thickened, dry, and hardened cells.

scorpioid Coiled like the tail of a scorpion.

scour Areas eroded by running water; to erode with running water.

scree An accumulation of loose stones or rock, typically at the base of a slope or cliff.

scrub A plant community dominated by shrubs or stunted trees.

sculptured With smooth curves.

scurfy Covered with small, coarse scales.

sediment Solid material created by erosion that is moved and deposited in a new location.

seed A ripened ovule. (p. 521)

CONIFER

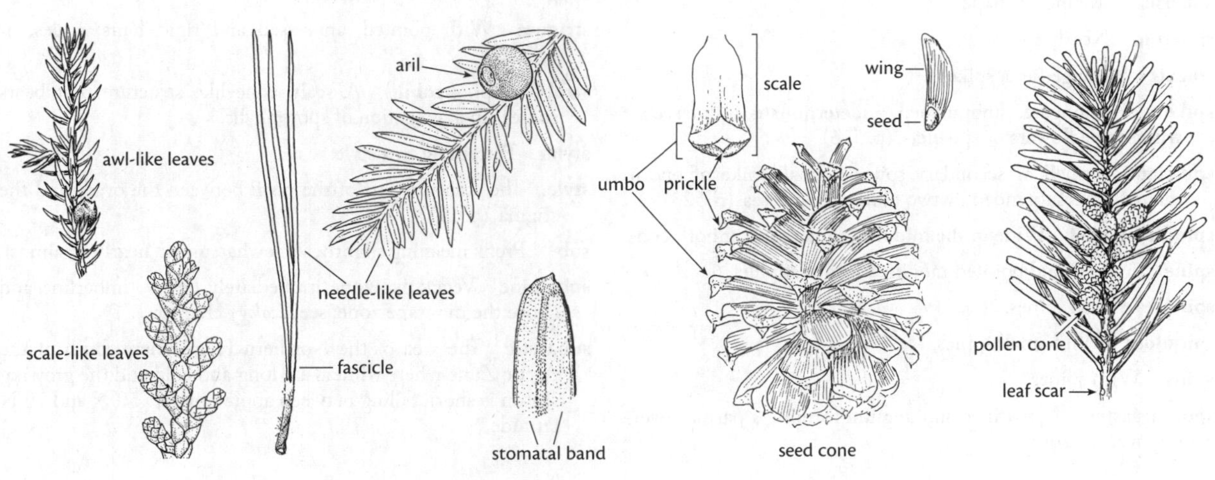

seep (alt. **seepage**) A place where ground water oozes up to the surface; see *Ecology* chapter.

segment Each free part of a divided whole.

sepal The outermost whorl of parts in a flower, often green and leaf-like. (p. 517)

septal Pertaining to a septum.

septate Internally divided by one or more partitions.

septicidal Dehiscing through the septa (compare to loculicidal).

septum (pl. **septa**) A partition or cross-wall.

serpentine Usually describes soil containing serpentinite, a dark green or brown mineral which inhibits the growth of most plants.

serrate With sharp teeth along the margin pointing toward the apex.

serrulate Serrate, but with very small teeth.

sessile Without a stalk.

setaceous Bristle-like.

sheath A tubular structure surrounding an organ or part.

sheathe To form a sheath.

shoot A young branch or stem.

shrub A woody plant, typically branched from the base.

silicified Impregnated with silica.

silky Covered with very fine, more or less appressed, lustrous hairs.

simple Composed of a single piece, not divided or branched.

sinuous Shallowly wavy, with many curves.

sinus The gap or recess between two lobes or segments.

slough Soft, muddy ground; backwater.

solitary Borne singly.

sorus (pl. **sori**) A group or arrangement of sporangia. (p. 519)

spadix (pl. **spadices**) A spike with small, sessile flowers crowded along a fleshy axis. (p. 516)

spathe A large bract subtending and sometimes enclosing an inflorescence.

spatulate (alt. **spathulate**) Spatula-like in shape, rounded at the tip and gradually tapering to the base.

species The fundamental unit of taxonomy; a group of plants of the same morphology, capable of interbreeding.

spherical Round in shape.

spheroid Nearly spherical.

spicate Arranged in a spike.

spike An elongate, unbranched, indeterminate inflorescence with sessile flowers or spikelets. (p. 516)

spikelet A small or secondary spike; a small spike of one to many florets subtended by two glumes in grasses. (p. 523)

spindle-shaped Wide in the middle and tapering at both ends.

spine A sharp and pointed modified leaf or stipule.

spinose With spines.

spinulose With small spines.

spiny With spines.

sporangiaster A modified sporangium forming a partial covering over the sorus.

sporangium (pl. **sporangia**) A case or body that contains the spores. (p. 519)

spore A vegetative, reproductive cell; the first cell of the gametophyte generation.

sporocarp A walled body enclosing one or more sporangia.

sporophore Fertile part of the leaf within some fern species.

sporophyll A leaf-like structure bearing one or more sporangia.

spray A slender shoot or branch with leaves, flowers, or fruits.

spreading Nearly horizontal.

spur A hollow, sac-like projection.

stalk A secondary stem, usually the supporting structure of an organ like a leaf, flower, or fruit.

stamen Male organ of the flower that bears pollen, composed of a stalk (filament) and pollen sacs (anthers). (p. 517)

staminate Having only stamens and lacking pistils (compare to pistillate). (p. 517)

staminode (pl. **staminodia**) A modified stamen that produces no pollen.

stem Central support of a plant.

steppe A large, semi-arid, treeless area of flat land; see *Ecology* chapter.

sterile Not producing seeds, pollen, or spores.

stigma (adj. **stigmatic**) The apical part of the pistil that is receptive to pollen. (p. 517)

stipe A stalk or support.

stipitate With a stipe.

stipular Pertaining to the stipule.

stipule One of a pair of leaf- or scale-like appendages inserted at or near the base of a petiole.

stolon A horizontal, elongate stem along the ground that can produce roots, stems, and leaves.

stoloniferous Bearing stolons.

stomata Small pores or apertures on the epidermis, allowing gas exchange between the tissue and outside air.

stomatal Relating to the stomata. (p. 521)

stramineous Straw-colored.

strap Straight, flat, and narrow.

striate With fine parallel lines or grooves.

strigillose Minutely strigose.

strigose With pointed, appressed and rigid hairs, scales, or bristles.

strobilus (pl. **strobili**) A scaly, cone-like structure that bears spores; an aggregation of sporophylls.

stylar Pertaining to a style.

style The elongate part of the pistil between the ovary and the stigma. (p. 517)

sub- Prefix meaning slightly, somewhat, under, nearly, or almost.

subalpine Vegetative zone immediately below timberline and above the montane zone; see *Ecology* chapter.

subarctic The area of the Northern Hemisphere south of the Arctic Zone where winters are long and cold and the growing season is short, falling between approximately 50°N and 70°N latitude.

submarginal Near the margin.

submerged Submersed.

submersed Under water.

subspecies A grouping within a species above the level of variety; sometimes a geographically isolated variant of a species.

subtend To be immediately below.

subterranean Underground.

subtidal Area below the low tide mark but still shallow and close to shore, usually submerged but exposed briefly during extreme low tide events.

subulate Awl-shaped, narrow and gradually tapering to a fine point. (p. 513)

subumbellate Nearly umbellate.

succulent Juicy or fleshy.

suffused Tinted.

sulcate With longitudinal grooves or furrows.

superior Situated above another organ or part; a superior ovary has other floral parts attached below its point of attachment (compare to inferior).

suture A seam or line formed at the junction of two margins.

swale A low tract of land that is usually moist or marshy.

swamp A wetland that has some woody vegetation and is flooded most or all of the year; see *Ecology* chapter (compare to marsh).

symbiotic Pertaining to a close relationship between species in which each species benefits.

sympatric Of two or more species occupying the same geographic area.

sympodial A growth axis that develops from a series of short lateral branches without a true central axis.

synandrium (pl. **synandria**) An androecium in which the anthers have been fused.

T

talus An accumulation of rock debris forming a slope, usually at the base of a cliff; see *Ecology* chapter.

tawny Tan in color.

taxon (pl. **taxa**) A unit of biological classification, such as a particular species, genus, or family.

temperate The area between the tropics and the Arctic in the Northern Hemisphere, and tropics and the Antarctic in the Southern Hemisphere.

tepal A segment of a perianth in which the sepals and petals are not well differentiated.

terete Cylindrical, round in cross section.

terminal At the end of a stem or axis.

ternate Arranged or divided into threes.

terrestrial Growing on the ground.

tertiary Third in order or level.

testa The seed coat.

tetrad A group of four.

tetrahedral Having four sides.

tetraploid Having four sets of chromosomes.

thalloid Resembling or pertaining to a thallus.

thallus (pl. **thalli**) A plant body not differentiated into leaves and stem.

throat The upper region of the grass leaf sheath, where it begins to spread apart and expose the stem; in orchids, the expanded portion of the corolla at the base of the column. (pp. 517, 523)

tiller A basal or subterranean shoot which is more or less erect.

timberline Elevation above which conditions prevent tree growth.

tomentose With a covering of short, matted or tangled, soft, woolly hairs.

toothed With teeth; with small points or lobes along the margin.

Grass

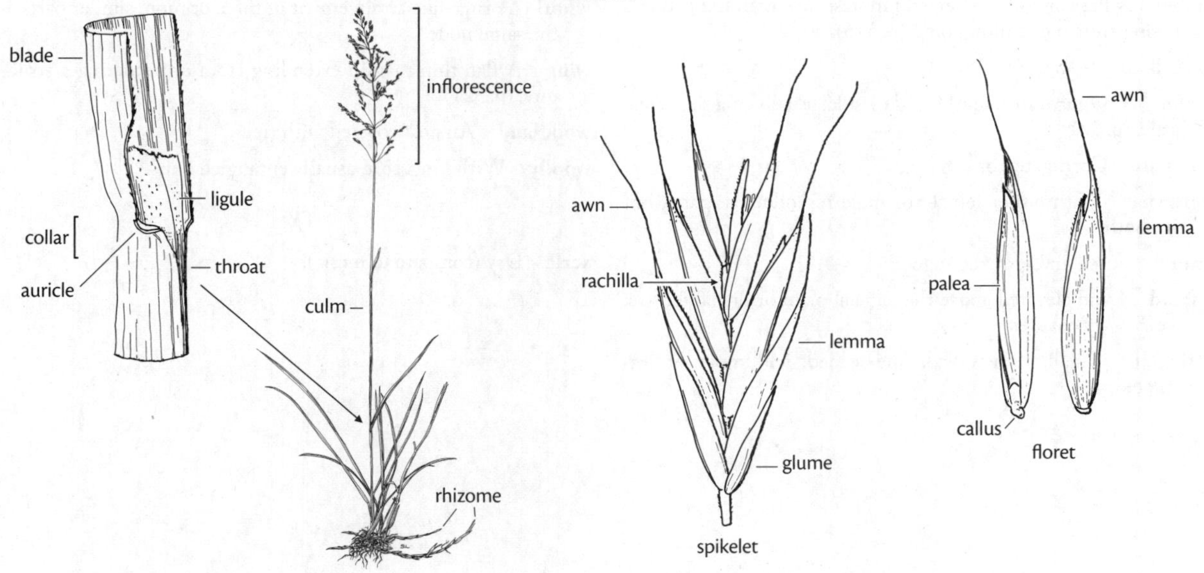

trailing Growing along the ground but not rooting.

translucent Letting light through, almost transparent.

transverse Lying across, set crosswise.

tree A large, woody plant with usually one main stem (trunk).

tribe A taxonomic grouping of closely related genera, a division of the family below the level of subfamily.

trifid Divided into three equal parts.

trigonous With three angles, triangular in cross-section.

trilete Of a spore with a radial, three-armed scar (compare to monolete).

triploid Having three sets of chromosomes.

trophophore Sterile part of the leaf within some fern species.

truncate With the base or apex flattened abruptly as though cut off. (p. 515)

trunk The erect, unbranched portion of a tree-like plant.

tuber A swollen, usually underground, part of a stem or root.

tubercle A small, wart-like swelling or projection.

tuberculate Bearing tubercles.

tuberous Producing tubers; tuber-like.

tuff A sedimentary rock formed from an accumulation of volcanic ash or dust and igneous rock.

tumid Swollen.

tunic Outer coating of a seed or bulb.

tunicate With sheathing, concentric layers; as in an onion bulb.

turbinate Top-shaped.

turgid Swollen, rigid and stiff by means of internal pressure.

turion A small, dormant shoot capable of overwintering and sometimes serving as a vegetative propagule.

tussock A tuft of grasses or sedges.

U

ultimate The final section or division of a structure.

ultramafic Substrates derived from igneous rock containing high amounts of magnesium and iron.

umbel A flat- or convex-topped inflorescence with the pedicels arising from a common point. (p. 516)

umbellate In umbels.

umbo A protuberance on the end or side, usually of a seed cone scale. (p. 521)

undulate Corrugated or wavy.

unisexual With either female or male reproductive parts, but not both.

united Fused and growing together.

upland Land elevated above the alluvial plain or fluvial terrace; see *Ecology* chapter.

utriculate Small, thin-walled, one-seeded, and more or less inflated.

V

valve One of the segments of a dehiscent fruit that separates from the other segments at maturity to effect dehiscence.

variegated Irregularly colored in patches.

variety A grouping within a species, below the level of subspecies, with a recognizable appearance but not usually genetically isolated from other varieties.

vascular Of tissue containing xylem and phloem, the conducting elements of the plant.

vegetative Without floral parts and/or sex organs.

vein Vessel by which water and nutrients are transported; often most easily seen in leaves. (p. 513)

velum The membranous flap of tissue partially covering the sporangium.

venation The arrangement of the veins, especially in leaves.

ventral On or pertaining to the upper or inward surface in relation to the axis; adaxial (compare to dorsal).

vernal In the spring; see *Ecology* chapter (vernal pool).

verrucose Warty.

verticil An arrangement of three or more similar parts at a single node; a whorl.

verticillate Arranged in whorls or verticils.

vespertine Opening or occurring in the evening.

vestigial Reduced to a rudiment and not functional.

villous Having long, soft hairs.

vine Trailing or climbing plant with a long, flexible stem, often supporting itself using tendrils.

viviparous Germinating or sprouting from a seed or bud while attached to the parent plant.

W

waif An unusual species found in the wild that is not native and persists only one or a few generations.

wetland An area of land where the seasonal or constant saturation with water determines the nature of plant and animal communities living there; see *Ecology* chapter.

whorl A ring-like arrangement of three or more similar parts at the same node.

wing A flat, thin margin extending from or bordering a structure. (p. 521)

woodland An area covered with trees.

woolly With long, soft, usually entangled hairs.

X

xeric Dry (compare to mesic).

REFERENCES

Abrams, L. & R.S. Ferris. 1923–1960. Illustrated flora of the Pacific states. Vols. 1–4. Stanford University Press, Stanford, CA.

Archibald, J.K., S.R. Kephart, K.E. Theiss, A.L. Petrosky & T.M. Culley. 2015. Multilocus phylogenetic inference in subfamily Chlorogaloideae and related genera of Agavaceae–informing questions in taxonomy at multiple ranks. Molecular Phylogenetics and Evolution 84:266–283.

Bailey, R.G. 1995. Description of the ecoregions of the United States. United States Forest Service, Washington, DC.

Baldwin, B.G., D.H. Goldman, D.J. Keil, R. Patterson, T.J. Rosatti & D.H. Wilken, eds. 2012. The Jepson manual: Vascular plants of California, 2nd ed. University of California Press, Berkeley, CA.

Ball, P.W. & D.E. Wujek. 2002. *Eriophorum*. In: Flora of North America Editorial Committee, eds. 1993+. Flora of North America north of Mexico. 18+ vols. New York and Oxford. Vol 23.

Ballantyne, O.H. 1983. Lilies of the Pacific coast. The Lily Yearbook of the North American Lily Society 36:11–26.

Balslev, H. 1996. Juncaceae. Flora Neotropica Monograph 68:1–168. New York Botanical Garden, Bronx, NY.

Barbour, M.G. & N.L. Christensen. 1993. Vegetation. In: Flora of North America Editorial Committee, eds. 1993+. Flora of North America north of Mexico. 18+ vols. New York and Oxford. Vol 1.

Beane, L. & A.L. Vollmer. 1955. Hunting lilies in California. The Royal Horticultural Society lily year-book 1955. Royal Horticultural Society, London.

Becking, R.W. 1986. *Hastingsia atropurpurea* (Liliaceae: Asphodeleae), a new species from southwestern Oregon. Madroño 33:175–181.

Binney, E.P. & G.E. Bradfield. 2000. An initial comparison of growth rates in the rare grass *Achnatherum hendersonii* and its common associate *Poa secunda.* Ecological Research 15:181.

Bishop, E.M. 2003. In search of ancient Oregon: A geological and natural history. Timber Press, Portland, OR.

Bogler, D.J., J.C. Pires & J. Francisco-Ortega. 2006. Phylogeny of Agavaceae based on *ndh*F, *rbc*L, and ITS sequences: Implications of molecular data for classification. Aliso 22:313–328.

Chambers, K.L. 1973. Floristic relationships of Onion Peak with Saddle Mountain, Clatsop County, Oregon. Madroño 22:105–160.

Christenhusz, M.J. & M.W. Chase. 2014. Trends and concepts in fern classification. Annals of Botany 113:571–594.

Christy, J.A. & E.R. Alverson. 2011. Historical vegetation of the Willamette Valley, Oregon, circa 1850. Northwest Science 85:93–107.

Consortium of Pacific Northwest Herbaria. 2013. Available at http://www.pnwherbaria.org. Accessed November 2014.

Cronquist, A., A.H. Holmgren, N.H. Holmgren, J.L. Reveal, P.K. Holmgren & R.C. Barneby. 1972–2013. Intermountain Flora: Vascular plants of the Intermountain West, U.S.A. Vols. 1–6. The New York Botanical Garden, Bronx, NY.

Daubenmire, R. 1978. Plant geography, with special reference to North America. Academic Press, NY.

Defenders of Wildlife. 1998. Oregon's living landscape. Defenders of Wildlife, West Linn, OR.

Duvall, M.R., G.H. Learn Jr., L.E. Eguiarte & M.T. Clegg. 1993. Phylogenetic analysis of *rbc*L sequences identifies *Acorus calamus* as the primal extant monocotyledon. Proceedings of the National Academy of Science U.S.A. 90:4641–4644.

Fishbein, M., S. Kephart, M. Wilder, K. Halpin & S. Datwyler. 2010. Phylogeny of *Camassia* (Agavaceae) inferred from plastid *rpl*16 intron and *trn*D-*trn*Y-*trn*E-*trn*T intergenic spacer DNA sequences: Implications for species delimitation. Systematic Botany 35:77–85.

Flora of North America Editorial Committee, eds. 1993–present. Flora of North America, 30 planned volumes. Oxford University Press, Inc., New York.

Franklin, J.F. & C.T. Dyrness. 1973. Natural vegetation of Oregon and Washington. General technical reports PNW-GTR-008. U.S. Department of Agriculture, Forest Service, Pacific Northwest Research Station. Portland, OR.

______. 1988. Natural vegetation of Oregon and Washington. Oregon State University Press, Corvallis, OR.

Franklin, J.F., F.C. Hall, C.T. Dyrness & C. Maser. 1972. Federal research natural areas in Oregon and Washington: A guidebook for scientists and educators. United States Forest Service, Portland, OR.

Garrett, S. & B. Youtie. 1992. The Painted Hills: Thirty million years of phytogeography. Kalmiopsis 2:21–31.

Gilkey, H.M. 1954. Tuberales. North American Flora, Series II 1:1–36.

Gilkey, H. M. & L.J. Dennis. 2001. Handbook of northwestern plants. Revised edition. Oregon State University Press, Corvallis, OR.

Gilmour, C.N., J.R. Starr & R.F.C. Naczi. 2013. *Calliscirpus*, a new genus for two narrow endemics of the California floristic province, *C. criniger* and *C. brachythrix* sp. nov. (Cyperaceae). Kew Bulletin 68:1–21.

Gould, F.W. 1942. A systematic treatment of the genus *Camassia*. American Midland Naturalist 28:712–742.

Grayum, M.H. 1987. A summary of evidence and arguments supporting the removal of *Acorus* from the Araceae. Taxon 36:723–729.

Halpin, K.M. & M. Fishbein. 2013. A chloroplast phylogeny of Agavaceae subfamily Chlorogaloideae: Implications for the tempo of evolution on serpentine soils. Systematic Botany 38:996–1011.

Harris, J.G. & M.W. Harris. 2001. Plant identification terminology: An illustrated glossary, 2nd ed. Spring Lake Publishing, Spring Lake, UT.

Haynes, R.R. & C.B. Hellquist. 2000. *Potamogeton.* In: Flora of North America Editorial Committee, eds. 1993+. Flora of North America north of Mexico. 18+ vols. New York and Oxford. Vol 22.

Hickman, J.C., ed. 1993. The Jepson manual: Higher plants of California. University of California Press, Berkeley, CA.

Hitchcock, C.L. & A. Cronquist. 1973. Flora of the Pacific Northwest. University of Washington Press, Seattle, WA.

Hitchcock, C.L., A. Cronquist., M. Ownbey & J.W. Thompson. 1955–1969. Vascular plants of the Pacific Northwest. Vols. 1–5. University of Washington Press, Seattle, WA.

Horn, E.L. 2006. Oregon's best wildflower hikes: Southwest region. Westcliffe Press, Englewood, CO.

Houck, M.C. & M.J. Cody. 2011. Wild in the city: Exploring the intertwine, the Portland-Vancouver region's network of parks, trails, and natural areas. 2nd ed. Oregon State University Press, Corvallis, OR.

Howell, T.J. 1897–1903. A flora of Northwest America. Self-published, Portland, OR.

Ito, Y., T. Ohi-Toma, J. Murata & N. Tanaka. 2010. Hybridization and polyploidy of an aquatic plant, *Ruppia* (Ruppiaceae), inferred from plastid and nuclear DNA phylogenies. American Journal of Botany 97:1156–1167.

James, M. & K. Baitis. 2003. Geologic interpretation of floodplain deposits of the southwestern Willamette Valley, T17SR4W, with some implications for restoration management practices. Unpublished report. James Geoenvironmental Services, Eugene, OR.

Jepson, W.L. 1925. Manual of the flowering plants of California. University of California Press, Berkeley, CA.

Jernstedt, J. 2002. *Chlorogalum*. In: Flora of North America Editorial Committee, eds. 1993+. Flora of North America north of Mexico. 18+ vols. New York and Oxford. Vol 26.

Jolley, R. 1988. Wildflowers of the Columbia River Gorge. Oregon Historical Society Press, Portland, OR.

Jones, P.N. 1992. Columbia River Gorge: A complete guide. The Mountaineers, Seattle, WA.

Kawano, S., M. Ihara, M. Suzuki & H.H. Iltis. 1967. Biosystematic studies on *Maianthemum* (Liliaceae) somatic chromosome number and morphology. Botanical Magazine 80:345–352.

Kawano, S. & H.H. Iltis. 1966. Cytotaxonomy of the genus *Smilacina* (Liliaceae). II. Chromosome morphology and evolutionary consideration of New World species. Cytologia 31:12–28.

Kim, S.C. & N.S. Lee. 2007. Generic delimitation and biogeography of *Maianthemum* and *Smilacina* (Ruscaceae sensu lato): Preliminary results based on partial 3' *mat*K gene and *trn*K 3' intron sequences of cpDNA. Plant Systematics and Evolution 265:1–12.

Kirschner, J., ed. 2002. Juncaceae. Species plantarum: Flora of the world. Australian Biological Resources Study, Canberra, Australia.

Kozloff, E.N. 2005. Plants of western Oregon, Washington & British Columbia. Timber Press, Portland, OR.

Kruckeberg, A.R. & R.M. Love, eds. In preparation. Plant hunters of the Pacific Northwest: Botanical explorers of Oregon, Washington, Montana, Idaho and British Columbia, 1790–1970. University of Washington Press, Seattle, WA.

Küchler, A.W. 1946. The broadleaf deciduous forests of the Pacific Northwest. Annals of the Association of American Geographers 36:122–147.

LaFrankie, J.V. 1986a. Morphology and taxonomy of the New World species of *Maianthemum* (Liliaceae). Journal of the Arnold Arboretum 67:371–439.

______. 1986b. Transfer of the species of *Smilacina* Desf. to *Maianthemum* Wiggers (Liliaceae). Taxon 35:584–589.

Lang, F.A. & P.F. Zika. 1997. A nomenclatural note on *Hastingsia bracteosa* and *Hastingsia atropurpurea* (Liliaceae). Madroño 44:189–192.

Les, D.H. & D.J. Crawford. 1999. *Landoltia* (Lemnaceae), a new genus of duckweeds. Novon 9:530–533.

Lint, H.L. 1977. A revision of *Juncus* subgenus *Genuini* (Juncaceae) in the Pacific states. Ph.D. dissertation. Oregon State University, Corvallis, OR.

Love, G.A. & R.M. Love. 1970. Ecological crisis: Readings for survival. Harcourt Brace Jovanovich. New York, NY.

Macfarlane, R.M. 1978. On the taxonomic status of *Fritillaria phaeanthera* Eastw. (Liliaceae). Madroño 25:93–100.

MacKinnon, S. 2014. Flowers of the Table Rocks. Self published. Medford, OR.

Majak, W., R.E. McDiarmid, J.W. Hall & A.L. Van Ryswyk. 1980. Seasonal variation in the cyanide potential of arrowgrass (*Triglochin maritima*). Canadian Journal of Plant Science. 60:1235–1241.

Malaby, S., ed. 2007. Common plants of the Upper Klamath Basin. Native Plant Society of Oregon: Klamath Basin Chapter & Rabe Consulting, Klamath Falls, OR.

Mansfield, D. 2000. Flora of Steens Mountain. Oregon State University Press, Corvallis, OR.

Mason, G. 1975. Guide to the plants of the Wallowa Mountains of northeastern Oregon. Museum of Natural History, University of Oregon, Eugene, OR.

McKain, M.R., N. Wickett, Y. Zhang, S. Ayyampalayam, W.R. McCombie, M.W. Chase, J.C. Pires, C.W. dePamphilis & J. Leebens-Mack. 2012. Phylogenomic analysis of transcriptome data elucidates co-occurrence of a paleopolyploid event and the origin of bimodal karyotypes in Agavoideae (Asparagaceae). American Journal of Botany 99:397–406.

Merriam-Webster Inc. 2000. Merriam-Webster's collegiate dictionary. 10th ed. Print.

Meyers, S.C., A. Liston & R.J. Meinke. 2006. The hybrid origin of *Fritillaria gentneri* Gilkey (Liliaceae). Presentation to Botany 2006, Chico, California.

Moerman, D.E. 1986. Medicinal plants of native America, technical report 19. University of Michigan Museum of Anthropology, Ann Arbor, MI.

Muth, G.J. 1976. The taxonomic and floristic relationships of *Quercus sadleriana* R. Br. Campst. to other chestnut oaks. PhD dissertation. University of California, Davis, CA.

Native Plant Society of Oregon. 1989 onwards. Kalmiopsis and NPSO Bulletin. Available at http://npsoregon.org/publications.html.

Natural Heritage Advisory Council. 1998. Oregon natural heritage plan. Natural Heritage Advisory Council, Salem, OR.

ORBIC (Oregon Biodiversity Information Center). 2013. Portland State University: Oregon Biodiversity Information Center. Institute for Natural Resources. Available at http://orbic.pdx.edu. Accessed November 2014.

Oregon Department of Agriculture. 2014. Oregon's threatened, endangered, and candidate plants. Available at http://www.oregon.gov/ODA/programs/PlantConservation/Pages/AboutPlants.aspx. Accessed November 2014.

Oregon Department of Fish and Wildlife. 2006. The Oregon conservation strategy. Oregon Department of Fish and Wildlife, Salem, OR.

Oregon Department of Geology and Mineral Industries. Oregon: A geologic history. Available at http://www.oregongeology.org/sub/publications/ims/ims–028/index.htm. Accessed December 2014.

Oregon Flora Project. 2012 onwards. Available at http://oregonflora.org.

Orr, E.L. & W.N. Orr. 2000. Geology of Oregon. Kendall/Hunt Publishing Company, Dubuque, IA.

______. 2012. Oregon geology. Oregon State University Press, Corvallis, OR.

Pater, D.E., S.A. Bryce, T.D. Thorson, J. Kagan, C. Chappell, J.M. Omernik, S.H. Azevedo & A.J. Woods. 1998. Ecoregions of Oregon. United States Geological Survey, Reston, VA.

Peck, M.E. 1961. A manual of the higher plants of Oregon, 2nd edition. Binfords & Mort Publishers, Metropolitan Printing Company, Portland, OR.

Pojar, J. & A. MacKinnon. 1994. Plants of coastal British Columbia, including Washington, Oregon, and Alaska. Lone Pine Publishing, Vancouver, British Columbia, Canada.

Ranker, T.A. & T. Hogan. 2002. *Camassia*. In: Flora of North America Editorial Committee, eds. 1993+. Flora of North America north of Mexico. 18+ vols. New York and Oxford. Vol 26.

Raven, P.H. & D.I. Axelrod. 1978. Origin and relationships of the California flora. University of California Publications in Botany 72:1–134.

Ross, R.A. & H.C. Chambers. 1988. Wildflowers of the western Cascades. Timber Press, Portland, OR.

Rudall, P.J., J.G. Conran & M.W. Chase. 2000. Systematics of Ruscaceae/Convallariaceae: A combined morphological and molecular investigation. Botanical Journal of the Linnean Society 134:73–92.

Schmitz, U., S. Köhler & A. Hussner. 2014. First records of American *Wolffia columbiana* in Europe – Clandestine replacement of native *Wolffia arrhiza*? BioInvasions Records 3:213–216.

Sen, S. 1975. Cytotaxonomy of Liliales. Feddes Repertorium 86: 255–305.

Skinner, M.W. 2002. *Lilium*. In: Flora of North America Editorial Committee, eds. 1993+. Flora of North America north of Mexico. 18+ vols. New York and Oxford. Vol 26.

Smith, S.G., J.J. Bruhl, M.S. Gonzalez-Elizondo & F.J. Menapace. 2002. *Eleocharis*. In: Flora of North America Editorial Committee, eds. 1993+. Flora of North America north of Mexico. 18+ vols. New York and Oxford. Vol 23.

Stevens, P.F. Angiosperm Phylogeny Website. Version 12. 2012 onwards. Available at mobot.org/MOBOT/research/APweb/. Accessed December 2014.

Sullivan, W.L. 2009. 100 hikes/travel guide: Oregon coast and coast range. 3rd ed. Navillus Press, Eugene, OR.

______. 2010. 100 hikes in southern Oregon. 3rd ed. Navillus Press, Eugene, OR.

______. 2012. 100 hikes/travel guide: Central Oregon Cascades. 4th ed. Navillus Press, Eugene, OR.

______. 2012. 100 hikes/travel guide: Eastern Oregon. 2nd ed. Navillus Press, Eugene, OR.

______. 2013. 100 hikes in northwest Oregon and southwest Washington. 4th ed. Navillus Press, Eugene, OR.

Sultany, M., S. Kephart & P. Eilers. 2007. Blue flower of tribal legend: "Skye blue petals resemble lakes of fine, clear water." Kalmiopsis 14:28–35.

Takhtajan, A.L. 1986. Floristic regions of the world. University of California Press, Berkeley, CA.

Thompson, S.A. 1995. Systematics and biology of the Araceae and Acoraceae of temperate North America. University of Illinois Press, Champaign, IL.

Thorne, R.F. 1993. Phytogeography. In: Flora of North America Editorial Committee, eds. 1993+. Flora of North America north of Mexico. 18+ vols. New York and Oxford. Vol 1.

Thorson, T.D., S.A. Bryce, D.A. Lammers, A.J. Woods, J.M. Omernik, J. Kagan, D.E. Pater & J.A. Comstock. 2003. Ecoregions of Oregon. United States Geological Survey, Reston, VA.

United States Fish and Wildlife Service. 2014. Federally listed, proposed, candidate, delisted species and species of concern under the jurisdiction of the Fish and Wildlife Service which may occur within Oregon. Available at http://www.fws.gov/oregonfwo/species. Accessed November 2014.

Uyeda, J. & S. Kephart. 2006. Detecting species boundaries and hybridization in *Camassia quamash* and *C. leichtlinii* (Agavaceae) using allozymes. Systematic Botany 31:642–655.

von Mering, S. & J.W. Kadereit. 2010. Phylogeny, systematics and recircumscription of Juncaginaceae - a cosmopolitan wetland family. In: O. Seberg, ed. Diversity, phylogeny, and evolution in the monocotyledons. Proceedings of the fourth international conference on the comparative biology of the monocotyledons. Aarhus University Press, Aarhus, Denmark.

Wagner, D.H. 1979. Systematics of *Polystichum* in western North America north of Mexico. Pteridologia 1:1–64.

Wilson, B.L., R. Brainerd, M. Huso, K. Kuykendall, D. Lytjen, B. Newhouse, N. Otting, S. Sundberg & P. Zika. 1999. Atlas of Oregon *Carex*. Native Plant Society of Oregon, Occasional Paper No. 1, Corvallis, OR.

Wilson, B.L., R. Brainerd, D. Lytjen, B. Newhouse & N. Otting. 2008. Field guide to the sedges of the Pacific Northwest. Oregon State University Press, Corvallis, OR.

Wuerthner, G. 2001. Oregon's best wildflower hikes: Northwest region. Westcliffe Press, Englewood, CO.

Yamashita, J. & M.N. Tamura. 2000. Molecular phylogeny of the Convallariaceae (Asparagales). In: K.L. Wilson & D.A. Morrison, eds. Monocots: Systematics and evolution. The Commonwealth Scientific and Industrial Research Organisation, Melbourne, Australia.

APPENDICES

Appendix 1

Taxa Not Fully Treated in Flora

The three tables presented here list pteridophytes, gymnosperms, and monocots that are not treated in this volume because, although there has been evidence suggesting Oregon occurrence, the listed taxa are not a viable part of the flora of Oregon. As such, the most current nomenclature may not be reflected for these excluded taxa.

Table 1: Excluded Taxa with Confirmed Oregon Vouchers

These taxa have confirmed Oregon voucher specimens, but either they are not naturalized, are known only from a single population, or are infrequently occurring hybrids.

Name	Reason Not Treated
ALISMATACEAE	
Sagittaria platyphylla (Engelm.) J.G. Sm.	recent waif
AMARYLLIDACEAE	
Allium membranaceum Ownbey ex Traub	recent waif
Amaryllis belladonna L.	recent waif
Leucojum aestivum L.	recent waif
Narcissus × incomparabilis Mill.	recent waif
Narcissus tazetta L.	escaping in natural areas
ARACEAE	
Dracunculus vulgaris Schott	recent waif
Pistia stratiotes L.	recent waif
ASPARAGACEAE	
Brodiaea minor (Benth.) S. Watson	known from a single OR population
Brodiaea nana Hoover	known from a single OR population
Dichelostemma × venustum (Greene) Hoover	occasional hybrid
Dichelostemma volubile (Kellogg) A. Heller	known from a single OR population
Hyacinthoides × variabilis P.D. Sell	escaping on land under cultivation
Ornithogalum nutans L.	escaping on land under cultivation
Triteleia ixioides (W.T. Aiton) Greene	known from a single OR population
Triteleia peduncularis Lindl.	known from a single OR population
ASPLENIACEAE	
Asplenium bulbiferum G. Forst.	historic waif
Asplenium trichomanes L. ssp. *quadrivalens* D.E. Mey.	known from a single OR population

Name	Reason Not Treated
ATHYRIACEAE	
Athyrium alpestre (Hoppe) Clairv.	known from a single OR population
COMMELINACEAE	
Murdannia keisak (Hassk.) Hand.-Mazz.	historic waif
CUPRESSACEAE	
Hesperocyparis macrocarpa (Hartw. ex Gordon) Bartel	recent waif
Hesperocyparis sargentii (Jeps.) Bartel	known from a single OR population
CYATHEACEAE	
Sphaeropteris cooperi (Hook. ex F. Muell.) R.M. Tryon	recent waif
CYPERACEAE	
Carex comans Berggr.	recent waif
Carex fissuricola Mack.	known from a single OR population
Carex pumila Thunb.	ballast
Cyperus aggregatus (Willd.) Endl.	ballast
Cyperus odoratus L.	ballast
Schoenoplectus × kuekenthalianus (Junge) D.H. Kent	recent waif
DRYOPTERIDACEAE	
Polystichum andersonii Hopkins × *Polystichum munitum* (Kaulf.) C. Presl	occasional hybrid
Polystichum californicum (D.C. Eaton) Diels × *Polystichum munitum* (Kaulf.) C. Presl	occasional hybrid
EQUISETACEAE	
Equisetum × nelsonii (A.A. Eaton) J.H. Schaffn.	occasional hybrid
HYDROCHARITACEAE	
Elodea bifoliata H. St. John	known from a single OR population
Najas guadalupensis (Spreng.) Magnus var. *guadalupensis*	known from a single OR population
IRIDACEAE	
Gladiolus communis L.	recent waif
Iris × thompsonii R.C. Foster	occasional hybrid
Romulea rosea (L.) Eckl. var. *australis* (Ewart) M.P. de Vos	escaping in natural areas
JUNCACEAE	
Juncus ensifolius Wickstr. × *Juncus nevadensis* S. Watson	occasional hybrid
JUNCAGINACEAE	
Triglochin palustris L.	recent waif
LILIACEAE	
Calochortus amabilis Purdy	known from a single OR population
Calochortus luteus Douglas ex Lindl.	known from a single OR population
Calochortus persistens Ownbey	known from a single OR population
Erythronium hendersonii S. Watson × *Erythronium oregonum* Applegate	occasional hybrid
Erythronium multiscapideum (Kellogg) A. Nelson & P.B. Kenn.	known from a single OR population

Name	Reason Not Treated
LILIACEAE	
Fritillaria camschatcensis (L.) Ker Gawl.	known from a single OR population
Fritillaria eastwoodiae R.M. Macfarlane	known from a single OR population
Fritillaria purdyi Eastw.	known from a single OR population
Tulipa sylvestris L.	escaping on land under cultivation
LYCOPODIACEAE	
Lycopodium alpinum L. × *Diphasiastrum sitchense* (Rupr.) Holub	occasional hybrid
PINACEAE	
Pinus attenuata Lemmon × *Pinus radiata* D. Don	escaping in natural areas
Pinus monophylla Torr. & Frém.	historic waif
Pinus muricata D. Don	known from a single OR population
Pinus pinaster Aiton	escaping in natural areas
Pinus radiata D. Don	escaping on land under cultivation
POACEAE	
× *Aegilotriticum sancti-andreae* (Degen) Soó	escaping on land under cultivation
× *Agropogon lutosus* (Poir.) P. Fourn.	historic waif
× *Leydeum piperi* (Bowden) Barkworth	occasional hybrid
× *Schedololium holmbergii* (Dorfl.) Soreng & Terrell	recent waif
Achnatherum × *bloomeri* (Bol.) Barkworth	occasional hybrid
Agropyron fragile (Roth) P. Candargy	historic waif
Agrostis × *murbeckii* Fouill. ex P. Fourn.	recent waif
Agrostis elliottiana Schult.	historic waif
Agrostis exarata Trin. × *Agrostis pallens* Trin.	occasional hybrid
Agrostis exarata Trin. × *Agrostis stolonifera* L.	historic waif
Agrostis hallii Vasey × *Agrostis pallens* Trin.	occasional hybrid
Amelichloa caudata (Trin.) Arriaga & Barkworth	ballast
Arthraxon hispidus (Thunb.) Makino	ballast
Avena fatua L. × *Avena sativa* L.	historic waif
Brachypodium distachyon (L.) P. Beauv.	ballast
Brachypodium pinnatum (L.) P. Beauv.	historic waif
Bromus madritensis L.	ballast
Calamagrostis nutkaensis (J. Presl) Steud. × *Calamagrostis stricta* (Timm) Koeler	occasional hybrid
Chloris radiata (L.) Sw.	ballast
Cladoraphis cyperoides (Thunb.) S.M. Phillips	ballast
Cynodon × *magennisii* Hurcombe	recent waif
Echinochloa colona (L.) Link	recent waif
Echinochloa crus-pavonis (Kunth) Schult. var. *crus-pavonis*	ballast
Eleusine tristachya (Lam.) Lam.	ballast
Elymus ciliaris (Trin.) Tzvelev	ballast
Elymus tsukushiensis Honda	ballast

Name	Reason Not Treated
POACEAE	
Leymus flavescens (Scribn. & J.G. Sm.) Pilg. × *Leymus triticoides* (Buckley) Pilg.	occasional hybrid
Lolium × *hybridum* Hausskn.	recent waif
Lolium rigidum Gaudin	recent waif
Phalaris caroliniana Walter	historic waif
Phalaris minor Retz.	historic waif
Phleum arenarium L.	ballast
Phleum paniculatum Huds.	ballast
Phleum subulatum (Savi) Aschers. & Graebn.	ballast
Polypogon australis Brongn.	ballast
Polypogon fugax Nees ex Steud.	historic waif
Secale cereale L. × *Secale strictum* (C. Presl) C. Presl	escaping on land under cultivation
Secale strictum (C. Presl) C. Presl	historic waif
Setaria parviflora (Poir.) Kerguélen	historic waif
Setaria viridis (L.) P. Beauv. var. *major* (Gaudin) Peterm.	historic waif
Sporobolus indicus (L.) R. Br.	ballast
Thinopyrum ponticum (Podp.) Z.W. Liu & R.R.-C. Wang × *Triticum aestivum* L.	escaping on land under cultivation
Thinopyrum pycnanthum (Gould) Barkworth	ballast
Urochloa mutica (Forssk.) T.Q. Nguyen	ballast
PONTEDERIACEAE	
Eichhornia crassipes (Mart.) Solms	historic waif
PTERIDACEAE	
Cheilanthes covillei Maxon	known from a single OR population
TAXACEAE	
Torreya californica Torr.	known from a single OR population
XANTHORRHOEACEAE	
Hemerocallis fulva (L.) L.	recent waif
Hemerocallis minor Mill.	recent waif
Kniphofia uvaria (L.) Oken	escaping on land under cultivation
Phormium tenax J.R. Forst. & G. Forst.	recent waif

Appendix 1

Table 2: Taxa Documented in Local Floras as Occurring in Oregon but Lacking a Voucher

One or more of the key references (see page 4) indicate these taxa occur in Oregon, in contradiction with their excluded status in the OFP Checklist. A voucher documenting the occurrence was sought; if one was located, it was determined by OFP to be misidentified or erroneously attributed to occur in Oregon. These taxa are not considered synonymous with any taxa in this flora.

Name	Reason Not Treated
AMARYLLIDACEAE	
Allium cepa L.	cultivated
Allium tribracteatum Torr.	no known voucher
ASPARAGACEAE	
Triteleia ixioides (W.T. Aiton) Greene ssp. *unifolia* L.W. Lenz	no known voucher
CUPRESSACEAE	
Juniperus communis L. var. *depressa* Pursh	no known voucher
CYPERACEAE	
Carex albonigra Mack.	no known voucher
Carex brunnescens (Pers.) Poir. ssp. *brunnescens*	former voucher was misidentified
Carex miserabilis Mack.	no known voucher
Carex prionophylla T. Holm	no known voucher
Carex tenera Dewey var. *tenera*	former voucher was misidentified
Eleocharis acicularis (L.) Roem. & Schult. var. *gracilescens* Svenson	no known voucher
Eleocharis atropurpurea (Retz.) J. Presl & K. Presl	no known voucher
Eleocharis coloradoensis (Britton) Gilly	poor voucher material; identification cannot be confirmed
Eleocharis parvula (Roem. & Schult.) Link ex Bluff, Nees & Schauer var. *anachaeta* (Torr.) Svenson	poor voucher material; identification cannot be confirmed
DRYOPTERIDACEAE	
Cyrtomium fortunei J. Sm. var. *fortunei*	cultivated
Dryopteris phegopteris (L.) C. Chr.	original voucher not collected in OR
EQUISETACEAE	
Equisetum × *litorale* Rupr.	former voucher was misidentified
Equisetum × *mackaii* (Newman) Brichan	no known voucher
Equisetum variegatum Schleich. ex F. Weber & D. Mohr var. *alaskanam* A.A. Eat.	no known voucher
IRIDACEAE	
Iris purdyi Eastw.	no known voucher
ISOËTACEAE	
Isoëtes lacustris L.	no known voucher

Name	Reason Not Treated
JUNCACEAE	
Juncus nodosus L.	former voucher was misidentified
Juncus xiphioides E. Mey.	no known voucher
LILIACEAE	
Fritillaria pluriflora Benth.	no known voucher
Lilium parvum Kellogg	former voucher was misidentified
Lilium rubescens S. Watson	former voucher was misidentified
ORCHIDACEAE	
Cypripedium calceolus L. var. *parviflorum* (Salsib.) Fernald	no known voucher
Cypripedium parviflorum Salisb. var. *pubescens* (Willd.) O.W. Knight	no known voucher
Habenaria orbiculata (Pursh) Torr.	no known voucher
Malaxis monophyllos (L.) Sw. ssp. *brachypoda* (A. Gray) Á. Löve & D. Löve	no known voucher
Malaxis monophyllos (L.) Sw. var. *brachypoda* (A. Gray) F. Morris & E.A. Eames	no known voucher
Platanthera leptopetala (Rydb.) R.M. Bateman	former voucher was misidentified
Platanthera michaelii (Greene) R.M. Bateman	no known voucher
Platanthera orbiculata (Pursh) Lindl.	no known voucher
POACEAE	
Achnatherum lettermanii (Vasey) Barkworth	former voucher was misidentified
Aegilops neglecta Req. ex Bertol.	no known voucher
Agropyron × *pseudorepens* Scribn. & J.G. Sm.	poor voucher material; identification cannot be confirmed
Agropyron caninum (L.) P. Beauv. var. *unilaterale* (Cassidy) Vasey	no known voucher
Agropyron scribneri Vasey	former voucher was misidentified
Agropyron semicostatum (Nees ex Steud.) Nees ex Boiss.	former voucher was misidentified
Agropyron subsecundum (Link) Hitchc.	no known voucher
Agropyron trachycaulum (Link) Malte ex H.F. Lewis var. *glaucum* (Pease & A.H. Moore) Malte	no known voucher
Agropyron trachycaulum (Link) Malte ex H.F. Lewis var. *unilaterale* Malte	no known voucher
Agrostis aequivalvis (Trin.) Trin.	former voucher was misidentified
Agrostis perennans (Walter) Tuck.	former voucher was misidentified
Alopecurus arundinaceus Poir.	no known voucher
Bouteloua curtipendula (Michx.) Torr.	no known voucher
Bouteloua gracilis (Kunth) Lag. ex Griffiths	no known voucher
Bromus berteroanus Colla	no known voucher
Bromus catharticus Vahl var. *elatus* (E. Desv.) Planchuelo	former voucher was misidentified
Bromus erectus Huds.	former voucher was misidentified
Bromus polyanthus Scribn.	no known voucher
Bromus porteri (J.M. Coult.) Nash	no known voucher
Bromus pumpellianus Scribn. ssp. *pumpellianus*	no known voucher
Bromus racemosus L.	former voucher was misidentified
Bromus trinii É. Desv.	no known voucher

Name	Reason Not Treated
POACEAE	
Calamagrostis bolanderi Thurb.	no known voucher
Cenchrus incertus M.A. Curtis	no known voucher
Cenchrus pauciflorus Benth.	no known voucher
Cenchrus spinifex Cav.	no known voucher
Crypsis vaginiflora (Forssk.) Opiz	no known voucher
Echinochloa crus-galli (L.) P. Beauv. var. *zelayensis* (Kunth) Hitchc.	no known voucher
Elymus hispidus (Opiz) Melderis	no known voucher
Elymus scribneri (Vasey) M.E. Jones	former voucher was misidentified
Elymus sierrae Gould	former voucher was misidentified
Elymus triticoides Buckley var. *simplex* (Scribn. & T.A. Williams) Hitchc.	no known voucher
Elymus virginicus L.	no known voucher
Eragrostis mexicana (Hornem.) Link ssp. *mexicana*	no known voucher
Festuca californica Vasey var. *parishii* (Piper) Hitchc.	no known voucher
Festuca minutiflora Rydb.	former voucher was misidentified
Festuca ovina L. var. *ovina*	cultivated
Festuca saximontana Rydb. var. *saximontana*	former voucher was misidentified
Hesperostipa comata (Trin. & Rupr.) Barkworth ssp. *intermedia* (Scribn. & Tweedy) Barkworth	former voucher was misidentified
Hordeum caespitosum Scribn.	former voucher was misidentified
Hordeum jubatum L. var. *caespitosum* (Scribn.) Hitchc.	former voucher was misidentified
Hordeum pusillum Nutt.	former voucher was misidentified
Koeleria gerardii (Vill.) Shinners	no known voucher
Koeleria phleoides (Vill.) Pers.	no known voucher
Lagurus ovatus L.	no known voucher
Leymus racemosus (Lam.) Tzvelev	former voucher was misidentified
Melica californica Scribn.	former voucher was misidentified
Muhlenbergia frondosa (Poir.) Fernald	former voucher was misidentified
Muhlenbergia montana (Nutt.) Hitchc.	no known voucher
Nassella chilensis (Trin.) E. Desv.	no known voucher
Oryzopsis micrantha (Trin. & Rupr.) Thurb.	no known voucher
Panicum capillare L. var. *occidentale* Rydb.	no known voucher
Panicum thermale Bol.	no known voucher
Panicum villosissimum Nash var. *pseudopubescens* (Nash) Fernald	no known voucher
Pennisetum setaceum (Forssk.) Chiov.	cultivated
Phalaris brachystachys Link	former voucher was misidentified
Phleum graecum Boiss. & Heldr.	no known voucher
Phyllostachys aurea Carrière ex Rivière & C. Rivière	no known voucher
Piptatherum micranthum (Trin. & Rupr.) Barkworth	no known voucher
Poa arctica R. Br.	no known voucher
Poa fendleriana (Steud.) Vasey ssp. *fendleriana*	former voucher was misidentified

Name	Reason Not Treated
POACEAE	
Poa grayana Vasey	former voucher was misidentified
Poa interior Rydb.	no known voucher
Poa kelloggii Vasey	former voucher was misidentified
Poa occidentalis Vasey	former voucher was misidentified
Poa pattersonii Vasey	no known voucher
Podagrostis aequivalvis (Trin.) Scribn. & Merr.	former voucher was misidentified
Setaria italica (L.) P. Beauv.	former voucher was misidentified
Spartina densiflora Brongn.	original voucher not collected in OR
Sphenopholis intermedia (Rydb.) Rydb.	former voucher was misidentified
Sporobolus asper (P. Beauv.) Kunth	no known voucher
Sporobolus neglectus Nash	no known voucher
Stipa comata Trin. & Rupr. var. *intermedia* Scribn. & Tweedy	former voucher was misidentified
Stipa divaricata Columbus & J.P. Sm.	no known voucher
Stipa lemmonii (Vasey) Scribn. var. *pubescens* Crampton	no known voucher
Stipa lettermanii Vasey	former voucher was misidentified
Stipa minor (Vasey) Scribn.	former voucher was misidentified
Stipa occidentalis Thurb. ex S. Watson var. *minor* (Vasey) C.L. Hitchc.	former voucher was misidentified
Stipa occidentalis Thurb. ex S. Watson var. *occidentalis*	no known voucher
Vulpia octoflora (Walter) Rydb. var. *glauca* (Nutt.) Fernald	no known voucher
Zea mays L.	cultivated
Zizania aquatica L.	no known voucher
POTAMOGETONACEAE	
Potamogeton vaginatus Turcz.	former voucher was misidentified
PTERIDACEAE	
Pellaea glabella Mett. ex Kuhn var. *simplex* Butters	no known voucher
Pellaea suksdorfiana Butters	no known voucher
SELAGINELLACEAE	
Selaginella densa Rydb. var. *standleyi* (Maxon) R.M. Tryon	no known voucher
THELYPTERIDACEAE	
Phegopteris connectilis (Michx.) Watt	original voucher not collected in OR
Thelypteris phegopteris (L.) Sloss.	original voucher not collected in OR
TYPHACEAE	
Typha × *glauca* Godr.	cultivated
Typha domingensis Pers.	no known voucher

Table 3: Misapplied and In Part Names

This table lists taxa not treated because they represent names misapplied to Oregon taxonomic concepts. A *misapplied* name is one that is accepted in a key reference but does not truly refer to the taxonomic description with which it is associated. This happens for two reasons. Either the name was applied in error or a name is used in a flora to describe a taxonomic concept that is subsequently determined to be new to science. Once the new concept is published with a new name, the initial name—although correctly applied to a different taxon—is henceforth considered *misapplied* to the new concept. Misapplied names are not true synonyms.

A few of the names listed are either misapplied to more than one Oregon taxon or are good synonyms that refer, *in part*, to more than one accepted name in this flora.

The name under the column heading **Old Name** shows the name used in one of the key local floras whereas the name(s) in the column **Accepted Name(s)** shows the name(s) recognized and described in this volume.

Old Name	Accepted Name(s)
MISAPPLIED	
ALISMATACEAE	
Alisma plantago-aquatica L.	*Alisma triviale*
ARACEAE	
Wolffia punctata Griseb.	*Wolffia borealis, W. brasiliensis*
CYPERACEAE	
Carex backii Boott	*Carex cordillerana*
Carex deweyana Schwein.	*Carex bolanderi, C. infirminervia, C. leptopoda*
Carex engelmannii L.H. Bailey	*Carex breweri*
Carex filifolia Nutt. var. *erostrata* Kük.	*Carex filifolia* var. *filifolia*
Carex garberi Fernald	*Carex hassei*
Carex haydenii Dewey	*Carex haydeniana*
Carex lanuginosa Michx.	*Carex pellita*
Carex lenticularis Michx. var. *lenticularis*	*Carex kelloggii* var. *kelloggii*
Carex macloviana d'Urv.	*Carex pachystachya*
Carex multicostata Mack.	*Carex pachycarpa*
Carex muricata L.	*Carex echinata* ssp. *echinata*
Carex norvegica Retz.	*Carex media*
Carex nova L.H. Bailey	*Carex pelocarpa*
Carex podocarpa R. Br.	*Carex paysonis*
Carex rostrata Stokes ex With.	*Carex utriculata*
Carex saximontana Mack.	*Carex cordillerana*
Cyperus bushii Britton	*Cyperus schweinitzii*
Cyperus lupulinus (Spreng.) Marcks ssp. *lupulinus*	*Cyperus schweinitzii*
Cyperus lupulinus (Spreng.) Marcks ssp. *macilentus* (Fernald) Marcks	*Cyperus schweinitzii*
Eleocharis mamillata H. Lindb.	*Eleocharis palustris*
Eleocharis uniglumis (Link) Schult.	*Eleocharis palustris*
Scirpus americanus Pers.	*Schoenoplectus americanus, S. pungens*
Scirpus lineatus Michx.	*Scirpus pendulus*
Scirpus pedicellatus Fernald	*Scirpus cyperinus*
Scirpus tuberosus Desf.	*Bolboschoenus glaucus*

Old Name	Accepted Name(s)
CYSTOPTERIDACEAE	
Gymnocarpium dryopteris (L.) Newman	*Gymnocarpium disjunctum*
DRYOPTERIDACEAE	
Dryopteris austriaca (Jacq.) Woyn. ex Schinz & Thell.	*Dryopteris expansa*
Dryopteris dilatata (Hoffm.) A. Gray	*Dryopteris expansa*
Polystichum mohrioides (Bory) C. Presl	*Polystichum lemmonii, P. scopulinum*
IRIDACEAE	
Sisyrinchium angustifolium Mill.	*Sisyrinchium bellum, S. halophilum, S. idahoense* var. *idahoense, S. idahoense* var. *occidentale, S. sarmentosum*
JUNCACEAE	
Juncus brachyphyllus Wiegand	*Juncus trilocularis*
Juncus drummondii E. Mey. var. *drummondii*	*Juncus drummondii*
Juncus dubius Engelm.	*Juncus nevadensis* var. *nevadensis*
Juncus falcatus E. Mey. ssp. *falcatus*	*Juncus falcatus* ssp. *sitchensis*
Juncus falcatus E. Mey. var. *falcatus*	*Juncus falcatus* ssp. *sitchensis*
Juncus phaeocephalus Engelm.	*Juncus nevadensis* var. *inventus*
Juncus phaeocephalus Engelm. var. *glomeratus* Engelm.	*Juncus nevadensis* var. *inventus*
Juncus phaeocephalus Engelm. var. *paniculatus* Engelm.	*Juncus nevadensis* var. *inventus*
Juncus phaeocephalus Engelm. var. *phaeocephalus*	*Juncus nevadensis* var. *inventus*
Juncus sphaerocarpus Nees	*Juncus bufonius* var. *occidentalis*
Juncus xiphioides E. Mey. var. *xiphioides*	*Juncus saximontanus*
Luzula campestris (L.) DC. var. *congesta* (Thuill.) E. Mey.	*Luzula multiflora*
Luzula campestris (L.) DC. var. *frigida* Buchenau	*Luzula multiflora*
Luzula divaricata S. Watson	*Luzula parviflora*
Luzula glabrata (Hoppe) Desv.	*Luzula hitchcockii*
Luzula multiflora (Ehrh.) Lej. ssp. *frigida* (Buchenau) Krecz.	*Luzula multiflora*
Luzula multiflora (Ehrh.) Lej. var. *congesta* (Thuill.) E. Mey.	*Luzula multiflora*
Luzula subcongesta (S. Watson) Jeps.	*Luzula multiflora*
Luzula wahlenbergii Rupr.	*Luzula piperi*
LILIACEAE	
Calochortus lobbii (Baker) Purdy	*Calochortus subalpinus*
Calochortus nuttallii Torr.	*Calochortus bruneaunis*
MELANTHIACEAE	
Trillium angustipetalum (Torr.) J.D. Freeman	*Trillium kurabayashii*
Trillium chloropetalum (Torr.) Howell	*Trillium albidum* ssp. *albidum, T. albidum* ssp. *parvifolium, T. kurabayashii*
OPHIOGLOSSACEAE	
Botrychium boreale Milde	*Botrychium pinnatum*
Botrychium lunaria (L.) Sw. var. *onondagense* (Underw.) House	*Botrychium lunaria, B. minganense*
Ophioglossum vulgatum L.	*Ophioglossum pusillum*

Old Name	Accepted Name(s)
ORCHIDACEAE	
Habenaria hyperborea (L.) R. Br.	*Platanthera huronensis*
Limnorchis viridiflora (Cham.) Rydb.	*Platanthera stricta*
Spiranthes romanzoffiana Cham. var. *romanzoffiana*	*Spiranthes romanzoffiana, S. stellata*
PINACEAE	
Abies concolor (Gordon & Glend.) Lindl. ex Hildebr.	*Abies concolor* × *Abies grandis*
Abies magnifica A. Murray bis	*Abies magnifica* × *Abies procera*
Abies magnifica A. Murray bis var. *shastensis* Lemmon	*Abies magnifica* × *Abies procera*
Larix lyallii Parl.	*Larix occidentalis*
Pinus contorta Douglas ex Loudon ssp. *murrayana* (Grev. & Balf.) Critchf.	*Pinus contorta* var. *latifolia*
Pinus contorta Douglas ex Loudon var. *murrayana* (Grev. & Balf.) Engelm.	*Pinus contorta* var. *latifolia*
Pinus ponderosa Douglas ex C. Lawson var. *scopulorum* Engelm.	*Pinus ponderosa* var. *ponderosa*
POACEAE	
Agropyron caninum (L.) P. Beauv. ssp. *caninum*	*Elymus ciliaris, E. tsukushiensis*
Agropyron caninum (L.) P. Beauv. var. *latiglume* (Scribn. & J.G. Sm.) Pease & A.H. Moore	*Elymus trachycaulus* ssp. *trachycaulus*
Agropyron pungens (Pers.) Roem. & Schult.	*Thinopyrum pycnanthum*
Agrostis alba L. var. *alba*	*Agrostis gigantea, A. stolonifera*
Agrostis hyemalis (Walter) Britton, Sterns & Poggenb.	*Agrostis scabra*
Anthoxanthum nitens (Weber) Y. Schouten & Veldkamp ssp. *nitens*	*Hierochloë odorata*
Aristida fendleriana Steud.	*Aristida purpurea* var. *longiseta*
Danthonia pilosa R. Br.	*Rytidosperma penicillatum*
Elymus caninus (L.) L.	*Elymus ciliaris, E. tsukushiensis*
Festuca idahoensis Elmer var. *oregona* (Hack. ex Beal) C.L. Hitchc.	*Festuca idahoensis, F. roemeri* var. *roemeri*
Festuca ovina L. var. *rydbergii* St.-Yves	*Festuca saximontana* ssp. *purpusiana*
Festuca perennis (L.) Columbus & J.P. Sm.	*Lolium multiflorum, L. perenne*
Festuca supina Schur	*Festuca saximontana* ssp. *purpusiana*
Gastridium ventricosum (Gouan) Schinz & Thell.	*Gastridium phleoides*
Muhlenbergia glomerata (Willd.) Trin.	*Muhlenbergia mexicana*
Muhlenbergia racemosa (Michx.) Britton, Sterns & Poggenb.	*Muhlenbergia mexicana*
Setaria glauca (L.) P. Beauv.	*Setaria pumila* ssp. *pumila*
Setaria lutescens F.T. Hubb.	*Setaria pumila* ssp. *pumila*
POTAMOGETONACEAE	
Potamogeton compressus L.	*Potamogeton zosteriformis*
Potamogeton lucens L.	*Potamogeton illinoensis*

Old Name	Accepted Name(s)
IN PART	
JUNCACEAE	
Juncus arcticus Willd. var. *balticus* (Willd.) Trautv.	*Juncus balticus* ssp. *ater*, *J. breweri*
ORCHIDACEAE	
Habenaria dilatata (Pursh) Hook.	*Platanthera dilatata*, *P. sparsiflora*
Habenaria elegans (Lindl.) Bol.	*Platanthera elegans*, *P. elongata*, *P. transversa*
Habenaria unalascensis (Spreng.) S. Watson	*Platanthera ephemerantha*, *P. unalascensis*
Habenaria unalascensis (Spreng.) S. Watson var. *elata* (Jeps.) Correll	*Platanthera elegans*, *P. elongata*, *P. transversa*
Piperia elegans (Lindl.) Rydb.	*Platanthera elegans*, *P. elongata*
POACEAE	
Agropyron caninum (L.) P. Beauv. ssp. *majus* (Vasey) C.L. Hitchc.	*Elymus trachycaulus* ssp. *trachycaulus*, *E. violaceus*
Agropyron trachycaulum (Link) Malte ex H.F. Lewis	*Elymus trachycaulus*, *E. violaceus*
Festuca elatior L.	*Schedonorus arundinaceus*, *S. pratensis*
Festuca howellii Hack. ex Beal (true synonym of *F. elmeri*, but misapplied for *F. viridula*)	*Festuca elmeri*, *F. viridula*
Festuca ovina L.	*Festuca brachyphylla*, *F. brachyphylla* ssp. *coloradoensis*, *F. filiformis*, *F. saximontana* ssp. *purpusiana*
Glyceria pauciflora J. Presl	*Torreyochloa erecta*, *T. pallida* var. *pauciflora*
Poa gracillima Vasey	*Poa* × *multnomae*, *P. secunda* ssp. *secunda*
Poa vaseyochloa Scribn.	*Poa leibergii*, *P. leptocoma*
Stipa columbiana Macoun	*Achnatherum lemmonii* ssp. *lemmonii*, *A. nelsonii*
Stipa occidentalis Thurb. ex S. Watson	*Achnatherum nelsonii* ssp. *nelsonii*, *A. occidentale*
POLYPODIACEAE	
Polypodium vulgare L.	*Polypodium glycyrrhiza*, *P. hesperium*
TOFIELDIACEAE	
Tofieldia glutinosa (Michx.) Pers.	*Tofieldia glutinosa*, *T. occidentalis*

Appendix 2

Taxa Endemic to Oregon

The 115 taxa listed here are naturally occurring only in the state of Oregon. For taxa treated in this Volume, we considered only native terminal taxa receiving a full floristic description and designated by the floristic author as occurring solely in Oregon. Additional criteria for inclusion in this list were congruence with the endemic status of ORBIC (2013) and with earlier research by OFP (Scott Sundberg, personal communication 2004). Any taxon whose concept (as it appears under any name for that concept as tracked by OFP) is treated in the most recently published flora of California (Baldwin et al. 2012) was excluded. Taxa occurring outside of Oregon, as indicated by *Flora of North America* (1993–present) and specimens mapped in the Consortium of Pacific Northwest Herbaria (2013), were also evaluated for exclusion. The nomenclatural status of dicots is based on existing floristic treatments for the *Flora of Oregon* or, if not available, on the OFP Checklist Version 4.1. Endemic taxa are indicated in the floristic treatments and in Appendix 3 with a ✤ symbol.

Monocots

ASPARAGACEAE

- *Camassia leichtlinii* (Baker) S. Watson ssp. *leichtlinii*
- *Camassia quamash* (Pursh) Greene ssp. *intermedia* Gould
- *Camassia quamash* (Pursh) Greene ssp. *walpolei* (Piper) Gould
- *Hastingsia bracteosa* S. Watson var. *atropurpurea* (Becking) F.A. Lang & Zika
- *Hastingsia bracteosa* S. Watson var. *bracteosa*

IRIDACEAE

- *Iris tenax* Douglas ex Lindl. var. *gormanii* (Piper) R.C. Foster
- *Iris tenuis* S. Watson

LILIACEAE

- *Calochortus coxii* M.R. Godfrey & Callahan
- *Calochortus howellii* S. Watson
- *Calochortus indecorus* Ownbey & M. Peck
- *Calochortus longebarbatus* S. Watson var. *peckii* Ownbey
- *Calochortus umpquaensis* Fredricks ssp. *confertus* Callahan
- *Calochortus umpquaensis* Fredricks ssp. *flavicomus* Callahan
- *Calochortus umpquaensis* Fredricks ssp. *umpquaensis*
- *Erythronium elegans* P.C. Hammond & K.L. Chambers
- *Scoliopus hallii* S. Watson

POACEAE

- *Achnatherum wallowaense* J.R. Maze & K.A. Robson
- *Muhlenbergia mexicana* (L.) Trin. var. *filiformis* (Torr.) Scribn.
- *Pleuropogon oregonus* Chase
- *Poa chambersii* Soreng

Dicots

APIACEAE

Lomatium cookii Kagan

Lomatium erythrocarpum Meinke & Constance

Lomatium greenmanii Mathias

Lomatium minus (Rose) Mathias & Constance

Lomatium ochocense Helliwell & Constance ex Helliwell

Lomatium oreganum (J.M. Coult. & Rose) J.M. Coult. & Rose

Perideridia erythrorhiza (Piper) T.I. Chuang & Constance

ARISTOLOCHIACEAE

Asarum wagneri K.L. Lu & Mesler

ASTERACEAE

Artemisia ludoviciana Nutt. ssp. *estesii* K.L. Chambers in prep.

Chaenactis nevii A. Gray

Erigeron cascadensis A. Heller

Erigeron decumbens Nutt.

Erigeron stanselliae K.L. Chambers

Eucephalus gormanii Piper

Lasthenia ornduffii R. Chan

Luina serpentina Cronquist

Microseris howellii A. Gray

Senecio ertterae T.M. Barkley

Stephanomeria malheurensis Gottlieb

BORAGINACEAE

Amsinckia carinata A. Nelson & J.F. Macbr.

Plagiobothrys figuratus (Piper) I.M. Johnst. ex Cronquist var. *corallicarpus* (Piper) K.L. Chambers

Plagiobothrys hirtus (Greene) I.M. Johnst.

Plagiobothrys lamprocarpus (Piper) I.M. Johnst.

BRASSICACEAE

Boechera hastatula (Greene) Al-Shehbaz

Boechera horizontalis (Greene) Windham & Al-Shehbaz

Cardamine holmgrenii Al-Shehbaz

Cardamine pattersonii L.F. Hend.

Draba cusickii B.L. Rob. ex O.E. Schulz var. *cusickii*

Draba cyclomorpha Payson

Noccaea fendleri (A. Gray) Holub ssp. *siskiyouense* (P. Holmgren) Al-Shehbaz & M. Koch

Physaria kingii (S. Watson) O'Kane & Al-Shehbaz ssp. *diversifolia* (Greene) O'Kane & Al-Shehbaz

Thelypodium eucosmum B.L. Rob.

Thelypodium howellii S. Watson ssp. *spectabilis* (M. Peck) Al-Shehbaz

Dicots

CARYOPHYLLACEAE

Eremogone pumicola (Coville & Leiberg) Ikonn.

Silene douglasii Hook. var. *oraria* (M. Peck) C.L. Hitchc. & Maguire

Silene scaposa B.L. Rob. var. *scaposa*

CRASSULACEAE

Sedum moranii R.T. Clausen

Sedum stenopetalum Pursh ssp. *ciliosum* (Howell) R.T. Clausen

ERICACEAE

Kalmiopsis fragrans Meinke & Kaye

Kalmiopsis leachiana (L.F. Hend.) Rehder

FABACEAE

Astragalus accidens S. Watson var. *accidens*

Astragalus applegatei M. Peck

Astragalus collinus (Hook.) Douglas ex G. Don var. *laurentii* (Rydb.) Barneby

Astragalus curvicarpus (A. Heller) J.F. Macbr. var. *brachycodon* (Barneby) Barneby

Astragalus curvicarpus (A. Heller) J.F. Macbr. var. *subglaber* (Rydb.) Barneby

Astragalus diaphanus Douglas ex Hook. var. *diurnus* (S. Watson) Barneby ex M. Peck

Astragalus howellii A. Gray

Astragalus misellus S. Watson var. *misellus*

Astragalus peckii Piper

Astragalus robbinsii (Oakes) A. Gray var. *alpiniformis* (Rydb.) Barneby ex C.L. Hitchc.

Astragalus tegetarioides M.E. Jones

Astragalus tyghensis M. Peck

Lupinus biddlei L.F. Hend. ex C.P. Sm.

Lupinus lepidus Douglas ex Lindl. var. *ashlandensis* (B.J. Cox) Isely

Lupinus lepidus Douglas ex Lindl. var. *cusickii* (S. Watson) C.L. Hitchc.

Sophora leachiana M. Peck

Trifolium eriocephalum (Piper) J.M. Gillett ssp. *cascadense* J.M. Gillett

GROSSULARIACEAE

Ribes erythrocarpum Coville & Leiberg

HYDROPHYLLACEAE

Phacelia bicolor Torr. ex S. Watson var. *leibergii* (Brand) A. Nelson & J.F. Macbr.

Phacelia capitata Kruckeb.

Phacelia peckii J.T. Howell

Phacelia verna Howell

Romanzoffia thompsonii Marttala

LIMNANTHACEAE

Limnanthes alba Hartw. ex Benth. ssp. *gracilis* (Howell) Morin

Limnanthes pumila Howell ssp. *grandiflora* (Arroyo) S.C. Meyers & K.L. Chambers

Limnanthes pumila Howell ssp. *pumila*

Dicots

MALVACEAE

Sidalcea cusickii Piper

OROBANCHACEAE

Castilleja chambersii J.M. Egger & Meinke

Castilleja chlorotica Piper

Castilleja fraterna Greenm.

Castilleja pilosa (S. Watson) Rydb. var. *steenensis* (Pennell) N.H. Holmgren

Castilleja rubida Piper

Castilleja xanthotricha Pennell

PHRYMACEAE

Diplacus cascadensis G.L. Nesom

Diplacus deschutesensis G.L. Nesom

PLANTAGINACEAE

Penstemon eriantherus Pursh var. *argillosus* M.E. Jones

Penstemon glaucinus Pennell

Penstemon peckii Pennell

Penstemon richardsonii Douglas ex Lindl. var. *curtiflorus* (D.D. Keck) Cronquist

Penstemon richardsonii Douglas ex Lindl. var. *dentatus* (D.D. Keck) Cronquist

Penstemon spatulatus Pennell

POLEMONIACEAE

Collomia mazama Coville

Navarretia willamettensis S.C. Spencer

POLYGONACEAE

Eriogonum chrysops Rydb.

Eriogonum cusickii M.E. Jones

Eriogonum scopulorum Reveal

Eriogonum villosissimum Reveal & D.A. York

Polygonum gabrielae Costea & Tardif

RANUNCULACEAE

Delphinium leucophaeum Greene

Delphinium pavonaceum Ewan

Ranunculus austrooreganus L.D. Benson

ROSACEAE

Fragaria cascadensis K.E. Hummer

Horkelia congesta Douglas ex Hook. ssp. *congesta*

Horkelia fusca Lindl. ssp. *filicoides* (Crum) D.D. Keck

SAXIFRAGACEAE

Micranthes hitchcockiana (Elvander) Brouillet & Gornall

Appendix 3

Taxa Found in a Single Ecoregion Within Oregon

Taxa with an Oregon distribution limited to a single ecoregion are listed here. All are accepted terminal taxa that are native or naturalized exotics; any hybrids listed have a status of "frequent." Geographic location is confirmed from five or more voucher specimens from the OSU Herbarium and other selected herbaria. Taxa endemic to the state of Oregon are included regardless of the voucher count and are indicated by the ❧ symbol. If a taxon is documented as occurring infrequently in an ecoregion adjacent to that in which it is predominantly found, it is indicated with an asterisk (*). The nomenclatural status of dicots is based on existing floristic treatments for the *Flora of Oregon* or, if not available, on the OFP Checklist Version 4.1.

Basin and Range (29)

GYMNOSPERMS

EPHEDRACEAE

Ephedra nevadensis

MONOCOTS

CYPERACEAE

Amphiscirpus nevadensis

POACEAE

Leucopoa kingii

DICOTS

AMARANTHACEAE

Allenrolfea occidentalis

Atriplex pusilla

APIACEAE

Lomatium roseanum

ASTERACEAE

Chaenactis xantiana

Cirsium eatonii var. *peckii*

BRASSICACEAE

**Caulanthus major* var. *nevadensis*

Draba cusickii var. *cusickii* ❧

Lepidium montanum var. *nevadense*

CLEOMACEAE

Cleomella plocasperma

CRASSULACEAE

Sedum debile

ELAEAGNACEAE

Shepherdia argentea

FABACEAE

Lupinus nevadensis

HYDROPHYLLACEAE

**Phacelia bicolor* var. *leibergii* ❧

MONTIACEAE

Claytonia nevadensis

OROBANCHACEAE

**Castilleja pilosa* var. *steenensis* ❧

PAPAVERACEAE

Canbya aurea

PLANTAGINACEAE

Penstemon davidsonii var. *praeteritus*

POLEMONIACEAE

Aliciella lottiae

POLYGONACEAE

Eriogonum cernuum

Eriogonum collinum

Eriogonum crosbyae var. *mystrium*

Eriogonum cusickii ❧

ROSACEAE

Ivesia rhypara var. *shellyi*

SALICACEAE

Salix glauca var. *villosa*

Salix planifolia var. *monica*

Salix sitchensis var. *angustifolia*

Blue and Wallowa Mountains (142)

GYMNOSPERMS

CUPRESSACEAE

Juniperus scopulorum

PINACEAE

Pinus flexilis

PTERIDOPHYTES

OPHIOGLOSSACEAE

Botrychium ascendens

Botrychium pedunculosum

*Botrychium pinnatum

PTERIDACEAE

Cheilanthes feei

MONOCOTS

AMARYLLIDACEAE

Allium fibrillum

Allium madidum

CYPERACEAE

Carex capillaris

Carex concinna

Carex gynocrates

Carex media

Carex paysonis

Carex scirpoidea ssp. *scirpoidea*

Kobresia simpliciuscula

JUNCACEAE

Luzula orestera

LILIACEAE

*Calochortus elegans var. elegans

*Calochortus eurycarpus

Calochortus longebarbatus var. *peckii*

MELANTHIACEAE

*Trillium petiolatum

ORCHIDACEAE

Platanthera huronensis

POACEAE

Achnatherum wallowaense

Elymus violaceus

Piptatheropsis exigua

Poa alpina ssp. *alpina*

DICOTS

APIACEAE

Ligusticum tenuifolium

*Lomatium ambiguum

Lomatium cusickii

Lomatium erythrocarpum

Lomatium greenmanii

Lomatium oreganum

Lomatium serpentinum

ASTERACEAE

Antennaria howellii ssp. *petaloidea*

Antennaria lanata

Arnica rydbergii

Askellia pygmaea ssp. *ramosa*

Chaenactis alpina

Ericameria nauseosa var. *nana*

Erigeron chrysopsidis var. *brevifolius*

Erigeron coulteri

Erigeron davisii

*Erigeron eatonii var. villosus

Erigeron simplex

Erigeron vagus

Eurybia conspicua

Eurybia integrifolia

*Helianthella uniflora var. douglasii

Hulsea algida

Luina serpentina

*Oreostemma alpigenum var. haydenii

Pyrrocoma radiata

Symphyotrichum cusickii

Tonestus lyallii

Townsendia montana

BETULACEAE

Betula papyrifera

BORAGINACEAE

Eritrichium nanum var. *elongatum*

BRASSICACEAE

Cardamine holmgrenii

Draba cyclomorpha

Draba oligosperma

Blue and Wallowa Mountains

Physaria kingii ssp. *diversifolia*

Physaria oregona

Smelowskia americana

**Thelypodium howellii* ssp. *spectabilis*

CARYOPHYLLACEAE

Silene acaulis var. *exscapa*

ERICACEAE

Cassiope mertensiana ssp. *gracilis*

FABACEAE

Astragalus alpinus var. *alpinus*

Astragalus arthuri

Astragalus australis var. *glabriusculus*

Astragalus canadensis var. *mortonii*

**Astragalus cusickii* var. *cusickii*

**Astragalus inflexus*

Astragalus kentrophyta var. *tegetarius*

Astragalus robbinsii var. *alpiniformis*

Astragalus sheldonii

Hedysarum boreale var. *mackenzii*

**Lathyrus nevadensis* var. *cusickii*

Lupinus arbustus × *Lupinus sericeus*

Lupinus sabinianus

**Lupinus sulphureus* ssp. *sulphureus*

Oxytropis campestris var. *cusickii*

Oxytropis viscida var. *viscida*

Trifolium douglasii

Trifolium latifolium

Trifolium longipes var. *reflexum*

GENTIANACEAE

Frasera albicaulis var. *idahoensis*

Swertia perennis

GROSSULARIACEAE

Ribes cereum var. *colubrinum*

Ribes oxyacanthoides ssp. *irriguum*

HYDROPHYLLACEAE

Hydrophyllum fendleri var. *fendleri*

MONTIACEAE

Lewisia columbiana var. *wallowensis*

OROBANCHACEAE

Castilleja chrysantha

Castilleja cusickii

Castilleja fraterna

Castilleja lutescens

**Castilleja rhexiifolia*

Castilleja rubida

Castilleja viscidula

Pedicularis bracteosa var. *pachyrhiza*

PAPAVERACEAE

Corydalis caseana ssp. *cusickii*

PHRYMACEAE

Erythranthe patula

PLANTAGINACEAE

**Penstemon attenuatus* var. *attenuatus*

Penstemon attenuatus var. *palustris*

Penstemon confertus

Penstemon eriantherus var. *redactus*

Penstemon fruticosus var. *serratus*

Penstemon globosus

Penstemon payettensis

Penstemon spatulatus

Penstemon triphyllus

**Penstemon venustus*

Penstemon wilcoxii

**Synthyris rubra*

Tonella floribunda

**Veronica cusickii*

Veronica verna

POLEMONIACEAE

Collomia debilis

Ipomopsis congesta ssp. *viridis*

Phlox colubrina

Phlox pulvinata

POLYGONACEAE

Aconogonon phytolaccifolium var. *glabrum*

Eriogonum flavum var. *piperi*

Eriogonum scopulorum

Polygonum gabrielae

PRIMULACEAE

Androsace septentrionalis

RANUNCULACEAE

Anemone parviflora

Anemone piperi

**Aquilegia flavescens*

Clematis hirsutissima var. *hirsutissima*

Clematis occidentalis var. *grosseserrata*

ROSACEAE

Dryas drummondii

Dryas hookeriana

**Physocarpus malvaceus*

Rubus bartonianus

Sanguisorba stipulata

SALICACEAE

Salix farriae

**Salix petrophila*

Salix vestita

SAXIFRAGACEAE

Heuchera grossulariifolia var. *grossulariifolia*

Micranthes occidentalis

Mitella stauropetala

Saxifraga bronchialis var. *austromontana*

Saxifraga oppositifolia

Cascades (53)

MONOCOTS

CYPERACEAE

Carex breweri

Carex halliana

Trichophorum cespitosum

LILIACEAE

Erythronium montanum

ORCHIDACEAE

Spiranthes stellata

POACEAE

Calamagrostis breweri

Poa × multnomae

DICOTS

ADOXACEAE

Viburnum edule

ARISTOLOCHIACEAE

**Asarum wagneri*

ASTERACEAE

Arnica nevadensis

Arnica viscosa

Erigeron cascadensis

Erigeron howellii

Erigeron oreganus

Eucephalus gormanii

Hieracium longiberbe

Microseris borealis

Oreostemma alpigenum var. *alpigenum*

Raillardella scaposa

Rainiera stricta

BRASSICACEAE

Boechera horizontalis

Draba aureola

Smelowskia ovalis

CARYOPHYLLACEAE

**Eremogone pumicola*

Silene douglasii var. *rupinae*

ERICACEAE

Cassiope mertensiana ssp. *mertensiana*

Kalmiopsis fragrans

Vaccinium deliciosum

FABACEAE

Lupinus latifolius var. *subalpinus*

GENTIANACEAE

Gentiana newberryi var. *newberryi*

GROSSULARIACEAE

Ribes erythrocarpum

Ribes watsonianum

LENTIBULARIACEAE

Utricularia ochroleuca

MENYANTHACEAE

Nephrophyllidium crista-galli ssp. *crista-galli*

OROBANCHACEAE

Castilleja parviflora var. *oreopola*

Castilleja rupicola

Castilleja suksdorfii

Cascades

PAPAVERACEAE

**Corydalis aquae-gelidae*

PHRYMACEAE

Erythranthe moniliformis

PLANTAGINACEAE

Synthyris missurica ssp. *stellata*

POLEMONIACEAE

Collomia mazama

POLYGONACEAE

Eriogonum umbellatum var. *sandbergii*

Eriogonum villosissimum

Polygonum cascadense

RANUNCULACEAE

Cimicifuga laciniata

ROSACEAE

**Fragaria cascadensis*

Horkelia fusca ssp. *filicoides*

Spiraea × hitchcockii

SALICACEAE

Salix planifolia var. *planifolia*

SAXIFRAGACEAE

Heuchera glabra

Leptarrhena pyrolifolia

Micranthes nelsoniana var. *cascadensis*

Micranthes tolmiei

Coast Range (11)

MONOCOTS

LILIACEAE

Erythronium elegans

DICOTS

ASTERACEAE

Erigeron peregrinus var. *peregrinus*

ASTERACEAE

Packera flettii

BALSAMINACEAE

Impatiens ecornuta

BRASSICACEAE

Cardamine pattersonii

ERICACEAE

Cladothamnus pyroliflorus

MONTIACEAE

Lewisia columbiana var. *rupicola*

PLANTAGINACEAE

Synthyris schizantha

PRIMULACEAE

Dodecatheon austrofrigidum

ROSACEAE

Filipendula occidentalis

SAXIFRAGACEAE

Micranthes hitchcockiana

Columbia Basin (15)

MONOCOTS

AMARYLLIDACEAE

Allium robinsonii

POACEAE

Elymus lanceolatus ssp. *psammophilus*

DICOTS

APIACEAE

Lomatium farinosum var. *hambleniae*

Lomatium laevigatum

ASTERACEAE

Artemisia campestris var. *scouleriana*

Balsamorhiza rosea

BRASSICACEAE

Physaria douglasii ssp. *douglasii*

FABACEAE

Astragalus collinus var. *laurentii* ⚘

Astragalus conjunctus var. *rickardii*

Astragalus sclerocarpus

Astragalus succumbens

Astragalus tweedyi

Astragalus tyghensis ⚘

Lupinus leucophyllus × *Lupinus sericeus*

PHRYMACEAE

Erythranthe jungermannioides

Eastern Cascade Slopes and Foothills (11)

MONOCOTS

LILIACEAE

**Calochortus longebarbatus* var. *longebarbatus*

DICOTS

APIACEAE

**Lomatium peckianum*

ASTERACEAE

Packera eurycephala var. *eurycephala*

BRASSICACEAE

Thelypodium brachycarpum

**Thelypodium howellii* ssp. *howellii*

CONVOLVULACEAE

Cuscuta cephalanthi

FABACEAE

Astragalus applegatei ⚘

OROBANCHACEAE

**Castilleja chlorotica* ⚘

PLANTAGINACEAE

Penstemon barrettiae

**Penstemon peckii* ⚘

POLYGONACEAE

Eriogonum umbellatum var. *glaberrimum*

Estuarine Coast (104)

PTERIDOPHYTES

POLYPODIACEAE

Polypodium calirhiza

**Polypodium scouleri*

GYMNOSPERMS

PINACEAE

Pinus contorta var. *contorta*

MONOCOTS

ASPARAGACEAE

Brodiaea terrestris

CYPERACEAE

Carex brevicaulis

Carex macrocephala

Carex pansa

Eriophorum chamissonis

Isolepis cernua

IRIDACEAE

Crocosmia × *crocosmiiflora*

JUNCACEAE

Juncus breweri

Juncus canadensis

Juncus falcatus ssp. *sitchensis*

Juncus gerardi ssp. *gerardi*

Juncus lescurii

**Juncus nevadensis* var. *inventus*

Juncus planifolius

JUNCAGINACEAE

Triglochin striata

Estuarine Coast

LILIACEAE

Lilium occidentale

POACEAE

Agrostis densiflora

Bromus maritimus

Danthonia decumbens

Leymus × vancouverensis

Poa confinis

Poa macrantha

Poa unilateralis ssp. *pachypholis*

Poa unilateralis ssp. *unilateralis*

Puccinellia nutkaensis

Puccinellia pumila

TOFIELDIACEAE

Triantha glutinosa

ZOSTERACEAE

Phyllospadix scouleri

Phyllospadix serrulatus

Phyllospadix torreyi

Zostera japonica

Zostera marina

DICOTS

AIZOACEAE

Carpobrotus chilensis

AMARANTHACEAE

Atriplex gmelinii var. *gmelinii*

Sarcocornia perennis

APIACEAE

Angelica hendersonii

Glehnia littoralis ssp. *leiocarpa*

Sanicula arctopoides

ASTERACEAE

Agoseris apargioides var. *maritima*

**Ambrosia chamissonis* var. *bipinnatisecta*

Ambrosia chamissonis var. *chamissonis*

Artemisia pycnocephala

Cotula coronopifolia

Erigeron glaucus

Eriophyllum staechadifolium

**Grindelia stricta* var. *stricta*

Hesperevax sparsiflora var. *brevifolia*

Jaumea carnosa

Lasthenia maritima

Lasthenia ornduffii

Microseris bigelovii

Senecio triangularis var. *angustifolius*

Solidago simplex var. *spathulata*

Tanacetum camphoratum

BERBERIDACEAE

Berberis pinnata ssp. *pinnata*

BORAGINACEAE

Amsinckia spectabilis var. *spectabilis*

Cryptantha leiocarpa

BRASSICACEAE

Cakile edentula ssp. *edentula*

Cakile maritima

Crambe maritima

Erysimum concinnum

CARYOPHYLLACEAE

Cardionema ramosissimum

Honckenya peploides var. *major*

Spergularia macrotheca var. *macrotheca*

Stellaria humifusa

CONVOLVULACEAE

Cuscuta pacifica

CRASSULACEAE

Dudleya farinosa

ERICACEAE

Empetrum nigrum

Vaccinium macrocarpon

FABACEAE

**Lathyrus japonicus* var. *glaber*

Lathyrus littoralis

**Lupinus littoralis* var. *littoralis*

HYDROPHYLLACEAE

Phacelia argentea

Romanzoffia tracyi

HYPERICACEAE

Hypericum mutilum ssp. *boreale*

MALVACEAE

Sidalcea hendersonii

NYCTAGINACEAE

Abronia latifolia

Abronia umbellata var. *breviflora*

ONAGRACEAE

Camissoniopsis cheiranthifolia ssp. *cheiranthifolia*

Oenothera wolfii

OROBANCHACEAE

Castilleja affinis ssp. *litoralis*

Castilleja ambigua ssp. *ambigua*

Castilleja mendocinensis

Chloropyron maritimum ssp. *palustre*

PHRYMACEAE

Diplacus aurantiacus

PLANTAGINACEAE

Plantago coronopus

Plantago macrocarpa

Plantago maritima var. *californica*

Plantago subnuda

PLUMBAGINACEAE

Armeria maritima var. *californica*

Limonium californicum

POLEMONIACEAE

Gilia capitata ssp. *chamissonis*

Gilia capitata ssp. *pacifica*

Gilia millefoliata

POLYGONACEAE

Eriogonum latifolium

Eriogonum nudum var. *paralinum*

Polygonum paronychia

PRIMULACEAE

Lysimachia terrestris

ROSACEAE

Fragaria chiloensis ssp. *lucida*

**Fragaria chiloensis* ssp. *pacifica*

SAXIFRAGACEAE

Heuchera pilosissima

Lava Plains (10)

DICOTS

APIACEAE

Lomatium ochocense

ASTERACEAE

Artemisia ludoviciana ssp. *estesii*

Chaenactis nevii

BRASSICACEAE

**Thelypodium eucosmum*

FABACEAE

Astragalus diaphanus var. *diurnus*

Astragalus misellus var. *misellus*

OROBANCHACEAE

Castilleja xanthotricha

PLANTAGINACEAE

**Penstemon eriantherus* var. *argillosus*

POLYGONACEAE

Eriogonum microthecum var. *microthecum*

ROSACEAE

Crataegus castlegarensis

Owyhee Uplands (34)

DICOTS

APIACEAE

Cymopterus acaulis var. *acaulis*

Lomatium bentonitum

Lomatium ravenii var. *paiutense*

ASTERACEAE

Artemisia packardiae

Artemisia papposa

Brickellia microphylla var. *microphylla*

Chaenactis cusickii

Owyhee Uplands

Erigeron latus

Senecio ertterae

BORAGINACEAE

Amsinckia carinata

**Cryptantha propria*

Hackelia cronquistii

Hackelia ophiobia

BRASSICACEAE

**Lepidium davisii*

FABACEAE

Astragalus cusickii var. *sterilis*

Astragalus kentrophyta var. *jessiae*

Astragalus mulfordiae

Astragalus nudisiliquus

Astragalus toanus

Trifolium leibergii

Trifolium owyheense

LOASACEAE

Mentzelia mollis

Mentzelia packardiae

PLANTAGINACEAE

Penstemon perpulcher

POLEMONIACEAE

Aliciella triodon

Collomia renacta

Gymnosteris nudicaulis

Langloisia setosissima ssp. *punctata*

POLYGONACEAE

**Eriogonum chrysops*

Eriogonum novonudum

Eriogonum salicornioides

Eriogonum watsonii

PRIMULACEAE

Primula cusickiana var. *cusickiana*

ROSACEAE

Ivesia rhypara var. *rhypara*

Siskiyou Mountains (179)

PTERIDOPHYTES

PTERIDACEAE

Adiantum jordanii

GYMNOSPERMS

CUPRESSACEAE

Hesperocyparis bakeri

PINACEAE

**Picea breweriana*

Pinus sabiniana

MONOCOTS

AMARYLLIDACEAE

Allium bolanderi var. *mirabile*

**Allium falcifolium*

ASPARAGACEAE

Camassia howellii

**Camassia leichtlinii* ssp. *leichtlinii*

Dichelostemma multiflorum

Hastingsia alba

Hastingsia bracteosa var. *atropurpurea*

Hastingsia bracteosa var. *bracteosa*

Triteleia crocea

CYPERACEAE

**Calliscirpus criniger*

Carex klamathensis

Carex nervina

Carex scabriuscula

Carex serratodens

Eleocharis parishii

IRIDACEAE

Iris bracteata

LILIACEAE

Calochortus coxii

Calochortus greenei

Calochortus howellii

Calochortus indecorus

Calochortus persistens

Calochortus umpquaensis ssp. *confertus*
Erythronium citrinum var. *citrinum*
**Erythronium hendersonii*
Fritillaria gentneri
Lilium bolanderi
Lilium pardalinum ssp. *vollmeri*
Lilium pardalinum ssp. *wigginsii*

MELANTHIACEAE

Toxicoscordion exaltatum

POACEAE

Calamagrostis koelerioides
Festuca elmeri
Poa pringlei
Poa rhizomata

SMILACACEAE

Smilax californica

DICOTS

APIACEAE

Lomatium cookii
Lomatium engelmannii
Lomatium howellii
**Sanicula peckiana*
Sanicula tuberosa
Tauschia glauca
Tauschia howellii
Tauschia kelloggii

ASTERACEAE

Achyrachaena mollis
**Arnica cernua*
Balsamorhiza sericea
Brickellia greenei
**Cacaliopsis nardosmia* ssp. *nardosmia*
Calycadenia fremontii
Centromadia fitchii
Cirsium ciliolatum
Erigeron bloomeri var. *nudatus*
Erigeron cervinus
Erigeron klamathensis
Erigeron petrophilus var. *petrophilus*
Erigeron stanselliae
Eriophyllum lanatum var. *lanceolatum*
Eucephalus glabratus
Grindelia hirsutula var. *davyi*
Helianthus bolanderi
Hemizonia congesta ssp. *clevelandii*
Microseris howellii
Microseris laciniata ssp. *siskiyouensis*
Packera hesperia
Pyrrocoma racemosa var. *congesta*
**Rudbeckia glaucescens*
Stephanomeria elata
Symphyotrichum hendersonii

BERBERIDACEAE

Berberis pumila
Vancouveria chrysantha

BORAGINACEAE

Lithospermum californicum
**Plagiobothrys figuratus* var. *corallicarpus*
Plagiobothrys fulvus var. *campestris*
Plagiobothrys hirtus
Plagiobothrys lamprocarpus
Plagiobothrys shastensis

BRASSICACEAE

**Arabis aculeolata*
Arabis mcdonaldiana
Arabis modesta
Arabis oregana
Boechera koehleri
Boechera subpinnatifida
Draba howellii
Streptanthus glandulosus ssp. *josephinensis*
Streptanthus howellii
Streptanthus tortuosus
Thysanocarpus radians

CARYOPHYLLACEAE

Minuartia californica
Minuartia howellii
Silene bolanderi
Silene grayi
Silene hookeri ssp. *serpentinicola*

CELASTRACEAE

Parnassia palustris

CRASSULACEAE

Sedum laxum ssp. *heckneri*

SISKIYOU MOUNTAINS

Sedum moranii
Sedum oblanceolatum
Sedum radiatum ssp. *depauperatum*
Sedum spathulifolium ssp. *purdyi*

ERICACEAE

Arctostaphylos × *oblongifolia*
Arctostaphylos canescens × *Arctostaphylos nevadensis*
Arctostaphylos viscida
Kalmiopsis leachiana
Leucothoë davisiae

FABACEAE

Astragalus accidens var. *accidens*
**Astragalus accidens* var. *hendersonii*
Astragalus californicus
Astragalus gambelianus
**Lathyrus delnorticus*
Lupinus breweri
Lupinus lepidus var. *ashlandensis*
Lupinus onustus
Lupinus tracyi
Sophora leachiana
Trifolium longipes var. *oreganum*
**Trifolium siskiyouense*

FAGACEAE

Quercus × *morehus*
**Quercus sadleriana*

GENTIANACEAE

Gentiana plurisetosa
Gentiana setigera

GROSSULARIACEAE

Ribes marshallii

HYDROPHYLLACEAE

Phacelia corymbosa
Phacelia egena
Phacelia leonis

LAMIACEAE

Monardella villosa ssp. *sheltonii*
Scutellaria tuberosa

LIMNANTHACEAE

Limnanthes alba ssp. *gracilis*
Limnanthes floccosa ssp. *floccosa*
Limnanthes pumila ssp. *grandiflora*
Limnanthes pumila ssp. *pumila*

MONTIACEAE

Lewisia cotyledon var. *howellii*
Lewisia leeana
Lewisia oppositifolia

ONAGRACEAE

Clarkia heterandra
Epilobium oreganum
Epilobium rigidum
Epilobium siskiyouense
Tetrapteron graciliflorum

OROBANCHACEAE

Castilleja brevilobata
Castilleja elata
Castilleja schizotricha
**Orthocarpus cuspidatus* ssp. *cuspidatus*
Pedicularis densiflora
Pedicularis howellii

PAPAVERACEAE

Dicentra formosa ssp. *oregana*
Eschscholzia caespitosa

PHRYMACEAE

Diplacus kelloggii

PLANTAGINACEAE

Antirrhinum vexillo-calyculatum ssp. *breweri*
Collinsia latifolia
Penstemon azureus var. *azureus*
Penstemon parvulus

POLEMONIACEAE

Allophyllum gilioides
Navarretia pubescens
Navarretia tagetina

POLYGONACEAE

Eriogonum diclinum
Eriogonum luteolum var. *luteolum*
Eriogonum pendulum
Eriogonum ternatum
Eriogonum umbellatum var. *goodmanii*

RANUNCULACEAE

Delphinium nudicaule

Ranunculus austrooreganus

RHAMNACEAE

Ceanothus arcuatus

ROSACEAE

Drymocallis rhomboidea

**Horkelia congesta* ssp. *nemorosa*

Horkelia daucifolia var. *caruifolia*

Horkelia hendersonii

Horkelia sericata

Horkelia tridentata var. *tridentata*

Rosa gymnocarpa var. *serpentina*

RUBIACEAE

Galium porrigens var. *tenue*

SALICACEAE

Salix delnortensis

SAXIFRAGACEAE

Lithophragma affine

Micranthes aprica

Saxifragopsis fragarioides

VIOLACEAE

Viola lobata ssp. *integrifolia*

Viola quercetorum

VISCACEAE

Arceuthobium abietinum ssp. *wiensii*

Willamette Valley (26)

GYMNOSPERMS

CUPRESSACEAE

Juniperus virginiana var. *virginiana*

MONOCOTS

CYPERACEAE

Carex pendula

Eleocharis engelmannii

POACEAE

Coleanthus subtilis

DICOTS

APIACEAE

Lomatium bradshawii

ASTERACEAE

Erigeron decumbens

Grindelia integrifolia

Pyrrocoma racemosa var. *racemosa*

BETULACEAE

Corylus avellana

CAMPANULACEAE

Howellia aquatilis

FABACEAE

**Lathyrus holochlorus*

Trifolium vesiculosum

Vicia lutea

MALVACEAE

Malva sylvestris

OROBANCHACEAE

Castilleja levisecta

PHRYMACEAE

Mazus pumilus

PLANTAGINACEAE

Veronica peregrina var. *peregrina*

POLEMONIACEAE

Navarretia willamettensis

RANUNCULACEAE

Delphinium leucophaeum

**Delphinium pavonaceum*

Ranunculus sardous

ROSACEAE

Cotoneaster divaricatus

Crataegus phaenopyrum

Geum urbanum

RUBIACEAE

Galium odoratum

Galium parisiense var. *leiocarpum*

Appendix 4

Threatened and Endangered Species

This appendix lists the 59 vascular plant taxa designated as endangered or threatened within Oregon and administratively protected by the Oregon Department of Agriculture Native Plant Conservation Program (2014). The status ascribed by the Oregon Biodiversity Information Center (ORBIC 2013) and the United States Fish and Wildlife Service (2014) is also given. The definitions of conservation status by each organization are given below:

Oregon Department of Agriculture (State)

Endangered: Any native plant species determined by the director to be in danger of extinction throughout all or any significant portion of its range; or any plant species listed as an endangered species pursuant to the federal Endangered Species Act of 1973.

Threatened: Any native plant species the director determines is likely to become endangered within the foreseeable future throughout all or any significant portion of its range; or any plant species listed as a threatened species pursuant to the federal Endangered Species Act of 1973.

Oregon Biodiversity Information Center (ORBIC)

List 1: Taxa that are threatened with extinction or presumed to be extinct throughout their entire range.

List 4: Taxa which are of conservation concern but are not currently threatened or endangered. This includes taxa which are very rare but are currently secure, as well as taxa which are declining in numbers or habitat but are still too common to be proposed as threatened or endangered. While these taxa may not currently need the same active management attention as threatened or endangered taxa, they do require continued monitoring.

United States Fish and Wildlife Service (Federal)

Endangered: Any species that is in danger of extinction throughout all or a significant portion of its range.

Threatened: Any species that is likely to become an endangered species within the foreseeable future throughout all or a significant portion of its range.

Candidate: Plants and animals for which the U.S. Fish and Wildlife Service has sufficient information on their biological status and threats to propose them as endangered or threatened under the Endangered Species Act, but for which development of a proposed listing regulation is precluded by other higher priority listing activities.

Scientific Name	Common Name	State	ORBIC	Federal
PTERIDOPHYTES				
OPHIOGLOSSACEAE				
Botrychium pumicola	pumice moonwort	Threatened	1	
MONOCOTS				
ASPARAGACEAE				
Hastingsia bracteosa	large-flowered rush lily	Threatened	1	
LILIACEAE				
Calochortus coxii	crinite mariposa lily	Endangered	1	
Calochortus howellii	Howell's mariposa	Threatened	1	
Calochortus indecorus	Sexton Mountain mariposa	Endangered	1	
Calochortus umpquaensis	Umpqua mariposa	Endangered	1	
Erythronium elegans	elegant fawn-lily	Threatened	1	
Fritillaria gentneri	Gentner's fritillary	Endangered	1	Endangered
Lilium occidentale	western lily	Endangered	1	Endangered
POACEAE				
Pleuropogon oregonus	Oregon semaphoregrass	Threatened	1	

Scientific Name	Common Name	State	ORBIC	Federal
DICOTS				
APIACEAE				
Lomatium bradshawii	Bradshaw's lomatium	Endangered	1	Endangered
Lomatium cookii	Cook's lomatium	Endangered	1	Endangered
Lomatium erythrocarpum	red-fruited lomatium	Endangered	1	
Lomatium greenmanii	Greenman's lomatium	Threatened	1	
ASTERACEAE				
Artemisia campestris var. *wormskioldii*	Columbia Islands sagewort	Endangered	1	Candidate
Erigeron decumbens	Pacific fleabane	Endangered	1	Endangered
Eucephalus vialis	wayside aster	Threatened	1	
Microseris howellii	Howell's microseris	Threatened	4	
Pyrrocoma radiata	Snake River goldenweed	Endangered	1	
Sericocarpus rigidus	rigid white-topped aster	Threatened	1	
Stephanomeria malheurensis	Malheur wirelettuce	Endangered	1	Endangered
BORAGINACEAE				
Amsinckia carinata	Malheur Valley fiddleneck	Threatened	1	
Hackelia cronquistii	Cronquist's stickseed	Threatened	1	
Plagiobothrys hirtus	hairy popcornflower	Endangered	1	Endangered
Plagiobothrys lamprocarpus	shiny-fruited plagiobothrys	Endangered	1	
BRASSICACEAE				
Arabis mcdonaldiana	McDonald's rockcress		1	Endangered
Lepidium davisii	Davis' peppergrass	Threatened	1	
Thelypodium eucosmum	arrowleaf thelypody	Threatened	1	
Thelypodium howellii ssp. *spectabilis*	Howell's spectacular thelypody	Endangered	1	Threatened
CAMPANULACEAE				
Howellia aquatilis	water howellia		1	Threatened
CARYOPHYLLACEAE				
Silene douglasii var. *oraria*	Cascade Head catchfly	Threatened	1	
Silene spaldingii	Spalding's silene	Endangered	1	Threatened
FABACEAE				
Astragalus applegatei	Applegate's milkvetch	Endangered	1	Endangered
Astragalus collinus var. *laurentii*	Laurent's milkvetch	Threatened	1	
Astragalus cusickii var. *sterilis*	sterile milkvetch	Threatened	1	
Astragalus diaphanus var. *diurnus*	John Day milkvetch	Threatened	1	
Astragalus mulfordiae	Mulford's milkvetch	Endangered	1	
Astragalus peckii	Peck's milkvetch	Threatened	1	
Astragalus tyghensis	Tygh Valley milkvetch	Threatened	1	
Lupinus lepidus var. *cusickii*	Cusick's lupine	Endangered	1	
Lupinus oreganus	Kincaid's lupine	Threatened	1	Threatened
Trifolium owyheense	Owyhee clover	Endangered	1	

Scientific Name	Common Name	State	ORBIC	Federal
HYDROPHYLLACEAE				
Phacelia argentea	silvery phacelia	Threatened	1	
LIMNANTHACEAE				
Limnanthes pumila ssp. *grandiflora*	big-flowered woolly meadowfoam	Endangered	1	Endangered
Limnanthes pumila ssp. *pumila*	dwarf woolly meadowfoam	Threatened	1	
LOASACEAE				
Mentzelia mollis	smooth mentzelia	Endangered	1	
Mentzelia packardiae	Packard's mentzelia	Threatened	1	
MALVACEAE				
Sidalcea nelsoniana	Nelson's checkermallow	Threatened	1	Threatened
NYCTAGINACEAE				
Abronia umbellata var. *breviflora*	pink sandverbena	Endangered	1	
Mirabilis macfarlanei	MacFarlane's four o'clock	Endangered	1	Threatened
ONAGRACEAE				
Oenothera wolfii	Wolf's evening primrose	Threatened	1	
OROBANCHACEAE				
Castilleja levisecta	golden paintbrush	Endangered	1	Threatened
Chloropyron maritimum ssp. *palustre*	saltmarsh birdbeak	Endangered	1	
PLANTAGINACEAE				
Gratiola heterosepala	Boggs Lake hedge hyssop	Threatened	1	
POLYGONACEAE				
Eriogonum chrysops	golden buckwheat	Threatened	1	
Eriogonum crosbyae	Crosby's buckwheat	Threatened	1	
RANUNCULACEAE				
Delphinium leucophaeum	pale larkspur	Endangered	1	
Delphinium pavonaceum	peacock larkspur	Endangered	1	
ROSACEAE				
Ivesia rhypara var. *rhypara*	grimy mousetail	Endangered	1	

Appendix 5

Native Taxa Not Collected in the Past 50 Years

The 159 taxa listed here denote all accepted native taxa that are not represented by a herbarium specimen from the past 50 years (since 1964). Collections from multiple herbaria are tracked in the OFP specimen database, as are all Oregon specimens in the Oregon State University Herbarium (OSC, ORE, and WILLU). Institutions affiliated with the herbarium acronyms are listed in the Abbreviations and Symbols table (page 8). Nothogenera, nothospecies, and unnamed hybrids are excluded from this list. The year of the most recent collection is given. Taxa designated with an asterisk (*) are assessed by ORBIC to be extirpated or extinct (ORBIC 2013). The nomenclatural status of dicots is based on existing floristic treatments for the *Flora of Oregon* or, if not available, on the OFP Checklist Version 4.1.

Taxon	Herbarium Code	Most Recent Collection Year
PTERIDOPHYTES		
ASPLENIACEAE		
Asplenium trichomanes ssp. *quadrivalens*	WTU	1929
GYMNOSPERMS		
PINACEAE		
Pinus ponderosa var. *benthamiana*	OSC	1958
MONOCOTS		
ALISMATACEAE		
Sagittaria montevidensis ssp. *calycina*	WILLU	1943
AMARYLLIDACEAE		
Allium peninsulare var. *peninsulare*	WILLU	1931
ARACEAE		
Lemna valdiviana	ORE	1899
CYPERACEAE		
**Bulbostylis capillaris*	ORE	1900
Carex comosa	OSC	1945
Eleocharis decumbens	OSC	1929
Eleocharis erythropoda	WILLU	1950
Lipocarpha aristulata	WILLU	1944
Schoenoplectus heterochaetus	WILLU	1896
IRIDACEAE		
Sisyrinchium halophilum	WILLU	1939
JUNCACEAE		
Juncus capillaris	ORE	1896
Luzula campestris ssp. *campestris*	ORE	1903
LILIACEAE		
Calochortus luteus	SOC	1939
**Lilium kelloggii*	OSC	1943

Taxon	Herbarium Code	Most Recent Collection Year
ORCHIDACEAE		
Corallorhiza wisteriana	OSC	1882
POACEAE		
× *Elyhordeum macounii*	ORE	1954
× *Elyleymus aristatus*	ORE	1957
× *Leydeum piperi*	US	1921
Achnatherum × *bloomeri*	OSC	1957
**Agrostis hendersonii*	ORE	1930
Coleanthus subtilis	OSC	1927
Elymus canadensis var. *canadensis*	OSC	1957
Elymus glaucus ssp. *virescens*	ORE	1945
Leymus × *vancouverensis*	WILLU	1939
Phalaris californica	WILLU	1926
Poa fendleriana ssp. *longiligula*	ORE	1959
Poa laxa	ORE	1900
Poa lettermanii	ORE	1958
Scolochloa festucacea	ORE	1950
Sphenopholis obtusata	ORE	1934
POTAMOGETONACEAE		
**Potamogeton fibrillosus*	ORE	1927
DICOTS		
AMARANTHACEAE		
Atriplex patula var. *patula*	ORE	1958
Chenopodium capitatum var. *capitatum*	OSC	1888
Chenopodium nevadense	WILLU	1962
APIACEAE		
Angelica tomentosa	WILLU	1939
**Cicuta bulbifera*	OSC	1916
Cicuta maculata var. *angustifolia*	OSC	1957
Conioselinum scopulorum	ORE	1961
Zizia aptera var. *occidentalis*	OSC	1903
ASTERACEAE		
**Artemisia campestris* var. *wormskioldii*	OSC	1955
Balsamorhiza × *terebinthacea*	ORE	1931
Balsamorhiza × *tomentosa*	ORE	1930
Cirsium cymosum var. *cymosum*	OSC	1956
Cirsium occidentale var. *venustum*	ORE	1940
Cirsium scariosum var. *robustum*	ORE	1941

Taxon	Herbarium Code	Most Recent Collection Year
ASTERACEAE		
Crepis runcinata ssp. *hispidulosa*	OSC	1953
Dieteria canescens var. *incana*	OSC	1962
Ericameria parryi var. *nevadensis*	OSC	1919
Eriophyllum lanatum var. *grandiflorum*	ORE	1939
Grindelia nana var. *discoidea*	OSC	1947
Helenium puberulum	US	1950
Helianthella californica var. *nevadensis*	ORE	1899
Helianthus bolanderi	OSC	1936
Malacothrix stebbinsii	ORE	1894
Matricaria occidentalis	WILLU	1953
Microseris douglasii ssp. *douglasii*	ORE	1889
Packera porteri	ORE	1899
Sphaeromeria cana	OSC	1954
Stylocline psilocarphoides	ORE	1943
Symphyotrichum ericoides var. *pansum*	OSC	1937
Symphyotrichum laeve var. *geyeri*	WILLU	1947
Wyethia × cusickii	OSC	1950
BETULACEAE		
Betula × sargentii	OSC	1925
BORAGINACEAE		
Amsinckia spectabilis var. *spectabilis*	OSC	1939
Plagiobothrys lamprocarpus	OSC	1921
Plagiobothrys reticulatus var. *reticulatus*	ORE	1947
BRASSICACEAE		
Arabis mcdonaldiana	ORE	1953
Athysanus unilateralis	ORE	1889
Boechera breweri ssp. *shastaensis*	WILLU	1904
Boechera cascadensis	MO	1936
Boechera covillei	WILLU	1935
Boechera horizontalis	ORE	1929
Cardamine holmgrenii	UTC	1946
Draba incerta var. *incerta*	WILLU	1934
Draba oligosperma	ORE	1963
Erysimum inconspicuum var. *inconspicuum*	ORE	1941
Lepidium acutidens	ORE	1888
Lepidium densiflorum var. *densiflorum*	ORE	1927
Lepidium dictyotum	ORE	1948
Physaria geyeri ssp. *geyeri*	OSC	1894
Thelypodium integrifolium ssp. *integrifolium*	OSC	1935

Taxon	Herbarium Code	Most Recent Collection Year
CAMPANULACEAE		
Triodanis biflora	OSC	1958
CARYOPHYLLACEAE		
Loeflingia squarrosa	WILLU	1942
Minuartia austromontana	WILLU	1944
Silene bridgesii	ORE	1887
Silene campanulata ssp. *campanulata*	WS	1949
Silene hookeri ssp. *serpentinicola*	ORE	1884
Stellaria borealis ssp. *borealis*	OSC	1915
CERATOPHYLLACEAE		
**Ceratophyllum echinatum*	OSC	1951
CONVOLVULACEAE		
Cressa truxillensis var. *vallicola*	WILLU	1927
Cuscuta cephalanthi	ORE	1953
Cuscuta indecora var. *neuropetala*	OSC	1949
CORNACEAE		
Cornus glabrata	ORE	1963
CRASSULACEAE		
**Sedella pumila*	GH	1835
FABACEAE		
Astragalus lemmonii	OSC	1957
Astragalus lentiginosus var. *floribundus*	RSA	1940
Trifolium eriocephalum ssp. *martinii*	ORE	1902
Trifolium siskiyouense	WILLU	1918
Vicia hassei	ORE	1889
FAGACEAE		
Quercus × *morehus*	OSC	1956
Quercus garryana var. *semota*	ORE	1928
GERANIACEAE		
**California macrophylla*	ORE	1889
GROSSULARIACEAE		
Ribes acerifolium	ORE	1933
Ribes divaricatum var. *pubiflorum*	OSC	1935
Ribes inerme var. *klamathense*	ORE	1935
Ribes watsonianum	ORE	1938

Taxon	Herbarium Code	Most Recent Collection Year
HYDROPHYLLACEAE		
Nemophila heterophylla	OSC	1948
Phacelia adenophora	ORE	1941
Phacelia bicolor var. *bicolor*	WILLU	1948
Phacelia egena	ORE	1952
**Phacelia malvifolia*	ORE	1884
LAMIACEAE		
Physostegia parviflora	OSC	1928
MONTIACEAE		
Lewisia × whiteae	ORE	1926
ONAGRACEAE		
Chylismiella pterosperma	ORE	1945
Clarkia purpurea ssp. *purpurea*	OSC	1961
Epilobium leptocarpum	WILLU	1944
Tetrapteron palmeri	OSC	1940
OROBANCHACEAE		
**Castilleja levisecta*	ORE	1938
Orobanche ludoviciana ssp. *ludoviciana*	ORE	1948
PAPAVERACEAE		
Dicentra pauciflora	CAS	1926
PLANTAGINACEAE		
Antirrhinum kingii	OSC	1948
Penstemon acuminatus var. *acuminatus*	OSC	1955
Penstemon glandulosus var. *chelanensis*	OSC	1932
Plantago eriopoda	OSC	1957
POLEMONIACEAE		
Allophyllum gilioides	ORE	1946
Collomia tenella	ORE	1957
Navarretia atractyloides	ORE	1927
Navarretia heterandra	WILLU	1952
Navarretia leptalea ssp. *leptalea*	WILLU	1936
POLYGONACEAE		
Chorizanthe brevicornu var. *spathulata*	OSC	1949
Eriogonum douglasii var. *meridionale*	ORE	1950
Eriogonum elatum var. *villosum*	OSC	1917
Eriogonum roseum	OSC	1909
Eriogonum umbellatum var. *argus*	OSC	1949

Taxon	Herbarium Code	Most Recent Collection Year
POLYGONACEAE		
Polygonum erectum	WILLU	1949
Polygonum gabrielae	OSC	1953
Rumex persicarioides	WILLU	1939
PRIMULACEAE		
Lysimachia ciliata	OSC	1938
RANUNCULACEAE		
Ranunculus orthorhynchus var. *bloomeri*	WILLU	1927
ROSACEAE		
× *Amelasorbus jackii*	WILLU	1915
Crataegus macracantha var. *occidentalis*	ORE	1933
Horkelia daucifolia var. *daucifolia*	US	1927
Horkelia tridentata var. *flavescens*	ORE	1933
Potentilla supina ssp. *paradoxa*	WILLU	1948
Rosa californica	OSC	1930
Rosa pisocarpa ssp. *ahartii*	ORE	1929
RUBIACEAE		
Galium californicum ssp. *californicum*	WILLU	1929
Galium muricatum	OSC	1950
Galium porrigens var. *porrigens*	ORE	1926
SALICACEAE		
Salix laevigata	WILLU	1940
Salix tracyi	ORE	1932
SAXIFRAGACEAE		
Heuchera grossulariifolia var. *tenuifolia*	ORE	1957
Heuchera rubescens var. *rydbergiana*	ORE	1887
VALERIANACEAE		
Plectritis ciliosa ssp. *insignis*	OSC	1936
VIOLACEAE		
Viola purpurea ssp. *dimorpha*	WTU	1927
VISCACEAE		
Arceuthobium monticola	WILLU	1918

INDEX

Scientific and common names for plants described in this volume and scientific names in Appendix 1 are indexed. Accepted family names of plants in the flora are in **ALL CAPS BOLD**, accepted generic and lower taxa names are in **bold**. Taxonomic synonyms, misapplied names, and taxa not fully treated in the flora are in *italics*. Common names are in SMALL CAPS, and common family names are in ALL CAPS. Page numbers of the main descriptions are in **bold**; illustrations are in regular; other entries such as appearances in keys are in light.

Concepts that are more narrowly circumscribed are indented and listed below their inclusive name. Page numbers are listed for names (scientific or common) as they appear in correspondence to a description. An inclusive name may occur only in the context of one of these narrower concepts, and therefore page numbers are given only for the narrower concept, whether an accepted name, a synonym, or a common name.

A

E

G

H

I

J

K

L

M

N

O

Q

R

S

T

U

V

W

X

Y

Z

Flora of Oregon
Volume 1

The main typeface is Adobe Caslon Pro. Cronos Pro is used for captions, appendices, and the index. The symbol ❧, used to indicate endemics, is an ornament from Minion Pro.

Text pages are acid-free 70 gsm New Classic White paper.

Color pages are acid-free 100 gsm matte art paper.

End papers are 180 gsm woodfree paper.

The dust jacket is 150 gsm gloss art paper with UV coating.

The binding cloth for the cover is BK-3 from Dong-A Co. with Murata Gleam metallic foil stamping.

The book was designed and produced on a Mac in Adobe InDesign®.

Pen & ink illustrations created for this volume were scanned and arranged in Adobe Photoshop®. Other illustrations were created in Adobe Illustrator®.

Maps were created in ArcGIS.

Printed and bound by Sung In Printing Co., Ltd., Korea.

Book design is by Tanya Harvey Design.